The NALCO Water Handbook

OTHER McGRAW-HILL HANDBOOKS OF INTEREST

American Institute of Physics *AMERICAN INSTITUTE OF PHYSICS HANDBOOK*
Baumeister and Avallone *MARKS' STANDARD HANDBOOK FOR MECHANICAL ENGINEERS*
Brady and Clauser *MATERIALS HANDBOOK*
Chopey and Hicks *HANDBOOK OF CHEMICAL ENGINEERING CALCULATIONS*
Considine *PROCESS INSTRUMENTS AND CONTROLS HANDBOOK*
Davidson *HANDBOOK OF WATER-SOLUBLE GUMS AND RESINS*
Dean *LANGE'S HANDBOOK OF CHEMISTRY*
Dean *HANDBOOK OF ORGANIC CHEMISTRY*
Grant *HACKH'S CHEMICAL DICTIONARY*
Harper *HANDBOOK OF PLASTICS AND ELASTOMERS*
Hicks *STANDARD HANDBOOK OF ENGINEERING CALCULATIONS*
Hopp and Hennig *HANDBOOK OF APPLIED CHEMISTRY*
Juran *QUALITY CONTROL HANDBOOK*
Lyman *HANDBOOK OF CHEMICAL PROPERTY ESTIMATION METHODS*
McLellan and Shand *GLASS ENGINEERING HANDBOOK*
Maynard *INDUSTRIAL ENGINEERING HANDBOOK*
Meyers *HANDBOOK OF CHEMICAL PRODUCTION PROCESSES*
Meyers *HANDBOOK OF PETROLEUM REFINING PROCESSES*
Rohsenow, Hartnett, and Ganic *HANDBOOK OF HEAT TRANSFER FUNDAMENTALS*
Rohsenow, Hartnett, and Ganic *HANDBOOK OF HEAT TRANSFER APPLICATIONS*
Rosaler and Rice *STANDARD HANDBOOK OF PLANT ENGINEERING*
Schwartz *COMPOSITE MATERIALS HANDBOOK*
Schweitzer *HANDBOOK OF SEPARATION TECHNIQUES FOR CHEMICAL ENGINEERS*
Tuma *ENGINEERING MATHEMATICS HANDBOOK*
Tuma *HANDBOOK OF PHYSICAL CALCULATIONS*

Nalco Chemical Company

Frank N. Kemmer *Editor*

The NALCO Water Handbook

Second Edition

McGraw-Hill Book Company

New York St. Louis San Francisco Auckland
Bogotá Hamburg London Madrid Mexico
Milan Montreal New Delhi Panama
Paris São Paulo Singapore
Sydney Tokyo Toronto

Library of Congress Cataloging-in-Publication Data

The NALCO water handbook.

Includes index.
1. Water—Handbooks, manuals, etc. I. Kemmer,
Frank N. II. Nalco Chemical Company.
QD169.W3N34 1987 628.1′62 87-4171
ISBN 0-07-045872-3

9 0 DOC/DOC 9 8 7

ISBN 0-07-045872-3

This book is printed on acid-free paper.

The editors for this book were Harold B. Crawford and Rita T.
Margolies and the production supervisor was Richard A. Ausburn.
It was set in Times Roman by University Graphics, Inc.

Printed and bound by R. R. Donnelley & Sons Company.

CONTENTS

Part 4 Specialized Water Treatment Technologies

PREFACE

Since the appearance of the first edition of this handbook in 1979, old frontiers of knowledge have been pushed back as new technologies have been commercialized to solve the mounting problems of pollution control, energy management, and production of ultra-high-purity water for a host of new applications. The second edition meets the new needs for such technology by a significant expansion of editorial material.

The original intent of the handbook is still germane today: to provide orientation in water technology for those who need this information in an understandable form for their jobs, but may not necessarily aspire to becoming specialists in water treatment. As stated in the first edition preface, these persons "may be in such diverse occupations as plant manager, architect, utilities superintendent, city engineer, legislator, or environmental administrator." The first edition has served as both textbook and reference for those studying these and related professions.

There are four parts to this handbook. Part 1 introduces the nature of water and explores the fundamentals of chemistry, biology, water resources, and water analysis. Part 2 presents, in detail, the unit operations of water conditioning. Part 3 gives an overview of important industrial and municipal uses of water, including recycling and disposal; the typical operations of major industries, institutions, and municipalities are described to exemplify prevailing industrial practices. Part 4 presents chapters on specialized water treatment technologies, such as treatment for cooling purposes, steam generation, and oil recovery. The first part has been published separately as a softcover book, *Water: The Universal Solvent.*

It is impossible for me, looking back on 48 years of my career in this field, to acknowledge all of the ideas of many friends and authors whose experiences, added to my own, may appear here. Fortunately, I am able to acknowledge some of my many associates at Nalco Chemical Company who contributed individual chapters in their specialized fields. These include P. A. Barker, J. A. Beardsley, G. A. Casedy, J. L. Cole, L. J. Domzalski, K. E. Fulks, C. R. Hoefs, J. C. Jennings, K. E. Lampert, G. F. McIntyre, K. Odland, M. T. Perin, J. E. Phelan, R. S. Robertson, and R. F. Tuka. Many others, too numerous to mention, have made valuable suggestions, which I greatly appreciate and continue to acknowledge in person.

During months of this handbook's gestation the assiduous attention of John McCallion to innumerable editorial details and Beverly Newmiller to the preparation of the manuscript and illustrations was indispensable in ensuring a safe delivery.

Frank N. Kemmer

The NALCO Water Handbook

P · A · R · T · 1

THE NATURE OF WATER

CHAPTER 1
THE WATER MOLECULE

Three-quarters of the surface of the earth is covered with water. While this is an impressive statistic, it is pale beside the spectacular photographs that have come to us from outer space. They reveal a beautiful blue planet bathed in water, partly hidden by a veil of vapor.

Life came into being in this water. As living things became more complex and specialized, they left the sea for the land, taking water with them as the major part of their bodies. On the Planet Earth, water is life.

A philosopher observed that the proper study of mankind is man; the water chemist paraphrases this: "The proper study of water is the water molecule." The formula for water—H_2O—by itself tells us only its composition and molecular weight. It does nothing to explain the remarkable properties that result from its unique molecular arrangement (see Figure 1.1). Two hydrogen atoms are located 105° apart, adjacent to the oxygen atom, so that the molecule is asymmetrical, positively charged on the hydrogen side and negatively charged on the oxygen side. For this reason, water is said to be dipolar. This causes the molecules to agglomerate, the hydrogen of one molecule attracting the oxygen of a neighboring molecule. The linking of molecules resulting from this attractive force is called *hydrogen bonding.*

One of the consequences of hydrogen bonding is that molecules of H_2O cannot leave the surface of a body of water as readily as they could without this intermolecular attraction. The energy required to rupture the hydrogen bond and liberate a molecule of H_2O to form vapor is much greater than for other common chemical compounds. Because of this fact, the water vapor—steam—has a high energy content and is an effective medium for transferring energy in industrial plant operations, buildings, and homes.

Water also releases more heat upon freezing than do other compounds. Furthermore, for each incremental change in temperature, water absorbs or releases more heat—i.e., has great heat capacity—than many substances, so it is an effective heat transfer medium.

The freezing of water is unusual compared to other liquids. Hydrogen bonding produces a crystal arrangement that causes ice to expand beyond its original liquid volume so that its density is less than that of the liquid and the ice floats. If this were not the case, lakes would freeze from the bottom up, and life as we know it could not exist.

Table 1.1 compares the boiling point and other heat properties of water with similar molecules, such as hydrogen sulfide, and with dissimilar compounds that are liquid at room temperature.

Because of the unusual structure of the water molecule, it is present in the

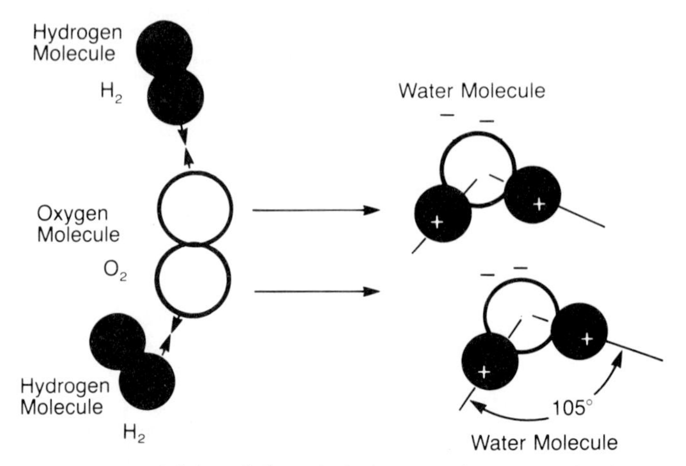

FIG. 1.1 The joining of diatomic hydrogen and oxygen molecules to produce water molecules of a polar nature.

TABLE 1.1 Thermal Properties of Water and Similar Compounds

Substance	Specific heat	Freezing point, °C	Boiling point, °C	Latent heat of evaporation, cal/g
H_2O	1.00	0	100	540
H_2S		−83	−62	132
Methanol	0.57	−98	65	263
Ethanol	0.54	−117	79	204
Benzene	0.39	6	80	94

FIG. 1.2 A steel needle, with a density about 7 times that of water, can be made to float because of water's high surface tension.

natural environment in all three states of matter, solid as ice, liquid as water, and gas as vapor. It is the only chemical compound having this unusual character.

In addition to its unusual heat properties, water has physical properties quite different from other liquids. Its high surface tension is easily demonstrated by the experiment of "floating" a needle on the surface of water in a glass (Figure 1.2). This high surface tension, due to hydrogen bonding, also causes water to rise in a capillary tube (Figure 1.3). This capillarity is partly responsible for the system of circulation developed by living plants through their roots and tissue systems.

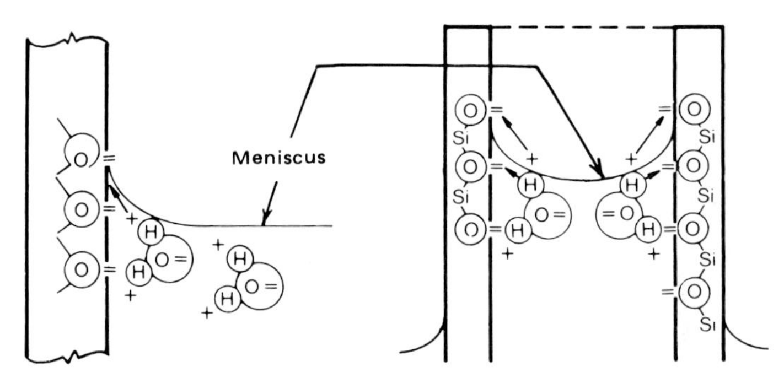

FIG. 1.3 A meniscus forms (left) when hydrogen atoms reach upward to wet oxide surfaces at the water line in a glass tube. The drawing at the right shows how hydrogen bonding of water to a thin glass tube causes the water in the tube to rise above the level of the surrounding water. Some liquids other than water do not wet a glass surface. They form an inverted meniscus.

Water is often called the universal solvent. Water molecules in contact with a crystal orient themselves to neutralize the attractive forces between the ions in the crystal structure. The liberated ions are then hydrated by these water molecules as shown in Figure 1.4, preventing them from recombining and recrystalizing. This solvency and hydration effect is shown quantitatively by water's relatively high dielectric constant.

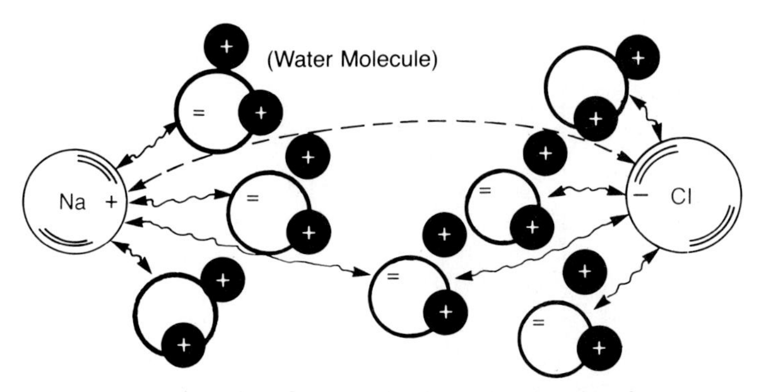

FIG. 1.4 The orientation of water molecules tends to keep ions from recombining and thus precipitating from solution. This accounts for water's capabilities as a solvent.

Water ionizes so very slightly, producing only 10^{-7} moles of hydrogen and 10^{-7} moles of hydroxyl ions per liter, that it is an insulator—it cannot conduct electrical current. As salts or other ionizing materials dissolve in water, electrical conductivity develops. The conductivity of naturally occurring waters provides a measure of their dissolved mineral content (Figure 1.5).

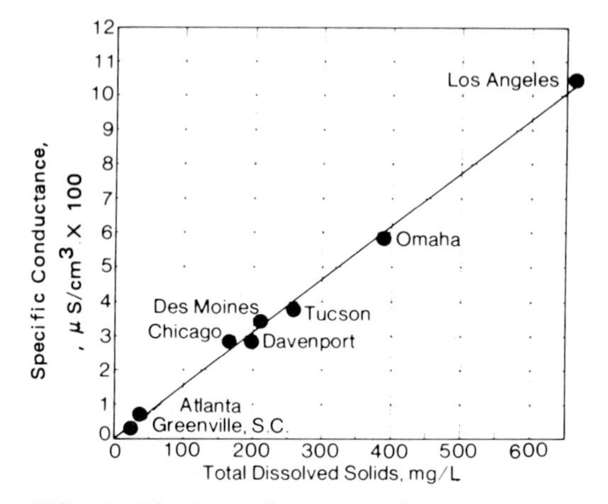

FIG. 1.5 Dissolved solids content of water can be estimated from its specific conductance. For most public water supplies, the conversion factor is 1.55μs conductance per milligrams per liter of total dissolved solids. For other kinds of water, e.g., wastewater and boiler water, the conversion factor must be established for each situation.

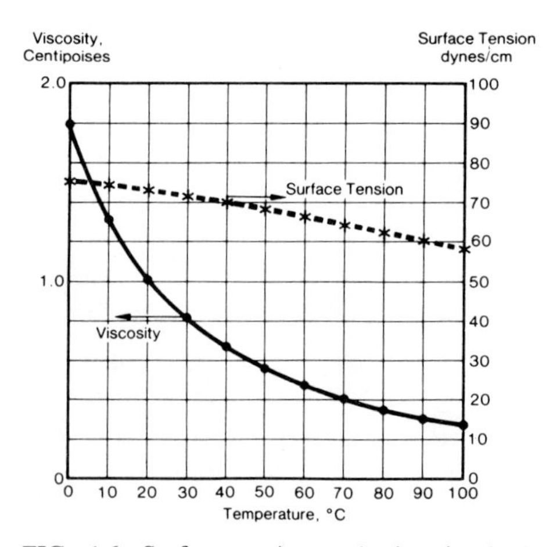

FIG. 1.6 Surface tension and viscosity both decrease as water is heated.

Another important phenomenon occurring in water solutions related to dissolved matter (solutes), rather than to water (solvent), is osmotic pressure. If two aqueous solutions are separated by a membrane, water will pass from the more dilute into the more concentrated one. This important process controls the performance of all living cells. It explains the effectiveness of food preservation by salting; the salt creates a strong solution, disrupting the cells of organisms that might cause food spoilage, as the water inside their bodies leaves in an attempt to dilute the external salt solution. In specially designed membrane cells, the osmotic flow of water across the membrane can be reversed by applying a sufficiently high pressure to the more concentrated solution. This process of "reverse osmosis" is a practical one for desalination of water.

Finally, viscosity is another property of water affecting its treatment and use. It is a measure of internal friction—the friction of one layer of molecules moving across another. As water temperature rises, this internal friction decreases. Because of the temperature effect, dissolved salts and gases can diffuse more rapidly through warmer water, chemical treatment is hastened, and the physical processes of sedimentation and degasification proceed faster. The effect of temperature on viscosity is shown in Figure 1.62

SUGGESTED READING

Boys, C. V.: *Soap Bubbles,* Doubleday, New York, 1959.

Buswell, A. M., and Rodebush, W. H.: "Water," *Sci. Am.,* April 1956, p. 76.

Carson, Rachel: *The Sea Around Us,* Oxford University Press, New York, 1951.

Day, John A., and Davis, Kenneth S.: *Water: The Mirror of Science,* Doubleday, New York, 1961.

King, Thompson: *Water, Miracle of Nature,* Macmillan, New York, 1953.

Leopold, Luna B., and Davis, Kenneth S.: *Water,* Life Science Library, Time-Life, New York, 1974.

CHAPTER 2

WATER SOURCES AND USES

Water has moved in ceaseless migration across the face of the earth since earliest time. Well-established currents such as the Gulf Stream and the Humboldt current continuously circulate in the seas, regulating the earth's climate and providing sustenance for the fisheries on which much of the world's population relies. Where stopped by land barricades, the sea, refusing to be held back completely, releases its water as vapor to condense and fall somewhere inland as rain to attack the land barrier. This continual evaporation and condensation is the *hydrologic cycle.*

Over the U.S. mainland, rainfall averages approximately 4250×10^9 gal/day (1.6×10^{10} m^3/day). Of this, about two-thirds returns to the atmosphere through evaporation directly from the surface of lakes and rivers and transpiration from plant foliage. This leaves approximately 1250×10^9 gal/day (0.46×10^{10} m^3/day) to flow across or through the earth to return to the sea (Table 2.1). Although

TABLE 2.1 Water Balance in the United States
[*Figures in billion gallons per day (bgd)*]*

Precipitation	4250
Evaporation and transpiration	3000
Runoff	1250
Withdrawal	310
Irrigation	142
Industry†	142
Municipal	26
Consumed‡	90
Returned to streams	220

* bgd × 44 = m^3/s.
† Principally utility cooling water (approximately 100 bgd).
‡ Principally irrigation loss to ground and evaporation.
Source: National Academy of Sciences—National Research Council Publ. 1000-B, 1962.

municipal usage of water seems a small fraction of this great volume, per capita consumption of water in the United States is very high—100 to 200 gal/day (0.38 to 0.76 m^3/day), probably because potable water is relatively inexpensive ($0.50 to $1.50 per 1000 gal). In Europe, municipal water costs are generally much higher, and per capita water consumption is only 25 to 35% of that in the United States. The cost of water for irrigation, the largest water use in the United States, is usually less than 10% of the cost of potable water, raising questions as to priorities in resources and the effectiveness of the national water policy.

Abundant supplies of water attracted the American colonists to settlements along rivers. Their waterfalls became the principal sources of power for early industry. Along the eastern seaboard, river waters were of excellent quality, ideal for textile and paper manufacture. As the frontiers moved inland, settlers found

Identification of Analyses Tabulated Below:

A. Delaware at Morrisville, PA
B. Edisto at Charleston, SC
C. Chattahoochee at Atlanta, GA
D. Ohio at Steubenville, Ohio
E. Tennessee at Decatur, AL
F. Mississippi at St. Louis, MO

Constituent	As	A	B	C	D	E	F
Calcium	$CaCO_3$	35	17	9	57	58	125
Magnesium	"	19	3	4	30	20	57
Sodium	"	4	10	5	14	18	66
Total Electrolyte	$CaCO_3$	58	30	18	101	96	248
Bicarbonate	$CaCO_3$	10	19	12	13	55	128
Carbonate	"	0	0	0	0	0	0
Hydroxyl	"	0	0	0	0	0	0
Sulfate	"	35	2	2	79	15	93
Chloride	"	10	8	3	8	21	23
Nitrate	"	3	1	1	1	4	4
Fluoride	"	NR	NR	0	TR	1	NR
M Alk.	$CaCO_3$	10	19	12	13	55	128
P Alk.	"	0	0	0	0	0	0
Carbon Dioxide	"	1	20	2	15	10	3
pH		7.1	6.1	6.9	6.4	7.0	7.9
Silica	SiO_2	4.7	5	10	5.6	5.6	13
Iron	Fe	0.1	0.2	0.1	TR	0.1	NR
Turbidity		NR	NR	26	88	50	NR
TDS		84	46	33	143	113	326
Color		7	54	4	2	30	19

pH in pH units; color in APHA units; all others in mg/l

In this and in subsequent analyses, the notation NR is used where data have not been reported on the original source analyses.

(*a*)

FIG. 2.1 (*a*) Selected river analyses. United States east of the Mississippi River.

that the salinity of the streams became higher, particularly where long rivers flowed over relatively soluble rock formations.

Figures 2.1*a* and *b* compare the chemical properties of some of the major rivers in the United States. Table 2.2 shows how water quality, using Colorado River data, is influenced from the headwaters to the discharge to the sea. The quality of such rivers is affected by rainfall, the geological nature of the watershed, conditions of evaporation, seasonal changes in stream flow, and by activities of human society. As an example of the latter, in column D of Figure 2.1*a*, the low alkalinity relative to total dissolved solids is attributed to industrial discharges upstream, chiefly from coal mines and steel mills. [These analyses date from the era prior

Identification of Analyses Tabulated Below:

G. Missouri at Great Falls, MT
H. Missouri at Kansas City
I. Neches at Evadale, TX
J. Colorado as delivered to Los Angeles, CA
K. Columbia at Wenatchee, WA
L. Sacramento at Sacramento, CA

Constituent	As	G	H	I	J	K	L
Calcium	$CaCO_3$	100	152	17	198	50	12
Magnesium	"	49	70	16	105	16	8
Sodium	"	47	86	23	220	25	5
Total Electrolyte	$CaCO_3$	196	308	56	523	91	25
Bicarbonate	$CaCO_3$	140	158	17	113	60	20
Carbonate	"	0	TR	0	7	0	0
Hydroxyl	"	0	0	0	0	0	0
Sulfate	"	36	120	13	302	22	2
Chloride	"	16	27	25	100	2	2
Nitrate	"	2	3	1	Nil	1	1
Fluoride	"	2	NR	0	1	TR	0
M Alk.	$CaCO_3$	140	158	17	120	60	20
P Alk.	"	0	TR	0	4	0	0
Carbon Dioxide	"	TR	NR	4	0	4	2
pH		8.4	8.3	6.7	8.4	7.7	7.2
Silica	SiO_2	19	12	16	8	9.6	9.2
Iron	Fe	TR	NR	0.6	Nil	TR	0
Turbidity		150	2000	35	NR	3	36
TDS		234	365	96	661	100	36
Color		NR	NR	60	NR	NR	15

pH in pH units; color in APHA units; all others in mg/l

(*b*)

FIG. 2.1 (*b*) United States west of the Mississippi River.

to the existence of the Environmental Protection Agency (EPA).] Water quality is affected by municipal sewage, even in the absence of industrial waste discharges, which typically shows an increase in total dissolved solids of 50 mg/L from water intake to outfall. (See Chapter 6.)

Identification of Analyses Tabulated Below:

A. Caroni (Eastern Venezuela) D. Athabasca (Canada)

B. Amazon (Northern Brazil) E. Kerteh (Malaysia)

C. Paraiba (Southern Brazil) F. Brantos (Indonesia)

Constituent	As	A	B	C	D	E	F
Calcium	$CaCO_3$	2	5	25	66	7	70
Magnesium	"	1	1	21	28	8	30
Sodium	"	2	13	10	18	3	43
Potassium		NR	NR	NR	1		
Total Electrolyte	$CaCO_3$	6	19-22	56	113	18	143
Bicarbonate	$CaCO_3$	3	8	35	80	3	100
Carbonate	"	0	0	0	0	0	0
Hydroxyl	"	0	0	0	0	0	0
Sulfate	"	Nil	5	8	18	1	15
Chloride	"	3	6	13	5	14	28
Nitrate	"	Nil	Tr	NR	10	TR	NR
M Alk.	$CaCO_3$	3	8	35	80	3	100
P Alk.	"	0	0	0	0	0	0
Carbon Dioxide		15	7	30	Nil	5	5
pH		5.8	6.5	6.5	7.9	6.2	7.6
Silica	SiO_2	6.7	3	1.5	1.3	10	45
Iron	Fe	0.6	0.5	0.5	NR	3	NR
Turbidity		30	3-4	15-20	67	Nil	70
TDS		NR	28	NR	152	78	250
Color		100	60-70	10	100	(Dark)	(Dark)

(*c*)

FIG. 2.1 (*c*) Selected rivers outside the United States.

It is not the province of this text to include orientation in geology, but a few basic facts about minerals can be helpful to understanding water chemistry. Of the elements of the earth's crust most readily soluble in water, calcium is the most prominent (Chapter 4), and the increase in hardness of the Colorado River in its journey to the sea (Table 2.2) is caused by the dissolution of calcareous (calcium-bearing) rock. The principal mineral contributing calcium to water is limestone ($CaCO_3$), an alkaline compound, and its source is the shells or skeletons of aquatic organisms (from tiny polyps to clams and larger shellfish) deposited in the bottom

Identification of Analyses Tabulated Below:

G. Tees, Durham — J. Dee, downstream

H. Tyne — K. Great Ouse

I. Dee, at headwaters in Wales — L. Severn, Gloucester

Constituent	As	G	H	I	J	K	L
Calcium	$CaCO_3$	74	77	8	78	357	136
Magnesium	"	30	33	1	21	25	45
Sodium	"	15	20	10	39	58	30
Total Electrolyte	$CaCO_3$	119	130	19	138	440	211
Bicarbonate	$CaCO_3$	77	80	3	62	242	71
Carbonate	"	0	0	0	0	0	0
Hydroxyl	"	0	0	0	0	0	0
Sulfate	"	24	32	2	38	144	70
Chloride	"	17	17	13	37	51	68
Nitrate	"	1	1	1	1	3	2
M Alk.	$CaCO_3$	77	80	3	62	242	71
P Alk.	"	0	0	0	0	0	0
Carbon Dioxide		3	3	5	3	Nil	3
pH		7.6	7.5	6.3	7.7	8.1	7.7
Silica	SiO_2	4	NR	1.5	8.0	NR	5
Iron	Fe	0.5	0.6	0.6	0.5	0.1	Nil
Turbidity							
TDS		NR	NR	NR	NR	NR	NR
Color		82	60	NR	NR	26	17
Conductivity		219	NR	NR	NR	665	440

(*c*)

FIG. 2.1 (*Cont.*)

of prehistoric seas and forced into a new shape by the pressure and heat of succeeding geological periods. The dissolution of limestone makes water alkaline.

Some areas were never submerged beneath prehistoric seas of recent geological eras and therefore have had no contact with limestone. Some portions of the immense Orinoco basin of Venezuela, a water-rich jungle, are in this category. Some rivers there contact chiefly siliceous rock, such as flint, quartz, and sandstone. The decaying vegetation produces humic acid, and in the absence of alkaline limestone, the acidity of the soils and rivers there creates an environment

Identification of Analyses Tabulated Below:

M. North Saskatchewan (Canada)
N. Saguenay (Canada)
O. Ottawa River (Canada)
P. Guadalquivir (S.W. Spain)
Q. Ebro (N.E. Spain)
R. Sava River (Yugoslavia)

Constituent	As	M	N	O	P	Q	R
Calcium	$CaCO_3$	107	12	20	248	350	119
Magnesium	"	54	4	8	179	110	39
Sodium	"	12	3	15	42	333	27
Potassium					NR	NR	
Total Electrolyte	$CaCO_3$	173	19	43	469	793	185
Bicarbonate	$CaCO_3$	110	10	18	76	180	150
Carbonate	"	10	0	0	0	0	0
Hydroxyl	"	0	0	0	0	0	0
Sulfate	"	48	1	18	198	290	30
Chloride	"	4	8	6	191	323	5
Nitrate	"	1	0	Nil	4	NR	ND
Fluoride				1			
M Alk.	$CaCO_3$	120	10	18	76	180	150
P Alk.	"	5	0	0	0	0	0
Carbon Dioxide		0	5	4	Tr	10	10-12
pH		8.4	6.8	7.1	8.0	7.6	7.4
Silica		3.3	6.8	4.7	20	4	3
Iron	Fe	Nil	0.4	0.3	NR	NR	0
Turbidity		5	0.8	8	NR	NR	NR
TDS		208	35	57	NR	NR	NR
Color		13	40	30	NR	NR	NR
Conductivity		–	44	72			385

FIG. 2.1 (*c*) (*Cont.*)

hostile to humans and the kinds of plants and animals they depend on for sustenance. (An analysis of the Caroni River illustrating this condition is given in Figure 2.1*c*.) Lakes suffering the effects of acid rain are usually in basins lacking limestone, often overlaid with soils high in humus; these lakes lack the alkaline reserve needed to neutralize the acidity of the rainwater runoff.

A variety of analyses of river waters from countries other than the United States are shown in Figure 2.1*c*. Rivers flowing through drainage basins having dense vegetation and substantial rainfall are generally rather highly colored the world over, as shown by columns A through H. In Venezuela and Brazil, the

Identification of Analyses Tabulated Below:

S	Tiber (Rome)	V	Nile (Cairo)
T	Po River (Casal Grasso)	W	Vaal (South Africa)
U	Po River (Moncalieri)	X	Orange (South Africa)

Constituent	As	S	T	U	V	W	X
Calcium	$CaCO_3$	275	92	159	68	104	68
Magnesium	"	130	52	66	44	40	26
Sodium	"	105	18	37	61	70	44
Total Electrolyte	$CaCO_3$	510	162	262	173	214	138
Bicarbonate	$CaCO_3$	285	120	182	112	102	66
Carbonate	"	0	0	0	24	0	0
Hydroxyl	"	0	0	0	0	0	0
Sulfate	"	80	30	58	10	67	42
Chloride	"	140	10	20	25	45	30
Nitrate	"	5	2	2	2	Nil	Nil
M Alk.	$CaCO_3$	285	120	182	136	102	66
P Alk.	"	0	0	0	12	0	0
Carbon Dioxide		15	15	18	0	Tr	8
pH		7.5	7.1	7.3	8.5	8.0	7.4
Silica		6	6.5	7.6	7.6	12	8
Iron	Fe	NR	0.1	NR	NR	NR	NR
Turbidity		NR	18	21	30	15	10
TDS		575	170	270	210	278	200
Color		NR	NR	NR	NR	2	Tr

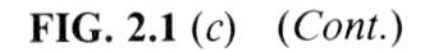

FIG. 2.1 (*c*) (*Cont.*)

mineralization of the highly colored Caroni and Amazon rivers is limited by the nature of the lithosphere, especially the absence of limestone, and the heavy rainfall. (The Amazon drains such an enormous, rain-drenched area that it discharges about 20% of the worldwide flow of freshwater to the oceans.) The influence of municipal and industrial use of river water in adding to its mineral con-

Identification of Analyses Tabulated Below:

Y. Seine River, above Paris

Z. Seine River, 100 km below Paris

Constituent	As	A	B	C	D	E	F
Calcium	$CaCO_3$	238	244				
Magnesium	"	12	60				
Sodium	"	20	43				
Potassium	"	3	9				
Ammonia	"	0	0				
Total Electrolyte	$CaCO_3$	273	356				
Bicarbonate	$CaCO_3$	192	154				
Carbonate	"	0	0				
Hydroxyl	"	0	0				
Sulfate	"	26	100				
Chloride	"	37	95				
Nitrate	"	18	7				
Fluoride	"	0	0				
M Alk.	$CaCO_3$	192	154				
P Alk.	"	0	0				
Carbon Dioxide	"	5	30				
pH		7.85	6.90				
Silica	SiO_2	5.4	2.6				
Iron	Fe	0.0	0.1				
Turbidity		–	–				
TDS		–	–				
Color		–	–				
Conductivity, μm		426	–				

FIG. 2.1 (*c*) (*Cont.*)

TABLE 2.2 Colorado River—1964–1965

Location	Total dissolved solids, mg/L	Total hardness mg/L as $CaCO_3$	pH
Hot Sulfur Springs, CO	92	56	7.4
Glenwood Springs, CO	292	159	7.6
Cameo, CO	380	181	7.7
Cisco, UT	530	276	7.7
Lees Ferry, AZ	609	300	7.6
Grand Canyon, AZ	655	307	7.8
Parker Dam, AZ-CA—3/15/65	753	360	7.8

Source: USGS Water Supply Paper, 1965.

tent is shown in columns I and J (the Dee River in southwest Great Britain) and columns P and Q (rivers of Spain seriously affected by a high degree of reuse and lack of dilution during the May-to-October drought).

EFFECTS OF RAINFALL

The sudden dilution of a river by heavy rainfall can be a disruptive factor in a water treatment plant. The location of a river water intake should be carefully chosen with this problem in mind. In the operation of a treatment plant, it is common practice to adjust the chemical dosages according to effluent water quality. However, there are many water supplies so variable that it is necessary to base changes in chemical treatment on raw water characteristics, rather than on finished water quality. This imposes a hardship on the treatment plant operators, requiring their constant attention to analysis and control.

Tides create another important influence on surface water quality in that they slow, or actually reverse, normal river flow. This is particularly pronounced during periods of low rainfall. The change in water quality between high and low tide sometimes justifies the installation of raw water supply reservoirs to receive water at low tide when the river flows unimpeded and quality is at its best. Plants so equipped stop pumping at high tide when saline bay waters move upstream into the upper channel. An example of the effect of tides and seasonal runoff is given in Table 2.3, showing the enormous variations in the Delaware River near Wilmington, Delaware.

OTHER SEASONAL CHANGES

Another characteristic of surface waters is seasonal temperature changes. This complicates treatment, particularly affecting the coagulation process in the winter. Low temperatures also create problems with air-binding of filters due to the increased solubility of gases and higher water viscosity. This binding causes pressure drop through the filter beds to increase, releasing gas and disrupting flow.

Another effect of temperature change occurs in water-cooled systems of industrial plants where heat exchange equipment is usually designed for the least favorable condition—the higher summer temperatures of surface waters. In winter, when the temperature is low, the flow must often be restricted to prevent overcooling. Lower water velocities may allow silting in heat transfer equipment, which can lead to corrosion and to pressure loss when higher cooling rates are needed.

Because rivers are warmer in the summer, designers take this into account in most water-dependent systems. But a complication arises in that many wastewaters contain heat from plant processes, and this added heat compounds the natural rise of the summer, sometimes producing an effluent warm enough to create an unhealthy condition for aquatic life. Pollution discharges not only add to the heat load of the river, but they usually also add to the oxygen demand and may have a pronounced influence on the oxygen content of the river water.

When pollutants are biodegradable, bacterial activity in the stream increases with pollution load, tending to reduce the dissolved oxygen level in the stream, (see Chapter 5), but there are offsetting factors. The principal one is the presence of algae in the stream; algae will produce oxygen by photosynthesis in daylight—

TABLE 2.3 Delaware River at Memorial Bridge near Wilmington, Delaware
Specific conductivity, uS/cm at 25°C October 1969–September 1970, showing daily variations and day with maximum and minimum reading

Date	Maximum	Minimum	Mean
October 16	6880	2660	4420
October 7	6220	1400	3540
November 2	8820	3500	5650
November 26	2460	420	1370
December 9	6940	1100	3610
December 27	240	100	145
January 23	5440	960	2580
January 17	2700	100	1470
February 27	980	180	493
February 24	560	100	235
March 21	3240	200	1450
March 27	1700	100	
April 23	1240	220	289
April 5	120	100	105
May 24	4520	920	
May 12	870	170	613
June 22	3900	200	1110
June 1	1740	180	905
July 20	6380	860	2930
July 7	2500	100	1380
August 19	8920	1900	5030
August 2	4060	560	2060
September 20	8300	3500	5630
September 1	5940	1740	3820

Source: USGS Water Supply Paper 2151.

often causing supersaturation on bright, sunny days—with a falloff at night as the process is restricted. This diurnal cycle affects not only dissolved oxygen, but also carbon dioxide, and thus pH. This can have a strong influence on the coagulation of a water supply in municipal and industrial water treatment plants.

It is unusual to find high levels of dissolved iron in surface waters, except where the water supply is highly colored and has a relatively low pH. In this case, the iron usually is complexed by the organic matter causing the color. An exception to this would be where there is acid mine drainage into the river; in this case, the iron is introduced into the water in the reduced (ferrous) condition and a lack of dissolved oxygen prevents its oxidation to the less soluble ferric state.

When a river is dammed, the water quality may be considerably different from that of the flowing stream. The impoundment behind the dam then takes on the characteristics of a lake. In deep impoundments, it is common to find stratification, with oxygen depletion in the bottom, stagnant zone, and development of significant levels of iron and manganese in the bottom water, even though the surface remains free of these heavy metals. Concentration gradients are sometimes found in impounded lakes, indicated by an increase in conductivity with depth. For example, Lake Mead, the impoundment of the Colorado River above Hoover Dam on the Nevada-Arizona border, has a conductivity of 900 μS at the surface; at 50 ft of depth, this figure begins to increase, reaching 1150 μS at 300 ft where it levels off; at the bottom—about 460 ft—the conductivity abruptly increases to almost 1500 μS.

From flowing streams and rivers, water may diffuse into underground aquifers when the surrounding water table is low; or, water may feed into the river from these aquifers when the water table is high. This, too, influences chemical composition, particularly iron and manganese in certain streams.

Identification of Analyses Tabulated Below:

A. Greenville, SC
B. New York City (Catskill)
C. Boston
D. Frederick, MD
E. Little Rock, AR
F. Colorado Springs, CO

Constituent	As	A	B	C	D	E	F
Calcium	$CaCO_3$	3	12	10	2	6	15
Magnesium	"	2	7	3	1	4	4
Sodium	"	4	4	1	6	1	11
Total Electrolyte	$CaCO_3$	9	23	14	9	11	30
Bicarbonate	$CaCO_3$	6	8	5	4	8	15
Carbonate	"	0	0	0	0	0	0
Hydroxyl	"	0	0	0	0	0	0
Sulfate	"	2	11	6	3	0	4
Chloride	"	1	4	3	2	3	4
Nitrate	"	0	0	0	0	0	1
Fluoride	"	0	0	0	0	0	6
M Alk.	$CaCO_3$	6	8	5	4	8	15
P Alk.	"	0	0	0	0	0	0
Carbon Dioxide	"	6	2	5	8	2	2
pH		6.2	6.9	6.3	6.1	7.1	7.4
Silica	SiO_2	7.8	2.5	0.9	3.0	3.5	8.4
Iron	Fe	0	0	0.1	0	NR	0.4
Manganese	Mn	0	0	0	0	NR	0.0
Turbidity		NR	2	1	1	10	NR
TDS		17	34	33	12	25	33
Color		6	1	7	3	35	NR

pH in pH units; color in APHA units; all others in mg/l

FIG. 2.2 Impoundments in watersheds supplied principally by precipitation.

In the impoundment of major rivers, such as the Columbia, the Colorado, and those in the Tennessee Valley system, the mineral content of the lakes behind the dams is similar to that in the river, as would be expected. On the other hand, natural or artificial impoundment of streams in smaller watershed areas of abundant rainfall produces water supplies of very low mineral content. This accounts for the excellent quality of such municipal waters as Greenville, South Carolina, New York City, and Boston, as shown by Figure 2.2.

WHERE RIVER MEETS OCEAN

The quality of water in estuaries where rivers meet the sea is unpredictable, depending on river flow, tidal conditions, the size of the basin or bay, and the presence or absence of land formations that restrict the flow to the sea. In large basins, such as Albemarle Sound on the east coast, the water, although saline, is of relatively uniform composition because of mixing of the shallow water by wind; but in smaller bays, the quality will change with the tides and the flow of the river. It is one of the miracles of nature that aquatic life adapts to these changes and flourishes in such tidal areas.

Many of the major rivers of the world finally reach the ocean through deltas, which have some of the characteristics of an estuary. In many instances, the flow of these rivers is so great that the dilution of the ocean can be measured for miles out to sea. Some plants have used the brackish waters of tidal basins as cooling tower makeup, since the water is low enough in dissolved solids to be concentrated by evaporation without severe scaling problems, just as fresh waters are.

TABLE 2.4 Composition of Ocean Waters
(U.S. Naval Oceanographic Office, 1966)

Location	Temperature, °C	Salinity, parts per thousand
North Atlantic		
Central water	4–17	35.1–36.2
Bottom water	1–3	34.8–34.9
South Atlantic		
Central water	5–16	34.3–35.6
Bottom water	0–2	34.5–34.9
Mediterranean	6–10	35.3–36.4
Red Sea	9	35.5
Indian Ocean, central water	6–15	34.5–35.4
North Pacific		
Eastern	10–16	34.0–34.6
Western	7–16	34.1–34.6
Subartic	2–10	33.5–34.4
South Pacific		
Eastern	9–16	34.3–35.1
Western	7–16	34.5–35.5
Subantarctic	3–7	34.1–34.6

The use of brackish water for this purpose permits installation of much smaller pipelines than would be needed if once-through cooling were practiced. One example is a chemical plant on the island of Trinidad, which is able to use water from the bay side of the island as cooling tower makeup. This water is diluted enough by the immense flow of the Orinoco River so that its salinity is lower than typical ocean water, permitting it to concentrate by evaporation without causing scale problems.

Out at sea, where surface waters have become part of the circulation system, the composition of the ocean water is remarkably uniform, as shown in Table 2.4. There are, of course, local changes in salinity, as mentioned earlier, caused by upwelling of subsurface waters into the ocean, the flow of mighty rivers into the sea, or the melting of glaciers and the polar ice caps. Even though it isn't usable by land animals, seawater is a valuable source of water for industry and is widely used for cooling.

LAKES AS RESERVOIRS

Lakes are a major source of fresh water. They are of particular significance in North America, where Canada and the United States share, in the Great Lakes Basin, what is usually considered to be the largest freshwater supply in the world. In the Soviet Union, a single body of water, Lake Baikal in Siberia, contains about the same volume of fresh water as the entire Great Lakes system, 5500 mi^3 (2.3×10^4 km^3). Lake Baikal is about 5000 ft (1525 m) deep, with a surface area of about 11,000 mi^2 (2.8×10^6 ha), compared with 95,000 mi^2 for the Great Lakes. Together, the Great Lakes and Lake Baikal contain 40% of the world's presently available fresh water. In addition to sharing the Great Lakes with the United States, Canada is dotted with countless smaller lakes, carved by ice-age glaciers, holding another 15% of the presently available fresh water of the world.

Of the Great Lakes, Lake Superior is the largest, with a total area of about 32,000 mi^2 and a maximum depth of approximately 1300 ft. It is also "the different lake," with a significantly lower dissolved solids content than the other Great Lakes due to differences in geologic formations of the lake bed and to temperature (Figure 2.3). The analyses of Lake Michigan and Lake Erie are shown in Figure 2.3. Even though these two lakes are different in configuration, area, and depth, their chemical compositions are similar. This concentration of mineral matter is maintained fairly uniformly until the water of the Great Lakes is finally discharged from Lake Ontario into the St. Lawrence River.

The composition of lake water changes seasonally and sometimes even daily with weather conditions. Although the major dissolved mineral constituents may not be greatly affected by seasons and weather, such factors as dissolved oxygen, temperature, suspended solids, turbidity, and carbon dioxide will change because of biological activity. Another factor that creates change is the seasonal turnover that occurs in most lakes in the United States during the spring and fall.

SEASONAL TURNOVER

An excellent classification of lakes is given in *The Microscopy of Drinking Water* by Whipple and Fair. Using as their basis for categorizing lakes the nature of the

Identification of Analyses Tabulated Below:

A. Lake Superior at Duluth
B. Lake Michigan at Chicago
C. Lake Erie at Erie, PA
D. Seneca Lake, Geneva, New York
E. Lake Coeur d'Alene, Idaho
F. Clear Lake, West Palm Beach, FL

Constituent	As	A	B	C	D	E	F
Calcium	$CaCO_3$	35	80	90	100	18	68
Magnesium	"	9	41	33	41	10	11
Sodium	"	6	19	20	135	2	23
Total Electrolyte	$CaCO_3$	50	140	143	276	30	102
Bicarbonate	$CaCO_3$	42	113	91	97	19	50
Carbonate	"	0	0	0	0	0	0
Hydroxyl	"	0	0	0	0	0	0
Sulfate	"	1	18	25	35	10	12
Chloride	"	5	9	27	142	1	40
Nitrate	"	2	NR	Nil	2	Nil	NR
Fluoride	"	TR	TR	TR	TR	TR	1
M Alk.	$CaCO_3$	42	113	91	97	19	50
P Alk.	"	0	TR	0	0	0	0
Carbon Dioxide	"	4	Nil	5	5	4	2
pH		7.4	8.2	7.5	7.6	7.0	7.7
Silica	SiO_2	3.3	2.3	1.2	1.9	10	2.2
Iron	Fe	0.2	0.1	TR	TR	NR	0.6
Turbidity		1	12	Nil	NR	7	18
TDS		54	171	172	323	58	173
Color		1	3	2	1	NR	25

pH in pH units; color in APHA units; all others in mg/l

FIG. 2.3 Selected lake analyses.

thermocline (temperature barrier) at different times of the year, they have defined three types of lakes: polar, temperate, and tropical, with those in the United States being chiefly in the second category.

In temperate lakes over 200 ft (60 m) deep (first order), the bottom water is at the temperature of maximum density, 39.2°F (4°C), year round. After the ice cover thaws in spring, the surface water gradually warms from 32 to 39.2°F (0 to 4°C), within which temperature range its density increases; vertical circulation commences as the surface reaches the bottom water temperature. After this *spring turnover,* the surface rises above 39.2°F and its density decreases as it continues to warm through the summer. The temperature gradient in the summer is illustrated by Figure 2.4.

The fall and winter cooling again reduces the surface temperature to its maximum density, and the *fall turnover* occurs.

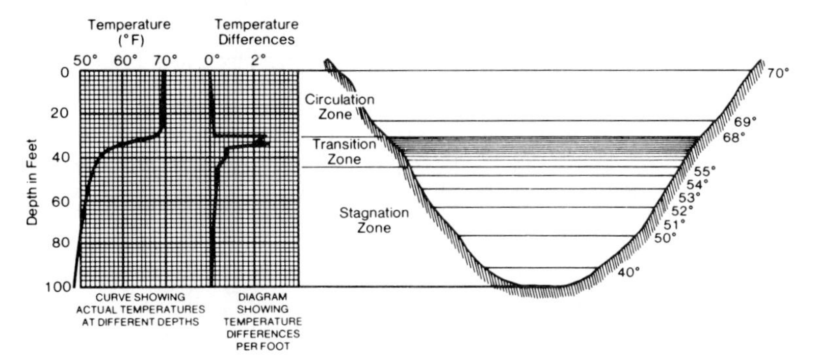

FIG. 2.4 In summer, lake water temperature drops with increasing depth. *(Adapted from Whipple and Fair, 1933.)*

During the period of summer stagnation, vertical circulation is induced by wind, but this force is not strong enough to cause mixing of the bottom water, the hypolimnion. The circulation pattern induced by wind is illustrated in Figure 2.5.

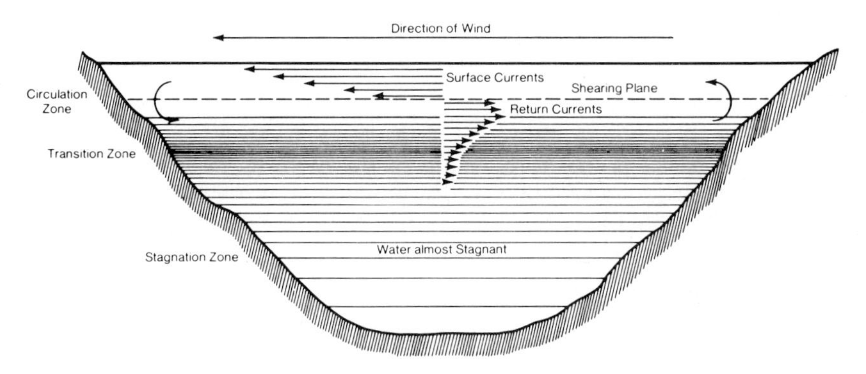

FIG. 2.5 Wind-induced lake water circulation. *(Adapted from Whipple and Fair, 1933.)*

A similar turnover occurs in lakes of the second order, in the approximate range of 25 to 200 ft (7.5 to 60 m) deep. In the second-order lake, the bottom temperature changes measurably but is never far from that of maximum density. Figure 2.6 illustrates surface and bottom temperatures of a 60-ft-deep lake, with turnover occurring at the points where the temperature curves cross.

In lakes of less than about 25 ft deep (third order), vertical circulation is induced almost exclusively by wind action, rather than density differences, so that there is little difference between surface and bottom temperatures.

An understanding of these characteristics of lakes is necessary to properly locate water intakes and discharges. It aids in anticipating the changes in water treatment needed to meet changes in composition caused by turnover and wind-induced circulation.

Another characteristic of lake waters, usually seasonal and related to biological

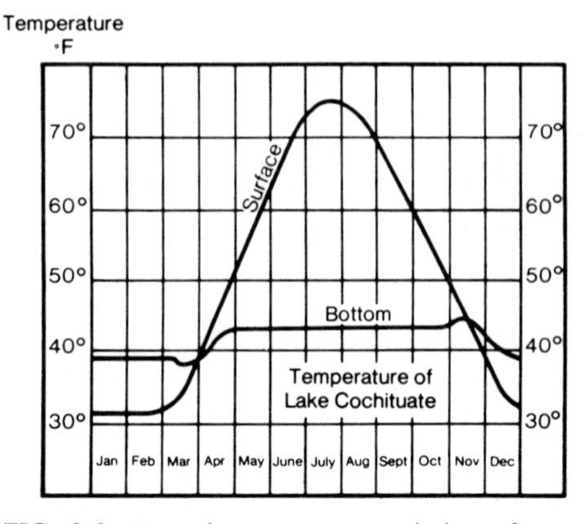

FIG. 2.6 Annual temperature variation of surface and bottom water in a 60-ft-deep lake. *(Adapted from Whipple and Fair, 1933.)*

activity, is taste and odor. Except where these may be introduced by wastewater discharges, they are usually attributed to organic matter, such as essential oils produced by the algae growth.

SUBSURFACE WATER

Underground reservoirs constitute a major source of fresh water. In terms of storage capacity, underground aquifers worldwide contain over 90% of the total fresh water available for man's use (Figure 2.7). Much of this is too deep to be exploited economically. In the United States, where the Great Lakes contain such a large

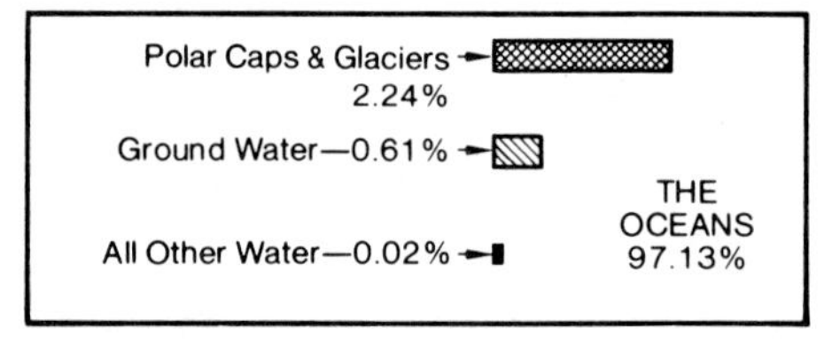

FIG. 2.7 More than 97% of the earth's water is in the oceans. All the rivers of the world contain only 0.02% of the world's total water. (There is some dispute among experts on the exact quantity of groundwater.) *(From "Water of the World," U.S. Dept. of the Interior/Geological Survey.)*

volume of fresh water, the proportion of surface water to groundwater is much higher. But even here, it is estimated that at least 50% of total available freshwater storage is in underground aquifers (Figure 2.8).

Over 60 bgd (2600 m^3/s) of water is withdrawn from wells for irrigation of the U.S. mainland (Figure 2.9). This well water usage is almost twice the total water

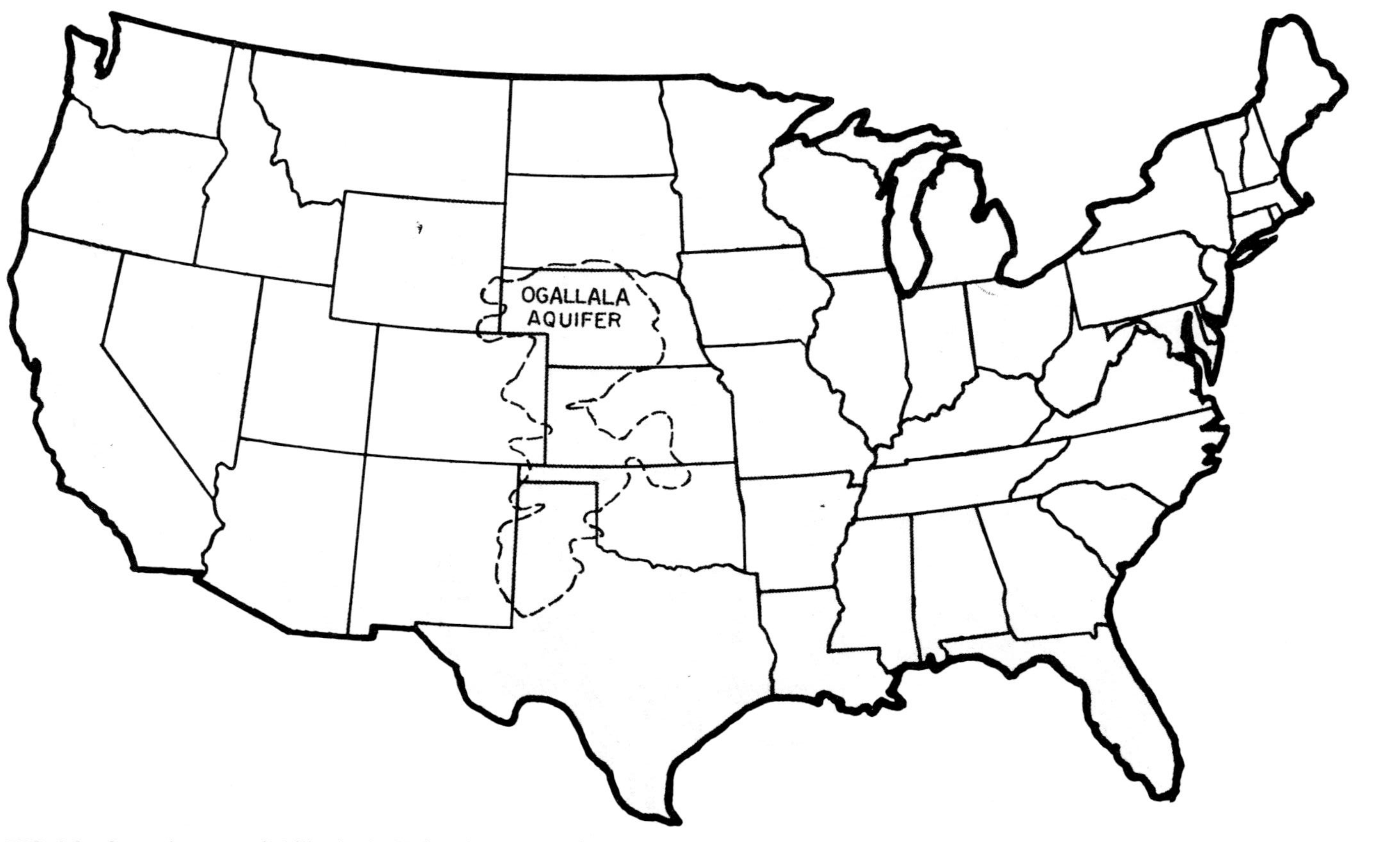

FIG. 2.8 Groundwater availability in the United States. Supplies are relatively scarce in the western third of the country, but plentiful in the east and middle west. The Ogallala aquifer, believed to be the world's largest (shown in outline) has a capacity of 650×10^{12} gal ($2.47 \times 10^{12} m^3$) of water. A typical recharge rate is about 10% per year, and west of the Mississippi, this is below current withdrawal rates. *(From Chemical Week by special permission. Copyright 1977 by McGraw-Hill, Inc.)*

used by municipalities throughout the United States. Over 80% of the municipalities in the United States depend on well water, although less than 30% of the total volume of water treated for municipal use is from this source. The largest well field in the United States supplies 72 mgd (3.1 m^3/s) to the city of Tacoma. Industry draws approximately 12 bdg (520 m^3/s) from wells.

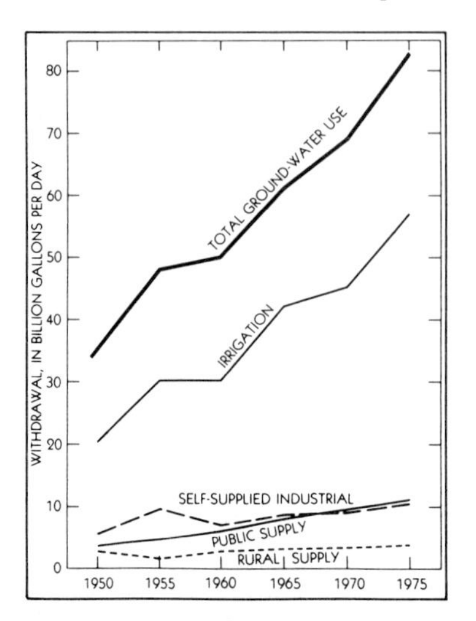

FIG. 2.9 The rate of withdrawal of groundwater for irrigation in the United States has increased rapidly since 1950. *(With permission from The Johnson Drillers Journal, July-August 1979.)*

WELL WATERS CONSTANT

Underground water usually moves very slowly. Its flow is measured in feet per year; compare this with surface streams, where velocities are in the feet-per-second range. Because of this slow movement, the composition of any one well is usually quite constant. Although shallow wells may vary seasonally in temperature, most wells are also constant in temperature, usually in the range of 50 to 60°F (10 to 16°C). Since the water has passed through miles of porous rock formation, it is invariably clear if the well has been properly developed to keep fine sand from entering the casing.

Since the composition is related to the chemistry of the geological formations through which the water has passed, waters from wells drilled into different strata have different characteristics. Some aquifers are so large that they may cover several states in total area, and wells drilled into that particular aquifer produce water of similar composition. For that reason, with good geological information it is possible to make some generalizations about the composition of well waters in different parts of the United States. Selected analyses are shown in Figure 2.10.

As water filters through the ground, soil organisms consume dissolved oxygen and produce carbon dioxide, one of the principal corrosive agents in dissolving the minerals from geological structures. It is common to find iron and manganese in waters that are devoid of oxygen if they have been in contact with iron-bearing minerals. Shallow wells containing oxygen are generally free of iron.

Extensive records on the composition of well waters are available from the U.S. Geological Survey and also from most state agencies regulating the use of water as a natural resource. Most well drillers also have records of subsurface supplies, including both chemical composition and water yield.

The water produced with oil, called connate water or oil-field brine, is unique and creates unusual problems in handling and treatment for reuse or disposal. It is usually more concentrated than seawater, often exceeding 100,000 mg/L in total salinity; this fact and the reduction in temperature and pressure as the brine travels upward from great depths cause difficult problems of scale and corrosion control. Several analyses are shown by Table 2.5.

In many areas of the country, particularly the west and southwest, underground water is being mined. This mining has resulted in net loss of water, which has steadily reduced the level of the water table and has contributed to land subsidence. In some areas, such as Long Island, heavy withdrawal has resulted in intrusion of seawater inland. Here, underground injection of highly treated wastewater is being practiced to provide an intermediate barrier to hold back the saltwater from the freshwater wells. Injection water of this kind must be free of pollution, since the pollution of an underground aquifer can be very serious, so much time being required to displace the pollutant once it has gotten into the aquifer.

Augmentation of a groundwater source by artificial recharge has also been practiced where a nearby source of surface water has been available for this purpose. The aquifer becomes, in effect, a storage reservoir. As an example, an industrial plant in southern New Jersey had for many years relied on wells for plant

Identification of Analyses Tabulated Below:

A. Manhattan, Kansas — D. Phoenix, Arizona

B. Amarillo, Texas, Well No. 1 — E. Norman, Oklahoma

C. Baytown, Texas, Well No. 1 — F. Richland, Washington

Constituent	As	A	B	C	D	E	F
Calcium	$CaCO_3$	282	145	25	192	(5)	125
Magnesium	"	86	255	14	230	(3)	65
Sodium	"	65	84	(593)	319	462	38
Total Electrolyte	$CaCO_3$	433	484	632	741	470	228
Bicarbonate	$CaCO_3$	352	366	384	220	295	169
Carbonate	"	0	0	0	0	57	0
Hydroxyl	"	0	0	0	0	0	0
Sulfate	"	54	86	(Nil)	128	98	45
Chloride	"	25	24	245	312	18	13
Nitrate	"	1	3	0	(79)	1	1
Fluoride	"	1	(5)	3	2	1	Nil
M Alk.	$CaCO_3$	352	366	384	220	352	169
P Alk.	"	0	0	0	0	(29)	0
Carbon Dioxide	"	35	20	20	5	0	8
pH		7.3	7.6	7.6	7.9	(9.1)	7.7
Silica	SiO_2	31	(61)	26	27	9.7	(55)
Iron	Fe	(6.9)	Nil	0.3	Nil	Nil	Nil
Manganese	Mn	(1.4)	NR	0	NR	NR	0
Turbidity							
TDS		488	530	733	887	550	307
Color		2	NR	10	3	5	5

pH in pH units; color in APHA units; all others in mg/l

(*a*)

FIG. 2.10 (*a*) Selected well waters of the United States (unusual values are circled).

Identification of Analyses Tabulated Below:

G. Camden, NJ, Well No. 1 J. Ft. Lauderdale, FL

H. Camden, NJ, Well No. 7 K. Bastrop, LA

I. Camden, NJ, Well No. 9N L. Dallas, TX, Well No. 39

Constituent	As	G	H	I	J	K	L
Calcium	$CaCO_3$	60	18	42	230	4	15
Magnesium	"	38	19	41	8	2	9
Sodium	"	42	28	31	22	616	838
Total Electrolyte	$CaCO_3$	140	65	114	260	622	862
Bicarbonate	$CaCO_3$	50	1	91	235	356	452
Carbonate	"	0	0	0	0	0	TR
Hydroxyl	"	0	0	0	0	0	0
Sulfate	"	38	42	12	TR	2	270
Chloride	"	50	18	10	25	262	134
Nitrate	"	1	4	1	NR	1	1
Fluoride	"	1	0	0	NR	1	5
M Alk.	$CaCO_3$	50	1	91	235	356	452
P Alk.	"	0	0	0	0	0	TR
Carbon Dioxide	"	80	80	30	10	5	Nil
pH		6.0	4.6	6.8	7.7	8.0	8.2
Silica	SiO_2	13	6.1	3.9	22	11	22
Iron	Fe	1.2	TR	21	1.6	TR	TR
Manganese	Mn	0.3	TR	5.6	NR	NR	0
Turbidity							
TDS		181	103	118	314	697	1040
Color		3	2	3	59	35	0

pH in pH units; color in APHA units; all others in mg/l

(*b*)

FIG. 2.10 (*b*) Selected well waters of the United States (unusual values are circled).

use. A long period of drought in the 1960s so drastically reduced the water table that the plant faced a water supply crisis. An injection well was constructed to recharge the aquifer from the Delaware River at times of optimum quality, permitting the plant to continue using existing pumping facilities. The city of Los Angeles reclaims storm water and segregated wastewater of selected quality by gravity recharge of underground reservoir from water collected by the Los Angeles River. The natural riverbed has been paved with concrete to prevent haphazard loss of collected water to the ground, and a collapsible dam has been constructed at the river's end. Collected waters are spread over the old, natural river delta for percolation through the original gravel riverbed into the underground reservoirs. During a heavy storm, the dam is momentarily and deliberately collapsed to flush out collected solids, then reinflated. This arrangement provides for recovery of most of the storm water in this arid metropolis.

Identification of Analyses Tabulated Below:

A. Vienna (Austria) (1)
B. Rijeka (Yugoslovia) (2)
C. Canary Islands (3)
D. Tarragona (Spain)
E. Ontario (Eastern Canada) (4)
F. Saskatchewan (Central Canada) (4)

Constituent	As	A	B	C	D	E	F
Calcium	$CaCO_3$	160	136	50	557	180	352
Magnesium	"	120	27	135	672	80	232
Sodium	"	6	39	760	1891	22	750
Potassium	"	NR	NR	NR	NR	NR	23
Total Electrolyte	$CaCO_3$	286	202	445	3120	282	1357
Bicarbonate	$CaCO_3$	220	180	144	181	210	380
Carbonate	"	0	0	6	0	0	0
Hydroxyl	"	0	0	0	0	0	0
Sulfate	"	40	10	48	510	41	307
Chloride	"	26	12	251	2429	21	670
Nitrate	"	NR	NR	NR	NR	10	NR
M Alk.	$CaCO_3$	220	180	150	181	210	380
P Alk.	"	0	0	3	0	0	0
Carbon Dioxide		10	8-10	0	Nil	15	40
pH		7.7	7.5	8.4	7.9	7.5	73
Silica		5-10	1.5	30	13	7.3	14
Iron	Fe	NR	Nil	0.8	0.1	Nil	2.6
Turbidity		Nil	Nil	NR	NR	0	2
TDS		NR	NR	NR	NR	300	1650
Color		NR	NR	NR	NR	5	5

Notes: (1) Lateral wells near the Danube River
(2) Low magnesium to calcium ratio
(3) Low calcium to magnesium ratio typical of diluted seawater
(4) A significant difference between eastern and western wells.

(*c*)

FIG. 2.10 (*c*) Selected well waters of other countries.

TABLE 2.5 Examples of Connate Water

Constituent	A	B	C	D
Calcium, mg/L Ca	8,300	2,630	1,500	630
Magnesium, mg/L Mg	260	690	500	40
Sodium, mg/L Na	56,250	16,800	9,150	5,640
Bicarbonate, mg/L HCO_3	50	315	1,000	500
Sulfate, mg/L SO_4	180	2,880	2,000	120
Chloride, mg/L Cl	98,300	30,500	17,800	8,350
TDS, mg/L	166,652	54,072	32,329	15,417

SOURCES AND PRACTICES IN OTHER COUNTRIES

Just as in the United States, water resources in other countries are characterized in quantity and quality by rainfall, the nature of the lithosphere, residence time in contact with soluble minerals, and social and industrial influences.

Often, the well waters of Mexico are high in silica content, like well waters of the southwestern United States. The surface waters of northern Spain, where annual rainfall is approximately 60 in, are like waters of the Pacific northwest, while the arid sections of that country produce waters of much higher salinity, as in the southwestern United States. Almost 75% of Spain, the central Meseta plateau, receives less than 18 in of annual rainfall; the Guadalquivir River, which originates in the plateau, is as still as a mill pond for 6 months of drought, while evaporation, groundwater flow, and wastewater discharges raise its salinity.

Well waters of central Europe are like those of the U.S. midwest and are as variable in quality based on location, the nature of the geological formations the water has passed through, and depth. An analysis of an industrial well outside Vienna is shown in Figure 2.10*c*. A well water analysis from a refinery on the Adriatic coast of Yugoslovia is unusual in its low sulfate and chloride level, a low ratio of magnesium to calcium, and a low silica level. The Ilova River in the same area also has a low concentration of sulfate and chloride relative to total dissolved solids. It is obvious from these and other analyses in this chapter that prediction of a water analysis is not possible and that, throughout the world, each source is unique.

Practices of storing water resources vary from place to place, much dependent on land values. In the Netherlands, where recovery of land from the sea is a slow and costly process, advantage has been taken even of sand dunes as storage reservoirs for water. A water source of growing importance is sewage, both domestic and industrial. Reuse of municipal sewage by industry is no longer a rarity; each instance presents unique problems, but the need for water in arid places has given economic incentive to practical solutions. In the United States, reuse of domestic sewage as barrier water, as described earlier, is the only example where a small flow of treated sewage may eventually return to potable wells. But in South Africa, where the total amount of available water is only 20 bgd—almost equally divided between irrigation and all other uses—research has been conducted on direct return of a portion of highly treated sewage to municipal water plant intakes. The city of Windhoek in Namibia has practiced such recycle for over a decade; this may become a growing practice as South Africa prepares to offset a 5% deficit in natural water supplies facing it in the year 2000.

THREATS TO WATER SOURCES: THE EFFECTS OF ACIDIC RAINFALL

The oxides of sulfur and nitrogen discharged from utility stations, ore smelters, and internal combustion engines—in diesel locomotives, automobiles, and trucks—react with water in the atmosphere to form sulfuric and nitric acids. These acidify the rainfall downwind from industrial populous areas. The pH profile (see Chapter 4 for an explanation of pH and acidity) of rainfall over the United States is shown in Figure 2.11. A similar pattern prevails over Europe, where the acidity has affected most prominently the Scandanavian countries.

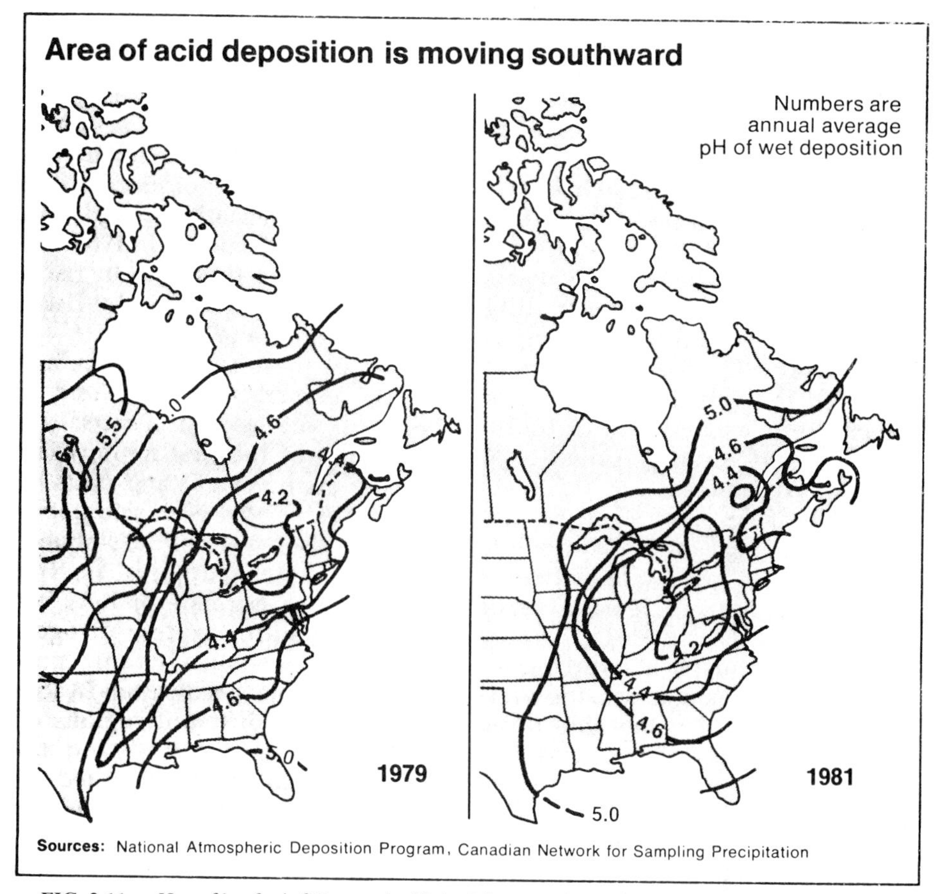

FIG. 2.11 pH profile of rainfall over the United States. *(Courtesy of Chem. Eng. News June 20, 1983, p. 27.)*

If the lakes and reservoirs receiving this rainfall are already low in natural alkalinity, as results from the absence of limestone in the area, they may be swamped by the acid rainfall, resulting in an acidic lake water. This condition can lead to a sterile aquatic environment unless the acidity is neutralized by application of lime, soda ash, or other alkalies. In general, the lakes and reservoirs affected by acid rain are contained in basins having a granite bedrock. These waters have little alkalinity, as shown by Figure 2.2, and are themselves quite corrosive even without the added input of acidic rain. Water supplies having such low buffer capacity will always be vulnerable to the influence of human activity, and as water sources they are invariably temperamental. In the United States, the major contributor (45%) to the estimated production of 25 million tons per year of nitrogen oxides is transportation; the major contributor (65%) of the estimated 29 million tons per year of sulfur oxides is the electric utility industry. These inputs suggest that lakes having a historical background alkalinity of less than about 20 ppm are vulnerable to the ravages of acid rain. Preliminary findings in New York state

TABLE 2.6 Annual Water Use by Types of Industry
(1977 Census of Manufactures estimated for 1978)

Industry	Gross*	Water usage, bgd		% Change†
		Intake	Discharge	
Primary metal manufacture	17.7	9.26	8.55	−34
Chemicals and allied products	34.1	11.8	10.7	None
Paper and allied products	28.4	5.36	4.82	−24
Petroleum and coal products	22.4	3.20	2.63	−19
Food	3.79	2.02	1.77	−10
Transportation equipment	6.65	0.64	0.60	None
Machinery	1.08	0.45	0.43	−43
Textile mill products	1.05	0.44	0.40	None
All industries	121.5	35.5	31.9	−17

* The difference between gross usage and intake represents the success in conserving water by in-plant reuse. The National Resources Council has estimated that by the year 2000, industry will have an intake 62% below that of 1975 while gross usage will increase by about 150%.

† Change from discharge data of 1973.

show that liming such lakes may effectively protect their ecology. However, even though the water sources may someday be protected in this fashion, the acidity of the rainfall itself will need correction for protection of soil, trees and plants, crops, and concrete and steel structures. It is one of our major pollution problems.

Table 2.1 lists the distribution of the water resources of the United States and the rates of withdrawal and consumption. That portion of this fresh water used by industry, not including utility cooling water, is shown in Table 2.6. Most of this water is drawn from the Great Lakes and the river systems shown in Figure 2.12.

THREATS TO WATER SOURCES: THE EFFECTS OF LANDFILL LEACHATE ON GROUNDWATER QUALITY

The random disposal of both domestic (municipal) and industrial wastes as solids and packaged liquids on dumps or landfill areas has proceeded uncontrolled and unchallenged for many years. In industrialized countries, the enormity of the risk this has created for goundwater contamination has only recently come to light as a major environmental issue.

In the United States, the Environmental Protection Agency is identifying hundreds of major sites that must be cleared, decontaminated, and restored to a nonhazardous status. Many of these sites owe their hazardous nature to the presence of toxic chemicals, usually leaking from damaged or partially drained shipping drums. Obviously, the great varieties of drummed chemicals means that each site has its own peculiarities. Water-soluble materials are usually carried into the soil by storm water. In some cases, soil bacteria may digest organic chemicals with a high BOD:COD ratio. In the aerobic zone, the by-product is mostly CO_2; if the leachate reaches the anaerobic zone, the by-products are ammonia, CO_2, methane, and residual nondigestible organic materials.

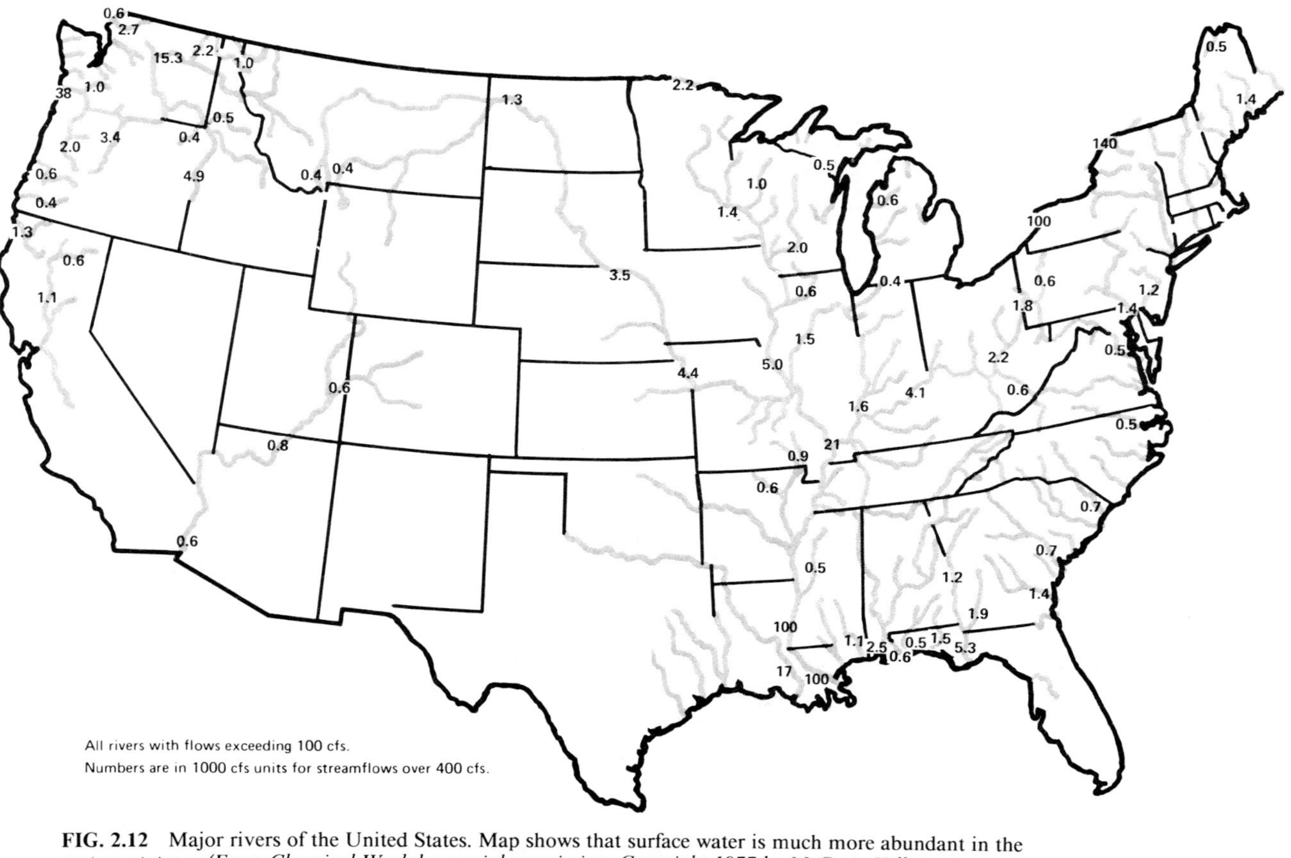

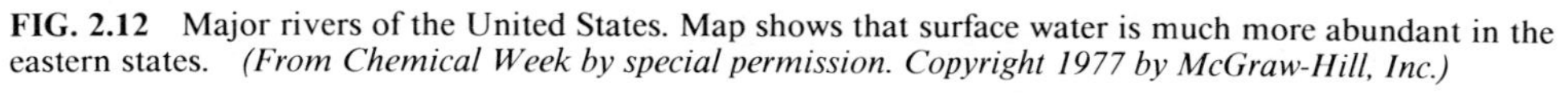
FIG. 2.12 Major rivers of the United States. Map shows that surface water is much more abundant in the eastern states. *(From Chemical Week by special permission. Copyright 1977 by McGraw-Hill, Inc.)*

NALCO

CONSULTING SERVICES

WATER TREATMENT PROCESS DATA

File: LEACHATE FROM AQUIFER 80-100 FT BELOW MUNICIPAL TRASH DISPOSAL SITE (EAST COAST - UNITED STATES)   Date: 10/1/86

Identification of Analyses Tabulated Below:

A. Model from several analyses   D. ______
B. ______   E. ______
C. ______   F. ______

Constituent	As	A	B	C	D	E	F
Calcium	$CaCO_3$	480					
Magnesium	"	170					
Sodium	"	450					
Potassium	"	150					
Ammonium	"	250					
Total Cations	$CaCO_3$	1500					
Bicarbonate	$CaCO_3$	1100					
Carbonate	"	0					
Hydroxyl	"	0					
Sulfate	"	40					
Chloride	"	350					
Nitrate & Nitrite	"	5					
Fluoride	"	5					
Phosphate	"	NIL					
Total Anions	$CaCO_3$						
M. Alk.	$CaCO_3$	1100					
P. Alk.	"	0					
Carbon Dioxide (calc.)	"	275					
pH		6.7					
Silica		12					
Iron (total)	Fe	48					
Iron (soluble)		1					
BOD	O_2	19					
TOC	C	82					
Dissolved O_2		0					
Extractibles	O & G	63					
Redox potential		−150					
Odor	Hydrocarbon & sulfide						
Turbidity		48					
TDS		1560					
Color	APHA	75					
Suspended solids	—	122					

Chemical	¢/lb.	Per One Thousand Gallons Outlet					
		Lbs.	Lbs.	Lbs.	Lbs.	Lbs.	Lbs.
		Lbs.	Lbs.	Lbs.	Lbs.	Lbs.	Lbs.
		Lbs.	Lbs.	Lbs.	Lbs.	Lbs.	Lbs.
		Lbs.	Lbs.	Lbs.	Lbs.	Lbs.	Lbs.
Total Cost Per 1000 Gal.		¢	¢	¢	¢	¢	¢

FIG. 2.13 Analysis of a landfill leachate used as an industrial water source.

In many cases, the soil adsorbs some toxic substances like PCBs and DDT until saturated, and then these materials continue downward until they reach and contaminate the water table. The soil must then be decontaminated.

Figure 2.13 shows the analysis of leachate from a test well sample taken below a municipal landfill; its character is essentially that of an anaerobically digested sewage. In this particular case, the leachate has gathered in an aquifer directly above the municipality's drinking water source, separated by a clay layer. To eliminate this potential threat to the potable supply, the municipality must withdraw the leachate and treat it for use as industrial water and cycle it to extinction in an evaporative cooling tower and sidestream demineralizing system.

It is always difficult to plan such a treatment system. Special procedures are required for sampling, and samples from different cores show different analyses. Once pumping has commenced, it may be that the quality will slowly improve, but that depends on the volume and shape of the leachate reservoir, rainfall, and the shape of the cone of depression at the withdrawal well.

SUGGESTED READING

AWWA Research Foundation: *Water Reuse Highlights—A Summary of Wastewater Reclamation and Reuse Information,* Denver, Colo., January 1978.

Ember, Lois R.: "Acid Pollutants: Hitchhikers Ride the Wind," *Chem. Eng. News,* September 14, 1981.

Espena, J. M.: "The Case for Artificial Recharge," *The Johnson Drillers J.,* Jan–Feb 1980.

Freeze, R. Alan, and Cherry, John A.: *Groundwater,* Prentice-Hall, Englewood Cliffs, N.J., June 1979.

Murray, C. R., and Reeves, E. B.: "Groundwater Withdrawal," *USGS Circular 765,* 1977.

Nordell, Eskel: *Water Treatment for Industrial and Other Uses,* Reinhold, New York, 1961.

Penman, H. L.: "The Water Cycle," *Sci. Am.,* September 1970, p. 149.

Rogers, Peter: "The Future of Water," *The Atlantic,* July 1983, p. 80.

Rudd, R. T.: "Water Pollution Control in South Africa," *J. Water Pollut. Control Fed.,* **51.3** 453 (1979).

Todd, David K., ed.: *The Water Encyclopedia,* Water Information Center, Port Washington, N.Y., 1970.

U.S. Department of Commerce: "Water Use in Manufacturing," *1972 Census of Manufactures,* MC 72(SR)4.

U.S. Geological Survey: "The Industrial Utility of Public Water Supplies in the United States," Papers 1299 and 1300, 1952.

Whipple, George C.: *The Microscopy of Drinking Water,* 4th ed. (Fair, G. M., and Whipple, M. C., eds.), Wiley, New York, 1933.

Wolman, M. G.: "The Nation's Rivers," *J. Water Pollut. Control Fed.,* **44** 715, 1972.

CHAPTER 3
BASIC CHEMISTRY

Comparing the atom to the solar system is useful in developing an understanding of basic chemical principles. The sun is likened to the nucleus of the atom and the planets to the electrons moving in orbits about the nucleus. The physicist, who would probably find the analogy simplistic, is concerned mostly with activity within the nucleus, whereas the chemist deals principally with the activity of the planetary electrons. Chemical reactions between atoms or molecules involve only these electrons; there are no changes in the nuclei.

The nucleus is made up of protons, having a positive electrical charge, and neutrons, without any charge. The positive charges in the nucleus are balanced by the negatively charged electrons orbiting it. Since the mass of an electron is less than 0.02% of the mass of a proton, for practical purposes the weight of the atom may be considered to be entirely in the nucleus.

Atoms are identified by name, atomic number, and atomic weight. The atomic number is the number of electrons in orbit about the nucleus, and, therefore, the number of protons in the nucleus. The weight of the atom is the sum of the protons and neutrons in the nucleus. The names and symbols are of historical interest. Some are ancient history; the symbol for lead is Pb from the Latin *plumbum,* from which is derived the word "plumber"; others are recent, such as fermium, atomic number 100, named for Enrico Fermi.

PERIODIC CHART OF ELEMENTS

Over 100 years ago, studying the elements known at that time, the Russian chemist Mendeleev tabulated the atomic weights of different elements in increasing order, and observed a repetition, or periodicity of certain significant physical properties of these elements. He constructed from this a periodic chart of sufficient accuracy to enable him to predict the properties of elements which had not in his time been discovered. A modern version of his periodic chart, modified to accent the elements of particular interest to water chemists, is reproduced in Figure 3.1.

At one time the atomic weights developed by chemists differed from those developed by nuclear physicists. Those arrived at by the chemists had been based on a system of logic which assigned to oxygen an atomic weight of 16 as a starting point. Oxygen combines with some metals and other elements in more than one proportion: for example, there are two oxides of carbon, CO_2 and CO. From this kind of information and the measured weights of the materials found in chemical

GROUP →

PERIOD	IA	IIA		IIIA	IVA	VA	VIA	VIIA	0
1	1.01 **H** 1 (+1, -1)								He
2	6.94 **Li** 3 (+1)	9.01 **Be** 4 (+2)		10.81 **B** 5 (+3)	12.01 **C** 6 (+2, +4, -4)	14.01 **N** 7 (+1, -1, +2, -2, +3, -3, +4, +5)	16.00 **O** 8 (-2)	19.00 **F** 9 (-1)	Ne
3	23.00 **Na** 11 (+1)	24.31 **Mg** 12 (+2)		26.98 **Al** 13 (+3)	28.09 **Si** 14 (+2, +4, -4)	30.97 **P** 15 (+3, +5, -3)	32.06 **S** 16 (+4, +6, -2)	35.45 **Cl** 17 (+1, +5, +7, -1)	Ar
4	39.10 **K** 19 (+1)	40.08 **Ca** 20 (+2)	TRANSITION ELEMENTS	Ga	Ge	74.92 **As** 33 (+3, +5, -3)	78.96 **Se** 34 (+4, +6, -2)	79.90 **Br** 35 (+1, +5, -1)	Kr
5	Rb	87.62 **Sr** 38 (+2)		In	118.69 **Sn** 50 (+2, +4)	121.75 **Sb** 51 (+3, +5, -3)	Te	126.90 **I** 53 (+1, +5, +7, -1)	Xe
6	Cs	137.34 **Ba** 56 (+2)		Tl	207.19 **Pb** 82 (+2, +4)	Bi			

TRANSITION ELEMENTS

PERIOD	IB	IIB	IIIB	IVB	VB	VIB	VIIB	VIIIB
4				47.90 **Ti** 22 (+2, +3, +4)	50.94 **V** 23 (+2, +3, +4, +5)	52.00 **Cr** 24 (+2, +3, +6)	54.94 **Mn** 25 (+2, +4, +7)	58.85 **Fe** 26 (+2, +3)
4	63.55 **Cu** 29 (+1, +2)	65.37 **Zn** 30 (+2)				Mo		58.93 **Co** 27 (+2, +3); 58.71 **Ni** 28 (+2, +3)
5	107.87 **Ag** 47 (+1)	112.40 **Cd** 48 (+2)						
6	Au	200.59 **Hg** 80 (+1, +2)						

FIG. 3.1 Periodic chart of elements. The elements of interest to the water chemist are in boldface; rare elements are omitted. Atomic weights are shown in the upper left corners; atomic numbers below in boldface. Valences are shown to the right of the chemical symbol. The symbols are identified throughout the text.

reactions, using the assumption of oxygen having an atomic weight of 16 made it possible to calculate the atomic weight of other elements.

The physicist uses the mass of the proton in the nucleus as unity. Scientists have discovered that there are several varieties of many of the elements, called isotopes, attributed to differences in the nucleus. An isotope of an element is an atom having the same structure as the element—the same electrons orbiting the nucleus, and the same protons in the nucleus, but having more or fewer neutrons. For example, isotopes of hydrogen are deuterium, having one neutron in addition to the proton in the nucleus, and tritium, having two neutrons in addition to the proton. These have weights of 2 and 3 relative to hydrogen at one. This basically accounts for the differences in numbers arrived at by the physicist and the chemist, since the latter was working with elements found in nature, and therefore, containing the element itself plus its naturally occurring isotopes.

The isotope most familiar to the lay person is probably ^{235}U, which is found in natural uranium (atomic weight 238.03). ^{235}U is fissionable and became well known as the basic material used in the first atomic bomb. Because ^{235}U and ^{238}U behave identically in chemical reactions, they must be separated by extremely sophisticated physical processes based on their very small difference in atomic weight, 235 versus 238.

FROM ATOMS TO MOLECULES

Atoms can combine to form molecules. The molecules may be as simple as the association of two hydrogen atoms to form molecular hydrogen, H_2. Molecules made up of a number of different atoms are called compounds. These vary from the simple union of sodium and chlorine to form sodium chloride, NaCl, having a molecular weight of 58.5 (the sum of the sodium atomic weight and the chlorine atomic weight) to extremely elaborate combinations of carbon, hydrogen, nitrogen, and phosphorus found in such structures as the DNA (deoxyribonucleic acid) molecule, the fundamental unit in all cells, which carries the genetic code. DNA has a molecular weight in the thousands. In water treatment, artificial molecules having molecular weights on the order of 5 million are used to coagulate suspended solids in water.

The atom is defined as the smallest unit of matter retaining the characteristics of an element; a molecule is the smallest unit of matter retaining the properties of a compound.

Since the weights of atoms and molecules are relative, and because the units themselves are unimaginably small, the chemist works with units called *moles.* A mole is an arbitrary unit, given in grams; the number of grams in a mole of a given substance is determined by adding the atomic weights of the constituents. For example, calcium carbonate ($CaCO_3$), a compound familiar to water chemists, has a molecular weight of 100, so a mole of calcium carbonate weighs 100 g.

This is also called the *gram–molecular weight.* A mole of any substance contains exactly the same number of molecules as a mole of any other substance. This is Avogadro's number, 6.02×10^{23}. When dealing with gases, a mole of any gas, containing 6.02×10^{23} molecules, always occupies 22.4 liters at 0°Celsius and 760 mmHg total pressure.

IMPURITIES IN WATER

Practically speaking, no chemical is pure, whether naturally occurring or artificial. Most industrial chemicals have a level of impurity usually measured in percentage, or parts per hundred. The water chemist seldom deals with water sources having percentage levels of impurity, except for seawater (about 3% dissolved mineral impurities), connate waters (produced with some crude oils, sometimes containing 20 to 30% dissolved salts), brackish waters, and certain industrial wastewaters. Since the water chemist is usually working with fresh water, the impurity level is measured in parts per million (ppm), 10,000 ppm equaling 1%; a sample of water from Lake Michigan, containing about 150 ppm, has an impurity level of only 0.015%. Since a liter of distilled water weighs 1000 g or 1,000,000 mg, it is apparent that 1 mg of impurity in a liter represents 1 ppm. However, a liter of seawater weights about 1032 g, so 1 mg of impurity in seawater is less than 1 ppm. Because the density of the water may be quite high, the use of milligrams per liter is more precise than parts per million, although for practical purposes, they are identical when dealing with fresh water. This book will use milligrams per liter, the preferred notation.

Anything in water that is not H_2O is a contaminant or impurity. All water is impure, and it is the principal job of the water chemist to define these impurities, set specifications for each impurity acceptable for the intended use of the water after treatment, and devise economical treatment methods to reach the quality limits that have been set. It is important to recognize that the terms impurity, contamination, and pollution are subjective ones. In this book, a contaminant is considered a pollutant when its concentration reaches a level that may be harmful either to aquatic life or to the public health if the water is for potable purposes. Dissolved impurities in water are broadly classified as inorganic salts (dissolved from minerals in the geological formations containing the water source) and organic matter (related to aquatic life and vegetative cover of the watershed). In most fresh waters, dissolved matter is largely inorganic.

To the water chemist, the molecular weight of a mineral is usually of less use in calculations than its equivalent weight, because most minerals dissolved in water are ionized. To understand the definition and usage of equivalent weight, the behavior of ionic solutions—called electrolytes—should be examined.

ELECTROLYTES

Water is known as the universal solvent. When it dissolves a mineral, new materials are produced from the atoms released by the mineral. These are fundamental particles called ions. The breakup of a chemical compound by dissolution in water forms cations, which are positively charged, and anions, which are negatively charged. Generally, an increase in water temperature causes an increase in the solubility of most salts. Important and notable exceptions are $CaCO_3$, $CaSO_4$, $MgCO_3$, and $Mg(OH)_2$, all of which become less soluble as the temperature increases.

There are a number of ways in which this phenomenon of ionization can be demonstrated. The simplest is an experiment in which an electric light is connected to an electrical circuit having two separated poles inserted in a beaker of water (Figure 3.2). In pure water, the lamp does not light when the switch is on because pure water is an insulator, not a conductor. Salt is added to the water in small increments, and gradually the light brightens, the intensity being propor-

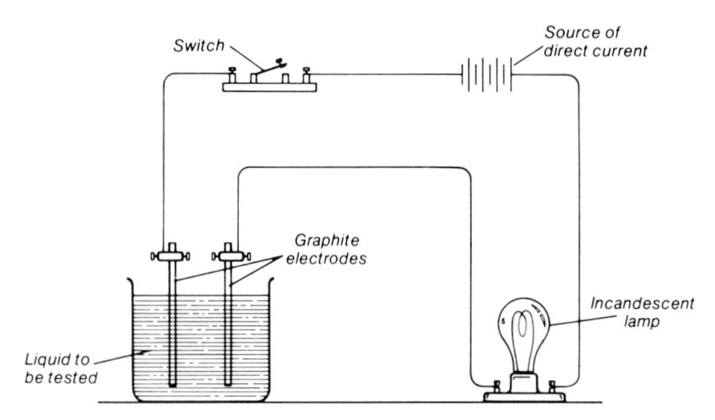

FIG. 3.2 Pure water is nonconductive, but addition of an ionizing salt allows current to pass.

tional to the amount of salt added. In pure water, the circuit is open; current begins to flow only after sodium and chloride ions from the salt are present to transport electrons through the solution. If crystals of sugar are added to the water instead of salt, nothing happens; salt is an electrolyte and sugar is not.

It was observed many years ago that impurities in water lower its freezing point. Upon careful investigation it was discovered that 1 mole/L of a nonelectrolyte such as sugar or alcohol dissolved in water lowers the freezing point by 1.86°C. It was found that 1 mole/L of sodium chloride lowered the freezing point by almost twice the value found for sugar and alcohol, and 1 mole/L of sodium sulfate produced almost 3 times this depression. This is due to the production of two ions, sodium and chloride, from one molecule of sodium chloride salt, and three ions, two sodiums and one sulfate, from one molecule of sodium sulfate (Na_2SO_4).

IONS AND ELECTRIC CURRENT

When sodium chloride dissolves in water to produce sodium and chloride ions, the sodium loses the single electron in its outer orbit to the chloride ion. The sodium, having lost a negative charge, becomes positively charged by one unit, and the chloride, having retained the electron, becomes negatively charged by one unit. If two electrodes are immersed in the solution of sodium chloride and direct current is applied, sodium will move toward the negatively charged electrode (which is called a cathode), and the chloride ion will move to the positively charged electrode (called the anode). It is because of this direction of movement that the positive sodium ion is called a cation and the negative chloride ion an anion.

Those elements shown in columns 1A and 1B of the periodic chart become cations with a charge of +1 when dissolved in water. Similarly, the elements in columns 2A and 2B become cations with a charge of +2. On the opposite side of the periodic table, elements in column 6A become anions with a charge of −2, and those in column 7A become anions with a charge of −1. These charges are called the valences of the ions; the common ions in water are mono-, di-, or trivalent.

Ions made up of several different atoms are called radicals. Common radicals in water chemistry include ammonium, NH_4^+, a cation; and nitrate, NO_3^-, sulfate, SO_4^{2-}, and phosphate, PO_4^{3-}, which are anions.

Elements that form cations in aqueous solution may be broadly classified as alkali metals (e.g., sodium, potassium), alkaline earth metals (e.g., magnesium, calcium), and heavy metals (e.g., iron, manganese).

In their elemental form, each of these can displace hydrogen from aqueous solutions. Sodium and potassium are so active that they will react with water itself to liberate hydrogen:

$$2Na + 2H_2O \rightarrow 2H_2\uparrow + 2NaOH \tag{1}$$

Magnesium will not react with cold water, but it will react with acid, which is a solution of the hydrogen ion:

$$Mg + 2HCl \rightarrow H_2\uparrow + MgCl_2 \tag{2}$$

ELECTROMOTIVE SERIES

These cations can be classified, then, according to their reactivity in an order known as the electromotive series, shown in Table 3.1. Not only will the more reactive metals displace hydrogen from solution, but they will even displace metals below them in the series. For example, if a strip of iron is placed in a solution of copper sulfate, the copper will plate out on the strip as the iron goes into solution:

$$Fe + CuSO_4 \rightarrow Cu + FeSO_4 \tag{3}$$

In comparing reactions (1) and (2), to produce one molecule of hydrogen requires two atoms of sodium but only one atom of magnesium, because the sodium has a charge of +1, while the magnesium ion has a charge of +2. This provides one basis for the concept of equivalent weight: The equivalent weight of a cation is that weight which will replace 1.0 g of hydrogen from aqueous solution. In most reactions, equivalent weight is the molecular weight of a substance divided by its valence.

When direct current is passed through an aqueous solution of an electrolyte (the process of electrolysis), a unit of electricity called the *faraday,* equal to 96,500 C (ampere-seconds), causes 1 eq of cations to react at the cathode and 1 eq of anions at the anode. In the electrolysis of HCl:

$$2HCl \rightarrow H_2\uparrow + Cl_2\uparrow \tag{4}$$

In this reaction, 1 F releases 1.0 g of hydrogen gas at the cathode and 35.45 g of chlorine gas at the anode. The molecular weights and equivalent weights of some common electrolytes are shown in Table 3.2 (see also Chapter 4).

TABLE 3.1 Electromotive Series of Elements

Element	Half-cell reaction	Volage, E*
Potassium	$K^+ + e \rightarrow K$	−2.93
Calcium	$Ca^{2+} + 2e \rightarrow Ca$	−2.87
Sodium	$Na^+ + e \rightarrow Na$	−2.71
Magnesium	$Mg^{2+} + 2e \rightarrow Mg$	−2.37
Aluminum	$Al^{3+} + 3e \rightarrow Al$	−1.66
Zinc	$Zn^{2+} + 2e \rightarrow Zn$	−0.76
Iron	$Fe^{2+} + 2e \rightarrow Fe$	−0.44
Nickel	$Ni^{2+} + 2e \rightarrow Ni$	−0.25
Tin	$Sn^{2+} + 2e \rightarrow Sn$	−0.14
Lead	$Pb^{2+} + 2e \rightarrow Pb$	−0.13
Hydrogen	$2H^+ + 2e \rightarrow H_2$	−0.00
Copper	$Cu^{2+} + 2e \rightarrow Cu$	0.34
Mercury	$Hg_2^{2+} + 2e \rightarrow 2Hg$	0.79
Silver	$Ag^+ + e \rightarrow Ag$	0.80
Gold	$Au^{3+} + 3e \rightarrow Au$	1.50

* The voltage is that developed by an electrode of the element immersed in a molal solution (1 mole dissolved in 1000 g of water) of its ions shown in the half-cell reaction.

TABLE 3.2 Weights of Some Common Compounds Used in Water Treatment

Common name	Formula	Mol. wt.*	Equiv. wt.
Table salt	$NaCl$	58.5	58.5
Caustic soda	$NaOH$	40	40
Salt cake	Na_2SO_4	142	71
Soda ash	Na_2CO_3	106	53
Limestone	$CaCO_3$	100	50
Quicklime	CaO	56	28
Slaked lime	$Ca(OH)_2$	74	37
Gypsum	$CaSO_4 \cdot 2H_2O$	172.2	86.1
Muriatic acid	HCl	36.5	36.5
Sulfuric acid	H_2SO_4	98	49
Aqua ammonia	NH_4OH	35	35

* Also called formula weight.

COLLOIDAL SYSTEMS

Some types of matter can be dispersed in water even though not truly soluble. This dispersion is accomplished by breaking down the material into an extremely small size, at the upper end of the size range for ions and molecules. Particles of this size are called colloids.

The surface of almost all matter—glass, steel, plastic—has a residue of electric charges. This can lead to the development of high surface voltage, as demonstrated on a small scale by the discharge of a spark of static electricity on a cold, dry day, and on a large scale by lightning. As matter is reduced in size, the ratio of surface charge to mass increases exponentially. Assume that a cube of sand measuring 1.0 mm on each side is reduced to colloidal size of 100 nm (1 nm = 10^{-6} mm); this would produce 10^{12} colloidal particles with a total surface 10,000 times larger than the original grain with a correspondingly larger surface charge.

It is this high surface charge, which is negative for silica (sand) in the example given, which causes colloidal particles to repel one another, thus maintaining the stability of the dispersion.

OXIDATION AND REDUCTION

The periodic chart shows that a number of elements have more than one valence. This is of great importance to the water chemist; an example is iron, whose oxide is quite soluble in water when its valence is +2 and almost completely insoluble when its valence is +3. If the valence of a material becomes more positive or less negative, it is said to be oxidized. Reduction occurs when valence becomes more negative or less positive—for example, in the change of iron +3 to iron +2, the iron is said to have been reduced. Another example of this is the chromate radical, CrO_4^{2-}, in which chromium has a valence of +6; when this is reduced, the chromium is changed in valence to +3.

In this example, an effective corrosion inhibitor, chromate, becomes a heavy metal cation with no useful properties, and it precipitates as $Cr(OH)_3$ to produce suspended matter.

Certain water treatment chemicals, such as chlorine gas, are effective because they are strong oxidizers. When chlorine is present in its elemental form, it has zero valence. If the chlorine attacks carbon in its elemental state, the carbon changes from zero valence to +4 and forms CO_2 gas, and the chlorine is reduced from zero valence state to the chloride anion, Cl^-. In reactions of this kind, one material is oxidized and the other is reduced, with a net change of zero in the electrical balance of the system, since there are as many positive charges produced as negative charges.

SOLVENT ACTION OF WATER

The solvent action of water varies from one mineral to another. Table 3.3 shows the solubility of some common compounds in water.

TABLE 3.3 Solubility of Some Inorganic Compounds in Water at 20°C

Cation	Compound (solid phase in parentheses)	Solubility, wt %
H^+	HCl	41.9
	HNO_3	*
	H_2SO_4	*
Na^+	$NaCl$	26.5
	$NaOH$	52.1
	$NaNO_3$	46.7
	$Na_2CO_3(\cdot 10H_2O)$	17.7
	$Na_2SO_4(\cdot 10H_2O)$	16.2
	$Na_3PO_4(\cdot 12H_2O)$	9.1
K^+	KCl	25.4
	KOH	52.8
	KNO_3	24.0
	$K_2CO_3(\cdot 2H_2O)$	53.5
Mg^{2+}	$MgCl_2(\cdot 6H_2O)$	35.3
	$Mg(OH)_2$	Nil
	$MgCO_3$	Nil
	$MgSO_4(\cdot 7H_2O)$	26.2
Ca^{2+}	$CaCl_2(\cdot 6H_2O)$	42.7
	$Ca(OH)_2$	0.17
	$CaCO_3$	Nil
	$CaSO_4(\cdot 2H_2O)$	0.20
	$Ca_3(PO_4)_2$	Nil
Fe^{2+}	$FeCl_2(\cdot 4H_2O)$	40.8
	$Fe(OH)_2$	Nil
Fe^{3+}	$FeCl_3$	47.9
	$Fe(OH)_3$	Nil

* Infinitely soluble.

Like minerals, gases and organic matter are soluble in water. Henry's law states that the amount of gas dissolved in water is directly proportional to the pressure of the gas above the water surface. In any gas system the pressure percent equals volume percent equals mole percent.

$$\frac{P_g}{P_t} = \frac{V_g}{V_t} = \frac{M_g}{M_t}$$

The atmosphere is approximately 20% oxygen and 80% nitrogen. At atmospheric pressure of 760 mmHg, then, the partial pressure of oxygen is 152 mm and the partial pressure of nitrogen is 608 mm. At these partial pressures, the approximate concentration of oxygen in water is about 10 mg/L and nitrogen is about 15 mg/L. Even though 4 times as many nitrogen molecules are bombarding the water as oxygen molecules, less than twice as many go into solution. If the total gas pressure at the water surface would double, so would the concentration of oxygen and nitrogen in the water solution. Unlike most mineral salts, which become more soluble at higher temperatures, gases decrease in solubility as the temperature rises.

ORGANIC CHEMICALS

Organic matter is a broad category that includes both natural and synthetic molecules containing carbon, and usually hydrogen. All living matter is made up of organic molecules. At one time, organic chemistry was considered to involve those carbon-containing compounds produced by living things. But most organic materials can now be made synthetically, so their relation to living things is not a necessary condition, although it is still true, of course, that life processes are involved with organic chemicals. Organics can be extremely soluble in water (as is the case with alcohol and sugar) or may be quite insoluble (as are most plastics). Even when dissolved in water at less than 1 mg/L, certain organic compounds may still cause serious physiological effects. Modern technology makes it possible to analyze such toxic organic materials as pesticides at concentrations in the part-per-billion (ppb) range. To put this in perspective, analyzing 1 ppb (1 ug of contaminant per liter of water) is comparable to seeing a bottle cap on the earth's equator from an orbiting satellite.

Just as elements combine to form radicals in inorganic chemistry, so do elements combine with carbon to form functional groups with their own individual properties. The major kinds of organic functional groups and compounds of importance to the water chemist are listed in Table 3-4.

PREDICTING SOLUBILITIES

There are a few rules of thumb to use in determining the solubility of any chemical compound—inorganic or organic—in water: the salts of sodium, potassium, and ammonium are highly soluble; mineral acids (H_2SO_4, HCl) are soluble; with the exception of certain heavy metal cations (Pb, Ag), most halides (Cl, Br, I) are soluble, except for fluoride; most carbonates, hydroxides, and phosphates are only slightly soluble, with the exception of those associated with Na^+, K^+, and NH_4^+.

TABLE 3.4 A Summary of Important Organic Groups and Compounds

---

1. Compounds of C and H

CH_4: methane; CH_3-: the methyl group

H_3C-CH_3: ethane; CH_3-CH_2-: the ethyl group

$H_3C-CH_2-CH_3$: propane; $CH_3-CH_2-CH_2-$: the propyl group

$H_3C-CH_2-CH_2-CH_3$: butane, a straight chain compound, since the carbons are linear (joined in a single line).

CH_3

\

$CH_3-CH_3-CH_3$: isobutane, a branched chain compound having the same empirical formula as butane; an isomer of butane.

$CH_3-CH_2-CH_2-CH_2-CH_2-CH_3$: hexane, another saturated compound in the straight chain series. The straight chain and branched chain isomers are called aliphatic compounds.

$H_2C=CH_2$: ethylene; an unsaturated compound, so-called because additional H could be added. The suffix "ene" and the double bond (=) in the structural formula denote unsaturation.

$HC\equiv CH$: acetylene; a compound having an even greater degree of unsaturation, indicated by a triple bond ($\equiv$).

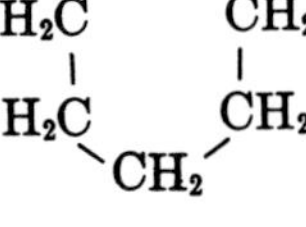

cyclohexane, a cyclic compound, since the carbon chain forms a closed configuration.

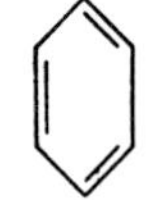

benzene, an unsaturated cyclic compound, also called aromatic. The structural formula is often shown as a simple hexagon.

NOTE: Formulas may be simplified by a number of shorthand notations, e.g., hexane, $CH_3(CH_2)_4CH_3$.

2. Functional groups containing C and O

    O
    ||
a. $-C-$: a carbonyl group present in ketones, acids, and aldehydes

                 O
                 ||
Example: $CH_3-C-CH_2CH_3$ is methyl ethyl ketone.

    | |
b. $-$ C $-$O$-$C$-$: an ether
    | |

Example: $CH_3-O-CH_2CH_3$ is methyl ethyl ether.

General example: $R-O-R'$, where R and R′ are alkyl groups such as methyl and ethyl; or aryl groups, such as phenyl or benzyl.

TABLE 3.4 A Summary of Important Organic Groups and Compounds (*Continued*)

3. Functional groups containing C, H, and O

   a. $-\overset{\overset{\displaystyle O}{\|}}{C}-OH$: an acid; also written $-COOH$, called the carboxyl group
      Example: CH_3-CH_2-COOH is propionic acid.

   b. $-\overset{|}{\underset{|}{C}}-OH$: an alcohol. $-OH$ is a hydroxyl group, often identified by the suffix, "ol"

      Example 1: CH_3-OH is methyl alcohol, or methanol.
      Example 2: $CH_3CH_2CH_2OH$ *n*-propyl alcohol (or propanol) where "*n*" means normal, or straight chain.
      $CH_3CHOHCH_3$ isopropyl alcohol where "iso" means isomer, or a compound having the same molecular weight as the normal compound but a different structure.
      Example 3: OH is phenol, a cyclic alcohol. A simplified formula is C_6H_5OH.

      General example: $R-OH$, where R is an alkyl group such as methyl, ethyl, etc.

      NOTE: 1. The alcohol is analogous to a base in inorganic chemistry. It reacts with organic acids to form esters such as stearyl glyceride (fat), the reaction product of the alcohol glycerine with a long-chain acid, stearic acid.
      2. Alcohols also react with caustic to produce alcoholates, such as CH_3-CH_2-ONa, sodium ethylate.

   c. $-\overset{\overset{\displaystyle O}{\|}}{C}-H$: an aldehyde

      Example: $CH_3-CH_2-CH_2-CHO$ butyraldehyde
      General example: $R-CHO$

4. Functional groups containing C, H, and N
   R, R′, and R″ represent alkyl (straight chain) and aryl (cyclic) organic groups such as methyl, ethyl, phenyl, etc.
   $R-NH_2$: a primary amine, a basic material similar to ammonia (NH_3)
   $R-NH-R'$: a secondary amine
   $R-\overset{\overset{\displaystyle R'}{|}}{N}-R''$: a tertiary amine

$\left[R-\overset{\overset{\displaystyle R'}{|}}{\underset{\underset{\displaystyle R''}{|}}{N}}-R'''\right]^{+}-OH^{-}$: a quaternary amine, a strong organic base

NOTE: Being basic, amines react with inorganic acids to form amine salts and with organic acids to form amides:

$$R-NH_2 + HCl \rightarrow R-NH_2 \cdot HCl$$

$$R-NH_2 + R'COOH \rightarrow R-NH-CO-R' + H_2O$$

TABLE 3.4 A Summary of Important Organic Groups and Compounds (*Continued*)

5. Polymers

Polymers are large molecules constructed of aggregations of individual units called monomers.

Example 1: $CH_2{=}CH_2$, ethylene monomer. These units join to form ethylene polymer:

$-CH_2-CH_2-CH_2-CH_2-CH_2-CH_2$ etc.

This polymerized ethylene is called polyethylene. It is a homopolymer, because all units are the same.

Example 2: acrylamide monomer: $CH_2{=}CH-CONH_2$

The building unit is $[CH_2{=}CH]$ with $O{=}C-NH_2$ attached to CH

which joins others to form polyacrylamide:

$-CH_2-CH — CH_2-CH — CH_2-CH-$ (each CH bearing, respectively, $O{=}C-NH_2$, $O{=}C-NH_2$, $O{=}CH-NH_2$)

This is also a homopolymer because all units are the same.

Example 3: Two different monomers may be polymerized, forming a copolymer. For example, acrylamide can be copolymerized with acrylic acid ($CH_2{=}CH-COOH$):

$-CH_2-CH — CH_2-CH —$ (each CH bearing, respectively, $O{=}C-NH_2$, $O{=}C-OH$)

This is called a polyacrylate. It is an anionic polymer, since the acrylate unit is anionic in nature, with the H^+ ion being an exchangeable cation. A great variety of monomers can be copolymerized in this fashion. By selecting monomers carefully, and combining them in the right proportions, molecular weight and strength of anionic or cationic charge in the final polymer can be controlled.

A simplified solubility chart is given in Table 3.5. Solubility data are determined only by laboratory investigation. For those materials that are very slightly soluble, which this text arbitrarily defines as being less than 2000 mg/L (the approximate solubility of $CaSO_4$), the *solubility product* is a useful tool.

The solubility product makes it possible to calculate the residual solubility of a chemical after chemical treatment to remove it from water by precipitation. According to the solubility product concept, the concentration of cations multiplied by the concentration of anions gives a product which is a constant at a given temperature. For calcium carbonate, a simple example, this expression would be:

$$[Ca^{2+}] \times [CO_3^{2-}] = K_s$$

In this example, the concentration of both the calcium and the carbonate is given in mole per liter (or moles·liter^{-1}). The solubility product in this case is expressed in moles^2·liters^{-2}. In a reaction where more than a single cation and anion are formed, the expression is more involved, as with magnesium hydroxide:

$$Mg(OH)_2 \rightleftharpoons Mg^{2+} + 2\,OH^- \qquad (5)$$

$$K_s = [Mg^{2+}] \times [OH^-]^2$$

in which case the units are moles^3·liters^{-3}.

TABLE 3.5 Simplified Solubility Chart

Anion → ↓ Cation	F^-	Cl^-	Br^-	I^-	HCO_3^-	OH^-	NO_3^-	CO_3^{2-}	SO_4^{2-}	S^{2-}	CrO_4^{2-}	PO_4^{3-}
Na^+	S	S	S	S	S	S	S	S	S	S	S	S
K^+	S	S	S	S	S	S	S	S	S	S	S	S
NH_4^+	S	S	S	S	S	S	S	S	S	S	S	S
H^+	S	S	S	S	CO_2	H_2O	S	CO_2	S	H_2S	S	S
Ca^{2+}	I	S	S	S	SS	VSS	S	I	VSS	X	S	I
Mg^{2+}	VSS	S	S	S	S	I	S	VSS	S	X	S	I
Ba^{2+}	VSS	S	S	S	VSS	S	S	VSS	I	X	I	I
Sr^{2+}	VSS	S	S	S	VSS	SS	S	I	VSS	X	VSS	I
Zn^{2+}	S	S	S	S	VSS	I	S	I	S	I	VSS	I
Fe^{2+}	SS	S	S	S	SS	VSS	S	VSS	S	I	X	I
Fe^{3+}	SS	S	S	S	I	I	S	I	S	X	X	I
Al^{3+}	S	S	S	S	X	I	S	X	S	X	X	I
Ag^{1+}	I	I	I	I	I	I	S	VSS	S	I	I	I
Pb^{2+}	VSS	S	SS	VSS	I	VSS	S	I	I	I	I	I
Hg^{1+}	I	I	I	I	I	I	S	S	VSS	I	VSS	I
Hg^{2+}	SS	S	S	I	I	I	S	I	VSS	I	SS	I
Cu^{2+}	SS	S	S	VSS	I	I	S	I	S	I	I	I

S—Soluble, over 5000 mg/L.
SS—Slightly soluble, 2000–5000 mg/L.
VSS—Very slightly soluble, 20–2000 mg/L.
I—Insoluble, less than 20 mg/L.
X—Not a compound.

In the case of $CaCO_3$, this mathematical expression is the equation for a hyperbola. The solubility of calcium carbonate is found at that point where the calcium equals the carbonate concentrations on the curve (Figure 3.3). It will be seen, then, that to reduce the concentration of calcium in water, the introduction of

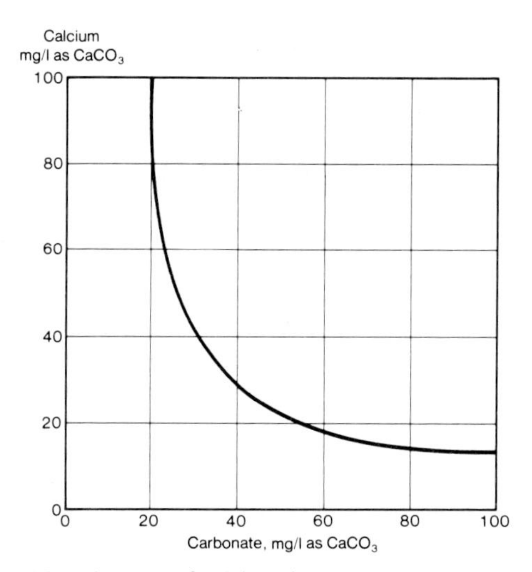

FIG. 3.3 Plot of calcium ions versus carbonate ions in saturated $CaCO_3$ solutions forms a hyperbola.

TABLE 3.6 Solubility Products at 25°C of Typical Hydroxides and Carbonates

Reaction	Solubility product, K_{sp}*
$Fe(OH)_2 \rightleftharpoons Fe^{2+} + 2OH^-$	1×10^{-14}
$Fe(OH)_3 \rightleftharpoons Fe^{3+} + 3OH^-$	2×10^{-39}
$Al(OH)_3 + 3H^+ \rightleftharpoons Al^{3+} + 3H_2O$	1×10^{-9}
$ZnO + 2H \rightleftharpoons Zn^{2+} + H_2O$	6.7×10^{-12}
$Mn(OH)_2 \rightleftharpoons Mn^{2+} + 2OH^-$	1.6×10^{-13}
$Ca(OH)_2 \rightleftharpoons Ca^{2+} + 2OH^-$	3.7×10^{-6}
$CaCO_3 \rightleftharpoons Ca^{2+} + CO_3^{2-}$	5×10^{-9}
$Mg(OH)_2 \rightleftharpoons Mg^{2+} + 2OH^-$	1×10^{-11}
$MgCO_3 \rightleftharpoons Mg^{2+} + CO_3^{2-}$	1×10^{-5}

* From concentrations in moles/liter.

Source: Stumm, W., and Morgan, J. J.: *Aquatic Chemistry*, John Wiley, New York, 1970.

extra carbonate ions—for example, by adding sodium carbonate—will reduce the calcium ions. Solubility products of significance in water chemistry are shown in Table 3.6.

FOREIGN IONS INTERFERE

The precipitation of most inorganic solids from water is so affected by the formation of ion pairs and complexes that the solubility product is of limited value in predicting actual treatment results. Because of the great variety of ions in most water systems, it is seldom that a pure precipitate, such as $CaCO_3$, is formed. The inclusion of foreign ions in the precipitate and the formation of a variety of ion pairs in the solution, in effect, increase the solubility of $CaCO_3$.

Even if pure $CaCO_3$ precipitates, it may form two kinds of crystals, calcite (the more stable) and aragonite, which have different solubilities.

Just as the water molecule can form chains (Chapter 1), so can ions associate themselves with one another. For example, calcium may be present in water as Ca^{2+}, or it may exist as $Ca(HCO_3)^+$, or $Ca(OH)^+$. Magnesium may be present, also, in a variety of ionic forms, e.g., Mg^{2+} and $MgHCO_3^+$.

The water molecule itself may take part in these reactions, usually associating with heavy metal cations in a ratio of 4:1 or 6:1. For example, iron in water may be present as $Fe(H_2O)_6^{3+}$ as well as Fe^{3+}.

In the presence of ammonia, copper is complexed, so that in addition to the presence of copper ions, Cu^{2+}, there is also present a whole family of cuprammonium complexes from $CuNH_3^{2+}$ to $Cu(NH_3)_5^{2+}$.

The presence of such ion pairs in solution increases the solubility of the precipitated salt. Equally as complicated is the nature of the precipitate. In the softening of water with lime, the following reactions are usually expected:

$$Ca(HCO_3)_2 + Ca(OH)_2 \rightarrow 2CaCO_3\downarrow + 2H_2O \qquad (6)$$

$$Mg(HCO_3)_2 + 2Ca(OH)_2 \rightarrow Mg(OH)_2\downarrow + 2CaCO_3\downarrow + 2H_2O \qquad (7)$$

In reaction (6), the calcium may be precipitated with trace amounts of other cations, such as strontium, producing a crystalline form more soluble than calcite.

The solubility of $CaCO_3$ determined for freshly precipitated material produced by reacting Na_2CO_3 with $CaSO_4$, for example, is higher than that obtained when calcite is ground and stirred into distilled water. The same effect is observed with almost all slightly soluble materials. The solubility of the precipitate may be expected to diminish with time and approach that of the pure solid as it ages and dehydrates.

In reaction (7), depending on temperature and pH, the magnesium may precipitate as a mixture of brucite [$Mg(OH)_2$], magnesite ($MgCO_3$), nesquehonite ($MgCO_3 \cdot 3H_2O$), or hydromagnesite [$Mg_4(CO_3)_3(OH)_2 \cdot 3H_2O$]. Each of these compounds has its own solubility.

OTHER ANOMALIES

Another factor affecting solubility is the presence of organic matter dissolved in water. It has been amply demonstrated that the residual calcium from lime softening of sewage at ambient temperatures is 2 to 3 times the solubility in fresh water, presumably due to organic complexes. If the same sewage is heated over about 150°F (66°C), the reactions with lime are about the same as obtained with fresh water. Inexplicably, the same effect is not observed with magnesium precipitation.

One example of complex effects on precipitation is found in the coagulation of water with alum [$Al_2(SO_4)_3 \cdot 14H_2O$], where the following reaction would be expected:

$$Al_2(SO_4)_3 + 6H_2O \rightarrow 2Al(OH)_3\downarrow + 3H_2SO_4 \qquad (8)$$

However, analysis of the precipitated floc will show the presence of sulfate in the solids, indicating that some SO_4^{2-} anions have replaced OH^- anions in the lattice of the precipitate.

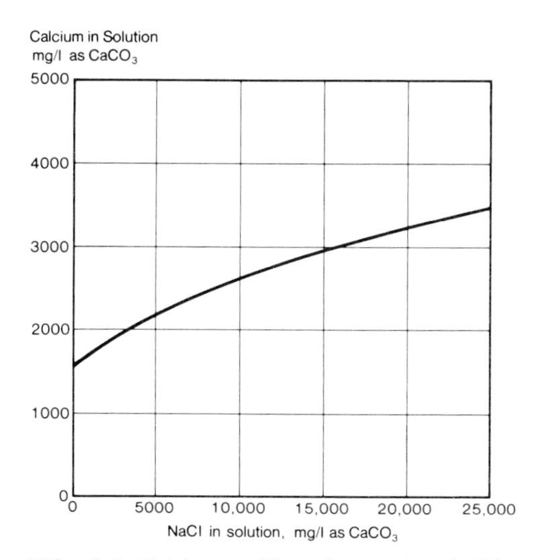

FIG. 3.4 Calcium sulfate (gypsum) solubility increases with increasing sodium chloride.

A final factor affecting solubility of these slightly soluble compounds is the so-called ionic strength of the aqueous solution. The higher the concentration of foreign ions not included in the precipitate, the more soluble is the precipitate. This means that $CaSO_4$, for example, is more soluble in seawater containing about 30,000 mg/L NaCl, and in oil-field brines, containing up to 150,000 mg/L NaCl, than in fresh water containing 100 mg/L dissolved solids (see Figure 3.4).

The solubility product, then, is seen to be a useful concept when used qualitatively or developed empirically for a specific system, but of limited value when taken directly from a handbook. (In fact, such data from different handbooks frequently disagree.)

EQUILIBRIUM

Another concept closely related to the solubility product is the *equilibrium constant,* also called the dissociation constant.

TABLE 3.7 Ionization of Electrolytes

Reaction	Degree of ionization*
$NaCl \rightleftharpoons Na^+ + Cl^-$	0.85
$Na_2CO_3 \rightleftharpoons 2Na^+ + CO_3^{2-}$	0.70
$HCl \rightleftharpoons H^+ + Cl^-$	0.92
$H_2CO_3 \rightleftharpoons H^+ + HCO_3^-$	0.0017
$NaOH \rightleftharpoons Na^+ + OH^-$	0.91
$NH_4OH \rightleftharpoons NH_4^+ + OH^-$	0.013

* In a 0.1 *N* solution at 18°C.

Note: From this tabulation is seen that some electrolytes are highly ionized, and are called strong salts, acids or bases; others are very slightly ionized, and are called weak electrolytes.

Not all chemical reactions go to completion because the products of the reaction exert a restraining effect on the reactants. The chemist shows this by the following conventions:

$$2H_2 + O_2 \rightarrow 2H_2O\uparrow \tag{9}$$

This reaction goes to completion—sometimes explosively. The yield arrow shows this reaction going in only one direction.

$$3Fe + 4H_2O \rightleftharpoons Fe_3O_4 + 4H_2\uparrow \tag{10}$$

In this reaction, if the hydrogen generated in the process is continually removed, the reaction proceeds to the right; if the system is contained, however, an equilibrium is reached, in which case all four materials, Fe, H_2O, Fe_3O_4, and H_2, are found in the container. If extra hydrogen is then introduced into the system, the reaction will move to the left.

Such an equilibrium can be expressed mathematically. In the reaction

$$A + BC \leftrightharpoons AB + C \tag{11}$$

the equilibrium constant is given as

$$K = \frac{[AB] \times [C]}{[A] \times [BC]}$$

where the bracketed values are expressed in concentration units, usually moles per liter.

When the equilibrium is related to the ionization of an electrolyte in water, the degree of ionization can be calculated, as shown for the reactions in Table 3.7.

PRACTICAL VALUE OF EQUILIBRIUM CONSTANTS

Strong electrolytes, such as sodium chloride, completely ionize in fresh water, so that the equilibrium constant is of no value in calculations involving this particular salt. However, calculations involving weak electrolytes rely on the equilibrium constant to show the distribution of the reacting materials and products both in their ionic and nonionized forms. This provides data useful in selecting chemical processes for removal of various contaminants from water. For example, the equilibrium constants for gases that ionize in water, such as hydrogen sulfide and ammonia, make it possible to calculate the optimum pH values for removal of these materials from water in their gaseous form. Typical dissociation constants are shown in Table 3.8.

TABLE 3.8 Dissociation Constants of Weak Acids and Bases

Calcium hydroxide	$Ca(OH)_2$	3.74×10^{-3}
Phosphoric acid	H_3PO_4	7.5×10^{-3}
Lead hydroxide	$Pb(OH)_2$	9.6×10^{-4}
Ammonium hydroxide	NH_4OH	1.8×10^{-5}
Acetic acid	CH_3COOH	1.8×10^{-5}
Carbonic acid	H_2CO_3	4.3×10^{-7}
Hypochlorous acid	$HClO$	3.5×10^{-8}
Boric acid	H_3BO_3	5.8×10^{-10}

SOLID REACTANTS

There are two distinctive mechanisms by which materials dissolved in water react with solids; these are adsorption and ion exchange.

Adsorption is the adhesion of a layer of molecules or collodial particles to the surfaces of a solid, which is usually porous. The gas mask is a common illustration

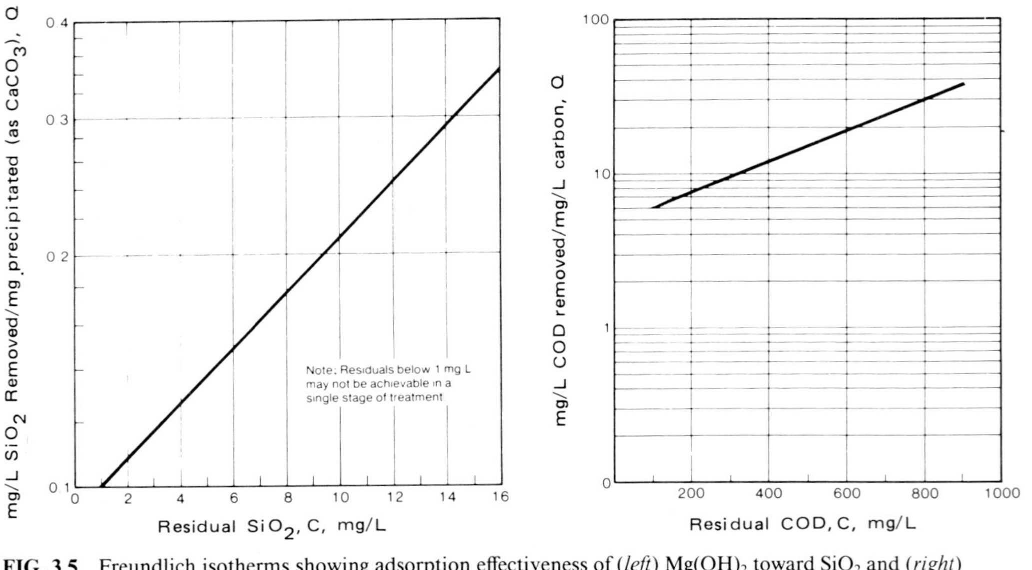

FIG. 3.5 Freundlich isotherms showing adsorption effectiveness of (*left*) $Mg(OH)_2$ toward SiO_2 and (*right*) activated carbon toward organics (COD). The formula for these isotherms is $Q = kC^{1/N}$.

of the use of this process. The mask contains a cannister of adsorbent material, usually activated carbon, capable of removing and storing hazardous or toxic gases so that the person wearing the mask can safely breathe even though in a contaminated atmosphere. In water systems, a similar activated carbon removes organic molecules which cause taste and odor problems.

Another example of the process is the adsorption of SiO_2, a negatively charged colloid, on freshly precipitated $Mg(OH)_2$.

The amount of adsorbent required for effective removal cannot be universally determined for all waters from a single equation. However, for any given system, experimental data are easily plotted on a semilog graph, producing a straight line—the *Freudlich isotherm* (Figure 3.5), so that an equation unique to that system can then be used for dosage adjustment.

Ion exchange is the process of removing unwanted ions from a solution in an equivalent exchange for preferred ions supplied by a solid having a special structure to do this. The solid, an ion exchange material—also called a zeolite—must be recharged periodically with the preferred ions. In this regeneration, the accumulated unwanted ions are flushed to waste.

The removal of calcium ions (hardness) by sodium zeolite, a common process, is shown by the reaction:

$$Ca^{2+} + Na_2X \rightarrow CaX + 2Na^+ \quad (12)$$

the letter X representing the cation exchange solid.

Many clays have ion exchange properties. This is an important aspect of soil chemistry and plant nutrition. One of these clays, clinoptilolite, is used to remove ammonia from wastewater. However, most ion exchangers are synthetic organic materials formulated for specific applications.

CHELATION

A final concept of great importance to water chemistry is the process of chelation. A chelating agent is a molecule—usually organic—which is soluble in water and can undergo reactions with metal ions to hold them in solution. So chelation is analogous to ion exchange, with the difference being the solubility of the exchange material: in chelation, the exchange material is very soluble in water as a large molecule; in ion exchange, the exchange material is insoluble in water and is a large solid particle.

The chemical equation for chelation is similar to that for ion exchange. For example, a common chelant is the sodium salt of ethylene diamine tetraacetic acid (EDTA).* Added to water, this chelant reacts with calcium ions to keep them in solution, preventing the formation of $CaCO_3$ scale:

$$Ca^{2+} + Na_4EDTA \rightarrow CaNa_2EDTA + 2Na^+ \quad (13)$$

There are a number of natural organic materials in water which have chelating ability, such as humic acid and lignin. Because of their chelating abilities, some organic materials interfere with certain water softening processes.

---

* *Note:* The above is an oversimplification of a complex field of chemistry—coordination chemistry. The organic portion of the EDTA compound forms bonds (ligands) with the central cation (Ca^{2+} or Mg^{2+}) other than ionic bonds, so the reaction shown is only part of the chelation process.

Somewhat related to chelation is the process of sequestration, also called "threshold treatment." A variety of phosphate compounds called polyphosphates are used in this process to prevent formation of iron, manganese, and $CaCO_3$ deposits. Condensed phosphates are produced by dehydration of one or more orthophosphate (PO_4^{3-}) compounds. If monosodium phosphate is used, the condensed product is sodium metaphosphate:

$$NaH_2PO_4 - H_2O = NaPO_3$$

With disodium phosphate, sodium pyrophosphate is formed:

$$2Na_2HPO_4 - H_2O = Na_4P_2O_7$$

The kinds of condensed phosphates formed can be varied by using different mixtures of orthophosphates. The condensed phosphates form chains containing the P—O—P group, as shown below for tripolyphosphate:

```
         O       O       O
         ||      ||      ||
Na—O—P —O—P —O—P —O—Na
         ||      ||      ||
        ONa     ONa     ONa
```

Proper selection of raw materials can produce long chains, called polyphosphates, which have strong electrostatic charges and which resist hydration (reversion) to the original orthophosphate form. The noncrystalline (glassy) phosphates with an $Na_2O:P_2O_5$ ratio of about 1.0 or slightly over are most stable and useful for threshold treatment. The most commonly used product is called hexametaphosphate $(NaPO_3)_6$.

Water containing a high concentration of calcium that is in a condition to precipitate, treated with only 0.5 mg/L of polyphosphate, can be kept from depositing $CaCO_3$ scale. The amount of polyphosphate required for effective scale control is far less than that required for softening on a stoichiometric basis, hence the name "threshold treatment." Polyphosphates can also hold Fe and Mn in solution (sequestration) in an environment where they would otherwise precipitate, e.g., in the presence of oxygen or chlorine at a pH over 8.

Other properties of polyphosphates are useful in water treatment. When $CaCO_3$ precipitates in the presence of polyphosphates, the usual calcite crystal form is distorted, and the scale structure is weak. Because they are strongly charged, polyphosphates adsorb on silt particles and help to keep them from settling because the individual particles repel one another.

The value of polyphosphates is destroyed if they revert to orthophosphate. Solutions of glassy polyphosphates remain stable for months, but in the water system they may be reverted by low pH, high temperature, and the presence of the oxides of certain heavy metals, including iron.

SUGGESTED READING

Degering, E. F., et al., eds.: *Organic Chemistry,* College Outline Series. Barnes & Noble, New York, 1955.

Gelender, M., and Geffner, S. L.: *Review Text in Chemistry,* Amsco School Publications, New York, 1964.

Moore, F. J., and Hall, W. T.: *A History of Chemistry,* McGraw-Hill, New York, 1931.

CHAPTER 4
WATER CHEMISTRY AND INTERPRETATION OF WATER ANALYSES

The common denominator of the majority of water problems is hardness. Hardness is one of the folk terms inherited from the past with origins in household use of water for washing. It was found that some waters were hard to use in doing the family laundry. More soap was needed to produce suds in these waters—so much so that many houses had a rain barrel or a cistern to collect soft rainwater for washing. Indeed, this relation between hardness and suds was so fundamental that the chemist devised a standard solution of soap, which was used for many years to determine the hardness of water. Thus, tradition defines hardness as the soap-consuming capacity of water. For practical purposes, it is the calcium and magnesium content of water, although heavy metals such as iron and manganese also consume soap.

Hardness, then, is the solution in water of both calcium and magnesium as cations, independent of the nature of the anions present. It has usually been expressed in terms of calcium carbonate, $CaCO_3$. This is a fortuitous choice because the molecular weight of $CaCO_3$ is 100 and its equivalent weight is 50, providing a convenient unit of exchange for expressing all ions in water, rather than showing each with its own equivalent weight. This is comparable to having the dollar as a convenient unit of exchange in international currency, rather than dealing in a mixture of marks, pounds, pesos, and francs.

The water analyses used for illustration in this book are based on the calcium carbonate equivalent concept, a widely used, but not universal, form for reporting a water analysis. There are several other forms also used in the United States: (1) most analyses of the U.S. Geological Survey and other government agencies report ions as they actually exist, usually in parts per million or milligrams per liter; (2) some chemists report ionic constituents in equivalents per million, arrived at by taking the concentration of each ion in parts per million as shown in the typical USGS report, and dividing this by equivalent weight.

Table 4.1 compares these three methods of reporting a water analysis. In the third column the sum of all the anions determined by analysis is 4.94. This slightly exceeds the total cations as determined by analysis, 4.91. Since the water must be electrically neutral, the sum of the cations should equal the sum of the anions; however, it is not unusual to find a modest discrepancy, usually because some minor constituents—perhaps ammonia (a cation) in this case—haven't been reported, or because of limitations of individual ion tests.

TABLE 4.1 Comparison of Water Analysis Report Methods—Mississippi River at Vicksburg

Component	mg/L*	epm†	mg/L as $CaCO_3$‡
Calcium	46	2.30	115
Magnesium	14	1.15	57
Sodium	32	1.39	75§
Potassium	2.7	0.07	NR
Total cations	NR	4.91	247
Bicarbonate	154	2.52	126
Sulfate	67	1.40	70
Chloride	34	0.96	48
Nitrate	3.6	0.06	3
Total anions	NR	4.94	247
Total hardness	172		
Dissolved solids	304		
Conductance (μS)	483		
pH¶	7.5		
Silica (as SiO_2)	8.3		
Iron (as Fe)	0.03		
Manganese (as Mn)	0.0		
Color¶	15		

NR = Not reported.
* USGS method (Water Supply Paper 1299), originally in ppm.
† Equivalents per million; in the above example, for calcium, epm = 46/20 (equivalent weight) = 2.30.
‡ $CaCO_3$ equivalent form, mg/L as $CaCO_3$.
§ By difference.
¶ pH in pH units, color in APHA units.

In the ionic method of reporting results, silica, iron, and manganese are not commonly reported in ionic form. These materials are usually colloidal. Strictly speaking, in a deep well water devoid of oxygen, iron and manganese are soluble in the reduced forms (Fe^{2+} and Mn^{2+}) and should be reported as cations if their concentrations are a significant portion of the total. Silica may be an anion at concentrations over 50 mg/L.

The common convention in using the $CaCO_3$ equivalent form is to overlook the analysis of sodium and potassium and to report these ions together as sodium, determined as the difference between the sum of the total anions and the hardness of the water. This maintains the electrical neutrality of the system. (If ammonia is determined, it should be reported as the cation, and the sodium content determined by the difference reduced accordingly.)

A convenient method for converting one analytical form to another is shown in Table 4.2. In this book, concentrations will normally be expressed in milligrams per liter.

One of the minor shortcomings of the $CaCO_3$ equivalent form is that the actual weight of minerals dissolved in the water (electrolytes plus inorganic colloidal substances) is different from the electrolyte expressed as $CaCO_3$ plus these same colloids. Often it is useful to know the actual weight of total dissolved minerals,

TABLE 4.2 Calcium Carbonate ($CaCO_3$) Equivalent of Common Substances

	Formula	Molecular weight	Equivalent weight	Substance to $CaCO_3$ equivalent (multiply by)	$CaCO_3$ equivalent to substance (multiply by)
	Compounds				
Aluminum sulfate (anhydrous)	$Al_2(SO_4)_3$	342.1	57.0	0.88	1.14
Aluminum sulfate (hydrated)	$Al_2(SO_4)_3 \cdot 14H_2O$*	600.0	100.0	0.5	2.0
Aluminum hydroxide	$Al(OH)_3$	78.0	26.0	1.92	0.52
Aluminum oxide (Alumina)	Al_2O_3	101.9	17.0	2.94	0.34
Sodium aluminate	$Na_2Al_2O_4$	163.9	27.3	1.83	0.55
Barium sulfate	$BaSO_4$	233.4	116.7	0.43	2.33
Calcium bicarbonate	$Ca(HCO_3)_2$	162.1	81.1	0.62	1.62
Calcium carbonate	$CaCO_3$	100.1	50.0	1.00	1.00
Calcium chloride	$CaCl_2$	111.0	55.5	0.90	1.11
Calcium hydroxide	$Ca(OH)_2$	74.1	37.1	1.35	0.74
Calcium oxide	CaO	56.1	28.0	1.79	0.56
Calcium sulfate (anhydrous)	$CaSO_4$	136.1	68.1	0.74	1.36
Calcium sulfate (gypsum)	$CaSO_4 \cdot 2H_2O$	172.2	86.1	0.58	1.72
Calcium phosphate	$Ca_3(PO_4)_2$	310.3	51.7	0.97	1.03
Ferric sulfate	$Fe_2(SO_4)_3$	399.9	66.7	0.75	1.33
Ferrous sulfate (anhydrous)	$FeSO_4$	151.9	76.0	0.66	1.52
Magnesium oxide	MgO	40.3	20.2	2.48	0.40
Magnesium bicarbonate	$Mg(HCO_3)_2$	146.3	73.2	0.68	1.46
Magnesium carbonate	$MgCO_3$	84.3	42.2	1.19	0.84
Magnesium chloride	$MgCl_2$	95.2	47.6	1.05	0.95
Magnesium hydroxide	$Mg(OH)_2$	58.3	29.2	1.71	0.58
Magnesium phosphate	$Mg_3(PO_4)_2$	262.9	43.8	1.14	0.88
Magnesium sulfate (anhydrous)	$MgSO_4$	120.4	60.2	0.83	1.20
Magnesium sulfate (epsom salts)	$MgSO_4 \cdot 7H_2O$	246.5	123.3	0.41	2.47
Manganese chloride	$MnCl_2$	125.8	62.9	0.80	1.26
Manganese hydroxide	$Mn(OH)_2$	89.0	44.4	1.13	0.89
Potassium iodide	KI	166.0	166.0	0.30	3.32
Silver chloride	$AgCl$	143.3	143.3	0.35	2.87
Silver nitrate	$AgNO_3$	169.9	169.9	0.29	3.40
Silica	SiO_2	60.1	60.1	0.83	1.20
Sodium bicarbonate	$NaHCO_3$	84.0	84.0	0.60	1.68
Sodium carbonate	Na_2CO_3	106.0	53.0	0.94	1.06
Sodium chloride	$NaCl$	58.5	58.5	0.85	1.17
Sodium hydroxide	$NaOH$	40.0	40.0	1.25	0.80
Sodium nitrate	$NaNO_3$	85.0	85.0	0.59	1.70
Trisodium phos.	$Na_3PO_4 \cdot 12H_2O$	380.2	126.7	0.40	2.53
Trisodium phos. (anhydrous)	Na_3PO_4	164.0	54.7	0.91	1.09
Disodium phos.	$Na_2HPO_4 \cdot 12H_2O$	358.2	119.4	0.42	2.39
Disodium phos. (anhydrous)	Na_2HPO_4	142.0	47.3	1.06	0.95
Monosodium phos.	$NaH_2PO_4 \cdot H_2O$	138.1	46.0	1.09	0.92
Monosodium phos. (anhydrous)	NaH_2PO_4	120.0	40.0	1.25	0.80
Sodium metaphosphate	$NaPO_3$	102.0	34.0	1.47	0.68
Sodium sulfate	Na_2SO_4	142.1	71.0	0.70	1.42
Sodium sulfite	Na_2SO_2	126.1	63.0	0.79	1.26

TABLE 4.2 Calcium Carbonate ($CaCO_3$) Equivalent of Common Substances (*Continued*)

	Formula	Molecular weight	Equivalent weight	Substance to $CaCO_3$ equivalent (multiply by)	$CaCO_3$ equivalent to substance (multiply by)
		Positive ions			
Aluminum	Al^{3+}	27.0	9.0	5.56	0.18
Ammonium	NH_4^+	18.0	18.0	2.78	0.36
Barium	Ba^{2+}	137.4	68.7	0.73	1.37
Calcium	Ca^{2+}	40.1	20.0	2.50	0.40
Copper	Cu^{2+}	63.6	31.8	1.57	0.64
Hydrogen	H^+	1.0	1.0	50.0	0.02
Ferric iron	Fe^{3+}	55.8	18.6	2.69	0.37
Ferrous iron	Fe^{2+}	55.8	27.9	1.79	0.56
Magnesium	Mg^{2+}	24.3	12.2	4.10	0.24
Manganese	Mn^{2+}	54.9	27.5	1.82	0.55
Potassium	K^+	39.1	39.1	1.28	0.78
Sodium	Na^+	23.0	23.0	2.18	0.46
Strontium	Sr^{2+}	87.6	43.8	1.14	0.88
Zinc	Zn^{2+}	65.4	32.7	1.53	0.65
		Negative ions			
Bicarbonate	HCO_3^-	61.0	61.0	0.82	1.22
Carbonate	CO_3^{2-}	60.0	30.0	1.67	0.60
Chloride	Cl^-	35.5	35.5	1.41	0.71
Chromate	CrO_4^{2-}	116.0	58.0	0.86	1.16
Fluroide	F^-	19.0	19.0	2.63	0.38
Iodide	I^-	126.9	126.9	0.39	2.54
Hydroxyl	OH^-	17.0	17.0	2.94	0.34
Nitrate	NO_3^-	62.0	62.0	0.81	1.24
Phosphate (tribasic)	PO_4^{3-}	95.0	31.7	1.58	0.63
Phosphate (dibasic)	HPO_4^{2-}	96.0	48.0	1.04	0.96
Phosphate (monobasic)	$H_2PO_4^-$	97.0	97.0	0.52	1.94
Sulfate	SO_4^{2-}	96.1	48.0	1.04	0.96
Bisulfate	HSO_4^-	97.1	97.1	0.52	1.94
Sulfite	SO_3^{2-}	80.1	40.0	1.25	0.80
Bisulfite	HSO_3^-	81.1	81.1	0.62	1.62
Sulfide	S^{2-}	32.1	16.0	3.13	0.32

* Empirical formula of commercial alum.

which can be obtained by recalculating the $CaCO_3$ equivalent analysis back to ionic form. In the example given in Table 4.1, the total number of ions in the "mg/L" column equals 353 mg/L. This exceeds the electrolytes (total number of cations or anions) of 247 mg/L as $CaCO_3$, and even the reported total dissolved solids (by evaporation) of 304 mg/L. The reason for this is the nature of the bicarbonate ions (equiv wt 61) in the water; this unstable ion breaks down under heat to form carbonate (equiv wt 30) when the water sample is evaporated to dryness at 100°C, which is the standard method of analyzing for total dissolved solids. Therefore, since the usual purpose of the conversion of an analysis to total mineral solids is to compare this to total dissolved solids after evaporation, the bicarbonate should be calculated as carbonate, its actual form in the evaporating dish. In this same example, Table 4.1, then, the calculated total mineral solids is 276 mg/L electrolytes plus 8.3 mg/L silica and iron, or 284 mg/L. The difference between total dissolved solids (304 mg/L) and total mineral solids (284 mg/L) is commonly ascribed to volatile and organic matter (V&O), which in this case is 20 mg/L.

Wastewaters usually contain a number of ions that may not be shown in this illustration. Composition depends on the kind of plant operations through which the water has passed before becoming waste. For example, heavy metals such as zinc and copper may be present in the waste from plating operations; chromate may be present as an anion in the blowdown from evaporative cooling systems; fluoride may be present as an anion or an anionic complex in wastewater from glass manufacturing.

ALKALINITY AND ACIDITY

Probably the most fundamental concept in the approach to understanding water chemistry is the acidity-alkalinity relationship. The first step in grasping this is an understanding of the dissociation of the water molecule itself into hydrogen ions and hydroxyl ions, according to equation:

$$H_2O \rightleftharpoons H^+ + OH^- \qquad K = [H^+] \times [OH^-] = 10^{-14} \tag{1}$$

(For the sake of simplicity, this book will use the hydrogen ion, H^+, while recognizing that it actually exists in a hydrated form, the hydronium ion—H_3O^+.)

THE pH NOTATION

Since the dissociation constant is so very small, 10^{-14}, at neutrality where there are the same number of hydrogen and hydroxyl ions there are only 10^{-7} moles/L of each. This is equal to only 10^{-4} mmoles/L, corresponding to an actual concentration of only 0.0001 mg/L H^+ ion, equivalent to 0.005 mg/L as $CaCO_3$. Because we are dealing with such small numbers in the dissociation of water into its ions, it is more convenient to substitute an expression involving the power of 10. This expression is defined as pH, in which the relationship is:

$$pH = \log \frac{1}{[H^+]} = -\log [H^+]$$

The dissociation constant, K, changes with temperature, and this must be taken into account in interpreting data involving H^+ and OH^- ions. For example, many water treatment operations are carried out at high temperatures, and samples from the system are usually cooled prior to analysis. The H^+ and OH^- concentrations measured on the cooled sample, even though different from those in the hot system, are usually used for control purposes. But a physical chemist needing to know the conditions prevailing in the hot system must use the dissociation constant for the temperature in that system.

The hydrogen ion concentration can be measured with a pH meter. It can also be titrated (see glossary) when the concentration becomes large enough to be detectable by chemical analysis. Since pH is a logarithmic function, the hydrogen ion concentration increases by a factor of 10 for each unit of pH reduction.

When the pH drops below approximately 5, the hydrogen ion begins to reach milligrams per liter levels, concentrated enough to be determined by titration, using the correct organic dye indicator. The chemical indicator originally selected by the water chemist for this purpose was methyl orange, changing color at pH 4.2 to 4.4. The color change of this indicator was so subtle—orange on the alkaline side to salmon pink on the acid side—that researchers looked for a substitute to give a more pronounced color change. The one they developed produces a blue color on the alkaline side and red on the acid side, with gray at the endpoint. Even though this special indicator has replaced methyl orange, the water chemist still defines alkalinity as *methyl orange alkalinity* ("M" alkalinity) which exists above the approximate pH range of 4.2 to 4.4. M *acidity* is strong mineral acidity that exists below this pH range. An approximate relationship between pH value and mineral acidity is shown in Table 4.3.

TABLE 4.3 Mineral Acidity* vs. pH

H^+, mg/L as $CaCO_3$	pH
2–3	4.3
4–5	4.0
6–7	3.9
8–9	3.8
10–11	3.7
12–13	3.6
14–16	3.5
17–20	3.4
21–25	3.3
26–30	3.2
31–40	3.1
41–50	3.0

* Mineral acidity is the presence of H^+ ion in mg/L concentrations.

The measure of pH, then, by a pH meter (or by a suitable colorimetric comparator) can be very valuable in determining the hydrogen and hydroxyl ion concentrations in the pH range above 4.2 to 4.4. It can also be used below this pH in the absence of reagents for performing an actual acidity titration.

USE OF METER FOR $[OH^-]$

Because the dissociation constant of water is approximately 10^{-14}, and pH has been defined as $-\log [H^+]$, the hydroxyl ion concentration can also be determined by a pH meter since the following relationships hold:

$$pOH = 14 - pH$$

$$pOH = \log \frac{1}{[OH^-]}$$

TABLE 4.4 Caustic Alkalinity vs. pH

OH, mg/L as $CaCO_3$	pH
2–3	9.7
4–5	10.0
6–7	10.1
8–9	10.2
10–11	10.3
12–13	10.4
14–16	10.5
17–20	10.6
21–25	10.7
26–30	10.8
31–40	10.9
41–50	11.0

As the pH of a water solution is increased and exceeds about 9.6 to 9.8, a measurable concentration of hydroxyl ions begins to appear. The hydroxyl alkalinity (caustic or OH alkalinity) can be determined either by using a pH meter (or the equivalent in a colorimetric comparator) or by titration. The relationship between hydroxyl alkalinity and pH is shown in Table 4.4.

An understanding of these concepts is necessary to put these acidity-alkalinity relationships into perspective. To the theoretical chemist, a pH of 7 is considered neutral; to the water chemist, a pH of 7 in itself means very little. He must also know how much total alkalinity and how much free or combined CO_2 may be present. For the water chemist, then, the dividing point between acidity and alkalinity is not pH 7.0, but rather the M alkalinity endpoint, corresponding to a pH of approximately 4.4.

The water chemist is also concerned with P alkalinity (phenolphthalein alkalinity), which exists when the pH is over a range of 8.2 to 8.4, corresponding to the change in phenolphthalein indicator from a colorless condition below 8.2 to pink or red above 8.4. In most natural water supplies, the pH is less than 8.2, so there is no P alkalinity. Very few natural waters have a pH below about 5.0, so it is seldom that strong mineral acids are found in fresh water. The pH range between the M endpoint and the P endpoint defines the alkaline range in which bicarbonate alkalinity exists and weak acids may be present, the most prominent of which is carbonic acid—carbon dioxide in solution.

THE IMPORTANCE OF CO_2

The atmosphere is a mixture of gases containing about 79% N_2 and 21% O_2 by volume. However, it also includes 0.04% CO_2, extremely important to the balance of life on the planet. Carbon dioxide is produced by the combustion of fuel. Even before human activities made a significant contribution to the CO_2 content of the atmosphere, the respiration of animal life—which is also a fuel consuming process—introduced CO_2 into the air. Plants containing chlorophyll utilize carbon dioxide in building cellular material such as carbohydrates. This reaction is called photosynthesis because photons of energy from the sun are needed for the reaction.

$$6CO_2 + 6H_2O \xrightarrow{\text{chlorophyll}} (CH_2O)_6 + 6O_2 \quad \text{or} \quad C_6H_{12}O_6 \text{ (glucose)} + 6O_2 \tag{2}$$

The chlorophyll catalyzes the reaction between water and CO_2. Since the solubility of CO_2 in water is less than 2000 mg/L at ambient air temperatures in a CO_2-saturated atmosphere, at the normal atmospheric level of 0.04% CO_2 less than 1 mg/L CO_2 dissolves in rainwater. However, once the rainwater penetrates the mantle of soil, it is exposed to CO_2 gas levels a hundredfold greater than in the atmosphere, created by the respiration of soil organisms as they convert organic food into its products of combustion. Well waters, then, which have percolated through this CO_2-rich zone may contain from 10 to several 100 mg/L dissolved CO_2.

When CO_2 dissolves in water, it reacts with water to form carbonic acid, which dissociates into the hydrogen ion and the bicarbonate ion according to the reaction:

$$CO_2 + H_2O \rightleftharpoons H_2CO_3 \rightleftharpoons H^+ + HCO_3^- \tag{3}$$

If distilled water is completely saturated with CO_2, approximately 1600 mg/L dissolves in the water, and the pH is decreased to approximately 4. The equivalent amount of a strong acid like H_2SO_4 would decrease the pH to about 2.5, illustrating the weak character of carbonic acid.

THE SOURCE OF ALKALINITY

The alkalinity of most natural water supplies is caused by dissolved bicarbonate (HCO_3^-) salts. The following equations show how water containing CO_2 from the atmosphere and from respiration of soil organisms dissolves magnesium and calcium from a common mineral, dolomite ($CaCO_3 \cdot MgCO_3$), to produce hardness and alkalinity in ground water:

$$H_2O + CO_2 + MgCO_3 \rightarrow Mg(HCO_3)_2 \rightleftharpoons Mg^{2+} + 2(HCO_3^-) \tag{4}$$

$$H_2O + CO_2 + CaCO_3 \rightarrow Ca(HCO_3)_2 \rightleftharpoons Ca^{2+} + 2(HCO_3^-) \tag{5}$$

In the pH range of 4.4 to 8.2, there remains a balance between excess CO_2 and bicarbonate ions which is measured by pH value as shown in Figure 4.1.

This shows that a water containing 1 mg/L CO_2 and 10 mg/L alkalinity has the same pH as one containing 10 mg/L CO_2 and 100 mg/L alkalinity. In the first case, the addition of 1 mg/L CO_2 would produce a large change in pH, whereas in the second case, the same 1 mg/L CO_2 addition would not produce a noticeable change. The alkalinity moderates or *buffers* the pH change, and alkalinity is known as a buffer.

Water attacks and dissolves many other minerals from the lithosphere in addition to dolomite. Figure 4.2 shows the distribution of the elements in the earth's crust. The three most prominent are present in the form of oxides, SiO_2, Al_2O_3, and Fe_2O_3, which are only very slightly soluble in water. The next most common elements, calcium, sodium, potassium, and magnesium, are seen to be the most prominent cations present in most freshwater supplies, as illustrated by the several water analyses included in this and earlier chapters.

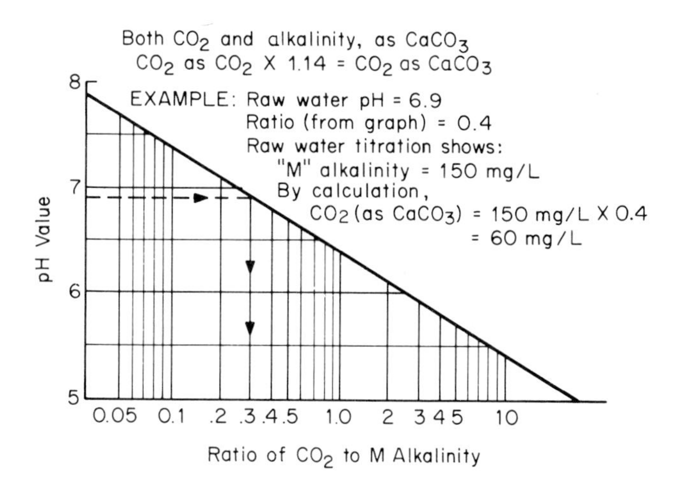

FIG. 4.1 Approximate relationship of carbon dioxide, alkalinity, and pH value.

When rock is attacked by wind and water, damaged by cracking and splitting, penetrated by plant roots, and weakened by alternating cycles of freezing and thawing, the by-product is clay. The analysis of most clays is similar to the composition of the earth's crust (Figure 4.2). Primary clays, those found where debris

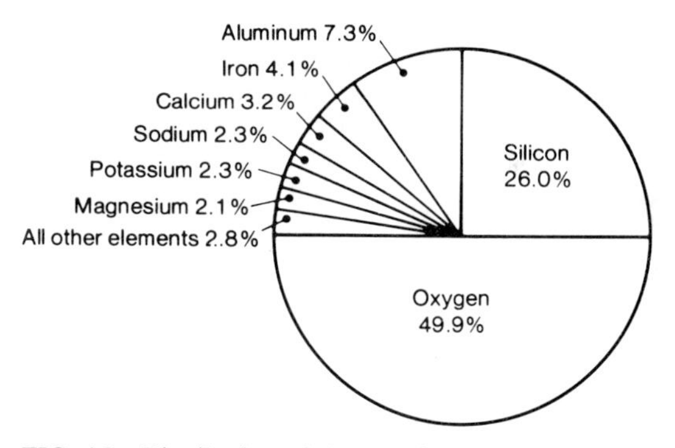

FIG. 4.2 Distribution of elements in the earth's crust.

accumulates from the exfoliation of weathered rock, have a somewhat different composition from secondary clays, those transported by water and deposited in slow-moving rivers or deltas. These latter clays usually contain organic matter that has deposited with the mineral components, and such clay may have less sodium or potassium and more iron than the native primary clays at their sources. These secondary clays, while in suspension, have an important influence on the processes subsequently used for treating surface water supplies. Equally important, in past geologic ages, the laying down of alternate layers of clay, sand and gravel, limestone, and clay provided the means for collecting and storing underground fresh water (which percolated into the gravel strata) and for pro-

TABLE 4.5 Principal Minerals Making Up the Crust of the Earth

Class	Mineral group	Compounds contained	% of earth's crust (approx)
1	Feldspars—orthoclase, microcline, albite, anorthite	Al_2O_3, SiO_2, Na_2O, K_2O, CaO, MgO	60
2	Ferromagnesians—olivine, pyroxene, amphibole	FeO, MgO, SiO_2, CaO_2, Al_2O_3	17
3	Quartz	SiO_2	12
4	Micas—biotite, vermiculite	SiO_2, Al_2O_3, MgO, FeO	3–4
5	Titania—rutile, ilmenite	TiO_2, FeO	1–2
6	Residual—calcite, dolomite, salt gypsum	Various—$CaCO_3$, MgO, NaCl, $CaSO_4$	5–6

Adapted from (a) Rhodes, Daniel: *Clays and Glazes for the Potter;* (b) Zim, H. S., Shaffer, P. R., and Perlman, R.: *Rocks and Minerals, a Guide to Familiar Minerals, Gems, Ores, and Rocks;* (c) Stumm, W., and Morgan, J. J.: *Aquatic Chemistry.* (See Suggested Reading.)

tecting the aquifer from contamination by brackish water or brine from prehistoric seas confined by a similar "sandwich" at another horizon.

The composition of the earth's crust shown elementally by Figure 4.2 can also be expressed by the proportions of minerals found in it, as shown by Table 4.5. The first five classes of minerals are very slowly soluble. For example, a prominent feldspar, orthoclase, has the formula $K_2O \cdot Al_2O_3 \cdot 6SiO_2$. Most potassium salts are very soluble in water, but the potassium (as K_2O) in this crystal structure is trapped by the strong attraction of the other elements present. It is slowly replaced by a process of ion exchange over a long period of time, and, when this happens, the by-product is $Al_2O_3 \cdot 2SiO_2 \cdot 2H_2O$, a clay known as kaolin. The last class of minerals comprise those that are more easily dissolved and are therefore the most important contributors to the mineral content of water. Of these, calcite ($CaCO_3$, one of the forms of limestone) is critically important because it is responsible for the alkaline nature of practically all natural water supplies.

CO_3^{2-}/HCO_3^- DISTRIBUTION

Most natural waters contain bicarbonate alkalinity and are at a pH less than about 8.2 to 8.4. Above this pH, CO_2 ceases to exist in measurable quantities and the carbonate ion begins to make itself known. The equilibrium reaction is as follows:

$$HCO_3^- \rightleftharpoons CO_3^{2-} + H^+ \qquad (6)$$

Many lime-treated waters have a pH above the phenolphthalein endpoint—that is, they contain P alkalinity. In the range of pH 8.2 to 9.6, the bicarbonate and carbonate ions exist together in the absence of measurable carbon dioxide or hydroxyl ions. The distribution between the carbonate and bicarbonate ions can

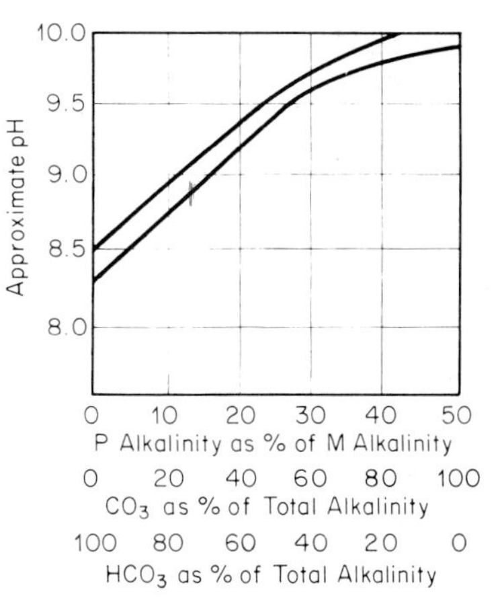

FIG. 4.3 Variation of the carbonate/bicarbonate ion distribution with pH, based on use of P and M titrations to determine CO_3^{2-} and HCO_3^-.

be determined approximately by the relationships shown in Figure 4.3. They can also be calculated after measuring P and M alkalinities according to the following equations.

$$CO_3 = 2 \times P$$

$$HCO_3 = M - CO_3 = M - 2P$$

The distribution of all forms of CO_2-related ions is shown by Figure 4.4. In this figure, free CO_2 is CO_2 gas; HCO_3^- referred to as halfbound CO_2, and CO_3^{2-} as bound CO_2.

As the pH increases above 9.6, the hydroxyl alkalinity becomes measurable.

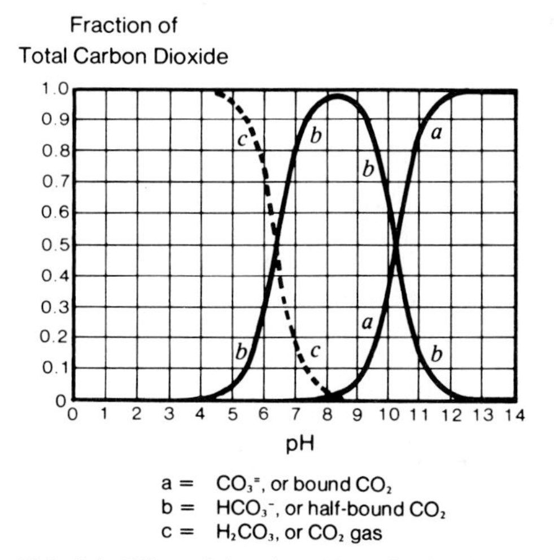

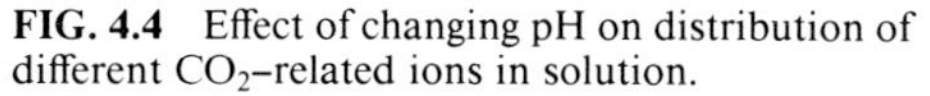

FIG. 4.4 Effect of changing pH on distribution of different CO_2–related ions in solution.

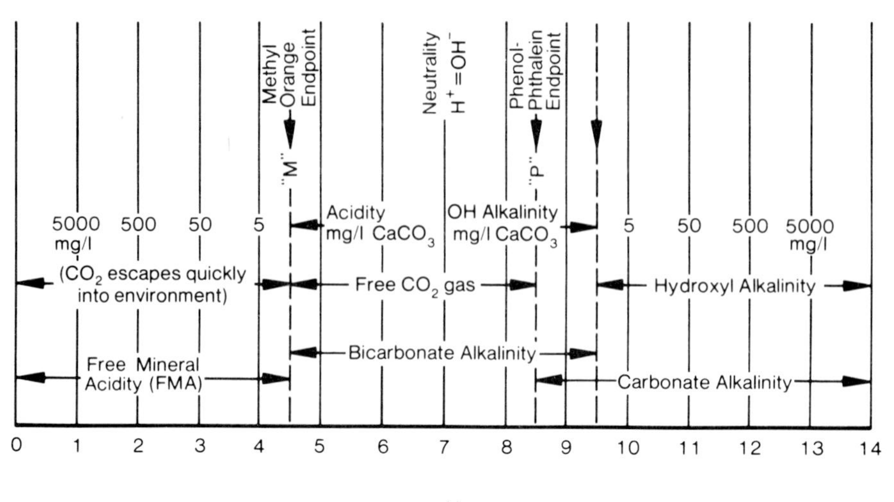

FIG 4.5 Acidity and various types of alkalinity and their pH ranges.

Its concentration can be determined by pH meter, as shown in Table 4.4, or the distribution between carbonate and hydroxyl can be calculated as follows:

$$CO_3 = 2(M - P)$$

$$OH^- = 2P - M$$

$$HCO_3^- = O$$

These are very useful approximations for interpreting most water analyses. The various regimes of acidity and alkalinity are illustrated by Figure 4.5.

INTERFERING IONS

These approximations are subject to some interferences, especially in contaminated waters. Common interferences are ammonia and alkaline anions other than carbonate and bicarbonate, such as sulfide, phosphate, silicate, and borate.

These P and M relationships have a long history of use in water treatment. Most plant data relating to results of lime softening, for example, are based on these convenient calculations. They trace their history to titration curves of Na_2CO_3 and NaOH added to water and neutralized with acid (Figure 4.6).

These relationships are not consistent, however, with the theoretical distribution between HCO_3^- and CO_3^{2-} based on dissociation constants. The theoretical distribution between these ions (Figure 4.4) indicates that at a pH as high as 10.3, for example, where OH^- alkalinity is about 10 mg/L as $CaCO_3$, there are equal quantities of HCO_3^- and CO_3^{2-} present.

Although inconsistent with this physical chemistry theory, in dealing with results of lime softening where pH exceeds 10 and alkalinity is less than 50 mg/L, this book uses the P and M relationships as a convenience and because so much

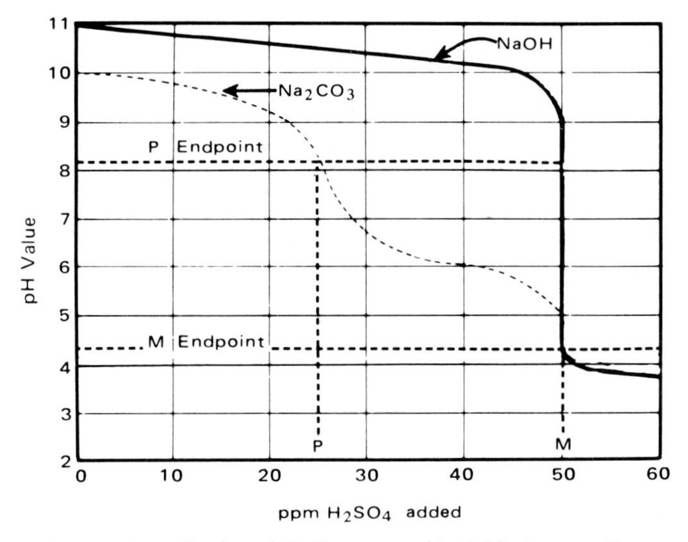

FIG. 4.6 Neutralization of alkaline water with H_2SO_4 (assume 50 mg/L M alkalinity).

field data have been developed from them. This practice tends to overstate carbonate alkalinity because it indicates no HCO_3^- alkalinity exists in the presence of OH^-. For example, at M = 50 mg/L and P = 25 mg/L, CO_3^{2-} = 50 mg/L at pH 9.8 (Figure 4.3); however, Figure 4.4 shows CO_3^{2-} to be only 25% of total alkalinity, or 12.5 mg/L at pH 9.8. This accounts for part of the discrepancy

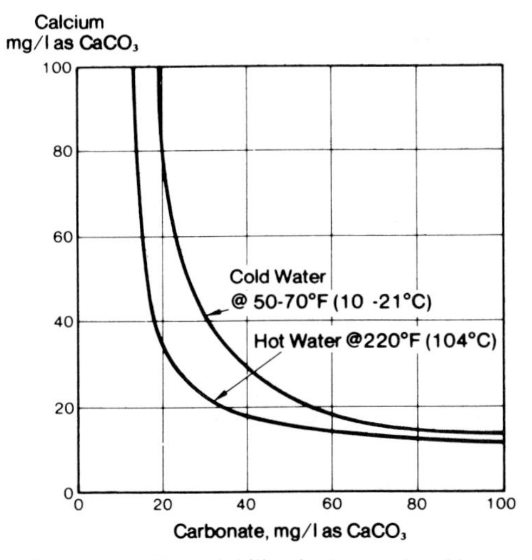

FIG. 4.7 $CaCO_3$ solubility in hot and cold systems—empirical data based on 60 to 90 min reaction time and settling time.

observed between theoretical $CaCO_3$ solubility and that actually achieved in practice by lime softening where the results are interpreted from the P and M relationships.

Examination of these acidity-alkalinity relationships in water reveals that the solubility of minerals such as calcium carbonate and magnesium hydroxide is more complicated to predict than the solubility product concept suggests. For example, in the case of calcium precipitating from water as calcium carbonate, it is now apparent that there are numerous equilibrium reactions occurring in the water, affecting the interpretation of simple chemical data. For that reason, water chemists usually rely on empirical data based on reported plant experiences to estimate results of a precipitation reaction.

Figure 4.7 shows data generally used for predicting calcium carbonate solubility, specifically related to precipitation in a lime-softening operation. The difference in results between hot process [above approximately 212°F (100°C)] and cold process (river water temperatures) is considerably beyond what would be predicted by simple solubility data, which show a decrease of only about 2 mg/L solubility at the higher temperature. The empirical temperature effect must be caused by other factors, such as rate of reaction [which, as a rule of thumb, doubles for each 18°F (10°C) temperature increase], the nature of the crystalline or amorphous precipitate, and the possible effect of coagulating organic materials at higher temperature.

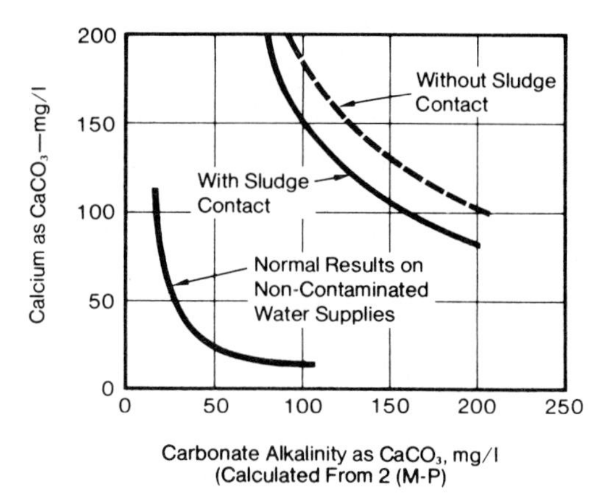

FIG. 4.8 Cold lime softening of sewage (upper two curves) compared with noncontaminated water.

As an example of this, in cold lime softening of municipal sewage, residual calcium carbonate is invariably higher than when the same process is used in the treatment of fresh water. This is illustrated in Figure 4.8. If, however, the sewage is heated above approximately 160°F (71°C), then results approach those achieved in fresh water.

EFFECTS OF IMPURITIES

If magnesium is precipitated with calcium carbonate, the residual calcium in solution may be increased. The inclusion of other impurities, such as strontium, has also been shown to increase the solubility of calcium carbonate. It would seem, then, that empirical data from a variety of plants should be used cautiously in estimating what to expect in a given situation; if the calculation involves lime treatment of an unknown water supply and the contamination levels are not fully defined, there is no substitute for an actual bench test to determine the response of that water to lime treatment.

Magnesium solubility is as difficult to predict with accuracy as calcium solubility, and for the same geneal reasons. The solubility of magnesium hydroxide is reported to be in the range of 30 to 40 mg/L at 77°F (25°C). Because of the difference in solubility product relationships,

$$K_{Ca} = [Ca^{2+}] \times [CO_3^{2-}] \qquad K_{Mg} = [Mg^{2+}] \times [OH^-]^2$$

magnesium concentration is affected more by changes in hydroxyl concentration than calcium is by changes in carbonate concentration. Any increase in carbonate produces a corresponding decrease in calcium, but an increase in hydroxyl concentration affects the magnesium concentration in a squared relationship. Thus a doubling of OH^- will cause Mg^{2+} to drop to one-fourth its original value.

When water is softened with lime—in the pH range of 9.5 to 10.5—magnesium precipitates as the hydroxide and the precipitate is positively charged. In the same pH range, $CaCO_3$ precipitate is negatively charged. Furthermore, silica in the system is usually present as a strong, negatively charged colloid. Sodium aluminate may be introduced in the system as a strong anionic complex. All of this may account for the coprecipitation of magnesium with calcium, the strong adsorption of silica on $Mg(OH)_2$ precipitate, and the low residual magnesium often achieved by the treatment of water with sodium aluminate in addition to lime.

In conventional softening operations, then, it is usual to show that lime treatment of fresh water produces a $CaCO_3$ solubility of approximately 35 mg/L cold and 25 mg/L hot, and a residual magnesium hydroxide solubility of approximately 35 mg/L cold and 2 to 3 mg/L hot.

Calcium carbonate and magnesium hydroxide are precipitated in the lime-softening operation; however, these can also precipitate from the unstable water—that is, a water containing these materials in a supersaturated condition—if anything is done to the system to upset the equilibrium. This could be an increase in temperature, a decrease in pressure, turbulence, or contact with surfaces which seed the chemical precipitation. The most common product of instability of fresh water is $CaCO_3$; in seawater the usual precipitate is $Mg(OH)_2$ or one of its complex carbonate hydroxide salts.

$CaCO_3$ STABILITY INDEXES

A water can be evaluated for its $CaCO_3$ stability by testing or by theoretical calculation from known data. The *marble test,* although infrequently used for control purposes, is a valuable test to illustrate the principles of calcium carbonate stability. If an unstable water is supersaturated, then the addition of finely powdered marble ($CaCO_3$) to a sample of the water should cause calcium carbonate to precipitate and coat the marble. The effect is measured by a reduction in both the hardness and alkalinity of the water, measured after the suspension has settled and the water has been filtered. There is also a decrease in pH of the marble-contacted water, which is said to have a *positive saturation index.*

On the other hand, if the water sample is undersaturated with respect to $CaCO_3$—that is, if it is an aggressive, corrosive water—then some of the fine marble added to the sample dissolves, increasing the hardness, alkalinity, and pH. This water is said to have a *negative saturation index.*

Working with the $CaCO_3$ equilibrium values, incorporating the dissociation factors for carbonic acid, bicarbonate, and carbonate, and based on the theoretical solubility of $CaCO_3$ at different temperatures, as affected by water salinity, Langelier developed a method for predicting the saturation pH (called pH_s) of any

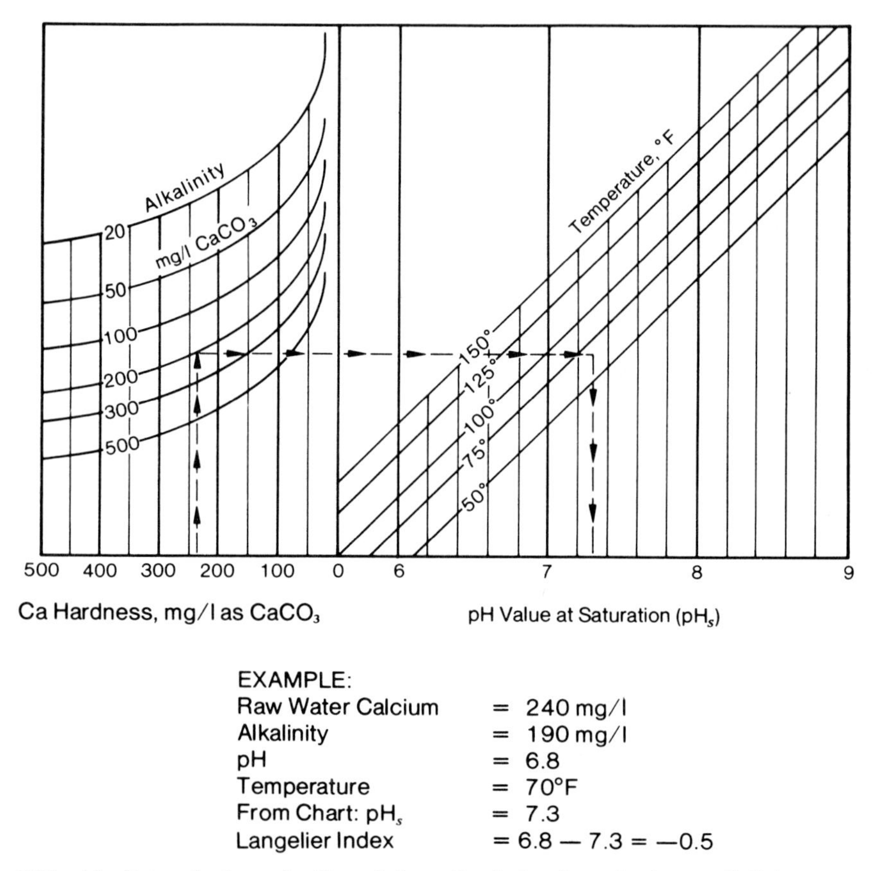

FIG. 4.9 Determination of pH_s and Langelier index from hardness, alkalinity, and temperature.

water. If the actual pH of the water is below the calculated level (pH_s), the water has a negative Langelier index and will dissolve $CaCO_3$. This is generally also interpreted to indicate that the water may be corrosive to steel if oxygen is present. If the measured pH exceeds pH_s, the Langelier index is positive, and being supersaturated with $CaCO_3$, the water is likely to form scale. The greater the deviation of actual pH from pH_s, the more pronounced is the instability. The saturation pH_s and the Langelier index can be determined from a water analysis by reference to Figure 4.9.

$$\text{Langelier index} = pH - pH_s$$

Based on studies of reported conditions of scaling and corrosion in a variety of municipal systems, Ryznar modified the Langelier index to more reliably predict the likelihood of scale forming or corrosion occurring, based on pH and Ph_s, using his *Ryznar index* or *stability index.*

$$\text{Stability index (SI)} = 2pH_s - pH$$

In using this index, a water solution is considered to be corrosive when the stability index exceeds approximately 6.0 and to be scale-forming when the index is less than 6.0

The Langelier index is most useful in predicting events in a bulk system (low velocity of flow), such as a lime softener, filter, or reservoir. The Ryznar index is empirical and applies only to flowing systems, where the environment at the pipe wall is quite different from that of the bulk water. (See Figures 20.1, 20.4, and 20.6) If corrosion is occurring, oxidation-reduction reactions create different conditions at the cathode and anode from those of the flowing water; if suspended solids are present, the velocity of flow has an important effect on potential deposit formation from sedimentation that will then influence the aqueous environment wetting the pipe wall. As a general rule, the Langelier index is most useful in bulk systems and the Ryznar index in flowing systems where the velocity is greater than about 2 ft/s (0.6 m/s), or sufficient to prevent sedimentation.

The Langelier index has been modified by Stiff and Davis for oil-field brines, where the high salinity affects ionic strength and influences $CaCO_3$ solubility. This index is:

$$SDI = pH - pCa - pAlk - K$$

where pH is used as measured,

$$pCa = \log I/[Ca]$$

$$pAlk = \log I/[Alk]$$

$$K = \text{constant based on total ionic strength and temperature}$$

Ionic strength of the solution is calculated as in Table 4.6; K is then calculated from temperature using Figure 4.10; pCa and pAlk are obtained by reference to Figure 4.11.

Such data developed for oil-field brine may become useful in wastewater applications such as treatment of open recirculating cooling water systems for zero discharge. Similar indexes have been developed for other commonly occurring minerals depositing in distribution systems and heat exchangers. Notable among these indexes are those for calcium phosphate and magnesium hydroxide. These are of much more limited value to the water chemist than the Langelier or Ryznar index, chiefly because of the variety of complexes which form and the different solubility values for each of the complexes involved in the system.

The effect of ionic strength is very pronounced in the illustration of $CaSO_4$ solubility in a concentrating recirculating cooling water, shown in Figure 4.12.

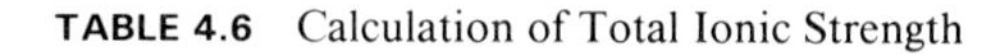

TABLE 4.6 Calculation of Total Ionic Strength

1. Calculate all ions in mg/L as $CaCO_3$.
2. Multiply monovalent ions by 1×10^{-5}, and divalent ions by 2×10^{-5}.
3. Add the sum of factored ions to determine total ionic strength, μ (all ions in mg/L as $CaCO_3$) Example:

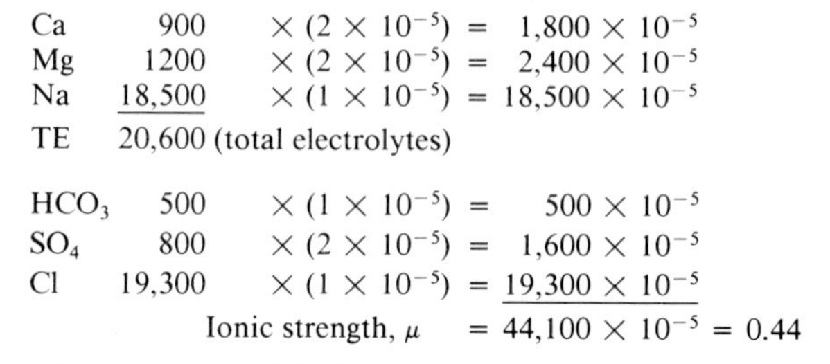

Ca	900	$\times (2 \times 10^{-5})$ =	$1{,}800 \times 10^{-5}$
Mg	1200	$\times (2 \times 10^{-5})$ =	$2{,}400 \times 10^{-5}$
Na	18,500	$\times (1 \times 10^{-5})$ =	$18{,}500 \times 10^{-5}$
TE	20,600 (total electrolytes)		
HCO_3	500	$\times (1 \times 10^{-5})$ =	500×10^{-5}
SO_4	800	$\times (2 \times 10^{-5})$ =	$1{,}600 \times 10^{-5}$
Cl	19,300	$\times (1 \times 10^{-5})$ =	$19{,}300 \times 10^{-5}$
		Ionic strength, μ =	$44{,}100 \times 10^{-5} = 0.44$

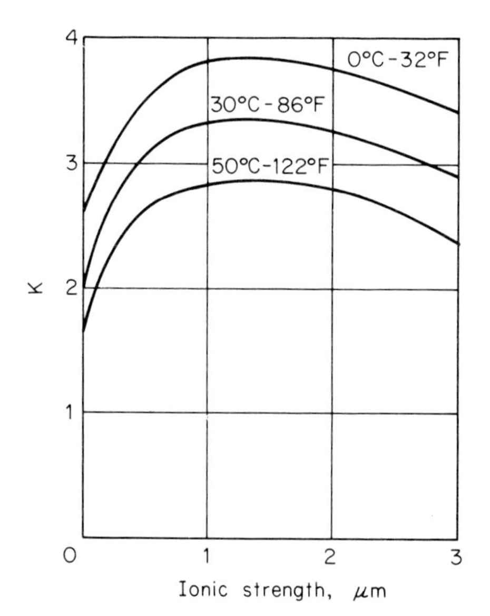

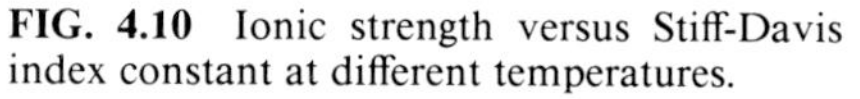

FIG. 4.10 Ionic strength versus Stiff-Davis index constant at different temperatures.

The predicted solubility of Ca with increasing SO_4 concentration is shown by curve A; actually, the Ca concentration reaches a minimum of about 900 mg/L as $CaCO_3$, which is maintained even at SO_4 concentrations in excess of 25,000 mg/L.

DISSOLVED GASES

Further consideration of the Langelier index provides instruction on the nature and effects of gases dissolved in water. Carbon dioxide, for example, dissolves to

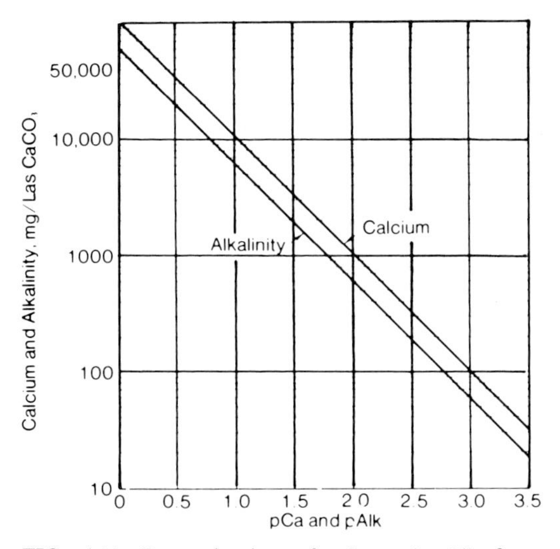

FIG. 4.11 Determination of pCa and pAlk from concentrations.

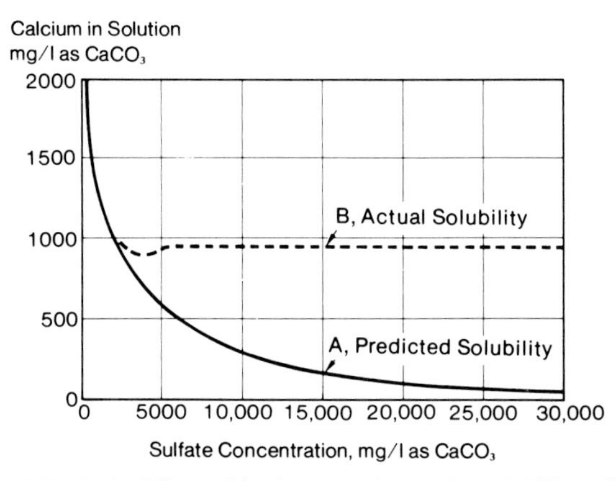

FIG. 4.12 Effect of ionic strength on the solubility of $CaSO_4$ in a recirculating cooling water.

form carbonic acid, which ionizes to produce H^+ and HCO_3^- ions. Other ionizing gases include sulfur dioxide, hydrogen sulfide, and hydrocyanic acid, which form weak acids when dissolved in water. Because of this effect, the addition of a strong acid to water containing these gases will force the equilibrium to the left, essentially eliminating the ionized portion so that all of the gas is in molecular form and free to escape from the water as gas molecules. For example, with CO_2:

$$CO_2 + H_2O \leftrightharpoons H_2CO_3 \leftrightharpoons H^+ + HCO_3 \quad (7)$$

The addition of H^+ ions by addition of H_2SO_4 would force the reaction to the left.

On the other end of the scale is ammonia, which dissolves and ionizes to form a weak base according to the equation:

$$NH_3 + H_2O \rightleftharpoons NH_4OH \rightleftharpoons NH_4^+ + OH^- \quad (8)$$

As opposed to acidic gases, which are converted to the molecular form by the addition of a strong acid, in this case the addition of a strong alkali, such as sodium hydroxide, will force the reaction to the left and produce molecular ammonia, which is then free to escape as a gas molecule (Figure 4.13).

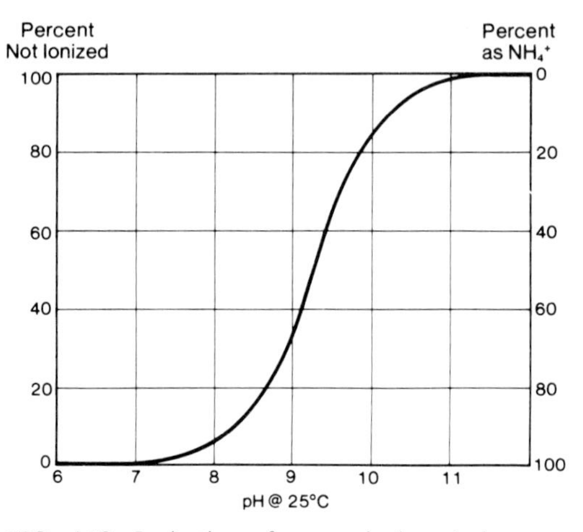

FIG. 4.13 Ionization of ammonia in solution as a function of pH.

There are other gases that dissolve in water, but do not ionize. Among these are oxygen, nitrogen, and methane. The most prominent of these in its effect on water systems is oxygen. If a water solution has a negative Langelier index (a positive stability index), it is considered agressive and corrosion is expected. This corrosion, however, is usually caused by dissolved oxygen which is free to attack the bare metal surface when it is wetted by a water which cannot form a protective calcium carbonate film or scale. An example of this process is condensate in a steam plant: in a tight system free of oxygen, plain carbon steel piping is successfully used to handle this condensate without undue corrosion problems; if oxygen finds its way into the system through valve or pump leaks, the condensate becomes extremely aggressive and corrosion occurs, unless the metal can be protected by a film-forming corrosion inhibitor.

MINERALS AND CONDUCTANCE

An important consideration in water chemistry is electrical conductivity. The higher the mineral content of the water, the higher its conductivity. This has several important consequences. First, the higher the conductivity, the more freely can electrical current flow through the water, and the more rapid is the corrosion rate if other conditions favor corrosion. Second, the higher the conductivity, the less completely ionized are the minerals dissolved in the water, as the ions are packed more closely together and collide more frequently. This decreases the activity coefficient, or freedom, increasing the solubility of $CaCO_3$ and other slightly soluble materials. As a result, $CaCO_3$ is more soluble in seawater than in fresh water under equal conditions of pH, alkalinity, and temperature, as explained in the earlier discussion of the Stiff-David stability index.

There are many factors involved in the corrosion process, of which pH, conductivity, temperature, and dissolved oxygen are usually foremost. Other important factors in the corrosion mechanism include the presence of dissimilar metals in the system, with the more anodic becoming corroded, and dissimilarities in the metallurgical structure of a single metal in the system. An example of the latter is zinc in brass, whereby one portion of the metal surface, the zinc crystals, become anodic relative to an adjoining cathodic portion, the copper.

All of these corrosion mechanisms may be magnified by the presence of suspended matter, particularly if velocities are low and the material deposits to form a concentration cell, or the velocities are exceedingly high and a combination of erosion and corrosion magnifies the individual effects of each to produce a very rapid rate of attack.

SOLID MATTER

Solid matter occurs in most waters as suspended solids and colloidal matter. The concentration of suspended solids is determined by filtration, the collected solids on the filter membrane being dried and weighed. Those suspended solids which are large and heavy are called settleable solids, and these may be determined volumetrically in a settling cone as a simple control test (Figure 4.14) or weighed. The solids remaining with supernatant water above the settled matter are fine and called *turbidity.*

Turbidity in water is measured by the effect of the fine suspended particles on a light beam (Figure 4.15). Light-interference analytical methods are classified as nephelometric, and one system of turbidity measurement uses *nephelometric turbidity units* (NTU). The original nephelometric method used a standardized candle, providing results in *Jackson turbidity units* (JTU), named for the man who developed the standard candle. Turbidity standards prepared with formazin for comparator tube determinations have given rise to a third turbidity unit, the FTU.

The JTU is measured with a transmitted light beam, while the NTU is measured by light scattering, so there is no comparison between the two units that applies to all waters. In the case of turbidity standards prepared from 325 mesh diatomaceous earth, a reading of 100 NTU is equivalent to about 40 JTU.

Colloidal matter in water is sometimes beneficial and sometimes harmful. Beneficial colloids are those that provide a dispersant effect by acting as protective colloids. In a suspension of clay or silt in water, the smallest particles usually carry a negative charge. If these negative charges are neutralized, the particles are coagulated; however, if additional negative charges are introduced into the system deliberately, sedimentation may be prevented. Sodium silicate probably works in this way; so too do some kinds of organic materials in water, such as lignins and tannins, which occur in highly colored water sources. Silica in itself can be troublesome, forming a very hard scale when it deposits on heat transfer surfaces. So

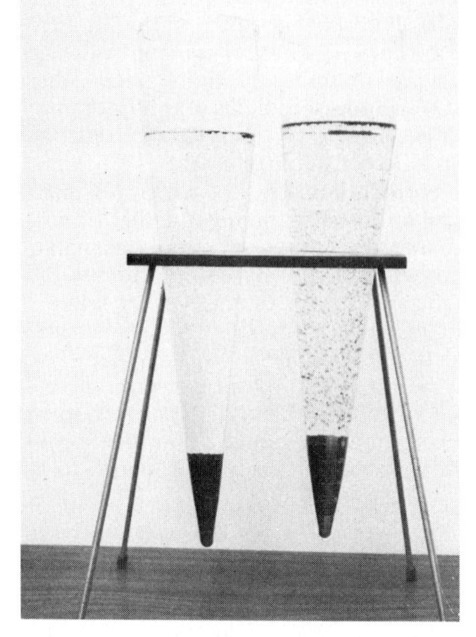

FIG. 4.14 Volumetric determination of settleable solids in Imhoff cones.

FIG. 4.15 Turbidity in this water scatters Tyndall beam as it passes through cell.

adsorbents are used for the removal of these colloidal materials from water when they exceed acceptable concentration limits. Many of the heavy metals are present in colloidal form and are removed from water by coagulation, filtration, adsorption, or a combination of these methods.

ORGANICS IN WATER

While tannin and lignin are generally present as colloidal suspensions in water, many organic compounds are actually soluble in water. There is an astounding variety of organic compounds in water, put there both by nature and humankind. It is unusual for a water analysis to show all of this organic matter broken down by individual molecular constituents, because of the great variety. If the analyst should know what may be present [for example, polychlorinated biphenyls (PCBs), or total trihalomethane (TTHM)], it can be found. But in the absence of this kind of information, a battery of nonspecific tests must be relied on. Those tests shown in Table 4.7 are evidence of carbonaceous materials, but cannot distinguish between them. They are reasonably simple, inexpensive control methods. Interpretation is usually difficult; it is not at all unusual to find an increase in one test, such as BOD, with a concurrent decrease in another test, such as TOC.

Figure 4.16 shows the diversity of change in color, TOC, and BOD in the Hackensack River. Variations in color, LOI and COD for the Hudson River are shown in Figure 4.17. Finally, Table 4.8 shows an increase in BOD with a decrease in COD along the Connecticut River at progressive downstream sampling stations.

TABLE 4.7 Organic Matter in Water, Nonspecific Tests

Test	Description
BOD	**Biochemical oxygen demand.** Measures the ability of common bacteria to digest organic matter, usually in a 5-day incubation at 20°C, by analyzing depletion of oxygen. This measures biodegradable organic matter. Expressed as O_2.
CCE	**Carbon-chloroform extract.** Organics adsorbed on an activated carbon cartridge are extracted from the carbon by chloroform. The extract is weighed or further analyzed.
CAE	**Carbon-alcohol extract.** Organics adsorbed on an activated carbon cartridge are extracted from the carbon by ethyl alcohol. (This is performed after the chloroform extraction of the same column.)
COD	**Chemical oxygen demand.** Measures the ability of hot chromic acid solution to oxidize organic matter. This analyzes both biodegradable and nonbiodegradable (refractory) organic matter. Expressed as O_2.
Color	**Color.** This is a rough measure of tannin, lignin, and other humic matter in surface waters and certain wastes, such as kraft pulping wastes. Reported in APHA units, related to platinum standard.
IDOD	**Immediate dissolved oxygen demand.** Measures the presence of strong reducing matter in wastes which would have an immediate effect in reducing the oxygen level of receiving streams. Determined by measuring reduction in oxygen 15 min after dilution of sample with an O_2-saturated water. Expressed as O_2.
LOI	**Loss on ignition.** In this determination, the water sample is first evaporated to dryness, then weighed. The solids are then heated to a dull red heat and reweighed. The difference in the two weights is loss on ignition. The firing step burns off organic matter, but it also affects mineral composition, as in the breakdown of carbonates to oxides, so the loss is not all organic.
O_2 consumed	**Oxygen consumed from permanganate.** Measures the susceptibility of organic matter in sample to oxidation by $KMnO_4$. Not as strongly oxidizing as chromic acid. Preceded the COD test and still a useful control test. Reported as O_2.
Solvent extractables	**Extractables.** Measures the organic matter directly extractable from water, usually using hexane, although CCl_4 and $CHCl_3$ are also used.
TOC	**Total organic carbon.** Measures the CO_2 produced from organics when a water sample is atomized into a combustion chamber. The CO_2 equivalent to the alkalinity may be removed first, or this equivalent may be subtracted from the total CO_2 to determine organic carbon. Reported as C.

A great deal of experience is required, involving collection of considerable data for analysis, to decide on the significant tests for organic matter for a given water treatment control application. Some help can be gained in recognizing that the COD test itself involves a powerful oxidizing agent, chromic acid; therefore, the test is comparable to combustion of the organic matter in an aqueous medium. On this basis, some theoretical relationships between the several organic indexes can be drawn, as shown in Table 4.9.

A computer regression analysis of BOD, TOC, and COD data of wastes from nine standard industrial categories (SICs) of plants was undertaken to determine

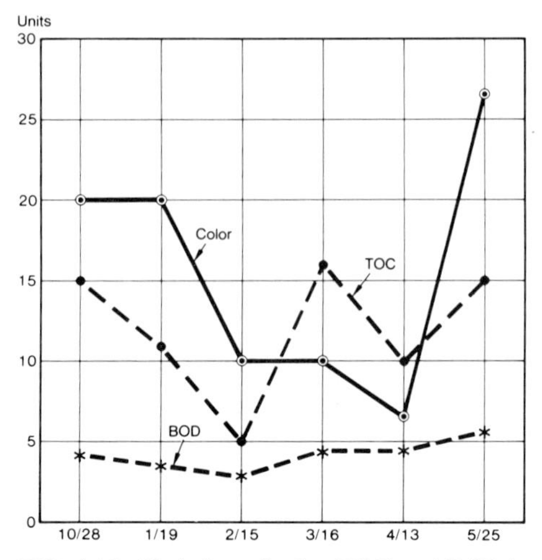

FIG. 4.16 Variation of color, TOC, and BOD in the Hackensack River.

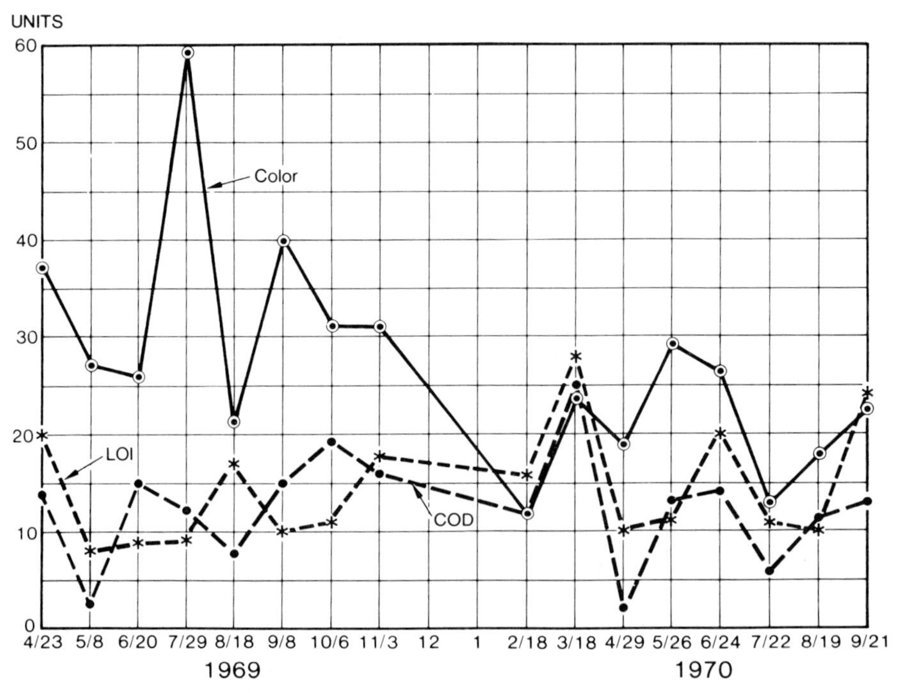

FIG. 4.17 Variation of color, LOI, and COD in the Hudson River.

TABLE 4.8 Connecticut River Organic Matter

Station	COD	BOD	Other
Wilder, Vt. (9/1/70)	19	0.8	Color—32
Northfield (9/1/70)	15	1.4	Color—12
Thompsonville (9/1/70)	9	2.8	LOI—19

Source: USGS Paper 2151.

TABLE 4.9 Organic Matter in Water—Theoretical Relationships of Several Indicators

1. Total organic carbon (TOC), as carbon (C)
   a. If present as a carbohydrate, $(CH_2O)_n$, TOC × 2.5 = organic matter
   b. If present as a biomass,* TOC × 2.8 = organic matter
2. Chemical oxygen demand (COD) as O_2†
   a. If present as carbohydrate, $(CH_2O)_n$, COD = 2.67 × TOC as O_2
   b. If present as biomass, COD = 3.6 × TOC, as O_2
   c. If present as hydrocarbon, $(-CH_2-)_n$, COD = 3.4 × TOC, as O_2
3. Biochemical oxygen demand $(BOD)_5$, as O_2
   a. If present as carbohydrate or biomass, BOD is approximately 0.7–0.8 × COD
   b. Other organic materials show lower ratios of BOD/COD as follows:
4. Total organic nitrogen, as N
   If present in biomass, N × 16 = organic matter

* For algal biomass, a relationship of $C_{106}H_{263}O_{110}N_{16}P$ has been suggested; for bacterial matter, the relationship $C_{60}H_{87}O_{23}N_{12}P$ has been used. These correspond to a C:H:O relationship of 1:2.5:1 and 1:1.45:0.4 respectively compared to 1:2:1 for carbohydrates.

† The ratio of COD/TOC varies considerably from sample to sample, even on a common waste stream from a single plant. Figure 4.18 plots COD and TOC on a chemical plant waste. Although the plot shows an average ratio of 3.0, individual samples range from 2 to 5.

*This ratio is a rough measure of the ability of bacteria to digest the organic matter present (biodegradability).

4. Total organic nitrogen, as N
   If present in biomass, N × 16 = organic matter

* For algal biomass, a relationship of $C_{106}H_{263}O_{110}N_{16}P$ has been suggested; for bacterial matter, the relationship $C_{60}H_{87}O_{23}N_{12}P$ has been used. These correspond to a C:H:O relationship of 1:2.5:1 and 1:1.45:0.4 respectively compared to 1:2:1 for carbohydrates.

† The ratio of COD/TOC varies considerably from sample to sample, even on a common waste stream from a single plant. Figure 4.18 plots COD and TOC on a chemical plant waste. Although the plot shows an average ratio of 3.0, individual samples range from 2 to 5.

how these values related to each other. In general, the BOD-TOC and TOC-COD relationships were linear. The BOD-TOC relationship tended to be geometric. However, the linear data (Figure 4.18) and the geometric data (Figure 4.19) are quite scattered.

Finding a specific organic molecular species in water usually requires sophisticated procedures and equipment and is generally too expensive to be justified

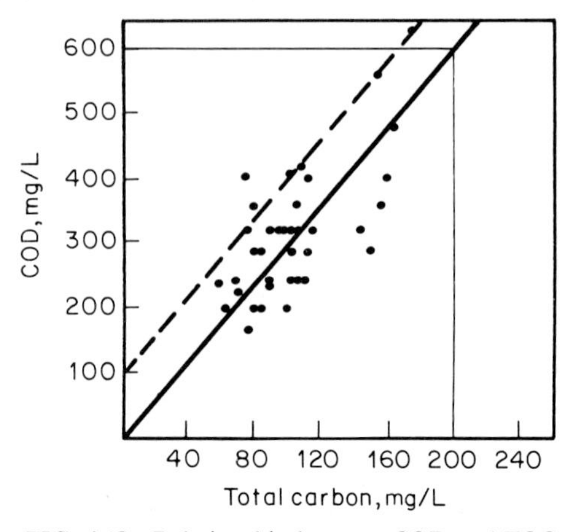

FIG. 4.18 Relationship between COD and TOC in chemical plant's waste (solid line). Some wastes have a background noncarbonaceous COD, as shown by dashed line; others have different slopes.

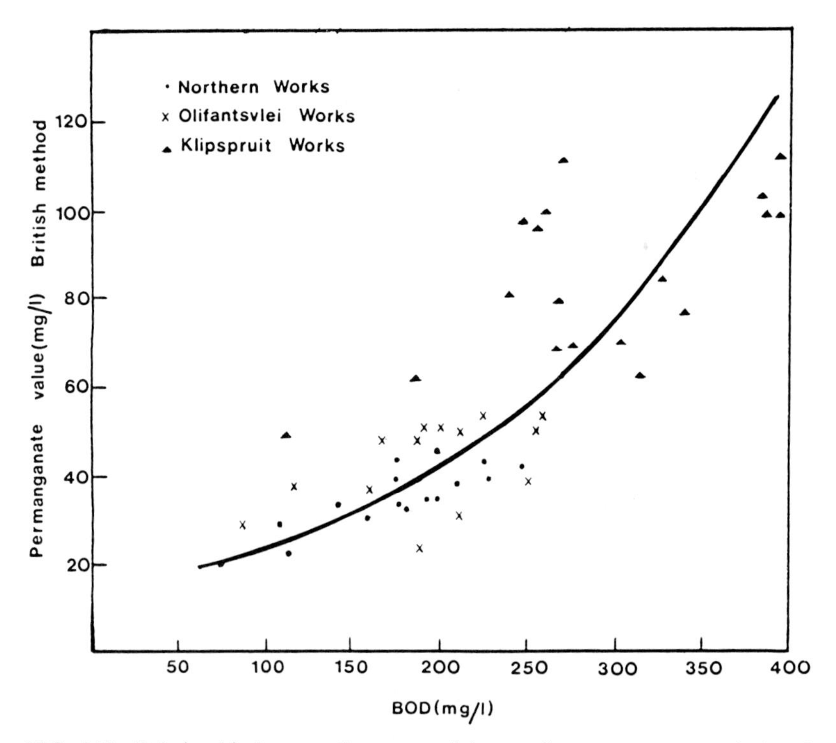

FIG. 4.19 Relationship between O_2 consumed (reported as permanganate value) and BOD in testing effluent from a primary basin in a sewage plant. A special permanganate solution was used. [*From Osborn, D. W. "Evolution of Wastewater Purificattion Process in Johannesburg, South Africa," J. Water Pollut. Control Fed.,* ***51:*** *3(472) (March, 1979). Copyright Water Pollution Control Federation, reprinted with permission.*]

TABLE 4.10 Some Problems Caused by Water Impurities

	Process water		Boiler water			Cooling water		
	Industries affected	How affected	Deposits	Corrosion	Other	Deposits	Corrosion	Other
Hardness (Ca, Mg)	All Paper and textile Laundries	Scale and deposits Deposits on fibers Soap scum on fabrics	M*	—	—	M	—	—
Alkalinity	Paper, textile, beverages	Destroys acid reagents, dyes, alum, flavors	—	M(1)	—	M	—	—
Dissolved solids	Electronics, utilities	Adds to cost of making high-purity water	—	—	High blow-down	—	C	High blow-down
Suspended solids	All	Deposits, wear	M	—	—	M	M	—
Dissolved oxygen	All	Major cause of corrosion	—	M	—	—	M	—
Carbon dioxide	All (3)	May flash at well pump and cause scaling	—	M(1)	—	—	—	Affects pH
Iron and manganese	All Paper, textile	Deposits and discoloration Stains fibers	M	—	—	M	—	—
Organic matter	Foods, beverages All	Tastes and odors Food for bacteria Fouls ion-exchange resins	—	—	May cause foaming	—	—	May cause foaming
Silica	—	—	C	—	M(2)	C	—	—
Microorganisms	All	Produce slimes and odors	—	—	—	M	—	—

Code: M, major factor in problem; C, contributor to problem; — no significant effect. (1) in steam condensate system; (2) in steam turbine; (3) deep well water supply.

as a routine control; it may be done only for specific purposes where toxic materials may be suspected and the analysis is essential for environmental control purposes. Even in this case, the analysis is usually performed on a daily composite, hourly grab sample, or other scheduled samples, since continuous analysis is seldom practical.

INTERPRETATION: PROBLEMS CAUSED BY IMPURITIES

The orientation in basic water chemistry presented here has included an overview of some of the major impurities present in water and has suggested the nature of the problems they cause. Table 4.10 briefly summarizes the most common problems arising from the presence of such major impurities as hardness, iron, and

TABLE 4.11 A Summary of Some Results Obtained from Water Treatment Processes

Substances in water	Process used for removal	Chemical used
Hardness	Precipitation	Lime, soda ash, caustic, phosphate,
	Ion exchange	Salt, acid
Alkalinity	Precipitation	Lime, gypsum
	Ion exchange	Acid, salt
	Neutralization	Acid
Carbon dioxide	Precipitation	Lime
	Ion exchange	Caustic
	Neutralization	Lime, caustic
	Degasification	None
Dissolved solids	Reverse osmosis	None
	Reduction by removal of separate components adding to dissolved solids	
Suspended solids	Coagulation, flocculation, sedimentation	Alum, aluminate, coagulant aids
Iron and manganese	Oxidation and filtration	Chlorine, lime
	Precipitation	Lime, chlorine, air
	Ion exchange	Salt, acid
Silica	Precipitation	Iron, salts, magnesia
	Ion exchange	Caustic
Organic matter	Clarification	Alum, aluminate
	Oxidation	Chlorine
	Adsorption	Activated carbon
Oxygen	Degasification	None
	Reduction	Sulfite, hydrazine
Microrganisms	Clarification	Various, coagulants, coagulant aids
	Sterilization	Chlorine, sterilants, heat

microorganisms. Subsequent sections of this book cover these in much more detail. Referring to these same impurities, Table 4.11 focuses on some of the processes and chemicals most generally used to control them for avoidance of problems of water use both in industry and by municipalities.

SUGGESTED READING

AWWA: *Water Quality and Treatment,* McGraw-Hill, New York, 1971.

Blanchard: D. C.: *From Raindrops to Volcanoes,* Anchor Books, Doubleday, New York, 1967.

Cowan, J. C., and Weintritt, D. G.: *Water-Formed Scale Deposits,* Gulf, Houston, 1976.

Day, A. D. and Davis, K. S.: *Water, The Mirror of Science,* Anchor Books, Doubleday, New York, 1961.

Deevey, E. S. Jr.: "Mineral Cycles," *Sci. Am.,* September 1970, p. 149.

Henderson, J. J.: *The Fitness of the Environment,* Beacon Press, Boston, 1958.

Hill, David R., and Spiegel, Stuart J.: "BOD, TOC, and COD in Industrial Wastes," *Industrial Wastes,* **21** November/December 1979.

Miheuc, E. L., and Luthy, R. G.: *Estimate of Calcium Carbonate Scaling Technology in Blast-Furnace Recycle Water,* Intl. Water Conf. Paper IWC-82-16, October 1982.

Nordell, Eskel: *Water Treatment for Industrial and Other Uses,* Reinhold, New York, 1961.

Nancollas, G. H., and Reddy, M. M.: "Crystal Growth Kinetics Minerals Encountered in Water Treatment Processes," in *Aqueous-Environmental Chemistry of Metals* (Ruben J., ed.), Ann Arbor Science, Ann Arbor, Mich., 1976.

Rapport, S., Wright, H. et al., eds.: *The Crust of the Earth,* New American Library, New York, 1955.

Rhodes, Daniel: *Clays and Glazes for the Potter,* Chilton, Philadelphia, 1971.

Rubin, Alan J.: *Chemistry of Wastewater Technology,* Ann Arbor Science Publishers, Ann Arbor, Mich., 1978.

Stumm, W., and Morgan, J. J.: *Aquatic Chemistry,* Wiley, New York, 1970.

Weber, W. J., Jr., et al.: *Physicochemical Processes for Water Quality Control,* Wiley, New York, 1972.

Zim, H. S., Shaffer, P. R., and Perlman, R.: *Rocks and Minerals,* Simon & Schuster, New York, 1957.

CHAPTER 5
AQUATIC BIOLOGY

Scientists picture the primordial earth as a planet washed by a hot sea and bathed in an atmosphere containing water vapor, ammonia, methane, and hydrogen. Testing this theory, Stanley Miller at the University of Chicago duplicated these conditions in the laboratory. He distilled seawater in a special apparatus, passed the vapor with ammonia, methane, and hydrogen through an electrical discharge at frequent intervals, and condensed the "rain" to return to the boiling seawater. Within a week the seawater had turned red. Analysis showed that it contained amino acids, which are the building blocks of protein substances.

Whether this is what really happened early in the earth's history is not important; the experiment demonstrated that the basic ingredients of life could have been made in some such fashion, setting the stage for life to come into existence in the sea. The saline fluids in most living things may be an inheritance from such early beginnings.

A HEALTHY AQUATIC ENVIRONMENT

There are both physical and chemical conditions that define a healthy environment for aquatic organisms. Among the physical conditions are temperature, pressure, osmotic pressure, light, and turbidity.

Although a few organisms can live in hot water, most prefer a moderate temperature. Because there are so many varieties of acquatic life, it is impossible to generalize about an optimum temperature for a healthy aquatic environment. Among bacteria, some prefer warm water, some cold, and others grow best at intermediate temperatures (Figure 5.1). The majority of aquatic organisms have become acclimated to the prevailing ranges of temperatures of surface waters and underground aquifers. But the demand of life to exist and propagate itself is so strong that old ideas of fixed limits to the environmental regimes of temperature, pressure, pH, and salinity are being overturned as biologists discover new life forms in seemingly hostile surroundings. Bacteria have recently been found adjacent to ocean-floor vents living in water at 250°C, kept liquid by the immense pressure of the ocean above—a challenge to previous understandings of pasteurization and denaturation of protein structures by heat.

Freshwater fish populations are broadly classed as game fish and rough fish depending on preference for cold water or warm water, respectively. This is the extent to which generalizations can be made about water temperature without focusing on specific organisms.

The influence of pressure is also specific for different organisms. Pressure is determined by the water depth, but the preferred depth for an organism is not always directly related to pressure. It may be more closely related to the depth of sunlight penetration, temperature, dissolved oxygen content, and other factors indirectly influenced by depth.

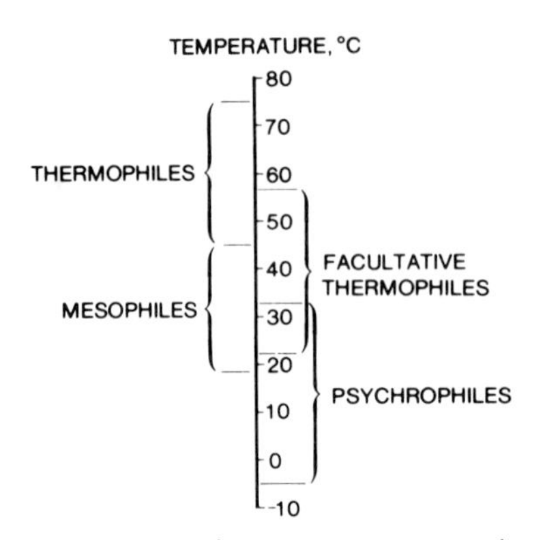

FIG. 5.1 Approximate temperature regimes of common bacteria.

Osmotic pressure is an important consideration because it affects the movement of water into or from the cell. Osmosis is the process by which water molecules move through a permeable membrane separating a weak solution from a strong solution in a direction to dilute the stronger solution. A cell that has reached equilibrium with seawater may rupture because of the inrush of water molecules if it is suddenly exposed to fresh water. Since osmotic pressure is influenced by solution concentration, the salinity of the aquatic environment is an important factor controlling the biota able to live in that environment. Solutions causing swelling of the cell are hypotonic; those producing shrinkage are hypertonic; and those in equilibrium with the cell fluid are isotonic.

Another important physical consideration is water turbidity. Suspended solids and turbidity have important consequences, including obstruction of sunlight, blanketing of the bottom causing a shift in the population of bottom organisms, and abrasion or clogging of gills and other organs of larger aquatic species.

THE ROLE OF CHEMISTRY

The chemical nature of the aquatic environment plays a major role in establishing and supporting the aquatic population. But just as the chemical conditions affect the biota, so too do the organisms affect their chemical environment.

The primary chemical consideration is food. A class of organisms called autotrophs, including algae and certain bacteria, are able to synthesize their own food

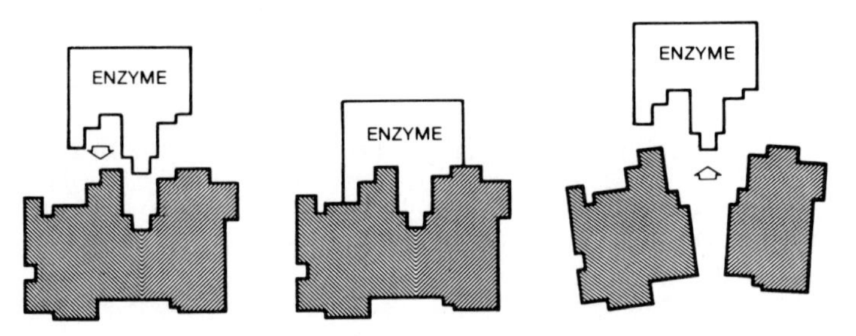

FIG. 5.2 The exact mechanism by which enzymes break down large molecules is uncertain, but it is believed to be related to a matching of molecular shapes.

from carbon dioxide and water. The more complex carbon-containing organic matter, such as polysaccharides, becomes food for other types of microorganisms. Some of these organic molecules are too large to pass through the cell membrane. Bacteria have developed a process for taking care of this problem: they produce enzymes which break down the large organic molecule into smaller components which can then be transported across the cell membrane and used as food (Figure 5.2). Larger organisms use algae and bacteria as their food supply, and in turn become the food for higher forms of life. This progression is known as the food chain.

Those organisms that feed on algae and other forms of plant life are herbivores; those that prey on protozoa and larger animals are carnivores. Some organisms serve a useful purpose in feeding on decaying matter; these are known as saprophytes.

SIX IMPORTANT ELEMENTS

In the process of metabolism, all cells must reproduce themselves and form new cellular matter. The most common elements required for biosynthesis in the order of their occurrence are C, H, O, N, and S. A sixth element, P, is of considerable importance also, especially as it is present in sewage and is considered a good nutrient for algae. It is also a major component of agricultural fertilizers, so it is present in storm runoff from farmland. There is usually an ample supply of C, H, O, and S in various forms in the water to encourage growth of aquatic life. In the deliberate culture of bacteria, such as in the biological digestion process for waste treatment, there may be insufficient N and P in the water for healthy growth. In that case, these elements are added to produce the ratio represented by the empirical formulas sometimes used to represent some of the major constituents in protoplasm, $C_{106}H_{263}O_{110}N_{16}P$ in the case of algae and $C_{60}H_{87}O_{23}N_{12}P$ for aerobic bacteria. In addition to these major nutrients, there are countless minor nutrients, some of which are generally required by almost all organisms, and some of which are specific for certain organisms. One example is magnesium: it is required by all organisms for cell growth; but it also has a specific use in the chlorophyll molecule, which is required by plant life; a second is iron, which is generally needed in small amounts by all cells, but is specifically needed for the hemoglobin found in the blood stream of animals. Aquatic organisms concentrate different minerals in different organs of their bodies. For example, the respiratory system of the sea cucumber contains a high concentration of vanadium; the distinctive flavor of scallops is due to the presence of copper in its hemocyanin, analogous to the iron in hemoglobin.

Low pH in water has a primary effect on aquatic organisms because it inhibits enzymatic activity. Some bacteria can live in acid conditions, as illustrated by the bacterial activity responsible for the acidity in coal mine drainage. Other organisms prefer alkaline conditions and can exist at fairly high pH levels; certain microbes have adapted to the hostile environments of the saline lakes of Kenya, having pH values as high as 11. Algae faced with the need to obtain CO_2 will break down bicarbonate to carbonate, in the absence of CO_2 gas, often producing a pH above 9. However, most organisms seem most comfortable in a pH range of about 6.5 to 8.5.

The pH of the aquatic environment may have secondary, indirect effects. For example, if ammonia is present, pH determines the ratio of the ionized portion to the portion present as a gas. Since nonionized ammonia may be toxic to certain aquatic organisms, pH can have an indirect influence on their health.

SOME NEED OXYGEN, OTHERS DON'T

Oxygen influences both microscopic life and larger life forms. The bacteria that require oxygen, and that produce CO_2 as a by-product, are known as aerobic organisms. Those that can live without oxygen are called anaerobes. Some, called facultative organisms, can adapt to either situation. In a deep lake, the oxygen at the surface will support aerobic organisms, and the bottom layers devoid of oxygen may support a population of anaerobic bacteria. The action of anaerobic bacteria is called fermentation, and the by-products are carbon dioxide, methane, hydrogen, ammonia, and hydrogen sulfide. In swamps, where the continual accumulation of dying vegetation leads to fermentation of the bottom, organic-rich layers, the production of methane may be sufficient to produce the phenomenon known as swamp fire. The increasing concentration of methane in the atmosphere, which is of current concern to environmentalists, is chiefly from anaerobic digestion.

Unlike bacteria, algae are single-cell organisms that contain pockets of chlorophyll that the cell uses as plants do for the photosynthesis of carbohydrate from CO_2 and H_2O [reaction (2), Chapter 4].

In most surface water, there is a symbiotic relationship between oxygen-producing algae and oxygen-utilizing bacteria occurring together in slime masses, although individual colonies of these microbes also thrive separately. The amount of oxygen produced by the algae is directly affected by sunlight, so the oxygen content of most surface waters increases during the day and falls off at night.

The oxygen content of surface waters has a strong influence on the species of fish in the aquatic population. Trout and other game fish usually require high oxygen levels and prefer lower temperatures; rough fish, such as carp and catfish, can survive at oxygen levels as low as 2 mg/L and can also tolerate warmer water.

FIG. 5.3 Algae bloom occurs when phosphorus, nitrogen, and nutrients are abundant. Bacteria feeding on the algae multiply rapidly, producing debris that speeds up the eutrophication process.

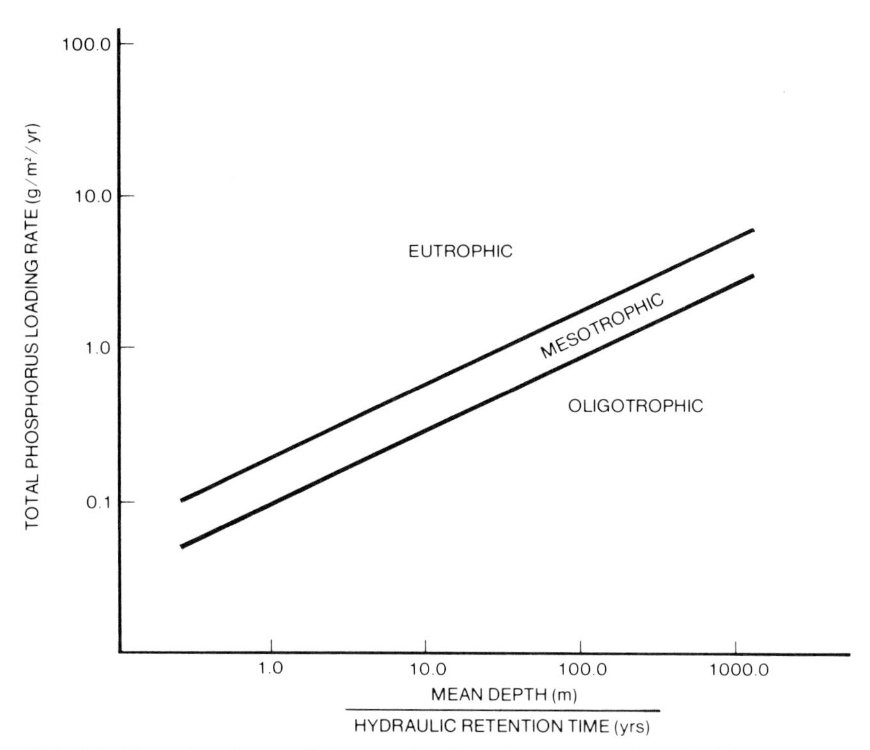

FIG. 5.4 How phosphorus affects eutrophication of a lake, as determined by its geometry. *(From National Water Quality Inventory, EPA 440/9-75-014.)*

Although the presence of algae may be beneficial in the production of dissolved oxygen, too much produces blooms, which are extremely troublesome and can lead to eutrophication (slow death) of the body of water (Figure 5.3). When algae have reached the end of their growing season, they die and provide a rich source of organic material for bacteria. With an ample food supply, the bacteria may initially grow at an exponential rate, consuming dissolved oxygen in the process. The process slows down as food is consumed and waste products accumulate. Even large bodies of water, e.g., Lake Erie, may become completely devoid of dissolved oxygen in certain areas where this process has occurred. Since the growth of the algae that starts the process of eutrophication is dependent on nitrogen and phosphorus as fertilizers, the concentration of these nutrients entering the body of water has a strong influence on the rate of eutrophication. A relationship between the input of nutrients and the geometry of the lake has been hypothesized (Figure 5.4). Even those waters classed as oligotrophic (abundant in oxygen, deficient in plant nutrients) are destined to become filled with debris eventually, even though the rate of organic production is very low. The largest lake in the world, Lake Baikal in Siberia, located on a geologic fault, has a water depth of about 5000 feet (1525 m); beneath this is an accumulation of another 20,000 feet (6000 m) of sediment.

A final chemical factor influencing the aquatic biota is toxic matter in a variety of forms. Many toxic materials are produced by nature; for example, there are many toxic varieties of algae. Other toxic materials may be present because of municipal sewage or industrial waste discharges. Some elements that are required in trace amounts for healthy aquatic growth are toxic at higher concentrations.

The aquatic organisms of concern to both water users and naturalists are (a) microorganisms, which are visible only under the microscope; (b) macroorganisms, which are visible to the unaided eye, but generally rather small; and (c) higher life forms.

MICROBIAL ANALYSES

In considering the various fields of aquatic biology, it is found that there are two principal kinds of analyses used to identify microorganisms: the first is that used by the municipal water treatment plants for control purposes to assure that the finished water is free of pathogenic organisms. Coliform bacteria are used as indicator organisms in this test. Although these are not pathogenic, they are found in the intestinal tract of all warm-blooded animals, so their presence is a warning that pathogenic organisms may also be present. A more positive indication of a health hazard is the further identification of some of these as fecal coliform bacteria. The effluent is tested at a frequency specified by law and must meet minimum counts for coliform organisms on a statistically significant number of samples. These records are submitted to the control agency, usually at the local level.

TABLE 5.1 Odors Produced by Algae

Species	Moderate	Abundant
Anabaena	Grassy, Nasturtium	Septic
Asterionella	Geranium, spicy	Fishy
Dinobryon	Violet	Fishy
Melosira	Geranium	Musty
Oscillatoria	Grassy	Musty
Stephanodiscus	Geranium	Fishy
Synura	Cucumber	Fishy
Tabellaria	Geranium	Fishy
Volvox	Fishy	Fishy

In addition to determining coliform organisms, the municipal plant may on its own initiative analyze for specific organisms based on its past operating experience with biologically related problems. For example, the spring turnover in a lake may introduce filter-clogging organisms at the water intake: a regular evaluation for the organisms which experience has found in the past to cause filter problems is a helpful control measure. Later in the year, certain algal blooms may produce tastes or odors in the finished water, and analysis for these specific algae can forewarn the plant of the time when activated carbon may be needed for taste or odor control (Table 5.1).

SAMPLING AND ANALYSIS

Microbiological analyses conducted to pinpoint problems in industrial water systems, the second type of analysis, are considerably more complex. Figures 5.5 to

5.8 show such analyses for a river, a recirculating cooling water system, a paper mill system, and a waste plant discharge.

In reviewing these analyses, it is apparent that there is space for sample identification that calls for observation by the person taking the sample, and these examples illustrate some shortcomings in the sampler's performance. Sample B42292 should have been completely identified, even to the extent of the location of the sample point on the river and the time of the day of sampling; certainly, the river should have been named. The appearance "clear liquid" is important, but this could be expanded to include color and transient characteristics such as temperature, pH, and odor at the time of sampling. Usually the sample is drawn from a pipe tap. Of course, there is always a background of microbes in the air seeking a good site for colonization, including the valve and nipple at the sample pipe tap. The same tap, then, must be sterilized by flame or alcohol and thoroughly flushed before the sample is finally drawn into a sterile container.

The time between sampling and analysis is significant, and any comparison between analysis of the same sample point at different times must take this into account. Comparing the analysis of a sample that took 5 days in transit to another that was sent by overnight delivery requires judgment that comes only with experience.

In this type of analysis, although total count is a very significant number, by itself it means little unless samples collected at regularly scheduled times from the same location are analyzed and compared for the levels of populations and the species (organisms) normally encountered. The particular analyses shown break down the microbiological life into three categories: bacteria, fungi, and algae. In each category, the specific organisms sought are those that experience has shown to be the most troublesome in industrial systems.

The microbiologist who performs these analyses is dealing with single-cell organisms, most of which can be seen only under the microscope. In the case of bacteria, those common to the aqueous environment are generally in the size range of 0.5 to 20 μm (1 μm is about 0.00004 in).

There are larger single-cell organisms such as fungi and algae, perhaps 10 times larger. Another example is the human ovum, barely visible to the unaided eye and about 100 μm in diameter. This largest of the cells of the human body must hold a large reserve of nourishment for growth. The cell is fertilized by the smallest of human cells, the sperm—only about 10^{-5} times the size of the ovum, or on the order of 1 nm, or 10 Å (Angstrom units). The sperm is essentially a compact package containing 23 chromosomes with a simple appendage for propulsion. Since the chromosome is about the same order of magnitude as a virus, a chemical complex of protein and nucleic acid, these analogies suggest that the human ovum is a rough model for the common bacteria found in water systems and the sperm an approximate model for viruses and phages that invade microbes' cells. Their size is of considerable significance in that the organisms larger than 0.5 μm can be filtered from water in a typical sand filter, but the viruses cannot. [In treatment of potable water, viruses, like colloids, can be coagulated and flocculated, and then can be removed by filtration (see Chapters 8 and 9)].

There are thousands of species of microbes, so it is fortunate that methods of identification other than visual appearance alone have developed over the years. The term *plate count* is used to express the number of microbe colonies that develop on a selective nutrient gel contained in a specially designed covered dish after a fixed period of incubation at a controlled temperature. For example, when analyzing for total count—a measure of a variety of bacteria living together—the sample is added to a gel prepared from nutrient agar and the sample is incubated for 24 h at 37°C (body temperature). Bacteria growing in this fashion reproduce by fission, each cell splitting into two new cells approximately at 15-min intervals, so that the enormous numbers theoretically produced after 24 h (2^{96}) create fairly large, readily visible colonies that can be counted. By using a variety of nutrients, gels, dyes, and pH environments that encourage the growth of one species and exclude others, the microbiologist is able to make the census listed on Figures 5.5 through 5.8.

NALCO

REPORT OF MICROBIOLOGICAL ANALYSIS

Company [redacted]

Address [redacted]

Sample Marked RIVER WATER

Analysis No. 018047
Sampling Date 04/21/77
Date Rec'd by NALCO 04/25/77

Physical Appearance

CLEAR LIQUID

(All counts express quantity of organisms per ML of sample)

BACTERIA

AEROBIC SLIME FORMING	
Non-Sporeforming	
Flavobacterium	NEG IN 1/10
Mucoids (S)	NEG IN 1/10
Aerobacter	NEG IN 1/10
Pseudomonas	2,500
Sporeforming	
B. subtilis	
B. cereus	
B. megatherium	
B. mycoides	
ANAEROBIC CORROSIVE	
Desulfovibrio	20
Clostridia	
IRON DEPOSITING	
Sphaerotilus	NONE
Gallionella	
OTHER BACTERIA	NONE
TOTAL COUNT	5,000

FUNGI

MOLDS	
Aspergillus	1
Penicillium	
Trichoderma	
Alternaria	
FUSARIUM	3
YEASTS	
Torula	NONE
Monilia	
Saccharomyces	
Rhodotorula	

ALGAE

BLUE GREEN	NONE
Oscillatoria	
GREEN	NONE
Chlorococcus	
DIATOMS	NONE

OTHER ORGANISMS

NONE

REMARKS:

Ronald Christensen
Head, Microbiological Laboratory

NALCO CHEMICAL COMPANY
6216 W. 66TH PLACE ▫ CHICAGO, ILLINOIS 60638

FIG. 5.5 Microbial analysis of a river water.

At the head of the list are slime-forming aerobic bacteria. The most common of these is *Pseudomonas,* a non–spore-forming variety reproducing by fission. The spore-formers secrete a thick coating that encapsulates the cell's nucleus and helps it survive changes in temperature or food supply and other shocks from the environment. The next major classification includes those common anaerobic (oxygen-free environment) organisms that can cause corrosion in metal piping sys-

NALCO

REPORT OF MICROBIOLOGICAL ANALYSIS

Company [redacted]
Address [redacted]

Analysis No. 017593
Sampling Date 04/13/77
Date Rec'd by NALCO 04/15/77

Sample Marked COOLING TOWER BASIN

Physical Appearance
THIN GREEN MATERIAL

(All counts express quantity of organisms per ML of sample)

BACTERIA

AEROBIC SLIME FORMING		
Non-Sporeforming		
Flavobacterium		300,000
Mucoids	(S)	100,000
Aerobacter		NEG IN 1/1000
Pseudomonas		10,000,000
Sporeforming		
B. subtilis		
B. cereus		1,500
B. megatherium		200
B. mycoides		
ANAEROBIC CORROSIVE		
Desulfovibrio		10,000
Clostridia		500
IRON DEPOSITING		
Sphaerotilus		NONE
Gallionella		
OTHER BACTERIA		NONE
TOTAL COUNT		47,000,000

FUNGI

MOLDS	*1
Aspergillus	
Penicillium	
Trichoderma	
Alternaria	
PAECILOMYCES	100
YEASTS	
Torula	
Monilia	6,000
Saccharomyces	
Rhodotorula	

ALGAE

BLUE GREEN	NONE
Oscillatoria	
GREEN	NONE
Chlorococcus	
DIATOMS	NONE

OTHER ORGANISMS

NONE

REMARKS:

*1 FUSARIUM	1,000
*1 MUCOR	100

Ronald Christensen
Head, Microbiological Laboratory

NALCO CHEMICAL COMPANY
6216 W. 66TH PLACE ▫ CHICAGO, ILLINOIS 60638

FIG. 5.6 Microbial analysis of a recirculating cooling water.

tems and equipment, especially beneath deposits where they are shielded from contact with dissolved oxygen. The next troublesome category contains those bacteria responsible for depositing iron oxide in piping systems, leading to clogging and deterioration.

The fungi are categorized as molds and yeasts, and these are of importance because of their ability to attack cooling tower lumber, deteriorate paper or textile

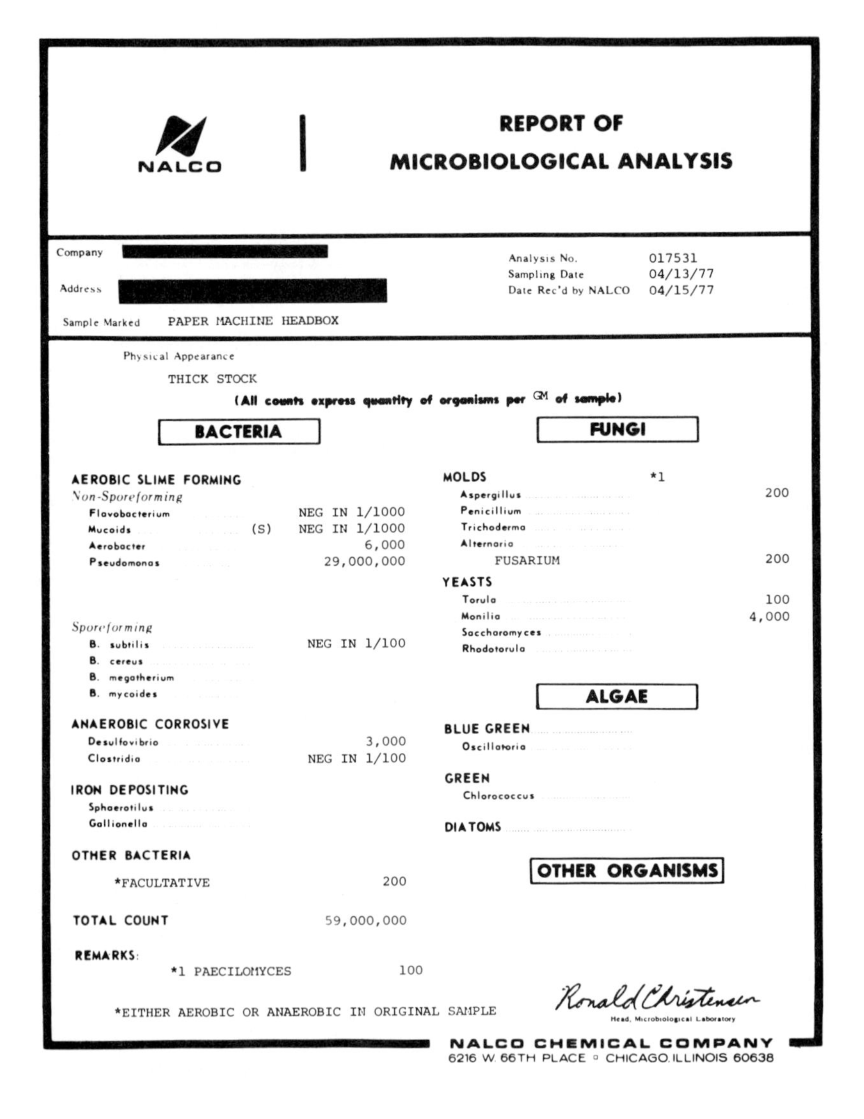

NALCO

REPORT OF MICROBIOLOGICAL ANALYSIS

Company [redacted]
Address [redacted]
Sample Marked PAPER MACHINE HEADBOX

Analysis No. 017531
Sampling Date 04/13/77
Date Rec'd by NALCO 04/15/77

Physical Appearance
THICK STOCK

(All counts express quantity of organisms per GM of sample)

BACTERIA

AEROBIC SLIME FORMING	
Non-Sporeforming	
Flavobacterium	NEG IN 1/1000
Mucoids (S)	NEG IN 1/1000
Aerobacter	6,000
Pseudomonas	29,000,000
Sporeforming	
B. subtilis	NEG IN 1/100
B. cereus	
B. megatherium	
B. mycoides	
ANAEROBIC CORROSIVE	
Desulfovibrio	3,000
Clostridia	NEG IN 1/100
IRON DEPOSITING	
Sphaerotilus	
Gallionella	
OTHER BACTERIA	
*FACULTATIVE	200
TOTAL COUNT	59,000,000

FUNGI

MOLDS *1	
Aspergillus	200
Penicillium	
Trichoderma	
Alternaria	
FUSARIUM	200
YEASTS	
Torula	100
Monilia	4,000
Saccharomyces	
Rhodotorula	

ALGAE

BLUE GREEN	
Oscillatoria	
GREEN	
Chlorococcus	
DIATOMS	

OTHER ORGANISMS

REMARKS:
*1 PAECILOMYCES 100

*EITHER AEROBIC OR ANAEROBIC IN ORIGINAL SAMPLE

Ronald Christensen
Head, Microbiological Laboratory

NALCO CHEMICAL COMPANY
6216 W. 66TH PLACE ▫ CHICAGO, ILLINOIS 60638

FIG. 5.7 Microbial analysis of a liquid sample from a paper machine headbox.

products, or affect use of water in food production. The algae, as mentioned earlier, produce tastes and odors and interfere with flocculation and sedimentation processes for clarification of water. Like their larger plant relatives, they have certain blooming seasons during the year. An example of this is shown in Figure 5.9.

Identification of species of algae is done with the microscope, since plate count techniques are applicable only to bacteria and fungi. This means that the analyst

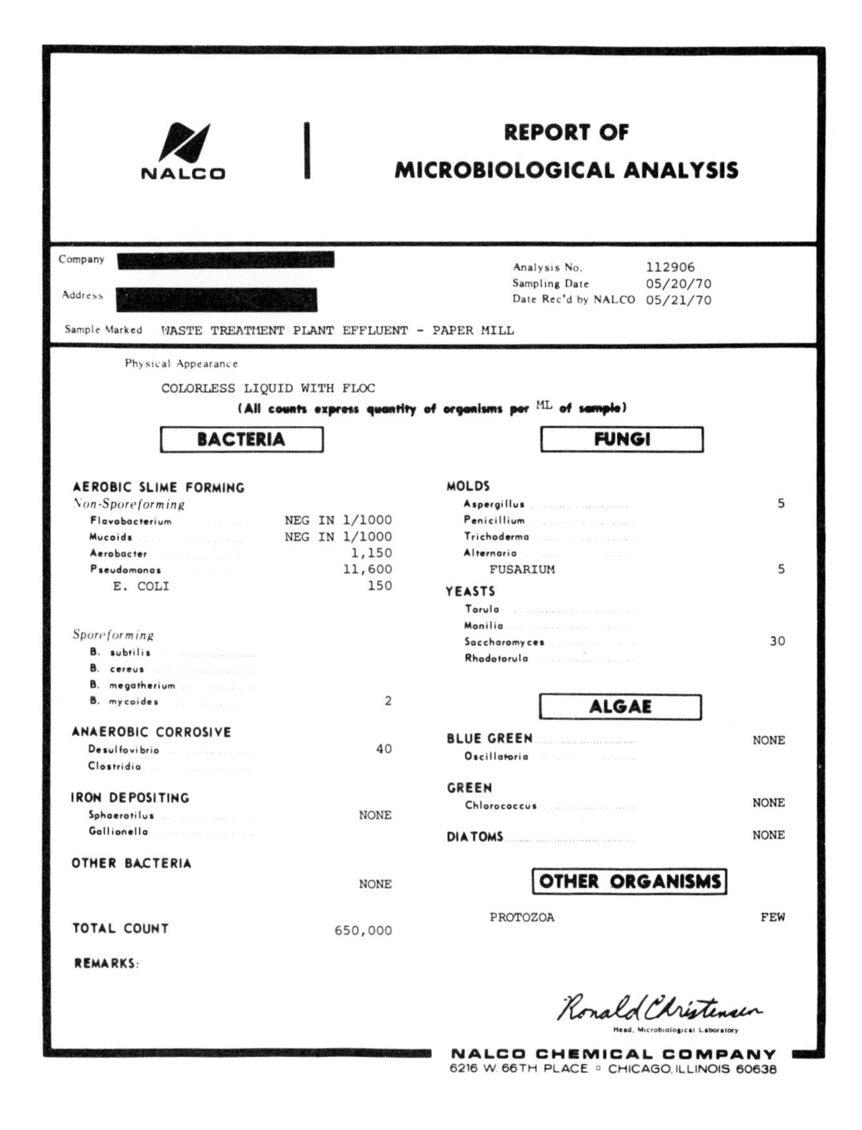
NALCO

REPORT OF MICROBIOLOGICAL ANALYSIS

Company
Address

Analysis No. 112906
Sampling Date 05/20/70
Date Rec'd by NALCO 05/21/70

Sample Marked WASTE TREATMENT PLANT EFFLUENT - PAPER MILL

Physical Appearance
COLORLESS LIQUID WITH FLOC

(All counts express quantity of organisms per ML of sample)

BACTERIA

AEROBIC SLIME FORMING	
Non-Sporeforming	
Flavobacterium	NEG IN 1/1000
Mucoids	NEG IN 1/1000
Aerobacter	1,150
Pseudomonas	11,600
E. COLI	150
Sporeforming	
B. subtilis	
B. cereus	
B. megatherium	
B. mycoides	2
ANAEROBIC CORROSIVE	
Desulfovibrio	40
Clostridia	
IRON DEPOSITING	
Sphaerotilus	NONE
Gallionella	
OTHER BACTERIA	NONE
TOTAL COUNT	650,000

FUNGI

MOLDS	
Aspergillus	5
Penicillium	
Trichoderma	
Alternaria	
FUSARIUM	5
YEASTS	
Torula	
Monilia	
Saccharomyces	30
Rhodotorula	

ALGAE

BLUE GREEN	NONE
Oscillatoria	
GREEN	
Chlorococcus	NONE
DIATOMS	NONE

OTHER ORGANISMS

PROTOZOA	FEW

REMARKS:

Ronald Christensen
Head, Microbiological Laboratory

NALCO CHEMICAL COMPANY
6216 W. 66TH PLACE ▫ CHICAGO, ILLINOIS 60638

FIG. 5.8 Microbial analysis of a waste treatment plant discharge.

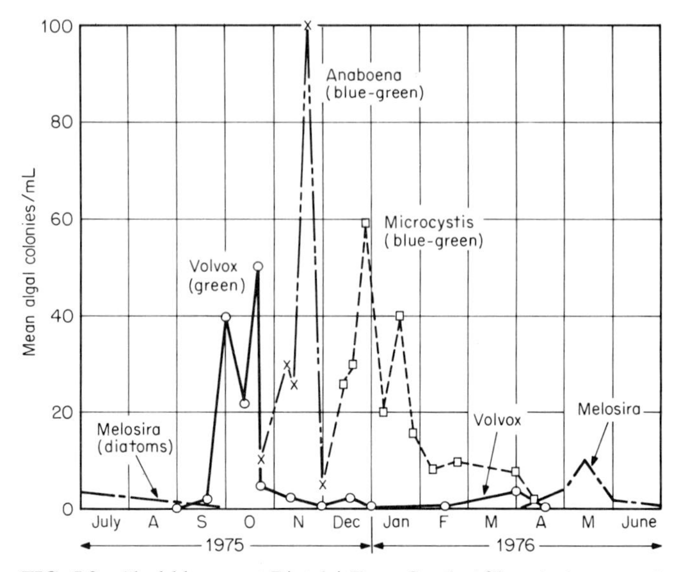

FIG. 5.9 Algal blooms at Rietulei Dam, South Africa, during annual cycle. Peak counts of blue-green algae developed at about 25°C temperature peaks, November through January. [*From Ashton, P. J.: "Nitrogen Fixation in a Nitrogen Limited Impoundment," J. Water Pollut. Control, Fed.,* ***51*** *(3) March 1979.*]

must be able to recognize the large varieties of blue-green and green algae and the delicately patterned diatoms that are part of this family of microorganisms. Some common forms are shown in Figure 5.10.

The analysis shown in Figure 5.6 reports a total count of 47,000,000/mL. This population density in itself will produce a turbid water—usually a haze begins to develop at a level of about 10,000,000/mL. A liter of this sample (Figure 5.6) would contain only about 10 mg of bacteria on a dry basis, equivalent to a suspended solids concentration of 10 mg/L.

When organisms other than those shown on the standard analytical form are found, these are noted in the appropriate location. For example, *Escherichia* are reported as aerobic, non-spore-forming bacteria; the sulfur-producing bacteria, *Beggiatoa,* is reported under the column for "Other bacteria" when it is identified. Organisms other than bacteria, fungi, and algae include the protozoa and rotifers that are often found grazing on the floc containing other microorganisms. *Actinomyces* are sometimes identified by special techniques, as these can contribute significantly to tastes and odors in finished water.

When a wastewater is analyzed, it is sometimes necessary to determine fecal coliform and fecal streptococci to establish the need for final disinfection. If the sample is taken from a stream below a wastewater discharge, a common variety of bacteria called *Sphaerotilus* is often found. This is usually classed as an iron-depositing bacteria, but iron is not needed for its growth.

If a bioassay is being made of a stream to assess its health, a census must be made of many larger organisms. Macroorganisms—visible to the unaided eye—include a variety of worms, crustacea, and larvae. A healthy stream contains a large number of species with relatively low populations. These provide food for the higher forms of life, such as fish, crawfish, and shrimp.

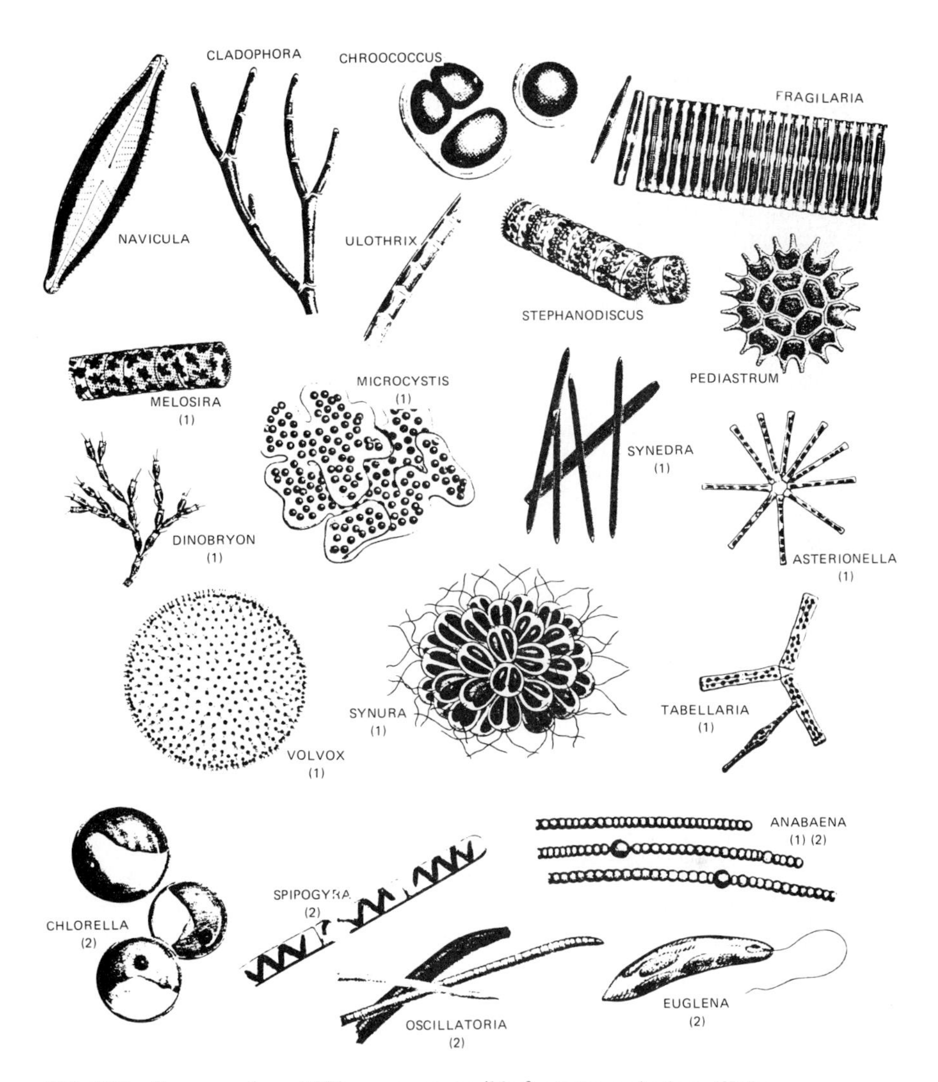

FIG. 5.10 Common algae. (1)These are responsible for tastes and odors; (2) these are common to polluted water as well as fresh water. *(Adapted from Palmer, Tarzwell, and Walter, 1955.)*

THREE LIVING ZONES

Three well-defined habitats must be examined in making a complete bioassay of a stream or lake: the littoral area just off the shoreline; the limnetic zone, or open water beyond the shore; and the benthic zone at the bottom. There is an endless variety of organisms indigenous to each of these zones, and a periodic evaluation of their populations is needed to assess the health of the lake and the effect of municipal or industrial discharges.

If such discharges have not been properly treated, the receiving body of water becomes polluted. In fact, by definition, pollution refers to contaminants that degrade the body of water, either wiping out desirable populations or hastening eutrophication. The mere presence of contaminants, which exist in all waters, does not constitute pollution.

Of the indicators of pollution, certainly dissolved oxygen level is the foremost. A reduction of dissolved oxygen is always apparent in a receiving stream immediately below a wastewater discharge. Figure 5.11 shows the oxygen content at each sampling location along a stream. The data are typical of the effect of a wastewater discharge on a receiving stream. If the stream is severely polluted,

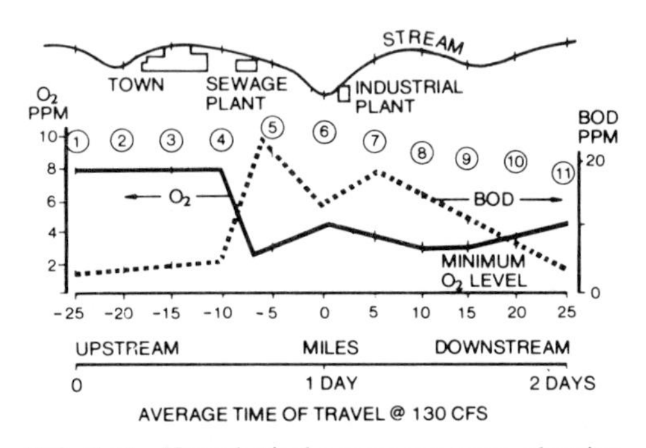

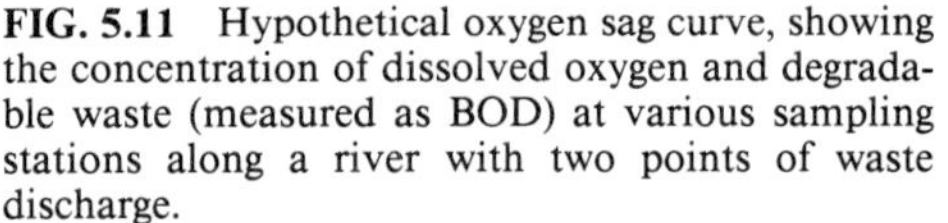

FIG. 5.11 Hypothetical oxygen sag curve, showing the concentration of dissolved oxygen and degradable waste (measured as BOD) at various sampling stations along a river with two points of waste discharge.

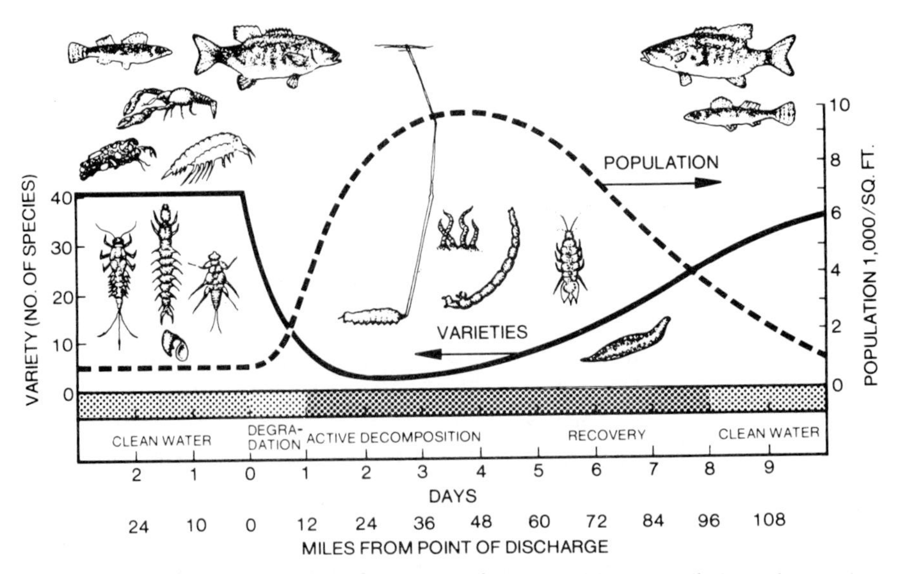

FIG. 5.12 Changes in population of macroorganisms caused by waste discharge into a clean stream. *(From Ingram, Mackenthum, and Bartsch*, 1966.)

oxygen may disappear entirely. There will be a zone of degradation below the point of discharge where the water may become septic and foul odors develop. Depending on the type of exposure of the stream to the atmosphere—that is, whether the stream is turbulent or limpid—oxygen will be redissolved in the water, and the stream will rapidly or slowly recover.

Correlating closely to the oxygen sag curve is the change in stream biota below the point of discharge: the number of species of organisms drops precipitately, and the population of undesirable organisms, such as sludge worms, increases rapidly. Farther down stream, as the river begins to recover, species of desirable organisms begin to reappear, as shown by Figure 5.12.

CHEMICAL INDICATORS

Certain chemical tests, too, correlate with the oxygen sag curve, as shown by Figure 5.13. The BOD values shown are a measure of food for bacteria (Chapter 4). When food is introduced, bacteria begin to multiply rapidly at an exponential rate, resulting in the reduction of dissolved oxygen in the water. As food is consumed, as indicated by a reduction in BOD, the bacterial populations die off and

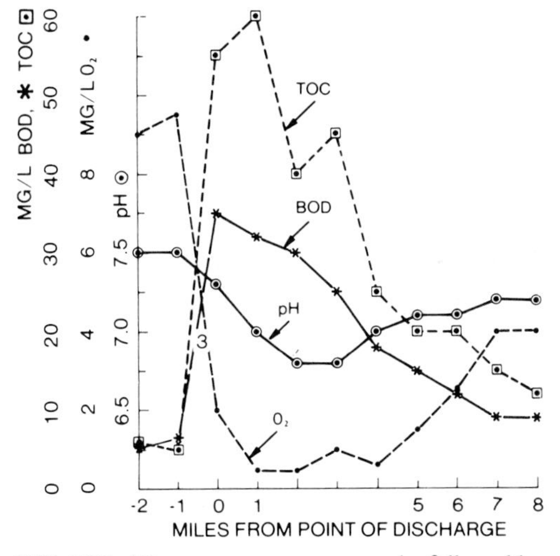

FIG. 5.13 The oxygen sag curve may be followed by a variety of analyses showing effects of biological activity.

become food for protozoa, which in turn die off and become food for rotifers and crustaceans, as shown by Figure 5.14. This marks the return of the stream to a healthy condition.

Because pollution of a stream is quickly shown by these effects on the biota, periodic bioassays of a river above and below the point of wastewater discharge to determine variety and density of aquatic species is a practical means of mon-

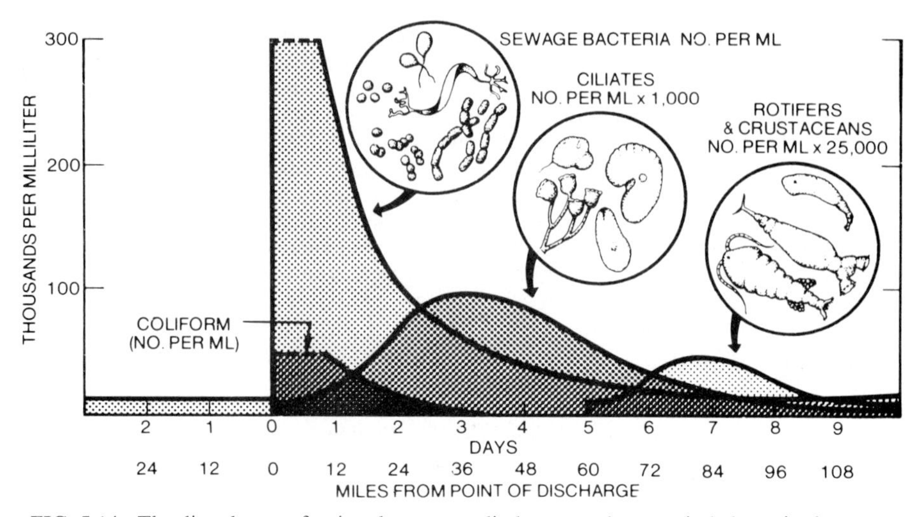

FIG. 5.14 The disturbance of a river by a waste discharge produces an imbalance in the population distribution. The types of species found indicate the condition of the stream at each sampling point. *(From Ingram, Mackenthum, and Bartsch, 1966.)*

itoring the performance of a waste treatment plant. Other bioassay methods are being developed that rely on the reactions of fish exposed to an effluent stream, usually at a controlled dilution rate. It is quite possible that a waste treatment plant may perform within the specifications of the effluent guidelines established by federal regulations, and yet the effluent may be damaging to the stream because of the presence of toxic materials for which there may be no analysis. For that reason, biological studies to evaluate the suitability of an effluent for mixing into the receiving stream are valuable. Biological monitoring upstream and downstream of a plant outfall is a valuable program for assessing the effect of the discharge on the receiving stream. Certain indicator organisms, such as diatoms, may be useful in measuring the health of the aquatic environment in this monitoring program. Population counts and varieties of species are noted, and the data are used in calculating a *diversity index* that becomes a guide in assessing the effects of the plant discharge.

It is seen, then, that gross pollution of a stream has severe biological implications, but that a well-treated effluent may have little or no effect on the receiving stream, depending to a large extent on the ability of the stream to recover through the natural process of reaeration.

BACTERIA CAN HELP

In most municipal sewage plants, microbiology can be applied as a beneficial science for the destruction of pollutants in wastewater. In such a process, the wastewater is collected and held under controlled conditions, similar to a balanced aquarium, and bacteria are introduced with the proper application of nutrients to break down and digest the organic wastes. Under such controlled conditions, the final discharge does not impose an excessive oxygen demand on the receiving stream.

While bacteria can be beneficial in the purification of wastewater, when they are uncontrolled in industrial water systems, microbiological organisms produce an endless variety of problems. Aside from diseases such as typhoid and dysentery, which are common threats from exposure to waters that have not been disinfected, the potential for equipment damage and product damage due to microbiological activity is great. Slime masses are often part of deposits found in water distribution systems and water-using equipment. Corrosion is often accelerated because of biological activity; this is particularly true when deposits form and anaerobic bacteria are shielded from the flowing water stream by these deposits, permitting them to corrode steel piping by the H_2S produced through their metabolism. However, proper identification of the organisms entering a system and found in deposits enables the water engineer to develop a chemical program for controlling these problems by application of biocides or dispersants, or a combination of both.

SUGGESTED READING

American Society of Testing Materials: "Evaluating Acute Toxicity of Industrial Wastewater to Fresh Water Fishes," ASTM Standards, Part 23, D1345-59, 1970.

Edmonson, W. T. et al., eds.: *Fresh Water Biology,* Wiley, New York, 1959.

Gaufin, A. R., "Use of Aquatic Invertebrates in the Assessment of Water Quality," in *Biological Methods for Assessment of Water Quality,* STP 528, American Society for Testing Materials, 1973.

Ingram, W. M., Mackenthun, K. M., and Bartsch, A. F.: *Biological Field Investigative Data for Water Pollution Surveys,* U.S. Department of Interior, U.S. Government Printing Office, 1966.

Kushner, D. J. et al.: *Microbial Life in Extreme Environments,* Academic, New York, 1978.

Mackenthun, K. M., and Ingram, W. M.: *Biological Associated Problems in Fresh Water Environments,* U.S. Department of Interior, U.S. Government Printing Office, 1967.

Maitland, Peter S.: *Biology of Fresh Waters,* Wiley, New York, 1978.

Mitchell, Ralph: *Water Pollution Microbiology,* Wiley, New York, 1971–72.

Mowat, Anne: "Measurement of Metal Toxicity by BOD," *J. Water Pollut. Control Fed.,* p. 853, May 1976.

Palmer, C. M., and Tarzwell, C. M.: "Algae of Importance in Water Supplies," *Public Works,* p. 107, June 1955.

Whipple, George C.: *The Microscopy of Drinking Water,* 4th ed., rev. (Fair, G. M., and Whipple, M. C., eds.), Wiley, New York, 1933.

CHAPTER 6

WATER CONTAMINANTS: OCCURRENCE AND TREATMENT

The introduction of contaminants into water supplies has been shown to be related to rainfall, the geologic nature of the watershed or underground aquifer, and the activities of nature and the human population. Water contaminants to be examined in more detail fall in two categories: dissolved matter (Table 6.1) and nonsoluble constituents (Table 6.2). Dissolved gases are included in discussions of the biological cycles affecting water quality.

As shown in Table 6.1, soluble materials in water are arbitrarily assigned to five classifications, the first four of which are based on concentration levels, with the last covering those materials usually transient because continuing reactions in the aquatic environment change their concentrations.

Many materials are transient because of biological activity. The change in CO_2 and O_2 content with sunlight is one example. Equilibrium between NH_3, N_2, NO_2^-, and NO_3^- is another, discussed later in this chapter as part of the nitrogen cycle. (See Class 2, Secondary Constituents.)

There are also longer term processes by which nature cycles matter through living organisms, which in turn modify the environment and leave their records in the rocks. This chapter examines the sources of contaminants in water, many of which are minerals created by living things. Perhaps the best known are the chalk cliffs of Dover and the coral atolls of the Pacific, both composed of $CaCO_3$.

Discussing these atolls in his essay on formation of mineral deposits, C. C. Furnace says, "To the casual observer, it would seem that the polyp has built these great masses of land out of nothing; but, of course, it cannot do that any more than man can. It has taken calcium compounds from very dilute solutions of sea water and built up a shell of calcium compounds to protect itself. In this process of following its preordained metabolic rite, it has concentrated calcium by several thousandfold in the form of an insoluble compound. Insignificant as the coral polyp may appear, it is one of the most important creatures in changing the character of the earth's surface."

Many other natural cycles have been operating over countless geologic ages to produce deposits of sulfur, iron, manganese, silica, and phosphate, to name only a few. In the village of Batsto, New Jersey, the early American colonists set up the first blast furnace in the New World. Their source of iron ore was "bog iron"—pure iron oxide precipitated from artesian water by iron-depositing bacteria.

So the presence of many of the mineral constituents in water supplies may simply represent the return to the aquatic environment of a loan made by the earth to living organisms long ago.

TABLE 6.1 Soluble Material in Water Supplies

Class 1	Primary constituents—generally over 5 mg/L		
	Bicarbonate	Magnesium	Sodium
	Calcium	Organic matter	Sulfate
	Chloride	Silica	Total dissolved solids
Class 2	Secondary constituents—generally over 0.1 mg/L		
	Ammonia	Iron	Potassium
	Borate	Nitrate	Strontium
	Fluoride		
Class 3	Tertiary constituents—generally over 0.01 mg/L		
	Aluminum	Copper	Phosphate
	Arsenic	Lead	Zinc
	Barium	Lithium	
	Bromide	Manganese	
Class 4	Trace constituents—generally less than 0.01 mg/L		
	Antimony	Cobalt	Tin
	Cadmium	Mercury	Titanium
	Chromium	Nickel	
Class 5	Transient constituents		
	Acidity-alkalinity		
	Biological cycles		
	Carbon cycle constituents		
	Organic C/CH_4/CO/CO_2/$(CH_2O)_n$/C-tissue		
	Oxygen cycle		
	O_2/CO_2		
	Nitrogen cycle constituents		
	Organic N/NH_3/NO_2^-/NO_3^-/N^0/amino acids		
	Sulfur cycle constituents		
	Organic S/HS^-/SO_3^{2-}/SO_4^{2-}/S^0		
	Redox reactions		
	Oxidizing materials		
	From the natural environment—O_2, S		
	Treatment residues—Cl_2, CrO_4^{2-}		
	Reducing materials		
	From the natural environment—		
	Organics, Fe^{2+}, Mn^{2+}, HS^-		
	Treatment residues—		
	Organics, Fe^{2+}, SO_2, SO_3^{2-}		
	Radionuclides		

TABLE 6.2 Nonsoluble Constituents in Water Supplies

Class 1—Solids
Floating
Settleable
Suspended
Class 2—Microbial organisms
Algae
Bacteria
Fungi
Viruses

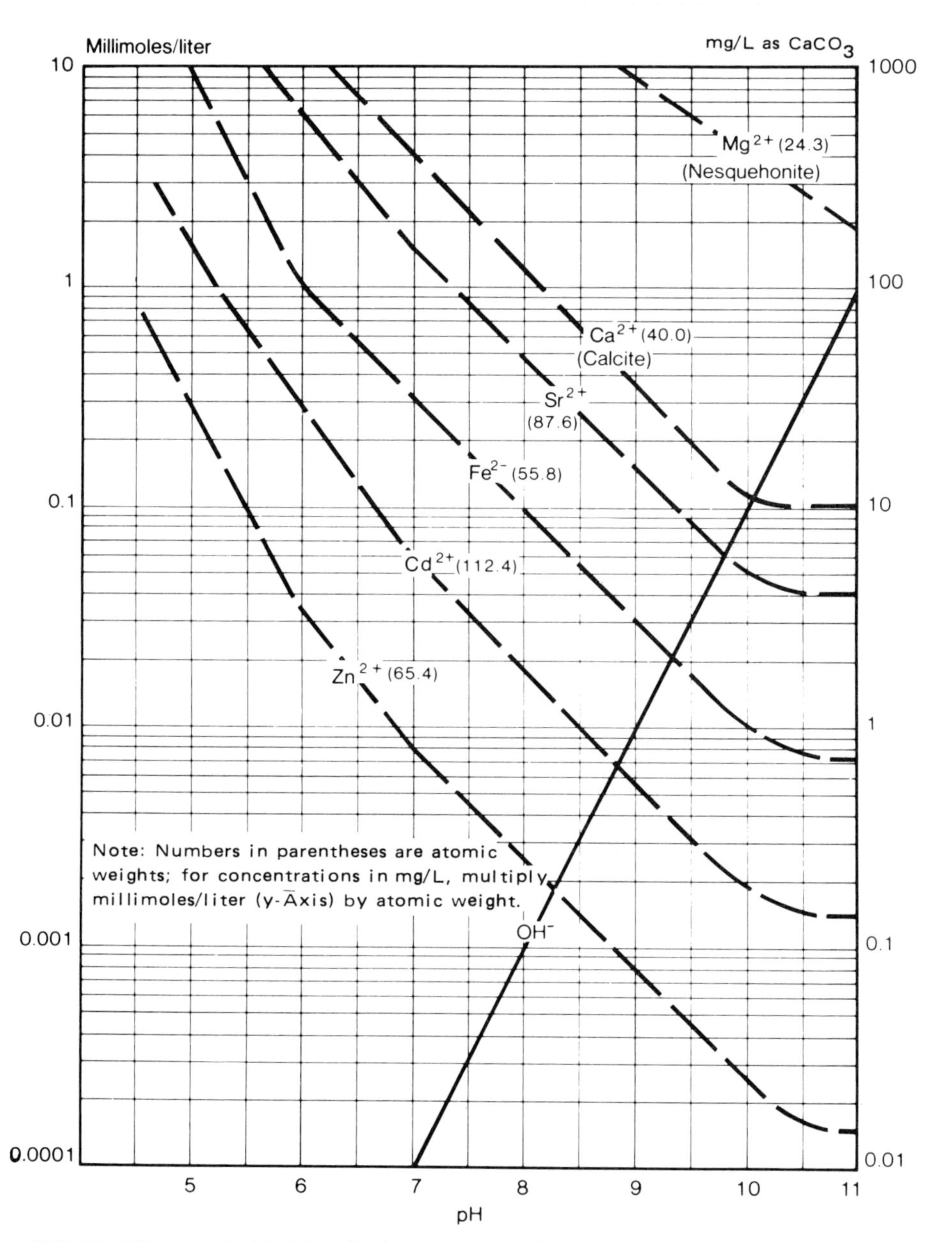

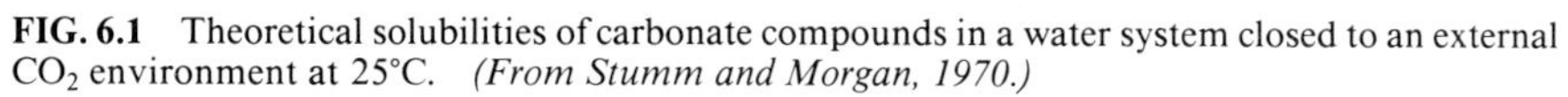

FIG. 6.1 Theoretical solubilities of carbonate compounds in a water system closed to an external CO_2 environment at 25°C. *(From Stumm and Morgan, 1970.)*

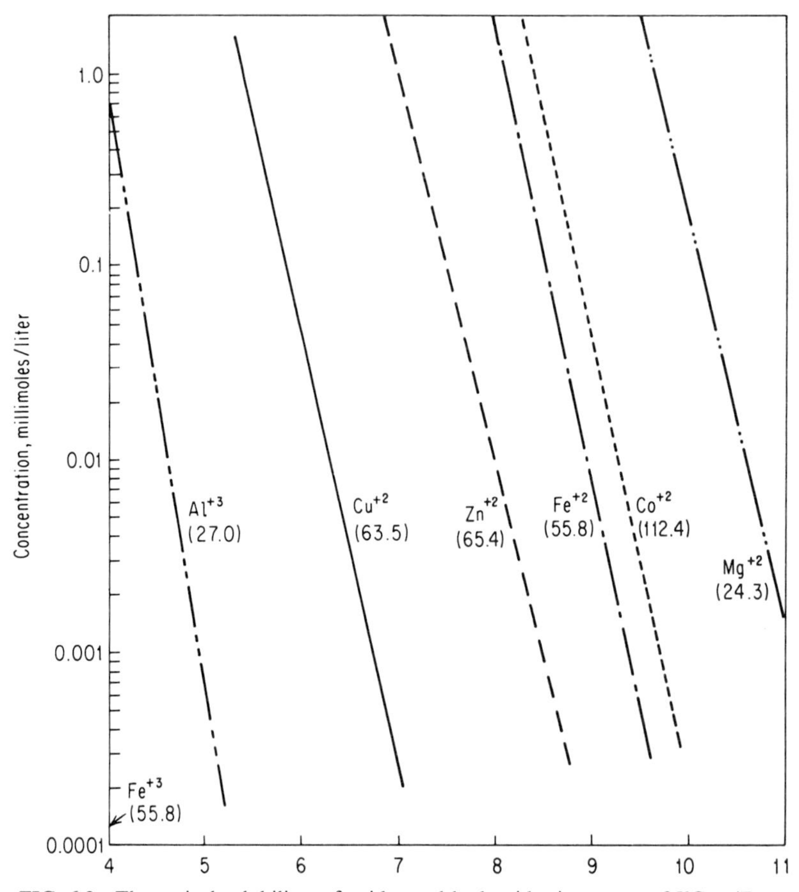

FIG. 6.2 Theoretical solubilites of oxides and hydroxides in water at 25°C. *(From Stumm and Morgan, 1970.)*

As an aid to appreciating the solubilities of the constituents being examined, and thus the limitations of their occurrence in natural water supplies and the residuals which may be reached in precipitation processes, Figures 6.1, 6.2 and 6.3 present solubility characteristics in appropriate locations in the text.2

CLASS 1—PRIMARY CONSTITUENTS

This category includes dissolved solids generally exceeding 5 mg/L, and often several orders of magnitude above this level.

Bicarbonate (HCO_3^-—Molecular Weight 61)

The bicarbonate ion is the principal alkaline constituent of almost all water supplies. It is generally found in the range of 5 to 500 mg/L, as $CaCO_3$. Its introduc-

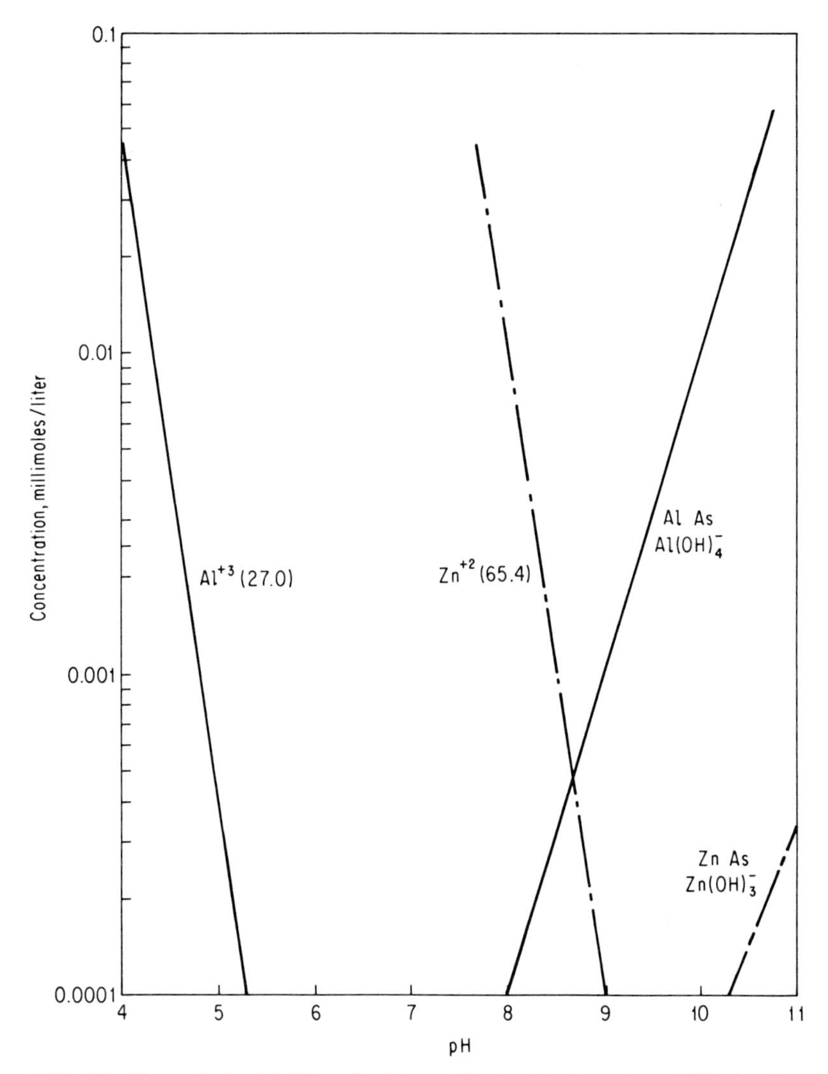

FIG. 6.3 Theoretical solubilities of oxides and hydroxides in water at 25°C, showing amphoteric nature of aluminum and zinc. *(From Stumm and Morgan, 1970.)*

tion into the water by the dissolving action of bacterially produced CO_2 on carbonate-containing minerals has been explained elsewhere. Normal activities of the human population also introduce alkaline materials into water, evidenced by a typical increase of alkalinity of sewage plant effluent of 100 to 150 mg/L above the alkalinity of the municipal water supply. Much of this is due to the alkalinity of industrial and domestic detergents.

Alkalinity in drinking water supplies seldom exceeds 300 mg/L. The control of alkalinity is important in many industrial applications because of its significance in the calcium carbonate stability index. Alkalinity control is important in both concentrated boiler water and cooling water in evaporative cooling systems.

Makeup for these systems must often be treated for alkalinity reduction either by lime softening or direct acid addition. Alkalinity is objectionable in certain other industries, such as the beverage industry, where it neutralizes the acidity of fruit flavors, and in textile operations, where it interferes with acid dyeing.

Calcium (Ca^{2+}—Atomic Weight 40; Group 11A, Alkaline Earth Metal, Figures 6.1 and 6.3)

Calcium is the major component of hardness in water and usually is in the range of 5 to 500 mg/L, as $CaCO_3$, (2 to 200 mg/L as Ca). It is present in many minerals, principally limestone and gypsum. Limestone deposits are often the residue of the fossils of tiny aquatic organisms, such as polyps, that have taken calcium from the seawater in which they lived, and used it for their skeletons. This is but one of many cycles in nature whereby some component of the environment is continually withdrawn by living things and eventually returned directly or indirectly.

Calcium removed from water in softening operations is later returned to the environment, often to the watershed, by way of a precipitate or a brine which is the by-product of the softening reaction. Calcium is a major factor in determining stability index. Calcium reduction is often required in treating cooling tower makeup. Complete removal is required for many industrial operations, particularly for boiler makeup, textile finishing operations, and cleaning and rinsing in metal finishing operations.

Calcium hardness can be reduced to a level of 35 mg/L as $CaCO_3$ by cold lime-soda softening and to less than 25 mg/L by hot lime-soda softening. It is reduced to less than 1 mg/L by cation exchange methods.

Chloride (Cl^-—Atomic Weight 35.5; Group VIIA, Halide)

Since almost all chloride salts are highly soluble in water, chloride is common in freshwater supplies, ranging from 10 to 100 mg/L. Seawater contains over 30,000 mg/L as NaCl, and certain underground brine wells may actually be saturated, approximately 25% NaCl. Many geologic formations were once sedimentary rocks in the sea, so it is not surprising that they contain residues of chlorides that are continually leaching into freshwater sources. The chloride content of sewage is typically 20 to 50 mg/L above the concentration of the municipal water supply, accounting in part for the gradual increase in salinity of rivers as they proceed from the headwaters to the sea.

Anion exchange is the only chemical process capable of removing chlorides from water; however, physical processes such as evaporation and reverse osmosis can separate a feedwater into two streams, one with a reduced chloride and the other with an increased chloride content.

The recommended upper limit for chloride in drinking waters is 250 mg/L, based entirely on taste, not on any known physiological hazards.

Magnesium (Mg^{2+}—Atomic Weight 24.3; Group IIA, Alkaline Earth Metal, Figs. 6.1 and 6.2)

The magnesium hardness of a water is usually about one-third of the total hardness, the remaining two-thirds being calcium hardness. Magnesium typically ranges from 10 to 50 mg/L (about 40 to 200 mg/L as $CaCO_3$). In seawater, magnesium concentration is about 5 times that of calcium on an equivalent basis. The production of magnesium hydroxide from seawater is the starting point in the manufacture of magnesium. Magnesium is a prominent component of many minerals, including dolomite, magnesite, and numerous varieties of clay.

Since magnesium carbonate is appreciably more soluble than calcium carbonate, it is seldom a major component in scale except in seawater evaporators. However, it must be removed along with calcium where soft water is required for boiler makeup or for process applications. It may be removed by lime softening to a residual of 30 to 50 mg/L as $CaCO_3$ cold, or 1 to 2 mg/L as $CaCO_3$ hot. It is also reduced by ion exchange to less than 1 mg/L as $CaCO_3$.

Organic Matter (Carbon, C^{4+}—Atomic Weight 12; Group IVA, Nonmetal)

Since organic material makes up a significant part of the soil and because it is used by aquatic organisms to build their bodies and produce food, it is inevitable that water-soluble organic products of metabolism should be present in all water supplies. There is not much information available on specific organic compounds in most water sources. (See Table 4.7.) There are literally hundreds of thousands of known organic compounds, many of which might somehow find their way into the hydrologic cycle. A complete "organic analysis" of water is impossible. However, one of the by-products of space-age technology has been the development of new instruments for organic analysis (see Chapter 7). With these instruments, the analyst can develop methods of analysis for organic materials of interest—especially those considered by the EPA to be toxic or carcinogenic, such as PCBs (polychlorinated biphenyls) and TTHMs (total trihalomethanes). But unless such specific organic compounds are requested at the time a sample is presented, the analyst uses indirect measures of organics instead (e.g., COD, TOC).

Many waters have a yellowish or tea color due to decayed vegetation leached from the watershed by runoff. These organic materials are broadly classified as humic substances, further categorized as humic acid (a water-soluble compound), fulvic acid (alkali-soluble material), and humin (high molecular weight, water insoluble matter). These organic compounds are molecules having many functional groups containing oxygen and hydrogen atoms in various proportions, so that when organic matter is reported as carbon, as it is in the TOC determination, it is probable that the molecular weight of these humic organic molecules is 2.0 to 2.5 times greater than the value reported as carbon. A survey of 80 municipal supplies in the United States showed an average total organic carbon content in the finished water of 2.2 mg/L, as C, so the organic matter was probably on the order of 5 mg/L.

There are a variety of indexes for measuring the gross organic content of water, and there is generally no correlation between them. The organic matter at a sampling station on the Mississippi River as determined by these indexes is shown in Figure 6.4. Because some of the functional groups in humic compounds have ion exchange properties, they tend to chelate heavy metals. In spite of this, there is no correlation whatever between the color of a water and its total heavy metal concentration. A study of the Rhine River showed that humic substances comprised from 25% (at 1000 m^3/s) to 42% (at 3500 m^3/s) of the dissolved organic matter; sulfonic acids ranged from 41% at 1000 m^3/s to 17% at 3500 m^3/s; a third category, chloro-organics, ranged from 12% at low flow to 5% at high flow. These are refractory, or nonbiodegradable, classes of organic matter. The significance of

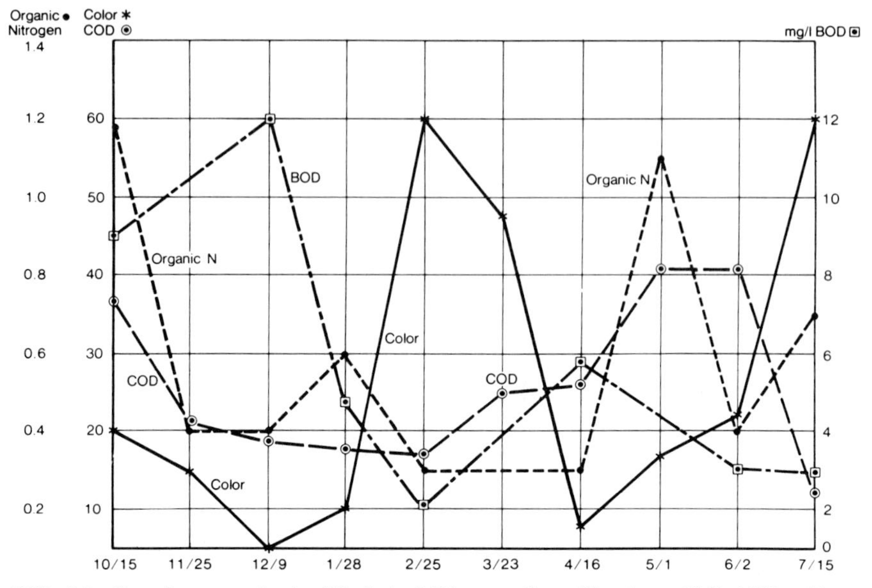

FIG. 6.4 Organic matter in the Mississippi River at Cape Girardeau, 1969–1970. *(From USGS Water Supply Paper 2156.)*

this information is simply that each investigator has his or her own purpose in studying organic matter in water and selects the most practical categories to study and the simplest methods of analysis; there is usually no purpose to identifying 30 to 40 specific organic compounds in water—a rather costly procedure—if rough indexes, such as TOC or humic substances will suffice for the study.

Some organic materials are truly soluble, but much of it—certainly the humic matter—is present in colloidal form and can generally be removed by coagulation. Alum coagulation at a pH of 5.5 to 6.0 typically reduces color to less than 5 APHA units. Organic matter such as is found in domestic sewage often inhibits calcium carbonate precipitation. If the natural color exceeds about 50, it must be partially removed for lime softening to occur. Organic matter may be removed by activated carbon treatment, widely practiced in municipal treatment plants when organic matter causes objectionable tastes or odors in the finished water. Generally these tastes and odors are produced by algae, each species having its characteristic odor or taste just as with land plants. Also like land plants, algae produce organic compounds which may be toxic if enough is ingested by fish or animals.

Certain organic materials in water polluted by agricultural runoff (e.g., pesticide residues) or by industrial wastes in concentrations far below 1 mg/L still exert a significant effect on the biota of the receiving stream. Even when the effect is not dramatic, as with fish kill, it may have long-term consequences, such as affecting reproduction or disrupting the food chain.

Organic matter is objectionable in municipal water chiefly for aesthetic reasons. It can be troublesome in industrial supplies by interfering with treatment processes. It is a major factor in the fouling of anion exchange resins, degrading effluent quality of demineralized water, and requiring early replacement of resin.

Silica (SiO_2—Molecular Weight 60; Oxide of Silicon, Group IV, Nonmetal)

Silica is present in almost all minerals, and is found in fresh water in a range of 1 to 100 mg/L. The skeletons of diatoms are pure silica, so the silica content of surface waters may be affected by seasonal diatom blooms. Silica is considered to be colloidal because its reaction with adsorbents like MgO and Fe $(OH)_3$ show characteristics similar to typical colloids. At high concentrations—over 50 mg/L—the adsorption isotherms (Chapter 3) no longer apply, and it appears that chemical precipitation occurs instead. There is probably an equilibrium between the silica in colloidal form and the bisilicate ($HSiO_3^-$) anion. Because of this complexity, it is difficult to predict the conditions under which silica can be kept in solution as water concentrates during evaporation.

The term "colloidal silica" is loosely used by water chemists and can be confusing. Very little research has been done to categorize the size distribution of the silica micelles (polymeric groups). It is very clear to the water analyst that this needs investigation. The analyst uses a colorimetric test that develops a blue color to measure silica concentration. Sometimes, particularly with demineralizing systems, there is evidence that some of the silica in water does not produce the blue color needed for detection, and this slips through the demineralizer without reaction. It seems that some of the silica micelles are too large to react with the chemical test reagents and the ion exchange resin, and in this case, the analyst may report that colloidal silica is present. A more accurate statement would be that inert (or nonreactive) silica is present, since for all practical purposes all of the silica is colloidal, although of differing sizes.

Silica is objectionable at high concentration in cooling tower makeup because of this uncertainty about its solubility limits.

It is objectionable in boiler feedwater makeup not only because it may form a scale in the boiler itself, but also because it volatilizes at high temperatures and redeposits on turbine blades. Treatment processes that remove silica are: adsorption on magnesium precipitates in the lime softening operation; adsorption on ferric hydroxide in coagulation processes using iron salts; and anion exchange in the demineralization process.

Sodium (Na^+—Atomic Weight 23; Group IA, Alkali Metal)

All sodium salts are highly soluble in water, although certain complexes in minerals are not. The high chloride content of brines and seawater is usually associated with the sodium ion. In fresh waters, its range is usually 10 to 100 mg/L (about 20 to 200 mg/L as $CaCO_3$). Sodium is present in certain types of clay and feldspar. There is an increase of sodium in municipal sewage of 40 to 70 mg/L in excess of the municipal water supply. Its concentration is not limited by Federal Drinking Water Standards, so persons on low sodium diets may require special sources of potable water. The only chemical process for removing sodium is cation exchange in the hydrogen cycle. Evaporation and reverse osmosis also reduce sodium, producing a product stream low in sodium and a spent brine high in sodium.

Sulfate (SO_4^{2-}—Molecular Weight 96; Oxide of Sulfur, Group VIA, Nonmetal)

Sulfate dissolves in water from certain minerals, especially gypsum, or appears from the oxidation of sulfide minerals. Its typical range is 5 to 200 mg/L. The suggested upper limit in potable water is 250 mg/L, based on taste and its potential cathartic effect. Because calcium sulfate is relatively insoluble—less than 2000 mg/L—sulfate may be objectionable in concentrating water high in calcium, as in an evaporative system. High sulfate levels may be reduced measurably by massive lime or lime-aluminate treatment, or in rare cases by precipitation with barium carbonate. It may also be reduced by anion exchange. In the coagulation of water with alum, sulfate is introduced at a rate of 1 mg/L SO_4 for each 2 mg/L alum added, while an equivalent amount of alkalinity is neutralized.

Total Dissolved Solids

Since this is the sum of all materials dissolved in the water, it has many mineral sources. Its usual range is 25 to 5000 mg/L. The suggested limit for public water supplies, based on potability, is 500 mg/L. The principal effect of dissolved solids on industrial processes is to limit the extent to which a water can be concentrated before it must be discarded. High concentrations affect the taste of beverages. The related electrical conductivity tends to accelerate corrosion processes. A reduction in dissolved solids is achieved by a reduction in the individual components.

CLASS 2—SECONDARY CONSTITUENTS

These are generally present in concentrations greater than 0.1 mg/L and occasionally in the range of 1 to 10 mg/L.

Ammonia [NH_3—Molecular Weight 17, Usually Expressed as N (Nitrogen); Atomic Weight 14, Group VA, Nonmetal]

Ammonia gas is extremely soluble in water, reacting with water to produce ammonium hydroxide. Since this ionizes in water to form $NH_4^+ + OH^-$, at high pH, free ammonia gas is present in a nonionized form. At the pH of most water supplies, ammonia is completely ionized.

$$NH_3 + H_2O \rightleftharpoons NH_4OH \rightleftharpoons NH_4^+ + OH^- \qquad (1)$$

(The addition of excess OH^- drives the reaction to the left.)

Ammonia is one of the transient constituents in water, as it is part of the nitrogen cycle and is influenced by biological activity. As is seen in the illustration of the nitrogen cycle, Figure 6.5, ammonia is the natural product of decay of organic nitrogen compounds. These compounds first originate as plant protein matter, which may be transformed into animal protein. The return of this protein material to the environment through death of the organism or through waste elimination produces the organic nitrogen compounds in the environment which then decay to produce ammonia.

Because this biological process also occurs in sewage treatment plants, ammonia is a common constituent of municipal sewage plant effluent, in which its usual concentration is 10 to 20 mg/L. It also finds its way into surface supplies from the agricultural runoff in areas where ammonia is applied to land as a fertilizer. Ani-

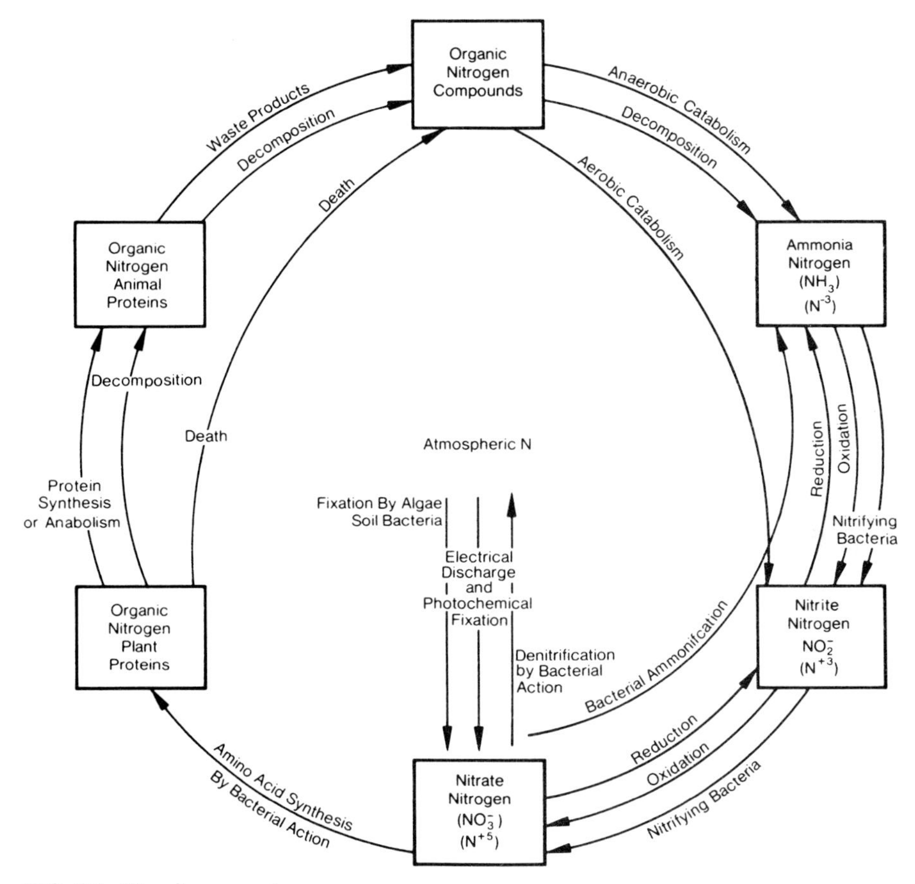

FIG. 6.5 The nitrogen cycle.

mal feed lots also contribute ammonia, which may run off into surface streams or find its way into underground aquifers.

The effect of sewage plant effluent on the ammonia content of a receiving stream is shown in Figure 6.6. Ammonia is oxidized by bacterial action first to nitrite and then to nitrate, so the concentration is continually being affected by the input from decay of organic nitrogen compounds and by the output, which is the uptake by bacteria to convert ammonia to nitrate.

The typical concentration range in most surface supplies is from 0.1 to 1.0 mg/L, expressed as N. It is not usually present in well waters, having been converted to nitrate by soil bacteria. Certain industrial discharges, such as coke plant wastes, are high in ammonia and account for the ammonia content of some surface waters.

The concentration of ammonia is not restricted by drinking water standards. Ammonia is corrosive to copper alloys, so it is of concern in cooling systems and in boiler feedwater.

Ammonia is often deliberately added as the nitrogen source for biological waste treatment systems. This is because the bacteria require nitrogen to produce

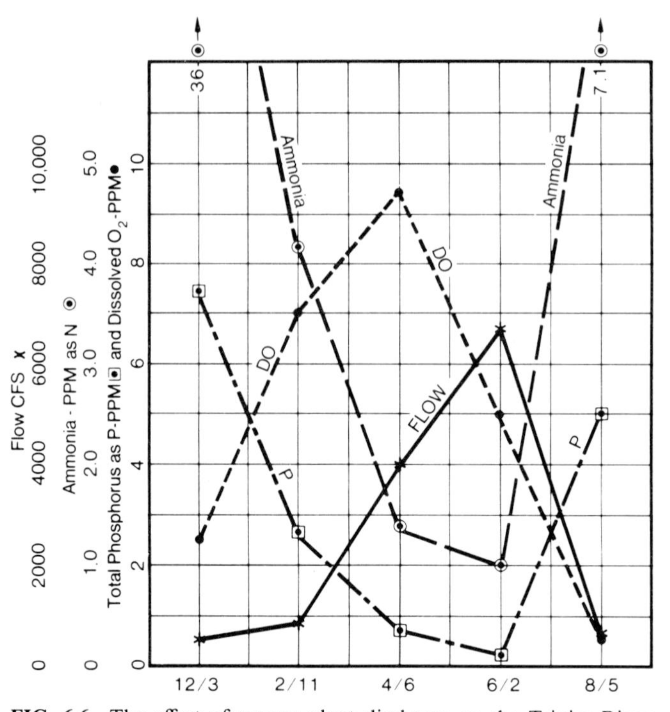

FIG. 6.6 The effect of sewage plant discharge on the Trinity River below Dallas, Texas, 1969-70. *(From USGS Water Supply Paper 2157.)*

protein substances (Figure 6.5), so the nitrogen is usually applied at the ratio of one part of nitrogen per 20 parts of food, measured as BOD.

Ammonia can be removed by degasification, by cation exchange on the hydrogen cycle, and by adsorption by certain clays, such as clinoptilolite. It is also reduced in concentration by biological activity, as noted above.

Borate [$B(OH)_4^-$, Compound of Boron, Atomic Weight 10.8; Group IIIA, Nonmetal]

Most of the world's boron is contained in seawater, at 5 mg/L B. Pure supplies of sodium borate occur in arid regions where inland seas have evaporated to dryness, especially in volcanic areas. Boron is frequently present in freshwater supplies from these same geologic areas.

It is present in water as nonionized boric acid, $B(OH)_3$. At high pH (over 10), most of it is present as the borate anion, $B(OH)_4^-$. It has little known significance in water chemistry. Its concentration is not limited in municipal waters by potable water standards. It can be damaging to citrus crops if present in irrigation water and the irrigation methods tend to concentrate the material in the soil. Although boron is in the same group on the periodic chart as aluminum, it behaves more like silica in aqueous systems; it can be removed by anion exchange and by adsorption.

Fluoride (F^-—Atomic Weight 19; Group VIIA, Halide)

Fluoride is a common constituent of many minerals, including apatite and mica. It is common practice to add fluoride to municipal water to provide a residual of 1.5 to 2.5 mg/L, which is beneficial for the control of dental caries. Concentrations above approximately 5 mg/L are detrimental, however, usually causing mottled, brittle tooth structure. Because of this, the concentration is limited by drinking water standards. High concentrations are present in wastewaters from glass manufacture, steel manufacture, and foundry operations. Lime precipitation can reduce this to 10 to 20 mg/L. Fluoride is also reduced by anion exchange and by adsorption on calcium phosphate and magnesium hydroxide. Fluoride forms a number of complexes, so residuals in fluoride wastes should be analyzed chemically and not by the use of a fluoride electrode.

Iron (Fe^{2+} and Fe^{3+}—Atomic Weight 55.9; Group VIII, Transition Element, Figures 6.1, 6.2, and 6.3)

Iron is found in many igneous rocks and in clay minerals. In the absence of oxygen, iron is quite soluble in the reduced state (as seen in the analysis of well waters containing iron). When oxidized in a pH range of 7 to 8.5, iron is almost completely insoluble, and the concentration can be readily reduced to less than 0.3 mg/L, the maximum set by drinking water standards. Because iron is so insoluble when oxidized completely, the actual residual iron after treatment is determined by how well the colloidal iron has been coagulated and filtered from the water.

Because iron is a product of corrosion in steel piping systems, often the iron found in water from a distribution system is from this source and does not represent iron left from the treatment process in the water treatment plant.

Nitrate [NO_3^-—Molecular Weight, 62, Usually Expressed as N (Nitrogen), Atomic Weight 14; Group VA, Nonmetal]

Nitrate, like ammonia, comes into water via the nitrogen cycle, rather than through dissolving minerals. Its concentration is limited by drinking water standards to 45 mg/L for physiological reasons. There are no reported uses of water where nitrate is a restrictive factor. There is an increase of total nitrogen in sewage plant effluent in the range of 20 to 40 mg/L as N above the level in the water supply. A great deal of this is ammonia, but some is nitrate. The only chemical process that removes nitrate is anion exchange; nitrate can be converted to nitrogen in a biological system by the action of the nitrifying bacteria. The nitrate content of well water is usually appreciably higher than surface water.

Potassium (K^+—Atomic Weight 39.1; Group IA, Alkaline Metal)

Potassium is closely related to sodium—so much so that it is seldom analyzed as a separate constituent in water analysis. Its occurrence is less widespread in nature, and for that reason it is found at lower concentrations than sodium. It has no significance in public water supplies or in water used for industrial purposes. As with sodium it can be removed chemically only by cation exchange, or by physical processes such as evaporation and reverse osmosis. Potassium salts are highly soluble in water (Table 3.5); but as a common constituent of clays, potassium is kept from dissolving by the nature of the structure of clay. For that reason, when water-formed deposits contain significant levels of potassium, it is probably caused by silt, in which case the deposit would also be high in Al_2O_3 and SiO_2.

Strontium (Sr^{2+}—Atomic Weight 87.6; Group IIA, Alkaline Earth Metal, Figure 6.1)

Strontium is in the same family as calcium and magnesium. Although its solubility in the presence of bicarbonate is significant (about half that of calcium), its occurrence is usually restricted to geologic formations where lead ores occur, and therefore its concentration in water is typically quite low. It is completely removed by any process used for calcium removal. If not removed by softening, in a scaling water it will be a contributor to the scale problem.

CLASS 3—TERTIARY CONSTITUENTS

This group includes materials genrerally found at concentrations exceeding 0.01 mg/L.

Aluminum (Al^{3+}—Atomic Weight 27; Group IIIA, Metal)

Although aluminum constitutes a high percentage of the earth's crust as a common component of a wide variety of minerals and clays, its solubility in water is so low that it is seldom a cause for concern either in municipal supplies or indus-

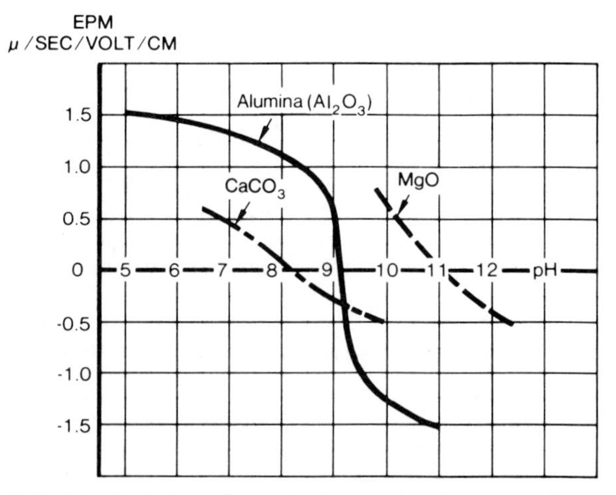

FIG. 6.7 Variation of particle charge with the nature of the particle and pH. *(From Stumm and Morgan, 1970.)*

trial water systems. However, in industrial systems, the carryover of alum floc from a clarifier may cause deposit problems, particularly in cooling systems where phosphate may be applied as a stabilizing treatment. Aluminum found in treated water systems is usually there because of colloidal residues (alumina, Al_2O_3) from the coagulation of the water if alum or aluminate is used as the coagulant. If the residuals are objectionable, they can be removed by improved filtration practices.

As shown by the solubility curves (Figure 6.3) aluminum is amphoteric, being present as Al^{3+} or lower valence hydroxyl forms at low pH and the aluminate anion at higher pH values. As might be expected from this amphoteric nature, alumina particles are positively charged at low pH and negatively charged at high pH, as indicated by Figure 6.7. The effectiveness of alum in precipitating negatively charged colloids, such as clay particles from water, is more likely related to the charge on the precipitated alumina than the charge on the aluminum ion itself, since the aluminum ion is not soluble in the typical coagulation pH range of 5 to 7. Its strong negative charge at pH 10.0 to 10.5 helps explain the effectiveness of sodium aluminate in precipitating magnesium hardness, which is positively charged at this pH.

Arsenic (As—Atomic Weight 74.9; Group VA, Nonmetal)

The solubility of arsenic in water is so low that its presence is usually an indicator of either mining or metallurgical operations in the watershed or runoff from agricultural areas where arsenical materials have been used as industrial poisons. If in colloidal form, it would be removed by conventional water treatment processes. Federal regulations limit the content in public water supplies to a maximum of 0.1 mg/L total arsenic. If the material is present in organic form, it may be removed by oxidation of the organic material and subsequent coagulation, or by an adsorption process, such as passage through granular activated carbon.

Barium (Ba^{2+}—Atomic Weight 137.3; Group IIA, Alkaline Earth Metal)

In natural waters containing bicarbonate and sulfate, the solubility of barium is less than 0.1 mg/L, and it is seldom found at concentrations exceeding 0.05 mg/L. Removal to low residuals can be expected in conventional lime treatment processes. There are instances of barium being added to water for the specific purpose of sulfate reduction. The reaction is hindered because the barium reagent itself is so insoluble that considerable time is needed for the reactions to occur; furthermore, sulfate deposition on the surface of the barium reagent makes the process inefficient. Barium is limited in drinking water to a maximum concentration of 1 mg/L.

Bromide (Br^-—Atomic Weight 79.9; Group VIIA, Halide)

Bromine is found in seawater at about 65 mg/L as the bromide ion; some connate waters produced with oil contain several hundred milligrams per liter and are the source of commercial bromine. Over 0.05 mg/L in fresh water may indicate the presence of industrial wastes, possibly from the use of bromo-organo compounds as biocides or pesticides.

Copper (Cu^{2+}—Atomic Weight 63.5; Group IB Metal)

Copper may be present in water from contact with copper-bearing minerals or mineral wastes from copper production. It is more likely, however, that the copper found in water will be a product of corrosion of copper or copper alloy piping or fittings, or may have been added deliberately to a water supply reservoir for algae control, as copper sulfate. When copper sulfate is added for algae control, because its solubility is limited, organic chelating materials may be added to the copper sulfate formulation to keep the copper from precipitating and, therefore, maintain its effectiveness. Drinking water regulations limit the municipal water supply concentration to 1 mg/L maximum. At higher concentrations, the water has an astringent taste. If a water supply is corrosive to copper, the first drawing or tapping of the supply from piping which has been idle overnight may contain relatively high concentrations, and ingestion of this water may cause immediate vomiting. In industrial supplies, the presence of copper can be objectionable as it is corrosive to aluminum. Copper is essential to certain aquatic organisms, being present in hemocyanin in shellfish, the equivalent of hemoglobin in humans.

Lead (Pb^{2+}—Atomic Weight 207.2; Group IV, Metal)

The presence of lead in fresh water usually indicates contamination from metallurgical wastes or from lead-containing industrial poisons, such as lead arsenate. However, lead may also appear in water as a result of corrosion of lead-bearing alloys, such as solder. Being amphoteric, lead is attacked in the presence of caustic alkalinity.

The limitation on lead in drinking water has been established as 0.05 mg/L, which should be readily achieved with good filtration practice. In wastewaters where lead may be complexed with organic matter, it may be solubilized, and oxidation of the organic may be required for complete lead removal.

Lithium (Li^{+}—Atomic Weight 6.9; Group IA Alkali Metal)

This alkaline earth element is rare in nature and seldom analyzed in water. There are no records of experience indicating that this material is of concern either in industrial or municipal water supplies. However, lithium salts are used in psychotherapy to combat depression, so there may be a concentration level in water that has a psychotropic effect. Lithium salts have a wide variety of uses, but the industrial consumption is so low that it is not likely to be a significant factor in the wastewaters from industries using these products.

Manganese (Mn^{2+}, Mn^{4+}—Atomic Weight 54.9; Group VIIB, Metal)

Manganese is present in many soils and sediments as well as in metamorphic rocks. In water free of oxygen, it is readily dissolved in the manganous (Mn^{2+}) state and may be found in deep well waters at concentrations as high as 2 to 3 mg/L. It is also found with iron in acid mine drainage. Wastewaters from metallurgical and mining operations frequently contain manganese.

It is an elusive material to deal with because of the great variety of complexes it can form depending on the oxidation state, pH, bicarbonate-carbonate-OH equilibria, and the presence of other materials, particularly iron.

It is limited to 0.05 mg/L maximum by drinking water regulations, because higher concentrations cause manganese deposits and staining of plumbing fixtures and clothing. However, concentrations even less than this can cause similar effects, as it may accumulate in the distribution system as a deposit, to be released in higher concentrations later if the environment should change, such as by change in pH, CO_2 content, oxidation potential, or alkalinity.

In industrial systems it is as objectionable as iron, particularly in textile manufacture or the manufacture of bleached pulp, since small amounts of deposited manganese can slough off to cause stained products which must be rejected. Reduction to levels as low as 0.01 mg/L are required for certain textile finishing operations.

In the oxidized state, manganese is quite insoluble, and can be lowered in concentration, even in an alum coagulation process by superchlorination with adequate filtration, even at a pH as low as 6.5. However, the conventional process for removal of manganese by itself is oxidation plus elevation of the pH to approximately 9 to 9.5, with retention of approximately 30 min in a reaction vessel before filtration. Filters that have gained a coating of manganese oxides can work very effectively, but may slough manganese if the aquatic environment is radically changed. Manganese is also precipitated by the continuous application of potassium permanganate ahead of a manganese form of zeolite.

Organic materials can chelate manganese much as they chelate iron, so

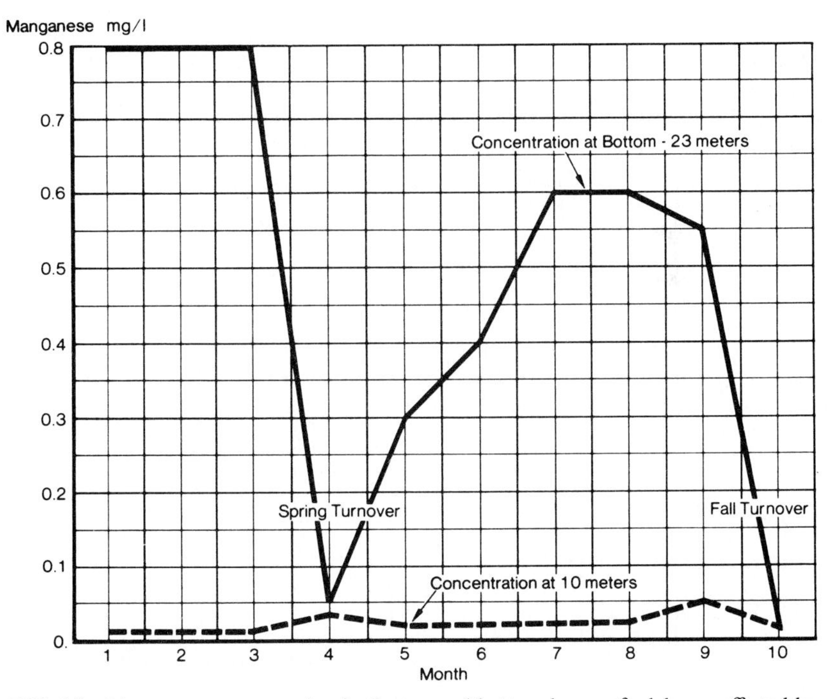

FIG. 6.8 Manganese concentration in the top and bottom layers of a lake, as affected by seasonal changes. *(From "Chemistry of Manganese in Lake Mendota, Wisconsin," Environ. Sc. Tech., December 1968.)*

destruction of the organic matter is often a necessary part of the manganese removal process.

Because manganese accumulates in sediments, it is common to find high levels of manganese in deep water where none may be apparent at the surface. This should be studied in designing the proper intake structure for a plant water supply. An example of this is illustrated in Figure 6.8, showing the manganese concentrations in a lake at various times of the year and at different depths in the lake.

Phosphate (PO_4^{3-}, Molecular Weight 95; Compound of Phosphorus P—Atomic Weight 31)

Phosphorus is found in many common minerals such as apatite, in the form of phosphate (PO_4^{3-}equivalent weight 31.7). Since phosphate compounds are widely used in fertilizers and detergents, it is common to find phosphate in silt from agricultural runoff, with fairly high concentrations being found in municipal wastewater, usually in the range 15 to 30 mg/L as PO_4 (about 5 to 10 mg/L P). Since phosphate is commonly blamed as the primary cause of excessive algal growths, which lead to eutrophication of lakes and streams, a reduction of phosphate is being brought about by legislation restricting the amount of phosphate in detergents and also requiring treatment of municipal sewage for phosphate removal.

Phosphate may be present in water as HPO_4^{2-} and $H_2PO_4^-$, as well as the higher pH form, PO_4^{3-}. The distribution as affected by pH is shown in Figure 6.9.

Phosphate can be reduced to very low levels by treatment with alum, sodium aluminate, or ferric chloride, with a formation of insoluble aluminum phosphate and iron phosphate. It can also be precipitated with lime at a pH over 10 to pro-

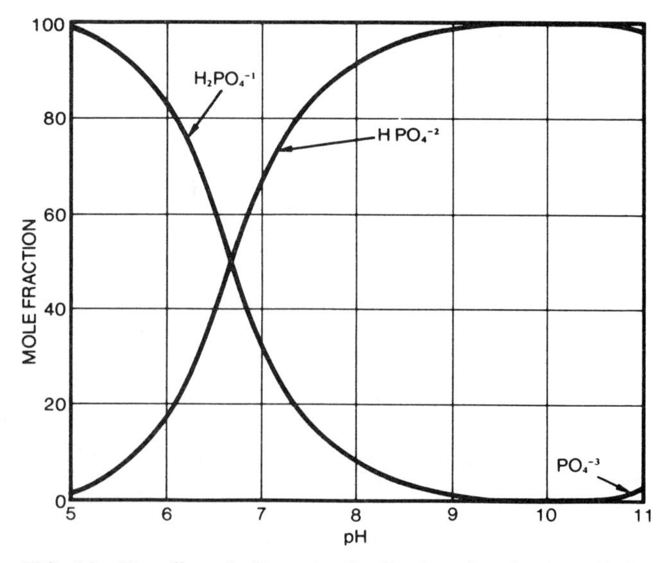

FIG. 6.9 The effect of pH on the distribution of various phosphate species.

duce residuals less than 2 to 3 mg/L in the form of hydroxyapatite; in a hot process system, the residuals would be less than 0.5 mg/L.

These phosphate precipitates are often colloidal, and filtration is required to achieve the low residuals specified.

Zinc (Zn^{2+}—Atomic Weight 63.4; Group IIB, Metal)

Zinc is a Group IIB metal, behaving quite like calcium in solution, although of considerably lower solubility in natural waters with a neutral pH and having bicarbonate alkalinity. (See Figure 6.3 for its solubility characteristics.) Zinc is seldom found at concentrations over 1 mg/L, with a typical concentration being approximately 0.05 mg/L. Because it tends to have an astringent taste, its concentration in public water supplies is limited to 5 mg/L maximum.

Zinc may be present in water because of waste discharges from mining, metallurgical, or metal finishing operations. It may also appear because of corrosion of galvanized steel piping. It is often included in proprietary corrosion inhibitors where its effect on steel piping is similar to that of galvanizing.

Zinc would be removed in lime softening operations to residuals well below 0.1 mg/L. It can also be removed by cation exchange on either the sodium or the hydrogen cycle.

CLASS 4—TRACE CONSTITUENTS

Materials in this group (Table 6.3) are generally found at concentrations less than 0.01 mg/L.

CLASS 5—TRANSIENT CONSTITUENTS

This class includes constituents which change in concentration or activity not by dilution, dissolution, or precipitation, but rather by changes in the aquatic environment which disturb the equilibrium. These changes may come about from biological activity, oxidation-reduction potential, or radioactive decay.

Acidity-Alkalinity

The typical domain of almost all natural waters is characterized by a pH range of 6 to 8, the presence of bicarbonate alkalinity, and some CO_2 dissolved in the water. All waters in contact with limestone, dolomite, or geologic formations including these minerals tend to reach this equilibrium: it is the end result of the chemical reactions that cause the weathering of rocks and of the oxidation-reduction reactions which are mediated by aquatic organisms. Because of this, the few exceptional streams that contain free mineral acidity (i.e., have a pH below about 4.5) usually dissipate this condition by accelerated weathering of the alkaline components of the rocks they contact. Likewise, when the pH exceeds 8 and carbonate alkalinity begins to appear, this is brought into balance by reaction with carbon dioxide from the atmosphere or from respiration of aquatic life.

TABLE 6.3 Class 4—Trace Constituents

Constituent	Occurrence*	Behavior†
Antimony (Sb, group VA, at. wt. 74.9)	Leaching of metallurgical slags; colloidal hydrous oxides	Insoluble in aqueous systems containing $HCO_3/CO_3/OH$. Filterable.
Cadmium (Cd, Group IIB, at. wt. 112.4)	Plating wastes; limited solubility as Cd^{2+}. Note: Restricted in potable water supplies to 0.01 mg/L maximum.	Behaves as Ca and Zn (see Figure 6.3). Can be precipitated as carbonate or removed by cation exchange.
Chromium (Cr, group VIB, at. wt. 52)	Plating wastes, cooling tower blowdown. Soluble as CrO_4^{2-} (Cr^{6+}); insoluble as Cr^{3+}. Note: Restricted in potable water supplies to 0.05 mg/L maximum as Cr.	As CrO_4^{2+}, oxidizing agent. Can be reduced by SO_2 to Cr^{3+} or removed by anion exchange. As Cr^{3+}, colloidal hydrous oxide at neutral pH, filterable.
Cobalt (Co, group VIII, at. wt. 59)	Present in copper- and nickel-bearing ore tailings; ceramic wastes.	Behaves as iron.
Cyanide (CN^-, eq. wt. 26)	Wastes from plating shops, coke plants, blast furnaces, petroleum refining. Note: Restricted in potable water supplies to 0.2 mg/L maximum.	Behaves as Cl^-, NO_3^-, highly soluble. Can be oxidized with Cl_2 to CNO^- and $N_2 + CO_2$. Reduced in activated sludge biodigestion. Removable by anion exchange. Forms complexes with Cd, Fe.
Mercury (Hg, group IIB, at. wt. 200.6)	Wastes from electrolytic NaOH production, leaching of coal ash. Note: Restricted in potable water supplies to 0.002 mg/L maximum.	May be methylated by bacterial activity and taken up by the aquatic food chain. Removed by closing plant loop or by reduction and filtration.
Nickel (Ni, group VIII, at. wt. 58.7)	Plating wastes; electric furnace slag or dust; ore tailings.	Ni^{2+} behaves as iron, Fe^{2+}. Seldom present as Ni^{3+}. Precipitates as hydroxide and basic carbonate. Sulfide precipitates are insoluble. Removable by cation exchange.
Tin (Sn, group IVA, at. wt. 118.7)	Tinplate waste.	Converted to hydrous oxide colloid at neutral pH or in HCO_3/CO_3 environment.
Titanium (Ti, group IVB, at. wt. 47.9)	Ilmenite in well formation.	Colloidal TiO_2. Removed by filtration.

* Most constituents in class 4 are introduced by industrial waste discharges, and are more generally found in surface than in well waters.

Uncontrolled discharge of industrial wastes could wipe out this natural buffering effect of equilibrium between the aquatic environment, the atmosphere, and the lithosphere.

Since legislation now prohibits such discharge, the only circumstances that can cause waters to fall outside the natural conditions are accidental spills of large volumes of strong chemicals, seepage of acid mine drainage into a stream, or acid rain from air pollution. The mine seepage may not be controllable because of inability to locate the source or the point of entry into the stream.

Acid rain is caused by the dissolution of acidic gases from the environment, chiefly the oxides of sulfur (SO_2, SO_3), and perhaps aggravated by nitrogen oxides (NO_x). The most prominent source appears to be residues of sulfur from coal-fired boiler plants. When the acid rain falls on alkaline rock or into rivers or lakes in limestone basins, there may be enough reserve alkalinity in the rock or dissolved in the water to neutralize the acidity. But often the rain falls in forested areas where vegetative litter has developed a soil high in humus. There is no natural alkalinity then to counteract the acidity of the rainfall, so the runoff is acidic and if the drainage leads to a lake in a granite basin (or a formation free of limestone), the lake itself will become acid and normal aquatic life will disappear.

Areas of the United States and Canada affected by acid rain are shown in Figure 2.11. Possibly at some time in the past, alkaline components in industrial gas discharges (e.g., alkaline fly ash, cement kiln, or lime kiln dust) alleviated some of this problem, and it is likely that application of limestone to some lakes affected by rain will be both effective and economically justified. Reduction of sulfur oxides from boiler plant flue gas often utilizes lime as the neutralizing agent. But complete elimination of these sulfur oxides is impossible in the boiler stack. So alkali treatment of acid-affected lakes may prove more economical and practical than further treatment at the boiler stack as further addition of alkali there runs into diminishing returns. However, this does not correct other deleterious effects of acid rain, such as damage to the flora of forests and croplands and etching and corrosion of buildings and other structures.

Carbon Cycle Constituents (Carbon, C—Atomic Weight 12.0; Group IVA, Nonmetal, see Figure 6.10)

Carbon is one of the primary elements of living matter, as shown by the generalized formula for biomass. It has been hypothesized that earth's primitive environment contained carbon in the form of methane plus ammonia, water, and hydrogen gases. Methane is one of the carbon compounds present in the carbon cycle (Figure 6.10), produced by fermentation of larger organic molecules. Carbon dioxide and bicarbonate-carbonate alkalinity are also prominent in the cycle. These reactions are proceeding in the aquatic environment in the carbon cycle, Figure 6.10.

Methane, a major component of natural gas, is produced by anaerobic decomposition of organic chemical compounds. Methane is given off by anaerobic decomposition of organic sediments in marsh gases, and the concentration may become so high that the swamp gas may ignite. It is more common to find methane in well waters in areas where natural gas is produced than in surface waters.

Oxygen Cycle Constituents

The most common carbon-containing gas is carbon dioxide, discussed in earlier chapters. The carbon dioxide content of surface waters is greatly influenced by

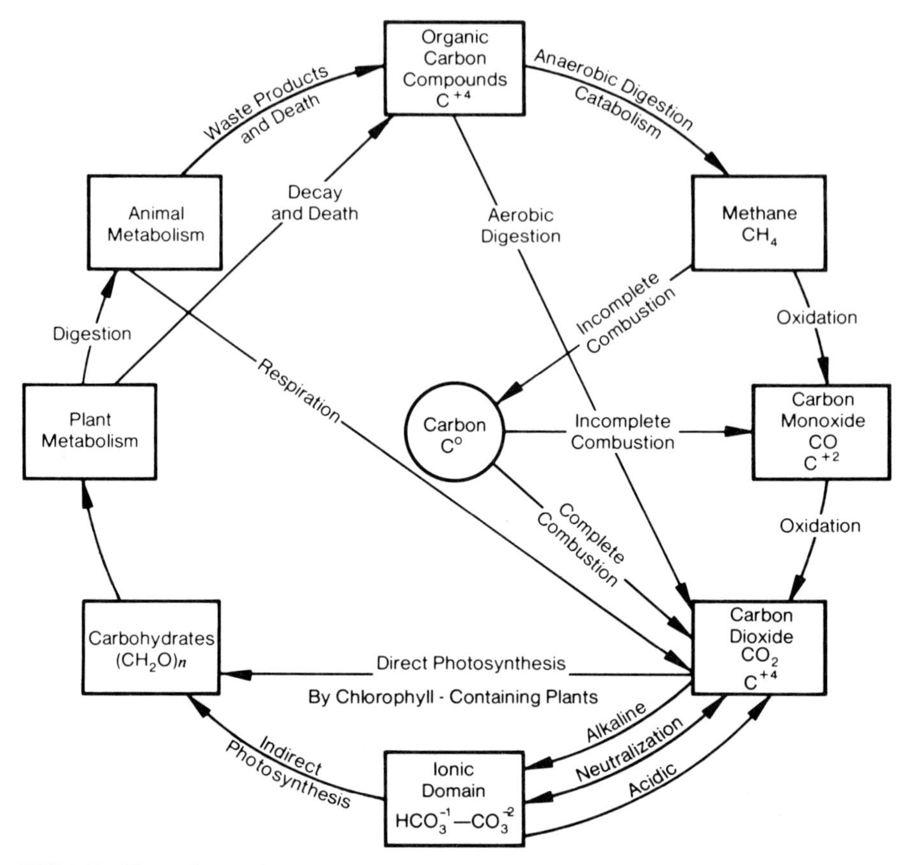

FIG 6.10 The carbon cycle.

bacterial and algal symbiotic existence, illustrated by the oxygen cycle, Figure 6.11. During bright sunlight, the photosynthetic reactions proceed so rapidly that the water may actually become supersaturated with oxygen, beyond the capacity of the bacteria to utilize. If algae require more carbon dioxide than is available from bacterial respiration, they assimilate carbon dioxide from the bicarbonate alkalinity, producing a trace of carbonate alkalinity. For that reason, carbon dioxide and oxygen are variable in most surface supplies, as affected by sunlight and the photosynthetic process (Figure 6.12).

The plant tissue built up by photosynthesis is eventually metabolized by larger aquatic organisms that produce organic compounds and discharge them in their wastes. Organic compounds are also produced by the death and decay of the aquatic plants and fish life. Organic matter thus produced becomes food for bacteria and is returned to the cycle as methane by anaerobes and as carbon dioxide by aerobic bacteria.

Deep well water often contains over 25 mg/L CO_2, and may be saturated with this gas at the hydrostatic pressure and temperature in the water table. A drop in pressure as water flows across the well screen may cause the CO_2 to come out of solution, disrupting the equilibrium and depositing $CaCO_3$ scale.

The theoretical solubility of oxygen exposed to the atmosphere is dependent

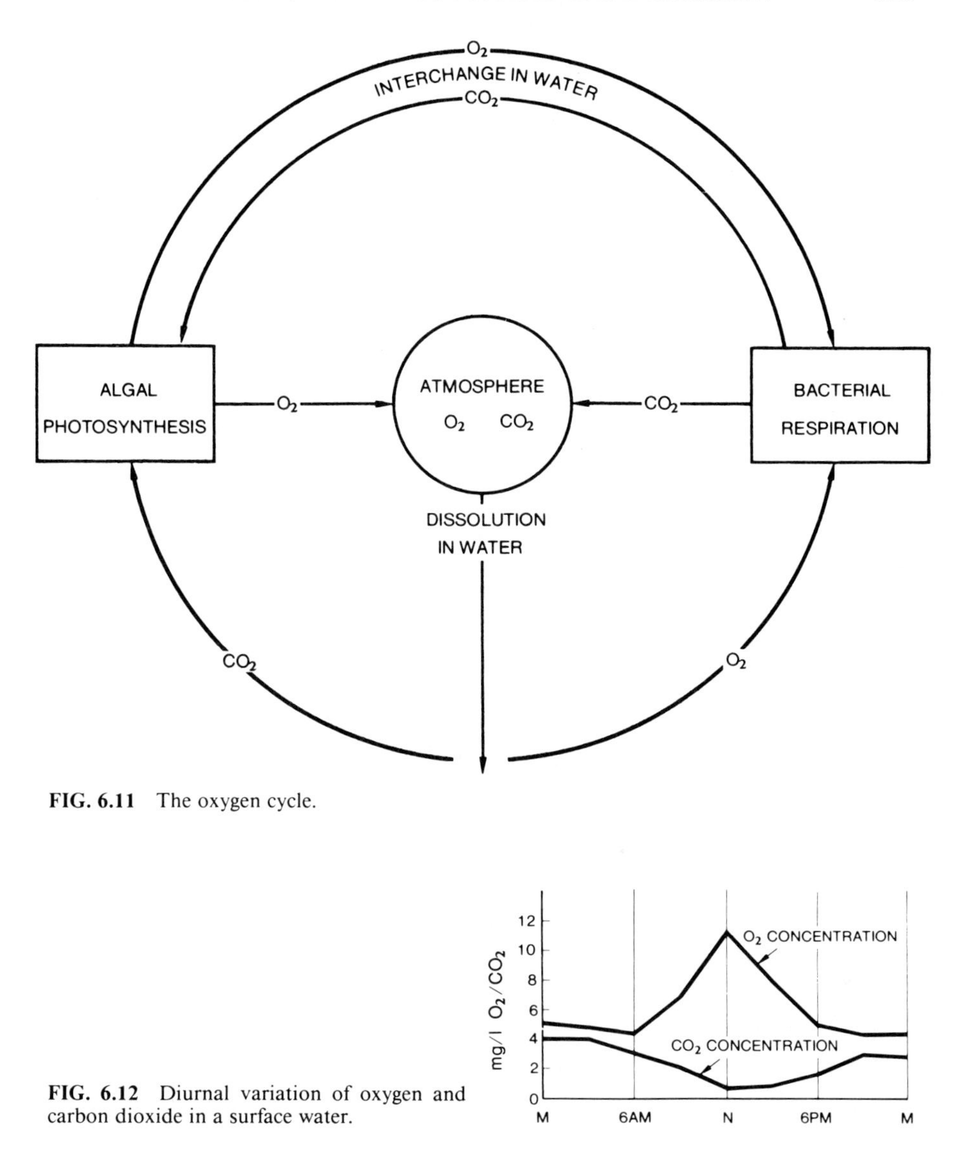

FIG. 6.11 The oxygen cycle.

FIG. 6.12 Diurnal variation of oxygen and carbon dioxide in a surface water.

on the temperature (Figure 6.13). Excess oxygen concentration is due to photosynthesis, and a deficiency is usually caused by bacterial activity or reducing agents.

Nitrogen Cycle Constituents (Figure 6.5)

Constituents of the nitrogen cycle have been discussed earlier. As was true with the carbon cycle, the nitrogen cycle is involved with life in the aquatic environment.

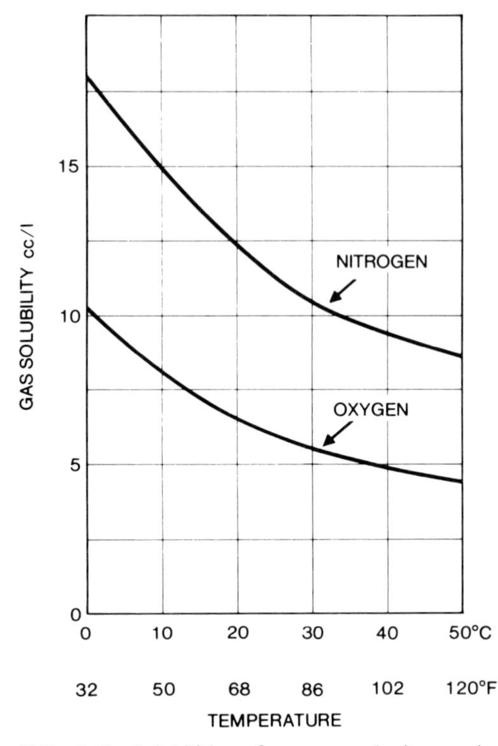

FIG. 6.13 Solubilities of oxygen and nitrogen in water at various temperatures.

Sulfur Cycle (Figure 6.14)

Because sulfur is in the same family as oxygen, there are many compounds where sulfur replaces oxygen in a compound with similar properties. For example, ethanol (CH_3CH_2OH) and ethyl mercaptan (CH_3CH_2SH) are analogous compounds with only the sulfur and oxygen atoms interchanged.

Certain bacteria can metabolize the sulfur atom in hydrogen sulfide, just as algae and other plants can metabolize oxygen from water in photosynthesis to produce free oxygen and carbohydrate. The by-product of the bacterial process of splitting H_2S is free sulfur. The corresponding chemical equations are:

$$CO_2 + 2\ H_2O \xrightarrow{\text{Algal photosynthesis}} CH_2O + O_2 + H_2O \tag{2}$$

$$CO_2 + 2\ H_2S \xrightarrow{\text{Bacterial action}} CH_2O + 2\ S + H_2O \tag{3}$$

Hydrogen sulfide, which is present in some deep well waters and some stagnant surface waters, is generally produced by the anaerobic decomposition of organic

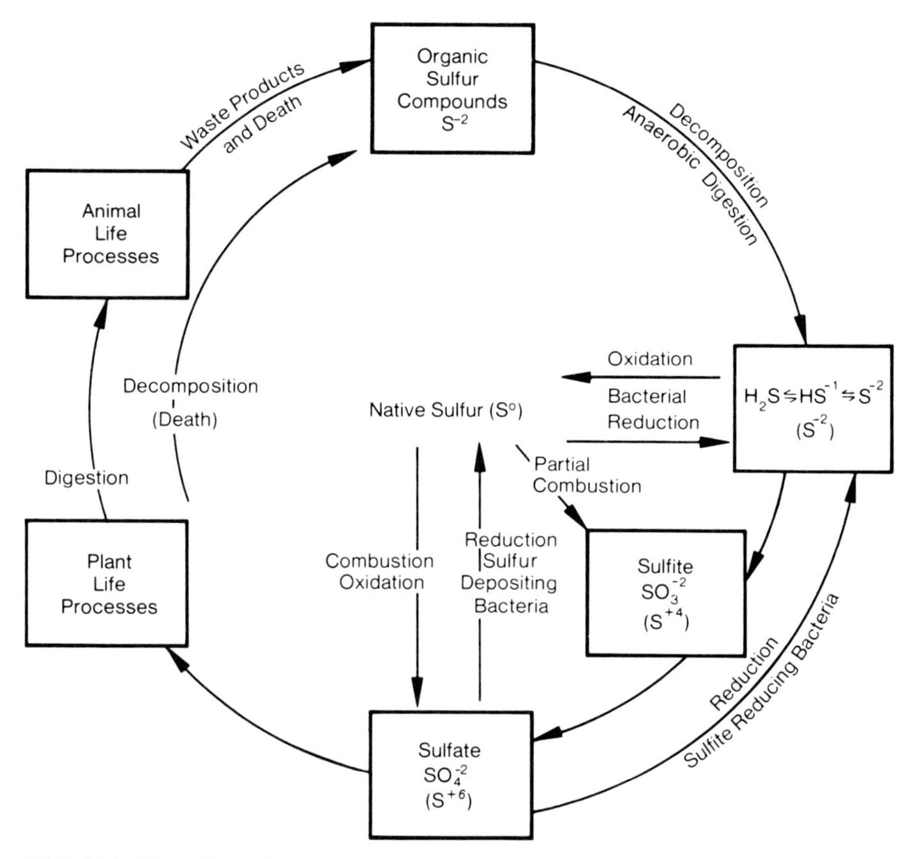

FIG. 6.14 The sulfur cycle.

compounds containing sulfur or by sulfate-reducing bacteria capable of converting sulfate to sulfide.*

All of these biological processes that occur in nature can be put to work under controlled conditions by the water specialist, to digest and eliminate undesirable organic wastes or their by-products.

* In each of these cycles involving biological activity, the changes occurring in surface water are much more pronounced than those occurring in deep well waters. The surface supplies are constantly inoculated by microbes from the air and from the soil and supplied with solar energy. On the other hand, well water usually represents a dead end in the cycle, with the water generally having a long residence time in the aquifer, with the likelihood that most organisms have been filtered by the porous formation so that all of the microbes may have been removed from the strata at the well screen. Because of this, it is common to find that the constituents of the carbon cycle, particularly CO_2, those of the nitrogen cycle, particularly NO_3^-, and those of the sulfur cycle, particularly SO_4^{2-} and HS^-, are relatively constant in well waters. For the same reason, the concentrations of CO_2, NO_3^-, and HS^- are generally higher in deep well water than in surface supplies.

Oxidation-Reduction Potential

Some materials in water are transient because they tend to oxidize or reduce other constituents, either through biological activity, as illustrated by the several biological cycles, or directly. The presence of these materials has already been considered in the text covering the biological cycles and the constituents taking part in these cycles.

The only significant oxidizing materials encountered in the natural environment that can participate in chemical reactions without needing a biological route are oxygen, which is prevalent in surface supplies, and sulfur, which is encountered by subsurface waters in contact with native sulfur, a special situation. There are numerous oxidizing materials that can appear as residues in treated wastewaters—among these are free chlorine and chromate.

The common reducing materials are organic material, ferrous iron (Fe^{2+}), manganese (Mn^{2+}), and bisulfide (HS^-) from the natural environment; reducing materials which may be added as treatment by-products or waste residues include a wide variety of organic matter, ferrous iron from such operations as steel pickling, and sulfite, present in certain kinds of pulp mill wastes.

Radionuclides

Water itself is not radioactive, but may contain elements that are. These enter the water cycle as wastes from nuclear power plants, fallout from nuclear blasts, or the by-products of metallurgical processing of radioactive materials. In very rare cases, well waters may contain radionuclides—common radioactive elements and their isotopes—as natural contaminants.

The water chemist normally deals with concentrations at part per million or milligrams per liter levels. One mole of any substance represents its molecular weight, so that the concentration of 1 mole/L as calcium carbonate represents 100 g/L as $CaCO_3$. Since a mole contains 6×10^{23} molecules, 100 g/L contains this number of molecules, and 100 mg/L contains 6×10^{20} molecules. Even at the very low concentration of 1 ppb (0.001 mg/L), then, there are still 6×10^{15} molecules of the substance as calcium carbonate in a liter of water. Concentrations of radioactive substances at this level would be extremely dangerous.

In chemical change, one element reacts with another in a process involving the electrons surrounding the nucleus, with the nucleus remaining unchanged. Energy may be given off or may be absorbed in the form of heat. With nuclear reactions, on the other hand, only the nucleus is affected, and the products of reaction may include nuclear particles (such as protons, neutrons, or electrons) and energy, including heat and electromagnetic radiation. The typical by-products of nuclear reactions, then, are alpha particles, electrons, and electromagnetic radiations. The general characteristics of these emanations are shown in Table 6.4.

TABLE 6.4 Types of Radiation

Alpha	Helium nuclei, charge +2, mass 4	Can penetrate air, but is stopped by solids.
Beta	Electrons, charge −1, mass 0	About 1000 times more penetrating than alpha radiation. Can penetrate 2–3 mm into solids.
Gamma	Similar to x-rays	Can penetrate air or solids about ten times deeper than beta radiation.

Some nuclear reactions are extremely rapid, occurring in a split second, but others may require 1000 years. Because these reactions may never go to completion, the life of a material participating in a nuclear reaction (called a radionuclide) is expressed as the time required for one-half its energy to be dissipated, its half-life. Each nuclear disintegration can be measured by the particles given off. The unit of measurement is the curie, defined as 3.7×10^{10} disintegrations per second. The levels of radioactive disintegration of concern to the water technologist are so far below this that they are expressed as micromicrocuries, or *picocuries* (pCi), a pCi being equal to 2.2 disintegrations per minute.

A variety of radionuclides may be found in waters leaving a nuclear power plant. The government license to operate such a plant specifies the particular radionuclides which must be identified individually. This requires sophisticated and painstaking analytical techniques. In municipal supplies, it is not required that individual species be identified because the radiation levels are so low, but an analysis is required for alpha radiation, which is assumed to be contributed by ^{226}Ra, beta radiation (assumed to be contributed by strontium, ^{90}Sr) and gross beta activity. Table 6.5 shows the relative relationship between occupational levels and

TABLE 6.5 Radionuclide Contamination in Water

	^{226}Ra	^{90}Sr
Occupational levels (body exposure)		
pCi/L	4×10^{1}	8×10^{2}
ppb	4×10^{-5}	4×10^{-6}
Potable levels (ingestion)		
pCi/L	3	10^{1}
ppb	3×10^{-6}	5×10^{-8}

potable water levels and also relates the radiation intensity to the concentration of ^{226}Ra or ^{90}Sr in the water. It is obvious that the concentration levels are far below those considered significant when dealing with nonradioactive contaminants in water supplies.

The heavier radionuclides are generally insoluble and may be removed by coagulation and filtration; the soluble constituents may be removed by ion exchange. These processes become quite complicated when the radionuclides are present in very low concentrations and the treatment process may first have to react with more common contamination, such as calcium and magnesium, at concentrations far above those of the radionuclides before the radionuclides are reduced to acceptable levels.

METHODS OF ANALYSIS

It is not in the scope of this handbook to include methods of analysis for each of the constituents presented in this chapter. These methods are covered in manuals devoted exclusively to such analyses (e.g., Standard Methods for Analysis of Water and Wastewater, ASTM Standards). However, included in Chapter 7 is a summary of the methods of analysis used, detection limits, methods of sample preservation, and other facts of importance.

SUGGESTED READING

Davies, S. N., and DeWiest, R. C. M.: *Hydrogeology,* Wiley, New York, 1966.

Faust, S. J., and Hunter, J. V.: *Organic Compounds in the Aquatic Environment,* Dekker, New York, 1961.

McKee, J. E., and Wolf, H. W.: *Water Quality Criteria,* State Water Quality Control Board, Resources Agency of California, 1963.

Stumm, W., and Morgan, J. J.: *Aquatic Chemistry,* Wiley, New York, 1970.

Todd, Dand K.: *The Water Encyclopedia,* Water Information Center, Port Washington, N.Y., 1970.

U.S. Environmental Protection Agency: *National Water Quality Inventory,* Report to Congress, EPA-440/9-75-014, 1975.

U.S. Environmental Protection Agency: *Quality Criteria for Water,* EPA-440/9-76-023, 1976.

U.S. Environmental Protection Agency: *Recommended Uniform Effluent Concentration,* U.S. Government Printing Office, 1973.

William, S. L.: "Sources and Distribution of Trace Metals in Aquatic Environments," in *Aqueous-Environmental Chemistry of Metals* (Ruben, J., ed.), Ann Arbor Science, Ann Arbor, Mich., 1976.

CHAPTER 7

WATER GAUGING, SAMPLING, AND ANALYSIS

Knowing the rate of flow of water and the concentrations of substances it carries is necessary to the solution of many problems in water supply, utilization, and ultimate disposal. Development of such information requires:

1. Measurement, or gauging, of water flow rates
2. Collection of representative samples of the water
3. Examination and analysis of the samples

For purposes of establishing user charges, as well as operating treatment facilities and proportioning treatment chemicals, most plants meter incoming water supplies. The total flow as well as the flow to individual processing areas within a plant may be measured.

After collection and treatment, water is usually distributed through completely filled mains and pipes under the pressure (head) produced at the intake. The rate of flow, then, depends principally upon the difference in pressure existing between the inlet and outlet, and the size of the pipe. The pressure difference establishes the velocity of water flowing through the line. The rate of flow equals the cross-sectional area of the pipe multiplied by the velocity of the water movement through it, or

$$Q = AV$$

where Q = volume of flow past a given point per unit of time
A = cross-sectional area of the pipe or conduit
V = velocity of fluid travel in terms of distance per unit of time

Most water distribution systems are sized for a velocity of 3 to 6 ft/s (0.9 to 1.8 m/s) or for a pressure loss of 1 to 5 lb/in^2 per 100 ft of piping. In this range, flow is turbulent and fine particulates usually stay in suspension. There is variation in the velocity: pressure loss relationship between small and large diameter piping, as shown by Table 7.1. Increased roughness of piping with age, as corrosion or scale develops, also affects this ratio.

Pipe fittings and valves introduce additional pressure loss, as shown by Table 7.2, which provides a basis for estimating overall pressure loss and pumping requirements for water distribution systems. Wastewater collection piping is usually larger than freshwater distribution systems because it frequently must carry off its flow by gravity alone.

TABLE 7.1 Selecting Pipe Size for Water Systems

Range of flow gal/min	Nominal pipe size, in (std wt)	Velocity range ft/s	Head loss* in feet
1–2	½	1–2	2–10
2–5	¾	1–3	2–10
4–9	1	1.5–3.5	2–10
8–19	1¼	1.7–4.0	2–10
12–30	1½	1.9–4.7	2–10
23–55	2	2.4–5.3	2–10
38–85	2½	2.5–5.7	2–10
65–150	3	2.9–6.5	2–10
130–310	4	3.3–7.8	2–10
240–570	5	3.9–9.3	2–10
390–920	6	4.3–10.4	2–10
800–1300	8	5.1–8.3	2–5
1400–2400	10	5.7–9.8	2–5
2300–3700	12	6.5–10.0	2–5

* Based on clean pipe, coefficient of 100; head loss is given in feet of water per 100 ft of pipe. To convert to lb/in^2: feet of water/2.3 = lb/in^2.

FLOW MEASUREMENT

Some water meters, called displacement meters, measure flow by responding to the volume of water throughput. A familiar example is the disk type meter widely used in homes and buildings. In this design, the displacement of the disk, usually by a rotary motion, is directly proportional to the volume of water passing through the meter body. The disk rotation drives a gear train actuating a totalizer comparable to the mileage indicator in an automobile. Another displacement meter is the turbine type flow transmitter (Figure 7.1), in which water causes a multiblade rotor to spin at a velocity proportional to flow rate.

Using Pressure Changes

A different method, which measures flow indirectly, responds to pressure differential manifestations of water velocity. Because water is incompressible, the quantity passing one point in a full conduit in a unit of time is exactly the same as that passing any other point, regardless of line size restrictions or changes occurring along the conduit. Since the volumetric flow is constant, the velocity must increase where flow passes through a restriction. The pressure at the restriction decreases in proportion to such velocity increase. This relationship is the basis for operation of widely used differential pressure type measuring devices where the restriction is a thin-plate orifice, flow nozzle, or venturi tube. Such devices are carefully designed to be installed in the path of flow to create a velocity

TABLE 7.2 Useful Formulas for Estimating Capacity of Water Distribution Systems

1. Velocity (ft/s) through steel pipe:
   $V = 0.41 \times$ gpm/d (d = pipe diameter in inches)
2. Friction losses (in feet of water) through steel pipe:
   a. Friction length (L) = pipe length (ft) + extra for fittings.
      Extra length for fittings:

Gate valve	=	$0.3 \times$ pipe size (d)
Elbow	=	$2.7 \times$ pipe size
Side outlet tee	=	2 elbows
Globe valve	=	5 elbows

   Example: Estimate friction length (L) of a distribution system having 350 ft of 4-in pipe with 5 gate valves and 10 elbows

$$
\begin{aligned}
L = \text{length of pipe} &= 350 \text{ ft pipe} \\
+\,5 \times 0.3 \times 4 &= 6 \\
+\,10 \times 2.7 \times 4 &= \underline{108} \\
&\ 664 \text{ ft total}
\end{aligned}
$$

   b. Approximate head loss due to friction (H):
      $H = 0.1\,(\text{gpm})^2/(d)^5 \times (L/100)$
      Example: Estimate the head loss through the system above at 250 gal/min:

$$
\begin{aligned}
H &= 0.1\,(250)^2/(4)^5 \times (664/100) \\
&= 0.1\,(62{,}500/1024) \times 6.64 \\
&= 0.1 \times 61 \times 6.64 = 41 \text{ ft} \\
&\ 41/2.3 = 17.7 \text{ lb/in}^2
\end{aligned}
$$

3. Horsepower for pumping (hp):

$$\text{hp} = \frac{(\text{gpm}) \times \text{head loss}}{3960 \times \text{efficiency}}$$

Example: Estimate the horsepower required by the above system using a pump having an efficiency of 85%.

$$\text{hp} = \frac{250 \times 41}{3960 \times 0.85} = 3.05$$

Use a 5-hp motor to drive this pump.

increase and a corresponding pressure decrease which can be accurately measured. The formula for calculating flow through the thin-plate orifice is:

	English	**Metric**
	$Q = 11.782d^2f\sqrt{H}$	$= 0.0125d^2f\sqrt{H}$
where	Q = gal/min	m^3/min
	d = orifice diam., in	cm
	H = head loss, ft of H_2O	m
	f = 1, unless d is more than 50% of pipe diameter D	

The factor f is calculated as follows:

$$f = \sqrt{\frac{1}{1 - (d/D)^4}}$$

FIG. 7.1 The turbine meter is widely used in industrial plants to measure the consumption of municipal water, where measurement of flow rate is not essential. *(Courtesy Neptune International Corporation.)*

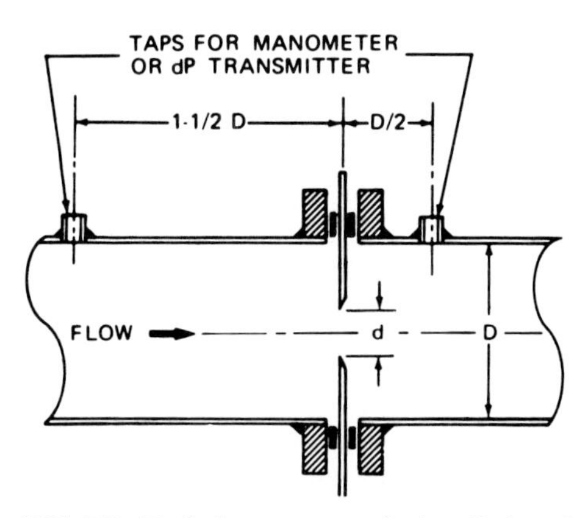

FIG. 7.2 Typical arrangement for installation of an orifice to meter flow rate as well as volume.

Figure 7.2 shows an enclosed orifice installation and the correct location of the pipe taps for measuring head loss. Flow through an orifice based on head loss can be determined from the graph in Figure 7.3.

With the enclosed orifice, either a mercury manometer type meter body or a diaphragm-type differential converter (Figure 7.4) is connected by means of suitable pipe taps to measure the differential pressure.

Thin-plate orifices are the least expensive, easiest to install, and the most widely used of the differential pressure type devices for flow measurement. But

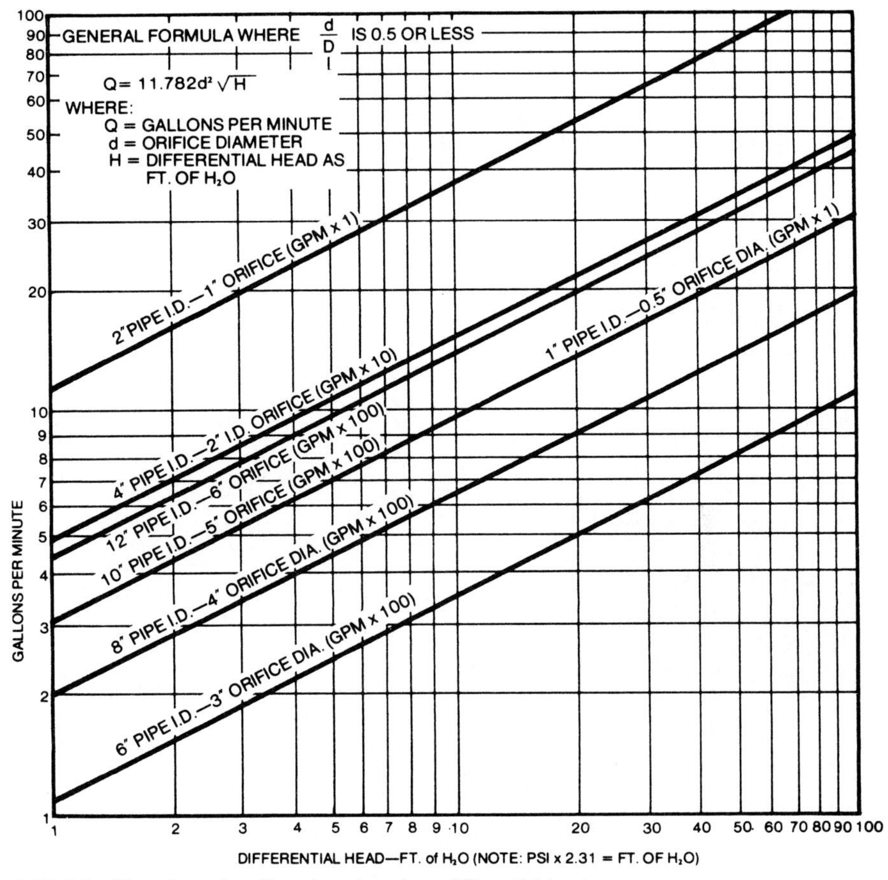

FIG. 7.3 Flow through orifice plates based on differential head.

an undesirable feature, as illustrated in Figure 7.5, is that they produce a relatively high permanent pressure loss. The venturi tube (Figure 7.6) is designed to reduce this loss to a negligible level. Venturi tubes offer the additional advantages of freedom from accumulation of entrapped solids. But, they are expensive compared to other devices. Flow nozzles fall between venturi tubes and thin-plate orifices in both cost and permanent head loss.

Another differential pressure type flow measuring device, providing inexpensive and accurate measurement of relatively small rates of flow for clean water, is the rotameter, Figure 7.7. Transparent rotameter tubes have a tapered inner bore increasing in diameter from bottom to top of the tube. A metal bob in the tube comes to an equilibrium position at a point where the annular flow around it produces just the velocity and pressure difference needed to support its weight. The bob position alongside the calibrated graduations on the rotameter tube directly indicates rate of flow.

Some supply water metering methods do not require the installation of inline primary devices. For example, there is always a frictional loss of pressure proportional to water velocity across any pipe fitting in a line flowing full under pressure. Pipe elbows can be calibrated, though with considerably less accuracy than orifice plates or venturis, for use with flow measuring devices.

For very large supply lines, where the installation of permanent inline primary devices might be impractical or costly, pitot tubes (Figure 7.8) are useful for gauging flows. These are inserted into the line through special pressure fittings and

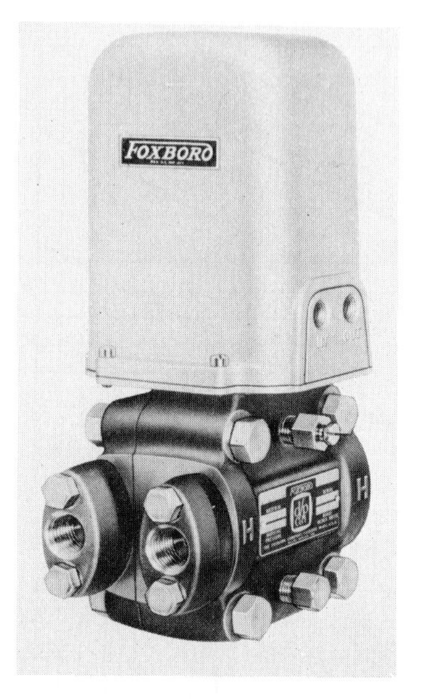

FIG. 7.4 Differential pressure cell used to convert head loss across a metering element into a signal to activate a recording and integrating device. *(Courtesy The Foxboro Company.)*

positioned to measure the velocity head of the flowing water. The velocity head is the difference between the total head, sensed by the probe aperture facing upstream, and the static head in the line. Because velocities in a pipe are usually greater at the center of the pipe than along the walls, flow calculations are based

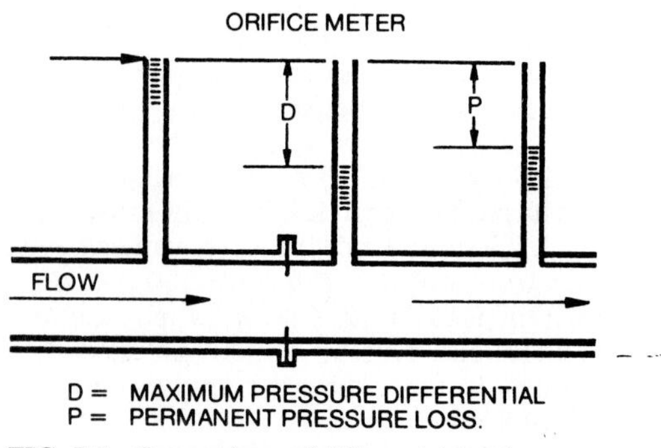

FIG. 7.5 Comparison of differential head to permanent head loss for flow through an orifice.

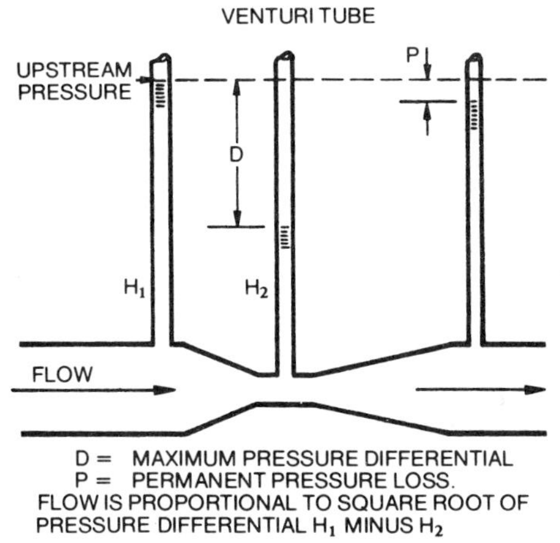

D = MAXIMUM PRESSURE DIFFERENTIAL
P = PERMANENT PRESSURE LOSS.

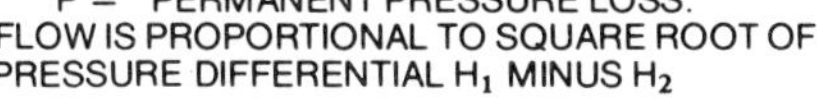

FLOW IS PROPORTIONAL TO SQUARE ROOT OF
PRESSURE DIFFERENTIAL H_1 MINUS H_2

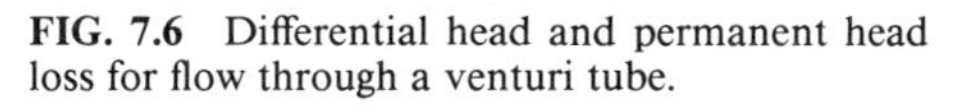

FIG. 7.6 Differential head and permanent head loss for flow through a venturi tube.

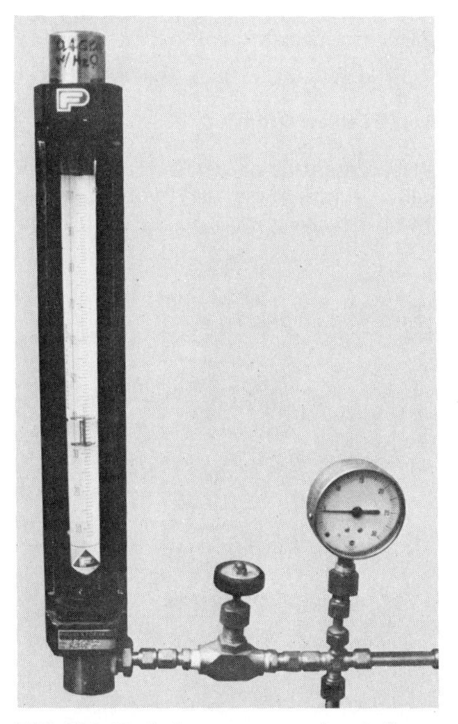

FIG. 7.7 Typical rotameter used to indicate rate of flow. An output signal can be provided for remote indication and recording. *(Courtesy Fischer & Porter Co.)*

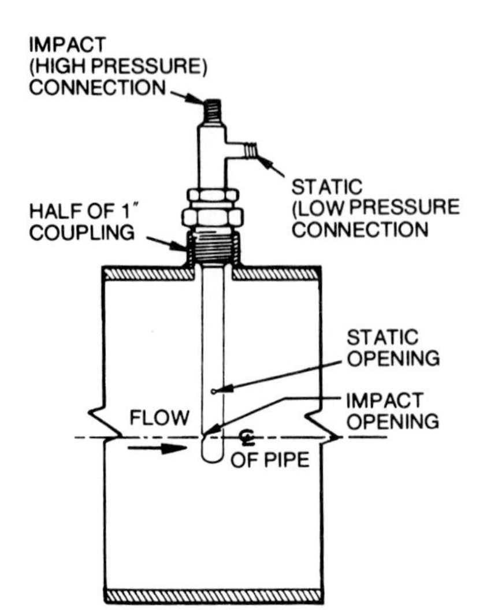

FIG. 7.8 Pitot tube installation, showing connections for differential head measurement. *(Courtesy The Foxboro Company.)*

on the average of ten velocities measured at specific locations as the pitot traverses the diameter of the pipe. This average velocity multiplied by the cross-sectional area of the pipe produces the volumetric flow.

Methods Not Related to Pressure Drop

There are two flow measurement devices with measuring elements external to the stream; they are designed that way so they won't obstruct the flow or become clogged with solids that might be present in the liquid. One of these is the mag-

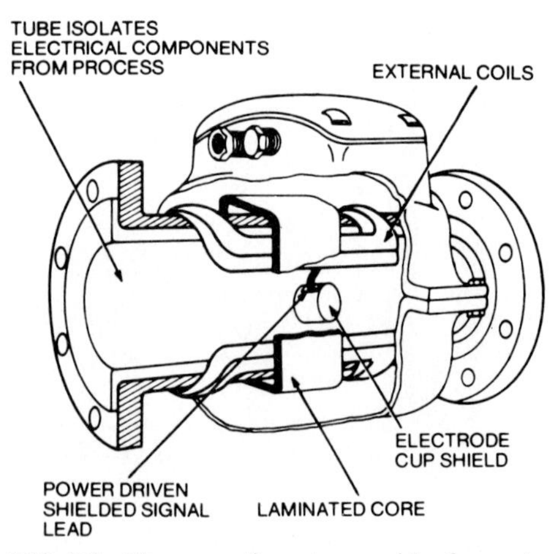

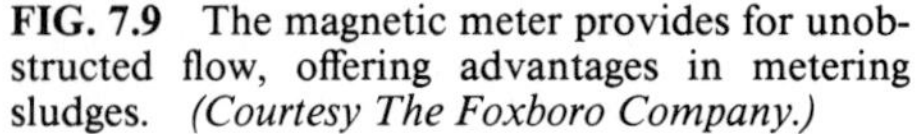
FIG. 7.9 The magnetic meter provides for unobstructed flow, offering advantages in metering sludges. *(Courtesy The Foxboro Company.)*

netic and the other the ultrasonic flowmeter. In the case of the magnetic flowmeter (Figure 7.9), the principle of operation is based on Faraday's law of induction: a conductor moving through a magnetic field produces a voltage proportional to the velocity of the conductor. The water flowing through the pipe is the conductor, and an electrically energized coil produces the magnetic field. The induced voltage, proportional to velocity of flow, is measured by electrodes in contact with the water and transmitted to a meter that converts the signal to flow readout, taking into account the pipe diameter as a constant in initial calibration of the meter.

In the case of the ultrasonic flowmeter, the sensing device is strapped onto the outside of the pipe: metal, plastic, glass, or lined metal—this can be either a permanent installation or a temporary one, as in the situation where the meter is moved from one location to another in making a plant survey of water use. The sensing element contains an ultrasonic transmitter and a receiver. The ultrasonic waves may be projected and received by two basic methods:

1. *Travel time difference technique:* With this method, the meter is provided with two elements strapped to opposite sides of the pipe wall, one being at a set distance downstream of the other. Each element has a transmitter and a receiver, so that one beam is projected downstream, and the other upstream; the difference in travel time of these two beams is proportional to flow. High suspended solids or gas bubbles interfere with performance.
2. *The doppler technique.* With this meter there is one element on one side of the pipe wall. The ultrasonic beam is projected at an angle downstream at a controlled frequency. Solids or bubbles in suspension reflect the echo back to the element at a lower frequency (the doppler effect), and the change in frequency is proportional to flow. This technique requires suspended solids or bubbles; it is not applicable to clear liquids. A typical ultrasonic meter is shown in Figure 7.10. This is a travel time differential unit. Single units are available to handle either travel time or doppler techniques to accommodate changes in suspended solids concentration in the measured stream.

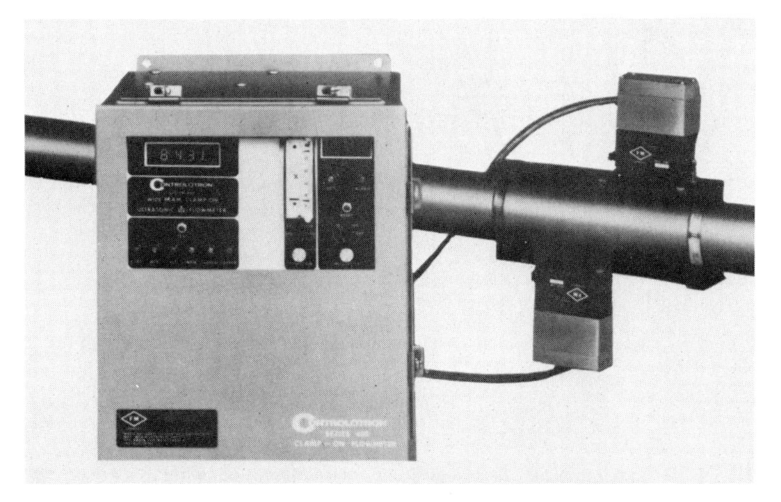

FIG. 7.10 Ultrasonic flowmeter. *(Courtesy Controlotron Corporation.)*

TABLE 7.3 Methods of Wastewater Flow Measurement

A. For open channels
   1. Weirs
   2. Flumes
   3. Float timing
B. For discharge from open-end pipes
   1. Discharge timing method
   2. Pipe cap orifice method
   3. Coordinate measurement
   4. California pipe method
C. Manning Method for Partially Filled Sewers
D. Indirect methods
   1. Chemical concentration
   2. Temperature measurement

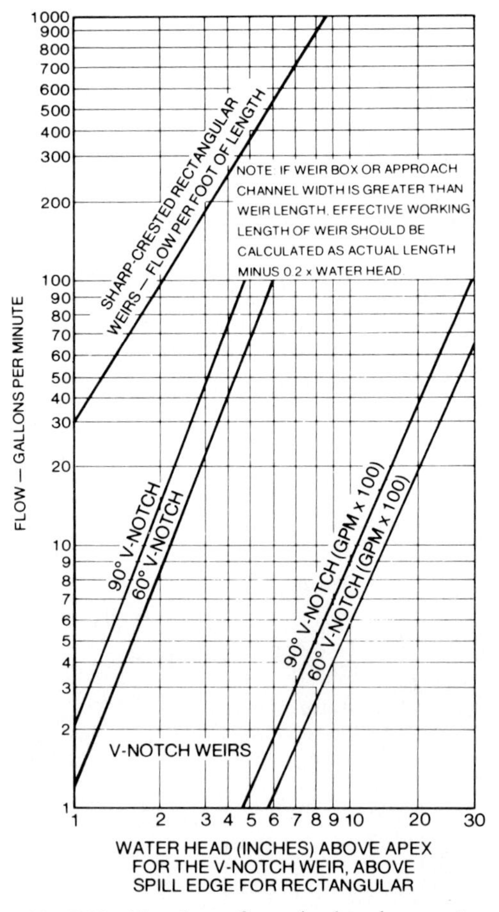

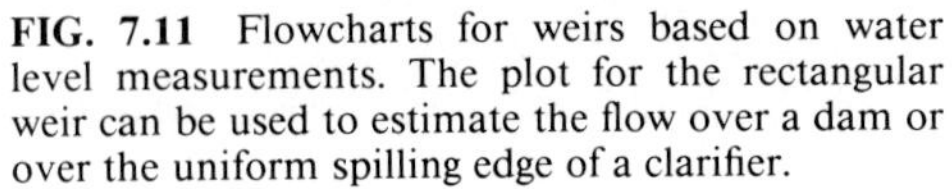

FIG. 7.11 Flowcharts for weirs based on water level measurements. The plot for the rectangular weir can be used to estimate the flow over a dam or over the uniform spilling edge of a clarifier.

MEASURING WASTEWATER FLOWS

Measuring these flows of wastewater often challenges the ingenuity of the expert. The stream may flow through subterranean sewers that are difficult to access. A careful survey of the sewerage system may reveal junction boxes and manholes where some type of measuring device may be installed. Wherever sewers, pipelines, and trenches are accessible, accurate measurement of flow is possible. Some of the devices and procedures used are listed in Table 7.3.

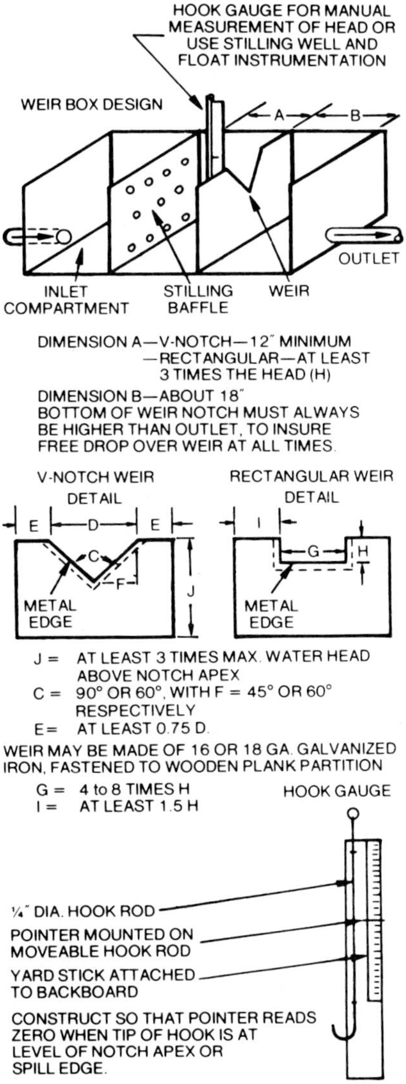

Weirs are used frequently because of their simplicity and relative ease of use in wastewater flow measurement.

Commonly used weirs are either the rectangular or V-notch type. Rectangular weirs are applicable for flows of 100 to 4000 gal/min (0.4 to 15.1 m^3/min). V-notch weirs are more accurate at lower flows and are widely used for measurements in the 20 to 400 gal/min range (0.08 to 15.1 m^3/min).

Flow is proportional to the head of water above the lowest point of the weir opening, as shown in Figure 7.11. Head may be measured manually with a gauge, or automatically by float mechanisms. Typical design of a weir box is shown in Figure 7.12.

In the use of weirs for wastewater flow measurement, care must be taken to avoid the accumulation of solids in the bottom of the box behind the weir.

Parshall flumes are used for measurement of flows in ditches and open channels where only very small head drops are available. These provide good accuracy over wide flow ranges. Under normal freeflow conditions, for example, a Parshall flume with 9 in (22.9 cm) throat width will accurately measure flows ranging from 45 to 2800 gal/min (0.17 to 10.6 m^3/min). A further advantage is that this flume design offers no traps for debris to settle.

FIG. 7.12 Construction of a weir box.

Figure 7.13 illustrates typical Parshall flume design. Prefabricated flume liners can be purchased in almost any size. The liner can be used as the inside form for a concrete structure, with the liner left permanently in place. For temporary use, wood construction may be employed.

Flow is proportional to the head (depth) of water in the converging section of the flume, measured two-thirds of the lengths of the converging section upstream

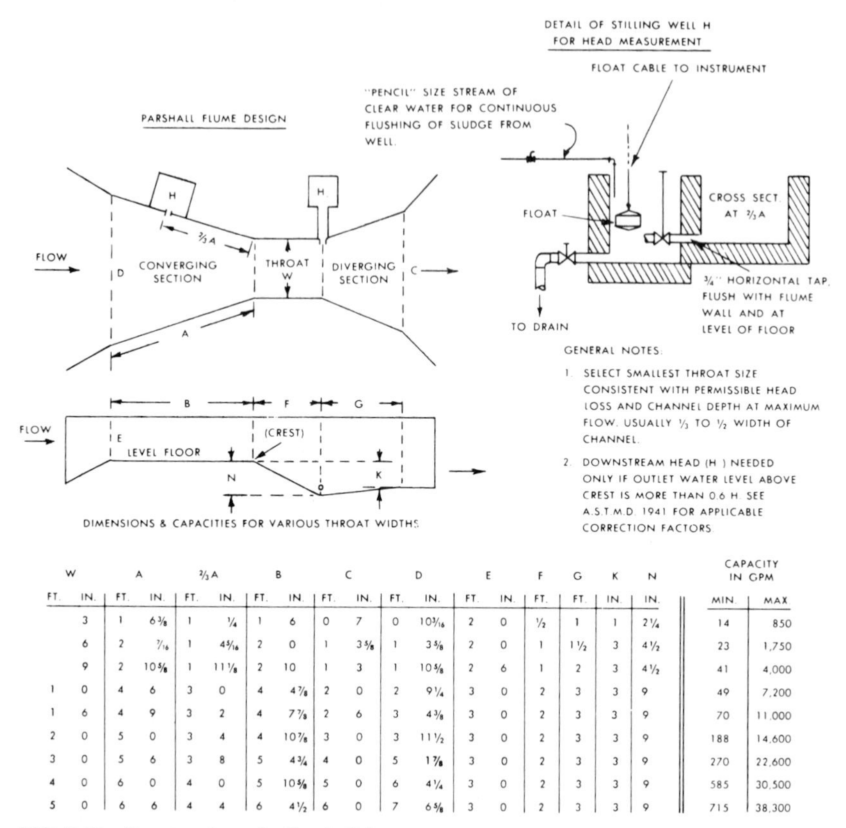

W		A		2/3 A		B		C		D		E		F	G	K	N	Capacity in GPM	
FT.	IN.	FT.	IN.	FT.	IN.	FT.	IN.	FT.	IN.	FT.	IN.	FT.	IN.	FT.	FT.	IN.	IN.	MIN.	MAX.
	3	1	6 3/8	1	1/4	1	6	0	7	0	10 3/16	2	0	1/2	1	1	2 1/4	14	850
	6	2	7/16	1	4 5/16	2	0	1	3 5/8	1	3 5/8	2	0	1	1 1/2	3	4 1/2	23	1,750
	9	2	10 5/8	1	11 1/8	2	10	1	3	1	10 5/8	2	6	1	2	3	4 1/2	41	4,000
1	0	4	6	3	0	4	4 7/8	2	0	2	9 1/4	3	0	2	3	3	9	49	7,200
1	6	4	9	3	2	4	7 7/8	2	6	3	4 3/8	3	0	2	3	3	9	70	11,000
2	0	5	0	3	4	4	10 7/8	3	0	3	11 1/2	3	0	2	3	3	9	188	14,600
3	0	5	6	3	8	5	4 3/4	4	0	5	1 7/8	3	0	2	3	3	9	270	22,600
4	0	6	0	4	0	5	10 5/8	5	0	6	4 1/4	3	0	2	3	3	9	585	30,500
5	0	6	6	4	4	6	4 1/2	6	0	7	6 5/8	3	0	2	3	3	9	715	38,300

FIG. 7.13 Construction of a Parshall flume.

of the throat. Flow versus head measurement, for a 9-in throat width flume, is shown in Figure 7.14.

There are a number of other flume designs including easily installed prefabricated fiberglass flumes such as illustrated in Figure 7.15.

For automatic measurement of head at flumes and weirs of most varieties and sizes, relatively inexpensive instruments with suitable charts and flow totalizers are readily available. Typical float actuated instruments are pictured in Figure 7.16.

For measurement of flows in larger channels, receiving streams, and rivers, a useful procedure for approximating flows is to calculate the average cross-sectional area of the stream over a selected segment of known length; then to drop a suitable float into the water upstream of the selected test segment, and measure the time required for the float to travel the known length of the test segment.

To minimize the effects of wind, the float should be almost submerged. For such surface floats, if nothing is known of velocity distribution from top to bottom of the stream, mean velocity is usually assumed to be 0.8 times the surface velocity. Therefore, flow may be calculated as follows:

$$\text{Flow (ft}^3\text{/s)} = 0.8 \text{ float velocity (ft/s)} \times \text{stream cross section (ft}^2\text{)}$$

$$\text{Flow (m}^3\text{/s)} = 0.8 \times \text{m/s} \times \text{m}^2$$

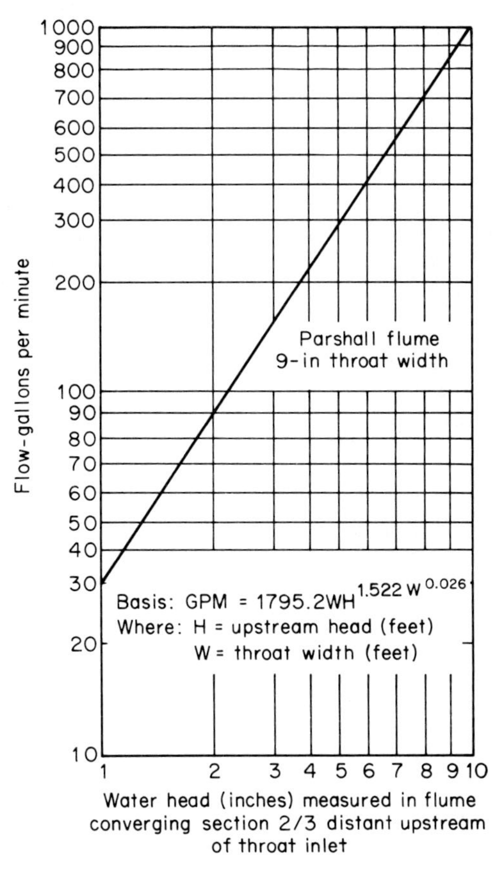

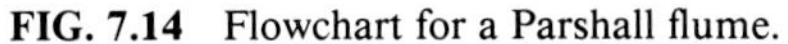

FIG. 7.14 Flowchart for a Parshall flume.

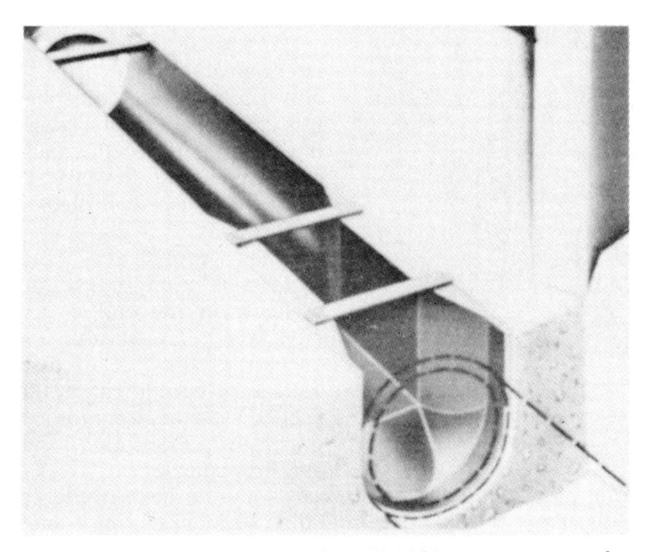

FIG. 7.15 Installation of a prefabricated flume. *(Courtesy of F. B. Leopold Co.)*

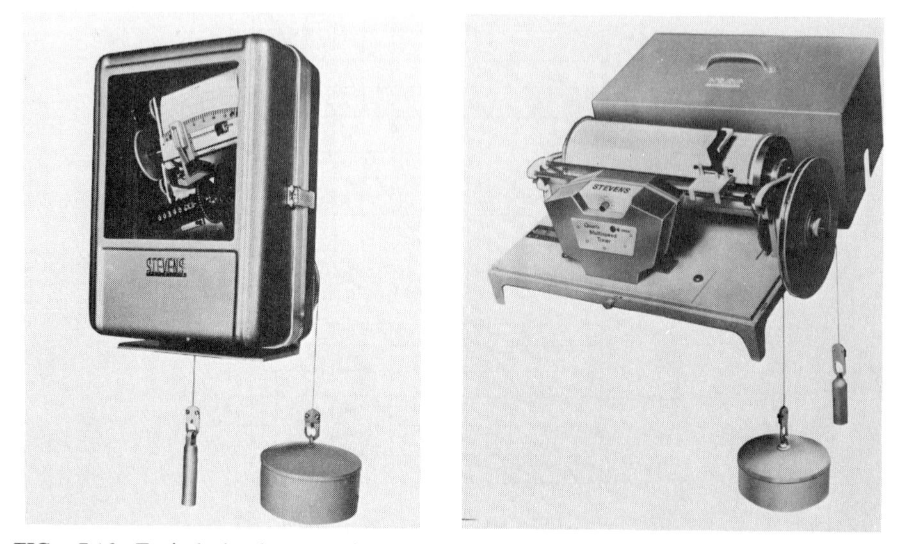

FIG. 7.16 Typical level measuring devices for measuring flows over weirs and flumes. *(Courtesy Leupold & Stevens, Inc.)*

Open Pipe Discharges. For open pipe discharge the simplest method is the discharge timing method. This is adaptable to small flows usually under 30 gal/min (0.1 m^3/min) where it is possible to catch the discharge in a container of known capacity. The container is put under the waste discharge and the filling is timed. This is repeated at least three times.

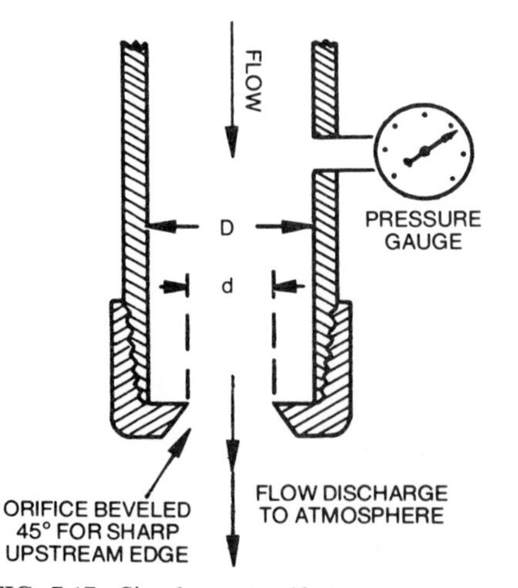

FIG. 7.17 Simple open orifice cap measurement of flow to a sewer or open receiver.

For larger volumes of free flow from vertical pipes into open sumps, another simple flow measurement technique is the use of a drilled and beveled pipe cap orifice, Figure 7.17.

Flow is a function of the pressure differential across the orifice. Since the flow discharges to atmosphere, flow for this device is simply proportional to the square root of the gauge pressure measured. The flowchart previously given for thin-plate orifices can be used (Figure 7.3).

The coordinate measurement method is used to estimate flow of waste from the end of a horizontal pipe, where it is possible to observe all or part of the freefall discharge. No devices or equipment installations are needed with this technique. Figure 7.18 illustrates coordinate measurement and flow calculation.

The California pipe method is a relatively accurate measurement which utilizes open-end discharge from a horizontal pipe segment having a length equivalent to six pipe diameters. For example, a 4-in diameter pipe used for this method

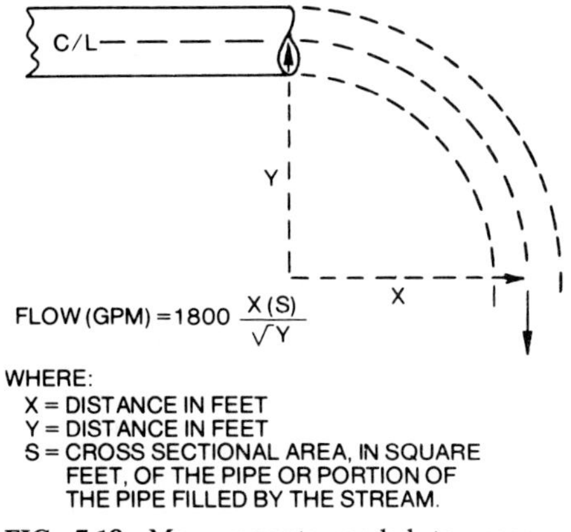

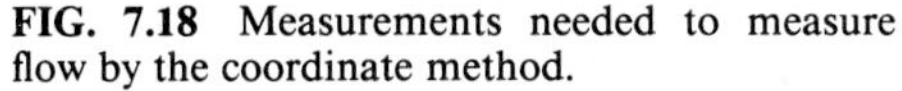

FIG. 7.18 Measurements needed to measure flow by the coordinate method.

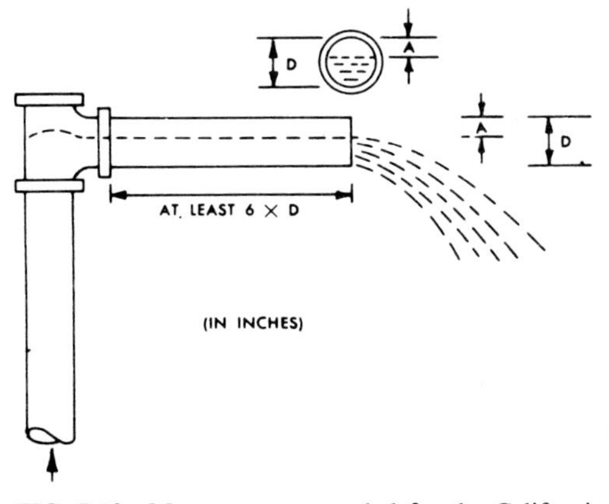

FIG. 7.19 Measurements needed for the California pipe method.

should be at least 24 in long. Figure 7.19 illustrates this method of flow measurement. Figure 7.20 plots flow versus measured water depth and pipe diameter.

Gravity Flow in Partially Filled Sewers—The Manning Formula

Since its initial development in the nineteenth century, the Manning formula has been used extensively to measure flow in a variety of open channels, ranging from natural streambeds to open flumes and sloped sewer lines. If it can be checked by a referee method at a selected flow, the formula is particularly useful because if this can be done once, the channel itself can thereafter be used as the primary metering device with flow varying in a definite relationship to the depth of water in the channel.

Without such a referee method (such as the dye tracer method, described in the next section), the Manning formula is only an approximation of flow because

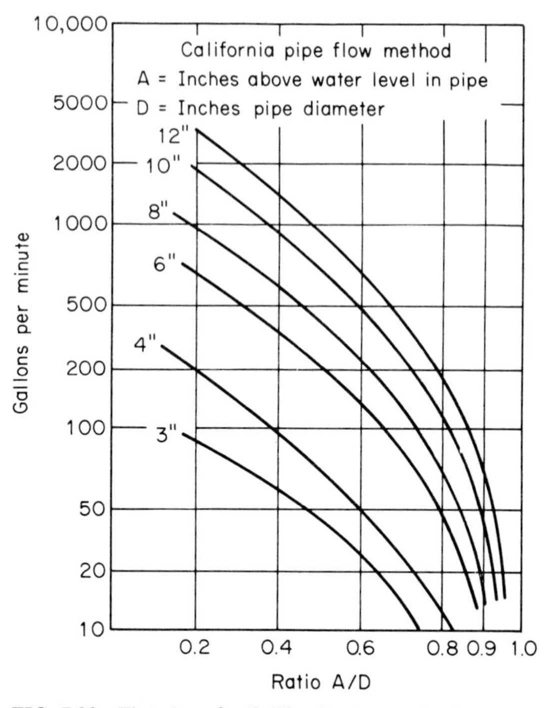

FIG. 7.20 Flowchart for California pipe method.

it is dependent on such unknown factors as the roughness of the channel surfaces, the presence of sediment and debris on the bottom of the channel, and dips or bends in the channel, which may have developed by subsidence or shifting of the soil surrounding the channel.

Figure 7.21 shows the basic data needed to use the Manning formula to measure flow in a partially filled, sloped sewer line. A section of line should be selected that is free of such obstructions as irregular manholes, which radically change the shape of the channel, and bends or dips in the pipes. Such a section should be at least 200 ft (61 m) long.

Table 7.4 shows the factors for calculating the area of flow, A, and hydraulic radius, R, from the diameter and the ratio of water depth to pipe diameter.

Using Figure 7.21 and Table 7.4, the Manning formula can then be applied to calculate flow as follows:

$$Q = 670 \times \frac{A\, R^{2/3}\, S^{1/2}}{n}$$

where Q = flow, gal/min
A = cross-sectional area of flowing stream, ft^2
R = hydraulic radius, ft (the ratio of area to wetted perimeter of the pipe)
S = slope, ft/ft or percent/100
n = Manning coefficient

n, the roughness coefficient, can only be approximated (from Table 7.5), unless there has been a referee method used to determine flow, in which case n can actually be measured and used with confidence thereafter.

The most frustrating aspect of the Manning coefficient is that it is not a constant in the equation, varying as the water depth changes. The value for n may increase as much as 15% when a steel sewer pipe runs half-full, and decrease 15% at very high and very low water depths.

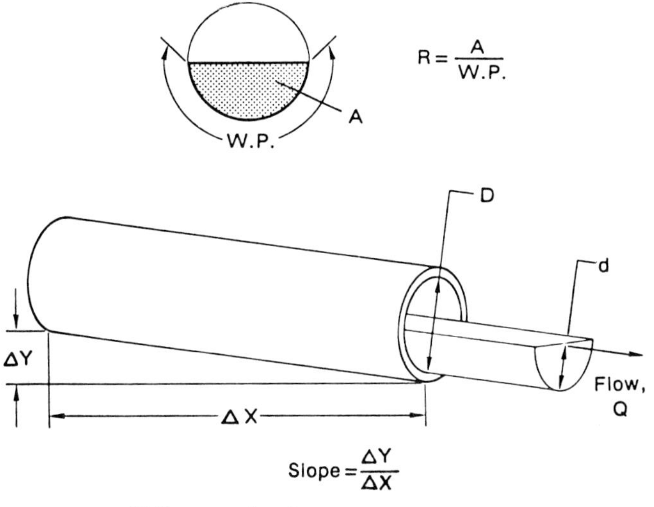

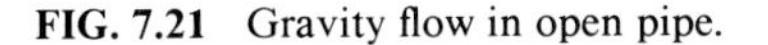
FIG. 7.21 Gravity flow in open pipe.

As an example of the application of the Manning formula, assume a slime-coated 36-in-diameter sewer line with a slope of 1 ft per 100 ft ($S = 0.01$) carrying wastewater at a depth of 27 in. From Table 7.5, with a d/D ratio of 0.75, $A = 5.7$ ft^2, and $R = 0.9$ ft.

$$\text{Flow} = \frac{670 \times 5.7 \times 0.9^{2/3} \times 0.1^{1/2}}{0.014}$$

$$= 26{,}000 \text{ gal/min}$$

Chemical Techniques

The chemical concentration technique, which is applicable to either open or closed sewer systems, employs fixed-rate injection of a suitable chemical into the flowing stream and testing downstream for the resultant concentration of that chemical. Flow may then be approximated as follows:

$$Q = \frac{2000\ I}{C_1 - C_2}$$

where Q = flow, gal/min ($\times$ 3.785 = L/min)
I = chemical injected, lb/h ($\times$ 0.454 = kg/h)
C_1 = upstream concentration of chemical (same chemical as I), mg/L
C_2 = downstream concentration of chemical, mg/L
2000 = conversion factor

TABLE 7.4 Area of Flow (A) and Hydraulic Radius (R), for Various Flow Depths (d)

d/D	A/D^2	R/D	d/D	A/D^2	R/D
0.01	0.0013	0.0066	0.51	0.4027	0.2531
0.02	0.0037	0.0132	0.52	0.4127	0.2561
0.03	0.0069	0.0197	0.53	0.4227	0.2591
0.04	0.0105	0.0262	0.54	0.4327	0.2620
0.05	0.0147	0.0326	0.55	0.4426	0.2649
0.06	0.0192	0.0389	0.56	0.4526	0.2676
0.07	0.0242	0.0451	0.57	0.4625	0.2703
0.08	0.0294	0.0513	0.58	0.4723	0.2728
0.09	0.0350	0.0574	0.59	0.4822	0.2753
0.10	0.0409	0.0635	0.60	0.4920	0.2776
0.11	0.0470	0.0695	0.61	0.5018	0.2797
0.12	0.0534	0.0754	0.62	0.5115	0.2818
0.13	0.0600	0.0813	0.63	0.5212	0.2839
0.14	0.0668	0.0871	0.64	0.5308	0.2860
0.15	0.0739	0.0929	0.65	0.5404	0.2881
0.16	0.0811	0.0986	0.66	0.5499	0.2899
0.17	0.0885	0.1042	0.67	0.5594	0.2917
0.18	0.0961	0.1097	0.68	0.5687	0.2935
0.19	0.1039	0.1152	0.69	0.5780	0.2950
0.20	0.1118	0.1206	0.70	0.5872	0.2962
0.21	0.1199	0.1259	0.71	0.5964	0.2973
0.22	0.1281	0.1312	0.72	0.6054	0.2984
0.23	0.1365	0.1364	0.73	0.6143	0.2995
0.24	0.1449	0.1416	0.74	0.6231	0.3006
0.25	0.1535	0.1466	0.75	0.6318	0.3017
0.26	0.1623	0.1516	0.76	0.6404	0.3025
0.27	0.1711	0.1566	0.77	0.6489	0.3032
0.28	0.1800	0.1614	0.78	0.6573	0.3037
0.29	0.1890	0.1662	0.79	0.6655	0.3040
0.30	0.1982	0.1709	0.80	0.6736	0.3042
0.31	0.2074	0.1755	0.81	0.6815	0.3044
0.32	0.2167	0.1801	0.82	0.6893	0.3043
0.33	0.2260	0.1848	0.83	0.6969	0.3041
0.34	0.2355	0.1891	0.84	0.7043	0.3038
0.35	0.2450	0.1935	0.85	0.7115	0.3033
0.36	0.2546	0.1978	0.86	0.7186	0.3026
0.37	0.2642	0.2020	0.87	0.7254	0.3017
0.38	0.2739	0.2061	0.88	0.7320	0.3008
0.39	0.2836	0.2102	0.89	0.7384	0.2996
0.40	0.2934	0.2142	0.90	0.7445	0.2980
0.41	0.3032	0.2181	0.91	0.7504	0.2963
0.42	0.3130	0.2220	0.92	0.7560	0.2944
0.43	0.3229	0.2257	0.93	0.7612	0.2922
0.44	0.3328	0.2294	0.94	0.7662	0.2896
0.45	0.3428	0.2331	0.95	0.7707	0.2864
0.46	0.3527	0.2366	0.96	0.7749	0.2830
0.47	0.3627	0.2400	0.97	0.7785	0.2787
0.48	0.3727	0.2434	0.98	0.7816	0.2735
0.49	0.3827	0.2467	0.99	0.7841	0.2665
0.50	0.3927	0.2500	1.00	0.7854	0.2500

TABLE 7-5 Approximation of Manning Coefficient

Type of construction	Range of n values
Welded steel pipe	0.010–0.014
Spiral steel pipe	0.013–0.017
Corrugated storm drain	0.021–0.030
Concrete sewer with manholes	0.013–0.017
Slime-coated sewers	0.012–0.016
Clay drainage tile	0.011–0.017

The chemical selected for injection must be readily soluble in the flowing stream, should be compatible and nonreactive with other ions in the stream, and should be easy to analyze with reasonable accuracy. Care should be taken to allow adequate mixing and distribution of the chemical in the stream ahead of point of downstream sampling.

Start with a chemical injection rate calculated to give a 100 mg/L analysis increase for the estimated flow. Using common salt, for example, approximately 5 lb/h injection per 100 gal/min flow would be required for 100 mg/L increase in NaCl reading.

Other Chemical Injection Techniques

Several modifications of the salt dilution method may be used if the background concentration of NaCl in the wastewater varies so much it masks the increment added by salt injection. Instead of injecting NaCl, the investigator may select a fluorescent dye or an element absent from the wastewater (e.g., lithium or a radioactive element). Fluorescent dye is the easiest to see, provided the wastewater is not already highly colored.

In the fluorescent dye tracer method, a range of dilutions of the dye (e.g., 1:1000, 1:10,000, or 1:50,000) are made and the fluorescence of diluted samples and of a blank containing no dye is measured by a fluorometer. Dye concentration in milligrams per liter versus fluorescence of the diluted samples is then plotted to provide data for determining dye concentration during the period of study while dye is being injected. A dye is selected which is not absorbed on typical wastewater solids (e.g., Rhodamine WT*), and a dye solution of standardized strength is injected into the wastewater flow at a point where mixing is adequate. The test procedure is then identical to the salt dilution method.

Another technique, applicable to occasional flow measurement in laterals not equipped with meters, is the measurement of the velocity of an additive injected into the line. The injected additive may cause a measurable increase in conductivity, such as by sodium chloride addition, or an increase in color, such as by dye injection. The salt or dye is injected in a slug into the line upstream, and the travel of the salt or dye cloud is carefully timed between downstream sampling points a known distance apart. The conductivity chart example shown in Figure 7.22 illustrates the measurement of passage time between the recorded median conductivity values of a dissolved salt slug flowing past two sampling points. For dye injection, similar charts can be obtained through the use of commercially available recording fluorometers.

* DuPont Company.

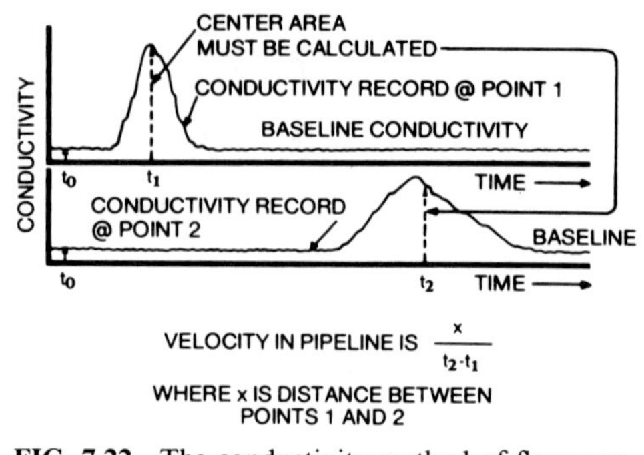

FIG. 7.22 The conductivity method of flow measurement requires care in slugging and measuring chemical concentrations. This provides only a single measurement for each application, but is useful in establishing flow data in inaccessible sewers.

Where none of the preceding methods can be used, it may be possible to approximate supply water flow on the basis of line pressure readings and pump curves, draw-down from a supply tank, or by electrical amperage draw of the supply pump motor. Amperage readings can be converted to approximate flow discharge values from the pump curve normally supplied by the manufacturer. Newer plants provide permanent facilities for measuring wastewater flow as well as water supply. Many older plants lack such devices. In either case, wastewater flow is often quite different from raw water supply in that it may be collected in open sewers or flumes instead of closed piping.

Estimating Individual Flows

There is a simple method used for measuring the flows in tributary streams when the flow in the combined sewer is known.

A common variable, such as conductivity or temperature, of all streams is measured. The flow rates in the tributaries may then be calculated as a ratio of the tributary conductivities or temperatures from the following equations. The ratio then does not carry any unit label.

$$Q_t = Q_1 + Q_2$$

$$Q_t\lambda_t = Q_1\lambda_1 + Q_2\lambda_2$$

$$\text{or } Q_tT_t = Q_1T_1 + Q_2T_2$$

where Q_t = total flow rate
Q_1 = first tributary flow rate
Q_2 = second tributary flow rate
λ = conductivity
T = temperature

Substituting and solving for the ratio of Q_1 to Q_2,

$$\frac{Q_1}{Q_2} = \frac{\lambda_t - \lambda_2}{\lambda_1 - \lambda_t} \quad \text{or} \quad \frac{(T_t - T_2)}{(T_1 - T_t)}$$

For example:

Main stream flow (Q_t) = 498 gal/min (1.89 m^3/min)
Main stream conductivity (λ_t) = 1120 μS
Tributary 1 conductivity (λ_1) = 1500 μS
Tributary 2 conductivity (λ_2) = 340 μS

$$\frac{Q_1}{Q_2} = \frac{1120 - 340}{1500 - 1120} = 2.05$$

The flow of tributary 1 is calculated as follows:

$$Q_1 = \frac{Q_1}{Q_1 + Q_2} \times Q_t$$

If $Q_2 = x$, then $Q_1 = 2.05x$ and

$$\frac{Q_1}{Q_1 + Q_2} = \frac{2.05x}{2.05x + x} = \frac{2.05}{3.05}$$

$$Q_1 = \frac{2.05}{3.05} \times 498 = 335 \text{ gal/min } (1.27 \text{ m}^3\text{/min})$$

$$Q_2 = 498 - 335 = 163 \text{ gal/min } (0.62 \text{ m}^3\text{/min})$$

SAMPLING

The success of an entire water project may depend on the collection of a representative sample. In most water supply systems, this is rarely a problem. The composition of the supply water is seldom subject to either continuous or sudden, instantaneous variation. Even a surface water diluted by a spring downpour will usually change gradually enough so that the water treatment plant operator can cope with it. The operator must be aware of past climatological history and must schedule sampling based on the meteorologist's predictions of rainfall.

Some supplies fluctuate because the town or plant draws on several wells of different composition, and these wells may cut in and out depending on demand. Here, too, it is usually possible to anticipate when the changeover in wells may occur, so that sampling can be scheduled accordingly.

Suspended Solids

With these relatively constant quality water streams, the presence of suspended solids may be one of the factors requiring special care in sampling. In the distribution pipe, suspended solids have a tendency to travel along the pipe wall, so that a sample pipe with its opening in the wall of the distribution pipe may trap more suspended solids than would be representative of the water flowing through the pipe. So, where suspended solids content is important, as in industrial condensate systems, the sampling line should protrude into the flow of the stream. A similar situation develops in open distribution flumes if floating matter is present. Flushing of lines and rinsing of containers with sample are always necessary to help assure drawing a representative sample once the correct provisions have been made for the sampling line.

Sampling wastewater correctly in an industrial complex often requires ingenuity. First, something must be known about the sewer system to select the best sample location. For example, if the sewer pipe is not completely filled, there may be three strata in the flowing water: the upper contains floating matter, the center may be free of floating matter and sediment, and the lower contains solids that have settled in spite of the water velocity through the pipe. Truly representative sampling could conceivably require turbulent mixing upstream of the sampling location. Usually a compromise is made because mixing may be impossible: the

TABLE 7.6 Checkerboard Sampling

(Note: 1 mL of sample collected for each gpm of stream flow)

	Station, flow rate at time of sampling, gal/min						Composite
Time	1	2	3	4	5	Total	(hourly plant,* mL)
9 A.M.	15	105	70	0	180	370	370
10	2	90	55	75	175	397	397
11	8	90	100	60	175	433	433
12 noon	6	90	150	0	205	445	445
1 P.M.	10	95	45	10	160	320	320
2	10	105	45	25	175	360	360
3	1	100	50	35	170	356	356
4	5	110	150	30	165	460	460
5	6	160	160	10	170	446	446
6	10	105	105	50	200	460	460
7	15	110	110	55	210	480	480
8	2	90	90	70	190	442	442
9	3	75	75	70	195	428	428
10	8	60	60	15	170	343	343
11	10	105	105	30	160	405	405
12 midnight	12	120	120	40	165	437	437
Average	7.7	96	93	36	179	417	417
Daily composite,† mL	123	1535	1390	575	2865	6582	6582‡

* Hourly plant composites (H-series, or horizontal) are taken by withdrawing from each bottle a volume proportional to flow at the time of sampling [for example, 1 mL/gal/min)] and combining. Each individual sample is examined for odor and color and checked for pH, TDS, or both. The composite may then be analyzed for other constituents, for example, oil. If the composite is too small, use 2 mL/(gal/min) or more to make a larger composite.

† Daily composite at each sample station by the same technique. These may be more extensively analyzed for BOD, COD, mineral components, special toxicants (for example, cyanide from a refinery overhead), etc.

‡ The hourly composites may be combined as a daily total plant composite using the residue after tests have been run and combining on a proportional basis. For example, if only 100 mL has been used for analysis, measure out ½ of original volume of each hourly composite, combine, and use as total plant composite for one day.

person doing the sampling must decide what data are the most significant for plant control purposes.

Assess Variability First

A first step in deciding on a sampling schedule is to determine an instrumental profile of the stream to be sampled. One procedure is to install automatic recording conductivity, pH, and temperature instruments at the point to be sampled—to operate over at least a 24-h period. If the results show wide fluctuations, then the frequency of these fluctuations can be used as a guide for sampling intervals. There may be some periodicity in the readings, indicating a repetitive operation within the plant; if so, this should be investigated. If the readings remain relatively steady, samples can be widely spaced.

In the absence of continuous recording of conductivity, pH, and temperature, manual samples should be taken following a checkerboard schedule, illustrated by Table 7.6. With this program, which requires data on water flow rate, samples are obtained every hour for a full operating day.

Each sample should be checked for conductivity and appearance. Those found to vary widely from the norm should be checked for pH and temperature. A measured volume of each sample should then be taken in proportion to flow rate past the point of sampling at the time of sampling. If a number of sewers flow into a main sewer, then the checkerboard schedule shows the measured volume of each sample to be taken in proportion to the flow to make an hourly plant composite. A more complete analysis is then made on the composite. Following this procedure for the whole day, a graph of analytical data plotted against time of sampling provides a clue for checking plant records to see which plant operations may have caused the deviations from normal conductivity, pH, or temperature. Table 7.6 shows a checkerboard sampling program with five sampling points on tributary sewers leading to a common collection main. For such a program to be successful, the flow rate through each lateral must be known either by measurement or by reliable estimate.

Automatic Samplers

A variety of samplers are available, the choice depending on the water source to be sampled. Samplers for surface waters can obtain samples from whatever depth may be desired. A complete program for surface waters would include sampling at a number of elevations from surface to bottom (Figure 7.23).

Wastewater samplers have special characteristics depending on the nature of the waste. Some of these devices, such as that illustrated in Figure 7.24, are designed for placement in a calibrated flume, such as a Parshall flume: the sampler is shaped so that the volume delivered on each cycle to a sample container is proportional to the flow.

Other samplers, such as those illustrated in Figure 7.25, operate on a timed sequence, and the identification of each sample must then be referred to a flowmeter chart so that a flow-proportional composite can be prepared from the individual samples.

Electrical, mechanical, and pneumatic samplers are available. Great care must be taken in the selection of the proper instrument for sampling based on knowledge of the operating conditions within the plant and the nature of the wastewater

FIG. 7.23 The sampler used for surface water studies is tripped to close at predetermined depth.

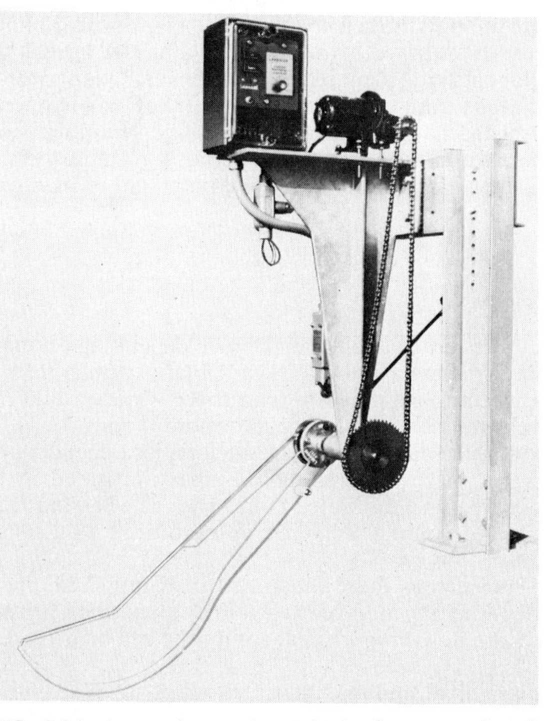

FIG. 7.24 A sampler used to obtain flow-proportioned samples from a flume. *(Courtesy Lakeside Equipment Corporation.)*

FIG. 7.25 A programmed sampler obtaining intermittent sewage samples.

being sampled. There is certainly nothing more frustrating to a water survey team than to set up a sampling device and to return 24 h later to find the sample container empty because the device has failed through lack of power or by clogging of the water channels or ports.

ANALYSIS

Once the sample has been taken, the person responsible should make certain observations and measurements before sending it to the laboratory. These observations should include appearance (color, haziness, turbidity, appearance of floc, oil on surface), odor, and measurements that are valid only on a fresh sample (temperature, dissolved oxygen, and pH).

There are many reference books on water analysis procedures, some of which include simplified control tests as well as standard, or referee, methods.

Table 7.7 lists typical methods of analysis used for control purposes and in field studies. Most control laboratories in municipal and industrial plants are equipped to conduct these analyses. In doing survey work or in providing technical service, most water specialists have miniaturized analytical equipment so that many of the critical tests shown in Table 7.7—tests used to control water treatment programs—can be conducted using portable test kits (Figure 7.26). Examples of the success that has been achieved in miniaturizing equipment for portability are the spectrophotometer, shown in Figure 7.27, and the membrane technique for bacteriologic counts shown in Figure 7.28.

The list in Table 7.7 does not include specialized test equipment such as corrosion meters, corrosion coupons, steam quality meters, and pH, conductivity, and temperature recorders.

Many of these analytical control procedures are modifications of standard methods acceptable to regulatory agencies. When a sample is taken for record purposes, standard methods must be used, and the sample is usually sent to a

TABLE 7.7 Summary of Methods for Chemical Analysis of Water and Wastewater

Analysis	Code	Definition	Method	Normal detection limit, mg/L	Sample size, mL	Container*	Preservation
Acidity—total	AP	Acidity to phenolphthalein. Reported as $CaCO_3$	Titration with standard base to phenolphthalein endpoint	2	50	P	Keep sample cool
Acidity—free mineral	AM	Acidity to methyl orange. Reported as $CaCO_3$	Titration with standard base to methyl orange endpoint, or potentiometric to pH 4.5	2	50	P	Keep sample cool
Alkalinity—total	M	Alkalinity to methyl orange. Reported as $CaCO_3$	Titration with standard acid to methyl orange endpoint, or potentiometric to pH 4.5	2	50	P	Keep sample cool
Alkalinity—hydroxide	O	Alkalinity to phenolphthalein after precipitation of carbonate. Reported as $CaCO_3$	Precipitation of carbonate with $BaCl_2$. Titration with standard acid to phenolphthalein endpoint	2	50	P	Keep sample cool
Alkalinity—phenolphthalein	P	Alkalinity to phenolphthalein. Reported as $CaCO_3$	Titration with standard acid to phenolphthalein endpoint	2	50	P	Keep sample cool
Aluminum (Al)—total	AL	Soluble and insoluble aluminum	Atomic absorption spectroscopy	0.2	200	G	5 mL 10% HNO_3 per liter

Aluminum (Al)—soluble	AL2	Soluble aluminum	Filtration—atomic absorption spectroscopy	0.2	200	G	None or filter sample thru 0.4 μ millipore and add 5 mL 10% HNO_3 per liter
Ammonia (NH_3)	NH3	Evolvable ammonia nitrogen	Distillation at pH 9.5 followed by Nessierization or titration with standard acid	0.1	100	G	10 mL 10% H_2SO_4 per liter
Antimony (Sb)—total	SB	Soluble and insoluble antimony	Atomic absorption spectroscopy	1	50	G	5 mL 10% HNO_3 per liter
Arsenic (As)—total	AS	Soluble and insoluble arsenic	Atomic absorption spectroscopy—flameless (graphite furace) or flame (hydride generation)	0.002	100	G	5 mL 10% HNO_3 per liter
Barium (Ba)—total	BA	Soluble and insoluble barium	Atomic absorption spectroscopy	0.5	50	G	5 mL 10% HNO_3 per liter
Beryllium (Be)—total	BE	Soluble and insoluble beryllium	Atomic absorption spectroscopy	0.1	50	G	5 mL 10% HNO_3 per liter
Biochemical oxygen demand (BOD)	BOD	Biochemical oxygen demand	Dilution—5 day incubation—dissolved oxygen determination using probe method	5	1000	G	10 mL 10% H_2SO_4 per liter or refrigeration
Bismuth (Bi)—total	BI	Soluble and insoluble bismuth	Atomic absorption spectroscopy	1	50	G	5 mL 10% HNO_3 per liter

TABLE 7.7 Summary of Methods for Chemical Analysis of Water and Wastewater (*Continued*)

Analysis	Code	Definition	Method	Normal detection limit, mg/L	Sample size, mL	Container*	Preservation
Boron (B)	B	Soluble and insoluble boron	Colorimetric (circumin) or mannitol titration	0.1	100	P	None
Bromide (Br)	BR	Bromide ion concentration	Conversion to bromate—titration with sodium thiosulfate	2	1000	P	None
Cadmium (Cd)—total	CD	Soluble and insoluble cadmium	Atomic absorption spectroscopy	0.01	50	G	5 mL 10% HNO_3 per liter
Calcium (Ca)—total	CA	Soluble and insoluble calcium	Atomic absorption spectroscopy	0.08	50	P	5 mL 10% HNO_3 per liter
Calcium (Ca)—soluble	CA2	Soluble calcium	Filtration—atomic absorption	0.08	50	P	None or filter sample thru 0.45 μm millipore and add 5 mL 10% HNO_3 per liter
Carbon (C)—total	TC	Total carbon	Total carbon analyzer—I.R. detector	1	100	G	None
Carbon (C)—total organic	TOC	Total organic carbon	Total organic carbon analyzer—I.R. detector	1	100	G	10 mL 10% H_2SO_4 per liter
Cesium (Cs)—total	CS	Soluble and insoluble cesium	Atomic absorption spectroscopy	1.0	50	G	5 mL 10% HNO_3 per liter

Chloride (C)	CL	Soluble chloride ion concentration	Titration with silver nitrate—chromate indicator	1	50	P	None
Chromium (Cr)—total	CR	Soluble and insoluble chromium	Atomic absorption spectroscopy	0.01	50	G	5 mL 10% HNO_3 per liter
Chromium (Cr)—total	CR5	Soluble and insoluble chromium for high purity systems	Atomic absorption spectroscopy—flameless (graphite furnace)	0.005	50	G	5 mL 10% HNO_3 per liter
Chromium (Cr)—hexavalent	CR6	Hexavalent chromium	Colorimetric (diphenylcarbazide) or titration with thiosulfate	0.01	50	G	None
Chromium (Cr)—trivalent	CR3	Trivalent chromium	Atomic absorption spectroscopy after separation from hexavalent chromium	0.01	50	G	None
Cobalt (Co)—total	CO	Soluble and insoluble cobalt	Atomic absorption spectroscopy	0.1	50	G	5 mL 10% HNO_3 per liter
Color—APHA	Q2	APHA color	Visual comparison against APHA standards	1 unit	75	G	None
Color—boiler	Q	Organic boiler color	Visual comparison against Nalco boiler water color standards on filtered samples	1 unit	50	G	None
Chemical oxygen demand (O_2)	COD	Determination of organic and oxidizable inorganic material	Oxidation with potassium dichromate—titration with ferrous ammonium sulfate	4	100	G	10 mL 10% H_2SO_4 per liter

TABLE 7.7 Summary of Methods for Chemical Analysis of Water and Wastewater (*Continued*)

Analysis	Code	Definition	Method	Normal detection limit, mg/L	Sample size, mL	Container*	Preservation
Conductivity	CON	Determination of specific conductance—ambient temperature	Conductivity meter	1 micro mho/cm	50	P	Keep sample cool
Copper (Cu)—total	CU	Soluble and insoluble copper	Atomic absorption spectroscopy	0.01	50	G	5 mL 10% HNO_3 per liter
Copper (Cu)—soluble	CU2	Soluble copper	Filtration—atomic absorption spectroscopy	0.01	50	G	None or filter sample thru 0.45 μm millipore and add 5 mL 10% HNO_3 per liter
Cyanide (CN)—free and combined	CN	Free and combined cyanide	Distillation—colorimetric (pyridine barbituric acid) or titration with silver nitrate	0.01	1000	G	Raise pH to above 12 with 1 *N* NaOH
Cyanide (CN)—free	CN2	Free cyanide	Distillation—colorimetric (pyridine barbituric acid) or titration with silver nitrate	0.01	1000	G	Raise pH to above 12 with 1 *N* NaOH
Fluoride (F)—free and combined	F	Free and combined fluoride ion concentration	Distillation—ion—selective electrode	0.05	500	P	None
Fluoride (F)—free	F2	Free fluoride ion concentration	Ion selective electrode	0.05	100	P	None

Gold (Au)—total	AU	Soluble and insoluble gold	Atomic absorption spectroscopy	0.2	50	G	5 mL 10% HNO_3 per liter
Hydrocarbons—chlorinated solvents	HC	Determination of methylene chloride, chloroform, 1, 2-dichloroethane, 1.1.1-trichloroethane, 1.1.2-trichloroethene, 1.2-dichloropropane 1.1.2.2–tetrachloroethene, 1.1.2.2–tetrachloroethane O-dichlorobenzene	Gas chromatographic determination using a flame ionization detector	10	100	G	Cool, 4 °C
Iodide (I)	I	Iodide ion concentration	Conversion to iodate—titration with sodium thiosulfate	2	1000	P	None
Iron (Fe)—total	FE	Soluble and insoluble iron	Atomic absorption spectroscopy	0.1	50	G	5 mL 10% HNO_3 per liter
Iron (Fe)—soluble	FE2	Soluble iron	Filtration—atomic absorption spectroscopy	0.1	50	G	None or filter sample thru 0.45 μm millipore and add 5 mL 10% HNO_3 per liter
Iron (Fe)—total	FES	Soluble and insoluble iron for high purity systems	Colorimetric—ferrozine method	0.01	50	Special	5 mL 10% HNO_3 per liter

TABLE 7.7 Summary of Methods for Chemical Analysis of Water and Wastewater (*Continued*)

Analysis	Code	Definition	Method	Normal detection limit, mg/L	Sample size, mL	Container*	Preservation
Lead (Pb)—total	PB	Soluble and insoluble lead	Atomic absorption spectroscopy	0.05	100	G	5 mL 10% HNO_3 per liter
Lithium (Li)—total	LI	Soluble and insoluble lithium	Atomic absorption spectroscopy	0.01	50	P	None
Magnesium (Mg)—total	MG	Soluble and insoluble magnesium	Atomic absorption spectroscopy	0.02	50	P	5 mL 10% HNO_3 per liter
Magnesium (Mg)—soluble	MG2	Soluble magnesium	Filtration—atomic absorption spectroscopy	0.02	50	P	None or filter sample thru 0.45 μm millipore and add 5 mL 10% HNO_3 per liter
Manganese (Mn)—total	MN	Soluble and insoluble manganese	Atomic absorption spectroscopy	0.05	50	G	5 mL 10% HNO_3 per liter
Mercury (Hg)—total	HG	Soluble and insoluble mercury	Flameless atomic absorption spectroscopy (cold vapor technique)	0.0001	150	G	5 mL 10% HNO_3 per liter
Molybdenum (Mo)—total	MO	Soluble and insoluble molybdenum	Atomic absorption spectroscopy	1	50	G	5 mL 10% HNO_3 per liter
Nickel (Ni)—total	NI	Soluble and insoluble nickel	Atomic absorption spectroscopy	0.1	50	G	5 mL 10% HNO_3 per liter
Nitrate (NO_3)	NO3	Nitrate ion concentration	Cadmium reduction—colorimetric (diazotization)	1	100	G	10 mL 10% H_2SO_4 per liter
Nitrite (NO_2)	NO2	Nitrite ion concentration	Colorimetric (diazotization)	0.1	100	G or P	10 mL 0.4% $HgCl_2$ per liter

Nitrogen (N)—Kjeldahl	KJ	Total free and combined ammonia nitrogen	Kjeldahl digestion—distillation—Nesslerization or titration with standard acid	0.1	50	G	10 mL 10% H_2SO_4 per liter
Nitrogen (N)—organic	NO	Total combined ammonia nitrogen	Evolution of free ammonia—Kjeldahl digestion—distillation—Nesslerization or titration with standard acid	0.1	50	G	10 mL 10% H_2SO_4 per liter
Oil and grease—freon	OIL	Freon extractable—nonvolatile at 70°C	Acidification—liquid—liquid extraction with freon—gravimetric determination	1	1000	G	10 mL 10% H_2SO_4 per liter
Oil and grease—hexane	HE	Hexane extractable—nonvolatile at 70°C	Acidification—liquid—liquid extraction with hexane—gravimetric determination	1	1000	G	10 mL 10% H_2SO_4 per liter
Oil and grease—soxhlet extraction—hexane	SHE	Hexane extractable—nonvolatile at 70°C	Acidification—soxhlet extraction with hexane—gravimetric determination	1	1000	G	10 mL 10% H_2SO_4 per liter
Pesticides—chlorinated	DDT	Determination of DDT, DDD, DDE, and other common commercial toxicants	Extraction—gas chromatographic determination using an electron capture detector	0.001	1000	G with foil lined caps	None

TABLE 7.7 Summary of Methods for Chemical Analysis of Water and Wastewater (*Continued*)

Analysis	Code	Definition	Method	Normal detection limit, mg/L	Sample size, mL	Container*	Preservation
pH	pH	Measurement of hydrogen ion activity in aqueous solution	pH meter with glass electrode	0.1 pH unit	50	G or P	None
Phenols	PHE	Phenols reacting with 4-amino-antipyrine	Distillation—colorimetric (4-aminio-antipyrine)	0.001	1000	G	5 mL 20% $CuSO_4$ Lower pH to less than 4 with 10% H_3PO_4. Keep sample cool. Ship immediately.
Phosphorus (P)—total	D4	Total organic and inorganic phosphorus	Oxidation with persulfate—reversion—colorimetric	0.03	100	P	10 mL 0.4% $HgCl_2$ per liter
Phosphorus (P)—total inorganic inorganic	D2	Soluble and insoluble ortho and polyphosphate	Reversion in acid—colorimetric	0.03	100	P	10 mL 0.4% $HgCl_2$ per liter
Phosphorus (P)—soluble inorganic	D2S	Soluble ortho- and polyphosphate	Filtration—reversion—colorimetric	0.03	100	P	10 mL 0.4% $HgCl_2$ per liter
Phosphorus (P)—ortho	D	Soluble ortho phosphate	Colorimetric	0.03	25	P	10 mL 0.4% $HgCl_2$ per liter
Platinum (Pt)—total	PT	Soluble and insoluble platinum	Atomic absorption spectroscopy	2	50	G	5 mL 10% HNO_3 per liter

Polychlorinated biphenyls (PCB)	PCB	Determination against reference PCB Aroctor 1221, 1242, 1248, 1254, and 1260	Extraction—gas chromatographic determination using an electron capture detector	0.001	1000	G with foil lined caps	None
Potassium (K)—total	K	Soluble and insoluble potassium	Atomic absorption spectroscopy	0.1	50	P	None
Selenium (Se)—total	SE	Soluble and insoluble selenium	Atomic absorption spectroscopy—flameless (graphite furnace) or flame (hydride generation)	0.002	100	G	5 mL 10% HNO_3 per liter
Silica (SiO_2)	SI	Molybdate reactive silica	Colorimetric (ammonium molybdate with 1-amino-2-naphthal-4-sulfonic acid reduction)	0.1	25	P	None
Silica (SiO_2)	SI5	Molybdate reactive silica for high purity systems	Colorimetric (ammonium molybdate with 1-amino—2-naphthol 4-sulfonic acid reduction)	0.001	100	P	None
Silver (Ag)—total	AG	Soluble and insoluble silver	Atomic absorption spectroscopy	0.01	50	G	5 mL 10% HNO_3 per liter
Sodium (Na)—total	NA	Soluble and insoluble sodium	Atomic absorption spectroscopy	0.1	50	P	None
Solids—total	TS	Total filterable and nonfilterable residue—nonvolatile at 180°C	Gravimetric determination—dried 180°C	5	200	G	None

TABLE 7.7 Summary of Methods for Chemical Analysis of Water and Wastewater (*Continued*)

Analysis	Code	Definition	Method	Normal detection limit, mg/L	Sample size, mL	Container*	Preservation
Solids—total dissolved	TDS	Total filterable residue—nonvolatile at 180°C	Millipore filtration—gravimetric determination on filtrate dried at 180°C	5	200	G	None
Solids—total fixed	TSF	Total residue remaining after ignition at 550°C	Gravimetric determination—ignition at 550°C	5	200	G	None
Solids—total suspended	SS	Total nonfilterable residue—nonvolatile at 105°C	Millipore filtration—gravimetric determination—dried at 105°C	5	200	G	None
Solids—volatile suspended	SSV	Total nonfilterable residue volatile at 550°C	Millipore filtration—gravimetric determination—ignition at 550°C	5	200	G	None
Solids—total volatile	TSV	Total residue volatile at 550°C	Difference between total solids (TS) and total fixed solids (TSF)	5	200	G	None
Solids—settleable	SET	Settleable solids at 5 min and 1 h	Imhoff cone	1	1000	G	None
Strontium (Sr)—total	SR	Soluble and insoluble strontium	Atomic absorption spectroscopy	0.1	50	G	5 mL 10% HNO_3 per liter
Sulfate (SO_4)	SO4	Soluble sulfate	Turbidometric	2	50	P	None
Sulfide (s)	H2S	Sulfide ion concentration	Evolution—iodometric titration or colorimetric (methylene blue)	0.1	1000	G	5 mL 24% Zn $(C_2H_3O_2)_2$ per liter

Surfactants—anionic	ABS	Methylene blue active substances	Colorimetric (methylene blue dye and chloroform extraction)	0.01	200	G	None
Thallium (TI)—total	TL	Soluble and insoluble thallium	Atomic absorption spectroscopy	0.1	50	G	5 mL 10% HNO_3 per liter
Thiosulfate	THI	Thiosulfate ion concentration	Iodometric	1	200	G	None
Tin (Sn)—total	SN	Soluble and insoluble tin	Atomic absorption spectroscopy	1	50	G	5 mL 10% HNO_3 per liter
Titanium (Ti)—total	TI	Soluble and insoluble titanium	Atomic absorption spectroscopy	1	50	G	5 mL 10% HNO_3 per liter
Turbidity	JTU	Light scatter reported in nephelometric turbidity units	Nephelometric	0.1 NTU	50	G or P	None
Vanadium (V)—total	V	Soluble and insoluble vanadium	Atomic absorption spectroscopy	1	50	G	5 mL 10% HNO_3 per liter
Zinc (Zn)—total	ZN	Soluble and insoluble zinc	Atomic absorption spectroscopy	0.01	50	G	5 mL 10% HNO_3 per liter
Zinc (Zn)—soluble	ZN2	Soluble zinc	Filtration—atomic absorption spectroscopy	0.01	50	G	None or filter sample thru 0.45 μm millipore and add 5 mL 10% HNO_3 per liter
Zirconium (Zr)—total	ZR	Soluble and insoluble zirconium	Atomic absorption spectroscopy	5	50	G	5 mL 10% HNO_3 per liter

* P, polyethylene; G, glass.

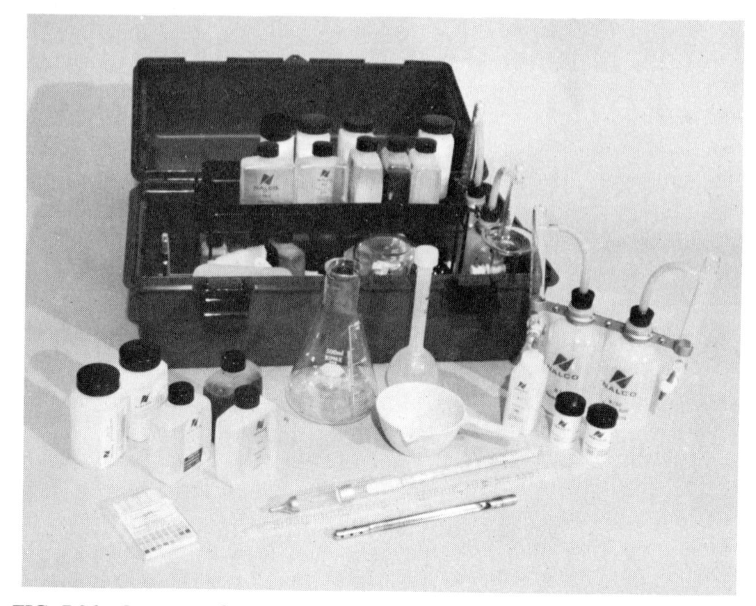

FIG. 7.26 One type of portable laboratory used for field and plant surveys and for monitoring services.

FIG. 7.27 A miniature spectrophotometer used for a wide variety of analyses in the laboratory or in field studies. *(Courtesy Bausch & Lomb.)*

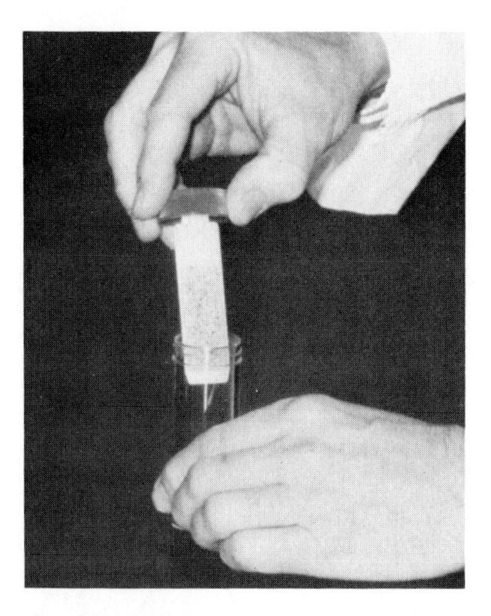

FIG. 7.28 Microbes are collected and cultured by pretreated elements which are easily incubated. The sampler has a fine grid printed on it to facilitate counting of colonies. *(Courtesy Millipore Corporation.)*

water laboratory fully equipped to follow these procedures. Figure 7.29 shows the application of special equipment available for analysis of water and water-formed deposits in such a laboratory.

Chemical analysts continually face new challenges in analyzing water supplies and waste discharges. Not only is there a growing need to measure a broader spectrum of water constituents, but also lower concentrations are being measured. Instrumental techniques have become essential to meet the need for rapid, accurate identification of water impurities.

Analyses for the inorganics found in water and wastewater have been established for many years. However, technological advances in analytical chemistry are continually creating new and more sensitive techniques and instruments for measuring these inorganic constituents. A good example of this is inductively coupled argon plasma (ICAP). This excitation source, when coupled to a direct reading optical emission spectrometer, provides a highly sensitive technique for quantitative determination of 50 or more elements simultaneously. These advances, coupled with the deep concern over the health effects of certain of these inorganic materials at very low concentrations in drinking water and foods, create a constantly changing state-of-the-art in analytical chemistry.

Emphasis on identifying and measuring organic constituents in water is also increasing. Because of physiological concerns, the volume of activity in organic analysis of water continues to grow. Recent research and the development of more sophisticated and powerful tools for organic analysis and molecular characterization have resulted in the generation of large amounts of valuable information. Additionally, research in these areas continues at a very high level in academic, government, industrial, and medical laboratories.

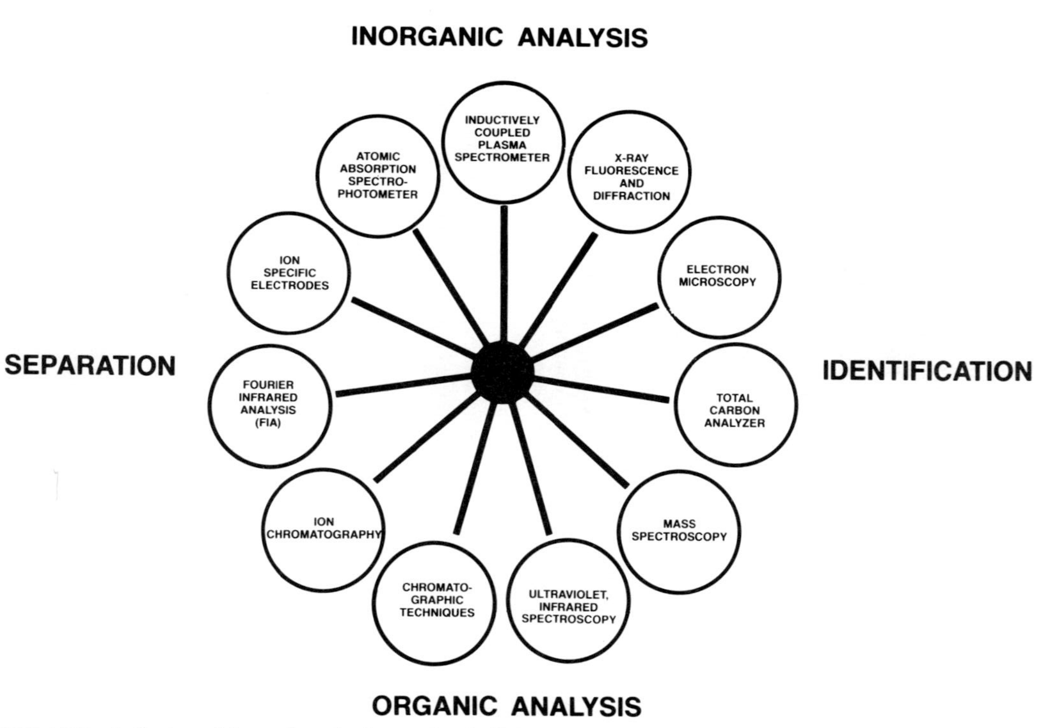

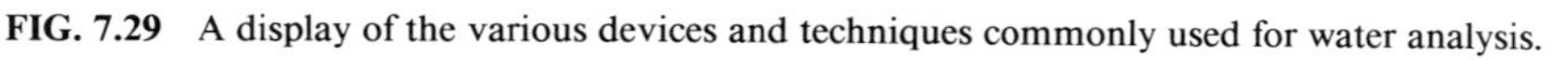

FIG. 7.29 A display of the various devices and techniques commonly used for water analysis.

Brief descriptions of some of the instruments used for water analysis along with types of impurities and concentration ranges for each are given below. This section is divided into inorganic and organic analysis instrumentation groupings depending on the principal use of each instrument; but most have applications in both types of analysis.

Inorganic Analysis

Atomic Absorption Spectroscopy (Figure 7.30). Atomic absorption is a technique for measuring specific metals in both aqueous and nonaqueous systems. It is uniquely suited to analyze water because of its sensitivity and its ability to detect a wide diversity of cations and metallic constituents.

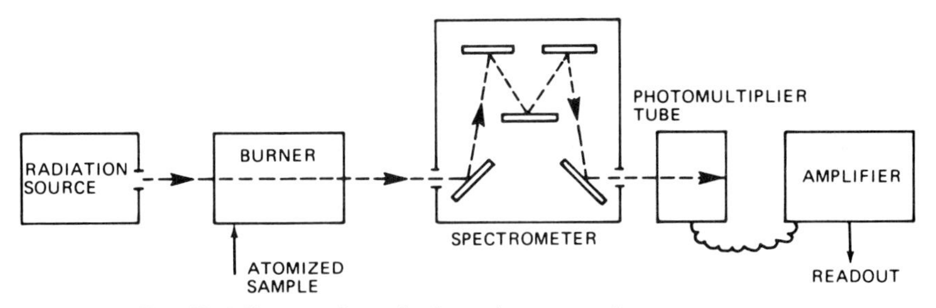

FIG. 7.30 Simplified diagram of atomic absorption spectrophotometer.

The operation of the instrument is simple and rapid. About 68 different elements can be determined with varying degrees of sensitivity in the milligram per liter range and below. Results are accurate to within 5% of true values. The technique is relatively interference-free compared to wet chemical procedures and it is much faster. The use of this technique has continued to grow as the need for measuring heavy metals in various streams has become more important.

Typical analyses performed by atomic absorption spectrophotometers include:

1. Raw water—trace metals and hardness
2. Brines and salts—heavy metals and other trace impurities
3. Plant effluents—pollutants such as Cr, Sn, Mn, Pb, and Hg
4. Treated water—hardness (to 0.1 ppm), Fe, and Cu

Optical Emission Spectrography (Figure 7.31). The optical emission spectrograph is similar in principle to the atomic absorption spectroscope. It is a tool for qualitative and quantitative elemental analysis of solids, liquids, or gaseous materials. It has been used for many years for measuring low or trace level element concentrations in various matrices in manufacturing control and research applications. It has been used extensively for quality control analyses of metals in the steel, aluminum, and other metal industries. Improvements in instrument design and availability of analytical techniques for solution analysis by optical emission make this technique useful in the analysis of water samples.

The use of inductively coupled argon plasma (ICAP) as an excitation source for optical emission in recent years has dramatically increased the application of

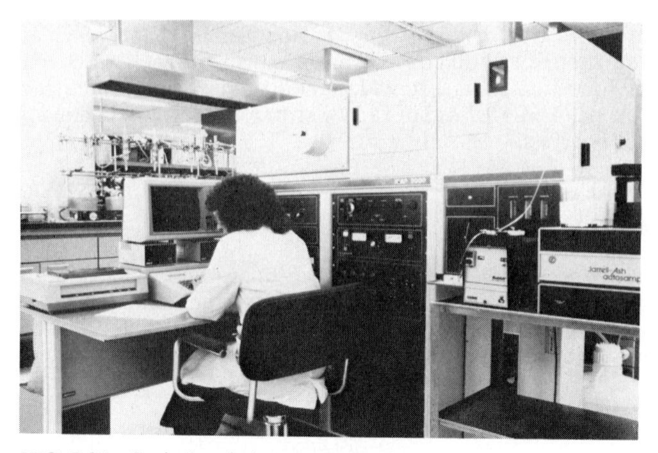

FIG. 7.31 Optical emission spectrograph.

this technique for water analysis. ICAP provides a very stable and reproducible excitation source and gives sensitivities in the microgram per liter range (parts per billion). The ICAP, coupled to a direct-reading spectrograph, can be used to measure quantitatively up to 50 elements simultaneously in 1 min or less. Additionally, most commercially available instruments are available with a scanning photomultiplier tube to permit the use of the spectrograph to measure other elements not in the direct reading program.

Ion Selective Electrodes (Figure 7.32). A variety of ion selective electrodes have proved valuable in water analysis through the direct potentiometric measurement of specific ions in aqueous systems. They can also be used as endpoint detectors for potentiometric titrations. These electrodes respond only to ionic materials, so to use them, sample preparation is sometimes necessary. The cyanide electrode, for example, requires distillation of the sample to break down complex metal cyanides that are not detectable with the electrode.

Electrodes specific for anions are especially useful, since previous chemical procedures for analyzing anions were tedious. Electrodes are available for such

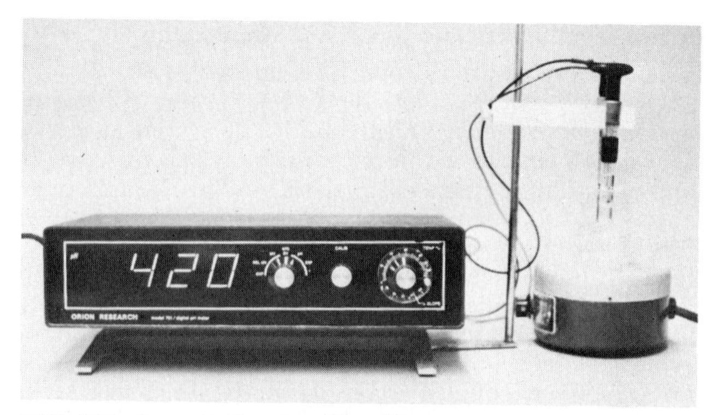

FIG. 7.32 Ion selective electrode with output measuring device.

anions as sulfide, fluoride, cyanide, and nitrate. These make it possible to monitor specific anions continuously in a water system, as is commonly done with pH instrumentation.

The ion selective electrodes work like pH electrodes, depending on a potential developed across a thin membrane by the difference in the concentrations of the ion to be measured on each side of the ionically conducting thin layer. The concentration within the electrode is fixed and the potential varies with the concentration of ions in the sample.

Ion Chromatography. The use of the ion chromatograph has become widespread in recent years as a tool for water analysis. In this technique, the sample is injected into an ion exchange chromatographic column where the various components of the sample are separated by an eluting solvent. To effect this separation in elution volumes, the solvent system and support media must be selected from a variety of available materials for the particular analytical program. In this way the components are separated according to their relative affinity for the column media or the solvent system. As the components are eluted they are converted to a highly conductive form, and then they pass through a very sensitive conductivity bridge. The ion chromatograph provides a sensitive selective technique for many of the compounds in natural waters, and can be adapted to a variety of other analyses in aqueous systems.

X-Ray Fluorescence and Diffraction (Figures 7.30, 7.34). X-ray flourescence has been used extensively for the analysis of water-formed deposits and corrosion products from various industrial systems. When the fluorescence technique is supplemented by several chemical methods of analysis, it is possible to obtain

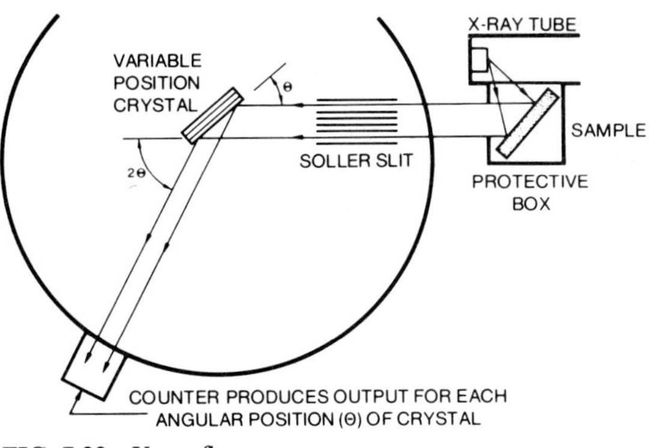

FIG. 7.33 X-ray fluorescence.

complete identification of these deposits. In wastewater analyses, the fluorescence method is used to characterize particulate matter filtered from the sample.

X-ray diffraction is used to identify crystalline materials in water-formed deposits. Based on the elemental analysis obtained by fluorescence techniques, one of several crystalline structures may exist, and identifying the structure may provide a clue to the mechanism of deposit formation. X-ray diffraction has limited usefulness in identifying organic materials.

In x-ray fluorescence, the sample, irradiated with high intensity x-rays, fluoresces to produce the x-ray line spectra of the elements present in the sample. The spectra are dispersed within the instrument to allow readout of these lines correlated to a particular analysis program. In x-ray diffraction, the crystalline

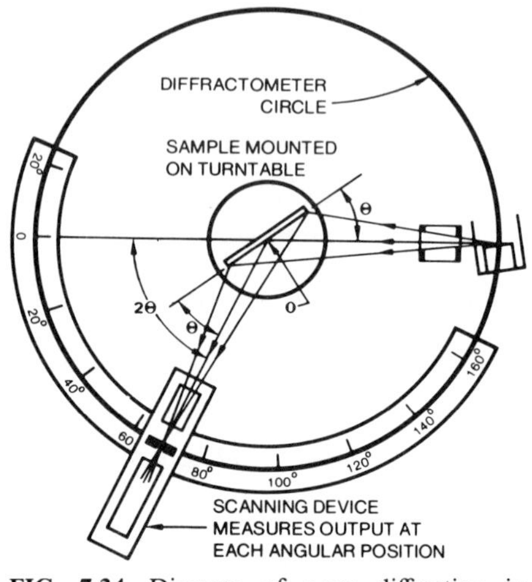

FIG. 7.34 Diagram of x-ray diffraction instrument.

material acts as a three-dimensional grating for the x-rays. The diffracted x-rays produce a pattern formed by the various sets of parallel planes which define the crystal lattice.

Flame Emission. Flame emission spectroscopy provides a sensitive, rapid method of analysis for the alkali metals in water. The method is less sensitive and more prone to interferences than atomic absorption spectroscopy, but it takes only about half the time.

Visible and Ultraviolet Spectrophotometry. These techniques, especially using the visible region of the spectrum, continue in wide use today after many years of service. They are invaluable tools for measuring many inorganic constituents in water. They also have some applications in organic analyses, primarily in the ultraviolet region.

Microscopic Techniques. The analyst uses various microscopic techniques to characterize particulate material in water. These particulates must first be separated by filtration or other means. Light microscopy has been a much used tool for making particulate identification. Modern microscopes routinely provide magnification of 2000×. An experienced microscopist can often identify the chemical composition of particulate material by simple microscopic observation. He may add certain reagents to the specimen under observation to verify the presence of certain elements.

Recent improvements in electron microscopy have advanced the use of microscopic techniques in water chemistry. Both the transmission electron microscope and the scanning electron microscope are used in water analyses.

Transmission Electron Microscope (TEM) (Figure 7.35). Transmission electron microscopy (TEM) is used for the direct observation of small particulates stable in a high vacuum. Form and structure (morphology) can be established for rigid surfaces such as ceramics, metals, plastics, fibers, and other organic and inorganic solids separated from water. The magnification range of the TEM is 220× to 1,000,000× with a resolution of 2Å.

FIG. 7.35 A view of the components of a transmission electron microscope.

Scanning Electron Microscopy (SEM). The scanning electron microscope can be used for morphological characterization of particulates such as clays, spores, pollen, and bottom sediment. Particle size determination in the range of 0.05 to 100 μm can be made. The magnification range is from 20× to 200,000× with a resolution of 100Å. This technique has been used extensively for characterizing very small particulates in water (Figure 7.36).

Both the SEM and the TEM can be equipped with x-ray detectors to determine the elements present in the portion of the specimen in the beam of the microscope (Figure 7.37).

Miscellaneous Instrumental and Wet Chemical Analyses. There continues to be a great need for the "wet chemistry" methods of water analysis. In many cases they are the only ones acceptable to regulatory agencies. Chemical oxygen demand (COD), biochemical oxygen demand (BOD), and microorganism counts are examples. Although improved wet techniques have been developed, the methods are still tedious and time-consuming.

Many of the anions in water are analyzed by classical chemical methods because faster, more economical techniques are not yet available. In many cases,

FIG. 7.36 Diatoms and silt observed by the scanning electron microscope at 1000× magnification.

automated devices have been developed to facilitate these wet-chemistry procedures (Figure 7.38).

Many instrumental techniques require chemical pretreatment of the sample to solubilize the constituent to be analyzed; to concentrate or dilute to the range of the instrument; or to handle chemical or instrumental interferences.

Total Carbon Analyzer (Figure 7.39). The total carbon analyzer is used for the quantitative determination of carbonaceous material in aqueous media. The instrument accepts only a very small sample (approximately 20 μm), so it is necessary that the carbon-containing materials be either dissolved or dispersed homogeneously. The instrument can be operated over several concentration ranges, such as 0 to 100 or 0 to 1000 mg/L. The sensitivity and accuracy are approximately 1 to 2% of the range covered, providing a capability to measure 1 to 2 mg/L of total carbon.

Most carbon analyzers measure both organic and inorganic carbon. To analyze only the organic carbon, it is necessary to remove the carbonate before injecting the sample into the instrument. Where chemical or biochemical oxygen demand tests (COD or BOD) are necessary to monitor organic pollutants, it may be possible to establish a correlation between total organic carbon (TOC) analysis and the COD or BOD values. The carbon analyzer requires 5 to 10 min per sample, including pretreatment time. With the COD or BOD tests, considerably more operator time is required and results are not available for 3 to 4 h in the case of the COD test, and 5 days for the BOD test.

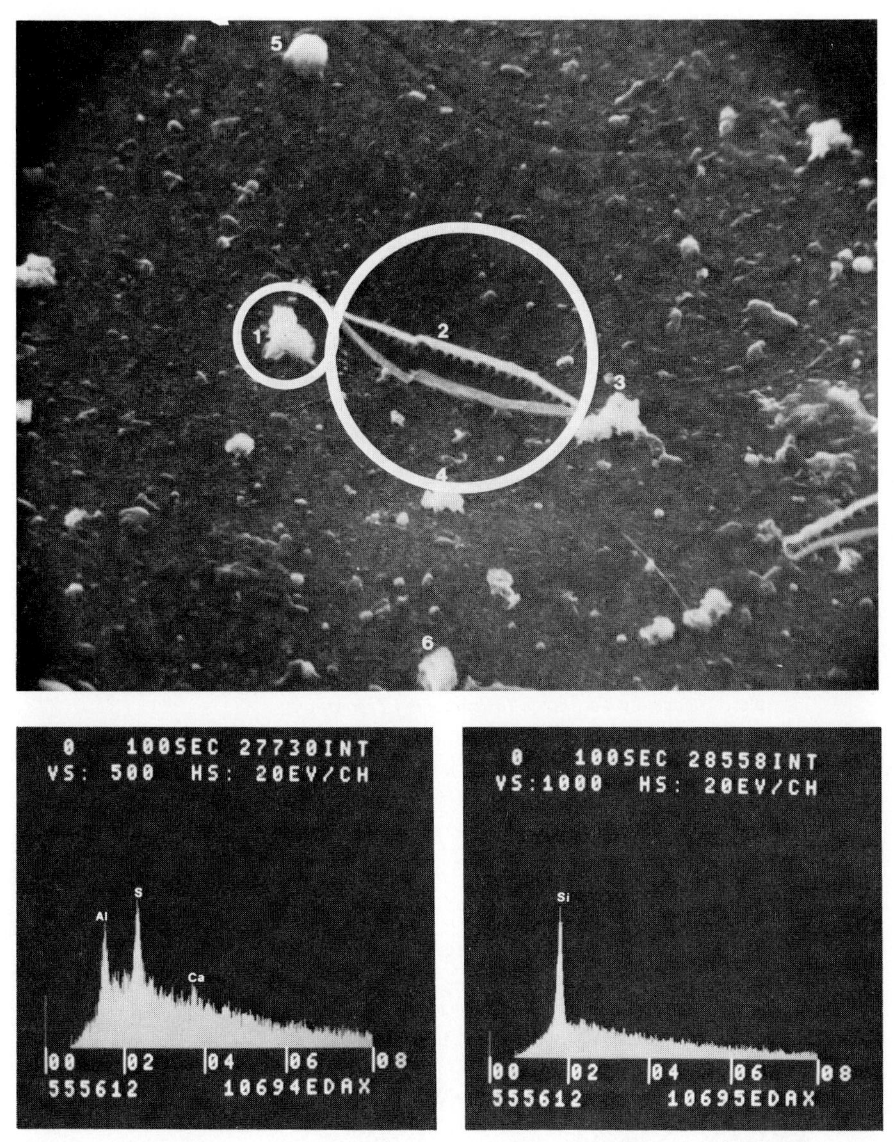

FIG. 7.37 Observation of particulates in a wastewater sample by SEM, with simultaneous analysis of individual particles 1 and 2, identified as a precipitate containing calcium, sulfur, and aluminum, and a diatom.

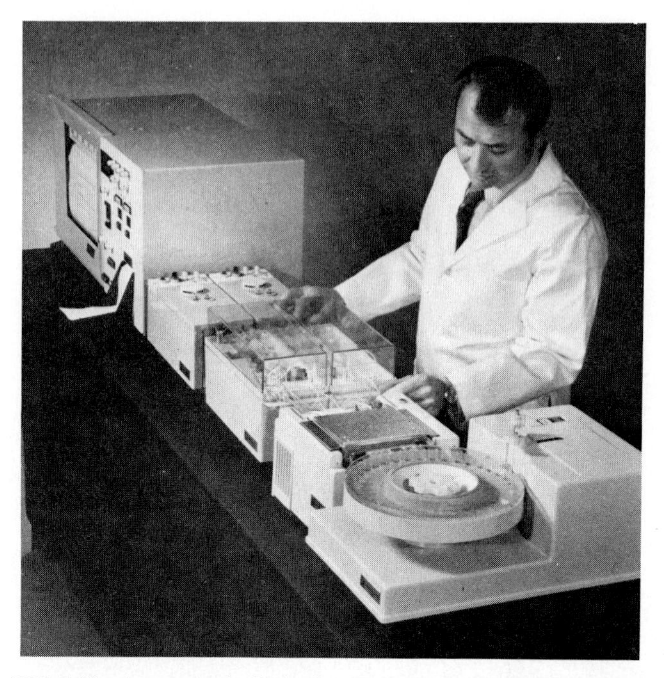

FIG. 7.38 An automatic device for wet analysis of a battery of samples. *(Courtesy Technicon Instruments Corporation.)*

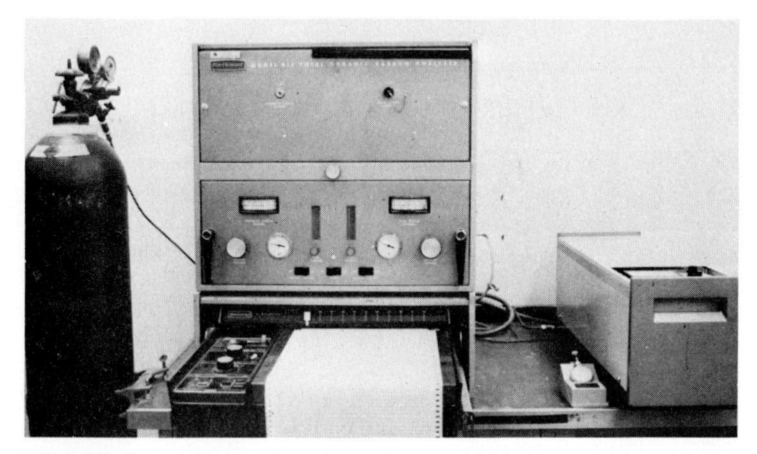

FIG. 7.39 Total carbon analyzer in use for determining organic matter in water.

Organic Analysis

Infrared Spectroscopy (Figure 7.40). Infrared spectroscopy (IR) is used to detect the presence of most organic functional groups, such as $-COOH$, $-CHO$, $-CH_3$, and $-OH$. Essentially all forms of organic materials that can be separated from water, wastewater, and water-formed deposits can be characterized by this method after appropriate sample preparation. This technique is qualitative, since the separated fraction is usually a mixture. The sensitivity for various organic functional groups varies greatly, but for ideal cases is as low as a few percent of the fraction analyzed. Detailed characterization of complex mixtures requires information from other types of analyses and chemical separations, such as elemental data and thin layer chromatopography.

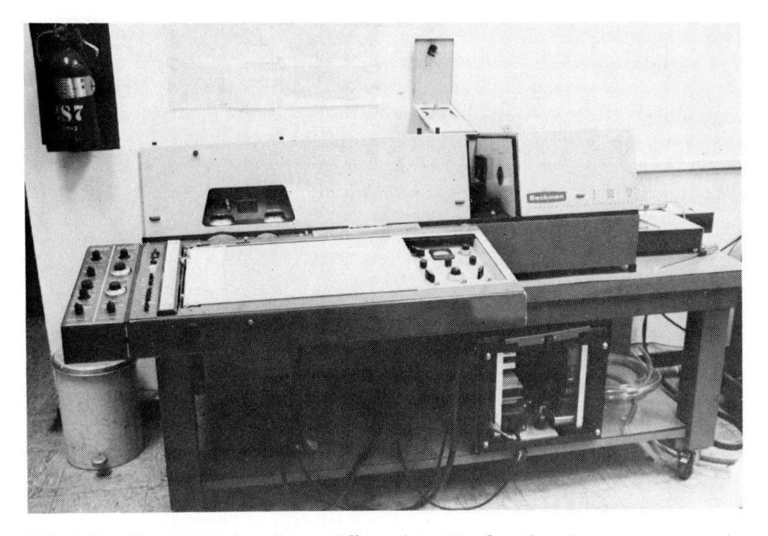

FIG. 7.40 The infrared analyzer differentiates the functional groups on organic molecules extracted from water.

An example of an IR spectrum is shown in Figure 7.41. The information shown led to the selection of additional analytical tools to more fully identify the organic materials present. The identification of fatty bisamide and polyethylene glycol ester pinpointed the production areas where process pollution was occurring.

IR is typically used on solvent extracts and, therefore, gives information on the least polar, nonvolatile components of a sample. New procedures are being developed for more effective separation of additional types of organic materials to be characterized by IR.

Nuclear Magnetic Resonance (Figure 7.42). The IR spectrum provides a fingerprint by which to compare organic matter in a water sample to patterns of known compounds. The nuclear magnetic resonance (NMR) spectroscope supplements the IR analysis by differentiating structural elements in the organic molecules, using radio frequency energy.

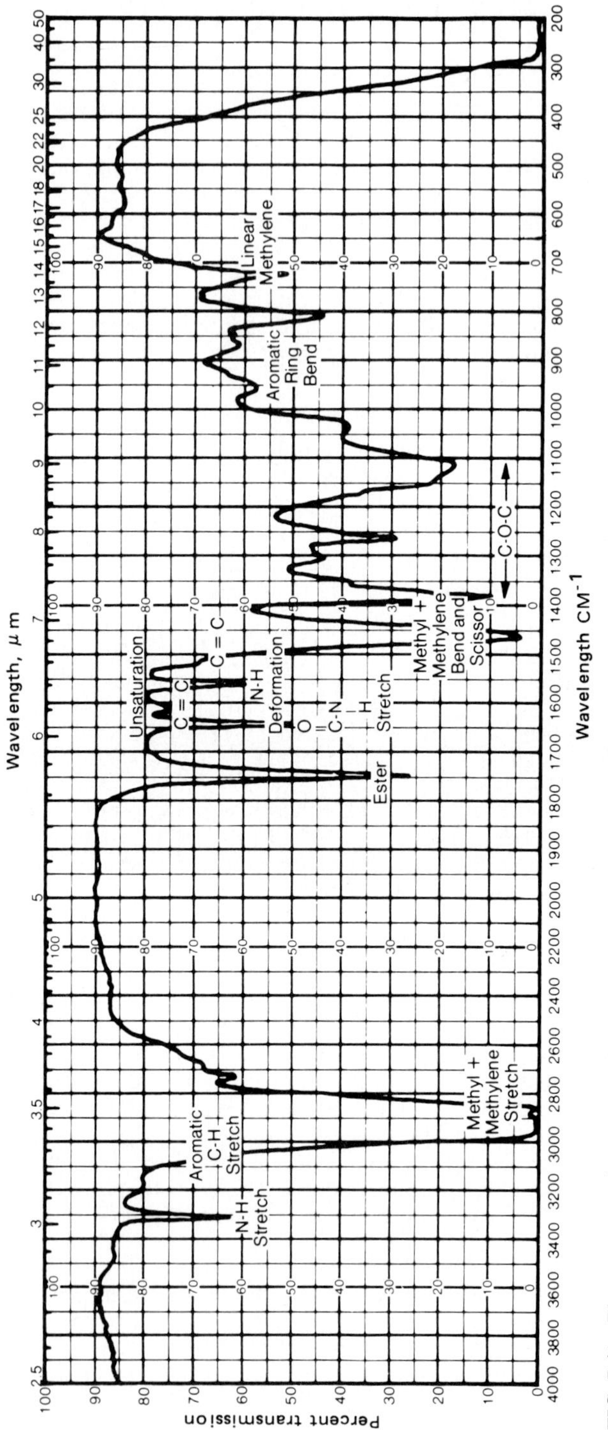

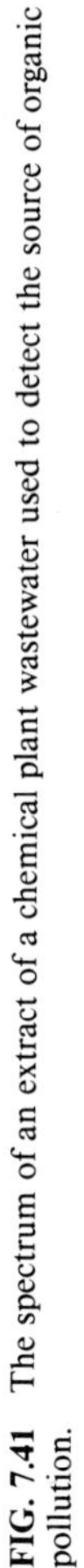

FIG. 7.41 The spectrum of an extract of a chemical plant wastewater used to detect the source of organic pollution.

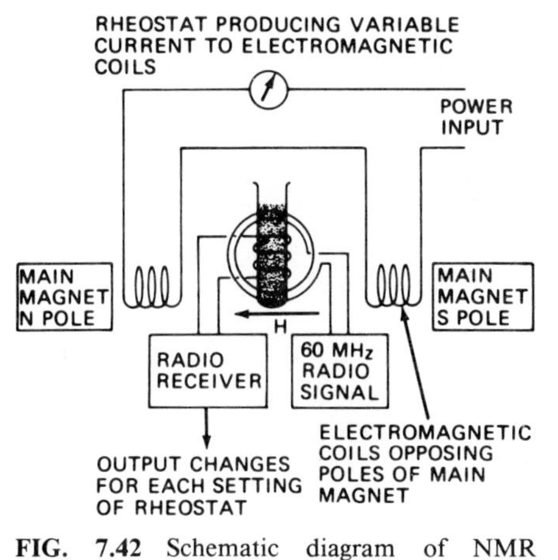

FIG. 7.42 Schematic diagram of NMR analyzer.

Liquid Chromatography (Figure 7.43). Liquid chromatography (LC) is a technique used for the separation and determination of a specific material in a liquid sample.

A fixed volume of sample is injected at the head of a separative column packed with a bed of adsorbent material. A mobile liquid is then forced through the column to cause the components to be differentiated into separate bands in the bed. A detector at the effluent of the separative column then analyzes these bands (analytes) as they leave the column. Several different types of separation columns using different mechanisms of analyte retention, and different mobile liquids can be selected depending on the particular separation objective. Many types of detectors have been used; the most widely used sensitive detector is the ultraviolet spectrophotometer.

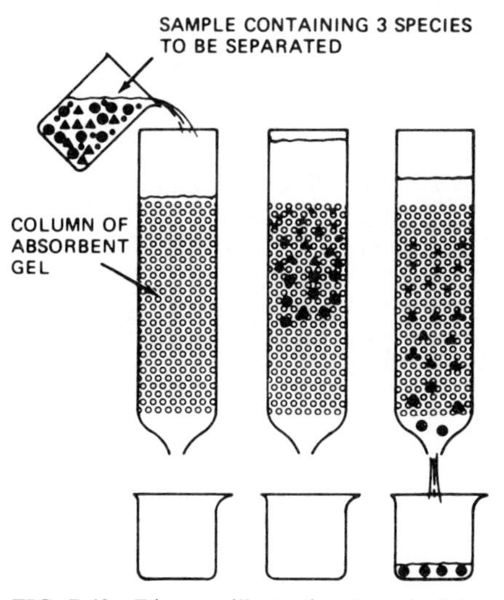

FIG. 7.43 Diagram illustrating the principles of liquid chromatography.

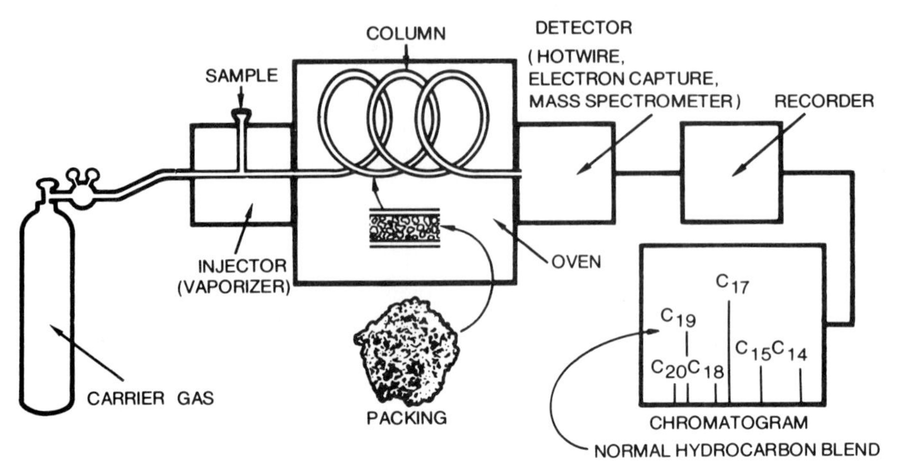

FIG. 7.44 Simplified diagram of a gas chromatograph.

Gas Chromatography. Gas chromatography (GC) (Figure 7.44) is used for the determination of volatile materials in water samples. Some of these can be measured directly; others may require separation or special preparation. Where the analyte is not sufficiently volatile, it may be extracted from water and then treated with chemical reagents to convert it into a volatile compound. Gas chromatography is being used extensively to determine concentrations of various organic compounds in water, especially drinking water. It provides a convenient method of testing for pesticides or halogenated hydrocarbons that may be present as a result of runoff or chlorination of naturally occurring organic materials during water treatment.

To fractionate a sample, it is vaporized and swept by a stream of inert carrier gas into the chromatographic column. The components separate into bands as the carrier gas passes through the column packing. The separated components are measured at the end of the column by various types of detectors. Examples of materials analyzed by the GC technique include pesticides and chlorinated solvents.

The mass spectrometer is one of the detectors used with GC for the identification of the structures of compounds. Gases are ideal samples for the mass spectrometer (MS). The MS operates by bombarding gaseous molecules with energetic electrically charged particles. The energy transferred to the gas molecules fragments them into smaller charged particles in a reproducible manner. The mass and relative abundances of these charged fragments is determined by a scanning instrument which produces the mass spectrum related to the analyte.

SUGGESTED READING

APHA, AWWA, and WPCF: *Standard Methods for the Examination of Water and Wastewater,* 1971.

American Society for Testing and Materials: *Annual Book of ASTM Standards,* Part 23: "Water and Atmospheric Analyses," 1973.

Envirex Inc.: *Handbook for Sampling and Sample Preservation of Water and Wastewater,* U.S. Department of Commerce, National Technical Information Service, PB-259 946, September 1976.

McCrone, W. C., and Delly, J. S.: *The Particle Atlas,* 2d ed, Ann Arbor Science, Ann Arbor, Mich. 1973.

U.S. Department of the Interior, Bureau of Reclamation: *Water Measurement Manual,* U.S. Government Printing Office, 1967.

P · A · R · T · 2

UNIT OPERATIONS OF WATER TREATMENT

CHAPTER 8
COAGULATION AND FLOCCULATION

The processes of coagulation and flocculation are employed to separate suspended solids from water whenever their natural subsidence rates are too slow to provide effective clarification. Water clarification, lime softening, sludge thickening, and dewatering depend on correct application of the theories of coagulation and flocculation for their success.

Taking surface water clarification as an example, turbid raw water contains suspended matter—both settleable solids, particles large enough to settle quiescently, and dispersed solids, particles which will not readily settle. A significant portion of these nonsettleable solids may be colloidal. Each particle is stabilized by negative electric charges on its surface, causing it to repel neighboring particles, just as magnetic poles repel each other. Since this prevents these charged particles from colliding to form larger masses, called flocs, they do not settle. Coagulation is the destabilization of these colloids by neutralizing the forces that keep them apart. This is generally accomplished by adding chemical coagulants and applying mixing energy. Aluminum salts, iron salts, or polyelectrolytes are the chemicals usually used.

Figure 8.1 illustrates how these chemicals reduce the electric charges on colloidal surfaces, allowing the colloidal particles to agglomerate into flocs. These initially small flocs join, creating larger, settleable agglomerates. The destabilization step is coagulation (charge neutralization); the floc-building stage is flocculation.

The terms coagulation and flocculation are often used interchangeably; however, when seen as two different mechanisms they can provide a better understanding of clarification and dewatering.

COAGULATION

Colloidal species encountered in raw water and wastewater include clay, silica, iron and other heavy metals, color, and organic solids such as the debris of dead organisms. Colloids may also be produced in precipitation processes such as lime softening. Oil in wastewater is frequently colloidal.

Among the wide variety of colloidal materials in water, there is a broad distribution of particle sizes. Table 8.1 shows how particle size affects the tendency of particles to settle in quiet water. Colloids always require coagulation to achieve

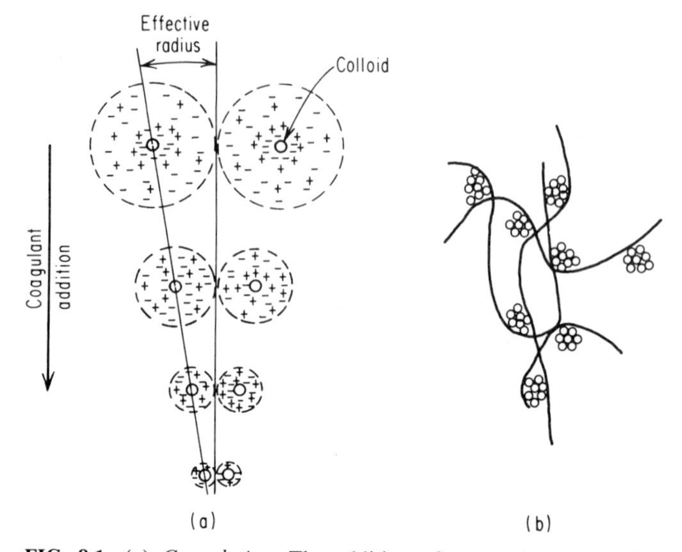

FIG. 8.1 (*a*) *Coagulation*: The addition of a coagulant neutralizes charges, collapsing the "cloud" surrounding the colloids so they can agglomerate. (*b*) *Flocculation*: The bridging of the flocculant chemical between agglomerated colloidal particles forms large settleable flocs.

an effective size and settling rate; but even larger particles, which are not truly colloidal and would settle if given enough time, may require coagulation to form larger, faster settling floc.

When insufficient settling time is available in a treatment plant to remove syspended solids, coagulation and flocculation may cause them to grow in size and settle rapidly enough to overcome the physical limitation of the plant design.

Colloids are categorized as hydrophobic (water hating) or hydrophilic (water loving). Hydrophobic colloids do not react with water; most natural clays are hydrophobic. Hydrophilic colloids react with water; the organics causing color are hydrophilic. Of importance in water treatment, the hydrophiloc colloids may

TABLE 8.1 Sedimentation of Small Particles of Silica of 2.65 Sp. Gr.

Typical	mm	μm	Surface area (total)	Settling time, 1 m fall
Gravel	10.	10,000	3.14 cm^2	1 s
Coarse sand	1.	1,000	31.4 cm^2	10 s
Fine sand	0.1	100	314 cm^2	125 s
Silt	0.01	10	0.314 m^2	108 min
Bacteria	0.001	1.	3.14 m^2	180 hr
Colloidal matter	0.0001	0.1	31.4 m^2	755 days

NOTE: Particles larger than 100 μm are visible to the naked eye and are considered to be settleable solids. In the range of 10–100 μm, they are considered to be turbid. Below 10 μm they are considered colloidal. Particles larger than 0.1 μm are visible by light microscope; below 0.1 μm, the electron microscope is used for detection.

chemically react with the coagulant used in the treatment process. So hydrophilic colloids require more coagulant than hydrophobic, which do not chemically react with the coagulant. For example, removal of color from a water having an APHA color of 50 requires higher coagulant dosages than removal of 50 JTU turbidity.

Several theories have been advanced to describe the colloidal particle and the forces surrounding it. For practical purposes the determination of the nature and strength of the particle charge is all that is needed to define the colloidal system. The particle charge strength, illustrated as the layer surrounding the colloid in Figure 8.1, affects how closely colloids can approach.

ZETA POTENTIAL

Zeta potential is a measurement of this force. For colloids in natural water sources in a pH range of 5 to 8, the zeta potential is generally -14 to -30 mV; the more negative the number, the stronger the particle charge. As the zeta potential diminishes, the particles can approach one another more closely, increasing the likelihood of collision. In a conventional clarification system at a pH of 6 to 8, coagulants provide the positive charges to reduce the negative zeta potential. Coagulation usually occurs at a zeta potential which is still slightly negative, so complete charge neutralization is not usually required. If too much coagulant is added, the particle surface will become positively charged (a positive zeta potential), and the particle will be redispersed.

Coagulants may be required in high pH water treatment systems, such as in lime softening. Calcium carbonate particles also carry a negative charge, and cationic coagulants may be useful in reducing residual colloidal hardness. Magnesium hydroxide, on the other hand, carries a positive charge until the pH exceeds 11; so in lime or lime-soda softening processes where both $CaCO_3$ and $Mg(OH)_2$ precipitate, the oppositely charged particles coprecipitate. This coprecipitation in past geologic periods produced the mineral dolomite, $CaCO_3 \cdot MgCO_3$. The coagulation and flocculation of materials other than silt and color, which are the common targets of a water clarification program, are discussed in the chapter on precipitation processes.

Zeta potential is determined indirectly from data obtained in observing particle motion under a microscope. Figure 8.2 *a* and *b* shows typical instruments employed for this determination. Zeta potential measurements have been used successfully to monitor plant coagulant dosages. However, for selecting the best coagulant, zeta potential readings alone are not reliable. Observation of results in a jar test remain the best method of coagulant selection.

Somehwat related to zeta potential in a qualitative way is streaming current, an electric current produced when colloidal particles are trapped in a capillary tube or a confined space with water flowing past them at high velocity. The adsorbed charges, or counter-ions, are stripped from the surface of the colloid and pass along with the water until the velocity is dissipated. The flow of ions constitutes an electric current that is measurable by an instrument called the *streaming current detector* (SCD). As is true with zeta potential, if coagulants have neutralized the charge on the colloids, the current is reduced to zero. Streaming current detectors require frequent maintenance to control plugging of the close clearances in the water passage spaces. In many cases, the head of the detector (a piston or plunger reciprocating in a closed cup) can be kept clear by an ultrasonic cleaning device. The SCD can produce an output signal to control the coagulation process.

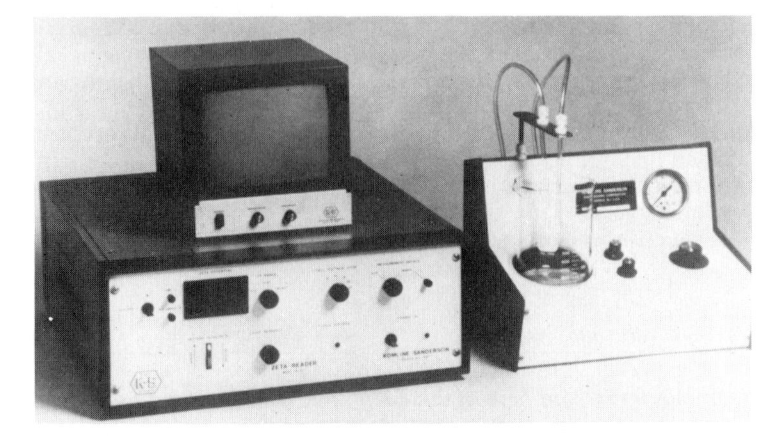

(*a*)

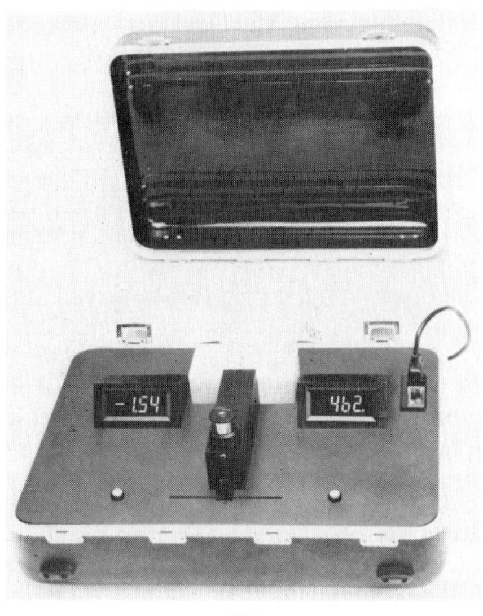

(*b*)

FIG. 8.2 (*a*) The "Zeta-Reader," an instrument for continuously observing the mobility of particles in water and measuring zeta potential. *(Courtesy of Komline-Sanderson Engineering Corporation.)* (*b*) The Mobility Meter is a bench-type instrument for observation and measurement of average mobility. This is translatable to zeta potential by a standard factor. *(Courtesy of Paper Chemistry Laboratory, Inc.)*

One of the variables that needs study in each system is the time factor; the speed or rate of charge neutralization varies with the type of colloid present and with temperature. Therefore, a sample taken immediately after coagulant addition is usually inadequate. Typically, an equilibration time of 5 to 10 min is needed before putting the sample through the SCD.

Mixing is required to supplement coagulant addition to destroy stability in the colloidal system. For particles to agglomerate they must collide, and mixing promotes collision. Brownian movement, the random motion imparted to small particles by bombardment by individual water molecules, is always present as a natural mixing force. However, additional mixing energy is almost always required. High intensity mixing, which distributes the coagulant and promotes rapid collisions, is most effective. The frequency and number of particle collisions are also important in coagulation. In a low turbidity water, the addition of solids such as clay or the recycle of previously settled solids may be required to increase the number of particle collisions.

FLOCCULATION

The floc formed by the agglomeration of several colloids may not be large enough to settle or dewater at the desired rate. A flocculant gathers together floc particles in a net, bridging from one surface to another and binding the individual particles into large agglomerates as shown in Figure 8.3. Alum, iron salts, and high molecular weight polymers are common flocculants. Flocculation is promoted by slow mixing, which brings the flocs gently together; too high a mixing velocity tears them apart, and they seldom re-form to their optimum size and strength. Flocculation not only increases the size of floc particles, but it also affects the physical nature of the floc. Sludges and slurries, when flocculated, dewater at faster rates on sand beds and in mechanical dewatering equipment because of the less gelatinous structure of the floc.

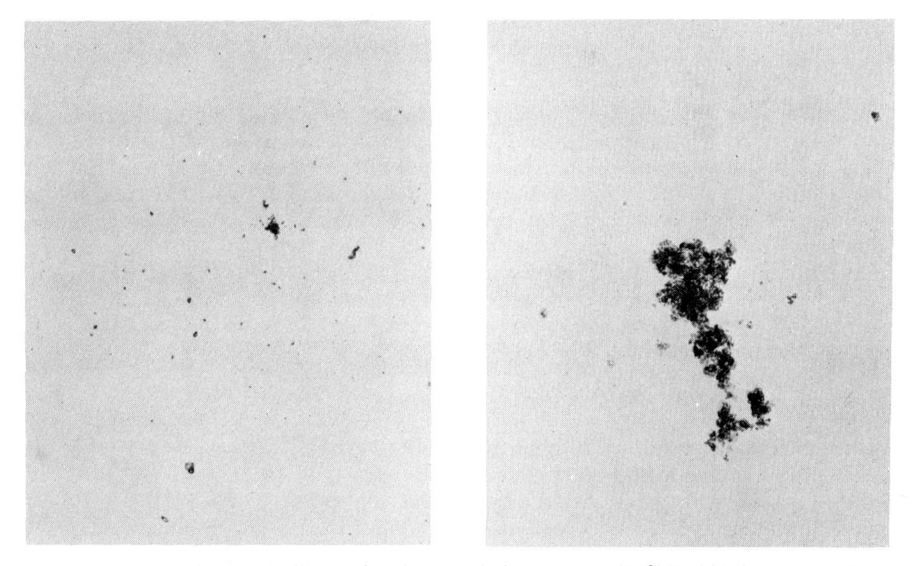

FIG. 8.3 Photomicrographs illustrating the coagulation process. *(Left)* Turbid river water, showing fine dispersion of tiny solid particles. *(Right)* Same water treated with coagulant. Particles are "collected" in the floc. (130×)

TABLE 8.2 Constraints between Coagulation and Flocculation

Variable condition	Coagulation	Flocculation
Nature of solids	Numerous, fine particulates	Scattered, large gels
Type of chemical applied	Low molecular weight charge neutralizer	High molecular weight particle binder
Energy requirement	Rapid mixing	Slow stirring
Velocity gradient	High	Low
Time in process	Seconds	Minutes

It is apparent that the processes of (a) charge neturalization, or coagulation, and (b) floc building, or flocculation, are so different that each system containing the chemically treated solids being processed has its own physical constraints. These are outlined in Table 8.2.

In an attempt to develop a mathematical procedure to express some of these variables, hydraulic engineers have examined this problem of fluid mechanics and have developed the concepts of velocity gradient and shear rate, or "G factor."

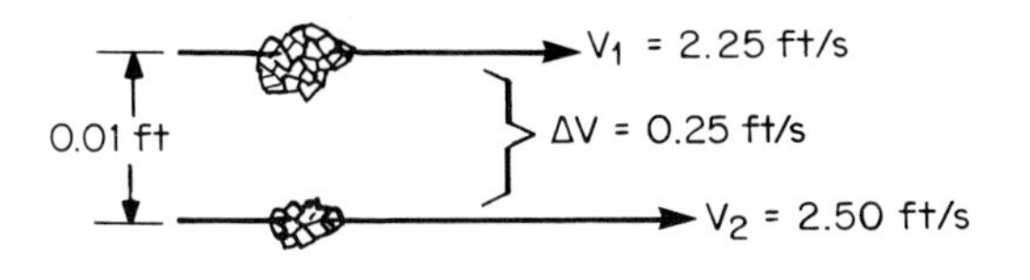

G – FACTOR (shear rate) = 0.25 ft/s / 0.01 ft = 25 s⁻¹

FIG. 8.4 Illustration of velocity gradient and shear rate.

Figure 8.4 illustrates the basis of these concepts. In this illustration, the differential velocity between two particles 0.01 ft apart is 0.25 ft/s, so the shear rate G is 25 s^{-1}. Obviously, it is impractical to measure the G factor in this way, but fortunately further development of the mathematical model shows that shear rate is also related to rate of energy input (power) per unit volume (equivalent to detention time in the process) and water viscosity. The latter has a direct bearing on the frequency of particle collisions and explains, in part, the strong influence of water temperature on both coagulation and flocculation. The formula is:

$$G\ factor = \sqrt{\frac{P/V}{\mu}}\ \text{in s}^{-1}$$

where P is power input, in ft lb/s (hp × 550) (watts)
V is volume in the process, ft^3 (m^3)
μ is viscosity, in (lb) (s)/ft^2 (viscosity in centipoises $\times 2.1 \times 10^{-5}$ = (lb)(s)/ft^2)

The G factor usually recommended for most coagulation units is about 900 s^{-1} for a 30-s mixing time, varying inversely with time. The required mixing time is usually established by bench tests, as described later. The recommended G factor for flocculation is lower, varying from about 50 for a cold, colored water carrying a very fragile floc, to about 200 for a solids contact lime softener on warm river water. Again, the G factor for flocculation must be determined by bench testing, and this should lead to a design of flocculator which can be varied in speed and power input as river conditions change and lead to fluctuations in solids concentation and sensitivity of floc to shear.

COAGULATION AND FLOCCULATION CHEMICALS

Historically, metal coagulants (alum and iron salts) have been most widely used in water clarification. These products function both as coagulants and flocculants. When added to water, they form positively charged species in the typical pH range for clarification, about 6–7. This hydrolysis reaction produces insoluble gelatinous aluminum or ferric hydroxide.

$$Al_2(SO_4)_3 + 6H_2O \rightleftharpoons 2Al(OH)_3\downarrow + 3H_2SO_4 \quad (1)$$

$$FeCl_3 + 3H_2O \rightleftharpoons Fe(OH)_3\downarrow + 3\,HCl \quad (2)$$

Note that the by-products are hydroxide precipitates and mineral acids; the latter react with alkalinity in the water, reducing pH and producing a second by-product, CO_2. Sometimes the gaseous CO_2 by-product interferes with the coagulation process by coming out of solution, adsorbing on the hydrous precipitate, and causing floc flotation rather than settling.

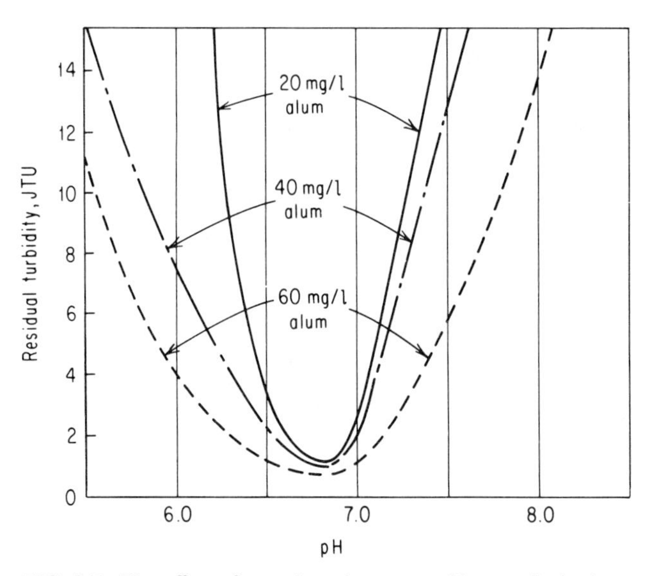

FIG. 8.5 The effect of coagulant dosage on pH range limitations. The optimum pH remains almost constant, but the pH range becomes less restrictive as coagulant dosage increases.

TABLE 8.3 Properties of Common Coagulants

Common name	Formula	Equiv. weight	pH at 1%	Availability
Alum	$Al_2(SO_4)_3 \cdot 14H_2O$	100	3.4	Lump—17% Al_2O_3 Liquid—8.5% Al_2O_3
Lime	$Ca(OH)_2$	40	12	Lump—as CaO Powder—93–95% Slurry—15–20%
Ferric chloride	$FeCl_3 \cdot 6H_2O$	91	3–4	Lump—20% Fe Liquid—20% Fe
Ferric sulfate	$Fe_2(SO_4)_2 \cdot 3H_2O$	51.5	3–4	Granular—18.5% Fe
Copperas	$FeSO_4 \cdot 7H_2O$	139	3–4	Granular—20% Fe
Sodium aluminate	$Na_2Al_2O_4$	100	11–12	Flake—46% Al_2O_3 Liquid—26% Al_2O_3

Polyaluminum chloride, a product in wide use in Japan, avoids the problem of alkalinity reduction. When the material hydrolyzes, the floc formed incorporates the chloride ion into the floc structure so it is not available to produce acid, reduce alkalinity, and form by-product CO_2. Even if there are no suspended solids in the water initially, the metal coagulants form flocs which enmesh the destabilized colloids. However, the voluminous sludges produced by the addition of metal coagulants create disposal problems because they are usually difficult to dewater. This is why alum and iron salts are not often used to improve efficiency of centrifuges, filter presses, and other dewatering devices.

Metal coagulants are particularly sensitive to pH and alkalinity. If pH is not in the proper range, clarification is poor, and iron or aluminum may be solubilized and cause problems to the water user. The lower the coagulant dosage, the more sensitive the floc is to pH changes (Figure 8.5). Table 8.3 lists some important properties of common coagulants.

The introduction of activated silica in the 1940s significantly improved the performance of alum and iron salts as coagulants and flocculants in water clarification. The subsequent development of a variety of organic polymers called polyelectrolytes in the next decade was an even more spectacular contribution to water treatment technology.

Polyelectrolytes are large water-soluble organic molecules made up of small building blocks called monomers, repeated in a long chain. They usually incorporate in their structures ion exchange sites which give the molecule an ionic charge. Those having a positive charge are cationic, and those with a negative charge are anionic. These molecules react with colloidal material in the water by neutralizing charge or by bridging (tying together) individual particles to form a visible, insoluble precipitate or floc.

TAILORING POLYELECTROLYTES

The performance of these materials can be modified to suit the nature of the colloidal matter to be removed from the water. These modifications include variations in molecular weight and ion exchange capacity. These materials can also be produced without an ionic charge; these are called nonionic polymers. Although they are not, strictly speaking, polyelectrolytes, nonionic polymers exhibit many of the same flocculating properties in solution, and are generally considered as part of the general polyelectrolyte family of compounds.

Although most polyelectrolytes are synthetic organic materials, nature also produces an endless variety of such materials. Some of these are chemically processed to improve performance and are commercially available.

The cationic polyelectrolytes are either polyamines or quaternary amines. In water, a polyamine hydrolyzes as follows:

$$R_2NH + H_2O \rightarrow R_2NH \cdot H^+ + OH^-$$

Because of the hydrolysis to yield OH^-, at high pH, the reaction is forced to the left, and the polymer becomes nonionic. This is illustrated by Figure 8.6, which shows loss in exchange capacity for a specific polyamine as pH increases.

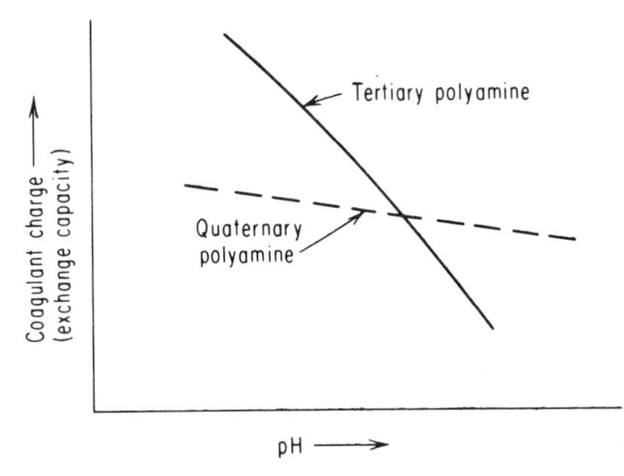

FIG. 8.6 Generalized plot showing loss of cationic strength for tertiary polyamines as pH increases, and relative pH independence of quaternary amine coagulants.

In contrast, the quaternary polymers are but slightly affected by pH, remaining positively charged over a broad pH range (Figure 8.6).

The anionic polymers incorporate a carboxyl group (—COOH) in their structure. These ionize in the following manner:

$$R{-}COOH \rightleftharpoons R{-}COO^- + H^+$$

The hydrogen ion forces the reaction to the left, so anionics become nonionic at low pH.

The ionic nature of polyelectrolytes is only one factor determining the performance of these materials as coagulants and flocculants. Other factors, such as the polar nature of nonionic bonds in the molecule, molecular size, and molecular geometry, also play a large part and may in some cases overshadow the effects of charge and charge density. Hence, high molecular weight nonionic polymers are effective flocculants in many systems because of their ability to attract and hold colloidal particles at polar sites on the molecule. Furthermore, because of their molecular size, they can bridge together many small particles. Less sludge is generated by organic polymers than by inorganic salts, since they do not add weight or chemically combine with other ions in the water to form a precipitate. Organic polymers do not affect the pH of the water and generally do not require pH adjustment for effective use.

So, as a general rule, cationics are designed to work at lower pH values, anionics at higher. Nonionics and quaternaries are only slightly infleunced by pH. The general rule should not be interpreted to mean that anionic polymers do not work at low pH value; it simply means they are no longer ionic. They may produce good results in flocculating solids at low pH simply because of their nonionic bonds. The same applies to cationics; even though they are not charged at high pH, they may act as effective coagulants because of their polar groups.

Organic polymers overcome many of the problems inherent in the use of alum or iron salts. These polymers are long chain organic molecules made up of small building blocks, called monomers, repeated along the chain. Depending on the selection of monomers and processing methods, a wide variety of polymers can be made of various configurations and molecular weights. Molecular weight is proportional to polymer chain length; the wide selection of structures and molecular weights makes it possible to design a polymer specifically for a given coagulation or flocculation problem, but this is seldom practical for economic reasons.

Organic polymers used in water treatment are of two major types, coagulants and flocculants. Coagulants are positively charged molecules of relatively low molecular weight. Although they exhibit some tendency for bridging, they are not particularly effective flocculants. Flocculant polymers have much higher molecular weights, providing long bridges between small flocs to enhance particle growth. Flocculants are either cationic, anionic, or nonionic. The flocculant that

TABLE 8.4 Some Characteristics of Organic Polymers

Class	Mol. wt. ranges	Form and availability
1. Cationic coagulants Polyamines Polyquaternaries PolyDADMAC Epi-DMA	Below 100,000	All are available as aqueous solutions
2. Cationic flocculants Copolymers of: Acrylamide and DMAEM Acrylamide and DADMAC Mannich amines	Over 1,000,000	Powders or emulsions
3. Nonionic flocculants Polyacrylamides	Over 1,000,000	Powders or emulsions
4. Anionic flocculants Polyacrylates Copolymers of acrylamide and acrylate	Over 1,000,000	Powders or emulsions

NOTE: DADMAC: diallyl-dimethyl ammonium chloride; Epi: epichlorhydrin; DMA: dimethylamine; DMAEM: dimethyl-aminoethyl-methacrylate.

works best in any system can be determined only through laboratory screening and in-plant testing. Polymer flocculants, unlike coagulants, are not selected for neutralization. Table 8.4 lists some characteristics of commonly used organic coagulants and flocculants used in water treatment.

Unlike inorganic salts, polymers do not produce voluminous, gelatinous floc. In applications where additional solids improve results, inorganic coagulants or clay may be required to supplement the use of polymers. Polymers do not affect pH, nor is their performance as sensitive to the pH of the treated water as metal coagulants.

ACTIVATED SILICA

Some inorganic compounds can be polymerized in water to form inorganic polymer flocculants. Activated silica (sometimes identified as $^{-}SiO_2^{-}$) is an example. When sodium silicate, which contains alkali, is diluted to 1.5 to 2.0% and then partially neutralized (usually with chlorine or sodium bicarbonate), the silica becomes colloidal and then begins to slowly polymerize. After aging for 15 to 30 min, the solution is diluted to about 0.5 to 1.0% SiO_2 arresting further polymerization, producing activated silica. Although this preparation procedure is complicated, this is a very effective flocculant for such applications as assisting alum treatment for color removal and improving the softening of organic-containing waters, such as some of the colored well waters in Florida.

APPLICATIONS OF COAGULATION AND FLOCCULATION

A sample of turbid water in a graduated cone separates into two layers, the settleable and the colloidal solids (Figure 8.7.) In raw water clarification a coagulant is almost always used since the colloidal haze must be removed to produce the low turbidity demanded by most water-using processes. In wastewater clarification, a coagulant is required only where the suspended solids create a problem in meeting effluent guidelines; here a flocculant may be required to speed the settling rate.

Two types of laboratory tests are used to select the best chemical and approximate dosage level required for clarification: (1) the jar test, and (2) the cylinder test. The jar test is used when the stream to be clarified has less than approximately 5000 mg/L suspended solids. Raw water clarification, settling of biological solids, and most primary waste streams are in this category. The cylinder test is used for heavy slurry streams where suspended solids exceed approximately 5000 mg/L. Coal and mineral processing wastes and the sludge resulting from a primary clarification are examples of heavy slurries.

The jar test simulates the types of mixing and settling conditions found in a clarification plant. The laboratory unit for running these tests (Figure 8.8) allows up to six individual tests to be run simultaneously. The jar tester has a variable speed motor that allows control of the mixing energy in the jars.

Clarification results are sensitive to chemical dosage, mixing energy, and length of mixing. Figure 8.9 shows a typical sequence in jar testing where a colloidal haze is removed. The coagulant is added with high energy to disperse it in the water and promote increased frequency of collisions. The duration may be

FIG. 8.7 The solid particles in the left cone are a conglomeration of materials of various particle sizes, identified as suspended solids. After settling for 30 min, two fractions are obtained (right cone): settleable solids, expressed as milliliters per liter, and turbidity, expressed in nephelometric units.

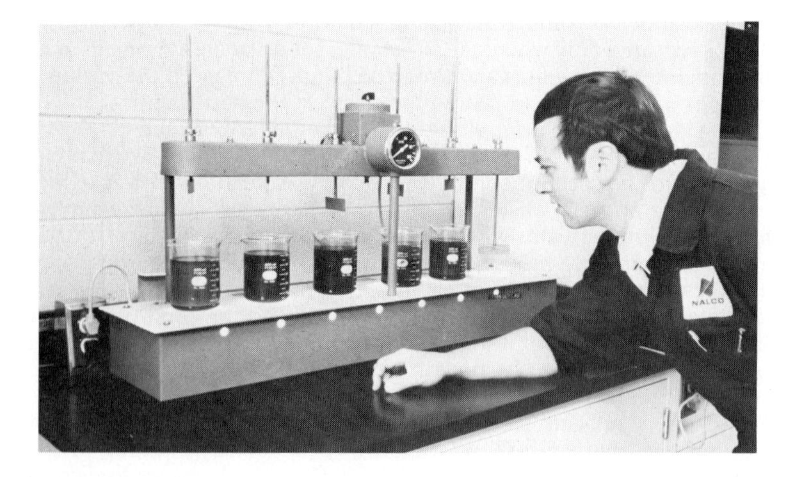

FIG. 8.8 This type of gang-stirrer is widely used for jar testing both as a research tool and a plant control device.

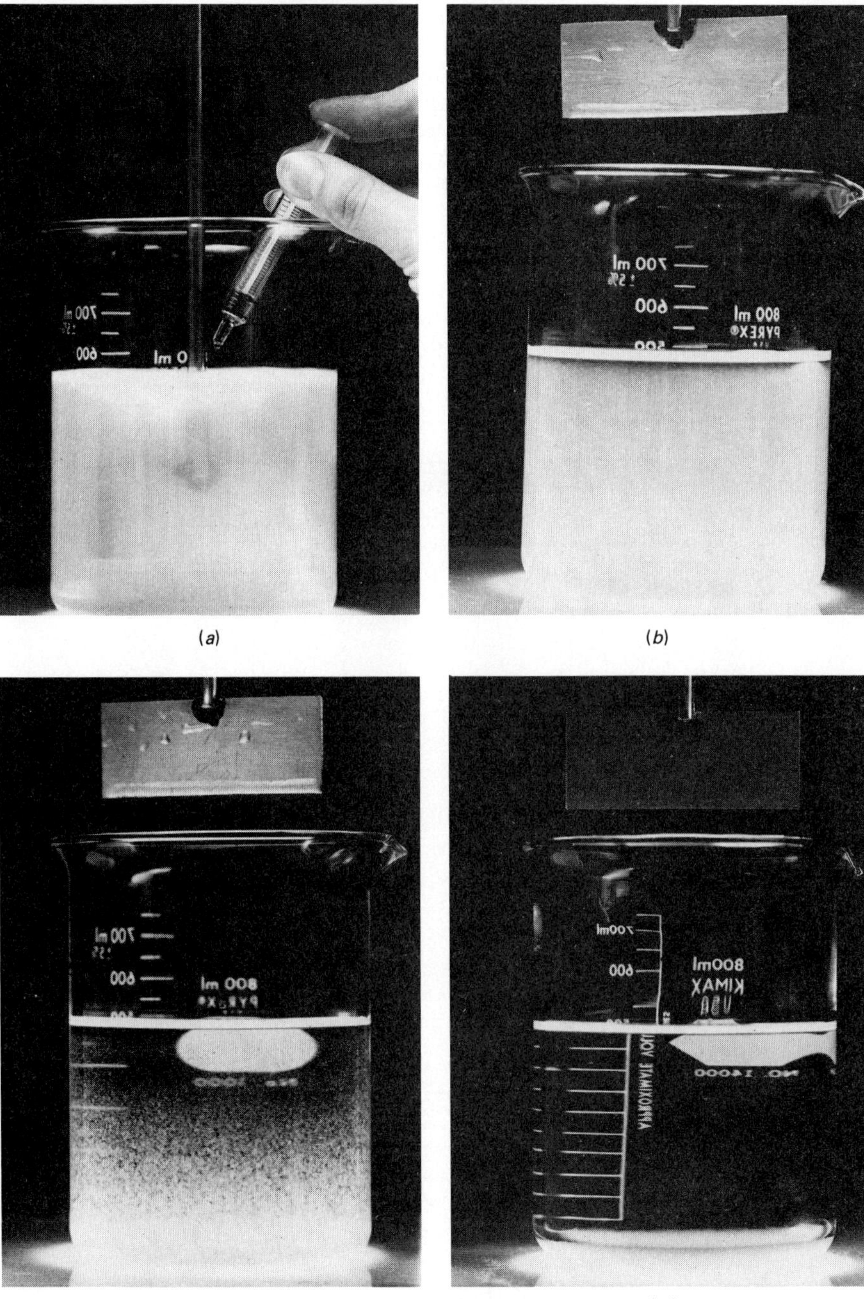

FIG. 8.9 (*a*) The coagulant is measured into a sample of turbid water with a high degree of mixing. (*b*) After coagulant addition, particle growth occurs because of charge neutralization. Additional coagulant or a high molecular flocculant may then be added. (*c*) After flocculation, at a very low stirring speed—10 to 15 r/min, for example—the sample is examined after an established time period. Note the fine, pinpoint floc which has escaped entrapment by the larger floc. (*d*) The supernatant is examined and tested after 5 to 10 min settling time, and the nature and volume of the floc may be recorded. In some cases, the floc is reclaimed for use in the next series of jar tests.

short, less than 1 min. The actual mixing time is refined as the test regimen proceeds—in essence, defining the optimum G factor. A polymer flocculant, if required, is added during the last few seconds of the rapid mix. In the slow mix period which follows, floc building proceeds until the floc becomes so big that shear forces finally overcome the bridging forces, breaking the floc apart. This limits the size of the floc. After slow mixing for an optimum period of time, found only by repeated tests (usually 5 to 20 min), the jars are allowed to settle for 5 to 10 minutes.

Jars with different chemicals or the same chemical at different dosages are run side by side and the results compared. Floc settling rate, final clarity or suspended solids, and volume of sludge produced (if measurable) are contrasted between jars. Although clarity can be judged by the eye, the more accurate standard mea-

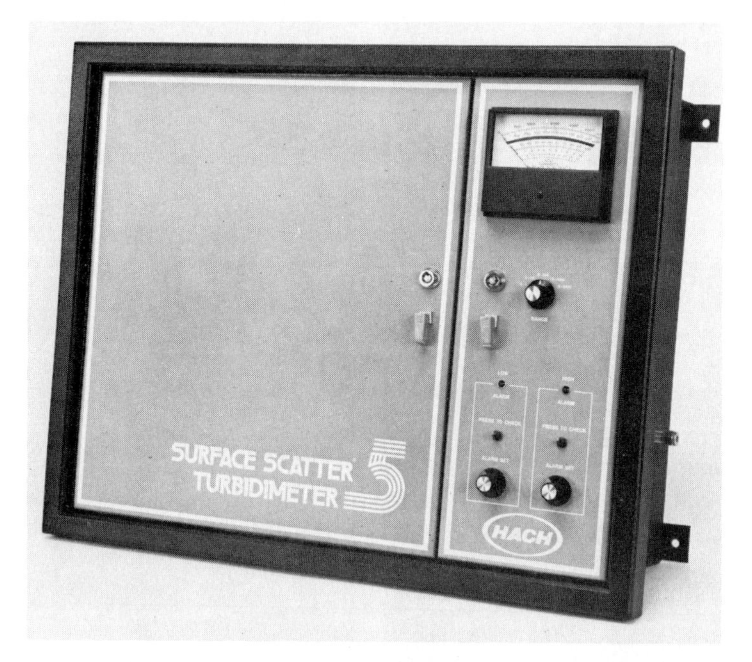

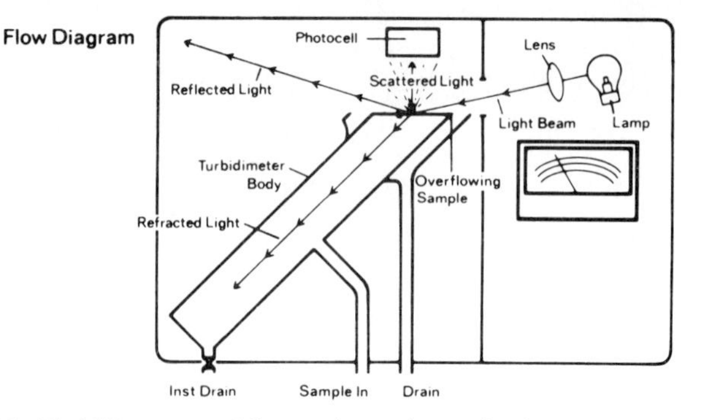

FIG. 8.10 Turbidimeter used for continuously monitoring a raw water stream to assist in the control of coagulant dosage. *(Courtesy of Hach Company.)*

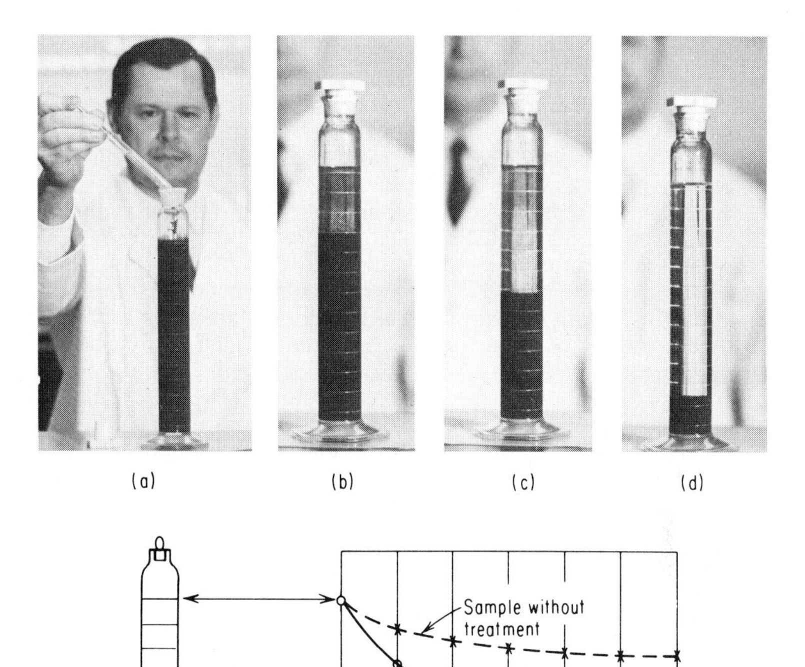

FIG. 8.11 Cylinder tests on a coal slurry. After chemical treatment, the 500-mL cylinder is inverted several times to mix and induce flocculation, and the drop in solids interface is measured at fixed time intervals, as shown above.

surement is made with a turbidimeter (Figure 8.10). Other quality tests such as pH, BOD, color, COD, and soluble metals, are run on the settled water to establish performance standards.

The cylinder test, designed to indicate how fast the suspended solids will settle, employs a 500-mL stoppered graduated cylinder, stopwatch, and labware for dosing the chemical being evaluated. The slurry sample is placed in the cylinder, chemical is added, and the cylinder gently inverted several times. Mixing is much less severe than in the jar test because solids are present at much higher levels, so that frequent collisions can occur at the lower mixing energy. After mixing, the cylinder is set upright, and the interface between the water and the settling solids observed. Time and solids level are recorded, and the data are plotted on a graph. As in the jar test, a number of analytical tests can be run on the clear water; however, rapid settling rate is usually the goal. By running coagulants and flocculants at different dosages and comparing settling rates, the most effective products are selected. Figure 8.11 illustrates the results of cylinder tests conducted on a coal slurry.

COLOR REMOVAL

The selection of an effective chemical program for the removal of color from water is accomplished by jar testing as in the case of suspended solids removal, but there are pronounced differences in results: The floc produced from coagulated organic matter is fragile, so it is very important that the jar test device be operated in a manner that duplicates mixing energy and flocculation shear corresponding to what would prevail in full-scale equipment.

For the most part, color in water is a mixture of colloidal organic compounds that represent breakdown products of high molecular weight substances produced by living cells. (See Chapters 4 and 6.) These materials are analagous to the polyelectrolytes used in water treatment—as a matter of fact, natural organics like starch have been used both as dispersants and flocculants since the earliest days of water treatment science. They are variously identified as humic acid (a polymer containing phenolic groups), polysaccharides (polymers similar to sugar and cellulose), polypeptides (protein polymers), and lignins and tannins (other relatives of cellulose). For the most part, these substances are anionic or nonionic polymers. It is not surprising, then, that they are coagulated by cationic materials, and the amount of coagulant needed is directly proportional to the color.

Alum is commonly selected as the first coagulant to be evaluated in the jar test. After the alum demand has been satisfied, excess alum produces a floc that ties the coagulated particles together. The pH range is extremely narrow, usually about 4.8 to 5.5, and variation in pH usually disperses the floc and creates a haze. Most natural colored waters are low in alkalinity, so the alum used for coagulation often destroys the natural alkalinity, and the addition of an alkali may be needed for pH adjustment. After coagulation and formation of the alum floc, an anionic polymer is usually used to strengthen the floc and to hasten sedimentation. A complicating factor is temperature; many colored waters are found in Canada and the northern United States, and the selected program must be effective at 32°F, where viscosity greatly increases shear forces and hinders sedimentation; this complicates the jar test procedure. Another potential complication is the usual need for pH correction of the finished water to render it less corrosive than water at pH 5.5. The color matter in water behaves much like an acid-base indicator, and an increase in pH usually results in a color increase—which may not be a serious problem in most cases.

Certain of the cationic polyelectrolytes are useful for the partial replacement of alum in the color removal process, permitting treatment at a higher pH and reducing the destruction of alkalinity by the high alum dosage otherwise required.

TABLE 8.5 Chemical Treatment of Colored Water
(Florida swamp drainage at 400 APHA color)

Variable	Conventional alum program	Alum-polyamine program
Alum dosage, mg/L	55	35
Aluminate dosage, mg/L	40	30
Nonionic polymer, mg/L	0.5	none
Polyamine, mg/L	none	5
Final results		
pH	5.2	6.5
Color	5–10	5–10

Table 8.5 compares the results of a conventional alum program to alum-polyamine treatment.

Wastewaters containing color, such as pulp/paper mill discharges, are sometimes even more difficult to treat than natural water sources. Experience and ingenuity are needed to screen potential coagulants; this is an area of water treatment that is still more an art than a science. An example of this was a study of a textile wastewater where the color could not be removed by alum treatment followed by pH correction with alkali—yet it could be treated by aluminate, followed by pH correction by acid. A theory was developed to explain this—but only after the investigator had worked out the solution by trial and error.

PLANT DESIGN

The flowsheet of a surface water clarification plant shows how the principles of coagulation and flocculation apply to actual plant design. Generally, the lower the suspended solids in the process stream, or the higher the required effluent clarity, the more critical is mixing to the final results. Surface water is relatively low in suspended solids and removal to a low concentration of residual solids is usually required. For this reason many water plants are designed with flash mixing and flocculation mixing. The jar test protocol of rapid and slow mixing, which works best for raw water clarification, is duplicated on the plant scale. Flash mixing is accomplished in several ways: in-line hydraulic mixing (Figure 8.12), and high-speed mixing in a small mixing basin (Figure 8.13). The coagulant is added at or

FIG. 8.12 High energy in-line mixer used for optimizing the effectiveness of polymer coagulants in water treatment systems. *(Courtesy of Mixing Equipment Company.)*

FIG. 8.13 Flash mixers are designed to disperse chemicals throughout the water instantaneously, prior to flocculation. *(Courtesy of FMC Corporation, Material Handling Division.)*

FIG. 8.14 Reel-type paddle flocculator in a municipal water treatment plant. *(Courtesy of Envirex, a Rexnord Company.)*

FIG. 8.15 Turbine-type flocculators installed in a large water treatment plant. *(Courtesy of Envirex, a Rexnord Company.)*

before the flash mix. Mixing can also be accomplished by (1) hydraulic jumps in open channels, (2) venturi flumes, and (3) pipelines with tortuous baffles. However, these do not maintain the necessary G factor at low flows, so are somewhat limited in application.

Flocculation mixing occurs in gently stirred compartments. Two common flocculator designs are the horizontal reel (Figure 8.14) and the turbine mixer (Figure 8.15). Variable speed motors may be provided to allow variation in mixing energy. Some plants employ hydraulic flocculation mixing, but this has limited use because its effectiveness falls off at reduced flows.

Silt or color removal in raw water is done using two basic treatment schemes: conventional clarification or direct filtration. The most common plant operation is conventional: rapid mix, slow mix, sedimentation, and filtration (Figure 8.16). Historically, alum has been most widely used, because the optimum pH is often below 6.0. It is fed at the head of the plant, sometimes in conjunction with alkali for pH control. Iron salts are seldom used. Because alum floc is light, a polymer flocculant is usually required to reduce the carryover of floc from the settling basin to the filters.

Polymer coagulants often are used to either replace or reduce inorganic salts. The polymer coagulant is added at the flash mix. In some water, alum can be replaced only if clay is fed to add enough to the raw water to ensure high collision frequency and to add weight to the floc. An alternate method is to return sludge from the bottom of the settling basin to the rapid mix.

When the color or turbidity is very low in the raw water, direct filtration is often practiced in water clarification plants. In a direct filtration plant, water passes through flash mixing, sometimes a flocculator, and then directly to the filters (Figure 8.17). There are so few solids in the water that the filters do not plug excessively. A polymer is used as the primary coagulant in this process, since inorganic salts add solids which could clog the filter.

FIG. 8.16 Conventional-design water treatment plant showing reel-type flocculation units, rectangular sedimentation basins in parallel, with common wall construction and sludge collection flights. Final filtration is not shown. *(Courtesy of FMC Corporation, Material Handling Division.)*

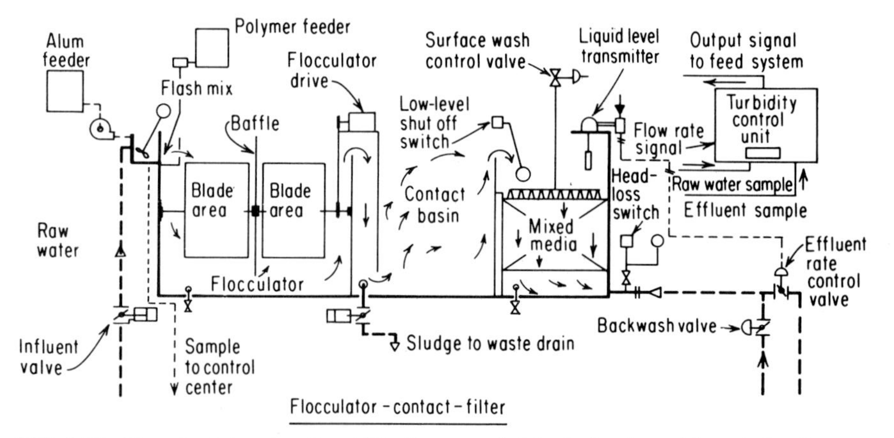

FIG. 8.17 Flow sheet showing direct filtration preceded by flash mix and flocculation for preparation of water for a small municipality. *(Courtesy of Neptune Microfloc, Inc.)*

Wastewater streams usually have higher solids than raw water and the required suspended solids removal may not be as stringent. Generally, for wastewater clarification, hydraulic mixing has been widely used in the past; but newer plants are being designed with mechanical mixing similar to raw water clarification plants to improve removal of suspended solids.

SUGGESTED READING

AWWA: *Water Quality and Treatment,* McGraw-Hill, New York, 1971.

Water Treatment Handbook, Infilco-Degrémont, Inc., Halsted Press, New York, distributors, 1979.

Nordell, Eskel: *Water Treatment for Industrial and Other Uses,* Reinhold, New York, 1961.

Vold, M. J., and R. D. Vol,: *Colloid Chemistry: The Science of Large Molecules, Small Particles, and Surfaces,* Reinhold, New York, 1964.

CHAPTER 9

SOLIDS/LIQUIDS SEPARATION

Solids/liquids (S/L) separation in water treatment includes the processes for removal of suspended solids from water by sedimentation, straining, flotation, and filtration; it also includes solids thickening and dewatering by gravity, sedimentation, flotation, centrifugation, and filtration, processes that remove water from sludge or liquids/solids (L/S) separation. Suspended solids are defined as those captured by filtration through a glass wool mat or a 0.45-μm filter membrane. Those solids passing through are considered to be colloidal or dissolved.

REMOVAL OF SOLIDS FROM WATER

Selection of the specific process or combined processes for removal of suspended solids from water depends on the character of the solids, their concentration, and the required filtrate clarity. For example, very large and heavy solids can be removed by a simple bar screen or strainer. Fine solids may require both sedimentation and filtration, usually aided by chemical treatment. The approximate relationship of particle size to the S/L separation devices used in water treatment is shown by Figure 9.1.

Straining includes such conventional devices as bar screens (Figure 9.2), traveling trash screens, and microstrainers. In some plants, instead of using screens, a device called a comminutor grinds the gross solids so they will settle and not interfere with the sedimentation equipment.

In some applications, microstraining can be used instead of granular media filtration for solids reduction. Microstraining has been used for many years for algae removal in the United Kingdom and is used as a tertiary polishing step in some wastewater treatment plants in the United States.

A typical microstraining system is shown in Figure 9.3. The unit consists of a motor-driven rotating drum mounted horizontally in a rectangular pit or vat. The rigid drum support structure has either a stainless-steel or plastic (polyester) woven screen covering fastened to it. Mesh size is normally in the 15- to 60-μm range. Sometimes a pleated configuration is used to increase surface area.

Feed passes from the inside to the outside of the drum, depositing solids on the inner surface. Water jets on top of the screen dislodge collected solids into a waste hopper. Where biological growth is a problem, the units may be equipped with uv lights. The peripheral drum speed is usually adjustable. Filtration rates are in the 10 to 30 gal/min/ft^2 range. Pressure drop through the screen is 3 to 6 in H_2O, while head loss through the complete system is in the 12 to 18 in H_2O range.

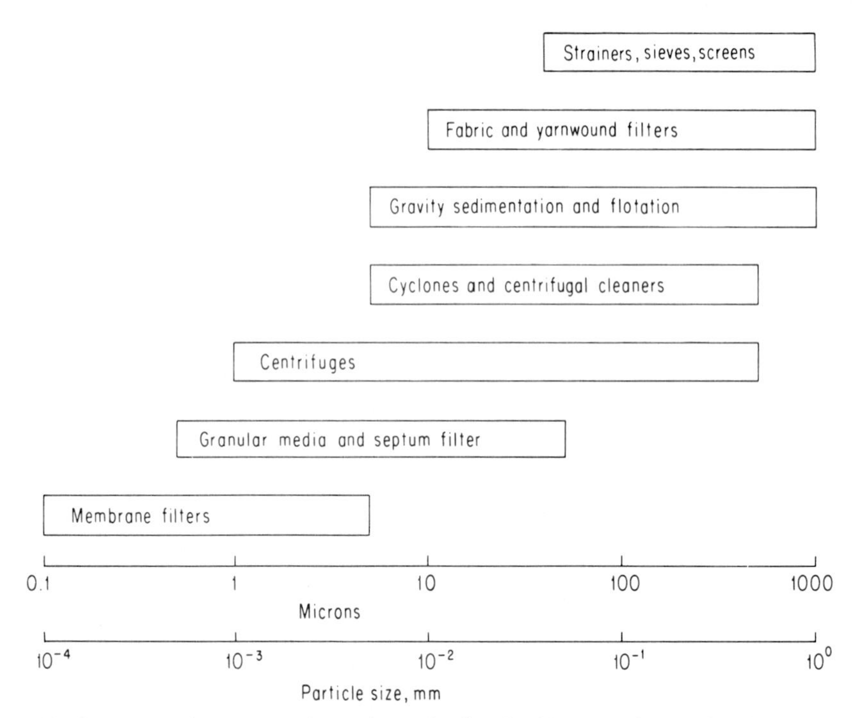

FIG. 9.1 Approximate operating regions of solids/liquids separation devices in treating water.

FIG. 9.2 Bar screen with automatic cleaning mechanism. *(Courtesy of Envirex, a Rexnord Company.)*

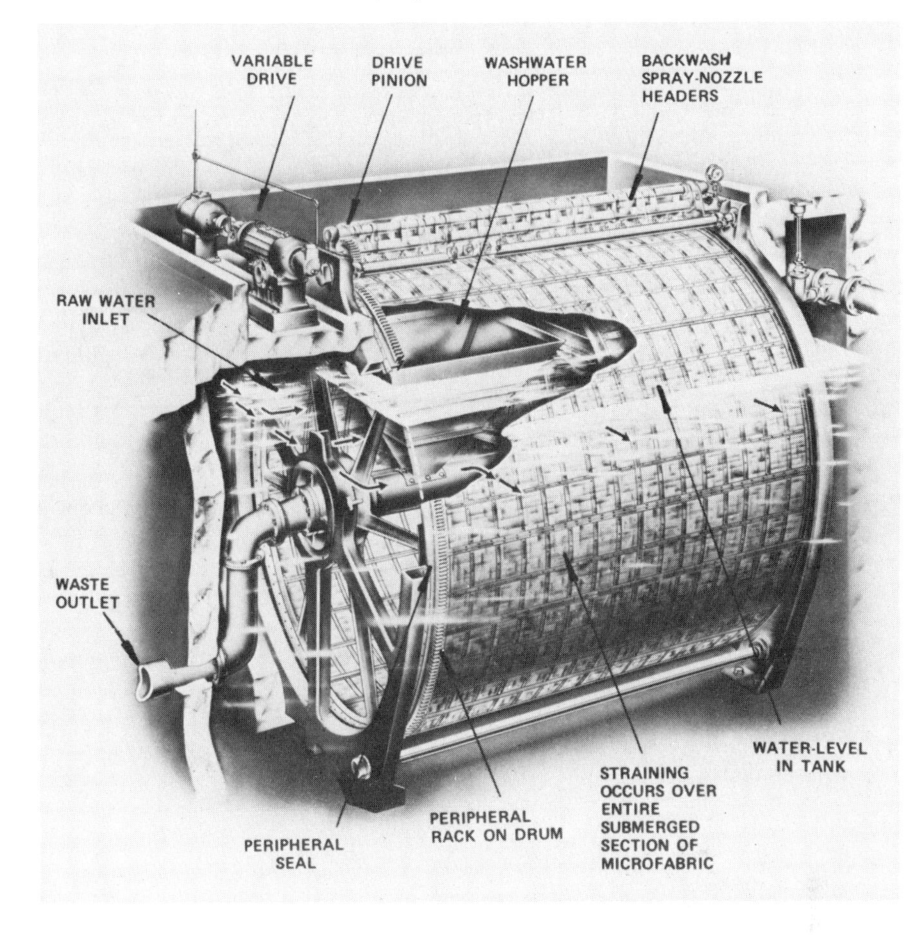

FIG. 9.3 Microstrainer used for removal of fine suspended solids from storm water or wastewater. *(Courtesy of Cochrane Division, the Crane Company.)*

Microstrainer effluent is used to backwash and to remove solids. If grease, algae, or slime are present, hot water wash and industrial cleaners may be required at regular intervals to dislodge these materials.

Sedimentation

Sedimentation is the removal of suspended solids from water by gravitational settling. Flotation is also a gravity separation, but is treated as a separate process. To produce sedimentation, the velocity of the water must be reduced to a point where solids will settle by gravity if the detention time in the sedimentation vessel is great enough. The effect of overflow rate on settling is shown in Figure 9.4.

The settling rate of particles is affected by their size, shape, and density as well as by the liquid they are settling through. As a particle settles, it accelerates until

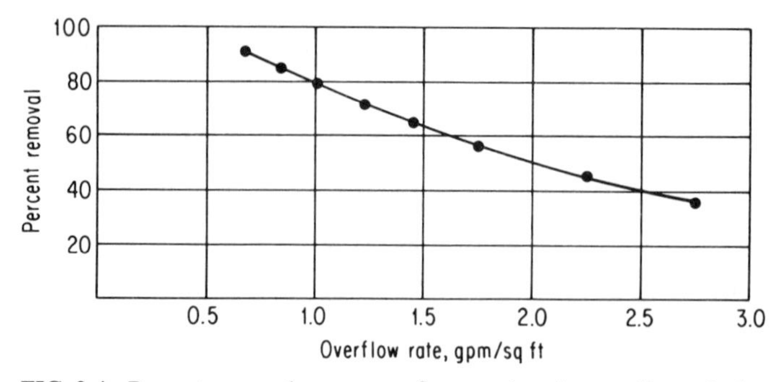

FIG. 9.4 Percent removal versus overflow rate based on settling velocity data for a specific system.

the frictional drag of its surface against the liquid equals the weight of the particle in the suspending fluid. The relationship governing particle settling is given by the following equation:

$$F \propto g \frac{(S_1 - S_2)}{V}$$

where F = impelling force
g = gravitational constant
V = volume of the particle
S_1 = density of the particle
S_2 = density of the fluid

Hindered Settling

When particles settle through a liquid in free fall, the liquid displaced by the particles moves upward and the space between the particles is so large that the counterflow of water does not interpose friction. When the particles approach the bottom of the vessel and begin to form a liquid/solid interface, their free-fall velocity is arrested. The collected solids, or sludge, now slowly compact in a process known as hindered settling. In hindered settling, the particles are spaced so closely that the friction produced by the velocity of the water being displaced interferes with particle movement. Figure 9.5 illustrates the change from free-fall to hindered settling. As sedimentation continues, the particles reach a previously established dense sludge layer; settling then becomes even slower because of the apparent increase in density in the liquid through which the particles are settling (Figure 9.6).

Settling rate is also affected by water temperature. Raising the temperature from 32 to 85°F (0 to 29°C) doubles the settling rate for a given discrete particle because both the density and viscosity of the water are reduced.

In clarification, the major objective of the sedimentation is a clear effluent water, rather than a dense underflow sludge. Clarification is used for raw water preparation and for wastewater treatment. Many process applications also use clarification, e.g., separation of fines from coal preparation tailings.

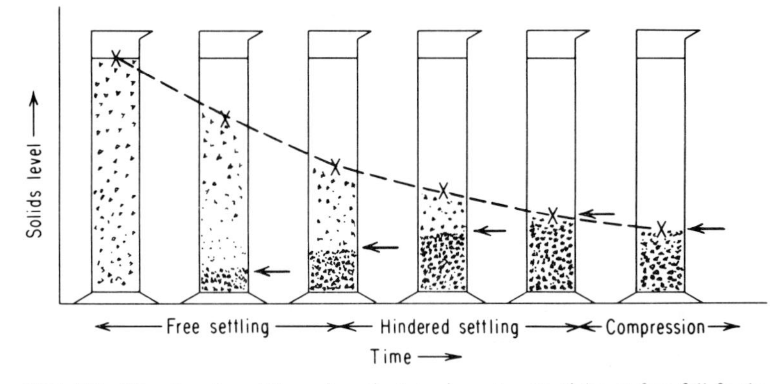

FIG. 9.5 The steps in settling of particulates in water: Particles at first fall freely through the water. As they come closer together, their rate of sedimentation is restricted, and settled sludge volume increases. In the final stages, compaction or compression becomes very slow.

FIG. 9.6 (*a*) Hindered settling is reached as particulates become so close to each other that the passages between them restrict the ability of water to escape from the sludge. (*b*) Compaction occurs naturally, but slowly, by gravity and by dehydration of the particulates; it is aided by gently moving the sludge to develop crevices behind the moving pickets or scraper blades for water release.

Gravity Clarifier Design

There are three major types of gravity clarifiers: plain sedimentation, solids contact units, and inclined plane settlers. There are several designs of plain sedimentation basins (clarifiers): center feed (the most common), rectangular, and peripheral feed. The center feed clarifier has four distinct sections, each with its own function (Figure 9.7).

The inlet section of the center feed clarifier provides a smooth transition from the high velocities of the influent pipe to the low uniform velocity required in the settling zone. This velocity change must be carefully controlled to avoid turbulence, short-circuiting, and carryover.

The quiescent settling zone must be large enough to reduce the net upward flow of water to a velocity below the subsidence rate of the solids. The outlet zone provides a transition from the low velocity of the settling zone to the relatively high overflow velocities, which are typically limited to values less than 12 to 15 gal/min per lineal foot of weir or launder.

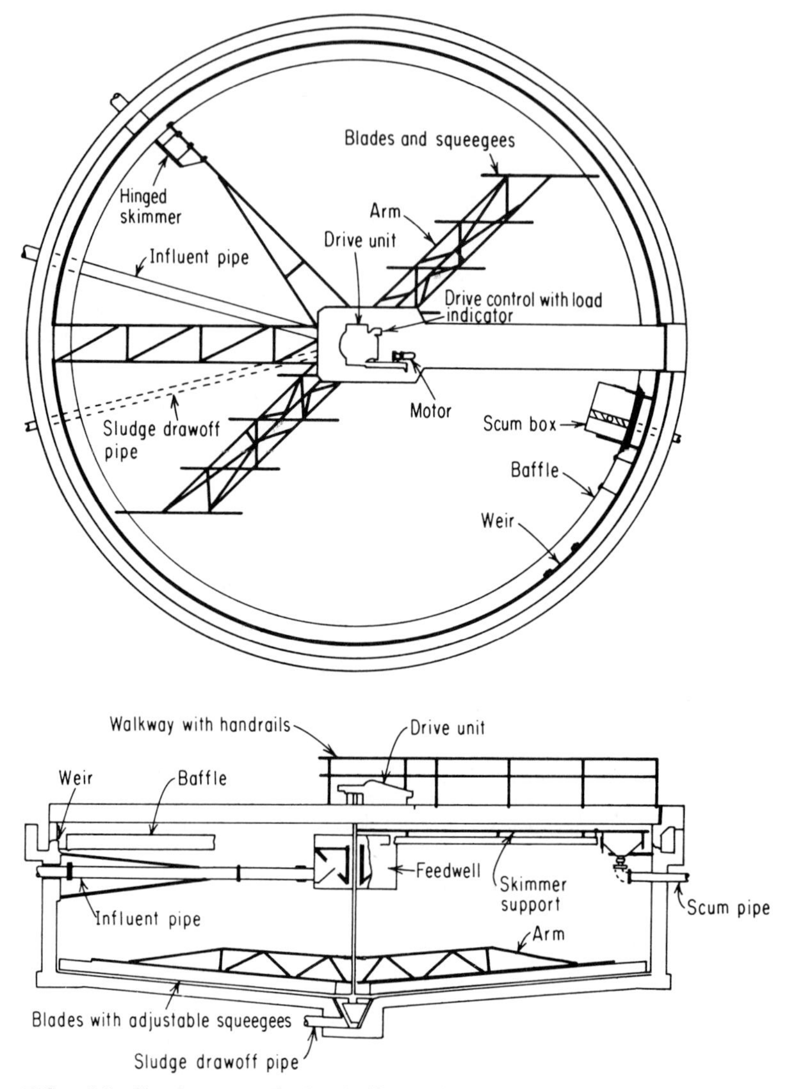

FIG. 9.7 Simple centerfeed clarifier with scum skimmer and sludge scraper. *(Courtesy of Envirotech.)*

The fourth section, the sludge zone, must effectively settle, compact, and collect solids and remove this sludge from the clarifier without disturbing the sedimentation zone above. The bottom of a circular clarifier is normally sloped 5 to 8 degrees to the center of the unit where sludge is collected in a hopper for removal. Usually, mechanically driven sludge scrapers plow or rake the sludge down the sloping bottom to the sludge hopper. Some of the collected sludge may be returned to the feedwell for seeding, if chemical treatment is applied.

The rectangular basin is somewhat like a section taken through a center feed clarifier with the inlet at one end, and the outlet at the other. A typical rectangular

FIG. 9.8 Side-by-side rectangular clarifiers with common wall, each with a traveling bridge sludge collector. *(Courtesy of Walker Process Division, Chicago Bridge & Iron Company.)*

clarifier has a length to width ratio of approximately 4:1. Sludge removal in rectangular sedimentation tanks is normally accomplished by a dual purpose flight system. The flights first skim the surface for removal of floating matter and then travel along the bottom to convey the sludge to a discharge hopper. However, the surface skimming is not a common feature, being used almost exclusively for wastewater, rather than raw water treatment. This flight system must move slowly to avoid turbulence, which could interfere with settling. An advantage of this type of clarifier is that common walls can be used between multiple units to reduce construction costs (Figure 9.8).

The peripheral feed (rim feed) clarifier (Figure 9.9) attempts to use the entire volume of the circular clarifier basin for sedimentation. Water enters the lower

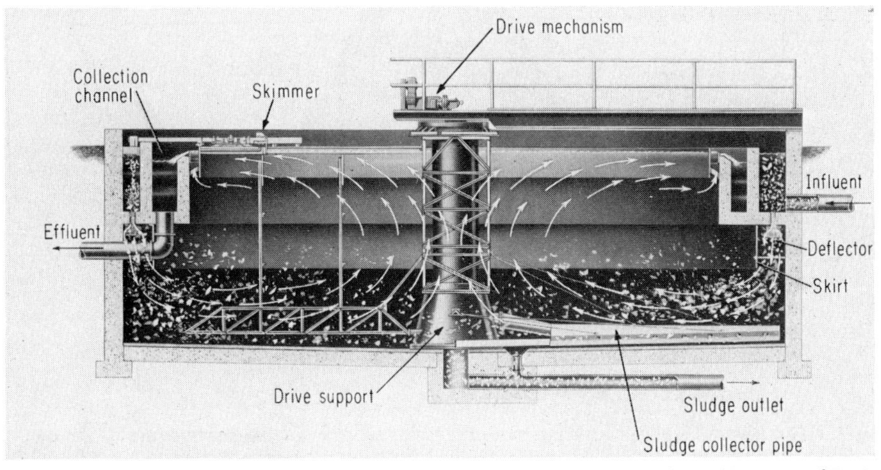

FIG. 9.9 Peripheral feed clarifier with sludge pipe and scum removal device. *(Courtesy of Envirex, a Rexnord Company.)*

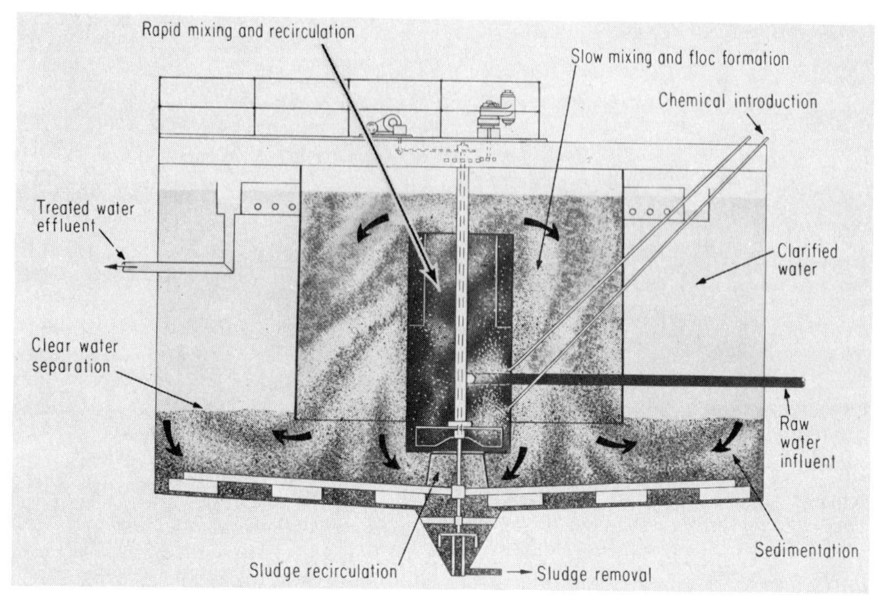

FIG. 9.10 Slurry recirculation design clarifier. *(Courtesy of Ecodyne Corporation.)*

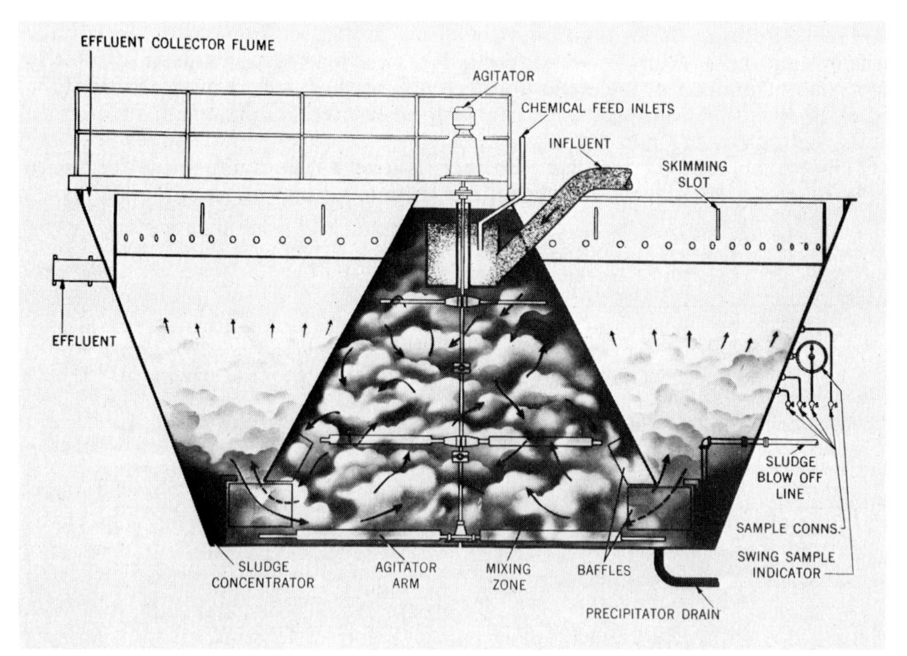

FIG. 9.11 Sludge blanket clarifier providing increasing area for the water rising in the outer annulus, resulting in reducing velocity to match sludge settling rate. *(Courtesy of the Permutit Company.)*

section at the periphery at extremely low velocities providing immediate sedimentation of large particles. The velocity accelerates toward the center, then drops as the flow is reversed and redirected to a peripheral overflow weir. Since the flow pattern depends entirely on hydraulics, this type of clarifier is sensitive to temperature changes and load fluctuations. Recirculation of sludge is very difficult in the peripheral feed design.

A second major category of clarifier is the solids contact unit, available in two basic types: the slurry recirculation clarifier and the sludge blanket clarifier (Figures 9.10 and 9.11). Both of these combine chemical mixing, flocculation, and clarification in a single unit. In the mixing zone of a solids contact clarifier, the solids concentration may be as much as 100 times that of the simple clarifier. This high solids level greatly increases the rate of chemical destabilization reactions and particle growth. Because of these features, the solids contact units are usually used in lime softening. In the slurry recirculation unit, the high floc volume is maintained by recirculation from the flocculation zone to the clarification zone. In the blanket-type clarifier, the floc solids are maintained in a fluidized blanket through which the water must flow. Because of the increased solids in a solids contact unit, clarifier size may be reduced. The even distribution of the inlet flow and the vertical flow pattern of this type clarifier provides better performance than standard horizontal flow clarifiers. In passing through the sludge blanket, the larger floc settles to the bottom by gravity and the remaining fine floc is removed by straining and adsorption.

Variable speed mixers are used to control flocculation and solids concentration in the reaction zone. The solids concentration in the reaction zone is maintained by bleeding solids out of the system to balance those coming in with the raw water and the solids produced by chemical reaction. Sludge removal can be accomplished either by a sludge blowoff pipe as in Figure 9.12, or by a conventional rake and pump system as in Figure 9.13. Balancing the solids budget—solids input versus output—is the most difficult aspect of controlling a sludge blanket unit.

FIG. 9.12 Sludge collection pipe (arrow) for periodic removal of sludge to a collection hopper. *(Courtesy of Envirex, a Rexnord Company.)*

FIG. 9.13 Circular sludge collectors, with rakes designed for corner cleaning. *(Courtesy of FMC Corporation.)*

Control of Flow Pattern

Two major problems with gravity clarifiers are short-circuiting and random eddy currents. These are related in that both can be induced by changes in flow, inlet composition, temperature, and specific gravity. They are both aggravated by localized sludge deposits, which block the normal flow pattern.

It is quite obvious that in the conventional circular clarifier, water must be relatively stagnant in a significant proportion of the total volume in detention. Notably, there is no flow in the annular space below the peripheral overflow launder. The actual detention can be determined by measuring effluent chloride concentrations or conductivity at timed intervals after injecting a measured slug of salt into the feed. The results of this should be discussed with the equipment designer if short-circuiting is suspected.

Eddy currents are usually readily observed, showing up as an apparent boiling of the sludge. Often these disturbances can be traced to weather conditions, such as high winds or bright sunlight which either heats the sludge unevenly or encourages algal production of O_2.

Floc separators have been a solution to these problems in many gravity clarifiers. These units, made up in modules for installation in a variety of clarifier designs, add just enough frictional resistance to flow to even out the hydraulic pattern and eliminate the problems of short-circuiting and eddy currents. Figure 9.14 shows a typical installation of floc barriers in a sludge-blanket type clarifier. Figure 9.15 shows the design of one kind of floc barrier module.

In most gravity clarifiers, the mean water depth through which sludge particles must fall is on the order of 5 ft or more. The time required for the sludge to fall this distance is a critical factor in limiting the clarifier capacity. Two similar modifications to the standard design of gravity clarifiers reduce the distance of fall

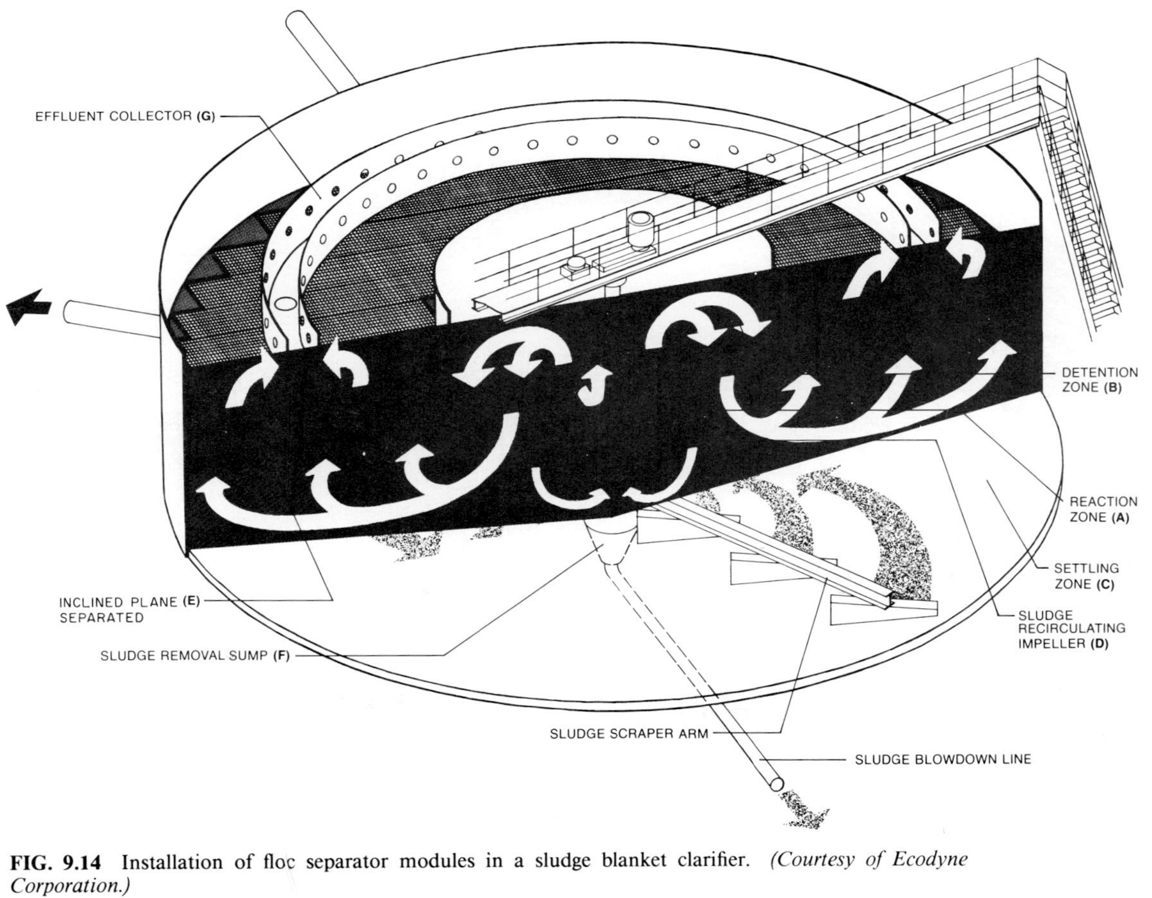

FIG. 9.14 Installation of floc separator modules in a sludge blanket clarifier. *(Courtesy of Ecodyne Corporation.)*

FIG. 9.15 Plastic module of inclined plates simulating inclined tubes. These modules help equalize water distribution. *(Courtesy of Neptune Microfloc, Inc.)*

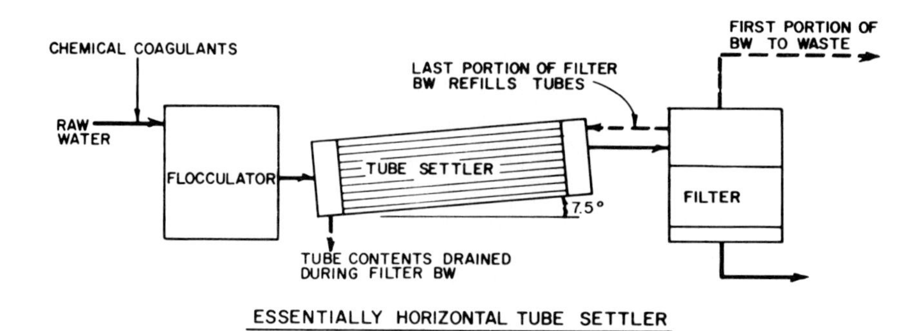

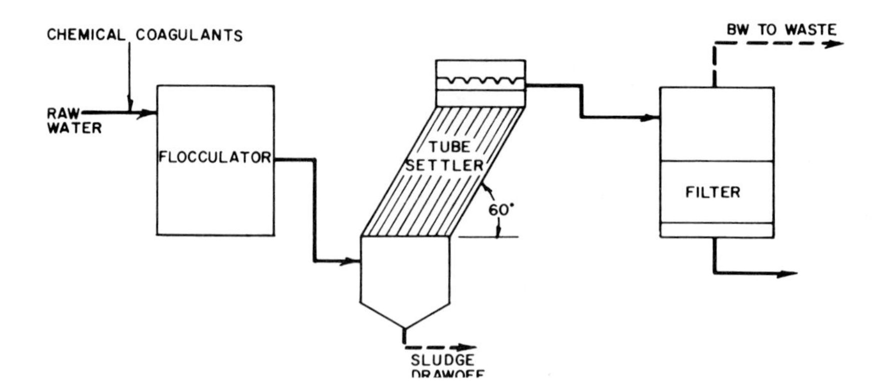

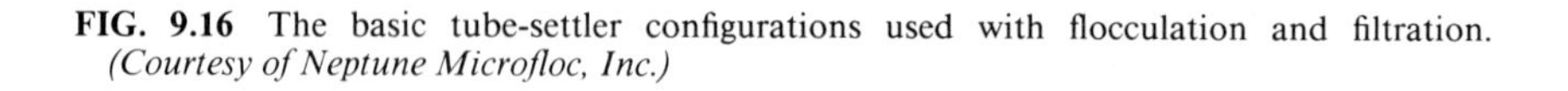

FIG. 9.16 The basic tube-settler configurations used with flocculation and filtration. *(Courtesy of Neptune Microfloc, Inc.)*

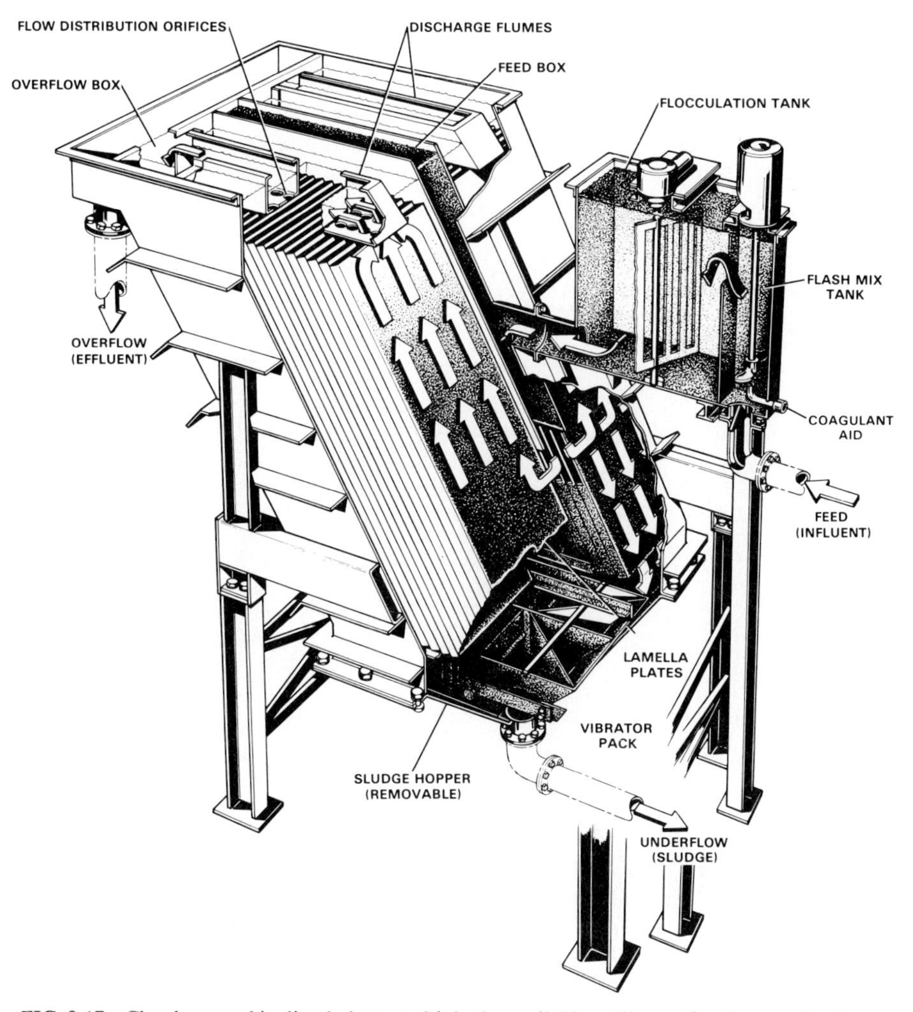

FIG. 9.17 Closely spaced inclined plates multiply the available settling surface in a small volume and reduce installation space. *(Courtesy of Parkson Corporation.)*

from feet to inches, increasing the effective rise rate and radically reducing space requirements for clarification. These are the tube-settler and the inclined plane settlers.

The so-called tube-settler may in fact be a series of inclined tubes, somewhat like a heat exchanger bundle connected at the inlet to a flocculation chamber and at the outlet to a clear well (Figure 9.16). The angle of inclination is varied to suit the required duty. It is affected by the concentration and nature of the solids, and by the ensuing processes of water filtration and sludge thickening. The tube-settler may also be a vessel packed with floc separators or with parallel inclined plates.

The inclined plate separator (Figure 9.17) is more intricate, but the same principle applies in that the sludge particles have a very short settling distance, and the accumulated sludge is induced to flocculate and concentrate as it rolls down the inclined surface. These units are ideally suited for localized treatment of individual waste streams in cramped locations. An example is the installation of these separators for treating chemical plant wastes (Figure 9.18).

A final example of a gravity clarifier, which is a modification of the plain sedimentation unit, is the rectangular drag tank (Figure 9.19). This is designed for

FIG. 9.18 Installation of an inclined plate settler in a chemical plant, clarifying wastewater. *(Courtesy of Parkson Corporation.)*

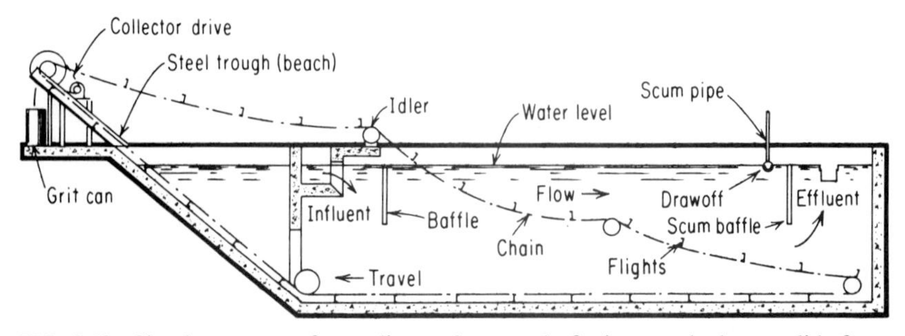

FIG. 9.19 Simple separator for settling and removal of gritty, nonhydrous solids from wastewater. Some dewatering occurs on the incline above the water line. *(Courtesy of FMC Corporation.)*

FIG. 9.20 Steel mill scale pit, provided with oil skimmer and sludge collector. *(Courtesy of FMC Corporation.)*

removal of dense solids, such as granulated slag from a foundry cupola. As the solids are dragged from the vessel by flights moving up the beach, water drains off, producing a relatively dry mass. Fragile solids are broken down by the movement of the flights up the beach, so the drag tank is limited in the type of solids that can be separated. The detention is usually short, and overflow clarity rather poor, even when chemicals are used for coagulation and flocculation. However, the drag tank can be modified, such as by providing an hour or so of detention and preflocculating the feed, to deliver clear water, as in handling scale pit solids in a steel mill (Figure 9.20).

Flotation Clarification

Solids can also be removed from water using an air-flotation clarifier such as the one shown in Figure 9.21. In this unit, light solids are floated to the surface by air bubbles and skimmed off while heavier solids are settled and removed in the normal fashion.

Air flotation has been used for many years in the mining industry for concentrating mineral ores, and in the paper industry for treating white water for fiber recovery and water clarification. The use of dissolved air flotation has broadened to include treatment of oily waste from refineries, petrochemical plants, steel rolling mills, automotive plants, and railroad terminals. In these industries, the oil

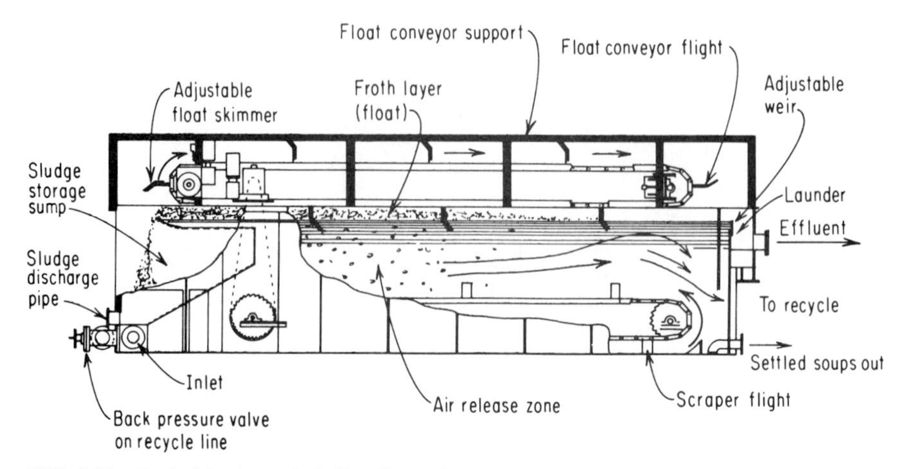

FIG. 9.21 Typical horizontal air flotation unit.

in the waste may coat solid particles, giving them a tendency to float rather than settle. In these applications, the air flotation clarifier is often preceded by an American Petroleum Institute (API) separator for the removal of free oil (Figure 9.22).

Another important application of dissolved air flotation is food industry waste treatment. Meat and seafood processing plants, canneries, and wineries have significantly reduced BOD and suspended solids using air flotation equipment.

In flotation clarification, the waste flow is usually pressurized and supersaturated with air. When the pressure is released, air comes out of solution, forming microbubbles, which float the solids to the surface. In some cases, instead of pressurizing the influent, a portion of the effluent is recycled through an air-saturation tank to meet the feed stream, as discussed later.

In treating wastes containing solids which tend to float, air flotation may be so effective that it may reduce clarification time to 15 to 20 min of detention time, compared to the several hours typical of gravity sedimentation. As with gravity clarifiers, it is often necessary to add coagulant or flocculant chemicals such as ferric chloride and alum to flotation units to aid in floc formation, using lime if needed for pH adjustment. Polyelectrolytes have been gaining popularity for this

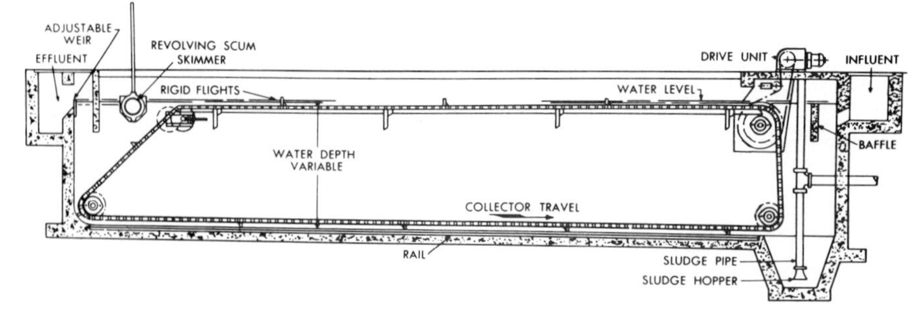

FIG. 9.22 Typical design of API separator. *(Courtesy of Envirex, Inc., a Rexnord Company.)*

application, and in almost all cases, they increase the efficiency of a flotation clarifier. Cylinder tests, pilot plant tests, or both are used to select the best chemical program.

Theory

The amount of air that can be dissolved in water is determined by Henry's law, which states that for nonionizing gases of low solubility, the volume dissolved in water varies with absolute pressure. At 75 lb/in^2 abs (5 bars), for example, 5 times as much air can be dissolved in water as at atmospheric pressure. The quantity of gas that will theoretically be released from solution when pressure is reduced to atmospheric is:

$$G_R = G_A\left(\frac{P_A}{14.7} - 1\right)$$

where G_R = gas released, mg/L
G_A = gas solubility at atmospheric pressure, mg/L (see Table 9.1)
P_A = absolute pressure in saturation tank, lb/in^2 abs

The above must be corrected for the efficiency of gas absorption in the saturation vessel, which is influenced by mixing and detention time. The efficiency varies in the range of 40 to 60%, so the gas released would typically be about half of that determined by the above formula.

TABLE 9.1 Gas Solubility at Atmospheric Pressure in mL/L

Temperature	O_2	N_2	Air
0°C (32°F)	10.3	18.0	28.3
10°C (50°F)	8.0	15.0	23.0
20°C (68°F)	6.5	12.3	18.8
30°C (86°F)	5.5	10.5	16.0
40°C (104°F)	4.9	9.2	14.1
50°C (122°F)	4.5	8.5	13.0

The air bubbles formed in a DAF unit normally carry a slight negative charge. Depending on the type of particulate matter and the degree of agglomeration of the solids, the air bubbles can attach themselves by any of the following mechanisms:

1. Simple adhesion of the air bubble to the solid surface. This can occur either through collision or by formation of the air bubble on the particle surface.
2. Trapping of air bubbles under sludge floc, such that the waste particle "takes a ride" to the surface. Sometimes referred to as "screening," this implies that there need be no real attachment of air bubbles to sludge particles to accomplish flotation.
3. Incorporation of air bubbles into floc structures. This is believed to be the most efficient mode of air usage because there is less chance of floc separation from

the air bubble. This process is encouraged by the use of polyelectrolytes, which, when applied correctly, will cause the flocculation of sludge particles at the sites where air bubbles are coming out of solution.

Since the net specific gravity of the air-solid or air-liquid particles is less than that of water, they rise to the surface. There, they consolidate to form a float, which can be removed by mechanical skimmers. The clear subnatant is withdrawn from the bottom of the unit. Figure 9.21 shows a cross section of a typical horizontal unit.

Usually the size of a flotation unit is selected on the basis of solids loading on the bottom, expressed as pounds per hour per square foot of floor area. Depending on the nature of the solids, the floor loading ranges from 0.5 to 5.0 lb/h/ft^2 (2.5 to 25 kg/h/m^2).

The hydraulic load and inlet solids concentration must be balanced to arrive at an acceptable floor loading. For instance, a unit designed to handle a floor loading of 2 lb/h/ft^2 can handle a hydraulic loading of 0.8 gal/min/ft^2 (0.33 m^3/min/^{2}) at 0.5% solids (5000 mg/L) or 1.6/gal/min/ft^2 at 0.25% solids (2500 mg/L) with about equal efficiency. A lower flow rate should be maintained as a safety factor to allow for fluctuations in concentrations. A high effluent solids concentration may be the result of an overloaded unit; when this happens the unit feed should

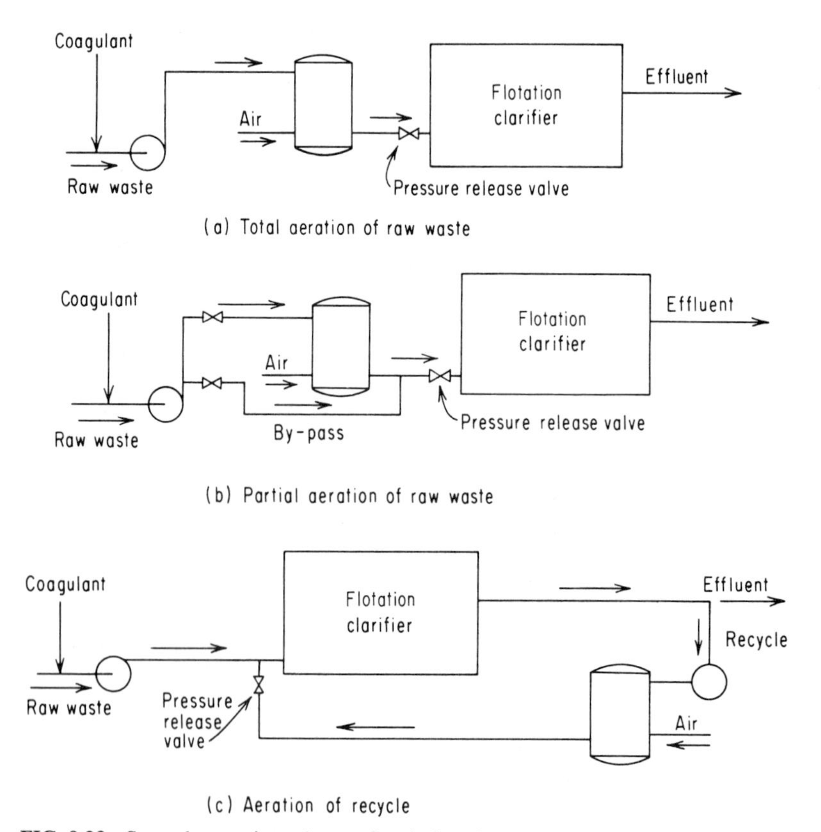

FIG. 9.23 Several operating schemes for air flotation clarification.

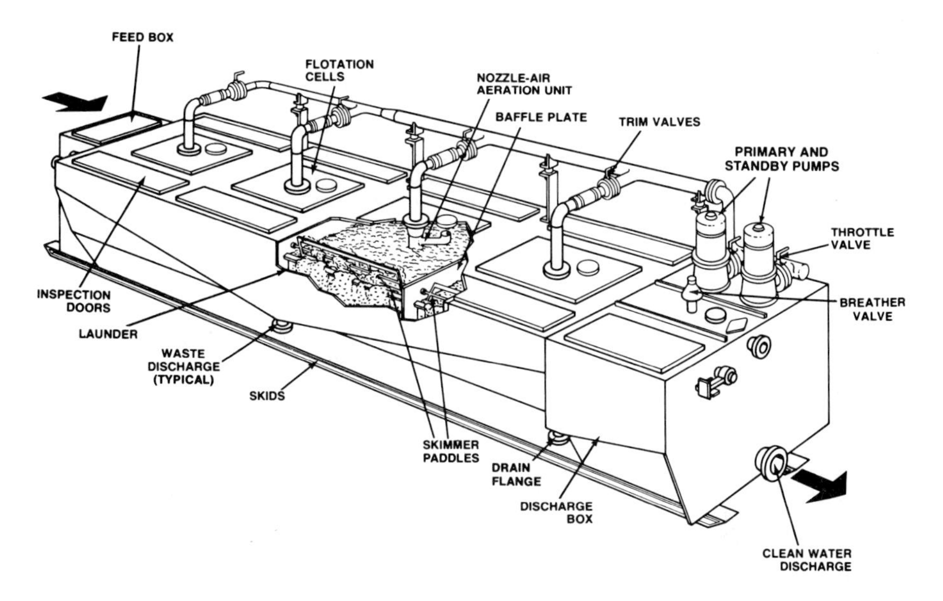

FIG. 9.24 Dispersed air flotation, typically used for clarification of oily waste. *(Courtesy of Wemco Division, Envirotech Corporation.)*

be closed and effluent recycled to allow the tank to clear. If this is a persistent problem, it may be possible to increase the amount of dissolved air by increasing the pressure or the flow rate of the pressurized stream. It may be necessary to decrease the unit feed stream to compensate for the overload situation. If the unit is shut down on a daily basis, clear water should be recycled routinely prior to shutdown to remove suspended and floated material.

Types of Flotation Systems

There are three basic types of dissolved air flotation systems in use (Figure 9.23). In direct aeration, the entire waste stream is pressurized and aerated. In this case, the material to be separated must be able to withstand the shearing forces in the be closed and effluent recycled to allow the tank to clear. If this is a persistent problem, it may be possible to increase the amount of dissolved air by increasing the pressure or the flow rate of the pressurized stream. It may be necessary to decrease the unit feed stream to compensate for the overload situation. If the unit is shut down on a daily basis, clear water should be recycled routinely prior to shutdown to remove suspended and floated material.

Effluent recycle is recommended where fragile floc is formed. This floc would be destroyed by the intense mixing which occurs in the pressurization system. Gas is dissolved in the recycle stream. This stream is then combined with the feed stream at a point where the pressure is released. Mixing of these streams prior to entering the flotation zone results in intimate contact of the precipitated gas and suspended solids to effect efficient flotation. Effluent recirculation is required when light flocculent solids such as biological or hydroxide sludges are to be thickened. Flotation areas must be large when effluent recirculation is employed, since hydraulic loading is based on both feed and recycle flows.

Flotation is also practiced by application of dispersed air into a vessel containing water with oily or solid particulates. Air is mechanically entrained and dispersed through the liquid as fine bubbles in contrast to release of dissolved gas from solution. The dispersed air flotation design is especially suitable for treating oily wastewater (Figure 9.24). It is widely used in water-flooding for crude oil recovery, where natural gas is used in place of air.

Filtration

Granular Media Filtration. Granular media filtration is generally applicable for removal of suspended solids in the 5 to 50 mg/L range where an effluent of less than 1 JTU (Jackson turbidity unit) is required. Sand filters have been used for many years as a final polishing step in municipal and industrial water plants where the clarifier effluent contains 5 to 20 mg/L of suspended solids. In areas where a very low turbidity raw water source is available, e.g., the gravel-bed rivers of the Rocky Mountains that carry snow melt, both industrial and municipal plants use granular media filtration, with minimal chemical treatment, as the only treatment process for solids removal. Granular media filters are also being used to filter cooling water sidestreams to reduce suspended solids buildup where effluent clarity is not critical. Granular media filters may handle suspended solids up to 1000 mg/L and provide about 90% removal.

A number of mechanisms are involved in solids removal by filtration, some physical and others chemical. These filtration mechanisms include adsorption and straining.

Adsorption is dependent on the physical characteristics of the suspended solids and the filter media. It is a function of filter media grain size and such floc properties as size, shear strength, and adhesiveness. Adsorption is also affected by the chemical characteristics of the suspended solids, the water, and the filter media. The amount of surface exposed for adsorption is enormous—about 3000 to 5000 ft^2 per cubic foot of media. Straining, which occurs in all granular media filters, is the major factor controlling the length of filter runs. A major objective of good filter design is to minimize straining since it leads to rapid head loss. This occurs because straining causes cake formation on the surface of the filter bed (particularly on sand filters), with the deposited cake then acting as the filter media. The filter media in essence become finer as the cake forms, and head loss increases exponentially with time.

Of the several types of filtration media used to remove suspended solids, the most common is silica sand, but crushed anthracite is also widely used. When a single medium such as silica sand is used, it will classify in the filtration vessel according to size, the smallest particles rising to the top. When water flows downward through the sand, which is the traditional path, solids form a mat on the surface, and filtration typically occurs in the top few inches. The sand is cleaned by upward washing with water or with water and air (backwash), and this hydraulically classifies the bed, keeping the finest material on top. If the sand could be loaded into a filter with the larger grains at the top and the smaller at the bottom, this coarse to fine grading would allow in-depth penetration. The increased solids storage would allow longer filter runs. However, since backwashing fluidizes the bed, the washed sand would again return to a fine to coarse grading.

If a single medium bed is used, the only path to coarse to fine filtration is upflow. Water is applied into the bottom of the bed. Solids can penetrate the coarser grain medium, resulting in deeper bed filtration. Backwashing occurs in the same direction as the filtration. The bed is classified fine at the top to coarse at the bottom. Upflow filters operate at up to 5 gal/min/ft^2. Some more sophisticated designs combine upflow and downflow filtration and provide extra facilities for bed cleaning, resulting in a system that competes with larger clarifiers for treatment of turbid river waters.

Typical single medium filters operate downflow at 2 gal/min/ft^2 of bed area in potable water service, and up to 3 gal/min/ft^2 in industrial filtration. The filter bed is 24 to 30 in deep, supported on several courses of graded gravel (Figure 9.25*a* and *b*).

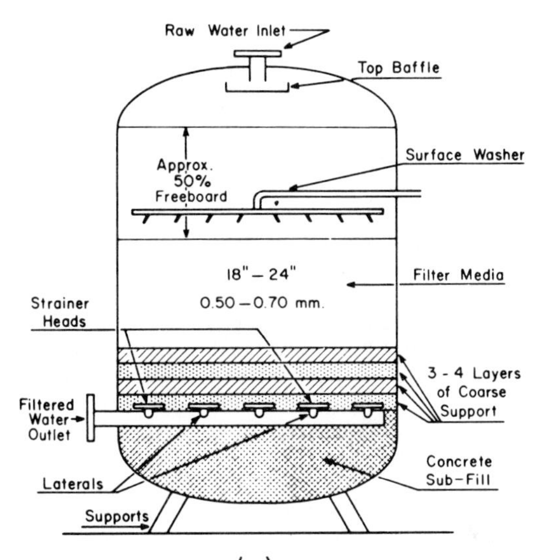

(a)

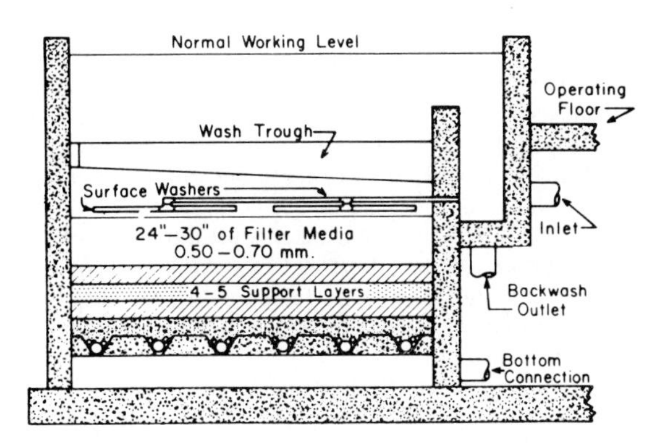

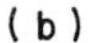
(b)

FIG. 9.25 Schematic details of conventional, municipal-type filtration units. (*a*) Pressure filter, vertical cylindrical design, fabricated of steel. Usually limited to 10 ft 0 in diameter. (*b*) Gravity filter, usually of concrete construction; used in larger municipal and industrial plants.

Silica sand normally has a grain size of 0.5 to 0.8 mm. Anthracite is usually about 0.7 mm. Smaller grains filter better, but filter runs are short. Larger grains allow longer filter runs, but if the flow is too high, hydraulic breakthrough will occur. A coarse filter media will produce acceptable effluent and reasonable filter runs if its depth is increased.

Multimedia Filter Beds

A stacked media bed or two layers (dual media) is one answer to providing coarse to fine filtration in a downflow pattern. The two materials selected have different grain sizes and different specific gravities. Normally, ground anthracite is used in conjunction with silica sand. The anthracite grains with a specific gravity of 1.6 and a grain size of 1 mm settle slower than sand with a specific gravity of 2.65 and a grain size of 0.5 mm, so the coarse anthracite rests on top of the fine sand after backwashing. In a typical dual media bed, 20 in of anthracite is placed above 10 in of sand. The coarse anthracite allows deeper bed penetration and provides longer filter runs at higher filter rates. The finer sand polishes the effluent. Under normal conditions, this dual media can produce acceptable effluent at flow rates up to 5 gal/min per square foot of bed area.

Just as coarse-to-fine dual media is more effective than a single medium filter, further improvement can be gained by introducing a third, smaller, heavier media under the sand. Garnet with a specific gravity of 4.5 and a very fine grain size settling faster than the silica sand can be used as the bottom layer. A typical multimedia contains 18 in of 1.0 to 1.5 mm anthracite, 8 in of 0.5 mm silica sand, and 4 in of 0.2 to 0.4 mm garnet. This filter operates at higher flow rates and provides deeper penetration and longer filter runs than a single or dual media filter.

To design a filter for maximum performance, the first consideration is the desired quality of effluent. The selection of filter design required to produce an effluent of 0.1 JTU is different from that required to produce 1.0 JTU.

Flow rate through a filter is critical, since it limits the throughput and dictates the number of filters required. Generally, as flow rate increases, penetration into the filter increases. The flow rate is limited by the head available and the media size. As the media starts to load with solids, the net velocity at a given flow rate increases until shear forces tear the solids apart and they escape into the effluent. Most filters are designed to be backwashed before this breakthrough occurs at a point determined by head loss. Typically, single media filters are backwashed when the head loss reaches about 10 ft. In deep bed filtration, a terminal head loss of 15 to 20 ft is tolerable.

The gradual increase of head loss across a granular filter as solids accumulate in the bed has been used as the means to actuate backwash of the filter bed. This has led to development of the automatic-backwash filter (Figure 9.26) to permit reliable operation of a battery of such filters in remote locations where operator attention may be infrequent.

In a finer grain media, since solids removal is primarily accomplished in the first few inches, increasing bed depth is of little value except for improving hydraulic distribution. But in coarse filters where penetration is wanted, the coarser the media the deeper must be the bed for equivalent effluent quality.

Water temperature affects filter performance due to viscosity. At 32°F, water viscosity is 44% higher than at 72°F. Backwashing, on the other hand, improves with cold water, since increased viscosity more effectively scours the bed to

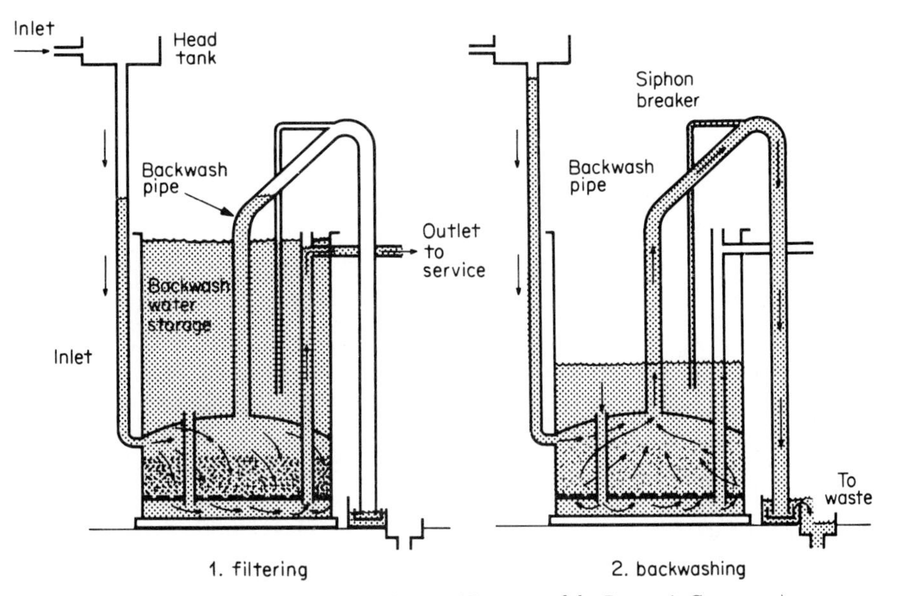

FIG. 9.26 Automatic backwash gravity filter. *(Courtesy of the Permutit Company.)*

remove solids. Floc formation is much slower at low temperatures so the filterability at a given plant may vary seasonally. In the summer, floc may stay on the surface, but penetrate deeply into the filter in the winter.

The best method of determining filter media selection for a chemical coagulation/flocculation program is by operation of a pilot test column. Chemicals can be fed directly to the column or into a separate flash mix tank ahead of the column. Various laboratory tests have been used to determine filterability, but none are as accurate as the pilot test column.

Granular media filters have been used for treatment of oil-bearing waters. An example is the use of anthracite-bed filters for removal of oil from industrial plant condensates. In this case, a slurry of aluminum hydrate floc is formed by reacting alum with sodium aluminate; a portion of this (the precoat) is applied to the filter bed at a rate of about 0.2 lb per cubic foot of bed volume, with the effluent discharged to waste during this application, followed by a short rinse to eliminate by-product salts; the balance of the slurry is then fed directly to the incoming oily condensate (the body feed) at a rate of about 2 to 3 parts of floc per part of oil. This is a modification of the usual feed of a coagulant directly to the feed stream for charge neutralization.

Significant advances in engineering design have produced sophisticated filtration systems that compete with sedimentation and flotation devices for removal of solids from water even at high suspended solids concentrations. A design of a continuous filter with a moving, recycling sand bed is shown in Figure 9.27. This type unit has been used in such diverse applications as the direct filtration of river water and the removal of oily solids from scale-pit waters in steel mills. Performance is improved by the application of low dosages of polyelectrolyte to the feed stream. Pilot plant testing is required in many potential applications to tie down all of the cost and performance data needed in choosing between direct filtration, sedimentation, flotation, or a combination of these.

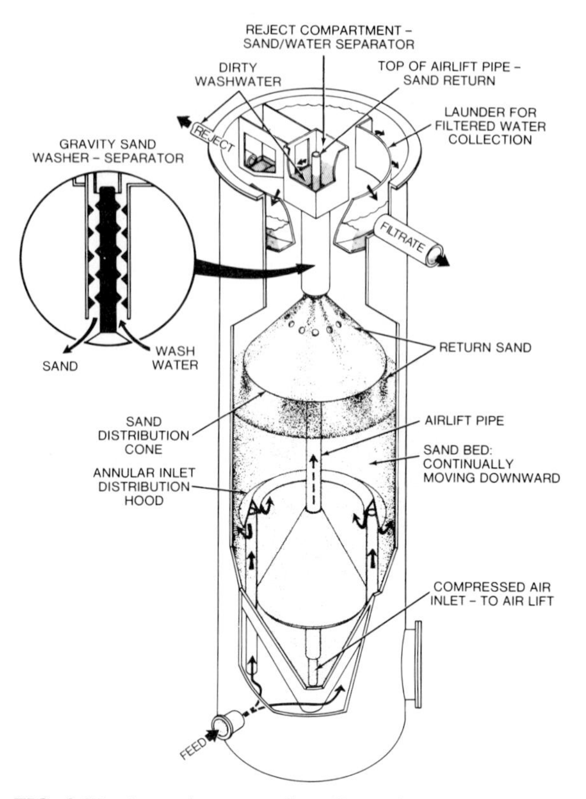

FIG. 9.27 A continuous upflow filter with a recycling sand filter media and continuous cleaning of a portion of the media. *(Courtesy of Parkson Corporation.)*

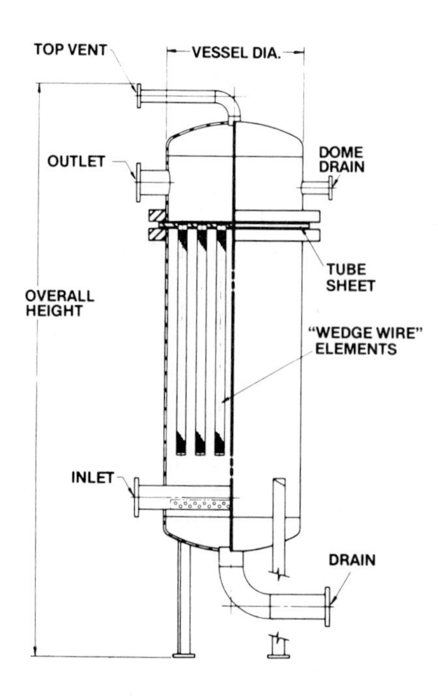

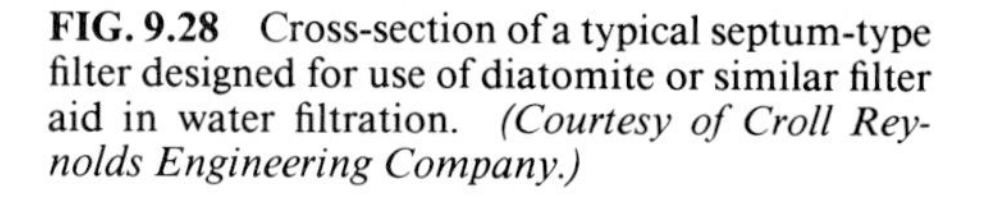

FIG. 9.28 Cross-section of a typical septum-type filter designed for use of diatomite or similar filter aid in water filtration. *(Courtesy of Croll Reynolds Engineering Company.)*

Septum Filters. Where suspended solids concentrations are very low, septum filtration can be used. These filters are often referred to as DE (diatomaceous earth) filters since this material is usually used as a filter precoat, although other filter aids can be used. The septum filter (Figure 9.28) relies on a thin layer of precoat applied as a slurry to a porous septum to produce a filtering surface to strain the suspended solids. In most cases water being filtered is pumped through the filter under pressure; in special designs where low head loss is possible the water may be pulled through using vacuum. As the filter becomes plugged, head loss increases and the solids, including precoat, have to be removed by reversing the flow through the unit. A new precoat is then applied and filtration is resumed. Usually in addition to the precoat, a body feed of filter aid is used. This body feed is simply additional filter aid added to the influent to extend filter runs by continually providing a fresh filter surface. Because the filter aid has a different shape (morphology) than the solids in the water, the heterogeneous mixture is more permeable than the solids alone.

A relatively high ratio of filter aid to suspended solids is required to operate septum filters making operating costs fairly high. Therefore, these units are not as common as granular media filters in most industrial systems. DE filters are often used for applications such as municipal swimming pools, and they are excellent for removal of oil from industrial plant condensate.

Septum filters can be cleaned of accumulated solids by air-bumping, a procedure requiring little or no water and producing a thick slurry or cake of accumulated solids. This simplifies solids disposal, and reduces backwash water requirements. They can also be fitted into a relatively small space, compared to granular media filters.

While diatomaceous earth—the fossil remains of diatoms, a type of algae having a silica skeleton—is the commonly used filter aid (see Figure 9.29), mixtures of DE and asbestos are often used. At the high temperatures encountered in filtration of oily condensate, silica dissolves from the DE filter cake, so Solka-Floc, a cellulosic product, is used to avoid this problem if the condensate is to be fed to a boiler.

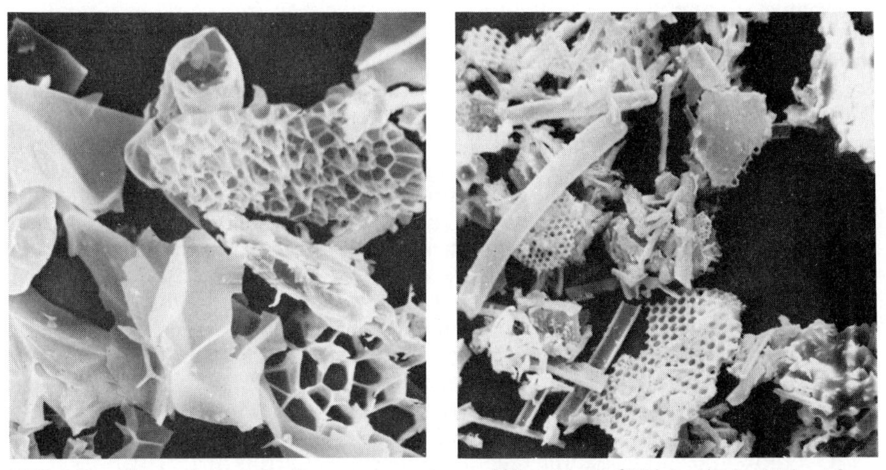

FIG. 9.29 Photomicrograph of two common types of filter aid. *(Left)* Diatomite is obtained from natural deposits of the siliceous skeletons of diatoms, a variety of algae. *(Right)* Perlite, a mineral of volcanic origin, is processed at high temperature to produce a variety of forms of glassy slivers. (About 500×.) Special grades of cellulose and carbon are also used in water filtration.

REMOVAL OF WATER FROM SLUDGE

Thickening

Thickening is a solids/liquids separation method used to increase the solids content of a slurry prior to dewatering. Thickening normally follows a clarification process where the suspended solids have been separated from the liquid. In clarification, feed solids are normally in the 10 to 1000 mg/L range, while influent to a thickener is usually in the 0.5 to 10% range.

The purpose of thickening is to increase the solids of the underflow thereby reducing sludge volume and cost of subsequent handling; in clarification, the purpose is to remove solids and produce a clear effluent. The clarity of water leaving a thickener is not as critical as the density of the underflow, since the effluent water normally is recycled back to the head of the plant. The thickened sludge must remain liquid to the extent that it can be pumped to subsequent dewatering operations. In municipal waste treatment plants, where digestion follows thickening, improved digestion and conservation of digester space is achieved through thickening.

In the clarifier, solids have separated from the water primarily by free-fall. The solids collected in the lower region have encountered the effects of hindered settling. In the thickener, there is no free-fall; the process of hindered settling controls the design and the final compaction of sludge.

Gravity thickening and flotation thickening are the two major methods.

Gravity Thickening. Gravity thickening is often used in municipal plants for primary sludges and in industrial plants for chemical sludges. A typical thickening operation in a steel mill will double the solids concentration. The gravity process works well where the specific gravity of the solids is much greater than that of the liquid.

A gravity thickener is constructed much like a clarifier: usually it is circular with a side wall depth of approximately 10 ft and with the floor sloping toward the center (Figure 9.30). The floor angle is greater than in a clarifier, normally 8 to 10 degrees. As in a clarifier, the sludge is moved to a well by a rake assembly and then pumped out by a positive displacement pump. In a gravity thickener, because the process of hindered settling controls solids compaction, the sludge rake arm has a dual purpose; besides raking the solids to the sludge well, the arm is constructed like a picket fence to gently muddle the slurry, dislodging interstitial water from the sludge and preventing bridging of the solids.

As the sludge blanket gets deeper, up to about 3 ft, the density of the solids increases, after which there is little advantage in increasing sludge depth. When thickening municipal sludges, close attention must be paid to the length of time sludge is in the thickener, since it can become septic and produce gas bubbles that may upset the system. This is particularly true with thickening biological secondary sludge. If septic sludge is encountered, chlorine may be added to the feed to the thickener. The SVR (sludge volume ratio), which is used to monitor sludge age, is the volume of sludge in the blanket divided by the daily volume of sludge pumped from the thickener. This gives the retention time, which is normally between 0.5 and 2 days.

Overflow and solids loading rates are important controls in gravity thickening; if thickener performance is not satisfactory, the operator can alter these rates to improve solids capture. Overflow rates for gravity thickeners range from 400 to

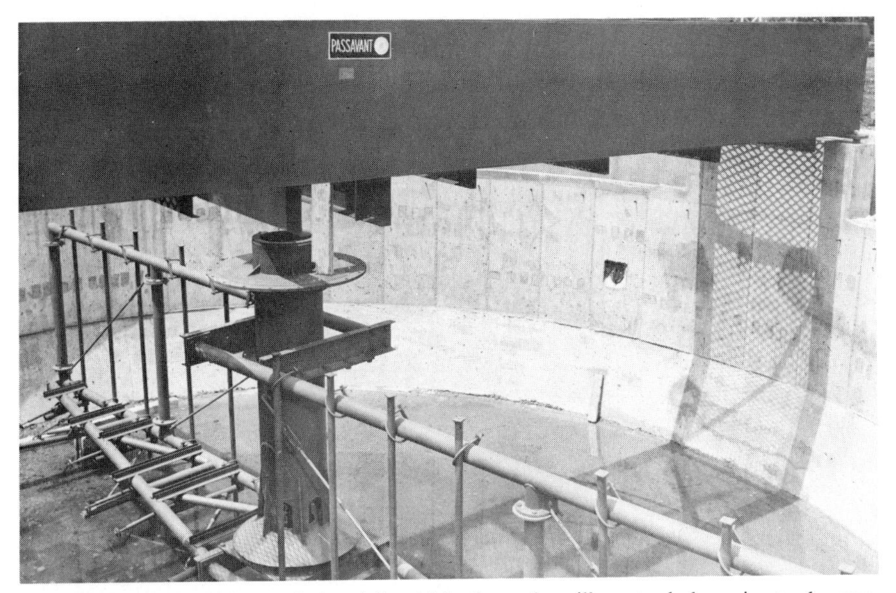

FIG. 9.30 Sludge thickener designed for thickening pulp mill waste sludge prior to dewatering. *(Courtesy of Passavant Corporation.)*

800 gal/day per square foot of surface area. The solids loading rate, expressed in pounds of solids per day per square foot, depends on the type of sludge being thickened.

Chemical Treatment. Chemicals may aid gravity thickening. Salts of iron and aluminum have little effect; in some cases they improve the overflow clarity, but do not provide increased loading. Polymer flocculants are effective aids to gravity thickening, forming larger, heavier floc particles, which settle faster and form a denser sludge. Depending on the system involved, these polymers can be cationic, nonionic, or anionic in character. Effective dosages of polymers usually are in the range of 2 to 20 lb per ton of sludge solids on a dry solids basis. Two test methods are used to determine the best chemical program: these are a simple cyclinder settling test and a stirred thickening test shown in Figure 9.31, where the stirrer rotates at only 0.1 to 1.0 r/min. The sludge is mixed with chemical, placed in the thickening apparatus, and stirred. Effectiveness of treatment is determined by measuring the density of sludge samples removed from the bottom of the beaker and comparing it to the concentration of unthickened sludge. An alternate method is to measure the percentage of supernate in each sample of settled sludge representing untreated and treated conditions.

When polymer is used in gravity thickening, special attention should be given to the application point and to the dilution water rate. Since older gravity thickeners were not designed specifically to use polymers, feed taps are often not readily available. Suitable feed points would be either directly ahead of the thickener to the feed pump or into the sludge feed line. Dilution water needs to be adjusted for optimum dispersion of the polymer into the sludge without defeating the goal of water removal.

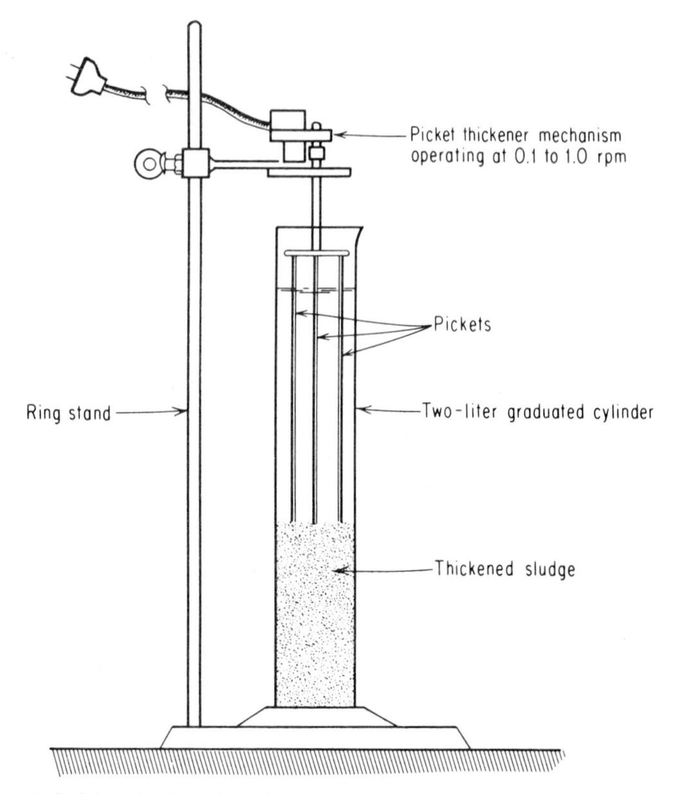

FIG. 9.31 Sludge thickening test apparatus. Note: Results are appreciably influenced by sludge depth, a greater depth hindering compaction rate. The test time varies from minutes to hours, and is best judged by the plot of sludge level versus time to determine the time required to fall 1 to 2 in. *(Courtesy of Eimco Division, Envirotech Corporation.)*

Thickening by Flotation. An alternate to gravity thickening is flotation thickening. This is usually more effective than gravity thickening when the solids being thickened have a specific gravity near or less than that of the liquid from which they are being removed. Because of its high solids loading, a DAF (dissolved air flotation) thickener normally occupies one-third or less of the space required for a gravity thickener.

Flotation may be either dispersed air or dissolved air. In dispersed air flotation, bubbles larger than 100 μm are generated by mechanical shearing devices acting on air injected into the water in a flotation cell (Figure 9.24). In the dissolved air flotation process, discussed earlier as a clarification process, the slurry being thickened or a portion of the recycle flow is supersaturated with air under pressure. When the pressure is released, air is precipitated as small bubbles in the 10- to 100-μm size range. The air bubbles attach to the solids, increasing the bouyancy of the particles and cause them to rise to the surface and concentrate.

Biological suspended solids are usually difficult to settle; however, with the addition of dissolved air bubbles and polymer, these particles have a tendency to float.

The DAF thickening process is often augmented by the addition of chemical aids. Chemicals that have been used include inorganics, such as ferric chloride and lime, and organic polyelectrolytes. Of the polyelectrolytes used, the most effective have been either moderate molecular weight polyamines or very high molecular weight flocculants. In most cases, cationic flocculants are used. They are particularly effective in flocculating biological solids. Introduction of the polymer into the line at a point where bubbles are precipitating and contacting the solids normally produces the best results.

Methods of evaluating flotation aids vary with the type of slurry being thickened. A standard cylinder settling test is often used to determine which polymers will form stable floc. Floatability of a waste sludge can easily be determined using the apparatus shown in Figure 9.32. More sophisticated pilot plant test equipment, such as is shown in Figure 9.33, can also be used to develop chemical programs. This equipment can closely duplicate the conditions found in actual DAF thickeners.

FIG. 9.32 Bench apparatus for air flotation studies. *(Courtesy of Infilco Degrement Incorporated.)*

In summary, DAF can be an effective method of thickening materials that have a tendency to float. The use of DAF to concentrate sludge offers some advantages over gravity thickening. This is especially true in the case of concentration of activated biological sludge, which is troublesome to concentrate by gravity. Gravity thickening of waste-activated sludge will seldom yield concentrated sludge of more than 2% solids. Dissolved air flotation of the same sludge will normally yield greater than 4% solids in the float. During the concentration operation, since air is used, the sludge will remain fresh and not become septic as it can if left in a gravity thickener.

Like other thickeners, flotation devices are designed for specific solids loading rates and overflow rates. Solids loading rates without chemical addition range from 1 to 2 lb of dry solids per hour per square foot; chemical addition allows the load to be doubled. The air to solids ratio of a DAF is critical since it affects the rise rate of the sludge. The air to solids ratio required for a particular sludge is a function of the sludge characteristics such as SVI. Typically, the air to solids ratio varies from 0.02 to 0.05.

The DAF column test can be used to simulate the thickening process on a small scale. This apparatus can be used to measure the floatability of a particular sludge and to evaluate various chemical flotation aids that improve performance, either solids capture or solids density or both.

Centrifugation. Solids concentration or thickening can also be accomplished by centrifugation. The three widely used types of centrifuges are basket, solid bowl, and disk-nozzle, but the basket and solid bowl centrifuges are more commonly used in dewatering of sludges.

FIG. 9.33 Pilot plant for evaluation of dissolved air flotation results on specific wastewater problems. *(Courtesy of Komline-Sanderson Engineering Corporation.)*

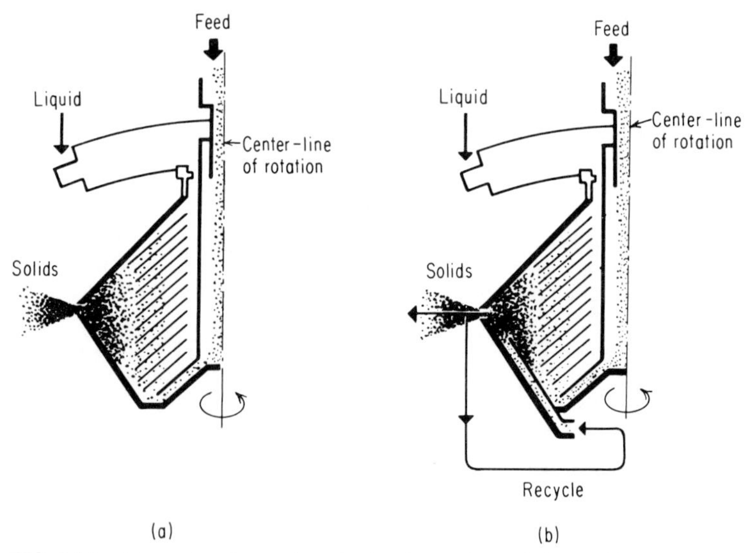

FIG. 9.34 Schematic of several methods of operating a disk centrifuge: (*a*) basic scheme of operation; (*b*) with recycle of solids.

The disk-nozzle centrifuge (Figure 9.34), while primarily a liquid/liquid separation unit, can be used to thicken slurries for further dewatering. It is suitable for thickening a slurry with very fine uniform particle size since it creates greater centrifugal force than a solid bowl centrifuge.

Slurry is fed into the center of the machine at the top and then directed to an area on the outside of the disks (Figure 9.34*a*). The disks are stacked in the centrifuge so that they are 0.10 to 0.25 in apart. The solids settle as the feed is forced through this narrow space. Since the distance is so narrow, particles do not have far to travel. The settled solids slide down the underside of the plates and out of the bowl wall where compaction takes place prior to discharge through nozzles in the periphery. The centrate passes under the sludge and is discharged from the center of the centrifuge. Since the disk-nozzle centrifuge has such close tolerances it is subject to frequent plugging when coarse solids are encountered. For this reason coarse solids are often screened from the fluid before it enters the centrifuge.

DEWATERING—L/S SEPARATION

Dewatering is normally the final step in a liquids/solids separation. The goal is to produce a cake of such density and strength as to permit hauling to a final disposal site as a solid waste. It usually follows clarification and thickening operations. In waste treatment, the dewatering method is often dictated by the nature of the solids being dewatered and the final method of solids disposal. If sludge is being incinerated, as in the case of a raw or biological sewage sludge, it is necessary to extract as much water as possible to minimize the requirement for auxiliary fuel for incineration. If solids are being used in a land reclamation program or as landfill, it may not be necessary to dewater to such an extent.

Centrifugation. Centrifugation has long been used for dewatering as well as for thickening, discussed earlier. Selection of the proper centrifuge is important since design characteristics can be tailored to meet specific application needs. There are several inherent advantages in centrifugation that make it attractive for many dewatering applications. Among the important advantages are compact design, high throughput, and relative simplicity of operation. Auxiliary equipment is very simple.

Solid Bowl Centrifuge. The type of centrifuge normally used in dewatering applications as opposed to thickening is the solid bowl unit. These are widely used in municipal sewage plants, paper mills, steel mills, textile mills, and refineries. They are also used extensively in mining operations, such as processing coal and refuse. In more moderate climates, they can be installed and operated outdoors.

There are three solid bowl designs: conical, cylindrical, and conical-cylindrical. The conical bowl achieves maximum solids dryness, but at the expense of centrate clarity by employing a large beach area over a small centrate pool volume. In comparison, the cylindrical bowl has a deep centrate pool throughout its entire length and provides good centrate clarity, but relatively wet cake.

The conical-cylindrical design (Figure 9.35) is the most commonly used solid bowl centrifuge. It is flexible in its ability to shift the balance of cake dryness and centrate quality over a broad range by changing pool depth depending upon the desired performance criteria. The conical-cylindrical solid bowl centrifuge consists of a rotating unit comprising a bowl and a conveyor joining through a special system of gears which causes the bowl and conveyor to rotate in the same direction, but at slightly different speeds. Most solid bowl centrifuges operate at 1500

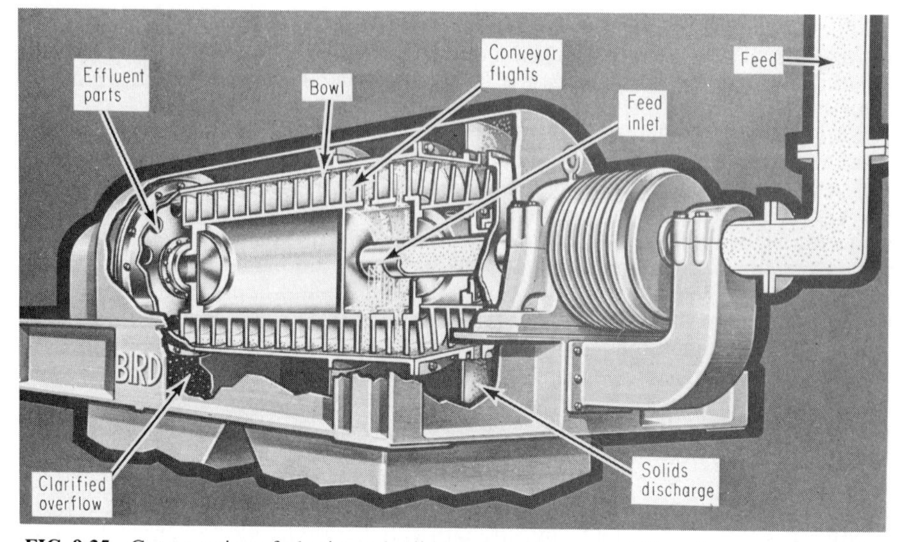

FIG. 9.35 Cutaway view of a horizontal solid-bowl continuous centrifuge showing typical design features. *(Courtesy of Bird Machine Company, Inc.)*

to 2500 r/min. The conveyor normally operates at 50 to 150 r/min lower. The conical section at one end of the bowl forms a dewatering beach over which the conveyor pushes the sludge to outlet ports. The clarified supernatant liquid is allowed to escape over weirs or is removed by a skimmer. If centrate and sludge cake leave the same end of the bowl, the centrifuge is called a concurrent type; if they leave the machine at opposite ends, the centrifuge is a countercurrent type.

The slurry to be dewatered enters the rotating bowl through a feed pipe extending into the hollow shaft of the rotating screw conveyor and is distributed through ports into a pool within the rotating bowl (Figure 9.36). The slurry pool forms a concentric ring of liquid on the inner wall of the bowl. The solids are settled by centrifugal forces to the peripheral surface of the rotating bowl. The depth of the

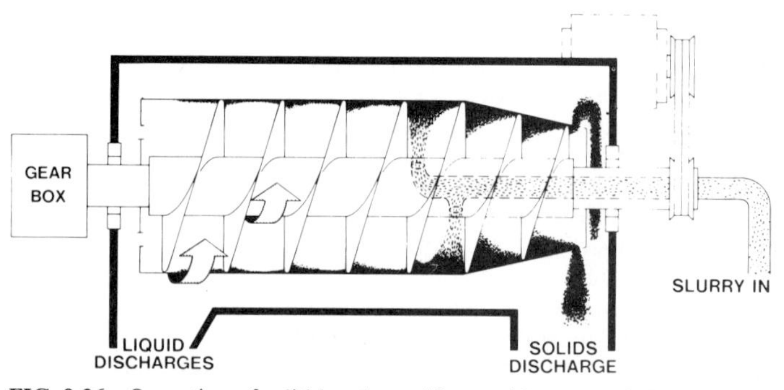

FIG. 9.36 Operation of solid-bowl centrifuge. *(Courtesy of Sharples Division, Pennwalt Corporation.)*

bowl and its volume can be varied by adjustment of the overflow weir plates or a skimmer. The overflowing liquid is removed by pumping or by gravity. The helical rotating conveyor pushes the solids to the conical section where the solids are forced out of the water and free water drains from the solids back into the pool.

Slurry feed pumps should be of the nonpulsating type. A progressing cavity pump is normally recommended, since it affords positive displacement without surging. Sludge feed and chemical feed pumps should be interlocked to stop automatically if the centrifuging or cake conveying operation is interrupted. A wash water system is normally included to provide a means of flushing the machine when it is to be shut down. This prevents the accumulation of solids that may cause vibrations because of imbalance. In general, those sludges that separate more readily and concentrate to the greatest thickness by plain sedimentation can be dewatered most efficiently by centrifuge.

Theory. The solid bowl centrifuge can be compared with a horizontal clarifier as illustrated in Figure 9.37. Both processes are applications of Stokes' law:

$$V = \frac{2R^2(S_1 - S_2)g}{9\mu}$$

where V = settling velocity
R = particle radius
S_1 = specific gravity of particle
S_2 = specific gravity of liquid
g = acceleration due to gravity
μ = viscosity

The important difference between the processes of sedimentation and centrifugation is that centrifugation increases the G factor from 1.0 to 2000 to 3000. Another difference is this high gravitational force of the centrifuge reduces the required detention so that the particles have only to settle a very short distance—inches instead of feet.

There are several design variables to be considered when choosing a centrifuge for a specific dewatering application, the most important being bowl design, rotational speed, and conveyor (scroll) speed.

As the bowl length increases, retention time increases, resulting in drier cake and clearer centrates. However, power requirements also increase, since they are proportional to the volume contained in the centrifuge. As with bowl length, increasing the diameter of a centrifuge results in drier cakes and clearer filtrate at the same feed rate. Besides requiring higher power input, a larger diameter machine requires more critical balancing, since the center of gravity moves farther from the center.

As might be expected, higher bowl speeds produce drier cakes and clearer centrates, but also increase the wear on the machine. When centrifuging abrasive materials, low-speed centrifuges are preferred. As the conveyor speed differential increases, machine throughput can increase, but at the expense of a wetter cake and dirtier centrate. While bowl speed and scroll speed differential are treated as a design variable, it is possible by changing pulleys and gear boxes to alter them on existing machines, so they are also an operating variable. On some newer machines the scroll speed differential can be varied while the centrifuge is operating.

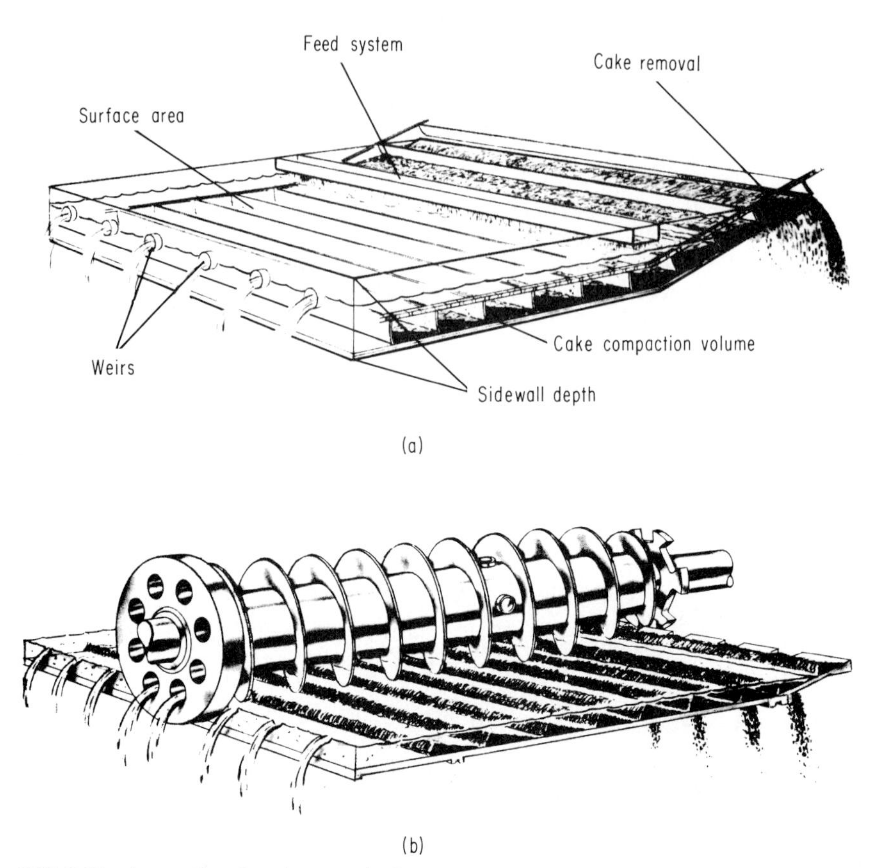

FIG. 9.37 A centrifuge has the same basic characteristics as a clarifier. It's a clarifier whose base is wrapped around a center line so it can be rotated to generate elevated *g*'s. *(Courtesy of Sharples Division, Pennwalt Corporation.)*

In operating a centrifuge for slurry dewatering, the following variables must be balanced to obtain the desired results:

1. *Sludge flow:* The sludge feed rate is one of the more important variables. The rate of sludge feed affects both clarity and cake dryness. As sludge flow increases, machine throughput increases, but at the expense of centrate clarity and cake dryness (Figure 9.38). Extremely high flows cause machine flooding. Very low flow rates produce extremely dry cakes causing increased wear of the conveyor scroll. Sludge should be pumped to the machine at the highest rate that will not cause the carryover of excessive fines in the centrate.
2. *Chemical conditioning:* Polyelectrolytes have been used with success to increase the recovery efficiency of centrifuges. They may also increase throughput and widen the spectrum of sludge characteristics that can be processed. The most active polymers are of high molecular weight, either anionic, nonionic, or cationic, depending on the sludge being dewatered. The degree of solids recovery and cake dryness can be regulated over a wide range depending on the amount of chemical used (Figure 9.39). The use of flocculation aids

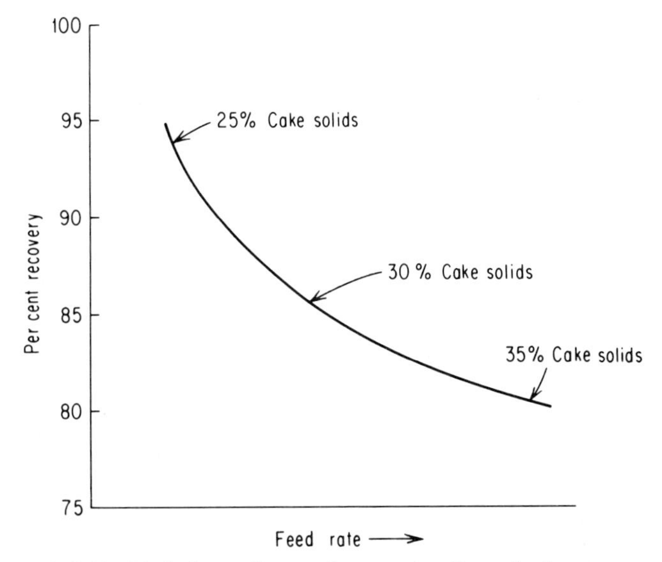

FIG. 9.38 Typical centrifuge performance on digested primary sewage solids.

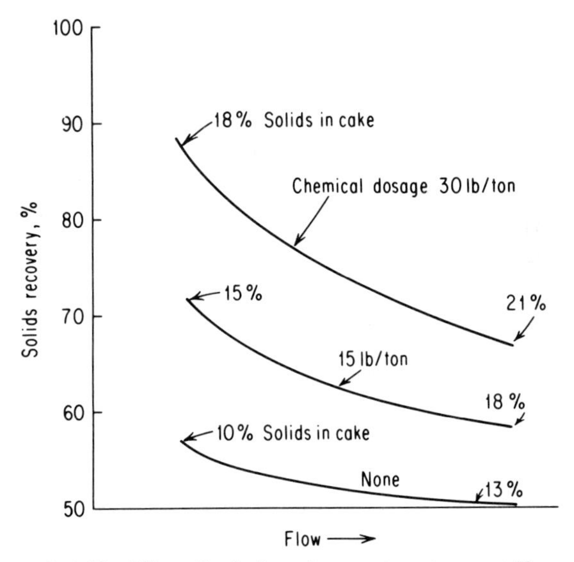

FIG. 9.39 Effect of polyelectrolyte treatment on centrifuge performance.

sometimes produces wetter sludge because of the increased capture of fine solids. The variables that affect the efficiency of chemical conditioning are dosage, dilution, and chemical injection point. Cylinder settling tests should be run to select the proper polymer and its approximate dosage. Exact dosages can only be determined in the centrifuge. If a centrifuge has been designed with more than one polymer application point, all points should be evaluated. The purpose of an internal feed point is to add polymer to the pool after the heavy, easily settleable solids have been removed. However, it is sometimes advantageous to add polymer to the total sludge stream, thereby using the heavier solids to help in bridging and flocculating the finer ones. Since in either case conditioning time is relatively short, it is important to dilute polymer to a level that will provide good distribution into the sludge without overloading the machine with dilution water.

3. *Pool depth:* Pool depth determines pool volume and affects both clarification and cake dryness. Lowering the pool exposes more drainable deck area, increases the dewatering time, and produces a drier cake; conversely, increasing pool depth increases clarification by increasing detention time. However, just as in sedimentation, too deep a pool prevents a particle from reaching the sediment zone in time to avoid being discharged with the effluent. Theory indicates that the thinnest possible layer in the bowl or in a sedimentation basin gives maximum clarification. In practice, this is not true because of practical inefficiencies. Thin layers mean high linear liquid velocities and the resuspension of nonsettling fine materials. Also, at minimum pool depths the moving conveyor tends to redisperse solids as does turbulence in the feed zone. Consequently, it is normally found that a medium pool depth gives optimum performance. Increasing the pool depth beyond the optimum depth does not contribute to clarity as the retention time and the distance the particle must settle are both directly proportional to the pool depth. The shallower the pool in the centrifuge, the longer the drainage time for solids on the beach out of the pool and generally the drier the settled solids (Figure 9.40). Fine particles, however, compact to a limiting dryness at the bowl wall under the pool and capillary forces may prevent appreciable drainage of liquid from the interstices of the sludge on the beach. In this case, a shallow pool is no advantage, and in fact a deep pool often improves conveying and removal of this type of sludge solids.

All conditions of operating variables should be evaluated to fine tune operating parameters for a given centrifuge application. Gradual changes should be made, allowing 10 to 15 min between changes.

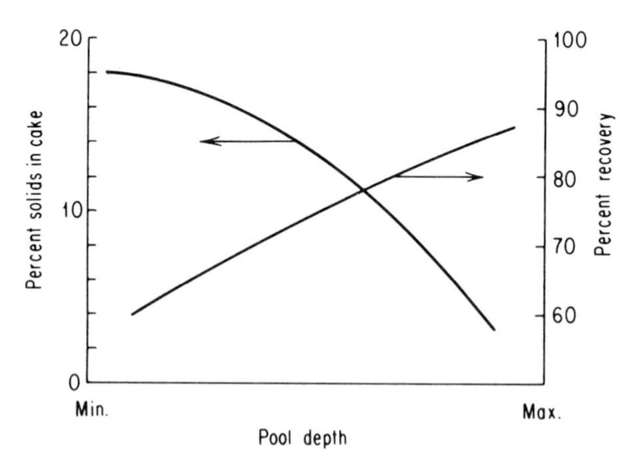

FIG. 9.40 The effect of pool depth on centrifuge performance. The deeper pool limits the area available for cake dewatering, but increases detention.

Basket Centrifuge. A basket centrifuge may also be used for slurry dewatering. In a basket centrifuge (Figure 9.41) dewatering is a batch process. The slurry enters through a feed line from the top to a point near the bottom of the rotating drum. If the drum has solid (imperforate) walls, the centrate is collected by a skimmer or by simply overflowing a weir at the top of the bowl. The perforate bowl centrifuge has a drum with holes similar to that of an automatic washing machine. This type of centrifuge uses a filter medium placed on the inside of the drum, and water, under centrifugal force, passes through the cake and the medium to the outside of the drum from which it drains. After the machine is loaded, speed is increased to effect separation. After the centrate is removed, the

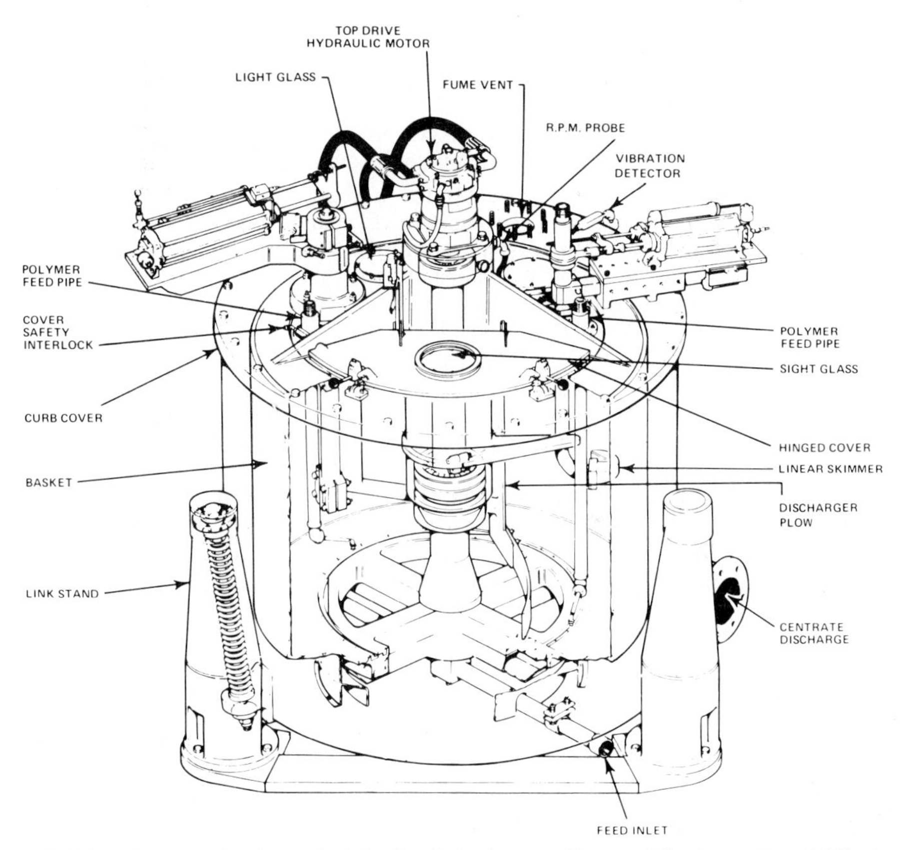

FIG. 9.41 Cutaway showing typical details of a basket centrifuge used for dewatering of difficult sludges, and sludges that are highly abrasive. *(Courtesy of the DeLaval Separator Company.)*

centrifuge is slowed down for unloading. A knife or "plow" is then moved into the bowl to cut out the cake. This plow is hydraulically operated and is adjusted so as not to cut into the bowl wall.

Basket centrifuges are used where solids recovery and cake dryness are most important. Since perforate bowl machines can easily be equipped with wash sprays, they are often used where cakes need to be purified and washed as in the pharmaceutical industry.

System Evaluation. To determine whether a program is successful, some performance criteria must be established. The desired results, whether it be dry cake or clear centrate, will determine the weighting of the following criteria.

Recovery: Centrifuge dewatering performance is normally stated in terms of recovery, expressed as the percentage of suspended solids in the sludge feed that ends up in the discharged sludge cake. Recovery is computed using the following equation:

$$\% \text{ efficiency} = \frac{C_i - C_o}{C_i} \times 100$$

where C_i is feed solids concentration in, mg/L and C_o is centrate solids concentration out, mg/L.

Note that this formula is not corrected for dissolved solids, which must be taken into account when centrifuging a chemical process liquor, such as a 10% brine. Where the water slurry contains less than 500 mg/L and the feed solids exceed 2%, the above formula is adequate for most plant calculations.

Yield: The amount of cake throughput expressed in dry pounds of solids per hour is the centrifuge yield. This is not an efficiency factor since the machine could conceivably have a high yield, but a poor centrate. Yield is determined by multiplying the pounds of wet cake per hour by the percent solids (on a dry basis) in the wet cake. For example, 2100 lb of wet cake per hour at 30% solids is equal to 700 lb solids per hour, dry basis.

Centrate quality: Centrate quality refers to the amount of suspended solids in the centrate and is directly related to recovery. Typically, solids found in the centrate are extremely fine. Since the centrate is normally settled and recycled through the system, running with a poor centrate will eventually build fines in the system and lower the recovery rate. Getting 100% recovery, on the other hand, is uneconomical and is not necessary to meet real plant needs.

Cake solids: The desired degree of cake dryness will depend on final disposition of the solids. If solids are to be incinerated, low moisture is desired. As drier solids entail higher dewatering costs, higher scroll wear, and lower yield, cake solids should not be higher than realistically needed for final disposal.

Vacuum Filtration. Rotary vacuum filters have been used for many years by industry and municipalities to dewater waste sludges. They are also used in many process applications such as the pulping operations in paper mills. The mining industry also uses vacuum filters extensively in coal recovery and in ore processing. Solids content of the filter cake may vary considerably depending on the slurry being dewatered. Inorganic slurries may dewater to 70% solids, whereas a biological sludge may only dewater to a 20% solids level.

A rotary vacuum filter consists of a perforated cylindrical drum rotating in a vat containing the suspension to be dewatered. Vacuum is applied to the interior and liquids/solids separation is accomplished by drawing the liquid through a filter medium, leaving solids on the medium for separate collection. As the drum rotates, this medium provides a continuous filter going through the following sequence (Figure 9.42).

1. *Cake formation:* During this step, the drum is submerged in a vat containing slurry and vacuum is applied. The first solids that collect on the medium act as a filter for subsequent cake formation. For this reason, it is desirable that the cake be formed gradually.

2. *Liquid extraction or drying:* A vacuum is also applied during the liquid extraction or drying step. Water is drawn out of the filter cake through the filter media to the inside of the drum and from there into a filtrate receiver.
3. *Filter cake removal:* Removal of the filter cake from the drum can be accomplished by one of several methods discussed later.
4. *Media washing:* This final step in the cycle of vacuum filter operation can be accomplished on a continuous or intermittent basis. High pressure water sprays are normally used. These sprays dislodge the particles that could build up and clog the medium. It often becomes necessary to acid wash the filter media on a routine basis. This is particularly true when ferric chloride and lime are used as conditioning chemicals.

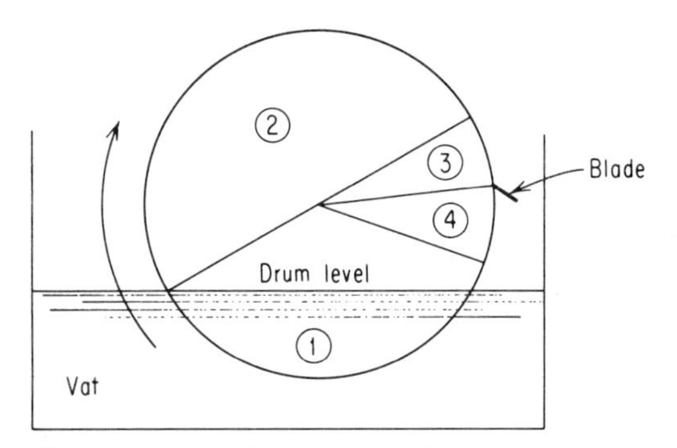

FIG. 9.42 Operating sequence of rotary-drum vacuum filter.

Types of Vacuum Filters. There are several variations of rotary vacuum filters. The drum filter (Figure 9.43) uses a "doctor blade" to scrape off the cake for discharge. It is sometimes supplemented by a pressure blowback in the discharge zone. Air pressure from the inside of the drum just ahead of the doctor blade loosens the cake for subsequent removal by the doctor blade. The filter medium never leaves the drum surface on this type filter. The filter medium may be either a stainless-steel mesh or a synthetic or natural fiber fabric. The more common materials are nylon, polypropylene, and polyethylene. These media come in a wide range of porosities to meet specific application needs. A filter medium should be selected with porosity small enough to obtain an acceptable filtrate, but large enough to avoid plugging. Consideration also should be given to the compatibility of the medium with the slurry being filtered.

A second type of vacuum filter is the belt filter (Figure 9.44) in which the filter medium is a continuous belt that leaves the drum surface and travels over a short radius discharge roll. This sharp change in curvature normally causes the sludge cake to fall from the media. In some cases, it is also necessary to use a doctor blade to ensure maximum production. The advantage of this filter over a drum filter is that the media are washed on both sides.

The coil type filter (Figure 9.45) employs two layers of stainless-steel coil springs arranged in a corduroy fashion around the drum. These layered springs act as the filter medium. On completion of the dewatering cycle the two layers of springs leave the drum and are separated from each other in such a manner that the filtered sludge cake is lifted off the lower layer of coil springs and discharged from the upper layer with the aid of a positioned tine bar. The two layers of

FIG. 9.43 Assembly of a typical rotary-drum vacuum filter. *(Courtesy of Dorr-Oliver, Inc.)*

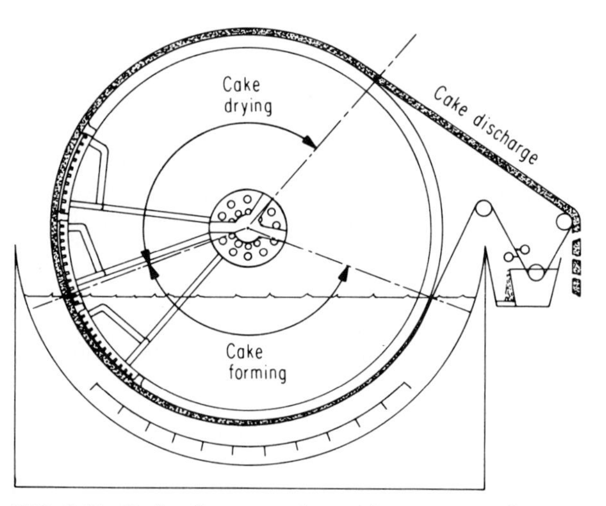

FIG. 9.44 Design features and operating sequence of rotary-drum belt-type vacuum filter.

springs are then washed separately by sprays and returned to the drum by grooved aligning rolls. Because of its positive filter cake release system, the coil filter is often selected for dewatering of sludges with poor release tendencies.

A fourth type of vacuum filter, the continuous disk filter (Figure 9.46), operates on the same general principle as other vacuum filters except that the filtering sur-

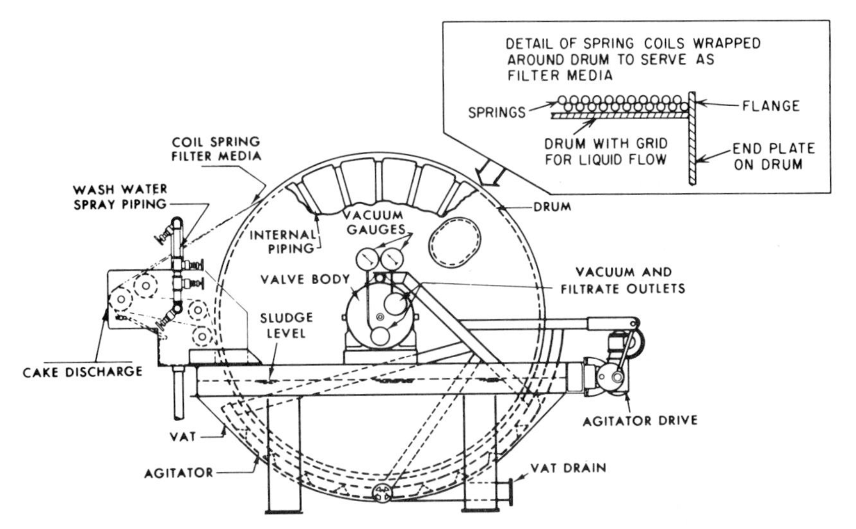

FIG. 9.45 Cross-section of a coil filter. *(Courtesy of Komline-Sanderson Engineering Corporation.)*

FIG. 9.46 Disk-type vacuum filter used for solids which form a thick cake. *(Courtesy of Dorr-Oliver Incorporated.)*

face is on both sides of several disks mounted perpendicularly to a rotating horizontal center shaft. Filtrate is collected from each disk and cake is removed by scrapers.

Auxiliary Equipment. There are a number of auxiliary devices required for operation of rotary filters. These include vacuum pump, vacuum receiver, filtrate pump, sludge pump, and chemical conditioning equipment.

Normally each filter has its own vacuum pump. Vacuum pumps may be either reciprocating positive displacement, centrifugal, or rotary positive types. Reciprocating pumps are the most efficient for high vacuum of 20 to 25 in of mercury. The required capacity for a vacuum pump is determined in part by the porosity of the filter medium. If the filter medium is changed to one more porous, this may result in the vacuum pump being undersized. In operation this condition may be evidenced by an inability of the vacuum system to prevent backflowing of filtrate as the drum turns out of the sludge in the vat. Of course, poor performance of the vacuum system can also result from normal wear or poor maintenance.

Each vacuum filter must be supplied with a vacuum receiver between the filter valve and the vacuum pump. The receiver acts as a separator of air and filtrate, as well as a reservoir for the filtrate pump suction.

Sludge feed pumps can be of the piston, diaphragm, or progressing cavity type. They should be installed so that their capacity is constant at a given setting and should be interlocked to a level sensing device to stop at a predetermined high liquid level in the filter vat, and to automatically restart when the filter has caught up. A separate pump should be provided for each filter conditioning tank combination. Pumps should have stroke counters or some other flow totalizing device.

Chemical conditioning equipment includes chemical storage tanks, sludge conditioning tanks, and chemical preparation and feed systems. Sludge conditioning containers may be horizontal tanks, vertical tanks, or horizontal rotating drums. The sludge should flow by gravity from the conditioning tank to the filter vat to minimize floc damage. To provide optimum sludge mixing and flocculation under varying conditions, sludge conditioning tanks should be provided with variable speed mixer drives, removable weirs, and multiple points of chemical application. Rates of rotation of mixers generally should be adjustable to between 10 and 60 r/min, depending on the design of the stirrers.

There are several variables to be balanced to operate a vacuum filter efficiently. Since each sludge being filtered has its own dewatering characteristics, the variables must be tuned to give the desired results. Changes should be made in small increments and the system given at least ½ h to react and reach a new equilibrium. If the filter vat is filled with improperly conditioned sludge, it is best to drain the pan and start over.

Sludge conditioning: Polyelectrolytes, ferric chloride, and lime are conditioning chemicals used to coagulate and flocculate solids in suspension. The larger floc particles filter more easily and have less tendency to blind the filters. With the use of iron and lime, deposits may accumulate on the filter medium and necessitate periodic acid cleaning. To properly coagulate and flocculate a sludge for vacuum filtration, controlled mixing is required. Conditioning tanks normally are sized to provide retention time of 30 to 60 s. With too little mixing, polymer will not be properly dispersed; too much mixing breaks the floc. The proper amount of mixing for a particular sludge must be determined empirically.

Vat agitation: The conditioned sludge must be agitated in the filter pan to eliminate settling. Organic sludges without heavy solids usually require only intermittent agitation.

Vat level: Conditioned sludge in the filter pan should be set at a level which, when coupled with drum cycle time, gives the desired cake thickness. Higher vat levels give thicker and somewhat wetter cakes than low levels if other variables are kept constant.

Cycle times: The drum speed should be adjusted to give the desired cake thickness and moisture content. For most applications, a higher speed resulting in a thinner, drier cake (¼ to ½ in) is better than a low speed producing thick (1 in) cakes.

Vacuum: Normally vacuum filters are run with as much vacuum as can be attained. Since usually only one vacuum pump is used per filter, it is necessary to balance vacuum between the cake-forming and drying sections. Higher pickup vacuum will produce wetter cakes and greater sludge dewatering rates.

Filter media: A variety of filter media is available, providing a selection of size and weave. The coarser media normally give drier cakes, but at the cost of fines escaping into the filtrate. Filter leaf tests can be used to determine the proper medium for a particular application (Figure 9.47).

Sludge density: Thicker sludges will filter and release from the medium more easily than thin sludges. Fillers such as sawdust, diatomaceous earth, and fly ash are sometimes added to thicken thin sludges. Where sludges are incinerated, recycled ash fed to the thickener ahead of the filter may increase solids and improve performance.

Sludge particle size: By achieving high recovery (capture) performance, the level of recycled fines accumulating in the plant can be minimized. If the level of fines in the plant becomes too high, filter blinding can occur. Fines can also upset other plant processes, such as clarification and BOD removal, by overloading their solids separation capacity.

System Evaluation. To determine whether a program is successful some performance benchmarks must be established. Desired results, such as a dry cake, a clear filtrate, or high production can be determined by the following criteria:

1. *Filter yield:* The most common measure of filter performance is expressed as pounds of dry solids per square foot per hour. Typical filter loadings are 3 to 10 $lb/ft^2/h$ for a municipal plant and 8 to 10 $lb/ft^2/h$ for an industrial plant. Some industrial filters dewatering inorganic sludges are rated in the 40 to 300 $lb/ft^2/h$ range. Filter yield measures only the rate of production in terms of dry solids equivalent, not filter efficiency.
2. *Filter efficiency:* Since the vacuum filter is a device used for separating cake solids from a thinner sludge, the actual efficiency of the process is determined by the percentage of feed solids recovered in the filter cake. This is calculated by the same formula as used for the centrifuge. Solids removal efficiencies in vacuum filters have ranged from 85% for coarse mesh or coiled media to 99% with close-weave long-nap media. Since recycled filtrate solids increase the load on the plant treatment unit, they should be kept to a minimum or balanced within cost-effectiveness for overall plant operations.
3. *Filter cake quality:* Filter cake moisture content varies with the type of slurry handled and filter time in drum submersion. Good filter performance does not always mean producing the driest possible cake. Cake moisture should be adjusted to obtain only as dry a cake as is required for the intended end use. Anything more is inefficient and wasteful of filter capacity.
4. *Filtrate quality:* Suspended solids content is the most important indicator of filtrate quality. This depends on the filter medium (fine or coarse), the quality of filter feed slurry (percent fines), and the amount of vacuum. Generally, filtrate suspended solids should be kept as low as possible, especially when the

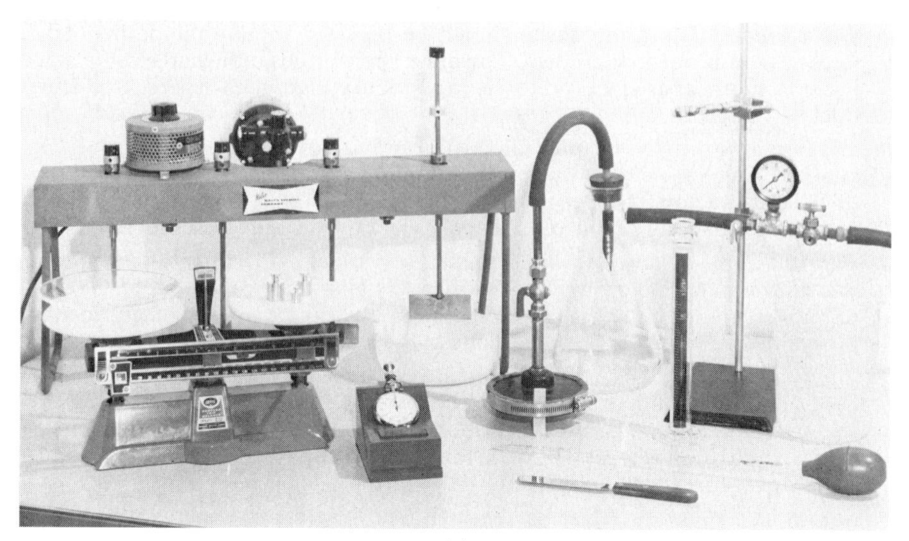

(a)

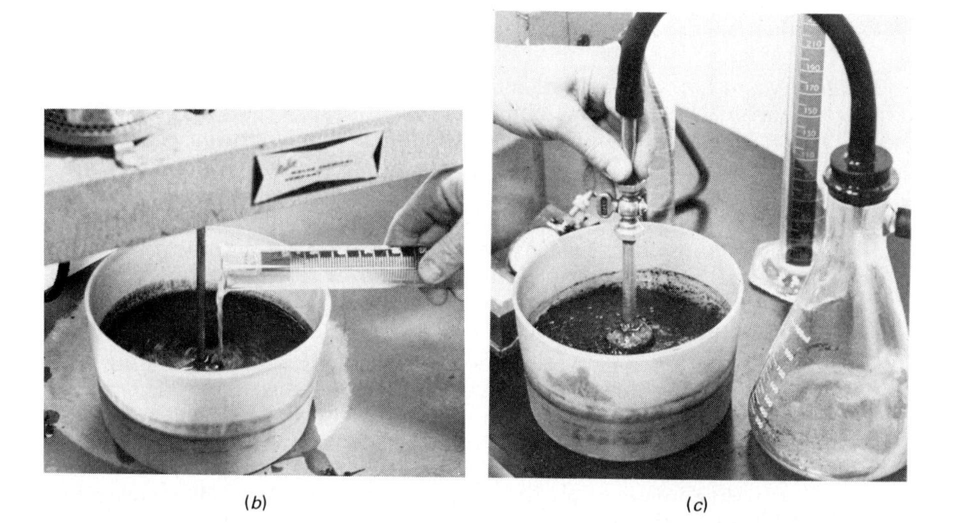

(b) (c)

FIG. 9.47 (*a*) Apparatus required for complete evaluation of vacuum filtration for sludge dewatering. (*b*) Chemical is added to the sludge with carefully controlled stirring to flocculate the sludge particles. (*c*) The test filter pad is submerged in the flocculated sludge for a measured time under a controlled vacuum with filtrate being collected.

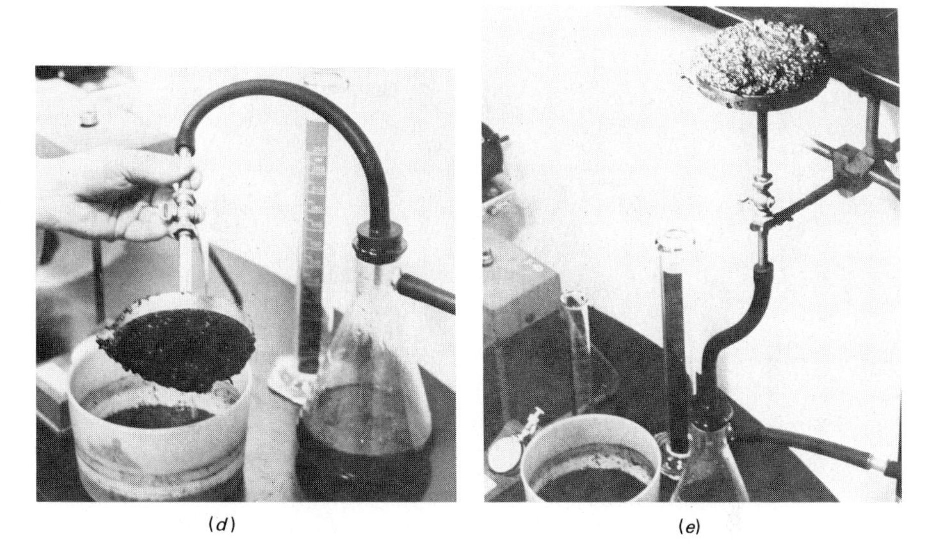

(d) (e)

FIG. 9.47 (*d*) The filter pad is removed from the pan with a motion simulating the withdrawal of the drum surface from the vat. (*e*) The vacuum is shut off after the measured time, during which the filtrate continues to flow into the collecting flask.

filtrate is settled and returned to a thickener in advance of the filter. A high percentage of solids in the filtrate can result in a buildup of recirculating solids, interfering with thickener and filter operation. Overtreating to a water-clear filtrate is not necessary since a small amount of fines will not generally cause such upsets.

Vacuum filters lend themselves to a wide variety of dewatering operations. In most cases, thickening of the slurry prior to vacuum filtration will increase filter yield substantially. The Buchner funnel test is useful in evaluation and selection of a chemical program to aid vacuum filtration. The filter leaf test can also be a valuable tool in program selection, as well as in the actual media selection and determination of filter yields in a vacuum filter installation.

Plate-and-Frame Filter Presses. In addition to centrifuge and vacuum filter dewatering, sludges can also be dewatered by pressure filtration. One method of pressure filtration is the plate-and-frame press. The press consists of vertical plates held in a frame and pressed together between a fixed and a moving end (Figure 9.48). Each plate is fitted with a medium, normally woven monofilament polypropylene. Water passes through the cloth and out the press. The solids are collected on the surface of the cloth.

Slurry is fed to the press until the flow rate drops radically. Most presses operate at pressures of 100 to 250 lb/in^2 gage. When filtrate flow stops, the pressure is relieved and the unit opened. Usually, presses are equipped with an automatic opening gear which separates the plates so that dry sludge cake can be discharged.

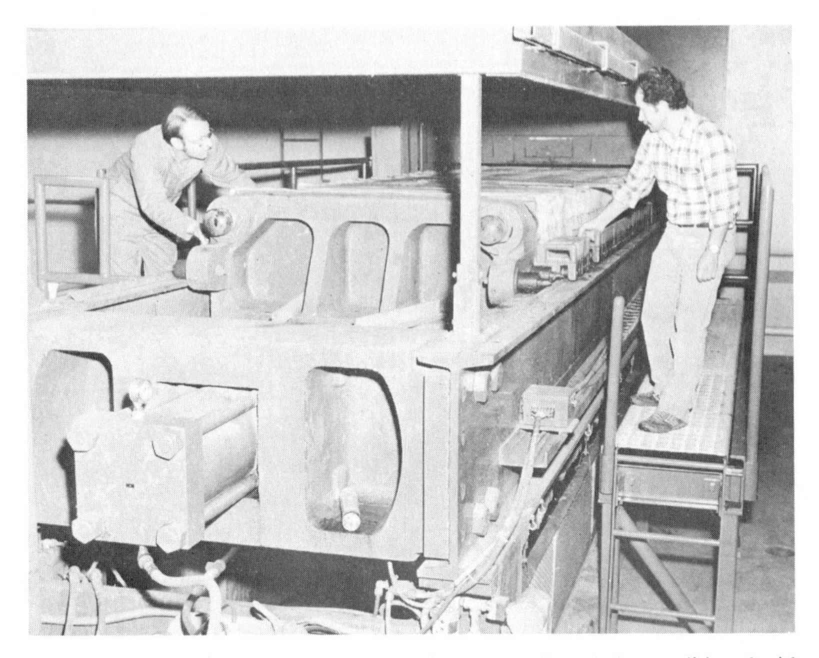

FIG. 9.48 A plate-and-filter press dewatering sewage plant sludge conditioned with polymer. Polymer increased throughput and eliminated 400 lb per ton inorganic ash previously produced by lime/ferric chloride treatment.

When the plates are separated, the cake normally falls out. Presses are commonly installed above floor level so that a hopper can be placed under the press to collect dry cake. Because of the high pressures involved, plate-and-frame presses produce much drier cakes than either centrifuges or vacuum filters. However, plate-and-frame presses are not as widely used since they operate on a batch basis rather than continuously.

Chemical conditioning aids are normally used to shorten filter time and produce drier cakes. When it is available, fly ash may be used. It is not unusual to use as much as a ton of fly ash to dewater a ton of sludge. Polyelectrolytes can be used to replace part, and in many cases, all of the fly ash. Ferric chloride and lime have also been used to condition sludges prior to plate and frame presses.

Belt Filter Press. Belt filter press dewatering is accomplished in three stages: (a) chemical conditioning of the feed sludge, (b) gravity drainage of water from the sludge through a fine-mesh endless-belt screen to yield a nonfluid mass, and (c) final compression of this nonfluid sludge. Compression is accomplished by squeezing the sludge slurry between two belts on rollers, first in a low-pressure zone and then in a higher pressure shear zone. Several typical belt press designs are shown in Figure 9.49. Some belt presses are equipped with mixers or flocculating drums ahead of the gravity drainage section to promote good conditioning of the sludge with chemicals.

Good chemical conditioning, essential to successful and consistent performance of the belt press, can be evaluated using the free-drainage test. This test selects the correct chemical and its proper dosage to produce acceptable filtrate clarity and initial drainage (dewatering) capability. This test is demonstrated in Figure 9.50.

After conditioning, free interstitial water drains from the flocculated sludge through the support mesh as the belt moves through the gravity drainage section. Typically, 1 to 2 min of drainage is required to reduce sludge volume by about 50% and increase solids concentration of the sludge to 6 to 10%.

The compression stage of the belt press begins as the sludge is squeezed by the carrying belt and cover belt coming together. Pressure can be applied in various ways depending on the belt press equipment selected for the application. Some alternative pressure application methods for compression dewatering are shown in Figure 9.51. As the belts move into the higher pressure zone, the sludge cake is squeezed between the belts and subjected to flexing in opposite directions as it passes over various rollers. This action causes further water release and greater compaction of the sludge.

There are four major operating variables that affect the performance of the belt press:

1. *Chemical conditioning:* Chemical dosage can be varied to adjust the percent solids and solids recovery in the final cake. If underconditioned, the sludge will not drain well in the gravity section, resulting in extrusion of sludge in the compression section or overflow of sludge from the drainage section. Overconditioned sludge drains so rapidly that solids cannot be uniformly distributed across the media. Both under- and overconditioning can produce blinding of the belt.

2. *Sludge flow:* The sludge feed rate affects both clarity of filtrate and cake dryness. Machine throughput increases with increasing sludge flow, but at the expense of filtrate quality and cake dryness. Polymer dosages should be increased or decreased automatically with increasing or decreasing sludge feed rates.

3. *Belt speed:* Too high a belt speed can result in sludge extruding from the sides of the belt in the pressure section and inadequate gravity drainage in the drainage section. Belt speed should be adjusted as necessary to the recommended range established by the equipment manufacturer.

4. *Pressure setting:* Too high a pressure setting in the press section hinders release of the sludge cake from the belt; too low a pressure causes a wet sludge cake. Pressure setting must be adjusted to prevailing conditions to produce best operating results.

Belt press performance is judged on the basis of the sludge cake yield, filtrate clarity, percent solids recovery (efficiency), and sludge cake density or moisture content. The efficiency is calculated as shown earlier for the centrifuge.

The amount of sludge cake throughput in dry pounds of solids per hour is the

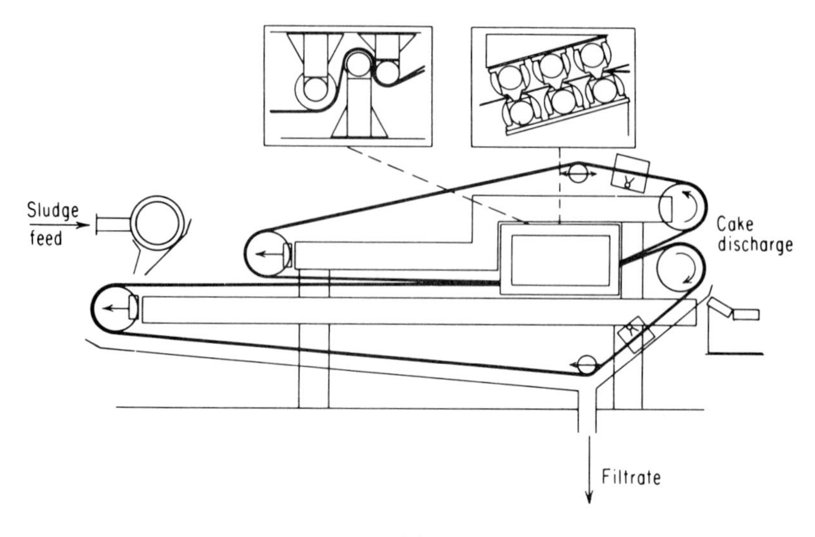

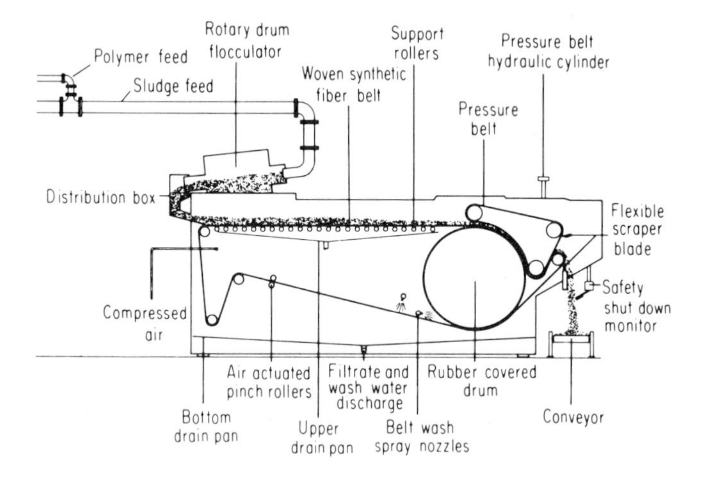

FIG. 9.49 Some designs of belt presses for sludge dewatering. (*a*) The Andritz machine was adapted from a device used for pulp dewatering. (*Courtesy of Arus-Andritz.*) (*b*) This design was developed for waste treatment plant sludges. *(Courtesy of Infilco Degremont, Inc.)*

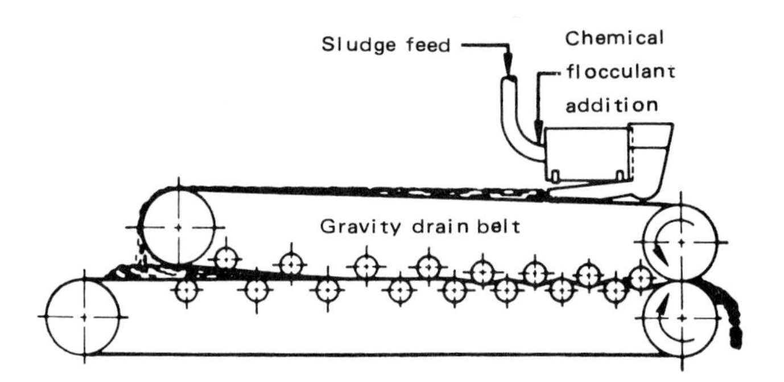

(c)

(d)

FIG. 9.49 (c) Schematic diagram of Carter press. (d) In this photo, conditioned sewage sludge is shown leaving the gravity drainage section and entering the pressure section of the belt filter press. *(Courtesy of Ralph B. Carter Company.)*

FIG. 9.50 A drainage test used to evaluate performance of chemicals applied to sludge to improve dewatering on belt presses. (*a*) Pour test loading into pipe, (*b*) lift pipe straight up, (*c*) a stable sludge cake, (*d*) fair to runny sludge cake.

yield. Yield is determined by multiplying the wet sludge cake processed in pounds per hour by the percent solids in the cake to give dry sludge solids processed.

The degree of cake dryness is a critical measure of machine performance: the drier the cake, the smaller the volume of sludge requiring final disposal. This may show up as a lower cost of hauling to a disposal site or a lower cost of incineration if this is the ultimate disposal method.

Drying Beds. Drying beds are often used to dewater waste sludges, where land is available and climatic conditions are favorable. In early designs, sand was used as the filtering medium. The typical sand bed has tile underdrains covered by 12

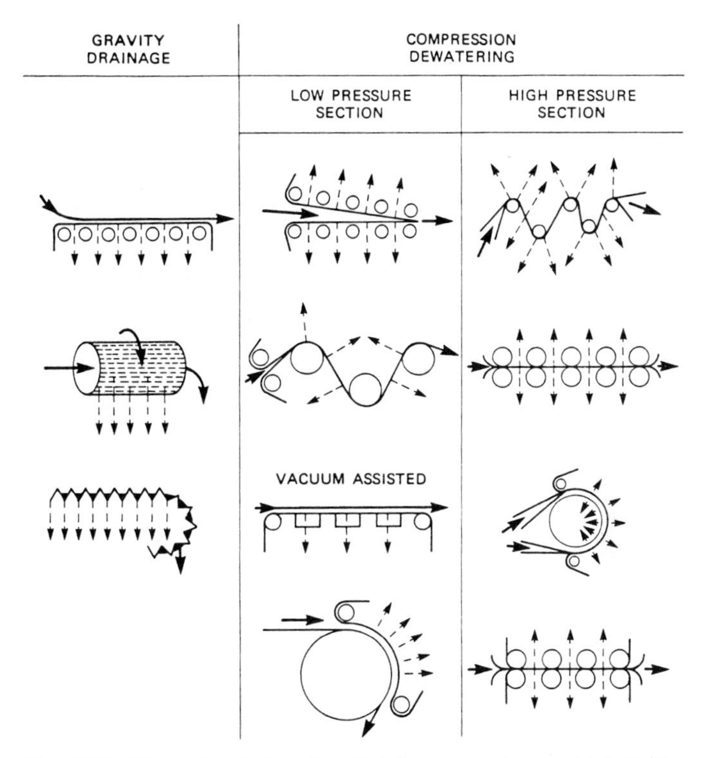

FIG. 9.51 Alternative designs for obtaining water releases with belt filter presses.

in of graded gravel, then topped with 6 to 9 in of 0.3 to 1.2 mm sand, with a uniformity coefficient less than 5.0 (Figure 9.52).

Under the right circumstances, the drying bed provides advantages over mechanical dewatering devices such as low operating costs due to low energy requirements, little maintenance, and minimal operator attention. They have the ability to handle variable sludges and to produce drier cakes than most mechanical devices. There are, however, limitations to their use, such as high land requirements and weather dependency. Since drying beds are outdoors, odor may be a problem if the sludge is high in organic matter.

For effective dewatering on drying beds, the sludge must have physical properties which allow drainage of contained water without blinding of the filtering medium. Water drainage should be rapid and relatively complete, reducing to a minimum the residue that must be evaporated. The sludge applied to the drying bed should be as thick as possible to reduce the required drying time. Chemicals that improve the release of free water have proven effective in speeding drainage and significantly reducing drying time.

Drying beds have been popular for dewatering municipal sewage sludges, including primary, aerobically or anaerobically digested, and mixtures of primary with biological sludges. Many other types of sludges are dewatered on drying beds, including water plant clarification sludges, softening sludges, and industrial sludges.

FIG. 9.52 Conventional sand drying bed with tile underdrain system, shown here during construction.

Drying bed designs generally fall into the following categories: impervious bottom beds, impervious bottom with drainage strips, sand beds, and sand beds with multiple cement tracks to provide complete mechanical unloading.

The simple impervious bottom beds and those with drainage strips are relatively easy to install and maintain, but they require longer drying times due to the reduced water drainage. Originally, sand beds were designed with clay tile as the underdrain system. The clay tile tended to break easily, ruling out the use of mechanical equipment for dry sludge removal. In addition, sludge removal with mechanical loaders has caused relatively high sand losses, which can be an important consideration if sand is scarce.

Some plants have been very successful with resilient plastic underdrain tile and specially designed buckets to skim off the dry sludge from the sand. A relatively new bed using alternating sand and cement strips throughout the entire bed, coupled with a drive-down ramp across the full width of the bed, allows a front endloader to make parallel passes when removing the dried sludge. The cement strips spaced at tire track distances prevent the bucket from digging into the sand media. This bed appears to offer the advantage of hard surface support along with open sand drainage. The use of polymer will significantly increase the dewatering speed and therefore reduce the area required.

Where sand beds are already in place, polymer addition will allow more complete drying and reduce the volume of dried sludge to be hauled away. Surface loading rates can often be two or more times as great as on beds not using polymer dewatering chemicals.

Sludge application will largely depend on local plant needs and the type of bed used. Normally, sludges are applied from 6 to 10 in deep, depending on local conditions and the amount of sludge available for drying. Addition of new sludge on top of partially dried sludge is not recommended. An exception to this may be thin digester supernate which is sent to the drying bed first in order to minimize its impact on the primary clarifier. In this case chemical must be used in order to prevent blinding of the drying bed by multiple sludge applications.

MISCELLANEOUS DEVICES AND PROCESSES

An entire volume could be devoted to the subject of solids/liquids separation devices. Only the most familiar have been discussed in this chapter. To illustrate how special problems can produce special solutions, four examples of specific solids/liquids separator designs are presented to conclude this chapter.

Cyclone Separators (Figure 9.53)

Often difficulties are encountered from the presence of very fine sand in a well water. This may be caused by improper completion of the well or failure of the well screen. For some purposes, the fine sand may be an acceptable contaminant,

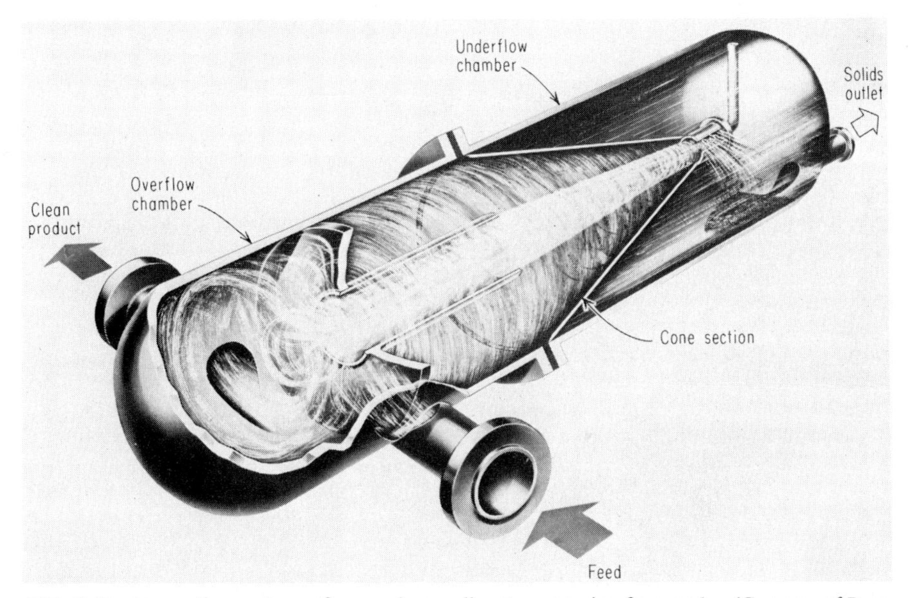

FIG. 9.53 Desanding cyclone often used on well waters carrying fine sand. *(Courtesy of Dorr-Oliver Incorporated.)*

but it may cause erosion or may plug filters and ion exchange beds. The hydraulic cyclone is a common solution to this problem, separating the sand from water by centrifugal force. These cyclones are also used extensively in the pulp/paper industry as pulp cleaners. The heavy silt, rust particles, or other debris is rather easily separated from lightweight pulp fibers by the centrifugal forces in the cyclones.

Strainers (Figure 9.54)

An alternative to cyclone separators is often a mechanical device to strain particulates from water. These units may handle particles in the size range of 10 to 200

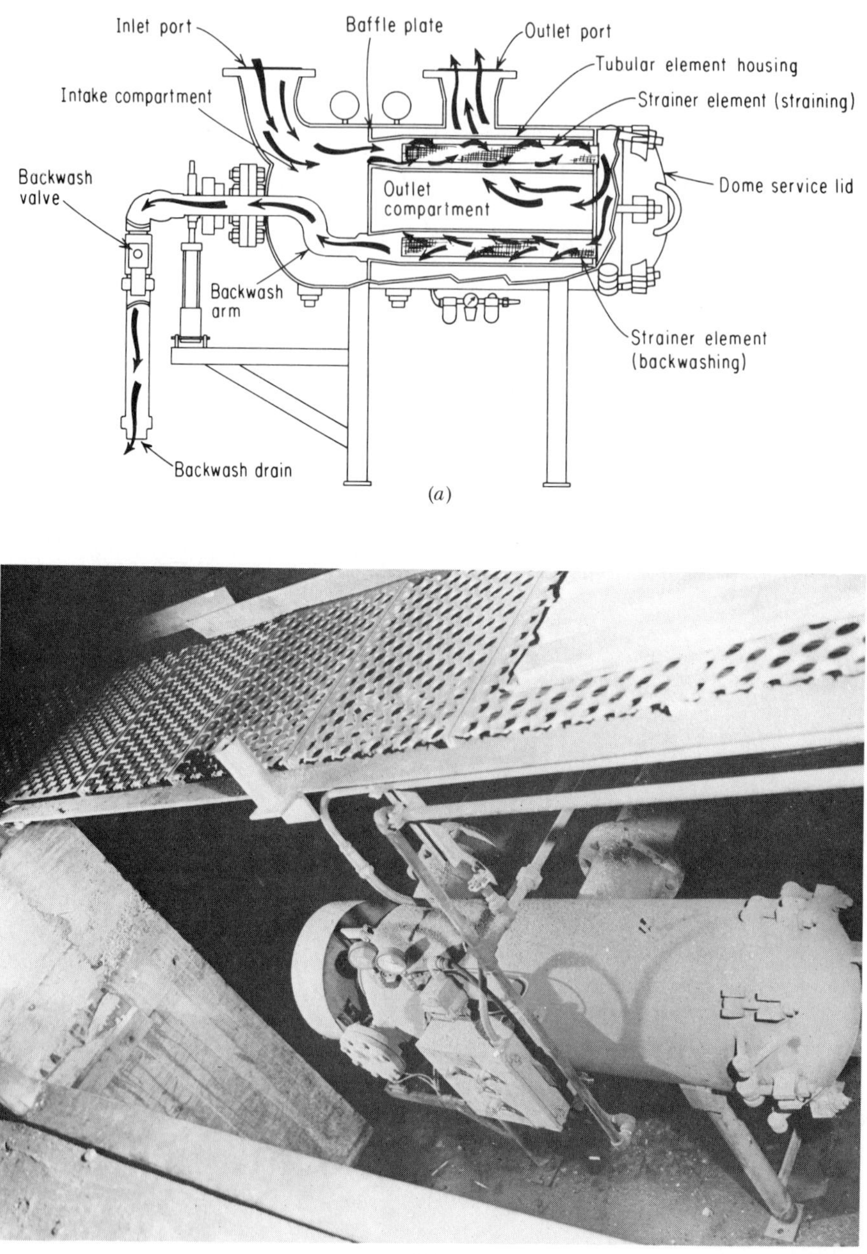

FIG. 9.54 (*a*) Cutaway showing internal details of an automatic backwash strainer. (*b*) Automatic backwash strainer in service on scale pit water in a steel mill, recirculating water for roll cooling. *(Courtesy of Ronninger-Petter Division, Dover Corporation.)*

μm. They are used where smaller suspended solids would not harm the process (as in roll-cooling in a steel mill), but larger particles could clog piping, distributors, or spray nozzles. Many designs include motor drives for rotation of the sieve openings for cleaning on a programmed basis.

Tubular Cartridge Filters (Figure 9.55)

Cartridge filters have been used in the chemical process industries for many years in such diverse applications as removal of oversize pigment from paints and removal of haze from beverages and pharmaceuticals. These devices are effective in removing particles from 0.5 to 500 μm from liquids typically containing less than 100 mg/L suspended solids.

These filters are effective in a variety of water treatment applications, including (a) cleanup of wastewater for recycle; (b) protection of pump glands and compressor jackets from scoring or scaling; (c) cleanup of water used in spray nozzles in a variety of industries; (d) safeguard against loss of resin in ion exchange systems; (e) pretreatment ahead of fine filters or membrane systems; and (f) protection of intermittently operated steam traps in remote locations.

There are two basic types of cartridge filters: the surface type and depth type, shown schematically in Figure 9.55. The surface type has a paper, mesh, or sintered septum of a specified pore size, with the solids being collected essentially on the surface. Table 9.2 can be used to relate strainer mesh to the pore sizes available for these units, but it is important to recognize that the physical measurement

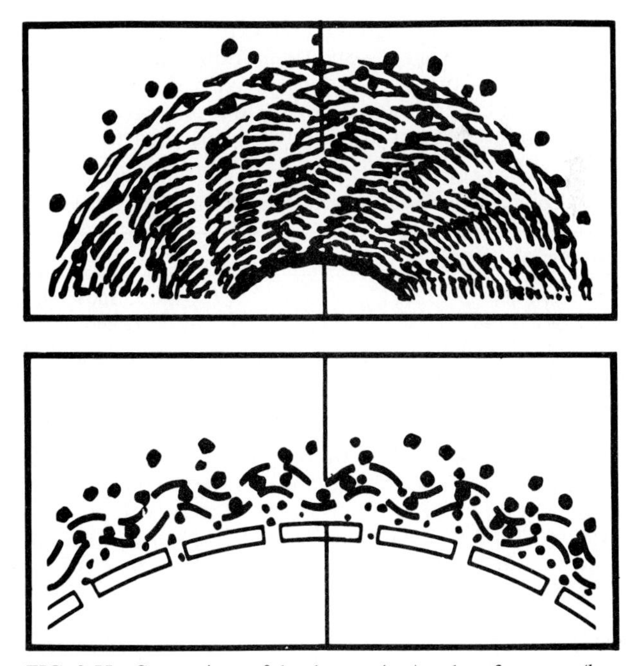

FIG. 9.55 Comparison of depth-type *(top)* and surface-type *(bottom)* cartridge cross sections. *(Courtesy of Brunswick Technetics.)*

TABLE 9.2 Strainer Size versus Size of Particle Passing Through

Mesh size*	Particle size†		Category of solids
	Inches	Micrometers	
10	0.079	2000	↑
20	0.033	840	
30	0.023	590	Settleable, solids,
40	0.017	420	visible to naked eye
60	0.012	250	↓↑
80	0.007	180	Turbidity, visible
100	0.006	150	through light
200	0.0027	70	microscope
270	0.0020	50	
325	0.0016	40	↓

* Mesh is the number of openings per unit of length, which is inches in this case.

† The opening depends both on mesh and the size of wire forming the screen. In this case U.S. Standard Screen dimensions are given. Commercial screens may not match established screen standards. For example, if stainless-steel wire is used, because it is stronger than most other materials, the wire may be smaller and the opening correspondingly larger than standard screen dimensions.

of pore size is only roughly related to the size of particles removed. For example, a filter rated at 10 μm may remove particles less than 1 μm in size after coarser solids have collected on the surface to act as a precoat, improving capture efficiency. This is particularly likely to happen in a recirculating system. On the other hand, the same 10-μm filter may pass a 20-μm (or longer) particle that has a width of only 2 to 3 μm. The depth-type filter, typically, of yarn wound on a tubular support frame, has a much higher solids retention capacity, because the solids penetrate into the depth of the yarn, but at a penalty of a higher pressure drop—about 5 to 6 times that of the surface-type cartridge of the same flow rating. Table 9.3 compares performance factors of these two types of cartridge filters.

TABLE 9.3 Comparative Performance Data for Cartridge Filters

Surface type	Depth type
1. Usually cleanable	1. Disposable
2. Low pressure drop	2. High pressure drop
3. High capture efficiency	3. High solids capacity
4. Low pumping cost	4. Low cartridge cost

Cartridge filters are usually taken out of service at a differential pressure of about 35 lb/in^2. Since the flow through these filters is laminar, pressure drop increases linearly with flow and viscosity and inversely with pore size. So a 1-μm filter may have only 20 to 30% the capacity of a 5-μm filter, and a 5-μm filter, in turn, only about 30% of the capacity of a 20-μm filter. As a result, filtration of 1-μm particles costs 8 to 10 times as much as filtration of 20-μm particles. To optimize cost, coarse filters are used for pretreatment ahead of fine filters.

A cartridge replacement in a typical filter installation is shown in Figure 9.56. A variety of cartridge element designed for submicron filtration of high purity water is shown in Figure 9.57.

FIG. 9.56 Cartridge replacement in a typical filter installation. *(Courtesy of Brunswick Technetics.)*

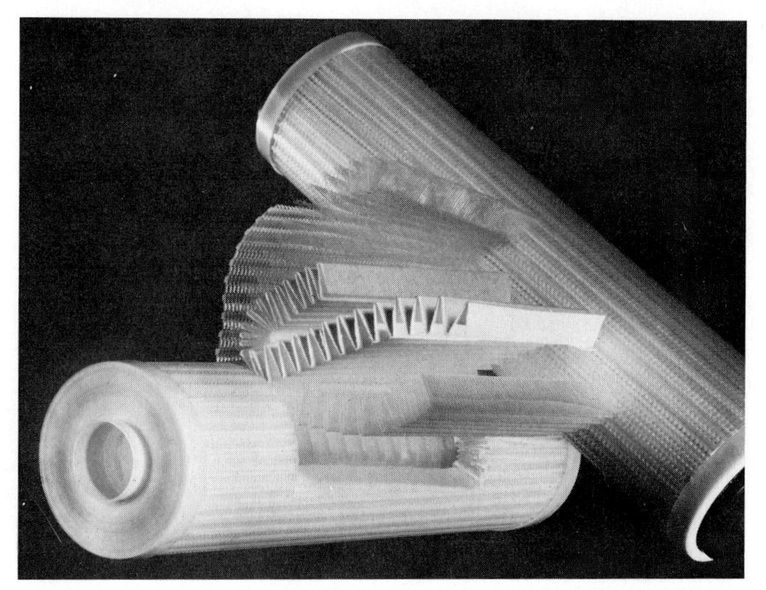

FIG. 9.57 A variety of cartridge element design for submicron filtration of high purity water. *(Courtesy of Brunswick Technetics.)*

Cross-Flow Sieves (Figure 9.58)

This type of solids/liquid separator combines the techniques of screening with initial separation caused by a change in direction of fluid flow. The screening slots are at right angles to flow, and the incline angle of the screening device is selected to give complete drainage of water through the slots before the flow reaches the base of the incline. This device is used for separation of fibrous materials from water, producing a dewatered pulp and water which may be recycled or disposed of more readily free of the bulk of solids.

Electroosmosis

Because colloidal particles are charged, they will migrate when an electric field is applied to the liquid phase. Most colloidal particles in aqueous solutions are negatively charged, so they move to the anode when a direct current potential is applied. Observed under the microscope, this is how zeta potential is determined. On an industrial scale, this principle is used to apply paint to an electrically charged surface; this is how water-based paints are applied to auto bodies.

FIG. 9.58 Inclined (side-hill) sieves, installed as trash screens in a municipal sewage plant handling a mixture of domestic and industrial wastes. *(Courtesy of C-E Bauer.)*

As a method of dewatering or densifying sludges, this method has not been commercialized, usually because the high current required makes the process uneconomical. However, a modification of this process, called electroendoosmosis, has been used to densify clay residues from phosphate mining tailings.

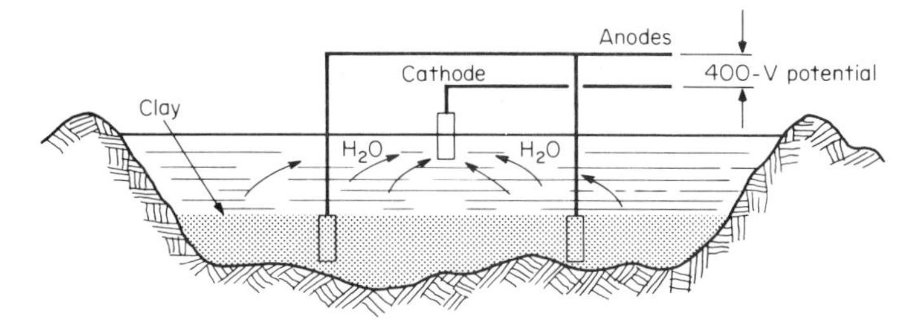

FIG. 9.59 Application of electroendoosmosis to dewatering of phosphate slimes.

These particles are so highly charged that they seldom settle to a density of more than about 30% solids in the tailings ponds, even after years of quiescence. In electroendoosmosis (Figure 9.59), water migrates out of the mass of negatively charged clay sludge and moves to the cathode, and the space it vacates is replaced with clay particles. In one Tennessee mine, a single cathode surrounded by eight anodes, 40 ft long, operating at a 400-V potential, draws about 40 A of direct current, so that power consumption is quite small. As the areas around the steel anodes are densified, the anodes are relocated to other areas of lower solids density. This same principle has been used in the construction industry on a much smaller scale to stabilize cuts through clay or soil banks during highway or railroad construction.

SUGGESTED READING

AWWA: *Water Quality and Treatment,* McGraw-Hill, New York, 1971.

Cartridge Filtration Guide, Brunswick Technetics Division, 1984.

"Electrical Separation Tested in Phosphate Settling Ponds," *Chem. Eng. News,* January 30, 1984.

Fair, G. M., Geyer, J. C., and Okun, D. A.: *Elements of Water Supply and Wastewater Treatment,* Wiley, New York, 1981.

New York State Department of Health: "Manual of Instruction for Sewage Treatment Plant Operators," Health Education Service, Albany, undated.

U.S. Environmental Protection Agency: "Process Design Manual for Sludge Treatment and Disposal," Municipal Environmental Research Laboratory, Office of Research and Development, September 1979.

Water Pollution Control Federation: *WPCF Manual of Practice No. 8,* "Sewage Treatment Plant Design," Washington, D.C., 1963.

Water Treatment Handbook, Infilco-Degrémont, Inc., Halsted Press, New York, distributors, 1979.

CHAPTER 10
PRECIPITATION

Once water has been taken from its source, where it may have been in a state of equilibrium, it is often exposed to pumping, aeration, and heating, any of which may upset its stability and lead to corrosion or scaling. Whether a particular water will tend to corrode metal or form a $CaCO_3$ scale can be roughly predicted by its stability index, which can be calculated from the solubility product of calcium carbonate and the concentrations of certain ions in the water. The same principles used in predicting water stability apply to all precipitation processes.

The precipitation process makes use of the solubility product of a compound containing an ion or radical that is considered detrimental, and that should, therefore, be removed before the water is put to use. The reduction of calcium ion concentration by precipitation as calcium carbonate is one example of this.

Adsorption is a process with some similarities to precipitation. The choice of an adsorbent and the degree of removal that can be achieved may be only roughly determined; data for estimating purposes may be found in technical literature, and these are useful as a guide to bench evaluation of the selected process. An example of adsorption is removal of silica from water on magnesium hydroxide precipitate.

Temperature is an important factor in both precipitation and adsorption reactions. The solubility product is affected by temperature; knowing the solubility characteristics of the desired precipitate will influence the selection of treatment equipment. For example, if preheating water to a higher temperature produces improved results over those expected at ambient temperature, then heat exchangers may be justified. Temperature also influences the rates of all chemical reactions, and heating may make it possible to select smaller reaction or sedimentation vessels for the process.

One of the fundamental principles of precipitation is that the size of a precipitate increases if the chemical reaction is encouraged to occur on previously precipitated particles. If a small crystal and a large crystal of the same substance are placed in a saturated solution of the substance in a beaker, the small crystal will slowly disappear as the larger one grows; if a crystal of salt is introduced into a supersaturated, clear salt solution, it will grow as salt comes out of solution on the surface of the seed crystal in preference to forming individual crystal nuclei. Because of these reactions, most precipitation processes in water treatment are conducted by introducing precipitating chemicals into the water in the presence of previously precipitated sludge.

SOFTENING BY PRECIPITATION

Lime softening, the most widely used precipitation process, serves well to illustrate the importance of four key variables in precipitation: (1) solubility, (2) particle charge, (3) temperature, and (4) time. Lime softening is the reduction of hardness by the application of hydrated lime to water to precipitate $CaCO_3$, $Mg(OH)_2$, or both.

At first glance, it may appear paradoxical that lime, a compound of calcium, can be added to water to remove calcium; the explanation is that the hydroxyl radical is the reactive component of lime, converting CO_2 and HCO_3^- to CO_3^{2-}, causing $CaCO_3$ to precipitate, as shown by reactions (1), (2), and (3).

$$Ca(OH)_2 \rightleftharpoons Ca^{2+} + 2OH^- \qquad (1)$$

$$Ca(OH)_2 + 2CO_2 \rightarrow Ca(HCO_3)_2 \qquad (2)$$

or, in ionic form:

$$2OH^- + 2CO_2 \rightarrow 2HCO_3^-$$

$$Ca(OH)_2 + Ca(HCO_3)_2 \rightarrow 2CaCO_3 \downarrow + 2H_2O \qquad (3)$$

in ionic form:

$$2OH^- + 2HCO_3^- \rightarrow 2CO_3^{2-} + 2H_2O$$

Other hydroxide compounds (NaOH, KOH) could also be used, but these cannot usually compete against the low cost of lime except in special circumstances where they may be available as by-products.

Most softening reactions are carried out at a pH of about 10. At this pH, $CaCO_3$ usually carries a negative, and $Mg(OH)_2$ a positive, charge. If these charges are not neutralized, colloidal hardness may resist flocculation and carry over into the effluent. Cationic coagulants may be needed when the bulk of the precipitate is negatively charged $CaCO_3$, as in partial lime softening. Sodium aluminate is frequently used as an anionic coagulant when low magnesium residuals [positively charged $Mg(OH)_2$] are needed in complete lime softening. The coagulant may be supplemented by an anionic or a nonionic flocculant.

The water chemist distinguishes between cold process lime softening, usually carried out in the range of 40 to 90°F (4 to 32°C), and hot process, at 215 to 230°F (102 to 110°C). Results achievable at intermediate temperatures are often of interest. Obtaining data in the intermediate range may require bench testing, as very few plants have operated in this range. At about 120°F (49°C), silica removal, which is negligible in cold process softening at temperatures close to freezing, increases. At about 140 to 160°F (60 to 71°C), lime softening of sewage, which is rather incomplete cold, begins to approach hot process results. These phenomena illustrate the importance of laboratory testing under actual operating conditions to obtain reliable data before a plant is designed and built.

PARTIAL LIME SOFTENING

The most prevalent precipitation process in water treatment is the reduction of calcium hardness by its precipitation as $CaCO_3$. Since the alkalinity of most waters is in the bicarbonate form, and because there is usually CO_2 present, the precipitation of calcium carbonate requires the conversion of CO_2 and bicarbonate to carbonate, as shown by reactions (1), (2), and (3).

As was pointed out previously, $CaCO_3$ precipitation is not as simple a reaction as it appears to be because of the formation of ion pairs and because of interferences, such as the stabilization of colloidal calcium carbonate preventing its precipitation in the normal reaction time. Because of this, the theoretical solubility curve is of limited value in predicting the actual results that might be achieved in a precipitation reaction such as partial lime softening. For estimating purposes, the solubility as shown by Figure 10.1 is commonly used in the water treatment industry. This assumes the solubility of calcium carbonate at ambient river water or well water temperatures to be about 35 mg/L as $CaCO_3$. (Obviously, if the water supply is already less than 35 mg/L in calcium hardness, lime softening for calcium reduction is of no value.) It also assumes about 60 min detention in the reaction tank.

In predicting the results of partial lime softening, there are two cases to be considered, one in which the calcium hardness of the raw water exceeds its alkalinity and the other in which the alkalinity exceeds the calcium hardness.

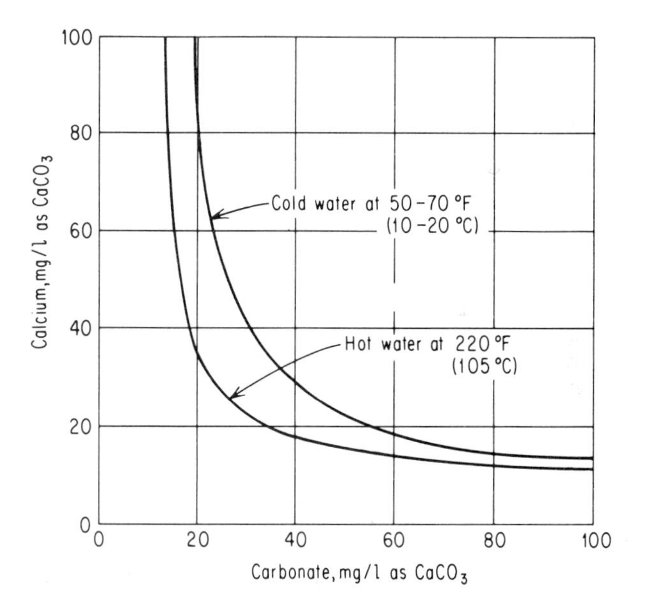

FIG. 10.1 The data used for estimating calcium carbonate residuals in conventional lime-softening plants with 60- to 90-min detention.

Where the calcium exceeds the alkalinity (Figure 10.2), the results of treatment are calculated by establishing first that there has been no change in any anions except alkalinity, which is converted to 35 mg/L CO_3, as $CaCO_3$. In the cation portion of the analysis, the magnesium is shown as being reduced by about 10%. The sodium remains unchanged, and the calcium is then calculated by difference. Figure 10.2 shows an analysis of a raw water having a calcium hardness exceeding the alkalinity, with results after partial lime softening, following this method of

Identification of Analyses Tabulated Below:

A. Raw water @ 60°F D. ________

B. Estimated results E. ________

C. ________ F. ________

Constituent	As	A	B	C	D	E	F
Calcium	$CaCO_3$	125	65				
Magnesium	"	65	60				
Sodium	"	20	20				
Total Electrolyte	$CaCO_3$	210	145				
Bicarbonate	$CaCO_3$	100	0				
Carbonate	"	0	35				
Hydroxyl	"	0	0				
Sulfate	"	60	60				
Chloride	"	50	50				
Nitrate	"						
M Alk.	$CaCO_3$	100	35				
P Alk.	"	0	18				
Carbon Dioxide		10	10				
pH		7.3	10.0				
Silica		10	10				
Iron	Fe	0.2	Nil				
Turbidity							
TDS		230	165				
Color							

FIG. 10.2 Partial cold lime softening results where Ca > M alkalinity in the raw water analysis.

calculation. The approximation of pH is taken from alkalinity relationships discussed in an earlier chapter.

Where the alkalinity exceeds the calcium (as shown in Figure 10.3), the cation section of the analysis is calculated first: the calcium is shown as 35 mg/L; the magnesium is reduced by 10%, and the sodium is unchanged. In the anion section of the analysis, except for alkalinity, the remaining anions are unchanged. The total alkalinity is then calculated by difference. Of the total alkalinity, at least 35 mg/L must be present as carbonate. The balance may be bicarbonate, or as much of this may be converted to carbonate as wished, depending on the desired stability index and pH of the system. Figure 10.3 shows a raw water having an alkalinity in excess of calcium hardness and shows the results of treatment, with several examples of varying degrees of conversion of bicarbonate to carbonate.

Identification of Analyses Tabulated Below:

A. Raw water @ 60°F

B. Estimated results with complete conversion to CO_3

C. Estimated results with partial conversion to CO_3

D.

Constituent	As	A	B	C	D	E	F
Calcium	$CaCO_3$	100	35	35			
Magnesium	"	50	45	45			
Sodium	"	75	75	75			
Total Electrolyte	$CaCO_3$	225	155	155			
Bicarbonate	$CaCO_3$	125	0	20			
Carbonate	"	0	55	35			
Hydroxyl	"	0	0	0			
Sulfate	"	50	50	50			
Chloride	"	50	50	50			
Nitrate	"						
M Alk.	$CaCO_3$	125	55	55			
P Alk.	"	0	28	18			
Carbon Dioxide		0	0	0			
pH		8.3	10.0	9.8			
Silica		8	8	8			
Iron	Fe	0.3	Nil	Nil			
Turbidity							
TDS		255	185	185			
Color							

FIG. 10.3 Partial cold lime softening results where M alkalinity > Ca in the raw water analysis.

With experience in using such calculations and in reviewing actual plant results, the water treatment engineer can improve accuracy in prediction of results of partial lime softening. For example, it is apparent that if the calcium is greatly in excess of alkalinity in the raw water, the shape of the solubility product curve is such that the carbonate would be lowered below 35 mg/L. Carbonate alkalinities as low as 20 mg/L have been observed in partial lime softening of well water at 55°F (13°C) having an excess calcium of more than 200 mg/L.

The reason for showing a magnesium reduction despite the fact that this may not be one of the goals of the treatment program is that it is impossible to instantaneously mix a slurry of lime into a large body of water; at the point of lime introduction the water is massively overtreated with lime, and it is inevitable that some magnesium will precipitate because of this. The 10% reduction is arbitrary

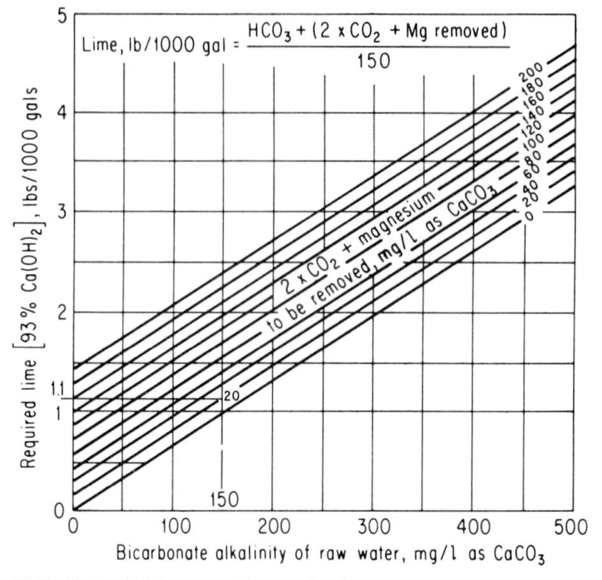

FIG. 10.4 Cold process lime softening.

and empirical, based on a ratio of Ca:Mg of 2:1; if this ratio is reduced, magnesium reduction increases.

$$Mg^{2+} + Ca(OH)_2 \rightarrow Mg(OH)_2 \downarrow + Ca^{2+} \quad (4)$$

In this process, the lime requirement is based on the CO_2, alkalinity converted to carbonate, and magnesium reduction. A simplified chart for determining lime requirement is shown in Figure 10.4.

COMPLETE LIME SOFTENING

Sometimes the residual calcium hardness after partial lime softening may still be higher than the municipality or industrial operation requires. If so, additional calcium reduction is achieved by adding soda ash. The reaction is as follows:

$$Ca^{2+} + Na_2CO_3 \rightarrow CaCO_3 \downarrow + 2Na^+ \quad (5)$$

As calcium is precipitated as $CaCO_3$, it is replaced by the sodium from the soda ash. Using the same example as shown earlier for partial lime softening, Figure 10.2, the further reduction of calcium hardness by soda ash addition in several stages is shown in Figure 10.5. The amount of soda ash required is simply calculated on the basis of the additional calcium hardness reduction wanted. This addition can continue until the calcium reaches a level of about 35 mg/L; beyond this, excess soda ash has only a partial, rather than a direct, effect in reducing calcium further. Figure 10.5 shows the results of excess soda ash addition on calcium hardness, alkalinity, and dissolved solids.

Identification of Analyses Tabulated Below:							
A. Raw water @ 60°F				D. "C" after an additional 30 mg/l soda ash*			
B. Estimated results with lime				E.			
C. "B" after addition of 30 mg/l soda ash*				F.			
Constituent	**As**	**A**	**B**	**C**	**D**	**E**	**F**
Calcium	$CaCO_3$	125	65	35	20		
Magnesium	"	65	60	60	60		
Sodium	"	20	20	50	80		
Total Electrolyte	$CaCO_3$	210	145	145	160		
Bicarbonate	$CaCO_3$	100	0	0	0		
Carbonate	"	0	35	35	50		
Hydroxyl	"	0	0	0	0		
Sulfate	"	60	60	60	60		
Chloride	"	50	50	50	50		
Nitrate	"						
M Alk.	$CaCO_3$	100	35	35	50		
P Alk.	"	0	18	18	25		
Carbon Dioxide		10	0	0	0		
pH		7.3	10	10	10		
Silica		10	10	10	10		
Iron	Fe	0.2	Nil	Nil	Nil		
Turbidity							
TDS		230	165	165	180		
Color							

*Expressed as $CaCO_3$ ≡ 32 mg/l Na_2CO_3

FIG. 10.5 Cold lime–soda softening for calcium reduction.

If reduction of magnesium is desired along with calcium reduction, additional lime must be added beyond that required for partial softening to react with all of the magnesium and to provide an excess hydroxide alkalinity. The procedure for estimating results again depends on which of two categories the water falls into: in the first, the total hardness exceeds total alkalinity, and in the second the reverse is true.

Where total hardness exceeds alkalinity, the anions are calculated first. There is no change in anions other than alkalinity. Since magnesium reduction is desired, an excess hydroxide alkalinity of 20 mg/L is usually selected. The carbonate alkalinity is shown as 35 mg/L, and there is no bicarbonate alkalinity. In the cation section, the sodium is unchanged and the magnesium is reduced to 20 mg/L. The calcium is then determined by difference. If this calcium level is still

Identification of Analyses Tabulated Below:	
A. Raw well water @ 55°F	D.
B. After lime treatment	E.
C. After lime-soda treatment	F.

Constituent	As	A	B	C	D	E	F
Calcium	$CaCO_3$	175	110	20			
Magnesium	"	100	20	20			
Sodium	"	25	25	130			
Total Electrolyte	$CaCO_3$	300	155	170			
Bicarbonate	$CaCO_3$	200	0	0			
Carbonate	"	0	35	50			
Hydroxyl	"	0	20	20			
Sulfate	"	60	60	60			
Chloride	"	40	40	40			
Nitrate	"						
M Alk.	$CaCO_3$	200	55	70			
P Alk.	"	0	38	45			
Carbon Dioxide		20	0	0			
pH		7.3	10.6	10.6			
Silica		15	12	12			
Iron	Fe	0.5	Nil	Nil			
Turbidity							
TDS		350	205	220			
Color							

FIG. 10.6 Complete cold lime–soda softening where raw water hardness exceeds alkalinity. The 20% reduction in silica can be improved by longer detention, higher solids in the recirculation zone, temperature, or a combination of these.

higher than desired, soda ash is added to reduce it, and each increment of calcium reduction results in an incremental sodium increase.

The lime is again calculated as that required for CO_2, bicarbonate conversion, and magnesium reduction plus an excess of 20 mg/L. The soda ash dosage is calculated by the desired calcium reduction in a stepwise fashion. Figure 10.6 shows complete cold process lime soda softening of a well water where the hardness exceeds the alkalinity.

At one time, batch lime–soda softening was a common process in treating water for steam locomotives. Hardness levels considerably below those shown in Figure 10.6 were achieved because of the long detention time. Batch softeners are still in use in some steel mills today producing final hardness of less than 20 mg/L.

When the alkalinity exceeds the hardness, the cation section is calculated first, and soda ash is never used. The sodium is shown unchanged, the calcium is reduced to 35 mg/L, and the magnesium reduced to 20 mg/L. The anions other than alkalinity remain unchanged, the hydroxide alkalinity is shown as 20 mg/L, and the balance is carbonate. If this carbonate alkalinity is too high, it can be reduced by the addition of gypsum, $CaSO_4$, each increment of gypsum producing an equivalent reduction of alkalinity, according to the following reaction:

$$CO_3^{2-} + CaSO_4 \rightarrow CaCO_3 \downarrow + SO_4^{2-} \qquad (6)$$

In all of these cold process reactions, it is extremely important that previously precipitated sludge be returned to the reaction chamber for mixing with raw water and treatment chemicals. A typical design of a cold process lime softening precipitation unit is shown in Figure 10.7. Another design, shown in Figure 10.8(*a*), is a unit in which the reactions occur on a bed of calcium carbonate granules. This is more compact than the first design, but it requires sophisticated chemical feed equipment so that the chemical feed rate is instantaneously adjusted to changes in water flow rate, since there is no provision for recirculation within the reaction vessel itself. The detention time is only about 12 to 15 min. The precipitation of $CaCO_3$ from the lime reaction occurs directly on the granules, but neither $Mg(OH)_2$, $Fe(OH)_3$, nor raw water suspended solids are trapped, so these remain in the effluent. This type unit has had a long history of use in Europe on hard well waters. It is sometimes used as the first stage of treatment, with the effluent going to a conventional sludge-contact softener for final treatment, including removal of suspended solids.

In all of these softening systems, a check of the Langelier index shows that the treated water is supersaturated with respect to $CaCO_3$. After-precipitation will occur, hastened if the water is heated, unless it is further treated or stabilized. The commonly used treatment is pH reduction, addition of a polyphosphate or an equivalent sequestering agent, or both. Recarbonation is usually the preferred method of pH reduction, especially in municipal softening plants, although acid is a common substitute. Recarbonation, however, can often do more harm than good if the softened water contains OH alkalinity, because this will be converted to carbonate in reacting with CO_2, further supersaturating the water with $CaCO_3$.

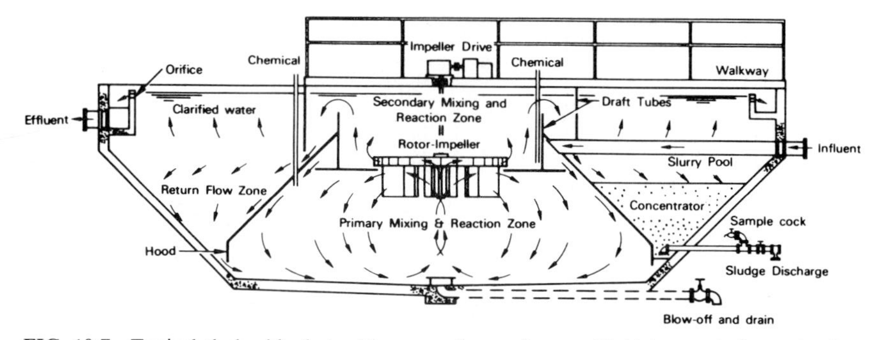

FIG. 10.7 Typical sludge-blanket cold process lime softener with high-rate sludge recirculation. *(Courtesy of Infilco Degremont Incorporated.)*

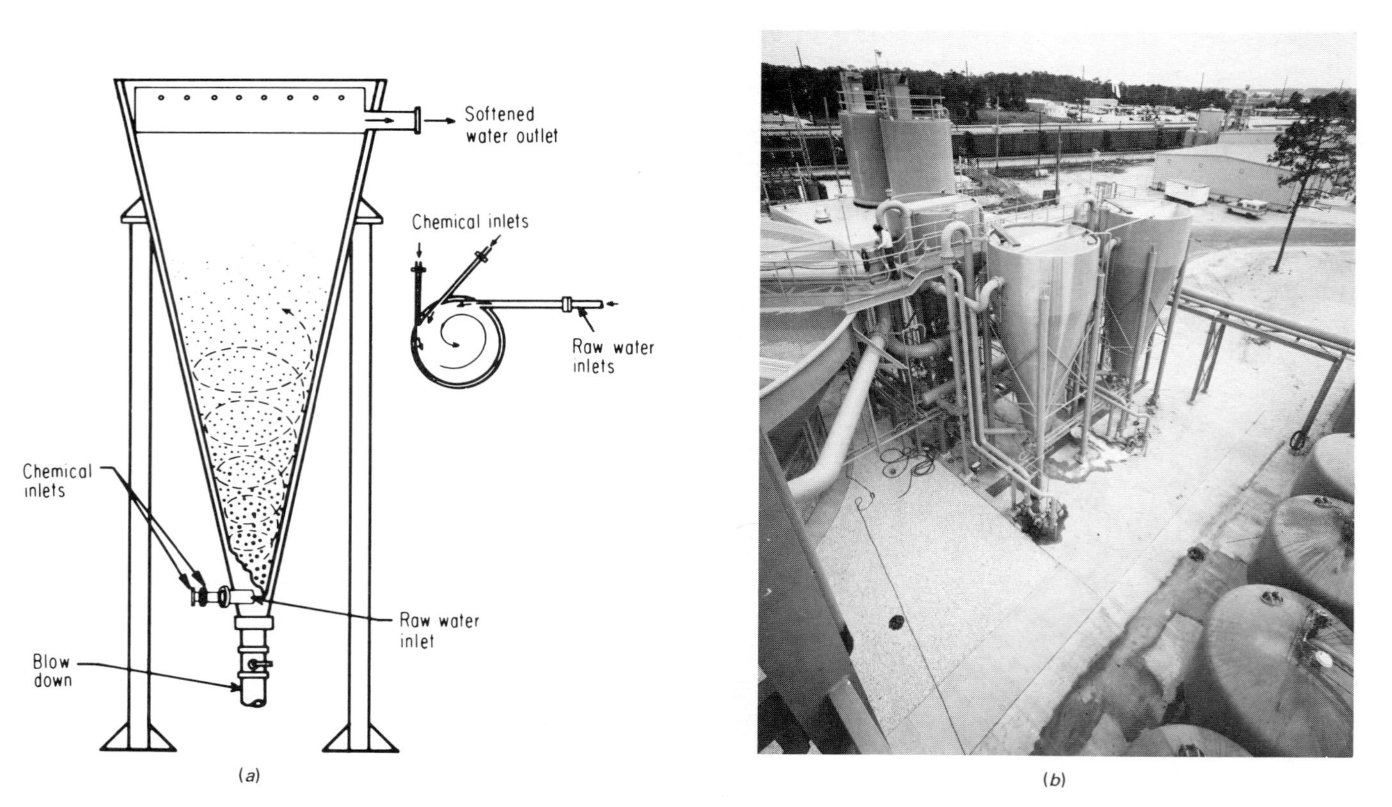

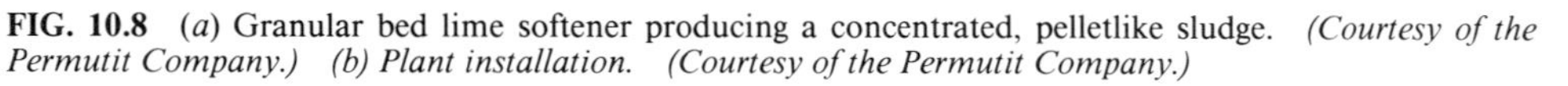

FIG. 10.8 (*a*) Granular bed lime softener producing a concentrated, pelletlike sludge. *(Courtesy of the Permutit Company.)* *(b) Plant installation.* *(Courtesy of the Permutit Company.)*

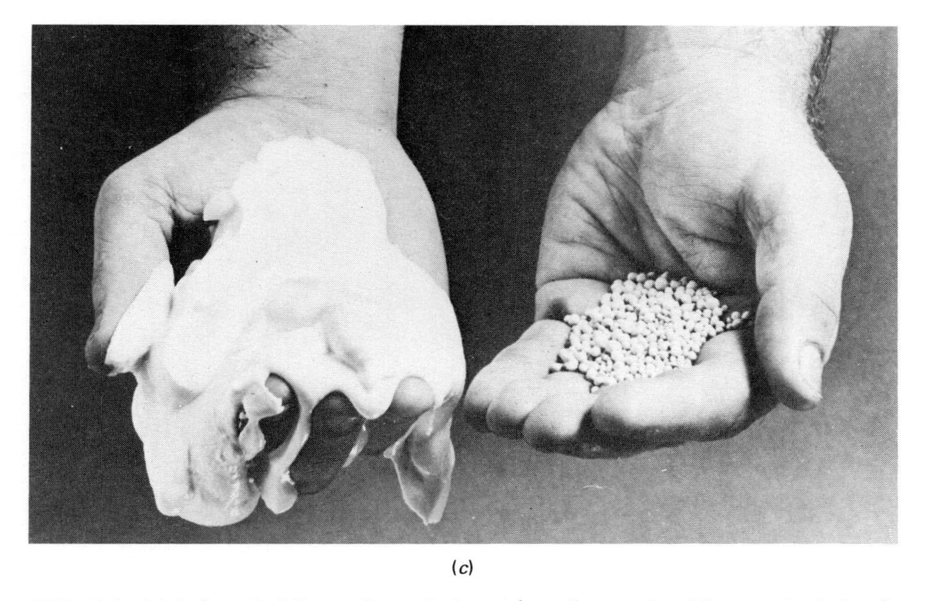

(*c*)

FIG. 10.8 (*c*) *Left,* typical lime softener sludge; *right,* pellets produced by granular bed softener. *(Courtesy of the Permutit Company.)*

This often results in filter plugging, scaling of piping, deposits in low-flow areas, or all three. (See Chapter 13, Neutralization.)

MORE COMPLETE SOFTENING AT HIGHER TEMPERATURES

Hot process softening is somewhat different in that the solubility of calcium and magnesium are both lower at elevated temperatures, and the rate of reaction is considerably increased. The $CaCO_3$ precipitates as calcite in the cold process and aragonite in the hot process. The precipitates settle much faster in hot water,

Identification of Analyses Tabulated Below:

A. Raw well water @ 55°F D.

B. After treatment* E.

C. F.

Constituent	As	A	B	C	D	E	F
Calcium	$CaCO_3$	175	15				
Magnesium	"	100	2				
Sodium	"	25	133				
Total Electrolyte	$CaCO_3$	300	150				
Bicarbonate	$CaCO_3$	200	0				
Carbonate	"	0	40				
Hydroxyl	"	0	10				
Sulfate	"	60	60				
Chloride	"	40	40				
Nitrate	"						
M Alk.	$CaCO_3$	200	40				
P Alk.	"	0	25				
Carbon Dioxide		20	0				
pH		7.3	10.3				
Silica		15	1-2				
Iron	Fe	0.5	0				
Turbidity							
TDS		350	185				
Color							
* Not corrected for about 15% steam dilution							

FIG. 10.9 Hot process lime–soda softening at 220°F.

whose density and viscosity are appreciably less than cold water's. There are two basic hot process systems, lime-soda and lime-zeolite.

The lime-soda process is similar to the cold process system, but the results are different, as shown in Figure 10.9. The amount of lime required is less because usually none is needed for CO_2 reaction, since most CO_2 is eliminated by heating the water and spraying it into the reaction vessel. Some high CO_2 well waters are difficult to degas, and may require excess lime for the residual CO_2. Silica is appreciably reduced by adsorption on the magnesium hydroxide precipitate, and the residual hydroxide alkalinity needed for magnesium precipitation is also considerably reduced.

There are two basic designs of hot process softeners applicable either to the lime-soda process or the lime-zeolite process. The first (Figure 10.10) is a sludge blanket design, particularly effective where silica reduction is important, but somewhat sensitive to load fluctuations. The second (Figure 10.11) is called a downflow design, and depends on a recirculation pump entirely for providing sludge contact; it is generally not quite as effective for silica reduction as the sludge blanket unit, but is less susceptible to upset caused by load fluctuations.

With the lime-zeolite design, complete hardness removal is achieved and lower alkalinity levels can be produced because soda ash is not required for reducing calcium hardness. Figure 10.12 shows the results of treating a water high in excess hardness. If the water has an alkalinity in excess of hardness, alkalinity reduction can be achieved either by acid feed ahead of the softener or by the addition of gypsum to the reaction zone. Figure 10.13 illustrates a complete lime-zeolite system.

SILICA REMOVAL

Although silica can be removed by adsorption on iron floc in the coagulation process, efficiency is low. For that reason, where silica removal is required, it is usually done in a hot or warm process system. The silica is adsorbed on magnesium hydroxide precipitate, which is either formed from the lime softening reaction or added by using dolomitic lime as the water-softening reagent. Typically, the equipment used is similar to the hot process softeners shown in Figures 10.10 and 10.11. Since the process can be made to operate effectively at temperatures as low as 120°F (49°C), these systems can be modified to maintain this temperature within rather close limits, which is necessary to avoid thermal currents that would upset the sedimentation process. The results anticipated by precipitation of magnesium hydroxide for hot process adsorption of silica are shown by Figure 10.14.

Rules of thumb can often be misleading, and the frequent statement that silica removal is poor in cold process softening is in that category; it assumes that (1) in winter, the water will be close to 32°F (0°C); (2) the detention time in the softener is only about 60 min; and (3) the precipitation of magnesium is only in the range of 20 to 30%.

Under proper conditions, silica removal can be quite effective in the cold process, as shown in the treatment of a 70°F (21°C) well water in Mississippi with a softener having effective solids recirculation and substantial magnesium precipitation (Figure 10.15). Silica residuals are shown for three conditions of operation. In some plants having high CO_2 concentrations in the raw water, silica removal is enhanced by installing a magnesium dissolving basin ahead of the softener. A

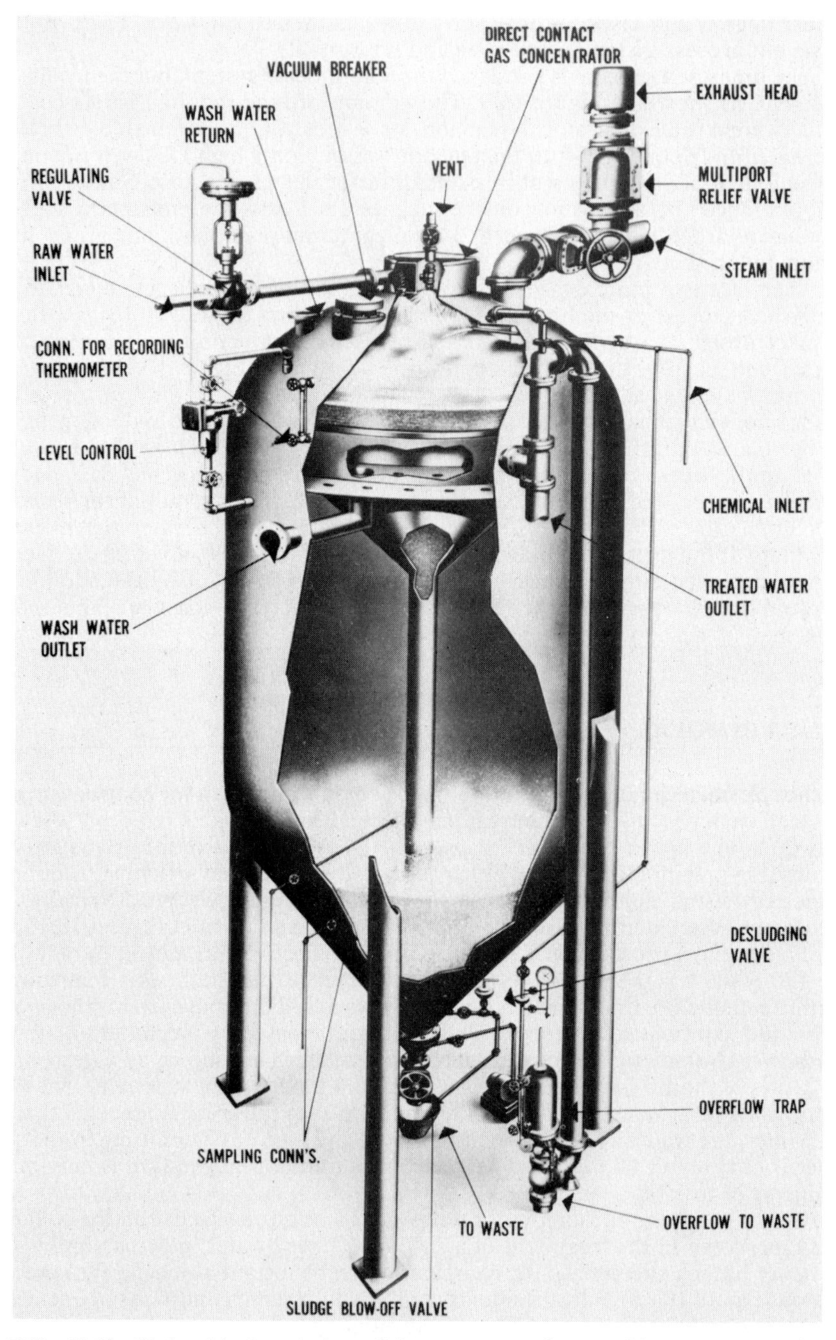

FIG. 10.10 Sludge blanket design of hot process softener with accessory equipment. *(Courtesy of Cochrane Division, the Crane Company.)*

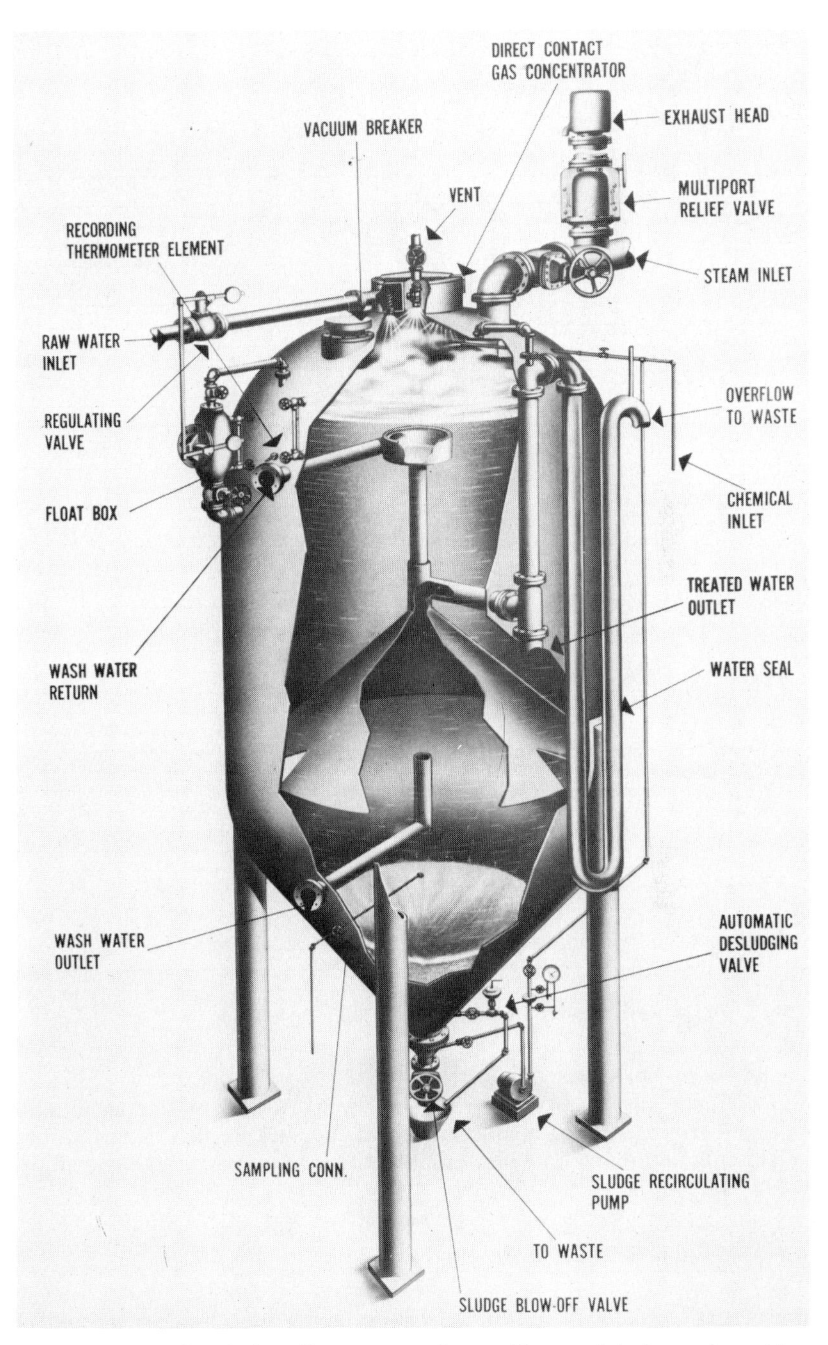

FIG. 10.11 Downflow design of hot process softener with control devices and provision for sludge recirculation. *(Courtesy of Cochrane Division, the Crane Company.)*

Identification of Analyses Tabulated Below:

A. Raw well water @ 55°F D.

B. After lime treatment* E.

C. "B" after zeolite treatment* F.

Constituent	As	A	B	C	D	E	F
Calcium	$CaCO_3$	175	98	Nil			
Magnesium	"	100	2	Nil			
Sodium	"	25	25	125			
Total Electrolyte	$CaCO_3$	300	125	125			
Bicarbonate	$CaCO_3$	200	0	0			
Carbonate	"	0	15	15			
Hydroxyl	"	0	10	10			
Sulfate	"	60	60	60			
Chloride	"	40	40	40			
Nitrate	"						
M Alk.	$CaCO_3$	200	25	25			
P Alk.	"	0	18	18			
Carbon Dioxide		20	0	0			
pH		7.3	10.3	10.3			
Silica		15	1-2	1-2			
Iron	Fe	0.5	0	0			
Turbidity							
TDS		350	160	160			
Color							

* Not corrected for approximately 15% steam dilution

FIG. 10.12 Hot process lime–zeolite treatment at 220°F.

portion of the sludge from the softener is recycled to this mixing basin and the CO_2 selectively redissolves magnesium from the sludge, increasing the magnesium subsequently reprecipitated in the softener to improve silica adsorption.

HEAVY METALS REMOVAL

Heavy metals are usually removed from water by precipitation, although ion exchange and adsorption are also used. Iron is a typical example of a heavy metal requiring removal, because it is a common constituent of well water and must be

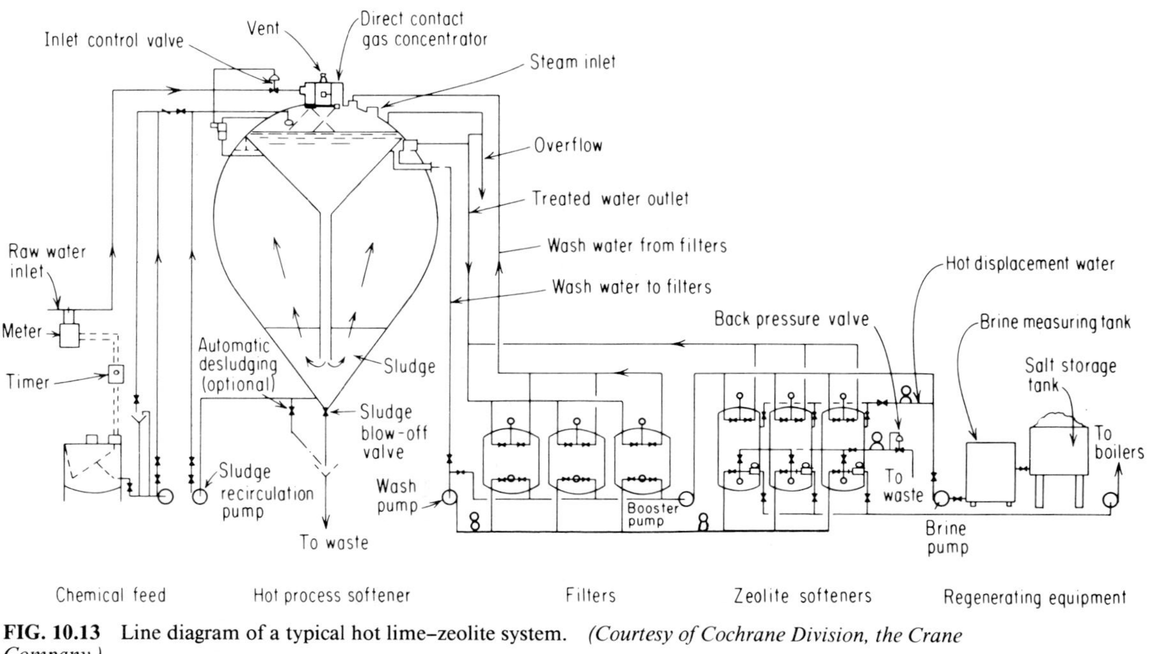

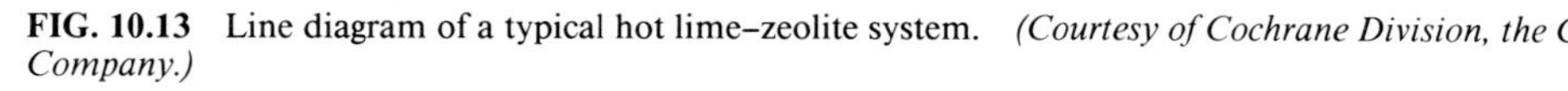

FIG. 10.13 Line diagram of a typical hot lime–zeolite system. *(Courtesy of Cochrane Division, the Crane Company.)*

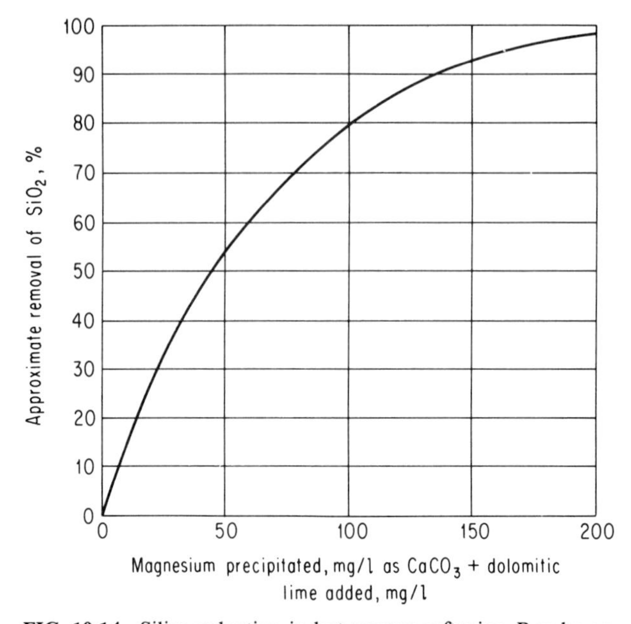

FIG. 10.14 Silica reduction in hot process softening. Results are affected by time, temperature, sludge density, the amount of Mg precipitated from ionic form versus particulate MgO added, and interferences from organic matter (color). The high silica levels (100 to 150 mg/L) found in arid regions (e.g., northern Mexico and the southwestern United States) are removed much more effectively by magnesium precipitation (almost a stoichiometric reaction) than are the lower concentrations typical of many lakes and rivers (5 to 20 mg/L). As this chart indicates, it is surprising to find that it is as difficult to reduce SiO_2 from 10 mg/L to a residual of 1 mg/L as to reduce it from an initial level of 100 mg/L down to 10 mg/L.

removed from potable supplies. It is also frequently found in wastewaters requiring treatment before discharge.

The hydroxides of heavy metals are usually insoluble, so lime is commonly used for precipitating them. However, sometimes the carbonates, phosphates, or sulfides are less soluble than the hydroxides, so precipitation in these forms must also be considered. There are probably situations where economics justify partial precipitation with lime or soda ash to the solubility level of the hydroxide or carbonate followed by a secondary treatment with phosphate or sulfide for reduction to the specified limits (Figure 10.16). So, the choice of the reactant is the first consideration in the precipitation of heavy metals. Since solubility is affected by temperature, this becomes a second consideration.

A third important factor in precipitation of heavy metals is the valence state of the metal in the water. For example, ferrous iron is considerably more soluble than ferric iron. Because of this, treatment of the water with an oxidizing agent such as chlorine or potassium permanganate to convert ferrous iron to the ferric state is an essential part of the iron removal process. Another example is chromium, whose hexavalent form, chromate, CrO_4^{2-} ($Cr^{6+} + O_4^{8-}$), is considerably

WATER TREATMENT PROCESS DATA

FIGURE 10.15

File: Partial Cold Lime Softening in Slurry-Type solids contact unit with silica removal with magnesium precipitation

Date: February 16, 1983

Identification of Analyses Tabulated Below:

A. Water from comb. wells
B. Accelator effluent - lime softened @ 65°F
C.
D.
E.
F.

Constituent	As	A	B	C	D	E	F
Calcium	$CaCO_3$	256	70				
Magnesium	"	168	32				
Sodium	"	114	122				
(Hardness)		(424)	(102)				
Total Electrolyte	$CaCO_3$	538	224				
Bicarbonate	$CaCO_3$	366	0				
Carbonate	"	0	16				
Hydroxyl	"	0	36				
Sulfate	"	Nil	Nil				
Chloride	"	172	172				
Nitrate	"	0	0				
M Alk.	$CaCO_3$	366	52				
P Alk.	"	0	42				
Carbon Dioxide							
pH		7.0	10.6				
Silica		28	7				
Iron	Fe						
Turbidity		Nil	Nil				
TDS							
Color							

NALCO CHEMICAL COMPANY
2901 BUTTERFIELD ROAD – OAK BROOK, ILLINOIS 60521

Form 402-PC Printed in U.S.A. WG 3/72

FIG. 10.15 Partial cold lime softening in slurry-type solids contact unit with silica removal with magnesium precipitation.

more soluble than the trivalent form, Cr^{3+}. In this case, the chromate (in which Cr is present as Cr^{6+}) must be reduced, usually with SO_2 at a low pH, for removal of chromium (as Cr^{3+}) by a precipitation process.

A fourth aspect of the precipitation process is the zeta potential of the initial heavy metal colloidal precipitate. In many plants where heavy metals are being removed, one of the principal problems in reaching the desired effluent limits is the colloidal state of the precipitated materials—they have not been properly neutralized, coagulated, and flocculated.

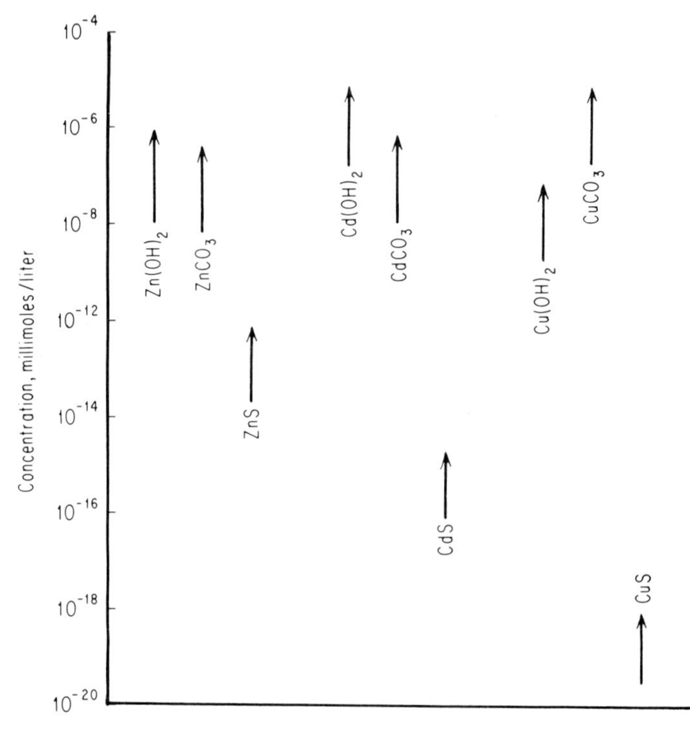

FIG. 10.16 Comparison of sulfide solubilities of certain cations to their hydroxides and carbonates.

A final aspect of heavy metals precipitation is the possible formation of complex ions, which are common when dealing with wastewaters containing ammonia, fluoride, or cyanide along with the heavy metals. For example, iron may be complexed as the ferrocyanide ion, which is rather soluble, and will remain in solution unless the complex can be broken by chemical treatment.

Because of these important aspects in the precipitation of heavy metals, there is no way to predict the best solution of a specific problem without undergoing a series of bench tests to evaluate the alternatives available. The removal of iron from water is an interesting example of this.

In the precipitation of iron from water, the first step is the oxidation to the ferric condition, as illustrated by the following reaction:

$$2Fe^{2+} + Cl_2^0 \rightarrow 2Fe^{3+} + 2Cl^- \quad (7)$$

Sometimes air can be used successfully to oxidize iron, but most frequently chlorine or potassium permanganate is required. The chlorine may be applied as chlorine gas or a calcium hypochlorite. Unless there is past experience with a specific water supply, each of these oxidation reagents must be evaluated to select the best process.

The pH of the water must be adjusted to an optimum value, determined not only by the solubility of the precipitate, but also by its charge. It may be necessary to determine the zeta potential in selecting the optimum pH for treatment. There are distinctive differences in the chemical used for pH adjustment also. For example, lime is usually more effective than caustic soda at the same pH, and this may very well be attributed to the charge on the particles and charge neutralization.

Once the iron has been oxidized and precipitated, the volume of sludge produced must be examined to decide whether the treated water can then be clarified by direct filtration or will require treatment through a sedimentation tank prior to filtration. Generally where the iron is less than 5 mg/L, the oxidized water can be fed directly to a filter. This is usually a mixed media filter provided with air scour devices so that the bed can be kept clean. A typical iron removal plant of this type used in municipal service is illustrated by Figure 10.17.

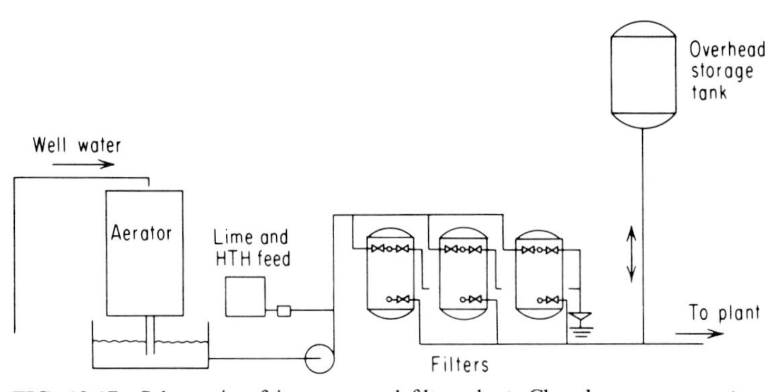

FIG. 10.17 Schematic of iron-removal filter plant. Closed pressure aerator may be used in place of the open tower shown. Where an overhead storage tank is used for town or plant supply this can be used to supply backwash water.

Manganese-treated zeolite is often used as the filtration medium in filters of this type. Where this is done, permanganate may be fed to the water ahead of these filters, and lime may be fed for pH correction as well. Those who have worked with this process claim that the oxides of manganese produced by this reaction are catalytic to the oxidation of iron within the filter bed. A permanganate-treated water supply using manganese zeolite as the filter medium is illustrated by Figure 10.18.

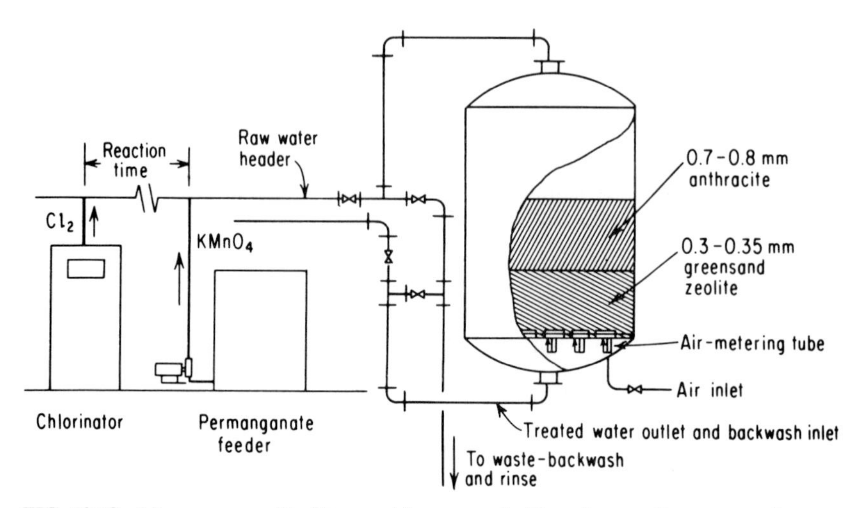

FIG. 10.18 Manganese zeolite filter used for removal of iron from well water supplies.

One of the complications of the iron removal process is the fact that some iron-bearing waters also contain sulfide. This adds to the demand for chlorine or other oxidizing agents.

MISCELLANEOUS PROCESSES

Using lime softening or iron removal type equipment, precipitation processes are often used to remove manganese from raw water supplies, phosphate from municipal sewage, fluorides from industrial wastes, and copper from metallurgical wastes. Each system is unique and requires bench testing to determine the best reagent for the process, optimum pH, proper temperature conditions, and the nature of the sludge expected, which will influence the selection of solids/liquids separation devices.

Sulfates are present at high concentrations in a variety of industrial wastes. If discharged into cement sewers, concentrations in excess of about 500 mg/L sulfate may deteriorate the cement, so reduction of sulfate to below this level may be required to protect the sewer system. Precipitation with lime to yield $CaSO_4$ is sometimes effective, but because the solubility of $CaSO_4$ is on the order of 1800 mg/L, the residual sulfate may be too high. Precipitation as a Ca:A1:SO_4:OH complex is effective in reducing the sulfate well below 500 mg/L. Sodium aluminate and lime, combining with some of the calcium ions already present in the wastewater, are effective agents to be used in this process at a pH over 9.5 to 10.0.

Sulfates can be removed by anion exchange (Chapter 12); when the exchanger is regenerated with salt, the by-product is a concentrated sodium sulfate solution, which must then be processed by precipitation for disposal. However, the precipitation treatment of the smaller volume of Na_2SO_4 brine on a batch basis may be easier than treatment of the lower sulfate concentration of the raw wastewater on a continuous basis.

Fluorides can be precipitated by lime at a pH of about 10 to a residual in the range of 10 to 20 mg/L F. Alum or aluminate used with the lime produces a fluoride complex, and the residual F, as such, is negligible by the usual tests for the fluoride ion. However, the complex fluoride can be detected if the sample is evaporated to dryness and analyzed. Reduction to low fluoride levels requires a second stage of treatment—adsorption by activated alumina, calcium phosphate, or magnesium hydroxide.

A modification of the conventional lime-soda softening process has been applied to treatment of cooling tower blowdown for its recovery (see Chapter 38).

TABLE 10.1 Side-Stream Treatment of Cooling Water; Two-Stage Softening in Slurry-Type Units

Variable	Circulating water	Treated water
Calcium, mg/L as $CaCO_3$	811	129
Magnesium, mg/L as $CaCO_3$	480	15
Silica, mg/L as SiO_2	137	7

Source: Abstracted from Grobmyer, Edge, and Hancock. American Power Conference, April 1983.

In some cases, the blowdown is treated directly for reduction of calcium, magnesium, and silica; in other cases, it is more advantageous to combine the blowdown with raw water makeup and then process the blend. A complicating factor is the presence of dispersants in most concentrated cooling waters, deliberately added to inhibit scale and deposit formation, and these have an inhibiting effect on some of the softening reactions. Data from a unique two-stage pilot plant treating a side stream from a cooling water cycle are shown in Table 10.1. In this process, lime is added to pH of about 10.8 in the first stage, followed by recarbonation and soda ash addition to the second stage.

BENCH AND PILOT PLANT TESTING

Each special application, as in these several illustrations, must be bench tested—and often carried through to pilot plant evaluations—because so many wastewaters contain interfering or inhibiting additives.

The test procedures usually involve a modification of the jar test where there may be provision for heating if high temperature is required for minimum solubility. In doing these tests, the residuals are usually determined both on the supernatant sample and also on a filtrate after the supernatant has passed through a 0.45-μm membrane filter. The volume of sludge produced is measured with an Imhoff cone to help establish the type of solids/liquids separation equipment required. If the jar test does not produce good results, more sophisticated equipment, such as a zeta meter, may be needed to more precisely determine the nature of the particle being removed.

In all of these processes, accurate and reliable chemical feeding is essential. The pH may be critical, so the chemicals used for pH control must be proportioned to the flow of water and be corrected by a signal showing deviations from the control pH setting.

Although there are instruments available for recording effluent hardness (Figure 10.19), where a precipitation process is selected for removal of heavy metals there may not be an instrument available for a direct measurement of the heavy

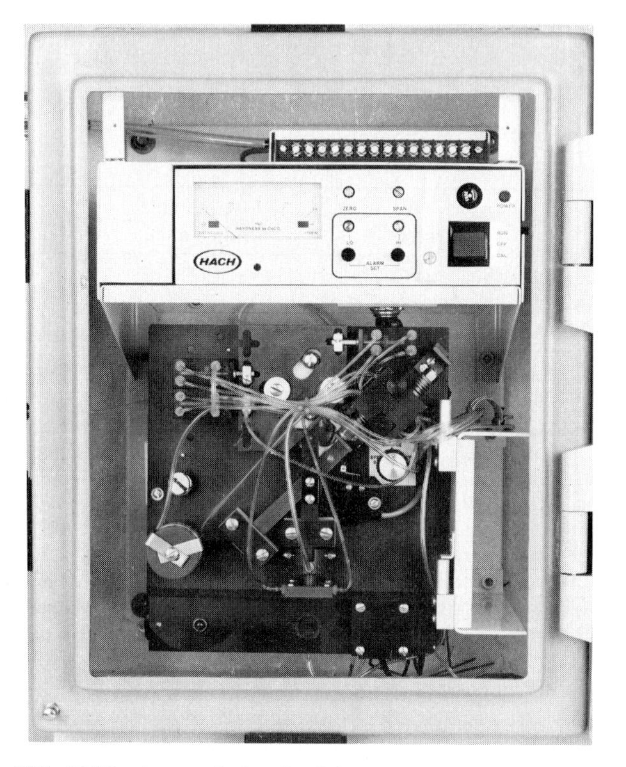

FIG. 10.19 Automatic low-level hardness analyzer. (*Courtesy Hach Chemical Company.*)

metal after treatment. This imposes special limitations on the plant, often requiring that the treated effluent be held temporarily in an inspection basin. The operator analyzes the final discharge and the precipitator effluents, and if specifications are not met, he must correct the treatment before the plant effluent is discharged from the retention basin to a receiving stream or sewer.

SUGGESTED READING

AWWA: *Water Quality and Treatment,* McGraw-Hill, New York, 1971.

Babber, N. R.: "Sodium Bicarbonate Helps Metal Plant Meet Federal Standards," *Industrial Wastes,* January–February 1978.

Christoe, J. R.: "Removal of Sulfate from Industrial Wastewaters," *J. Water Pollut. Control Fed.,* **48**(12) 2807 (December 1976).

Grobmyer, W. P., Edge, H. D., Jr., and Hancock, Fred: "Water Supply Pilot Testing Key to Design Optimization," Proceedings of the American Power Conference, April 1983.

Nancollas, G. H., and Reddy, M. M.: "Crystal Growth Kinetics of Minerals Encountered in Water Treatment Processes," in *Aqueous-Environmental Chemistry of Metals* (Rubin, J., ed.), Ann Arbor Science, Ann Arbor, Mich., 1976.

Neil, R. E., and O'Connell, R. T.: "A New Approach to Softening Plant Sludge Reduction," Proceedings American Water Works Association, June 1976, Paper 5897-676-3C.

"Treating Lead and Fluoride Wastes," *Environ. Sci Tech.* **6**(4): 321 (April 1972).

Water Treatment Handbook, Infilco-Degrémont, Inc., Halsted Press, New York, distributors, 1979.

CHAPTER 11
EMULSION BREAKING

An emulsion is an intimate mixture of two liquid phases, such as oil and water, in which the liquids are mutually insoluble and where either phase may be dispersed in the other. In water chemistry, two types of emulsions are commonly found, oily wastewater (oil emulsified in water or O/W emulsions) and waste oil emulsions (water emulsified in oil or W/O emulsions).

Oily waste and waste oil emulsions can usually be differentiated visually. O/W emulsion appears to be just oily, dirty water; a drop of the emulsion added to water disperses (Figure 11.1*a*). A W/O emulsion is usually thick and viscous; a drop of this emulsion added to water does not disperse (Figure 11.1*b*).

OIL-IN-WATER EMULSIONS

An oily waste emulsion, in which oil is dispersed in the water phase, may contain any of various types of oil in a wide range of concentrations. These oils are defined as substances that can be extracted from water by hexane, carbon tetrachloride, chloroform, or fluorocarbons. In addition to oils, typical contaminants of these emulsions may be solids, silt, metal particles, emulsifiers, cleaners, soaps, solvents, and other residues. The types of oils found in these emulsions will depend on the industry. They may be fats, lubricants, cutting fluids, heavy hydrocarbons such as tars, grease, crude oils, and diesel oils, and also light hydrocarbons including gasoline, kerosene, and jet fuel. Their concentration in the wastewater may vary from only a few parts per million to as much as 5 to 10% by volume.

A stable O/W emulsion is a colloidal system of electrically charged oil droplets surrounded by an ionic environment. Violent mixing and shearing of oily wastewater in transfer pumps disperses these minute oil droplets throughout the water. Emulsion stability is maintained by a combination of physical and chemical mechanisms. These emulsions are similar in behavior to the colloidal systems encountered in swamps (color) and rivers (silt).

One such stabilizing mechanism, ionization, is brought about by the addition of surface-active agents, such as organic materials or cleaners, which aid in maintaining a stable colloidal system. These molecules usually carry an electric charge and seek out the oil/water interface of the emulsified droplet. Here, the accumulated charges cause the emulsion to be stabilized through repulsion of the commonly charged droplets. Neutral (nonionic) surfactants can also stabilize an emulsion, since these molecules are bifunctional: one end is soluble in water, and the other end in hydrocarbon, so the molecule bridges the interface and stabilizes it.

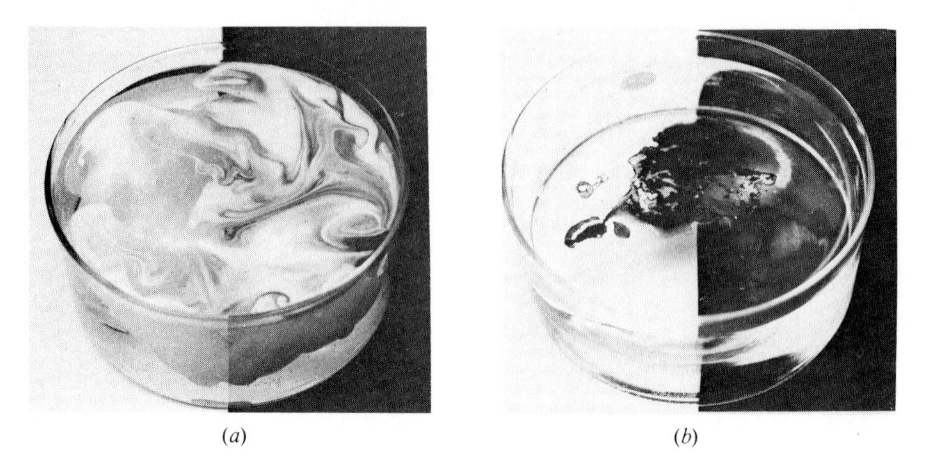

(*a*) (*b*)

FIG. 11.1 (*a*) If a few drops of oily water (O/W emulsion) are added to water, they disperse in the water. This is a test to distinguish O/W from W/O emulsions. (*b*) A few drops of this emulsion added to water remained as a separate phase floating on the surface, showing it to be a W/O emulsion.

Fine, solid particles may stabilize an emulsion if they are of correct size and abundance. In this case, stabilization occurs because the solid particles adsorbed at the oil/water interface tend to reinforce the interfacial film. The dispersed droplets cannot coalesce because of the interference or blocking effect caused by the solids (Figure 11.2).

Emulsions can also be stabilized by friction between the oil and water phases created by vigorous mechanical or physical agitation. Static electric charges developed by this action tend to collect at the oil/water interface.

An emulsifier is usually a complex molecule, often having a hydrophilic (water-loving) group at one end and a lyophilic (oil-loving) group at the other (Figure 11.3). Emulsifiers disperse oil droplets in the water phase because they have an affinity for both water and oil that enables them to overcome the natural forces of coalescence.

Most emulsifiers are surfactants having either anionic or nonionic polar groups. Petroleum sulfonates and sulfonated fatty acids are common anionic emulsifiers, and ethoxylated alkyl phenols are common nonionic emulsifiers. Examples of naturally occurring surfactants are organic sulfur compounds, various simple esters, and metal complexes. Alkaline cleaners containing surfactants

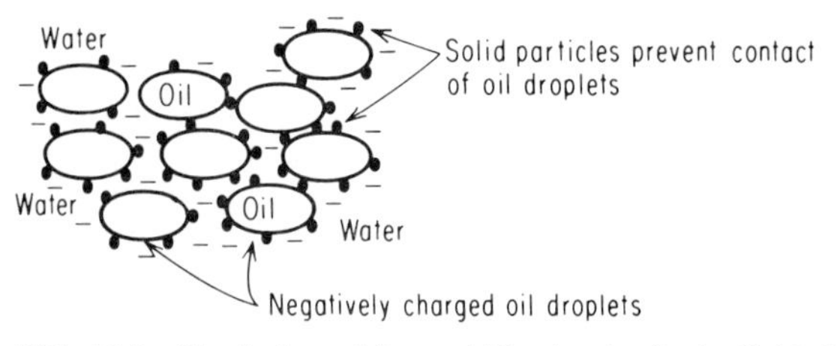

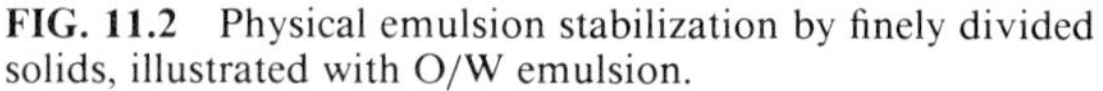

FIG. 11.2 Physical emulsion stabilization by finely divided solids, illustrated with O/W emulsion.

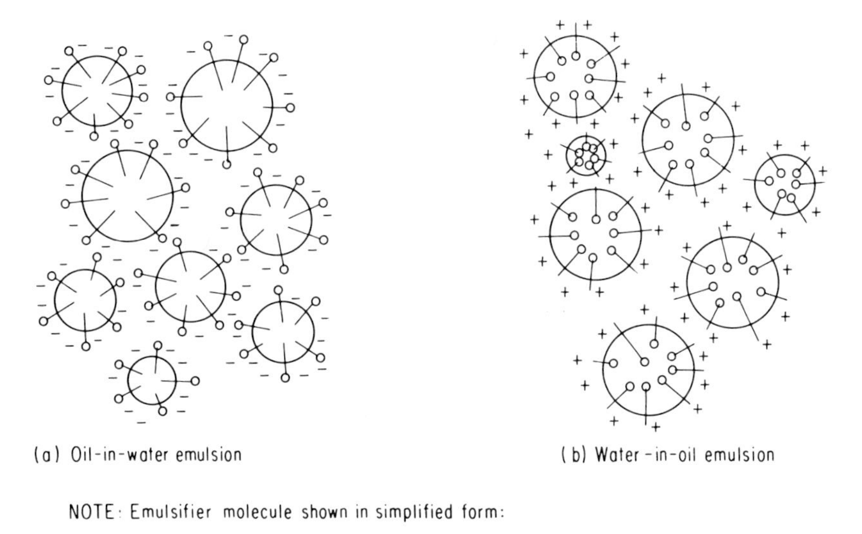

FIG. 11.3 Chemical stabilization of emulsions by surfactants having hydrophilic and lipophilic groups.

that emulsify free oil are present in many wastewaters. Table 11.1 lists a variety of emulsifiers for both O/W and W/O systems.

Breaking Oil-in-Water Emulsions

Emulsions may be broken by chemical, electrolytic, or physical methods. The breaking of an emulsion is also called resolution, since the result is to separate the

TABLE 11.1 Emulsifying Agents

Oil-in-water type	Water-in-oil type
1. Formation—when soaps are colloidally dispersed in water phase	1. Formation—when soaps are precipitated from aqueous phase
2. Ionic emulsifiers	2. Ionic emulsifiers
a. Sodium, potassium soaps and sulfides	*a.* Multivalent metal soaps
b. Sodium naphthenes and cresylates	*b.* Sulfide ion plus carbon particles
c. Precipitated sulfides plus surfactants	*c.* Multivalent metal oxides
d. Organic amines	*d.* Mercaptans
	e. Naphthenic or cresylic acids
3. Electrolytes which favor stability	3. Electrolytes which favor stability
a. Salts of univalent cations	Salts of di- and trivalent cations
b. Salts of di- and trivalent cations	

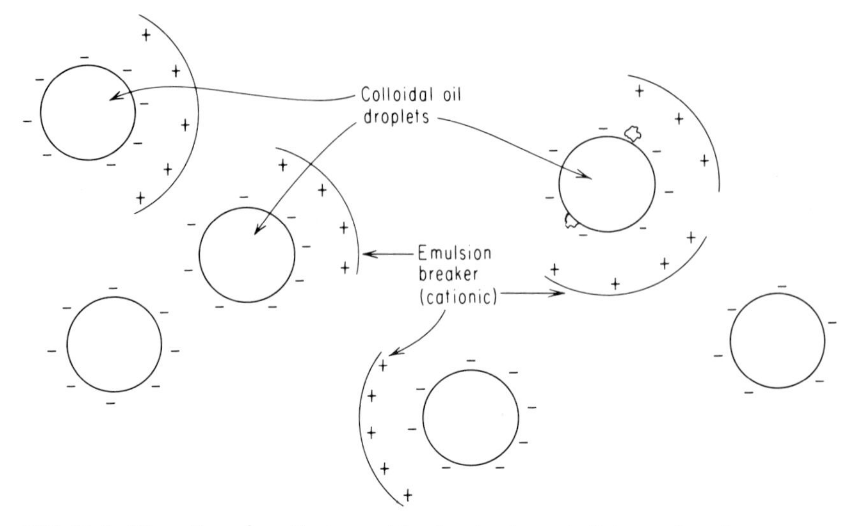

FIG. 11.4 The action of a cationic emulsion breaker in neutralizing surface charges on a colloidal oil droplet in oily wastewater.

original mixture into its parts. Chemicals are commonly used for the treatment of oily wastewaters, and are also used to enhance mechanical treatment. In breaking emulsions, the stabilizing factors must be neutralized to allow the emulsified droplets to coalesce. The accumulated electric charges on the emulsified droplet are neutralized by introducing a charge opposite to that of the droplet (Figure 11.4). Chemical emulsion breakers provide this opposite charge. The dielectric characteristics of water and oil cause emulsified oil droplets to carry negative charges. Therefore, to destabilize an oil-in-water emulsion, a cationic (positive charge) emulsion breaker should be used.

The resolution of an O/W emulsion should ideally yield an oil layer and a water layer. However, such a clear resolution is seldom achieved: there is often a scum, called a rag, at the interface where solids and neutralized emulsifier collect.

The treatment of oily wastewater is normally divided into two steps.

1. *Coagulation:* This is destruction of the emulsifying properties of the surface-active agent or neutralization of the charged oil droplet.
2. *Flocculation:* This is agglomeration of the neutralized droplets into large, separable globules.

Traditionally, sulfuric acid has been used in oily waste treatment plants as the first step in emulsion breaking. Acid converts the carboxyl ion in surfactants to carboxylic acid, allowing the oil droplets to agglomerate. Chemical coagulating agents, such as salts of iron or aluminum, can be used in place of acid, with the additional benefit that these aid in agglomeration of the oil droplets. However, the aluminum or iron forms hydroxide sludges that are difficult to dewater. Acids generally break emulsions more effectively than coagulant salts, but the resultant acidic wastewater must be neutralized after oil/water separation.

FIG 11.5 The use of organic emulsion breakers in place of alum or salts which form hydrous flocs greatly reduces sludge volume. The oil can be extracted from the sludge.

Organic demulsifiers are extremely effective emulsion breaking agents, giving more consistent results and producing better effluent quality than an inorganic program. In many treatment plants, organic emulsion breakers have replaced traditional alum treatment for exactly those reasons. In addition to yielding a better quality effluent, organic emulsion breakers often require lower dosages than a corresponding inorganic treatment. Organic emulsion breakers reduce the amount of sludge generated in a treatment program by as much as 50 to 75% (Figure 11.5).

Table 11.2 lists emulsion breakers used in both O/W and W/O treatment programs.

TABLE 11.2 Types of Emulsion Breakers

Main type	Description	Charge	Used for
Inorganic	Polyvalent metal salts such as alum, $AlCl_3$, $FeCl_3$, $Fe_2(SO_4)_3$	Cationic	O/W
	Mineral acids such as H_2SO_4, HCl, HNO_3	Cationic	O/W and W/O
	Adsorbents (adding solids)—pulverized clay, lime	None	O/W
Organic	Polyamines, polyacrylates and their substituted copolymers	Cationic	O/W
	Alkyl substituted benzene sulfonic acids and their salts	Anionic	W/O
	Alkyl phenolic resins, substituted polyalcohols	Nonionic	W/O

Treatment Methods

Oil separation and removal can be divided into two processes, (a) gravity separation of free, nonemulsified oil, and (b) chemical treatment and separation of emulsified oil. Oily waste is typically a combination of free, nonemulsified oil; stable emulsified oil; and insoluble solids, such as grit, metal fines, carbon, paint pigments, and corrosion products. Free oil and the insoluble solids are physically removed by gravity separation using American Petroleum Institute (API) or corregated plate interceptor (CPI) separators.

The primary function of a gravity separator is to remove free oil and settleable solids from wastewaters to enhance subsequent emulsion-breaking treatment and optimize chemical utilization, since oil and solids consume emulsion-breaking chemicals unnecessarily. Such separators cannot, of course, remove soluble impurities nor break emulsions. Gravity separators depend on density differences to provide the buoyant force that causes the droplets of free oil to rise to the surface. Theoretically, oil droplets rise linearly as predicted by Stokes' law; in practice, turbulence and short circuiting usually disturb the separation pattern.

The API separator consists of a rectangular trough or basin in which the wastewater flows horizontally while free oil rises to the surface. (See Figure 9.22.) Oil collecting on the surface is skimmed off to a recovery circuit.

The CPI separator consists of inclined packs of 12 to 48 corrugated plates mounted parallel to each other at distances of ¾ to 1½ (1.9 to 3.8 cm). As wastewater flows between the plates, the lighter oil globules float up the incline into the concave upper corrugations where they coalesce into larger masses which move along the plates to weep holes or to the trailing edge, and then to a floating layer at the surface (Figure 11.6). A modification of this design, shown in Figure 11.7, has been found effective on oil droplets of 0.85 sp gr as small as 5 μm in diameter.

After treatment of the oily wastewater for removal of free oil and solids, emulsion-breaking chemicals are applied to destabilize the colloidal oil, and the treated stream is then subjected to a second separation process, most commonly air flotation, which has been used for many years in the treatment of industrial wastes.

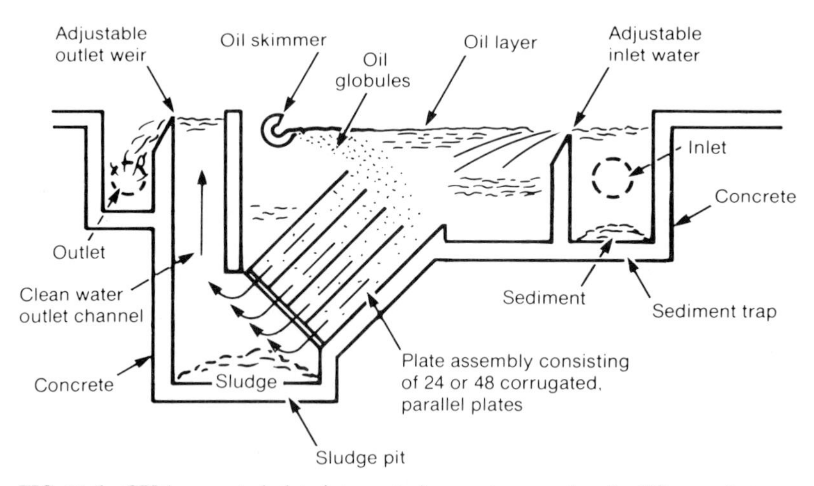

FIG. 11.6 CPI (corrugated plate interceptor) separator uses density difference between oil and water to separate free nonemulsified oil from wastewater.

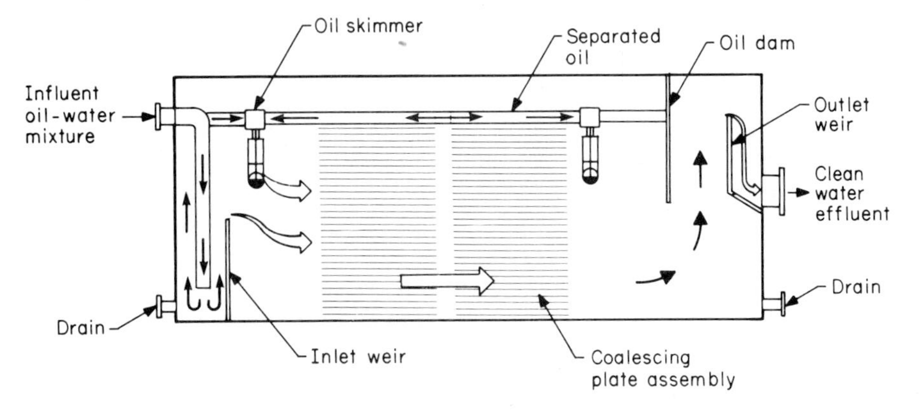

(*a*)

(*b*)

FIG. 11.7 Compact coalescing plate separator. (*a*) Line diagram showing details; (*b*) plant installation. *(Courtesy of Fram Industrial Filter Corporation.)*

(See Chapter 9 for discussion of air flotation devices.) Two basic methods of air flotation are used for O/W treatment: (a) dissolved air flotation (DAF) and (b) induced, or dispersed, air flotation (IAF).

Dissolved air flotation uses pressurized water, supersaturated with air, to produce bubbles of 30 to 120 μm upon release of pressure in the separation vessel. This causes flotation of the coagulated oil, solids, or both, as the fine bubbles are incorporated in the chemically produced floc. Induced air, on the other hand, utilizes mechanical agitation to aspirate or entrain air into the water undergoing treatment, resulting in larger bubbles, up to 1000 μm in size, for attachment to the flocs to bring them to the surface as a froth.

FIG. 11.8 The float produced by dissolved air flotation of an oily wastewater (O/W) emulsion. *(Courtesy of Infilco Degremont Incorporated.)*

Of the two processes, dissolved air flotation has been the more prominent in the United States, particularly in the refining industry. There are three basic flow sheets for the DAF process, (a) total pressurization, (b) partial pressurization, and (c) recycle pressurization, the latter being the preferred process in about 80% of oily wastewater treatment systems. As the supersaturated recycle water meets the chemically treated raw waste stream, the release of pressure causes the air to come out of solution on nucleation sites created by colloidal, neutralized particles. The common chemical treatment program is to apply an emulsion breaking chemical to the suction side of the supply pump, with addition of an inorganic coagulant such as alum to the discharge side; a high molecular weight flocculant is applied to the recycle stream so that the floc formation occurs as the air comes out of solution on the nucleation sites.

The DAF system usually includes the following (see Figs 9.23 and 11.8):

1. The DAF flotation vessel, a rectangular or vertical cylindrical unit with baffles and skimmers.
2. Recycle hardware, including the recycle pump, air compressor, and saturation tank.
3. Rapid mix and flocculation devices where required as a separate entity because of low temperature (slowing floc formation and separation velocity), fragile floc, or both.

As with the DAF unit, froth formation is produced in the IAF unit by chemical treatments normally added to the influent. The treated waste sometimes goes through a static mixer to hasten reaction and allow floc to form in the first cell of the series. The treatment scheme is different because the IAF depends on adherence of impurities to the bubble surface. Because the IAF can be designed to operate at higher hydraulic loading than DAF units, the larger bubbles rise faster. These units are often selected because of space savings and lower installed cost. Table 11.3 compares some of the performance characteristics of IAF and DAF units in oily waste treatment.

Induced air flotation systems usually have a lower installed cost per unit of throughput compared to DAF units. However, as the table shows, this is achieved at the expense of effluent clarity and collected sludge density. However, there are many applications where scalping the bulk of oil and solids is the major goal, as in side-stream treatment of recirculating cooling water in rolling mills, where the presence of some oil is acceptable, but a certain volume must be continuously removed to stay within acceptable limits in the bulk of the recirculated water.

Ultrafiltration consists of forcing an oily emulsion to pass through very small pores (less than 0.005 μm) in a membrane. Only water and dissolved low molecular weight materials can pass through the pore structure of the membrane, leaving a concentrate of the emulsified oil droplets and suspended particles. Plugging does not occur as it can in ordinary filtration because particulates are much larger than the pores and cannot enter the membrane structure.

Activated carbon adsorption has been used to clean up wastewater containing lesser amounts of soluble and emulsified organic contaminants (less than 100 mg/L). This is usually employed as a polishing step.

Coalescers are used where oily water may contain free oil and oil in a weakly emulsified state; examples are tanker ballast and oil field brines. There are a variety of designs, ranging from simple, baffled vessels where coalescence is induced by eddy currents, to rather sophisticated devices using membranes which allow water to pass through but reject oil. Coalescing devices for treating oily water, such as oil-contaminated condensate, include terry-cloth filters (a variety of cartridge filter) and coarse sand pressure filters. Of course, the coalescer by itself is

TABLE 11.3 Comparison of DAF and IAF Flotation Units

Dissolved air flotation	Induced air flotation
1. Air is dissolved into water. Air saturation equipment is required.	1. Air is entrained and mechanically mixed into water. Requires high-energy mixing.
2. Flotation of impurities due to bubbles enmeshed into flocs of solids and oil.	2. Surface interaction at air-water interface between air bubbles and impurities causes separation.
3. Air bubble size of 30–120 μm produces slow rise rate.	3. Air bubbles of up to 1000 μm produce rapid rise rate.
4. Lower loading rates because particles rise slower; longer detention time.	4. Rapid kinetics, therefore shorter detention time.
5. Low turbulence and longer detention time reduces floc carrythrough.	5. High energy and shear tends to increase floc carrythrough.
6. Chemical program focuses on coagulation and flocculation.	6. Chemical program focuses on air-water interface reactions.
7. Higher density of solids and oil in skimmed float.	7. Lower density of solids and oil in collected froth.

seldom adequate; it must be followed by a gravity separator or a filter, as is used in the treatment of oily condensate.

WATER-IN-OIL EMULSIONS

Water-in-oil emulsions are viscous, concentrated substances formed when oil comes into contact with water and solids. Metal particles and other solids may be coated with surfactants in such a manner that they are preferentially wetted by the oil rather than the water (Figure 11.9). In circumstances where agitation is

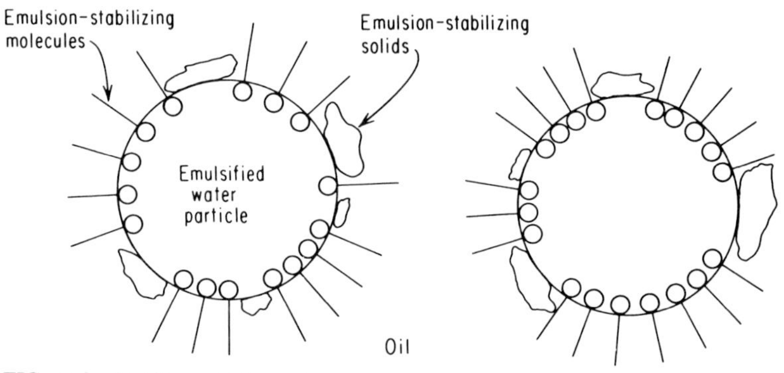

FIG. 11.9 Stabilized W/O emulsion.

present, the water becomes dispersed in the oil as a fine emulsion, and these small water droplets together with the oil-coated solids maintain the stability of the emulsion. Many types of water-in-oil or waste oil emulsions are found. Table 11.4 classifies some of the more common waste oils according to their relative oil, water, and solids content.

TABLE 11.4 Classification of Waste Oils

	Percent oil	Percent water	Percent solids
Refining—crude	90–95	5–10	0–5
Machining—cutting, grinding oil	50–80	20–50	0–20
Waste treatment sludges	40–50	40–50	5–10
Steel mills—rolling oils	80–95	5–20	0–5

W/O Emulsion Stabilizers

Among commonly occurring substances that promote or stabilize W/O emulsions are soaps, sulfonated oils, asphaltic residues, waxes, salt, finely divided coke, sulfides, and mercaptans. Finely divided solids varying in size from colloidal to 100 μm and larger are particularly effective in stabilizing these emulsions.

The presence of hydrophobic and hydrophilic colloidal solids at the oil/water interface may result in the larger particles aggregating around the dispersed droplets. The presence of these large particles throughout the continuous phase tends to form a complex matrix consisting of solids, oil-in-water, and water-in-oil emulsions existing simultaneously as a stable mixture in which either oil or water may be the continuous phase.

Breaking Water-in-Oil Emulsions

Water-in-oil emulsions can be broken by chemical or physical methods, including heating, centrifugation, and vacuum precoat filtration. Centrifugation breaks oil emulsions by separating the oil and water phases under the influence of centrifugal force (Figure 11.10). Filtration of waste oil emulsions can be accomplished

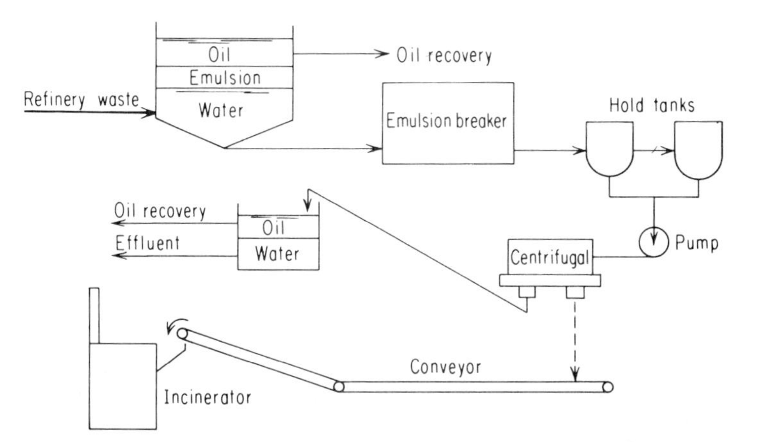

FIG. 11.10 The application of a centrifuge in treatment of waste oil (slop oil) in a refinery. *(Courtesy of Bird Machine Company.)*

through high rate sand filters (HRF) or diatomaceous earth (DE) filters (Figure 11.11). The operation of these pieces of equipment must be carefully controlled to provide the highest quality of upgraded oil.

Chemical treatment of a waste oil emulsion is directed toward destabilizing the dispersed water droplets and solids or destroying emulsifying agents. Acidification may be effective in breaking W/O emulsions if the acid dissolves some of the solid materials and thus reduces surface tension.

The newest method involves treatment of a W/O emulsion with a demulsifying agent containing both hydrophobic and hydrophilic groups that is able to form a water wettable adsorption complex. The mechanism of W/O emulsion breaking can be best explained by visualizing the displacement of the original emulsifying agent from the interface by a more surface-active demulsifying material. This process can be enhanced by heating to reduce viscosity and increase the solubility and rate of diffusion of the emulsifying agent in the oil phase. Because water droplets in oil tend to be positively charged, these types of emulsions are typically treated with an anionic (negative charge) organic emulsion breaker. Sometimes a combination of acid and organic demulsifying agent provides the best results.

FIG. 11.11 Diatomaceous earth filter of the pressure leaf design used for treating emulsions in oil field flooding operations. *(Courtesy of Industrial Filter/Pump Manufacturing Company.)*

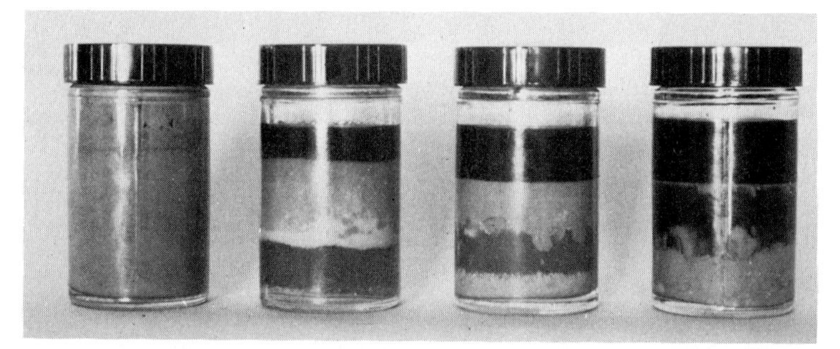

FIG. 11.12 Breaking a waste oil emulsion with chemical emulsion breaker, showing the effect of increasing chemical dosage on maximizing oil recovery and solids separation.

In all cases, the treatment reagents must be thoroughly mixed into the W/O emulsion to provide intimate contact with the emulsified water droplets. Heating the emulsion to 120 to 180°F (49 to 82°C) often produces rapid separation. Adequate settling time must be allowed to provide for optimum resolution of the oil, water, and solids phases (Figure 11.12).

INDUSTRIAL SOURCES OF EMULSIONS AND TREATMENT PRACTICES

An industrial activity that requires the use of process oils and water may be susceptible to emulsion formation at any point in its system. The three major industries producing oily waste are petroleum refining, metals manufacturing and machining, and food processing.

Hydrocarbon Processing Industry

Petroleum refining operations generate both O/W and W/O emulsions, present in drainage water, spills, separator skimmings, tank bottoms, and various oil recovery traps.

In the petrochemical industry quench waters generated in ethylene and other olefin manufacturing operations may contain mixtures of heavy, middle, and light hydrocarbons.

In the processing of crude oil, desalting is used initially to remove corrosive salts carried by the oil. In desalting, water is mixed with crude oil to wash it free of these salts. This forms a water-in-oil emulsion that is resolved in an electrostatic device, usually assisted by demulsifying chemicals. Frequently, however, not all of the emulsion is resolved, and the residue must be separated for further treatment. Often this emulsion is dumped into an oily sewer along with O/W and W/O emulsions formed in crude fractionation, thermal cracking, catalytic cracking, hydrocracking, and other processes. All of these emulsions contain too much water to justify processing, so they are usually dumped into an oily sewer system.

In the oily sewer, these emulsions are stabilized. Oil spills from process and general oil runoff are sent to this sewer along with oily wastewater, storm water, dirt, and debris. These components are mixed by pumps and flow turbulence in the sewer to form additional emulsions, or to further stabilize those already formed. The oils making up these emulsions are typically mixtures of light and heavy hydrocarbons.

The oily wastewaters collected from various points throughout the plant are delivered in a sewage system to an API separator. The skimmings from the API separator, which are usually good oil with low levels of solids and water present, may be sent back to the desalter or the coker. The underflow from the separator goes on to dissolved air flotation units (DAF). The underflow (the water) from the DAF unit may flow to a bio-oxidation pond and then to a clarifier before the water is ultimately sent to a river or sewer system (Figure 11.13). Emulsifying agents are not usually found in these O/W emulsions.

Waste oil treatment in refineries may be concerned with two different types of sludges or emulsions: skimmings from the API separator and the DAF units are typically sent to an oily sludge holding tank; the second type of waste oil emulsion in a refinery, leakage from processes, is often collected in various types of traps and drains throughout the plant. This oil may be of high enough quality that it can be blended with crude feedstock and returned to a process desalter or the coker unit.

A common measure of oil quality is the percent bottom sediment and water (BS&W). (Crude feed to a desalter typically has no greater than 5% BS&W.) If the reclaimed oil quality is too poor for use as desalter feed blend, the emulsion may

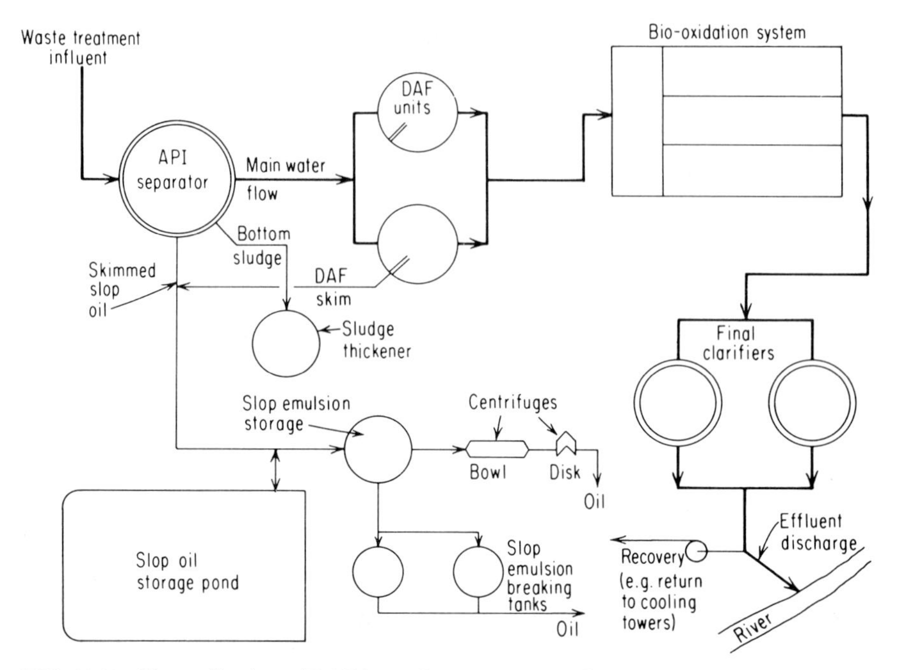

FIG. 11.13 The application of DAF in a refinery wastewater flow sheet.

be stored in a separate container or may be mixed with the API and DAF skimmings. These tanks typically have heating and recirculation or other type of mixing capabilities. Often, the fugitive waste oil emulsions can be separated solely by heat and a quiescent settling period. The API and DAF skimmings typically do not respond to this treatment, requiring the addition of a chemical W/O emulsion breaker. In either case, emulsion breakers can be used to effect a more rapid, complete separation of the oil, water, and solids phases. Treatment with an organic emulsion breaker involves feeding the chemical to the line while the treatment tank is filled. This is followed by a quiescent period ranging from one to several days. The separated layers can then be drained. The solids which are not oil-wet can be trucked to an approved landfill. The water, which may contain residual emulsified oil, is recycled back to the head of the oily wastewater treatment system and the recovered oil may be used according to individual plant needs.

Basic Metals Industry

Oily wastes generated in steel mills include both emulsified and nonemulsified, or floating, oils. Oily wastes from hot rolling mills contain primarily lubricating and hydraulic pressure fluids. Wastes from cold strip operations contain rolling oils that were used to lubricate the steel sheet and reduce rust. Oil-in-water emulsions are sprayed on the metal during rolling to act as coolants.

Wastewaters from these plants typically have a wide variety of contaminants, and the objectives of treatment include removal of not only oil, but also solids and metal fines. In a typical steel mill waste treatment scheme (Figure 11.14), the

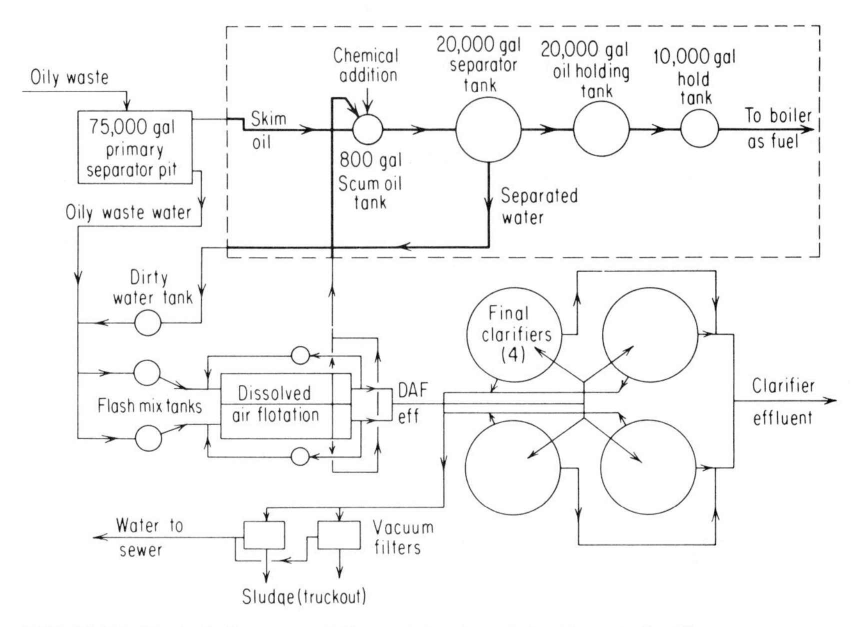

FIG. 11.14 Typical oil recovery/oily waste treatment plant in a steel mill.

plant wastewater enters an equalization tank, where the detention time is a few hours. The water then flows to an air flotation unit, where it may be treated with organic or inorganic coagulants. The skimmings from the flotation unit may be pumped to a storage tank for batch treatment with waste oil emulsion breakers. The underflow from the flotation unit usually passes to further treatment or is recycled.

Waste oil treatment in the primary metal industry is usually more complicated than in a refinery. Chemical emulsion breaking treatment programs often require the addition of acid to dissolve metal fines. In batch treating these emulsions, it is important that the chemical be thoroughly mixed into the emulsion and that heat be applied throughout the treatment and settling stages. After the settling period, the solids and water layers can be disposed of in the same way that refinery rag layers are handled. The oil layer is high quality and may be suitable for use as a boiler fuel blend or may be sold to a re-refiner or oil reclaimer.

Automotive and Machining Industries

Metalworking and metal parts manufacturing plants generate waste streams containing lubricating and cutting oils, lapping and deburring compounds, and grinding and other specialty fluids.

The wastewater generated in these plants contains a wide variety of oils and surfactants, so may require complicated treatment. The wastewater from several different operations often arrives at the treatment plant in separate streams or may be blended in the sewer system (Figure 11.15). All wastes flow to an equalization tank and then through several chemical feed and mix tanks prior to entering a flotation unit. The flotation skimmings are collected in storage tanks and

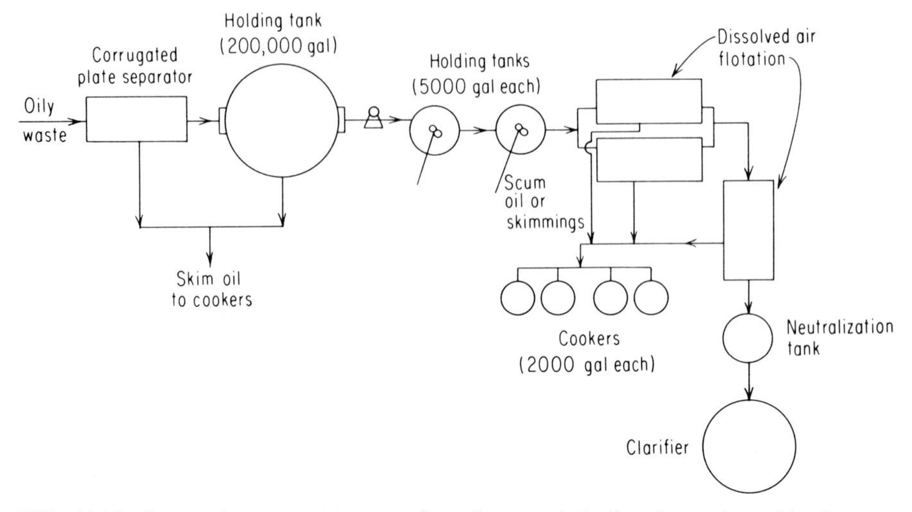

FIG. 11.15 Automotive waste treatment flow sheet, typical of engine and machined parts operation.

the underflow from the air flotation unit may then pass to final treatment or recycle.

The flotation skimmings can be pumped to lead-lined cooking tanks where they are treated with acid, an organic waste oil emulsion breaker, and a high molecular weight polymer for oil recovery. The rag layer from this treatment is sold to an oil scavenger and the good, recovered oil may be burned as boiler fuel or reused in various plant processes.

Meat and Food Processing

Rendering plants, creameries, bakeries, breweries, and canneries generate emulsions containing natural fats and oils from animal processing and oils from packing and container manufacture. Both of these operations occur on the same premises, and the waste streams may be mixed.

Textiles

Wastewaters from cotton and wool manufacturing plants contain oils and greases from the scouring, desizing, and finishing operations. Oily wastewater in the synthetic fibers industry is produced during desizing and scouring. Finishing oils are used to reduce friction and snagging of fibers on spinning machines. These types of oils may also be associated with machine lubricating oils in the plant oily wastewater.

Miscellaneous Industries

Many other industries have processes which may generate oil or oily wastewater such as paints, surface coatings, and adhesives; oils, fats and waxes; soaps and detergents; dyes and inks; and certain processes of the leather industry. All have potential emulsion problems that can be solved by one or more emulsion-breaking techniques.

RECOVERY OF SEPARATED OIL

Whether the oil separated from waste had originally been present as the continuous or the dispersed phase, a final step is required to segregate this oil fraction and recover its value as fuel or product.

Overflow skimmers, sometimes simply a horizontal pipe with a wide slot cut lengthwise in the top that can be rotated until the slot is partially submerged, are commonly used in API and CPI separators. Inevitably, some water is withdrawn with the oil; the skimmings are drained to a manometer-type separator (Figure 11.16) and the oil fraction is separated for recovery. An alternative is to use a mechanical separation device comprising a continuous belt or tubing partially immersed in the floating oil layer; the belt or tubing is fabricated of a material that is selectively oil-wetted, so its surface becomes coated with oil that is then removed by a doctor blade into an oil receiver (Figure 11.17).

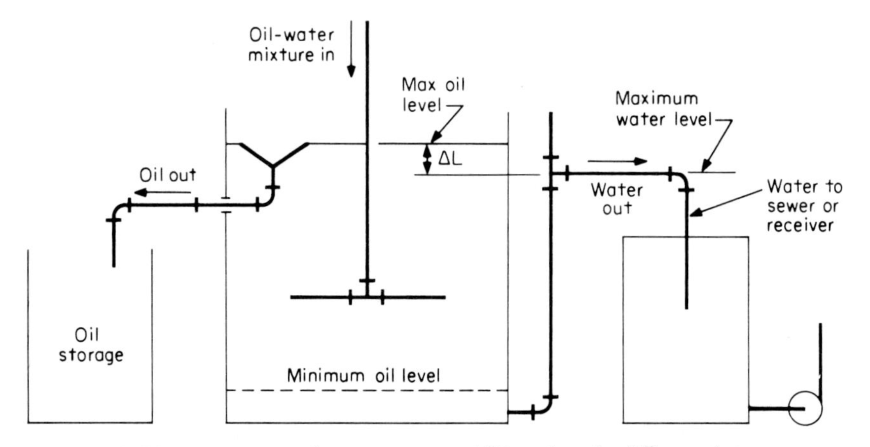

FIG. 11.16 Manometer-type oil-water separator ΔL based on the difference between specific gravity of oil and specific gravity of water.

FIG. 11.17 Free oil is often collected from water surfaces in a separation tank by means of an endless belt or loop made of oil-wettab plastic. *(Courtesy of Oil Skimmers Inc.)*

If the oil emulsion has been broken by heavy chemical treatment, especially if a coagulant like alum has been used, the oily product is pastelike in consistency and requires additional chemical treatment to free the oil for recovery. Although hot acid treatment alone has been successfully used for this, special corrosion-resistant equipment and protective devices for personnel are required. High-temperature reaction with organic surface-active reagents at a controlled pH has been used as an economical replacement for strong acid treatment. In some cases, the plant having this problem has an incinerator for other wastes, and the heating value of the pastelike oily matter will support combustion and allow it to be burned in the incinerator despite the presence of chemical conditioners and imbibed water.

CHAPTER 12
ION EXCHANGE

Ion exchange removes unwanted ions from a raw water by transferring them to a solid material, called an ion exchanger, which accepts them while giving back an equivalent number of a desirable species stored on the ion exchanger skeleton. The ion exchanger has a limited capacity for storage of ions on its skeleton, called its *exchange capacity;* because of this, the ion exchanger eventually becomes depleted of its desirable ions and saturated with unwanted ions. It is then washed with a strong regenerating solution containing the desirable species of ions, and these then replace the accumulated undesirable ions, returning the exchange material to a usable condition. This operation is a cyclic chemical process, and the complete cycle usually includes backwashing, regeneration, rinsing, and service.

The earliest ion exchangers were inorganic sodium aluminosilicates, some of which were manufactured synthetically and others made by processing natural greensand, which is a mineral called zeolite, into more stable, higher capacity forms. Even though these zeolites now have only limited use for water treatment, the name has persisted, and even synthetic organic ion exchangers are often called zeolites.

The ion exchangers used in water conditioning are skeletonlike structures having many ion exchange sites, as shown in Figure 12.1. The insoluble plastic skeleton is an enormously large ion that is electrically charged to hold ions of opposite charge. As such, the ion exchanger is related to the polyelectrolytes used for coagulation and flocculation (Chapter 8), but deliberately made so high in molecular weight as to be essentially insoluble. Exchangers with negatively charged sites are cation exchangers because they take up positively charged ions. Anion exchangers have positively charged sites and, consequently, take up negative ions. The plastic structure is porous and permeable, so the entire ion exchange particle participates in the process.

EXCHANGER CAPACITY

Typical exchangers are in the form of beads, having an approximate size of 20 to 50 mesh (0.8 to 0.3 mm), as illustrated by Figure 12.2. The chemist expresses the strength of a chemical solution in terms of normality; a 1.0 N solution contains 1 g-eq of electrolyte in a liter of solution. Ion exchange beads can be considered as solid solutions; the typical cation exchanger has a normality of approximately 2.0, and the typical strong-base anion exchanger a normality of approximately 1.3.

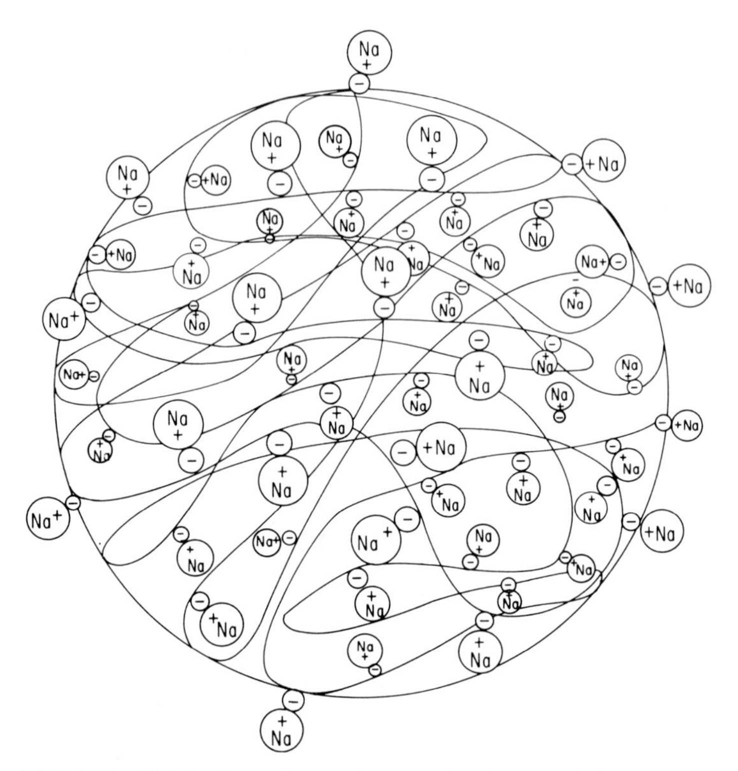

FIG. 12.1 Model of a cation exchanger, showing negatively charged exchange sites on the skeleton holding sodium ions like grapes on a vine.

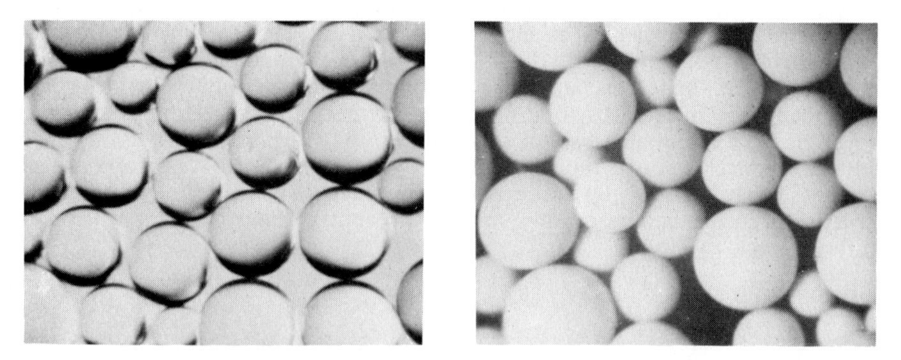

FIG. 12.2 Typical commercial exchangers are of two general structures, gel type *(left)* and macroporous type *(right)*. These are magnifications.

Capacity is also expressed as milliequivalents per milliliter (meq/mL), which is the same as normality; milliequivalents per dry gram (meq/g); and kilograins per cubic foot (kgr/ft^3).

In the early history of zeolite softening, it was common to express water hardness in grains per gallon (gr/gal) (1 gr/gal = 17.1 mg/L). Because of this common usage, the capacity of zeolites was rated in kilograins exchange capacity per cubic foot of zeolite, an unfortunately cumbersome term. The factor for conversion of resin normality to kilograins per cubic foot (kgr/ft^3) is approximately 22, so that a cation exchanger with a capacity of 2.0 meq/mL has an exchange capacity of approximately 44 kgr/ft^3.

Using normality as a basis for expressing the exchange of an ion exchange material and also for expressing the concentration of electrolytes in water, it is very easy to use the classic expression,

$$V_x N_x = V_w N_w$$

where V_x = volume of exchange material
N_x = normality of the exchanger
V_w = volume of water processed per cycle
N_w = normality of exchangable electrolytes in the water

For example, if a water having a total electrolyte content of 200 mg/L as $CaCO_3$ has a hardness of 150 mg/L, the normality of the exchangable electrolytes if this water is to be softened is 150/50, or 3.0 meq/L, or 0.003 N. If this is softened through a cation exchanger with a normality of 2.0, then the theoretical volumetric ratio would be

$$\frac{V_w}{V_x} = \frac{N_x}{N_w} = \frac{2.00}{0.003} = 667 \text{ volumes of water per volume of bed}$$

This is illustrated by Figure 12.3. To make the graph useful, the percentage of total capacity (in normality or kgr/ft^3) that is actually available under optimum conditions of regeneration must be established. This is covered in a later section of the chapter.

Most commercial ion exchangers are synthetic plastic materials, such as copolymers of styrene and divinyl benzene (Figure 12.4). There is a fine balance between producing a loosely cross-linked polymer that has free access to water for rapid reaction but is slightly soluble, and a tightly cross-linked resin, which would be insoluble, but more difficult to use because of restricted rates of exchange both in exhaustion and regeneration. Water treatment ion exchangers are essentially insoluble and can be expected to last for 5 to 10 years.

To produce cation exchangers, the plastic is reacted with sulfuric acid. Sulfonic groups attach to each nucleus in the skeleton to provide an exchange site. This produces a strong electrolyte, for which a typical reaction with cations in water is shown below:

$$Na^+ + R\cdot SO_3\cdot H \rightarrow R\cdot SO_3\cdot Na + H^+ \quad (1)$$

In this equation, the resin structure is represented by R. The usual convention omits showing the active group ($-SO_3\cdot H$) and simply uses the total exchanger molecule, which may be shown as Z (Na_2Z, from the historical usage of the word "zeolite") or X, which is used in this text for the cation exchanger. The above equation is, then, more usually written:

$$2Na^+ + H_2X \rightarrow Na_2X + 2H^+ \quad (2)$$

the X being considered a divalent cation exchange unit. In this example, sodium ions in water are being exchanged for hydrogen ions on the exchanger, a common exchange reaction. This is known as hydrogen cycle operation. On depletion of the hydrogen ion inventory of the exchange resin, called "exhaustion", the exchanger is regenerated by an acid wash:

$$2HCl + Na_2X \rightarrow H_2X + 2NaCl \quad (3)$$

Other examples of cation exchange include the following:
Hardness removal, with Na-form exchanger, Na_2X:

$$Ca^{2+} + Na_2X \rightarrow CaX + 2Na^+ \quad (4)$$

This is known as sodium cycle operation.

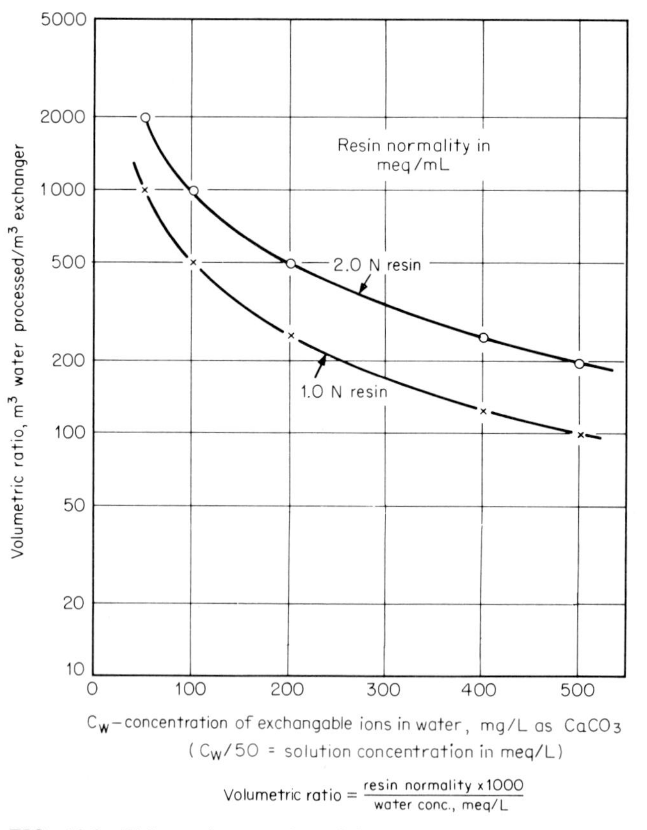

FIG. 12.3 Volumetric capacity of ion exchange resins at 100% utilization.

(1) Basic monomers

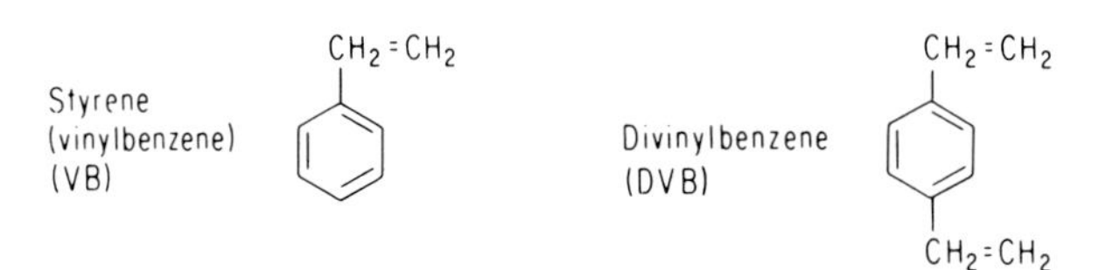

(2) Polymerization of styrene to polystyrene

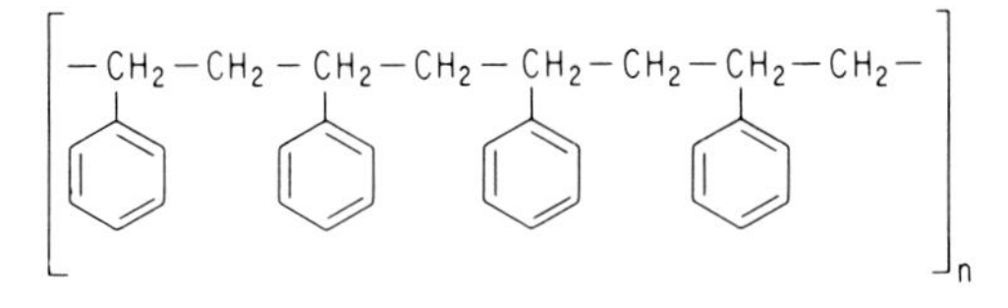

(3) Copolymerization of styrene and divinylbenzene, showing the role of the DVB in cross-linking the polystyrene chains.

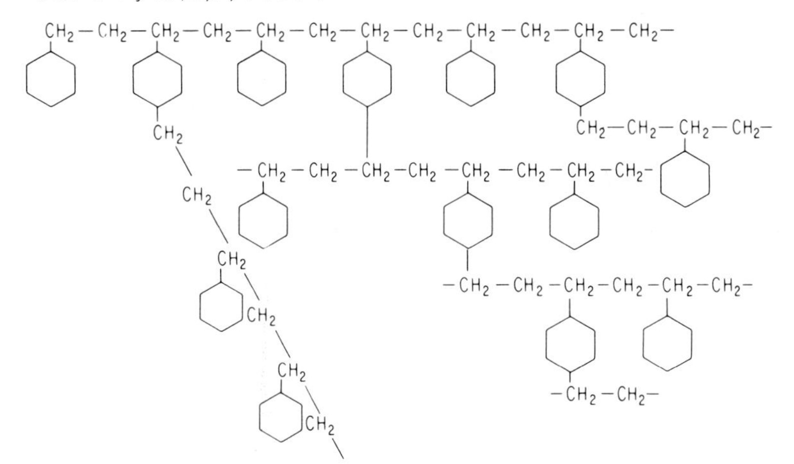

FIG. 12.4 The production of a stable ion exchange resin by copolymerization.

Iron removal with Na_2X:

$$Fe^{2+} + Na_2X \rightarrow FeX + 2Na^+ \tag{5}$$

(*Note:* Of the two species of iron in water, only Fe^{2+} is soluble, requiring an oxygen-free, reducing environment. The ferrous iron is ion exchangable. However, if oxygen is present either in the water or in the regeneration solution (e.g., brine), the Fe^{2+} will be oxidized to insoluble Fe^{3+}, which will precipitate and foul the ion exchanger.)

Nickel recovery from plating waste by H_2X:

$$Ni^{2+} + H_2X \rightarrow NiX + 2H^+ \tag{6}$$

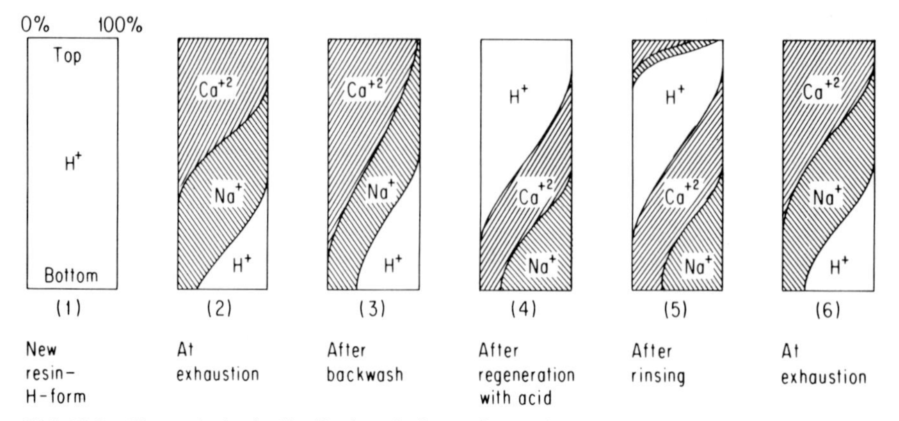

FIG 12.5 Change in ionic distribution during cation exchange—H-cycle, with downward flow of water and acid (cocurrent regeneration).

TABLE 12.1 General Order of Ion Selectivity in Water below 1000 mg/L TDS

Cations	Anions
Fe^{3+}	CrO_4^{2-}*
Al^{3+}	SO_4^{2-}*
Pb^{2+}	SO_3^{2-}*
Ba^{2+}	HPO_4^{2-}*
Sr^{2+}	CNS^-
Cd^{2+}	CNO^-
Zn^{2+}	NO_3^-
Cu^{2+}	NO_2^-
Fe^{2+}	Br^-
Mn^{2+}	Cl^-
Ca^{2+}	CN^-
Mg^{2+}	HCO_3^-
K^+	$HSiO_3^-$
NH_4^+	OH^-
Na^+	F^-
H^+	
Li^+	

* These ions may be displaced as they are protonated at low pH to: $HCrO_4^-$, HSO_3^-, $H_2PO_4^-$

Notes: Changes in position may occur between products of different manufacture or having slightly different skeletons or exchange groups. In general, selectivity is affected by:

(*a*) Ionic valence: 3 > 2 > 1.

(*b*) Atomic number: Ba > Sr > Ca > Mg in Group IIA.

(*c*) Hydrated ionic radius: the larger the radius, the lower the selectivity and exchange capacity.

A more realistic example which illustrates an ion exchanger property called selectivity is hydrogen cycle (H-form exchange) processing of a typical water containing a variety of ions:

$$\left.\begin{array}{l} Ca^{2+} \\ Mg^{2+} \\ Fe^{2+} \\ 2Na^{+} \\ 2NH;^{+} \end{array}\right\} + H_2X \rightarrow \left\{\begin{array}{l} CaX \\ MgX \\ FeX \\ Na_2X \\ (NH_4)_2X \end{array}\right. + 2H^{+} \tag{7}$$

If this ion exchange process continues to exhaustion, the first ions to appear in the effluent will be NH_4^+ and Na^+, and if the exhausted bed is then analyzed, the distribution of ions would be as shown by Figure 12.5. This is due to a selectivity or preference of the cation exchanger for certain ions over others. A typical selectivity list is shown in Table 12.1.

The order of selectivity shown applies to ions in typical water of less than 1000 mg/L TDS. The selectivity is different at high concentrations. For example, in the sodium cycle, the exchanger has a preference for Ca^{2+} over Na^+ at 1000 mg/L; but at 100,000 mg/L its preference is for Na^+ over Ca^{2+} (see Figure 12.6). This is fortunate, as it enhances the efficiency of regeneration with brine.

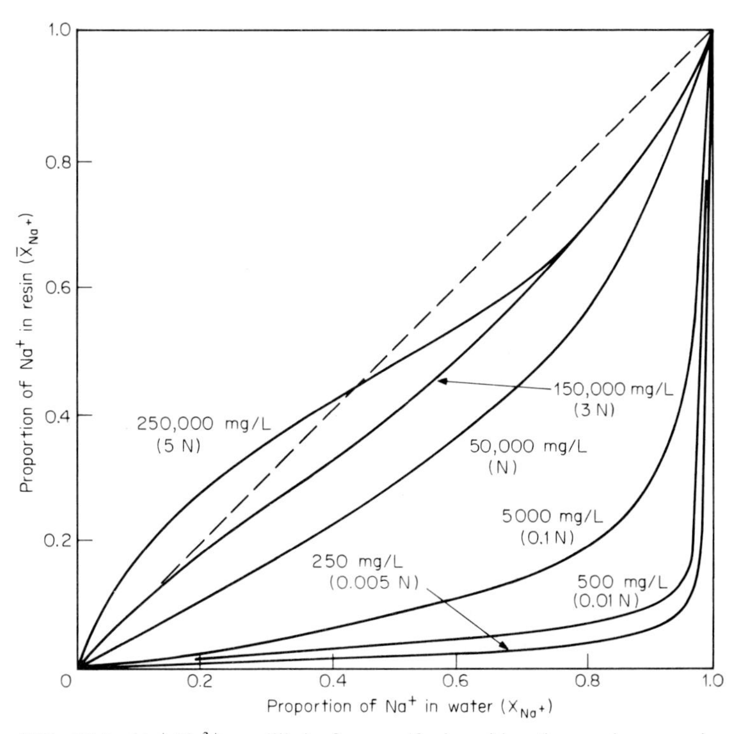

FIG. 12.6 Na^+/Ca^{2+} equilibria for a sulfonic acid cation exchange resin. *(Courtesy of T. V. Arden, Surrey, England.)*

Returning to the examination of the nature of cation exchangers, there are several materials in addition to the styrene-based material of commercial importance. One is processed greensand—the original zeolite—which is still used effectively for removal of iron and manganese. It can operate only in the sodium cycle over a limited pH range. Another, of increasing importance, is a resin having the carboxyl group ($-COOH$, with the hydrogen ion exchangeable) as the exchange site. This is called a weak cation exchanger, discussed later in a consideration of hydrogen exchange operations. This material is highly selective for H^+ and divalent cations with hindered ability to exchange Na^+. A third material is a true synthetic zeolite (aluminosilicate) that can be tailored for specific applications (e.g., ammonia removal) and for adsorption of specific molecules (molecular sieves). Other plastic materials are also used, such as acrylic and phenolic resins.

ANION EXCHANGERS

Anion exchangers may be produced from a variety of resinous or plastic skeletons, including the same styrene–divinyl benzene copolymer as is used for cation exchangers. As with cation exchangers, two general varieties are used commercially, weak base and strong base exchangers. The functional group of an anion exchanger is an amine, the organic equivalent of ammonia. Weak base exchangers contain a secondary or tertiary amine group, $RR'-NH$ or $RR'-N-R'$, which can adsorb strong acids. Strong base exchangers contain a quarternary amine, $R\,R'\,R''\,R'''N^+ \cdot Cl^-$, which can exchange all anions. The most common quarternary resin has the formula, $-R \cdot N(CH_3)_3 \cdot Cl$, or structurally

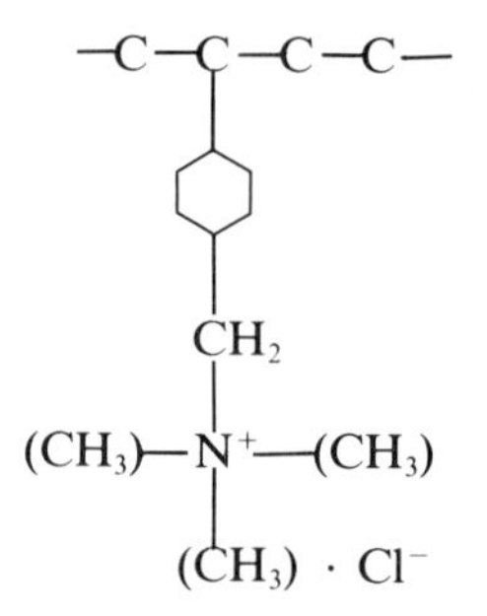

Weak-base anion exchangers are able to remove only strong mineral acids—HCl, H_2SO_4, HNO_3—with practically no exchange capacity for weak acids—such as CO_2, SiO_2, and organic acids. The typical reaction is shown as an adsorption, rather than ion exchange process:

$$HCl + A \rightarrow A \cdot HCl \tag{8}$$

(In this text, the letter A represents an anion exchanger.) The exhausted resin is very efficiently regenerated with any alkali, which simply neutralizes the adsorbed acid and releases it as a neutral salt.

Strong-base exchangers are true ion exchange materials. Several typical reactions are:

Dealkalization, with Cl-form exchanger, ACl_2:

$$2HCO_3^- + ACl_2 \rightarrow A(HCO_3)_2 + 2Cl^- \tag{9}$$

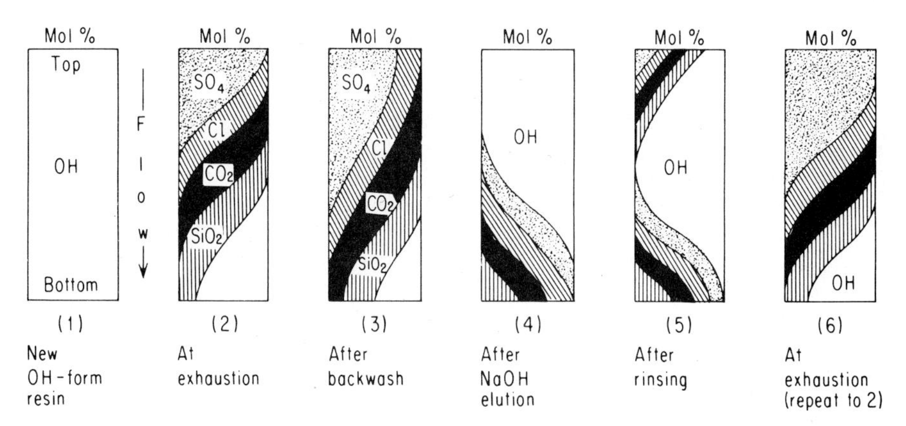

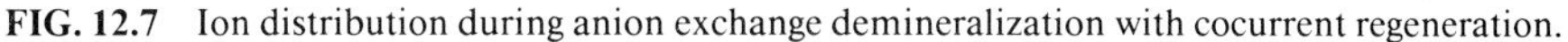

FIG. 12.7 Ion distribution during anion exchange demineralization with cocurrent regeneration.

Sulfate removal, with Cl-form exchanger:

$$SO_4^{2-} + ACl_2 \rightarrow ASO_4 + 2Cl^- \tag{10}$$

The most common use for strong base exchangers is complete anion removal from hydrogen exchanger effluent, to produce demineralized water:

$$\left.\begin{array}{l} H_2SO_4 \\ 2HCl \\ \\ 2H_2CO_3 \\ 2H_2SiO_3 \end{array}\right\} + A(OH)_2 \rightarrow \left\{\begin{array}{l} A \cdot SO_4 \\ A \cdot Cl_2 \\ \\ A \cdot (HCO_3)_2 \\ A \cdot (HSiO_3)_2 \end{array}\right. + 2H_2O \tag{11}$$

The exchange resin is regenerated with NaOH. In this form, the exchanger can also convert neutral salts to bases, known as salt-splitting.

$$2NaCl + A(OH)_2 \rightarrow ACl_2 + 2NaOH \tag{12}$$

The selectivity of the anion exchanger, also shown in Table 12.1, indicates that silica will leak through the bed first, and the exhausted bed will have the composition shown by Figure 12.7.

PRACTICAL ION EXCHANGE

In the commercial application of ion exchange to water treatment, there are four important principles:

1. Most ion exchange units are simple vessels containing a bed of ion exchange resin operated downflow on a cyclic basis: *(a)* a unit is operated to a predetermined leakage level, where it is considered to be exhausted; *(b)* the unit is then regenerated, first by upflow cleaning (backwash) and then by chemical elution, downflow; *(c)* the resin bed is then rinsed downflow. Because both water and regenerant flow in the same direction, the water leaving the unit is in contact with the resin having the highest level of contaminating ions, so quality and efficiency both suffer.

2. The ion exchange bed has a considerably higher capacity than is used, because uneconomical excesses of regenerating chemical would be required to convert the resin entirely to the desired ion form. For example, the cation resin may have a capacity of 2 *N* (about 44 kgr/ft^3), but only about half of this (20 to 22 kgr/ft^3) is used for sodium cycle softening. Therefore, there is always a high concentration of contaminating ions, calcium in this case, on the resin with a potential for spoiling the treated water quality.
3. Because of cyclic operation with cocurrent flow of water and regenerant, chemical utilization in regenerating ion exchange resins is usually poor. This drawback is most pronounced with strong resins—sulfonic-type cation exchangers and quarternary ammonium anion exchangers. For example, in the sodium-cycle, if the utilized capacity is 21 kgr (about 3 lb) as $CaCO_3$ per cubic foot (48 kg/m^3) of resin, the salt required is theoretically only $58.5/50 \times 3 = 3.5$ lb NaCl/ft^3. The actual salt consumption is typically 6 to 10 lb NaCl ft^3, so the efficiency is about 30 to 50%. Acid efficiency for hydrogen exchange using a sulfonic-type resin and H_2SO_4 for regeneration, and caustic efficiency for regeneration of strong base anion resins are even poorer, about 20 to 40%.
   However, weak cation resins (carboxylic type) and weak anion resins (amine type) can be operated at close to 100% chemical efficiency.
4. Most ion exchange materials used in water treatment are in the size range of 20 to 50 mesh, or about 0.5 mm effective size. This makes an ion exchange bed a very effective filter, a characteristic having both advantages and disadvantages. This filtering ability is combined with ion exchange properties in designing industrial condensate polishing systems using ion exchange beds. But the filtering ability also leads to fouling and unpredictable operating runs. Sometimes this is caused by the accumulation of high microbial populations in an exchanger bed, even when operating on a chlorinated municipal water supply.

AN EVOLVING SCIENCE

Because of these limitations, in spite of added capital cost and increased sophistication of equipment, ion exchange designs are changing to reduce chemical cost, improve effluent quality, and reduce pollution created by excessive chemical dosages. These improvements are realized by:

1. Countercurrent regeneration techniques
2. Multiple stages of treatment
3. Multiple stages of regeneration
4. Use of weak exchangers wherever possible
5. Pretreatment of water prior to ion exchange (e.g., by lime softening, reverse osmosis) to reduce the ionic load on the ion exchange system
6. Pretreatment to reduce particulates or soluble organic foulants (e.g., by filtration, adsorption)

In subsequent sections, leakage will be mentioned frequently. The term "leakage" implies a slipping of some of the influent ions into the effluent. In fact, unwanted ions reach the effluent by two different processes (see Figure 12.8*a*).

1. At the beginning of the softening run (using the Na_2X process as an example) there is appreciable calcium left in the bottom of the resin bed. Water entering at the top gets completely softened, the only cation being sodium. This softened water, in effect a very dilute brine, regenerates the calcium from the bottom of the bed. So at the beginning of the run the hardness in the effluent is due to calcium residue on the exchanger from the previous run.

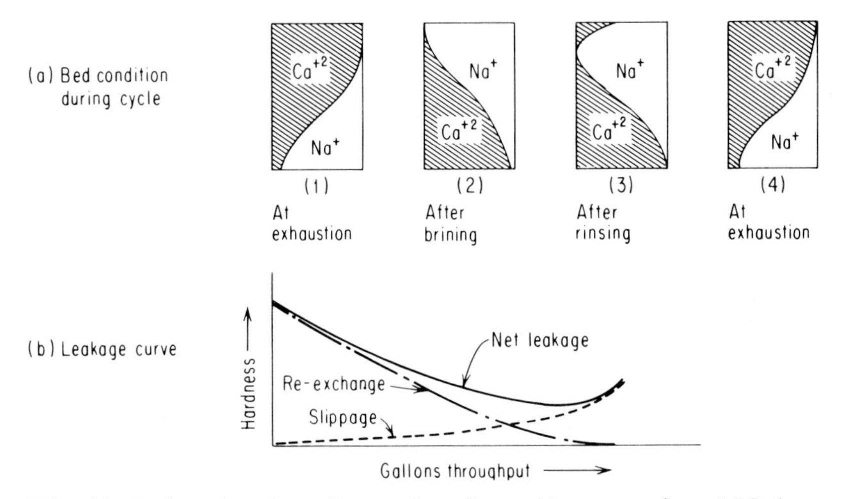

FIG. 12.8 Leakage through a sodium zeolite softener with cocurrent flows. (*a*) Bed condition during cycle; (*b*) leakage curve.

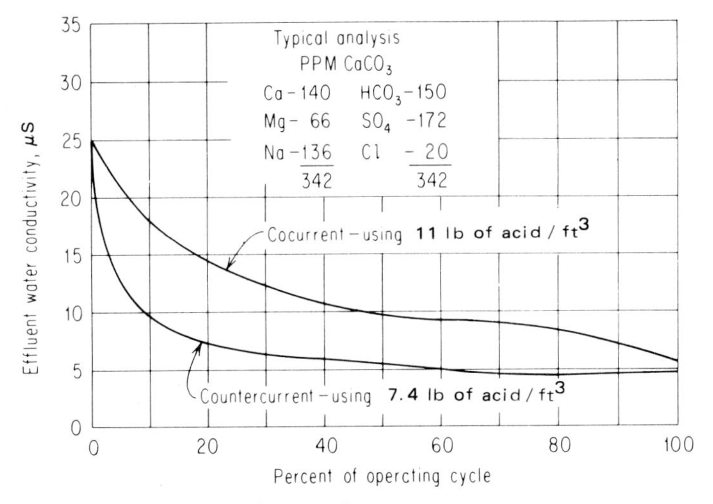

FIG. 12.9 Comparison of results of cocurrent versus countercurrent regeneration of cation exchange resin. *(Thompson, J., and Reents, A. C.: "Counterflow Regeneration," Proceedings of 27th International Water Conference, November 1966.)*

2. As the bed becomes exhausted, the calcium contamination at the bottom of the bed decreases and the quality continually improves until true leakage—slippage of incoming Ca^{2+} through the bed into the effluent—begins to appear. The net effect is the leakage curve shown by Figure 12.8*b*.

This same kind of effect is observed with all exchange processes having cocurrent exhaustion and regeneration. If the bed is regenerated countercurrent to the water flow, this effect is eliminated and leakage is then due to actual ion slippage. The improvement of both quality and efficiency by counterflow regeneration is strikingly illustrated by Figure 12.9, showing a hydrogen cycle operation with downward service flow and upflow regeneration with sulfuric acid.

A final backwash is another procedure for minimizing leakage.

Slippage is easy to appreciate as one consequence of the limited residence time of the water in the ion exchange bed, typically about 1 min.

In an idle unit, there is migration of ions between resin particles. This causes a change of water quality when flow commences again. This migration is also observed during backwash, illustrated in steps 2 and 3 of Figure 12.5.

SODIUM CYCLE EXCHANGE (ZEOLITE SOFTENING)

Zeolite softening is the oldest and simplest of the ion exchange processes. It removes hardness from water, including iron and manganese if these constituents can be kept in the reduced ionic form. When the ion exchange bed is saturated with the hardness constituents, the exchanger is regenerated with sodium chloride brine.

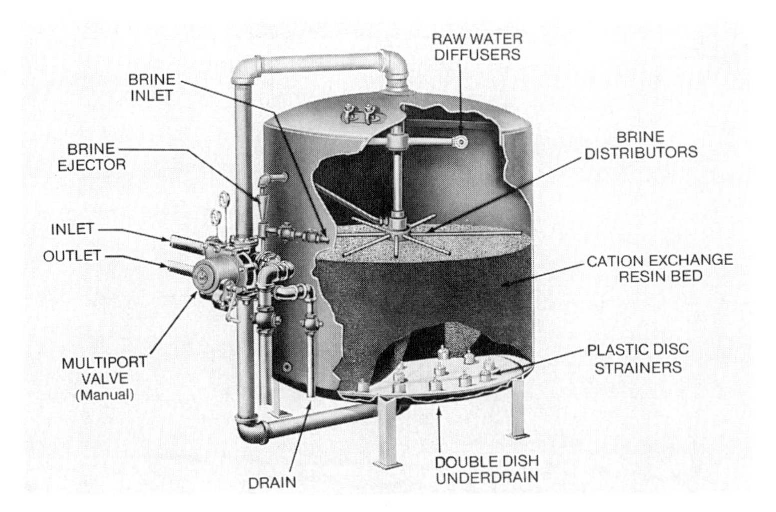

FIG. 12.10 Cross section of an ion exchange unit, showing design details of internals arranged for cocurrent regeneration. *(Courtesy of The Permutit Company.)*

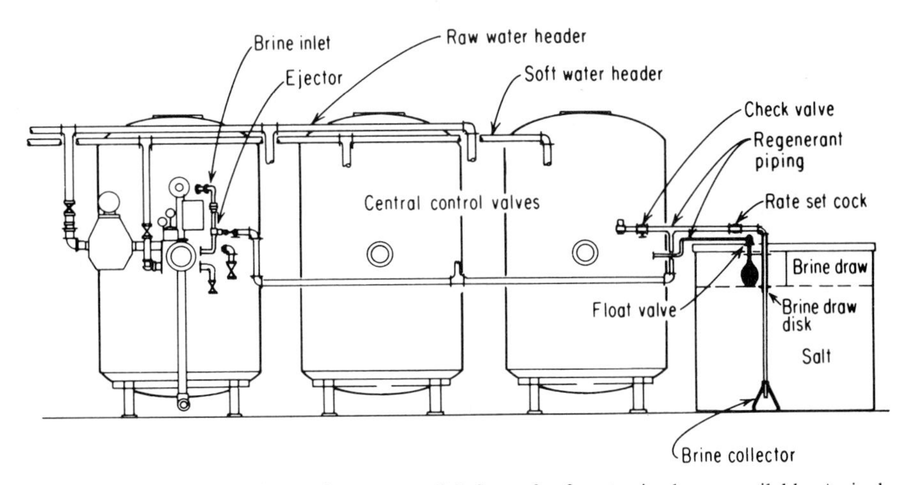

FIG. 12.11 With a multiple-unit system, a full flow of soft water is always available. A single regeneration system serves the battery of units, and individual regenerations are staggered to equalize water quality.

During the processing step—the softening run—the hardness of the raw water may vary and the flow rates through the system may change, but the bed continues to operate effectively to produce a soft water.

The equipment design is simple, consisting of a steel shell holding the ion exchange bed and provided with piping and valves to permit the essential operations of softening, backwash, brining, and rinsing (Figure 12.10). Instrumentation may be provided to help the operator anticipate exhaustion of the bed, so that the bed can be regenerated before the effluent quality depreciates beyond acceptable limits. Almost all sodium zeolite units include a meter for measuring water flow, usually equipped with an alarm so that when the gallonage corresponding to the capacity of the bed has been reached, the operator will be warned of the need to regenerate. More sophisticated units, required by higher quality limits, include automatic hardness analyzers.

If storage capacity is provided for softened water, a single unit may be adequate for the plant needs. In larger plants it is common to have more than one unit, so that one can be taken out of service for regeneration without interrupting the flow of soft water.

A diagram of a multiple unit system is shown in Figure 12.11. This system uses steel tanks to saturate the brine solution and then to measure it for application to the ion exchange bed. Larger systems, where salt can be more conveniently purchased in truckload quantities than in bags, are designed to have a large concrete salt-saturating basin located adjacent to a track or roadway where brine can be brought to the plant in bulk and simply dumped into the underground salt storage and brine saturation basin. A typical design of the salt storage facility for these larger installations is shown in Figure 12.12.

Figures 12.13 and 12.14 illustrate several typical installations, the first being a simple zeolite softener treating city water as boiler makeup, and the second a sodium zeolite softener especially designed for high-temperature service and installed in a plant to follow a hot process lime softener and filter system.

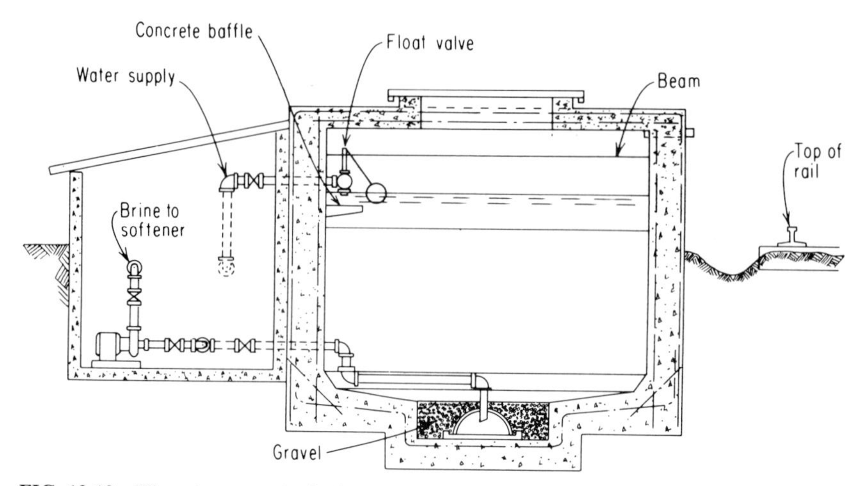

FIG. 12.12 Wet salt storage basin for bulk truckload or carload deliveries of sale. *(Courtesy of Cochrane Division, the Crane Company.)*

FIG. 12.13 Multiple-unit sodium zeolite installation for continuous delivery of softened water. *(Courtesy of The Permutit Company.)*

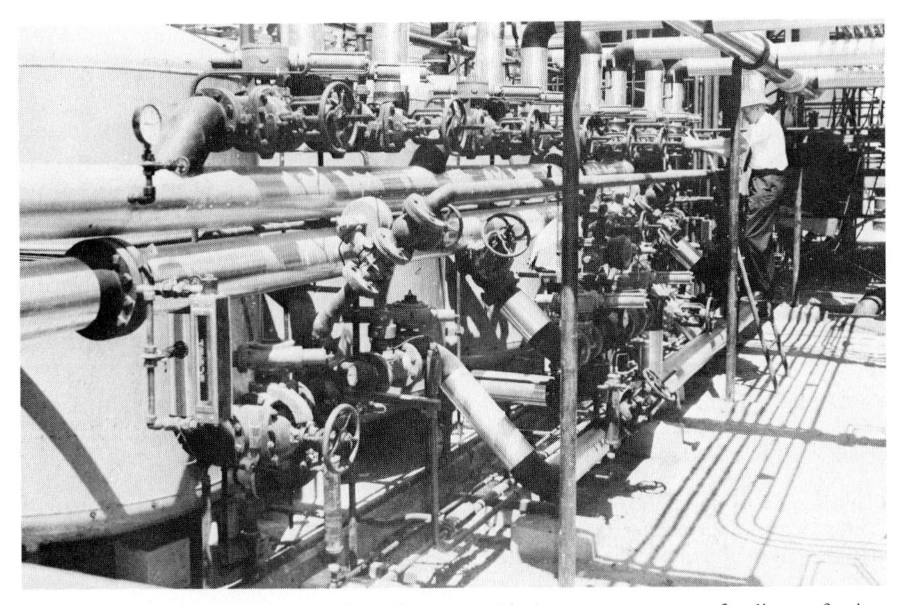

FIG. 12.14 Battery of sodium zeolite softeners used in hot process system after lime softening. Designed for automatic operation. *(Courtesy of Cochrane Division, the Crane Company.)*

The chemical reactions for the sodium zeolite process are shown below:
Softening (exhaustion):

$$Ca^{2+} + Na_2X \rightarrow CaX + 2NA^+ \tag{13}$$

$$Mg^{2+} + Na_2X \rightarrow MgX + 2Na^+ \tag{14}$$

Note: The selectivity of the exchanger favors Ca^{2+} over Mg^{2+}. Therefore, Mg hardness predominates at the point of breakthrough.

Regeneration (brining):

$$2NaCl + CaX \rightarrow Na_2X + CaCl_2 \tag{15}$$

$$2NaCl + MgX \rightarrow Na_2X + MgCl_2 \tag{16}$$

The amount of salt applied to the individual zeolite bed for regeneration is determined by the acceptable effluent quality limits and the capacity the plant wishes to achieve to properly schedule regeneration of the system based on available labor. The greater the salt dosage, the lower is the hardness leakage and the higher the capacity. However, the higher salt dosage also results in poorer chemical economy and greater quantities of spent brine to be handled through the waste disposal system. (See Table 12.2.)

As mentioned briefly earlier, countercurrent regeneration is one technique used to improve both effluent quality and regenerant efficiency; this will be covered in more detail in a later section of this chapter. Another technique is multiple-stage regenerant application at high chemical dosages. For example, salt could be applied for regeneration of a sodium exchanger at 15 lb/ft^3. The initial dis-

TABLE 12.2 Typical Sodium Zeolite Performance—Cocurrent Regeneration

Salt applied, lb/ft^3	6	8	10	15
Approx. capacity, kgr/ft^3	20	23	25	30
Salt rate, lb/kgr	0.30	0.35	0.40	0.50
Salt efficiency, %	56	48	42	33
mg/L salt/mg/L hardness	2.1	2.5	2.8	3.5
Hardness leakage, mg/L, at				
100 mg/L total electrolyte	0.1–0.3	0.1–0.2	0.1	0.1
250 mg/L total electrolyte	0.5–2	0.5–1.0	0.3–0.5	0.1–0.2
500 mg/L total electrolyte	3–6	2–4	1–3	0.5–1.0
1000 mg/L total electrolyte	12–25	8–15	5–10	2–4

TABLE 12.3 Sample Calculation of Sodium Exchange System

Water Analysis: See Figure 12.15. Total hardness = 200 mg/L, allowable leakage = 3 mg/L, TE = 300 mg/L.

Plant Requirements: 1,200,000 gal/day = average flow of 1,200,000/1440 = 833 gal/min; must accommodate extended peak flow of 1200 gal/min.

1. Calculation of required size to handle 1200 gal/min. (See Table 12.4.) Use three 10 ft 0 in diameter units.
2. Calculation of required resin volume.
   a. Required capacity in kilograins:
   Hardness = 200 mg/L; 200/17.1 = 11.7 gr/gal,
   1,200,000 × 11.7/1000 = 14,035 kgr/day
   Schedule all 3 units to be regenerated each shift.
   ∴ Required capacity = 14,035/3 = 4678 kgr total
   4678/3 units = 1560 kgr/unit.

   b. Allowable leakage = 3 mg/L:
   This leakage can be achieved at 6 lb NaCl/ft^3. (Table 12.2)
   (Actual leakage at 6 lb/ft^3 will be only 1.5 mg/L hardness)
   Capacity at 6 lb/ft^3 = 20 kgr/ft^3 (Table 12.2)
   ∴ Required resin = 1560/20 = 78 ft^3 minimum per unit.

   c. Bed depth: Area of 10 ft diameter unit = 78.5 ft^2. Therefore, bed would be approximately 1 ft deep. To achieve the desired leakage, minimum bed depth is 30 in. Therefore, each unit requires:

   $$2.5 \text{ ft deep} \times 78.5 \text{ ft}^2 \text{ area} = 195 \text{ ft}^3$$

   d. The actual service run will be:

   $$\frac{196 \text{ ft}^3 \times 3 \text{ units} \times 20 \text{ kgr/ft}^3}{14{,}035 \text{ kgr/day}} = 0.84 \text{ day}$$

   $$0.84 \times 24 = 20 \text{ h}$$

   or one unit will be regenerated every 6 h, 40 min.
3. Salt consumption at 6 lb salt—20 kgr/ft^3
   6/20 = 0.3 lb/kgr (2.1 mg/L salt/mg/L hardness), (theoretical = 0.167 lb/kgr)
   14,035 kgr/day × 0.3 lb NaCl/kgr = 4200 lb NaCl
   This requires bulk deliveries for practical handling by plant labor.

charge of spent brine will be high in Ca and Mg hardness, but after passage of a brine volume equivalent to perhaps 5 to 6 lb/ft^3 the hardness concentration of the discharge usually drops sharply; at this point, the subsequent brine discharge can be diverted from sewer to a reclaim tank to be used as the first step of the next regeneration cycle. This same principle has been applied to acid and caustic regenerations in water treatment and has been used extensively in metallurgical applications of ion exchange, such as uranium recovery.

An example is worked out completely in Table 12.3 to give a perspective on the relative size of the equipment, the choice between salt dosage, capacity, and leakage, and the volumes of waste produced. Note that NaCl utilization of 0.3 lb/kgr compared to a theoretical (stoichiometric) of 0.167 corresponds to an efficiency of only about 56%.

Identification of Analyses Tabulated Below:

A. Raw Water   D.

B. Softened Water   E.

C.   F.

Constituent	As	A	B	C	D	E	F
Calcium	$CaCO_3$	150	2				
Magnesium	"	50	1				
Sodium	"	100	297				
Total Electrolyte	$CaCO_3$	300	300				
Bicarbonate	$CaCO_3$	200	200				
Carbonate	"	0	0				
Hydroxyl	"	0	0				
Sulfate	"	50	50				
Chloride	"	50	50				
Nitrate	"	0	0				
					N O		
				C	H A N G	E	
M Alk.	$CaCO_3$	200	200				
P Alk.	"	0	0				
Carbon Dioxide		20	20				
pH		7.4	7.4				
Silica		15	15				
Iron	Fe	0.2	Nil				
Turbidity		Nil	Nil				
TDS *		350	370				
Color		Nil	Nil				

* Note the increase in TDS due to higher equivalent weight of sodium (23) compared to Ca (20) and Mg (12.2)

FIG. 12.15 Expected results of simple sodium zeolite treatment.

TABLE 12.4 Ion Exchange Flows and Dimensions

Diameter	Area, ft²	Min. resin, ft³ at 30 in	Backwash, gal/min			Service flow, gal/min max.*		
			†	‡	§	1 unit	2 units	3 units
2 ft–0 in	3.14	8	9	19	38	19	25	50
2 ft–6 in	4.91	12	15	30	59	30	40	80
3 ft–0 in	7.07	18	21	42	85	42	56	112
3 ft–6 in	9.62	24	29	58	116	58	77	154
4 ft–0 in	12.6	32	38	75	150	75	100	200
4 ft–6 in	15.9	40	48	95	190	95	130	260
5 ft–0 in	19.6	49	59	118	235	118	160	320
5 ft–6 in	23.8	60	71	143	285	143	190	380
6 ft–0 in	28.3	71	85	170	340	170	230	460
6 ft–6 in	33.2	83	100	200	400	200	270	540
7 ft–0 in	38.5	96	115	230	460	230	310	620
7 ft–6 in	44.2	110	132	265	530	265	350	700
8 ft–0 in	50.3	126	150	300	600	300	400	800
8 ft–6 in	56.8	142	170	340	680	340	450	900
9 ft–0 in	63.6	160	190	380	760	380	500	1000
9 ft–6 in	70.9	176	213	425	850	425	570	1140
10 ft–0 in	78.5	196	235	470	940	470	630	1260
11 ft–0 in	95	240	285	570	1140	570	750	1500

* Service flow is based on rates of 6 gal/min/ft², single unit maximum rate, and 8 gal/min/ft² short term rate, on multiple units when one unit is out of service for regeneration. To provide a continuous supply of treated water, most plants use two units, so that while one is being regenerated, the other continues to provide finished water. Plants having variable demand store finished water to eliminate surges in flow, minimizing equipment size. Treated water storage permits smaller plants to install only a single unit, relying on stored water to provide requirements during regeneration. This is risky, since it provides no margin for error or for normal maintenance.

† Backwash rate of 3 gal/min/ft² for anion resins at 70°F.

‡ Backwash rate of 6 gal/min/ft² for cation resins at 70°F.

§ Backwash rate of 12 gal/min/ft² for cation resins at 220°F.

The zeolite softener simply removes hardness when used for treatment of a raw water supply. A typical analysis showing raw water and treated water is given in Figure 12.15. However, the system is also used for cleanup of process condensates in industrial operations, such as paper mill condensate, chemical plant condensate, and central heating station condensate, where the units are operated at very high flow rates and remove not only traces of hardness that enter the system from leakage of heat exchangers, but also particulate matter, such as corrosion products.

The wastes from regeneration of an ion exchange system are convenient in that they are in solution form. However, they increase the dissolved solids content of the effluent from the waste treatment plant through which they are processed. If this saline water is a problem, it can be treated with lime and soda ash to precipitate the hardness, producing a brine which can then be reused. This process, which converts the liquid saline waste to a solid waste, has been proposed for installations in the southwest where the spent brine may lower the quality of the receiving streams and where the climatic conditions are favorable for drying of solid waste.

HYDROGEN CYCLE EXCHANGE

The development of ion exchange materials that could be regenerated with acid to exchange hydrogen ions for cations in water provided the first practical chemical process for the removal of sodium, potassium, and ammonia, all of whose salts are extremely soluble. The generalized reactions for hydrogen cycle exchange have been given earlier. The following specific reactions and data shown in Figure 12.16 illustrate the process more completely. The effluent is acid, since it contains carbon dioxide equivalent to the incoming alkalinity, and hydrogen which has been exchanged for substantially all of the cations in the influent, except for leakage:

$$Ca(HCO_3)_2 + H_2X \rightarrow CaX + 2H_2O + 2CO_2 \quad (17)$$

$$CaSO_4 + H_2X \rightarrow CaX + H_2SO_4 \quad (18)$$

The strong acidity of the treated water must be corrected to provide a water satisfactory for practically any conventional use. The carbon dioxide is easily removed by degasification, since the equilibrium carbon dioxide concentration corresponding to the typical composition of the atmosphere is less than 1 mg/L. Therefore, the acidity may be neutralized by the addition of alkali, or by anion exchange.

There are several choices for alkalinity neutralization. If the finished water need not be completely softened, the hydrogen exchange effluent can be blended with raw water and the blend degasified, to provide whatever alkalinity is needed in the finished water. Column C of Figure 12.16 gives the analysis of a blend of raw water with hydrogen exchange effluent to produce an alkalinity of 50 mg/L. If the blended water must be completely softened, a sodium zeolite softener can supply the alkalinity for neutralization of the acidity of the hydrogen exchanger effluent, and the analysis will then be that shown by column D. In both cases the proportions of hydrogen exchanger effluent to be blended is calculated by the following formula:

$$\text{Percent } H_2X = \frac{\text{alkalinity reduction}}{\text{total electrolyte}} \times 100\%$$

In the example given in Figure 12.16 the blend is:

$$\text{Percent } H_2X = \frac{200 - 50}{300} \times 100 = 50\%$$

$$\text{Percent raw water or } Na_2X = 50\%$$

Another way of neutralizing the acidity in the hydrogen exchanger effluent is to feed an alkaline chemical such as NaOH into the effluent to neutralize the hydrogen ion and to reach the desired alkalinity. This treatment produces the same effluent shown by column D. In this example, the caustic soda dosage would be equal to the 95 mg/L hydrogen ion neutralized plus 50 mg/L desired alkalinity, or 145 mg/L as $CaCO_3$, equivalent to 116 mg/L NaOH.

The major advantage of these variations of the hydrogen cycle process over direct acidification to reduce alkalinity is a reduction of total solids equivalent to alkalinity reduction, as shown by Figure 12.16.

By far the most common process is the blending of hydrogen exchanger and

Identification of Analyses Tabulated Below:	
A. Raw Water	D. H_2X-Na_2X Blend or H_2X + NaOH
B. H_2X Effluent	E. Neutralization
C. H_2X + Raw Water Blend	F.

Constituent	As	A	B	C	D	E	F
Calcium	$CaCO_3$	150	Nil	75	Nil		
Magnesium	"	50	Nil	25	Nil		
Sodium	"	100	5	50	150		
Hydrogen		0	95	0	0		
Total Electrolyte	$CaCO_3$	300	100	150	150		
Bicarbonate	$CaCO_3$	200	0	50	50		
Carbonate	"	0	0	0	0		
Hydroxyl	"	0	0	0	0		
Sulfate	"	50	50	50	50		
Chloride	"	50	50	50	50		
Nitrate	"	0	0	0	0		
M Alk.	$CaCO_3$	200	0	50	50		
P Alk.	"	0	0	0	0		
Carbon Dioxide		20	220	5*	5*		
pH		7.4	2.7	7.4	7.4		
Silica		15	15	15	15		
Iron	Fe	0.2	Nil	Nil	Nil		
Turbidity		Nil	Nil	Nil	Nil		
TDS		350	150	200	200		
Color		Nil	Nil	Nil	Nil		
* After degasification							

FIG. 12.16 Expected results of split-stream treatment.

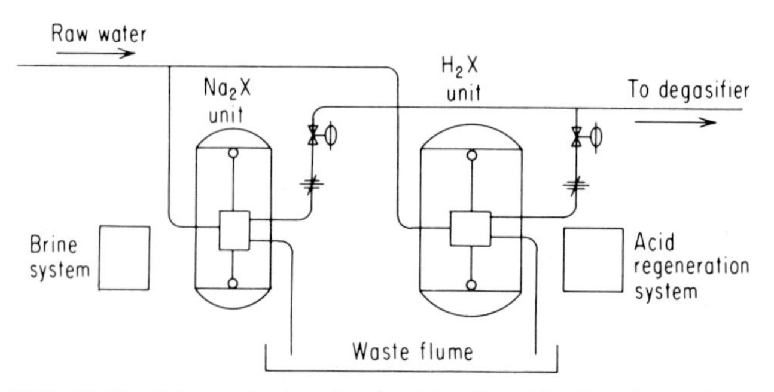

FIG. 12.17 Schematic showing the blending of acid effluent from H_2X unit with alkaline sodium zeolite effluent through adjustable rate controllers designed to maintain correct blend as flow demand varies.

FIG. 12.18 Split-stream H_2X–Na_2X exchangers softening and dealkalizing a municipal water. The exchangers are fitted for automatic regeneration. *(Courtesy of The Permutit Company.)*

sodium exchanger effluents, known as split-stream softening (Figure 12.17). A typical split-stream system is shown in Figure 12.18.

Acid Regeneration of Cation Exchangers

In the systems just described, a strong cation resin is usually used with sulfuric acid regeneration. The approximate capacities realized with cocurrent sulfuric acid regeneration of such an exchanger are shown in Table 12.5. The regeneration efficiency is rather poor when compared to theoretical, 0.143 lb/kgr (1 kg/kg).

Part of the inefficiency is due to the need to keep the concentration of H_2SO_4 low enough to avoid precipitation of calcium sulfate in the bed. In an ideal regeneration, the H_2SO_4 strength should be as high as possible while having a waste acid discharge that is clear when sampled but shows $CaSO_4$ precipitation in the same sample after it has stood for 3 to 5 min. A second major factor in low sulfuric acid efficiency is the incomplete ionization of the acid at the applied concentration:

$$H_2SO_4 \rightleftharpoons H^+ + HSO_4^- \rightleftharpoons 2H^+ + SO_4^{2-} \tag{19}$$

In the range of 2 to 8%, appreciable acid is still in the HSO_4^- form, limiting the mass of H^+ ions available for ion exchange.

The development of weak cation (carboxylic) exchangers (discussed later) has diminished the popularity of the split-stream dealkalizer. A layered bed with acid-regenerated carboxylic resin above a salt-regenerated strong cation resin provides advantages in efficiency and reduction of wastes. But the split-stream offers flexibility in dealing with a water supply of variable composition, since the ratio of Na_2X flow to H_2X flow can be adjusted.

Another alkalinity reduction process utilizes a strong cation exchanger regenerated first with salt followed by a "starvation" dosage of acid, creating both Na_2X

TABLE 12.5 Strong Cation Resin Performance H-Cycle, H_2SO_4 Regeneration with Cocurrent Regeneration

Acid applied, lb/ft³	6	8
Approx. capacity, kgr/lb/ft³		
1. At TE/M ratio = 3.0	12	13.5
lb/kgr	0.5	0.6
acid efficiency	29%	24%
parts H_2SO_4/part cations	3.5	4.2
2. At TE/M ratio = 2.0	13.5	16
lb/kgr	0.45	0.5
acid efficiency	32%	29%
parts H_2SO_4/part cations	3.2	3.5
3. At TE/M ratio = 1.5	15	17
lb/kgr	0.4	0.47
acid efficiency	36%	30%
parts H_2SO_4/part cations	2.8	3.3

Notes:
1. For split-stream dealkalization, use 6 lb H_2SO_4/ft^3.
2. For demineralization:
   a. 6 lb H_2SO_4/ft^3 for mixed bed.
   b. For 2-bed systems, use 6 lb/ft³, where Na < 50 mg/L, 8 lb/ft³ otherwise.

and H_2X sites in the same bed and realizing 100% acid efficiency. This cannot be used on waters of variable composition.

The spent regenerant requires careful attention for disposal since it contains a large amount of unused acid plus calcium sulfate at concentrations far above its solubility, creating a potential sludge problem. The use of hydrochloric acid avoids the calcium sulfate precipitation, but in most areas hydrochloric acid is too expensive compared to sulfuric acid and special materials of construction are required. HCl vapors must be properly vented, too.

As shown by Table 12.5, an increase in the amount of acid applied for regeneration produces an increase in capacity, but at the expense of efficiency. A second effect of higher regeneration level is a reduction in leakage. In the hydrogen cycle treatment scheme shown by Figure 12.16 sodium leakage is not critical, since the acid effluent is neutralized by a sodium-based alkali. Therefore it is practical to use only 6 lb/ft³ (96 kg/m³) regeneration level for better acid efficiency. However, as more fully developed later in the text regarding demineralization, the higher dosage of acid may be needed to meet the quality goals of demineralizing systems.

Using the water analysis of Figure 12.16, data on acid dosage and resulting capacity and sodium leakage with cocurrent regeneration are plotted in Figure 12.19.

Using Weak Cation Exchangers

The utilization of regenerant acid can be considerably improved by incorporating a carboxylic exchanger into the system. In such a system, the carboxylic exchanger bed is sized to react with the alkalinity or the hardness of the incoming water, whichever is the lesser. The strong cation exchanger is then sized to react with the balance of the cations. The carboxylic unit then operates at close to theoretical

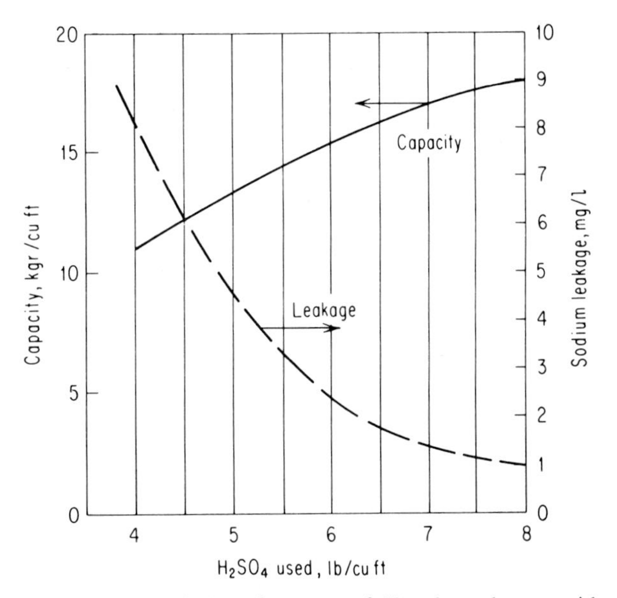

FIG. 12.19 Typical performance of H-cycle exchanger with cocurrent flows, sulfonic acid resin.

efficiency, approximately 0.15 lb/kgr (1.0 kg/kg). The distinctive performance of a carboxylic exchanger is strikingly illustrated by the comparison of acid-base titration curves for carboxylic and sulfonic exchangers, Figure 12.20.

Carboxylic exchangers typically have a capacity as high as 60 kgr/ft^3 (2.8 *N*), but the operating capacity is usually in the range of 20 to 40 kgr/ft^3 (0.9 to 1.8 *N*). This compares with a range of only 12 to 18 kgr/ft^3 (0.6 to 0.9 *N*) for the strong cation exchanger regenerated with H_2SO_4.

The results of treatment whether using a strong cation resin or a combination of a strong cation resin and a carboxylic resin are the same. The advantage of the carboxylic resin is a reduction in cost by improved acid efficiency and reduced

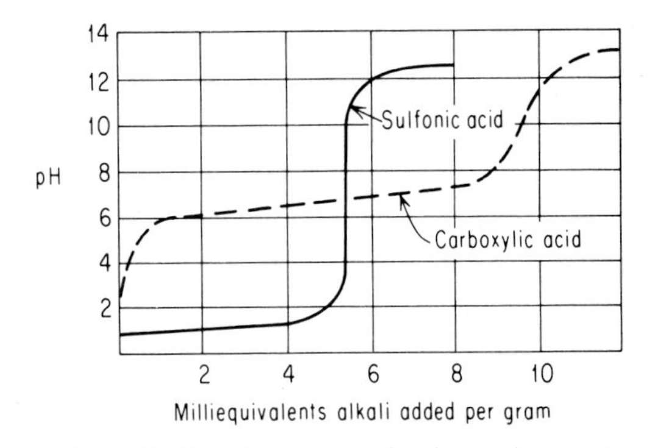

FIG. 12.20 Titration curves of cation exchange resins. *(Courtesy of Rohm & Haas.)*

TABLE 12.6 Comparison of Cocurrent to Countercurrent Regeneration of Cation Exchanger Regenerated with H_2SO_4

Operating factor	Cocurrent	Countercurrent
Regeneration:		
Acid, lb/ft^3	6.0	3.7
Concentration	2 and 4%	2%
Capacity:		
kgr/ft^3	7.7	14.5
lb H_2SO_4/kgr	0.78	0.26
Regeneration:		
Acid consumption, as a percent of theoretical	520%	170%

waste treatment load through reduction of excess acid. Against this chemical saving must be balanced the cost of the more sophisticated design required by the use of two resins. Some systems incorporate layered beds of carboxylic and sulfonic resins in the same vessel; these are susceptible to fouling because the carboxylic resin is light and difficult to clean and should be provided with a separate regeneration system to avoid calcium sulfate deposition.

Countercurrent Regeneration of H_2X Units

Another choice is available to improve results, and that is to design the exchanger for countercurrent regeneration.

An excellent example of this is given in Table 12.6. This compares side-by-side performance of two cation exchangers in a Canadian plant, where these units are part of a demineralizing train. One was operated by conventional cocurrent acid regeneration, and the other with countercurrent. The units were operated to produce equal low levels of sodium leakage, using H_2SO_4 for regeneration. The units each contained the same kind of strong cation resin. The remarkable increase in capacity and regenerant efficiency showed up not only on these cation units, but also on the anion units, which were likewise converted and compared. A second example, where the exchanger was operated on a relatively high solids water, is shown by Figure 12.9.

The countercurrent system is a little more sophisticated than cocurrent design, but improved quality and increase in efficiency make this attractive.

ANION EXCHANGE PROCESSES

The nature of anion exchange was briefly introduced earlier, with reference to both the strong- and weak-base anion exchangers. The first reaction shown for the strong-base exchanger involved the removal of alkalinity from water by the exchanger in substitution for chloride ions. This is a simple and practical process for the reduction of alkalinity without the use of acids. Since the exchanger is regenerated with sodium chloride brine, chemical handling is very simple. The chloride-form of the anion exchanger removes sulfate as well as bicarbonate, and because sulfate removal is not usually necessary, this is an added cost to the operation.

In some water treatment operations, the reduction of sulfate by itself may be important, and this can be accomplished in the same fashion. The removal of sulfate is somewhat selective as compared to the reduction of bicarbonate, so the efficiency of this operation is higher than that for the dealkalization process.

Although most plants soften water before dealkalization through an anion exchanger, this is not strictly necessary. For example, cooling tower makeup could be dealkalized through the anion exchanger without prior softening.

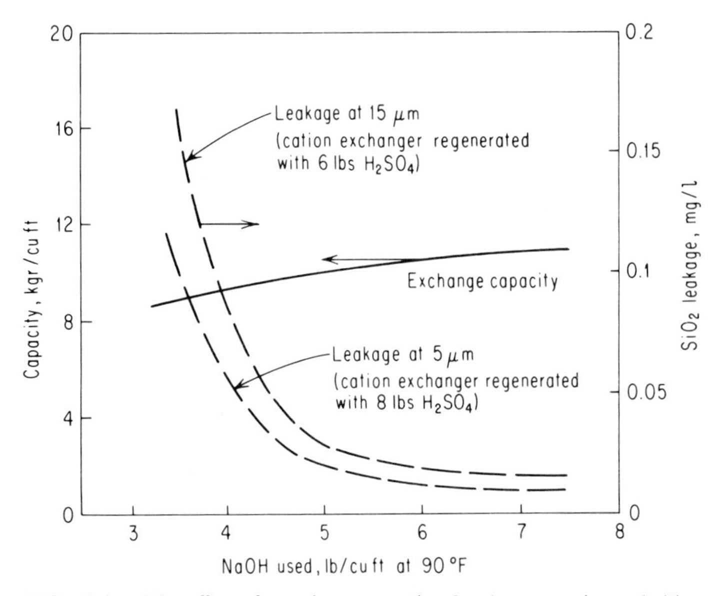

FIG. 12.21 The effect of caustic regeneration level on capacity and silica leakage in a two-bed, cocurrent demineralizer with degasifier (analysis of Figure 12.16).

By far the most common form in which anion exchangers are used in water treatment is the hydroxide form, with caustic soda being used for regeneration; and the most common process using this form of strong-base resin is demineralization, where the anion exchanger follows the hydrogen-form cation exchanger. In this process, the anions in the cation effluent are exchanged for hydroxide; if there are cations present, such as sodium, the effluent will contain sodium hydroxide, and this is the most prominent factor affecting the quality of finished demineralized water. When demineralized water is used for boiler makeup, silica leakage is of primary interest, and the amount of sodium leakage that occurs proportionately affects silica leakage. Using the water analysis of Figure 12.16 and the sodium leakage through the cation exchange shown by Figure 12.18, the quality of demineralized water that can be produced by a strongly basic anion exchanger is shown in Figure 12.21 for two different levels of acid regeneration of the cation exchanger and varying levels of caustic used for regeneration of the anion exchanger.

A simplified chart of typical performance capacities for strongly basic anion exchangers is listed in Table 12.7.

TABLE 12.7 Strong Base Exchange Capacities in Demineralizing Service (Regenerated with NaOH at 95°F, with 90-min Contact Time and Cocurrent Flow)

NaOH, lb/ft³	Cap., kgr/ft³	lb/kgr	Final SiO_2, mg/L
3.5	10	0.35	0.6
5.0	12	0.42	0.3

Note: SiO_2 leakage is based on anion loading of 100% SiO_2, water at 75°F, and final conductivity of 3 μS. Prorate to suit actual conditions:

1. If SiO_2 in influent to anion unit is only 10% of the total anions, final SiO_2 will be only 10% of value shown.
2. If water temperature is 90°F, SiO_2 leakage will increase 50%; at 50°F, SiO_2 leakage will decrease 50%.
3. If conductivity of finished water exceeds 3 μS, SiO_2 leakage will increase about 20% for each 5 μS increase.
4. In using these data, estimate the amount of water produced from capacity and total anion loading, which is the sum of all anions entering the exchange unit. If there is a degasifier in the system for the reduction of CO_2 ahead of the strong base resin, this reduces the load on the strong base resin.

Process Refinements

If the quality of water as shown by these illustrations is not adequate for the intended service, alternative methods of treatment must be used. The most obvious possibility is the installation of a second stage demineralizer to polish the water produced from the first stage. Another alternative is to use a mixed bed demineralizer either for the primary service or for polishing; in this type of unit, the cation and anion resins are intimately mixed after regeneration, in effect providing hundreds of stages of demineralization (Figure 12.22).

Other choices include improvements in design, such as counterflow regeneration, and reduction of the loading on the ion exchange system by pretreatment.

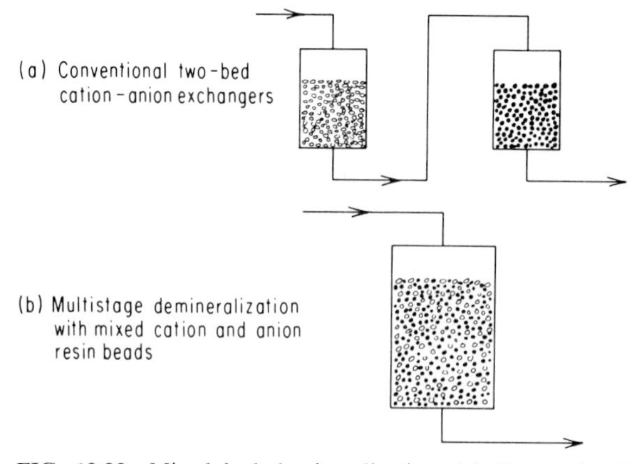

FIG. 12.22 Mixed bed demineralization. (*a*) Conventional two-bed cation-anion exchangers. (*b*) Multistage demineralization with mixed cation and anion resin beads.

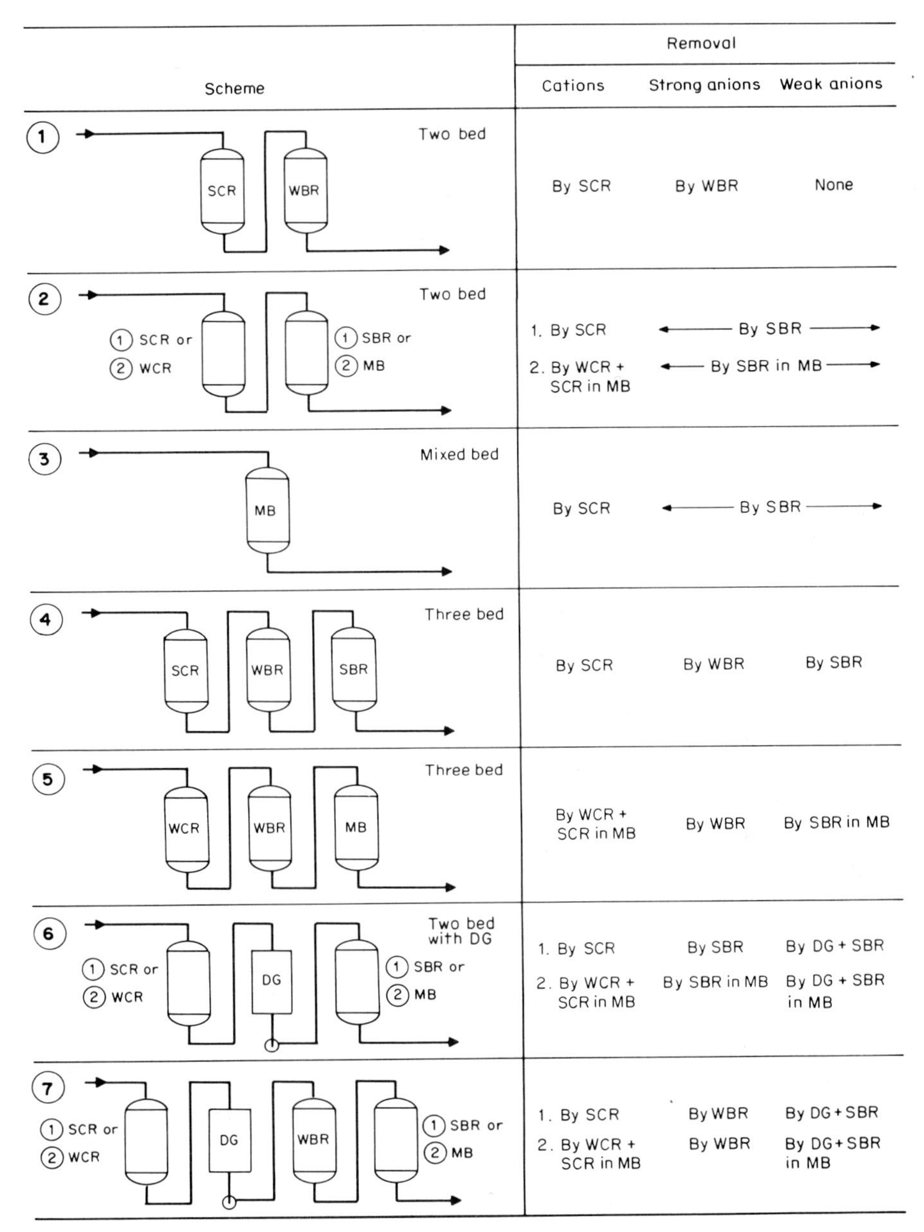

Scheme	Removal		
	Cations	Strong anions	Weak anions
(1) Two bed: SCR, WBR	By SCR	By WBR	None
(2) Two bed: (1) SCR or (2) WCR; (1) SBR or (2) MB	1. By SCR 2. By WCR + SCR in MB	← By SBR → ← By SBR in MB →	
(3) Mixed bed: MB	By SCR	← By SBR →	
(4) Three bed: SCR, WBR, SBR	By SCR	By WBR	By SBR
(5) Three bed: WCR, WBR, MB	By WCR + SCR in MB	By WBR	By SBR in MB
(6) Two bed with DG: (1) SCR or (2) WCR; DG; (1) SBR or (2) MB	1. By SCR 2. By WCR + SCR in MB	By SBR By SBR in MB	By DG + SBR By DG + SBR in MB
(7) (1) SCR or (2) WCR; DG; WBR; (1) SBR or (2) MB	1. By SCR 2. By WCR + SCR in MB	By WBR By WBR	By DG + SBR By DG + SBR in MB

SCR – strong cation resin; WCR – weak cation resin
SBR – strong base resin; WBR – weak base resin
DG – degasifier; MB – mixed bed

FIG. 12.23 Basic types of demineralizers and resins used.

TABLE 12.8 Demineralizer Unit Selection Chart*

Flow rate, gal/min†	Amount of impurity to be removed, mg/L†			Resins required			Units to be used				
	FMA	CO_2	SiO_2	C	WB	SB		C	DG	A	MB
1. Any	Any	None	None	x	x	—	(1)‡	x	—	x	—
							(2)	—	—	—	x
2. Any	Any	Any	None	x	x	—	(1)	x	x	x	—
							(2)	—	x	—	x
2. 0–20	Any	Any	Any	x	—	x	(1)	x	—	x	—
							(2)	—	—	—	x
4. 20–50	Any	0–50	Any	x	—	x	(1)	x	—	x	—
							(2)	—	—	—	x
5. Over 100	0–100	Over 100	Any	x	—	x	(1)	x	x	x	—
							(2)	x	x	—	x
6. Over 100	Over 200	Over 100	Any	x	x	x	(1)	x	x	x	—
							(2)	x	x	x	x

* C = cation, A = anion, DG = degasifier, MB = mixed bed, WB = weak base, SB = strong base, FMA = free mineral acidity (SO_4 + Cl + NO_3).

† Intermediate ranges (50–100 gal/min flow, 50–100 mg/L CO_2, 100–200 mg/L FMA) require careful evaluation for best balance in capital and operating costs.

‡ Numbers in parentheses: (1) multibed plant, (2) mixed bed for 1 μS effluent.

One of the common procedures for reducing loading on the anion exchanger is the removal of carbon dioxide by degasification so that it need not be removed by ion exchange.

Weakly basic exchangers can be installed in demineralizing systems both to relieve the load on the strong anion exchanger and also to improve the economy of operation. Weakly basic exchangers remove only the strong mineral acids, such as hydrochloric and sulfuric acid, and the alkali used for regeneration neutralizes these adsorbed acids so that the regeneration is close to 100% efficient.

There are numerous possibilities for the design engineer in putting together individual units of a demineralizing system to suit the type of raw water being processed, the economics of plant operation, and the ultimate quality demanded by the process use. These are illustrated by Figure 12.23. Some general rules for the selection of individual components of a demineralizing system based chiefly on economics of operation are shown in Table 12.8.

TABLE 12.9 Typical Weak Base Exchange Capacities

Chemical	Dosage, lb/ft³	Capacity,* kgr/ft³	Usage, lb/kgr
NH_3	1.5	20.0	0.08
NaOH	3	21.3	0.14
Na_2CO_3	6.6	19.6	0.34

* Capacity based on a 1:1 ratio of HCl to H_2SO_4.

Because the weakly basic exchanger is regenerated simply by neutralization of the adsorbed mineral acids, a variety of alkalies can be used for this operation. The approximate capacity of these exchangers and the kinds of regenerants that may be used are illustrated by Table 12.9.

Waste Treatment Considerations

A prominent factor in the selection of any ion exchange process is the waste created by regeneration of the ion exchange bed. The waste may be reusable, as in the discharge from the unit during backwashing, or it may be troublesome, as in the case of spent acid from regeneration of a cation exchanger. The proper treatment of these wastes requires an evaluation of each step in the regeneration process to determine the amount and flow rate of the waste from the regeneration step and the concentration of contaminants.

As an example of the importance of studying each step in the regeneration process, when a sodium zeolite system is installed to follow a hot process softener, there is essentially no undesirable material in the backwash from the zeolite, so this entire flow is reclaimed by returning the zeolite backwash to the head of the filters to reclaim both the heat and the water. There may be useful chemical values even in spent regenerants; spent caustic from regeneration of a strong anion exchanger may be used to regenerate the weakly basic anion exchanger. And, as mentioned earlier, multiple-stage elution may prove a realistic way to reduce net effluent salinity while maintaining required quality limits for the produced water. An elution study will show the optimum point for diversion of spent regenerant to a reclaim tank to be used as the initial stage of subsequent chemical regenerant applications.

And finally, careful attention should be given to recovery of discharge from the rinsing operation; cation exchanger rinse following acid regeneration has been used as cooling tower makeup to provide alkalinity reduction. Hot zeolite rinse water can be returned to the filter inlet when its hardness reaches the normal inlet hardness level. Spent caustic from anion exchanger regeneration has been used as salt-cake makeup in kraft pulping operations. Under the right circumstances, this could be combined with the spent sulfuric acid.

If no use can be made of the spent regenerants, they must be balanced to produce a neutral effluent that will not upset the overall plant waste treatment system. This may have an influence on such decisions as whether to install a degasifier for CO_2 removal and at which efficiency to operate the exchangers. For example, there may be no need to improve the efficiency of anion exchanger regeneration with caustic if the spent acid from regeneration of the cation exchanger is going to require extra alkali for neutralization.

There is a variety to select from in choosing the best ion exchange unit for each particular process. The simplest and most common ion exchange unit is a vertical cylindrical vessel designed for downflow operation and making provision for backwash, chemical injection, rinsing, and normal service operations. Figure 12.10 shows a cutaway of a typical unit for this four-step cycle. In this system, the chemical regenerant flows downward through the ion exchange bed, and normal flow of process water is also downward.

If the system requires more efficient chemical utilization, countercurrent regeneration can be used. In this case, the service flow may be upward through a vessel that is completely packed with ion exchange resin. The chemical injection is then downward, or countercurrent, followed by rinse in the same direction. This arrangement has no provision for resin backwashing in place, so periodically the resin must be withdrawn from the unit and cleaned in an external vessel for return to the system. External cleaning may be required every 50 to 100 cycles; but to achieve this frequency it is important that the water supply to the system be clear.

Countercurrent operation can also be achieved with a more conventional design that provides head room for backwashing, as shown by Figure 12.24. In this case, the service flow is downward, with regeneration flow upward. To prevent the bed from expanding during upflow regeneration, a counterflow of water may be required to balance the hydraulic pressure in the system.

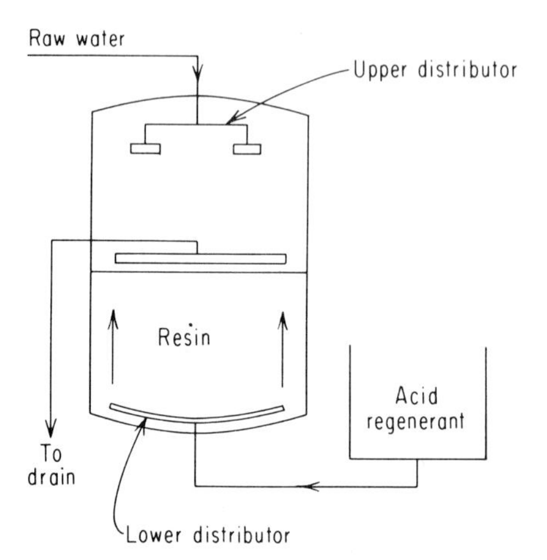

FIG. 12.24 A method of countercurrent regeneration. The regenerant is introduced upflow, with the resin bed held in place by a downward flow of water. *(Courtesy of Illinois Water Treatment Company.)*

The mixed bed demineralizer operates with a charge of cation and anion resin in the same vessel, with sequences shown in Table 12.10. A typical internal construction is shown in Figure 12.25. Units of this kind are used for polishing demineralized water produced in a two-bed system, although they are sometimes used as a complete unit by themselves if the water is of relatively good quality so that organic fouling will not be a problem. Mixed bed demineralizers are frequently used in purification of utility plant condensate, in which case the units operate at very high flow rates because of the high purity of the water being treated. This saves on resin investment. The resin is withdrawn from the processing unit for regeneration in external facilities, so the design of the process unit is quite simple. A typical condensate polisher with external regeneration is shown in Figure 12.26.

Earlier it was mentioned that ion exchange is a cyclic operation. One path toward improved chemical efficiency has been the design of a continuous ion exchange system. One such design which has been commercially successful is

TABLE 12.10 Sequence of Steps in Regenerating a Mixed Bed Demineralizer

1. Backwash to separate resins with upper layer of anion resin and lower of cation resin.
2. Settle bed.
3. Commence regeneration, with acid upflow through cation bed, caustic downflow through anion bed, and spent liquor withdrawn at interface.
4. Discontinue chemical feed and flush feedlines.
5. Displace chemical from both resin sections.
6. Fast rinse cation section while continuing displacement of anion section.
7. Rinse downflow through both sections.
8. Drain water from freeboard to 3–5 in above bed.
9. Air mix from bottom to fluidize bed.
10. While air mixing, dewater bed through drawoff at the interface collector.
11. Refill unit slowly through bottom distributor.
12. Final rinse downflow until acceptable quality is produced.

FIG. 12.25 View of internals of a mixed-bed demineralizer looking downward, showing reinforcing needed to prevent distributor damage from resin movement. *(Courtesy of Illinois Water Treatment Company.)*

FIG. 12.26 Condensate polisher designed for cleanup of 6000 gal/min of utility station condensate. The resin is regenerated by sluicing from the exchangers to separate cleanup vessels. *(Courtesy of Illinois Water Treatment Company.)*

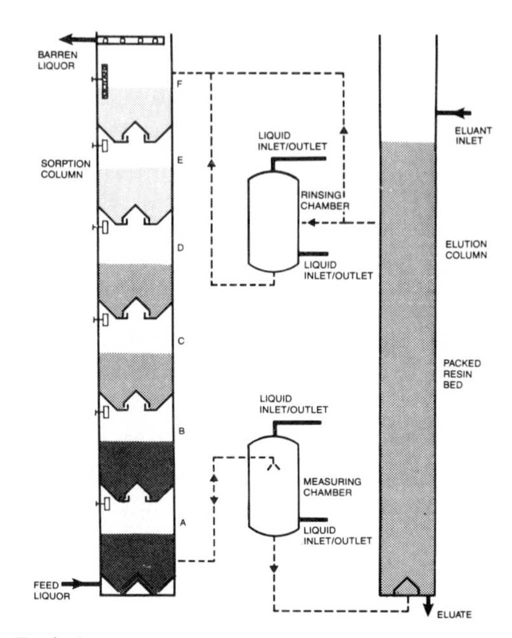

FIG. 12.27 Typical arrangement of a continuous ion exchange system. *(Courtesy of Illinois Water Teatment Company.)*

illustrated by Figure 12.27. One reason for its slow acceptance has been the improvement in chemical regeneration efficiency through operating conventional units in a countercurrent manner. The continuous ion exchange units are considerably more sophisticated than the conventional units, and early designs were often unreliable and costly as a result of resin bead breakage.

Powdered Resins

Another unique type of ion exchange system involves the depositing of powdered ion exchange resins on the surface of a septum-type filter. The powdered resins are a mixture of regenerated cation and anion resins, so they have some capacity for pickup of ionic materials, but they also perform as excellent filters, lending themselves ideally to the filtration of utility plant condensate. Such a unit, illustrated by Figure 12.28, offers cost savings over conventional ion exchange beds with external regeneration, particularly when handling nuclear plant condensate where spent resins must be disposed of because of radioactive contamination. The volume to be handled is generally much less with the powdered resin system than with the packed columns.

Finally, in the preparation of ultrapure water for such services as transistor manufacture, special ion exchange designs are used where the ion exchange material is in cartridge form and is expendable. A high purity system of this kind might include a primary demineralizer followed by an evaporator, with the distillate being processed through ion exchange cartridges and filtered through membrane filters at the point of use. The many facets of ion exchange technology cannot be explored in a general text on water treatment such as this. The production of ultrahigh purity water and selective ion exchangers are only two aspects of interest to water treatment specialists that require their own field of literature. But one final item of special interest that deserves mention is the availability of an ion

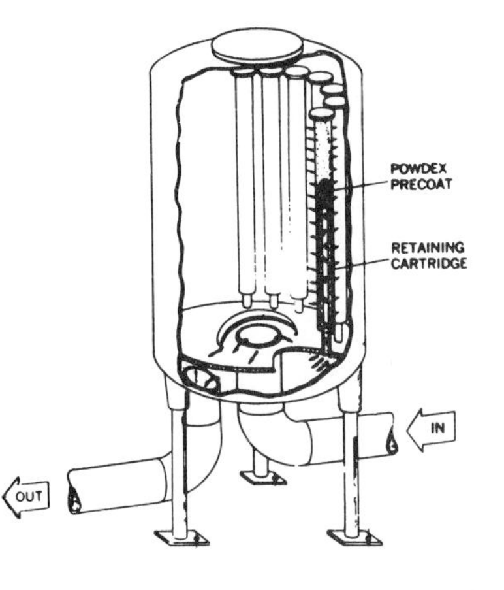

FIG. 12.28 Powdered resin filtration system. Cutaway showing internals of condensate polishing unit *(top)*. Typical installation of filter unit, skid mounted *(bottom)*. *(Courtesy of Graver Water Conditioning Company.)*

exchange material having both cation and anion exchange sites on the same polymer backbone. These sites are weak electrolytes, so they may be regenerated to their weak acid and weak base forms with hot water.

The formula for such an ion exchanger may be represented as

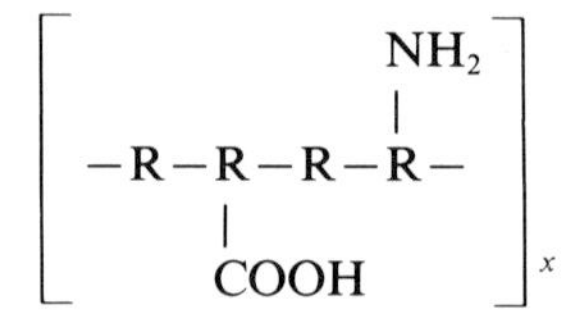

and the exchange reaction with NaCl in solution is:

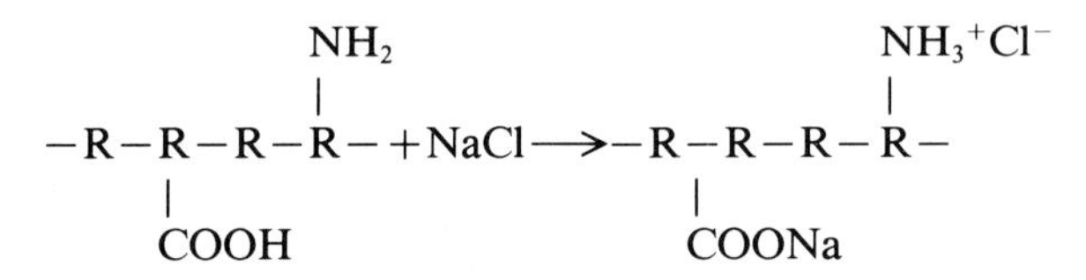

When the resin is regenerated with hot water, the NH_2 group becomes basic, and the carboxylic group returns to the acid form:

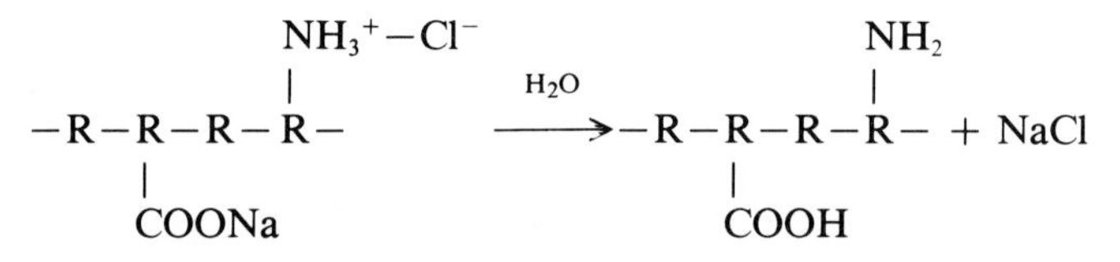

A schematic diagram showing how such a resin may be used in a continuous ion exchange system for water desalination is shown in Figure 12.29. This process may make it possible to use low-grade industrial heat sources beneficially for the first time. This process should become competitive with other desalination processes with which it is comparable in yield and production of reject brine.

EVALUATION OF ION EXCHANGE RESIN PERFORMANCE

Although water is known as the universal solvent, industrial grade ion exchange materials are relatively impervious to it. However, certain changes appear in the ion exchange material during its lifetime, as it cycles from regenerated to exhausted form, and users of ion exchange materials should budget an annual replacement of about 10% for cation resin and 20 to 30% for anion resin.

Capacity Variance Normal

A single ion exchange unit operating on a constant quality influent will show variations in capacity from one run to another, basically because it is difficult to achieve perfect flow distribution with a flat wave front through the ion exchange bed. Deviations of 5 to 10% each cycle are not unusual. A steady falloff of capacity, however, is a matter of real concern because it may lower effluent quality,

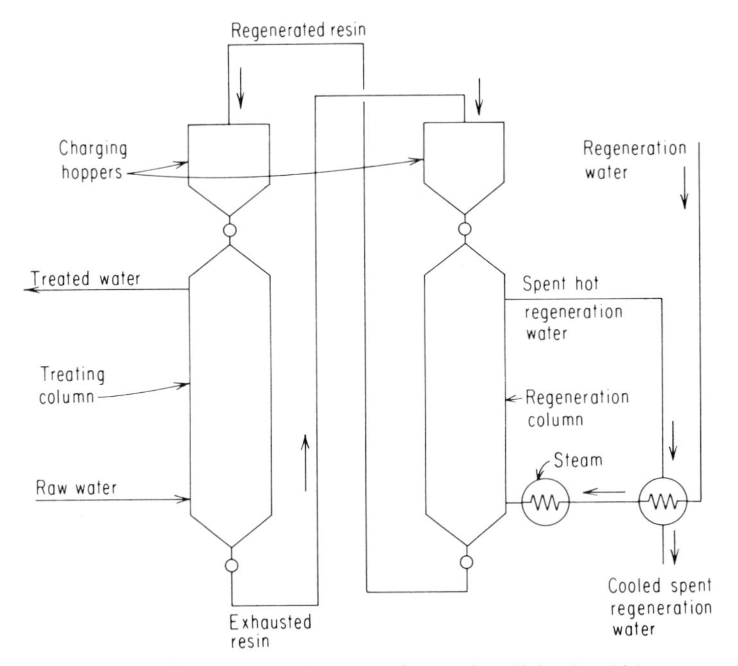

FIG. 12.29 Bifunctional resin system for demineralizing brackish water, using heat for regeneration.

decrease chemical utilization efficiency, and make it difficult for the operator to properly control the system.

A simple reason for loss of capacity may be material lost because of excessive backwash, resin flotation induced by the presence of gas or vapor in the water during backwash, or mechanical failure of an internal distributor. So, observation of the bed and a periodic measurement of bed level, always done after backwashing and draining at the end of the exhaustion period, is the first step in investigating capacity loss. If the bed is not flat, the causes of disturbances that create the unevenness in the bed should be investigated. While the ion exchange bed is open for inspection, a core sample should be taken for examination. A 1-in diameter copper tube or conduit will extract ample material for examination.

Most capacity problems can be attributed to dirty ion exchange beds, and this is usually caused by improper backwash. For adequate cleaning, the bed should be expanded until it is very close to the backwash offtake in the top of the ion exchange unit, and this requires adjustment of the backwash as water temperature changes. In effect, dirt moves at a higher velocity through the bed than it does when it reaches the bed surface and passes into the freeboard space above the expanded bed. This is why it is important that the expanded bed be very close to the level of the backwash offtake.

Even when flow rate is adequate, backwash time may be inadequate. This is particularly true of anion resins, which are commonly backwashed at a flow rate of about 3 gal/min/ft^2 (0.12 m^3/min/m^2) corresponding to a rise rate of only about 5 in/min (12.7 cm/min). If the freeboard is 30 in (0.76 m), which is common, it will require at least 6 min before the surface of the bed has expanded to the backwash offtake, and additional time must be allowed for the solids accumulated within the bed to be washed free and discharged from the unit.

The second most common cause of erratic capacity is channeling, which can be detected fairly reliably by the surface appearance. If the surface is uneven, it is helpful to probe the surface if there is a supporting bed beneath the ion exchange resin, to see if the supporting bed itself has been upset. The level of the supporting bed can be readily felt by probing with a stiff wire or rod. Some ion exchange units are equipped with underdrain systems having strainer nozzles, and the plugging of these will cause an uneven pattern of distribution; corrosion may actually destroy the threads of these nozzles and cause them to break off, allowing a full flow of water to come through the opening. Another common cause of channeling is low flow; this should be maintained at a minimum 2 gal/min/ft^2 even if this requires recycle or intermittent operation.

Regenerant Concentration Important

Incorrect chemical application is a third important cause of poor resin performance. There is an optimum concentration for each application, so the first step in investigating this potential cause of trouble is to determine the incoming concentration of regenerating chemical. If this is correct, then there may be some problem in the regenerant distribution system or in excessive dilution of the chemical between the distributor and the bed surface as it passes through the freeboard. To promote pistonlike flow during regeneration, the regenerant distributor should be no more than 3 in above bed level.

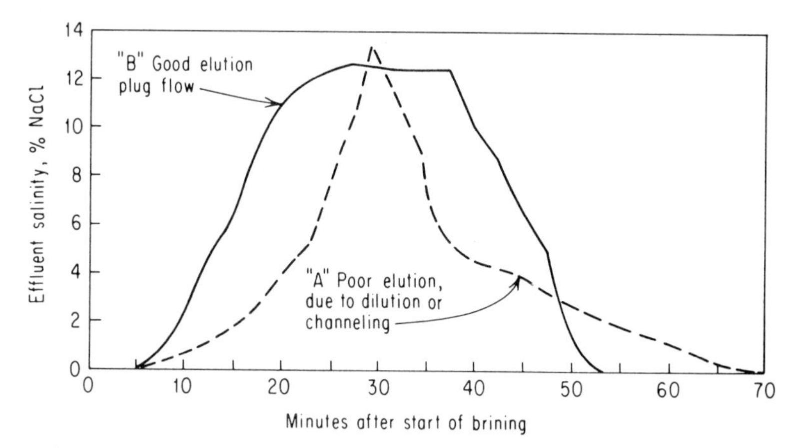

FIG. 12.30 Evaluating regenerant distribution by elution studies. Optimum cocurrent regeneration of a sodium exchanger requires the brine discharge to exceed 8% for about 30 min.

Detection of regeneration problems is relatively easy: follow an elution study during the chemical injection period. Samples are taken from the bottom collector at regular intervals, the solution strength is measured, and the strength is plotted versus time on a graph. Figure 12.30 shows poor elution characteristics, with an uneven wave front passing down through the bed during regeneration. This is caused by excessive dilution and departure from piston displacement characteristics. Curve B shows the same unit after correction of the regeneration procedure.

If none of these factors proves to be the cause of faulty performance, it may be that the effluent quality is simply being spoiled by bypass of raw water through a multiport valve or the individual valve system directly into the effluent. For example, backwash is normally accomplished by using raw water, and the backwash is fed to the bottom of the unit to the same point where effluent is normally withdrawn. A leak through the backwash control unit could cause raw water to appear in the effluent. If there is a multiport valve in the system, this leakage can be verified by taking a sample from the effluent of the multiport valve and comparing this to a sample taken from the bottom header of the ion exchange unit before it approaches the multiport valve connection.

Finally, resin characteristics may have changed to produce poor results. The change may be a physical one, with the development of excessive fines; it may be caused by the accumulation of fouling materials on the surface of the ion exchange resin, with metal oxides commonly found on the cation resin and organic material commonly found within the pores of the anion resin. Or, the resin structure itself may have been attacked by the water environment, most commonly by the presence of oxidizing agents.

Troubleshooting Approaches

Sometimes, the cause of trouble can be observed by using a low-power magnifying glass and comparing the resin sample taken from the unit with a sample of fresh material retained from the initial resin shipment. This examination may reveal the presence of excessive fines, fractured particles, or debris in the resin mix or coating the resin particles. Illustrations of these situations are shown in Figures 12.31 through 12.34. If visual observation does not provide the clue, then a complete resin analysis should be undertaken. Three examples of resin analyses are shown in Tables 12.11 through 12.13.

The cation resin shown in Table 12.11 was sampled from a sodium zeolite softener, and the analysis is compared to the analysis of new material.

FIG. 12.31 Gel-type resin containing a large amount of loose precipitate, indicating improper backwash. This condition is responsible for most problems with ion exchange systems.

FIG. 12.32 Badly cracked gel-type resin. This may be caused by internal precipitate or osmotic shock, the latter being rather common in hot zeolite units if pressure is suddenly released.

FIG. 12.33 Inorganically fouled gel-type resin showing both external and internal fouling. Barium and strontium frequently contribute to this problem in units regenerated with H_2SO_4.

FIG. 12.34 External organic and inorganic fouling of gel-type resin.

A review of the analysis shows that there are significant changes from new material. Loss in total capacity is caused by breakage of cross-linkages in the polymer with consequent swelling, as shown by increased water content. Increase in water retention from a typical 53 to about 60.5% causes a drop in total capacity from 40 kgr/ft^3 to 28 kgr/ft^3. Since dry weight capacity (measured in milliequivalents per gram of oven-dried resin) is almost equal to typical new resin, there has been no loss of ion exchange sites.

The shift of the wet screen analysis from 20 mesh toward 50 mesh indicates either poor sampling (most of the sample from the bed surface) or gradual production of fines. The large proportion of 50 mesh suggests a gradual increase in pressure drop which may be affecting performance.

Since the increase in water retention is caused by attack of the resin structure at the cross-links, the water supply should be examined to determine the cause of attack. In a hot lime–zeolite system, a usual cause is dissolved oxygen. Catalyzed sodium sulfite may be applied to lime-treated water ahead of the filters and zeolite softeners, and also to the regenerant brine. This resin will continue to perform well to a water retention capacity of about 70%; removing the cause of attack will delay the time when this point is reached. Care should be exercised in backwashing, since the lower density, swollen resin expands more during backwash than heavier, fresh resin.

Anion resins are generally much more sensitive to degradation than cation resins, particularly through irreversible fouling with organic material. The organic material may have been present in the original water supply, but it is often the accumulation of breakdown products from attack of the cation resin preceding it.

Table 12.12 shows the analysis of a strongly basic anion resin from a demineralizer system where the loss of the salt-splitting capacity is very evident. (Salt-splitting is the ability of OH-form anion resin to convert a solution of NaCl to NaOH.)

This sample shows measurable loss of exchange sites and severe loss of salt-splitting capacity. Even if the water being treated has a low ratio of weak acids to total acids, this reduction of salt-splitting capacity is undoubtedly producing short service runs. The resin appears to have reached the end of its useful life. The water

TABLE 12.11 Cation Resin Evaluation—Styrene-Sulfonic Acid Type

Characteristic	Sample	Typical new
Total capacity (T.C.):		
meq/mL	1.27	1.8
meq/g	4.96	4.9
kgr/ft^3	28	40
Water retention capacity, %	60.5	53
Wet screen analysis:		
% retained on 16 mesh	1.2	4–8
% retained on 20 mesh	14.1	35–45
% retained on 30 mesh	20.3	35–45
% retained on 35 mesh	9.5	7–10
% retained on 40 mesh	11.3	3–8
% retained on 50 mesh	19.1	2–5
% through 50 mesh	24.5	0–1

TABLE 12.12 Anion Resin Evaluation

Characteristic	Sample	Typical
Total capacity (T.C.):		
meq/mL	1.17	1.4
meq/g	2.43	3.0
kgr/ft^3	25	30
Salt-splitting capacity, % of T. C.	17	95–100
Water retention capacity, %	50.0	36–42
Wet screen analysis		
% retained on 16 mesh	0.2	1–3
% retained on 20 mesh	0.2	20–30
% retained on 30 mesh	18.6	40–50
% retained on 35 mesh	58.8	15–25
% retained on 40 mesh	14.4	5–10
% retained on 50 mesh	4.9	2–5
% through 50 mesh	2.7	0–1

TABLE 12.13 Anion Resin Evaluation

Characteristic	Sample	Typical
Total capacity (T.C.):		
meq/mL	1.19	1.3
meq/g	3.57	3.2
kgr/ft^3	26	28
Salt-splitting capacity, % of T.C.	75	95–100
Water retention capacity, %	N.D.	44–48
Wet screen analysis		
% retained on 16 mesh	0.2	1–3
% retained on 20 mesh	19.4	20–30
% retained on 30 mesh	49.1	40–50
% retained on 35 mesh	16.9	15–25
% retained on 40 mesh	10.0	5–10
% retained on 50 mesh	4.0	2–5
% through 50 mesh	0.4	0–1

TABLE 12.14 Ion Exchange Evaluation—Sulfonic Resin, Sodium Form

Total exchange capacity*		Water retention capacity, %	62.3
% of original, when completely cleaned	71	Microscopic examination	
meq/mL (wet)	1.34	Bead size	
kgr/ft³	29.2	% normal	50
Salt splitting capacity*		% large	—
% of original, when completely cleaned	—	% small	50
meq†/mL (wet)		Broken beads	
		%	50
Small column capacity*			
Weak acid resin, % of original, when completely cleaned	—		
kgr/ft³	—		
Weak base resin, % of original, when completely cleaned	—		
kgr/ft³	—		
Metal contaminants analysis			
Elements (g/ft³) as such			
Iron (Fe)	20	Magnesium (Mg)	—
Silica (Si)	26	Calcium (Ca)	—
Aluminum (Al)	21	Barium (Ba)	1
Manganese (Mn)	2	Strontium (Sr)	11
Chromium (Cr)	—	Zinc (Zn)	—
Copper (Cu)	—	Chlorine (Cl)	3
Nickel (Ni)	—	Phosphorus (P)	2
Condition of resin			
Chemical	Fair		
Physical	Very poor		

* meq/mL × 21.8 = kgr/ft³.
† Ultimate capacity (not operating capacity).

supply should be examined to determine whether this type of anion resin is applicable to this plant. If so, plant records should be examined to determine how long this resin has been in service. After replacement with new resin, a program of annual resin evaluation should be instituted to plan for periodic resin additions or replacement to maintain uniform operating results.

Although screen analysis shows a shift from 20 mesh, there is little evidence of fines. Operating problems are due to loss of resin capacity rather than to hydraulic problems caused by change in particle size.

The analysis of another type of strongly basic anion resin is shown in Table 12.13. In this analysis, reduction in total capacity is insignificant. Loss in salt-splitting capacity, while measurable, should cause little change in operating results, since exchange capacity is in the 10 to 12 kgr range, and salt-splitting capacity is still approximately 7.5 to 9.0 kgr. The higher the percentage of weak acids in water, the more pronounced will be the reduction in exchange capacity as salt-splitting capacity is lost. Analysis shows the resin is in good physical and chemical condition; the plant should be performing up to capacity.

Various forms of analyses are used to evaluate the performance and characteristics of ion exchange materials. These tables list the screen analysis in detail, but this is seldom required. Sometimes of equal importance is the composition of materials coating the resin surface or found loose in the resin bed. Tables 12.14 and 12.15 are representative analyses of cation resins including an assay of the crud contaminating the resin. In neither case was a sample of the original material provided, but observation indicated that it was the typical sulfonated styrene cation exchanger used in most industrial applications.

In the case of Table 12.14, the analysis shows the cation resin to be in poor physical condition, but still having an acceptable chemical composition. There has been an increase in water retention capacity from a normal of 50 to 53%, showing some decross-linking, but this has not yet reached serious proportions.

The resin sample was observed to be mixed with large amounts of loosely held particulate matter and fouled with organic substances. The contaminants on the resin include large amounts of silica, moderate amounts of iron, aluminum, and strontium, plus small quantities of manganese, barium, chlorine, and phosphorus. The accumulation of particulate matter within the resin bed suggests that higher backwash rates may be required to improve performance. Tests of the resin indicated that capacity loss could not be restored by any chemical cleaning procedures, and the loss was caused by the contaminants found on the resin. In this case, the resin has undergone extensive physical degradation and it cannot be cleaned, so replacement is recommended.

In the next example, Table 12.15, the cation resin is in fairly good physical condition, but very poor chemical condition. This cation resin is also mixed with

TABLE 12.15 Ion Exchange Evaluation—Softener

Total exchange capacity*		Water retention capacity, %	71.8
% of original, when completely cleaned	48	Microscopic examination	
meq†/mL (wet)	0.91	Bead size	
kilograins/ft^3	19.8	% normal	80
		% large	—
		% small	20
		Broken beads	
		%	20

Metal contaminants analysis

Elements (g/cu ft) as such			
Iron (Fe)	270	Magnesium (Mg)	—
Silica (Si)	6	Calcium (Ca)	—
Aluminum (Al)	—	Barium (Ba)	—
Manganese (Mn)	—	Strontium (Sr)	4
Chromium (Cr)	1	Zinc (Zn)	—
Copper (Cu)	20	Chlorine (Cl)	120
Nickel (Ni)	—	Phosphorus (P)	74
		Lead (Pb)	2
Condition of resin			
Chemical	Very poor		
Physical	Fair		

* meq/mL × 21.8 = kgr/ft^3.
† Ultimate capacity (not operating capacity).

large amounts of loose particulates, and is contaminated with extremely high concentrations of iron, large amounts of chlorine and phosphorus, and moderate amounts of silica and copper.

The water retention capacity has increased to 71.8%, at which point the material is extremely light, difficult to contain during backwash, and subject to early failure at an unpredictable time. The capacity is so low that the resin is essentially of marginal value and should be replaced at an early date. The loss in cross-linkage should be investigated; this particular system is operating on a municipal supply, and free chlorine in the supply water should be measured.

It should not be surprising that foreign matter should be attracted to ion exchangers, since exchangers are so closely related to polyelectrolytes used in coagulation and flocculation. Often it is difficult to find the source of the foulant because its concentration may be below the limits of detection, and it becomes identifiable only after it has accumulated on the resin, sometimes after millions of gallons have been processed per cubic foot of resin. A great deal of experience has been gained in cleanup of inorganic materials, but the prevention and alleviation of organic fouling of anion resins remains, after over 30 years' experience with these materials, an uncertain program at best.

It is common practice to protect demineralizers operating on low-solids surface waters by activated carbon prefiltration. However, even this is not a guarantee of full protection. Nor is there much help to be gained in analyzing the raw water for organic compounds. Most researchers seem to agree that certain reactive organic materials diffuse into the gel structure of strongly basic anion exchangers, and then polymerize into molecules so large that they are trapped in the gel, unable to diffuse out again. On the basis of this idea, the current trend is toward (1) maximum reduction of organic matter by pretreatment and (2) application of a cleaning agent during each backwash to disperse the organic matter and reduce

TABLE 12.16 Ion Exchangers Available in the United States

Rohm & Haas Amberlite	Dow Chemical Dowex	Sybron Corp. Ionac	Rohm & Haas Duolite
	Cation		
IR-120 plus	HCR-S	C-249	C-20
IR-122	HGR	C-250	C-20×10
IR-130C	HCR-W2	C-298	C-225
IR-132C	HGR-W2	C-299	C-225×10
IR-200,252	MSC-1	CFP-110	C-26
IRC-50,84	MWC-1	CC,CNN	C-464,433
	Anion		
IRA-400	SBR	ASB-1	A-109
IRA-402	SBR-P	ASB-1P	A-101D
IRA-410	SAR	ASB-2	A-104
IRA-900	MSA-1	A-641	A-161
IRA-401S	Dowex 11	—	A-143
IRA-910	MSA-2	A-651	A-162
IRA-93,94	MWA-1	AFP-328,329	A-368,378
IRA-47	WGR-1,2	A-305	A-340,A-30B

its opportunity to diffuse into the gel structure. In addition, periodic treatment with strong brine causes the anion resin to shrink, expelling water that carries with it some of the organic matter, like squeezing a sponge. Semiannual examination of resin then provides the data to assess the success of the program or the need for a new treatment procedure.

In the previous tables, the identification of the resins has not been noted on the analyses. Although there are variations in certain physical and chemical properties depending on manufacturing techniques, the general performance specifications for the products of the several manufacturers in the United States are quite similar. Table 12.16 provides a basis for identifying comparable categories of ion exchangers as produced by the four major U.S. manufacturers, with no attempt made to distinguish between them as to bead strength, particle size, uniformity coefficient, and other properties that affect performance and resin life.

SPECIAL ION EXCHANGE APPLICATIONS

The versatility of ion exchange has led to many specialized applications of the process, from the small-scale use of ion exchangers in chromatographic separations in laboratory equipment (Figure 12.35) to large scale uses in special process applications in such industries as mining, nuclear wastes handling, and metal finishing. Specific examples are given here as a way of suggesting potential water-treatment processes for the future.

1. *Shallow-bed, short-cycle operation:* In valuable metal-ion recovery from metal-finishing operations, resin inventory and equipment cost are minimized by employing shallow resin beds with frequent regenerations. The regeneration of a cation exchange unit of this design can be completed in less than 5 min. Leakage is a minor problem in this application as the processed water is recovered for rinsing and bath makeup. Lower leakages can be achieved when the process is applied to conventional water treatment applications.
2. *Resin-in-pulp operation (Figure 12.36):* The normality of strong cation resin of the nuclear sulfonic acid type is about 2.0 *N* and its void volume is about 30%, so it cannot be used to treat solutions stronger than about 0.2 *N* in conventional equipment, because the resin is scarcely flooded by the liquor before its capacity has been used up. However, if the resin is moved in baskets and "dunked" in tanks of liquor to be processed, moving counter to the flow of liquor from tank to tank, the resin can concentrate and then unload (in a similar dunking operation in regenerant tanks) valuable minerals at relatively high solution strengths. This process has been used for uranium recovery.
3. *The use of ion-selective resins; the softening of highly saline waters:* Although the major use for carboxylic exchangers is for water dealkalization, the high selectivity of these resins for calcium and magnesium over sodium makes them very useful for softening of highly saline waters, usually in a polishing operation following conventional salt-regenerated sulfonic-acid strong cation resins. In this case, the carboxylic resin is first regenerated with sulfuric acid to remove the accumulated calcium and magnesium, and the resin is then converted to the sodium form by neutralization with caustic, NaOH. Used as polishers, these units can reduce hardness to less than about 0.2 mg/L even at a salinity as high as 5000 mg/L. These systems are used to provide minimal

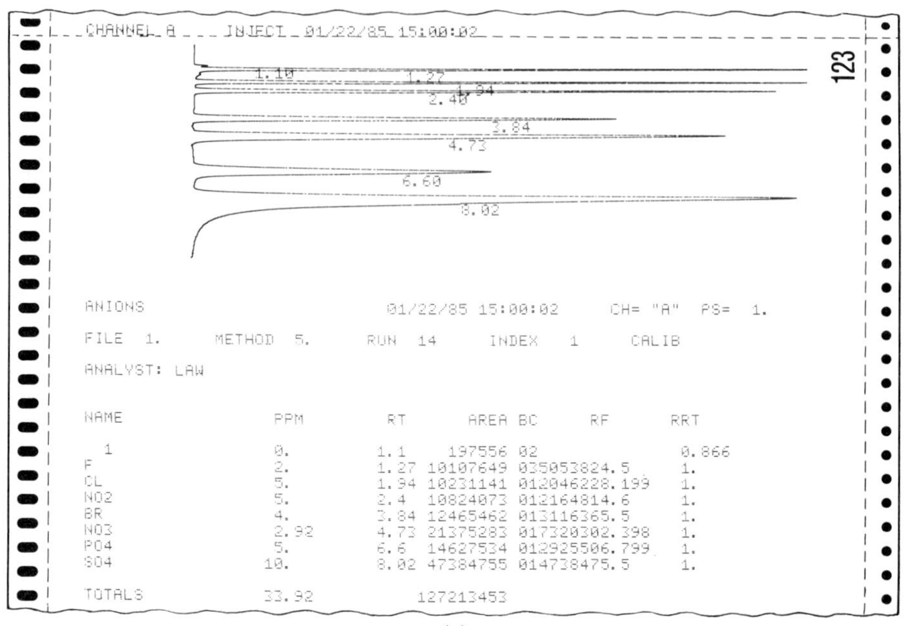

NAME	PPM	RT	AREA	BC RF	RRT
1	0.	1.1	197556	02	0.866
F	2.	1.27	10107649	035053824.5	1.
CL	5.	1.94	10231141	012046228.199	1.
NO2	5.	2.4	10824073	012164814.6	1.
BR	4.	3.84	12465462	013116365.5	1.
NO3	2.92	4.73	21375283	017320302.398	1.
PO4	5.	6.6	14627534	012925506.799	1.
SO4	10.	8.02	47384755	014738475.5	1.
TOTALS	33.92		127213453		

(*a*)

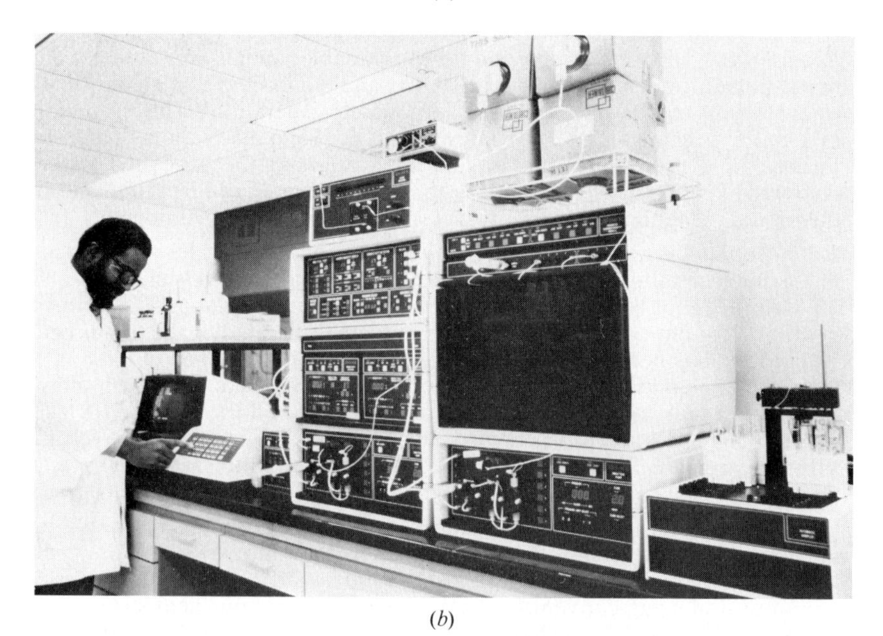

(*b*)

FIG. 12.35 Ion exchange columns are used analytically on a miniature scale to separate ions on the basis of the selectivity of the ion exchange material for each ionic component of the sample. (*a*) The apparatus shown here can analyze 60 samples in 30 h, an operation that formerly required about 2 weeks by wet chemistry methods. (*b*) The peaks and the areas under them are integrated to determine the ionic concentrations as the ions (anions) captured when the sample was passed through the ion exchange columns are eluted during the analysis step.

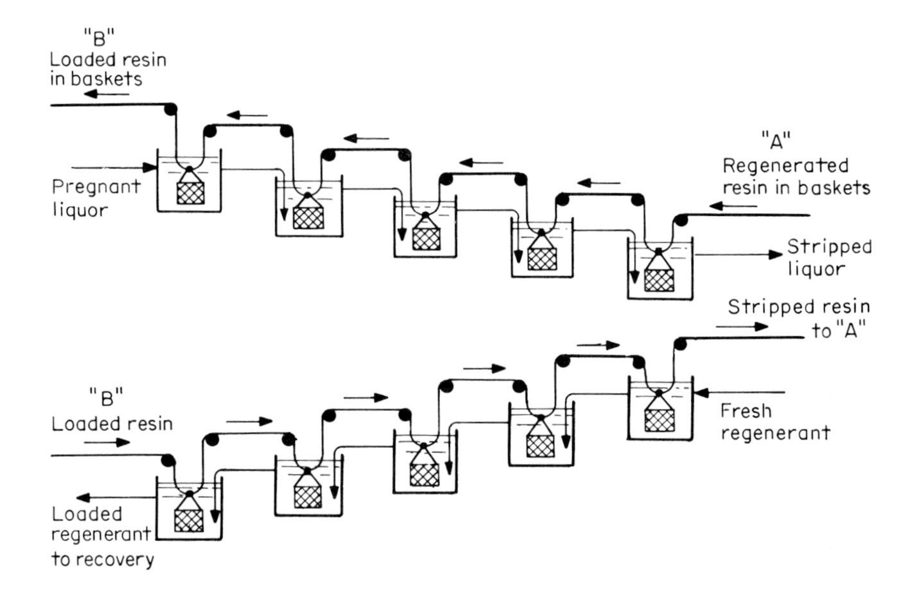

FIG. 12.36 Resin-in-pulp system. Here, the high concentration of the pregnant liquor is close to the normality of the ion exchange resin, so the resin is moved physically counter to the flow of pregnant liquor and regenerant.

hardness makeup for high-pressure steam generators used in steam flooding of oil fields, where the produced brine is the only available water for the boiler makeup (see Chapter 43). There are a variety of other ion-selective resins used for special applications other than softening, such as removal of boron from waste geothermal brines.

SUGGESTED READING

Arden, T. V.: *Water Purification by Ion Exchange,* Butterworths, London, 1968.

Crits, G. J.: "Technology of Mixed Beds," 22d Annual Liberty Bell Corrosion Course, 1984.

Harper, H. H., and Drummonds, D. L.: "Makeup Demineralization with an Organic Scavenger/Monobed System," Proceedings 45th International Water Conference, Pittsburgh, October 1984.

Kunin, Robert: *Elements of Ion Exchange,* Reinhold, New York, 1960.

Miller, W. S., and Yeugar, M. B.: "Factors Affecting Counterflow Ion Exchange Quality," *Ind. Water Eng.,* Vol. 18. No. 2 March/April, 1981.

CHAPTER 13
NEUTRALIZATION

Acid-base reactions are among the most prevalent of the chemical processes used in water conditioning. It is fortunate that commercially available acids and bases are relatively low in cost. Typical costs of acids and bases used in water treatment are shown in Table 13.1, which also indicates the relative value of these materials on an equivalent basis.

In addition to these chemicals, there are chemical wastes having residual acid or base values that can be used for neutralization in waste treatment. For example, pickle liquor from the steel industry is useful both for its acid value and for the iron it contains as a coagulant; sludge from lime–soda water softening is a good source of alkali for neutralization of acid wastes.

Progress in the neutralization of an acid or base is easily measured by pH. However, the practicality of using pH to control the neutralization reaction is greatly influenced by the buffer capacity of the water. For example, it is much easier to neutralize an acidic waste containing hydrochloric acid to a pH of 7.5 with sodium carbonate than with caustic soda. This is illustrated by Figure 13.1.

Treating an acidic wastewater may be relatively easy because the allowable effluent pH range is fairly broad, usually 5 to 9. However, adjustment of the stability index of a water requires controlling pH to a rather narrow range, so the equipment for feeding acid or alkali to achieve a specified stability index in process water is generally more sophisticated than equipment used in a wastewater treatment plant.

EXAMPLES OF ACID-BASE REACTIONS

The correction of the stability index with lime is shown in Figure 13.2. This is a relatively common problem, although lime is not always used because it is more difficult to handle than liquid caustic or soda ash. When used in large quantities, it is purchased as CaO (see Table 45.1 for various forms of lime) and slaked to $Ca(OH)_2$. A typical feeding system is shown in Figure 13.2. Figure 13.3 shows incremental additions of lime to a water requiring stabilization. The first increment of 6 mg/L is added for neutralization of all of the CO_2; additional lime then converts HCO_3 to CO_3^{2-} in successive steps. The resulting pH and stability index at each increment is shown in the same tabulation. These data are plotted in Figure 13.4. Assuming the goal is a stability index of 7, the graph shows that this can be accomplished with approximately 13 mg/L lime as $CaCO_3$, resulting in a pH of approximately 9.1.

TABLE 13.1 Acid and Alkali Costs for Neutralization

Chemical	Specification	Cost,[a] ¢/lb	lb/lb $CaCO_3$	Equiv. cost,[b] ¢/lb $CaCO_3$ equivalent
Sulfuric acid [H_2SO_4 (oil of vitriol)]	66° Be—93%	4.5[c]	1.05	4.7
Hydrochloric acid [HCl (muriatic acid)]	20° Be—32%	3.0[c]	2.28	6.8
Limestone ($CaCO_3$)	200 mesh, 93%	1.2[c]	1.08	1.3
Quicklime (CaO)	Lump, 90%	2.4[d]	0.62	1.5
		1.75[c]	0.62	1.1
Hydrated lime [$Ca(OH)_2$][f]	Pulv, 93%	2.4[d]	0.80	1.9
Soda ash (Na_2CO_3)	Light, 50% Na_2O	8.5[d]	1.06	9.0
		6.8[c]	1.06	7.2
Caustic soda (NaOH)	Dry	28[d]	0.80	22.4
	Dry	24[e]	0.80	19.2
	50% liquid	18[a]	0.80	14.4
Sodium bicarbonate ($NaHCO_3$)	Dry, 99%	16.3[d]	1.68	27.4
Ammonia (NH_3)	Liquid	12[c]	0.34	4.1

[a] Cost is given in cents per pound of product as shipped, not including freight, except for 50% liquid caustic, where cost is ¢/lb NaOH.

[b] Cost per lb as $CaCO_3$ [(¢/lb) × (lb product/lb $CaCO_3$ equivalent)].

[c] Bulk cost in tank cars, tank trucks, or hopper cars.

[d] Cost in bags.

[e] Cost in drums.

[f] Available as a slurry in some locations.

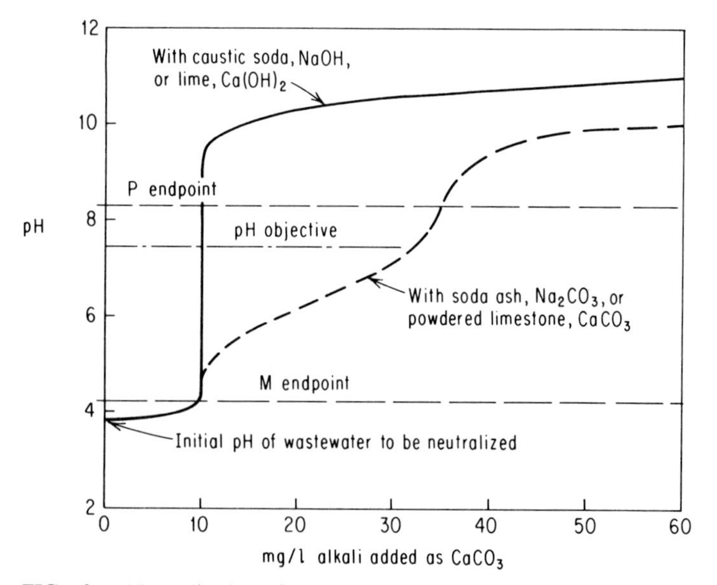

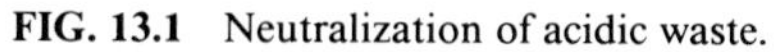

FIG. 13.1 Neutralization of acidic waste.

Identification of Analyses Tabulated Below:

A. Hypothetical Water
B. A + 6 mg/l Lime (as $CaCO_3$)
C. B + 4 mg/l Lime (as $CaCO_3$)
D. C + 5 mg/l Lime (as $CaCO_3$)
E. D + 5 mg/l Lime (as $CaCO_3$)
F.

Constituent	As	A	B	C	D	E	F
Calcium	$CaCO_3$	60	66	70	75	80	
Magnesium	"	20	20	20	20	20	
Sodium	"	10	10	10	10	10	
Total Electrolyte	$CaCO_3$	90	96	100	105	110	
Bicarbonate	$CaCO_3$	55	61	57	52	47	
Carbonate	"	0	0	8	18	28	
Hydroxyl	"	0	0	0	0	0	
Sulfate	"	30	30	30	30	30	
Chloride	"	5	5	5	5	5	
Nitrate	"	0	0	0	0	0	
M Alk.	$CaCO_3$	55	61	65	70	75	
P Alk.	"	0	0	4	9	14	
Carbon Dioxide		6	0	0	0	0	
pH		7.3	8.3	8.9	9.3	9.5	
Silica		5	5	5	5	5	
Iron	Fe	0.1	0.1	0.1	0.1	0.1	
Turbidity		0	0	0	0	0	
TDS		100	106	110	115	120	
Color		0					
pH_s @ 70° F		8.25	8.2	8.15	8.05	8.0	
S. I. @ 70° F		9.2	8.05	7.14	6.8	6.5	

FIG. 13.2 Correction of stability index with lime. Goal: stability index (S.I.) of 7.0.

The equivalent weight of commercial lime [93% $Ca(OH)_2$] is approximately 40, so that lime as calcium carbonate multiplied by 0.8 gives commercial lime requirements.

Another example of stability index correction is the reduction of the scaling tendencies of a lime-softened water, which usually remains supersaturated with calcium carbonate even after filtration. The common practice is to recarbonate by injection of CO_2 gas after the sedimentation basins and ahead of filters (see Figure 13.5). In some areas liquefied CO_2 is readily available and often is a cheaper reagent than natural gas, burned to produce CO_2 as shown in Figure 13.5. However, the same effect can be achieved by the feeding of sulfuric acid (see Figure 13.6).

(*a*)

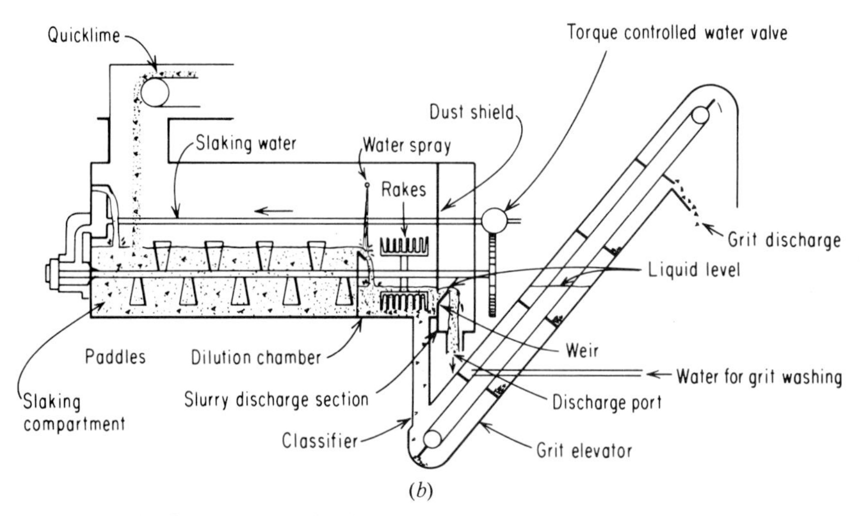

(*b*)

FIG. 13.3 Device for slaking and feeding large quantities of lime in water treatment plants. (*a*) Photo of slaker-feeder assembly. (*b*) Cutaway showing details of operation. *(Courtesy of Wallace & Tiernan Division, Pennwalt Corporation.)*

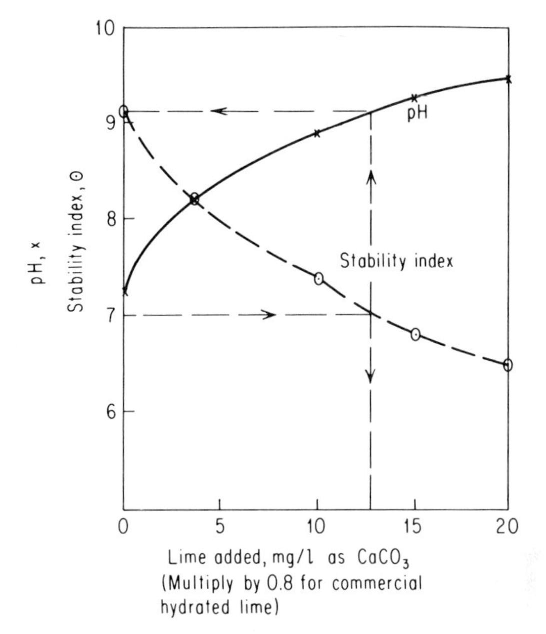

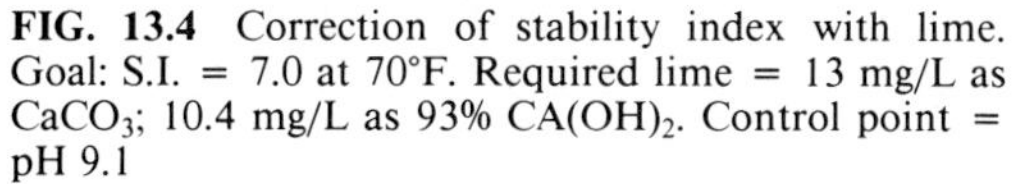

FIG. 13.4 Correction of stability index with lime. Goal: S.I. = 7.0 at 70°F. Required lime = 13 mg/L as $CaCO_3$; 10.4 mg/L as 93% $CA(OH)_2$. Control point = pH 9.1

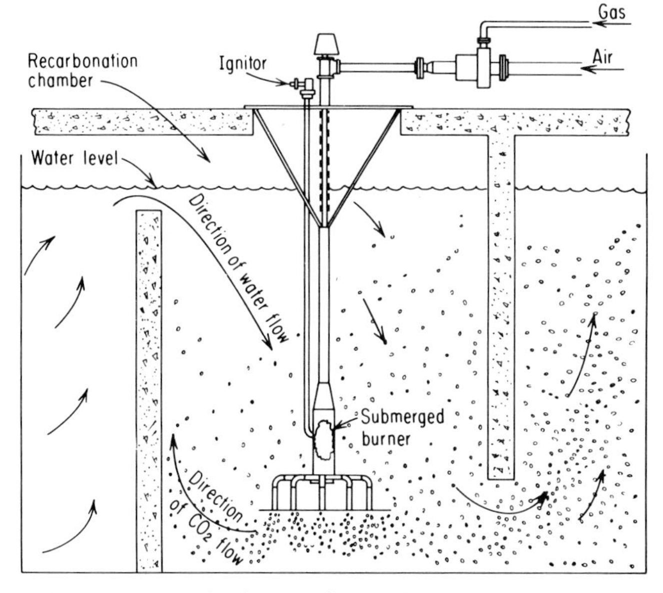

FIG. 13.5 Schematic diagram of water flow through recarbonation basin demonstrating how CO_2 bubbles are diffused by the submerged combustion burner at the bottom of the basin, and then further dispersed by the water flow, giving optimum absorption time. *(Courtesy of Ozark-Mahoning Company.)*

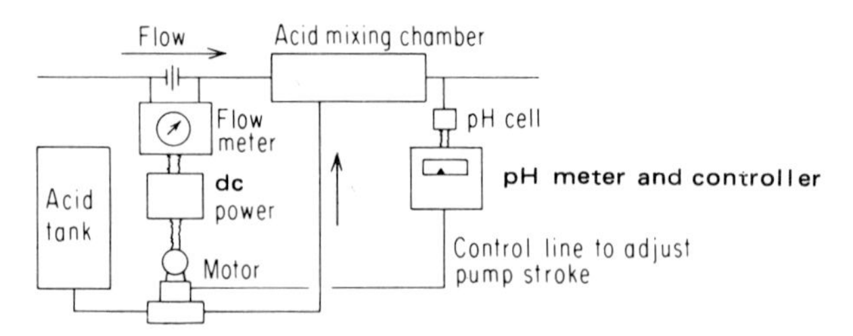

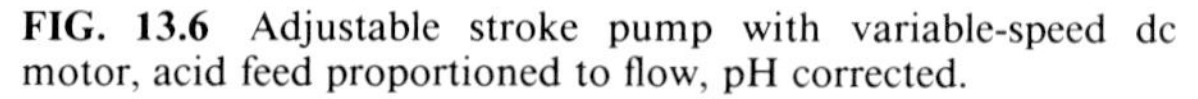

FIG. 13.6 Adjustable stroke pump with variable-speed dc motor, acid feed proportioned to flow, pH corrected.

Identification of Analyses Tabulated Below:

A. Hypothetical water
B. A + 10 mg/l CO_2 as $CaCO_3$
C. B + 10 mg/l CO_2 as $CaCO_3$
D. C + 10 mg/l CO_2 as $CaCO_3$
E. D + 10 mg/l CO_2 as $CaCO_3$
F. E + 10 mg/l CO_2 as $CaCO_3$

Constituent	As	A	B	C	D	E	F
Calcium	$CaCO_3$	140	140	140	140	140	140
Magnesium	"	60	60	60	60	60	60
Sodium	"	20	20	20	20	20	20
Total Electrolyte	$CaCO_3$	220	220	220	220	220	220
Bicarbonate	$CaCO_3$	0	0	10	20	30	40
Carbonate	"	40	50	40	30	20	10
Hydroxyl	"	10	0	0	0	0	0
Sulfate	"	70	70	70	70	70	70
Chloride	"	100	100	100	100	100	100
Nitrate	"						
M Alk.	$CaCO_3$	50	50	50	50	50	50
P Alk.	"	30	25	20	15	10	5
Carbon Dioxide		0	0	0	0	0	0
pH		10.3	10.0	9.8	9.7	9.6	9.5
Silica		10	10	10	10	10	10
Iron	Fe	0	0	0	0	0	0
Turbidity		0	0	0	0	0	0
TDS		250	250	250	250	250	250
Color		0	0	0	0	0	0
pH_s @ 70°F		7.95	7.95	7.95	7.95	7.95	7.95
S.I. @ 70°F		5.60	5.95	6.15	6.25	6.35	6.40

(*a*)

FIG. 13.7 (*a*) Correction of stability index with CO_2 (recarbonation of lime-softened water).

The results of sulfuric acid addition and recarbonation are markedly different, as shown by Figures 13.7*a* and *b*. These are compared in Figure 13.8. Here the safety features and buffer capacity of recarbonation are apparent when the goal is simply to increase the stability index above 6. As noted in Chapter 10, recarbonation can be troublesome if the water to be treated is in equilibrium with $CaCO_3$ and contains an excess of hydroxyl alkalinity. Before the CO_2 can produce bicarbonate alkalinity from CO_3^{2-}, it must first react with the OH^-, thus

$$2OH^- + CO_2 \rightarrow CO_3^{2-} + H_2O \tag{1}$$

and the increase in carbonate concentration will precipitate $CaCO_3$—often plugging the filter beds downstream. So there are aspects other than safety, expe-

Identification of Analyses Tabulated Below:

A. Hypothetical Water
B. + 5 mg/l H_2SO_4 as $CaCO_3$
C. B + 5 mg/l H_2SO_4 as $CaCO_3$
D. C + 5 mg/l H_2SO_4 as $CaCO_3$
E. D + 5 mg/l H_2SO_4 as $CaCO_3$
F. E + 5 mg/l H_2SO_4 as $CaCO_3$

Constituent	As	A	B	C	D	E	F
Calcium	$CaCO_3$	140	140	140	140	140	140
Magnesium	"	60	60	60	60	60	60
Sodium	"	20	20	20	20	20	20
Total Electrolyte	$CaCO_3$	220	220	220	220	220	220
Bicarbonate	$CaCO_3$	0	0	0	5	10	15
Carbonate	"	40	40	40	30	20	10
Hydroxyl	"	10	5	0	0	0	0
Sulfate	"	70	75	80	85	90	95
Chloride	"	100	100	100	100	100	100
Nitrate	"	0	0	0	0	0	0
M Alk.	$CaCO_3$	50	45	40	35	30	25
P Alk.	"	30	25	20	15	10	5
Carbon Dioxide		0	0	0	0		
pH		10.3	10.0	9.9	9.9	9.8	9.6
Silica		10	10	10	10	10	10
Iron	Fe	0	0	0	0	0	0
Turbidity		0	0	0	0	0	0
TDS		250	250	250	250	250	250
Color		0	0	0	0	0	0
pH_s @70° F		7.95	8.0	8.05	8.1	8.2	8.25
S. I. @ 70° F		5.6	6.0	6.2	6.35	6.6	6.9

(*b*)

FIG. 13.7 (*b*) Correction of stability index with H_2SO_4.

diency, and chemical cost involved in choosing between H_2SO_4 and CO_2 for correction of the stability index.

Simple changes in water alkalinity or acidity without regard to stability index correction are easily calculated: the reduction of alkalinity, such as application of acid to a sodium zeolite effluent prior to its use as boiler feedwater, requires the addition of 1 mg/L commercial 66° Baumé sulfuric acid for each milligram per liter of alkalinity reduction needed. An increase in alkalinity requires the addition of 0.8 mg/L caustic or lime for each milligram per liter alkalinity increase required. These changes in alkalinity are calculated on the basis of calcium carbonate equivalents and then converted to the equivalent weight of the commercial alkali or acid added to the system.

Of the numerous examples of acid-base neutralization reactions used in wastewater treatment, one deserving mention is pH control in anaerobic digesters. A common problem in these digesters is an upset in the biota, causing sudden production of acids, with a drop in pH, a loss of control, and a dangerous reduction in methane production. Sodium bicarbonate is frequently used in this application because its buffering action prevents the production of an equally damaging high pH if the reagent is overfed. In this case, the cost per unit of alkali value is of minor importance compared to the buffering.

An example of the reduction of alkalinity with acid in the treatment of boiler makeup is shown in Figure 13.9. In this case, the alkalinity is added to the makeup system ahead of the zeolite. There is some risk in such an operation because it requires excellent acid feed and control equipment, and protection of the piping between the injection point and the deaerator, since corrosion could introduce objectionable iron. However, often the gain in reduced blowdown, with its attendant heat savings, justifies the installation of sophisticated equipment that can accomplish these results using a basic system already in operation in the

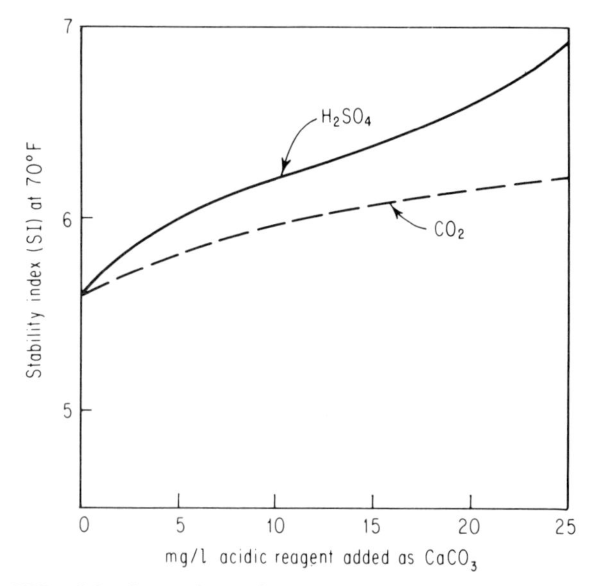

FIG. 13.8 Comparison of recarbonation to acidification of lime-softened effluent (data from Figure 13.7).

Identification of Analyses Tabulated Below:

A. Zeolite Softened Makeup (1) D. Concentrated 10 X in Boiler

B. After H_2SO_4 E.

C. "B" After Deaerator * (2) F.

Constituent	As	A	B	C	D	E	F
Calcium	$CaCO_3$	Nil	Nil	Nil	Nil		
Magnesium	"	Nil	Nil	Nil	Nil		
Sodium	"	125	125	125	1250		
Total Electrolyte	$CaCO_3$	125	125	125	1250		
Bicarbonate	$CaCO_3$	90	30	30	0		
Carbonate	"	0	0	0	50		
Hydroxyl	"	0	0	0	250		
Sulfate	"	15	75	75	750		
Chloride	"	20	20	20	200		
Nitrate	"	0	0	0	0		
M Alk.	$CaCO_3$	90	30	30	300		
P Alk.	"	0	0	0	275		
Carbon Dioxide		10	70	0	0		
pH		7.1	5.8	8.3	11.7		
Silica		3	3	3	30		
Iron	Fe	0	0	0	0		
Turbidity		0	0	0	0		
TDS		150	150	150	1500		
Color		0	0	0	0		

(1) If max. boiler alkalinity is 300 mg/1 max. concentration ratio is 300/90 = 3.3 ≡ 30% blowdown.
(2) After alkalinity reduction, max. conc. is 10 (column D) ≡ 10% blowdown.
* Not including steam dilution.

FIG. 13.9 Alkalinity reduction of boiler makeup. Although this example shows the acid applied to the zeolite effluent, it may be fed to the influent to buffer accidental overfeed. In either case, the final analyses (columns B, C, and D) are the same.

power plant. The risk is minimized by adding the acid ahead of rather than downstream of the zeolite softener, since the sodium zeolite could pick up hydrogen ions in the event of accidental overfeed of acid.

An operation of this kind, and even less sensitive situations, requires the installation of properly designed feed systems to avoid the consequences of over- or underfeeding acid. Figure 13.6 illustrates a system that can handle the chemical injection problem by proportioning the flow of chemical to the flow of water to be treated and correcting the basic dosage with a pH controller, which in this case adjusts the stroke of the pump. In this system, there must be adequate lag built into the mixing chamber in which the pH is measured so that the controller can

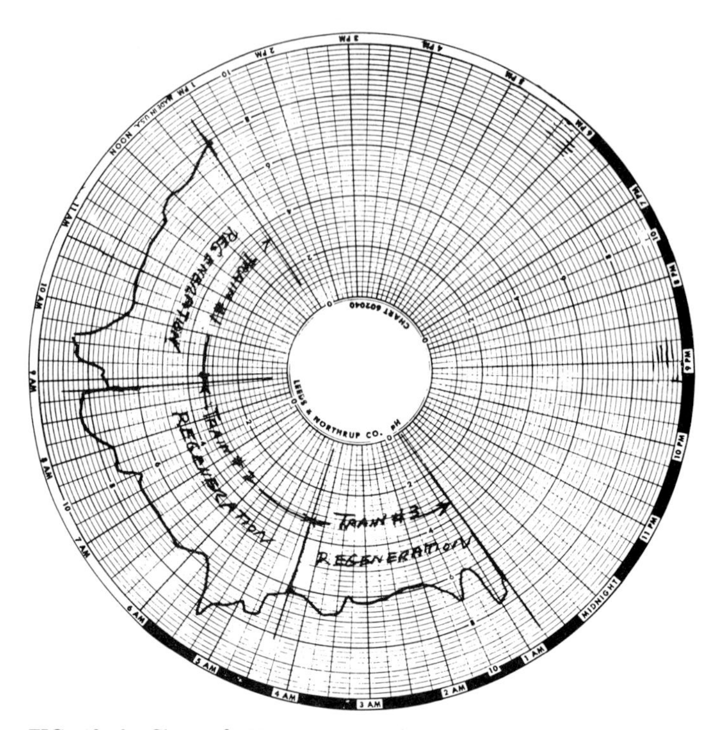

FIG. 13.10 Chart of pH recorder monitoring the effluent from an ion-exchange neutralizer. *(Courtesy of Cochrane Division, the Crane Company.)*

function properly without overriding or underriding while hunting for the pH setpoint.

Ion exchange has been applied to acid-base neutralization problems in a unique way: Weak base anion exchangers and carboxylic cation exchangers have the ability to pick up acid and release it as a neutral solution in treating a fluctuating wastewater which is periodically acidic and alkaline. If the excess free acid is balanced by the alkalinity, regeneration is unnecessary, and only periodic backwash is needed to keep the system operating. Although more expensive than a chemical feed system for the same purpose, it saves space and eliminates the maintenance costs of equipment and controls inherent in a chemical feeder. This type of system has been used to neutralize spent demineralizer regenerants. Figure 13.10 shows the pH record of the effluent of an ion exchange neutralizer in this kind of service. Although the pH fluctuates, it is kept within limits of 6 to 9 usually specified in municipal sewer codes.

ACID WATERS

By definition, in water chemistry, free mineral acidity is present when a water has a pH value below 4.4 to 4.6, the end point of the methyl orange titration. (See Table 4.3 and Figure 4.5.) It can be calculated from the definition of pH that a water with a pH of 4.0 contains about 5 mg/L of hydrogen ion as $CaCO_3$ (Table 4.3). So it may be assumed that this water could be neutralized to the methyl orange end point simply by the addition of 5 mg/L of lime or other alkali, as $CaCO_3$. However, this would be a completely erroneous assumption if the water contains heavy metals in solution, for these metal ions are also acids, just as the hydrogen ion is. One definition of an acid is a "proton donor," and the heavy metal ions act in this role in neutralizing alkalies:

$$Fe^{2+} + 2\ NaOH \rightarrow Fe(OH)_2 + 2\ Na^+ \quad (2)$$

$$Zn^{2+} + 2\ NaOH \rightarrow Zn(OH)_2 + 2\ Na^+ \quad (3)$$

$$Cu^{2+} + 2\ NaOH \rightarrow Cu(OH)_2 + 2\ Na^+ \quad (4)$$

Many mine waters are acidic and also contain metals in solution, as shown in Table 13.2. The alkali required to neutralize the hydrogen ion in this example is

TABLE 13.2 Alkali Demand of Mine Water at pH 4.0

	Concentration, mg/L	
Ions	As such	As $CaCO_3$
Hydrogen	0.1	5
Aluminum	26.1	145
Manganese	8.8	16
Zinc	2.2	3
Copper	0.3	1
Cobalt	2.0	4
Iron	0.5	1
Total acids	—	175

only 5 mg/L as $CaCO_3$; but the total alkali demand for both the H^+ ion and the metal ion "proton donors" is 175 mg/L as $CaCO_3$. An even larger dosage of alkali is required, of course, to raise the pH further to an acceptable level for heavy metals precipitation (usually above 8.5) and final discharge to a receiving stream.

CHAPTER 14
DEGASIFICATION

Water may contain a variety of gases in solution as well as minerals and organic matter. Because oxygen and nitrogen are the predominant gases in the atmosphere, surface water contains these two gases. Carbon dioxide is another common atmospheric gas; its concentration varies from place to place depending to a large extent on industrial activity.

There is an important distinction between oxygen and nitrogen on the one hand and carbon dioxide on the other: the former do not ionize in water, so these molecules exert a gas pressure in solution; carbon dioxide forms carbonic acid, which ionizes in water, so only the unreacted part can exert a gas pressure. At a pH below about 4.5, all of the CO_2 dissolved in water is present as a gas; above about 8.5, all of it is ionized. Other common ionizing gases are H_2S, HCN, and NH_3.

The products of respiration influence the composition of gases in water. In surface waters, algae produce oxygen in the presence of sunlight through the photosynthetic reaction of carbon dioxide and water to produce carbohydrates. Bacteria, on the other hand, utilize oxygen in their metabolism and produce CO_2. The action of soil bacteria frequently produces high levels of carbon dioxide in well waters, and deep well waters are usually devoid of oxygen because these bacteria have removed it from the percolating ground water. Anaerobic decomposition of organic debris on the bottom of swamps and shallow lakes frequently produces H_2S and CH_4. Methane is also found sometimes in well waters.

HENRY'S LAW

The amount of gas dissolved in water is directly proportional to the partial pressure of that gas in the vapor space above the water/gas interface. This is known as Henry's law. Another important consideration of gas solubility is Dalton's law, which states that the total pressure of a mixture of gases is made up of the individual pressures of those gases, and these are in direct relationship to their molar or volume ratios in the vapor space. For example, air normally contains about 80% nitrogen and 20% oxygen by volume and the standard atmospheric pressure is 760 mmHg. Dalton's law states that the partial pressure of O_2 in the atmosphere is 152 mm (0.20 × 760) and the partial pressure of N_2 is 608 mm (0.80 × 760).

Temperature is an important factor in gas solubility; solubility decreases as temperature increases. This is because the increasing temperature results in an increasing vapor pressure of the water itself, so that escaping water molecules at the liquid-gas surface force away other gas molecules.

A final factor important to the solution of gases in water is the diffusion of the gas molecules through water. The rate of diffusion increases with increasing temperature as molecular activity increases and the viscosity of the water decreases.

In the removal of gas from water, the mass transfer principles that apply to the process are analogous to the principles of heat transfer. The rate of gas transfer is directly proportional to three factors as shown by the following equation:

$$Q = Ka\,\Delta P$$

where Q = rate of gas transfer
K = gas transfer coefficient
a = exposed contact area
ΔP = log mean driving force (pressure units)

The gas transfer coefficient, K, is a function of both the specific gas being transferred and also the geometry of the system. As noted below, there are various shaped packing materials that can be used in packed towers (Figure 14.1); there is also variety in the designs of diffuser elements and atomizers. At the liquid-gas interface, there is assumed to be a liquid film on one side and a gas film on the other, and these films are affected by the nature of the liquid and the gas, with their combined effects influencing the gas transfer coefficient.

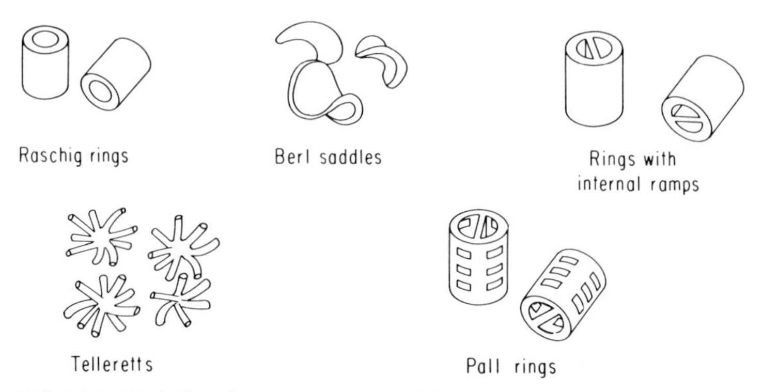

FIG. 14.1 Varieties of common tower packings.

The area factor, a, is the total surface area created in the degasification chamber by packing, by producing numerous fine water droplets, or by forming minute gas bubbles. In the case of tower packing, each size and shape of packing material—raschig rings, berl saddles, spirals—has a definite exposed area:volume relationship. This ratio increases as the size of the packing decreases, but so, too, does the pressure required to force stripping gas through the packing.

The gas driving force can be calculated readily for nonionizing gases. Ionizing gases often present a difficult problem, because the pH of the system establishes the degree of ionization, and the pH changes as the gas is removed.

DEGASIFICATION DESIGNS

The three types of degasification units commonly used in water treatment are shown in Figures 14.2, 14.3, and 14.4.

The first of these is a packed stripping tower. Water is introduced at the top, sometimes through a spray pipe, and allowed to percolate through the packing against a countercurrent of gas other than the gas being removed. Typically air is used as the stripping gas to remove dissolved gases such as carbon dioxide, ammonia, hydrogen sulfide, or methane.

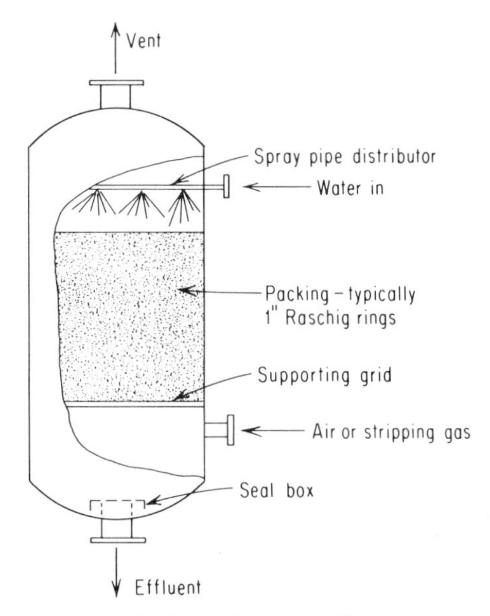

FIG. 14.2 Packed column degasifier.

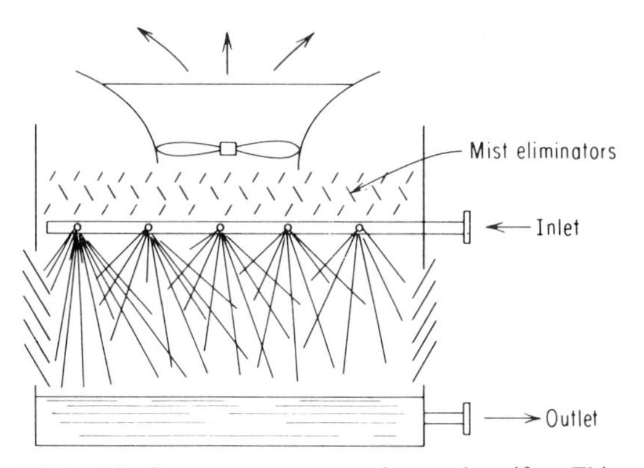

FIG. 14.3 Spray-type tower used as a degasifier. This type of design lends itself to ammonia stripping in treating wastewaters.

In the second type of degasifier, water is atomized into an inert gas, which is continually withdrawn as the stripped gas leaves the water surface.

In the third design, the stripping gas is introduced into the water through submerged diffusers and bubbles its way up through the water, carrying the stripped gas out of the aqueous environment into the atmosphere above.

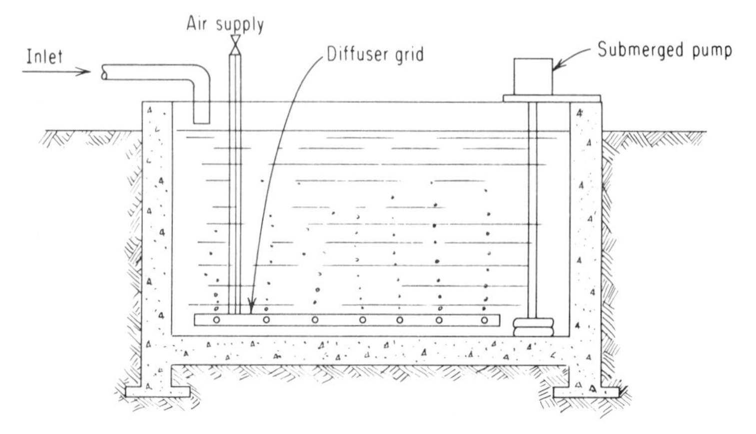

FIG. 14.4 Diffusers used for degasification of water in open, exposed areas.

The packed stripping tower is commonly used in refineries and paper mills for steam stripping of sour condensates. Figure 14.5 shows a refinery application, and Table 14.1 shows the analysis of a sour condensate before and after stripping. The steam consumption for this operation is very high, in the range of 0.5 to 1.0 lb/gal (0.06 to 0.12 kg/L). The basic reason is that the gases being removed are ionized, so that they exert only a fraction of the potential vapor pressure that would be available if pH could be properly adjusted. Unfortunately, the ammonia is best removed at a pH over 10 and the H_2S and HCN are best removed at a pH below 6; in the pH range of 7 to 8, these gases are highly ionized (Figure 14.6). It can be seen from this that a more practical system for stripping sour condensate might be a two-stage operation with alkali added for increasing the pH for ammonia removal in the first tower and acid added for reduction of pH for stripping of H_2S

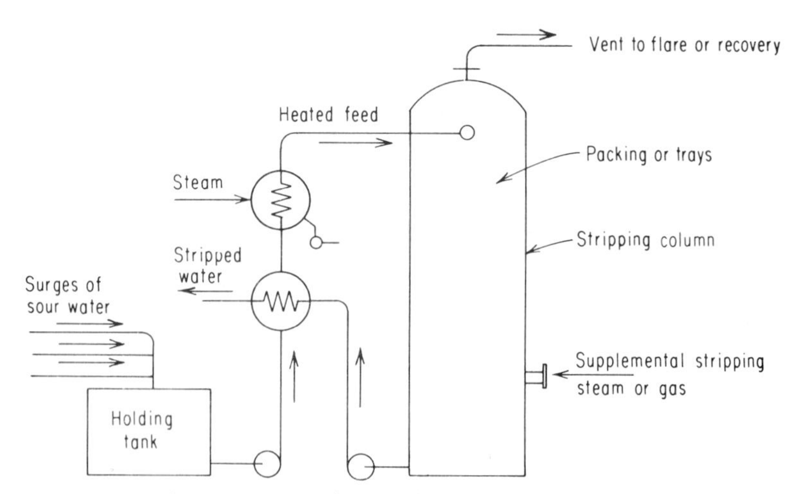

FIG. 14.5 Typical arrangement for stripping column operation processing refinery sour water.

TABLE 14.1 Refinery Sour Water Stripper

	In	Out
Sulfides, as HS, mg/L	1876	0
Ammonia, as NH_3, mg/L	1480	194
Cyanides, as CN, mg/L	24	2
Phenol, mg/L	102	65
pH	9–9.5	7.5
Temperature, °F	195	230

Physical design data: 60 gal/min flow, 3500 lb/h steam; tower diameter 3 ft 6 in, packing 12 ft of 1-in, raschig rings.
Source: API Manual 1969, Chapter 10, Table 10-1.

and HCN in the second tower. There are many other complex reations occurring in a system like this, for example, the formation of thiocyanates and ferrocyanides, which further reduce the potential for gas stripping.

In kraft pulp mills, condensate from the first stages of the multiple-effect black liquor evaporator is high in volatile organics, particularly mercaptan compounds and turpentine. These exert a BOD loading on the waste plant and produce offensive odors. Stripping columns have been installed in some kraft mills to process this foul condensate; the odorous gases are sent to the lime kiln for burning.

A similar design stripping column is used for the vacuum deaeration of cold water, as illustrated in Figure 14.7. In this kind of unit, vacuum is usually created by ejectors. Steam consumption can often be minimized by having two stages of stripping sections in the same column, with a steam jet ejector on each stage. The

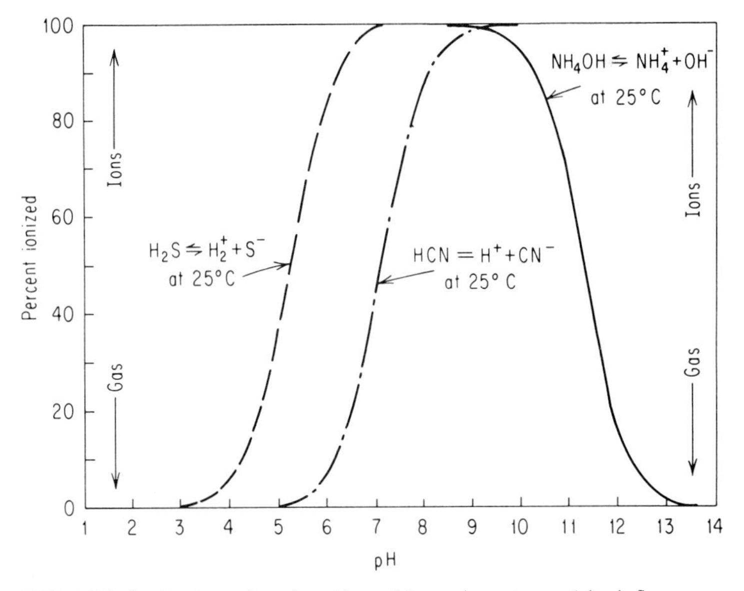

FIG. 14.6 Ionization of weak acids and bases in water and its influence on internal gas pressure.

FIG. 14.7 Vacuum deaerator at an oil production operation. *(Courtesy of L*A Water Conditioning Company.)*

vacuum produced is calculated to reduce the partial pressure of oxygen to less than 50% of that corresponding to the partial pressure equivalent to the desired effluent oxygen concentration. This type of tower usually uses raschig rings as packing.

Still a third variation in the design of stripping columns is shown in Figure 14.8, which is a stripper using air for stripping out carbon dioxide, methane, and hydrogen sulfide. This is of fiberglass construction, with raschig ring packing. This unit may also have wood stave construction to eliminate corrosion problems, and the packing itself consists of layers of wooden slats 3 to 4 in (7.5 to 10 cm) apart. The open spacing allows for a low pressure drop through either design of tower, so that the air blower is seldom required to operate in excess of 3 in H_2O discharge pressure. This provides economical gas removal.

Some stripping towers are constructed like cooling towers, where high ratios of stripping air to water (e.g., 200 ft^3 air per gallon of water) are needed, as in the removal of ammonia from sewage at high pH. (See Figure 36.5.)

DEAERATING HEATERS

A final design of packed column is one of the oldest devices in water treatment, the tray-type deaerating heater. In this unit, the packing is made up of a series of

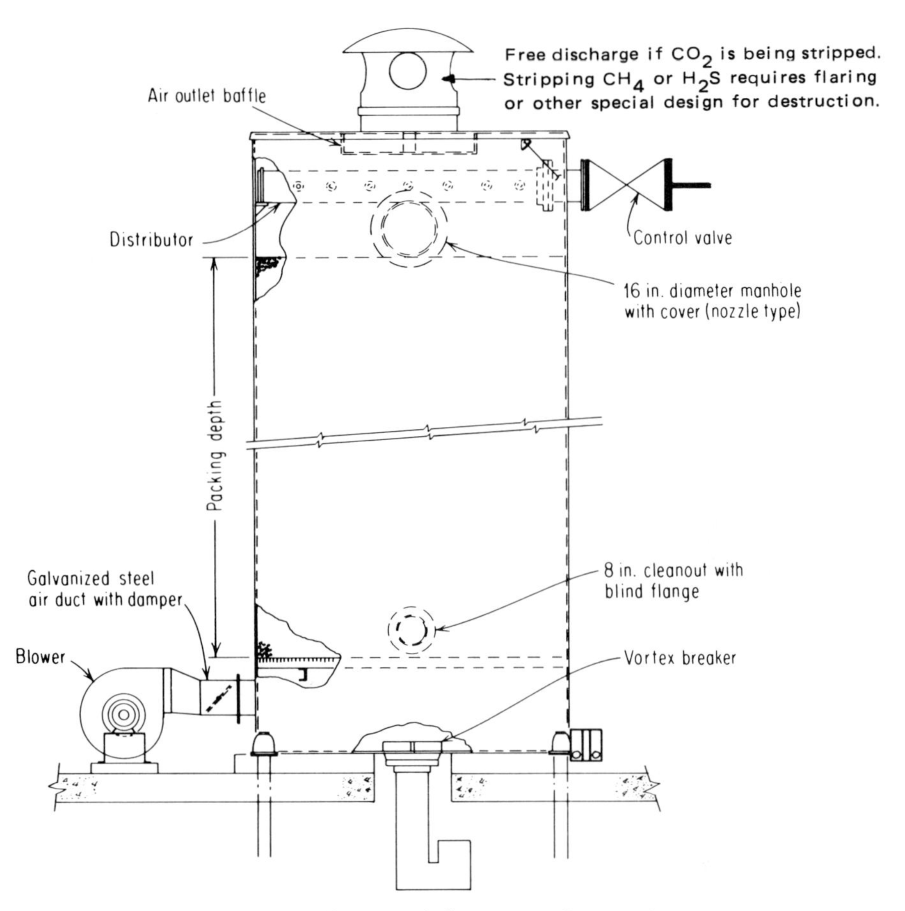

FIG. 14.8 FRP decarbonator. *(Courtesy of The Permutit Company.)*

stainless steel or cast iron slotted trays, and steam is used for both heating and stripping oxygen out of the water. This design is still very popular in boiler plants because it is more efficient thermally than the atomizing type deaerators described below (Figure 14.9).

Probably the major use for the atomizing-type stripper is as a deaerating heater in steam power plants. These units atomize the water by the energy of steam passing through an atomizing nozzle, and the steam both heats and strips the water of its dissolved oxygen content (Figure 14.10). Since the condensation of the steam in the vapor space creates the steam flow needed for atomization, this device requires a temperature difference of at least 50°F between the water entering the vessel for stripping and the steam. Because of the pressure drop across the atomizer, this device is less efficient than a tray-type deaerating heater.

The final design category, submerged diffusion, is seldom used unless certain space requirements favor it. The gas transfer coefficient is poor, and to be effective there must be some means for continually sweeping the surface of the basin by adequate ventilation to carry away the stripped gases.

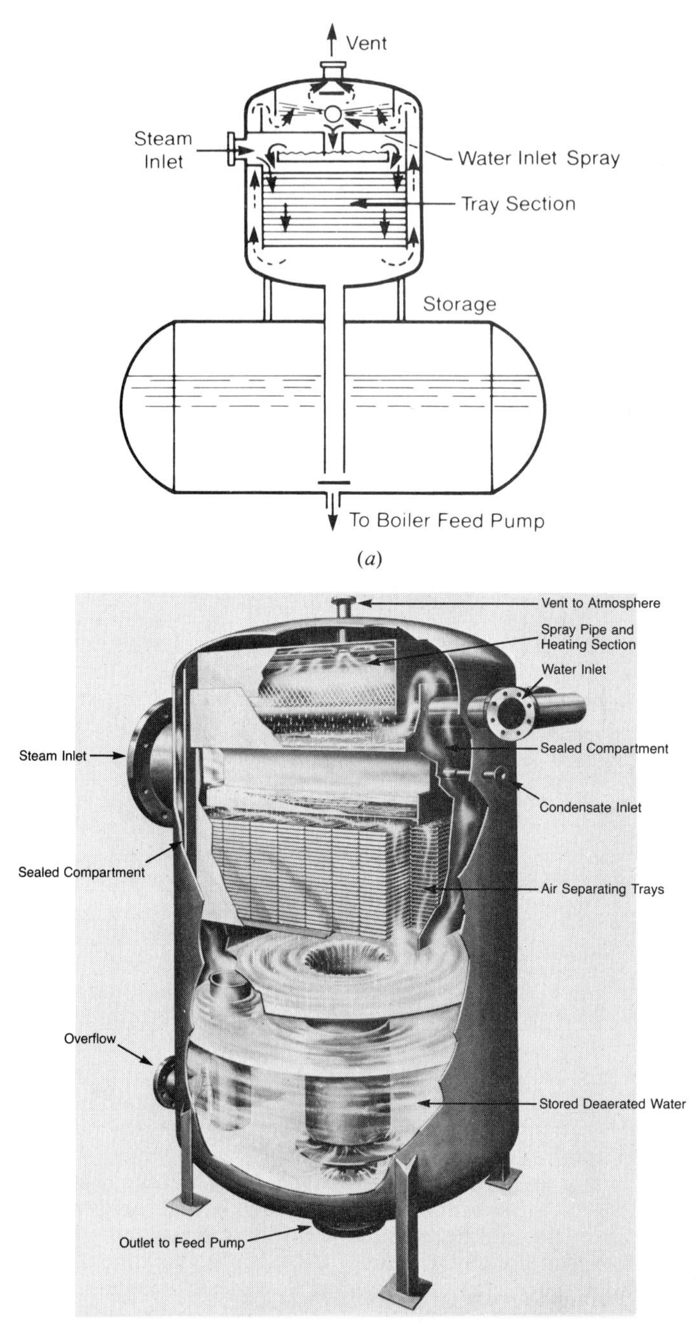

FIG. 14.9 (*a*) A spray-tray deaerator. The bulk of the oxygen is removed by the spray heater, with the remainder stripped by the total steam flow over the trays. *(Courtesy of Cochrane Division, the Crane Company.)* (*b*) Tray-type deaerating heater. *(Courtesy of Cochrane Division, the Crane Company.)*

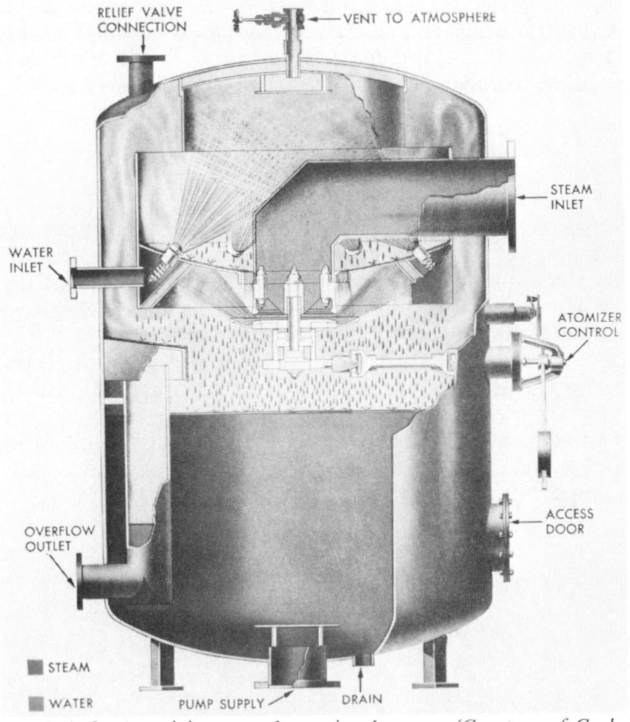

FIG. 14.10 Atomizing-type deaerating heater. *(Courtesy of Cochrane Division, the Crane Company.)*

CHEMICAL REMOVAL

While most gas removal in water treatment is done by mechanical action, chemical destruction of these gases is an equally important process. Ion exchange is also effective.

Even at high temperatures, there is some residual dissolved oxygen after water has been deaerated in a power plant deaerator. Sodium sulfite or hydrazine can be fed to the effluent to react against the possibility of accidental oxygen leakage into the system, as shown in the following chemical reactions:

$$2Na_2SO_3 + O_2 \rightarrow 2Na_2SO_4 \quad (1)$$

$$N_2H_4 + O_2 \rightarrow 2H_2O + N_2 \quad (2)$$

Because of the high temperatures found in the feedwater system, oxygen corrosion can be rapid. Catalysts are usually included in the scavenger formulation to assure fast, complete oxygen destruction.

While carbon dioxide can be stripped from water in the free gas form, it can also be neutralized by the application of lime, caustic, or sodium carbonate. The carbon dioxide that escapes with the steam can be neutralized by the application of amines to the system. In this case, the amine forms a bicarbonate compound, and actually develops some alkalinity to the condensate.

Carbon dioxide is also removed in the anion exchanger of a demineralizer.

Ammonia can be destroyed chemically by chlorination. The initial reaction forms chloramine, and until this material has been completely broken down, the application of chlorine does not produce a free chlorine residual.

$$NH_3 + Cl_2 \rightarrow NH_2Cl + HCl \quad (3)$$

$$2NH_3 + 3Cl_2 \rightarrow N_2 + 6HCl \quad (4)$$

When the ammonia has been completely reacted, then the residual appears: this is called the break point. Ammonia cannot always be destroyed in this fashion, however, since many wastewaters contain organic materials that react with chlorine in preference to the ammonia, so that these may be destroyed before the excess chlorine is available for reacting with the ammonia.

Ammonia can be removed by cation exchange. A type of clay, clinoptilolite, is specific for removal of ammonia in preference to other cations. This has been used in several wastewater treatment applications.

Finally, hydrogen sulfide may be destroyed chemically by oxidation with chlorine, or it may be precipitated by application of a heavy metal salt. The chlorination reactions are as follows:

$$H_2S + Cl_2 \rightarrow S + 2HCl \quad (5)$$

$$H_2S + 4Cl_2 + 4H_2O \rightarrow H_2SO_4 + 8HCl \quad (6)$$

CHAPTER 15
MEMBRANE SEPARATION

The use of semipermeable membranes is a comparatively recent addition to the technology of water renovation or purification. As design engineers strive to perfect these membranes and the techniques of their application, they can look to nature for examples of semipermeable membranes performing aqueous separations both in plant and animal tissues. In a sense, synthetic membrane devices copy what takes place in many natural systems: membranes are used by the roots and heartwood of trees and by the kidneys and small intestines of animals to transfer nutrients to cells and to remove waste products.

When membrane separation is used for purification, water passes through the membrane as a result of a driving force, or a combination of driving forces, leaving behind some portion of its original impurities as a concentrate. The type of membrane or barrier, the method of application of the driving forces, and the water characteristics determine the type of impurities removed and the efficiency of removal.

In the past, problems of membrane fouling or degradation and concentrate disposal have kept costs comparatively high and limited application of membrane systems to special situations, such as to cases where the impurities removed had commercial value. More recently, improvements in membranes and application technology have made membrane separation practical, and it is now commonly used in ultrapure water systems, in desalination of brackish water, and in effluent treatment.

MICROFILTRATION

In the simplest membrane separation process, the membrane acts as a porous barrier. Water is forced through the membrane by a pressure differential. The pores in the membrane may be rather large, generally in the range of 0.1 to 20 μm. When the membrane is used as an analytical device to detect suspended solids in a water sample, a pore size of 0.45 μm is usually used. The membrane can be made of many different materials, such as cellulose acetate and polyamide. The number of pores per unit surface area and their shape or configuration can vary greatly, influencing production rates and quality. The membrane chemistry and structure are also important factors.

In filtration, suspended solids are removed on the face of the membrane. If the suspended solids are slimy or easily compressed on the surface, the membrane may become clogged much like any other filter medium, and filtration rates will drop to impractical levels; the process must be stopped and the membrane may require replacement. Seldom can the membranes be cleaned by backflushing. Membranes used for filtration remove little colloidal or soluble material. The efficiency with which very fine suspended solids are removed depends primarily on the size and shape of the pores and the type of filter cake developed during filtration.

There may be more than sieve-type action taking place during microfiltration. The membrane may swell, changing its characteristics, and there may be chemical or electrochemical interactions between the membrane and certain soluble or colloidal substances in the water being filtered. Because of this, microfilters have two ratings, nominal and absolute. The nominal rating takes into account the filtration of particles even finer than the pores by the filtering action of the accumulated solids on the membrane surface. The absolute rating is removal of 100 percent of particles larger than pore size. The nominal rating must be established empirically for a specific membrane/liquid system.

As with other water filtration processes, filtration rate varies directly with water temperature (inversely with viscosity) and with the pressure differential across the membrane. Unlike other membrane processes, microfiltration recovers 100 percent of the filtrate as product water.

FIG. 15.1 Ultrapure water is needed in many areas of manufacture of electronic components. The final step in the treatment system shown here is membrane filtration following several stages of ion exchange. *(Courtesy of Bell Laboratories.)*

To make membrane filtration practical, the bulk of the suspended solids should be removed by conventional depth or septum filters (Chapter 9) before the solids reach the membrane unit. A typical example of membrane filtration is shown in Figure 15.1, as part of a total system producing ultrapure water for integrated circuit manufacture. The membrane is the last step in the process, removing crud that would be undetected by ordinary analysis, but which could ruin the integrated circuit. The membrane may process over 1000 L/cm^2 (250,000 gal/ft^2) before plugging in this application, where it would pass less than 1 L/cm^2 (250 gal/ft^2) if used to filter samples of many municipal water supplies.

ULTRAFILTRATION

As the pore sizes in a membrane becomes smaller, i.e., much below about 0.1 μm, the pressure differential required to produce acceptable flow rates increases substantially. Processes using small pore-size membranes and increased pressure, called ultrafiltration (UF), generally require pressure differentials greater than 20 lb/in^2 (1.4 kg/cm^2). The purpose of using a smaller pore size is to remove colloids and certain high molecular weight organic materials from water. However, this small pore size makes the membranes even more susceptible to clogging or blinding than microfiltration membranes. In certain configurations, UF membranes can be backflushed to solve this problem. UF membranes may be damaged permanently by excessive heat or pressure, causing flux to drop drastically.

More pronounced than in straight filtration, the character and form of the UF membrane is critical to the results obtained. To obtain suitable flow rates, the UF semipermeable membrane usually has an extremely thin skin incorporated on the surface of a more porous, thicker substrate. This type of membrane is said to be *anisotropic.* The skin may be less than 0.1 μm thick, while the substrate may be 25 to 50 μm in depth. Additional mechanical support is usually necessary. The skin and the substrate are often the same material, the skin being modified on the surface of the bulk membrane by treating the surface thermally, mechanically, chemically, or by some combination of these.

In addition to possible blinding of the UF membrane, concentration polarization may also effect flux rates. This phenomenon, occurring in the water layer adjacent to the membrane surface, is the result of a localized increase in the concentration of rejected impurities. This increases the solution density and viscosity at the membrane surface, reducing flow rate. A number of design techniques have been developed to reduce concentration polarization.

MATERIAL BALANCE IN MEMBRANE PROCESSES

In microfiltration and depth or septum filtration, the water flow path is perpendicular to the filtration surface, and all of the water to be processed passes through the barrier and is recovered. By contrast, in both ultrafiltration and reverse osmosis, the flow of water to be filtered is cross-flow, or parallel to the membrane surface, with only a portion of this peeling off the mainstream to pass through the membrane. The high rate of cross-flow is maintained either by recycle or by excessive wastage, and this procedure of maintaining high surface velocity effectively minimizes surface fouling by its scouring action and reduces concentration polarization effects. Because of this operating procedure, not all of the original liquid to be processed becomes filtrate. The ratio of permeate (the term applied to product water instead of "filtrate" where semipermeable membranes are used) to applied feed water is the *recovery ratio, R*. A schematic diagram of an ultrafiltration system is shown in Figure 15.2. The mass balance shown by this diagram

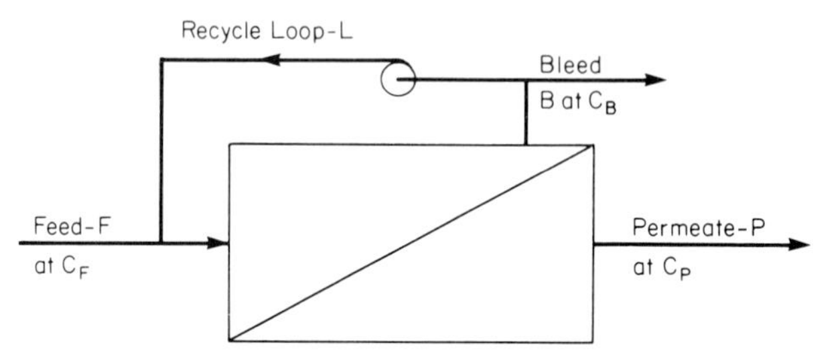

FIG. 15.2 Schematic of an ultrafiltration or reverse osmosis system with recycle loop to maintain high flow rate across the membrane surface. Legend: *C*, concentration of solute, mg/L; *F*, feed water flow, gal/min at C_F mg/L; *P*, permeate flow, gal/min at C_P; *B*, bleed, or blowoff, at C_B, mg/L; *L*, recycle rate, gal/min; *R*, recovery ratio = *P*/*F*; *CR*, concentration ratio = C_B/C_F.

applies both to ultrafiltration and reverse osmosis processes. The following relationships apply to this diagram:

1. Assuming complete rejection of a species in the feed stream, meaning that none of these species pass through the membrane:

$$CR = C_b/C_f = F/B \tag{1}$$

$$F = P(CR/(CR - 1)) \tag{2}$$

$$CR = (1/(1 - R)) \tag{3}$$

   In this example, complete rejection could apply to an organic material with a molecular weight of 10,000, a condition that could relate equally to ultrafiltration and reverse osmosis.

2. Where there is less than complete rejection, e.g., 80% rejection of nitrate ions in a reverse osmosis system (0.8 rejection factor, 0.2 passage factor), the concentration ratio for each species must be determined by analysis; but this will be less than *F*/*B*. It will be:

$$CR = (F/B) \times \text{rejection factor} \tag{4}$$

DESIGN FEATURES OF COMMERCIAL UNITS

There are four basic designs of ultrafiltration and reverse osmosis systems:

1. *Large tube construction (Figure 15.3):* Hollow porous tubes up to 1-in diameter are fastened at each end to tube sheet supports, and the membrane is cast

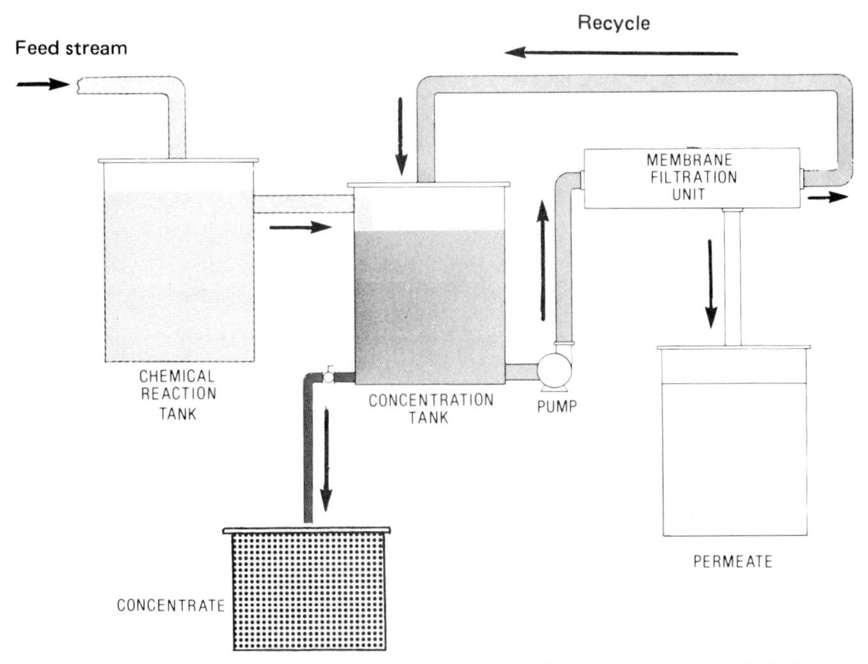

FIG. 15.3 Tubular design. Membrane tubes are housed in PVC modules. An individual membrane tube may be added or replaced without replacing an entire module.

on the inside surface of each tube. The permeate is collected in the shell of the vessel, and the concentrate discharges at the end of the tubes, where it may be discarded or collected for recycling. The tubes are individually removable, and may be cleaned physically and chemically. This design is used in chemical process and waste treatment applications, but seldom for treatment of fresh water or seawater.

2. *Flat plate construction (Figure 15.4):* In this design, the membrane is assembled somewhat after the fashion of a plate-and-frame press on a small scale. The membranes in one such design are cast on both sides of a flat, porous substrate, which collects the permeate, and a multiplicity of such "sandwiches" are assembled into cartridges with alternate spaces between for passage of the feed and recirculation. This design has been more applicable to food processing (e.g., in the dairy industry) than to water or wastewater treatment.
3. *Spiral-wound construction (Figure 15.5):* This is an ingenious extension of the flat-plate design. The membrane-porous collector-membrane sandwich, sealed at three edges, is overlain by a mesh spacer and roller around a collector tube with the open edge of the sandwich adjacent to the tube. The mesh spacer carries the feed solution, which permeates the membrane surfaces that constrain it on two sides, with excess concentrate discharging at the opposite end of the rolled cartridge.
4. *Hollow fiber membranes (Figure 15.6):* In this system, hundreds of hollow filaments about the size of human hair are collected in a bundle that is folded into a U-shape; the ends are set in plastic, but protrude so that they remain open and unobstructed, and this assembly of filaments and tube sheets is

FIG. 15.4 Flat plate construction, showing insertion of a cartridge into the pressure housing. *(Courtesy of Dorr-Oliver, Incorporated.)*

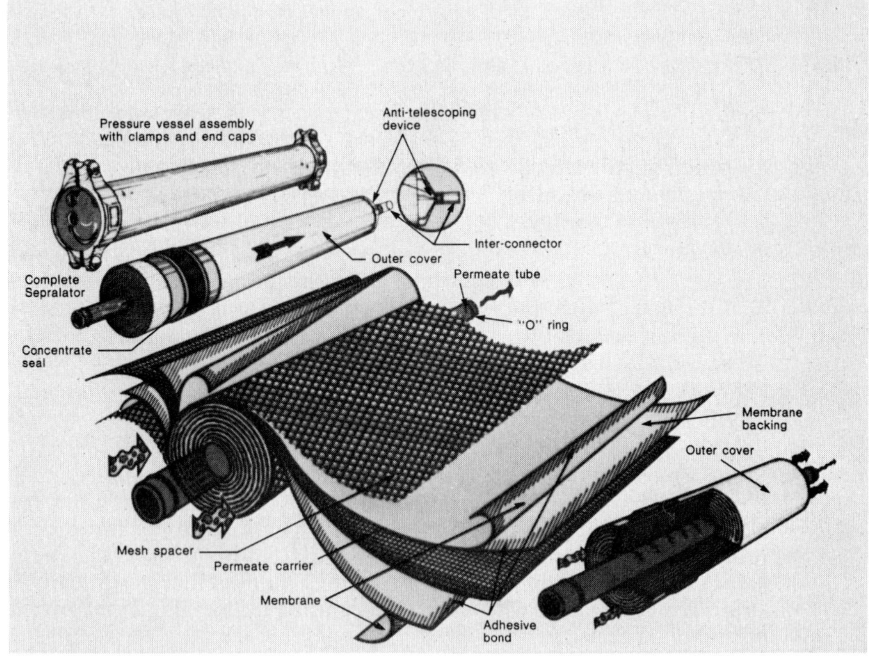

FIG. 15.5 Membrane assembly partially unrolled shows the distinct layers wrapped around permeate tube. Two leaves (complete membrane envelopes) are shown. Cutaway shows layers and the relationship to feed and concentrate flow. (Courtesy of Osmonics; Inc.)

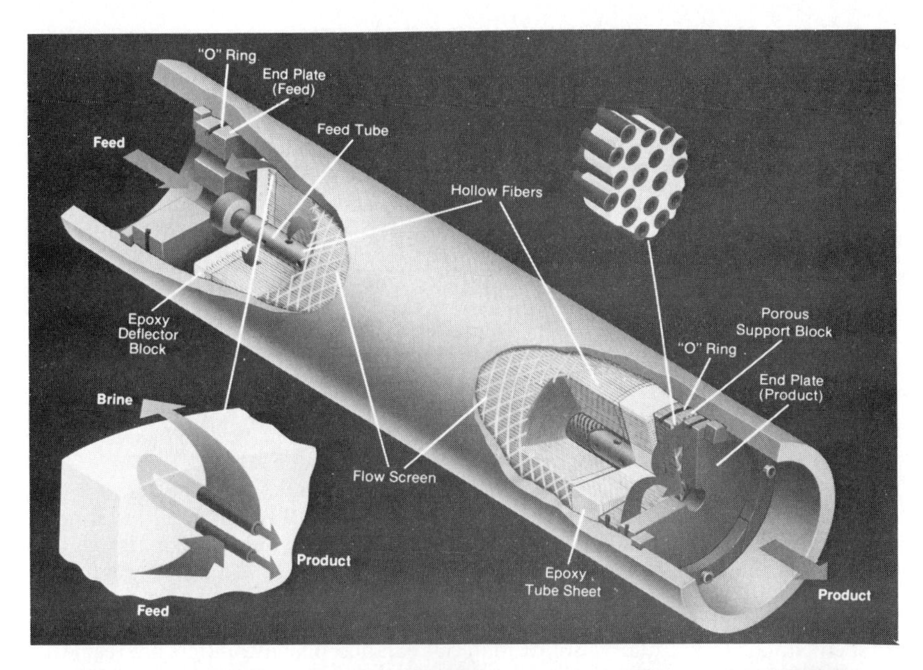

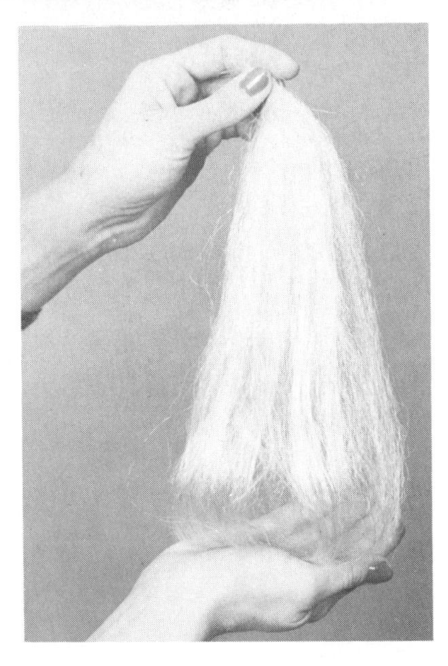

FIG. 15.6 Hollow fiber design. The two insets show the nature of the fibers both life-size and microscopically. (Courtesy of Dupont Company.)

mounted in a cylindrical pressure vessel. The pressure vessel has provision for pressurized feed inlet, saline outlet, and permeate outlet, suitably compartmented to prevent leakage of salines into the permeate. The large surface area of the fine filaments results in a total membrane surface of about 5000 ft^2 per cubic foot of cartridge volume. A number of cartridges are mounted in parallel in a battery to produce the required output; several batteries may be operated in series, then, to achieve the needed recovery ratio and rejection rate. This is a common feature of reverse osmosis systems as compared to ultrafiltration systems.

Figure 15.7 shows a typical UF installation. This unit is used in the automotive industry to continually filter a special electrodepositing paint from washwater, recovering the paint and preventing a buildup in salinity. It is also used to filter the paint itself from the dip tank to maintain density as paint is exhausted from the paint bath.

Figure 15.8 shows a process combining lime softening with ultrafiltration. The conventional lime softener (Chapter 10) having a 60-to 90-min detention time is replaced with a much smaller 10-to 15-min reaction vessel in which a high sludge density (about 15% sludge volume) is maintained. The total output is pumped through an ultrafiltration battery, with the permeate discharging as lime-softened water. The blowby is recirculated to the lime reactor, a portion being discarded as spent sludge to maintain a solids material balance. The ultrafilters are regularly backflushed, and also chemically cleaned, usually on a weekly basis.

REVERSE OSMOSIS

Osmosis is the process in which a solvent flows across a membrane separating a stronger from a weaker solution; the solvent flows in the direction that will reduce

FIG. 15.7 Ultrafiltration system being used to control density and recover paint solids. *(Courtesy of Abcor, Inc.)*

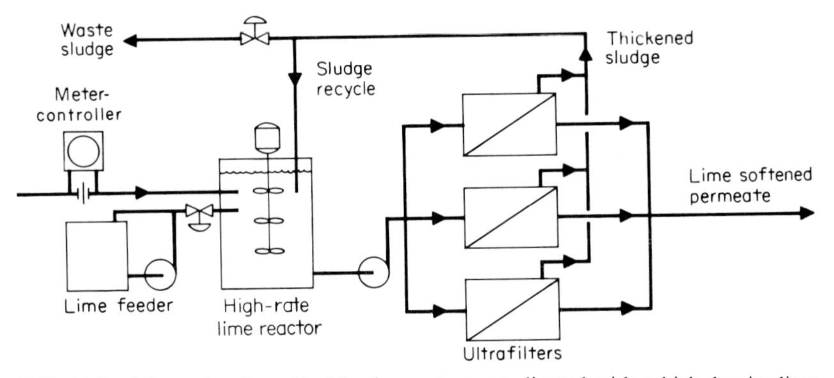

FIG. 15.8 Schematic of an ultrafiltration system coordinated with a high-density lime reactor for lime softening.

the concentration of the stronger solution. The flow of solvent between solution compartments can be observed as the liquid in the compartment of stronger solution increases in volume. If these compartments are fitted with standpipes, the flow will continue until the level in the stronger solution compartment is higher than that in the weak by a dimension which equals the osmotic pressure, as shown in Figure 15.9.

In reverse osmosis (RO), a driving force, the differential pressure across the membrane, causes water to flow from the stronger solution to the weaker. There-

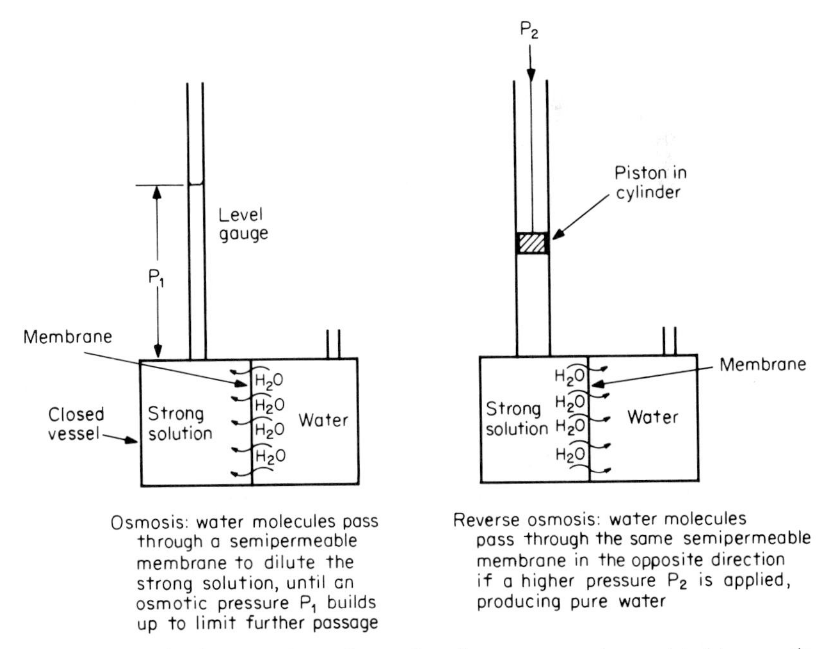

FIG. 15.9 A simple comparison of osmosis and reverse osmosis, as related to osmotic pressure.

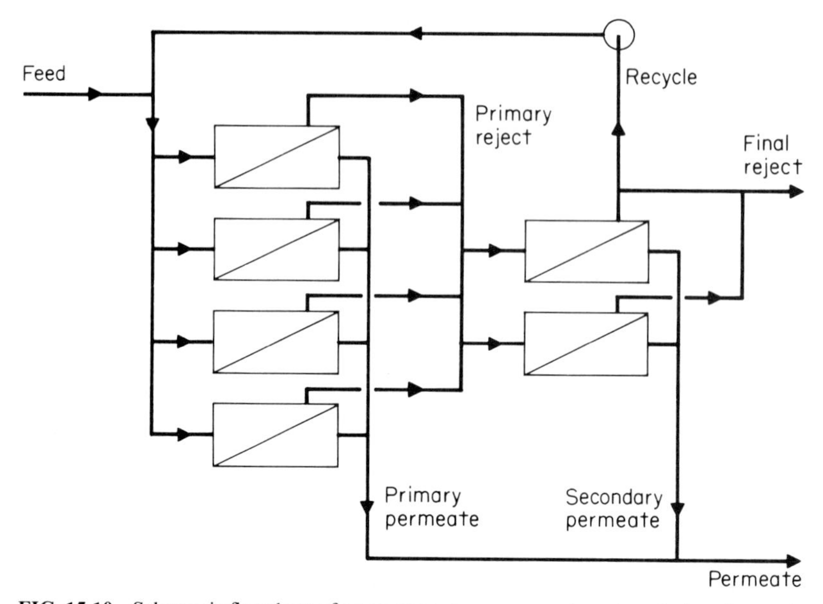

FIG. 15.10 Schematic flowsheet of a two-stage reverse osmosis system designed to optimize energy cost and recovery.

fore, the pressure required must exceed the osmotic pressure. The differential pressure (transmembrane pressure) is often greater than 300 lb/in² (21 kg/cm²), depending on concentration differences. It averages about 10 lb/in² for each 1000 mg/L difference in total dissolved solids. From this, it is apparent that a large portion of the operating cost is pumping energy; the other major cost factors are the cost of disposal of the reject stream (usually about 25% of the volume of feed stream), depreciation, and membrane replacement. To optimize these cost factors, most RO systems are two-stage units, as shown by Figure 15.10. There may be a temptation to reduce pressure as a potential energy-saving step, but it is very important to maintain design pressure drop (transmembrane pressure) because a reduction in this operating variable not only reduces production, but also allows more salt passage (the opposite of rejection). This is shown in Figure 15.11. If back-pressure builds up as a result of deliberate throttling, the same effect of increased salt passage (decreased rejection) also occurs. So, control of transmembrane pressure is critical to optimum performance.

Transmembrane pressure differences are influenced by the types of membranes used. In reverse osmosis, transport of water through the membrane is not the result of flow through definitive pores, at least not pores as commonly conceived. It is the result of diffusion, one molecule at a time, through vacancies in the molecular structure of the membrane material. The vacancies in amorphous polymers are in a state of flux, or are not fixed, while in crystalline materials the vacancies are voids in lattice structures and are essentially fixed in number, position, and size. RO membranes are made of amorphous polymers, but usually contain some crystalline or less amorphous regions.

The most widely used membrane materials are cellulose acetate, triacetate, and polyamide polymers. A condensed tabulation of their most important character-

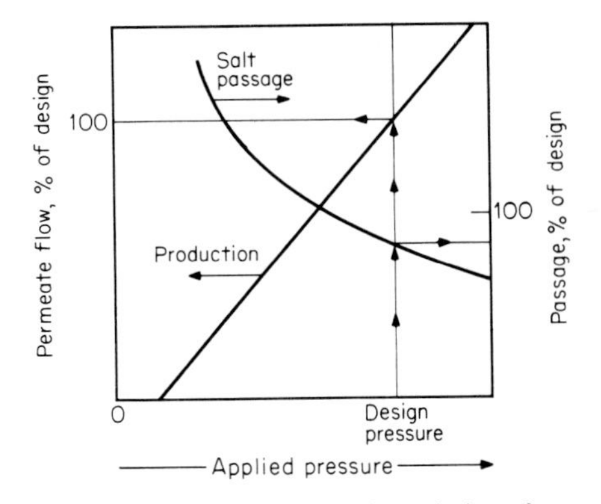

FIG. 15.11 The effect of deviating from design pressure on production and permeate quality.

istics is given in Table 15.1. Operating temperature must be carefully monitored, since failure occurs if the temperature limit is exceeded, and flow is restricted as the temperature drops (Figure 15.12). Polysulfone membranes are used at higher temperatures. For wider operating pH ranges, polyethylenimine and polyfurane membranes may be used.

There is an aging effect on RO membranes that results in surface compaction and flow constriction. This is hardly noticeable below 200 lb/in^2, but may amount to 10% loss of production at 400 lb/in^2 and as much as 40% after a year of operation. Therefore, it is important to allow for this compaction in designing an RO system so that plant needs can be met over the expected life of the membranes.

Reverse osmosis has also been called hyperfiltration, indicating its relationship to a high-pressure filtration process. However, it should not be confused with ultrafiltration, which uses lower applied pressures and different membranes, because ultrafiltration removes essentially no solutes of low to intermediate molecular weight, while reverse osmosis can remove even low molecular weight

TABLE 15.1 RO Membrane Characteristics
(Based on 90% rejection)

	Type membrane		
	Triacetate hollow fibers	Polyamide hollow fibers	Cellulose acetate spiral wound
Flux at 400 lb/in^2, gpd/ft^2	1.5	1.0	15–18
Back-pressure, lb/in^2	75	50	0
pH range	4–7.5	4–11	4–6.5
Maximum temperature, °F	100°	110°	100°
Cl_2, maximum mg/L	1.0	0.1	1.0
Bioresistance	Good	Excellent	Fair
Backflushing	Ineffective	Ineffective	Effective
Silt density index (SDI)	4.0	4.0	7.0

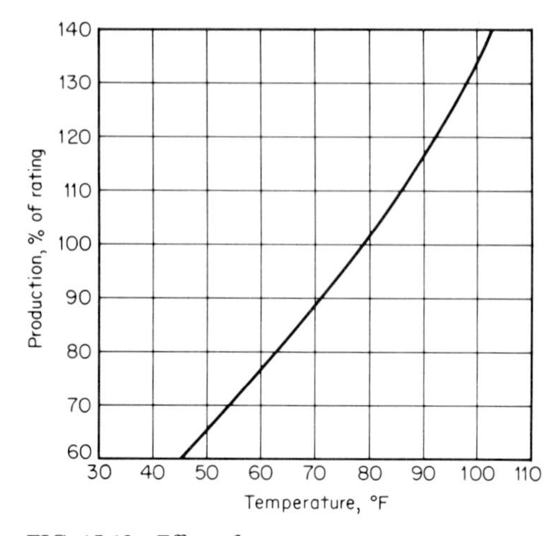

FIG. 15.12 Effect of temperature on permeate production at fixed pressure, TDS, and recovery.

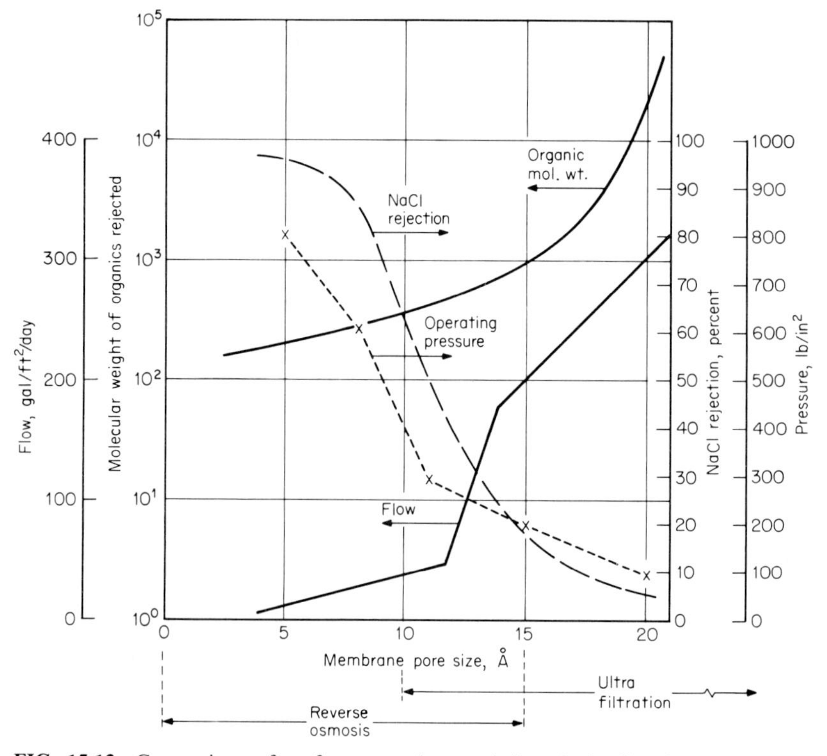

FIG. 15.13 Comparison of performance characteristics of ultrafiltration and reverse osmosis membranes. *(Abstracted from literature of Osmosis, Inc.)*

TABLE 15.2 Typical Passage of Ions* Across RO Membranes

Ions	% passage	% rejection
Ammonium	8	92
Sodium	5	95
Potassium	5	95
Magnesium	3	97
Strontium	3	97
Calcium	2	98
Nitrate	15	85
Bisilicate	10	90
Chloride	5	95
Fluoride	5	95
Bicarbonate	5	95
Sulfate	3	97
Phosphate	1	99

*Excluding heavy metal ions, which in fresh water are often colloidal compounds rather than ions and thus tend to foul the membrane.

Note: CO_2, O_2, and N_2 gases generally pass through readily and may actually be enriched in the permeate. The same is true of certain organics of smaller size than the pore diameter; for example, phenol is enriched in the permeate.

ionic species. The differences between the performance characteristics of ultrafiltration and reverse osmosis membranes are shown in Figure 15.13.

The salt removal or rejection obtained is primarily a characteristic of the semipermeable membrane, one rejecting a specific ion more effectively than another, as shown in Table 15.2; this rejection may be affected by the presence of other ionic species in the water being processed. In general the greater the water flux through a given type of membrane, the lower the salt rejection—the less salt retained in the concentrate and the more contaminated the product. Conversely, the higher the salt rejection rates, the lower the water flux through the membrane, using the same transmembrane pressure. These properties can be varied by changing the type of polymers used and methods of manufacturing and processing them. As with electrodialysis, potential for concentration polarization, scaling, and fouling are serious considerations in designing process units to take advantage of the various membrane materials available.

MEMBRANE MAINTENANCE

Because the membranes have such small pores, it is extremely important to provide a feed stream free of crud and stabilized to prevent precipitation or scaling reactions. The use of high velocity to scour the surface is a design feature that helps keep the surface from fouling, but it cannot solve the problem completely.

The susceptibility of the membrane to fouling by water-borne debris is measured by a special filtration technique that determines the silt density index (SDI). In this test, a sample of feed water at a constant pressure of 30 lb/in^2 is passed through a 0.45 μm membrane filter. The filtrate is collected in a 500-mL graduate, and the time for collection of the first 500-mL sample is recorded as t_{15}. From the data, the SDI is calculated as follows:

$$SDI = \left(\frac{t_{15} - t_0}{t_{15}}\right) \times 6.7$$

(The factor 6.7 applies to the 15-minute test period only.)

Usually spiral-wound membranes can tolerate a higher SDI than the hollow-fiber–type RO membranes, as shown by Table 15.2. If the SDI exceeds 10, prefiltration is always needed, but may be required for hollow-fiber membranes even when the SDI exceeds 4.

As indicated by Figure 15.2, the dissolved solids rejected by the RO membrane remain in the bleed, so obviously the concentration of the bleed, which occurs at the membrane surface, may produce precipitation. As a general rule, the Langelier index (Chapter 4) of the concentrate should be negative. In the desalination of brackish water or seawater, the Stiff-Davis index should be used in place of the Langelier index. It is rather common practice to feed acid to the incoming water to obtain a negative index, and often this is supplemented by application of a low dosage of polyphosphate. It is important to check the saline bleed for $CaSO_4$ stability as well as $CaCO_3$ stability. A method of evaluating the scaling potential of the bleed is given in ASTM Procedure D-3739-78. It is somewhat difficult to estimate the exact acid dosage required for this corrective treatment because the by-product CO_2 produced by the acid passes into the permeate rather than remaining in the reject as a pH-depressant.

Even with prefiltration and control of the Langelier index, fouling may still occur from such things as heavy metals (Fe, Mn), silica (SiO_2), and organic matter. The latter may provide food for microbes, so disinfection is frequently required. Some membranes cannot tolerate more than 0.1 mg/L chlorine, so nonoxidizing biocides must be used. Formaldehyde has sometimes been used for batch disinfection.

Chemical cleaning is an integral part of the membrane maintenance program. The procedures must be matched to the nature of the deposit and may include detergents, biocides, acids, and alkalies.

RO units are finding widespread application in desalination of brackish water for potable or industrial use, pretreatment of fresh water ahead of demineralizers to reduce chemical consumption and minimize production of strong wastes, and treatment of wastewater. There are now a number of municipal RO installations, one of the largest, in Riyadh, Saudi Arabia, having a capacity of about 12 mgd (31 m^3/min).

Results of RO treatment of a water supply to reduce the load on a demineralizer are shown in Table 15.3. Although the RO system is usually chosen for its economic benefit of reducing regeneration chemicals (Chapter 9), its added advantage of reducing organic matter and colloidal silica is becoming increasingly important and is often the prime reason for its selection as a demineralizer pretreatment unit. (See Chapter 20 on ultrapure water.)

The application of RO to wastewater treatment is illustrated by Figure 15.14, which shows an installation in a paper mill designed to reduce TDS in the mill discharge. Operating results from this unit are shown in Table 15.4. A 100-mgd (262 m^3/min) desalting plant at Yuma, Arizona, will include RO modules of several types for comparison of performance. The plant will desalt a portion of the Colorado River so that the salinity will be acceptable for agricultural uses on

TABLE 15.3 Typical RO Performance

	Constituent, mg/L	
	Raw	Finished
Hardness, as $CaCO_3$	380	20
Alkalinity, as $CaCO_3$	215	16
Total electrolyte, as $CaCO_3$	445	29
Silica, as SiO_2	25	3
pH	7.2	6.0
CO_2	25	25
CO, mg/L as O_2	6	0

downstream farms in Mexico. The largest seawater RO plant produces 3 mgd of potable water for the city of Key West, Fla. (Figure 15.15).

DIALYSIS

Dialysis is seldom used for the purification or renovation of water. However, this membrane-separation process can be used to reclaim strong chemical solutions, which eliminates a serious disposal problem. Thus, dialysis does have a use in water technology (Figure 15.16). In simple dialysis, the driving force is the concentration gradient across the membrane. The membrane has pores through

FIG. 15.14 RO system processing wastewater for recycle of concentrate and for use of permeate for effluent dilution. *(Courtesy of Green Bay Packaging Inc.)*

TABLE 15.4 Performance of Reverse Osmosis Unit Treating Pulp Paper Mill Wastewater

	February 25, 1976			July 21, 1976		
Constituent	Feed	Permeate	Concentrate	Feed	Permeate	Concentrate
Soluble solids*	54,400	372	66,900	47,600	246	61,355
Sodium*	6,500	62	8,000	6,000	58	7,750
BOD_5*	13,900	1,005	17,600	11,467	775	15,900
Color	112,500	65	146,800	83,000	52	108,000
pH	5.8	4.1	—	—	—	—
	(5.5–6.0)	(4.0–4.8)	—	—	—	—

* In milligrams per liter.

Note: Permeate is rejected from plant cycle, producing acceptable wastewater for discharge to municipal treatment plant with pH correction.

Source: William R. Nelson, Green Bay Packaging Inc., private communication.

which solutes can diffuse. Ions of large ionic radius diffuse through more slowly than those of smaller radius, so a separation of ionic species is possible. Solute or colloids too large to pass through the pores are retained in the concentrate. Some water may pass through in a direction opposite the flow of ions because of the osmotic pressure.

Even though almost any semipermeable membrane could be used in dialysis, the most commonly used membranes are made of a type of hydrated cellophane.

FIG. 15.15 At Key West, Fla., six banks of permeators process seawater with individual high-pressure pumps and controls to produce 0.5 mgd each of potable water. *(Courtesy of Water Services of America and DuPont.)*

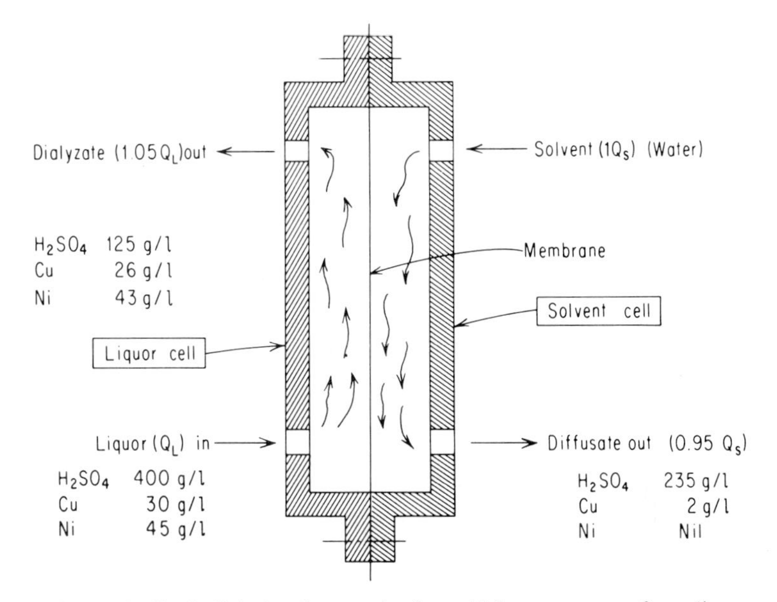

FIG. 15.16 Single dialysis cell separating free acid from a copper refinery liquor. *(From Chamberlin, N. S., and Vromen, B. H.: "Making Dialysis Part of Your Unit Operations," Chem. Eng., May 4, 1959.)*

This has been chemically altered to produce the surface properties and pore sizes needed for the specific separation. These membranes may also be stretched to change diffusion rates. Selectivity of a given membrane is largely related to its pore size. All of the problems discussed in connection with other membrane separation processes are also inherent in dialysis.

In the textile industry, dialyzers have been used to reclaim caustic from mercerizing baths; the NaOH passes through the membrane into the recovery water flow, and the organic residue is kept in the liquor for disposal.

ELECTRODIALYSIS

In electrodialysis (ED), the driving force is electrical. Semipermeable membranes having anion and cation exchange properties are alternately stacked in a press with narrow water passageways between them (Figure 15.17). When a direct current is applied to electrodes on either side of the stacked membranes, the anions migrate toward the anode and the cations toward the cathode. Since the cation exchange membrane allows only the passage of cations, and the anion exchange membranes the passage of anions, the alternation of membranes creates concentration and dilution in the alternate compartments of the stack, as shown in the figure. The flow rate through these narrow compartments or channels, the number of stacks used, and the amount of driving force (electric current and voltage applied) determine the amount of salt (cations plus anions) removed from the water (Figure 15.18).

To minimize current requirements, designers make these channels as narrow as possible. As the salt concentration decreases in the product water passageways,

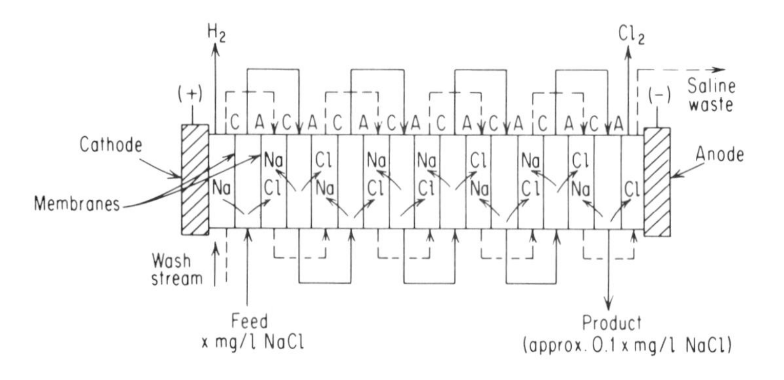

FIG. 15.17 Schematic drawing of a stack of permselective membranes assembled as an electrodialysis unit.

its electrical conductivity decreases, and the electric energy required to remove additional salt increases. If too much current is applied, some electrolysis (formation of H_2 and O_2 from water molecules) will occur, reducing overall efficiency.

Concentration polarization is even more limiting with electrodialysis than with other membrane separation processes. Concentration polarization causes the development of a membrane potential opposite to the applied potential. Concentration polarization and water-splitting, mentioned earlier, increase the tendency for scale formation as concentrations of calcium in the brine channels exceed the solubility of calcium carbonate or calcium sulfate, for example.

Calcium carbonate of magnesium hydroxide precipitates are most likely to form at the anion-exchange membrane-concentrate interface due to the higher pH (OH ions) there. The development of scale or deposits on the membrane surfaces reduces ion transfer and increases current requirements. A unique procedure called polarity reversal is effective in controlling this problem. Every 15 to 20 min, the electrical polarity of the stack is reversed, while at the same time the flows of feed, brine, and electrode water are reversed.

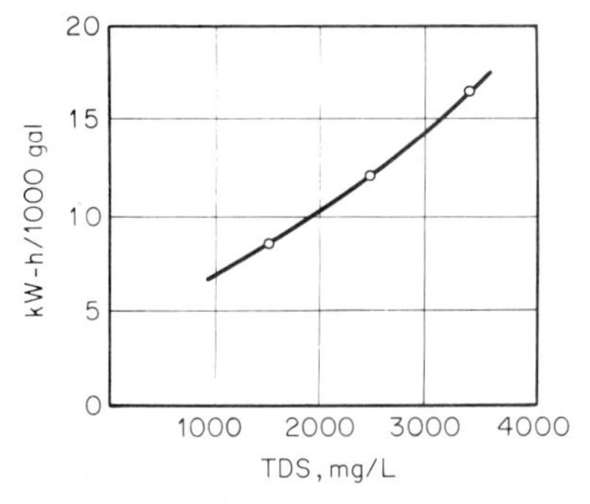

FIG. 15.18 Typical power consumption for electrodialysis as influenced by feed concentration (TDS, ionic) at 60°F with 83% TDS removal. *(Abstracted from literature of Ionics, Inc.)*

Improvements in membranes and stack designs have helped minimize the effects of concentration polarization. Methods of stabilizing scaling tendencies and cleaning stacks on the run to postpone shutdowns have been devised. Special cells have been designed for individual applications in attempts to increase reliability and reduce costs.

FIG. 15.19 Package electrodialysis plant, showing filter on feed, control system, and membrane stacks (left to right). *(Courtesy of Ionics, Inc.)*

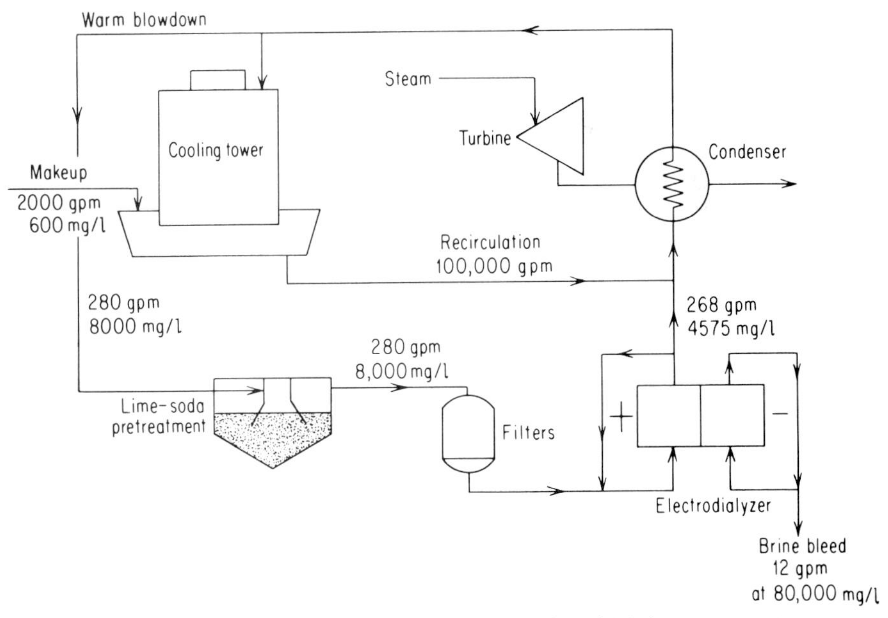

FIG. 15.20 A cooling-tower circuit with sidestream desalting with brine concentration. The figures above will vary depending on makeup water analysis, pretreatment capabilities, metallurgy restrictions of the cooling system, and operating considerations for the higher purity boiler water circuit.

15.19). One of the largest municipal plants produces over 3.5 mgd (10 m^3/min) for the city of Corfu in the Greek islands. Units are also installed in utility systems to reprocess cooling tower blowdown (Figure 15.20).

SUGGESTED READING

Argo, D. G., and Moutes, J. G.: "Wastewater Reclamation by Reverse Osmosis," *J. Water Pollut. Control Fed.,* **51,** (3), (590) (March 1979).

Birkett, J. D.: "Electrodialysis—an Overview," *Ind. Water Eng.,* September 1977.

"Calculation and Adjustment of the Langelier Saturation Index for Reverse Osmosis," ASTM D3739-78. *Annual Book of ASTM Standards,* Part 31. American Society for Testing and Materials Philadelphia. 1979.

Don Tang, T. L., Boroughs, R. D., and Chu, T. Y. J.: *Application of Membrane Technology to Power Generation Waters,* EPA Publication 600/7-80-063.

Katz, W. E.: *The Electrodialysis Reversal (EDR) Process,* International Conference on Desalination and Water Reuse, Tokyo, Nov.–Dec. 1977. Reprinted by Ionics, Inc., Watertown, Mass. 1986

"Mercury Recovery System Utilizes Ultra-filtration Technology to Minimize Metal Discharge," *Chem. Process,* Chicago, May 1982.

Milone, J. C., and Quinn, R.: "R-O Converts Ocean to Boiler Feedwater, *Power,* January 1980.

Southworth, F. C., and Applegate, L. E.: *The Role of Reverse Osmosis in Producing High Pressure Boiler Feed,* Proceedings of 38th American Power Conference, 1976.

Ultrafiltration Handbook, Romicon, Inc., 1983. Sales Publication of Romicon, Inc., Woburn, Mass.

CHAPTER 16
AERATION

Aeration is the mechanical process of providing intimate contact of air with water. As applied to water treatment, aeration transfers gas molecules, most notably oxygen, from air (gas phase) to water (liquid phase). While dissolving oxygen in water is most often the goal, aeration also includes removal of undesirable gases, such as CO_2 and methane from water, sometimes referred to as degasification.

Aeration is almost always accompanied by other processes or reactions, which may be physical, chemical, or biochemical in nature. Table 16.1 gives an overview of such processes with the common objectives and results of aeration of water and wastewater.

By far the greatest use for aeration equipment is in the field of biochemical oxidation of organic wastes, domestic or industrial. But, aeration is also used extensively for oxidation of inorganic impurities such as iron, manganese, and hydrogen sulfide, and for removal or oxidation of volatile impurities causing odor or bad taste. Aeration simply to increase oxygen content of water is sometimes carried out as the last step in a water or waste treatment plant.

Air is sometimes injected into water at high temperatures and pressures specifically to oxidize organics in wastewater. This process, known as wet oxidation, is discussed in Chapter 19.

MECHANISM OF AERATION

As a gas, oxygen is slightly soluble in water. At 20°C and sea level, saturation concentration is only 9.5 mg/L. At this concentration, it represents only 0.00095% of the weight of the water. But this concentration, when released as a gas, occupies 6.7 mL, or 0.67%, of the volume of the water which contained it. To correct gas solubility from weight to volume, use

$$\text{mL/L} = \frac{22.4}{\text{mol wt}} \times \text{mg/L}$$

As with most gases, solubility of oxygen in water is inversely proportional to temperature. Dissolved solids concentrations also affect dissolved oxygen (DO). Table 16.2 shows solubility of oxygen at various temperatures, elevations (which affect atmospheric pressure), and dissolved solids levels. As these data suggest, saturation values for oxygen in a particular water must be determined by aeration

TABLE 16.1 Aeration Processes in Water and Wastewater Treatment

Process	Simultaneous or subsequent reactions	Results	Examples
Aeration	None	Increase in dissolved oxygen content.	Oxygenation of streams. Post aeration of sewage plant effluent.
Degasification	Aeration	Increase in dissolved oxygen content; displacement of gaseous or volatile impurities.	Removal of CO_2, H_2S, methane, taste, and odor.
Aeration	Chemical oxidation	Oxidation of inorganic impurities; increase in dissolved oxygen content.	Removal of Fe, Mn, H_2S.
Aeration	Biochemical oxidation	Removal of organic impurities by biochemical digestion.	Sewage treatment BOD removal.

TABLE 16.2 Solubility of O_2 in Water as Influenced by Temperature, Elevation, and Salinity

Temperature		Elevation, ft				Salinity, mg/L*	
°F	°C	0	1000	2000	5000	400	2500
32	0	14.6	14.1	13.6	12.1	14.55	14.25
50	10	11.3	10.9	10.5	9.4	11.25	11.00
68	20	9.2	8.8	8.5	7.6	9.16	8.97
86	30	7.6	7.4	7.1	6.4	7.57	7.40

* These O_2 solubility values are at sea level.

tests. While it is not necessary to obtain data as extensive as those shown in Table 16.2, it is necessary to establish the ratio (beta factor) between oxygen saturation levels in a specific water and in pure water at the same temperature and pressure.

PRINCIPLES OF GAS TRANSFER

Aeration proceeds in three separate steps: (1) Air is brought into intimate contact with water by exposing a large surface area. This is created mechanically in the form of countless drops or small bubbles, depending on the type of aerator used. (2) Gas molecules pass across this surface into the liquid phase. The gas must pass through a thin barrier at the liquid surface, a liquid film, before it reaches the bulk of the liquid phase. The liquid underneath the film quickly becomes saturated with oxygen. (3) Gas molecules diffuse away from the liquid film into the bulk of the liquid until saturation is complete.

It is obvious that a large surface and turbulent conditions, which disrupt the liquid film and agitate the bulk liquid, increase transfer rates. The rate of oxygen transfer to the liquid body is expressed mathematically by the overall transfer equation:

$$\frac{dc}{dt} = K_La(C_s - C)$$

This says that the change of oxygen concentration with time, dc/dt, is equal to a transfer coefficient, K_La, times the oxygen deficiency in the liquid. Oxygen deficiency is the difference between saturation concentrations, C_s, and the prevailing oxygen concentration in the water, C, in milligrams per liter. The above equation is rearranged to

$$\frac{dc}{C_s - C} = K_La\, dt$$

and the integrated form is

$$-\ln (C_s - C) = K_Lat$$

Inserting limits gives the equation

$$\ln \frac{C_s - C_0}{C_s - C_t} = K_Lat$$

where C_s = oxygen concentration at saturation
C_0 = oxygen concentration at time zero
C_t = oxygen concentration at time t

A plot of $C_s - C_t$ versus time on a semilog graph will yield a straight line with a slope equal to K_La.

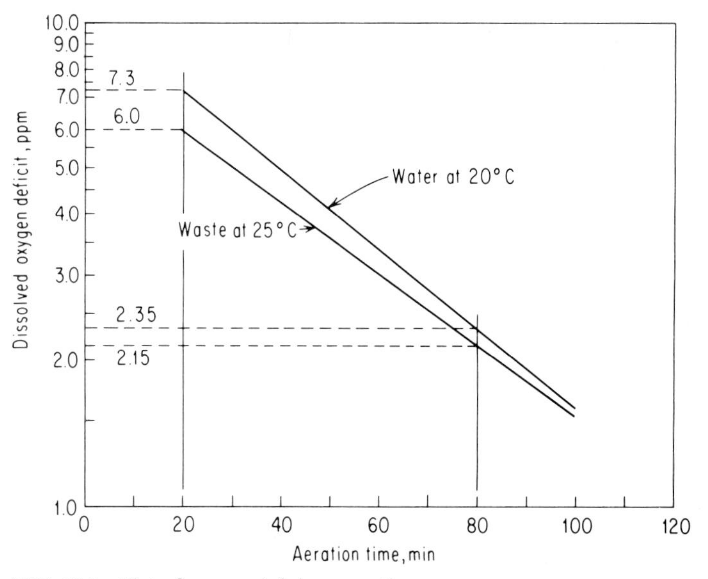

FIG. 16.1 Plot of oxygen deficiency vs. time.

TABLE 16.3 Aeration Test and Sample Calculations for Determination of Alpha Factor

Aeration equipment: diffused aerator; diffuser to be located 12 ft below surface
Temperature: 68°F, 20°C
Elevation: 1000 ft above sea level

Step 1:
Aeration Tests
Water
Dissolved oxygen readings, mg/L, and oxygen deficit versus time, minutes, based on saturated DO of 8.8 mg/L:

Time	0	20	40	60	80	100
DO	0	1.6	3.8	5.4	6.5	7.2
Deficit	8.8	7.2	5.0	3.4	2.3	1.6

Deficit plotted versus time on semilog graph (Figure 16.1)
Slope of straight line is equal to K_La for water

Waste
Waste aerated in same equipment as the water, at 25°C
Dissolved oxygen readings and deficit versus time, minutes assuming saturation at 6 h:

Time	0	20	40	60	80	100
DO	0	1.0	2.7	4	4.8	5.5
Deficit	7.0	6.0	4.3	3	2.2	1.5

Step 2:
Deficit plotted versus time on semilog graph, Figure 16.1

Step 3:
Slope of straight line is equal to K_La
To convert K_La_{25} to K_La_{20}, use the relationship

$$K_La_{25} = K_La_{20} \times 1.02^{(25-20)}$$

For water at 20°C,

$$K_La \text{ water } 20°\text{C} = \frac{\log 7.3 - \log 2.35}{1 \text{ h}} \times 2.3 = 1.132$$

For waste at 25°C,

$$K_La \text{ waste } 25°\text{C} = \frac{\log 6.0 - \log 2.15}{1 \text{ h}} \times 2.3 = 1.025$$

$$K_La \text{ waste } 20°\text{C} = \frac{1.025}{1.02^{(25-20)}} = \frac{1.025}{1.10} = 0.93$$

$$\text{Alpha factor} = \frac{K_La \text{ for waste}}{K_La \text{ for water}} = \frac{0.93}{1.132} = 0.82$$

Note: The effect of temperature on K_La depends on the type of aeration system. The relationship shown here applies to high ratios of air to liquid volume. With diffuser-type aerators in facultative lagoons, K_La may actually decrease with temperature rise.

Since the oxygen saturation concentrations are dependent on temperature, pressure, and dissolved solids as well as the nature of the solids, the rate of transfer of oxygen into water is also dependent on these variables. To use the equation for design purposes, it is necessary to perform aeration tests and to plot oxygen uptake versus time. When the differences between saturation levels and actual oxygen concentrations are plotted against time on semilog paper, the slope of the line becomes equal to the overall transfer coefficient, K_La (Figure 16.1). An example of such data collection and plotting is shown in Table 16.3.

The ratio of K_La for the wastewater and K_La for water, at the same pressure and temperature, is referred to as the *alpha factor*. This figure is very useful in sizing aeration equipment. An example of data collection and determination of the alpha factor is shown in Table 16.3.

EQUIPMENT

In freshwater treatment, gravity aerators and spray aerators have traditionally been used to remove iron, manganese, and hydrogen sulfide. Aeration techniques developed for wastewater treatment are applicable to fresh water, but capacity requirements are generally much greater.

Gravity or cascading aerators often resemble cooling towers, as the water is pumped over the top and allowed to cascade over wooden slats to a sump or basin.

The forced draft aerator is similar to a cooling tower in that air is forced upward against the counterflow of water over wooden slats. The coke-packed aerator, also a gravity flow unit, is popular in iron and manganese removal installations because of its simplicity.

In the area of wastewater treatment, greater capacities and higher efficiencies are required. The most widely used types of equipment are: diffused aerators, surface aerators (high and low speeds), and submerged turbine aerators.

In diffused or submerged aeration, air is forced through a diffuser, releasing small bubbles near the bottom of the aeration vessel. This establishes good contact between the oxygen and water at a pressure above atmospheric. It creates a large liquid-gas interface by producing small diameter bubbles through the small pores of the diffusers (Figure 16.2).

FIG. 16.2 Air diffusers in municipal sewage treatment plant. The diffuser assemblies may be retracted, as in basin on right, for maintenance and to prevent clogging when basin is out of service. *(Courtesy of FMC Corporation.)*

Efficiencies and power requirements for submerged aerators are closely related to the type of diffusers used. These can be porous, nonporous, or perforated pipe. The type of diffuser used is selected on the basis of oxygen transfer characteristics and maintenance requirements.

The porous type diffusers have found most use in municipal water treatment systems employing conventional activated sludge treatment. Oxygen transfer efficiencies as high as 10 to 12% may be achieved. The biggest problem is the tendency for clogging from the air side; clogging from the water side can also be a problem if the units are left in the liquor without air being bled through the system.

The nonporous diffusers do not clog as readily from the air side. However, oxygen transfer efficiencies (i.e., the ratio of oxygen absorbed to oxygen applied by pressurized air passing the diffuser) may amount to 4 to 8%. This means that power requirements may be as high or even higher than with porous diffusers.

FIG. 16.3 Details of construction of a fixed-position surface aerator, shaped to produce high-volume pumping and air entrainment. *(Courtesy of Infilco Degremont Inc.)*

Nonporous diffusers are usually constructed from nozzles or orifice plates made of metal or plastic.

The simplest type of diffuser is a perforated pipe laid along the bottom of the aeration vessel. This arrangement is used in ponds or narrow aeration tanks.

Mechanical surface type aerators have increased in popularity, especially in industrial waste treatment plants. While submerged aerators bring air into contact with the water, surface aerators operate in a reverse fashion, bringing the water into contact with the atmosphere. Such aerators actually lift large volumes of water above the surface and expose small liquid droplets to the atmosphere (Figure 16.3).

Basically, a surface aerator is a motor-driven impeller located at or beneath the liquid surface, mounted on fixed supports or on pontoons moored to the aeration basin. A draft tube is sometimes used to improve pumping capacity.

Two designs of impellers are used, updraft and plate-type vanes. The updraft impeller moves large volumes of water in the upward direction while the plate aerator causes the water to be scattered outward.

A combination surface-submerged aeration device is sometimes used in units designed for high loadings. These are furnished with a diffuser located at the bottom of the aeration tank with an impeller immediately above it. This combination can transfer more oxygen per unit volume than similar devices. It is usually applied to deep tanks, complicating maintenance because a steady bearing is required at the bottom.

Typical installations and flow patterns of submerged, surface, and combination aerators are shown in Figure 16.4.

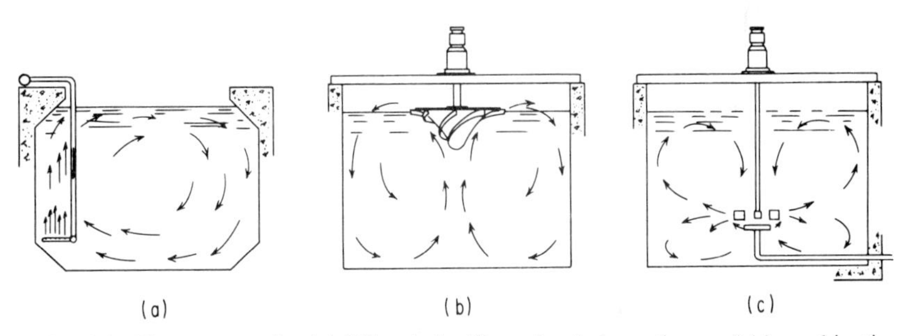

FIG. 16.4 Flow patterns for (*a*) diffused-air, (*b*) mechanical-aeration, and (*c*) combination turbine and diffuser-ring systems. *(From Chem. Eng., April 17, 1972, p. 97.)*

FIG. 16.5 Horizontal rotors with radial quills circulate and aerate wastewater in these oxidation ditches. *(Courtesy of Lakeside Equipment Company.)*

The brush aerator, also a surface unit, transfers oxygen by providing a liquid film on a wetted surface. This unit lends itself to aeration of streams or oxidation ditches. Mounted at a fixed level, it generates moderate turbulence and pumping action, with some spray of liquid from the rotating elements (Figure 16.5).

TABLE 16.4 Summary of Aerator Capacities

	Typical ratings, lb O_2/hp-h
Submerged aerators	
Porous diffusers	4
Nonporous diffusers	1.5
Surface aerators	
Plate	2.0–2.5
Turbine	3.0–3.5*
Propeller	2.5–3.5
Combination aerators	2–3

* Small units may be higher.

A summary of oxygen transfer rates of various types of aeration devices is shown in Table 16.4.

SIZING OF AERATION EQUIPMENT

Except in cases of degasification, aeration requirement is a function of degree of oxygen deficiency in the water and the oxygen-consuming reactions taking place. Of the examples given in Table 16.1, oxidation of iron and manganese requires much less oxygen on a pound-per-1000 gallons basis than biochemical oxidation of organic material (BOD). In addition, BOD concentrations in both industrial and municipal wastewater are generally much higher than the concentration of iron and manganese found in natural waters. Aeration capacity requirements for BOD removal, therefore, far exceed those of any other oxygen-consuming reaction.

The oxidation of both iron and manganese takes place most readily at a pH over 8. Theoretically, 1.0 mg/L O_2 oxidizes approximately 7 mg/L of either Fe^{2+} or Mn^{2+}. Oxidation of these metals requires increasing oxygen concentration to near saturation.

Oxygen requirements for biochemical oxidation processes (BOD removal) are well established for municipal applications. In cases where no industrial wastes are included with the sewage, aeration capacity is strictly a function of loadings and the type of treatment process to be applied. Some plants provide more or less aeration capacity depending on the extent to which the sludge is being aerated and the amount of denitrification required.

In industrial applications, as in municipal, aeration requirements are functions of BOD loadings and extent of sludge oxidation. However, other factors such as variations in loadings and waste composition may also demand greater aeration design capacities. Pilot studies are usually required.

Oxygen transfer rates of aeration devices are expressed in pounds of oxygen per horsepower per hour. Aerator horsepower requirements can be calculated once the loading, flow rates, and oxygen uptake requirements are known. Examples of such calculations are below.

Sizing Aeration Equipment

For simple aeration applications, horsepower requirements are expressed by:

$$\text{hp} = \frac{Q \times d \times L}{24 \times q}$$

where Q = flow, mgd
d = density of liquid, 8.34 lb/gal for water
L = loadings or oxygen demand in mg/L
q = oxygen transfer rate in lb O_2/hp-h

Example. For addition of 6 ppm dissolved oxygen to a water stream flowing at 2.5 mgd using an aerator device having a transfer rate equivalent to 2.0 lb oxygen/hp-h, power input requirements will be:

$$\text{hp} = \frac{2.5 \times 8.34 \times 6}{24 \times 2} = 2.6 \text{ hp}$$

The same equation may be used for biochemical oxidation of soluble BOD.

Example. For oxidation of 240 ppm BOD in a 15 mgd sewage plant, using aeration equipment rated at 2.5 lb O_2/hp-h, power input requirements will be:

$$\text{hp} = \frac{15 \times 8.34 \times 240}{24 \times 2.5} = 500 \text{ hp}$$

In practice, for reasons of sludge oxidation requirements and overall safety design factors, as much as 50% extra capacity above that required by the soluble BOD may be furnished. Similar considerations apply to industrial installations.

SUGGESTED READING

Boyle, W. C., Berthouex, P. M., and Rooney, T. C.: "Pitfalls in Parameter Estimation for Oxygen Transfer Data," *Proc. Am. Soc. Civil. Eng., J. Envir. Eng. Div.,* **100** (EE2), 391 (1974).

Chao, A. C., Galarraga, E., and Howe, R. H. L.: "Re-Evaluation of the Oxygenation Coefficient-Temperature Relationship for Waste Treatment and Stream Modeling," Proceedings of the 13th Atlantic Industrial Waste Conference, 1981, Ann Arbor Science, Ann Arbor, Mich.

Kalinske, A. A., Shell, G. L., and Lash, L. D.: "Hydraulics of Mechanical Surface Aerator," *Water Wastes Eng.,* **5**(4), 65 (1968).

Schmit, F. L., and Redman, D. T., "Oxygen Transfer Efficiency in Deep Tanks," *J. Water Pollut. Control Fed.,* **47,** 2586 (1975).

CHAPTER 17
ADSORPTION

In typical fresh waters, the bulk of the constituents in a filtered sample are dissolved minerals and organic compounds, the latter in both molecular and polymerized forms. There are, however, significant amounts of nonionic materials present as colloids. Among these are silica, insoluble metal oxides, and organic compounds, such as color and taste- or odor-producing substances.

COLLOIDAL MATTER

The division between colloidal matter and particles of larger size in water is arbitrary. The generally accepted upper boundary for a colloid is a diameter of 1.0 μm. Above this size, particles such as fine silt have a measurable settling rate and can be removed by sedimentation, though the period for reducing the concentration to one-half its original may be as long as a week or a month. Particles smaller than 1.0 μm are kept in suspension by the impact of molecules and ions dissolved in the water. The zigzag movement may be observed under the microscope, and is known as Brownian motion.

As the particle size is reduced below 1.0 μm, the ratio of surface to volume and correspondingly the electric charge per unit weight increase. This further stabilizes the colloidal matter, preventing sedimentation. The lower limit of colloid size may be considered to be about 1 to 10 nm (1 nm = 0.001 μm, or 10^{-6} mm), close to the dimensions of large molecules in solution. Colloids are further classed as hydrophobic and hydrophilic, the former resisting wetting by water, and the latter being strongly attracted to water and therefore difficult to adsorb.

Adsorption is the physical adhesion of molecules—particularly organic molecules—or colloids to the surfaces of a solid, an adsorbent, without chemical reaction. In some respects, adsorption is similar to coagulation and flocculation. One distinction is that adsorption generally uses an adsorbent solid processed especially for water treatment; in coagulation and flocculation, the adsorbent is produced in situ by the reaction of a chemical, such as alum, with water.

Adsorbents may be either finely powdered materials, applied to water in a clarifier or ahead of the filter, or 0.5 to 1.0-mm granules contained in a vessel similar to a pressure filter. By far the most common adsorbent used in water treatment is activated carbon, used both in powdered and granular forms. Other adsorbents include a variety of clays, magnesium oxide, bone char, and activated alumina. Special ion exchange resins are also used.

Since adsorption is a surface reaction, a measure of the effectiveness of an adsorbent is its surface area. For carbon, the total surface area is usually 600 to

1000 m^2/g. This surface is negatively charged. In spite of this, the negative charge of most colloids in water does not deter adsorption of high molecular weight materials, since the molecular structure itself becomes the controlling factor as molecular weight increases.

Adsorbents are porous, and the size of the pores is important. This is measured by iodine number, methylene blue number, or molasses number. The iodine number, a simple titration most commonly used, measures pores passing colloids larger than 1 nm. For activated carbon, this number has a range of 650 to 1000.

TASTE AND ODOR CONTROL

Some tastes and odors are caused by mineral constituents in the water. Examples are the salty taste apparent when chlorides are present at 500 mg/L or more, and the "rotten egg" odor caused by hydrogen sulfide in some well waters. However, most objectionable tastes and odors in potable water are caused by biological activity (Table 17.1). Many species of algae, diatoms, and actinomycetes produce organic by-products, such as essential oils, which can be observed by microscopic examination. The release of these materials into water, particularly when large populations of organisms die, produces objectionable tastes and odors. The released oils are negatively charged colloids.

Usually these effects are seasonal. They are controlled by the addition of powdered activated carbon either to the coagulation basin or immediately ahead of the filters in a typical municipal water plant. If the water supply contains considerable organic matter, the plant may need prechlorination; because chlorine oxidizes activated carbon, the selection of the best point for carbon application must be made carefully. In these cases, it is often best to apply the carbon ahead of the chlorine application point, if possible, allowing 10 to 15 min for adsorption to take place. If applied ahead of the filters, because carbon is so finely divided, it is important that the filters be in good condition so that the carbon will not pass into the effluent. This final application of carbon ahead of filters may remove residual chlorine, so final chlorination after the filters might be required. The carbon dosage may be as high as 50 mg/L if applied prior to clarification; it should not exceed 5 mg/L if applied ahead of filters.

TABLE 17.1 Sources of Tastes and Odors

Cause	Percent of cases	Treatment, % of plants reporting		
		Carbon	Chlorination	Aeration
Algae*	82	82	15.5	2.5
Decaying vegetation	67	85	13.8	1.2
Trade wastes	38	61	35.7	3.3
Other	23	85	13.2	1.8

* Relative frequency of occurrence:
Diatomaceae: Asterionella, Synedra
Protozoa: Synura, Dinobryon
Cyanophyceae: Anabaena, Aphanizomenon
Chlorophyceae: Volvox, Staurastrum
Fungi: Crenothrix

Source: Sigworth, E. A.: "Control of Odor and Taste in Water Supplies," *dJ. Am. Water Works Assoc.*, **49**(12), 1507 (1957).

The test for determining the optimum powdered carbon dosage for taste or odor control—or for organic removal—is performed much like a jar test. The procedure for measuring odor is quite elaborate if carried out by APHA Standard Methods; however, practical modifications are possible as long as the basic principles of dilution, avoidance of fatigue, isolation from background odors, and testing by a panel of several analysts are followed. Different amounts of carbon are

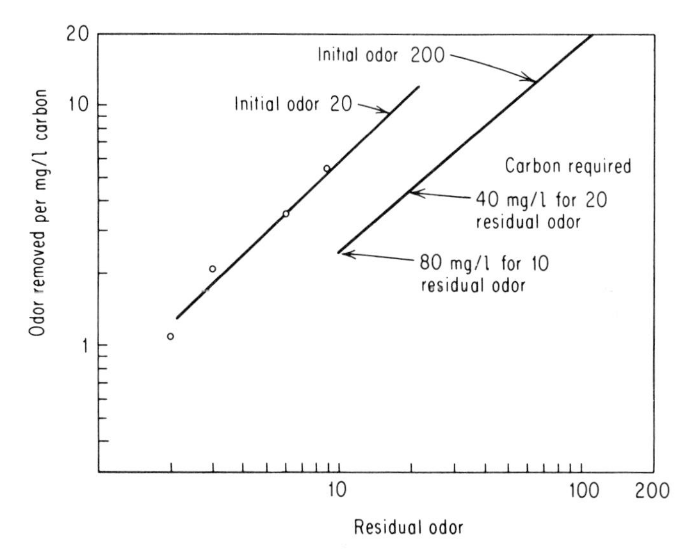

FIG. 17.1 Odor reduction with powdered activated carbon.

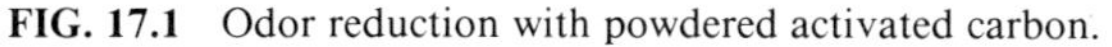

The equation expressing this data (the Freundlich isotherm) has the form

$$Q = K_F C_e^{1/n}$$

where Q = quantity removed per unit of carbon applied (e.g., mg/g C)
K_F = the Freundlich constant
C_e = the residual concentration, mg/L
$1/n$ = the Freundlich exponent

added to a series of jars, the jars are agitated for a controlled period corresponding to the detention that can be achieved in the plant system, and odors are determined on the treated samples after filtration. If the data, residual odor (X-axis) vs. odor removed (mg/L C) (Y-axis), are plotted on a log-log paper, a straight line results. Figure 17.1 is a typical plot.

REMOVAL OF ORGANIC MOLECULES AND COLLOIDS

If the organic content of the water is high, and relatively constant year-round, it becomes more economical to install pressure filters containing granular carbon or resin than to continuously apply the high dosages of powdered carbon needed to accomplish the same purpose. Typical granular carbon is about 8 to 30 mesh, having an effective size of 0.9 mm and a uniformity coefficient of 1.8.

TABLE 17.2 Typical Performance Characteristics of Activated Carbon Purifiers with Feed Concentration of About 1000 mg/L

Substance (nonpolar to polar)	% removal	Capacity, mg/g
Benzene (nonpolar)	95	80
Ethylbenzene	84	19
Butyl acetate	84	169
Ethyl acetate	51	100
Phenol	81	161
Methyl ethyl ketone	47	94
Acetone	22	43
Pyridine	47	95
Diethanol amine	28	57
Monoethanol amine	7	15
Acetaldehyde	12	22
Formaldehyde	9	18
Isopropyl alcohol	22	24
Methyl alcohol (polar)	4	7

The packed bed of adsorbent can be regenerated. If the organic material is volatile, the carbon bed may be regenerated by steaming. However, it is more conventional to remove the carbon and reprocess the material through a furnace. In large installations, the furnace is installed as part of the carbon installation; in smaller installations, the carbon may be discarded, or removed and returned to the manufacturer for reprocessing.

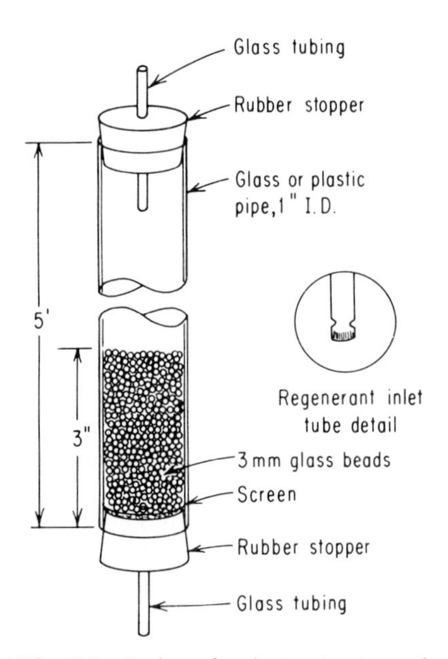

FIG. 17.2 Design of resin bead column for bench studies of adsorption process.

Packed beds of ion-exchange resin are regenerated with brine, causing the ion-exchange beads to shrink and expel the adsorbed organic matter.

Because adsorption reactions are relatively slow, particularly in cold water, packed beds are usually operated at lower flows than filters and ion-exchange systems, usually below 2 gal/min/ft^3 (0.27 m^3/min/m^3). Bed depths over 4 ft are common so that the retention time in the bed for adsorption may vary from 10 to 30 min. In evaluating the performance of granular beds of adsorbent, pilot plant studies are required. If organics are to be removed from a waste stream, the testing program may be extensive, since data must be obtained not only on the removal efficiency of the adsorbent itself, but also on the capability of the exhausted adsorbent to be regenerated and the relative performance of the various commercially available materials. Tests for commercial carbons used for color removal show variations in the K_F value from 6.9 to 13.4 and in $1/n$ exponent from 0.42 to 0.62.

Although adsorption may seem most often to be related to reaction of an adsorbent with colloidal substances, activated carbon and some other adsorbents react with molecules, too. This makes adsorption an effective waste treatment process in chemical, petrochemical, and other industries handling chemicals that may find their way into the plant sewer. Table 17.2 shows typical characteristics of granular activated carbon in processing specific organic compounds present in water.

When carbon is used, regeneration by reburning is a difficult process to evaluate by pilot plant; assitance by the carbon supplier may be needed. On the other hand, the testing of an ion-exchange-type adsorbent is simple, since regeneration is accomplished with brine.

The tests should be repeated for a long enough period of time to indicate whether the capacity of the resin is completely reclaimed each regeneration, and therefore to what extent and how frequently replacement of resin may be necessary. Tests of this kind are usually performed in glass tubes approximately 1 in (2.5 cm) in diameter, with a bed of 30 in (76 cm) of resin adsorbent. The tube is provided with end connections to permit normal downflow processing, backwashing for cleaning, and elution of accumulated organic material with brine. A typical test setup is shown in Figure 17.2.

A plot of organic removal by a packed column of resin, regenerated with brine, is shown in Figure 17.3.

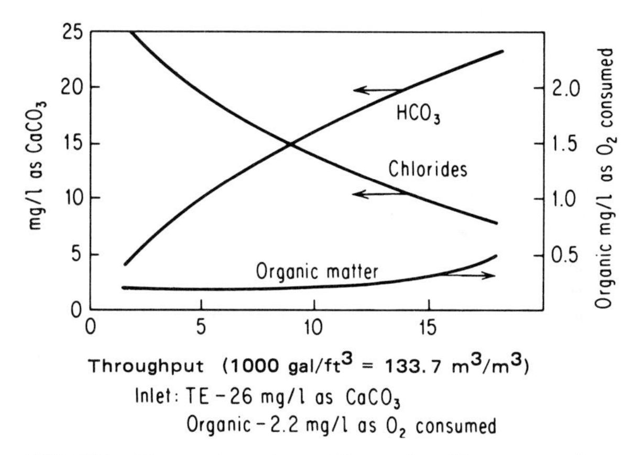

FIG. 17.3 Changes in water quality produced by passage of an organic-containing water through a brine-regenerated organic trap.

Where the material being removed is colloidal, the charge on the colloidal matter is influenced by pH, so careful control of pH may be needed in the adsorption process. It is possible for a sudden increase in pH to desorb the colloidal matter, rendering the effluent quality poorer than influent.

COLLOIDAL SILICA

Although some silica in water exists as the bisilicate ($HSiO_3^-$) anion, it is likely that it is for the most part present as a negatively charged colloid. It can be removed by anion exchange in a demineralizing system, suggesting that it is ionic; it can also be adsorbed on precipitated iron hydroxide and magnesium hydroxide, following an adsorption isotherm, which suggests that it is colloidal.

The most common adsorbent for silica when treating water at typical surface water temperatures in a clarifier is iron hydroxide produced when iron salts are used as a coagulant. Relatively high dosages are required for removal, and achieving these dosages produces high sulfate salts as a by-product. Nevertheless, in

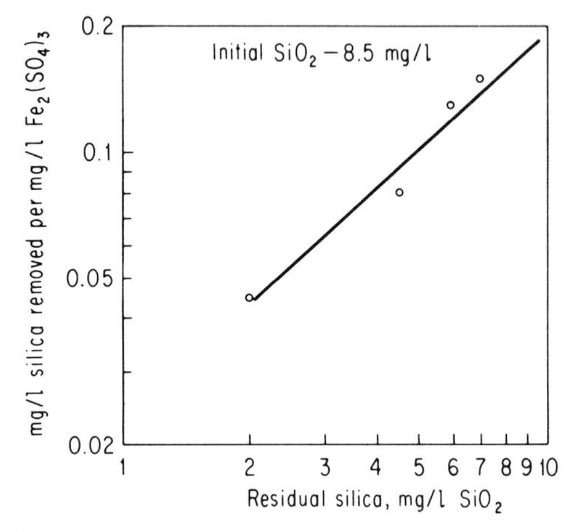

FIG. 17.4 Removal of silica by ferric sulfate at river water temperature.

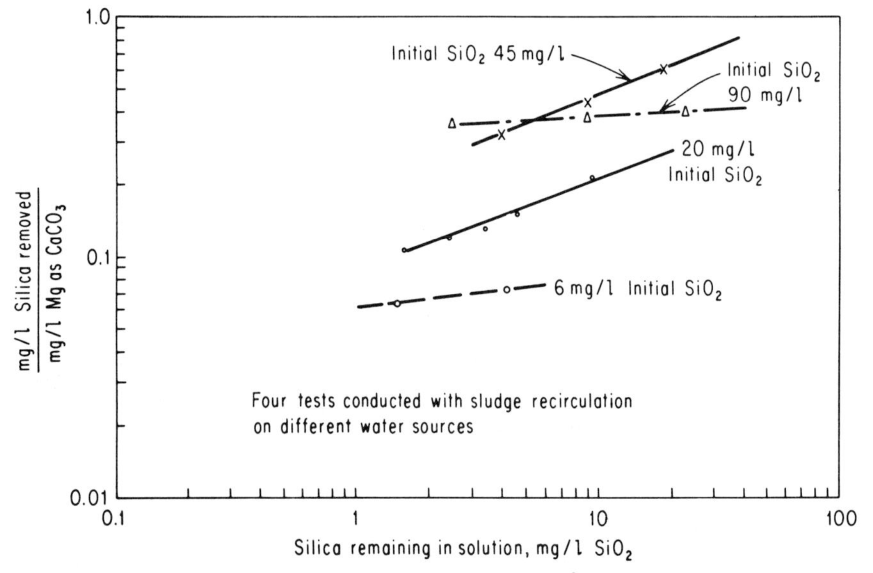

FIG. 17.5 Silica adsorbed by $Mg(OH)_2$ precipitated from Mg^{2+}, 100°C, pH 10.

some circumstances this is an effective process. Sodium aluminate has some effect at pH values over 8.5.

Silica is also removed in lime-softening processes. The magnesium hydroxide precipitated in the process is chiefly responsible for this. The process is inefficient at ambient surface water temperatures in a conventional lime softener with 1 h detention, but it becomes quite significant if the detention is increased to 4 h, the temperature increased to 120° to 140°F (44° to 60°C), or a favorable increase is made in both of these variables; it is very effective at typical hot-process softening temperatures in excess of 220°F (104°C). Contact time and density of the adsorbent sludge are important factors.

There is no way to predict with accuracy the effective dosage for silica removal, particularly where iron is used as the coagulant. This may be because there are other competing reactions. However, the jar test is reliable for anticipating the required chemical dosage for silica removal.

A typical isotherm showing the removal of silica at surface water temperatures with ferric sulfate is shown in Figure 17.4. A series of tests using magnesium as the absorbent are plotted in Figure 17.5. In all cases, the amount of magnesium required for precipitation is over twice what would be required if the reaction produced magnesium silicate.

APPLICATIONS OF CARBON FILTERS

Figure 17.6 shows carbon filters in the beverage industry following a lime softener. Accumulated solids require frequent backwash of the carbon filter beds. The density of carbon is only 25 to 30 lb/ft^3, so backwash rates must be kept low. This limits effective cleanup, so the water supply to the carbon bed should be clear.

Not only are the organic contaminants removed from water, but so also is residual chlorine, which is eliminated by the following chemical reaction:

$$2Cl_2 + 2H_2O + C \rightarrow CO_2 + 4HCl \quad (1)$$

In this chemical reaction, 1 mg/L Cl_2 reduces the alkalinity of the water by about 1.5 mg/L (as $CaCO_3$) by the production of HCl.

In the beverage industry, most bottling plants are under a franchise agreement that requires installation of water treatment facilities to reduce hardness and alkalinity, often involving cold-lime softening. Frequently high-chlorine dosages are applied to ensure complete sterilization. If the residual chlorine were not removed, it would destroy some of the organic materials used for flavoring. In the brewing and distillery industries, free chlorine would be objectionable in the fermentation process. So for these operations, the carbon filter is ideal for removing tastes and odors and for dechlorination. Because the water has been pretreated, the organic load is generally very low. The carbon may operate for 6 months to 2 years before exhaustion, and it is often more practical to replace the carbon bed with new material than to regenerate it.

Organic matter in water can cause permanent fouling of anion resin used in demineralizing service. The water supplied to demineralizers has generally received several steps of pretreatment, so that the application of carbon filters prior to demineralizing is very practical since the water is clear and the organic load is reasonably low. Since chlorine can damage ion-exchange resins, carbon has the same advantage in pretreating water for demineralization as in bottling plant operations, since it will dechlorinate the water as well as remove organics.

FIG. 17.6 This activated carbon filter *(right)* is used for the removal of organic matter and free chlorine. Two mixed media filters remove suspended solids to protect the carbon from fouling. *(Courtesy of Illinois Water Treatment Company.)*

The carbon filters are backwashed as needed, but regeneration is seldom practiced because of unfavorable economics. The exhaustion of the carbon can be determined by measurement of organic matter in the treated water.

WASTEWATER TREATMENT

The removal of organic matter from wastewater is in a completely different category from the earlier examples of carbon treatment, since the organic concentration may be quite high. Furthermore, it often comprises both molecular and colloidal organic matter. Most waste treatment plants in larger industries discharging directly to streams remove organic matter biologically in enormous aquariums in which bacterial cultures use dissolved organics as their food. However, bacteria have preferences in food, and those materials not readily digested—called refractory organics—may not be consumed at all. Carbon has the ability to adsorb many of the refractory materials which cannot be digested economically in a biological system. Powdered activated carbon is used in some biological digestion systems. The powdered carbon added to the water in aeration basins concentrates the food source so that bacterial colonies clinging to it can more efficiently convert the organic food to biomass or all the way to CO_2.

Because the organic content of wastewaters may be several orders of magnitude above that of fresh water, the economics of carbon treatment of wastewaters is vastly different from the treatment of fresh water. It is essential that the carbon system be designed for thermal regeneration. In such a system, the carbon may pick up 0.2 to 0.4 lb (kg) of organic matter for each pound (kg) of its own weight before needing regeneration. The adsorbed organics are usually burned off in a regeneration furnace. Approximately 5% of the carbon adsorbent is lost on each pass through the furnace. The performance and costs can be evaluated only by pilot plant operations (Figure 17.7).

When carbon beds are used for polishing the effluent from a biological system, the reduction by the biological system of the organic content to levels in the range

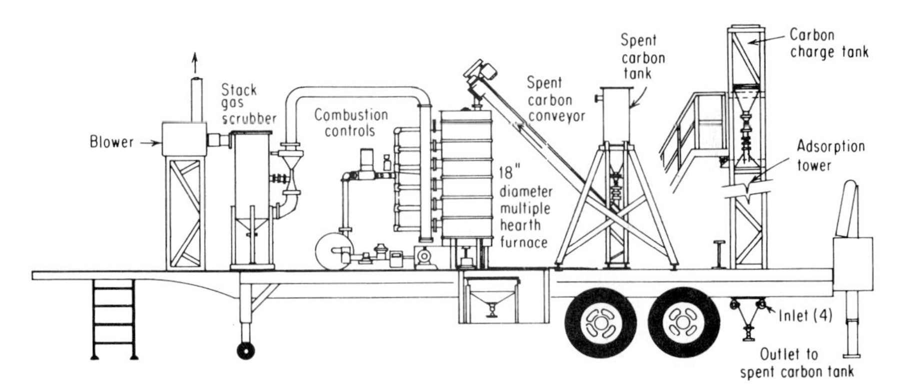

FIG. 17.7 A trailer-mounted activated carbon pilot plant used for wastewater treatment studies. *(Courtesy of Illinois Water Treatment Company.)*

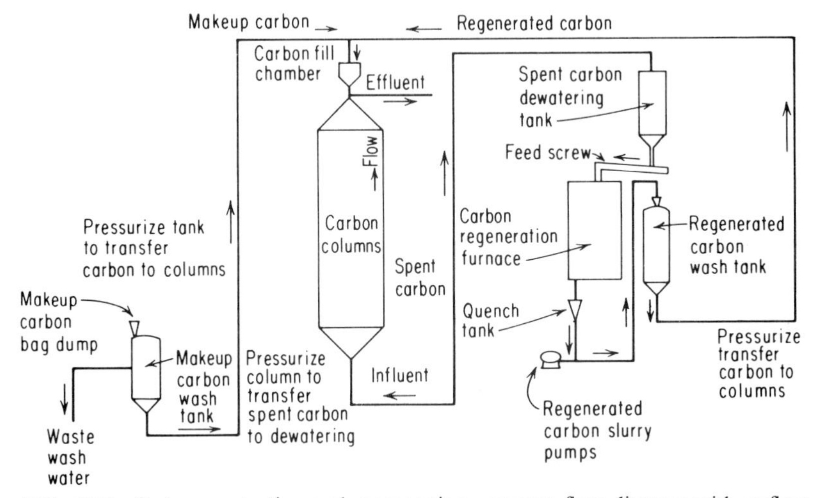

FIG. 17.8 Carbon contacting and regeneration—process flow diagram with upflow contactors.

of 10 to 20 mg/L organic as COD may permit the installation of conventional downflow carbon filters. However, where the carbon may be required to produce high-quality effluent from a fairly high concentration of organic matter in the influent, countercurrent operation is desirable so that the final effluent leaving the carbon bed will be in contact with the freshest carbon. A flow diagram illustrating such a countercurrent system is shown in Figure 17.8.

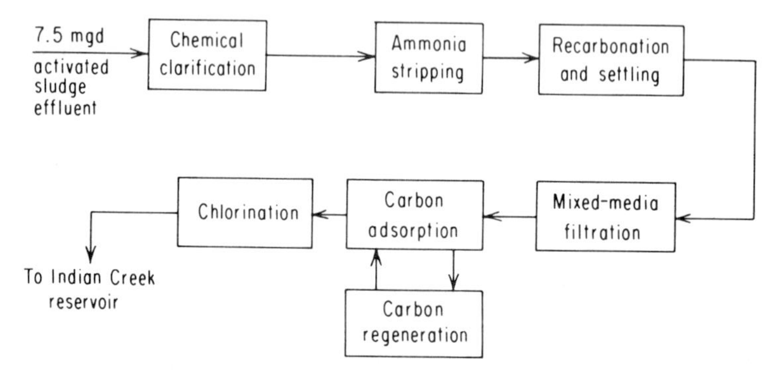

FIG. 17.9 Tertiary treatment schematic—South Lake Tahoe, Calif.

An activated carbon system has been operating at the South Lake Tahoe sewage plant since 1968 as part of the polishing system following conventional activated sludge sewage treatment. The carbon column is an upflow packed column having a total bed depth of 14 ft (4.2 m) and providing about 20 min residence time. The flow sheet for the plant is shown in Figure 17.9. With this system, the water quality at various stages of treatment is shown in Table 17.3.

TABLE 17.3 Water Quality at Various Stages of Treatment at South Lake Tahoe

Quality parameter	Raw wastewater	Effluent: Primary	Secondary	Chemical clarifier	Filter	Carbon	Chlorinated final
BOD (mg/L)	140	100	30		3	1	0.7
COD (mg/L)	280	220	70		25	10	10
SS (mg/L)	230	100	26	10	0	0	0
Turbidity (JTU)	250	150	15	10	0.3	0.3	0.3
MBAS* (mg/L)	7	6	2.0		0.5	0.10	0.10
Phosphorus (mg/L)	12	9	6	0.7	0.10	0.10	0.10
Coliform (MPN)†	50×10^6	15×10^6	2.5×10^6		50	50	<2.0

* Methylene blue active substance: detergents, surfactants.
† Most probable number: colonies/100 mL.

FIG. 17.10 A 2000 gal/min countercurrent activated carbon system treating refinery wastewater. Each column is 10 ft 0 in in diameter with an overall height of 65 ft 0 in and bed depth of 45 ft 0 in. A 5 ft 0 in diameter multiple hearth furnace regenerates spent carbon. *(Courtesy of Illinois Water Treatment Company.)*

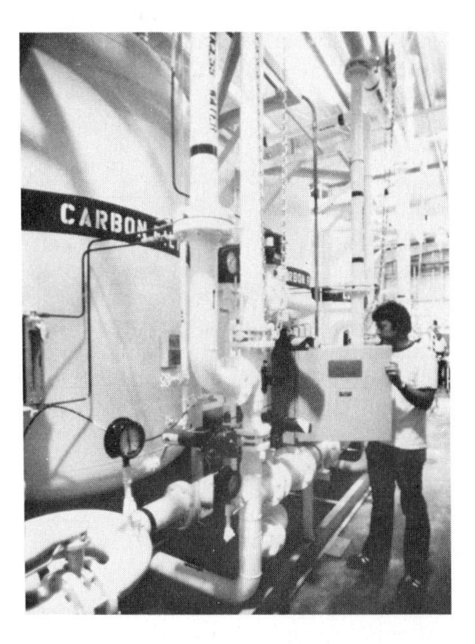

FIG. 17.11 A packaged carbon adsorption system with provision for carbon removal for regeneration at a separate servicing facility. *(Courtesy of Calgon Corporation.)*

In the operation of this system, approximately 200 lb (90 kg) of carbon are regenerated per million gallons (3785 m^3) of water treated. The organic material removed corresponds to a carbon loading of 0.4 lb as COD per pound of carbon regenerated. Activated carbon for organic removal is equally important in industrial waste treatment. The carbon is capable of removing many organic materials that are not biodegradable, as in refinery wastewaters (Figure 17.10). Large plants have their own complete facilities, including regeneration furnaces. Where this may not be economical, the carbon supplier may install the treatment units and provide regeneration as a service (Figure 17.11).

TABLE 17.4 Variations in Adsorption Rates Using Granular Activated Carbon for a Few EPA Priority Pollutants (Calgon Filtrasorb 300)

Priority pollutant	K_F	$1/n$
Dieldrin	606	0.51
Hexachlorobenzene	450	0.60
PCB-1221	242	0.70
2,4,6 Trichlorophenol	219	0.29
Phenol	21	0.54
Carbon Tetrachloride	11	0.83
Chloroform	2.6	0.73
Acrylonitrile	1.4	0.51
Benzene	1.0	1.6

Source: Weber, Walter J., Jr.: "Application of Adsorption to Wastewater Treatment," *Ind. Water Eng.*, July/August 1981.

Many of the EPA-selected priority pollutants can be removed from wastewater by activated carbon. To illustrate the extremes in adsorption rates, the Freundlich isotherm constants of a few of these are listed in Table 17.4, emphasizing the need for careful pilot plant studies to obtain design data for a carbon adsorption system.

RESIN COLUMNS

Special ion exchange resins have been developed to perform as adsorbents in the removal of organics from water. In the application described earlier, where resin is used to protect a demineralizer against fouling, the equipment utilizing the resin is almost identical to a common ion-exchange unit. The resin is sensitive to temperature change, and the slower rates of reaction at low temperatures require that the flow rates be selected for the coldest water anticipated. Similar units have been designed for the extremely high organic loading found in kraft pulp mill wastes, the resins being particularly designed for lignin and similar organics found in these wastewaters.

SLURRY SYSTEMS

The adsorption of silica by iron floc or magnesium hydroxide produced by lime softening requires the maintenance of a high sludge density for efficient performance. Of the varieties of coagulation basins available for water treatment, those that provide maximum contact of the water to be treated with dense sludge produce the best silica reduction.

Of the variety of miscellaneous adsorption processes, in addition to those discussed, fluoride removal is of some importance. Lime precipitation rarely reduces fluoride below 10 mg/L. Further reduction can be obtained by adsorption on magnesium oxide in a slurry-type contactor, or by passage of the fluoride-bearing waste through a column of hydroxyapatite. Experiences with these processes are limited, and bench tests must be run to obtain data for design and for estimation of costs.

SUGGESTED READING

AWWA: *Water Quality and Treatment,* McGraw-Hill, New York, 1971.

Baylis, J. R.: *Elimination of Taste and Odor in Water,* McGraw-Hill, New York, 1935.

Hassler, J. W.: *Activated Carbon,* Chemical Pub., New York, 1963.

Iler, Ralph K.: *The Chemistry of Silica,* Wiley, New York, 1979.

Lin, S. D.: "Sources of Tastes and Odors in Water," *Water Sewage Works,* **6,** 101–104 (1976); **7,** 64–67 (1976).

O'Brien, R. P., and Fisher, J. L.: "There Is An Answer to Groundwater Contamination," *Water Eng. Manag.,* May 1983.

Randtke, S. J., and Snoeyink, U. L.: "Evaluating GAC Adsorptive Capacity," *J. Am. Water Works Assoc.,* **75,** 406–413 (1983).

Robeck, G. G., Dostal, K. A., and Cohen, J. M.: "Effectiveness of Water Treatment Processes in Pesticide Removal," *J. Am. Water Works Assoc.,* **57,** 181–199 (1965).

Sigworth, E. A.: "Control of Taste and Odor in Water Supplies," *J. Am. Water Works Assoc.,* **49,** 1507 (1957).

Weber, W. J., Jr., and Morris, J. C.: "Equilibria and Capacities for Adsorption on Activated Carbon," *J. San. Eng. Div. Proc. Am. Soc. Civil Eng.,* **SA3,** 79–107 (1964).

CHAPTER 18
EVAPORATION AND FREEZING

Evaporation and freezing can be used to convert water (1) to a pure vapor that can be condensed or (2) to a pure solid that can be separated from a saline mother liquor and melted. Both processes leave saline residues containing essentially all the solute originally in the feed water.

EVAPORATORS

Evaporators are widely used in many water treatment operations, such as preparation of boiler feed water, concentration of diluted liquor, evaporation of seawater to produce fresh water, and concentration of waste liquors to reduce volume for further processing or disposal.

The typical evaporator is like a fire-tube boiler, with the flame replaced by steam or process vapor as the heat source; but, there are significant differences:

1. The evaporator has a much smaller temperature gradient across the heat transfer surfaces.
2. It usually holds less liquid.
3. Evaporator tubes are usually made of special metals (titanium) or alloys (stainless steel), whereas boiler tubes are made of steel.

To transfer heat efficiently at low temperature gradients, the evaporator surfaces must be kept free of deposits, which have an insulating effect. Correct chemical treatment and scheduled cleaning are important.

Every liquid exerts a vapor pressure, the magnitude of which is a measure of its volatility. High vapor pressure liquids evaporate readily, while those with low vapor pressures evaporate more slowly, requiring an increase in temperature to speed the rate. The kinetic energy of all molecules increases with increasing temperature. The rate of evaporation depends on the nature of the substance, the amount of heat energy applied to the liquid, and surface effects.

When a liquid reaches the temperature at which its vapor pressure equals atmospheric pressure, boiling occurs. This is the rapid evaporation from all parts of the liquid mass, with bubbles of vapor forming in the interior and rising to the surface. The pressure within these bubbles equals the vapor pressure of the liquid at that temperature, so the boiling point depends on the external pressure. For example, at sea level, pure water boils at 212°F (100°C) where its vapor pressure is 1 atm (1 bar), or 14.7 lb/in^2 (1.0 kg/cm^2). If the external pressure is reduced, as occurs at elevations above sea level or under vacuum, water boils at a lower temperature.

EFFECT OF SALT CONCENTRATION

Since evaporators may process liquids other than pure water, factors other than atmospheric pressure must also be considered. Soluble salts in the solution decrease the vapor pressure, elevating the boiling point. Therefore, as dilute liquor evaporates and becomes more concentrated, its boiling point rises. Figure 18.1 shows the boiling point elevation as the concentration of a salt increases in aqueous solution.

As water is evaporated from a solution and the liquid becomes more concentrated, it is possible to concentrate to the point where the solubility of the salts is

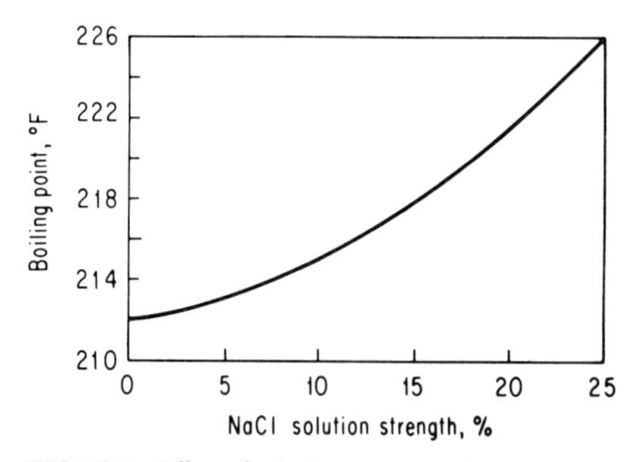

FIG. 18.1 Effect of solution concentration on atmospheric boiling point using NaCl as an example. For any specific concentration, the solution boiling points at several pressures plotted against the boiling points of water at these same pressures produce a straight line.

exceeded. This results in precipitation, usually as scale on the heat transfer surfaces. Where water is being evaporated, the scale may consist of salts of calcium, magnesium, and silica. This scale severely reduces the heat transfer rate, slowing evaporation and reducing thermal efficiency.

It takes a lot of heat to evaporate water. Raising the temperature of 1 lb of water 1°F requires 1 Btu; to change that 1 lb into vapor at atmospheric pressure requires 970 Btu. (It takes 1 cal to raise 1 g of water 1°C, 539 cal to vaporize 1 g at atmospheric pressure.) The high energy requirement for evaporation makes it important that the heat balance of a plant be controlled for maximum use of energy. A typical evaporator usually receives heat from live steam or from steam bled from a turbine.

EVAPORATOR DESIGN

There is a large variety of designs of evaporators, although the majority work on the principle of steam passing on the outside of a series of tubes with water or water solution, either confined or recirculated, flowing as a thin film over the inside of the tubular heating surface. The various types of evaporators are classified according to the way the water is vaporized:

1. *Boiling type:* Evaporators which heat water to the boiling point and evaporate it by applying an external heat source.
2. *Flash type:* Evaporators which superheat water by an external heat source and flash it into vapor.
3. *Compression type:* Evaporators which add energy to water vapor by compression and return this to the evaporator body as the heat source for boiling.

In the submerged tube boiling-type evaporators (Figure 18.2), steam enters a tubular element, boils water, and discharges water vapor from the evaporator shell. The heating elements are usually bundles of tubes of various configurations.

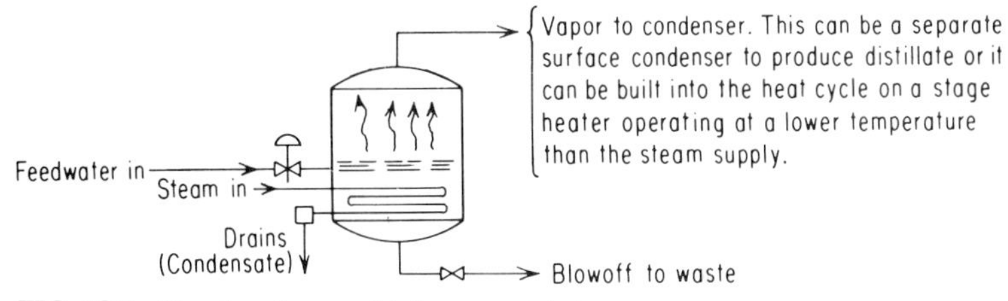

FIG. 18.2 Simple submerged-tube evaporator.

These may be completely submerged in the water, partially submerged, or arranged so that only a film of water flows across the surfaces. In each design, the space for vapor generation must be ample to avoid steam blanketing and to prevent fouling with baked on sludge.

As in a boiler, bleed-off regulates the solids concentration of the boiling liquid. Vapor-purifying devices trap entrained water droplets. This is particularly important where the aim of evaporation, as in most water treatment systems, is to produce high-quality distillate. The vapor purifiers are comparable to those in boilers. In addition to the conventional designs, bubble cap purifiers are sometimes used. These return part of the distillate to continually wash fresh vapor.

A boiling-type evaporator with proper disengaging area should produce distillate with less than 1 mg/L total dissolved solids. The quality is affected by the dissolved solids content of boiling water which may be entrained in the vapor discharge. Lower solids levels are attainable with more sophisticated vapor purifiers and conservatively designed evaporator elements. Vapor quality is affected by the CO_2 liberated from the bicarbonate alkalinity, just as in boiler operation.

CONDENSATION

Purified vapor leaving the evaporator is condensed in several ways:

1. In older utility systems where evaporators were used to provide high-quality makeup, the vapor was discharged through the deaerating heater and condensed by the boiler feedwater (Figure 18.3). (Modern utilities use demineralizers instead of evaporators to process makeup.)

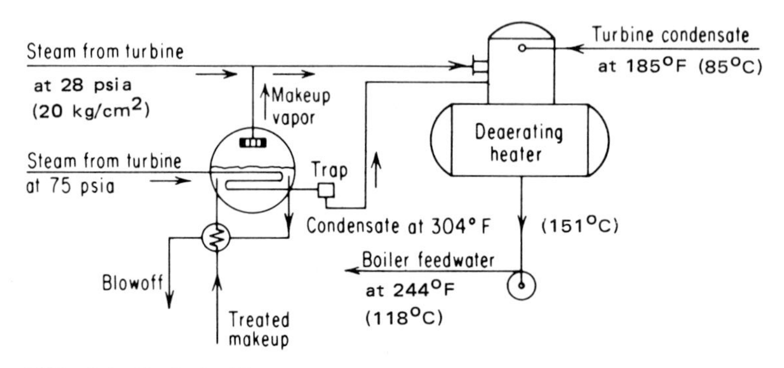

FIG. 18.3 Typical utility-type evaporator operation.

2. Vapor may be condensed by a surface condenser if the purified liquid phase is to be kept separated for some reason. Each pound of vapor becomes 1 lb of distillate in the condenser shell.

3. The vapor may be fed to the tube element of the second evaporator body, and the vapor from this second unit fed to a third, producing a multiple-effect evaporator (Figure 18.4). In this type of multiple-effect evaporation, vapor

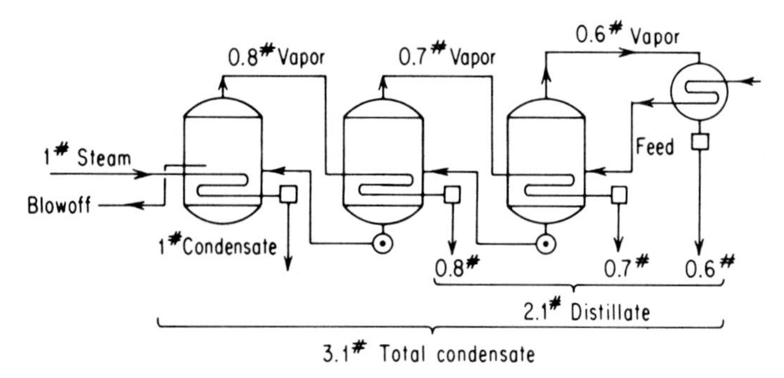

FIG. 18.4 Multiple-effect evaporator with condenser.

from the last unit is liquefied in a condenser. Each pound of fresh steam fed to the first stage theoretically produces 1 lb of condensate from each stage. In practice, however, a triple-effect evaporator produces about 3.1 lb of total condensate per pound of steam instead of 4.0, the total condensate including that produced by the fresh steam applied to the first effect.

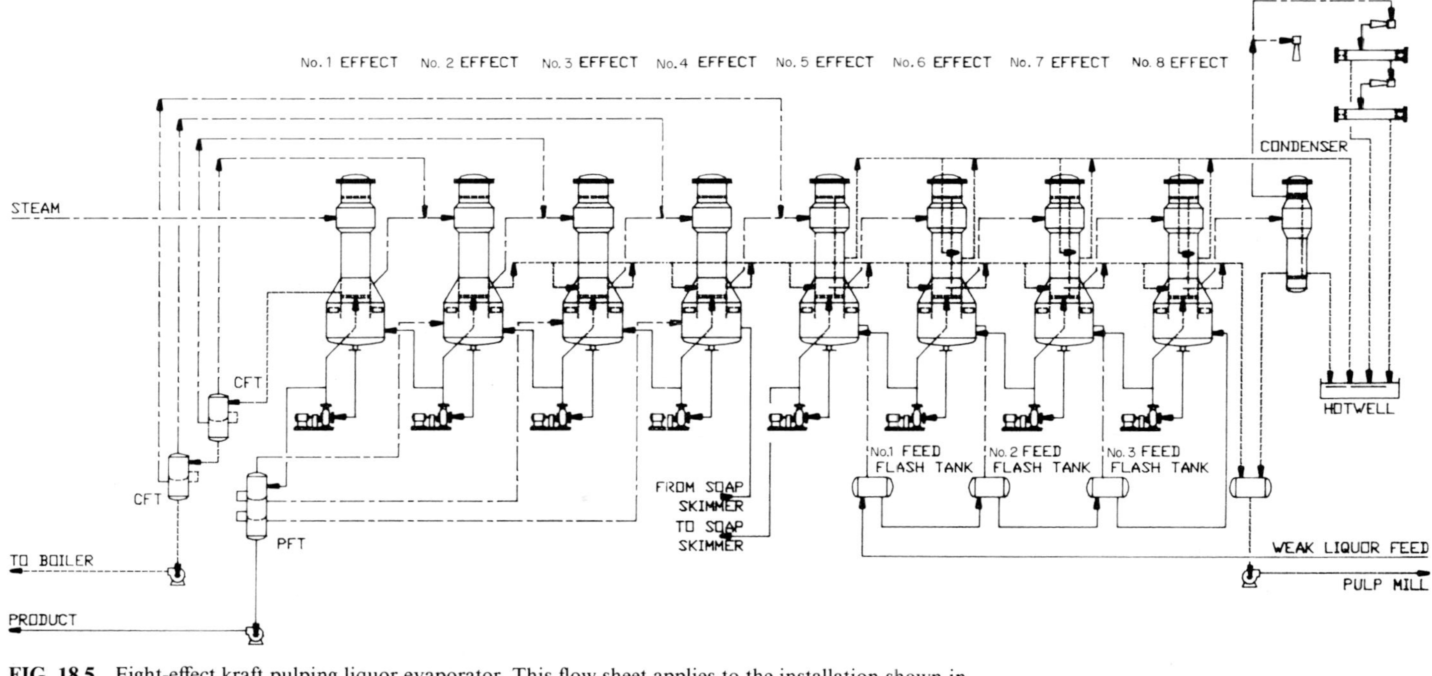

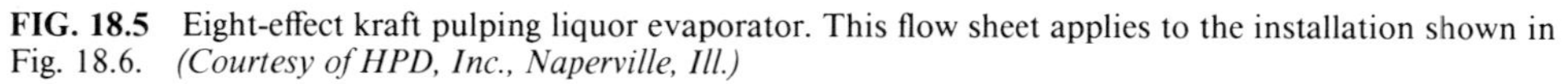

FIG. 18.5 Eight-effect kraft pulping liquor evaporator. This flow sheet applies to the installation shown in Fig. 18.6. *(Courtesy of HPD, Inc., Naperville, Ill.)*

TABLE 18.1 Typical Evaporator Controls Following Pretreatment

Reading	Range	Where and how maintained
TDS, mg/L	2500 max	In evaporator, by blowdown
SiO_2, mg/L	100 max	In evaporator, by blowdown
Hydroxide, mg/L $CaCO_3$	150–200	In evaporator, by blowdown or by NaOH feed
Dispersant and antifoam	Trace	In evaporator, by chemical feed
PO_4, mg/L	30–60	In evaporator, by phosphate-organic treatment
SO_3	30–60	In evaporator, by chemical feed
pH	8.2–8.6	In vapor, by amine treatment

Note: Where there is no pretreatment of evaporator makeup, chemicals should be fed for internal treatment just as for a low-pressure boiler, as shown by Table 18.2.

TABLE 18.2 Evaporator Controls—Internal Treatment Only

Reading	Range	Where and how maintained
TDS, mg/L	1500 max	In evaporator, by blowdown
SiO_2, mg/L	100 max	In evaporator, by blowdown
Hydroxide, mg/L $CaCO_3$	150–250	In evaporator, by blowdown or by chemical feed
Dispersant and antifoam	Trace	In evaporator, by selecting proper treatment combination
PO_4, mg/L	0	
pH	8.2–8.6	In vapor, amine treatment
SO_3, mg/L	If the evaporator is used intermittently, a residual SO_3 of 30–60 mg/L should be maintained to protect the evaporator shell from corrosion	

Multiple-effect evaporators are used principally for chemical process operations. Typical of such evaporators are the units found in the pulp industry for the concentration of sulfate black liquor (Figure 18.5).

The primary aim of evaporation in the power plant is to produce boiler makeup of high quality. The chemical treatment program must be designed to produce high-purity vapor while helping to maintain clean heat transfer surfaces. As in boiler water conditioning, the goals of chemical treatment are control of carryover, prevention of deposits, and elimination of corrosion. The makeup to most utility evaporators is pretreated to remove hardness, reduce alkalinity, and eliminate dissolved oxygen. The chemical treatment applied to the evaporator should be controlled to maintain the limits shown in Table 18.1.

Where there is no pretreatment of the evaporator makeup, then the chemical program should be the same as used for internal treatment of a low-pressure boiler, as shown in Table 18.2. Where the feed to the evaporator is brackish or seawater, it is often difficult to maintain a scale-free system even with a good internal treatment program. In such cases, the operation is programmed so that scale is allowed to build up on the heating tubes for a planned period, then the temperature of the system is suddenly dropped, creating a thermal shock that cracks the scale from the tube surface for removal from the bottom of the evaporator. Other evaporators are manufactured with bowed tubes which flex with temperature change, also resulting in scale-cracking and shedding.

MULTIPLE-EFFECT UNITS

Although multiple-effect evaporators are usually used for process operations, they have a definite tie-in to the utility system. In the pulp industry, five-, six-, or even seven-effect evaporators are used to concentrate water from the pulp washers for recovery of cooking chemicals. The black liquor may be concentrated to approximately 65% total solids, of which about half are organic materials. In this condition, the black liquor can be fired to a black liquor recovery furnace; the organic material supports its own combustion and smelts the cooking liquor salts to a recoverable form. Figure 18.6 shows a typical large installation of such a black

FIG. 18.6 Black liquor evaporator in a kraft pulp mill. Some of the condensate may be recovered for reuse, but much of it is contaminated with sulfur compounds. *(Courtesy of HPD Corporation.)*

liquor evaporator. Fresh steam fed to the first effect produces condensate that can be reused as boiler feed water. However, the condensate produced in subsequent stages is too contaminated by volatiles for such use, but may be used for brown stock washing, stock dilution, or other purposes. Some of this condensate is so foul that it must be stripped before it can be put into the sewer for treatment in the waste treatment plant. The vapors stripped from the foul condensate may be sent to the lime kiln where the organics responsible for the foul odors are burned.

The bauxite industry also uses multiple-effect evaporators for concentration of sodium aluminate liquors, producing more condensate than required for boiler makeup. Again the condensate is usually too contaminated to be directly usable as boiler feed water.

A final example is the beet sugar industry, where syrups are concentrated by evaporation, again producing an excess of condensate over boiler makeup requirements. These condensates are frequently contaminated with sugar, which is very detrimental to boiler operation, and ammonia, which is corrosive to systems that contain auxiliary equipment fabricated of copper alloys.

The advantage of multiple-effect in increasing yield per unit of energy can also be built into two quite different designs of evaporators which bear little resemblance to the standard multiple-effect evaporator. The first of these is the vapor compression still, which was developed initially for seawater evaporation aboard ship (Figure 18.7); the second is the multistage flash evaporator, which has

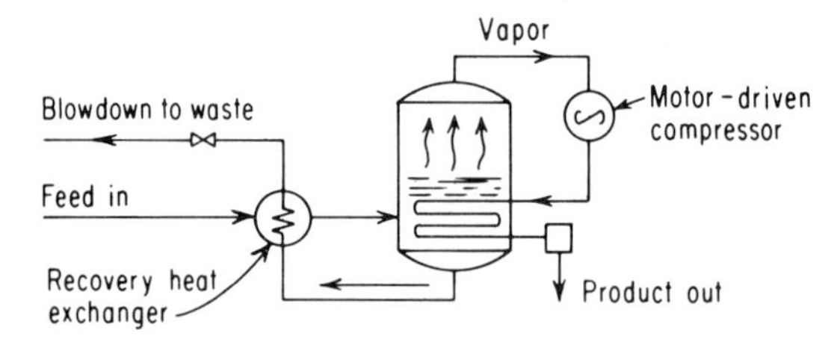

FIG. 18.7 Schematic of vapor compression still.

become popular for the production of potable water from brackish or seawater (Figure 18.8). Both of these designs work on low temperature differentials and must be kept free of deposits to maintain efficient heat transfer. Flash evaporators having a capacity of 7.5 mgd (20 m^3/min) have been installed for municipal water

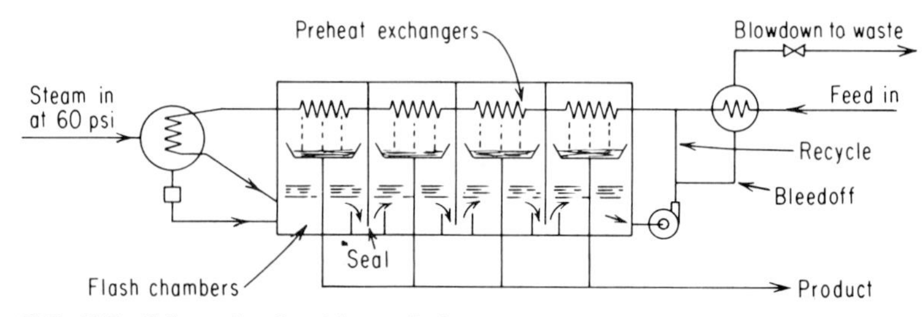

FIG. 18.8 Schematic of multistage flash evaporator.

supply in the Middle East, where energy costs are favorable for such an installation (Figure 18.9). Where seawater is being used as the feed, chemical treatment for prevention of calcium carbonate, calcium sulfate, and magnesium hydroxide scales is required. The treatment includes (a) reduction of alkalinity to minimize supersaturation of calcium and magnesium compounds, (b) application of scale-control agents, such as acrylates, polyphosphates, or combinations of these, (c) oxygen scavengers or other types of corrosion inhibitors, and (d) antifoams to protect the quality of the distillate. With this treatment it is possible to concen-

FIG. 18.9 Three flash evaporator modules, each with a capacity of 2.5 mgd (6.7 m^3/min), producing potable water for Al Khobar, Saudi Arabia. *(Courtesy of Aqua Chem Inc.)*

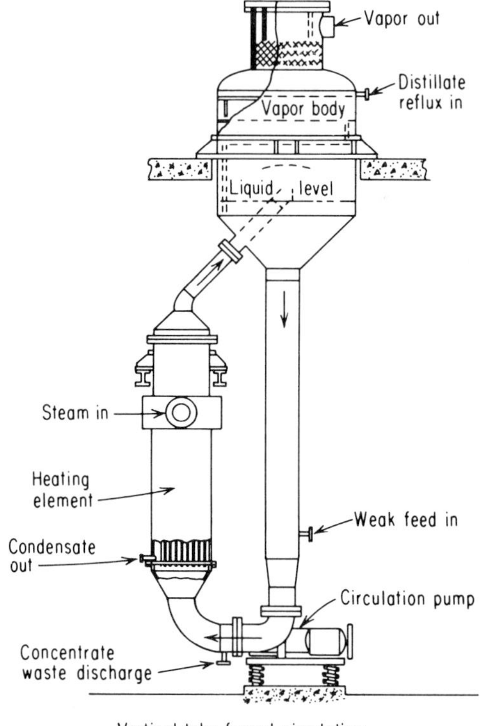

FIG. 18.10 A type of evaporator design used for concentration of radioactive wastes from nuclear fuel reprocessing. *(Courtesy of Unitech Division, Ecodyne Corporation.)*

FIG. 18.11 Compact evaporator designed for concentration and recovery of plating solutions from rinsewater. *(Courtesy of Industrial Filter & Pump Manufacturing Company.)*

trate seawater about 1.6 times. Large installations may find pretreatment of the seawater of value. Such pretreatment has been ruled out by the high cost of chemicals and the extra equipment, but the increasing cost of fuel may offset this in the future. Special alloys must be used throughout to counteract the corrosive effects of the concentrated seawater.

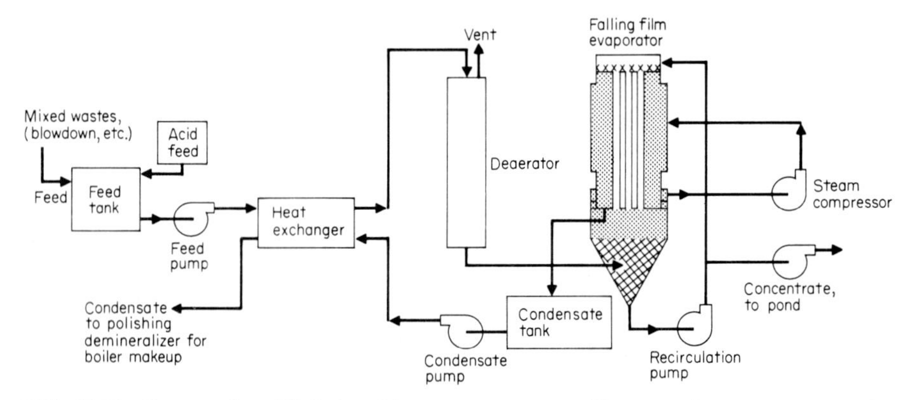

FIG. 18.12 Compression still designed to operate as a crystallizer as well as an evaporator in treating cooling tower blowdown and other wastes in a zero-discharge utility station. *(Courtesy of Resources Conservation Company.)*

TABLE 18.3 Chemical Characteristics of "Zero-Discharge" Evaporator System

Description	Feed: tower and scrubber	Liquor: evaporator concentrate	Product: distillate
pH, initial	8.6	6.6–6.8	6.8–7.4
after acid	6.2	—	—
Conductivity	6400	ND*	10–15
TDS, mg/L	4000	257,500	7
TS, mg/L	8400	324,900	7
SS, mg/L	4400	67,400	Nil

* Not determined

Evaporators are finding application in the concentration of wastes to minimize volume, simplify ultimate destruction, or both. Examples include the treatment of radwaste from nuclear power plant operations (Figure 18.10) and concentration of plating wastes (Figure 18.11).

In special cases where the EPA permit requires "zero discharge," a modified design of compression still is being used to concentrate combined wastes, such as cooling tower blowdown and flue gas scrubber effluent, to yield a saturated solution containing a crystal phase, so that the salinity can be removed as solids (Figure 18.12). Table 18.3 shows the concentration achieved at one utility station using this scheme. Treatment of the liquor to prevent scaling is essential.

FREEZING

As water begins to freeze in a container, the dendrites of ice that first form on the heat-extraction surface consist of fairly pure H_2O. The remaining water has concentrated the solute originally present in the makeup water. Before the advent of the home refrigerator, commercial ice plants manufactured ice in cans, and it was

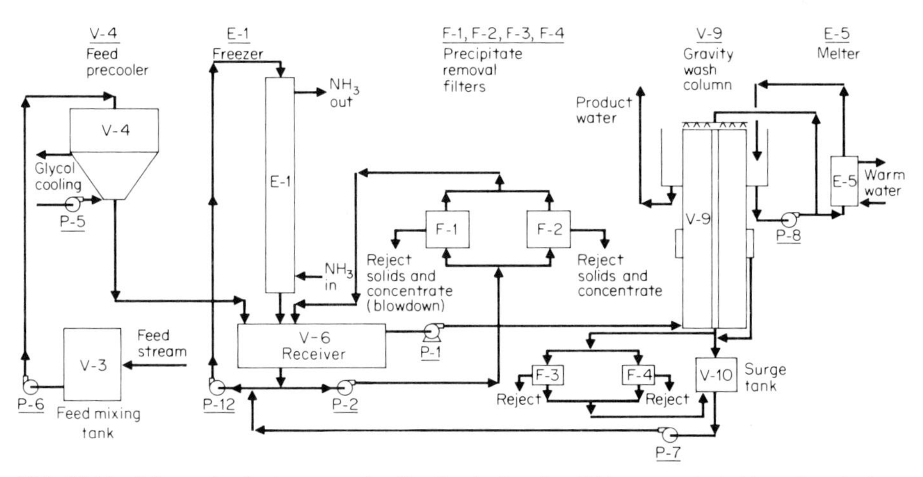

FIG. 18.13 Schematic of mine water desalination by freezing. This system is tied into the mine's air-conditioning load, improving the economics of the process. *(Courtesy of CBI Industries, Inc.)*

common practice to suck out the "core" of unfrozen water, containing most of the original dissolved solids, when this residue had concentrated about 10-fold. This improved the quality and strength of the finished cake of ice. If the core were not removed, the final solidification of the cake would include salt crystals mixed with ice crystals.

Various schemes have been proposed for producing pure H_2O as a solid, free of solute originally present in the feed water. These ideas are being seriously pursued because they hold some promise of energy savings over distillation processes; the evaporation of 1 lb of water requires about 1000 Btu (539 cal/g), compared with only 144 Btu (80 cal/g) to freeze it. Figure 18.13 is a schematic diagram of a pilot plant being used to desalinate mine drainage in South Africa by freezing. The feed system contains 9500 mg/L TDS, and product water contains 500 mg/L, with a water recovery rate of 90%. Product quality can be improved at the expense of a lower production rate.

Although freezing as an economical process of water desalination may be far from commercial use, its use as an energy storage scheme is being practiced to a limited extent. Water is frozen during nighttime off-peak electrical periods, when the cost of electricity is reduced; during the day, the ice is melted by cooling air to supplement the mechanical air-conditioning load in large buildings, hospitals, or other such facilities.

Finally, the potential use of natural ice as a source of potable water and as a means of cooling in arid regions has been seriously studied and proposed. Whether this is ever put into use will depend on future costs of energy.

CHAPTER 19
OXIDATION-REDUCTION

Fuel burned in air combines with oxygen, and if this fuel is carbon, the complete reaction is:

$$C + O_2 \rightarrow CO_2 \quad (1)$$

Although this illustration of combustion is an oxidation reaction where carbon combines with oxygen, not all oxidation reactions involve oxygen. In the above reaction it is apparent that there is a change in the valence of carbon from 0 to +4. There is also a change in the valence of each oxygen atom from 0 to −2. The broad interpretation of oxidation is therefore a chemical reaction in which there is an increase in valence (or a loss of electrons). To maintain the electric neutrality of a system every oxidation must be accompanied by a reduction. Thus, a reduction reaction can be defined as a reaction in which there is a decrease in valence (or a gain of electrons). By these definitions, the following chemical reactions can be identified as oxidation-reduction (redox) reactions:

$$Fe^0 + 2H^+ + 2Cl^- \rightarrow Fe^{2+} + 2Cl^- + H_2^0 \quad (2)$$

or

$$Fe + 2HCl \rightarrow FeCl_2 + H_2$$

$$2Cl^0 + 2H^+ + S^{2-} \rightarrow 2H^+ + 2Cl^- + S^0 \quad (3)$$

or

$$Cl_2 + H_2S \rightarrow 2HCl + S$$

$$2CrO_4^{2-} + 6Fe^{2+} + 16H^+ \rightarrow 2Cr^{3+} + 6Fe^{3+} + 8H_2O \quad (4)$$

or

$$2Na_2CrO_4 + 6FeSO_4 + 8H_2SO_4 \rightarrow$$
$$Cr_2(SO_4)_3 + 3Fe_2(SO_4)_3 + 2Na_2SO_4 + 8H_2O$$

where in the chromate radical (CrO_4), Cr has a valence of +6.

In the first reaction, iron is oxidized and the hydrogen ion is reduced to hydrogen gas; in the second, the sulfide ion is oxidized to free sulfur as the chlorine is reduced to the chloride ion; and in the third reaction, ferrous iron is oxidized to ferric iron as chromate radical is reduced with hexavalent chromium going to trivalent. Another way of showing the first reaction in its separate segments, called *half-cell reactions,* is as follows:

Oxidation: $$Fe^0 \rightarrow Fe^{2+} + 2e \quad (5)$$

Reduction: $$2e + 2H^+ + 2Cl^- \rightarrow H_2^0 + 2Cl^- \quad (6)$$

Redox reaction: $$2HCl + Fe \rightarrow FeCl_2 + H_2 \quad (7)$$

In each of these half-cell reactions, the material being oxidized is called a reducing agent, and the material being reduced is an oxidizing agent. The relative reactivity of the elements in terms of oxidation/reduction tendencies is illustrated by the electromotive series of the elements. The elements above hydrogen in this series will replace hydrogen from water or from acid solutions. For example, the series shows that sodium will react with water at room temperature to evolve hydrogen, and this can be shown as an oxidation-reduction (redox) reaction:

Oxidation: $$2Na^0 \rightarrow 2Na^+ + 2e \tag{8}$$

Reduction: $$2e + 2HOH \rightarrow H_2 + 2OH^- \tag{9}$$

Redox reaction: $$2Na + 2H_2O \rightarrow H_2 + 2NaOH \tag{10}$$

In this reaction, the element sodium (shown here as Na^0) as the reducing agent is oxidized to the sodium ion. Water, then, must be the oxidizing agent in this reaction.

Iron, which is much lower than sodium on the electromotive series, also reacts with water, though not so rapidly as to make the generation of hydrogen apparent. However, iron in an overheated superheater tube in a steam boiler will react readily with steam to produce hydrogen in measurable quantities. Although iron will not displace hydrogen from water readily at room temperature, it will very actively produce hydrogen in mineral acids. This is because the hydrogen ions are much more available in acid than they are in water.

Pursuing the electromotive series further, mineral acids should have very little effect on copper, since copper is below hydrogen in the series. However, hydrogen gas should reduce the copper in a copper sulfate solution to metallic copper. Because hydrogen and copper are so close in the electromotive series of elements, the reaction is not readily evident. However, iron and copper are relatively far apart, and the reaction of elemental iron with copper sulfate is easily demonstrated; an iron nail in a copper sulfate solution will soon be plated with copper, even at room temperature. These reactions all demonstrate the close relationship between oxidation-reduction reactions and the electromotive series.

The relative strengths of oxidizing and reducing agents are shown in Table 19.1. As an example, this table shows that ferric iron is most readily reduced by

TABLE 19.1 Relative Redox Potentials

Increasing strength of reductant ↑	Reductant	←(+e)— / —(−e)→	Oxidant	Increasing strength of oxidant ↓
	Na	⇌	Na^+	
	Mg	⇌	Mg^{2+}	
	Al	⇌	Al^{3+}	
	Zn	⇌	Zn^{2+}	
	Fe	⇌	Fe^{2+}	
	H_2	⇌	$2H^+$	
	Cu	⇌	Cu^{2+}	
	Fe^{2+}	⇌	Fe^{3+}	
	H_2O	⇌	O_2^0	
	I_2	⇌	IO_3^-	
	Mn^{2+}	⇌	MnO_2	
	Cl^-	⇌	Cl^0	

sodium, followed by magnesium, aluminum, and zinc. Conversely, aluminum is most readily oxidized by chlorine, followed by manganese dioxide, oxygen, and cupric ions.

Since oxidation and reduction reactions involve the transfer of electrons, it would be expected that oxidation and reduction reactions occur in an electrolytic cell. In the electrolysis of water, hydrogen is produced at the cathode and oxygen at the anode. Comparable reactions take place in all electrolytic cells—reduction at the cathode and oxidation at the anode. In the electrolysis of brine, the chloride ions lose electrons (oxidize) at the anode, becoming oxidized to chlorine gas; the sodium ions gain electrons (reduce) at the cathode and immediately react with water to produce hydrogen and caustic soda.

Strong oxidizing agents such as chlorine and chlorine dioxide can be hazardous; if they contact organic matter, such as paper or oil, the reaction is fast and violent. However, in many aqueous solutions, such as a surface water supply saturated with air, the concentration of oxidizing and reducing materials is so low that most redox reactions are extremely slow. The rates of these redox reactions are in no way comparable to the almost instantaneous reaction of acid with bases in water, or lime with carbon dioxide. Many oxidation-reduction reactions must be catalyzed; for example, the removal of dissolved oxygen from water by application of sodium sulfite is barely measurable until a catalyst is added to the water. Similarly, the oxidation of ferrous iron to the insoluble ferric condition is normally very slow, but may be catalyzed by the presence of oxides of manganese.

Even though they are slow there is appreciable energy available in some of the redox reactions that occur in water. It can be shown thermodynamically that the sulfate ion is strong enough to oxidize organic matter in water: This does not occur by a direct route, but nature carries out the reaction indirectly through microbes. This occurs under anaerobic conditions in well waters, and the result is the production of sulfides by anaerobic bacteria. Most of the oxidation-reduction reactions occurring in natural water systems are biologically mediated.

Because the oxidation and reduction equations for the overall redox reactions involve electron transfer, and since the voltage for these half-cell reactions can be determined, the progress of a redox reaction can be measured by proper instrumentation. The electrical system used for measuring voltage potential uses two electrodes, one being of a noble metal such as gold, and the other being a reference such as a hydrogen electrode or a calomel electrode. The voltage between these electrodes is measured, and as the oxidation-reduction reaction proceeds there is a change in voltage. For example, if a chromate solution is to be reduced using ferrous sulfate, the initial voltage is quite high because of the strong oxidation potential of chromate, and as ferrous sulfate is added the voltage drops, showing an inflection when the reaction is complete. The progress of the reaction is shown in Figure 19.1.

Redox recorders are used industrially to control the addition of oxidants or reductants to a water solution, or to follow the progress of the reaction in the system.

Using thermodynamic and physical chemical data, it is possible to predict the species of oxidizable and reducible matter in water based on redox potential and pH. Such a phase diagram is illustrated in Figure 19.2, showing the distribution of iron in an aqueous system at different pH values and different oxidation potentials.

There is another useful physical chemical notation representing the status of the oxidizing or reducing environment in an aqueous system for individual species that may be oxidized or reduced. This is the concept of pE, which is analogous to pH; like pH, it is a negative logarithm. The value of pE is minus the log

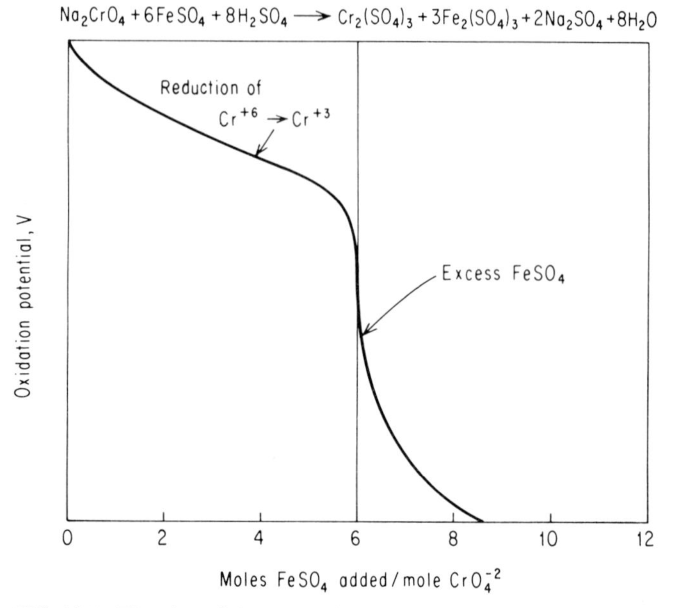

FIG. 19.1 Titration of chromate with $FeSO_4$ in acid solution.

of the electrons exchanged in the redox reaction. Although there are no free electrons in the aqueous environment, a number proportional to the number of moles of material being oxidized or reduced can be assigned to the electron transfer. If all oxidation and reduction equations are written with a single electron, an equilibrium can be established for which the constant can be determined. Mathematically, the relative electron activity, pE^0, at which the ionic species being examined are at an activity of unity, is shown as:

$$pE^0 = \log K$$

where K is the equilibrium constant.

Then

$$pE = pE^0 + \log(\text{product/reactant})$$

where pE represents a concentration of the species at other than unit activity.

Using ferric iron as an example, the half-cell reaction is as follows:

$$Fe^{3+} + e = Fe^{2+}$$

$$K = \frac{[Fe^{2+}]}{[Fe^{3+}][e]} = 3.4 \times 10^{13}$$

Using these data, $pE^0 = 12.53$, the level of electron activity at which ferrous and ferric ions are of equal concentration in an oxygen-free environment. When this system is exposed to the environment, the presence of oxygen causes an increase in pE to 16.49, and at this pE value the ferrous ion concentration is reduced to about 10^{-4} of the total iron concentration.

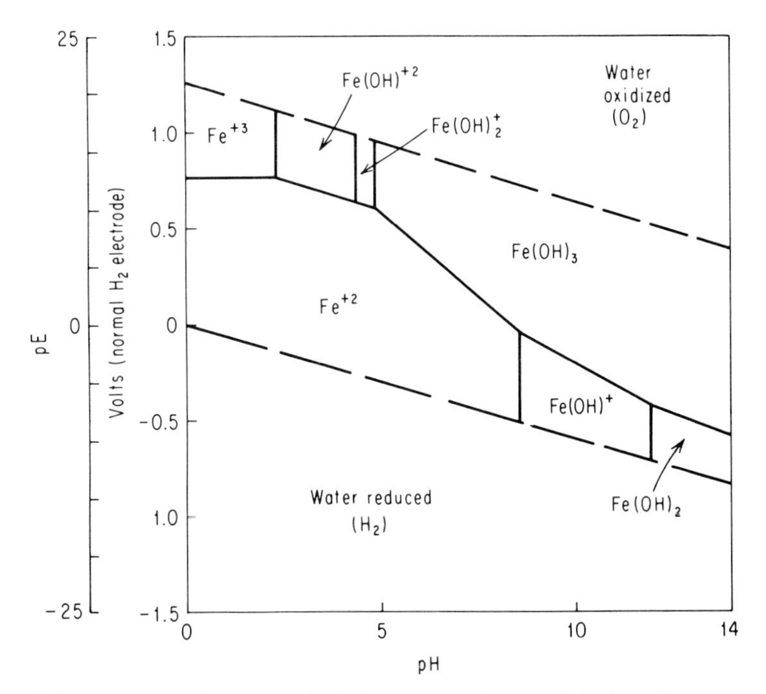

FIG. 19.2 Pourbaix diagram for Fe in water based on precipitation of hydroxide forms, at 2×10^{-7} *M* total Fe. [*From F. E. Clark, TAPPI J, 49 (10), 124A (October 1966).*]

Since there is a relationship between pE, the free energy involved in the electron transfer, the electrode potential, and the equilibrium constant for each redox reaction, it is possible to construct a phase diagram for most redox reactions in natural aqueous environments showing the distribution of species as established by pH and pE. In most natural aquatic environments, the pE can range from a strongly reducing value of -8 (corresponding to a voltage of approximately -0.3) to a strongly oxidizing value of 16 (corresponding to a voltage of approximately 1.0); below the lower limit water would be reduced to produce hydrogen, and above the upper limit water would be oxidized to produce oxygen.

It is important to recognize that although the environment may be strongly oxidizing or reducing, the voltage potential for the half-cell reaction does not influence the rate of reaction, and many such reactions in aqueous systems are very slow unless catalyzed.

The electromotive series as summarized in Table 19.1 shows that a substance can be defined as an oxidant or a reductant only in terms of other substances in the aqueous environment. Even though oxidation and reduction are accepted as relative terms, certain chemicals are normally described as oxidizing agents or reducing agents only if they are at the far ends of the series.

The most common oxidizing agents used in water treatment are air, chlorine and chlorine-releasing compounds, oxygen, ozone, hydrogen peroxide, potassium permanganate, chromate salts, and nitrate compounds. Following are five major water conditioning applications using these oxidizing agents.

IRON REMOVAL

Iron may be present in water (a) in colloidal form, (b) as the ferrous ion, or (c) as a chelated compound. The colloidal form can be removed by coagulation, flocculation and precipitation, or filtration, and so will not be dealt with here. The chelated form may require oxidation of the organic material composing the chelate so that the iron may then be precipitated for removal. In the ferrous form, iron may be found in certain well waters at concentrations as high as 25 mg/L under anaerobic conditions. The reducing conditions that permit iron to become soluble in well water frequently also produce sulfides. In the treatment of many municipal water supplies containing iron, the raw water is often aerated before being introduced into the clarifier, and this introduction of air is often adequate for the oxidation of soluble iron to the insoluble ferric state. One reason this process may be successful is that the water is then held for 60 to 90 min before being collected at the overflow weirs for filtration, allowing time for the iron to fully oxidize.

Where the amount of iron to be removed is relatively low, on the order of 2 to 3 mg/L, it may be possible to remove this iron directly on a filter, but this requires a more powerful oxidant than air and the presence of a catalyst, such as manganese, with an increase in pH to a minimum of 8.5. Even with this, the residence time in the filter itself may not be adequate for complete oxidation and iron removal, and a detention tank may be required for this purpose. The oxidation of iron with chlorine and potassium permanganate are shown by the following two reactions:

$$2Fe^{2+} + Cl_2^0 \rightarrow 2Fe^{3+} + 2Cl^- \qquad (11)$$

(In this reaction, 0.64 mg/L Cl_2 is required for each milligram per liter of Fe.)

$$MnO_4^- + 3Fe^{2+} + 2H_2O + 5OH^- \rightarrow MnO_2 + 3Fe(OH)_3 \qquad (12)$$

(In this reaction, 0.95 mg/L $KMnO_4$ is required for each milligram per liter of Fe.)

pH has a pronounced effect on iron removal. Although the oxidation potential of both chlorine and permanganate decrease as the pH increases, the rate of reaction increases significantly with increase in pH. Also, the solubility of the oxidized iron decreases as the pH increases.

MANGANESE REMOVAL

The removal of manganese from water is similar to iron removal. A variety of oxides of manganese is produced by chlorine and permanganate oxidation, but these are of little significance in the treatment technology. In most industrial water systems, manganese can be more troublesome than iron at much lower concentrations, so the final filtration process is critical to successful removal of manganese. The effectiveness of removal increases significantly as the pH is raised to about 9.0, and time is even more critical than with iron removal.

REMOVAL OF ORGANIC MATTER

Oxygen and air are seldom used directly for oxidation of organic material, but they are of course used in biologically mediated oxidation systems, such as in the activated sludge process. At high temperatures [about 300°F (150°C) and above] and high pressures, air or oxygen injected into a reaction vessel will oxidize appreciable organic matter (the so-called Zimmerman process). When chlorine is used for oxidation, the results may be disappointing in that chlorine forms substitution products, particularly in reacting with phenols, that may be more objectionable than the original organic molecule. Even when reacting with such a simple organic molecule as methane, the chlorine does not destroy the methane but produces methyl chloride according to the following reaction:

$$Cl_2 + CH_4 \rightarrow CH_3Cl + HCl \qquad (13)$$

The formation of chloroform and other trihalomethanes often occurs when organic-bearing wastewaters are chlorinated. Since these are considered to be carcinogenic, ozonation may be preferred to chlorination for final disinfection of certain types of sewage. On the other hand, ozone and permanganate may be used for complete destruction of phenol and simple organic molecules, as shown by the following equations:

$$14O_3 + \underset{\text{phenol}}{C_6H_5OH} \rightarrow 6CO_2 + 3H_2O + 14O_2 \qquad (14)$$

In this reaction, 7.2 mg/L ozone is required for each milligram per liter of phenol.

$$4KMnO_4 + \underset{\text{formaldehyde}}{3HCHO} \rightarrow 4MnO_2 + 2K_2CO_3 + CO_2 + 3H_2O \qquad (15)$$

In this reaction, 17.5 mg/L permanganate is required for each milligram per liter of organic carbon (TOC). Although these reactions take place, the rate is slow and therefore substantial residence time is needed in the reaction vessel. The rate can be increased substantially by exposure of the system to ultraviolet radiation.

Phenol can be oxidized by permanganate, but the reaction requires about 2 h at a pH of 7 for 90% completion; even with an increase in pH to 9 to 10, it still takes about 30 min for 90% reaction.

CYANIDE REMOVAL

Cyanide can be oxidized to the cyanate form under alkaline conditions both by chlorine and permanganate, according to the following reactions.

$$CN^- + Cl_2^0 + H_2O \rightarrow 2HCl + CNO^- \qquad (16)$$

In this reaction 2.75 mg/L chlorine is required for each milligram per liter of CN.

$$3CN^- + 2MnO_4^- + H_2O \rightarrow 2MnO_2 + 3CNO^- + 2OH^- \qquad (17)$$

In this reaction 4.1 mg/L $KMnO_4$ is required for each milligram per liter of CN. Both of these reactions are carried out at a pH greater than 10. In the case of chlorine, it is important to maintain this pH to avoid formation of toxic cyanogen chloride.

Although the cyanate is much less toxic than cyanide, complete destruction is often necessary, requiring further oxidation of the cyanate with chlorine or hypochlorite as shown in equation (18).

$$2CNO^- + 3OCl^- + H_2O \rightarrow N_2 + 3Cl^- + 2HCO_3^- \qquad (18)$$

Combined with the above reaction, complete destruction of CN to N_2 requires 6.9 mg/L Cl_2 per milligram per liter of CN.

The optimum pH for the second stage of this reaction is approximately 8.5. As in other oxidation reactions, time is an important factor to be studied; the cyanide reactions generally require over 10 min and an excess of oxidizing agent in order to provide complete destruction.

SULFIDE REMOVAL

In the aeration of sulfide-bearing waters to eliminate odor, some of the sulfide is converted to colloidal sulfur according to the following reaction.

$$2HS^- + O_2 \rightarrow 2S^0 + 2OH^- \quad (19)$$

Sulfides can also be oxidized by chlorine and permanganate according to the following reactions.

$$HS^- + Cl_2 + H_2O \rightarrow 2Cl^- + 2H^+ + S^0 + OH^- \quad (20a)$$

$$S^0 + 3Cl_2^0 + 4H_2O \rightarrow H_2SO_4 + 6HCl \quad (20b)$$

$$2MnO_4^- + 3(HS^-) + H_2O \rightarrow 3(S^0) + 2MnO_2 + 5OH^- \quad (21)$$

In all of the above examples, it is important to recognize that the oxidation was involved with specific individual species in the water. Since the aqueous system contains a variety of species, some of which may participate in the oxidation-reduction reactions, it is clear that predicting the best oxidation material for the reaction, optimum pH, required time for completion, and dosage of the oxidizing agent is impossible. In the final analysis, the examples above simply suggest potential oxidizing materials that might be used; bench testing is needed to evaluate these and others to obtain data to design an effective treatment process.

In each of the oxidation processes reviewed previously, chlorine was used in the examples as the oxidant. It is certainly the most widely used, and it is one of the common high-tonnage products of the chemical industry in the United States. Produced with caustic soda, it is relatively low in cost. It is widely used for disinfection of public water supplies and for control of microorganisms in industrial water systems. (Details of handling chlorine are given in Chapter 22.)

OTHER OXIDIZING AGENTS

There are sometimes objections to the use of chlorine, some chemical, and others regulatory, in nature. An example of a technical objection is the oxidation of cyanide in a coke-oven wastewater high in ammonia. Since the chlorine combines with ammonia to form chloramines (see Figure 22.7), it is unavailable to effectively oxidize cyanide, and the cost of cyanide conversion becomes prohibitive. A nontechnical objection is related to safety: Some cities have ordinances prohibiting transport of chlorine gas over bridges or through tunnels, or its storage in populous areas. Hypochlorite is commonly substituted in such situations. A comparison of the oxidizing power of various common water treatment chemicals is given in Table 19.2.

TABLE 19.2 Oxidation-Reduction Potential (ORP) of Common Oxidants

Chemical	Reaction	Volts at 25°C
Ozone	$O_3 + 2H^+ + 2e \rightarrow O_2 + H_2O$	2.87
Peroxide (acid)	$H_2O_2 + 2H^+ + 2e \rightarrow 2H_2O$	1.76
Permanganate	$MnO_4 + 8H^+ + 5e \rightarrow Mn^{2+} + 4H_2O$	1.49
Hypochlorite (acid)	$HOCl + H^+ + 2e \rightarrow Cl^- + H_2O$	1.49
Chlorine	$Cl_2 + 2e \rightarrow 2Cl^-$	1.39
Chlorine dioxide	ClO_2 (gas) $+ e \rightarrow ClO_2^-$	1.15
Bromine	$Br_2 + 2e \rightarrow 2Br^-$	1.07
Hypochlorite (basic)	$OCl^- + H_2O + 2e \rightarrow Cl^- + 20H^-$	0.90
Peroxide (basic)	$H_2O_2 + 2e \rightarrow 2OH^-$	0.87
Oxygen	$O_2 + 2H_2O + 4e \rightarrow 4OH^-$	0.40

Ozone has been used for many years for disinfection of public water supplies in Europe. It is gaining more acceptance in the United States, and improvements in equipment design are resulting in a more favorable cost-performance. Larger installations using over 500 lb/day (225 kg/day) can produce ozone at about 12 to 15 kw-h per pound of O_3 with atmospheric air, and at about half this energy consumption if oxygen is available for the conversion. The typical dosage for disinfection is 1 to 3 mg/L. It will deactivate most viruses at 0.5 mg/L in less than 5 min, where the same dosage of chlorine would require 1 h for deactivation.

REDUCING AGENTS

The commercially available reducing agents most commonly used in water treatment are sulfur dioxide, sodium sulfite, and ferrous sulfate or chloride, with occasional use of zinc hydrosulfite, borohydride compounds, hydrogen sulfide, and zinc dust. The most common reductions in water treatment are the scavenging of dissolved oxygen with sodium sulfite, reduction of excess chlorine with sodium sulfite, and the reduction of chromate to its trivalent form. Typical reactions for each of these processes are shown below.

$$O_2 + 2SO_3^{2-} \rightarrow 2SO_4^{2-} \tag{22}$$

In this reaction, 8 mg/L Na_2SO_3 is required for each milligram per liter of O_2.

$$Cl_2 + SO_3^{2-} + H_2O \rightarrow SO_4^{2-} + 2HCl \tag{23}$$

In this reaction, 1.8 mg/L Na_2SO_3 is required for each milligram per liter of chlorine.

$$2Na_2CrO_4 + 6FeSO_4 + 8H_2SO_4 \rightarrow Cr_2(SO_4)_3 + 3Fe_2(SO_4)_3 + 2Na_2SO_4 + 8H_2O \tag{24}$$

In this reaction 3.94 mg/L $FeSO_4$ are required per milligram per liter of chromate.

In commercial practice, the first reaction, elimination of dissolved oxygen from water, requires the use of a catalyst for practical treatment in a reasonable time; the reduction of chlorine need not be catalyzed.

The reduction of chromate is carried out at a pH of approximately 2.5 to 3.0 using more than twice the theoretical requirement of reducing agent for completion of the reaction in 15 to 20 min.

It is important to recognize that there are usually species present in water in addition to those represented by the above equations. For example, chromate is a common constituent in industrial recirculating cooling water systems, where it is used as a corrosion inhibitor. In the organic chemicals industry, leakage of organics into the cooling water often results in a reduction of chromate. Recognizing how difficult it is to reduce chromate at concentration levels of 10 to 20 mg/L with strong reducing agents under controlled conditions, it may be conjectured that the reduction of chromate in a cooling system may be a biologically mediated reaction.

Some wastewater problems require more powerful reducing agents than sulfites, hydrosulfites, or sulfides. Lithium borohydride is such a material. It has been applied to metallo-organic compounds, such as tetraethyl lead, in wastewaters, reducing the metallic fraction to free metal (e.g., lead), which is then filtered from the treated solution. Although borohydrides are costly, in some cases they provide the only practical solution to a difficult problem. Hydrogen is another effective reducing agent; it has been used in the primary loop of nuclear power plants where other chemicals cannot be used, and it serves the purpose of deactivating the oxygen formed from the breakdown of water under neutron bombardment.

SUGGESTED READING

Rice, R. G., and Browning, Myron, E.: "Ozone for Industrial Water and Wastewater Treatment; a Literature Survey," *EPA Document* 600/2-80-060. April 1980.

Stumm, W., and Morgan, J. J.: *Aquatic Chemistry,* Wiley, New York, 1970.

Weber, W. J., Jr., et al.: *Physicochemical Processes for Water Quality Control,* Wiley, New York, 1972.

CHAPTER 20
CORROSION CONTROL

Corrosion is nature's way of returning processed metals, such as steel, copper, and zinc, to their native states as chemical compounds or minerals. For example, iron in its natural state is an oxidized compound (i.e., Fe_2O_3, FeO, Fe_3O_4), but when processed into iron and steel it loses oxygen and becomes elemental iron (Fe^0). In the presence of water and oxygen, nature relentlessly attacks steel, reverting the elemental iron (Fe^0) back to an oxide, usually some combination of Fe_2O_3 and Fe_3O_4.

Although corrosion is a complicated process, it can be most easily comprehended as an electrochemical reaction involving three steps as shown in Figure 20.1:

1. Loss occurs from that part of the metal called the anodic area (anode). In this case, iron (Fe^0) is lost to the water solution and becomes oxidized to Fe^{2+} ion.
2. As a result of the formation of Fe^{2+}, two electrons are released to flow through the steel to the cathodic area (cathode).
3. Oxygen (O_2) in the water solution moves to the cathode and completes the electric circuit by using the electrons that flow to the cathode to form hydroxyl ions (OH^-) at the surface of the metal. Chemically, the reactions are as follows:

Anodic reaction:
$$Fe^0 \rightarrow Fe^{2+} + 2e^- \tag{1}$$

Cathodic reaction:
$$\tfrac{1}{2}O_2 + H_2O + 2e^- \rightarrow 2(OH^-) \tag{2}$$

In the absence of oxygen, hydrogen ion (H^+) participates in the reaction at the cathode instead of oxygen, and completes the electric circuit as follows:

$$2H^+ + 2e^- \rightarrow H_2 \uparrow \tag{3}$$

Every metal surface is covered with innumerable small anodes and cathodes as shown in Figure 20.1. These sites usually develop from: (1) surface irregularities from forming, extruding, and other metalworking operations; (2) stresses from welding, forming, or other work; or (3) compositional differences at the metal surface. In the case of steel, this may be caused by different microstructures; Figure 20.2 is an enlargement of a polished specimen of steel showing ferrite and cementite as two distinctly different components. In brass, the difference may be between the fine crystals of copper and zinc that make up this alloy. Inclusions on or below the metal surface may be the cause of anode-cathode couples. These impurities may have formed when the metal was molten or because impurities were pressed into the surface during the rolling, finishing, or shaping operations.

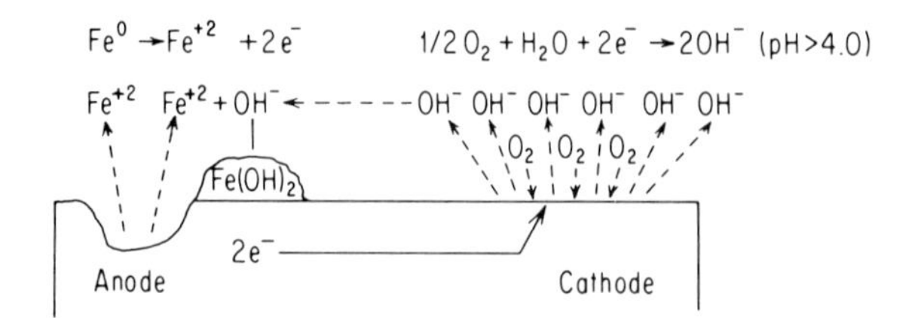

FIG. 20.1 Reactions occurring during the corrosion of steel in the presence of oxygen. The presence of OH^- at the cathode can be demonstrated by the pink coloration of phenolphthalein dye.

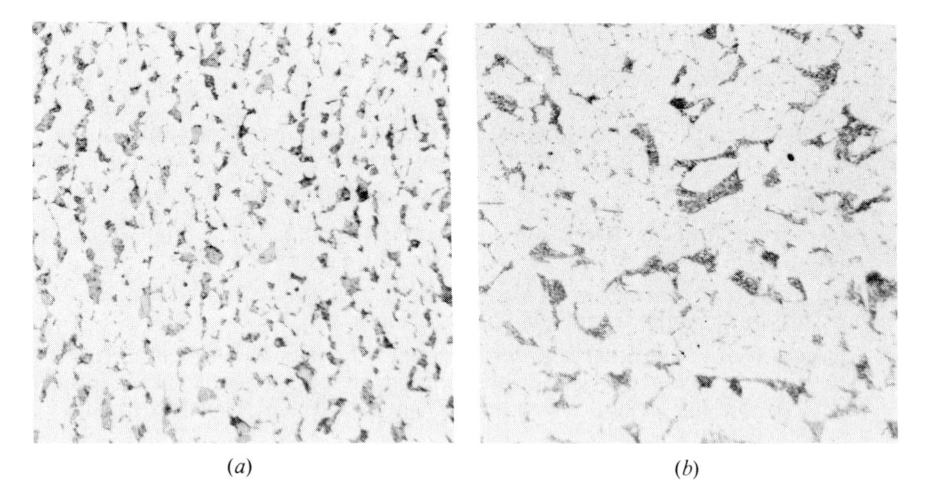

FIG. 20.2 Microstructure of low-carbon steel at 20× showing background matrix of ferrite (iron, light-colored) containing grains of pearlite (dark-colored). The pearlite consists of lamellae of ferrite and cementite (Fe_3C). (*a*) The pearlite is seen to be in layers, created by the extrusion of the metal through a die. (*b*) The larger pearlite grains in this specimen are randomly oriented, and the lamellae of ferrite and cementite can be seen.

CORROSION RATES

As noted above, three basic steps are necessary for corrosion to proceed. If any step is prevented from occurring, then corrosion stops. The slowest of the three steps determines the rate of the overall corrosion process. The cathodic reaction (step 3) is the slowest of the three steps involved in the corrosion of steel, so this reaction determines the rate. It is slow because of the difficulty oxygen encounters in diffusing through water. One factor in increasing corrosion, then, is increasing water temperature, which reduces its viscosity and speeds the diffusion of oxygen.

A large cathodic surface area relative to the anodic area allows more oxygen, water, and electrons to react, increasing the flow of electrons from the anode to corrode it more rapidly. Conversely, as the cathodic area becomes smaller relative to the anodic area, the corrosion rate decreases.

If, however, the anodic area is reduced with no corresponding decrease in the cathodic area, the same amount of metal will be lost, but from fewer or smaller anodic sites. The corrosion rate will not have been changed because the cathodic surface area has not been reduced, but each anodic site will be deeper. This is a simple explanation for the cause of pitting attack on metals. Pitting is more damaging than general loss over a large anodic area because it leads to rapid metal wall penetration and equipment failure. Figure 20.3 contrasts pitting metal loss

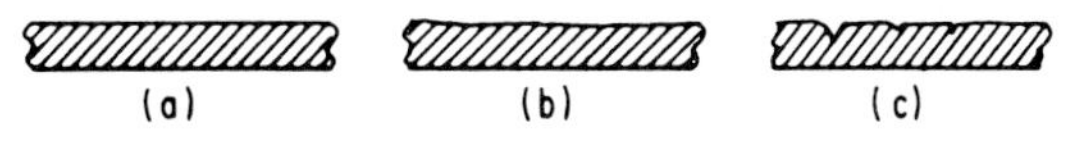

FIG. 20.3 (*a*) Surface of corrosion coupon is smoothed and cleaned prior to insertion in water line. (*b*) This coupon surface has been uniformly corroded, with some areas of localized attack. (*c*) This coupon surface has been pitted. The weight loss may be lower than (*b*), but failure (penetration) will occur earlier. Depth of pits may be measured by microscope or by a micrometer probe.

to general metal loss from corrosion coupons exposed to similar field environments in cooling systems.

Area ratios as they affect corrosion rate play a significant part in the selection of effective inhibitors to control corrosion. They are also a major consideration in designing equipment to minimize corrosion.

POLARIZATION-DEPOLARIZATION

As noted earlier, hydroxyl ions (OH^-), hydrogen gas (H_2), or both, are produced at the cathode as a result of the corrosion reaction. If these chemical reaction products remain at the cathode they produce a barrier (Figure 20.4) that slows the

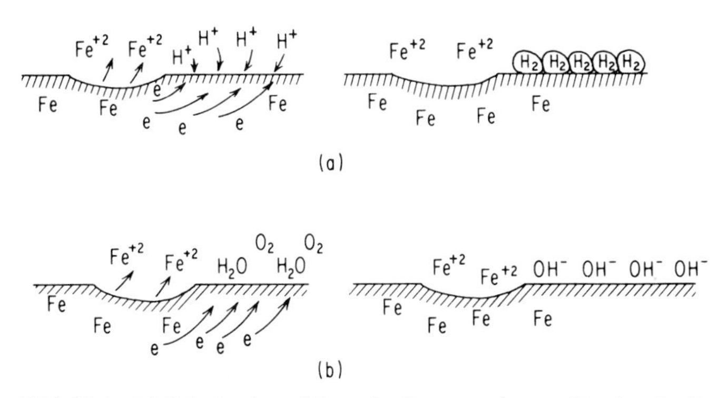

FIG. 20.4 (*a*) Polarization of the cathodic area at lower pH values by H_2 molecules. These may be depolarized by reaction with O_2. (*b*) Polarization of the cathode by an alkaline film highly concentrated in OH^- ions. These may be depolarized by chemical reaction with metal cations causing precipitation, by low pH, or by high-velocity water sweeping the surface.

movement of oxygen gas or hydrogen ions to the cathode. This barrier becomes a corrosion inhibitor because it insulates or physically separates oxygen in the water and the electrons at the metal surface. The formation of this physical barrier as a result of corrosion is called polarization. The removal or disruption of this barrier exposes the cathode, and corrosion resumes. This action, called depolarization or barrier removal, is enhanced by two factors:

1. Lowering the pH of the water. This increases the concentration of the hydrogen ions reacting with the hydroxyl ions to form water, thereby eliminating the hydroxyl barrier.
2. Increasing the water velocity into the turbulent flow region tends to sweep away hydroxyl ions and hydrogen from the surface of the cathode, thereby depolarizing it.

GALVANIC CORROSION

A special form of the general corrosion reaction is galvanic corrosion. This relatively common form of corrosion results when two dissimilar metals are connected and exposed to a water environment: one metal becomes cathodic and the other anodic, setting up a galvanic cell. For example, when copper and steel are connected in water, steel becomes the anode. It is said to be anodic to the copper, which is the cathode. The metal loss occurs at the anode, so the steel corrodes. The same principles of cathode-anode surface area ratios that apply to the general corrosion reaction apply to the galvanic cell: larger cathode—higher corrosion rate; smaller cathode—lower corrosion rate; large anode—general corrosion metal loss; small anode—pitting type attack. The galvanic couple corrosion rate is influenced by the types of metals that are connected. Table 20.1 lists a series of common metals and alloys frequently encountered in water systems. This is quite similar to the electromotive series of elements. The connection of two of these metals in a water environment (a galvanic couple) corrodes the more anodic. For metals close to one another in this series, the corrosion rate is less than for metals widely separated. In a galvanic couple, then, corrosion rate is dependent upon:

1. What metals are connected
2. Relative anodic to cathodic surface areas

Figure 20.5 shows extremely severe pitting and metal wall failure. This resulted principally from galvanic attack. The ultrapure aluminum tube experienced periodic copper plating from trace concentrations of copper in the circulating water which set up the severe galvanic attack. In this case, low flow velocities and inorganic and microbial deposits accentuated the problem.

CONCENTRATION CELL CORROSION

Just as the dissimilar metals noted above generate a galvanic current with a fixed concentration of water, a galvanic current can also be set up when a single metal is exposed to different concentrations (ionic strengths) of water solutions. The attack that occurs at the anode as a result of this mechanism, called *concentration cell corrosion,* takes place in the more concentrated solution. Concentration cells are the usual cause of the troublesome local etch or pitting type of metal loss.

TABLE 20.1 An Approximate Order of Galvanic Cell Corrosion

Electromotive series of elements	Galvanic series of metals and alloys		
Anodic			
Potassium			
Calcium			
Sodium			
Magnesium	Magnesium		
Aluminum	Aluminum		↑
Zinc	Zinc and Zn coatings		Al-2s
	Cadmium and Cd coatings		Aluminum alloys
Iron	Steel		Al-17s
	Cast iron		↓
Nickel		↓	
Tin	Tin	solders	
Lead	Lead		↑
	Ni resist	↑	60–40
			brasses
Hydrogen	Chromium		70–30
		↓	↓
	Nickel	Hastelloys	↑
	Inconel	↑	
Copper	Copper, bronze		
	Cupro-nickel, Monel		Stainless steels
Mercury			↓
Silver	Titanium		
Gold			
Cathodic			

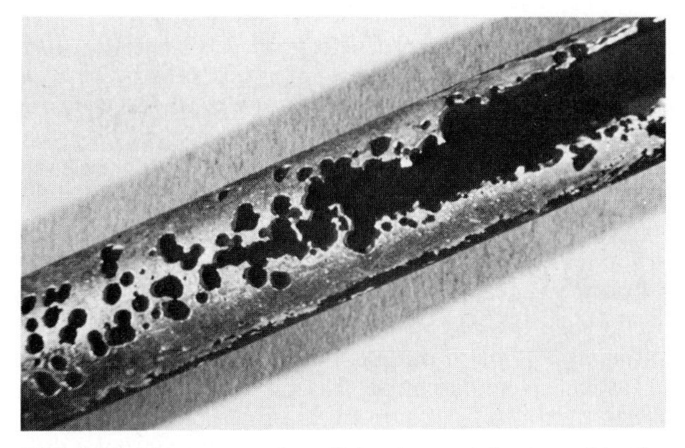

FIG. 20.5 Severe corrosion of aluminum piping caused by the presence of copper ions in the water in the piping system.

Concentration cell corrosion may severely shorten equipment life; equipment designed for years of service may fail in days. Generally, this type of corrosion occurs at any site where deposits, poor equipment design, or both, allows a localized concentration of a specific substance—such as NaCl or O_2—to be notably different from the amount found in the bulk of the water environment.

A typical sequence for the deposit-related concentration cell failure shown in Figure 20.6 is as follows: A deposit forms on a metal surface in a cooling water system. In a short time, the oxygen under the deposit is consumed by the normal

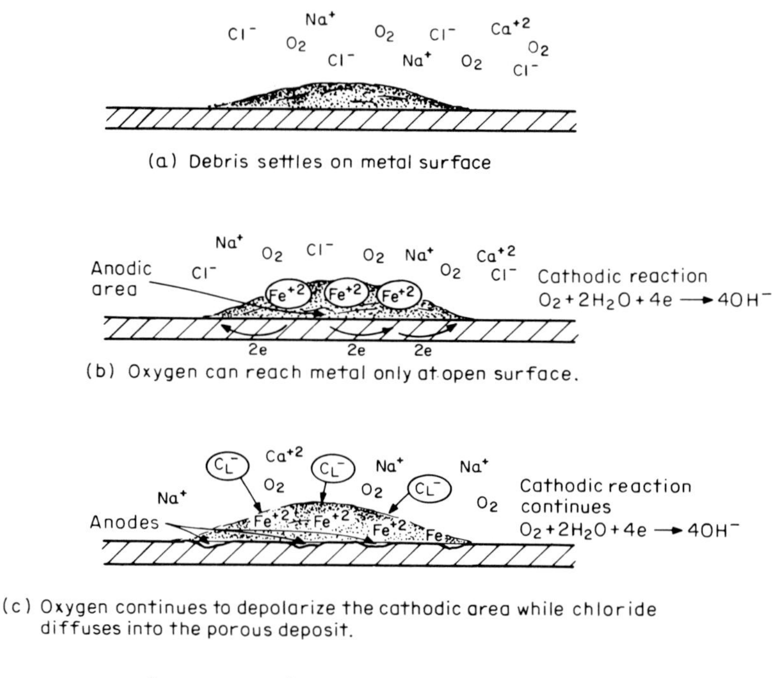
Na+
Cl−
O2
Ca+2
(a) Debris settles on metal surface
Anodic area
Fe+2
Cathodic reaction
O2 + 2H2O + 4e → 4OH−
2e
(b) Oxygen can reach metal only at open surface.
CL−
Anodes
Cathodic reaction continues
O2 + 2H2O + 4e → 4OH−
(c) Oxygen continues to depolarize the cathodic area while chloride diffuses into the porous deposit.

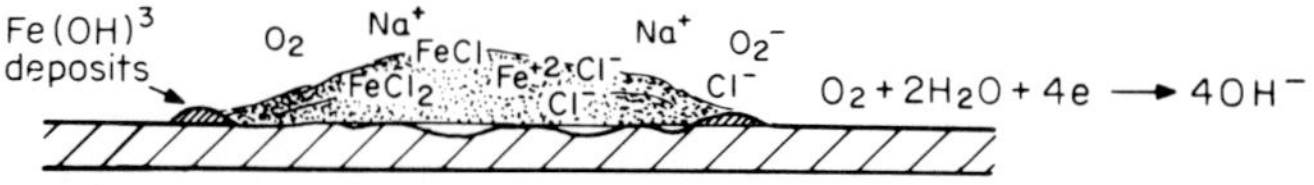
Fe(OH)3 deposits
O2
Na+
FeCl
FeCl2
Fe+2
Cl−
O2 + 2H2O + 4e → 4OH−
(d) The iron within the deposit remains soluble as Fe+2 in the absence of O2; and corrosion increases as ionic strength in the deposit increases.

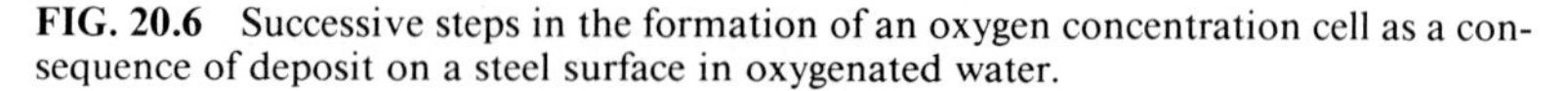

FIG. 20.6 Successive steps in the formation of an oxygen concentration cell as a consequence of deposit on a steel surface in oxygenated water.

corrosion reaction. As the oxygen concentration beneath the deposit becomes less than that in the bulk water surrounding the deposit because fresh O_2 is hindered from migrating through the deposit, the area under the deposit becomes anodic to the surrounding area. This unique concentration cell mechanism is called an oxygen differential cell. The corrosion that began as normal corrosion has now changed to a differential cell corrosion mechanism.

The rate of the differential cell reaction is proportional to the difference in concentration between the oxygen under the deposit and that found in the water around the deposit. In most cases, concentration cell corrosion will proceed much faster than the standard corrosion mechanism that originally occurred beneath the deposit and will result in a pitting-type metal loss.

An oxygen differential cell is rarely the sole cause of metal loss under a deposit. Coexisting with most oxygen differential cells are the corresponding chloride and sulfate concentration cells. These coexist because chloride and sulfate ions penetrate the deposit or crevice and concentrate. The deposit behaves like a semipermeable membrane (a Donnan membrane); as the iron ions (Fe^{2+}) leave the anodic surface, the $Cl^- + SO_4^{2-}$ anions diffuse through the deposit to maintain neutrality, resulting in a buildup of electrolyte under the deposit. This further accelerates the corrosion already taking place because of the oxygen differential cell.

Anaerobic microbes, which can reduce sulfate to sulfides under a deposit, create a very aggressive condition to further accentuate the metal loss.

In summary, deposits can lead to aggravated pitting attack initiated as normal corrosion, but compounded by various concentration cell corrosion mechanisms, which may include microbial involvement. Because these concentration cells and their corrosion products are shielded by deposits, not even the most effective inhibitors can get through to properly protect the metal surface. This emphasizes the importance of maintaining water systems free of deposits.

Other important factors which influence corrosion include the concentration of dissolved solids in water, dissolved gases, and temperature.

DISSOLVED SOLIDS

The influence of dissolved solids on corrosivity is very complex. Not only is the concentration important, but also the species of ions involved. For example, some dissolved solids (such as carbonate and bicarbonate) may reduce corrosion, while others (such as chloride and sulfate) may increase it by interfering with the protective film. Figure 20.7 shows that corrosivity does not increase at a linear rate with increasing total dissolved solids concentrations. In fact, over 5000 mg/L, a further increase in dissolved solids shows progressively less influence on corrosion rate of mild steel.

DISSOLVED GASES

CO_2 and O_2 are the major gases of concern in most industrial systems. Increasing the free CO_2 content in water reduces pH and tends to depolarize the cathodic surface area. In a relatively unbuffered condensate, this could produce a low pH and corrosive water. In a buffered water, such as a typical cooling water system, the impact of the increased CO_2 would be less.

As explained earlier, oxygen produces cathodic depolarization by removing the hydrogen produced at the cathode.

Hydrogen sulfide and ammonia are less frequently encountered than oxygen and CO_2, but both exert a strong influence on the corrosion of iron and copper alloys. Hydrogen sulfide is almost always ionized as bisulfide and sulfide ions, which tend to depolarize the anodic area. Ammonia increases the corrosion rate of copper and copper alloys by complexing the copper in the normally protective copper oxide or copper carbonate surface films.

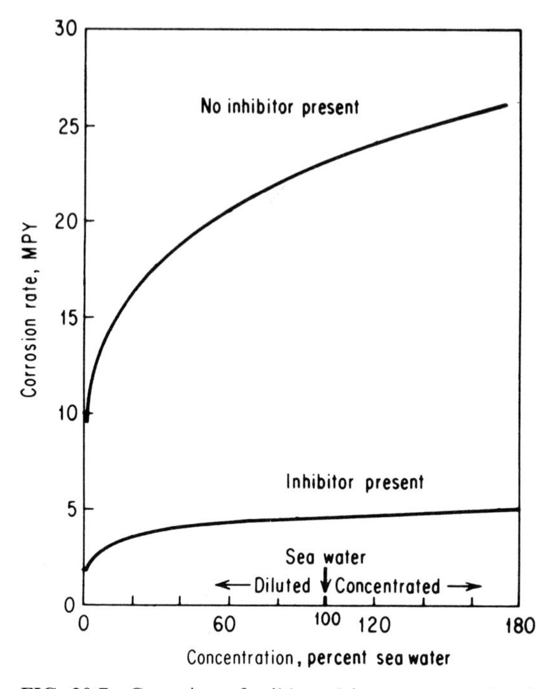

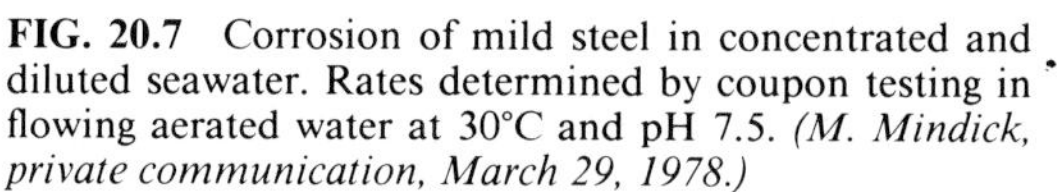

FIG. 20.7 Corrosion of mild steel in concentrated and diluted seawater. Rates determined by coupon testing in flowing aerated water at 30°C and pH 7.5. *(M. Mindick, private communication, March 29, 1978.)*

TEMPERATURE

As a general rule, each 15°F (8°C) increase in temperature doubles the rate of chemical reactions. Therefore, temperature increases the speed of corrosion because the cathodic reaction proceeds faster, although not at a rate predicted by the above rule of thumb. This is because the oxygen diffusion rate is also involved in the process. Figure 20.8 represents some of the actual increases in corrosion rate related to temperature.

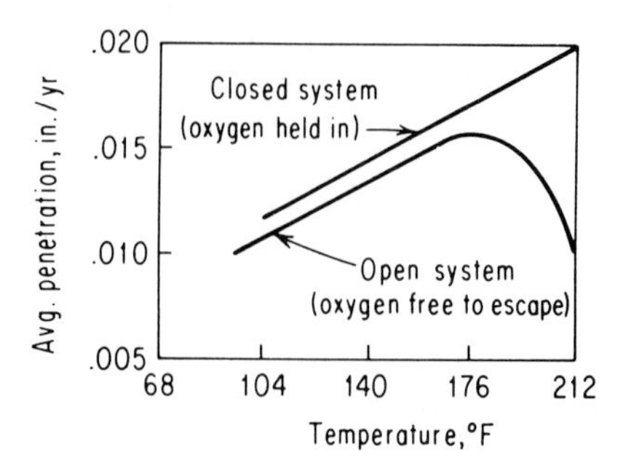

FIG. 20.8 High temperature boosts corrosion except where oxygen is free to escape. *(From "Corrosion Report," Power, December 1956.)*

STRESS CORROSION CRACKING

Metal under tensile stress in a corrosive environment may crack, a type of failure called stress corrosion cracking (Figure 20.9). The metal stress may be applied by any kind of external force which causes stretching or bending; it may also be due to internal stresses locked into the metal during fabrication by rolling, drawing, shaping, or welding the metal. The corrosive environment and the stress need not be of any specific minimum value. For stress corrosion cracking to take place, the combined effect of the stress and corrosion causes cracking either with a high stress in a mild corrosive environment or with a mild stress in a highly corrosive environment. At moderate stress and with a mild environment, there may be no failure.

If the combined effects are sufficient, generally both intergranular and transgranular cracks develop at right angles to the applied force. Highly corrosive waters tend to encourage randomly oriented cracking.

If the stresses are mechanically applied, this attack may be relieved by removing the applied stress. Where the stress is internal, relief is more difficult, since there is no practical way to locate and measure internal stresses in equipment at a plant site. A reliable means of reducing stress corrosion is to eliminate the corrosive environment or isolate the metal from water, such as by employing coatings. Application of a corrosion inhibitor may also help reduce this type of attack,

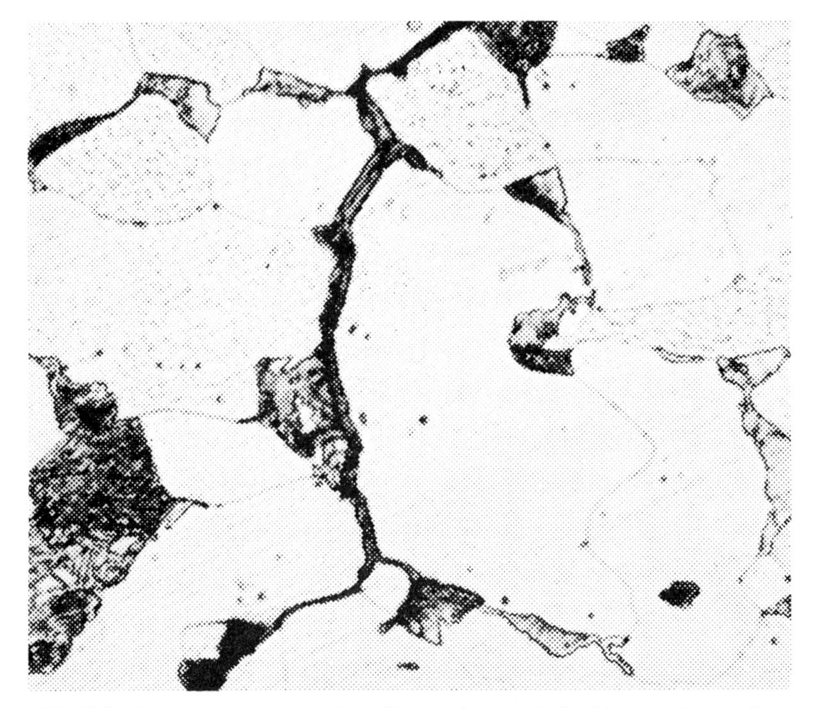

FIG. 20.9 Stress corrosion cracking of low-carbon steel. In this case, the specimen examined was taken from an embrittlement detector and illustrates caustic embrittlement. (500×)

but this approach is generally impractical because of the very high level of inhibitors required to effectively control the corrosion rate.

To a large extent, proper design and good fabrication procedures must be relied on to produce equipment, piping, or structures free of internal stresses. For example, heat treatment after fabrication (stress relief annealing) allows the crystalline structure of the metal to free itself of internal stresses. Even large process vessels are stress-relief-annealed to guard against stress corrosion cracking.

CAUSTIC EMBRITTLEMENT

Caustic embrittlement is a special type of stress corrosion cracking that sometimes occurs in boilers. At one time, this was a common cause of boiler failure, but improved fabrication practices and better water treatment have made it rare. Three concurrent conditions were found to cause caustic embrittlement:

1. A mechanism by which boiler water could concentrate to produce high concentrations of sodium hydroxide.
2. At the point of concentration, the boiler metal must be under high stress such as where the boiler tubes are rolled into the drum.
3. The boiler water must contain silica, which directs the attack to grain boundaries, leading to intercrystalline attack; and it must be of an embrittling nature. An embrittlement detector can be used to determine if the boiler water is embrittling in character.

CHLORIDE INDUCED STRESS CORROSION CRACKING

A specific type of stress corrosion cracking is induced by a chloride concentration cell, common with, but not limited to, stainless steels (Figure 20.10). To occur, it requires a chloride concentration and tensile stress focused together to cause both intergranular and transgranular branch-type cracking. This produces weakening of the metal and eventual failure. When this type of failure was first experienced, it was thought that chloride concentrations as low as 50 to 100 mg/L were responsible. But experience indicates that the concentration of chloride in the water contacting the stainless steel is not the critical factor. The main factor is the existence of conditions that allow chloride concentration cells to develop.

In the absence of concentration cells or stress, chloride levels in excess of 1000 mg/L have not caused stainless steel to crack. In fact, some desalination plants where chloride concentrations exceed 30,000 mg/L have not experienced failures, when properly annealed stainless-steel construction was used, and the system kept free of deposits. The key to preventing stress corrosion cracking is eliminating deposits and designing and fabricating stress-relieved equipment that does not allow concentration cells to occur.

CORROSION FATIGUE CRACKING

As the name implies, corrosion fatigue cracking is a result of a combination of both a corrosive environment and repeated working of a metal (Figure 20.11). The fatigue is brought on by the routine cyclic application of stress.

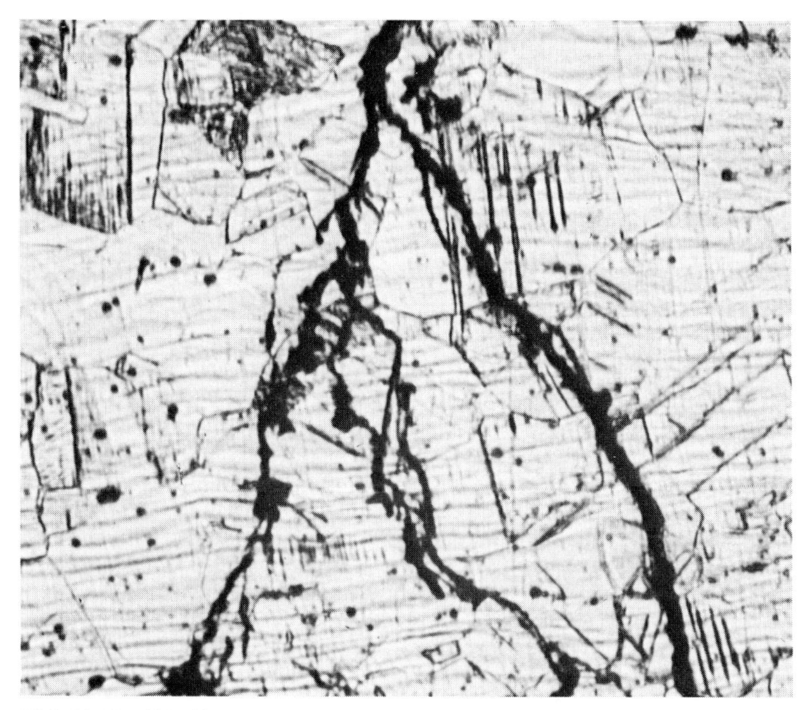

FIG. 20.10 Chloride stress corrosion cracking of type 316 stainless steel, showing the typical branching of the crack. (150×).

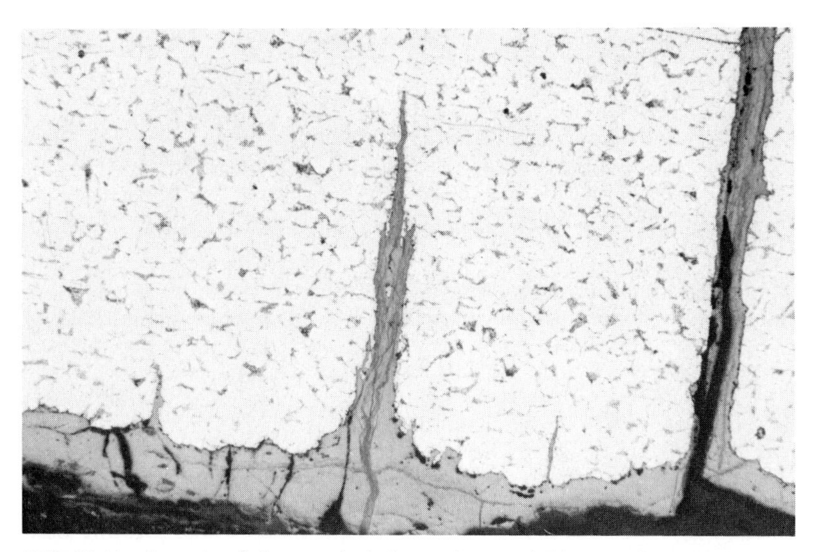

FIG. 20.11 Corrosion fatigue cracks in low-carbon steel. Blunt, wedge-shaped cracks are typical in this kind of failure. (150×)

This action occurring repeatedly in a corrosive environment eventually causes the metal to crack. Such failure can occur with any kind of corrosive environment and any type of metal.

The fatigue cracks are normally at right angles to the applied stress and the rate of propagation is dependent on the corrosivity of the water, the degree of stress, and the number of cycles that occur over a given time.

To minimize or prevent this type of attack, it is necessary to locate the source of stress and reduce the cyclic frequency or magnitude. Inhibitors are helpful in reducing the corrosivity at the metal-water interface, thereby minimizing one of the two forces that must be present for this transgranular, D-notched type of failure to propagate.

Corrosion fatigue cracking is more common than stress corrosion cracking.

TUBERCULATION

Tuberculation is the result of a series of circumstances that cause various corrosion processes to produce a unique nodule on steel surfaces. Figure 20.12 shows a cross section of a typical tubercle, with the majority of the mound composed of layers of various forms of iron oxide and corrosion products in laminar form. Initially, metal ions are produced at an anodic site. A high pH, caused by hydroxyl or carbonate ions, encourages iron to redeposit adjacent to the anodic area. This

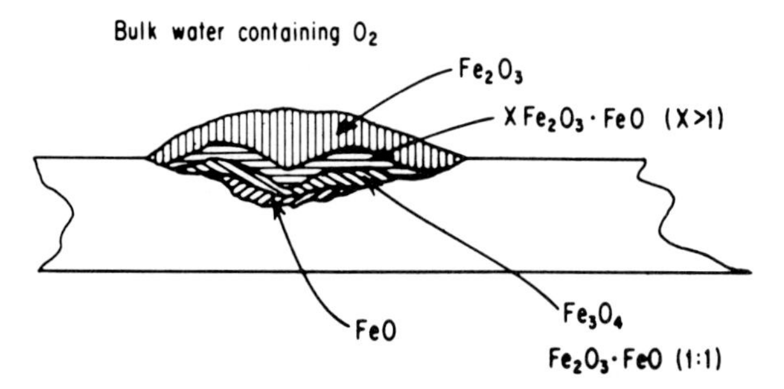

FIG. 20.12 Model of a corrosion tubercle, showing the forms of iron oxide found at various layers as influenced by oxidation-reduction potential.

mechanism continues until the original anodic area is pitted from metal loss, and the pit is filled with porous iron compounds forming a mound, since the by-products are more voluminous than the original metal. Within the tubercle, the aquatic environment is high in chlorides and sulfates and low in passivating oxygen. As a result, both oxygen differential cells and concentration cells form. Advanced tubercles may contain sulfides or acids.

This type of corrosion is common in systems not properly treated (Figure 20.13). Tubercles greatly increase the resistance to water flow and restrict carrying capacity. Large tubercles may break loose periodically and become lodged in critical passageways, such as heat exchanger water boxes or high-pressure descaling sprays in a hot rolling process.

FIG. 20.13 Tuberculation of a small-diameter pipe quickly blocks water flow and leads to perforation of the metal wall.

IMPINGEMENT ATTACK

Impingement attack is another type of selective corrosion involving both physical and chemical conditions, which produce a high rate of metal loss and penetration in a localized area. It occurs when a physical force is applied to the metal surface by suspended solids, gas bubbles, or the liquid itself, with sufficient force to wear away the natural or applied passivation film of the metal. This process occurs repeatedly and each occurrence results in the removal of successive metal oxidation layers.

The most easily identified characteristic of impingement attack is the "horseshoe walking upstream" (Figure 20.14). This pattern generally results when deposits on the metal surface create an eddy around an obstruction and debris or bubbles strike the metal around the deposit. A general metal loss or wall thinning indicates abrasive attack from suspended solids.

Cavitation is a special form of impingement attack most often found in pump impellers. This attack results from the collapse of air or vapor bubbles on the metal surface with sufficient force to produce rapid, local metal loss (Figure 20.15).

DEZINCIFICATION

Dezincification is a type of corrosion usually limited to brass. It takes two forms, general and plug-type, but the mechanism is believed to be the same. For example, dezincification occurs when zinc and copper are solubilized at the liquid-metal interface, with the zinc being carried off in the liquid medium while the copper replates. The replated copper is soft and lacks the mechanical strength of

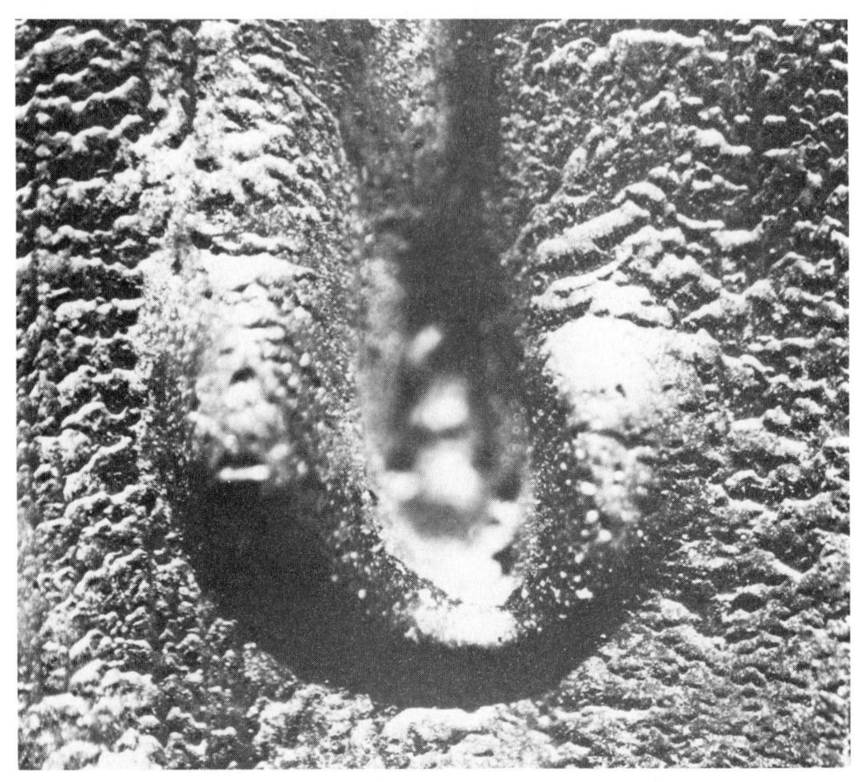

FIG. 20.14 Erosion-corrosion, or impingement attack, of brass. Soft metals and alloys are particularly susceptible to this type of corrosion, accelerated by silt or gas bubbles. (12×)

FIG. 20.15 Cavitation of a bronze pump impeller, common to the operation of a centrifugal pump with a starved suction.

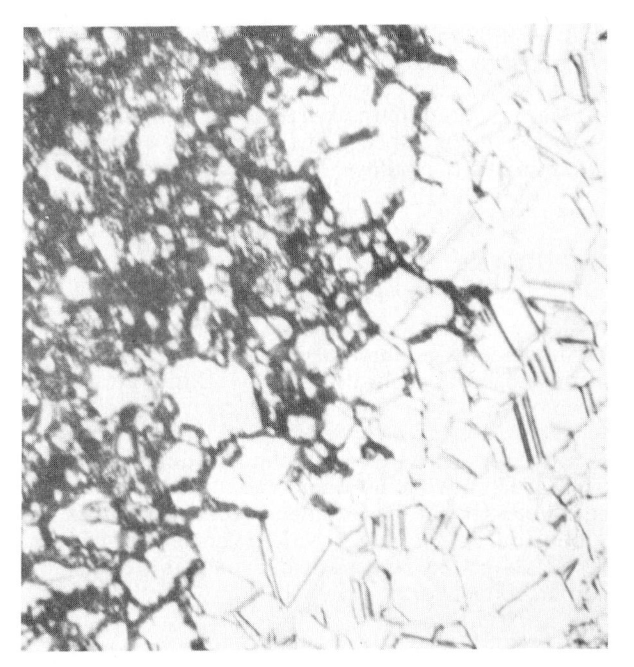

FIG. 20.16 Dezincification of brass, showing the selective leaching of zinc and the residue of copper redeposited in the cavity. (100×)

the original metal. The redeposition accounts for the copper color characteristic of dezincification (Figure 20.16).

General dezincification occurs wherever a large surface of the metal is affected, while plug-type is highly localized.

CORROSION INHIBITION

In water distribution systems, related water-using equipment, and boiler and condensate systems, complete corrosion protection of metal and alloys may be impractical. The goal is to control corrosion to tolerable levels by good design, selection of proper materials of construction, and effective water treatment. Levels of corrosion may be expressed as metal loss in mils per year (mpy), a mil being 0.001 in (0.0025 cm). (1 mpy = 0.025 mm/yr.) In a cooling system, an acceptable loss may be as much as 10 to 15 mpy (0.25 to 0.37 mm/yr); in a supercritical boiler it may be zero.

MATERIALS OF CONSTRUCTION

Where practical and economical, use of corrosion-resistant materials such as copper, stainless steel, copper-nickel alloys, concrete, and plastic may offer advantages over carbon steel. However, when taking this approach, it is important to thoroughly understand the total system. For example, substituting admiralty for mild steel in a heat exchanger in a system exposed to ammonia would be a mistake. Substituting a pure aluminum exchanger for aluminum alloy may reduce process-side corrosion at the expense of accelerating waterside corrosion. Galvanic couples from mixtures of alloys should be avoided.

COATINGS AND LININGS

Another practical way to prevent corrosion is to separate a metal from the water with coatings or linings. It is common to coat selected parts of some systems. For example, water boxes and tube sheets in many utilities are coated to minimize galvanic corrosion between these mild steel components and the Admiralty tubes generally employed in the surface condenser. Coatings and liners of various types are also used to prevent water line corrosion in feed tanks and emergency holding tanks. A special plastic sleeve is used in some utilities and industrial plants to prevent impingement attack at the entrance of condenser tubes. But as a general practice, most industrial systems are so widespread and complex that completely coating all pipes would be too expensive, and coatings are therefore usually limited to the examples given above.

INSULATION

As noted earlier, the joining of dissimilar metals can lead to galvanic corrosion. If this cannot be corrected by a substitution of materials of construction, and it is essential to use these dissimilar metals in the system, they may be insulated from

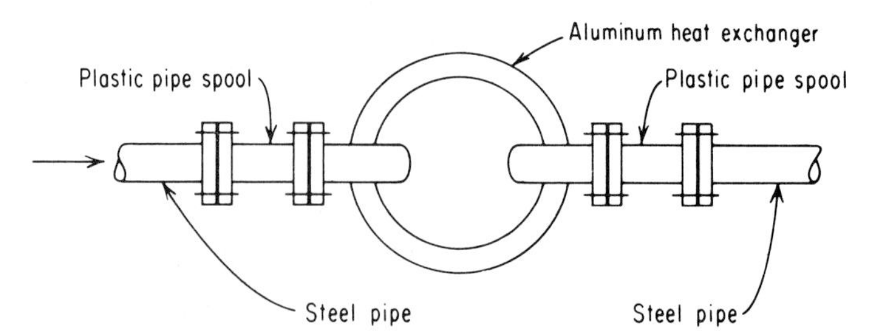

FIG. 20.17 A method of insulating aluminum equipment from steel piping to prevent galvanic attack.

one another. This can be accomplished by inserting nonconductive materials, such as plastic pipe, between them (Figure 20.17). When this is done, it may be necessary to put an electrical cable as a jumper around the isolated unit, because the system may be used as an electrical ground.

APPLIED CHEMICAL INHIBITORS

In spite of the fact that many options exist to minimize corrosion by improved design or better construction methods, because of economics, the majority of systems are designed and fabricated in such a way that a chemical inhibitor program is needed to control corrosion. This applies to all types of water systems: potable water distribution systems, cooling water systems, boiler water systems, process water, and effluent treatment plants.

In the following discussion, because the greatest use of corrosion inhibitors is in cooling water systems, these will serve as the basis for examples of techniques of corrosion protection. The techniques of controlling corrosion in other environments, such as steam/condensate systems or water distribution headers, are covered elsewhere in this text.

CORROSION INHIBITORS

As noted earlier, all the elements of a corrosion circuit must be completed for corrosion to proceed. This involves, among other things, an anodic and cathodic reaction. Therefore, any chemical applied to the water to stop the anodic reaction will stop corrosion; as a corollary, any material added to reduce the rate-determining cathodic reaction will reduce corrosion.

An effective corrosion control program usually depends on specific inhibitors to stop the anodic reaction, slow the cathodic reaction, or both. Typical inhibitors in use are shown by Table 20.2.

TABLE 20.2 Typical Corrosion Inhibitors

Principally anodic	Principally cathodic	Both anodic and cathodic
Chromate	Calcium carbonate	Organic filming amines
Orthophosphate	Polyphosphate	Phosphonates
Nitrite	Zinc	
Silicate		

Of the anodic inhibitors, chromate, once the most widely used, is a very strong inhibitor. Where many common corrosion inhibitors form a barrier layer on the metal surface, chromate reacts with the metal, forming a comparatively hard film of reduced chromate (Cr_2O_3) and alpha iron. The film formed in this manner is tightly adherent and long lasting. Unfortunately, chromates are toxic at these concentrations, so their use is prohibited in many countries for environmental reasons. Orthophosphate, also an anodic inhibitor, forms an iron phosphate film, but this is not as tightly adherent or as long lasting as chromate. Nevertheless, when properly established and maintained, orthophosphate can be an effective corrosion inhibitor. New polymer technology for calcium phosphate scale control has made orthophosphate based programs popular for chrome replacement. Ordinarily, nitrite and silicate receive less consideration than chromate in open recirculating cooling systems because of costs and technical limitations, including potential for deposits (glassy silica), nutrient effect on microbes, and limited effectiveness, particularly true of the silicate species. In addition, these inhibitors require closer control than chromates.

Anodic inhibitors carry a risk: if applied in insufficient quantities, they do not properly passivate all anodic sites. Under these circumstances, the few remaining anodic sites become the focal point of all the electron flow to the cathodic reaction area, and deep pitting can result. Therefore, it is important to maintain sufficient quantities of anodic inhibitor in the system at all times.

Cathodic inhibitors generally reduce the corrosion rate by forming a barrier or film at the cathode, restricting the hydrogen ion or oxygen migration to the cathodic surface to complete the corrosion reaction. Since the overall rate is under cathodic control, the corrosion rate is reduced proportionally to the reduction of cathodic surface area.

Cathodic species—zinc, polyphosphate, and calcium carbonate—are all considered safe compared to the risks of anodic inhibitors used alone.

ANODIC/CATHODIC INHIBITOR MIXTURES

When inhibitors were first introduced into water systems, they were frequently composed of single active components (e.g., chromate). Over the years, it has been found that some ingredients improve the performances of others, a chemical principle called synergism. For example, chromate used by itself requires 200 to 300 mg/L CrO_4 to prevent corrosion in an open recirculating system; but, chromate combined with zinc, various organic and inorganic phosphates, or molybdates, provides equal or better results at only 20 to 30 mg/L CrO_4. Some systems currently operate quite successfully with less than 10 mg/L CrO_4, a feat that would be impossible without effective supplements only recently developed. Currently, chromates are used alone only in unique situations.

ORGANIC FILMERS

These materials are typified by forming filming layers on metal surfaces to separate the water and metal. These include filming amines for condensate systems that work effectively only in significantly reduced oxygen environments and soluble oils that are generally limited to special applications in cooling water systems. These materials form and maintain a dynamic barrier between the water and metal phases to prevent corrosion. This film is generally substantially thicker than the films established with the proper application of inorganic inhibitors such as chromate or zinc. An inherent danger in the filmer approach is that a small break in the continuous film could allow the corrosive agent to be focused on the unprotected area resulting in rapid penetration of the metal.

Another series of inhibitors of note are those applied to reduce copper and copper-alloy corrosion. These include mercaptobenzothiazole, benzotriazole, and tolyltriazole.

These are organic compounds that react with the metal surface and form protective films on copper and copper alloys. They form an extremely thin film barrier that is not to be compared with the thicker soluble oil-type barrier. These materials are receiving more attention in modern treatments as more systems operate under increasingly higher TDS levels, a condition that encourages copper corrosion. Certain organophosphorous scale inhibitors used in cooling systems are aggressive to copper-type materials, requiring the addition of a copper inhibitor film when these are used.

CATHODIC PROTECTION

Sacrificial anodes reduce galvanic attack by providing a metal (usually zinc, but sometimes magnesium) that is higher on the galvanic series than either of the two metals that are found coupled together in a system. The sacrificial anode thereby becomes anodic to both metals and supplies electrons to these cathodic surfaces. Design and placement of these anodes is a science in itself. When properly employed, they can greatly reduce loss of steel from the tube sheet of exchangers employing copper tubes, for example. Sacrificial anodes have helped supplement chemical programs in a variety of cooling water and process water systems.

Impressed-current protection is a similar corrosion control technique, which reverses the corrosion cell's normal current flow by impressing a stronger current of opposite polarity. Direct current is applied to an anode—inert (platinum, graphite) or expendable (aluminum, cast iron)—reversing the galvanic flow and converting the steel from a corroding anode to a protected cathode. The method is very effective in protecting essential equipment such as elevated water storage tanks steel coagulation/flocculation vessels, or lime softeners (Figure 20.18).

WATER DISTRIBUTION SYSTEMS

The control of corrosion in water distribution systems presents some difficult problems in economics and water chemistry. In municipal water distribution systems, the choice of treatment chemicals is limited because the treated water must meet potable standards. Many industrial plants have the same problem, because drinking water is often tapped from the general mill supply. A secondary problem is the selection of a chemical for corrosion control that will not be harmful or undesirable in any of the uses of the water in the plant. And economics is an important consideration because the user resists the addition of any appreciable treatment costs to the 10 to 50 cents per 1,000 gallons that he may already be paying for the water.

Often where a water is corrosive, special materials of construction may be selected for the distribution system. These may include galvanized instead of ordinary steel piping; or it may be necessary to use special pipe such as thin-wall stainless-steel tubing, lined pipe, or nonmetallic pipe.

However, the majority of installations for municipal or mill water distribution use plain steel. This is subject to corrosion, chiefly because of dissolved oxygen, an unfavorable stability index, or both. Corrosion may be aggravated by deposition of suspended solids, which might develop from after-precipitation of a lime treatment or biological activity.

The first step in correcting a distribution system corrosion problem should be a review of the water analysis to see if a change in the stability index would be beneficial. In many cases the corrosion can be brought under control by carrying a positive Langelier index or a stability index below 6.

FIG 20.18 Cathodic protection system in place in a slurry-type clarifier. The anodes, which are hanging vertically, are almost expended.

Lime is usually fed to the system, since it has the dual effect of increasing calcium and alkalinity. One of the problems with lime, however, is that it may not be readily dissolved. Suspended lime in the system may settle at times of low flow or in areas of low velocity to create concentration cells. Even though it is more troublesome to do so, if proper feeding and dissolution of lime cannot be assured, then caustic soda should be fed and this may be supplemented with calcium chloride to achieve the overall effect of adding lime. If the plant has its own water treatment system, it may be possible to correct the stability index ahead of the final filters to avoid after-precipitation of undissolved lime.

If the water in the distribution system has been disinfected and carries a chlorine residual, this helps greatly in controlling corrosion by eliminating the additive effect of microbial activity to the corrosion process.

The conventional chemical treatments used in distribution systems include polyphosphates and silicates.

The exact mechanism by which polyphosphates control corrosion in distribution systems is not well known, but it probably includes dispersancy and the ability of polyphosphates to inhibit calcium carbonate precipitation (threshold treatment). Polyphosphates are usually applied at a dosage of about 2 mg/L. They are not objectionable in potable water, but they do sequester calcium so that polyphosphate-treated waters may not be completely reacted in ion exchange systems. In many cases this is not objectionable, but where high-purity demineralized water is required, the carrythrough of only 0.1 to 0.5 mg/L hardness could be objectionable.

Sodium silicates are less frequently used because the dosage required is usually in the range of 8 to 10 mg/L as SiO_2, requiring a feed of about 12 to 15 mg/L sodium silicate. Nevertheless, the treatments are effective, and although they may not be economical for treatment of the total water supply, they may be applied instead to points of use along the distribution system where the cost of the treatment can be justified. For example, in soft water areas where water is quite corrosive, such as along the northeastern seaboard, sodium silicate may be used to treat incoming municipal water at shopping centers and commercial buildings.

Zinc is effective in corrosion control of distribution systems just as in cooling water systems. It is usually fed with polyphosphates, but where the polyphosphate is objectionable it may be fed with orthophosphates. Since the orthophosphates are anodic inhibitors and because the level of treatment must be kept quite low to avoid calcium phosphate deposits, a treatment of this kind must be very carefully monitored for potential pitting-type attack. Such treatment may provide general corrosion protection, but could lead to localized pitting.

Sacrificial anodes are sometimes used at critical locations in a distribution system where the water is tapped off for once-through cooling of heat exchangers or process equipment. They may also be used in domestic hot water heaters in areas where the municipal supply is unusually aggressive, but it is not economical to treat the total flow.

Since oxygen is a prominent factor in corrosion of steel distribution systems, deaeration may be a practical process for corrosion control. Reduction of O_2 concentration to less than 0.5 mg/L will usually provide adequate protection. A number of industrial installations have used cold water deaerators (vacuum deaerators) for protection of long pipelines. The cost is usually justified not just on the basis of maintenance cost, but also in energy savings, since corrosion control maintains a smooth pipe wall surface, thus reducing pumping power consumption.

Dispersants, while not corrosion inhibitors per se, play a prominent role in controlling corrosion by preventing solids deposition and subsequent formation of oxygen concentration cells.

MONITORING RESULTS

Most water systems are in continuous use, so it is rare to have access to inspect the actual system to observe or measure corrosion and corrosion inhibition. A number of tools have been developed to measure corrosion by indirect means. Some of the more common of those employed today are outlined below:

Corrosion Coupons

These preweighed metal specimens (Figure 20.19) are normally put into a system for 30 to 90 days. Following removal, they are cleaned, reweighed, and observed. The metal loss (expressed in mpy) and the type of attack (general, pitting) is then determined and reported.

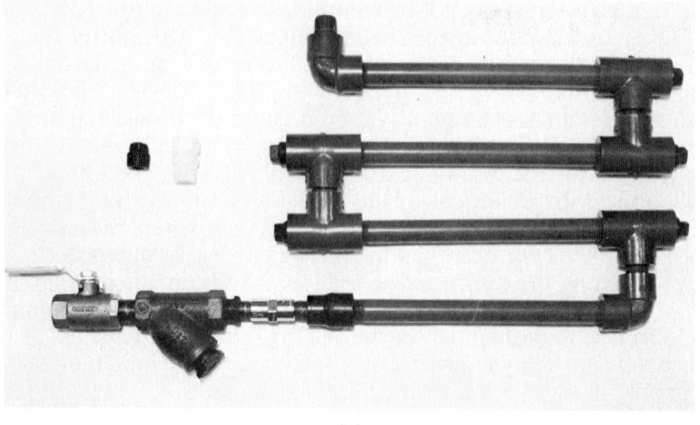

(*a*)

FIG. 20.19 (*a*) Installation of corrosion coupon rack. The control valve at left connects to the water line being sampled and is throttled to produce the correct velocity past the coupons, as measured at the overflow (top). The coupons are inserted at each tee.

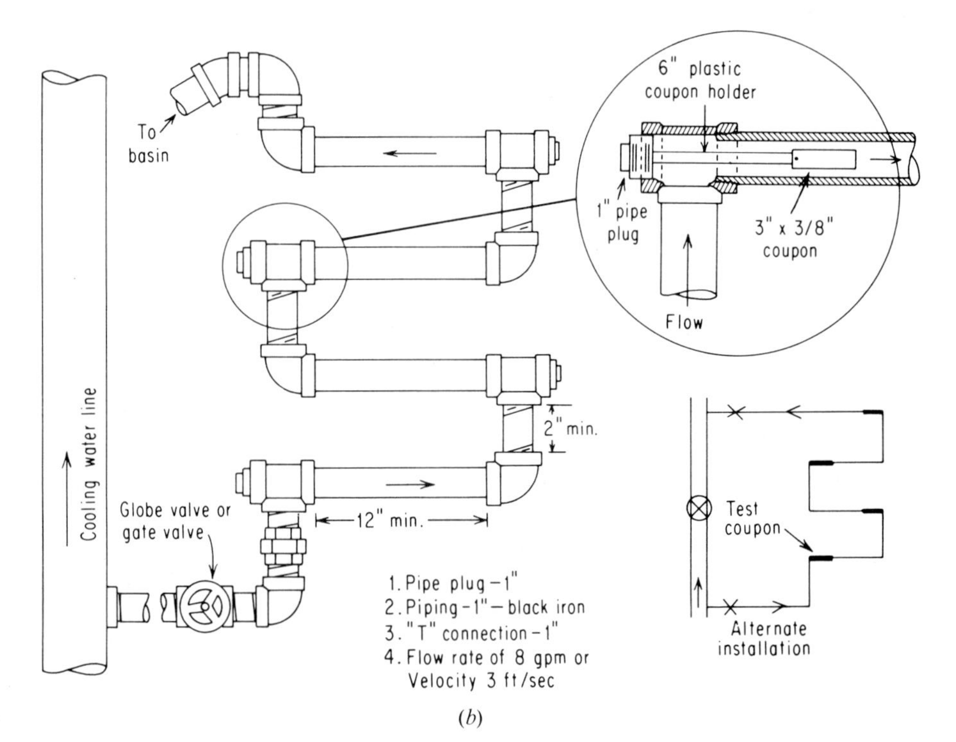

(*b*)

FIG.20.19 (*b*) Installation of corrosion coupons.

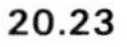

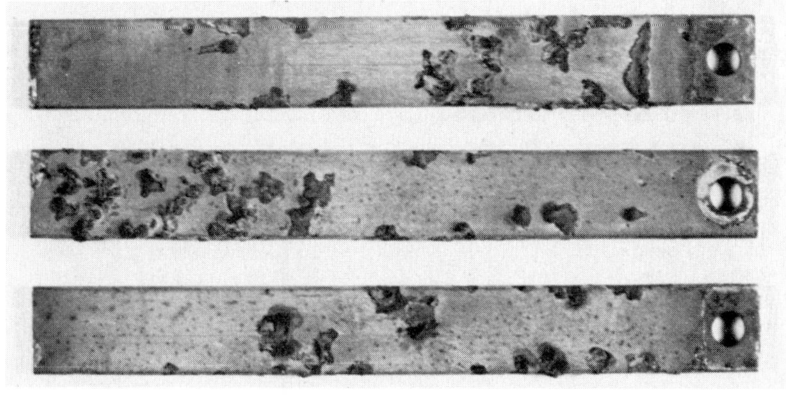

(*c*)

FIG. 20.19 (*c*) Typical coupons after exposure.

Occasionally, coupons undergo visual evaluation only. Coupon data vary widely based on coupon finish (e.g., polished versus sandblasted), location in the system, length of exposure, type of metallurgy, and type of pretreatment, if any. Corrosion coupons are excellent tools when properly used and evaluated. The outline diagram (Figure 20.20) shows a typical report of a coupon evaluation.

Corrosion Nipples

These are similar to coupons in concept, but are not preweighed and only visually evaluated. They offer an advantage over a poorly designed and installed corrosion coupon setup that does not take flow velocity into consideration. Generally, however, coupons have more utility because they provide more information if the coupon racks have been properly designed.

A nipple consists of a simple pipe that is installed on a bypass arrangement (Figure 20.21) so that the nipple can be easily removed for inspection at any time.

Corrosion Meters

Corrosion coupons provide long-term data. For more immediate results a corrosion meter can provide a readout in 24 h or less following initial system insertion. Like the coupon, the corrosion rate is indicated in mil per year. The meter works by measuring an electrical potential across electrodes made of the metal being evaluated.

When allowed to stabilize in the system, the meter illustrated in Figure 20.22 can provide instantaneous corrosion readings. This is useful for optimizing pH, TDS, inhibitor level, chlorine application, and other control variables. The more sophisticated meters provide a pitting tendency index, so both the amount of metal loss and type of loss is indicated.

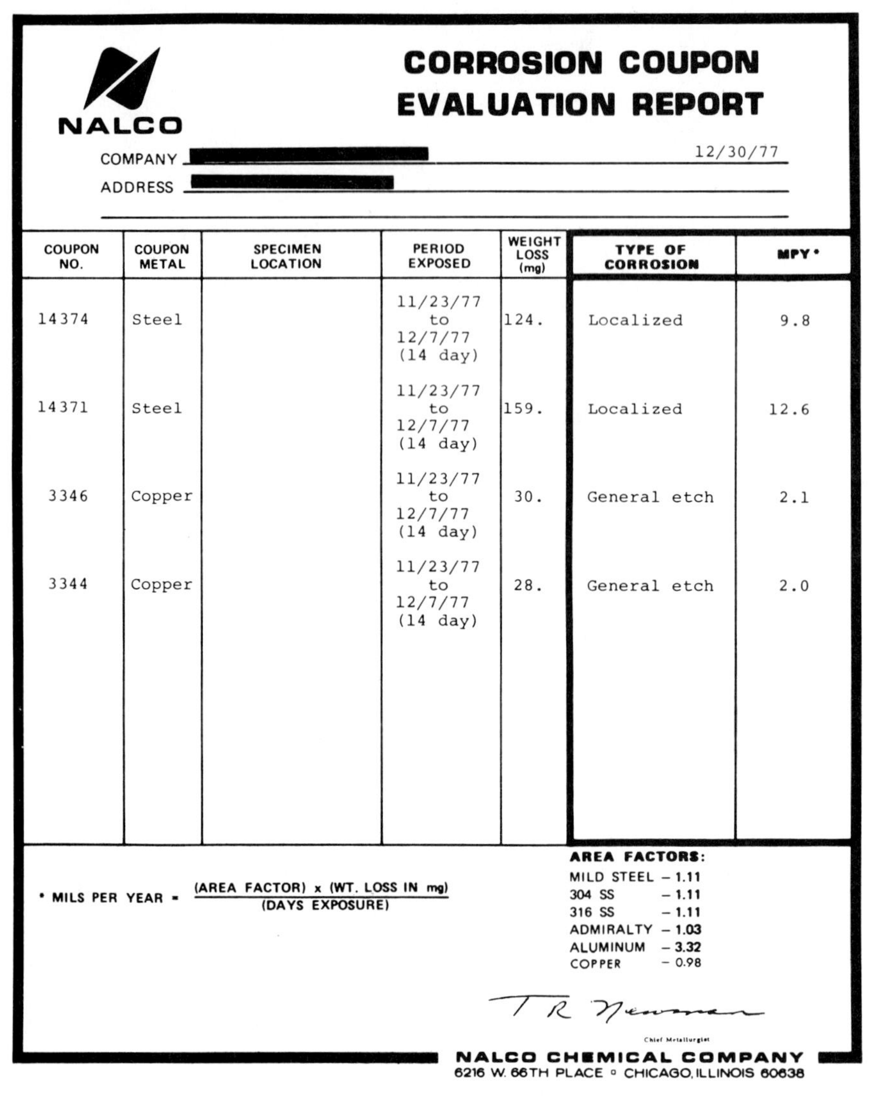

NALCO

CORROSION COUPON EVALUATION REPORT

COMPANY [redacted] 12/30/77

ADDRESS [redacted]

COUPON NO.	COUPON METAL	SPECIMEN LOCATION	PERIOD EXPOSED	WEIGHT LOSS (mg)	TYPE OF CORROSION	MPY*
14374	Steel		11/23/77 to 12/7/77 (14 day)	124.	Localized	9.8
14371	Steel		11/23/77 to 12/7/77 (14 day)	159.	Localized	12.6
3346	Copper		11/23/77 to 12/7/77 (14 day)	30.	General etch	2.1
3344	Copper		11/23/77 to 12/7/77 (14 day)	28.	General etch	2.0

$$\text{* MILS PER YEAR} = \frac{\text{(AREA FACTOR)} \times \text{(WT. LOSS IN mg)}}{\text{(DAYS EXPOSURE)}}$$

AREA FACTORS:

MILD STEEL – 1.11
304 SS – 1.11
316 SS – 1.11
ADMIRALTY – 1.03
ALUMINUM – 3.32
COPPER – 0.98

T R Newman
Chief Metallurgist

NALCO CHEMICAL COMPANY
6216 W. 66TH PLACE ▫ CHICAGO, ILLINOIS 60638

FIG. 20.20 Corrosion coupon evaluation report.

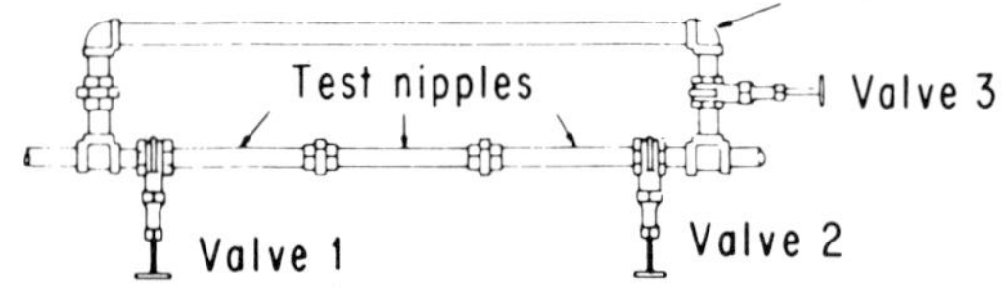

FIG. 20.21 The use of test nipples, installed in steam or condensate lines, permits both visual inspection of system conditions and a measure of corrosion.

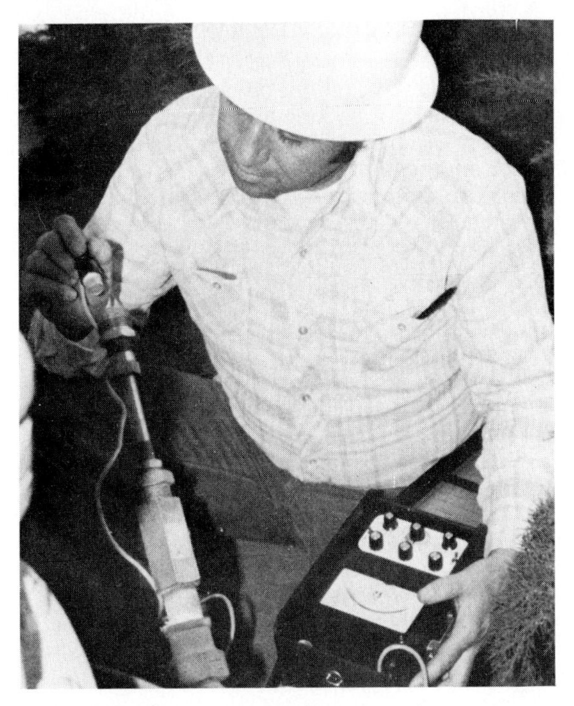

FIG. 20.22 Operator connecting a corrosion meter to a permanently installed probe to read corrosion rate of the probe material in the system. *(Courtesy of Rohrback Corporation.)*

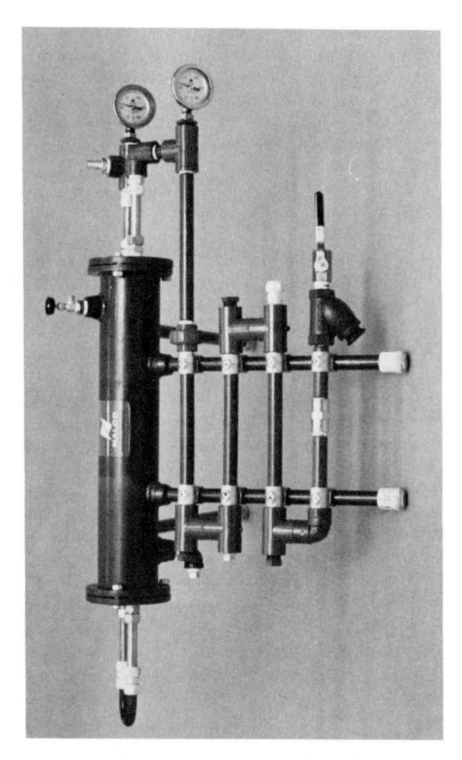

FIG. 20.23 A field test unit, including heat exchanger and coupon rack, used for evaluating corrosion inhibitors and the effect of temperature and flow rate on their performance.

Although meters provide some unique advantages over coupons, coupons continue to be employed to determine long-term effects of water and metal contact under fluctuating system conditions. The coupon and the meter are effective tools that complement one another.

Instantaneous corrosion monitoring has been applied to feed and control the level of inhibitor in cooling water systems. This technology has generally been limited to chromate inhibitors.

Field Evaluation Unit

Figure 20.23 shows a typical field evaluation unit. It has several key points. First, it incorporates a scientifically designed corrosion coupon rack. The flow (and therefore velocity) through the rack is selected and controlled. There are positions for both meter probes and corrosion coupons so that short-term and long-term corrosion rates can be effectively measured. Second, the unit duplicates heat transfer surfaces. Since temperature or heat flux can have a significant effect on the scaling, fouling, or corrosion rates being experienced in a system, monitoring this variable is important. A properly installed unit can provide an excellent readout and visual indication of what is happening in the operating equipment. The field evaluation unit effectively monitors corrosion on both heat transfer and nonheat transfer surfaces under conditions that closely duplicate flow velocities, metallurgy, and heat flux of the actual operating system.

CHAPTER 21
DEPOSIT CONTROL

Deposits are conglomerates that accumulate on water-wetted surfaces and interfere with system performance, either by gradually restricting flow or by interfering with heat transfer. Deposits include scale, foulants, or a combination of the two. Scale forms when the concentration of a dissolved mineral exceeds its solubility limit and the mineral precipitates. The most common example is calcium carbonate scale formed when the Langelier index or the stability index indicates a condition of $CaCO_3$ supersaturation (Chapter 4). A foulant is any substance present in the water in an insoluble form, such as silt, oil, process contamination, or biological masses.

Deposits are most often an accumulation of sediments or settled solids that drop out at some point in a system where the water velocity falls to a level too low to support the material in the stream. This definition is oversimplified, since deposits are usually found to contain—in addition to settled solids—scale, corrosion products, microbial slimes, reaction products, and oil or grease.

Deposits can occur in any type of water conduit, water-using device, or storage vessel. They occur in boilers as well as in distribution systems and heat exchange equipment. In this text, those deposits forming in industrial cooling systems and industrial and municipal distribution systems are concentrated on. The principles of deposit control in other systems can be extrapolated from this focus on the general fundamentals of deposit formation.

Unlike scale, a deposit is seldom composed of one substance; it is almost always an accumulation of materials. For example, silt from makeup water may begin to deposit in a low-flow section of a heat exchanger. If the water is on the border of $CaCO_3$ instability, the settling solids may act as the initiator for the scaling mechanism, further obstructing flow, and a deposit will form. This will then produce a combination of silt and scale, and the deposit may form even though the system is being treated for scale control by acid or some other chemical.

Once the deposit forms, dormant microbial organisms may become active, and if the deposit were then analyzed, all three constituents would be found. The difficulty in studying the analysis is deciding the likely order of events. The sequence could have started with microbial activity blocking the flow, causing the codeposition of the other two constituents. Analyzing deposit mechanisms and devising effective control actions demand a study of the total system. The problem may not be caused by suspended solids alone.

DEPOSIT SOURCES

The source of potential depositing material may be external or internal to the system. One of the most prominent external sources is the water supply itself, which may contain suspended solids, such as silt in a turbid surface water, soluble or precipitated iron, manganese, or carryover from a clarifier or other pretreatment unit.

A second source of depositing materials, particularly in an open recirculating cooling system with a cooling tower, is air. Cooling towers act as large air scrubbers, and the water is quite effective in capturing dust, microbes, and other debris from the large volume of air that it contacts. Because the amount of suspended solids in air is always changing, the suspended solids level in the circulating water also changes considerably from time to time; seasonal variations are common, depending on local environmental conditions. Atmospheric sources contribute significantly to deposits in arid sections of the country, near farm lands, and in areas of high prevailing winds.

Cooling towers also scrub industrial gases from air. Ammonia, hydrogen sulfide, sulfur dioxide, and other gases are readily absorbed into water, where they react chemically, changing the water characteristics. Sometimes this change can be so dramatic that scaling, corrosion, or microbial masses may suddenly obstruct a system in a matter of days.

Another significant external factor influencing deposits in industrial systems is leakage of process fluids into a water stream. This leakage may contribute directly or indirectly to the deposits. The most common effect is to provide food for microbial growth if the leaking contaminants are organic, such as oil or food substances.

Miscellaneous external sources include water used on pumps in the water circuit and lubricants applied to valves, pump glands, and bearings that leak into the water system through these seals.

Internal sources of deposit originate in the circulating water; chemical precipitation, formation of corrosion products, polymerization, and biological growth are examples of these.

Precipitation is usually induced by temperature change or a disturbance in the equilibrium of the system which causes the solubility product of some constituent in the system, such as calcium carbonate, to be exceeded. Steel corrosion products form iron oxides and hydroxides that tend to deposit in the system, particularly on hot surfaces. Polymerization of organic materials is a more nebulous effect; an example of this is the coagulation of proteins which occurs when the temperature of water reaches 140 to 150°F (60 to 65°C), the pasteurization range. There may be naturally occurring proteins in the system brought in with the raw water makeup, or the proteins may be by-products of the metabolism of microbial life in the system. The polymerization of other organic materials may also occur. Finally, biological activity may be encouraged by the nutrients and food substances present in the water.

DEPOSIT APPEARANCE

Because a deposit is frequently sedimentary, it may be found in the system as uncompacted silt in equipment or piping, or it may have hardened as additional chemical or biological activity developed after sedimentation of the initial mate-

FIG. 21.1 Laminar nature typical of most deposits, indicating variations in the environment. Analysis of each layer is sometimes required to determine all of the reasons for deposit formation.

rial. Where biological activity develops rapidly, the deposit may be predominantly organic material from cell substances, and it may be porous or slimy. The deposit may be modified by chemical reactions, particularly if high temperatures prevail, which could encourage chemical precipitation or corrosion reactions. At any rate, regardless of the nature of the deposit, because it is basically sedimentary, it usually has a laminar structure which tells something of its history (Figure 21.1).

DEPOSIT IDENTIFICATION

Evaluating the cause of a deposit in a water distribution system requires a complete examination of the water system involved. A flow diagram showing the water circuit, piping, and individual pieces of equipment in the system is required. Flow rates at critical points in the system, temperatures encountered (including seasonal variations), and materials of construction are also important data. The location of the deposit in the system should be indicated on the flow diagram.

Sensual observations are important, particularly those relating to the appearance of the equipment or piping where the deposit was found, the nature of the deposit itself, and samples of water from the system taken at the deposit location. The deposit may have an odor when first removed from the system which may disappear when it reaches the laboratory for analysis. The fresh deposit may feel slimy, yet may lose this characteristic if it dries before reaching the laboratory. So it is important that the person who is sampling record observations of the fresh sample.

Four types of analytical tests are used to supply chemical and biological data to supplement the physical data: a deposit analysis, a complete mineral analysis

NALCO

ANALYTICAL LABORATORY REPORT

FROM:

Analysis No.	329757
Date Sampled	4/22/73
Date Received	4/29/73
Date Pri...	5/03/73

Sample Marked: Water Delivery Line

>>>DEPOSIT ANALYSIS<<<

Physical Appearance: Layered Tan Solid

Inorganics in dried sample	
Iron Fe_2O_3	35
Silica (SiO_2)	30
Phosphorus (P_2O_5)	3
Alumina (Al_2O_3)	12
Calcium (CaO)	3
Sulfur (SO_3)	1
Carbonate (CO_3)	0
Loss at 200°C	16

The following elements were not determined:

Na Mg Zn Cl K Ti V Cr Mn Co Ni Cu Sr Sn Pb

Lab comments:

No iron present in top layers.

FIG. 21.2 Laminar deposit taken from water supply line. This is a composite analysis. Note that analysis of the top surface shows it to be free of iron.

of the water, and a microbial analysis of the system water and of the deposit (Figures 21.2 to 21.5).

The water analysis shown in Figure 21.3 indicates that this water is stable under prevailing river conditions at 50°F (10°C). However, if the water is used for cooling and is heated to 100°F (38°C), the stability index shows that $CaCO_3$ will precipitate, which may cement together the settled clay deposits. This indicates

NALCO

ANALYTICAL LABORATORY REPORT

FROM:

Analysis No.	C 557342
Date Sampled	4/22/73
Date Received	4/24/73
Date Printed	5/05/73

Sample Marked: River Water

>>>WATER ANALYSIS<<<

		PPM
CATIONS:		
	Calcium ($CaCO_3$) - Soluble	256
	Magnesium ($CaCO_3$) - Soluble	140
	Sodium ($CaCO_3$)	320
	Ammonia ($CaCO_3$)	2
ANIONS:		
	Bicarbonate ($CaCO_3$)	250
	Chloride ($CaCO_3$)	120
	Sulfate ($CaCO_3$)	200
	Nitrate ($CaCO_3$)	150
OTHERS:		
	pH (pH units)	8
	Alkalinity ($CaCO_3$) - Total	250
	Alkalinity ($CaCO_3$) - Phenophthalein	*ND (2.)
	Conductivity (Micromhos per cm)	1580
	Iron (Fe) - Soluble and insoluble	1.5
	Total dissolved solids (180°C)	865
	Suspended solids (105°C)	40
	Silica (SiO_2) - Soluble	12

* Not detected at concentration in parenthesis.

FIG. 21.3 Analysis of water from same system as Figure 21.2.

that either softening or a scale control chemical treatment should be considered to prevent problems of excessive deposit formation.

Taken together, a deposit analysis and a water analysis may indicate whether the suspended materials present in the water source can create sedimentary-type deposits. Confirmation of the role of raw water suspended solids in the formation of the deposit may be further verified by an analysis of the solids themselves.

NALCO

ANALYTICAL LABORATORY REPORT

From:

Sample Marked:
Condenser Deposit

Analysis No.	B 42384
Date Sampled	12/27/83
Date Received	12/29/83
Date Printed	1/ 4/84

>>> MICROBIOLOGICAL EVALUATION <<<

PHYSICAL APPEARANCE :	Brown Deposit
TOTAL AEROBIC BACTERIA	80,000,000
Aerobacter	<1000
Pigmented	20,000
Mucoids	<1000
Pseudomonas	35,000,000
Sporeformers	700
Others	44,979,300
TOTAL ANAEROBIC BACTERIA	13,000
Sulfate Reducers	10,000
Clostridia	3,000
IRON-DEPOSITING	
Gallionella	None
Sphaerotilus	None
TOTAL FUNGI	800
Molds	800
Yeast	<100
ALGAE	
Filamentous	None
Non-Filamentous	None
OTHER ORGANISMS :	None

Lab Comments:
(All Counts Express Colony Forming Organisms per ML of Sample.)

FIG. 21.4 Microbial analysis of water from supply line.

These analysis (Figures 21.6 and 21.7) indicate that this turbid water contains a high level of colloidal clay originating in the river supplying the plant. The settleability of these solids may be determined by the Imhoff cone to provide a basis for deciding whether clarification may be necessary.

Figure 21.8 shows a microbiologically contaminated water source. This level of contamination is common in river waters and some well systems fed by surface waters. The level of microbial contamination in this system indicates the need for

NALCO

ANALYTICAL LABORATORY REPORT

From:

Analysis No.	B 42385
Date Sampled	12/27/83
Date Received	12/29/83
Date Printed	1/ 4/84

Sample Marked:
River Make-Up

>>> MICROBIOLOGICAL EVALUATION <<<

PHYSICAL APPEARANCE :	Clear Liquid with Sediment
TOTAL AEROBIC BACTERIA	50,000
Aerobacter	2,000
Pigmented	5,000
Mucoids	<10
Pseudomonas	20,000
Sporeformers	None
Others	23,000
TOTAL ANAEROBIC BACTERIA	10
Sulfate Reducers	10
Clostridia	None
IRON-DEPOSITING	
Gallionella	None
Sphaerotilus	None
TOTAL FUNGI	<10
Molds	<10
Yeast	<10
ALGAE	
Filamentous	Few
Non-Filamentous	None
OTHER ORGANISMS :	None

Lab Comments:
(All Counts Express Colony Forming Organisms per ML of Sample.)

FIG. 21.5 Microbial analysis of deposit from condenser cooled by same water supply as Figure 21.3.

a continuous biocidal program to protect the system against fouling. For once-through water usage, a low chlorine dosage, and a specific biodispersant should be considered separately or in combination to maintain a clean system. Slime-forming bacteria present a heat transfer problem in all systems where they are found; bacterial slimes restrict heat transfer capacity much more severely than calcium carbonate scale.

NALCO

ANALYTICAL LABORATORY REPORT

FROM:

Analysis No.	P 8794
Date Sampled	10/1/81
Date Received	10/5/81
Date Printed	10/8/81

Sample Marked: River Water

>>>WATER ANALYSIS<<<

	PPM
CATIONS:	
CALCIUM ($CaCO_3$) - SOLUBLE	127
MAGNESIUM ($CaCO_3$) - SOLUBLE	84
ANIONS:	
BICARBONATE ALKALINITY ($CaCO_3$)	150
SILICA (SiO_2) - Soluble	17
OTHERS:	
pH (pH UNITS)	7.7
ALKALINITY ($CaCO_3$) - TOTAL	150
ALKALINITY ($CaCO_3$) - PHENOLPHTHALEIN	ND*
ALKALINITY ($CaCO_3$) - P-$BaCl_2$	ND*
CONDUCTIVITY (MICROMHOS PER cm)	350
TURBIDITY (NEPHELOMETRIC TURBIDITY UNITS)	47
CALCIUM ($CaCO_3$) - SOLUBLE AND INSOLUBLE	141
MAGNESIUM ($CaCO_3$) - SOLUBLE AND INSOLUBLE	60
IRON (Fe) - SOLUBLE AND INSOLUBLE	3.8
SUSPENDED SOLIDS	166

* NOT DETECTED (BELOW INDICATED LIMIT OF DETECTION)

FIG. 21.6 Analysis of a turbid river water.

In Figure 21.2 the deposit analyzed had a segmented or layered structure, the upper layers of this deposit indicating the high silt content of the system. The high iron content of the lower layers indicates active corrosion occurring below the silt.

Layering is often the result of long-term corrosion, wherein the strata represent changes in distribution of the various forms of iron, as in the development of a long-term corrosion tubercle. Examination of other layered deposits may indicate various periods of process contamination, variations in suspended solids, or lack of pH control of the water system.

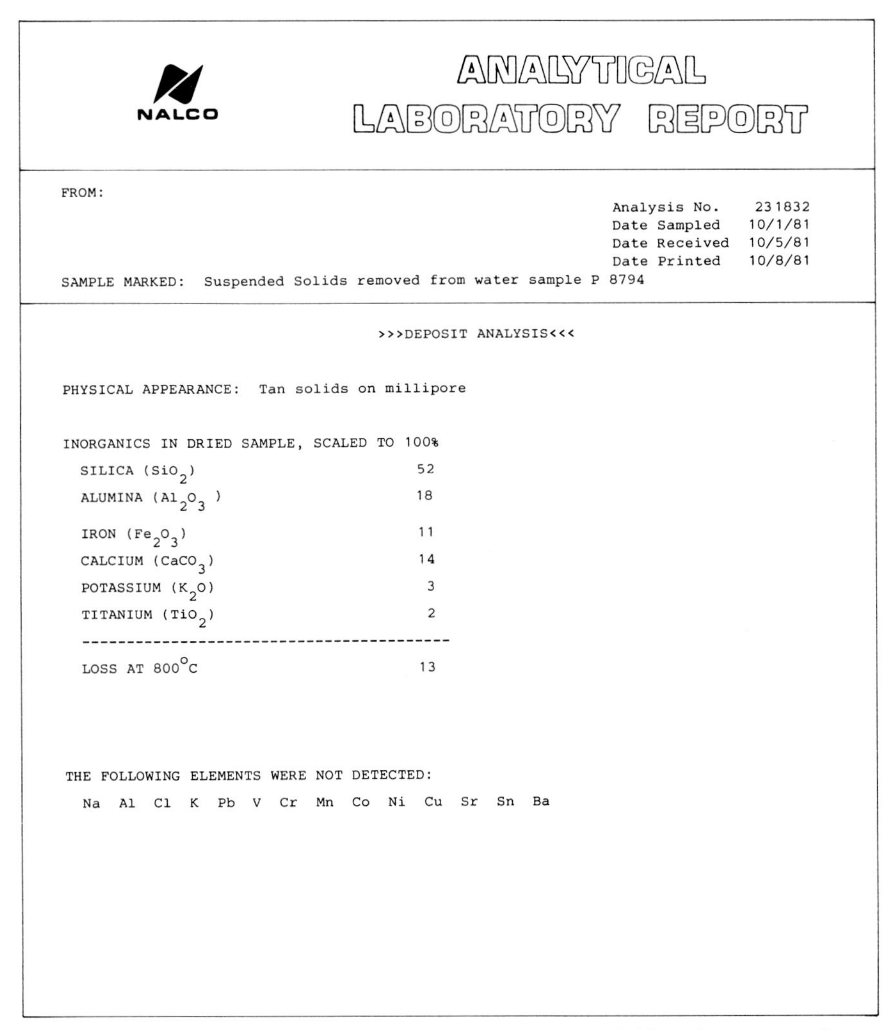

NALCO

ANALYTICAL LABORATORY REPORT

FROM:

Analysis No. 231832
Date Sampled 10/1/81
Date Received 10/5/81
Date Printed 10/8/81

SAMPLE MARKED: Suspended Solids removed from water sample P 8794

>>>DEPOSIT ANALYSIS<<<

PHYSICAL APPEARANCE: Tan solids on millipore

INORGANICS IN DRIED SAMPLE, SCALED TO 100%

SILICA (SiO_2)	52
ALUMINA (Al_2O_3)	18
IRON (Fe_2O_3)	11
CALCIUM ($CaCO_3$)	14
POTASSIUM (K_2O)	3
TITANIUM (TiO_2)	2
LOSS AT 800°C	13

THE FOLLOWING ELEMENTS WERE NOT DETECTED:

Na Al Cl K Pb V Cr Mn Co Ni Cu Sr Sn Ba

FIG. 21.7 Analysis of suspended solids from sample, Figure 21.6. Note the presence of potassium, normally water soluble, but bound in many forms of clay (Chapter 4).

Deposits can be analyzed physically as well as chemically. Microscopic examination of a deposit explores the appearance of individual layers. X-ray fluorescence shows the quantitative levels of specific atoms contained in a deposit. X-ray diffraction techniques determine the crystalline composition of materials within the deposit. A scanning electron microscope with an energy dispersive analysis of x-ray (EDAX) capability can identify elements and provide visual examination of individual deposit layers at high magnification. These examinations are not always necessary, but are valuable aids to solving difficult problems.

NALCO

ANALYTICAL LABORATORY REPORT

From:

Analysis No. B 42386
Date Sampled 12/27/83
Date Received 12/29/83
Date Printed 1/ 4/84

Sample Marked:
Illinois River Water

>>> MICROBIOLOGICAL EVALUATION <<<

PHYSICAL APPEARANCE :	Clear Liquid
TOTAL AEROBIC BACTERIA	320,000
Aerobacter	<10
Pigmented	<10
Mucoids	<10
Pseudomonas	80,000
Others	240,000
TOTAL ANAEROBIC BACTERIA	100
Sulfate Reducers	100
ESCHERICHIA COLI	40
IRON-DEPOSITING	
Gallionella	None
Sphaerotilus	None
TOTAL FUNGI	None
Molds	None
Yeast	None
ALGAE	
Filamentous	None
Non-Filamentous	None
OTHER ORGANISMS :	None

Lab Comments:
(All Counts Express Colony Forming Organisms per ML of Sample.)

FIG. 21.8 Microbial analysis of a contaminated water.

In instances where microbial activity has been a contributor to deposit formation, regularly scheduled sampling of water from the deposit area for microbiological analysis is necessary to assure that the disinfection program is protecting the system.

Deposit control programs work best when used for preventive rather than corrective purposes. All the tools noted earlier are employed for evaluating the problem. For example, if the survey shows that the concentrations and type of solids introduced into a system are likely to settle at the velocity encountered in the system, then either chemical treatment or a reduction of these solids is necessary, or both. This may take the form of complete treatment of the incoming makeup water. In a recirculating cooling system, it may require the inclusion of a slip-stream filter in the loop. If the change in the stability index resulting from a temperature change is such that it cannot be dealt with employing sequestering agents or dispersants, then the pretreatment system should provide not only for removal of suspended solids, but also softening of the water by lime treatment. Finally, sterilization of the makeup water, either with the clarification/softening operation or separately, should be practiced if microbial activity cannot be kept under control by treatment of the process water itself.

Preventive control may also require modification of the system. For example maintaining solids in suspension may require increasing the velocities through the water circuit by such means as recycling around the equipment with a pump, or agitating the water by injection of air into critical segments of the system at periodic intervals, called "air-rumbling."

TREATMENTS TO CONTROL SYSTEM DEPOSITS

Chemical treatment programs for the control of deposits fall into four general categories: (a) threshold inhibitors, (b) dispersants, (c) surface active agents, and (d) crystal modifiers.

Threshold Inhibitors

This first class of deposit control chemicals includes sequestering agents such as polyphosphates, organophosphorous compounds, and polymers (polyacrylates, for example). These exert a "threshold effect" (Chapter 3), reducing the potential for precipitation of calcium compounds, iron, and manganese. Threshold inhibition causes a delay in precipitation by the application of substoichiometric amounts of inhibitor; the amount of inhibitor is usually in the range of 1 to 5 mg/L, even in the treatment of very hard waters. This threshold dosage is possible because the chemical adsorbs only on the surface of the incipient precipitate, so that only a small fraction of the precipitating material consumes the active inhibitor.

Inorganic polyphosphates are used extensively for scale control primarily in potable water systems, particularly in municipal plants that have lime-softened their water supplies. Sequestration—another name for threshold treatment—is not practical in evaporative cooling systems because of the high dosage of polyphosphate required to form soluble complexes in the concentrated water.

Dispersants

The second group is made up of organic dispersants, including organophosphorous compounds and polyelectrolytes. If the applied dispersant is a charged molecule, such as a polyelectrolyte, it will disperse suspended solids by adsorbing to their surfaces, thus adding an electrostatic charge to each particle, causing mutual repulsion. In other words, dispersants act oppositely from coagulants, augmenting the charge of suspended solids rather than neutralizing them. (See Chapter 8.)

(*a*)

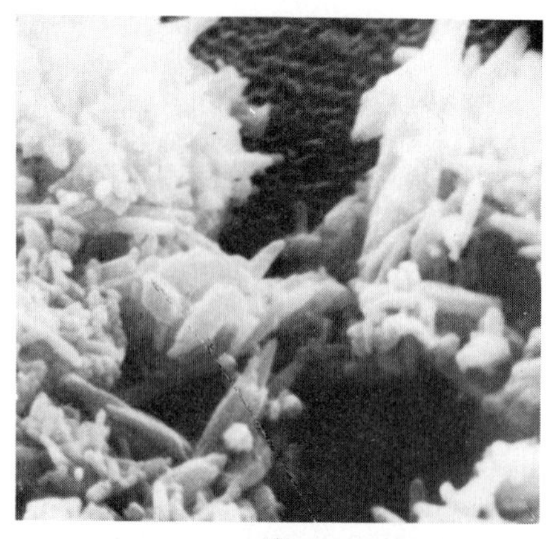

(*b*)

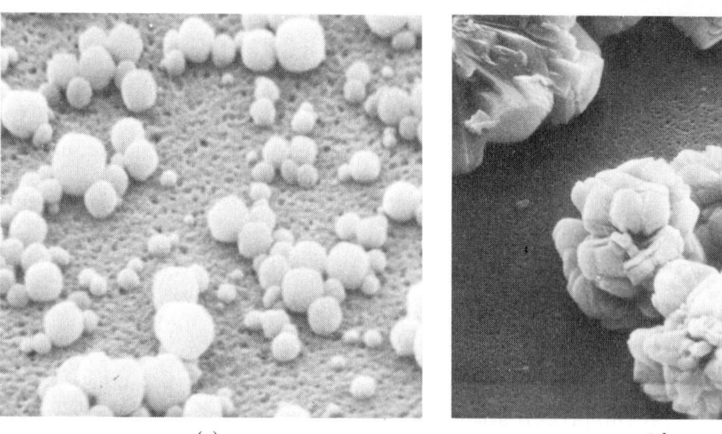

(*c*)

(*d*)

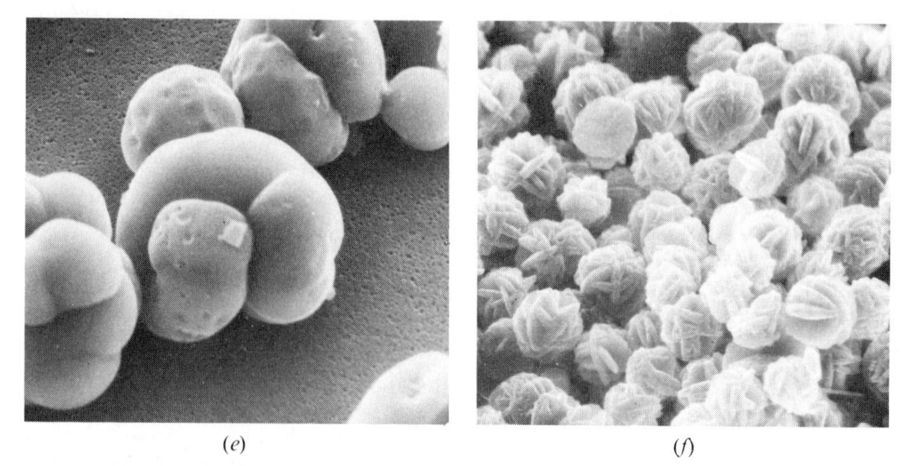

(*e*) (*f*)

FIG. 21.9 $CaCO_3$ crystal distortion caused by various organic treatment reagents. (*a*) $CaCO_3$ precipitated as calcite, the more common crystalline form at lower temperatures (2000×). (*b*) $CaCO_3$ precipitated as aragonite, the more common crystalline form at elevated temperatures (5000×). (*c*) Distortion created by polyacrylate treatment (2000×). (*d*) Changes in structure of precipitate produced. (*e*) Structure resulting from treatment with a sulfonated copolymer crystal modifier (5000×). (*f*) Changes produced by a blend of polyacrylate and phosphonate (2000×).

Other dispersants condition the surfaces of the suspended solids in other ways to keep them from coagulating and settling.

Surfactants

The third category of deposit control agents is surface-active chemicals. Those that penetrate and disperse biomasses are called *biodispersants*. Some of these are also biocides and help to kill the slime organisms. Frequently, chlorine is used in conjunction with biodispersants, although other kinds of biocides may also be applied.

Some surface-active agents are effective wetting agents and antifoulants which help fluidize solids and keep them moving with the flowing water. Still others are selected to emulsify hydrocarbons so that they can be eliminated from the system by conventional blowdown.

Crystal Modifiers

The fourth group of control agents includes those chemicals used to modify the crystal structure of scale (see Fig. 21.9). Since the presence of particulates induces precipitation of scale from a supersaturated solution, scale is often one of the fractions of a deposit. Although precipitation is not prevented by crystal modifiers, the resultant material that precipitates is structurally weak—more like a foulant than a scale. In this respect, these chemicals which distort or modify crystal growth may also be considered as deposit control agents. These chemicals are widely used in industrial water systems, especially in cooling systems and boilers. Natural organics such as lignin and tannin were the first organic treatments used to keep heat transfer surfaces clean by crystal modification. They function also as dispersants but are seldom used today except in special situations, such as when developing countries have their own sources of such natural organics. But, for general use their performance as scale inhibitors or dispersants is quite low in comparison with modern synthetic polymers.

SUGGESTED READING

Athey, R. D. Jr.: "Polymeric Organic Dispersants for Pigments," *J.*, **58**(9) (September 1975); 2d part, *Tappi, J.*, **58**(10) (October 1975).

Dubin, L.: "The Effect of Organophosphorus Compounds and Polymers on $CaCO_3$ Crystal Morphology," Proceedings National Association of Corrosion Engineers Annual Meeting, March 4, 1980.

Hatch, G. B., and Rice, O.: "Threshold Treatment on Water Systems," *Ind. Eng. Chem.* **37,** 710 (1945).

Lahann, R. W.: "A Chemical Model for Calcite Crystal Growth and Morphology Control," *J. Sediment. Petrol.* **48**(1), 337–344 (plus commentaries, pp. 345–347) (March 1978).

McCoy, J. W.: *The Chemical Treatment of Cooling Water,* Chemical Publ., New York, 1974.

Ralston, P. H., "Inhibiting Water Formed Deposits and Threshold Compositions," *Mater. Prot. Perform.*, **12**(6), 39–44 (1972).

Reitemeier, R. F. and Buehrer, T. F.: "The Inhibiting Action of Minute Amounts of Sodium

Hexametaphosphate on the Precipitation of Calcium Carbonate from Ammonical Solutions. I. Quantitative Studies of the Inhibition Process. II. Mechanism of the Process with Special Reference to the Formation of Calcium Carbonate Crystals," *J. Phys. Chem.*, **44,** 535–74 (1940).

Smallwood, P. V.: "Some Aspects of the Surface Chemistry of Calcite and Aragonite, Part I: An Electrokinetic Study," Part II: Crystal Growth," *Colloid Polym. Sci.*, **255,** 881–886 and 994–1000 (1977).

Williams, F. V. and Ruehrwein, R. A.: "Effect of Polyelectrolytes on the Precipitation of Calcium Carbonate," *J. Am. Chem. Soc.*, **79,** 4898–4900 (1957).

CHAPTER 22
CONTROL OF MICROBIAL ACTIVITY

All water treatment processes are affected by the presence of microbes. Many oxidation-reduction reactions are biologically mediated. In most cases, microbial effects are detrimental to the water-using process or system. However, certain industrial operations put microbes to work in a useful way: the activated sludge process uses microbes for digestion of organic wastes; microbes are used for the fermentation of beverages; microbial enzymes are useful for leather processing; and bacteria are used in the recovery of metal values—especially copper—from tailings, the residue of mineral beneficiation processes.

Disinfection of municipal drinking water and sterilization of food processing and hospital equipment are examples of applications of biocides (chemicals toxic to microbes) where the goal is to kill all microbes. However, in the treatment of nonpotable water, a complete kill is often costly and not always necessary. Cooling water in utilities, steel mills, refineries, and other industrial plants is treated to control microbe populations at levels that experience has proven to be tolerable to the system without complete sterilization. Papermaking systems, unlike cooling water systems, are designed to operate with large amounts of suspended solids, so the tolerable levels of microbe populations are considerably higher than for cooling water. The tolerable microbe count in a paper mill varies with the type of paper being made and machine operating conditions, such as pH and temperature.

Planning an effective microbial control program for a specific water treatment process requires an examination of:

1. The types of organisms present in the water system and the associated problems they can cause.
2. The population of each type of organism that may be tolerated before causing a significant problem.

Typical microbes encountered in water treatment and the problems they cause are summarized by Table 22.1.

Bacteria, the largest group of troublesome organisms, cause the most varied problems. They are usually classified in water treatment by the types of problems they cause: slime-forming bacteria, iron-depositors, sulfate-reducers, and nitrifying bacteria. Each group has its preferred environment and thrives in specific areas of a water system. Aerobic bacteria, for example, require oxygen, so they are found in aerated waters such as in a cooling tower basin or white water in a

TABLE 22.1 Typical Microorganisms and Their Associated Problems

Type of organism	Type of problem
A. Bacteria	
1. Slime-forming bacteria	Form dense, sticky slime with subsequent fouling. Water flows can be impeded and promotion of other organism growth occurs.
2. Spore-forming bacteria	Become inert when their environment becomes hostile to them. However, growth recurs whenever the environment becomes suitable again. Difficult to control if complete kill is required. However, most processes are not affected by spore formers when the organism is in the spore form.
3. Iron-depositing bacteria	Cause the oxidation and subsequent deposition of insoluble iron from soluble iron.
4. Nitrifying bacteria	Generate nitric acid from ammonia contamination. Can cause severe corrosion.
5. Sulfate-reducing bacteria	Generate sulfides from sulfates and can cause serious localized corrosion.
6. Anaerobic corrosive bacteria	Create corrosive localized environments by secreting corrosive wastes. They are always found underneath other deposits in oxygen deficient locations.
B. Fungi	
Yeasts and molds	Cause the degradation of wood in contact with the water system. Cause spots on paper products.
C. Algae	Grow in sunlit areas in dense fibrous mats. Can cause plugging of distribution holes on cooling tower decks or dense growths on reservoirs and evaporation ponds.
D. Protozoa	Grow in almost any water which is contaminated with bacteria; indicate poor disinfection.
E. Higher life forms	Clams and other shell fish plug inlet screens.

paper machine wire pit. Anaerobic bacteria, on the other hand, don't use oxygen and obtain their energy from reactions other than the oxidation of organic substances. The reduction of sulfur in sulfate to the sulfide ion is an example. Since anaerobes don't need oxygen, they are found in oxygen-deficient areas, such as under deposits, in crevices, and in sludges.

Iron-depositors occur in water high in ferrous iron, which they convert to insoluble ferric hydroxide and which becomes part of the mucilaginous sheath around the cell. These deposit and accelerate corrosion rates, which produces additional soluble iron, further increasing the population of iron-depositors in the system. The cycle accelerates until the whole system is plugged with iron deposits (Figure 22.1).

Nitrifying bacteria oxidize ammonia to nitrate. This nitrification reaction sometimes occurs in iron removal filters, accompanied by a reduction of oxygen and pH. These bacteria are often found in ammonia plants where leakage of

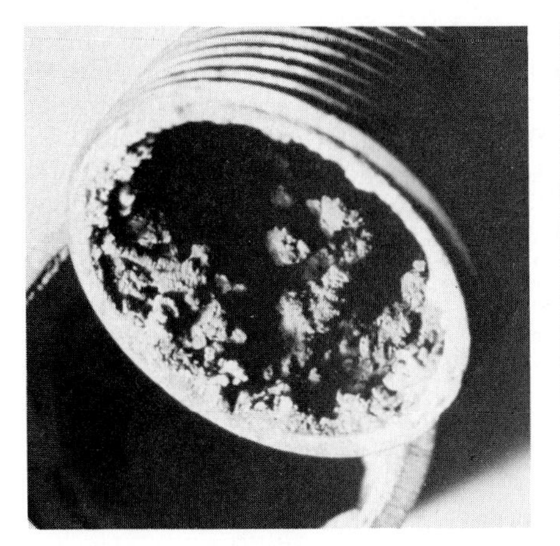

FIG. 22.1 Iron-depositing bacteria initiated the tuberculation attack on this steel distribution pipe.

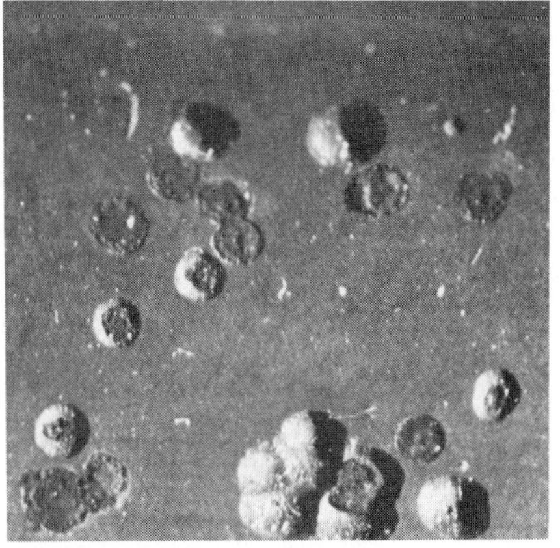

FIG. 22.2 The development of pits filled with voluminous products of corrosion from the attack of sulfate-reducing bacteria on steel pipe.

ammonia into cooling water encourages their growth. A pH drop caused by the conversion of ammonia to nitrate is often the clue to their presence.

Sulfate-reducing bacteria are found in many systems subject to deposit problems. The sulfides produced are corrosive to most metals used in water systems, including mild steel, stainless steel, and aluminum. Evidence of the sulfate-reducers is the unique pit etched on the metal surface, sometimes in the form of concentric rings (Figure 22.2).

Many bacteria secrete a mucilaginous substance that encapsulates the cell (Figure 22.3), shielding it from direct contact with water, so that the cell is protected from simple toxic biocides. Control of encapsulated bacteria usually requires both oxidation and dispersion of the protective sheath so that the biocide can reach the cell.

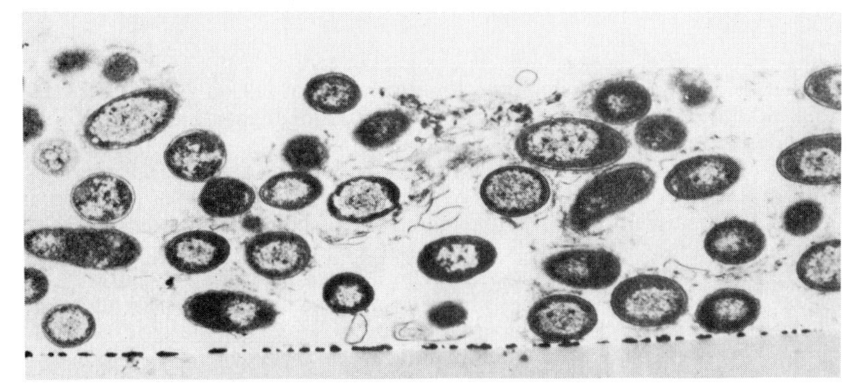

FIG. 22.3 A biofilm in development. Biofilm is the mucus-like coating produced by slime-forming bacteria. Organisms are *Pseudomonas aeruginosa* at 7000×.

Yeasts and molds can live on dead or inert organic matter. Fungi are often found on wooden structures, such as cooling tower fill and supporting members, and sometimes under bacterial or algal masses. Fungal attack of wood usually means permanent loss of strength of the wood structure, so protection of the wood requires control of fungi from the time the structure is put into service. Periodic testing of the wood to determine its resistance to fungal attack is an important maintenance step. Very thin sections of wood specimens taken from susceptible locations are examined to determine the extent of attack if there has been any (Figure 22.4).

With few exceptions, algae need sunlight to grow, so they are found on open, exposed areas, such as cooling tower decks or on the surface of reservoirs, ponds,

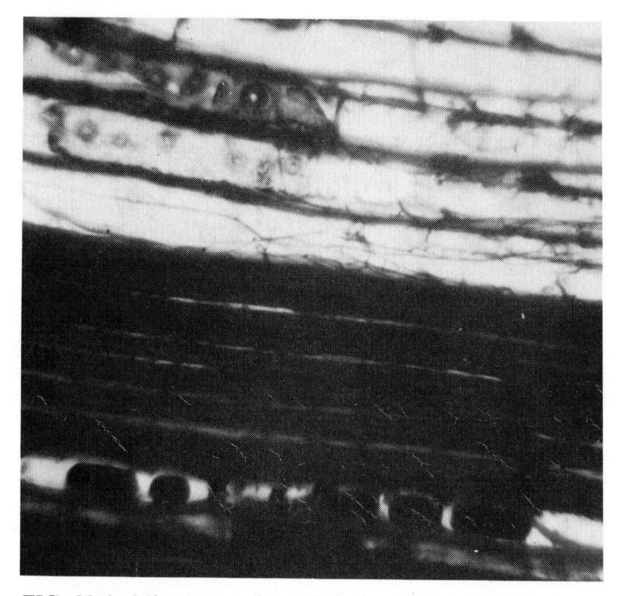

FIG. 22.4 Microtome of wood section from a cooling tower, showing fungal attack.

and lakes. Most algae grow in dense, fibrous mats that not only plug distribution piping and flumes, but also provide areas for subsequent growth of anaerobic bacteria under the algae deposits.

A century ago, a Danish biologist, Christian Gram, developed a method of staining bacterial cultures as a means of separating them into two broad categories as an aid to identification: those that retain a blue color produced by an iodine treatment are called Gram-positive; those not retaining the blue color and accepting a red dye following iodine treatment are Gram-negative. Most aquatic microbes are Gram-negative. They all are negatively charged colloids (have a negative zeta potential), a property not related to the Gram staining technique. Because they are negatively charged colloids, they are affected by cationic polymers and biocides.

PHYSICAL FACTORS AFFECTING MICROBE GROWTH

Many species of microbes indigenous to soil, water, and vertebrate organisms thrive in a rather broad temperature range of 10 to 45°C. Nature has produced select organisms that can live at temperatures as low as 0°C and as high as 100°C. Higher temperatures kill all common microbes, but scientists report finding life in hot springs and adjacent to ocean vents on the sea floor at temperatures of over 200°C.

Denaturation of proteins, which causes coagulation within the cell, occurs at temperatures below 70°C. Commercial pasteurization is a denaturation process. Milk is usually pasteurized at 63°C by holding that temperature for 30 min; if the temperature is raised to 72°C, pasteurization is completed in only 15 s. This process kills all disease-producing (pathogenic) organisms but does not produce sterile milk; some microbes remain to cause the milk to spoil in time. Most actively growing microbes of interest in water treatment technology are killed at 70°C in less than 5 min. Although pasteurization has a long history of success in food processing, it has never been reported in use for water disinfection, except in occasional emergencies where a community water supply may be contaminated and the public is warned to boil all drinking water until the crisis has passed.

Maintaining low temperatures is not an effective means of killing microbes. At 0 to 5°C, organisms become dormant. Freezing kills many cells, but those that survive are capable of complete recovery from the shock. One procedure used to preserve microbes involves freezing cells rapidly at −70°C and then removing the ice crystals as vapor (sublimation). This process is lyophilization.

Dry heat results in dehydration of all cellular matter and oxidation of intracellular constituents. Sterilization of laboratory media is usually carried out in an autoclave at 121°C. Sterilization of glassware is done with dry heat at 160°C for two hours.

Moisture is required for microorganisms to grow actively. Many species of pathogenic organisms are killed quickly by drying. However, organisms in the spore or cyst state can survive low moisture environments; and, if transported by wind or animals to a location where moisture levels suit them, they revive and form new colonies. To prevent attack by microbes, lumber and other vulnerable materials are dried to less than 20% moisture content.

Organisms containing chlorophyll are able to use the radiant energy of the sun or artificial lighting to convert CO_2 to carbohydrates, which they need for cell synthesis. However, not all radiant energy is useful to the cell and certain frequencies of radiation are harmful. Radiation is therefore one method of microbe control.

Short-wavelength forms of energy, such as gamma rays (0.01 to 1 Å) and x-rays (1 to 100 Å) are particularly useful. These create free hydrogen and hydroxyl radicals and some peroxides when they pass through the cell, causing cell damage or death. These forms of radiation are hazardous. Energy in the ultraviolet region (1000–4000 Å) is also useful for killing microbes. In this case, the energy is absorbed by the nucleic acids, creating chemical reactions that are lethal to the cell. This form of energy, however, has poor penetrating ability, so the use of ultraviolet light for disinfection requires a treatment unit of special design so that the energy does not have to penetrate deeply into the water. Ultraviolet sterilization uses about 0.2 kWh of electric energy per thousand gallons of water treated (0.05 kWh/m^3), so it is economically attractive in situations where the microorganisms are not shielded by large agglomerated masses or by suspended solids. This method of disinfection is widely used in ultrapure water systems.

Osmosis is the diffusion of water through a semipermeable membrane separating two solutions of different solute concentrations. The water flows in a direction to equalize the concentrations. When microbes are placed in 10 to 15% salt solutions or 50 to 70% sugar solutions, the water inside the cells is extracted by the surrounding medium. This dehydrates the cells so they are unable to grow or are killed. This technique is used commercially to preseve food. Bees use this principle to preserve honey, concentrating it by fanning the comb with their wings.

The interfaces between a liquid and a gas (such as surrounding a bubble of air in water), between two liquids (oil droplets in water), and between a solid and a liquid (sand grains in water) are characterized by unbalanced forces of attraction between the molecules of water at the surface and those in the fluid body. These forces are closely associated with the metabolic processes of the microbe. The cell must be able to accumulate nutrients at its surface for assimilation, and waste products must be eliminated from the cell and carried away. Therefore, the growth and well-being of a cell are influenced by surface forces in the surrounding aquatic environment. Substances having surface tension depressing effects (surfactants) tend to have a detrimental effect on microbes if the concentration is high enough. These materials can alter cell division, growth, and survival. Surfactants are often used to increase the effectiveness of biocides by dispersing cell colonies and protective sheaths to allow the toxicant to contact the cells. Some toxicants are themselves surface active (such as phenols and quaternaries) and tend to accumulate on cell surfaces by adsorption. This prevents entrance and utilization of food substances by the cell. Quaternaries will sometimes cause leakage of cellular material out of the cell wall by changes in surface tension at the membrane surface.

CHEMICAL FACTORS AFFECTING MICROBIAL GROWTH

Microbes have been found to exist in the broad pH range of 1 to 13. However, the most common microbes associated with water—algae and bacteria—usually maintain their internal pH at 7, so they prefer a neutral aquatic environment.

Generally, yeasts and molds favor depressed pH, in the range of 3 to 4. Dilute alum solution is sometimes contaminated by fungi, causing the plugging of alum feed lines and rotameters while at the same time the solution is free of bacterial growth.

Bacteria and fungi can both contribute to industrial problems over a pH range of 5 to 10. Other chemical factors, discussed in earlier chapters, include the presence of organic matter to serve as food for the microbes, and a supply of the common nutrients such as nitrogen and phosphorus required for cell metabolism.

One of the surprising facts of microbe life is that there is such a profusion and variety of forms that some can almost always be found that will resist damage by, or even thrive on, chemicals that are toxic to animal and plant life. For example, phenol was one of the early chemical biocides used for sterilization in medical practice, yet at low concentrations—up to 100 mg/L, which is sometimes found in coke plant wastes—it is readily digested in activated sludge waste treatment plants. Similarly, some bacteria thrive in wastewaters that contain herbicides, pesticides, cyanide, arsenic compounds, and a variety of other chemicals normally considered toxic.

METHODS FOR CONTROLLING MICROBIAL ACTIVITY

For practical reasons, in most industrial water systems, only limited use can be made of the physical conditions that inhibit or destroy microbial life. For example, heating water may control microbial activity, but if the water is used for cooling purposes, this is not useful. Radiation is sometimes used, but its adoption on a widespread basis would require the development of more efficient energy sources and better designs of equipment to expose the water to the radiant energy. Among chemical conditions that might be used for microbe control, pH is the only likely candidate for practical results. Even this is limited unless the system water can be kept at a pH over 10. However, pH does have important effects on the performance of biocides, as shown later when chlorine reactions are discussed.

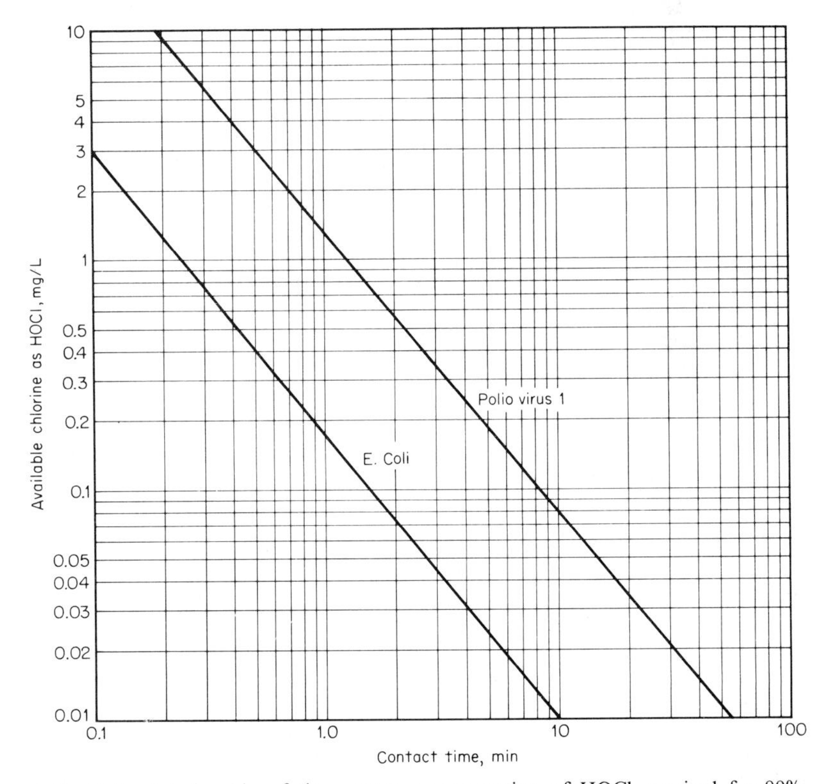

FIG. 22.5 Relationship of time versus concentration of HOCl required for 99% destruction or inactivation of a common bacteria *(Escherichia coli)* and a virus, at 0° to 6° C. *(From Water Pollution Control Federation Publication Deeds & Data, January 1980.)*

Since neither physical nor aquatic chemical conditions can be changed in a practical way to control microbial growth, toxic chemicals must be applied as biocides. The two commonly used types are oxidizing and nonoxidizing. Regardless of which type is used, there is a relationship for all chemical biocides that expresses effectiveness, measured as percent kill or inactivation, concentration of biocide applied to the water, and time of contact of the biocide with the organism or virus. This was first discovered by H. Chick in 1908 and further developed by H. E. Watson in the same year, and is now designated as the Chick-Watson law:

$$N/N_0 = \exp(-k'c^n t)$$

where N_0 represents the bacterial population at the time zero
N is the reduced population at time t, after biocide application
exp = base of natural logarithm system, 2.718
k' is a rate constant
c is the concentration of the biocide, mg/L
n is an empirical value
t is the time of contact, min

As an illustration of this law, Figure 22.5 shows the concentration versus time relationship for application of chlorine to water for destruction of *Escherichia coli,* the coliform bacteria used as an indicator organism in evaluating disinfection of municipal water supplies, and for inactivation of polio virus 1. This graph is typical of all biocidal charts, demonstrating the Chick-Watson law. In each case, the graph is specific for both the organism and the water supply, as there may be side reactions of the biocide with other constituents of the water.

OXIDIZING BIOCIDES (SEE ALSO CHAPTER 19)

Chlorine gas, the chemical biocide most commonly used in the United States, hydrolyzes rapidly when dissolved in water according to the following equation:

$$Cl_2 + H_2O \rightarrow H^+ + Cl^- + HOCl \qquad (1)$$

Hydrolysis occurs in less than a second at 65°F (18°C). Hypochlorous acid (HOCl) is the active ingredient formed by this reaction. This weak acid tends to undergo partial dissociation as follows:

$$HOCl \rightleftharpoons H^+ + OCl^- \qquad (2)$$

This reaction produces a hypochlorite ion and a hydrogen ion. Depending on pH and concentration, chlorine in water exists as free chlorine gas, hypochlorous acid, or hypochlorite ion. Figure 22.6 illustrates the distribution of these components at varying pH values. Above pH 7.5, hypochlorite ions predominate, and they are the exclusive form when the pH exceeds 9. The sum of the hypochlorous acid and hypochlorite ions is defined as free available chlorine. Hypochlorite salts such as calcium hypochlorite ionize in water to yield these two species, depending on pH.

$$Ca(OCl)_2 \rightleftharpoons Ca^{2+} + 2OCl^- \qquad (3)$$

$$2OCl^- + H_2O \rightleftharpoons HOCl + OCl^- + OH^- \qquad (4)$$

Thus, the same equilibria are established whether elemental chlorine or hypochlorite is used for chlorination.

Liquefied chlorine is available in bulk or in cylinders. A typical installation for feeding chlorine is shown in Figure 22.7*a* and *b*.

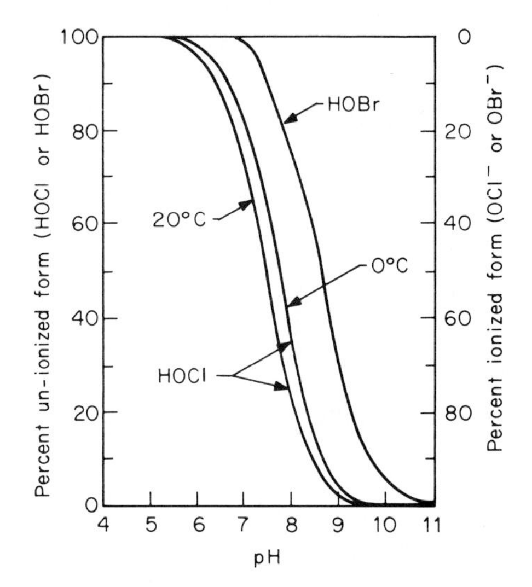

FIG. 22.6 Distribution of hypochlorite and hypobromite in water as affected by pH.

Chemicals other than chlorine gas that liberate hypochlorite ions are compared to one another in oxidizing power on the basis of "available chlorine." The available chlorine values of a number of disinfectants are shown in Table 22.2.

Chlorine is a strong oxidizing agent capable of reacting with many impurities in water including ammonia, amino acids, proteins, carbonaceous material, Fe^{2+}, Mn^{2+}, S^{2-}, and CN^-.

The amount of chlorine needed to react with these substances is called the chlorine demand. Chlorine reacts with ammonia to form three different chloramines:

$$HOCl + NH_3 \rightarrow NH_2Cl \text{ (monochloramine)} + H_2O \quad (5)$$

$$NH_2Cl + HOCl \rightarrow NHCl_2 \text{ (dichloramine)} + H_2O \quad (6)$$

$$NHCl_2 + HOCl \rightarrow NCl_3 \text{ (trichloramine)} + H_2O \quad (7)$$

These chloramine compounds also have biocidal properties; they are referred to as the combined residual chlorine. In general, the chloramines are slower acting than free residual chlorine, but have the advantage of being more effective at pH values above 10. Chloramines may also be more persistent in a water system.

Breakpoint chlorination is the addition of sufficient chlorine to satisfy the chlorine demand and produce free residual chlorine. When breakpoint chlorination is used, the ammonia nitrogen content is destroyed and the residual chlorine remaining will be almost entirely free available chlorine (Figure 22.8).

(*a*)

(*b*)

FIG. 22.7 (*a*) Typical chlorination system. In this plant, eight chlorinators each meter up to 8000 lb Cl_2/day to municipal sewage plant effluent. Each is supplied by a chlorine evaporator (in background, right). *(Courtesy of Wallace & Tiernan.)* (*b*) Ton cylinders of chlorine are mounted on a scale to monitor consumption by weight loss, and the discharge is converted to Cl_2 gas by the evaporator on the right. The chlorine supply room is isolated from operators in the control room. *(Courtesy of Wallace & Tiernan.)*

Chlorine also reacts with organic nitrogen in water. This is found in components of living cells, protein, polysaccharides, and amino acids. The toxicity of chlorine is thought to be derived not from the chlorine itself or its release of nascent oxygen, but rather from the reaction of the HOCl with the enzyme system of

TABLE 22.2 Available Chlorine of Chlorination Chemicals

Material	Percent available Cl_2
Chlorine gas (Cl_2)	100
Chlorine dioxide (ClO_2)	263
Hypochlorites (OCl)	
Calcium, HTH, $Ca(OCl)_2$	70
Sodium, NaOCl	
Industrial grade	12–15
Domestic grade	3–5
Lithium, LiOCl, laundry grade	35
Chlorinated isocyanuric acid $(CONCl)_3$	85

the cell. The superiority of HOCl over OCl may be due to the small molecular size and electrical neutrality of HOCl, which allow it to pass through the cell membrane.

On-site production of hypochlorite from seawater or brine is becoming popular as it limits the exposure of operating personnel to chlorine gas or hypochlorite compounds. An installation of such a system is shown in Figure 22.9.

Chlorine reacts with a variety of organic materials, and attention is being directed to the presence of chlorinated compounds in water thought to be pro-

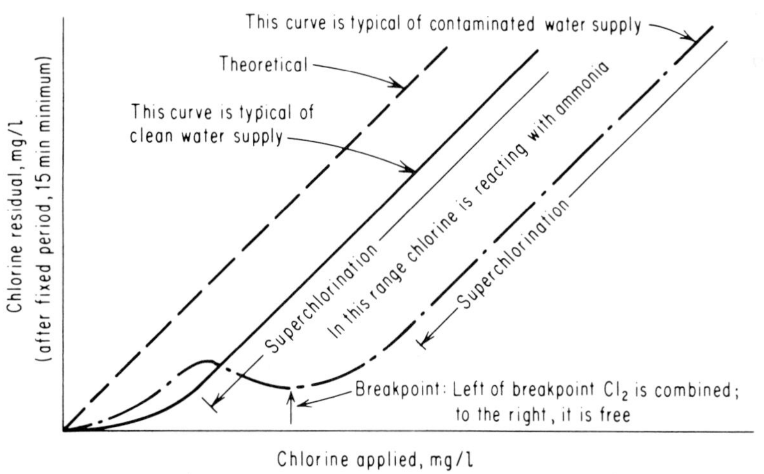

FIG. 22.8 Breakpoint chlorination curves showing reaction of Cl_2 with N-compounds.

duced by chlorination. Chloroform is one of these materials. Because of concern for the potentially adverse physiologic effects of these chlorinated compunds, regulatory agencies are severely restricting chlorine applications to large effluent flows. For example, treatment of utility station condenser water with chlorine may be limited to a total period of only 2 h/day at residuals averaging not over 0.2 mg/L Cl_2 (EPA limit mandated July 1, 1984).

FIG. 22.9 An electrolytic cell designed to produce chlorine from brine. This unit is generating Cl_2 from seawater for chlorination of sewage plant effluent. *(Courtesy of Electrolytic Systems Division, Diamond Shamrock.)*

For this reason, the electric utility industry has studied chlorine minimization techniques and alternatives to chlorination to meet the strict discharge limits without jeopardizing condenser performance. One such alternative is an activated bromide program: a chlorine-bromide mixture produces oxidant species that penetrate the biofilm; chlorine activates a bromide compound, according to the reaction

$$HOCl + Br^- \rightarrow HOBr + Cl^- \tag{8}$$

The Cl:Br ratio can be varied to produce a Br_2 residual, a Cl_2 residual, or a mixture, as conditions may vary and require a ratio change. With high ammonia levels, the bromamines that form degrade more rapidly than do chloramines, so they are less persistent in the environment. The combination reduces the total residual chlorine and aids in compliance.

Like chlorine, bromine residuals exist either as unionized or ionized species in water, as seen in Figure 22.6. At pH 7.5, 50% of the available chlorine is present as HOCl, with the balance as OCl^-. With hypobromous acid, at pH 7.5, over 90% of the oxidant is present as HOBr, the more active form, just as HOCl is more active than OCl^-, as mentioned earlier. These curves demonstrate why bromine residuals provide better biocidal performance than chlorine in systems operating at higher pH values. In an 8-month field study, a utility system treated with the activated bromide-chlorine blend required only about 10% of the oxidant used by a similar unit treated with chlorine alone, and in this period, the 0.2 mg/L chlorine residual specified by the EPA was never exceeded. (Note: The conventional total chlorine test reacts with both chlorine and bromine.)

Chlorine dioxide, ClO_2, is used to a limited extent in water treatment for the control of taste and odor problems and for the degradation of phenol. It is used extensively in the pulp and paper industry for bleaching. This compound must be generated from the reaction of chlorine with sodium chlorite as shown:

$$2NaClO_2 + Cl_2 \rightarrow 2ClO_2 + 2NaCl \quad (9)$$

Generally an excess of chlorine is used to drive the reaction to completion. For safety and to preserve its stability, the material is generated on-site. Since chlorine dioxide does not react with ammonia, it is useful in systems containing ammonia.

The next most common oxidizing biocide is ozone, O_3. This is in common use commercially throughout Europe in preference to chlorine and is finding growing acceptance in certain municipalities in the United States for disinfection of potable water. It is also used in certain waste treatment applications to avoid the residual chloramines that result from the usual chlorination of wastewater effluent. Ozone is produced on-site by an electric corona discharge through air or oxygen. A typical ozone generator is shown in Figure 22.10.

Oxidizing biocides such as chlorine, hypochlorites, and organochlorine materials will kill all organisms in the system quickly, if the free chlorine comes into direct contact with the organisms long enough and at a strong enough dosage

FIG. 22.10 A battery of three 110 lb/day (49 kg/day) ozonators installed in a municipal plant. These produce O_3 from air. *(Courtesy of Infilco Degremont, Inc.)*

level. They also retain their effectiveness because organisms cannot adapt to or become resistant to chlorine.

However, oxidizing biocides also react with contaminants like H_2S, NH_3, pulp lignins, wood sugars, and other organics. This increases the amount of chlorine required for biocidal effects. They are not persistent, and they decay quickly after the chemical feed stops. They do not penetrate slime masses, so they do not reach subdeposit microbes.

Thus oxidizing biocides require complementary treatments to improve their effectiveness. These include biodispersants to remove existing slime masses and to prevent organisms from settling on heat transfer surfaces; penetrants to permeate organic masses and to expose and kill subsurface organisms; and biocides for control of organisms in systems contaminated with H_2S, NH_3, and other reducing agents.

NONOXIDIZING BIOCIDES

Nonoxidizing biocides offer a way to control microbial activity in systems incompatible with chlorine, such as water systems high in organic matter or ammonia. With few exceptions (e.g. copper sulfate), they are organic chemicals. They provide the following features:

1. Activity independent of pH
2. Persistency
3. Control of organisms such as fungi, bacteria, and algae

Since all of these benefits are usually not available from a single penetrating biocide, individual ingredients are formulated into proprietary products designed to increase overall performance in very specific applications, e.g., paper machine systems, open cooling water systems, and process water in food plants. Slug feed is the preferred method of application.

Just as ammonia, sulfides, and reducing agents can interfere with the performance of oxidizing biocides, certain substances can reduce the effectiveness of nonoxidizing biocides; e.g., cationic biocides may react with anionic dispersants to cancel the effectiveness of both.

Organic Compounds

Methylene-bis-thiocyanate (MBT), $(SCN)-CH_2-(SCN)$, is a well-known organosulfur biocide. It is usually recommended for applications in paper mills and cooling systems where effluent limitations are strict, and where control of slime-forming bacteria is the main problem. Holding time and pH affect the half-life of MBT, which hydrolyzes in water to form less toxic substances.

Figure 22.11 illustrates the pH dependency of the half-life of MBT. At a pH of 11, it is destroyed in seconds.

A continuous feed of 1 mg/L to the makeup of a cooling system will control organisms entering with the water. If an alkaline cooling water treatment program is being used, the relatively higher pH will cause the MBT to hydrolyze faster. In most cooling systems, the molecule will eventually be destroyed if holding time is sufficiently long. Most of the degradation products are volatile and are stripped in the tower.

A study to assess the effect of MBT on biological sewage treatment programs indicated that 0.5 to 2.0 mg/L had no measurable effect on BOD and suspended solids removal. So it is feasible to treat a system with methylene-bis-thiocyanate to control microbes and still have an effluent amenable to biological treatment.

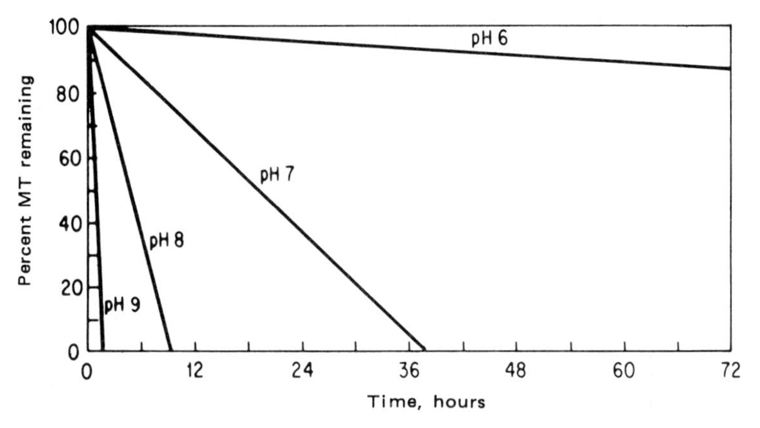

FIG. 22.11 Degradation of methylene-bis-thiocyanate at pH 6, 7, 8, and 9.

Dibromonitrilopropionamide (DBNPA) is also excellent where antibacterial activity and environmental acceptability are needed. DBPNA kills quickly, then decomposes to nontoxic compounds.

Chlorinated phenols are highly effective against most common organisms, especially fungi and algae. Tower fill is sometimes sprayed with chlorinated phenols to increase the lumber's resistance to fungal attack. These compounds are effective biocides when fed directly to cooling water. However, because of toxicity and danger to the environment, chlorinated phenols are banned in the United States and many other countries.

Organotriazine compounds, such as isothiazolones, are broad spectrum nonoxidizing biocides, particularly effective against bacteria, and active over a wide pH range. Isothiazolones are not inhibited by most organic and inorganic contaminants found in cooling waters, and are compatible with ionic and nonionic dispersants. They can be used with any of the oxidizing biocides.

A typical example of the organometallic group of biocides is the compound bis-tributyl tin oxide (TBTO), $(H_9C_4)_3{\equiv}Sn-O-Sn{\equiv}5(C_4H_9)_3$. This compound is used to control fungi and algae. It also has a tendency to adsorb on surfaces in the system, especially on wood, and therefore analyses usually do not show tin levels as high as would be expected from dosage calculations. This adsorption provides for residual algae and fungi control after treatment is discontinued. Although its toxicity is greatest on fungi and algae, it also provides good control of anaerobic corrosive bacteria.

CATIONIC BIOCIDES

If penetration of slime and algae masses is desired along with persistency, then amines and quaternaries (quats) are usually applied. The use of amines or quats with chlorination usually permits the dosage of chlorine to be reduced.

Many of these biocides are surface-active and therefore can disperse slime masses. This allows chlorine and other toxicants to contact organisms which, under normal circumstances, they would not harm. The wettability of organic slime masses is increased so that the toxicants can reach under the mass to get at anaerobic corrosive bacteria.

In a typical amine program in a cooling tower system, a once-per-month slug of an amine biocide at 75 mg/L for 48 h is used to supplement continuous chlorination at 0.5 to 1.0 mg/L free residual for 6 to 8 h/day.

METALLIC COMPOUNDS

The detection of mercury in aquatic organisms in the late 1960s—and especially in tuna and other edible fish—as a result of mercury discharges from chloralkali plants alerted the public to the potential threat of toxic metals in the environment. As a result of subsequent studies of the toxicity of heavy metals, some that had been used as biocides (mercuric salts) and others that were not used as poisons but had simply found their way into the environment indirectly (such as lead aklyls in gasoline), these compounds came under scrutiny and were withdrawn from the market or had their use greatly curtailed.

One of the most common of the metallics used as a biocide is copper in ionic form. It has been applied as copper sulfate to ponds and reservoirs for many years for control of algae at a concentration usually below 1 mg/L. Its solubility falls off rapidly as pH increases, so a chelating agent, such as citric acid, is often applied to improve the effectiveness of treatment. Since algae have their individual seasons for blooming, the copper sulfate is usually applied only seasonally, and its effective dosage is below the concentration permitted in potable water. The mechanism of action against algae is not clear, but copper ions form complexes with amines; the effectiveness of copper ions is probably a result of their reaction with the essential amino acids. Toxic metals can be transported across the cell membrane more readily in nonionic form, so a variety of neutral metallo-organic compounds have been developed to increase the toxicity of certain heavy metals as biocides, as discussed earlier (TBTO).

MICROBIAL MONITORING PROCEDURES

In systems where microbe populations may be maintained within a certain acceptable control range and complete sterilization is not required, frequent microbe analyses should be run for monitoring the program. Sometimes only total counts are run to indicate overall microbe population levels. However, in systems such as industrial recirculating cooling water loops, analyses of more specific organisms are required. Changes in total count do not always indicate changes in fungi, anaerobic bacteria, sulfate-reducing bacteria, or algae. Since these organisms can be troublesome, they should be monitored specifically.

In industrial waters, a microbial analysis usually expresses counts per 1-mL sample. Since a billion bacteria weigh on the order of 1 mg, a high count of 10,000,000/mL of sample represents only about 10 mg/L of suspended solids.

Since there are interfering substances that can reduce the effectiveness of certain biocides, a change in microbial population may be due to causes other than the inherent toxicity of the biocide. So, a review of plant conditions (temperature, chemical environment, sources of inoculation) should always accompany any sampling for microbe analysis.

COOLING WATER SYSTEMS

A typical report of those organisms most commonly found in industrial water systems is shown in Figure 22.12. This analysis is typical of the weekly analyses

NALCO

ANALYTICAL LABORATORY REPORT

From:

Analysis No. B 41616
Date Sampled 12/ 1/83
Date Received 12/14/83
Date Printed 12/21/83

Sample Marked:
Cooling Water - Ammonia Plant

>>> MICROBIOLOGICAL EVALUATION <<<

PHYSICAL APPEARANCE :	Yellow Liquid
TOTAL AEROBIC BACTERIA	500,000
Aerobacter	20,000
Pigmented	50,000
Mucoids	<1,000
Pseudomonas	70,000
Others	360,000
TOTAL ANAEROBIC BACTERIA	<10
Sulfate Reducers	<10
IRON-DEPOSITING	
Gallionella	None
Sphaerotilus	None
TOTAL FUNGI	<10
Molds	<10
Yeast	<10
ALGAE	
Filamentous	None
Non-Filamentous	None
OTHER ORGANISMS :	None

Lab Comments:
(All Counts Express Colony Forming Organisms per ML of Sample.)

FIG. 22.12 Analysis—a cooling system under control.

NALCO

ANALYTICAL LABORATORY REPORT

From:

Analysis No. B 41617
Date Sampled 12/ 1/83
Date Received 12/14/83
Date Printed 12/21/83

Sample Marked:
Ammonia Plant

>>> MICROBIOLOGICAL EVALUATION <<<

PHYSICAL APPEARANCE :	Yellow Liquid with Floc
TOTAL AEROBIC BACTERIA	14,000,000
Aerobacter	70,000
Pigmented	200,000
Mucoids	<1,000
Pseudomonas	400,000
Others	13,330,000
TOTAL ANAEROBIC BACTERIA	20
Sulfate Reducers	20
IRON-DEPOSITING	
Gallionella	None
Sphaerotilus	None
TOTAL FUNGI	<10
Molds	<10
Yeast	<10
ALGAE	
Filamentous	None
Non-Filamentous	None
OTHER ORGANISMS :	Few Protozoa

Lab Comments:
(All Counts Express Colony Forming Organisms per ML of Sample.)

FIG. 22.13 Analysis—the system in Fig 22.12 out of control because of an exchanger leak.

for microbe counts taken on a cooling system water in an ammonia plant, sampled when the system is under control. The notation "NEG IN 1/1000" means that no organisms were observed in a 1/1000 dilution of the sampled water. The total count is usually higher than the sum of individual aerobic slime-forming bacteria shown above it. This is because there are many more types of aerobic bacteria in cooling water than those specifically reported as troublesome.

Figure 22.13 demonstrates changes in microbe populations produced by an

NALCO

ANALYTICAL LABORATORY REPORT

From:

Analysis No.	B 41493
Date Sampled	12/ 6/83
Date Received	12/12/83
Date Printed	12/21/83

Sample Marked:
Well Water

>>> MICROBIOLOGICAL EVALUATION <<<

PHYSICAL APPEARANCE :	Liquid with Floc
TOTAL AEROBIC BACTERIA	None
Aerobacter	<10
Pigmented	None
Mucoids	None
Pseudomonas	None
Others	None
TOTAL ANAEROBIC BACTERIA	15
Sulfate Reducers	10
Clostridia	5
IRON-DEPOSITING	
Gallionella	None
Sphaerotilus	Few
TOTAL FUNGI	None
Molds	None
Yeast	None
ALGAE	
Filamentous	None
Non-Filamentous	None
OTHER ORGANISMS :	None

Lab Comments:
(All Counts Express Colony Forming Organisms per ML of Sample.)

FIG. 22.14 A well water with low counts can still produce troublesome iron growths.

ammonia leak into the cooling water system described above. This sudden influx of nutrient caused aerobic slime-formers to multiply. The resulting deposition of these aerobes increased the shelter for anaerobic corrosive bacteria underneath the deposits.

Using analyses such as these, an industrial plant can optimize biocide usages to minimize water treatment costs while preventing unnecessary shutdowns caused by rampant microbial contamination.

A common problem in once-through cooling water systems is iron fouling. In many instances the iron fouling is actually a result of contamination with iron-depositing bacteria such as *Sphaerotilus* or *Gallionella.* By routinely monitoring the makeup water with microbiological analyses, potential problems with iron depositors can be anticipated and treatment adjusted before the problem gets out of hand.

Figure 22.14 shows a typical makeup water analysis with iron-depositors present. A total count analysis would have indicated this system to be under good control and would have missed the potential for iron fouling from the iron-depositing bacteria.

PAPER MILL WATER SYSTEMS

In the manufacture of unbleached grades of paper or board, total counts of 30 to 40 million bacteria per milliliter can often be tolerated in the system without causing slime problems, because the solids are highly dispersed by residual lignin materials remaining in the system. By contrast, in the production of bleached grades, slimes and other microbial problems would be uncontrollable at such high total counts because of the absence of the dispersive lignins.

The high temperatures of linerboard production systems often allow the machines to run virtually slime-free even though total counts may approach the 30 million to 40 million range. Cylinder machines can tolerate extremely high levels of bacteria because the machines run slowly and are relatively immune to production interruptions from slime. They usually produce heavyweight, multiply board which can incorporate slime spots without damage to its properties.

Molds and yeasts are more responsible for slime problems in papermaking systems than are bacteria. Molds create more problems than yeasts because their threadlike branched form can trap fiber, filler material, and debris, and bind them into a tenacious deposit. Yeasts cause more problems than bacteria because they are considerably larger and have a tacky coating around the cell membrane.

Figure 22.15 shows an analysis of a deposit taken from a paper machine frame. The important features of this deposit are the high levels of yeasts, the presence of anaerobic corrosive bacteria, and the presence of coliform bacteria *(E. coli).* Deposits caused by aerobic bacteria require total counts many times higher than these, so this is not primarily a bacterial problem. The major culprit is yeast. Corrosion may be a problem in this system because of the high level of anaerobic corrosives associated with the deposits. Protozoa in the deposit are indicators that there are problems in treating the fresh water brought into the mill. Improper chlorination or filtration of the fresh water may be the cause.

STORING, HANDLING, AND FEEDING PRECAUTIONS

Because the purpose of biocides is to kill living organisms, the need for care in planning an effective but safe program to handle these toxic chemicals is obvious. Proper protective clothing, gloves, masks, and respirators must be worn by oper-

NALCO

ANALYTICAL LABORATORY REPORT

From:

Analysis No. B 41494
Date Sampled 12/ 6/83
Date Received 12/12/83
Date Printed 12/21/83

Sample Marked:
Tray Deposit (No. 1 Machine)

>>> MICROBIOLOGICAL EVALUATION <<<

PHYSICAL APPEARANCE :	Green Deposit
TOTAL AEROBIC BACTERIA	56,000,000
Aerobacter	50,000
Pigmented	1,000,000
Mucoids	<10,000
Pseudomonas	5,000,000
Sporeformers	<100
Others	49,950,000
TOTAL ANAEROBIC BACTERIA	10,200
Sulfate Reducers	10,000
Clostridia	200
IRON-DEPOSITING	
Gallionella	None
Sphaerotilus	None
TOTAL FUNGI	20,100
Molds	100
Yeast	20,000
ALGAE	
Filamentous	None
Non-Filamentous	None
OTHER ORGANISMS :	Many Protozoa

Lab Comments:
(All Counts Express Colony Forming Organisms per GRAM of Sample.)

Microscopic Examination - Many Bacteria, Moderate Fibers and Fines.

FIG. 22.15 Bioanalysis—paper machine frame.

ators responsible for charging tanks, adjusting feeders, and controlling the program by testing. Showers and eye baths must be readily accessible.

Bulk supply has advantages in minimizing exposure of personnel to biocides and eliminating the problem of package disposal. If products are obtained in drums, provision must be made for proper drum handling, piping hookup, and drum disposal. (Figure 22.15). Not only is the safety of plant personnel at risk, but also the safety of the environment. It is important to know the fate of the biocide selected when it reaches the waste treatment plant and eventually a receiving stream. Bench testing or pilot plant studies may be needed to assure the plant manager that the biocide program most effective in the plant will be compatible not only with discharge permit requirements, but also with the health of the receiving stream.

CHAPTER 23
BIOLOGICAL DIGESTION

Biological digestion is a process used to remove organic matter from municipal sewage and industrial water. In biological digestion, bacteria cultivated under controlled conditions utilize organic matter in the water as their food, producing sludge as a by-product and also products of respiration, such as CO_2 in aerobic and CH_4 in anaerobic systems. Since the process deals with living organisms, every factor influencing the growth and health of the culture must be considered, including an adequate food supply, the availability of nutrients, a temperate climate, and a relatively uniform environment free of pH and temperature shocks and similar disturbances. If the system is aerobic, oxygen must also be available for respiration.

There are many ways of characterizing organic matter in water. The carbonaceous fraction of the total organic matter can be determined (total organic carbon, TOC), and the susceptibility of the organic matter to oxidation by strong oxidizing agents, such as chromic acid, can also be measured (COD). However, the fact that the carbonaceous material can be oxidized is no assurance that the material is a suitable food for bacteria. The ability of bacteria to digest organic matter is measured by the BOD (biochemical oxygen demand) test. (See Chapter 4.)

The BOD test uses what is essentially a miniature incubator containing the organic-laden water, saturated with oxygen, a culture of aerobic bacteria, and suitable nutrients. The sample is incubated at 68°F (20°C), usually for 5 days (BOD_5). Additional information on the nature of the organics may be obtained by incubating a second sample for 20 days. At the end of the incubation period, the oxygen content is measured, and this value is related directly to the respiration of the bacteria and the reduction of organic matter during the test incubation. The organic matter may be metabolized to create new cells, or completely converted to CO_2. The aerobic bacteria used for this test are normally taken from an activated sludge plant, and contain a variety of organisms which have become acclimated to the particular sludge in which they were grown (Figure 23.1).

TREATING INDUSTRIAL ORGANICS

In working with a wastewater that contains organic matter from an industrial plant, it is important to acclimate the bacteria to the waste by operating an incubator for a period of about a week to obtain organisms that have adapted to the particular organics to be digested. It is common to find that a standard BOD test may indicate poor biodegradability of an industrial waste simply because the normal sewage organisms used in the test have not become adjusted to the waste.

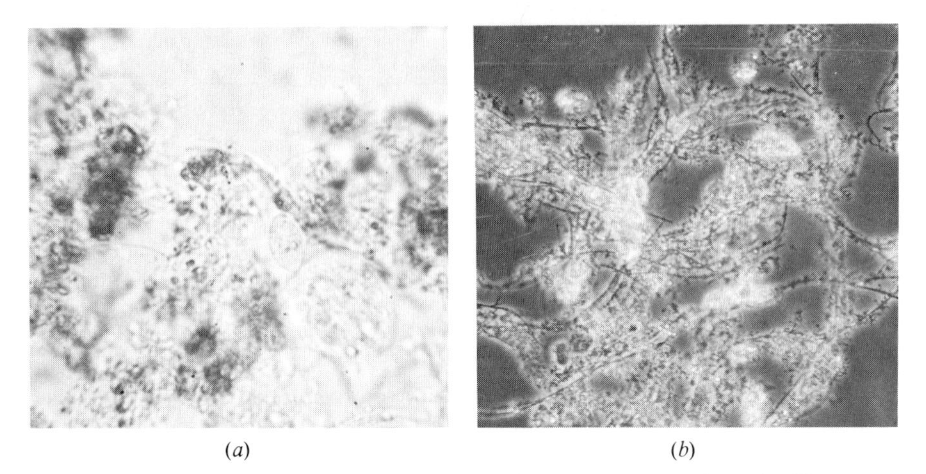

FIG. 23.1 (*a*) A variety of microbes develop in an activated sludge system, including filamentous types that interfere with sedimentation. (*b*) When the filamentous organisms (*Sphaerotilus* in this case) predominate, a condition known as sludge bulking develops.

The process of acclimatization results in the elimination of ineffective bacteria and the selection of those organisms that have the ability to digest organic matter that may actually be toxic to many organisms found in a municipal activated sludge.

The more closely the BOD result approaches the COD determination, the more readily is the organic matter in the waste digested by the selected bacteria. A ratio of 0.5, for example, is extremely favorable, but bacterial digestion is practical even when the ratio is as low as 0.1 to 0.2.

A number of industries have developed around biological processes, such as wine and antibiotics, which use fermentation to produce useful products. These industries use pure strains of microbes for a specific purpose. In water treatment processing by biological treatment, except for the seeding of the system during initial startup, the microbial populations that colonize and populate the system include a wide variety of species, many of which enter the system with air or water. The population of the typical activated sludge system is shown by the analysis of Figure 23.2. This population distribution will be maintained fairly uniformly as long as the system in which these organisms live is not upset beyond their ability to cope with changing conditions.

AVAILABLE PROCESSES

Although special organisms or strains are not usually required for water treatment processes, the method of digestion most suited to the organic material in the water will favor the organisms that are most effective in that system. The categories of organisms developing may be correlated to three operating conditions, aerobic, anaerobic, and facultative.

By far the largest category is aerobic, using organisms that require oxygen for their life processes. Facultative systems are selected for certain industrial waste

NALCO

ANALYTICAL LABORATORY REPORT

From:

Analysis No.	B 42387
Date Sampled	12/27/83
Date Received	12/29/83
Date Printed	1/ 4/84

Sample Marked:
Mixed Primary and Aeration Tank

>>> MICROBIOLOGICAL EVALUATION <<<

PHYSICAL APPEARANCE :	Liquid W Floc
TOTAL AEROBIC BACTERIA	700,000.
Aerobacter	5,000
Pigmented	<1,000.
Mucoids	1,000
Pseudomonas	100,000.
Sporeformers	2,500
Others	591,500.
TOTAL ANAEROBIC BACTERIA	2,000
Sulfate Reducers	1,000
Clostridia	1,000
ESCHERICHIA COLI	10,000
IRON-DEPOSITING	
Gallionella	None
Sphaerotilus	None
TOTAL FUNGI	1,220
Molds	20
Yeast	1,200
ALGAE	
Filamentous	Few
Non-Filamentous	None
OTHER ORGANISMS :	Few Protozoa

Note: Many Fungal Filaments

FIG. 23.2 Typical activated sludge organisms.

treatment operations and in many of the smaller municipalities. Anaerobic systems are usually confined to strong organic wastes, such as those coming from food processing plants, or accumulated organic sludges from the aerobic process.

An understanding of the growth pattern of bacteria is important to the design of a bacterial digestion system. Bacteria reproduce by fission, one cell splitting to produce two, two producing four, and so on. When there is adequate food supply, this reproduction rate occurs exponentially (Figure 23.3). During this logarithmic growth phase, the cells remain dispersed. The organic material may be synthesized into new cellular material, broken down completely to CO_2 and water, or there may be smaller organic molecules (metabolites) formed from the larger ones. These may eventually be completely digested to CO_2.

As the microbes flourish, the food supply begins to limit further growth, and in this declining growth phase, also shown in Figure 23.3, a change takes place in

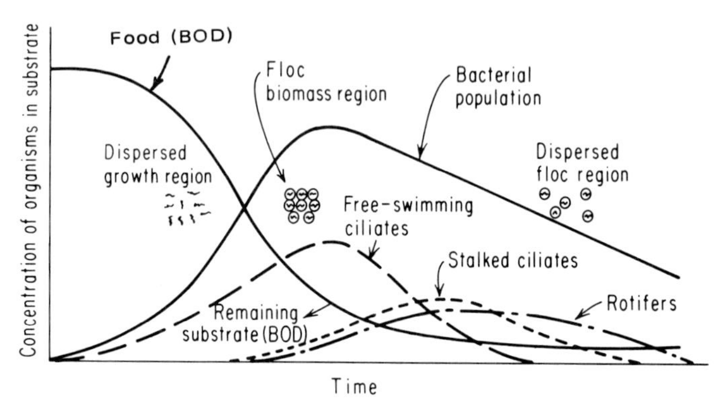

FIG. 23.3 Phases of microorganism development.

the microbial population: the colonies of bacteria begin to form settleable biological flocs. Unless additional food is introduced, a third phase, the endogenous growth phase, commences. In this phase, the bacteria become cannibalistic, feed on each other and on dead cellular matter, and once again become highly dispersed. The suspended solids in the system become difficult to settle.

This growth curve teaches that the ratio of food to bacterial population is of major importance in biological digestion processes. This ratio, called the food-to-microorganism or food/mass (F/M) ratio, can be measured and is an important factor in design and operation of aeration basins, such as activated sludge basins. The food is the BOD of the incoming wastewater; the suspended solids in the incubation basin are mostly attributed to the bacterial population, so the population density is obtained by multiplying the suspended solids concentration in milligrams per liter by the incubator volume. The F/M ratio is then expressed as follows:

$$\text{F/M} = B_i/C_s \times 1/D$$

where B_i = BOD of incoming wastewater, mg/L
C_s = suspended solids in the incubation basin, mg/L (also called mixed liquor suspended solids, MLSS)
D = detention in the incubator, in days

$$= \frac{\text{incubator volume, in gallons (m}^3\text{)}}{\text{flow, in gal/day (m}^3\text{/day)}}$$

Sometimes, instead of using total suspended solids in this ratio, the volatile portion of the suspended solids (MLVSS) may be considered more applicable, particularly if the incoming waste is handled without primary sedimentation to remove inorganic suspended solids. In an activated sludge system in a sewage plant, the volatile suspended solids are usually approximately 70% of the total solids in suspension.

In Figure 23.4, the F/M ratios are plotted to correspond to the approximate boundaries between the logarithmic growth phase, the declining growth phase, and endogenous respiration in aerated basins. In all of the varieties of activated

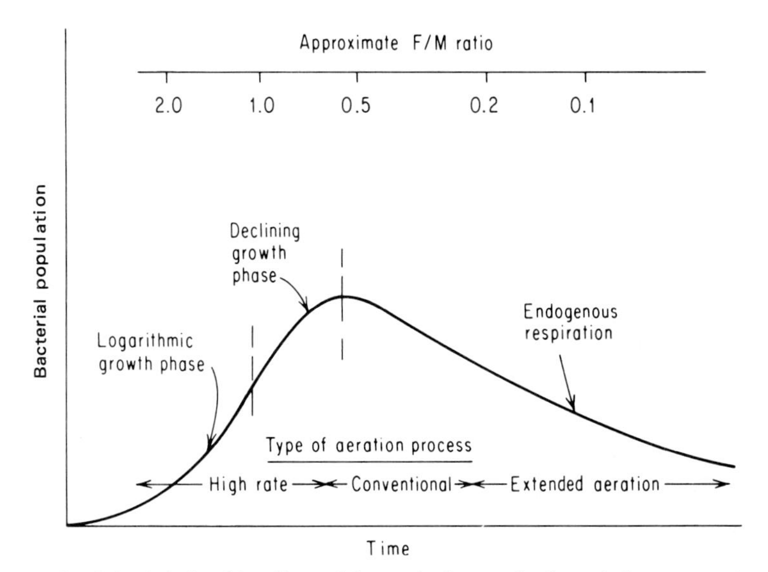

FIG. 23.4 Relationship of bacterial growth phase to food supply in an aerated system.

sludge processes, the aeration devices are usually designed to be capable of maintaining a minimum dissolved oxygen (DO) concentration of 2 mg/L.

TEMPERATURE EFFECTS

A major factor in biological digestion is water temperature, and the daily and seasonal changes that can be anticipated in the biological system. There are two forces at work over the operating temperature range: the first is the competition between psychrophilic, mesophilic, facultative, and thermophilic organisms, each having a preferred temperature range. Second, in the favorable operating range for each of these categories of organisms, the metabolic rate approximately doubles for every 20°F (11°C) temperature increase (Figure 23.5). This points up one of the most serious problems of biological digestion: the fact that the bacterial colonies are almost in a state of suspended animation in the winter if the plant must be shut down and the temperature of the incubation basin drops to 32°F (0°C).

The bacterial colonies favor a pH in the range of 5 to 9, and are seriously inhibited in their activities if the pH should drop below 5 or exceed 10. pH is only one

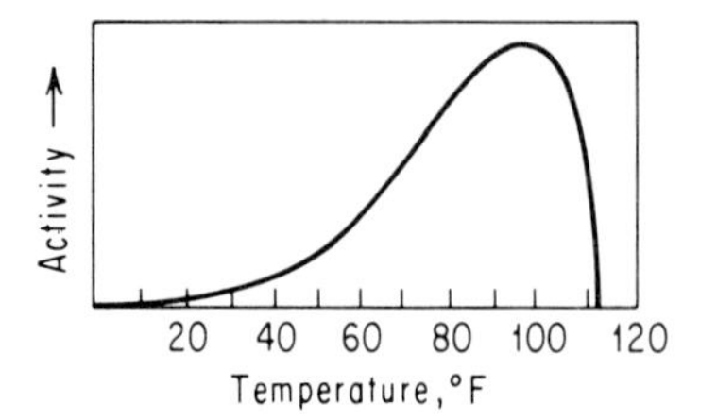

FIG. 23.5 Temperature change of 20°F alters typical activated sludge microbe metabolism rate by a factor of two up to 80°F. As 32°F approaches, rate nears zero. Above 100°F, activity decreases, and soon thermophilic organisms take over. *(Adapted from Power Special Report "Waste-Water Treatment," June 1967.)*

of the factors that must be evaluated to determine the variability of the waste stream to decide on the need for equalization, or storage capacity, prior to the biological digestion system to avoid upsets in the aquatic environment.

TRACE NUTRIENTS

Finally, the bacterial populations require nutrients in addition to carbonaceous food, as indicated by the empirical chemical formula for bacterial protoplasm, $C_{60}H_{87}O_{23}N_{12}O$. As a rule of thumb, the system requires approximately 5 parts of nitrogen as N and 1 part of phosphorus as P for every 100 parts of BOD in the wastewater. Many wastewaters, particularly municipal sewage, contain the required nitrogen, but industrial wastewater may require supplemental nutrients, usually added as ammonium salts and phosphoric acid.

The phenomenon called sludge bulking is thought to be caused by inadequate nutrient levels, although this is probably an oversimplification of a difficult problem. In sludge bulking, the sludge settles slowly because of the development of filamentous organisms in the biomass. The projecting filaments make the biological floc similar to thistledown, so that it moves with the water currents and resists settling. A bulking sludge produces a high sludge volume index (SVI) and is readily identified in microscopic examination. Other factors that are believed to promote bulking conditions may include low pH, which favors the growth of mold and yeasts; the present of high concentrations of carbohydrates; and low levels of dissolved oxygen.

BENCH TESTING

In working with a wastewater containing organic matter of an undefined nature, bench testing followed by pilot plant testing is absolutely essential for obtaining data needed to design the biological treatment system. Before this work can commence, the wastewater flow must be studied over a sufficient period of time to indicate the variations that can be expected in composition and temperature so that provisions can be made for equalization, which is usually required to eliminate variation both in flow rate and composition beyond the range acceptable for bacterial digestion. A sampling system should then be installed to withdraw a sample of the wastewater flow at regular intervals, deliver it to a miniature equalization basin, and produce a composite sample of wastewater that is representative of the total flow, and can be drawn on at any time for the bench studies. A water analysis must then be run on the composite sample to determine the mineral analysis and also the organic material as determined by BOD, COD, and TOC. The mineral analysis must include the determination of total nitrogen and phosphorus to determine whether there is a need for nutrient addition. Once this has been done, a miniature biological digestion system (incubator) is set up to operate on a batch basis (Figure 23.6). This consists of a 5-gal container with an

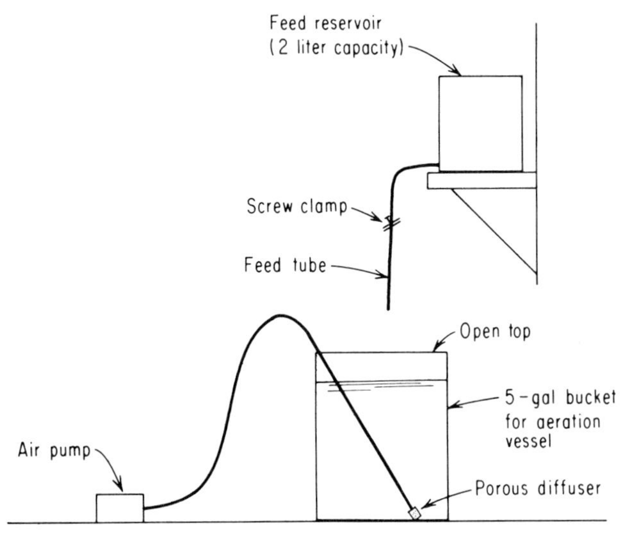

FIG. 23.6 Equipment for bench-scale biological treatment study.

aeration device, such as an aquarium aerator obtained from a pet supply store. The procedure for running the test is then carried out as outlined in Table 23.1. Feeding waste to the aeration vessel may be on a continuous basis if a suitable laboratory pump is available.

TABLE 23.1 Summary of Bench Scale Biological Treatability Procedure (See Figure 23.6)

Aeration volume:	15 L of wastewater
Feed portion volume:	800 mL
Feeding procedure:	Withdraw 1000 mL from aeration chamber. Settle for 30 min in a 1000-mL cylinder. Decant and discard 800 mL. Refill cylinder to 1000 mL with fresh waste and return to aeration vessel.

Phase	Feeding frequency	Average feed rate, mL/h	Detention time, h
Acclimatization	Once every 8 h	100	150
Sludge buildup	Once every 4 h	200	75
Rate studies	Once every 2 h*	400	37.5
Rate studies	Hourly	800	18.75
Rate studies	Once every 0.5 h	1600	9.38

* Feed may be semicontinuous or continuous. Withdrawal can be periodic.

Because the results of a BOD test are not known for 5 days, it is helpful to establish the ratio of BOD to COD or BOD to TOC prior to beginning the test, and then to use the COD or TOC test to measure the progress of the biological digestion experiment.

SLUDGE ACCLIMATIZATION

The common procedure for sludge acclimatization is to withdraw a sample from the incubator once a shift, at which time the sludge is settled and the sludge volume measured. This procedure is repeated until the sludge volume has stabilized or is increasing (consult Table 23.1 for feeding procedure). Sludge buildup is accomplished by increasing the feed rate after the period of sludge acclimatization. When sufficient quantities of sludge have been produced, rate studies can commence.

Rate studies can be made at sludge volumes of approximately 200 mL/L. Data such as BOD, TOC, COD, pH, temperature, and total dissolved solids are run on the raw waste and on the supernatant from settling tests. These tests are run for 1 to 2 weeks at constant feed rates to establish the degree of removal at various detention periods.

A test of this kind with long detention times simulates an extended aeration system with an extremely low F/M ratio. As such, it produces BOD reductions that are reproducible on a plant scale only with an extended aeration system. The test, in effect, indicates the maximum biodegradability that can be achieved and the minimum BOD levels that can be reached.

At the conclusion of the test, the mixed liquor in the incubator should be examined to establish the sludge volume index (the Mohlman index). This index is calculated as follows:

$$\text{SVI} = \frac{C_v}{C_s} \times 1000$$

where C_v is the settled volume (Imhoff cone test), in milligrams per liter and C_s is the amount of mixed liquor suspended solids, mg/L.

Therefore,

$$\text{SVI} = \text{mL/mg} \times 1000 = \text{mL/g}.$$

An SVI below 100 implies that the sludge will settle and compact well. If the SVI exceeds about 200, sediment action is slow, and chemical coagulation and flocculation will probably be required in the final clarifier.

Another measure of the performance of a biological digester is the sludge age, which is calculated as follows:

$$\text{Sludge age} = \frac{C_s}{C_i} \times D$$

where C_s = solids in aeration basin, mg/L
C_i = solids entering basin, mg/L
D = detention, days

For example, if C_s = 3000 mg/L, C_i = 100 mg/L, and D = 0.5 days, SA = 30 × 0.5 = 15 days.

There is no best value for any given process, but there is for each individual plant and this must be determined by experience.

A pilot plant must then be designed to treat this waste and evaluate the numerous combinations of units that can be put together for a complete biological digestion system. The following factors must be considered in choosing between the systems available, and some choices may be quickly discarded without trial or evaluation because they don't meet the needs of the plant:

1. Effluent limitations: These are established by the EPA under the NPDES (National Pollution Discharge Elimination System) program. Does the final BOD achieved by the test meet the requirements?
2. Reliability: This is also established by the NPDES permit, which spells out permissible deviations from average performance and the duration of such excursions.
3. Climatic conditions: Some systems may be rejected because of failure to perform in subzero weather, or because of turnover of the basin with high prevailing winds.
4. Location of residential areas and neighboring industries: This can be particularly significant with odor-bearing wastes.
5. Availability of land for lagooning or land application of liquid waste or sludge.
6. Space available for treatment units.
7. Energy required and available for pumping, aeration, sludge collection, and sludge disposal.
8. Equipment costs and operating costs.
9. Manpower for operation.

Before considering which of the biological systems may be best suited for the waste to be processed, it is important to evaluate the pretreatment steps that might reduce the load on the biological system and simplify its operation. Typical pretreatment steps include grit removal, screening, and primary sedimentation. If the suspended solids concentration is relatively low, below a range of 50 to 100 mg/L, pretreatment may be unnecessary and the waste may go directly to the biological digester.

AERATED SYSTEMS

By far the largest number of biological systems fall into the category of aerobic digesters, depending on aeration to support oxygen-consuming bacteria. If ample land is available, an aerated lagoon (Figure 23.7) may be the most practical incubator, usually having a detention time of about 30 to 60 days. This type of system is rather common in the treatment of sewage from small municipalities.

Aeration basins (Figure 23.8) are typically designed for a detention of 12 to 48 h, so the F/M ratio is quite low. The aeration devices are usually capable of delivering about 3 lb of oxygen per hour per horsepower applied (1.8 kg/kWh). At low F/M ratios, which is typical of these systems, the air required is usually in the range of 1200 to 1800 ft^3 per pound (75 to 110 m^3/kg) of BOD removed. Where surface aerators are used, the system operates as a completely mixed basin, and maintaining solids in suspension may require as much as 100 hP (75 kW) per million gallons of volume (3780 m^3).

FIG. 23.7 An aerated facultative lagoon for treatment of sewage from a community of approximately 5000 people. *(Courtesy of Hinde Engineering Company.)*

A modification of the aeration basin is the oxidation ditch (Figure 23.9). The activated sludge system (Figure 23.10) is much more compact, operates at a much higher loading (F/M ratio), and cannot produce quite as efficient reduction of organic matter as the aerated lagoon. Sludge production is significantly greater. This is the system typically used in larger municipal sewage plants, where it operates successfully under a wide range of climatic conditions. Air is usually supplied through subsurface diffusers at a rate of about 500 to 900 ft^3 per pound (30 to 55 m^3/kg) of BOD. In some plants, pure oxygen is supplied instead of air, further reducing the volume of the incubator.

FIG. 23.8 A typical aeration basin works with low F/M ratios and detention times in the 12 to 48 h range.

FIG. 23.9 Oxidation ditch processing 2.75 mgd of municipal sewage, with clarifiers in background. *(Courtesy Lakeside Engineering.)*

Sludge production is an important aspect of the biological digestion process. Referring to Figure 23.4, in a high rate aeration system, where there is so much food that the bacteria multiply exponentially, much of the food goes to production of new cells, so sludge (which is made up almost entirely of biomass) production is high. Conversely, when food is scarce, the bacteria become cannibalistic and cell production is low, so sludge production may actually cease as new cells reach an equilibrium with those consumed. An estimate of sludge production can thus be made on the basis of F/M ratio (Figure 23.11).

In the typical municipal sewage plant using activated sludge, shown in Figure 23.12, the waste is first passed through a grit chamber and then clarified in a primary sedimentation basin, where the BOD may be reduced by 35 to 40% and the suspended solids by about 50%. The sludge in the primary clarifiers may settle to a density in the range of 4 to 6%. The overflow from a primary clarifier enters the aeration basin, where it may remain under aeration for 6 to 8 h. The mixed liquor suspended solids level is typically in the range of 2500 to 4000 mg/L. The liquor from the aeration basins then flows to a final clarifier, where the biological sludge is settled, and the overflow is disinfected for discharge to the receiving stream. The biological solids in the secondary clarifier, being very light, flocculent colonies, often require chemical treatment for flocculation and sedimentation. The sludge in the secondary clarifier seldom compacts to more than 1.5 to 2.0% solids.

Using this example, assuming a BOD reduction in the secondary system of 200 mg/L, 8 h (0.33 day) detention and MLSS of 3000 mg/L, the F/M ratio is

$$\frac{200}{3000} \times \frac{1.0}{0.33} = 0.2$$

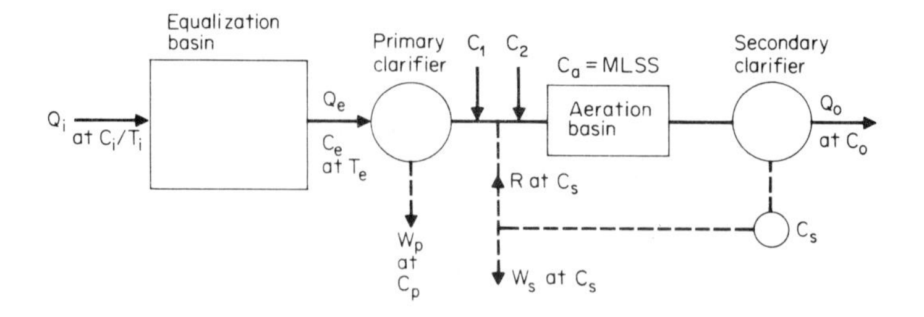

Q_i Influent flow, variable with time of day, rainfall, runoff, production schedule, etc.

C_i Influent solids concentration, variable as with Q_i

T_i Influent temperature, may be variable in an industrial plant with production; seasonally variable

Q_e Flow after equalization, usually a 3 to 4 h average

C_e Solids of Q_i, concentration after equalization

T_e Temperature after equalization

W_p Primary sludge flow, variable with C_o

C_p Solids density, usually fairly constant

C_i Primary clarifier effluent solids concentration, fairly constant

C_2 Solids concentration entering aeration, an operating variable

C_a Solids concentration in aeration, or mixed liquor suspended solids, an operating variable controlled by F/M ratio, R, and C_s

C_s Solids concentration in secondary clarifier blowoff, variable with C_a and SVI

R Return sludge flow, an operating variable adjusted for F/M ratio, SVI, optimum MLSS, and C_s

W_s Waste sludge flow, an operating variable to maintain solids balance and control sludge age

Q_o Effluent flow, usually equal to Q_e unless wasted sludge is not dewatered with return of filtrate

C_o Effluent solids, variable with flow, SVI, and MLSS

FIG. 23.10 Material balance around an aerobic treatment system.

At this ratio, the production of sludge is expected to be 0.3 lb/lb (0.3 kg/kg) BOD removed, equivalent to 60 mg/L (Figure 23.8).

Since the bacteria in the sludge are still active, they are returned to join the primary clarifier effluent entering the aeration basin. The excess production of sludge, which was calculated to be 0.3 lb/lb BOD removed, is drawn off for further processing; it may be joined with the primary sludge or dewatered separately for disposal. Modifications of the basic system include extended aeration, step aeration, and contact stabilization.

In an extended aeration plant (Figure 23.13), the waste stream is fed directly to the aeration basins without primary sedimentation, with the balance of the process being the same as the basic system.

With step aeration (Figure 23.14), secondary biological sludge is introduced at

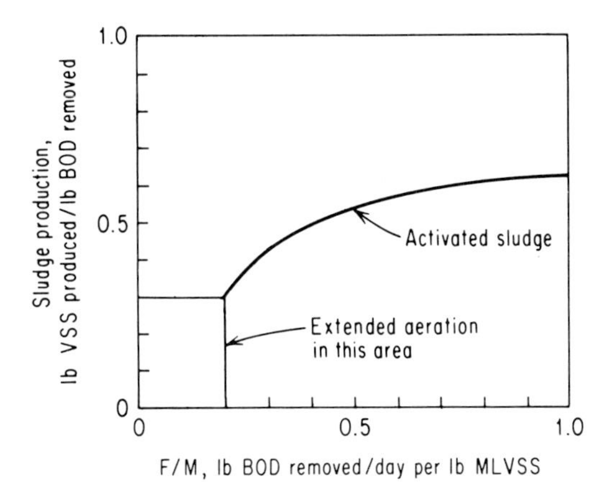

FIG. 23.11 Sludge produced in municipal activated sludge plants as influenced by F/M ratio. *(EPA Report 11010-EXQ-08/71, August, 1971).*

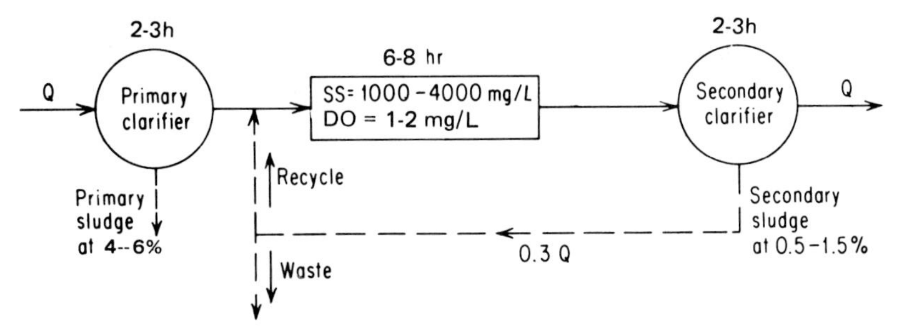

FIG. 23.12 Conventional activated sludge.

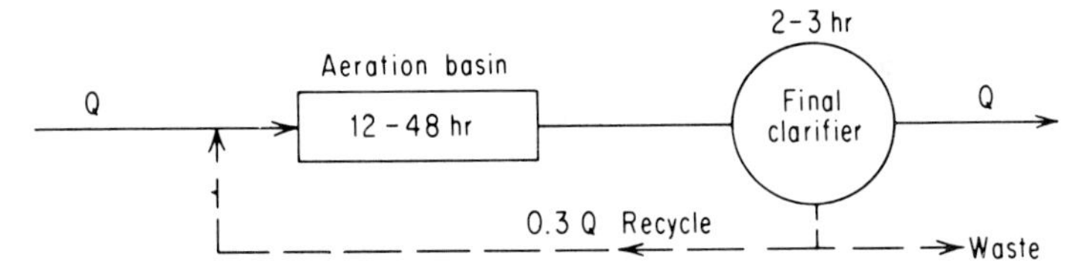

FIG. 23.13 Extended aeration.

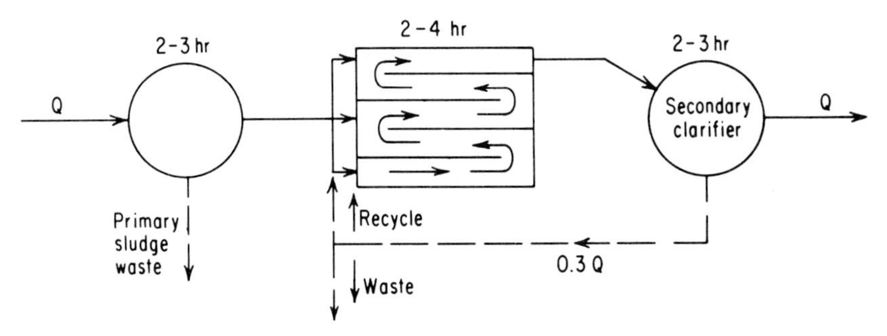

FIG. 23.14 Step aeration.

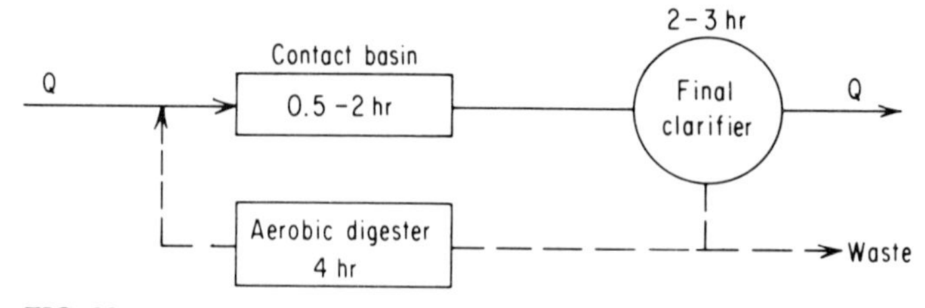

FIG. 23.15 Contact stabilization.

the head of the aeration basin, but the primary clarifier effluent is introduced in portions at several positions along the path through the aeration basin. In this design, the detention is usually reduced to about 3 to 4 h.

Contact stabilization (Figure 23.15) is a final modification of the activated sludge process, and here the biological solids are mixed with the clarified sewage for a period of only 30 to 60 min, in which time they act as flocculants for removal of suspended and colloidal matter and for adsorption of soluble organic materials. The collected sludge from the final clarifier is processed through a separate aeration basin in which the adsorbed and entrapped solids are then digested by the biomass.

The aerobic systems described here depend on air, comprising 80% nitrogen and 20% oxygen, as the source of oxygen. The large proportion of nitrogen does not aid the aerobic process biochemically but assists in the mixing process to keep the sludge suspended. Liquid oxygen has become a high-volume chemical commodity in industrialized countries, and its cost has been reduced to a level favoring its use in specially designed basins. These have mixers that produce finer bubbles than those produced by conventional aeration processes, increasing gas transfer efficiency to make the O_2 digestion process more economical. Oxygen has provided better performance than air in handling difficult wastes and in retrofitting older plants to meet the more stringent effluent standards required today. Another modification of the aerobic system used to process complex organic contaminants is the application of activated carbon to the aeration basin as a means of concentrating the organic matter on a surface more amenable to microbe growth.

BALANCING THE ACTIVATED SLUDGE PROCESS

There are so many variables in aerobic digestion systems that control of the process is difficult unless the plant is well designed, the operators well trained, and the performance carefully monitored by instrumentation and laboratory testing. The balance requires an evaluation of each of the components of the system:

1. *Clarifiers:* The primary and secondary clarifiers are vastly different in their characteristics because of differences in solids loading, as shown by Table 23.2. A major physical contrast between the primary and secondary units is the underflow rate of sludge withdrawal. A large portion of the secondary sludge underflow is returned to seed the feed to the aeration basin, to provide the F/M ratio needed for optimum digestion. However, if the food (F, or BOD) supply

TABLE 23.2 Comparison of Primary and Secondary Clarifiers

Variable	Primary	Secondary
Feed solids range, mg/L	50–500	1500–5000
Typical solids removal, %	50%	95%
Inlet-outlet, mg/L	200	2500
Solids in sludge, range	4–6%	0.5–1.5%
Typical, mg/L	50,000	15,000
Sludge concentration ratio	250	6
Sludge underflow, % of feed	0.4%	17%

decreases, the bacterial population must be reduced and the recycle rate reduced. This increases the sludge wastage. To reduce the hydraulic load on sludge processing units, in some plants, the waste is thickened. In other plants, some of the waste is diverted to the primary clarifier, where its flocculant biomass may assist in solids reduction.

A second variable that requires a controlled change of the underflow rate is the sludge volume index (SVI). If the SVI has been normally about 100 and then increases to 150 (approaching a bulking condition), the density of the underflow is drastically reduced, so the rate of sludge withdrawal must by increased to prevent excessive carryover into the effluent. But, as shown by Table 23.3, if the plant had been carrying 2300 mg/L MLSS with the SVI at 100 and a 0.3 return ratio, almost a twofold increase in the sludge return ratio (to 0.6) would be needed to raise the MLSS back to 2300 mg/L. If the food in the waste entering the aeration basin has not changed, the F/M ratio will increase with the reduced MLSS, producing more sludge. This may be beneficial, unless the new sludge also has a poor SVI. At this point, the adjustment of clarifier flows has gone about as far as possible toward correcting the upset (by increasing underflow and sludge return rate), and additional improvements require adjustment of the biology of the system.

TABLE 23.3 Effect of SVI on MLSS that Can Be Handled at Various Sludge Return Ratios

Assumes a system normally operating at 2300 MLSS, 0.3 return ratio, and 100 SVI

SVI, mL/g	Return ratio, % of plant flow	MLSS, mg/L
50	0.3	4615
	0.6	7500
100	0.3	2300
	0.6	3750
	0.9	4740
150	0.3	1540
	0.6	2500
	0.9	3160

Note: The relationships are as follows:

$$\frac{R}{Q_e} = \frac{M}{(10^6/\mathrm{SVI}) - M}$$

Until the SVI can be lowered to a satisfactory level, the plant must cope with the excess sludge underflow. This can be done by (a) chemical coagulation/flocculation to assist sludge settling and compaction in the final clarifier, (b) returning as much waste-activated sludge to the primary feed as possible, and (c) increasing the load on the thickener and dewatering devices, possibly by supplemental chemical coagulation and flocculation.

2. *Aeration basin:* Bringing the biological system back to health, with an SVI index in the range of 75 to 125, requires a careful comparison of all the control variables and indicator ratios prevailing during upset to normal or acceptable conditions. These include:

a. *Physical factors:* Most of these are uncontrollable.

Flow. Is there a stormwater overload? Is there an unusual tributary flow?

Temperature. Is there an unusual weather condition?

pH, conductivity, BOD, and suspended solids: Is there a clue in any of these measurements to an unusual dump?

b. *Biological factors:*

Appearance of sludge. Brown or gray?

Microscopic study of floc. Filamentous? Protozoa present?

Oxygen uptake rate. Is return sludge biologically active?

Dissolved O_2 in basin. Should be 1 to 2 mg/L.

Dissolved O_2 in final clarifier. Is digestion occurring there?

Aeration rate, cubic feet per minute, and cubic feet per pound of BOD.

F/M ratio. Is this in a flocculation range?

Sludge age, or cell residence time. Is sludge too old? Too young?

Soluble BOD/total BOD in raw waste.

Sulfides in primary system.

Nutrient balance. Ratio of N:P:BOD.

Toxic shock. Are protozoa present?

Microscopic evaluation of the sludge is essential to an understanding of the upset that has caused a high SVI. In addition to this, the oxygen uptake rate (OUR) and respiration rate must be determined. The test is done as follows.

1. Shake and aerate a sample of sludge in a large bottle for about one minute.
2. Transfer to BOD bottle, insert O_2 probe, and record O_2 concentrations for 10 min.

$$\text{OUR} = (\text{initial } O_2 - \text{final } O_2) \times 6, \text{ mg/L/h}$$

If there is no substantial uptake rate compared with values obtained during normal operations, there is essentially no biological activity. If there is a significant uptake rate, convert this to a respiration rate, using the MLSS concentration prevailing at the time of testing oxygen levels:

$$\text{Respiration rate} = \frac{\text{OUR} \times 1000}{\text{MLSS mg/L}}$$

TRICKLING FILTERS

Competing with the activated sludge process and its modifications is the biofilter or trickling filter (Figure 23.16). Though the trickling filter does not provide the high degree of organic removal afforded by the activated sludge process, it has advantages in space savings and in its ability to resist upsets. Newer modifications of the biofilter incorporate rotating disks (Figure 23.17). In either of these designs, a gelatinous biological mass forms on the filter surfaces, and this mass digests the organic matter in water passing over the surface. These units do not require air beyond that available to them by natural ventilation. Where the organic concentrations are high, the concentration ahead of the conventional trickling filter is reduced by recycling a portion of the biofilter effluent or the plant effluent. A variety of schemes are available, as shown by Figure 23.18.

The disk-type filters [called rotating biological contactors (RBCs)] do not require this recycle scheme. If loadings are high, disks can be operated in series, so the organic is reduced in two or three stages. The RBC filters are usually housed to minimize evaporative cooling and maintain as favorable an operating temperature as possible.

There is a continual sloughing of organic debris from the surface of the biofilter, requiring final clarification of the wastewater effluent for discharge to a receiving stream. This usually amounts to 0.1 to 0.3 lb solids per pound (kg/kg) BOD removed in a municipal sewage system. Handling industrial wastes, sludge production may be as high as 0.4 to 0.5 lb per pound BOD removed if the waste is high in carbohydrates, and as low as 0.05 to 0.1 lb per pound BOD removed if the organics are volatile acids or alcohols.

ANAEROBIC DIGESTION

Anaerobic processes are sometimes used for highly concentrated organic wastes. The two designs most prominently used are the anaerobic lagoon and the closed digester.

In the anaerobic lagoon, wastes simply enter the digestion basin without addition of air, and the oxygen content is quickly depleted, leading to fermentation. Very commonly, a thick crust forms on the lagoon, preventing oxidation and to some extent confining the odorous gases produced by anaerobic decomposition. In the anaerobic lagoon, carbonaceous material is converted to organic acids and alcohols, and later to methane. The proteinaceous material is converted to ammonia, and sulfates in the water are often reduced to sulfides, the principal source of odor from the system. In processing fruit or vegetable canning wastes, the fermentation may occur so rapidly that the basin does not crust over, because the solids are prevented from agglomerating by the effervescence of the gas produced.

Anaerobic digestion is generally slower than aerobic, and a typical food processing waste may require an anaerobic lagoon with a detention of as much as 30 days for effective operation. For the most part, however, the anaerobic lagoon has been limited in acceptance because of the potential for emission of sulfide odors.

The anaerobic digester is widely used in municipal sewage plants to stabilize the accumulated waste biological solids. These may be treated separately or mixed with primary sludge, depending on the volatile content of the sludge. A typical digester (Figure 23.19) can process these solids to yield a stabilized dewaterable

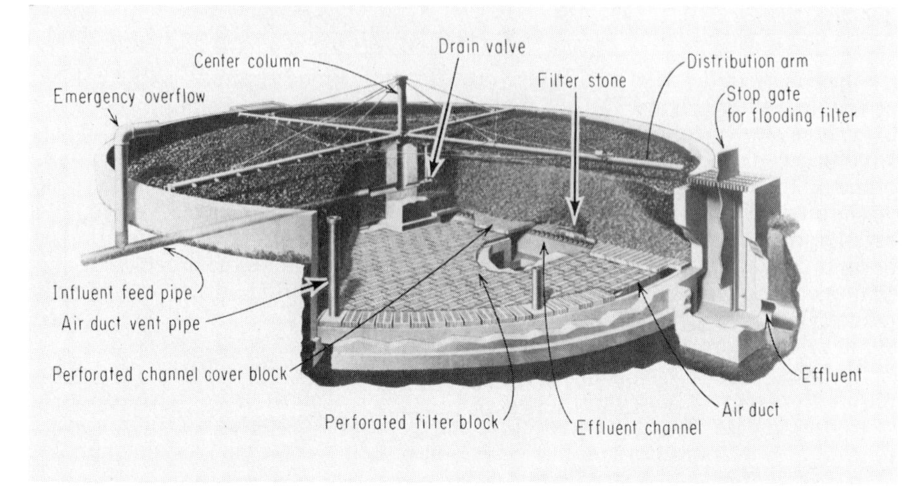

FIG. 23.16 Construction details of a trickling filter showing rotary distributor, underdrain, rock fill, and provision for air ventilation. *(Courtesy of FMC Corporation.)*

FIG. 23.17 In this 1.0-mgd sewage treatment plant, biological treatment takes place on these assemblies of disks mounted in basins through which pretreated sewage is flowing. A growth of biological gel develops on the rotating disks, providing the mechanism for digestion of organic wastes. Units are covered for odor control and to optimize temperature during winter operation. *(Courtesy of Walker Process Corporation.)*

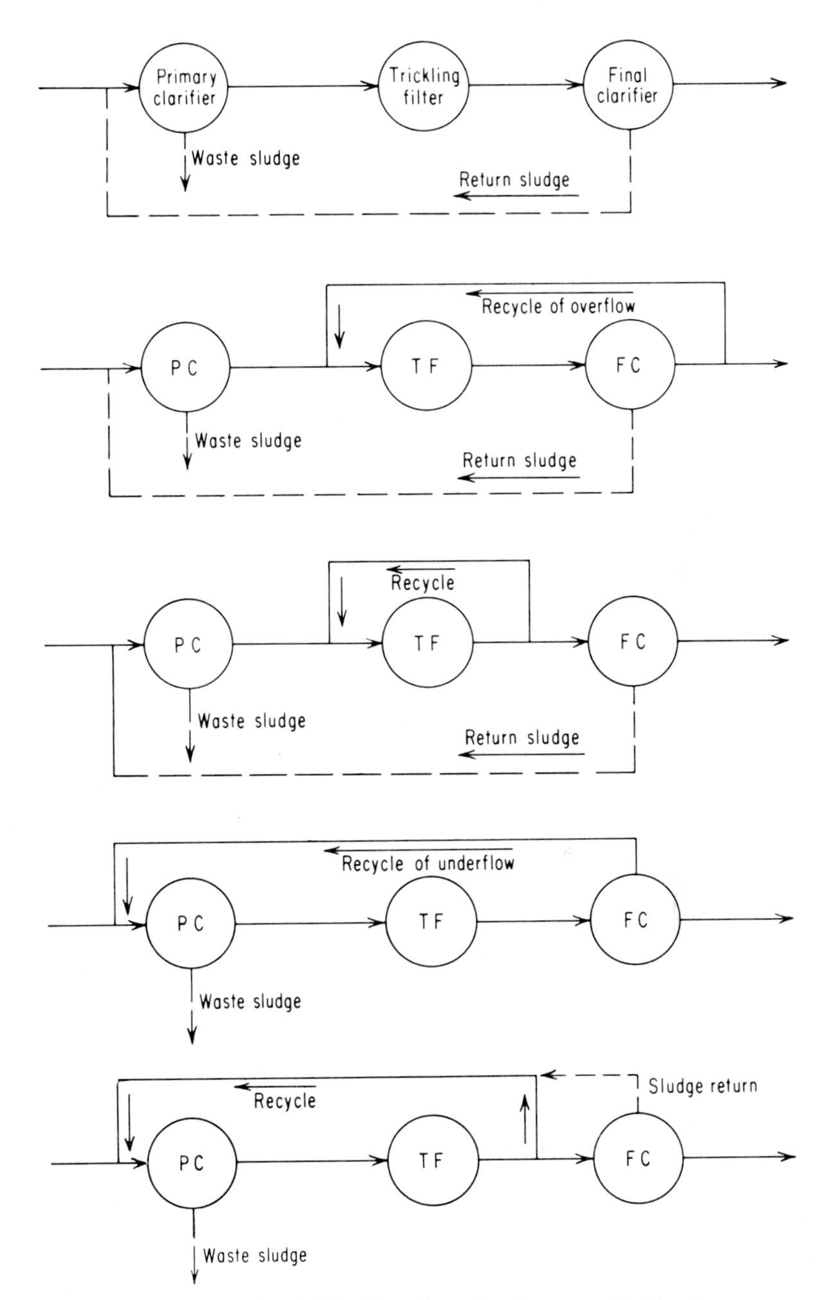

FIG. 23.18 Schemes for trickling filter plants based on strength of wastes.

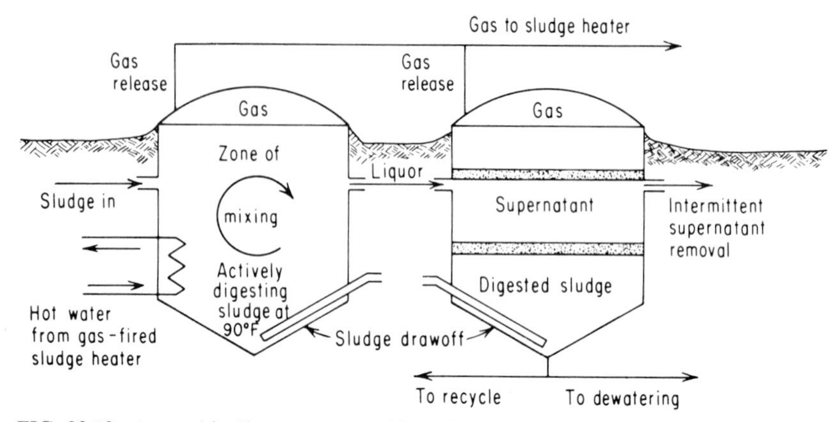

FIG. 23.19 Anaerobic digesters operated in series with gas used to heat first stage. Tanks are insulated by earthen backfill.

sludge at about 3 to 4% solids with acceptable odor. The anaerobic digestion produces 70% methane gas that is collected and burned to provide heat to the digester to maintain its temperature at about 90 to 95°F (32 to 35°C). The digester is insulated so that this temperature can be maintained for efficient biological activity during the digestion period of 15 to 20 days. The ammonia and CO_2 produced during digestion yield ammonium bicarbonate in a solution, producing a supernatant from the digester with an alkalinity in the range of 500 to 1500 mg/L. This supernatant is usually periodically withdrawn and fed to the aeration basins.

OTHER PROCESSES

Other methods of biological digestion include facultative lagoons, land irrigation, and cooling tower digestion. In the facultative lagoon, usually designed with a retention time in excess of 30 days, anaerobic digestion occurs in the settled sludge layer, and aerobic digestion occurs in the surface water. The lagoon may be 5 to 10 feet (1.5 to 3.0 m) deep. The effluent quality is subject to upsets by violent weather conditions and by the development of algal blooms in the summer. The algae play a prominent role in aerobic digestion, but they present difficulty in the production of a clear overflow from the lagoon.

With land irrigation, the soil bacteria become the active agents in the digestion of the organic matter if this is a spray-irrigation type of application. Liquid wastes have been successfully disposed of in this fashion on both forest and crop lands, and the use of wastewater for irrigation will play a prominent role in the future (Figure 23.20).

In some industrial plants, cooling towers have been used as a kind of combination biofilter and activated sludge system. This requires, of course, the maintenance of active biological populations in the cooling tower, and the goal of biological digestion may not always be consistent with the goal of maintaining clean heat transfer equipment in the plant process areas where the cooling water leaving the tower is used.

FIG. 23.20 Treated wastewater may be used for irrigation if it is free of heavy metals that might concentrate in food crops. Ridge and furrow irrigation, shown here, is common and relatively inexpensive. *(From EPA Publication 5012, Environmental Pollution Control Alternatives—Municipal Wastewater.)*

FIG. 23.21 Pilot plant designed for biological digestion studies on fresh wastewater at a plant site. This unit can be operated in a variety of modes, including extended aeration, step aeration, and contact stabilization.

In designing the biological digestion system, once the bench studies have been completed and the designer has reviewed the options open for biological treatment, a pilot plant must be set up to obtain the final design criteria for any given wastewater treatment problem. There are numerous handbooks providing rules of thumb that have been used in the past for design of biological treatment systems, but none of these can be relied on without pilot plant experience to determine the degradability of the organic materials in the wastewater under simulated plant conditions. An activated sludge pilot plant set up to investigate an industrial wastewater problem is shown in Figure 23.21. Pilot plants such as these must be designed for round-the-clock operation and require as careful attention for their control as a full-size wastewater treatment plant. The investment in pilot plant facilities and personnel is paid for in the efficiency and reliability of the final plant.

SUGGESTED READING

Bartsch, E. H., and Randall, C. W.: "Aerated Lagoons—A Report on the State of the Art," *J. Water Pollut. Control Fed.,* **43**(4), 699 (1971).

Chipperfield, P. N. J.: "Performance of Plastic Filter Media in Industrial and Domestic Waste Treatment," *J. Water Pollut. Control Fed.,* **39**(11), 1860 (November 1967).

DeWalle, F. B., Chian, E. S. K., and Small, E. M.: "Organic Matter Removal by Powdered Activated Carbon Added to Activated Sludge," *J. Water Pollut. Control Fed., 49*(4), 593 (April 1977).

Filion, M. P., Murphy, K. L., and Stephenson, J. P.: "Performance of Rotating Biological Contactor under Transient-Loading Conditions," J. Water Pollut Control Fed., *51*(7), 1925(July 1979).

Gagnon, G. A., Croudall, C. J., and Zanoni, A. E.: "Review and Evaluation of Aeration Tank Design Parameters," *J. Water Pollution Control Fed., 49*(5), 832 (May 1977).

Grimestad, D. E., and Wetegrove, R. L.: "Biological Process Troubleshooting," *Pollut. Eng., 14*(3), 25 (March 1982).

Kalinski, A. A.: "Comparison of Air and Oxygen Activated Sludge Systems," *J. Water Pollut. Control Fed.,* **48**(11), 2472 (November, 1976).

Kroeker, E. J., Schulte, D. D., Sparling, A. B., and Lapp, H. M.: "Anaerobic Treatment Process Stability," *J. Water Pollut. Control Fed.,* **51**(4), 718 (April 1979).

"Low Pressure Aeration Yields Effective Wastewater Treatment," *Ind. Wastes,* Nov/Dec. 1978.

Operator's Practical Guide to Activated Sludge, Parts 1 and 2. CRS Group Engineers, Inc., Houston, Texas, 1978.

Pipes, W. O.: "Microbiology of Activated Sludge Bulking," *Advances in Applied Microbiology.* Volume 24. Academic Press, New York, 1978.

Sherrard, J. H., and Schroeder, E. D.: "Stoichiometry of Industrial Biological Wastewater Treatment," *J. Water Pollut. Control Fed.,* **48**(4), 742 (1976).

Stone, R. W., Parker, D. S., and Cotteral, J. A.: "Upgrading Lagoon Effluent for Best Practicable Treatment," *J. Water Pollut. Control Fed.,* **47**(8), 2019 (1975).

Stover, Enos L., and McCartney, David E.: "BOD Results that Are Believable." *Water Eng. Manag.,* 37 (April 1984), page 37.

P · A · R · T · 3

USES OF WATER

CHAPTER 24
ALUMINUM INDUSTRY

Aluminum manufacture has four distinct phases of production from raw ore to finished products: mining, bauxite refining, reduction, and fabrication. Another category is secondary production by independent scrap recovery plants, but this is similar to recovery operations in the remelt section of the fabrication plant.

MINING

Bauxite is the ore used in the production of aluminum. It is a mixture of aluminum oxide trihydrate and monohydrate, with iron oxides, aluminum silicates, and titanium oxide impurities. The ore usually contains from 5 to 25% moisture, which is removed from imported ores to reduce shipping weight and improve handling. In the United States, most ore is imported, but some Arkansas bauxite is also processed. Lower grade aluminum minerals, such as kaolin and refuse from coal preparation, are being investigated in a variety of research programs attempting to extract alumina from such alternate sources.

Open pit mining accounts for almost all bauxite production. The overburden is stripped and the ore recovered by drag lines and shovels. The ore is then sized and sent to the bauxite plant where it is stored for blending and later use.

BAUXITE PROCESSING

To utilize the aluminum values in either domestic or imported ores, the alumina in these minerals must be extracted and purified. This is done in alumina refineries, which produce a finished product of calcined alumina, Al_2O_3, necessary for the electrolytic reduction process which is used to produce primary aluminum metal.

Of the nine basic alumina refineries operating in the United States, two are in Arkansas utilizing domestic bauxite. The others use ores imported from Australia, Jamaica, Africa, and other areas.

In an operation called the Bayer process, raw bauxite is digested at temperatures as high as 475°F (250°C) and at pressures up to 500 lb/in^2 (35 kg/cm^2), by caustic solutions to yield a slurry containing sodium aluminate ($NaAlO_2$) and suspended waste solids called red mud. A process flow sheet is shown in Figure 24.1. To allow efficient recovery and reduction of waste, the red mud is settled or fil-

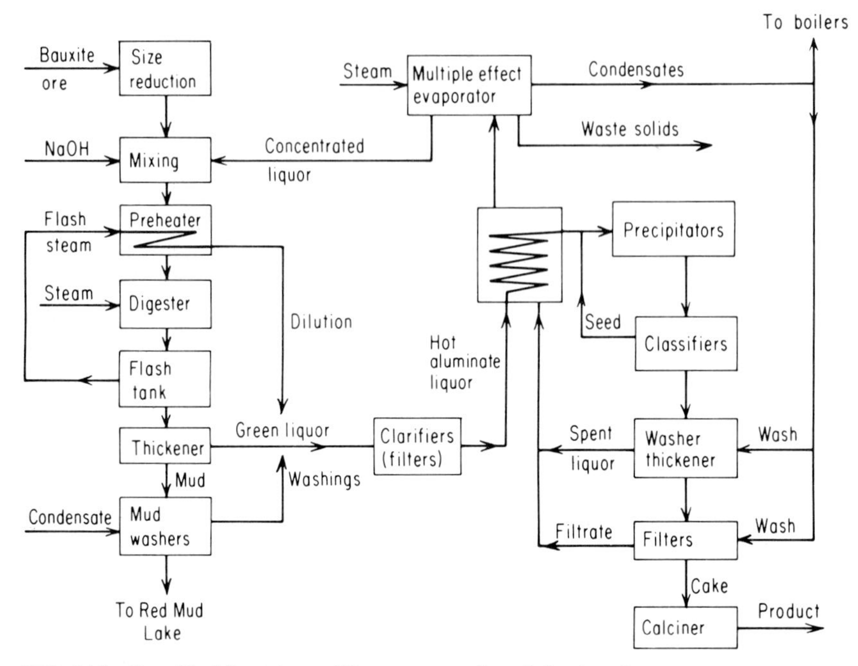

FIG. 24.1 Simplified flow sheet of Bayer process for refining bauxite ore.

tered from the dissolved liquor using natural and synthetic flocculants to produce a clear sodium aluminate solution.

The red mud from the first separation step is reslurried with weak liquor in a washing operation to extract more of the aluminum and alkali values. The most common practice uses a continuous countercurrent decantation arrangement with as many as seven washing stages to recover almost all significant alumina and alkali values. The red mud waste is finally discharged to a large mud lake (tailings pond).

After filtration, the pregnant liquor, rich in sodium aluminate, is cooled through heat exchangers to about 120 to 140°F (50 to 60°C) prior to entry into precipitation tanks. Here, seed crystals of previously precipitated alumina trihydrate are added and the tanks agitated for a day or two. This hydrolyzes the aluminate to yield alumina trihydrate (aluminum hydroxide), which grows on the seed material. This process is allowed to continue until about half of the alumina values are converted. Further precipitation may result in excessive dropout of impurities, such as silica. The alumina trihydrate slurry is then sent to classification and thickening tanks to separate the trihydrate from the spent liquor. The trihydrate is filtered and calcined to produce anhydrous alumina. Following the removal of the trihydrate crystals, spent liquor is concentrated by evaporators and returned to the beginning of the circuit.

Many Uses for Water

Water enters the process stream primarily as steam or condensate. A small amount of water may be introduced as purified makeup to prepare process reagents such as flocculants or cleaning solutions. Storm water and waste surface water may enter the process as return water from the mud lake. Noncontact cooling water is used for equipment cooling and finished alumina product cooling. This water may be from an outside source or from the mud lake reservoir (see Figure 24.2).

Once in the process stream, water is used again and again as a solvent, to transport heat or slurry, to wash away impurities, to create vacuum, to generate electricity, and to prepare chemical makeup solutions.

A major input of steam comes from the digestion tanks where the alumina values are dissolved from the ore. As pressure is relieved in flash tanks, the lower pressure steam is liberated with its heating potential. This is used to heat the spent liquor before it joins the fresh bauxite in the digester, thereby recycling heat in a closed loop. Other uses for the steam from the digester flash tanks are in the evaporators concentrating the spent liquor.

Wash-water streams vary in composition depending on the product purity required. Condensate is used to wash the precipitate on the trihydrate filters to remove residual impurities. This wash water or other condensate may be used to wash the thickened trihydrate before going to filtration dewatering. Excess condensate beyond that required in steam generation may be used in final red mud washing along with mud lake return water.

Mud lake return water consists of overflow from gravity thickening as well as stormwater inflows. The amount varies depending on local conditions. The water contains a low concentration of alumina and caustic values from mud leaching. This proceeds to the last mud thickener where it washes mud underflow from the previous thickener. Increased in alumina and caustic content, it joins the mud from the second previous washing, and so on through the circuit. As it progresses up through the wash train, the alumina and caustic values increase correspondingly. Finally it joins the partially cooled digester discharge where the final increase in alumina content is achieved. This countercurrent decantation process is designed to provide economical recovery of as much alumina and alkali as possible.

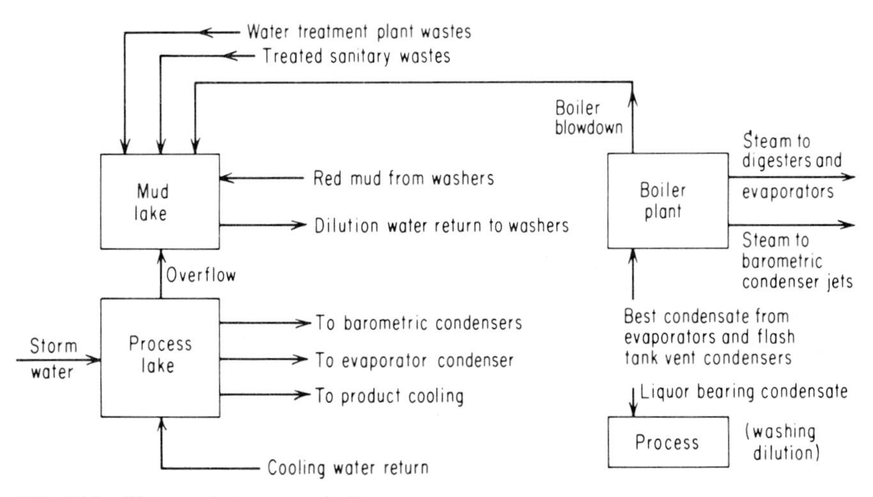

FIG. 24.2 Water and steam uses in Bayer process.

In addition to its use as an effective solvent, the water serves to transfer solids from unit to unit within the process. This includes not only the bauxite slurry, but the trihydrate and the red mud as well. It also provides a medium for wet grinding of the bauxite ore to the proper particle size for digestion. The use of water as a means of contact cooling of the overhead vapors and barometric condensers allows both the extraction of needed condensate water and also the concentration of process streams. Water use in these barometric condensers may reach 5000 to 10,000 gal/ton of product (20 to 40 m^3/t).

Because of process economics, the bulk of the unit process contact effluent streams becomes intermediate feed streams to the other units, as described. Noncontact streams represent a loss of heat rather than chemical values. These heat values are recovered wherever practical. Even though no significant discharges are planned, water may be lost from any of the following production areas:

1. Drying and calcining.
2. Red mud (evaporation and seepage from lakes or contained as part of the in-place mud content).
3. Evaporation of liquor streams.
4. Sand disposal.
5. Evaporation of washdown water streams.

The complete water use scheme, including the Bayer process circuit, is shown in Figure 24.2.

ALUMINUM REDUCTION

Aluminum is the most common metallic element in the earth's crust, comprising about 7.3% expressed in the elemental form, or about 15% in the commonly occurring oxide form, Al_2O_3. Unlike iron, the next most common element, aluminum was unknown to the ancients, and it was only a century ago that the metal became available in commercial quantity.

As indicated by its position in the electromotive series, aluminum is a reactive metal. Because of this, reduction of aluminum oxide by carbon in a furnace, a process which easily produces iron from its ore, is unsuccessful because the aluminum metal produced is vaporized and quickly returns to the oxide form when the reaction is reversed on cooling. Success in producing aluminum from the oxide was finally achieved in 1889 by the discovery of a unique electrochemical process. The process was discovered simultaneously by Hall in the United States, and Heroult in France.

Alumina produced by the Bayer process previously described is dissolved in a molten bath composed chiefly of cryolite ($3\,NaF \cdot AlF_3$), to which is added fluorspar (CaF_2) and aluminum fluoride (AlF_3). The molten mixture is prepared in an electrolytic cell, called a pot, illustrated by Figure 24.3. A series of carbon anodes project into the bath from above, and the amorphous carbon pot lining, with metal collector bars embedded in it, serves as cathode. The pot lining is encased in insulation and refractory brick surrounded by an outer steel shell. The exact reactions in the cell are not completely understood, but the important result is a reduction of the aluminum oxide by carbon, according to the following reactions:

$$Al_2O_3 + 3C \rightarrow 2Al + 3CO \uparrow \qquad (1)$$

$$2Al_2O_3 + 3C \rightarrow 4Al + 3CO_2 \uparrow \qquad (2)$$

FIG. 24.3 A potline in an aluminum reduction plant. Fumes are withdrawn from each cell and delivered to a scrubber. Fresh anodes are in the rack at the right. The crane moves the tapping device and crucible from cell to cell. *(Courtesy of Kaiser Aluminum and Chemical Corporation.)*

The oxidation occurring at the anode produces a mixture of about 25% CO and 75% CO_2, and this exiting gas mixture carries with it certain volatile fluoride compounds. Carbon is consumed at the rate of about 0.5 lb per lb of aluminum produced. Aluminum is produced by reduction at the cathode, and a pool of aluminum metal collects on the bottom of the cell. The passage of current through the cell generates sufficient heat to keep the bath molten and at a temperature of about 1650 to 1740°F (900 to 950°C).

Frequent Feeding Necessary

The heat is contained by the insulation of the pot lining material plus a crust which forms on the surface of the bath. Although aluminum oxide is the basic raw material, it makes up less than 8% of the bath, and must be continuously added. A supply of alumina is maintained on the crust where it heats and dehydrates. The crust is broken periodically and some of the alumina is stirred in to replenish what has been consumed by the reduction process.

Molten aluminum is periodically withdrawn (tapped) from the pot and cast into pigs or into any of the various types of ingots. The usual power consumption for aluminum reduction is in the range of 8 to 10 kWh per pound (4 kWh/kg) of aluminum produced, so that a pot line of perhaps 200 to 240 cells producing about 100,000 tons/year of aluminum would have a power requirement of about 250 MW. Because of this high electric power requirement, aluminum reduction plants are usually located near low-cost sources of power. Typically these plants use hydroelectric power or have contracted with nearby utility stations for a significant block of their electric output. Because direct current is needed for electrolytic cells, a substantial investment is required for electrical facilities to convert the alternating current available from public utilities to direct current.

Because of the high consumption of carbon electrodes, each plant has its own facilities for producing these from mixtures of petroleum coke and pitch. Rather

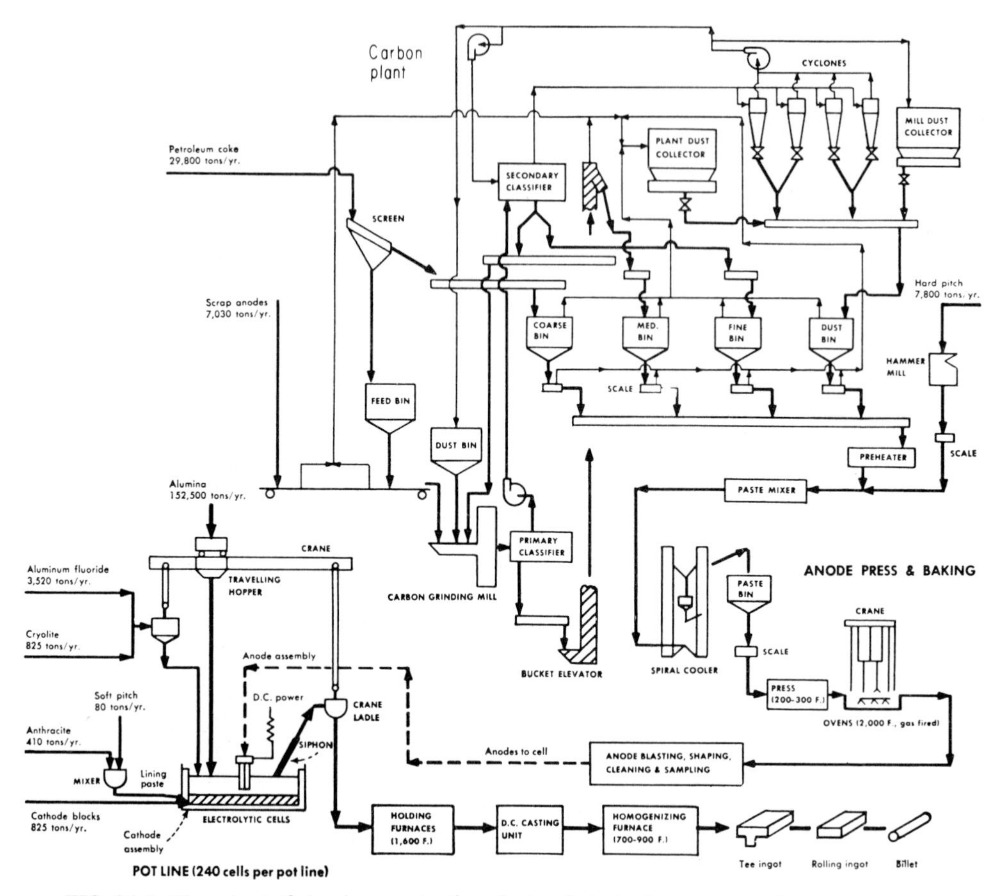

FIG. 24.4 Flow sheet of aluminum reduction plant and electrode processing plant.

extensive facilities are required for materials handling and processing of the electrodes. A flow sheet of an aluminum reduction plant and an electrode processing plant are shown in Figure 24.4.

Exhaust Gases

In the reduction process, as indicated by the chemical equation shown earlier, the consumption of the carbon electrodes produces a mixture of carbon monoxide and carbon dioxide. These gases leave the cell and carry with them certain volatile components from the bath, including fluoride compounds. Particulates from the surface of the pot are also entrained with the escaping gases. To control air emissions, there are hoods positioned over each cell and suction is applied to collect the gases. The gases are routed through wet scrubbers to remove both gaseous and particulate materials. Since the fluoride content of the gas stream is valuable, the scrubbing water is often fortified with reactive chemicals, such as sodium aluminate, to produce various mixtures of sodium and aluminum fluorides which can be returned to the bath after concentration and drying.

For air emission control in the electrode processing plant wet scrubbers are sometimes used, and the scrubber water becomes a significant part of the total waste treatment load. But it is more common to control these emissions with dry collectors.

Following the reduction of aluminum in the pot line, the metal is cast either by a continuous casting process or by conventional casting into molds on a casting floor. In continuous casting units, water is used for mold cooling. The requirements for high-quality water are much more stringent for the continuous casting operation than for conventional cooling operations. The cooling water may be reclaimed over evaporative cooling towers, and the systems can be troublesome because the water often contains oily mold-release agents. The materials provide food for microbes, making it difficult to keep the cooling system free of slime and other microbial masses.

After casting, the ingots are cooled by air or by water spray, and in some cases the system is designed to produce a very fine spray so that the water is evaporated completely to dryness and leaves no residual wastewater to be dealt with in a treatment system. If an excess of water is required, it will contain suspended solids and oil, and will require treatment for disposal. Air cooling, either alone or combined with a water spray, will undoubtedly find increasing use as a way of minimizing wastewater discharge.

There are various other cooling water applications in the reduction plant. These include water for cooling hydraulic oil, green anodes, and compressors. The plant receives electric power from a nearby utility at high voltage which must be reduced through transformers. The alternating current at reduced voltage is then converted by mercury arc rectifiers to direct current, which may be as high as 800 V, based on a loss of about 5 V at each pot, with the cells of the potline being connected in series. Cooling water is required for components of the rectifier. There is a transformer and rectifier unit for each pot line. Cooling water may also be required for the chlorine compressors, which are used to supply chlorine gas for degassing molten aluminum.

METAL FABRICATION

The aluminum ingot is converted to a wide variety of forms, including sheet, plate, structural shapes, rods and other bar stock, tubes, wire, and foil. Some of these can be extruded by heating the metal to a temperature at which it becomes plastic, but the bulk of production is done in the rolling mill. Since initial rolling is done on hot metal, the hot ingot may be sent directly to the rolling mill, but it is usually passed through furnaces to make certain that the temperature is uniform throughout. Cooling water is applied at the rolls to maintain the proper temperature for rolling, and lubricating oils (called rolling oils) are often applied to the system to facilitate rolling and prevent the development of surface imperfections otherwise caused by the friction of working through the rolls. The roll cooling water is recirculated, and the debris from the rolling operation settled, with facilities also being provided to reclaim the rolling oil wherever possible.

In addition to the cooling water used on the rolls, water may also be required for cooling furnace doors and for typical utilities operations such as compressor jacket cooling and bearing cooling.

Some fabricated shapes may be finished by special chemical treatment, such as the anodizing of aluminum sheet or tube by chromic acid treatment. These finishing operations may generate wastes that require separate treatment before being discharged into the final waste treatment plant.

Most reduction plants purchase their power from utilities and have no need for high-pressure boilers. Power may be generated in the bauxite plants where turbines exhaust into the steam line supplying the evaporators. But in the reduction plants, boilers are operated principally for space heating in the winter and for electrode paste heating, so the water treatment systems for these operations are usually quite simple.

In the secondary refinery, aluminum scrap is melted in induction furnaces in operations quite similar to recovery operations in the primary smelter itself where scrap must be reclaimed from butt ends and clippings. In secondary smelters, the nature of the impurities is unknown, so fluxing agents are added to eliminate them. The contaminants are removed as dross from the surface of the molten metal after casting.

SUGGESTED READING

Shreve, R. N.: *Chemical Process Industries,* 3d ed., McGraw-Hill, New York, 1967.

U.S. Environmental Protection Agency: *Development Document for Bauxite Refining,* EPA-440/1-74-019-c, March 1974.

U.S. Environmental Protection Agency: *Development Document for Primary Aluminum Smelting,* EPA-440/1-74-019-d, March 1974.

U.S. Environmental Protection Agency: *Development Document for Secondary Aluminum Smelting,* EPA-440/1-74-019-e, March 1974.

U.S. Environmental Protection Agency: *Economic Analysis of Effluent Guidelines, The Non-ferrous Metals Industry (Aluminum),* EPA-230/2-74-018 (NTIS-PB-239-161). August 1974.

CHAPTER 25
AUTOMOTIVE INDUSTRY

The automotive and machinery industries segment, classified by SIC codes 3400 to 3799, is composed of a wide variety of manufacturing operations, including (*a*) automotive (self-propelled) vehicles: farm implements, autos and trucks, construction equipment, and aircraft and aerospace; (*b*) machinery: appliances, electrical machinery, and fabricated metal products.

There are four basic operations in the manufacture of most of these products: casting (foundry operation, die casting, or investment casting), machining, stamping and fabricating, and final assembly. Although water consumption is relatively modest in each of these operations, water quality is important and aqueous wastes are quite concentrated.

Some plants may perform just one of the above operations, while others carry out the entire process from casting through assembly at a single integrated plant. Figure 25.1 shows the flow of material for the total operation of an automotive plant.

FOUNDRY OPERATIONS

In the foundry, parts such as crankshafts, engine blocks, and transmissions are cast. In a typical iron foundry, pig iron is purchased from a steel mill and melted in a furnace called a cupola, similar in design to a blast furnace. The iron is mixed in the furnace with a charge of coke and a flux or slag-forming material which may be limestone or fluorspar. Air is blown into the furnace at the tuyeres, as in the blast furnace, and the combustion of the coke melts the charge, with molten iron draining to the bottom and slag floating to the top.

The gases leaving the cupola are combustion products, basically CO_2 with perhaps a small amount of CO, plus a small amount of SO_2, if the coke was made from sulfur-bearing coal. When the charge is dumped into the burden, there is some breakup of the relatively weaker coke lumps, and there is an initial surge of coke fines into the exit combustion gases. There usually is iron oxide also broken loose from the bars of pig iron, so the discharge from the cupola is high in suspended solids.

To avoid creating an air pollution problem, the foundry may install a baghouse for dry collection of the dust, or a wet scrubber. If the latter is installed, this becomes the principal use of the water in the foundry operation. A typical foundry operation is shown in Figure 25.2.

Because the products of combustion are acidic, the pH of the scrubber water is generally quite low. Figure 25.3 shows a pH chart taken from a recorder sam-

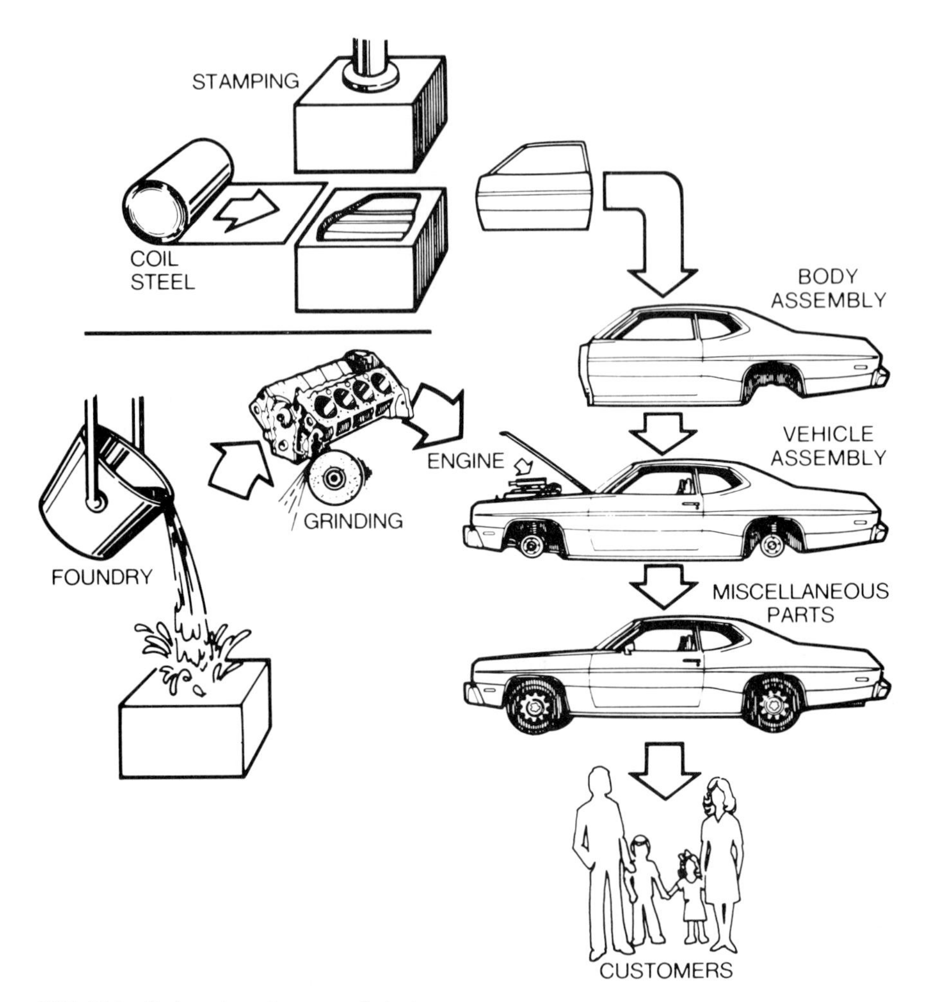

FIG. 25.1 Basic automotive manufacturing.

pling scrubber water. The effect of opening and closing the charging door is to dilute the stack gases, which is apparent on the strip chart.

A second use of water in the foundry is for cooling the cupola shell. This is usually done by direct spraying of the steel shell with water through a circumferential pipe at the top of the cupola. This water may be collected at the bottom of the cupola, pumped over a cooling tower and returned to the top of the cupola. Most foundries adjust pH and add a corrosion inhibitor to protect the cupola shell.

A third use for water is in the granulation of slag tapped from the cupola. The molten slag collects in the granulation tank, and the slag grains are removed by flights up a ramp and discarded into a totebox.

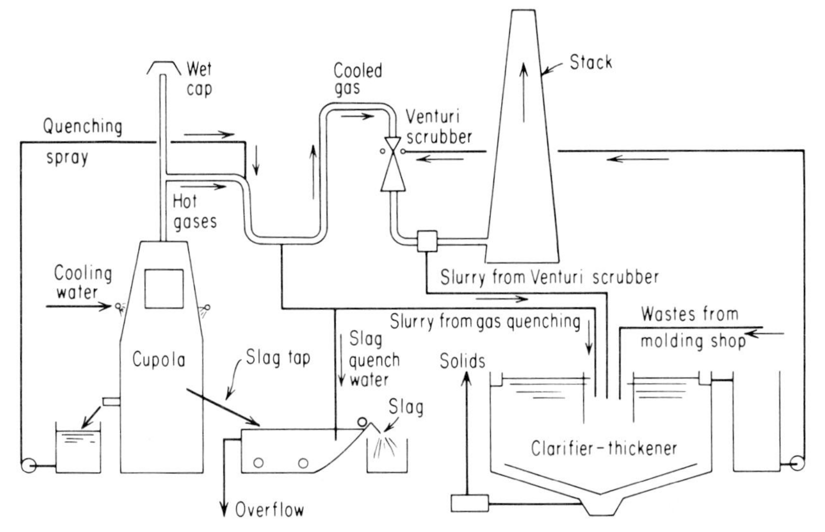

FIG. 25.2 Water circuit in a gray iron foundry with wet scrubbers.

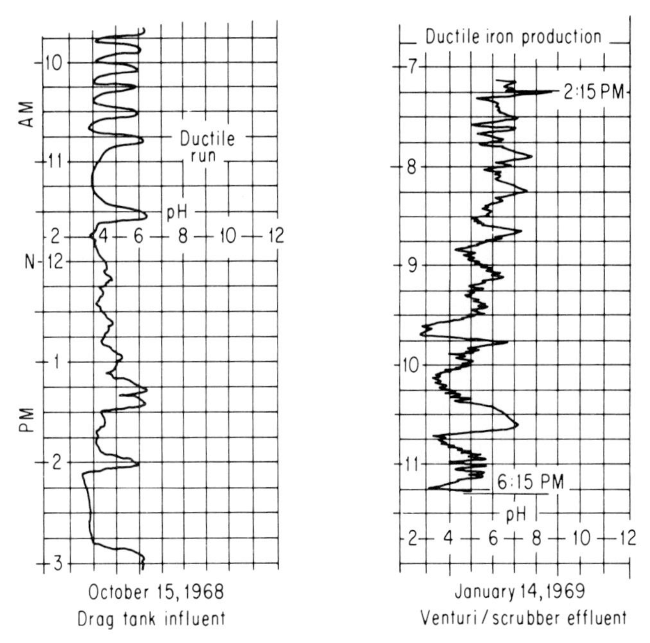

FIG. 25.3 Strip charts showing pH variations in Venturi scrubber effluent.

Most foundries encounter dust problems in the preparation of sand molds and the breakup of the sand from the finished casting, where the material is crushed for return to the molding room. A variety of chemicals may be mixed with the sand to produce the green mold, which must be cured before molten metal is poured into it. Phenolic compounds are sometimes used in preparing the mold, so phenol may be present in the foundry wastewater both from this source and also from the coke charged to the cupola. Oils may also be used in the preparation of the mold, and these volatilize during the baking of the mold and must be collected by a wet scrubber. Depending on the operations, there may be individual wet scrubbers at the cupola, the sand mold, and the shakeout room, or wastewater from these areas may be combined prior to treatment.

Table 25.1 shows the analysis of a foundry wastewater related to the installation illustrated in Figure 25.2. The oil content was quite high, but there was no evidence of free oil in the sample. This is a common occurrence where the bulk of suspended solids consists of carbon and iron oxide, as in foundry operations.

TABLE 25.1 Analysis of a Typical Foundry Wastewater

Constituent	mg/L
Total dissolved solids	770
Suspended solids	1900
Extractables (oil and grease)	290
Phenols*	1.5
pH	7.6

* Present in the coke and also in core binders.

After the treatment of this particular wastewater, the oil content was reduced to less than 50 mg/L, suspended solids to about 56 mg/L, and phenol to 0.5 mg/L. The collected solids were vacuum filtered for disposal, and the oil content of the dry cake was about 15%.

There is often some fluoride present in the wastewater, which may be introduced by volatilization into the cupola stack gas or may be dissolved from the slag in the granulation tank. Since most foundries discharge into a city sewer before the water eventually reaches a watercourse, the amount of fluoride is not usually so high as to require special treatment.

There is evaporation of the scrubber water, leading to concentration, so that chemical treatment is usually required for control of scale and corrosion in the recirculating system, and special alloys are often required for the circuits of a multiple-circuit washer where the pH may become very low.

MACHINING

Rough cast parts are sent to parts and accessories plants to be machined. Figure 25.4 shows the typical operations in a plant that machines parts for and assembles car engines. Major operations include machining, cleaning, engine testing (hydrotest), and waste treatment.

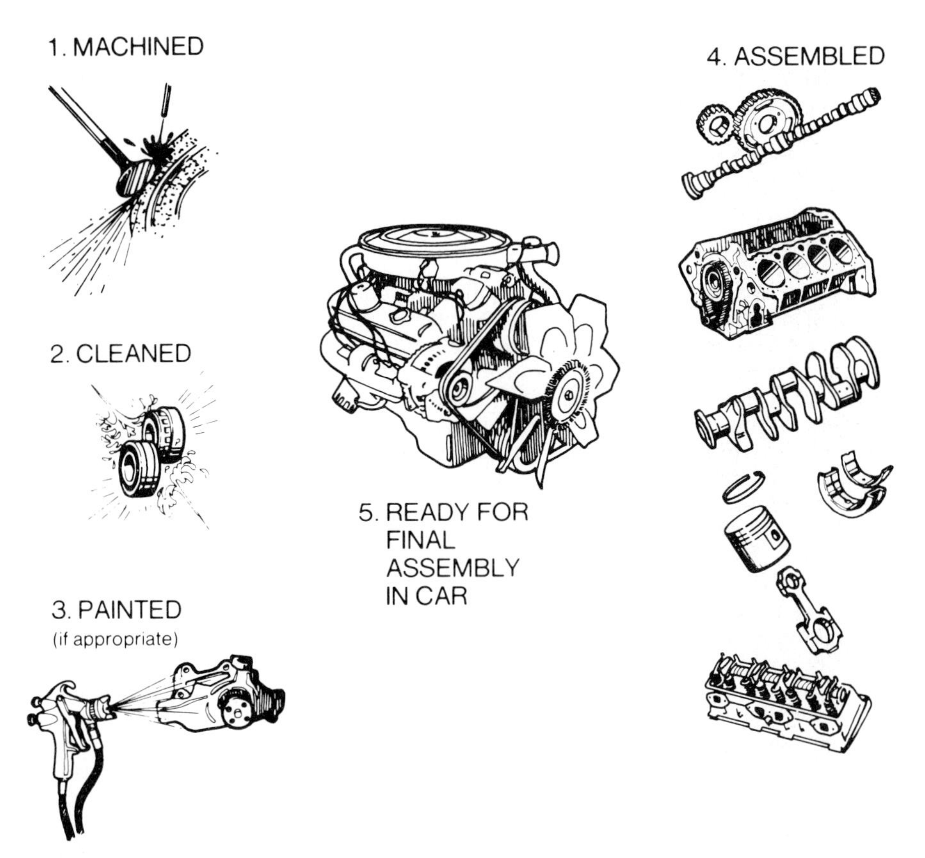

FIG. 25.4 Typical operations in an engine plant.

Rather than a single assembly line for the machining operation, a series of small lines is set up for each part and department. Each department performs its own particular machining operations on the subassemblies, and each of these may require a specific type of machine coolant, varying from a water-base material to a heavy sulfur oil. These coolants provide lubricity and cooling to the metal and tool used to shape it. Many of these oils contain emulsifiers that readily create O/W emulsions when mixed with water.

Soluble oil coolants are usually maintained at 5 to 15% oil concentration and stored in central sumps. The oil is recirculated through screens or filters to the machine tool and back to the sump. At the machine the soluble oil absorbs heat and picks up metal fines and hydraulic oil from leaks. Plants will normally dump the entire coolant system periodically because of bacterial buildup or because the emulsion is no longer stable. It is good practice to add biocide to control bacterial activity, but even this cannot extend coolant life indefinitely.

A recent development has been the introduction of synthetic and semisynthetic coolants. The synthetic coolants contain no oil, while the semisynthetics contain some oil but less than soluble oil coolants. These have some advantages over soluble oils on the machine; however, they are still in the developmental stage.

City water or clarified plant water is usually used to make up the coolants. If high hardness levels are found in the makeup water, soft water is sometimes advantageous because calcium ions tend to destabilize the emulsions, especially if the coolant concentrates. Removing the calcium and magnesium extends the stability of the soluble oils.

Parts Cleaning

The other major operation in the machinery plant is cleaning. The parts must be cleaned before and after machining to remove dirt, rust preventive, and coolant. Chemical cleaners are used, usually alkaline and detergent types. Anionic phosphate cleaners had been used extensively in the past, but have been greatly reduced. Nonionic surfactants have become quite popular. The cleaner usage is substantial since the cleaners are continually depleted and fresh chemicals are needed to maintain the strength necessary for effective performance.

In the machine shop, blowdown from the parts cleaners comprises roughly 80% of the flow to the waste treatment plant. Because a primary function of the cleaner is to remove coolant, blowdown usually contains 500 to 4000 mg/L oil. These wastes are emulsions stabilized by the surfactants in the coolant and in the cleaners. Other flows to the waste plant include soluble oil dumps, other cleaners such as floor cleaners, cooling water from a variety of sources (such as air conditioning systems), boiler blowdown, hydraulic oil leaks and spills, and storm water.

Occasionally, changing from soluble oil to a synthetic coolant has caused problems in the waste treatment plant when acid and alum were being used for emulsion breaking. Organic emulsion breakers have proved more effective in removing oil from the effluent under a wide variety of coolant selections. However, reduction of soluble BOD by alum or organic emulsion breakers is usually very limited. BOD loadings to the waste plant may, in fact, increase since the synthetics usually have a higher BOD than soluble oils.

Another factor that can have an impact on the quality of the waste treatment plant effluent is cleaner usage. Excessive cleaner usage, especially of certain strong nonionic cleaners, can increase dosages in the waste treatment plant or even make the waste virtually untreatable. Volume and type of cleaner should be balanced against the impact on both the cost and quality of the waste treatment plant operation.

TABLE 25.2 Raw Waste from Machining Operations

Constituent	mg/L
Hexane extraction	4000
P alkalinity (as $CaCO_3$)	424
Total alkalinity (as $CaCO_3$)	708
Suspended solids	300
Fe (total)	4
TDS	1500
pH	10.4
Ca hardness (as $CaCO_3$)	20
Total hardness (as $CaCO_3$)	26
Total PO_4	3.0

Table 25.2 shows a typical analysis of the untreated effluent from a machining plant. Basically the effluent is an oil-in-water emulsion. This analysis does not include free or floating oils.

Machining plants use cooling water for air conditioning, powerhouse diesel generators and compressors, hydraulic oil coolers, and furnace cooling if the plant is involved in heat treatments, such as annealing.

Each plant, depending on what it manufactures, may have its own unique uses for cooling water. For example, engine plants typically hydrotest the assembled engine on a dynamometer before shipment. A typical test system is shown in Figure 25.5. On the tube side or closed side, the heat exchanger takes the place of a radiator. Soluble oil-type products are used on the closed side if the engine is going to be drained before shipment. These products lay down a film that prevents flash corrosion. If the engine is not going to be drained, either soluble oil or conventional corrosion inhibitors can be used, depending on the manufacturer's restrictions.

A typical boiler plant generates an average of 2 million lb/day of 150 lb/in^2 steam, primarily for winter heat, but some steam is also used to heat various cleaner baths, so there is a light boiler load in the summer. If the plant has a

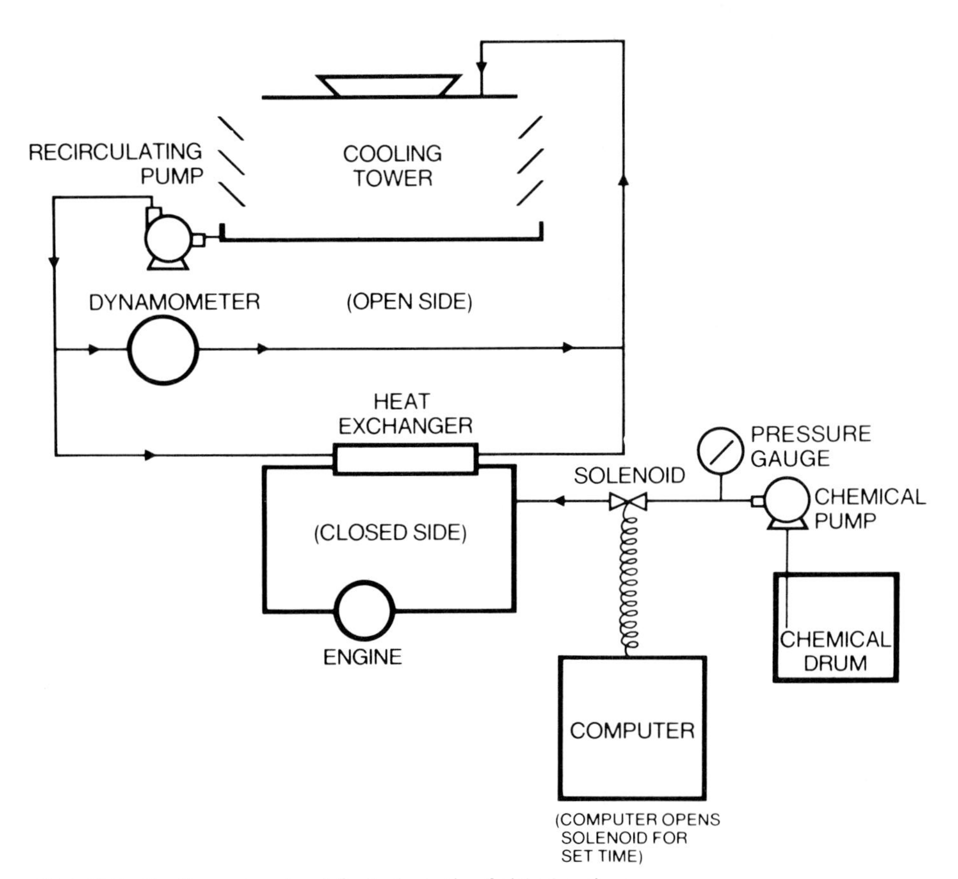

FIG. 25.5 Cooling water circuit for hydrotesting finished engines.

FIG. 25.6 A large steam-operated forging hammer. The dies and sow block have been removed for routine maintenance. *(Courtesy of Forging Industry Association.)*

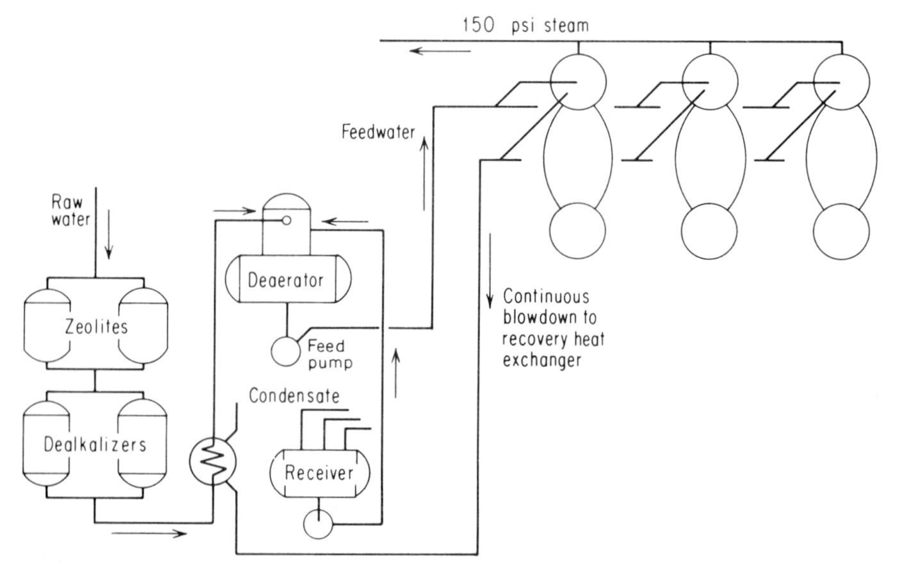

FIG. 25.7 The boiler house in an automotive plant is simple and reliable.

forging operation, steam may also be used to drive the hammers (Figure 25.6). Zeolites seem to be the most common method of softening, with average feedwater hardness being 1 to 2 mg/L. Figure 25.7 is a diagram of a typical boiler plant.

Water-Washed Paint Spray Booths

Some machinery plants (such as engine plants) have small spray booths where primer is sprayed on parts. These small booths experience the same maintenance problems as larger ones, such as those used to paint car bodies, and they require chemical treatment. They are not as important to production, so they are frequently poorly maintained.

Since spray booths are so widely used throughout the automotive and machinery industries, a description is given here to illustrate the type of water technology required for successful operation. Goals of the water treatment program are to (1) keep the water circuit free of deposits, (2) make the paint over-

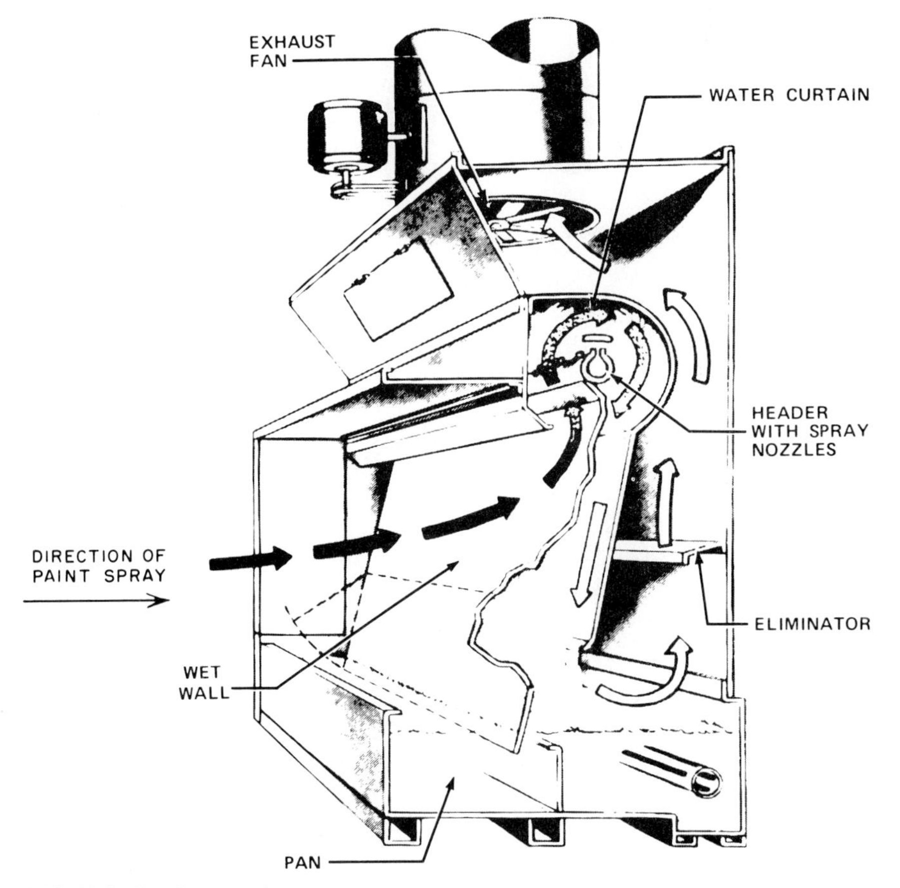

FIG. 25.8 Small parts paint spray booth.

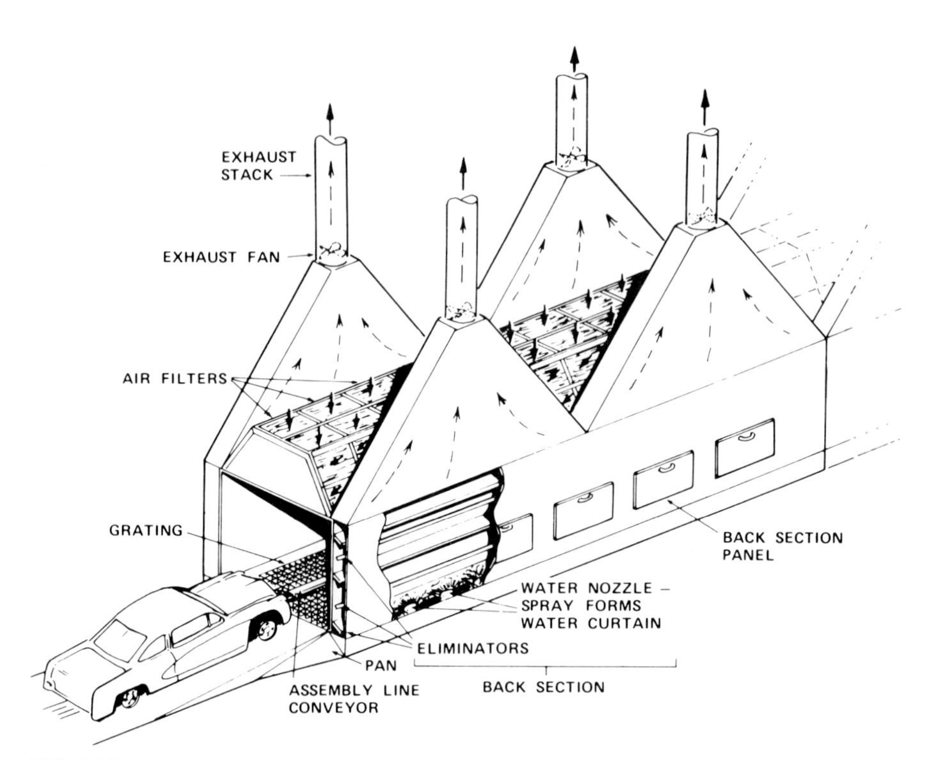

FIG. 25.9 Automotive paint spray booth.

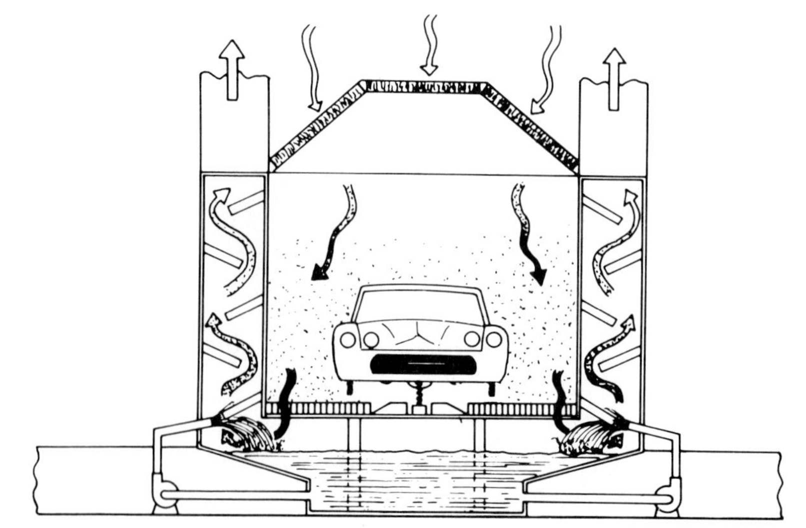

Fig. 25.10 Cross section of automobile paint spray booth.

spray collected in the water nonsticky and readily removable from the water, (3) minimize backsection deposits to prevent obstructing flow of paint-laden air from the booth, (4) minimize contaminants in the air discharged to the outside so it will not create an air pollution problem, and (5) minimize booth maintenance.

Figure 25.8 illustrates a typical paint spray unit for small parts, such as would be found in the machinery plants. The conveyor carrying the parts to be painted passes in front of the wet well, and the operator sprays paint on the parts as they pass by, with the excess or overspray being collected on the film of water flowing down the wet wall. The water must be chemically conditioned so that the pigment and the excess vehicle and solvent are killed and do not form a sticky mass that would be difficult to remove from the pan. Water is continually withdrawn and recirculated to the spray header, which provides scrubbing of the discharge of ventilating air before water drains back down the wet wall.

The large paint spray booths used in the automobile assembly shops are illustrated in Figures 25.9 and 25.10. Operators work inside these booths, applying paint to the car body. The floor of these units is grating supported above a water basin, and falling paint deposits on the water while other paint particles are carried by the air flow into the water curtain either on the wet walls or into the wall cavity, or back section, where additional sprays scrub the flow of air.

Detackifying Paint

Properly treated water will collect the pigment and organic components of the paint and condition these so that they are not tacky, producing a sludge which can be readily handled without sticking to the scrapers or flights used for its removal. The material which carries up into the back section is scrubbed out and killed by properly designed and treated water wash so that deposits do not form in this relatively constricted wall cavity. If the scrubber is not performing properly, particles penetrate the water curtain, build deposits in the back sections, and go out the stack as particulate emissions. In addition to the air pollution problem, extensive maintenance is required to remove the deposits in the back sections, on the eliminators, on the fan blades, and in the basin below the grating to keep the booth operating so that it provides safety to the operator and a clean discharge to the outside air.

Paint spray booths are in some respects like an air washer, in that the spray water may evaporate and cause concentration of dissolved solids, or water may be condensed from the air and there may be a continual dilution of the spray water. This means that a check on total dissolved solids, pH, and alkalinity is required periodically to keep the system under good control.

Recirculation volumes on washers of this type are high—about 10,000 gal/min for a typical body spray booth and 500 to 1000 gal/min for a parts spray booth. The water retention time in the system is relatively low, perhaps only 2 to 3 min. The basin is usually dumped at the end of the shift. The actual water makeup requirements are difficult to estimate because they depend on the evaporation rate and the frequency of dumping, which are unique for each installation.

STAMPING

In the stamping operation, the first step is to loosen mill scale from the metal surface in preparation for stamping by passing the strip steel through a flex roller

(Figure 25.11). This flex roller bends and flexes the steel through a series of rollers. Wash oil is brushed on during the flexing operation to remove the mill scale and dirt.

The flexed steel is cut to size and then stamped to the desired shape by large hydraulic presses. Drawing oil is sprayed on the dies to aid in stamping.

The stamped part is further cut and trimmed to the precise size; and the drawing oil, cutting oil, and dirt are removed in a parts washer, creating an oily wastewater. The part may then be welded or primed depending on the manufacturing requirements. The assembled part is then ready for shipment to the assembly plant. Finished products from stamping and fabrication plants include doors, floor pans, trunks, hoods, and fenders.

The flows from stamping and fabrication are usually quite low (50,000 gal/day) and consist of oily wastes from the parts washers, cooling water blowdown, and blowdown from the spray booths. The wastewater contains 100 to 500 mg/L oil and can be effectively treated with O/W emulsion breakers. Occasionally, wash oil, hydraulic oil, and drawing oil are dumped to the waste plant, but most plants try to segregate these relatively clean oils from the wastewater for recovery or sale.

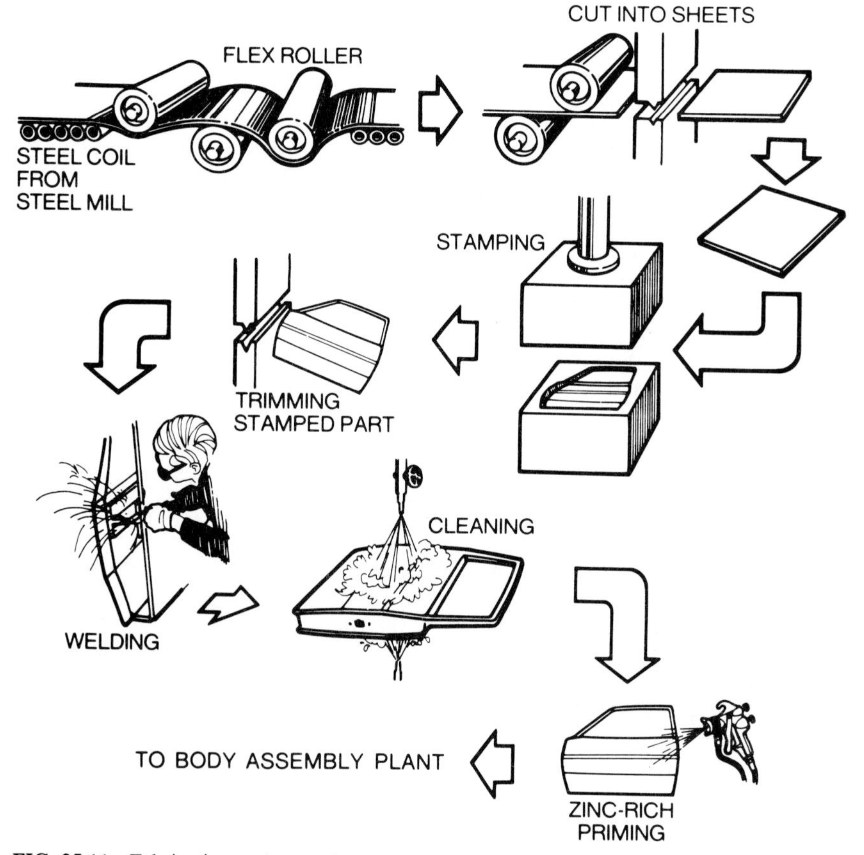

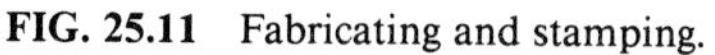
FIG. 25.11 Fabricating and stamping.

Plating Wastes

Where plating of parts is a major operation, waste treatment plants are designed to remove heavy metals. Traditional treatments of reduction and pH adjustment to remove the metal as its hydroxide are commonly used. A typical heavy metal removal plant is shown in Figure 25.12. Destruction of cyanide may also be required.

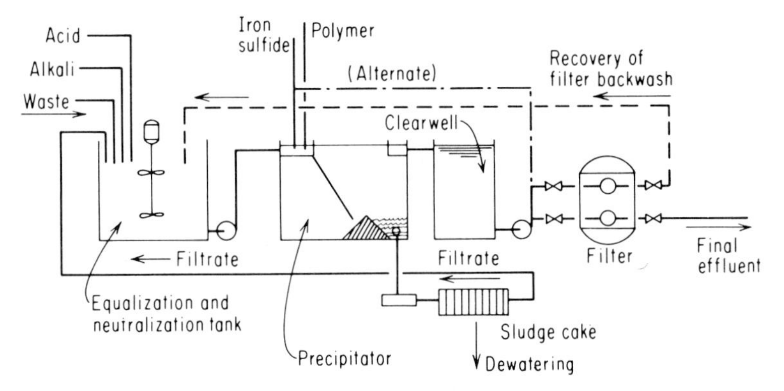

FIG. 25.12 Heavy metals precipitation in a metal finishing shop using sulfide precipitation. Freshly prepared iron sulfide slurry is normally fed to the neutralized waste at the precipitator inlet, but may be fed ahead of the filter where the nature of the waste better suits this alternative.

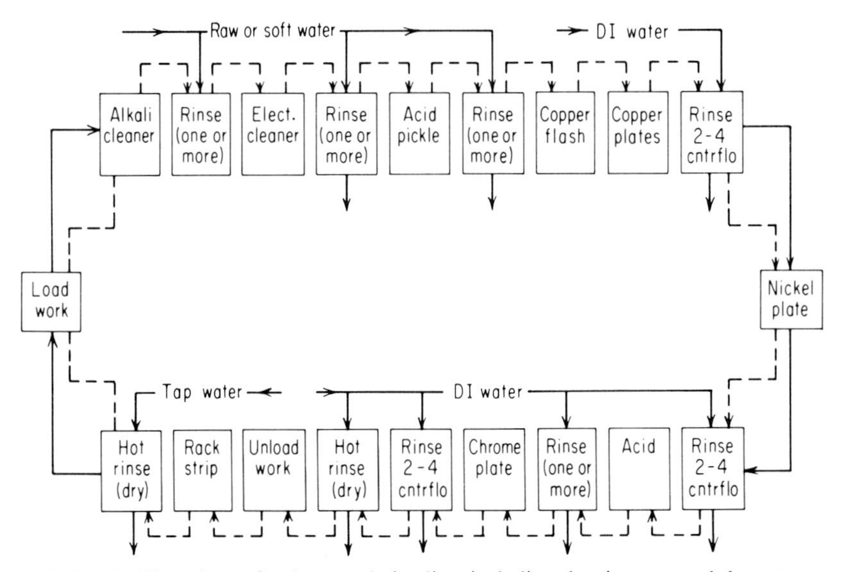

FIG. 25.13 Flow sheet of a chrome-plating line, including cleaning step and three stages of metal deposition.

Economics may justify heavy metal recovery, most frequently by an ion exchange operation. Nickel is easily reclaimed from rinse tanks following nickel plating on H-form cation resin, with the Ni-rich regenerant returned to the plating tank. Chromate is also recoverable, but not in a form useful in the plating operation.

Plating is as much an art as a science. Parts to be plated require careful cleaning and rinsing, often with soft water to avoid spotting. The plating baths are precisely controlled and may require demineralized water makeup. A typical plating line is shown in Figure 25.13.

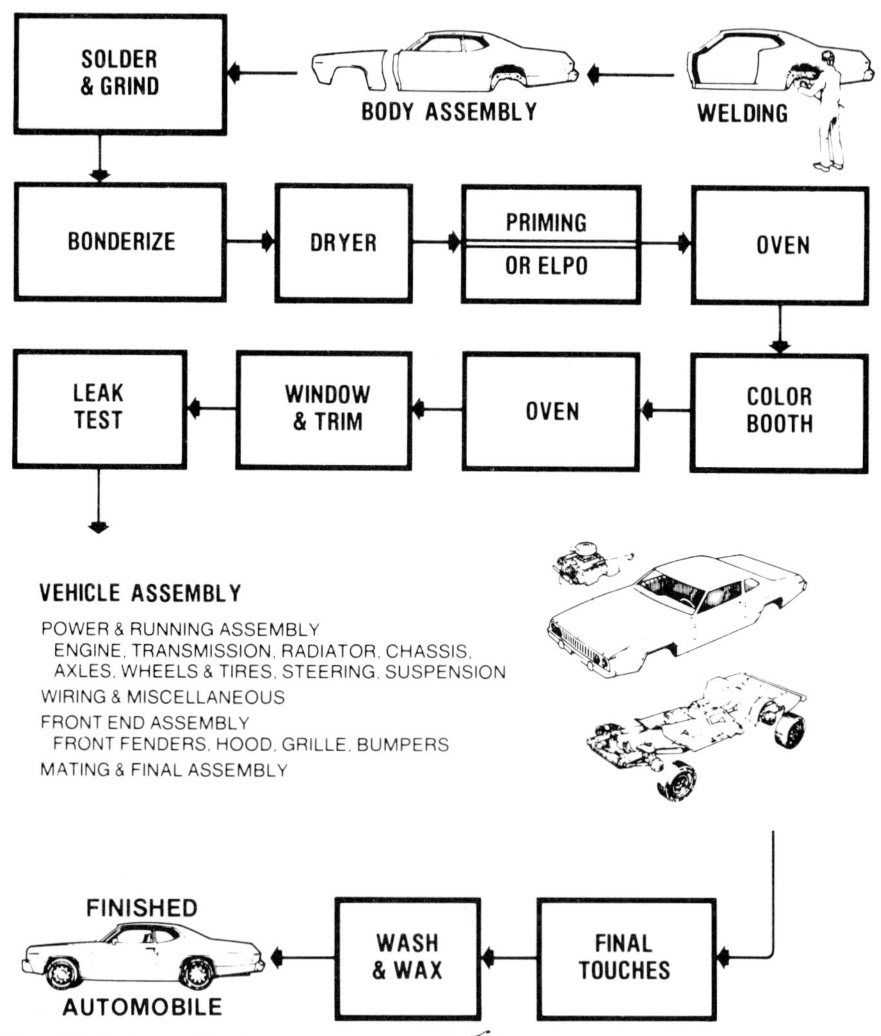

FIG. 25.14 Process flow in an automobile assembly plant.

Painting is increasingly important in the metal fabrication plants as the efforts to stop body corrosion increase. Zinc-rich primers are sprayed on the interiors of doors, fenders, and other areas that are enclosed after assembly to improve corrosion protection. This operation is carried out in the same type of spray booth as described earlier.

Utilities, such as steam and cooling water, are much the same in stamping and fabrication as described for the machining operation.

ASSEMBLY PLANTS

Figure 25.14 shows a typical flow diagram for an assembly plant. These plants receive all of the parts and subassemblies produced in supplier plants and assemble them into a finished car. Major operations include welding, bonderizing, painting, and assembly.

In the assembly plant, the sheet metal is welded to the frame and floor pan to construct the shell of the car. During the welding operation, the welder tips are cooled by recirculating water. The most common problem in this closed cooling system is fouling, plugging the tips with corrosion products. If the water flow is reduced by plugging, the tips may melt, causing a shutdown of the assembly line. Filtration and chemical treatment with soluble oil products and conventional corrosion inhibitors minimize this corrosion and plugging.

In the bonderizing step, a metal phosphate layer, such as zinc phosphate, is applied to the metal surfaces to provide corrosion protection and a surface for the paint finish. Figure 25.15 shows a typical phosphatizing operation consisting of cleaning, rinsing, phosphating, rinsing, and sealing with chromic acid. After bonderizing, many plants use a process to electrostatically deposit a primer coat of paint on the metal surface by dipping the electrically charged metal into a vat of

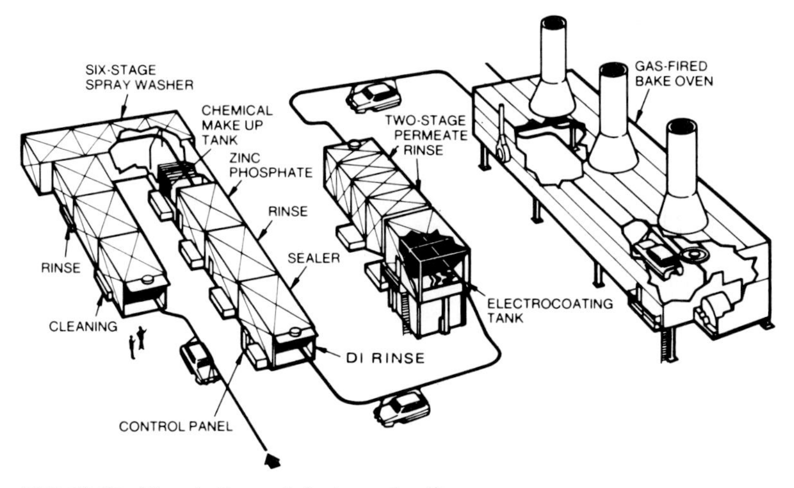

FIG. 25.15 Phosphating and electrocoating lines.

water-base paint. The oppositely charged paint particles are attracted evenly to the metal surface (see Figure 25.16). Makeup to this system is demineralized so that the bath conductivity can be controlled.

In this process, the bath tends to heat up. The paint is cooled below 90°F (32°C) through a heat exchanger by either a chiller or open recirculating cooling tower system. After the electrostatic dip, the painted car is baked in an oven to set the finish. The exhaust gases from the oven pass through heat exchangers to recover heat by warming the oven makeup air. These exchangers foul with paint residue and require frequent cleaning.

Periodically, concentrated rinse water from electrostatic coating is dumped to the waste plant. These dumps can upset the waste plant by introducing waste that is difficult to treat. If the pH of the paint is lowered to less than 4.0, gumballs form and plug the lines and pumps.

Most vehicle assembly plants have the type of wet spray booths described earlier. Their utility services are comparable to those found in the machining plants.

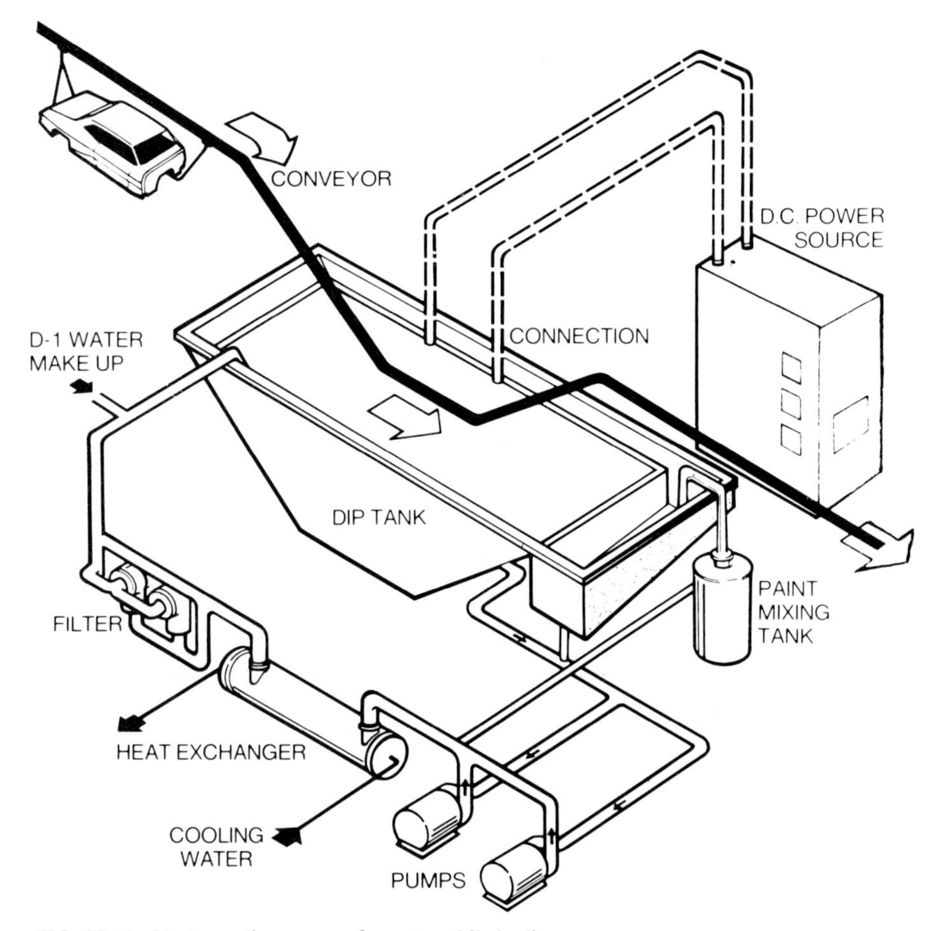

FIG. 25.16 Electrocoating process for automobile bodies.

TABLE 25.3 Summary of Water Uses

	Foundry	Machining	Fabricating and stamping	Assembly
Cooling	Cupola walls Cupola wet cap Slag granulator Comfort Compressors Furnace walls	Engine testing Oil coolers Comfort Compressors	Stamping Welder Comfort Compressors	Welder ELPO Comfort Compressors
Boilers	Makeup Wet scrubbers	Makeup Wet scrubbers	Makeup Wet scrubbers	Fuel Internal
Cleaning		Makeup Rinse	Makeup Rinse	Makeup Rinse
Spray booths		Makeup	Makeup	Makeup
Plating			Makeup Rinse	

There are two specific miscellaneous water uses in the assembly plant: (1) The assembled car is passed through a spray to test for leaks. Fluorescent dye is put in the water so a black light will clearly reveal leaks. Wetting agents are also added to make the test more severe and to prevent water spots. (2) The final process is a typical car wash operation. Various soaps are added to help clean the car.

Wastes from assembly plants are usually quite dilute. Typical contaminants are paint solids, oil, BOD, suspended solids, zinc, chrome, and phosphate. Coagulants and O/W emulsion breakers work well on these waters. Water comes from spray booths, stripping tanks, phosphatizing, wet sanding, car wash, cooling water, water leak test, and various parts washers. Typical flows range from 500,000 to 1,000,000 gal/day.

Table 25.3 summarizes water uses in the major automotive manufacturing steps.

CHAPTER 26
CHEMICAL INDUSTRY

The chemical elements can be combined in such a variety of ways that it is not surprising that the chemical industry is the most diverse of all industrial operations. Even when the number of elements that can take part in a reaction is limited, as with hydrogen, carbon, and oxygen, the combinations can produce an endless array of products.

Over 10,000 products are produced by the chemical industry; most are used to make other products of which the public is for the most part unaware. Even foods and beverages can be considered to be mixtures of chemical substances.

The government has broken down the chemical industry by its method of standard industrial classification coding. But there are other ways of categorizing it, including tonnage, dollar value, and the position the chemicals may occupy on the path between raw materials and consumer products. Using the latter method of classification, Table 26.1 shows the wide variety of chemical products, of which the fifth category covers those finally reaching the consumer market. Table 26.2 lists the major, high-production chemicals produced in the United States under the classification of inorganic and organic compounds.

PROCESS COOLING IS MAJOR H_2O USE

The industry is a major user of water, as shown by Table 26.3. The greatest use is for cooling process equipment. Many chemical reactions generate heat, and the reaction vessel is cooled so that the temperature is controlled at the desired limit and the reaction does not get out of control. Other chemical reactions require heat, and most chemical plants provide this with their own steam. If the steam is used only for low-temperature process heat, the plant may operate a low-pressure boiler; but if the plant requires a great deal of electric energy as well as heat, it is common to install high-pressure boilers to generate power through steam turbines, with the turbines exhausting into a process steam line, with perhaps some of the steam going directly to a turbine condenser. Many chemical plants are converting to cogeneration using high-pressure process steam to generate electricity for resale to the utility grid.

Cooling water is usually reclaimed by use of cooling towers, so the requirements for makeup water are minimized. Most heat exchangers use water on the tube side. However, in some chemical processes, it is more suitable to have water on the shell side with process liquor on the tube side. In this latter design, the water velocity is low compared to conventional practice, and suspended solids in the cooling water can create deposit problems.

TABLE 26.1 Varieties of Chemical Products

Class of chemicals	Examples
1. Heavy chemicals	Sulfur, sulfuric acid, lime, soda ash, ammonia, hydrochloric acid, caustic soda, methane, alum, chlorine, potash, salt, borax, phosphoric acid
2. Intermediates	Acrylic acid, acrylonitrile, cyclic compounds, unsaturated compounds, aldehydes, ketones, alcohols
3. Finished products—for fabrication or blending	Polyethylene, polyacrylamide, nylon, sodium phosphate, pigments, fertilizers, explosives, solvents, elastomers (natural and synthetic rubber)
4. Specialty products—for industrial and agricultural use	Corrosion inhibitors, coatings, herbicides, pesticides, food additives, inks, adhesives, dyes, cleaners
5. Consumer products	Fabrics, paints, drugs, cosmetics, detergents, toilet preparations, sterilizing agents, baking preparations, photographic products

Another large use of water is for process applications, which may include hydraulic conveying, hydraulic classification, washing, and equipment cleanup. Some water may become part of the finished product. Finally, provision is made in most chemical plants for fire protection, which usually requires water either applied directly, as a fog, or as a component of foam.

Chemical plant operations are classified into unit operations (such as evaporation, drying, crystallization) and unit processes (such as monomer production, sulfuric acid production, or catalytic cracking—all involving a number of unit operations). Because each chemical manufacturing plant differs so much from its neighbors, it is simpler to assess water and steam requirements for different classifications of chemical manufacturing plants.

In most chemical operations, there are processing steps involved in preparing the raw materials for the reaction that produces the finished product. The raw materials may be crushed and then conveyed by water to classification devices, such as screens, which may be operated either wet or dry. Soluble raw materials may be dissolved in water, or if gases, absorbed in water to facilitate the chemical reaction. The quality of water may be important; for example, it may contain iron, which could catalyze a reaction, discolor the finished product, or precipitate to produce a turbid product.

After preparation, the ingredients may be charged to a reactor. This may be done batchwise or continuously, depending on the volume of throughput, the value of the products being handled, and the economics of the alternative designs. If the reaction is endothermic, the reactor is usually heated with steam; if the reaction is exothermic, the reactor is designed so the mixture must be heated to a certain level before the reaction will occur, and thereafter the reaction may become exothermic and cooling will be required. Figure 26.1 illustrates a reaction vessel designed for both heating and cooling. The polymerization of organic materials is an example of a reaction requiring initial heating, followed by cooling to maintain a control temperature.

The chemical reaction does not always take place in such a sophisticated reactor; it often occurs in a simple sedimentation vessel similar to that used for lime-softening of water.

TABLE 26.2 United States Production of Chemicals with Output Exceeding One Billion (10^9) Pounds Per Year* (1983)

Inorganic		Organic	
Chemical	Output	Chemical	Output
Sulfuric acid†	69.5	Ethylene	28.6
Nitrogen	42.0	Propylene	14.0
Lime†	28.8	Urea	11.5
Oxygen	28.7	Ethylene dichloride	11.3
Ammonia	27.4	Benzene	9.5
Soidum hydroxide†	20.5	Ethyl benzene	7.9
Chlorine†	19.9	Toluene	7.1
Phosphoric acid	19.9	Styrene	7.0
Sodium carbonate†	16.9	Vinyl chloride	7.0
Nitric acid	14.8	Methanol	6.6
Ammonium nitrate	13.2	Terephthalic acid	5.7
Carbon dioxide†	7.2	Ethylene oxide	5.6
Hydrochloric acid	5.2	Xylene	5.6
Ammonium sulfate	3.9	Formaldehyde	5.4
Potash	3.0	Ethylene glycol	4.5
Aluminum sulfate†	1.9	*p*-Xylene	4.1
Calcium chloride	1.9	Cumene	3.3
Sodium sulfate	1.7	Acetic acid	2.8
Titanium dioxide	1.5	Phenol	2.6
Sodium silicate†	1.5	Carbon black	2.5
Sodium tripolyphosphate†	1.3	Butadiene	2.3
Total inorganics	331.1	Acrylonitrile	2.2
		Vinyl acetate	2.0
		Acetone	1.9
		Cyclohexane	1.7
		Propylene oxide	1.6
		Adipic acid	1.4
		Isopropyl alcohol	1.2
		Ethanol	1.1
		Total organics	168.0

* Abstracted from *Chem. Eng. News,* May 7, 1984.
† Widely used in water treatment.

TABLE 26.3 Water Use in Major* Chemical Process and Related Plants

	Annual usage, gal × 10^9			
Category	Gross	Intake	Recycle	Discharge
Organic chemicals	5716	2151	3880	1996
Inorganic chemicals	2279	965	1409	855
Synthetic plastics	1930	482	1481	455
Agricultural chemicals	1878	302	1653	233
Rubber and elastomers	623	187	516	168
Ceramics and glass products	419	207	267	182
Pharmaceuticals	241	90	160	87

* Plants using over 20,000,000 gal/year.

Source: 1977 Census of Manufactures, U.S. Department of Commerce Publ. MC77-SR-8, "Water Use in Manufacturing."

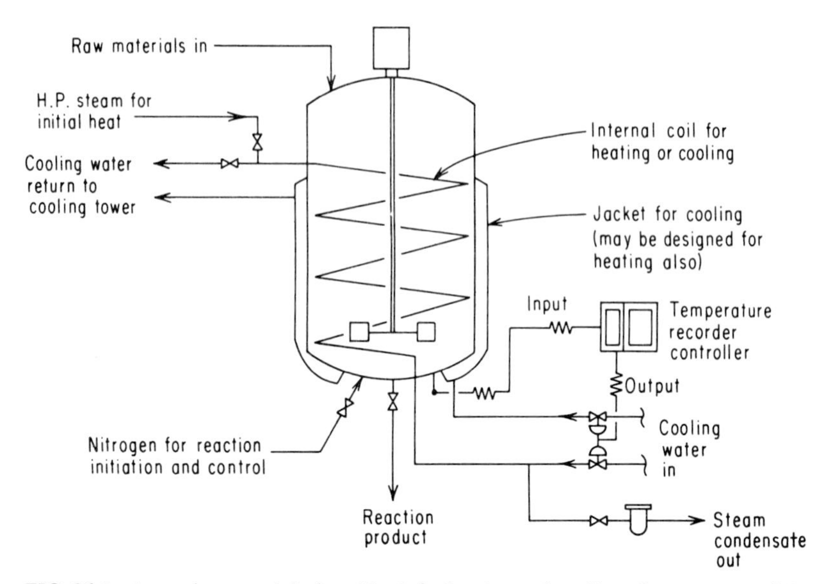

FIG. 26.1 A reaction vessel designed both for heating and cooling of process materials.

Once the chemical reaction has taken place, additional processing is usually required to remove unreacted raw materials or unwanted byproducts. When these impurities are solids, tyical solids/liquids separation processes are used. Depending on the nature of the solids to be removed, the choice may be sedimentation, filtration, or centrifugation. The solids removed are often washed for recovery of values in the liquor mixed with these solids (Figure 26.2).

If the impurities are not solids, then the final separation may require extraction with a solvent, which may be water if the process fluid is hydrocarbon. Or the separation may be achieved by an operation involving a phase change such as evaporation, distillation, crystallization, or drying. In all of these operations energy is applied to produce the separation, and cooling water may be used on condensers or reflux heat exchangers (Figure 26.3).

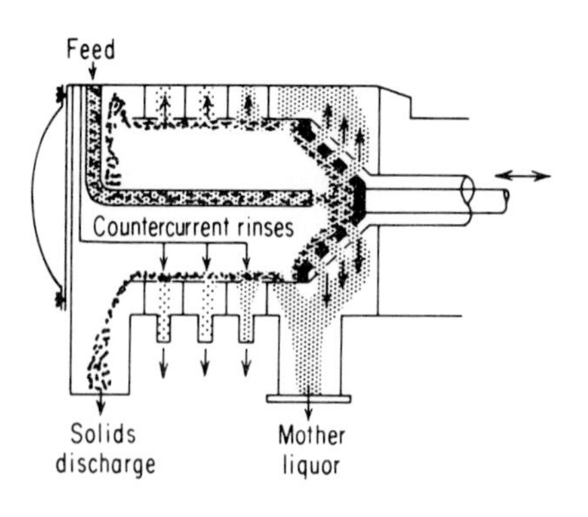

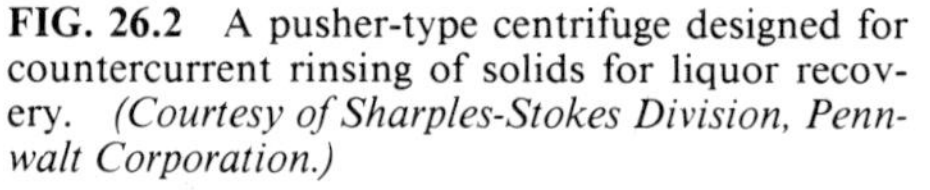
FIG. 26.2 A pusher-type centrifuge designed for countercurrent rinsing of solids for liquor recovery. *(Courtesy of Sharples-Stokes Division, Pennwalt Corporation.)*

FIG. 26.3 Energy changes accompany many chemical separation processes, requiring heating or cooling of the process liquor. In this large crystallizer, ammonium sulfate is produced at a rate of 600 tons/day. *(Courtesy of Whiting Corporation.)*

PROCESS FLOW SHEETS

Many of the heavy chemicals listed in Tables 26.1 and 26.2 are used in water treatment processes. They are used in various pretreatment processes (softening, dealkalization, coagulation, disinfection, etc.), and their costs are minimal; their use in pretreatment allows specialty water-conditioning chemicals needed for cooling systems, steam generators, and other special applications to perform at a reasonable maintenance cost. Without these less expensive pretreatment chemicals, the use of many of the more expensive specialty chemicals would not be cost-effective. The following brief look at the processes used for the manufacture of some of these chemicals provides an introduction to the chemical industry to those not familiar with it.

Some of the processes in the chemical industry seem more mechanical in nature than chemical. For example, *nitrogen* and *oxygen* are produced from air by a combination of compression, distillation, and refrigeration. Solutions such as brine—or even seawater—are evaporated, sometimes in solar evaporation ponds, to produce *salt.*

Other simple processes used in heavy chemical manufacture depend on a phase change: Native *sulfur* is melted underground by hot water and brought to the surface in molten form; *salt* and *sodium carbonate* are dissolved and brought from underground strata to the surface in solution form.

Simple processes that rely on combustion include the following.

Sulfuric acid: Sulfur is burned over a catalyst to produce SO_3, and this is dissolved in sulfuric acid and concentrated to market strength (98%) by recycle.

Lime: Limestone is heated in a kiln, producing *quicklime* (CaO), which is converted to *hydrated lime* by slaking with water. By-product *carbon dioxide* may be recovered from the kiln gases.

Phosphoric acid: In one process, phosphorus is burned to P_2O_5 which is dissolved in water. *Sodium phosphates* are produced by neutralizing this acid with sodium hydroxide.

The major water treatment chemicals manufactured electrically are *sodium hydroxide* and *chlorine,* made by electrolysis of brine, producing sodium at the cathode and chlorine at the anode. The sodium is then reacted with water to yield sodium hydroxide. Since they are coproducts, the price of one is affected by the market and price of the other.

Some heavy water treatment chemicals are by-products or waste products. *Ferric chloride,* a coagulant, is a by-product of the manufacture of titanium dioxide and a waste-product from HCl pickling of steel plate in the steel mill; *hydrated lime* is available in some areas as a by-product of acetylene manufacture; *hydrochloric acid* is often a by-product of the chlorination of organic compounds.

Alum (hydrated aluminum sulfate) is one of the water treatment chemicals produced in a multistage operation. The process flow sheet is shown in Figure 26.4. In the preparation of raw materials, the bauxite is crushed and ground in the dry form and fed into the first reaction vessel, where it meets concentrated sulfuric acid and is diluted with washings from later operations. The mixture is heated with steam so that the reaction can take place rapidly, iron is removed as an impurity, and the liquor is fed to a clarifier, where a flocculant is added. The underflow solids are pumped to a two-stage washing system, where the alum liquor is recovered by having these solids washed with condensate. The overflow liquor is further concentrated in an open evaporator, and the concentrate is then discharged onto a cooling floor where crystallization occurs. As seen in this flow sheet, the reaction takes place in relatively simple and compact vessels, and the balance of the equipment is for preparing the raw materials and finishing the product to meet the required specifications. Steam is required for the operation, and the condensate produced is used as the washwater in processing the waste solids.

Another example, where the preparation of raw materials and separation and purification of the final product is more extensive, is shown in Figure 26.5, relating to the manufacture of sodium chromate from chromite ore.

Beyond the confines of water treatment chemicals, a good example of the complexity of the petrochemical industry is the manufacture of synthetic rubber, illustrated by Figure 26.6. In this illustration, the raw materials are intermediates or specialty chemicals, the equipment for carrying out the reaction is quite complex, and the final processing steps are equally complicated. There is extensive use of steam and water throughout the process.

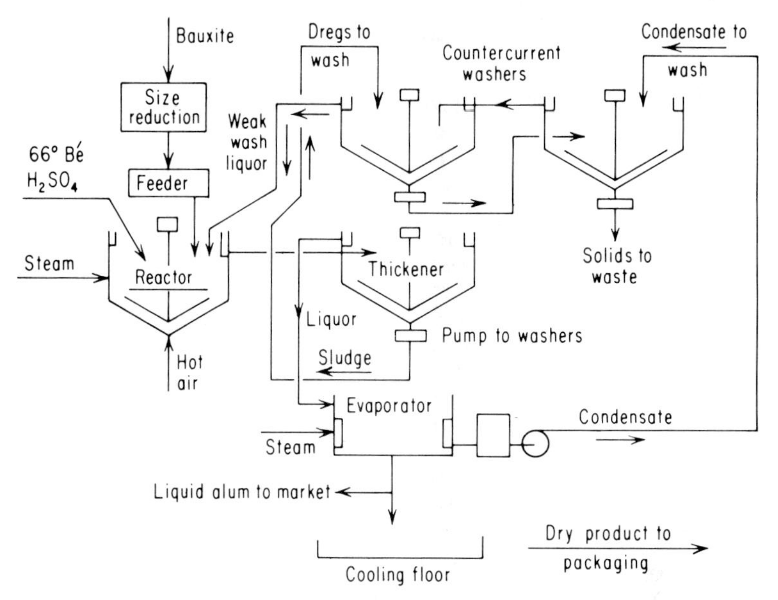

FIG. 26.4 Simplified flow sheet of alum manufacture uses several unit processes.

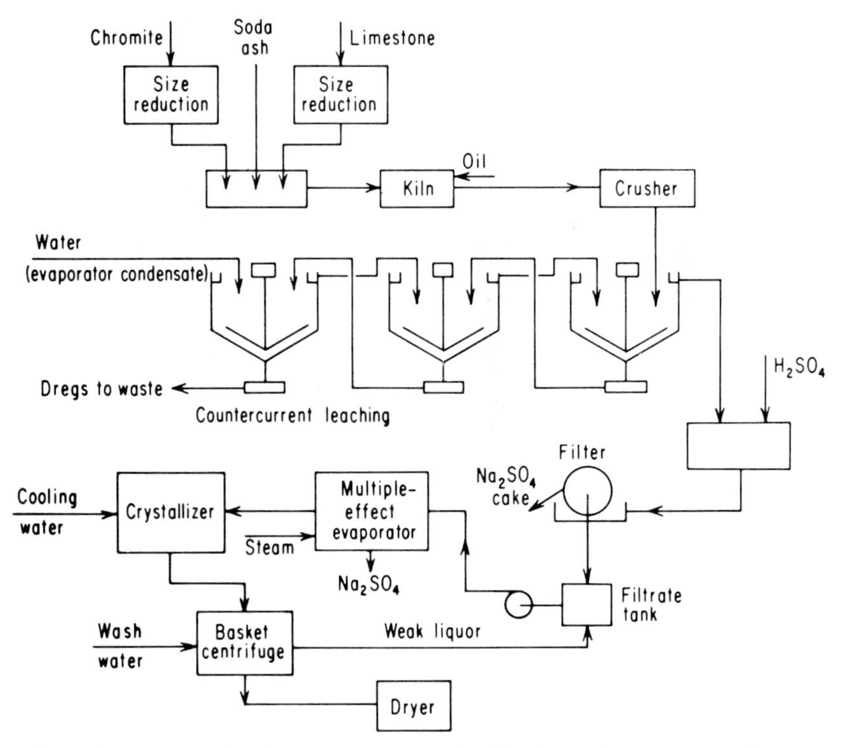

FIG. 26.5 A variety of unit processes are used in this scheme for production of sodium chromate and by-product salt cake from chromite ore.

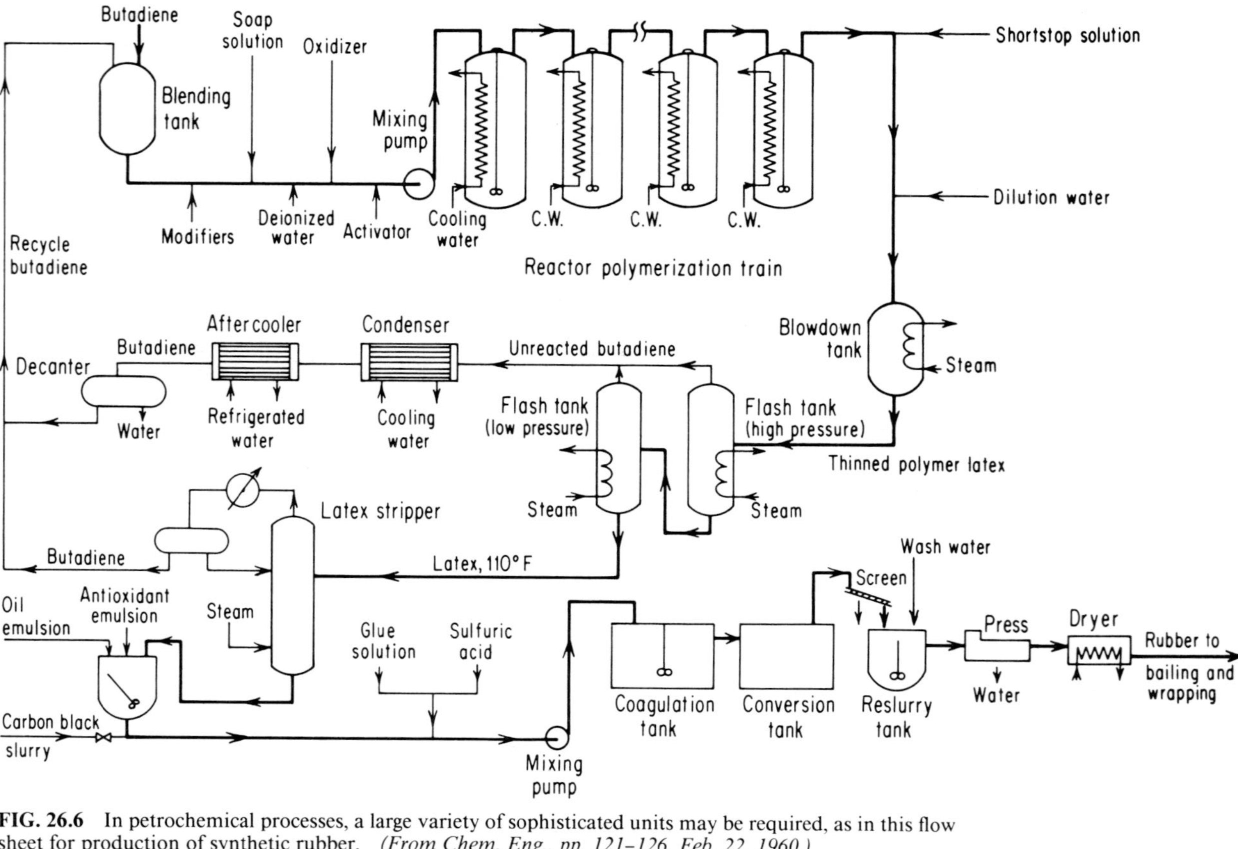

FIG. 26.6 In petrochemical processes, a large variety of sophisticated units may be required, as in this flow sheet for production of synthetic rubber. *(From Chem. Eng., pp. 121–126, Feb. 22, 1960.)*

ENVIRONMENTAL EFFECTS

The problems of water pollution control are particularly complicated in the chemical industry. The flowcharts have shown how widely water and steam are used in all kinds of operations, and each point of use represents a possibility for contamination with the raw material, intermediate, or finished product.

It is usually less expensive to design the process to prevent contamination and to install equipment to recycle and recover chemicals than it is to remove or neutralize the contamination after the fact in a waste treatment plant. For example, if the waste solids produced in the alum process, Figure 26.4, still contain too much liquor to be acceptable as landfill, a third washing stage may be added. In the removal of unreacted raw materials from a finished liquid organic compound, the installation of additional trays in the distillation column to recover the unreacted material may be much more practical than permitting the material to be discharged as a vapor or condensed in a barometric condenser, either of which represents a pollution problem.

As a general rule, if a chemical plant is producing inorganic chemicals, precipitation and solids/liquids separation equipment are usually used for the waste treatment operation; if handling organic chemicals, removal of organics by adsorption on activated carbon or by digestion in an activated sludge system is almost always required. The organic residues present a difficult problem in that

TABLE 26.4 Waste Profiles of Selected Organics in Water

*Based on oxygen demand**

Chemical	BOD	COD	Theoretical O_2 demand
Acids			
Acetic	0.34–0.88	1.01	1.07
Benzoic	1.37	1.95	1.97
Formic	0.15–0.27	—	0.35
Maleic	0.38	0.83	0.83
Alcohols			
n-Amyl alcohol	1.6	—	2.72
n-Butyl alcohol	1.5–2.0	1.90	2.59
Ethyl alcohol	1.0–1.5	2.0	2.1
Methyl alcohol	0.6–1.1	1.5	1.6
Phenol	1.6	—	2.4
Isopropyl alcohol	1.45	1.61	2.40
Aldehydes			
Formaldehyde	0.6–1.07	1.06	1.07
Furfural	0.77	—	1.66
Ketones—acetone	0.5–1.0	1.12	2.20
Ether			
Ethyl ether	0.03	—	2.59
Cyclic compounds			
Aniline	1.5	—	3.09
Benzene	0	0.25	3.07
Monochlorbenzene	0.03	0.41	2.06
Toluene	0	0.7	3.13
Xylene	0	—	3.16

* Given in g O_2/g of chemical

many of them are not easily digested by bacteria, and sometimes a combination of bacterial digestion plus activated carbon adsorption is required.

Table 26.4 lists a few organic compounds found in wastes from chemical plants where these materials are manufactured. For the organic acids and alcohols shown on this list, the BOD is usually over 50% of the COD or the theoretical oxygen demand, and such wastes are amenable to biological digestion in an activated sludge system. Benzene and its close relatives are usually impervious to biological attack and must be removed by stripping, adsorption, or both. So in many organic chemical plants, if a wide variety of compounds across the spectrum of active groups—alcohols, aldehydes, aromatics—is present, the waste may require both biological and activated carbon treatments.

In any case, the pollution control engineer must be aware of all raw materials, intermediates, and finished products produced, recognizing that they may be found in the effluent. He should be familiar with the toxicological properties and tolerance levels of each, and their potential effect on the receiving stream, the waste treatment plant, the land surface, and subsurface aquifers. Pollution control in the chemical industry is unique in that the industry is responsible not only for its own wastes, but also for the ultimate fate in the environment of products it sells to others. One of its most difficult problems is monitoring the end use of containers in which its products are supplied to its customers (see Chapter 40).

SUGGESTED READING

Shreve, R. H.: *Chemical Process Industries,* 3d ed., McGraw-Hill, New York, 1967.

U.S. Environmental Protection Agency: *Development Document for the Major Inorganic Products,* EPA-440/1-74-007-a, March 1974.

U.S. Environmental Protection Agency: *Development Document for the Major Organic Products,* EPA-440/1-74-009-a, March 1974.

U.S. Environmental Protection Agency, *Development Document for the Synthetic Resins,* EPA-440/1-74-010-a, March 1974.

CHAPTER 27
COAL PRODUCTS: COKE, PRODUCER GAS, AND SYNFUELS

Fossil fuels are those natural, combustible materials burned directly, such as natural gas, crude oil, peat, coal, or wood; or processed into by-product fuels, such as diesel fuel and gasoline, and having as their origin natural carbonaceous matter that can be traced to photosynthesis reactions. In a sense, these fuels are taken from natural, irreplaceable deposits of stored energy provided by the sun. Recent discoveries of hydrogen wells—mixtures of hydrogen and nitrogen—raise the question as to whether all fossil fuels are carbonaceous, but currently it is generally considered that they are.

The major use of coal in the United States is as a fuel for steam-electric plants, which currently burn about 550 million tons per year. The second largest consumer of coal is the steel industry, which cannot use the coal directly, but requires conversion to coke for reduction of iron ore to iron in the blast furnace. This industry uses about 70 million tons per year of coal in the production of its coke needs.

The major use for coal products in the future will probably be as raw materials for synfuels as reserves of oil and natural gas are depleted. Before the advent of cheap natural gas, made available throughout the United States through extensive pipeline systems developed during the 1950s, most urban centers had their own facilities for producing a combustible gas from coal and steam for domestic use. The two major products of the coal-steam reactions were called "water gas" and "producer gas."

In the *water-gas* process, coal was burned in the presence of steam to yield a mixture of CO, H_2, and CH_4:

$$C + H_2O \rightarrow H_2 + CO \tag{1}$$

$$C + 2H_2O \rightarrow CH_4 + O_2 \tag{2}$$

Periodically the coal charge had to be brought back up to a reactive temperature by injection of air:

$$C + O_2 \rightarrow CO_2 \tag{3}$$

So, this was a stepwise, regenerative process, with the heat of the CO_2 gas by-product returned to the combustion air.

The *producer-gas* reaction kept the coal up to reactive temperature continuously by controlled injection of air with the steam:

$$4C + 3H_2O + O_2 + 4N_2 \text{ (in the air)} \rightarrow CO + 2CO_2 + H_2 + CH_4 + 4N_2 \quad (4)$$

Because of the continuous injection of air, the heat content of producer gas was lower than for water gas because of the presence of inert nitrogen as a diluent.

With the availability of relatively economical oxygen from efficient air-separation plants, modern converters can produce CO, CO_2, CH_4, and H_2, by using oxygen for combustion, avoiding the diluent nitrogen, and yielding a product useful both for chemical and fuel synthesis. This is one of the major innovations of the projected synfuels development program in the United States.

Finally, coal can be hydrogenated in an oil slurry directly to yield a hydrocarbon product that can be used as the feedstock to a refinery or a petrochemical plant, producing either liquid fuels or by-products for the manufacture of petrochemicals.

An essential raw material for the iron blast furnace and the cast iron foundry is coke, which is produced from coal in the United States at a rate of 60 to 70 million tons per year. Coke is the source of both energy and carbon for chemical reduction in the blast furnace, in which it is used at a rate of approximately 0.5 ton per ton of finished iron. In foundry cupolas it is used as a fuel for melting pig iron.

Blast furnace coke is manufactured from carefully blended stocks of high and low volatility coals. The mix is ground, charged to a furnace, and heated for approximately 18 h at about 1800 to 2000°F (980 to 1100°C) to drive off moisture and volatile matter. At the end of the coking period, incandescent coke is pushed from the furnace into receiving cars, quenched with water, and sent to the blast furnace or to other industrial uses. The finished product is controlled by careful selection of raw materials and through the process steps to produce a carbonaceous product strong enough to avoid crushing in the blast furnace or cupola, and pure enough to avoid transmitting unwanted impurities to the product of the blast furnace or cupola.

Another variety of coke, produced by the petroleum refinery, is used for other purposes, such as electrode manufacture. This is considered a refinery operation and is not included here as part of the coke industry.

TABLE 27.1 Products of Coke Plant Operation
[*Based on 1 ton of coal (0.91 kkg)*]

Coke oven gas	10,000–12,000 ft^3 (283–340 m^3)
Coke	1400 lb (640 kg)
Ammonium sulfate	25 lb (11 kg)
Tar	10 gal (38 L)
Light oil	2–4 gal (7–15 L)

Because coal is a heterogenous mixture of amorphous carbon and a variety of organic chemicals and inorganic minerals (ash), the coking process yields not only coke but also a variety of by-products, as shown by Table 27.1. The integrated coke plant is usually broken down into two major areas, the coke plant itself, and the by-product plant, which recovers tar, ammonium sulfate, and light oils. The by-products plant refines the light oils to produce such valuable organic chemicals as benzene, toluene, xylene, and naphthalene.

COKE PLANT

A large inventory of high and low volatile coals is maintained at the plant site, partly for economic reasons and partly to permit blending of incoming coal to minimize variations in composition and to permit exercising quality control over the charge, and therefore the finished product. A typical coke plant processes 2000 to 4000 tons (1800 to 3600 t) of coal per day and may stockpile a 30- to 60-day supply. The several grades of coal are blended, crushed to about 50 mesh, and transferred to a storage bin for charging.

The coking operation is carried out in a battery of 10 to 100 individual ovens designed to provide relatively uniform production of finished products and to recover heat in a way that minimizes fuel requirements. Of the 10,000 to 12,000 ft^3 of gas produced per ton of coal charged (315 to 630 m^3/t) about 40 to 50% is returned as fuel to the coke ovens. Figure 27.1 is a simple schematic of a coke oven showing the oven itself and the regenerative brickwork used to reclaim heat and minimize fuel consumption.

The oven chamber is wider at the discharge end so that when the coke is pushed from the oven it will not bind. Typically, the oven is approximately 35 ft (10 m) long, 10 to 15 ft (3 to 4.5 m) high, 10 to 20 in (25 to 50 cm) at the pusher end, and about 2 to 3 in (5 to 7 cm) wider at the discharge end.

The blended, crushed coal charge is fed to the ovens by a charging car mounted on the coke oven battery and traveling from one oven to the other as needed (Figure 27.2). The cycle of charging, coking, and discharging is illustrated in Figure 27.3. To minimize air pollution during charging, new plants are being constructed to provide charging through a pipeline carrying preheated coal in a fluidized condition.

During the coking period, the coke oven gas is collected through ascension pipes at the top of the oven, scrubbed in the gas-collecting main with weak ammonia liquor to remove tar, and discharged to the by-products recovery operation. The heating gas is burned and the products of combustion pass through flues on both sides of the oven, finally passing through a regenerator. When the regenerator becomes hot, the flue gas is diverted to a second regenerator, and the first, heated regenerator is then used to preheat air to conserve fuel. The flue gas temperature is carefully controlled to avoid overheating the coke charge.

At the completion of coking, the oven is isolated from the gas main, and the incandescent coke is pushed into rail cars for prompt transfer to a quenching tower. Here, quench water showers the coke to cool it and prevent further loss by combustion.

BY-PRODUCT RECOVERY

The coke oven gas is then processed for recovery of valuable by-products as shown by the flow diagram in Figure 27.4. Depending on the capacity of the plant and the value of the by-products, each coke plant may decide on the extent to

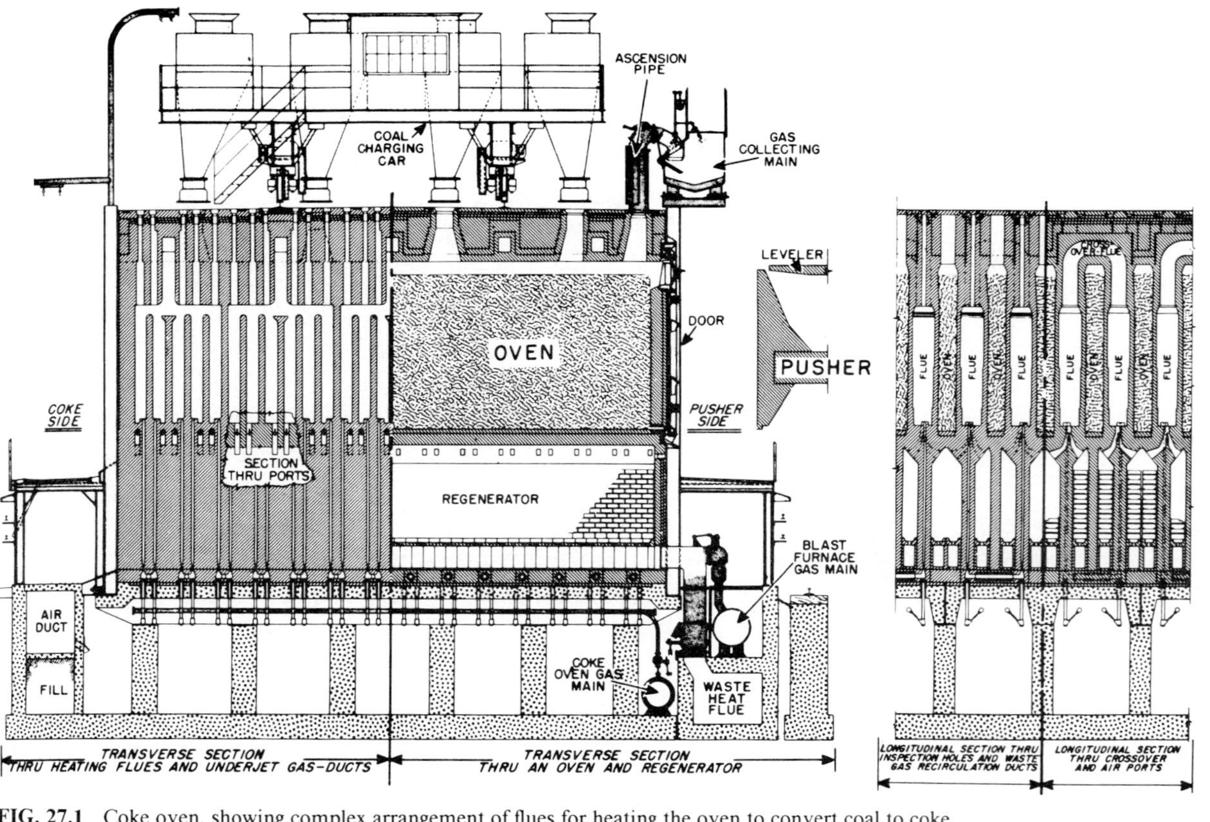

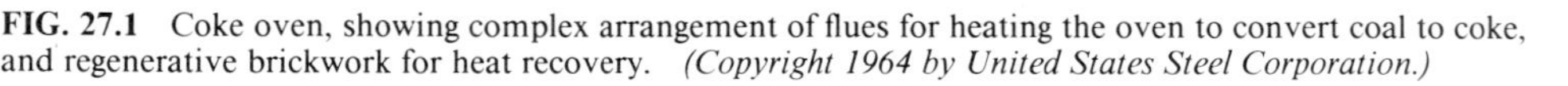

FIG. 27.1 Coke oven, showing complex arrangement of flues for heating the oven to convert coal to coke, and regenerative brickwork for heat recovery. *(Copyright 1964 by United States Steel Corporation.)*

which recovery is economical. The diagram shown represents complete recovery, but this may not be practiced in every plant.

The gas is first cooled by spraying with weak ammonia liquor. In a secondary loop, the weak ammonia liquor is then cooled either by heat exchangers or by an

FIG. 27.2 Car riding the top of the coke oven battery to charge coal to the ovens. *(Copyright 1964 by United States Steel Corporation.)*

open recirculation system incorporating a cooling tower. Since the spray cooling condenses the moisture originally in the coal, there is an increase in the volume of the weak ammonia liquor. Some liquor is bled off at a steady rate to maintain the material balance of the system. The gas is compressed and then passes to a tar precipitator, an electrostatic device for tar removal. The tar from the precipitator is combined with that decanted from the weak ammonia liquor and is a useful by-product, going into roofing pitch, asphalt shingles, and other products.

Following cooling, ammonia is removed from the coke oven gas by washing with a recirculating sulfuric acid solution. This liquor is sent to crystallizers for the production of ammonium sulfate. The gas is then sent through a final cooler where it is contacted with cool water to reduce the temperature to less than 68°F (20°C) so that the washwater will recover naphthalene from the gas. The cool gas is then washed with a wash oil, which extracts additional by-product hydrocarbons. Finally, the gas is sent to a scrubber for reduction of sulfur which may be required to meet air emission standards when it is used as fuel. The wash oil containing hydrocarbons may then be processed for separation of benzene, toluene, and xylene, as illustrated by the flow sheet in Figure 27.5.

UTILITY REQUIREMENTS

Because of the availability of excess by-product gas, the coke plant usually produces its own steam to drive turbines used for compressing the gas. The plant may also generate electric power for its own needs. Usually the turbine exhaust steam is used in the distillation required to separate benzene, toluene, and xylene.

Appreciable water is needed by the coke plant, principally for gas cooling. This may come to approximately 1000 to 2000 gal/ton (4 to 8 m^3/t) of product. In addition to gas cooling, water is used for coke quenching and for cooling of utility auxiliaries, such as air compressors and pump glands.

CHARGING, LEVELING AND PUSHING OPERATIONS
IN ONE COKING CYCLE OF A BY-PRODUCT COKE OVEN

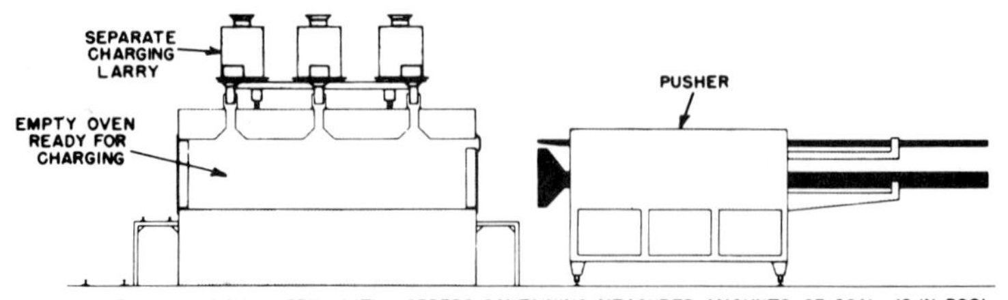

A. THE CHARGING LARRY, WITH HOPPERS CONTAINING MEASURED AMOUNTS OF COAL, IS IN POSITION OVER CHARGING HOLES FROM WHICH COVERS HAVE BEEN REMOVED. THE PUSHER HAS BEEN MOVED INTO POSITION.

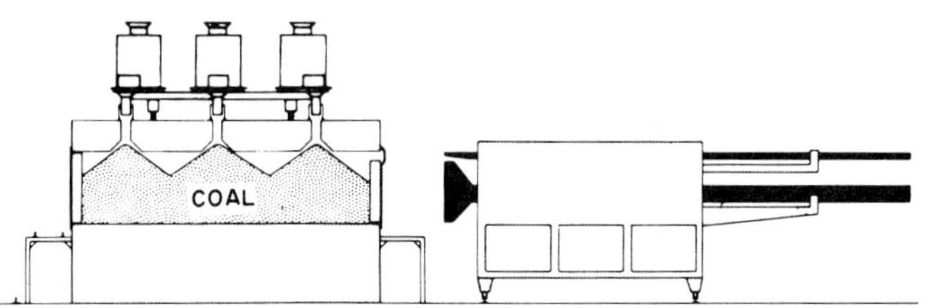

B. THE COAL FROM THE LARRY HOPPERS HAS DROPPED INTO THE OVEN CHAMBER, FORMING PEAKED PILES.

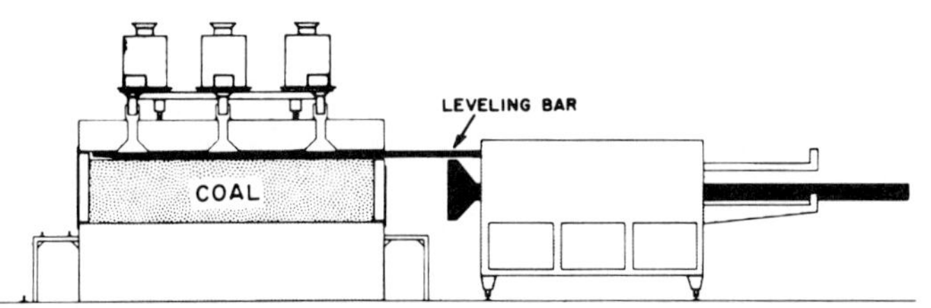

C. THE LEVELING DOOR AT THE TOP OF THE OVEN DOOR ON THE PUSHER SIDE HAS BEEN OPENED, AND THE LEVELING BAR ON THE PUSHER HAS BEEN MOVED BACK AND FORTH ACROSS THE PEAKED COAL PILES TO LEVEL THEM. THE BAR NEXT IS WITHDRAWN FROM THE OVEN, THE LEVELING DOOR AND CHARGING HOLES ARE CLOSED, AND THE COKING OPERATION BEGINS.

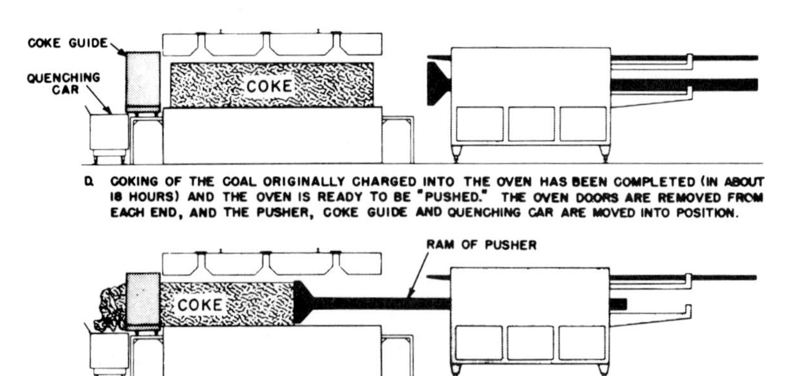

D. COKING OF THE COAL ORIGINALLY CHARGED INTO THE OVEN HAS BEEN COMPLETED (IN ABOUT 18 HOURS) AND THE OVEN IS READY TO BE "PUSHED." THE OVEN DOORS ARE REMOVED FROM EACH END, AND THE PUSHER, COKE GUIDE AND QUENCHING CAR ARE MOVED INTO POSITION.

E. THE RAM OF THE PUSHER ADVANCES TO PUSH THE INCANDESCENT COKE OUT OF THE OVEN, THROUGH THE COKE GUIDE AND INTO THE QUENCHING CAR.

FIG. 27.3 Cycle of coke oven production. *(Copyright 1964 by United States Steel Corporation.)*

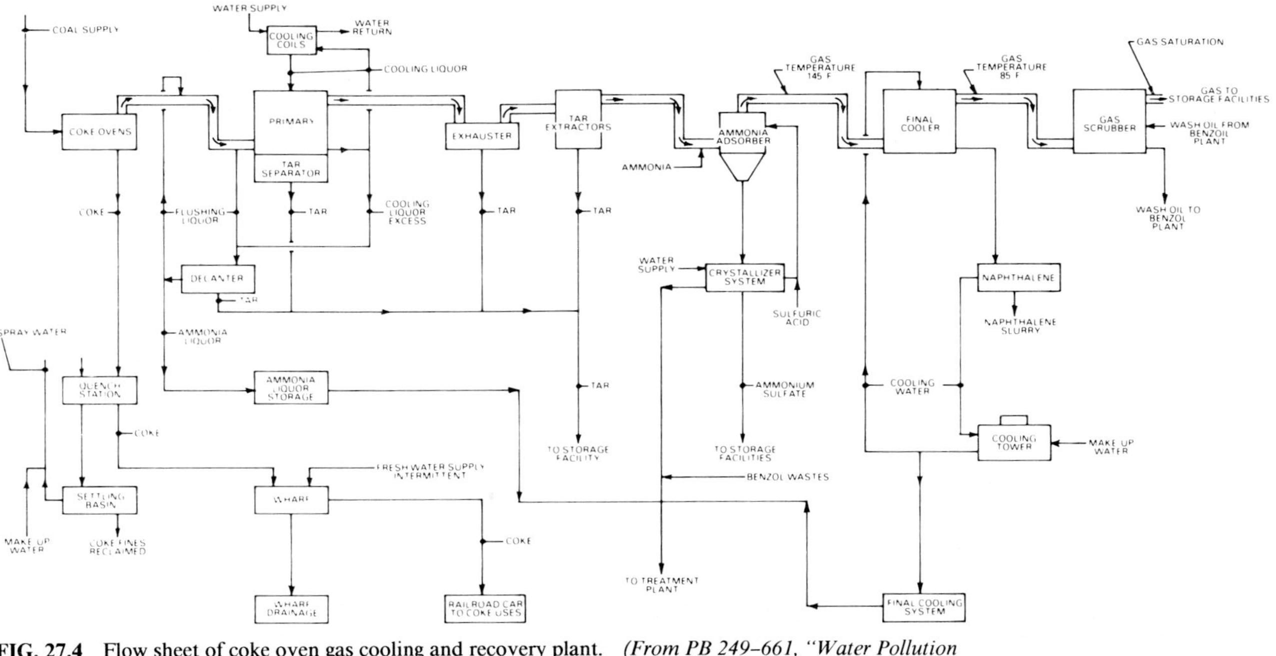

FIG. 27.4 Flow sheet of coke oven gas cooling and recovery plant. *(From PB 249–661, "Water Pollution Abatement Technology: Capabilities and Cost, Iron and Steel Industry," National Technical Information Service, Springfield, Va.)*

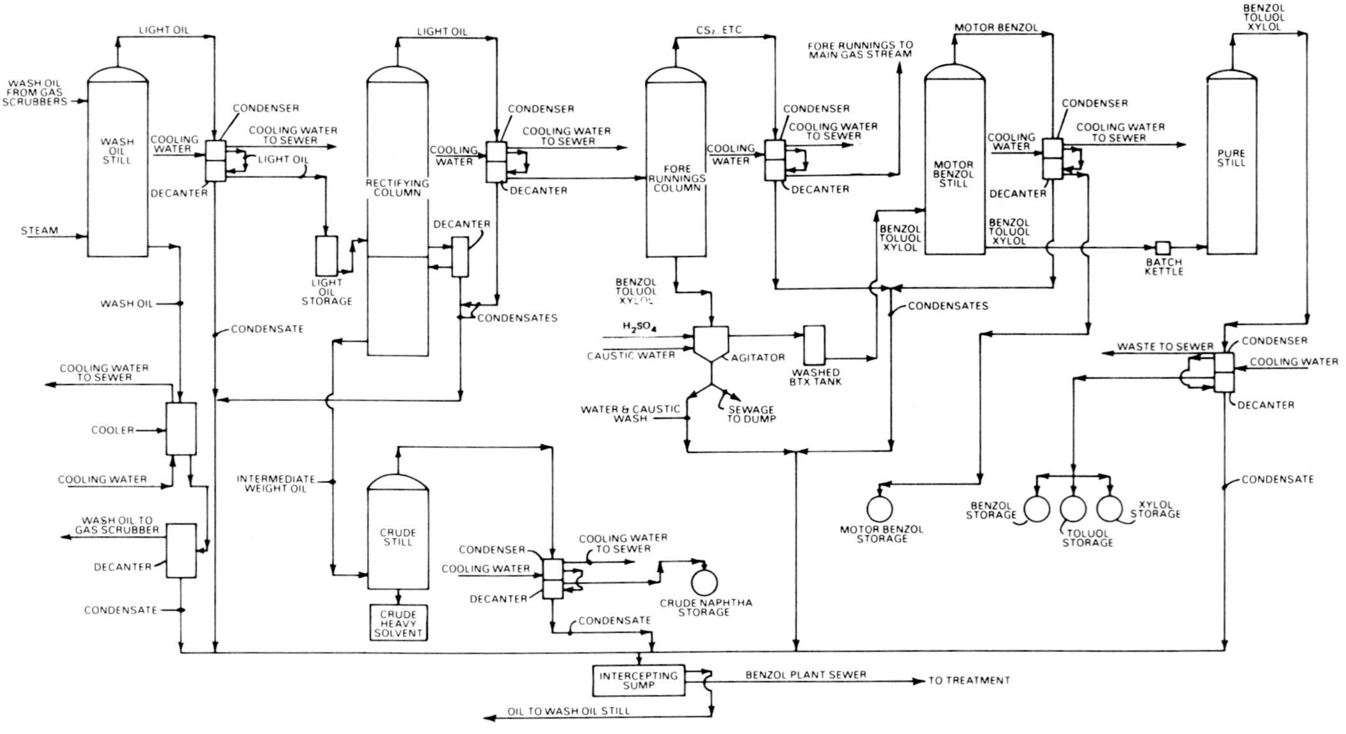

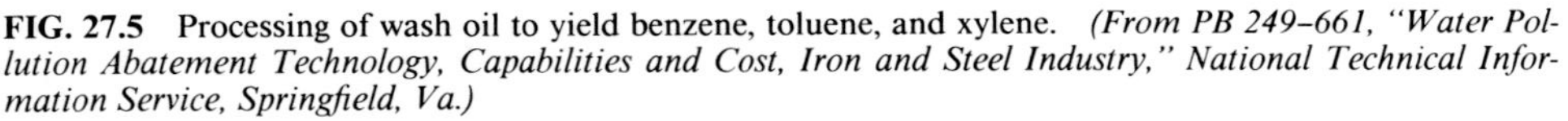

FIG. 27.5 Processing of wash oil to yield benzene, toluene, and xylene. *(From PB 249–661, "Water Pollution Abatement Technology, Capabilities and Cost, Iron and Steel Industry," National Technical Information Service, Springfield, Va.)*

As in many hydrocarbon processing plants, process water becomes contaminated not only with oil, but also with ammonia, phenol, cyanide, and thiocyanates. The major sources of wastewater are (*a*) the weak ammonia liquor from the primary coolers, (*b*) blowdown from cooling towers used either directly or indirectly for gas cooling, (*c*) contaminated condensate from steam distillation of some of the hydrocarbon fractions, (*d*) storage tank water draw, and (*e*) storm water. Strong wastes that may be segregated for separate disposal include acid and caustic used in light oil processing, spent soda ash solution from the sulfur removal tower, and excess liquors from the ammonium sulfate recovery process.

BIOLOGICAL TREATMENT

The weak ammonia liquor may be segregated for separate treatment through a biological system, which has proved to be an effective method for reduction of both phenol and cyanide.

Because of the relatively high requirement for quenching water, about 150 to 200 gal/ton (0.6 to 0.8 m^3/t) of coke, this has usually been used as a point for disposal of the stronger wastes, such as excess weak ammonia liquor, which would otherwise be difficult to process to a level acceptable for final discharge to a receiving stream. There is considerable heat in the coke pushed from the oven, and some plants reclaim this heat by using air cooling rather than water quenching. This preheats the air and reduces fuel consumption.

The problems of both air and water pollution control are formidable in the coke plant, being complicated by the nature of the raw materials, the design of the coke ovens, and the economics of by-product recovery. Each plant requires individual study to arrive at the best solution. Solving the problems will require revolutionary developments in technology, both in terms of pollution control facilities and also modification of the process itself. For example, coke oven fumes released during the pushing step are being captured by special emission control cars positioned on the rails at the oven being discharged (Figure 27.6). Where the coke plant is part of a steel mill, there may be opportunities for disposal of some wastes through steel mill operations such as slag quenching.

PRODUCER GAS AND SYNFUELS

The chemical reaction representing the manufacture of coke could be expressed in this way:

$$\begin{array}{llll} \text{Coal} \rightarrow & \text{gases:} + & \text{liquids:} & + \text{ solids:} \\ & CO & \text{tar} & \text{char} \\ & CH_4 & \text{benezene} & \text{(coke)} \\ & & \text{toluene} & \\ & H_2 & \text{xylene} & \\ & CO_2 & \text{naphthalene} & \end{array} \qquad (5)$$

In other coal-based processes, steam, air, oxygen, hydrogen, or hydrocarbons are introduced into the reactor (a special design of furnace, very different from a coke oven), depending on the products desired. There are two fundamental classifications of processes for coal conversion: coal gasification and coal liquefaction.

FIG. 27.6 Emissions from the coke oven discharged during the pushing operation are captured by this specially designed mobile unit that transports the coke to the quench tower. *(Courtesy of The Alliance Machine Company-Hartung Kuhn & Company.)*

Coal Gasification

As mentioned earlier, most gas burned in American homes prior to 1940 was producer gas; natural gas was used only in areas where it was readily available. A simple diagram of the gas producer is shown in Figure 27.7. The major reactions in the gas producer are:

$$C + O_2 \rightarrow CO_2 \text{ (exothermic)} \tag{6}$$

$$C + H_2O \rightarrow CO + H_2 \text{ (endothermic)} \tag{7}$$

$$C + 2H_2 \rightarrow CH_4 \text{ (exothermic)} \tag{8}$$

$$C + CO_2 \rightarrow 2CO \text{ (endothermic)} \tag{9}$$

The combustion of carbon [reaction (6)] provides the major energy source for the endothermic reactions, so the proportion of air to steam and carbon is a major control factor in the process. The pre-1950 producers usually yielded a low Btu gas, having about 150 Btu per standard cubic foot (1335 kcal/m^3), and operated at slightly over atmospheric pressure. The quality of the producer gas is affected by the characteristic of the coal: with high-grade coal, the producer gas might reach 180 Btu per standard cubic foot (1600 kcal/m^3) in heating value. A simplified flow diagram of the process is shown in Figure 27.8. Because the coal is gasified, the removal of sulfur, nitrogen, and carbon dioxide is relatively simple, as compared with their removal from liquid fuels. This is a fundamental advantage of coal gasification over direct coal liquefaction. Since air contains 80 percent nitrogen by volume, the nitrogen remaining in the finished gas dilutes the heating value of the producer gas.

If oxygen is used in place of air, the nitrogen is eliminated and the heating value increased to over 300 Btu per standard cubic foot (2670 kcal/m^3); this is called medium Btu gas.

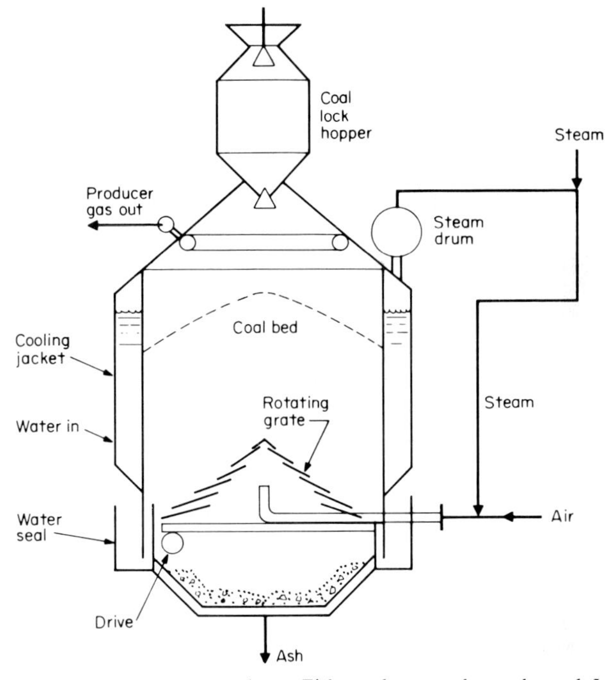

FIG. 27.7 Early gas producer. Either coke or coal was charged. In some designs, like this one, steam was produced in cooling the reactor jacket.

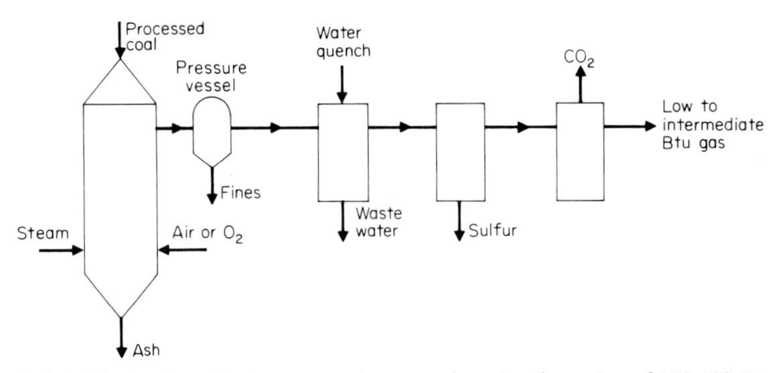

FIG. 27.8 Coal gasification to produce gas with a heating value of 150-300 Btu per standard cubic foot.

If the gas must be processed to provide the heating value of natural gas [about 1000 Btu per standard cubic foot (8900 kcal/m^3)], additional hydrogen must be provided. The ratio of atomic H to C in most coal is quite low, typically about 0.8, whereas the H:C ratio in oil and natural gas is over 1.6. So, the higher ratio required for high Btu pipeline gas must be obtained by additional process steps in reactors packed with catalysts to produce hydrogen by reaction (10), methane by reaction (11), or both:

$$CO + H_2O \xrightarrow{\text{catalyst}} H_2 + CO_2 \text{ (Shift Reaction)} \quad (10)$$

$$CO + 3H_2 \xrightarrow{\text{catalyst}} CH_4 + H_2O \text{ (Fischer-Tropsch)} \quad (11)$$

Hydrogen and methane may also be produced from char in special producers, based on reactions (7) and (8). The quality of pipeline syngas (SNG) depends on the nature of the coal used as feed, the size to which it is ground, the temperature

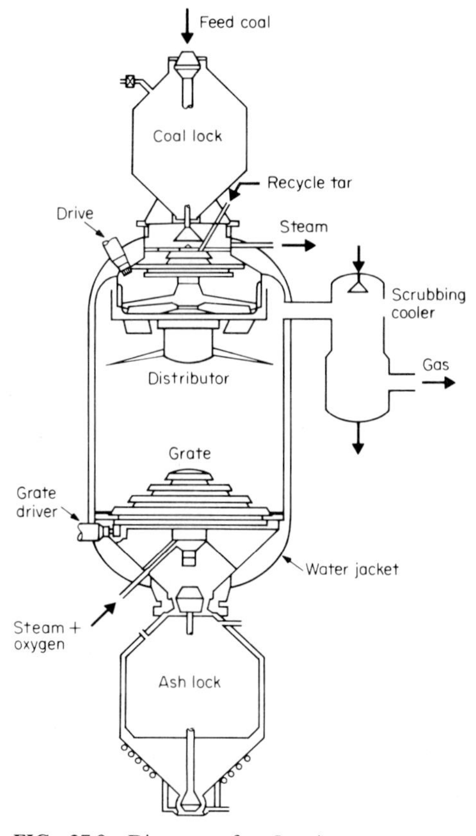

FIG. 27.9 Diagram of a Lurgi pressure gasifier. *(From the McGraw-Hill Encyclopedia of Science and Technology.)*

and pressure of operation, and the introduction of air or oxygen to the feed stream. Operating pressures range from 50 lb/in^2 gage for the COGAS producer to 1200 lb/in^2 gage (84 bars) for the Texaco design reactor. Worldwide, the most common design is the Lurgi producer, Figure 27.9, operating at 450 lb/in^2 gage (32 bars). The energy conversion efficiency of coal gasification processes ranges from 60 to 70 percent.

In general, the higher the operating temperature, the less complex the water pollution problems. The condensates from high-pressure plants are rather similar to refinery wastewaters, whereas those from low-pressure reactors are more complex and rather similar to coke plant wastes—high in ammonia, cyclic compounds, and more toxic organic compounds.

One of the significant factors in the synfuels development is that water is the source of hydrogen, and often the natural source of carbon—whether high-grade coal or lignite—occurs in water-short areas. For example, 12,000 gal of water are required for each million cubic feet (1600 m^3/million m^3) of SNG produced.

The SNG may be used either as high Btu pipeline gas or as feed to an indirect liquid fuels plant.

Coal Liquefaction

There are two categories of coal liquefaction processes, indirect and direct. The indirect process is so-called because the coal is first gasified, and the gas is then fed to converters to yield liquid products. Direct liquefaction processes put the coal solids in slurry form and hydrogenate the fine coal in high-pressure reactors.

Indirect Processes. There are two major indirect processes, one of which has been proved in large-scale operation at three plants operated by the South African Coal, Oil, and Gas Company (Sasol). The Sasol process is based on the Fischer-Tropsch reaction (11), using a catalyst that produces higher boiling hydrocarbons from producer-gas feed. The liquids are fractionated and processed in equipment similar to conventional petroleum refining units (Chapter 31). The second is the Mobil-M process: syngas high in hydrogen and methane is refined to remove sulfur and other contaminants, and this gas is catalytically converted to crude methanol, which is passed through a second catalytic converter to produce gasoline directly, with water by-product.

Since contaminants such as sulfur, carbon dioxide, and nitrogen are removed by conventional gas phase reactions, the indirect processes have an advantage of simplicity over direct liquefaction. The overall energy-conversion efficiency of indirect processes ranges from 40 to 60%, based on the ratio of gas to liquid of the total plant products.

Direct Processes. In the United States, there are currently three direct processes that are modifications of the original Bergius process developed by Germany 50 years ago and operated there during World War II to produce about 110,000 barrels per day (17,500 m^3/day) of liquid fuels from coal. All three processes slurry coal in a liquid hydrocarbon product stream, hydrogenate the slurry in reactors, and produce refinery feedstocks. The method of supplying hydrogen varies among the three processes [H-coal; solvent refined coal (SRC); and Exxon donor solvent (EDS)]. Currently, relatively small demonstration plants are being built in the United States to evaluate economics and environmental effects. However, even with the relatively high energy-conversion efficiencies projected for these direct

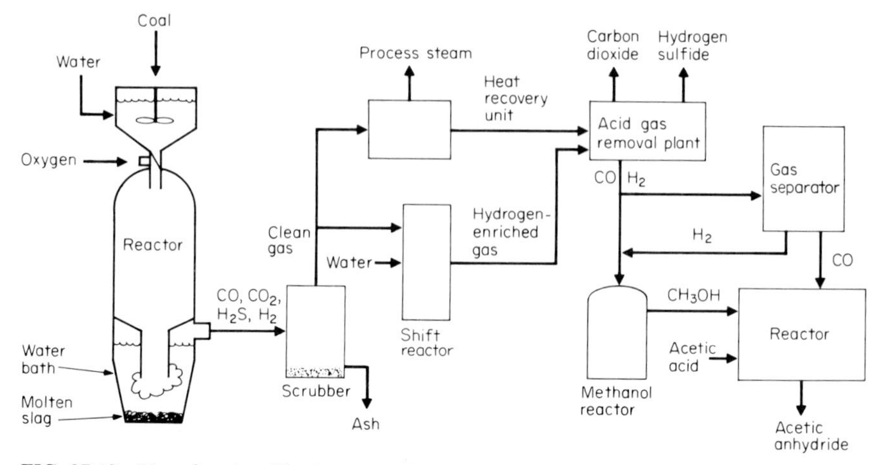

FIG. 27.10 Use of coal gasification to produce chemical intermediates. *(From Chem. Eng. News, Nov. 29, 1982.)*

liquefaction processes, the cost of synthetic oil production from coal exceeds the world price of crude oil.

Chemical Feedstocks. Although in the present economy, the United States cannot convert coal either to syngas or liquid hydrocarbons at costs competitive with natural gas or oil, there is growing interest in coal conversion to yield more valuable products. The flowsheet of Figure 27.10 illustrates a process plant being built by a chemical company in Tennessee to process 900 tons per day (816 t/day) of high sulfur Appalachian coal to produce about 750,000 pounds per day (341,000 kg/day) of acetic anhydride. It is probable that countries with economical sources of coal or lignite but having no crude oil or natural gas resources will be developing coal conversion plants for both fuels and chemical feedstocks at a rate far greater than is likely in the United States.

SUGGESTED READING

McGannon, H. E. (ed.): *The Making, Shaping and Treating of Steel,* 8th ed., United States Steel Corporation, 1964.

Parker, C. P., et al.: *Encyclopedia of Science and Technology,* Coal Gasification Systems: A Guide to Status, Applications and Economics. Electric Power Research Institute, EPRI AP-3109. Project 2207, Final Report June 1983. (Prepared by Synthetic Fuel Associates, Inc.) McGraw-Hill, New York, 1982.

Shreve, R. N.: *Chemical Process Industries,* 3d ed., McGraw-Hill, New York, 1967.

CHAPTER 28
FOOD PROCESSING INDUSTRY

Because of its intimate connection with the public health, the food industry has had a long history of surveillance of its activities by local, state, and federal agencies. The U.S. Congress passed the original Food and Drug Act in 1906. That act, with subsequent legislation, controls not only the chemicals that are directly added to food—salt, seasonings, and preservatives—but also such chemicals as sizing in food wrap that may indirectly become food additives by contacting food.

In addition to the close control of the FDA, additional surveillance is imposed on meat and poultry processing plants through the U.S. Department of Agriculture. No chemical can be brought into a meat or poultry processing plant unless approved by the USDA for its intended use, such as equipment cleaning or water treatment.

Because of this close regulation, the choices of chemicals used in water or wastewater treatment may be more limited in the food industry than in other major water-consuming industries.

There are many segments of the food processing industry, of which the major water-using categories include sugar cane and beet processing, beverage manufacturing, fruit and vegetable processing, meat and poultry, grain processing, fats and oils, and dairy products. The water consumptions of these segments are shown in Table 28.1.

Although there are wide variations in the process steps in each of these industry segments, there are a number of common unit operations. The distribution of water in the plant can be put into three categories: process water, cooling water, and boiler feed water. The percentage distribution varies considerably from a high of about 60% used for processing in the meat and poultry industry, to a low of only 15% in the sugar industry. However, 75% of the water used in the sugar industry is for cooling purposes (and later becomes process water), with only 25% being used in the meat and poultry industry. Most food processing plants generate steam for cooking or processing, and water used for boiler makeup ranges from about 6% of the total usage in fruit and vegetable processing, to about 15% in the fats and oils segment.

Process water uses include: washing of raw materials and process equipment; conveying products from one process area to another; dissolving or extracting; and addition to the finished product. Cooling water may be used to operate refrigeration equipment, to condense steam from evaporators or turbines, or to cool process equipment such as compressors, cookers, and engine jackets.

Steam may be generated for cooking, for heating evaporators, or for space heating. In some industries enough steam is required to justify installation of a turbine to extract power from the steam before it is sent to process (cogeneration). If the

TABLE 28.1 Water Usages in Food Processing*

Industry segment	Gross use	Water flows, mgd intake	Discharge
Sugar	1061	545	518
Beverages	797	275	226
Fruit and vegetables	620	348	324
Meat and poultry	458	296	288
Grain processing	386	218	199
Fats and oils	358	107	89
Dairy products	218	119	111

* 1972 Census of Manufactures, U.S. Department of Commerce Publication MC72 (SR-4), Water Use in Manufacturing.

steam can come into direct contact with food, there are strict limits on chemicals used for both steam and boiler water treatment, and on their maximum concentrations.

Knowing the processing operations in a food plant is helpful to an understanding of water use; water may be used sequentially for several purposes. For example, in the sugar industry, which has a very high requirement for condenser cooling water because of the evaporation and concentration of syrups, this cooling water is used for washing cane brought in from the fields before it is discharged, and it is categorized as cooling water rather than process water.

Because the problems of cooling water and boiler water treatment in the food industry are similar to other industries, this chapter will deal specifically with process water, with some consideration of its contamination and final treatment for disposal.

THE SUGAR INDUSTRY

There are many process steps in the sugar industry (some of which are similar to corn processing). As the largest food processing water user with a variety of process operations, the sugar processing industry offers a good example of water use in food processing.

Sugar (sucrose), a chemical classified as a disaccharide with the formula $C_{12}H_{22}O_{11}$, is derived from two major crops, sugar cane and sugar beets. Cane is cultivated in tropical and semitropical climates (e.g., Puerto Rico, Florida), while beets are raised in temperate climates (Idaho, California). In the United States, about 7 million tons of sugar are produced annually compared to world production of 75 million tons each year. In the United States the major sugar cane producers are Louisiana and Florida, with Hawaii and Texas also important contributors. Beet sugar production is principally in the western and northwestern states. Cane exceeds beet sugar production. Because these crops spoil rapidly, they cannot be stored and the production of sugar is seasonal, with a mill operating a production campaign of about 120 days geared to the plant harvest.

Sucrose, which constitutes up to 20% of the weight of the cane or beet, is readily degraded by bacterial action. The first step in the degradation is the production of invert sugars, fructose and glucose:

$$\underset{\text{sucrose}}{C_{12}H_{22}O_{11}} + H_2O \rightarrow \underset{\text{fructose}}{C_6H_{12}O_6} + \underset{\text{glucose}}{C_6H_{12}O_6} \quad (1)$$

The second step is the production of lactic acid ($C_3H_6O_4$) under the conditions prevailing in beet sugar manufacture, or dextran ($C_6H_{10}O_5$) under the conditions common in the cane sugar mills. Since this degradation is primarily caused by bacteria, it is important to maintain control of microbial organisms throughout the mill to avoid loss of production of sucrose. Microbial activity also causes processing difficulties, such as filter blinding, slime formation, and odors.

A basic flow diagram applying to cane and beet sugar processing is shown in Figure 28.1.

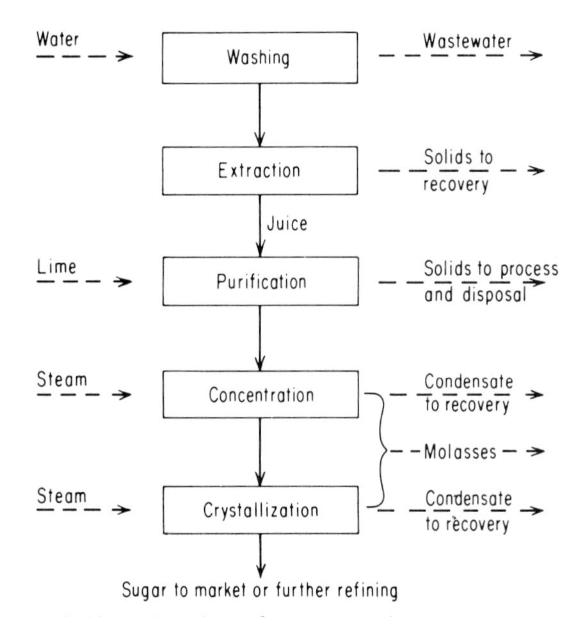

FIG. 28.1 Flow sheet of sugar processing.

As the crops arrive at the mill they contain soil and trash accumulated during the harvesting operation. In the case of cane sugar harvested by pushers, similar to bulldozers, the refuse may constitute as much as 10 to 25% of the weight of material delivered to the mill. This is not merely inert material; it represents a major source of bacterial inoculation since soil organisms are present with appreciable fecal matter from birds, rodents, and other small animals that live on the crop lands. Because of this, washing is a critical operation to the preparation of raw materials going to further extraction processes. On the other hand, washing should not be excessive, as this leads to loss of sucrose in the washwater.

After washing, sucrose is extracted from the raw material. In cane sugar mills, this is usually done by crushing and milling the washed, cut cane stalks, producing a juice containing approximately 12 to 15% sucrose. In the beet sugar industry, the beets are sliced into long, narrow pieces (cossettes), and the sucrose extracted by washing with water in diffusers at about 160°F (70°C). There is growing interest in the use of diffusers in place of crushers and mills in cane processing to reduce maintenance costs and improve yield.

The cane stalks are pressed after initial crushing and milling to reclaim as much sugar as possible, and the remaining solids (called bagasse) are usually

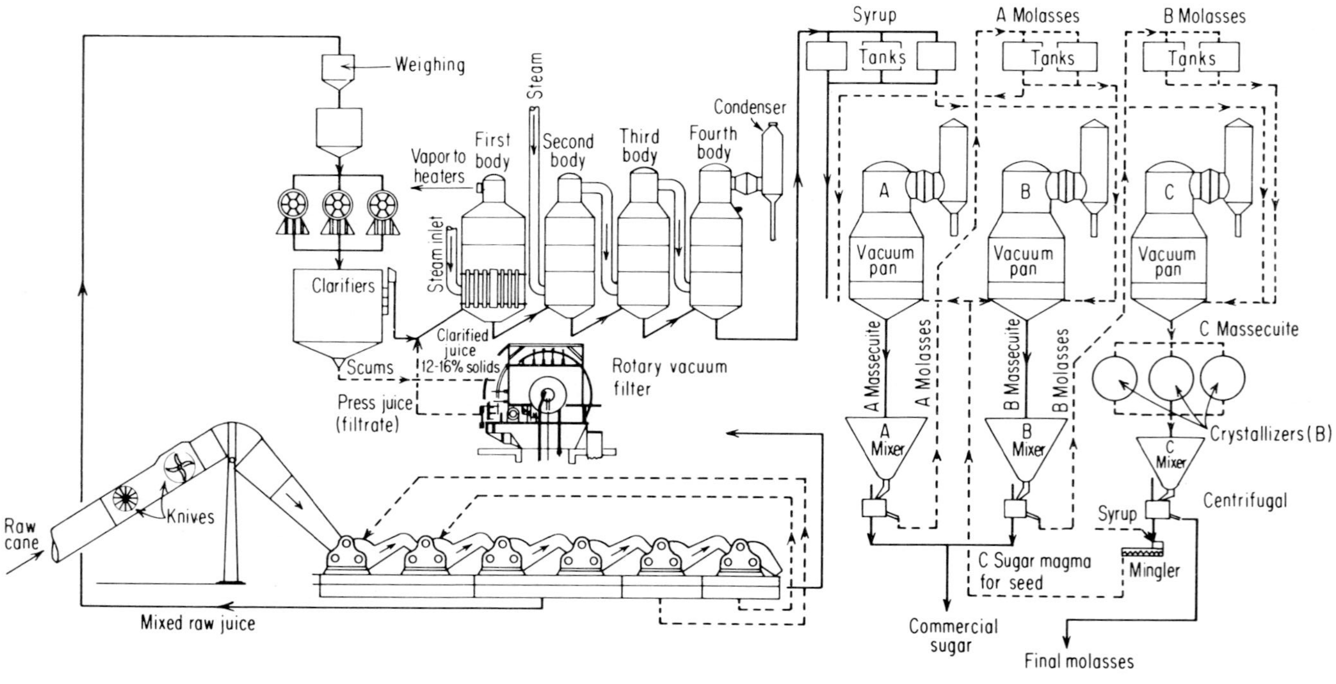

FIG. 28.2 Process flowchart of cane sugar mill. *(Courtesy of Rio Grande Valley Sugar Growers, Inc.)*

burned in boilers to generate steam. Bagasse may also be used as a raw material for such products as insulation board or acoustical tile. In the beet sugar industry, the beet pulp residue is quite high in protein, and it may be mixed with some of the plant production of molasses for cattle feed.

As with most other natural products, there are a variety of chemicals other than sucrose in the cane and beets. These must be removed to maximize the yield of sugar, minimize the production of molasses, and reduce taste-, color-, and odor-producing impurities. Lime is used to precipitate these impurities, and the lime mud is removed by conventional solids/liquid separation devices. The mud is washed to reclaim as much sugar as possible, and it may then be (*a*) reburned to produce fresh lime, (*b*) returned to the fields for its fertilizer value (it often contains significant phosphate), or (*c*) sent to landfill.

The purified juice must be concentrated to produce a thick syrup, the form of sugar often used by the beverage industry, or to produce a crystalline product. The juice is concentratd by evaporation. Since it contains calcium from the lime treatment, a common problem in the industry is the formation of scale in the pans (simple steam-jacketed evaporators) or in the multiple effect evaporators. Another common problem is foaming as the juices become concentrated during evaporation.

The flow sheet of a cane sugar mill producing raw sugar is shown in Figure 28.2. There is widespread use of steam throughout the plant, so the boiler house is a significant factor in economical production of sugar. Many sugar mills operate intermediate pressure boilers and produce electric power through turbo generators, taking extraction steam from the turbines for operation of pans, evaporators, and crystallizers, and in many cases taking part of the steam through the turbine to a condenser. Cooling water from both turbine and evaporator condensers may then be used as process water.

Because the operation of a multiple effect evaporator produces more water as condensate than it consumes as steam, there is usually an excess of condensate available as boiler feed water. This condensate often presents special problems in that it is likely to be high in ammonia, and it may periodically contain sucrose or invert sugar. The introduction of sugar into the boiler quickly produces an acid condition, so careful monitoring of the condensate system for sugar content is important to the protection of the boiler system.

The pollution control problems of the sugar mills are unusual because of the seasonal nature of the industry. The campaign is usually during the dry months, when streams cannot assimilate any excess organic loading. So the mills minimize discharge flows by recycling as completely as possible, and the wastewaters are treated biologically and impounded for solar evaporation or controlled discharge when stream flows return to acceptable rates.

THE BEVERAGE INDUSTRY

This segment of the food industry is another major consumer of water, some of which becomes part of the final product. The balance is used for washing of bottles and containers, cooling of compressors and refrigeration equipment, and makeup to boilers producing steam used for cooking, evaporation, heating of pasteurizers, and space heating.

The water used in the product must, of course, be potable; in addition, there are standards within the industry related to the effect of water quality on the taste of the finished beverage. In the soft drink industry, for example, it is common to lime soften the water for hardness and alkalinity reduction, since alkalinity destroys the flavor of acidic fruit extracts. In the lime softeners, breakpoint chlorination is also practiced. The finished water is filtered and then passed through activated carbon as a final precaution for the removal of chlorine and any residual tastes or odors. Most soft drink bottling plants have hot water boilers to provide the heat required for bottle and can washing.

Breweries and distilleries, on the other hand, operate their own steam plants because steam is required for cooking and for the operation of evaporators. In many of these plants, steam passes through turbines to generate power, exhausting to lower pressures for process operations. The unit operations in these plants are quite similar to those found in the chemical industry in principle, but special designs enable the equipment to be readily cleaned to prevent microbial contamination of the product and to avoid risks to the public health.

FIG. 28.3 A filter installation typical of the food and beverage industry practices. Note the use of special stainless-steel piping and fittings. *(Courtesy of Cross-Reynolds Engineering Company, Inc.)*

Special designs of piping and fittings, such as long-sweep pipe elbows, are used throughout the food industry because of this need for sanitation (Figure 28.3). Highly polished stainless-steel, monel, or chrome-plated steel eliminates scratches, nicks, and crevices which could offer a home for bacterial growths. The careful cleaning of equipment after each use creates a special problem of pollution control in that spent chemical cleaners, especially those containing biocides, often interfere with the performance of pollution control equipment.

Two water-using systems are unique to the food industry and require special attention to water quality: these are the bottle and container washing systems and the pasteurizer.

In the bottle washing operation, both cleaning and sterilization are required, so detergents and biocides are applied to match the severity of the problem. If the bottle washer is handling returnable bottles, since there is no way of knowing what might have been in the bottles when in the hands of the public, it is important to use effective cleaning chemicals. These are quite alkaline. Because of this, it is beneficial to have zeolite-softened water for washing and rinsing as this reduces the demand for detergents and also greatly facilitates the drainage of the bottle after rinsing for spot-free surfaces.

When strongly alkaline cleaners are used, these provide a biocidal effect that depends both on the length of time the chemical is in contact with the bottle, and the causticity of the cleaning solution. Even with this protection, however, chlorine is often applied to the final rinse water to ensure sterility.

In the pasteurizing operation, as practiced in breweries, the bottled product is moved through the pasteurizer, passing first through a chilling zone to halt the growth of specific spoilage organisms. A controlled temperature water bath then slowly brings the beverage to approximately 160°F (70°C) and holds it for the time required to ensure that the entire contents of the bottle have been pasteurized. Usually two heating stages are used to prevent thermal shock and bottle breakage. The bottle is then moved into a chilling compartment before leaving the pasteurizer for packaging. It is useful to have zeolite-softened water for this operation also, to avoid spotting the bottles or cans. The temperature in the pasteurizing section is maintained by circulating hot water; and the chilling sections (also staged to avoid thermal shock) may be tied into a cooling tower and supplemented with a closed chilled-water system (Figure 28.4), although once-through cooling water is still widely used.

In the event of bottle breakage, these water systems are inoculated with nutrients (the beverage), and microbial activity may quickly get out of hand. Biocides or chlorine-biodispersant treatments are used to keep this under control.

For the most part, beverage industry wastewaters are handled by municipal sewage systems. This may require the plant to install equilization facilities to unify the composition and flow rate. In-house handling of strong wastes, such as chemical cleaners, may also be necessary to make the equalization program effective.

A number of large distilleries provide their own independent waste treatment facilities, usually conventional biological treatment.

FRUIT AND VEGETABLE PROCESSING

Just as with fluming of sugar beets in the sugar industry, traditionally, water has been the media of in-plant conveyance of most fruits and vegetables in this industry segment. Not only has this choice been economical, but also it provides additional benefits in prewashing and cooling. However, because of the pollution load that results from fluming, new methods of conveyance (air, vacuum, and mechanical) are now becoming more common, so washing and rinsing, which may require as much as 50% of the total water used in process operations, is a separate

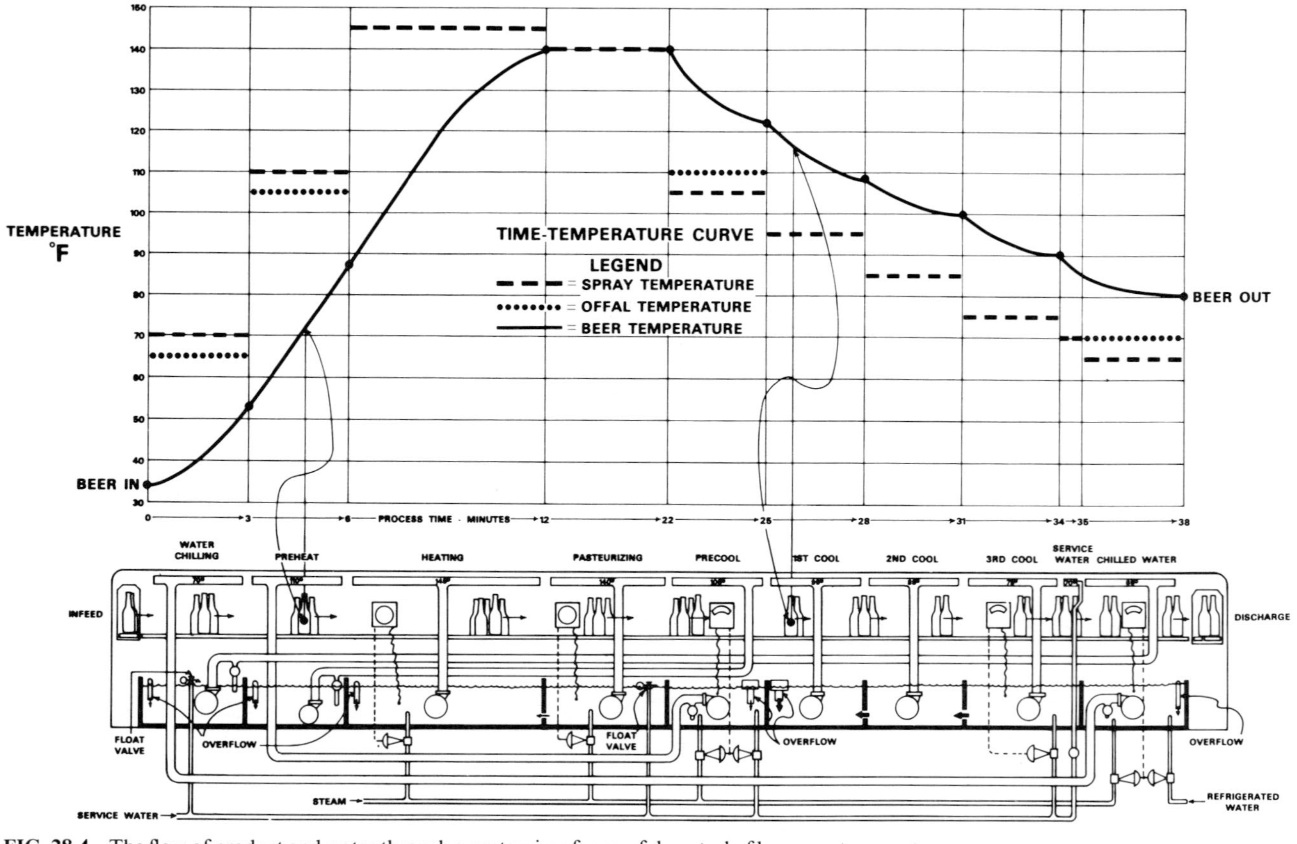

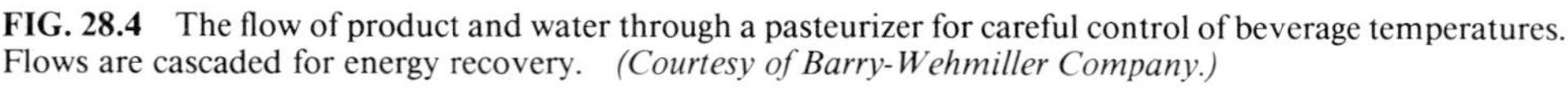

FIG. 28.4 The flow of product and water through a pasteurizer for careful control of beverage temperatures. Flows are cascaded for energy recovery. *(Courtesy of Barry-Wehmiller Company.)*

step. Grading and sizing are sometimes accomplished simultaneously with washing.

After the fruits and vegetables are washed, peels are removed in a variety of ways. Steaming or soaking in caustic solutions are the most common, but air and mechanical peeling are also used. Dry caustic peeling of potatoes is gaining acceptance within the industry as a means of greatly reducing pollution loads.

Water blanching is generally used for vegetable processing to remove air and to leach solubles before canning. Steam blanching of vegetables is usually used to destroy enzymes before freezing or dehydration. The blanching effluent stream contributes a signficant portion of the total pollution load in canning operations.

In most canneries, the cans are filled with uncooked food and then passed through a steam exhaust box to eliminate air preceding the can sealing operation. The food is then cooked in the cans by direct contact of steam in retorts, which are pressure cookers in which cans are stacked in racks. After cooking, large volumes of water are required to cool the cans as they pass through cooling water canals. For continuous production, horizontal rotary cookers may be used in place of retorts and cooling canals, with steam cooking in the first cylindrical shell and water cooling in the second. Even larger continuous units are available to process up to 50,000 cans on a 2-h cycle including preheat, steaming, cascade cooling, and final cooling.

A number of food products are cooked before they are bottled or canned, and many of these, such as catsup, require concentration through evaporation of water from the juice. So, there are many varieties of processing units used in the fruit and vegetable processing industry, including jacketed cookers and evaporators. All of this food processing equipment must be kept clean, usually by sanitizing with chlorinated water or detergents.

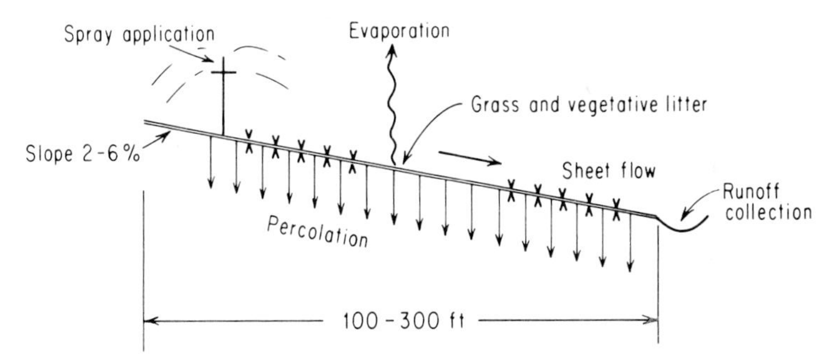

FIG. 28.5 Spray distribution of food processing wastes to sloped grassland with collection of treated runoff. This system is very efficient in temperate climates. *(Courtesy of U.S. Environmental Protection Agency.)*

Like most food industries, the fruit and vegetable processors are trying to adopt dry cleaning methods to reduce the pollutant loading on effluent water. This pollution loading varies enormously from one type of product to another. For example, in the processing of asparagus, the BOD and suspended solids are usually below 100 mg/L, whereas in the production of whole kernel corn, the BOD and suspended solids may be several thousand milligrams per liter. The solid wastes also vary considerably from one material to another.

Although many canneries are served by municipal sewage systems, many also operate their own waste treatment facilities. Those in farm areas where land is available have been successful in using spray irrigation as a means of disposal, often accomplishing as high as 99% BOD removal. One reason spray irrigation has been effective in the canning industry is because of the seasonal nature of its operations, permitting discharge during dry weather and avoiding the problems of freezing that would occur with a year-round operation of spray facilities in cold climates. A flow diagram of such a facility is shown in Figure 28.5.

Although activated sludge is an effective treatment process, strong waste from food processing may be fermented prior to final aerobic digestion to improve overall BOD reduction. This is especially useful in handling fruit processing wastes because anaerobic fermentation proceeds rapidly and greatly reduces the load on the aerobic polishing unit.

MEAT AND POULTRY

A generation ago, Carl Sandburg called Chicago "hog butcher for the world"; today its great stockyards are almost empty. Economics and environmental control problems have moved the meat industry closer to the animals' grazing lands and feedlots. There, the packinghouses purchase live animals, weigh them, and keep them for several hours or overnight in "cow palaces," or holding pens, for processing.

The diverse steps in meat processing are shown by the flow sheet in Figure 28.6. Poultry processing is illustrated by Figure 28.7.

In the first step of meat processing, animals are stunned by electrical, mechanical, or sometimes chemical means before slaughter. Large packinghouses can slaughter over 300 cattle or almost 1000 hogs per hour, while giant poultry processors can kill almost 500 broilers per minute. Water is used sparingly on the kill floor; there are two separate drains so that the blood can be reclaimed for further processing without being diluted by cleanup water.

Blood is either processed on-site or sold to renderers for processing. This consists of evaporating the water from the colloidal solids by "dry cooking" in a steam-jacketed vessel or by direct steam contact. The coagulated blood solids are then dewatered to about 57% moisture either by solids/liquid separation or further evaporation. The water from this process (serum water) carries a heavy pollutant load. Blood processing is a heavy contributor to air pollution also. The final product, blood meal, rich in amino acids, is sold as a nutrient to poultry and hog feeders.

Cattle hides are removed from carcasses, and these "green hides" are usually cured on-site by salting and stacking for 14 days or by soaking in a brine solution, which takes about 18 h. The brine solution is filtered for recycle to remove horn and flesh particles, hair and other solids, and discarded after a few days use. After curing and sorting, the hides are sent out for tanning.

Most hogs are processed without skinning; hair is removed by soaking in 140°F water, usually containing a surfactant, and then processing through a dehairer, a device using rubber hair scrapers with copious sprays of hot water. A gas singer burns remaining hair from the carcass before further processing.

Hair is usually further processed by a caustic soak followed by steam cooking to produce a high-protein animal feed supplement called hair meal.

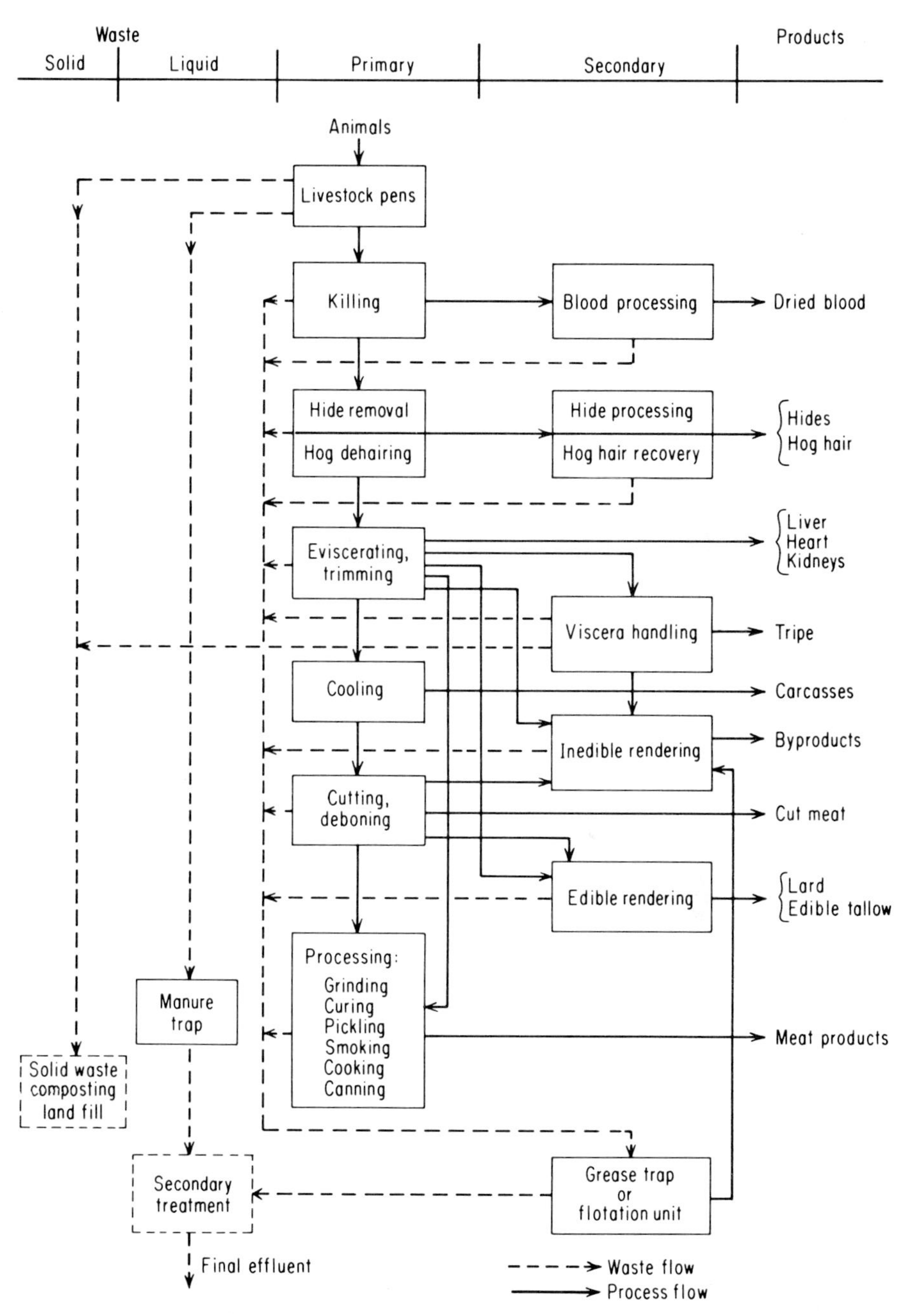

FIG. 28.6 Flowchart for a packinghouse. *(EPA Contract No. 68-01-0031.)*

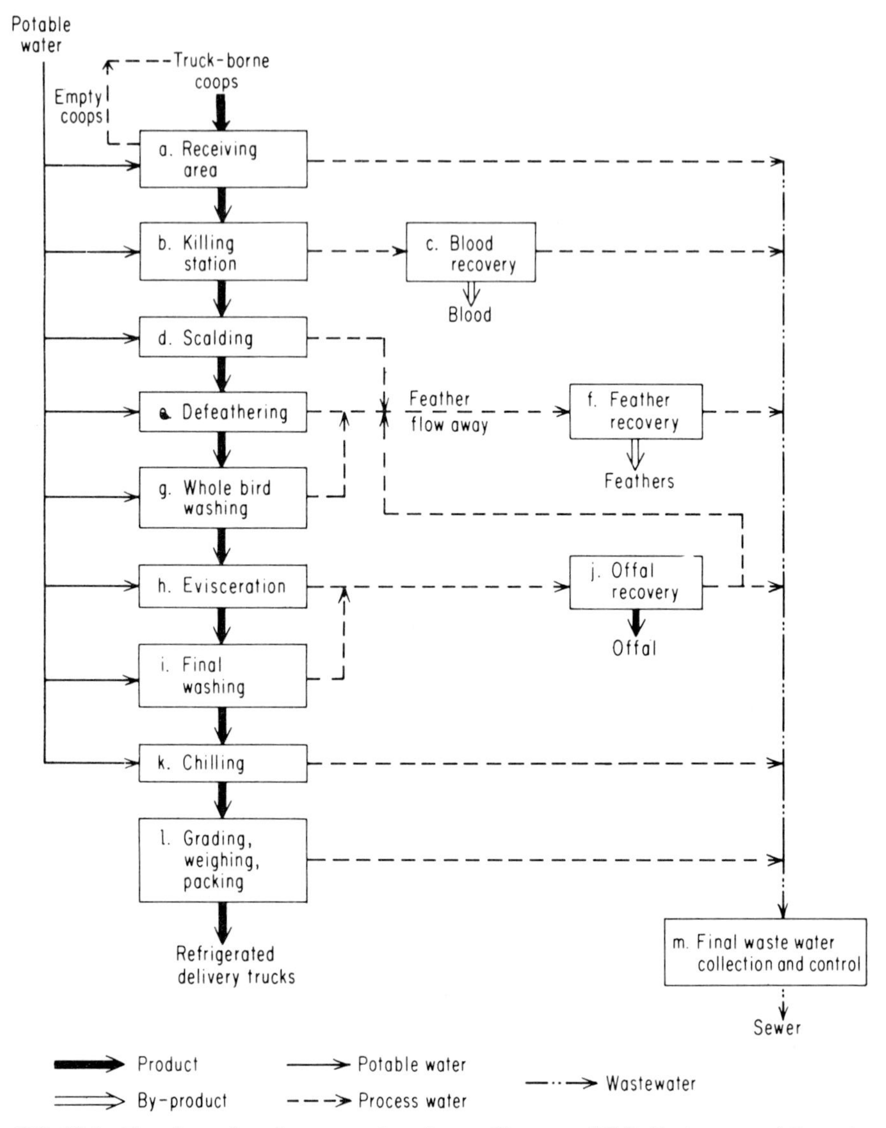

FIG. 28.7 Flowchart of poultry processing plant. *(Courtesy of U.S. Environmental Protection Agency.)*

Large volumes of water are used as the cattle and hog carcasses are cut and processed. Viscera are removed and conveyed by water to edible and inedible product processing.

Federally inspected and approved fat removed from the meat is rendered to make fine cosmetics and lard. This process of edible rendering is a "wet process." Live steam is injected into a pressure vessel and the fat is released to float on the

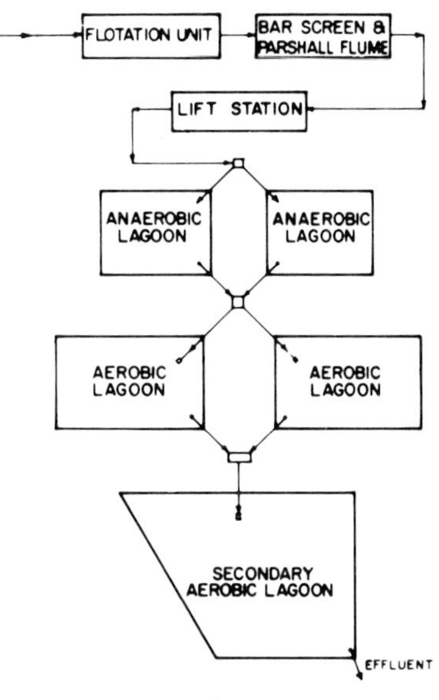

(*a*)

(*b*)

FIG. 28.8 Meat processing waste treatment. (*a*) Flow schematic; (*b*) aerial view. *(From EPA 625| 3-74-003, Water Treatment: Upgrading Meat Packing Facilities to Reduce Pollution, October 1973, U.S. Environmental Protection Agency, Technology Transfer.)*

condensate (stick water). The stick water is high in BOD and extractables, and constitutes a high loading on the waste treatment plant.

Most inedible rendering is accomplished dry. Ground offal is introduced into a steam-jacketed vessel, where water is evaporated under a slight vacuum. The cooked products are tallow and bone meal.

In basic operations, poultry and hog processing are similar. Scalding tanks are used to enhance feather removal; and the production of feather meal is similar to hair meal. Large amounts of water are used throughout the process.

Because of the high BOD of meat processing wastes, water pollution control is somewhat unusual in that anaerobic digestion is frequently used prior to aerobic digestion. A crust develops on the anaerobic lagoon which normally controls the odors often associated with anaerobic fermentation. Pretreatment by equalization, screening, and flotation is essential to good control of the operation. The digester loading is typically about 0.2 lb BOD/day per cubic foot of digester volume. The wastewater is usually warm, about 80 to 100°F (27 to 38°C), which is beneficial, as anaerobic digestion is greatly hindered at low temperatures. Detention may be 12 to 24 h. This combination of anaerobic-aerobic digestion followed by sedimentation usually results in 90% BOD reduction in the first stage and over 98% overall (Figure 28.8).

Many packing plants discharging to city sewage plants are required to provide pretreatment (equalization, screening, grease removal) to reduce the load and equalize the composition of wastes to avoid upsetting the municipal plant operations.

SUGGESTED READING

Lund, H. F., (ed.): *Industrial Pollution Control Handbook,* McGraw-Hill, New York, 1971.

Shreve, R. N.: *Chemical Process Industries,* 3d ed., McGraw-Hill, New York, 1967.

U.S. Environmental Protection Agency: *Development Document for Beet Sugar,* EPA-440/1-74-002-b, March 1974.

U.S. Environmental Protection Agency, *Development Document for Cane Sugar Refining,* EPA-440/1-74-002-c, March 1974.

U.S. Environmental Protection Agency: *Development Document for Red Meat Processing,* EPA-440/1-74-012-a, March 1974.

U.S. Environmental Protection Agency, *Development Document for the Apple, Citrus and Potato Processing Segment,* EPA-440/1-74-027-a, March 1974.

CHAPTER 29
MINING

Water uses in the mining and mineral processing industries may be divided into two parts. The first includes all uses of water and its treatment for removing the mineral from its surroundings, both in underground environments and also in surface stripping operations. The second deals with the processing of these recovered mineral ores. During these operations the beneficial mineral that has been mined is separated from valueless companion materials, normally by grinding and processing fine ore particles as a slurry. Water is used to prepare the slurry and to transport the mineral during this processing operation. Both aspects of water usage—in mining and in mineral processing—will be discussed here.

The most widespread use of water in underground mining operations is for dust control. Most automatic mining equipment includes an integral water spray system. Spray nozzles are supplied by a water hose coupling at the rear of the mining machine. The spray nozzles are set up to provide a water curtain around the area being mined when the mechanized drill, cutter, or continuous miner is in operation, thereby confining the dust at the face (Figure 29.1).

One of the major problems with such underground water systems is clogging of nozzles by scale and dirt particles in the water supply. Usually some type of strainer or filter immediately ahead of the mining equipment traps these particles to prevent their reaching the small nozzle orifices. Often, stabilization reagents are added to this water line to prevent scale formation if the water is hard or contains iron or manganese.

Another use for underground water is hydraulic mining. With this technique, water under high pressure is directed toward the face of the mine. The energy of the water stream breaks the mineral from the mine wall and washes it into a collection system. This type of mining may gain popularity as it becomes better understood. Currently its main use has been in the coal industry and to a limited extent the uranium industry. The cutting action of the water stream can be enhanced by coalescing the water into a very narrow stream, focusing its energy on the selected area of the mine face. Certain types of water-soluble polymers are used to give this cohesiveness to the water stream, thereby increasing mining efficiency.

UNDERGROUND WATER

In these examples of underground water use—dust control and hydraulic mining—it may be necessary to pump fresh water underground. However, in most instances there is already extensive underground seepage, and it is necessary to remove this from the mine to continue efficient operations. This water must often be pumped over appreciable distances from the mine floor to the surface for disposal. In many cases, it contains significant concentrations of dissolved solids leached from geologic formations. It may be acidic as a result of the types of minerals it has contacted. In the coal industry, reactions with forms of iron pyrites

FIG. 29.1 This continuous coal-mining machine is equipped with water spray nozzles to suppress dust at the working surface. *(Courtesy of Lee-Norse Company.)*

intimately mixed with the coal produce an acidic water, and treatment of acid mine drainage is a serious problem for this industry (Figure 29.2).

Bacterial action is believed to be responsible for the production of acid from the pyritic material, and the result is a water often at a pH below 3, containing up to 1000 mg/L Fe and with sulfate concentrations as high as 4000 mg/L. Not only is the cost of treatment a burden, but the by-product lime sludge is also a serious problem, often requiring large land areas for sludge impoundment.

It is often necessary to chemically treat underground water to prevent attack on pipes and pumps used to transport it to the surface. In many cases the water

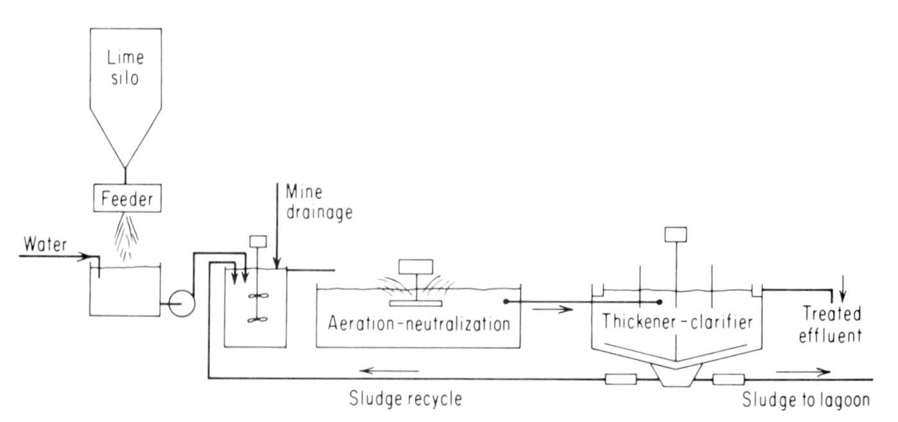

FIG. 29.2 Flow sheet of a treatment system to handle about 300 gal/min (1635 m^3/day) of coal mine drainage having an acidity of 1600 mg/L. Sludge recycle is essential to production of a dense, settleable precipitate.

has run through muddy areas and picked up a significant load of suspended solids, making it necessary to provide flocculation and coagulation of the solids prior to pumping the water to the surface.

Once at the surface, this drainage water is often added to the general water system for the mineral processing plant. Occasionally it requires further treatment to adjust pH or remove suspended solids precipitated upon exposure to the air. After pH adjustment with lime, most heavy metals precipitate. Settling basins are then required to clarify the water prior to discharge or recycling into the general plant water system.

Often in underground mining, a shaft or tunnel provides the approach to the ore body to be mined. As ore is withdrawn and the cavity (stope) becomes enlarged, it becomes necessary to regrade the floor of the stope at frequent intervals to make the mine face accessible to the miners and their machines and drills.

So, where the method of stope mining is practiced, tailings from the mineral processing plant are often used as backfill to provide the floor for the rising stope. These tailings compact in the stope, and water drains from beneath. This water is often the predominant portion of the mine drainage discussed earlier, which must be clarified before being pumped to the surface. Flocculant addition to the tailings often improves drainage and provides a more compact backfill in the stope area.

Very few minerals are mined and used directly without further treatment. The raw ore is either physically or chemically processed to remove the inert materials (gangue) from the desired mineral. This is normally done in a wet milling operation. In wet milling, the ore is crushed, slurried with water, and subjected to several techniques capable of separating the valued mineral from the gangue.

COAL PROCESSING

In the coal industry, various types of shale and clay are produced as a mixture with the coal. To increase the heating value of the coal and to reduce the hauling cost, a complex process of coal washing is normally used to reduce the total ash content. In this process the coal is graded to a certain size, usually less than 6 in, and then fed into a slurry bath in which the density of the media is closely controlled. The coal floats in this heavy media bath while the heavier rock sinks to the bottom.

Following this heavy media separation, all the floated material is again sized by vibrating screens for further purification. The smaller-sized fractions may be processed by shaking tables, hydrocyclones, or froth flotation. In each of these steps, coal is recovered and dried prior to shipment. The refuse is dewatered as much as possible by screens.

However, the final water effluent, after the larger materials have been removed, usually contains a significant concentration of very fine refuse in suspension. This might include some coal not recovered during the washing operations. Primarily, however, it contains sand and small pieces of rock and clay in a slurry of 3 to 15% solids. To close the coal preparation plant water system and minimize or eliminate effluent flow, it is necessary to dewater this refuse slurry as efficiently as possible. Normally a thickener is used to compact the solids into a mud of 30 to 40% solids by weight. The overflow from the thickener is normally of good enough quality to be reused in the coal washing plant.

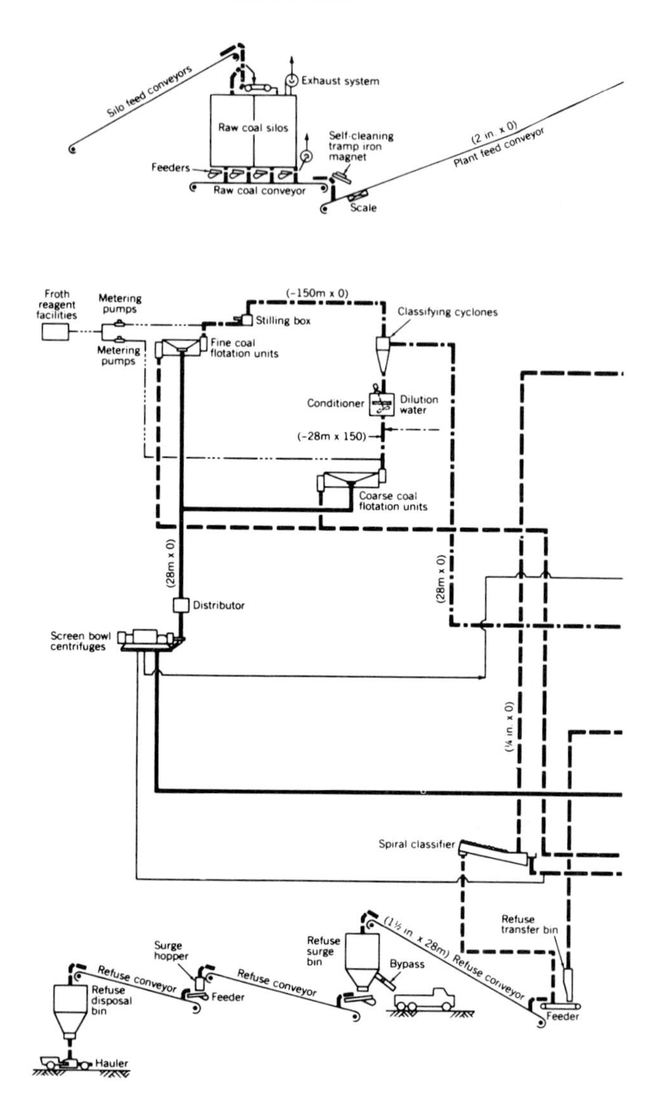

FIG. 29.3 Flow sheet of coal washing and drying plant. *(Reprinted from Coal Age, January 1976. Copyright McGraw-Hill, Inc.)*

FURTHER DEWATERING

The underflow slurry from the thickener is still pumpable but not yet suitable for final discharge. In the past, this slurry was discharged into tailings ponds and lagoons, and allowed to settle under its own gravity. Today, coal washing plants are closing their plant water systems entirely, requiring them to further dewater

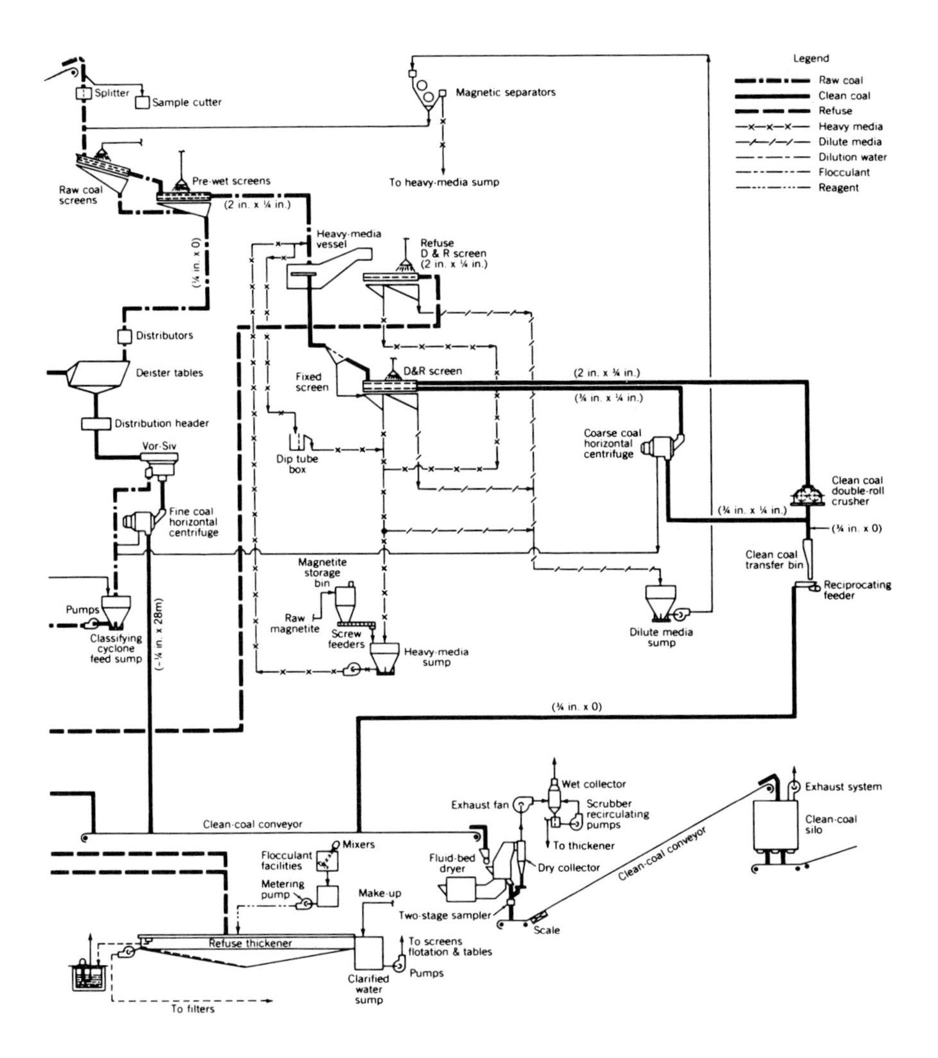

FIG. 29.3 (*Cont.*)

the refuse slurry to a dry, handleable form. The moisture must be reduced so that the refuse is dry enough to be trucked to a disposal site to be used as landfill.

Several techniques can accomplish this: vacuum filtration, centrifugation, and vibrating refuse screens. The key is to dewater the refuse slurry to the point where there is no subsequent water runoff when the dry solids are disposed of.

A complete water circuit for a modern coal washing plant is shown in Figure 29.3. Analyses of a typical makeup water and concentrated circuit waters are shown in Figure 29.4. There is a consistent increase in sodium, sulfate, and chloride. Lime is often added to the circuit to offset the development of acidity.

Identification of Analyses Tabulated Below:

A. Plant A--makeup
B. "A" recirculated water
C. Plant B--makeup
D. "B" recirculated water
E. Plant C--makeup
F. "C" recirculated water

Constituent	As	A	B	C	D	E	F
Calcium	$CaCO_3$	74	350	60	44	490	950
Magnesium	"	31	190	22	14	140	340
Sodium	"	644	1368	260	528	79	161
Total Electrolyte	$CaCO_3$	749	1908	342	586	709	1451
Bicarbonate	$CaCO_3$	170	18	88	118	220	0
Carbonate	"	0	0	0	0	0	0
Hydroxyl	"	0	0	0	0	0	0
Sulfate	"	9	290	230	410	450	1400
Chloride	"	570	1600	24	58	39	51
Nitrate	"						
M Alk.	$CaCO_3$	170	18	88	118	220	0
P Alk.	"	0	0	0	0	0	0
Carbon Dioxide							
pH		7.8	7.3	8.0	7.8	7.6	4.0
Silica	SiO_2	16	2	3	4	15	23
Iron	Fe	0.2	1.5	5	28	0.1	0.7
Turbidity							
TDS (Conductivity)		1500	3800	650	1200	1200	2200
Color							

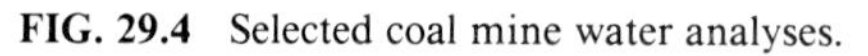

FIG. 29.4 Selected coal mine water analyses.

METAL-CONTAINING MINERALS

In the base metal industries, such as copper, lead-zinc, and iron ore processing, it is necessary to grind the mineral to such a fine size as to produce discrete particles of metal-rich minerals and quartz, feldspar, and other worthless materials. In the copper, lead, and zinc industries, the liberated metal-rich particle normally consists of a sulfide mineral, which is then amenable to separation and recovery by froth flotation techniques. In the iron ore or taconite industry, the iron is present as the mineral magnetite, $FeO \cdot Fe_2O_3$, which lends itself to separation by magnetic techniques. To recover minerals by either flotation or magnetic separation, the ore is ground in a wet condition to improve efficiency. Large volumes of water are used to slurry the dry ore.

During the hard rock grinding process, steel balls and rods are used to pulverize the mineral ore (Figure 29.5). In these mills, the grinding media are consumed at the rate of about 1 lb steel per ton of ore. This metal loss is due to physical abrasion and chemical corrosion. Chemical inhibitors can effectively reduce the rate of corrosion and subsequent wear of the steel balls and rods.

After separating the desirable mineral from the tailings, the final slurry is sent to a tailings thickener just as in the coal washing operation. Again, the tailings slurry is thickened in the thickener and the clarified water returned to the processing plant for reuse. The underflow slurry may be disposed of in large open tailings ponds where natural evaporation and seepage occur. If there is any significant breakout of free water in the tailings pond, this water is either discharged to a stream or recycled to the main plant water system.

The flow sheet for a copper mining operation is shown in Figure 29.6, including both the crushing and concentration steps.

In all thickening operations, the large demand for clear water and the size limitations of thickeners and settling ponds make it necessary to use synthetic flocculants and coagulants to aid in the settling process. These synthetic separation aids enable the plant to obtain all the desired recycled water necessary along with increased compaction and dewatering of the refuse solids. The size of thickeners and settling ponds is clearly illustrated in Figure 29.7, an aerial view of a copper mine in a mountainous area of Arizona.

However, no matter how efficient the thickening operation may be, a certain amount of process water always is lost to tailings with the slurry solids. Consequently it is necessary to add a quantity of fresh makeup water to maintain the balance of the plant water system.

FIG. 29.5 Fine copper ore is ground in these ball mills, each 16 ft 6 in diameter by 19 ft long. Each mill is driven by a 3000-hp motor. *(Courtesy of Duval Corporation, Stearns-Roger, Incorporated, and Ray Manley Photography.)*

In certain industries, such as copper, lead, and zinc where the sulfide minerals are floated at high pH, the addition of fresh makeup water containing significant calcium and magnesium hardness can result in unstable water (a low Ryznar index, high Langelier index). The refuse slurry which is clarified in the thickener contains sufficient alkalinity to react with the fresh calcium and magnesium added with the makeup water. Therefore, it is advantageous to add the makeup to the clarified recycled water in a vessel where precipitation of carbonates and hydroxides will not pose a major scaling problem. It has been found that addition of makeup directly to the thickener eliminates most problems.

However, it is also necessary to stabilize this combined water prior to sending it back to the plant. If this is not done, a significant buildup of scale can form in storage tanks as well as pipes and pump parts, causing problems and increasing plant maintenance. Analysis of makeup and recirculated process water from several copper mills is shown in Table 29.1.

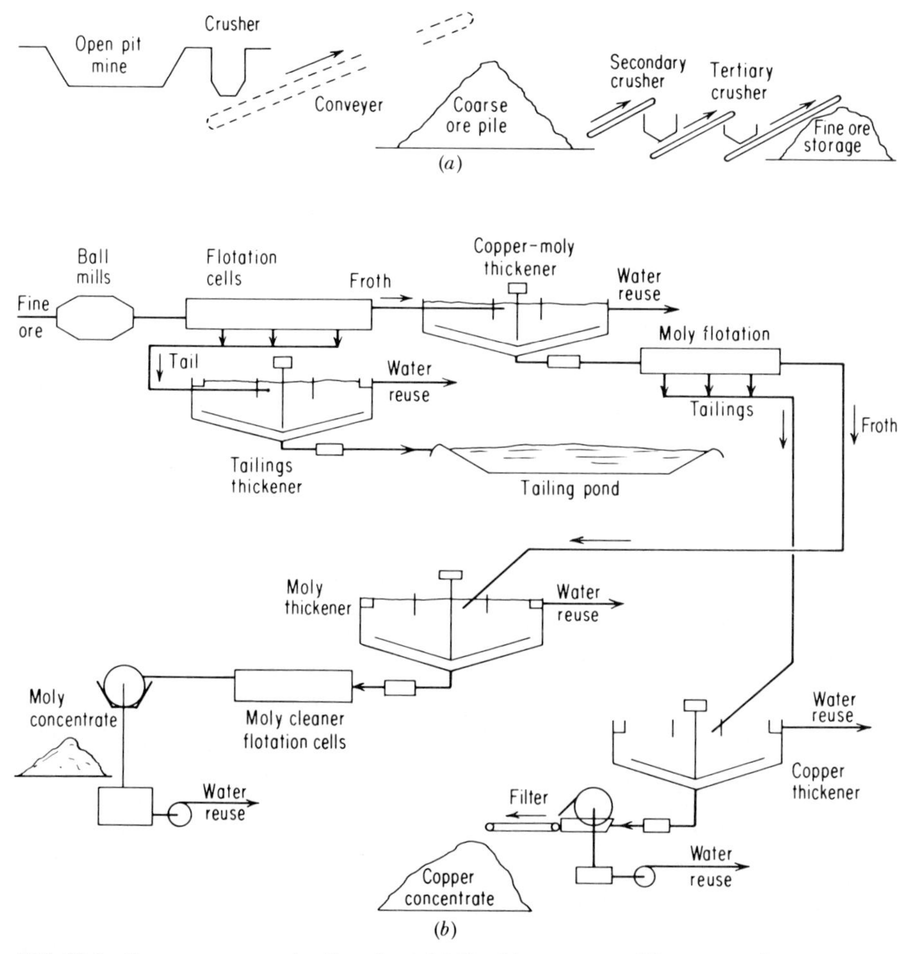

FIG. 29.6 Copper ore processing flow sheet. (*a*) Crushing sequence; (*b*) concentrating sequence.

FIG. 29.7 Aerial view of Duval copper-mining operation, Sahuarita, Arizona. *(Courtesy of Duval Corporation, Stearns-Roger, Incorporated, and Ray Manley Photography.)*

TABLE 29.1 Water Quality Comparison of Fresh Makeup Water and Recirculated Mill Process Water for Several Western Copper Concentrators

(Concentrations in milligrams per liter)

	Mill	Ca	P	M	O	pH	CO_3	HCO_3	TDS
A.	Makeup	65	4	136	0	8.7	8	128	300
	Clarified	1410	280	360	200	11.1	160	0	3000
B.	Makeup	60	20	115	0	8.4	40	75	270
	Clarified	1890	560	630	490	11.7	140	0	2000
C.	Makeup	—	12	60	0	8.1	24	36	580
	Clarified	—	214	254	164	11.5	90	0	1300

PHOSPHATE MINING

Phosphorus-bearing earth is mined at two major locations in the United States, Florida and the western mountain states, with minor production in Arkansas, Tennessee, and North Carolina. Ninety percent of the production is located in about 2000 square miles of Florida.

The phosphate industry ranks with coal, iron, and copper as a major tonnage processor of bulk material, surpassed only by stone, sand, and gravel. The ore is mined by both surface and underground methods, but the former accounts for almost 98% of annual tonnage. The overburden from surface mining exceeds 200 million tons annually, for a crude oil production of about 100 million tons. About two-thirds of the crude ore is discarded in beneficiation.

The overburden is essentially silica sand overgrown with vegetation. The underlying phosphorus-bearing matrix has a variable mineral composition, being a mixture of carbonate-fluorapatite (20 to 25%); quartz (30 to 35%); clays, chiefly montmorillonite (25 to 35%); and the balance feldspar, dolomite, and heavy minerals. Huge draglines separate the overburden from the matrix, and the matrix is then slurried by hydraulic guns and transported to the processing plant at about 30 to 35% solids.

At the beneficiation plant, the maximum phosphate values are extracted from the matrix. The processes are relatively simple, yet no two plants are exactly alike because of differences in the matrix screen analysis and the proportions of phosphate, clay, and sand in the matrix. The unit operations always include washing, feed preparation, and concentration (Figure 29.8). In the washing operation, particle size classifiers produce a product of about 1 in to +14 mesh with a BPL (bone phosphate of lime) value of 65 to 70%. This portion of the matrix is known as pebble product. The −14 mesh fraction moves to the second stage where fine clays are removed so that they do not interfere with the effectiveness of downstream chemical conditioning agents. This "desliming" process is accomplished by hydraulic sizing devices, such as hydrocyclones. The slimy, fine clay fraction is discharged to slime ponds. Because the mineral composition of the phosphate-bearing matrix is variable, the slime composition is also variable, usually containing 10 to 20% P_2O_5 and high concentrations of silica, iron, lime, and alumina.

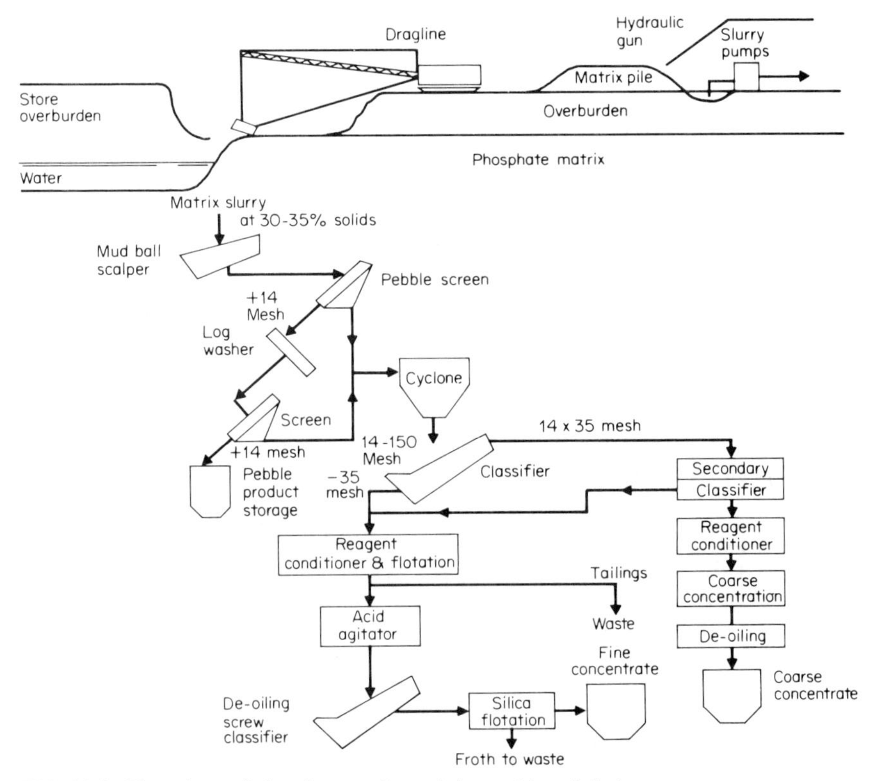

FIG. 29.8 Flow sheet of phosphate surface mining and beneficiation.

After desliming, the slurry is fractionated into 14 × 35 mesh and 35 × 150 mesh portions for further concentration by removal of silica. This is accomplished through the use of spirals, belt flotation, or froth flotation. Froth flotation is the most common practice and consists of:

1. Floating the phosphate with a tall oil fatty acid at an alkaline pH (float product: rougher concentrate; sink product: silica tailings).
2. De-oiling the rougher concentrate with sulfuric acid followed by rinsing.
3. Floating residual silica with an amine at a neutral or slightly alkaline pH (float product: silica tailings; sink product: final concentrate); the final concentrate generally has a BPL value of 70 to 74%.

Wastes generated by the mining and beneficiation steps include the overburden, clay slimes, and silica tailings. The mines usually reclaim about 75% of the mined acreage for citrus groves, timber stands, wildlife preserves, pasture land, and recreational use. The balance of the land is used for slime and tailings storage and for recirculation water reservoirs. Silica tailings can be used for slime pond dam construction; the clay slimes create a serious problem in that they cannot compact beyond about 30% solids by natural gravity sedimentation. In the original matrix, the clays are present at >60% solids, but swell as they absorb water during the slurrying process.

MINERAL LEACHING AND DISSOLVING

Other mineral industries use aqueous solutions to leach or dissolve the desired mineral from its ore. Included among these industries are uranium, bauxite for the recovery of alumina, soda ash for the recovery of sodium carbonate, potash for the recovery of potassium, and phosphate rock which processes naturally occurring phosphate ore to phosphoric acid for fertilizer intermediates. Aqueous solutions containing various leaching materials (acids, alkalies, or brines) dissolve the valuable mineral from the undesirable constituents in the ore. In most operations of this type, a series of countercurrent decantation (CCD) thickeners are used to recover as much soluble mineral as possible from the mud refuse. Flocculants are required in this process to compact the mud levels to the greatest extent and to yield as clear a pregnant liquor as possible for ultimate mineral recovery.

In addition to chemical leaching processes, there are three somewhat related mineral recovery operations: bacterial leaching, hot-water melting, and water dissolution.

Almost 20% of the copper recovered from tailings is put into solution by bacterial action, and the extract is then concentrated by conventional methods to recover the copper. As in any biological digestion process (Chapter 23), it is important to maintain a rather uniform aqueous environment, as the bacterial activity is upset by changes in temperature, pH, and nutrient levels; and the rate of activity depends on these factors and the food level, which in turn determine the time the bacteria must be detained in contact with the copper tailings to effectively dissolve the copper from the mineral substrate. It is theorized that many metallic compounds have been naturally concentrated in their ore bodies by biological processes, so it is likely that even more applications of bacterial leaching will materialize in the future.

The largest application of hot-water melting is the mining of native sulfur in the United States by the Frasch process (Figure 29.9). Deposits of pure sulfur are generally located 500 to 2000 ft (150 to 600 m) below the surface of the earth, often off-shore below the sea. Six-inch shafts are bored into the deposit, and a triple, concentric pipe system is inserted; the center pipe delivers compressed air to the deposit; the outside annulus delivers water heated to 340°F (171°C), above the melting point of sulfur; and the inner annulus collects the molten sulfur-water mixture, which is lifted to the surface in slugs by the bubbling compressed air.

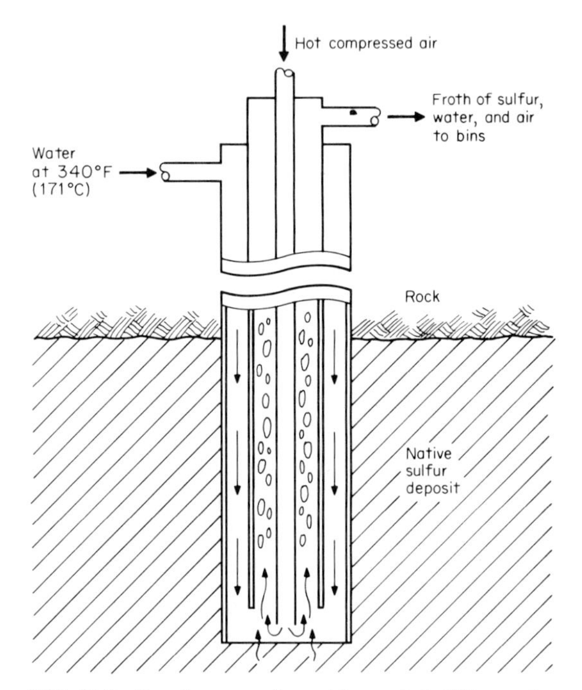

FIG. 29.9 Frasch process for melting and air-lifting molten sulfur from deposit to surface.

The mixture is fed into a bin, where the water drains off and the sulfur solidifies. The water required for the Frasch process is about 2000 gal per ton (8.34 m^3/t) of sulfur mined. If fresh water is used, it is usually lime-softened and deaerated to protect the extensive piping from scale and corrosion. If seawater is used, it is deaerated and stabilized. Since the water is heated to 340°F (171°C), a large boiler installation is required for the energy needs.

Water dissolution is used to mine both salt and sodium carbonate–bicarbonate mixtures. In both cases, the final product is usually a crystalline product, so the amount of water used must be minimized to reduce the energy cost of crystallization.

In several of these industries, process steam is required to supply the heat necessary for maximum leaching. Operations of this type require an effective boiler water treatment program to maintain maximum efficiencies. In those metal industries in which smelters are used to reduce the ore to its elemental form, waste heat boilers are used to convert heat from smelter gases into steam. These waste heat units gather additional heat energy by cooling the sides of the furnaces and other liquid metal transferring vessels. The steam generated by the process heat is then utilized for power generation as well as other process purposes.

Associated also with these heavily process-oriented industries are significant cooling water applications. For instance, in many of the metal sulfide smelting operations, acid plants have been built to recover the sulfur dioxide from waste gases. These recovery plants yield sulfuric acid, which can in turn be used to leach other types of metal mineral ores. Cooling water is a major problem for these satellite plants because of the salinity of the limited water sources in mining areas and the atmospheric contamination.

In summary, the mineral processing industry is continuing to close as many of its water systems as possible. To a great extent this industry has grown up with the philosophy of water recovery, having been forced for many years to reuse water because of its large water requirements for milling and mineral processing. Many of these plants are in water-short areas of the country. Consequently most mineral processing plants have always had thickeners to clarify their plant wastewaters (Figure 29.7). What is planned is to further dewater the thickener underflow slurries to produce dry, handleable solids as required for landfill, completely closing the system. Large tailings ponds are still used in the western states where enough land is available for long-term sedimentation. In these tailings ponds, the natural process of evaporation usually produces a dry, disposable refuse material.

The mineral processing industry needs all the clear water it can obtain for use in the processing of its ores. It will continue to improve its techniques for the recycling of its process waters. The primary objective is clarity. It is necessary to remove suspended solids to convert the wastewater to a usable quality for further mineral slurrying. Often pH control is necessary. Consequently, the addition of acid or alkali to the recycled water is necessary to bring the water into balance for scale and corrosion control. In most cases very little is done to reduce dissolved minerals in the effluent. The addition of fresh makeup water at a continuous rate momentarily dilutes the buildup of dissolved minerals. But in a closed system, there is a large increase in dissolved solids in the circuit (Table 29.1), and the only point of discharge of these solids is in the water lost with the final tailings discharge to landfill. The main water treatment process in the mineral processing industry is solids/liquids separation in the tailings system (Figure 29.10). The

FIG. 29.10 Tailings clarifier-thickeners at a large western copper concentrator.

effective use of synthetic polymers to improve solids/liquids separation efficiency enables the mineral processing industry to conserve water and minimize the strain on the environment.

SLURRY CONVEYING

One water-related operation in the mining industry that is advancing rapidly is hydraulic transport of minerals. Several pipelines are already in operation, the most notable being the Black Mesa coal pipeline feeding an electric generating station in Nevada. There are many indications that additional slurry pipelines will be needed to transport minerals such as coal and iron ore. The major concern in such slurry pipelines is that water is removed from the environment at one location and transported to another. Some slurry pipelines proposed for the more arid regions of Montana and Wyoming have encountered public opposition because they represent an export of water from the region. Perhaps this may be resolved by intermittent use of the pipeline to return water and refuse to their original source.

The pipeline system lends itself to some interesting water treatment problems. Of course, a chemical that could be added to the slurry while it is in the pipeline to increase the pumping efficiency could be of importance to the operation. Various materials which exhibit friction-reducing tendencies are being evaluated. In addition, there are usually periods when the pipeline is idle, normally filled with water. Under such conditions, corrosion begins to take place under the layer of solids on the bottom. Therefore, it is necessary to chemically treat these water slugs used to fill the gaps in the line between various quantities of the mineral slurries, to prevent this corrosive action. Because of the huge volumes of water involved, economics of this once-through system are of great concern.

CHAPTER 30

PULP AND PAPER INDUSTRY

Primitive people over countless ages communicated with each other or expressed themselves by audible sounds and visible symbols carved in wood or stone, impressed in clay, or painted on the walls of caves. As society progressed, these sounds and signs developed into language; symbols became alphabets, and records of important events could then be written on clay tablets. The Egyptians simplified record-keeping with the introduction of papyrus sheets, formed from the reeds of the papyrus plant, about 3000 B.C. The word paper is derived from papyrus.

During later periods of civilization, scribes kept handwritten records on parchment and hand-formed paper, chiefly for the ruling classes. The hope of written communication for the masses first came with the development of the printing press about 500 years ago. This created a demand for paper, so varieties of fibers and process methods were vigorously investigated for use in paper manufacture. In the early days of the American colonies, the fibers were obtained from rags; but by the nineteenth century, the demand for paper had increased so much that the quantity of rags available was insufficient to supply the need, and wood fibers began to find use in paper pulp.

The pulp and paper industry has since grown to produce not only paper for records, documents, and books, but also heavier grades for packaging, and plain and corrugated liner board for shipping containers, such as drums and cartons. Despite the relative light weight of the finished product, pulp and paper constitute one of the major tonnage products in the United States; it may soon challenge the output of the steel industry in annual production. The production of the major varieties of paper and board are shown in Table 30.1.

Today paper is almost entirely derived from wood, and since only about half of the weight of timber brought to the pulp mill is cellulose, researchers over the years have actively sought uses for the remaining 30 to 50% released during chemical pulping. They have developed a variety of by-products which include turpentine for the paint and coatings industry, tall oil for the manufacture of chemical intermediates, lignosulfonates as surface-active agents and dispersants, and such other products as yeast, vanillin, acetic acid, activated carbon, and alcohol. Table 30.2 shows the major categories of the components of wood.

Subsidiary operations in many pulp and paper mills include the manufacture of plywood, particle board, and chipboard, as well as the production of alpha cellulose, which is a dissolving pulp used in the manufacture of cellulose acetate.

Probably no industry has done as much in reclaiming waste products as the pulp and paper industry. Scrap paper is collected, sorted into grades for repro-

TABLE 30.1 Paper and Market Pulp Production, in 1000 Tons/Year Units

	1980	1984	1987 (est.)
A. Paper			
1. Newsprint	4,836	5,847	5,792
2. Printing and writing	16,798	19,770	21,953
3. Packaging and converting	6,072	5,991	5,769
4. Tissue	4,954	5,243	5,620
Total paper	32,659	36,851	39,134
B. Paperboard			
1. Unbleached kraft	15,967	17,790	18,362
2. Recycled board	8,617	8,702	8,989
3. Semichemical	4,946	5,219	5,393
Total paperboard	33,709	35,898	37,183
C. Construction paper and board	4,347	2,877	2,854
D. Machine-dried pulp	7,218	7,550	8,000

TABLE 30.2 Components of Wood Substance

Total wood substance	100%
Steam or solvent extractable	5%
Extractive free wood (including 0.5% inorganic)	95%
Mild oxidation yields soluble lignin	25%
Holocellulose (total polysaccharide)	70%
Dilute alkali extraction yields hemicelluloses	20%
Net wood cellulose	50%

cessing, and converted into new grades of tissue, book stock, container board, and other useful forms. Approximately 20% of the output of the paper industry is reused by secondary fibers processing plants.

WATER: A BASIC RAW MATERIAL

The manufacture of pulp and paper requires large volumes of water. Since its earliest days, the industry has located almost exclusively along major rivers. The first mills used water not only to make pulp and paper, but also for hydraulic power, by damming the stream to produce the head needed to drive water wheels which operated the grinding stones to convert wood to pulp. Many modern mills are located at the site of these early mills and continue to use hydraulic power for operation of grinders or of hydroelectric turbines.

The water required by a modern pulp/paper mill varies considerably with the pulping process, the availability of water, the bleaching sequence, and restrictions on wastewater discharge. Table 30.3 shows the water requirements for older and newer mills, based on the type of production.

The pulp and paper industry is a large user of water because the pulp is washed with water at several points in the process, and water is used to convey the pulp fibers from their initial production in the pulp mill through various refining operations, and finally to the paper machine, where it may be introduced to the forming wire at a slurry concentration which is 99% water and only 1% fiber (called 1% consistency in paper mill terminology). Whereas older mills used as much as 50,000 gal per ton of finished paper product, modern mills have closed up their systems to reduce the amount of water required and therefore the volume of waste

TABLE 30.3 Net Water Use for the Manufacture of Pulp and Paper Products

	Typical gal/ton*	New mills, gal/ton*
Pulp manufacturing process:		
Unbleached kraft	15,000–40,000	20,000
Kraft bleaching	15,000–35,000	20,000
Unbleached sulfite	15,000–50,000	25,000
Sulfite bleaching	30,000–50,000	40,000
Semichemical	8,000–40,000	10,000
Deinked	20,000–35,000	25,000
Groundwood	3,000–48,000	4,000
Soda pulp	60,000–80,000	65,000
Paper manufacture:		
Fine paper	8,000–40,000	10,000
Book or publication grades	10,000–35,000	12,000
Tissue	7,000–45,000	15,000
Kraft papers	2,000–10,000	5,000
Paperboard	2,000–15,000	8,000

* Gal/ton × 0.0042 = m^3/t.

to be treated. In a modern unbleached pulp/paper mill, producing linerboard, the consumption may be as low as 10,000 gal/ton (42 m^3/t), and in a bleached pulp/paper mill the requirement may be approximately 15 to 20,000 gal/ton (63 to 83 m^3/t). So, in a mill producing about 1000 tons/day (907 t/day), which is typical of southeastern U.S. kraft mills, the water treatment plant may be required to produce about 20 mgd (7570 m^3/day), of mill water for process and boiler makeup. Under certain circumstances, some smaller mills producing grades of product not harmed by high salinity have been able to close up and reduce fresh-water makeup to less than 5000 gal/ton (21 m^3/t).

Of the variety of pulping operations in use throughout the world, most fall into one of three categories: mechanical pulp, chemical pulp, or secondary fiber (reclaimed paper).

For most types of timber used to supply the pulp mill, the logs are cut to a convenient length and debarked in a device such as the barking drum shown in Figure 30.1. Since the bark is a significant portion of the raw material, the water used in the debarking operation is recycled and the bark removed, dewatered, pressed, and used as fuel for steam generation in specially designed bark boilers. Excess water from this operation is often high in BOD, because of the wood sugar extracted in the debarking process, and high in coliform organisms from the exposure of the tree to animal life both in the forest and in the wood yard.

FIG. 30.1 Bark is removed from cut wood in this rotating debarking drum.

GROUNDWOOD PULP

In the production of mechanical pulp, logs are loaded into the magazine of a grinder (Figure 30.2) in which they are pressed against a grinding wheel. Abrasion separates the bundles of fibers into individual strands, and except for a relatively small percentage of organic matter extracted during the grinding operation, the

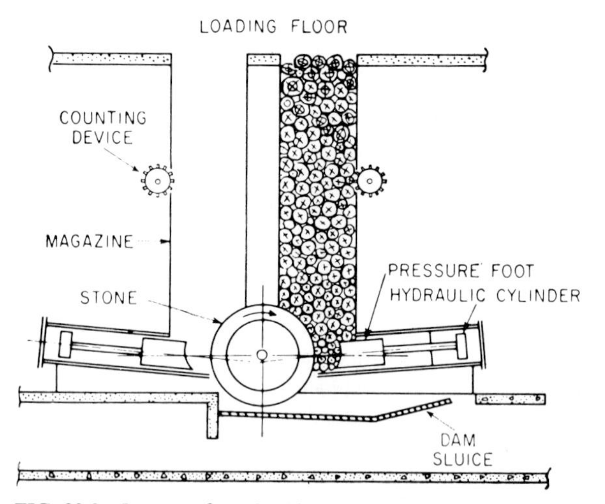

FIG. 30.2 Logs are forced against a grinding stone to produce groundwood pulp. *(From Pulp and Paper Science and Technology, vol. 1: Pulp, McGraw-Hill, 1962)*

lignin which holds the fibers together remains in the finished pulp, so that the pulp yield is quite high. The yield varies with the type of wood being ground but is generally on the order of 90%; that is, 90% of the weight of wood fed to the grinder becomes pulp available for paper manufacture.

Although the organic material extracted from the wood fiber during grinding is a small percentage of the fiber weight, it is a significant addition to the water; so the water wasted from the groundwood pulping operation is quite high in organic matter.

Paper produced from groundwood pulp is weak and has poor aging properties. For that reason it is used for manufacture of newsprint, where strength and aging properties are not critical factors. Even with newsprint, to provide the strength needed on a modern printing press, some chemical pulp, usually about 15%, is blended into the groundwood pulp to produce a satisfactory sheet.

THERMOMECHANICAL PULPING

Thermomechanical pulping (TMP) is a new process producing a mechanical pulp superior to groundwood in strength. Used in newsprint, it permits a reduction of chemical pulp in the furnish, at the same time producing a stronger sheet. But its major contribution to the industry is its ability to reduce chemical and water usage, with carrythrough benefits in waste treatment, while producing a pulp comparable to chemical pulp in many respects.

In the TMP process (Figure 30.3), wood chips are washed with recycled water,

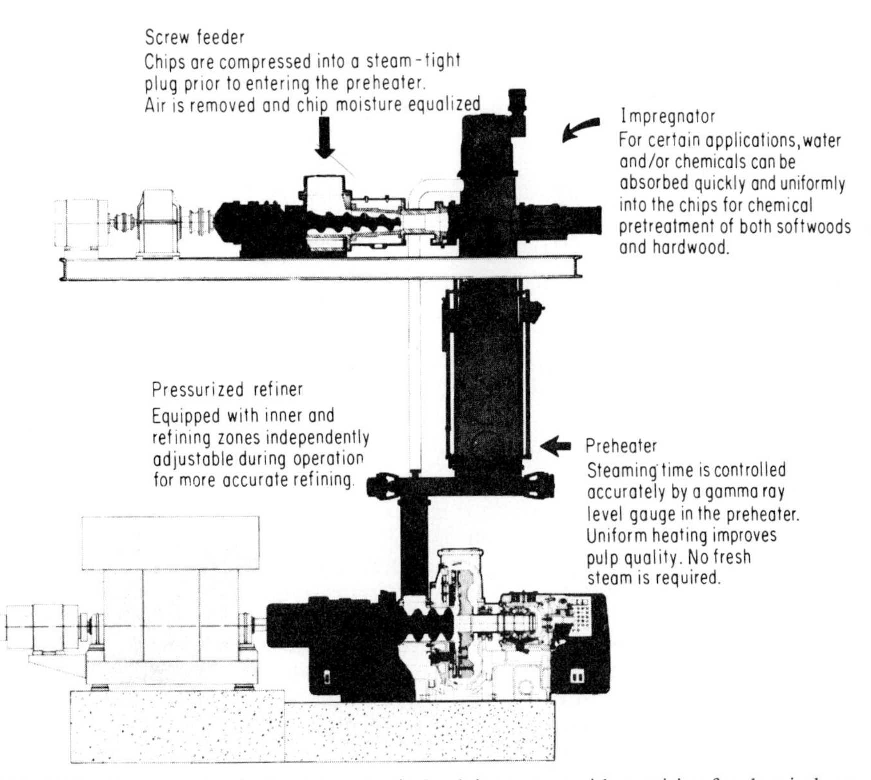
Screw feeder
Chips are compressed into a steam-tight plug prior to entering the preheater. Air is removed and chip moisture equalized
Impregnator
For certain applications, water and/or chemicals can be absorbed quickly and uniformly into the chips for chemical pretreatment of both softwoods and hardwood.
Pressurized refiner
Equipped with inner and refining zones independently adjustable during operation for more accurate refining.
Preheater
Steaming time is controlled accurately by a gamma ray level gauge in the preheater. Uniform heating improves pulp quality. No fresh steam is required.

FIG. 30.3 Components of a thermomechanical pulping system with provision for chemical pretreatment. *(Courtesy of American Defibrator Inc.)*

which may be white water or filtrate containing pulping chemical residues. This is macerated in a screw press to a homogeneous slush, and passed continuously through a steam-heated digester. This softens the fibers and permits them to separate later at boundaries which are relatively free of lignin. This improves refining, conducted in three stages following the digester. The refined pulp is screened, cleaned, and adjusted to the required consistency for bleaching. As with groundwood, hydrosulfite and hydrogen peroxide are the bleaching chemicals.

CHEMICAL PULPING

To improve the qualities of pulp fiber, early research workers developed chemical processes to dissolve the lignin and other organic materials holding the fibers together, releasing the fibers without extensive mechanical working. There is a continual development of new modifications to these basic chemical processes to meet modern needs for high-quality product and minimum wastewater contamination.

The basic chemical processes in use today are categorized as acid, neutral, or alkaline pulp processes, as shown by Table 30.4.

TABLE 30.4 Chemical Pulping Processes

pH environment	Acid	Neutral	Alkaline
Anionic species	SO_2, HSO_3^-	HSO_3^-, SO_3^{2-}, HCO_3^-, and CO_3^{2-}	HS^-, S^{2-}, and OH^-
Cationic species and process designation	(1) H^+—acid sulfite (2) Na^+—sodium base (3) NH_4^+—ammonium base (4) Ca^{2+}—calcium base (5) Mg^{2+}—magnesium base	Na^+ (1) Neutral sulfite (2) Neutral sulfite—semichemical (3) Chemiground*	Na^+ (1) Kraft (2) Kraft—semichemical*

* In these processes, liquor is used to soften the wood before grinding.

In all of the chemical pulping processes, a chemical solution is prepared and fed to a digester, a vessel in which it is mixed with wood that has been cut into chips to permit the liquor to penetrate effectively and produce a uniform pulp. The mixture is then cooked for a specified period at the optimum temperature for the particular process and the type of wood being pulped. Most digesters operate on a batch basis, but there is increasing use of continuous digestion processes.

At the conclusion of the digestion period, the contents of the digester are discharged into a blow tank. Because the liquor is at a high temperature, steam and volatile components from the wood flash off. These are condensed, and organic by-products may be produced from the condensed liquid. The quenching of the vapor provides heat for process water and also controls air pollution by concentrating and collecting sulfur compounds for burning.

RECOVERING PROCESS INGREDIENTS

The pulp from the blow tank is washed to remove the spent pulping liquor; the pulp then goes to further processing and the washwater containing the spent pulping chemicals is usually reclaimed. Because the liquor contains organic matter extracted from the wood, when concentrated by evaporation the organic material reaches a concentration level that permits the liquor to be sprayed into a furnace for burning, the combustion of organic matter providing the Btu's for evaporation of the liquor to dryness. The chemicals in the liquor become molten, and the molten products (smelt) collect at the bottom of the furnace for recovery. This recovery procedure does not apply to calcium-base sulfite cooking liquors. Since there is no practical way to recover the calcium-base liquor, the calcium-base sulfite process has practically disappeared. Magnesium-base sulfite liquors are recovered, but not exactly as described above.

Chemical recovery has been in operation for many years in the kraft process, because the process would not be economical without such recovery. The chemicals recovered are returned to the process through a chemical recovery operation which will be described later. The chemicals recovered from the NSSC process (neutral sulfite—semichemical) cannot be used directly again in the process, but can be used as makeup to a kraft mill. Because of this, some pulp mills have both kraft and a semichemical process; if the pulp mill is strictly an NSSC mill, it may recover its chemicals in a dry bead form, providing a product which can be supplied to a nearby kraft mill.

Because these pulping operations use sulfur-containing chemicals, many of which have a high odor level, sophisticated processing equipment is required to avoid leakage, which could result in air pollution. These sulfur compounds have a high oxygen demand, so it is equally important to avoid leakage of pulping liquors to the sewer, where they would impose a load on the waste treatment plant. Not only do these liquors have a high chemical oxygen demand (COD), but they also contain the organics extracted from the wood, and are therefore high in BOD.

The yield from chemical pulping is considerably less than from a groundwood operation, where the organic remains attached to the cellulose fiber. The extraction of lignin and hemicellulose is so effective in chemical pulping that the kraft process has a yield on the order of less than 50% compared with a semichemical pulping yield of about 70% and the groundwood pulp yield of about 90%.

THE KRAFT PROCESS

As an example of the chemical pulping operation, the kraft process, which produces about 75% of the total pulp output of the United States, is illustrated by Figure 30.4. In this process, chips and white liquor are measured into a digester, and the mixture is then brought up to about 350°F (177°C) either by direct steam injection or by recirculation of liquor through a steam-heated exchanger on the side of the digester. At the completion of the cooking period, which may take 2½ to 3 h, the contents of the digester are discharged to the blow tank, the vapors from which are used for heating water. In the process turpentine may be recovered, a significant by-product in pulping softwood. Noncondensible gases, such as H_2S and SO_2, are released for separate treatment to control sulfurous emissions.

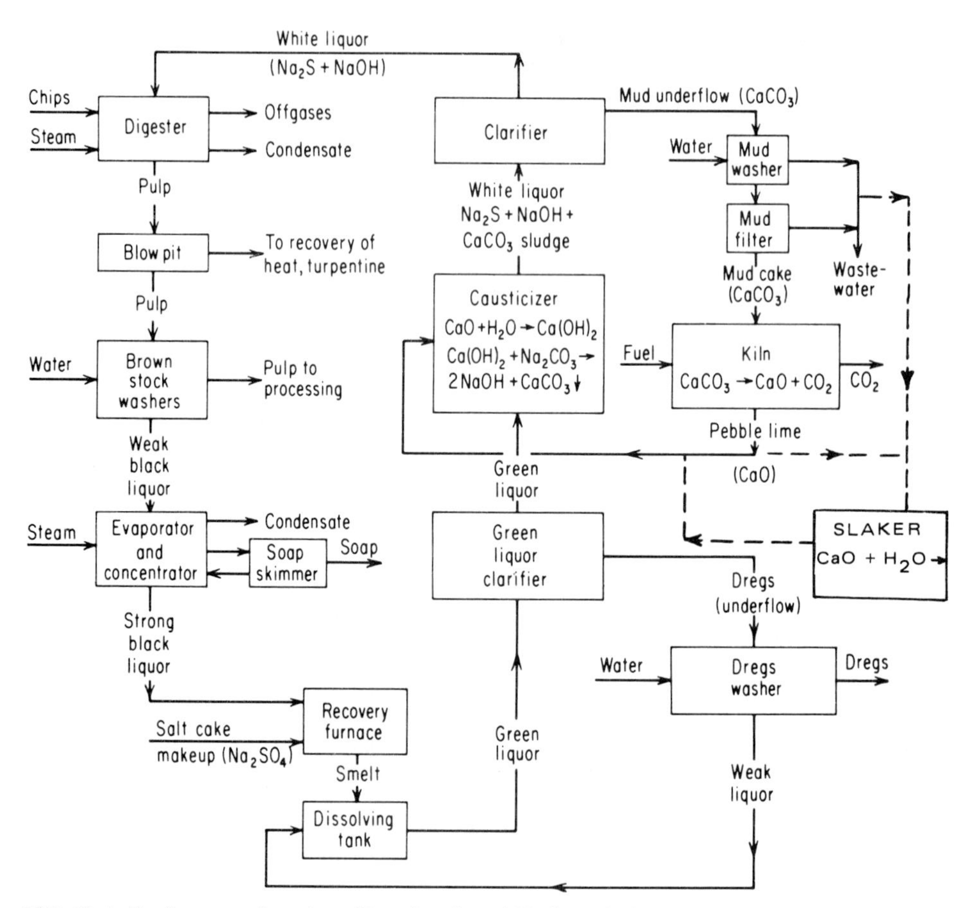

FIG. 30.4 Kraft process flow sheet. Note that the pebble lime (CaO) may go to a slaker or may be slaked in the causticizer.

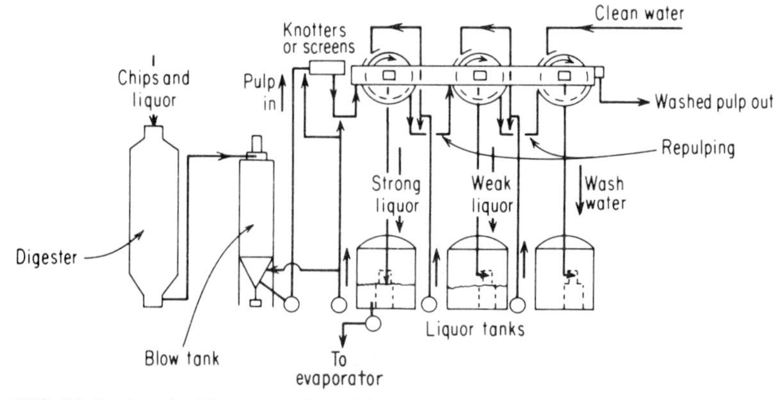

FIG. 30.5 A typical brown stock washing system.

The pulp is sent to the brown stock washers to remove the spent pulping liquor which may lower pulp quality. There are usually three stages of washing (Figure 30.5) with water flowing countercurrent to the pulp, picking up pulping chemical to become a weak black liquor containing approximately 10 to 12% total solids. This liquor is concentrated to over 55% total solids in an evaporator producing a strong black liquor so high in organic solids that it will burn when sprayed into the recovery furnace. Salt cake (the industry term for Na_2SO_4) is also added to the recovery furnace to provide makeup to the sulfur-bearing liquor to account for losses at various points in the circuit. In the recovery furnace, the sulfate is reduced to sulfide and some of the caustic is converted to carbonate by contact with the CO_2-containing gases produced from combustion of the organic matter. These materials become molten at the furnace temperatures, and the molten salts, called smelt, drain down the furnace tubes and collect at the bottom of the furnace. They are tapped off into the smelt dissolving tank.

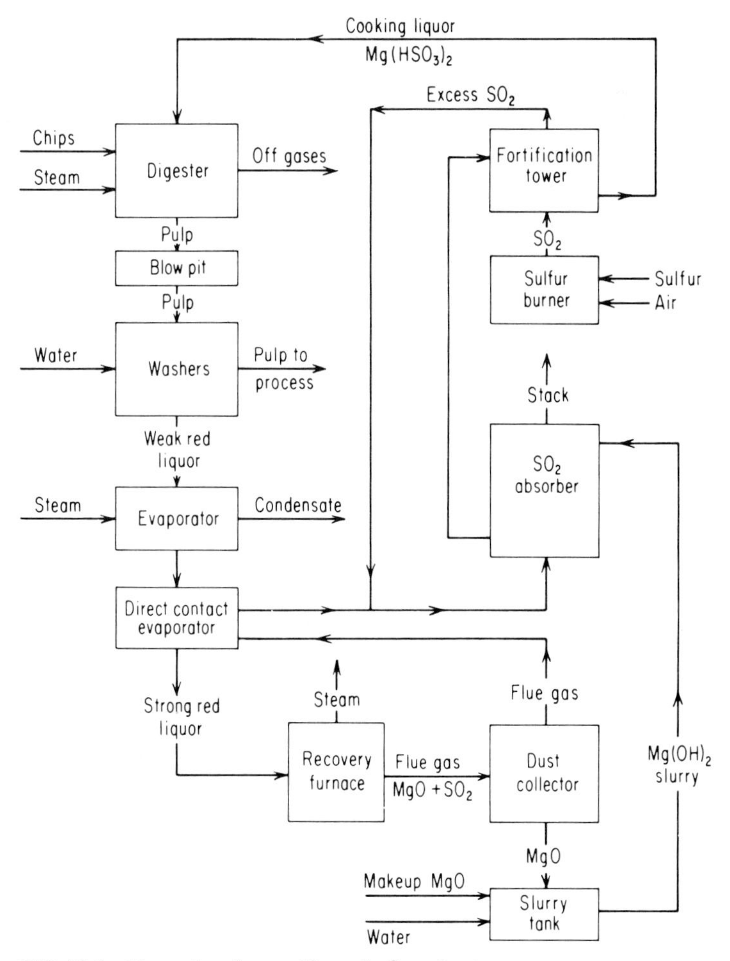

FIG. 30.6 Magnesium-base sulfite pulp flow sheet.

Recovered weak liquor quenches and dissolves the smelt, producing green liquor that must be clarified before it can be regenerated. The dregs from the clarifier contain valuable chemical, so they are washed and the weak liquor returned to the dissolving tank. The green liquor, which contains sodium carbonate, is then causticized by the addition of lime, converting the sodium carbonate to sodium hydroxide and producing calcium carbonate precipitate. The product of the liming operation is white liquor, which must be clarified before it can be returned to the digester. The mud produced from the clarifier is almost entirely calcium carbonate. This is washed and filtered, and the cake is fed to a kiln, which burns the calcium carbonate to calcium oxide for recycle.

The operation of a modern kraft mill is so tight that the amount of salt cake required for makeup to account for liquor losses is typically in the range of only 25 to 50 lb per ton (12 to 25 kg/t) of pulp. The condensates produced at the digester and evaporator vary in quality: those produced directly from the condensation of boiler house steam can usually be returned to the boiler plant, but even here, careful monitoring is required to make certain that leakage does not occur in the system to spoil the condensate and cause problems with steam generation. Other condensates may be used for pulp washing and other washing operations. The poorest grade of condensate from the kraft process, sour condensate, must be stripped of its volatile sulfur compounds, and it usually remains so badly contam-

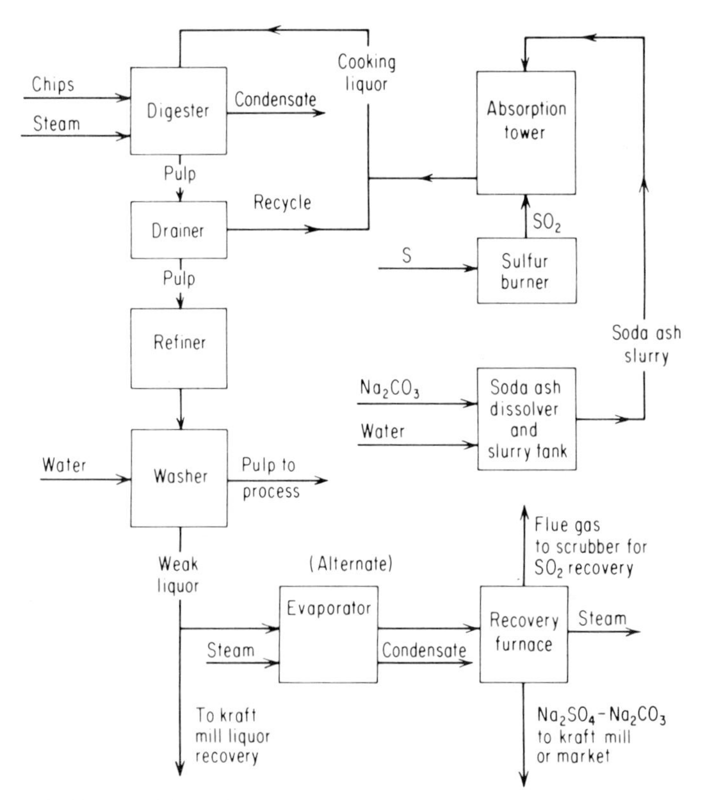

FIG. 30.7 NSSC pulping process.

inated that it must be sent to the waste treatment plant. It is a common source of problems in the operation of the waste treatment plant.

Using the kraft mill as a pattern, Figures 30.6 and 30.7 show the chemical pulping for the magnesium-base sulfite process and for the neutral sulfite-semi-chemical process, respectively. The sources of wastewater contamination in these circuits are similar to those found in the kraft mill.

PULP PROCESS UNITS

In the chemical pulping operation itself, the major items of equipment are the digester, the blow tank, vapor recovery, and pulp washers.

These are the basic units shown on the reference diagrams of chemical pulping operations. Contamination of steam condensate or water with pulping liquors can occur in the digester, except where direct injection of steam into the digester may be practiced. Two types of digesters are shown in Figures 30.8 and 30.9; the first is a batch-type digester in the kraft industry, where the liquor is externally circulated through a heat exchanger in which leakage of liquor into the condensate is always a potential threat. The second illustration shows a continuous digester, which also has external heat exchangers for heating of liquor or washwater with steam, again providing potential for leakage of contaminants into steam condensate.

The washers shown in Figure 30.5 were originally designed only to improve pulp quality by removing excess liquor. Operation of the washwater countercurrent to the pulp flow also allowed recovery of relatively strong liquor. With the attention today on minimizing wastewater discharge, the pulp washer also provides a potential point of reusing excess water from the paper mill or contaminated condensates from the pulp mill. In recycling these wastewaters to the washer, the original intent of washing to produce an acceptable pulp must be kept in mind; for example, certain pulp mill condensates are so badly contaminated with odorous sulfur compounds that they are unacceptable for pulp washing because they would impart this odor to the finished pulp.

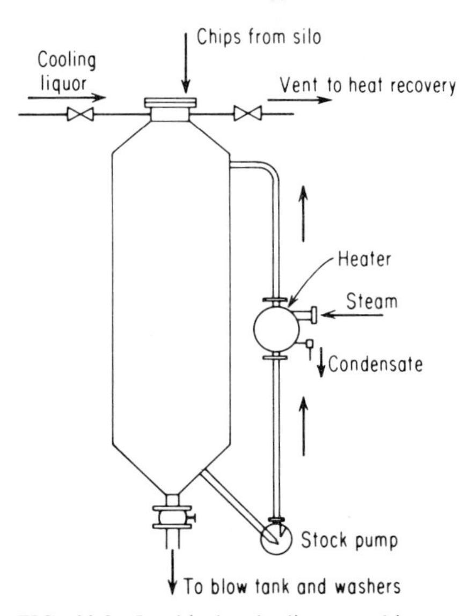

FIG. 30.8 In this batch digester, chips are cooked over a 60- to 90-min cycle, and the pulp product is drawn off for washing and further processing.

After the pulp has been washed and the liquor reclaimed, there are a variety of devices for purifying the pulp to the extent necessary for its end use in paper, board, or chemical manufacture. The major devices are thickeners, refiners, screens, and cleaners.

Some of these devices require that the pulp be diluted before it can be pro-

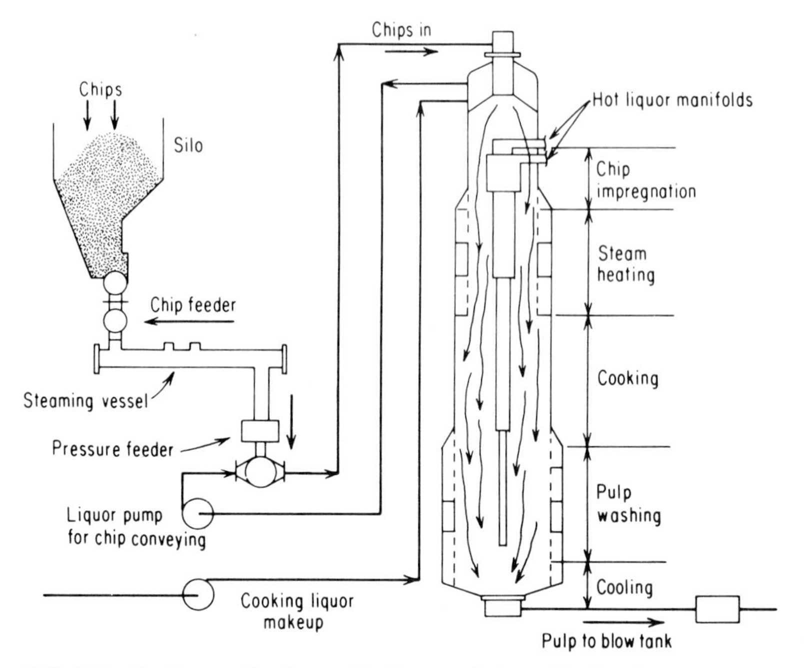

FIG. 30.9 Continuous digester used in the manufacture of kraft pulp.

cessed and others require that the pulp be thickened. This change in consistency may be done in storage vessels, called stock chests, or directly by in-line mixing if dilution is the effect to be achieved. However, where concentration of the pulp is required, the pulp must be passed over a thickener (Figure 30.10) for removal of water. The water recovered from this thickening operation is recycled to the most convenient point in the pulping system.

Refiners are somewhat related to grinders, providing mechanical force to abrade the pulp fibers and reduce them to individual strands of cellulose. The two most common types of refiners are disk refiners (Figure 30.11) and Jordans (Figure 30.12). Screens are used to classify pulp fibers to a certain size, with the oversize particles being returned for reprocessing; screens are also used for the removal of undesirable materials from the pulp, such as knots and noncellulose debris (Figure 30.13).

Centrifugal cleaners are hydraulic cyclones that separate pulp fibers both on the basis of density and size. The more dense or oversize materials are sent to the periphery of the cyclone, and the acceptable fiber fraction is withdrawn from the center of the cyclone. Very often three stages of centrifugal cleaning are involved (Figure 30.14). The rejects from the first stage go to the second, rejects from the second go to the third, and rejects from the third are wasted or returned to a point where their values may be reclaimed, such as in the barking drum where some waste cellulose particles may remain with the bark to be reclaimed in the bark boiler.

These basic units are put together in a pulp mill in different ways depending on the nature of the wood being pulped, the pulping process, and the type of paper

FIG. 30.10 Removal of excess water is often needed for pulp storage. This process is called deckering. These two types of deckers are commonly used for increasing consistency. *(Top)* This drum filter is thickening kraft fiber. *(Bottom)* Fiber dewatered through this vertical, slotted-screen decker is seen tumbling into the pulp discharge flume. *(Courtesy of Dorr-Oliver Incorporated.)*

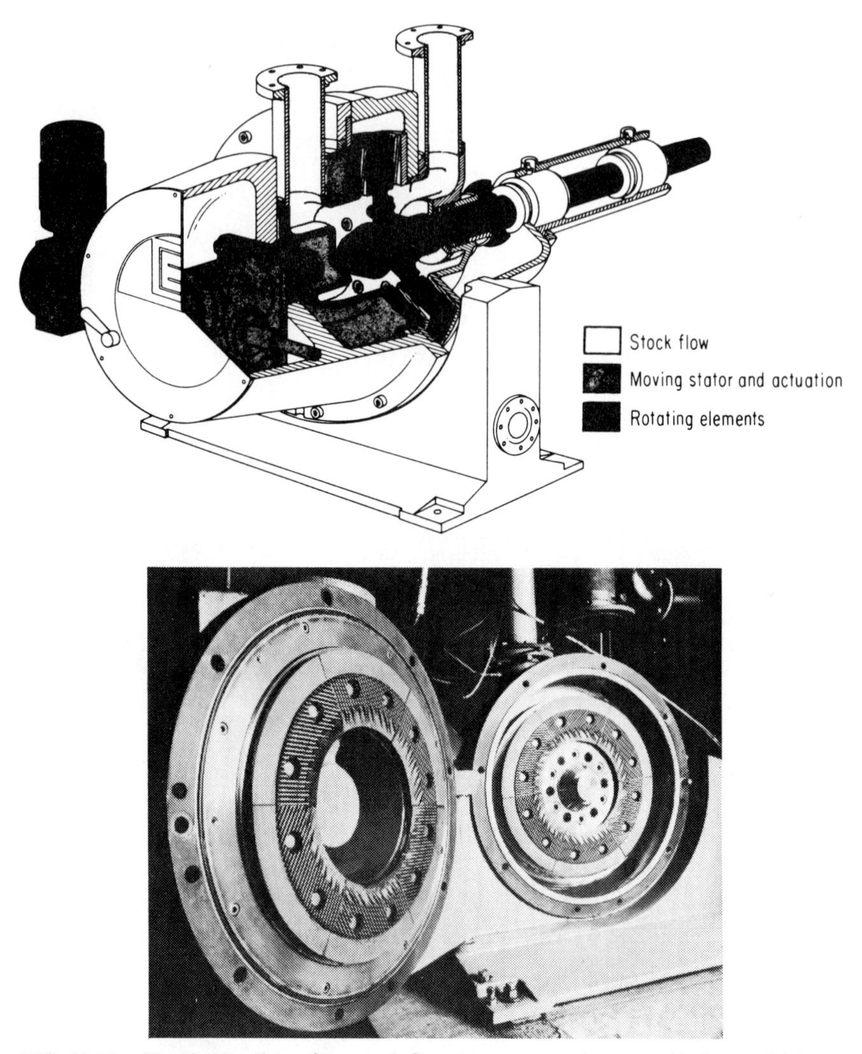

FIG. 30.11 *(Top)* In the disk refiner stock flows between rotating grooved plates, which separate bundles into individual fibers. *(Bottom)* Details of disk design. *(Courtesy of Bolton-Emerson.)*

or board being manufactured. Typical schematics for a groundwood mill, kraft mill, and semichemical mill are given in Figures 30.15, 30.16, and 30.17.

BLEACHING

If the final product is a bleached grade of paper or board, bleaching is done in the pulp mill. Chlorine and chlorine compounds are most commonly used for bleach-

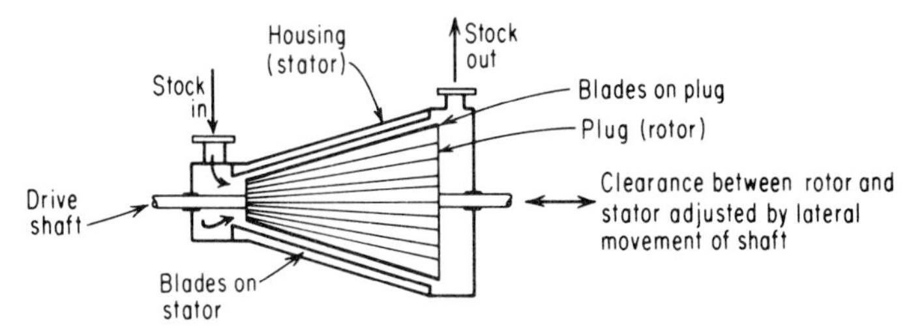

FIG. 30.12 In the Jordan refiner, stock is forced between a rotating plug and a stator, both of which have grooved surfaces or blades to break up fiber bundles.

FIG. 30.13 Rotary screens can handle large flows and classify fibers by their size. These screens are installed ahead of a bleached kraft foodboard machine. *(Courtesy of Black Clawson Company.)*

ing kraft pulp, but oxygen bleaching is assuming increasing importance in the industry. Typically, bleaching is accomplished in several stages, with the residues of the bleaching operation extracted from the pulp between these stages. A common example would be chlorination followed by caustic extraction as the first part of the sequence; this would be followed by hypochlorite bleaching, also followed by caustic extraction, with a final treatment with chlorine dioxide. In pulp mill terminology, this would be identified as CEHED bleaching. There is as yet no practical way to reclaim the materials extracted from the pulp during the bleaching operation, and these liquors constitute one of the major pollution problems in the pulp mill. This production of strong wastes from the bleaching operation is shown in a typical bleaching sequence diagram, Figure 30.18.

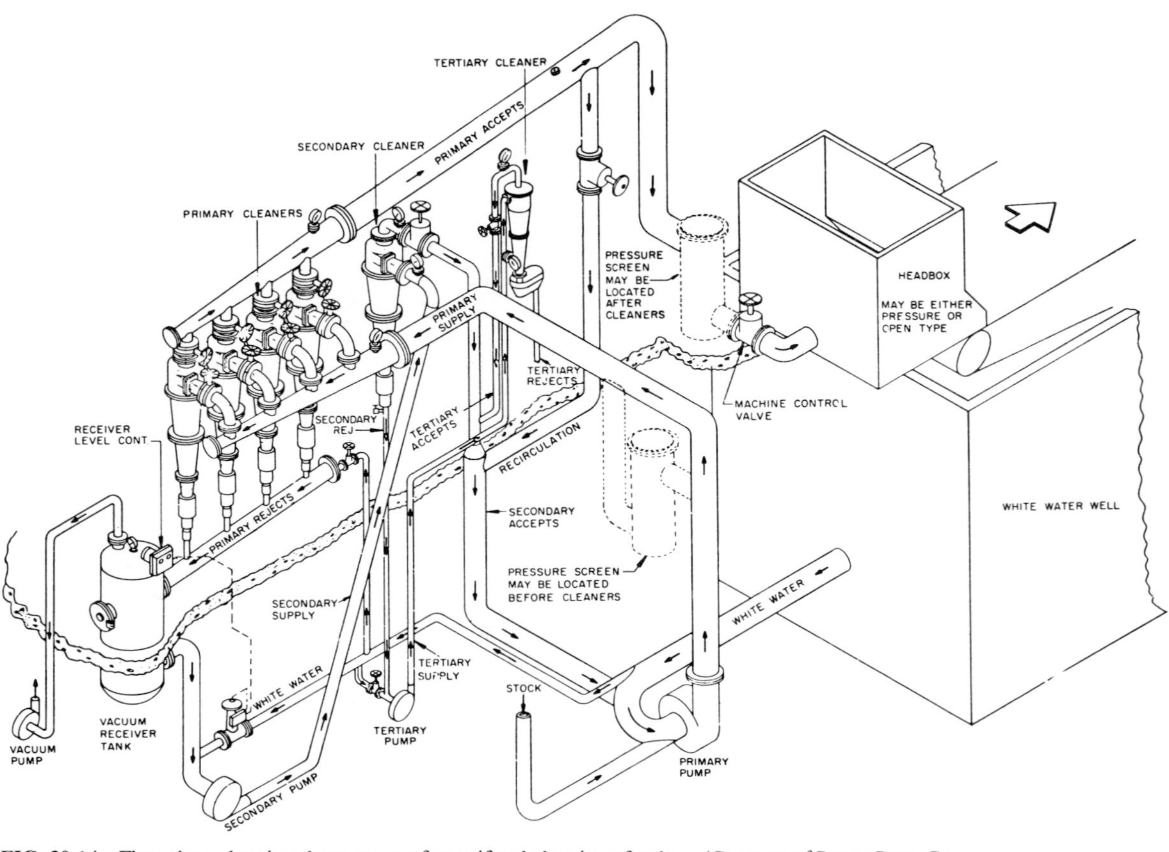

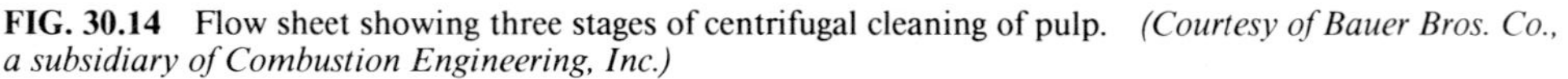

FIG. 30.14 Flow sheet showing three stages of centrifugal cleaning of pulp. *(Courtesy of Bauer Bros. Co., a subsidiary of Combustion Engineering, Inc.)*

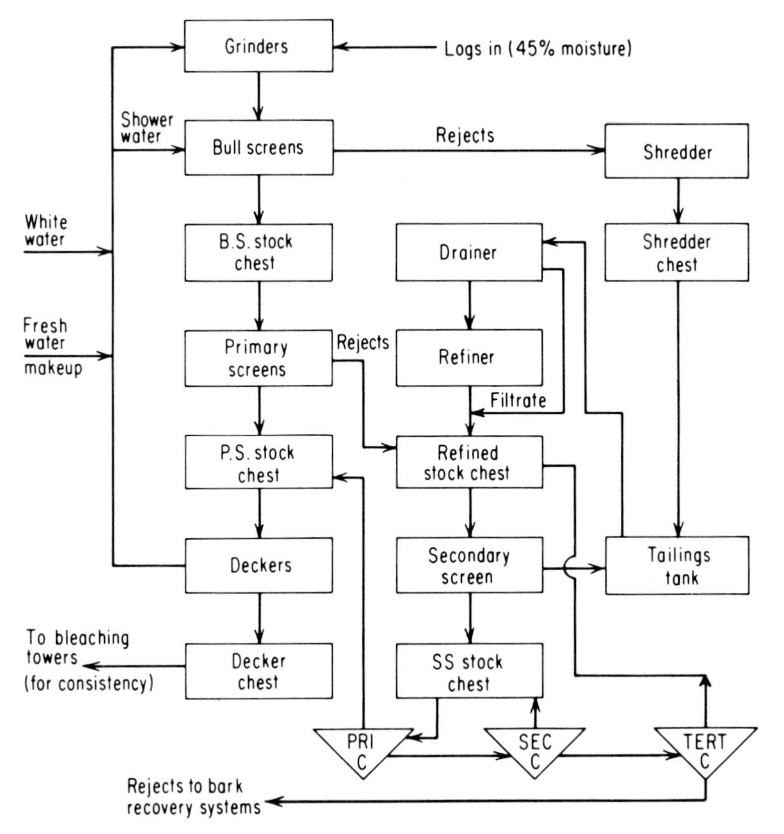

FIG. 30.15 Groundwood mill process units.

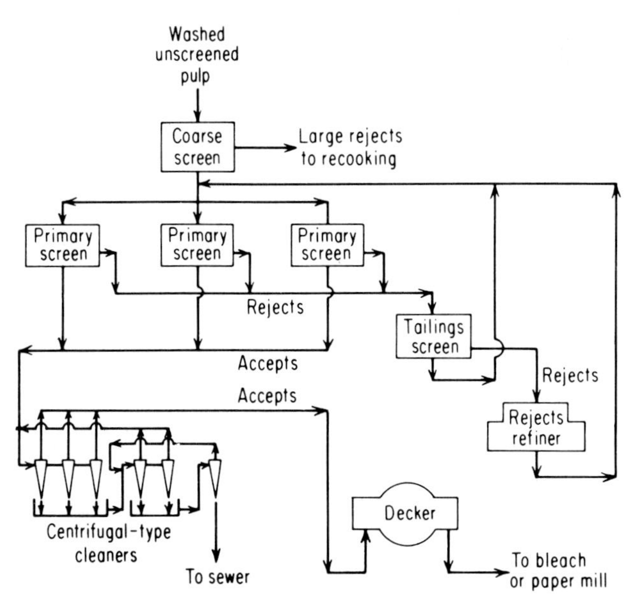

FIG. 30.16 Kraft mill process units.

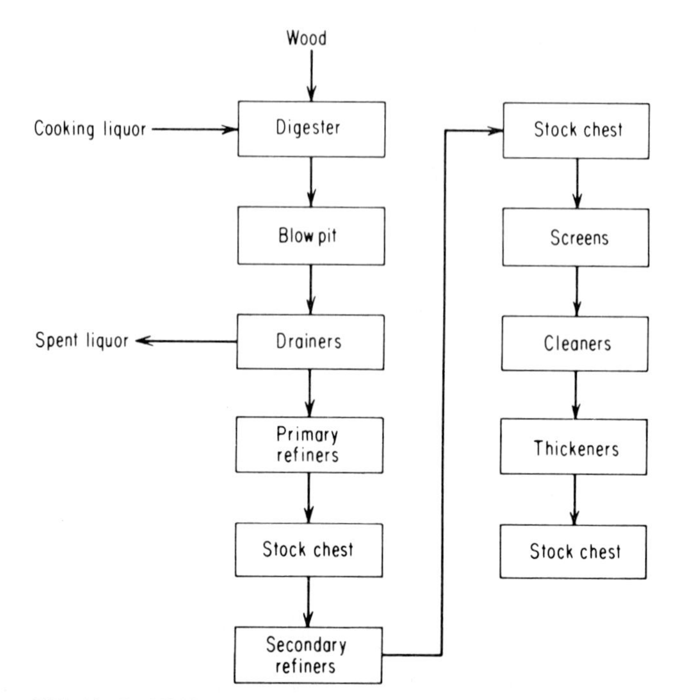

FIG. 30.17 NSSC process units.

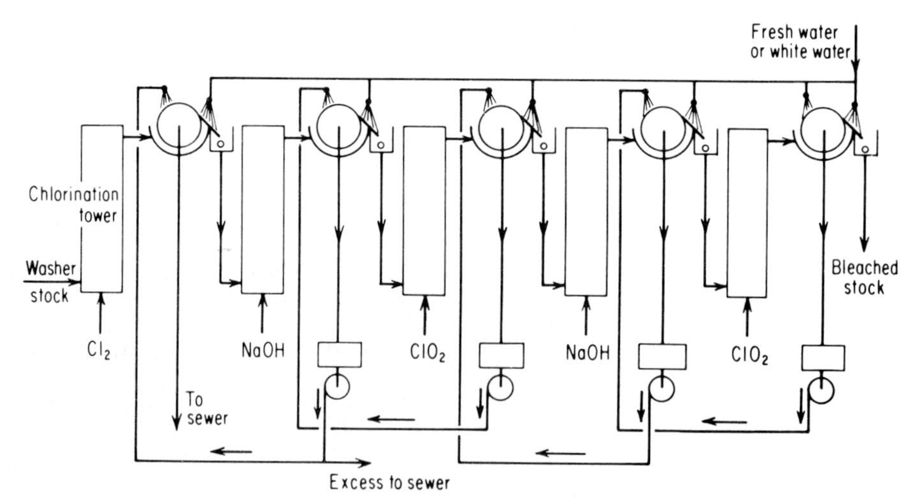

FIG. 30.18 Kraft mill pulp bleaching, CEHED sequence.

When oxygen is included in the sequence, it is the first stage and replaces chlorination and extraction, significantly reducing the concentration of wastes and the load on the waste treatment plant.

Other bleaching agents include zinc hydrosulfite, which is used for bleaching groundwood pulp, sulfur dioxide alone, or chlorination followed by sulfur dioxide.

PROBLEMS CREATED BY WATER

There are a variety of water-related problems in the pulp mill in addition to those already mentioned. These include contamination of steam condensate, discharge of concentrated wastes that cannot be recovered, and nonrecoverable water containing organic matter and other reducing chemicals that produce a load on the waste treatment plant. One of the principal problems is foaming, induced by the surface-active nature of some of the organic matter extracted from the wood. Chemicals for foam control are commonly required at the screens and washers to maintain production capacity.

A more difficult problem to control is the production of scale in evaporators, where the gradual increase in concentration of both inorganic and organic solids causes the solubility limit of calcium sulfate and other materials to be exceeded. It is common practice to boil out an evaporator with water on a scheduled basis to keep this under control. Sometimes the water used for boilout can be reclaimed if it is sufficiently concentrated to justify putting into the evaporator circuit; some of the water used for this operation, however, must be wasted, and this imposes a load on the waste treatment plant.

As in so many other industrial operations using water, corrosion is a constant threat; for the most part it is kept under control by the selection of suitable alloys or plastics, but even alloys may be subject to corrosion where microbial activity may cause slime deposits. Fortunately many of the circuits are relatively hot, which in itself prevents slime growths, but microbial activity frequently develops in the cooler areas of the pulp mill system.

SECONDARY FIBER RECOVERY

Waste paper and board make up a major raw material resource for the paper industry. These wastes are sorted, and the choice between the different grades and their prices determines how they are reprocessed and the type of finished sheet produced and its cost. There are about 40 to 50 grades varying in quality from clean clippings from a paper converter (such as envelope stock) to newsprint and used paper bags. The poorest grade is unsorted (Mixed Grade No. 2).

The raw stock is reduced to slurry by agitating with hot water in a mixing tank (Figure 30.19). Strings and bailing wire are separated at this point. If the stock is printed matter and the finished pulp must be equivalent to virgin pulp, it is deinked. This operation involves heating the stock to about 150°F (66°C) and adding chemical agents to release the ink from the fiber. Following this, the stock is screened and washed, either by flotation (Figure 30.20) or over side-hill washers. At this point, clay and other filler, which can amount to as much as 15% of the weight of the raw stock, are separated and removed from the system with the ink.

These separated wastes present a difficult disposal problem as these solids are hard to dewater. The dewatered cake can be used as landfill, which may still present a problem since land disposal sites near plants are becoming scarce.

The secondary fiber pulp may go directly to a cylinder machine for manufacture of paperboard. It may be bleached and dyed for the manufacture of tissue

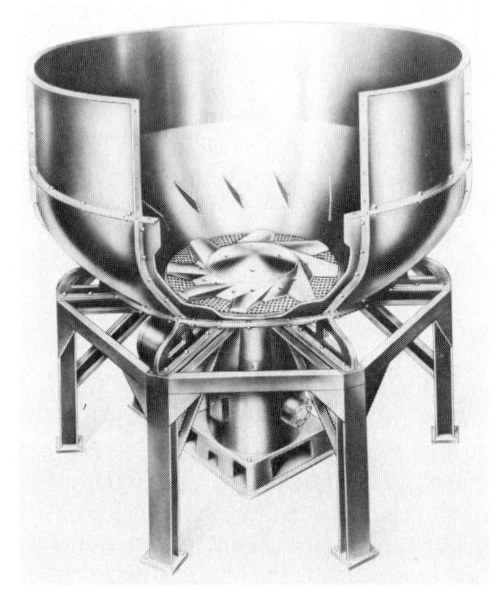

FIG. 30.19 Hydrapulper *(Courtesy of Black Clawson Company.)*

and toweling. Other products of secondary fiber processing include newsprint and writing papers. The same kinds of screens, thickeners, and cleaners may be used in processing reclaimed paper/board as are used in a mill processing virgin pulp, as described later.

WATER REMOVAL FROM PULP

The finished pulp is sent to the paper mill for conversion to paper or board. Dilution water, the majority of which comes from the paper machine wire pit, reduces the consistency of the pulp to less than 1%, and often below 0.5%, ahead of the paper machine. To convert this to the finished sheet, the 99.0 to 99.5% water must be reduced to produce a sheet usually containing less than 6% water.

The least costly way to remove water from the pulp is by drainage through a screen or wire, and this forms the basic part of most paper machine designs. Thereafter, water is removed at increasing increments of cost, first by vacuum, then by pressing, then by the blotting and pressing action of a felt blanket, and finally by evaporation as the sheet is carried over a multitude of stacks of steam-heated drums.

FIG. 30.20 Flotation by dispersed air separates ink sludge released from recoverable fiber by de-inking chemicals. *(Courtesy of Voith-Morden, Inc.)*

FOURDRINIER MACHINE

The most common paper machine is the fourdrinier, illustrated in Figure 30.21. The water that drains through the endless wire belt is collected in a pit and returned to the flow of incoming pulp stock to maintain the consistency at the feed point (the headbox) at a constant value. Sizing chemicals may be added to the stock at the headbox to produce certain desired sheet qualities. Since the water from the wire pit, called white water, is continuously recirculated, the addition of

FIG. 30.21 Fourdrinier paper machine installation looking from headbox toward dryer section.

soluble chemicals, or chemicals that form soluble by-products, results in a gradual increase in the dissolved solids in the white water. The white water also contains fibers which escape capture on the sheet. Excess white water is bled from the system, usually passed through a device for reclaiming fibers (called a saveall), and returned upstream in the process, often as far upstream as the pulp mill.

The sheet leaving the fourdrinier machine, called the wet web, may contain 2 to 4 lb of water per pound of fiber (33 to 20% consistency). It has very poor strength, so a very delicate balance of output from the fourdrinier machine and throughput at the press rolls is necessary to keep the wet web from breaking. The sheet leaves the felt section at a moisture content of only 20 to 40%, and the long stacks of drying rolls evaporate this moisture, delivering the sheet to a final calender stack for smoothing, and then to the reel as a finished product with a moisture content of less than 6%. The overall schematic diagram of the paper machine and the drying section is shown in Figure 30.22.

In some paper mills, the paper may be treated on a special size press located at some point midway between the ends of the drying rolls.

As in the pulp mill, sometimes centrifugal cleaners are installed on the paper machine system for removal of foreign solid matter from the paper stock. There are several kinds of savealls used for recovering fiber from excess white water, a common one being a dissolved air flotation clarifier. A unit of this type will typically reduce suspended fiber from a concentration of 200 to 300 mg/L to a residual of 25 to 50 mg/L producing a float with a consistency of approximately 5%.

Most mills producing white grades of paper size the sheet with alum and rosin, which works best in a pH range of 4 to 5. There is no buffer capacity in this type of water system, so excess white water from paper processing can create problems when it is reused in the pulp mill or sent to waste treatment. This process also builds up sulfate concentrations in the white water system. Many of these mills

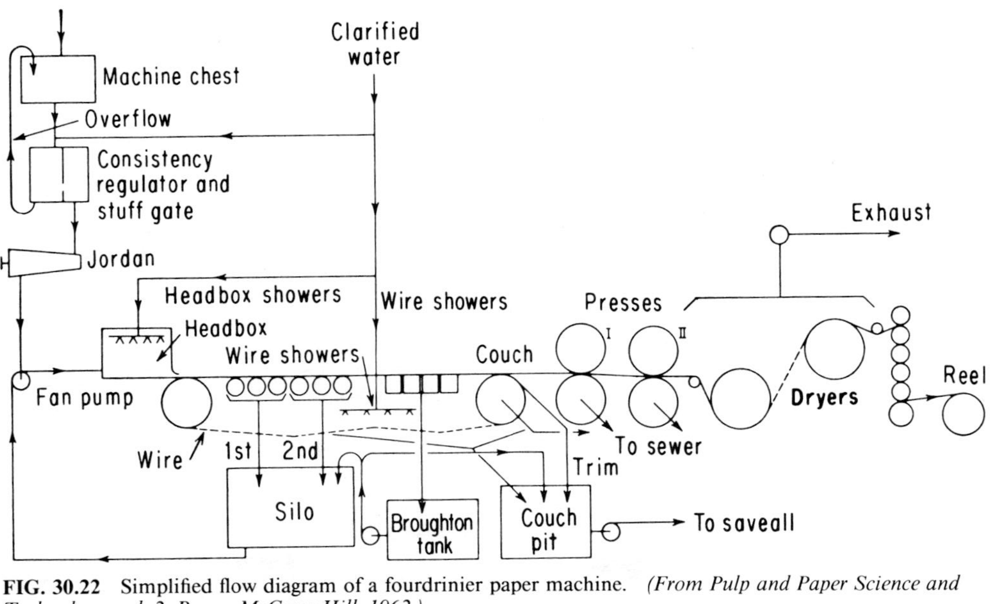

FIG. 30.22 Simplified flow diagram of a fourdrinier paper machine. *(From Pulp and Paper Science and Technology, vol. 2: Paper, McGraw-Hill, 1962.)*

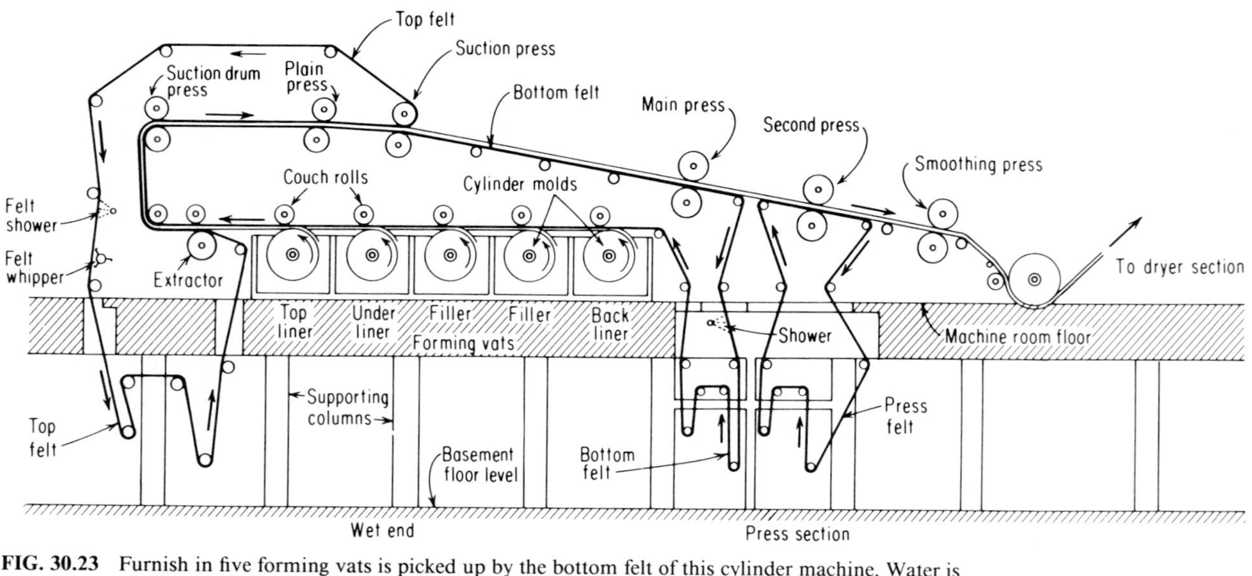

FIG. 30.23 Furnish in five forming vats is picked up by the bottom felt of this cylinder machine. Water is removed from the paperboard by suction and by pressing. The finished board is 5-ply, often distinguishable because of the variation in the pulps in each forming vat.

are now finding benefits to neutral-to-alkaline sizing, permitting them to close up the system further with minimal salinity increase and with a more favorable pH for corrosion control, recycle to the pulp mill, and final disposal.

CYLINDER MACHINE

Another common type of paper machine is the cylinder machine, commonly found in smaller mills and often used for the production of heavier grades of paper board. A typical cylinder machine is shown in Figure 30.23. The finished sheet leaving the cylinder machine is processed similarly to the sheet leaving the fourdrinier, by a battery of steam-heated dryer rolls. Heavier board may be dewatered on a single dryer drum, the Yankee dryer (Figure 30.24).

On both types of machines the sheet must be trimmed to an exact width depending on customer requirements, and the trim is collected for repulping. This material and broke, the term for excess product which cannot be handled at the reel, are fed to a beater (Figure 30.25), where they are mixed with white water to produce a pulp of a consistency which can be returned to the pulp mill or held in a stock chest until it can be returned to the paper machine system.

WATER-RELATED MILL PROBLEMS

Given the conditions of organic fiber, cellulose, residual organics extracted from the wood, and warm, oxygenated water in the white water circuit, bacterial

FIG. 30.24 Yankee dryer. *(Courtesy of AER Corporation.)*

growths become one of the major problems of paper manufacture. If these growths are not controlled, slimes form in the paper machine system and periodically break loose to enter the circuit, find their way through the headbox onto the paper machine, and develop as imperfections in the finished sheet. Equally important, these growths lead to corrosive attack even in stainless-steel alloy systems. Therefore, one of the major problems in paper manufacture is microbial control with chemicals that are effective and at the same time safe to handle.

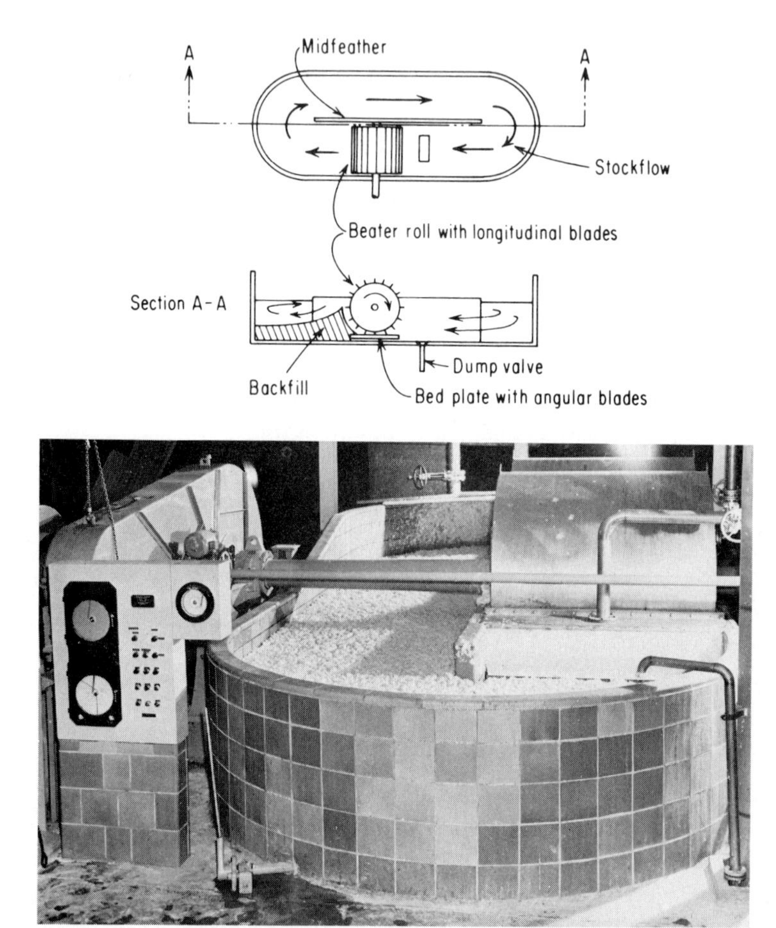

FIG. 30.25 *(Top)* The hollander beater was an early device for refining pulp, and is still used in many mills for repulping broke. *(Bottom)* Plant installation.

The trend toward tightening up the paper system to reduce water consumption has resulted in the buildup in the content of both inorganic and organic solids in the white water system, accentuating the problems of slime control and corrosion control. Figure 30.26 shows the results of corrosive attack of stainless steel caused by slime formation and the action of anaerobic bacteria at the slime/metal interface. Figure 30.27 shows sheet imperfections caused by bacterial colonies in the white water system. Foaming is another water-related problem that has been accentuated by the closing up of the paper mill white water circuit.

Because water is so important to the pulp and paper industry, it is common to find in modern pulp and paper mills a water treatment plant as sophisticated as

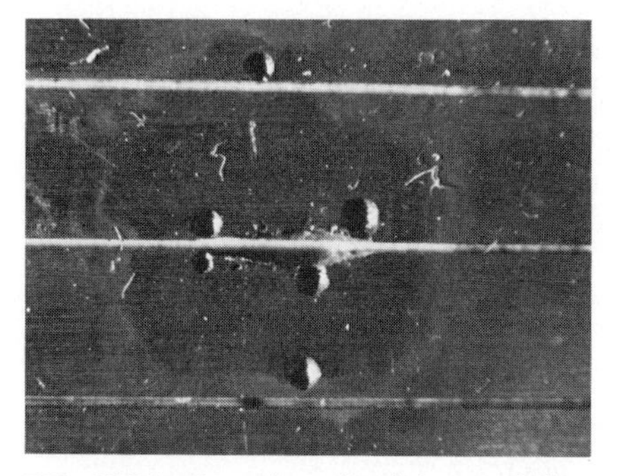

FIG. 30.26 Corrosion of stainless-steel sieve in a paper mill caused by sulfate-reducing bacteria.

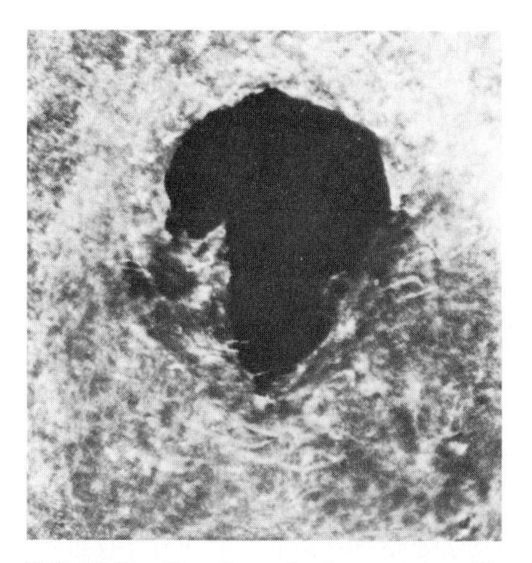

FIG. 30.27 Sheet imperfections are caused by sloughing of slime mass into machine system. In this photo, the removal of the slime mass during calendering left a hole in the sheet.

a municipal plant treating potable water. The quality of water required in the process varies with the grades of paper being produced and the specifications for these grades. These are generally summarized by Table 30.5. The standards shown in this table have been subject to many exceptions in past practice, and may be revised in the future as the industry learns what it can do with higher concentrations in the process water stream as mill systems are tightened to reduce discharge.

TABLE 30.5 Process Water Quality Standards

Substance, mg/L	Fine paper	Kraft paper (bleached)	Ground-wood papers	Soda and sulfate pulp
Turbidity as SiO_2	10	40	50	25
Color in platinum units	5	25	30	5
Total hardness as $CaCO_3$	100	100	200	100
Calcium hardness as $CaCO_3$	50	—	—	50
Alkalinity to methyl orange as $CaCO_3$	75	75	150	75
Iron as Fe	0.1	0.2	0.3	0.1
Manganese as Mn	0.05	0.1	0.1	0.05
Residual chlorine as Cl_2	2.0			
Silica (soluble) as SiO_2	20	50	50	20
Total dissolved solids	200	300	500	250
Free carbon dioxide as CO_2	10	10	10	10
Chlorides as Cl	—	200	75	75
Magnesium hardness as $CaCO_3$	—	—	—	50

Source: From TAPPI Monograph No. 18.

DESIGNING FOR THE FUTURE

Unfortunately, there are scant data to guide the designer of a new mill in the quality standards required for makeup water, which must be satisfactory not only in producing products that meet market standards, but also in conditioning the water to render it neither corrosive nor scale-forming under the wide variety of situations occurring in the pulp and paper mill.

For the most part, water treatment plants for new mills are designed based on past experiences and operating practices that may be quite different from the experiences the mill will encounter as pollution control restrictions become increasingly strict. One of the consequences of stricter discharge regulations may be a relaxation in product quality; for example, brightness has been associated with quality of fine paper, and there is little doubt that the brightness achievable today is beyond what is required for legibility of books and documents. This brightness is achieved at the expense of more severe chemical treatments in processing which add to the pollution load. A relaxation in brightness standards for paper products, then, would seem consistent with the goals of reducing the pollution load on the waste treatment plant.

Table 30.6 provides a general idea of the various uses of water in a typical pulp and paper mill and shows both volumes required and the wastes produced in a typical 500 ton/day (454 t/day) kraft mill with conventional chlorine and hypochlorite bleaching.

WASTE TREATMENT

The points of waste production will be apparent from a study of the flow sheets of the different pulp and paper manufacturing operations. In the present state of the art, most mills have tightened up their water systems to reduce discharge, but

TABLE 30.6 Effluents from a Typical 500 Ton/Day Kraft Mill (Built in 1960 to 1965 Period)

Process	Effluent, gal/ton*	Suspended solids, lb/ton†	BOD_5, lb/ton†
1. Debarking	2,640	31	7
2. Cooking and washing	264	—	9
3. Screening and cleaning	26,400	31	20
4. Bleaching			
a. Acid	15,800	6	22
b. Caustic	10,500	3	11
5. Pulp sheet formation	1,050	2	—
6. Evaporation	1,850	0.2	30
7. Causticizing	1,320	0.2	—
8. Recovery furnace	3,960	0.2	—
Total	83,784	73.6	99

* Gal/ton × 0.0042 = m^3/t.
† Lb/ton × 0.501 = kg/t.

have not done so to the extent that dissolved solids build up over 5000 mg/L. (There are actually a few mills having zero discharge; they produce a relatively low-grade paper which, of course, carries the solids out of the mill in its moisture content.) There is substantial flow of excess water carrying with it the rejects from pulp cleaning, excess pulping liquor which could not be reclaimed, dregs from the chemical recovery area, materials extracted from the pulp during bleaching, and excess clay and filler not retained by the sheet. Since a good portion of the wastes are organic materials extracted from the wood, the BOD of the combined wastes

TABLE 30.7 Sources of Solids in an Integrated Mill

Area	Suspended solids, mg/L	Dissolved solids,* mg/L
Woodroom	500–700	700–800
Pulp mill (bleach plant)		
Caustic extract	60–80	4000
Acid extract	60–80	1500
Total	200–250	1800
Recovery plant	3000	3000
Paper machines	400–700	300–400

* Increase over solids level in mill water.

is usually quite high—on the order of 50 to 150 lb per ton (25 to 75 kg/t) of product, or typically in the range of 100 to 300 mg/L. Suspended solids levels may be as low as 50% of the BOD values. Because these levels are not greatly different from municipal sewage plant loadings, the type of waste treatment facility installed in a pulp and paper mill is generally quite comparable to that found in municipal treatment plants. Table 30.7 shows the range of suspended solids and BOD loadings found in a variety of pulp and paper manufacturing operations. Figure 30.28 shows the layout of a bleached kraft mill using current technology

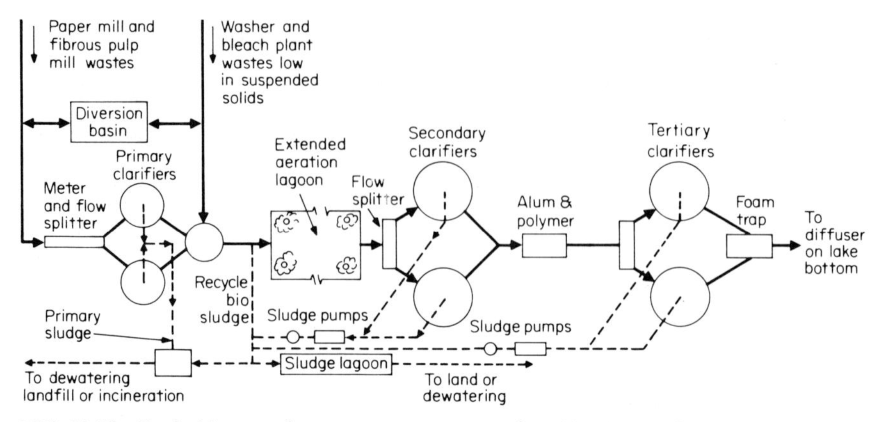

FIG. 30.28 Typical layout of waste treatment system for a bleached kraft mill.

for wastewater treatment for discharge to a nearby receiving stream. In this particular plant, the primary clarifiers handle only those wastes carrying over 50 mg/L suspended solids. After clarifying these wastes, the effluent is combined with the remaining low solids streams (such as bleach plant effluent) and treated in an extended aeration plant for biological reduction of organic solids. The digested waste is then processed through a final clarifier for discharge to the receiving water. Table 30.8 shows the analysis of the raw waste streams and the final effluent. The design of a treatment system of this type requires months of pilot plant study to determine the range of waste concentrations to be anticipated, to evaluate the performance of the aerobic digestion system under a broad range of temperatures and pH values and a study of the optimum procedures for dewatering and disposing of waste sludges.

Not only are large volumes of water needed for process, but also for the generation of steam. The pulp and paper industry is energy intensive, and the organic

TABLE 30.8 Integrated Pulp/Paper Mill Waste Treatment

	Inlet				
Constituent	From paper mill*	From pulp mill†	Lagoon feed	Secondary effluent‡	Final effluent
Suspended solids, mg/L	530	140	400	210	15
Alkalinity, mg/L as $CaCO_3$	224	(760)§	324	32	—
pH	7.6	2.3	7.2	6.3	—
BOD, mg/L as O_2	233	136	154	11	6
COD, mg/L as O_2	930	750	790	350	—
TOC, mg/L as C	270	230	260	110	—
Conductance, μS	1200	6400	1700	1700	—

* Fiber-containing streams.
† Lower suspended solids streams.
‡ Without chemical coagulation.
§ Acidity.

matter (such as bark and lignin) extracted from the wood fits into the overall planning and design of a steam-generation system to utilize these wastes as a fuel source. Because large volumes of steam are needed on the paper machines, particularly the drying rolls, the industry optimizes the heat balance by condensing turbine exhaust on the machines, making it economical for the industry to produce its own electric power.

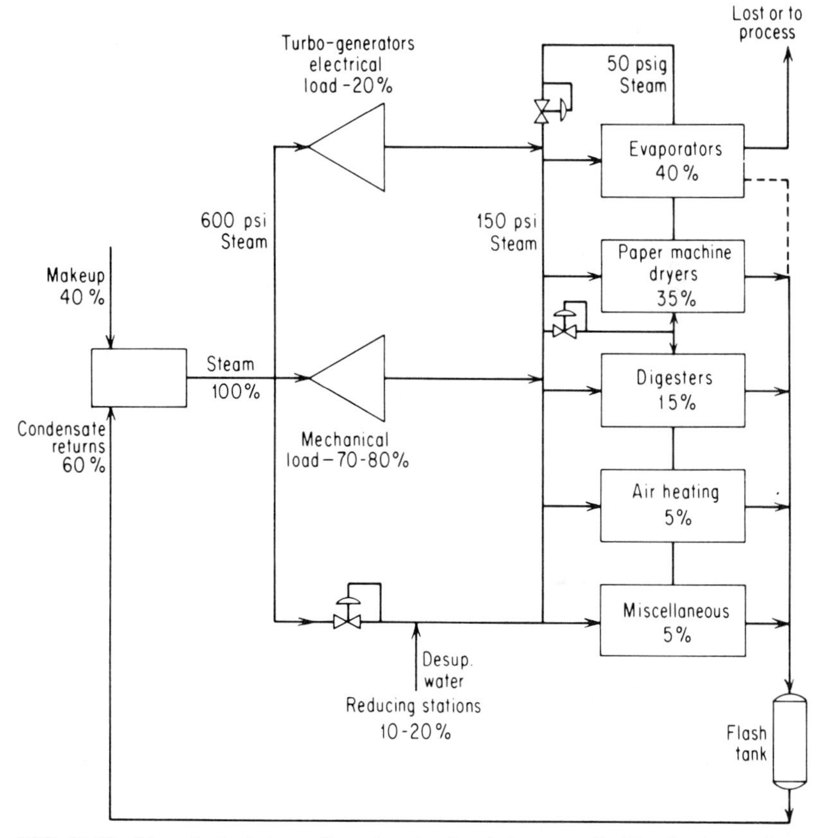

FIG. 30.29 Hypothetical steam flows in a kraft pulp/paper mill. Distribution varies somewhat seasonally as air and water temperatures change. Not shown are turbine condensers which are connected to the main generating turbines in most mills.

A flow diagram and heat balance of a steam generation plant in a typical kraft mill is shown in Figure 30.29. In an integrated kraft pulp/paper mill, a steam plant of this type will produce in the range of 5000 to 10,000 lb of steam per ton of production. The operation of the utilities system is so critical to the production process that the profit or loss of a mill requires efficient performance of the utility plant.

SUGGESTED READING

Lund, H. F. (ed.): *Industrial Pollution Control Handbook,* McGraw-Hill, New York, 1971.

Shreve, R. N.: *Chemical Process Industries,* 3d ed., McGraw-Hill, New York, 1967.

U.S. Environmental Protection Agency, *Development Document for Unbleached Kraft & Semichemical Pulp,* EPA-440/1-74-025-a, May 1974.

CHAPTER 31
PETROLEUM INDUSTRY

Petroleum refining is one of the major manufacturing industries in the United States, producing more than 75% of the petroleum products consumed in this country. Ninety percent of these products are fuels supplying over 46% of the country's energy needs. Plastics, solvents, asphalt, lubricants, and intermediate chemicals are among the more than 3000 products made from petroleum.

Petroleum refining is a continuous operation incorporating a number of interrelated process units. It is one of the largest "wet" processing industries in the United States.

Because it it used in the refinery for heating, cooling, and processing, water is an important factor in process operations. A refinery may draw water from a variety of sources, and the treatment processes used to condition this water vary accordingly. Little or no treatment may be required for some well waters, while other water sources may require an extensive plant incorporating clarification, softening, and filtration. The types of water treatment problems here are much the same as in other industrial applications. In general, the basic water flow scheme is illustrated in Figure 31.1.

After initial processing, the water is usually divided into several streams for use throughout the refinery. At most locations, this water can be used directly as cooling tower makeup, with little or no further pretreatment. However, an extensive treatment scheme is usually necessary to produce the high-quality water required for boiler feed.

Figure 31.2 shows a typical water-handling system designed for a larger refinery utilizing a surface water source as plant makeup.

Refinery processes are net consumers of heat, to the extent that 10 to 15% of the heat equivalent of the incoming crude is used in the refinery operations. This is provided by refinery by-products, including off-gases, residual oil, and coke, in many cases, supplemented by natural gas. A typical process heat balance is shown by Figure 31.3. This does not include heat recovered in process-heat boilers.

PROCESS OPERATIONS

The refining of petroleum products and petrochemicals involves two basic operations: physical change, or separation processes; and chemical change, or conversion processes. A refinery is a conglomerate of manufacturing plants, the number varying with the variety of products produced. The bulk of these products (kerosenes, fuel oils, lubricating oils, and waxes) are fractions originally present in and subsequently separated from the crude petroleum. Some of these are purified and

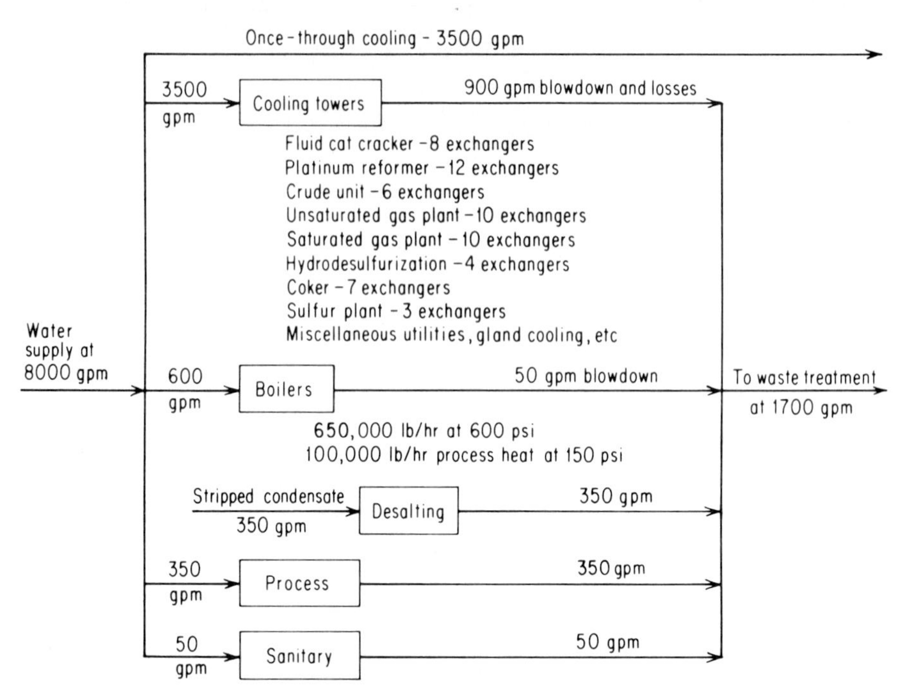

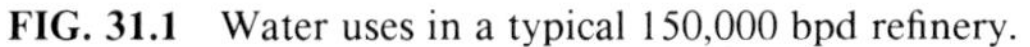

FIG. 31.1 Water uses in a typical 150,000 bpd refinery.

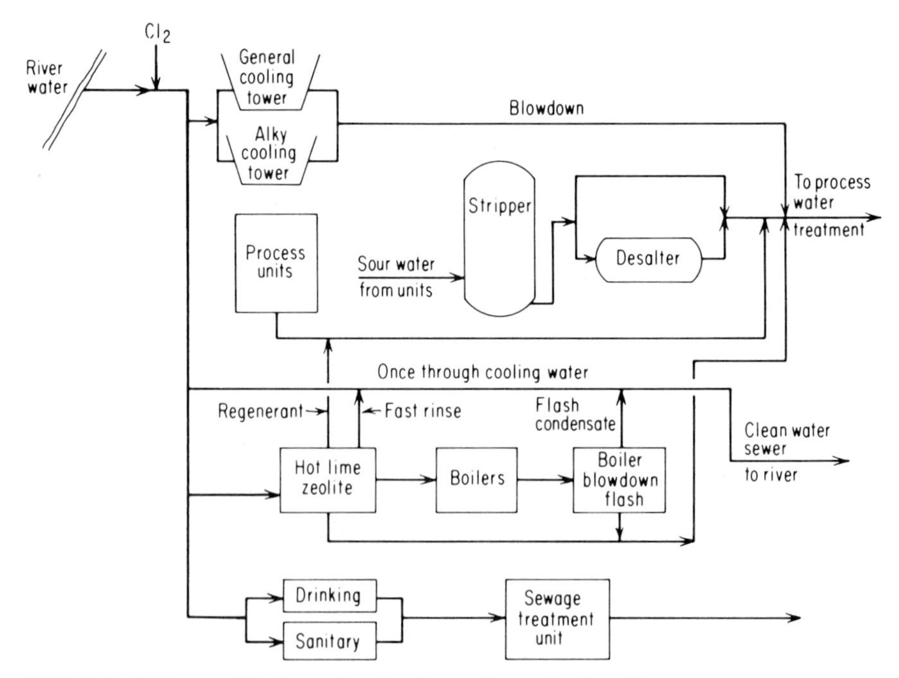

FIG. 31.2 Water conditioning system treating surface water for refinery usage.

supplemented with nonpetroleum materials to enchance their usefulness. A refinery with half a dozen processes, including distillation and cracking, can produce gasoline, kerosene, and fuel oils. The manufacture of solvents requires two or three more processes; lubricating oil production, at least five more; waxes another two or more. Asphalts, greases, coke, gear oils, liquefied petroleum gases, alkylate,

Source of energy	Percent of total heat
Crude oil	Nil
Distillate oils	1.6
Residual oil	9.0
LPG	1.3
Natural gas	36.2
Refinery gas	35.1
Coke	13.1
Steam (purchased)	1.1
Electricity (purchased)	2.6

Total heat requirement is approximately 674,000 Btu/bbl of crude processed.
SOURCE: From U.S. Bureau of Mines.

FIG. 31.3 Refinery process heat balance.

and all the other kinds of products that can be made would require as many as 50 different processes. Figure 31.4 is an overall flow diagram for a generalized refinery production scheme.

As previously mentioned, refining is basically concerned with separation and conversion processes, carried out by individual unit operations. Basic to most of them are furnaces, heaters, and heat exchangers, and distillation and extraction columns. Heat exchangers are typically of shell and tube design, often utilizing incoming hydrocarbon feedstock as a cooling medium for hot products. If additional cooling is required—trim cooling—this is done with water.

Distillation and extraction are usually the principal techniques in product separation. Distillation separates various hydrocarbon mixtures into components having different boiling points. In extraction, hydrocarbons are separated based on their different solubilities in a specific solvent. In some unit operations, filters are used to remove suspended contaminants from the hydrocarbon stream, such as catalyst fines and inorganic precipitates.

Distillation

Figure 31.5 illustrates a typical distillation column (pipe still). Preheated crude is charged into the bottom of a distillation column at a pressure slightly above atmospheric, and the vapors rise through the column, contracting a down flow stream (reflux). As a result, the lightest materials concentrate at the top of the column, the heaviest materials at the bottom, and intermediate materials in between. Desired products are withdrawn at appropriate points. Because the lighter products (as vapor) must pass through the heavier products (as liquid) and must be in equilibrium with them at each point in the column, each stream contains some very volatile, low-molecular-weight components (light ends).

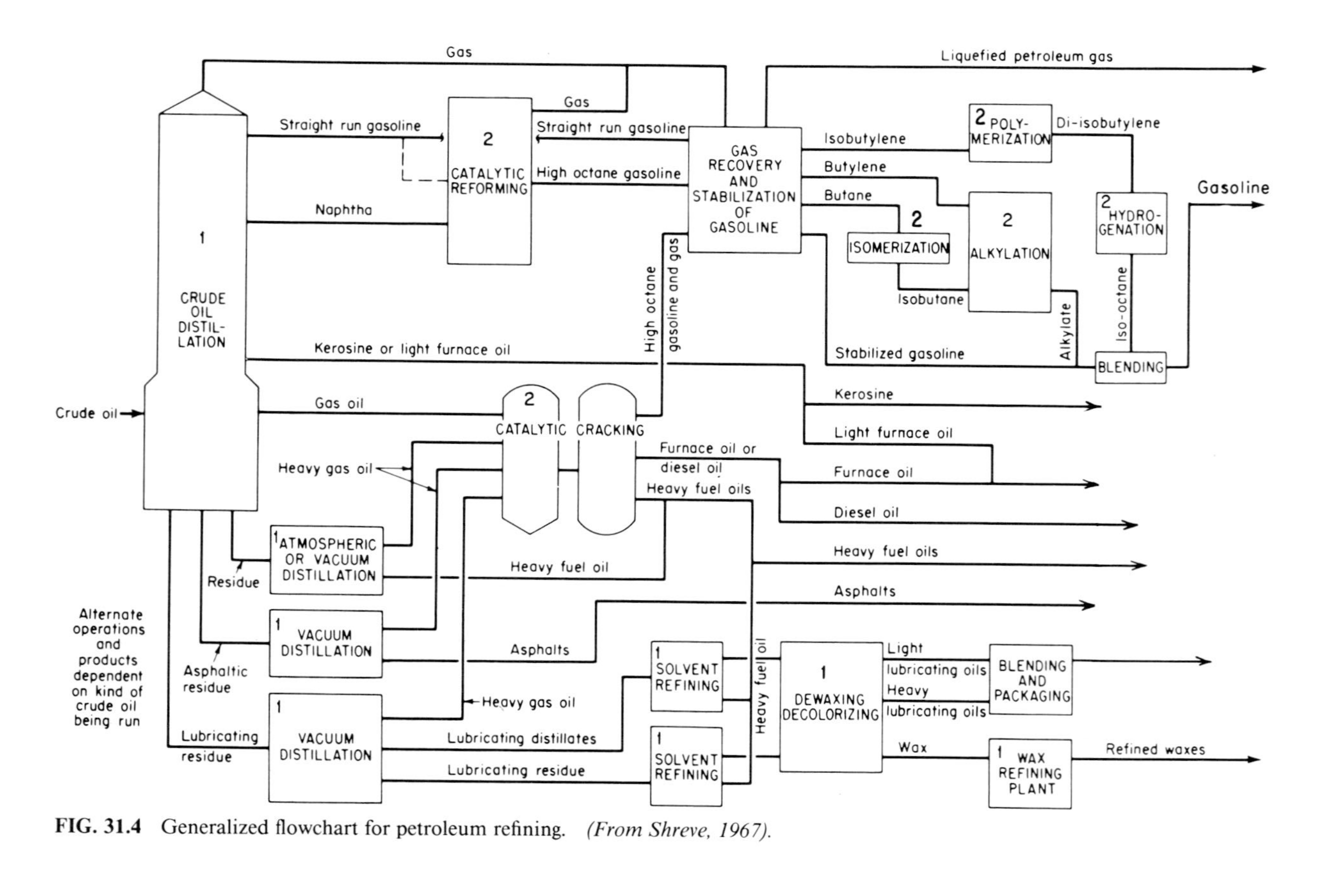

FIG. 31.4 Generalized flowchart for petroleum refining. *(From Shreve, 1967).*

As indicated in Figure 31.5, steam strippers are sometimes used to remove light ends from a sidestream. The sidestream is fed to the top of the stripper; countercurrent steam strips out the light ends and carries them back to the main column.

The wastes from crude oil fractionation, and in general from most distillation columns, come from three sources. The first is the water drawn from the overhead

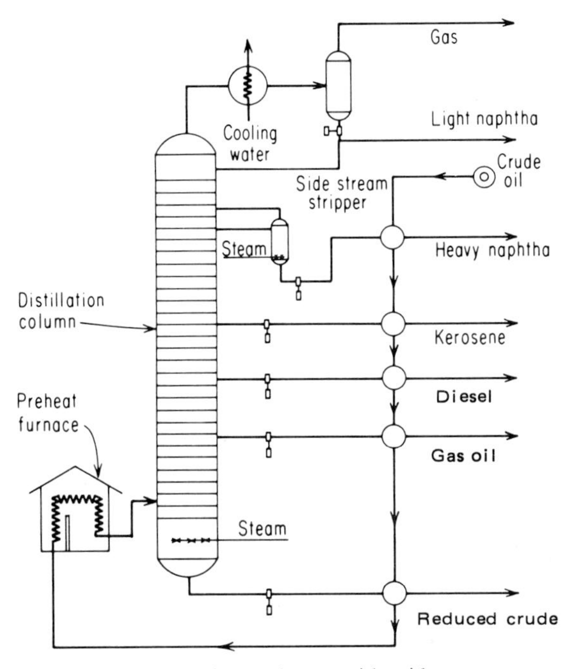

FIG. 31.5 Distillation column with sidestream steam stripper.

accumulator prior to recirculation or transfer of the hydrocarbons to another fractionator. The water that separates from the hydrocarbons in these accumulators is usually drawn off and discharged to the wastewater treatment system. This water can be a major source of sulfides, especially when sour crudes are being processed; it may also contain significant amounts of oil, chlorides, mercaptans, and phenols. A second significant waste source is discharge from oil sampling lines; this oil should be separable, but may form emulsions in the sewer. A third possible waste source is the stable oil emulsion formed in barometric condensers used to produce a vacuum in some distillation units (Figure 31.6).

In vacuum towers a steam jet ejector is the most widely used method for creating a vacuum. The steam and other vapors removed from the fractionator must be condensed, and the liquid removed prior to discharge of the vapor to the atmosphere.

The barometric condenser condenses the steam jet by a water spray in a closed chamber, and the water drains down the barometric leg. The organics, oils, and steam condensate are intimately mixed in a large volume of cooling water, which tends to form difficult-to-handle emulsions. Newer refineries use surface condensers instead of barometric condensers. These units consist of a series of shell and tube exchangers in which the condensibles are removed, and the water for cooling does not come into direct contact with the condensate (Figure 31.7). The water discharge from the distillation operation is sent to the refinery waste plant, sometimes through a separate gravity oil separation unit.

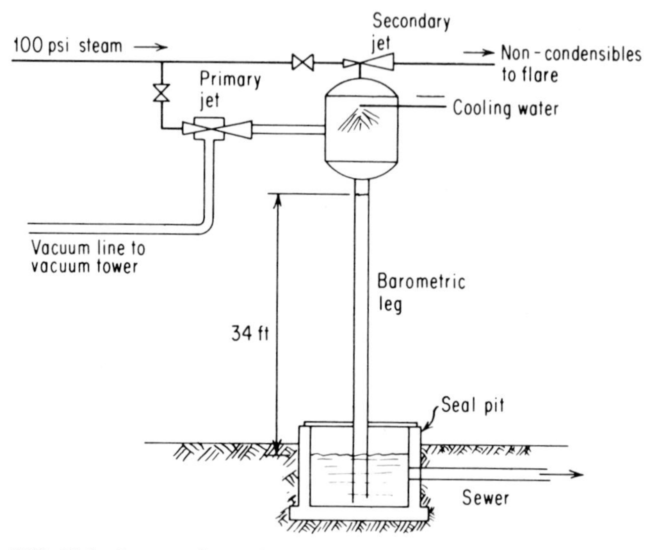

FIG. 31.6 Barometric condenser.

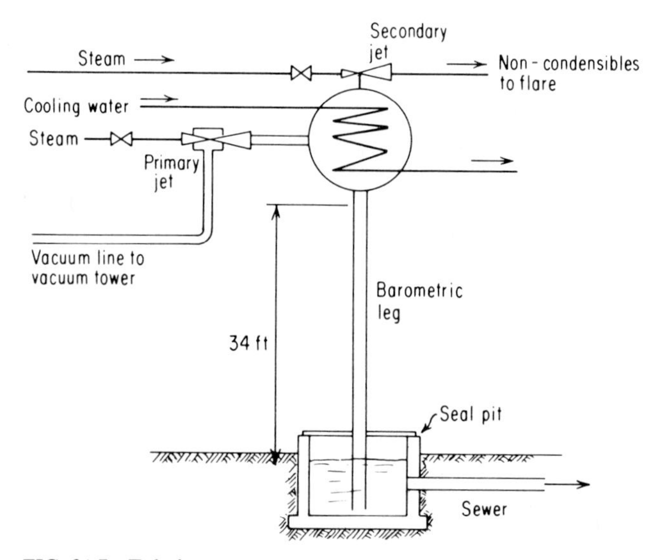

FIG. 31.7 Tubular process steam condenser.

Thermal Cracking and Related Subprocesses

In thermal cracking units, heavy oil fractions are broken down into lighter fractions by application of heat, but without the use of a catalyst. Production of gasoline is low, but middle distillates and stable fuel oils are high. Visbreaking and coking, the two major types of thermal cracking, maximize the production of catalytic cracking feedstocks, indirectly increasing gasoline production.

Oil feed is heated in a furnace to cracking temperatures and the cracked products are separated in a fractionator (Figure 31.8). The heat breaks the bonds holding the larger molecules together, and under certain conditions, some of the resulting smaller molecules may recombine into larger molecules again. The products of this second reaction may then decompose into smaller molecules depending on the time they are held at cracking temperatures.

Visbreaking is a mild form of thermal cracking; it causes little reduction in boiling point, but significantly lowers viscosity. The feed is heated, cracked slightly in a furnace, quenched with light gas oil, and flashed in the bottom of a fractionator. Gas, gasoline, and furnace oil fractions are drawn off, and the heavier fractions are recycled.

Residual oils may be cracked to form coke as well as the usual gaseous and liquid products. The most widely used process, known as delayed coking (Figure 31.9), accounts for about 75% of the total oil coking capacity.

Thermal cracking units require cooling water and steam on the fractionating towers used to separate products. Some towers employ steam-stripping of a side-stream to remove light ends, requiring an overhead condenser and accumulator system for product/wastewater separation. Wastewater usually contains various oil fractions and may be high in pH, BOD, COD, NH_3, phenol, and sulfides. Another important water use area is the high-pressure water sprays used for coke removal in delayed coking. Several refiners have instituted water recycle clarification systems to minimize water discharge from these coking units.

Catalytic Cracking

The fluid catalytic cracking process is the most widely used refining process (Figure 31.10). A large mass of finely powdered catalyst contacts the vaporized oil in the processing unit. The catalyst particles are of such a size that when aerated or "fluffed up" with air or hydrocarbon vapor, they behave like a fluid and can be moved through pipes and control valves.

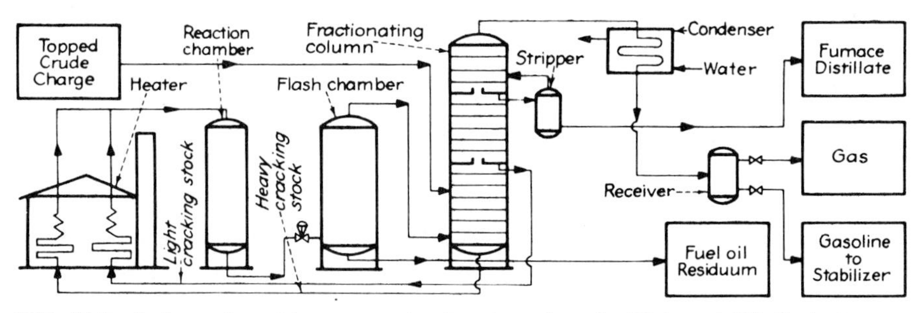

FIG. 31.8 A thermal cracking process treating topped crude (Universal Oil Products process). *(From Shreve, 1967.)*

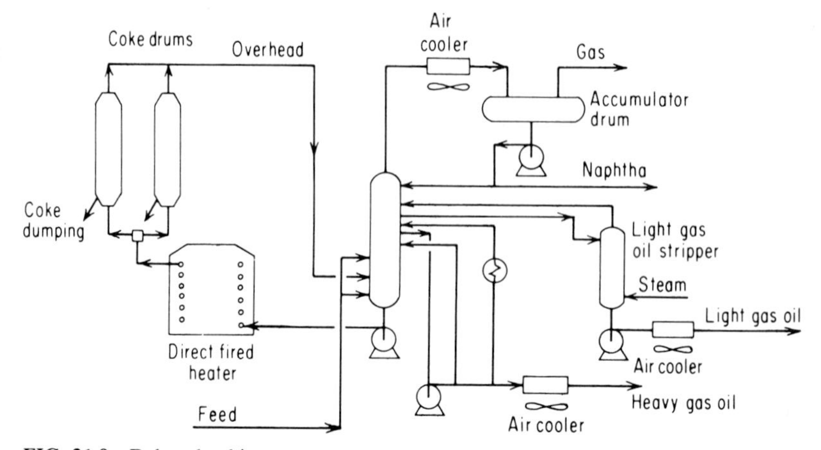

FIG. 31.9 Delayed coking process.

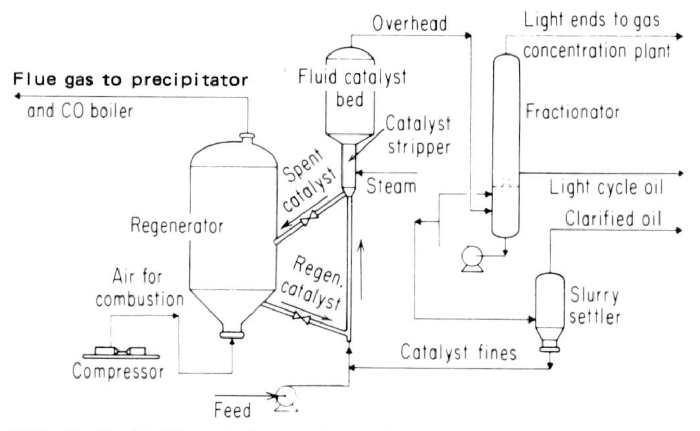

FIG. 31.10 Fluid catalytic cracking unit.

In the catalytic cracking process, feed and regenerated equilibrium catalyst flow together into the riser reactor. The cracked vapors from the reactor pass upward through a cyclone separator which removes entrained catalyst. The product vapors then enter a fractionator, where the desired products are removed and heavier fractions recycled to the reactor. Spent catalyst passes from the separation vessel, downward through a stream stripper, and into the regenerator where carbon deposits are burned off. The regenerated catalyst again mixes with the incoming charge stream to repeat the cycle. On units using a CO boiler, partial combustion in the regenerator is used. This incomplete combustion produces a preponderance of CO over CO_2. Significant amounts of hydrocarbons and other substances also remain unburned in the combustion gases. The CO and hydrocarbons in the exhaust make it a useful fuel. Most refineries burn this exhaust gas in specially designed carbon monoxide (CO) boilers to generate steam. The gas burned in the CO boiler carries a significant amount of residual catalyst fines, which can cause furnace deposits.

Today, most units operate in a complete combustion mode in the regenerator. This reduces carbon on regenerated catalyst, improves product yields, and helps reduce preheat fuel requirements.

Most fluid catalytic cracking units process vacuum and coker gas oils as feed. These feeds contain low levels of nickel and vanadium, which are serious catalyst poisons. "Heavy oil" units, which use part heavy oil for feed, may operate with more than 2000 ppm Ni and V on the equilibrium catalyst. This is made possible by using a metals passivator, which decreases the dehydrogenation poisoning activity of Ni and V.

The catalytic cracker is one of the largest producers of "sour water" in the refinery, coming from the steam strippers and overhead accumulators on the product fractionators. The major pollutants resulting from catalytic cracking operations are BOD, oil, sulfides, phenols, ammonia, and cyanide.

Catalytic Reforming

In catalytic reforming (Figure 31.11), the object is to convert straight chain into cyclic molecules (aromatics of high octane). So reforming is essential to production of high octane gasoline. Platforming, the most widely used reforming process, includes three sections: (1) in the reactor heater section, the charge plus recycle gas are heated and passed through reactors containing platinum catalyst; (2) the separator drum separates gas from liquid, the gas being compressed for recycling; and (3) the stabilizer section corrects the separated liquid to the desired vapor pressure.

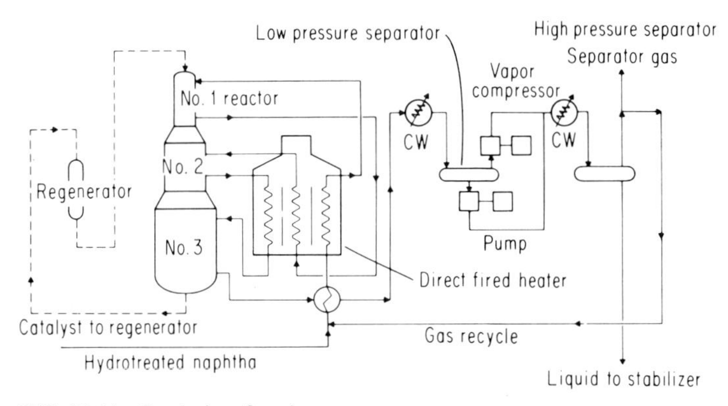

FIG. 31.11 Catalytic reforming process.

The predominant reforming reaction is dehydrogenation of naphthenes, or removal of hydrogen from the molecule. Important secondary reactions involve rearrangement of paraffin molecules. All of these result in a product with higher octane ratings that the reactants. Platinum and molybdenum are the most widely used catalysts, with platinum predominating because it gives better octane yields.

Because platinum catalysts are poisoned by arsenic, sulfur, and nitrogen compounds, feedstocks usually are treated with hydrogen gas (hydrotreated) before being charged to the reforming unit. This produces hydrogen compounds, such as H_2S, which can be removed from the hydrocarbon steam.

Reforming is a relatively clean process. The volume of wastewater flow and the pollutant concentrations are small.

Alkylation

The amalgamation of small hydrocarbon molecules to produce larger molecules is known as alkylation (Figure 31.12). In the refinery, this reaction is carried out between isobutane (an isoparaffin) and propylene, pentylenes, and in particular, butylenes (olefins). The product is call alkylate. The olefin-isobutane feed is combined with the fractionator recycle and charged to reactor (contactor) containing an acid catalyst at a controlled temperature. Aluminum chloride, sulfuric acid, and hydrofluoric acid are common acid catalysts. The content of the contactor are circulated at high velocities to expose a large surface area between the reacting hydrocarbons and the acid catalyst. Acid is separated from the hydrocarbons in a recovery section downstream and recirculated to the reactor. The hydrocarbon stream is washed with caustic and water before going to the fractionating sections. Isobutane is recirculated to the reactor feed, and alkylate is drawn from the bottom of the last fraction (debutanizer).

Acid may contaminate the cooling water should heat exchangers leak. Water drawn from the overhead accumulators contains varying amounts of oil, sulfides, and other contaminants, but they are not a major source of waste in this subprocess. The wastes from the reactor contain spent acids, which refineries may process to recover clean acids or may sell. Occasionally, some leakage to the sewer or cooling system does occur. The major contaminant entering the sewer from a sulfuric acid alkyation unit is spent caustic from neutralization of the hydrocarbon stream leaving the reactor.

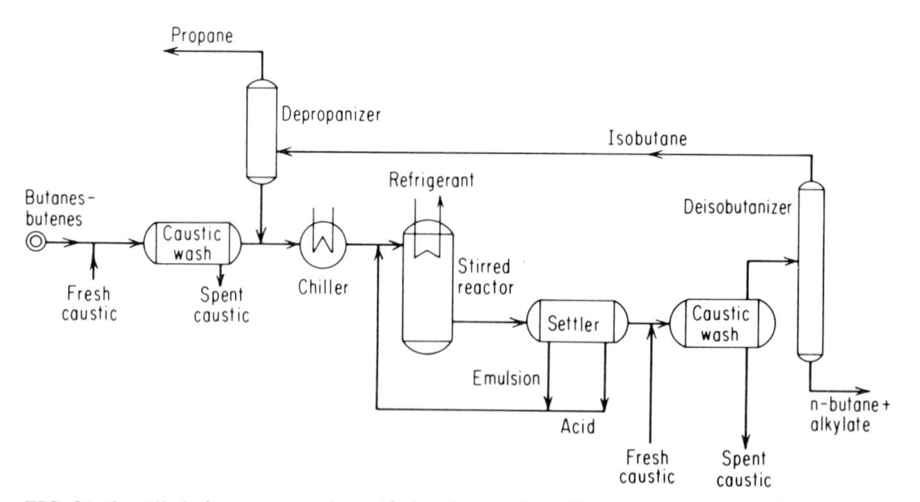

FIG. 31.12 Alkylation process using sulfuric acid. *(Adapted from Hengstebeck, R. J.: Petroleum Processing, McGraw-Hill, New York, 1959.)*

Hydrofluoric acid alkylation units do not have spent acid or spent caustic waste streams. Any leaks or spills that involve loss of fluoride constitute a serious and difficult waste problem.

Sulfuric alkylation units usually have a chilled-water refrigeration system with several compressors which have critical shell-side cooling water on inter- and after-coolers.

Hydrotreating

Hydrotreating is mild hydrogenation (Figure 31.13) that removes sulfur, nitrogen, oxygen, and halogens from a hydrocarbon feed and converts olefins (unsaturated hydrocarbons) to saturated hydrocarbons. Petroleum feedstocks ranging from light naphthas to lubricating oils are hydrogen treated. The major application of hydrotreating has been removing sulfur from feeds to catalytic reformers to prevent catalyst poisoning. Each of the different types of hydrotreaters, which vary in the selection of catalyst, incorporates a reactor and a separator. The oil feed, preheated to 400 to 700°F (204 to 371°C), passes through a fixed-bed reactor where it combines with hydrogen in the presence of a regenerable metal oxide catalyst at 200 to 500 lb/in^2 gage (14 to 35 kg/cm^2). The product stream is cooled before entering a separator where excess hydrogen gas is separated for use in other operations. After separation, the product is steam stripped for removal of residual hydrogen sulfide.

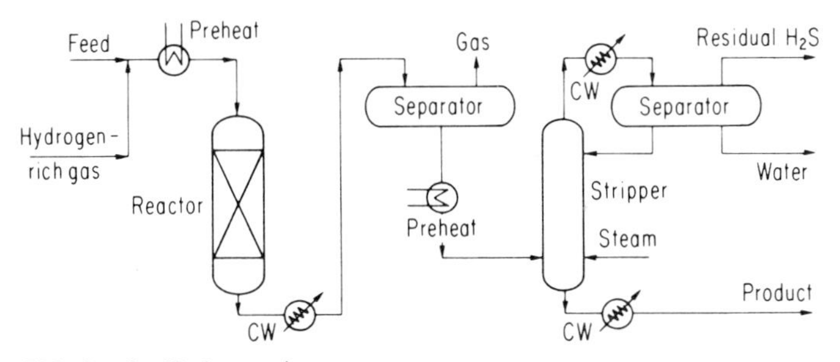

FIG. 31.13 Hydrotreating process.

Major wastewater streams come from overhead accumulators on fractionators and steam strippers, and sour water stripper bottoms. The major pollutants are sulfides and ammonia. Phenols may also be present, if the boiling range of the hydrocarbon feed is high enough.

UTILITY SYSTEMS

The steam-generating system is the heart of the refinery operation, since steam is a major source of energy in the refinery. It is used to drive pumps and compressors, to heat process streams, and to strip sour water. Few refineries generate electricity, but typically produce 10% of their motive power requirement. A typical water balance is shown by Figure 31.14.

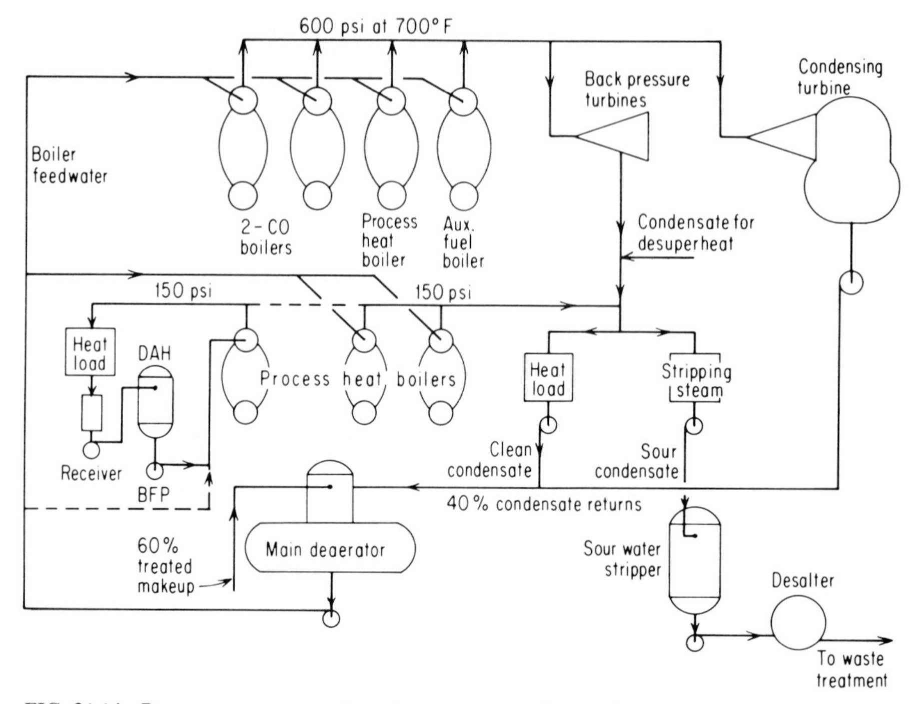

FIG. 31.14 Process steam uses and condensate recovery in a typical utility.

The pretreatment of boiler feed water is one of the most important steps in efficient boiler operation. The specific pretreatment scheme for a steam-generating system is dependent on such factors as boiler design, steam requirements, heat balance, outside power costs, and further expansion. Many refineries use hot process softeners, filters, and ion exchange trains.

Additional steam is generated in heat recovery and process-heat boilers. These boilers can be found throughout the refinery at the various process units. Process-heat boilers often resemble shell- and tube-heat exchangers in design. In most refineries, the steam generated by the process usually condenses as an acceptable quality condensate for return to the boiler that produced it. The major concern in these systems is prevention of corrosion in the condensate system, particularly at the point of initial steam condensation. In addition, depending on the pressure of the system, a condensate polishing unit may be installed to remove oil, corrosion products, or both, prior to returning this water to the boiler.

COOLING WATER SYSTEMS

As stated earlier, the cooling requirements of a refinery demand a large volume of water. It has been estimated that 80 to 85% of the total water requirements are for cooling if cooling towers are used for water conservation. The cooling systems in a refinery are similar to those found in many manufacturing plants, involving once-through, open recirculating, and closed cooling circuits. While each of these systems can be found in the refinery, the open recirculating system serves the greatest demand for process cooling. Once-through cooling is becoming rare except for coastal installations designed to use seawater.

REFINERY POLLUTION CONTROL—WASTE TREATMENT

Broadly speaking, a refinery wastewater system consists of:

1. A drainage and collection system.
2. Gravity-type oil-water separators and auxiliaries required to remove oil and sediment.
3. Treatment units or disposal facilities to handle segregated chemical solutions and other process wastes and to control the effects of pollutants that have toxic properties.

Figure 31.15 is a generalized list of wastewater sources; it also shows how they are segregated into the various sewer systems provided to optimize reuse and reduce overall treatment volumes to a minimum. The oil-free sewer collects wastewaters that have not contacted oil and that are not subject to any other contamination for which treatment must be provided.

Since these waters seldom contain significant oil contamination, they may bypass API separators and some may discharge directly into the refinery outfall

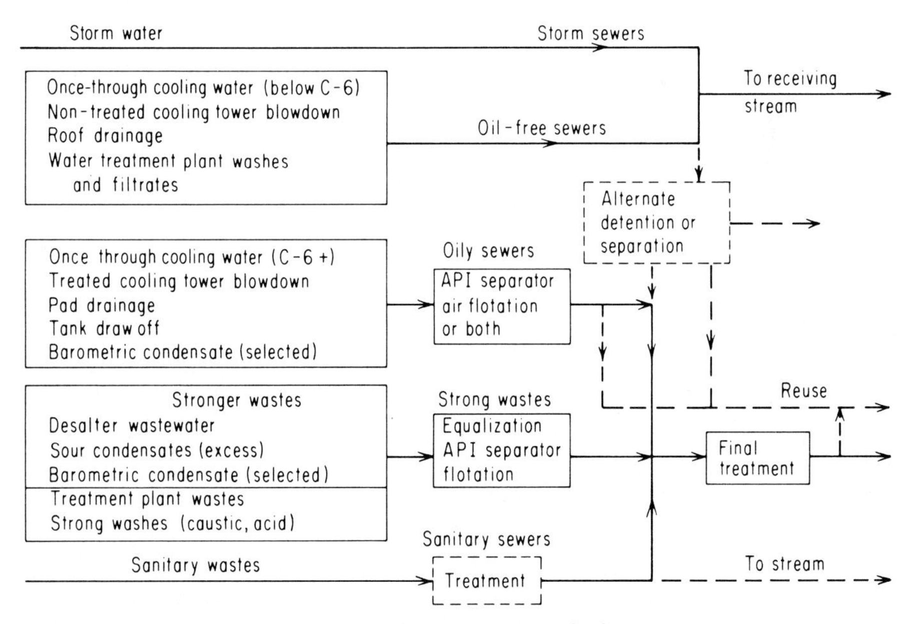

FIG. 31.15 Suggested scheme of collection and treatment of refinery wastewaters.

line. However, if collected in a common sewer, this flow is usually mixed with oily waste after API separators for a common treatment.

The oily cooling water sewer system is intended to handle waters that are expected to be subject to minor oil contamination from leaks in heat exchange equipment or from spills. In the absence of contamination by chemicals or fine solids which tend to cause emulsions, separation of the oil from the water can be readily accomplished. Barometric condenser cooling water subject to contamination by easily separable oil, but not containing emulsified oil, may also be included in this system.

The process water sewer system collects most wastewaters that come into direct contact with oil or that are subject to emulsified oil contamination or to chemicals limited by the plant discharge permit. Water from the process water sewer system is treated in an oil-water separator, and pollutants remaining after gravity separation are reduced by secondary treatment methods.

Separator skimmings, which are generally referred to as a slop oil, require treatment before they can be reused because they contain an excess of solids and water. Solids and water contents in excess of 1% generally interfere with processing.

The sanitary sewer system collects only raw sanitary sewage and conveys it to municipal sewers or to refinery treatment facilities. State or local regulations usually determine the sanitary disposal requirements. Raw sewage can be used for seeding refinery biological treatment units.

Special systems include those required for the separate collection and handling of certain wastes having physical or chemical properties that cause undesirable effects in the refinery drainage system, oil-water separators, or secondary treatment facilities. Spent solutions of acids and caustic, foul condensates, and degraded solvents are examples of such wastes.

Main wastewater treatment systems separate pollutants from the water by physical, chemical, or biological means. Primary treatment consists of physical, and often chemical, processes. Primary treatment separates the gross waste load of oil and suspended solids from the water. Secondary treatment removes much of the remaining organic and dissolved solid pollutants by biological treatment, which consumes and oxidizes organic matter.

There are a few physical, chemical, or biological methods known as tertiary treatment, including activated carbon adsorption and filtration. As pollution control regulations become more stringent, tertiary treatment methods will become more common.

Sour Water Stripping

Many wastewater effluents from petroleum refining processes originate from the use of steam within the processes. The subsequent condensation of the steam usually occurs simultaneously with the condensation of hydrocarbon liquids and in the presence of a hydrocarbon vapor phase that often contains H_2S, NH_3, HCN, phenols, and mercaptans. After separation from the hydrocarbon liquid, the condensed steam contains oil and a mixture of these contaminants. These wastewaters are typically called sour waters or foul waters because of the unpleasant odor characteristic of dissolved hydrogen sulfide.

The amounts of these contaminants in a sour water stream depend on the type of refining process from which the stream originated, as well as the feedstock to that process and the pressure level at which the steam was condensed within the process. The contaminant concentrations in typical sour waters will usually be 50 to 10,000 mg/L H_2S, 50 to 7,000 mg/L NH_3, and 10 to 700 mg/L phenolics.

The principal contaminants in sour waters are hydrogen sulfide and ammonia, ionized as HS^- and NH_4^+, respectively. These can be removed by single-step steam stripping, which is a simple form of distillation for the removal of dissolved gases or other volatile compounds from liquids. Stripping is rather inefficient and requires large volumes of steam because these ionized substances exert very little gas pressure unless the pH is adjusted. Stripping also removes phenolics to some extent, but the amount of phenolics removed may vary from 0 to 65%.

Effluent water from a sour water stripper is often used as makeup to the desalter. Refiners are investigating new schemes for incorporating portions of this water into refinery utilities systems.

Spent Caustic Treatment

Alkaline solutions are used to wash refinery gases and light products; the spent solutions, generally classified as sulfidic or phenolic, contain varying quantities of sulfides, sulfates, phenolates, naphthenates, sulfonates, mercaptides, and other organic and inorganic compounds. These compounds are often removed before the spent caustic solutions are added to refinery effluent. Spent caustics usually originate as batch dumps, and the batches may be combined and equalized before being treated or discharged to the general refinery wastewaters.

Spent caustic solutions can be treated by neutralization with spent acid or flue gas, although some phenolic caustics are sold untreated for their recoverable phenol value. Neutralization with spent acid is carried to a pH of 5 to ensure maximum liberation of hydrogen sulfide and acid oils.

In the treatment of spent caustic solutions by flue gas, hydroxides are converted to carbonates. Sulfides, mercaptides, phenolates, and other basic salts are converted to hydrogen sulfide, phenols, and mercaptans at the low pH conditions caused by the flue gas stripping. Phenols can be removed and used as a fuel or can be sold. Hydrogen sulfide and mercaptans are usually stripped and burned in a heater. Some sulfur is recovered from stripper gases. The treated solution will contain mixtures of carbonates, sulfates, sulfides, thiosulfates, and some phenolic compounds. Reaction time of 16 to 24 h is required for the neutralization of caustic solution with flue gas.

Shale Oil

Closely related to the petroleum refining industry in process technology is the developing synthetic fuels industry, with its two major segments, recovery of hydrocarbons from oil shale and conversion of coal to fuel gas, hydrocarbons, or a family of petrochemicals.

The processing of oil shale involves large-scale mining operations since the amount of oil in economically treatable shale is usually only 10 to 15% of weight of the native shale. At 15%, the production of 100,000 barrels of oil would produce a residue of about 100,000 tons of spent shale.

The release of oil is accomplished by heating the raw shale in a device like a kiln, although there is hope that in situ retorting may prove practical in the future. Since the heat capacity of the shale is high, this process consumes a great deal of energy, which is one reason for its rather slow development. The hydrocarbons released from the shale may be processed by fractionation and extraction to yield light ends and heavy fractions that may be used as refinery feedstock or fuel for retorting and steam generation.

Wastewater problems are similar to those of the refineries and coal-fired utility stations, the major one being the runoff from the large volumes of spent shale residue.

SUGGESTED READING

Lund, H. F., (ed.): *Industrial Pollution Control Handbook,* McGraw-Hill, New York, 1971.

Shreve, R. N.: *Chemical Process Industries,* 3d ed., McGraw-Hill, New York, 1967.

U.S. Environmental Protection Agency: *Development Document for the Petroleum Refining Industry,* EPA-440/1-74-014-a, April 1974.

CHAPTER 32
STEEL INDUSTRY

Archeologists have used the terms Stone Age, Bronze Age, and Iron Age to delineate successive stages of civilization. During the Bronze Age, humans had learned to use charcoal as a reducing agent, as well as a fuel, to convert mixed ores of copper and tin into molten bronze in clay furnaces. Iron ores were not so easily reduced, and the same process that yielded molten bronze from copper and tin minerals produced, instead, a spongy mass when iron ores were processed. The metalsmith had to learn to beat this mass into a solid, and in the process most of the slag fell away, but some was retained in the iron to produce what is known as wrought iron. Some metalworkers learned how to reheat this mass in the forge so that it dissolved some carbon from the charcoal and was thereby converted to steel. Such secrets were carefully guarded as they provided the balance of power in warfare and in commercial trade.

It was not until the late Middle Ages that relatively large-scale production of iron was achieved with the development of the blast furnace. But the iron produced by the blast furnace (pig iron) contained a high carbon content; it was excellent for casting into shapes in molds, but it wasn't malleable. Processes were developed to convert pig iron into more malleable wrought iron or steel, and today over 95% of the steel produced still follows the two-step process of blast furnace followed by the converter that burns carbon out of molten iron to produce steel.

However, in developing countries that have fossil fuel reserves, direct reduction processes are being developed that turn back the clock of history and produce "sponge iron." This product is mixed with scrap steel and the charge is melted in electric furnaces to produce steel. In the United States, "mini-mills" have sprung up to compete with the large integrated steel mills, and their major source of raw material is steel scrap. Almost one-third of U.S. steel production now comes from these electric furnaces, creating a demand for scrap steel. This has raised its price to the point that sponge iron (or DRI, for "direct reduced iron") will be required for future electric furnace charging with the likelihood that DRI processes will continue to grow.

Nationwide, in both the large integrated mills and the smaller mini-mills, iron and steel making use more water than any other manufacturing industry. A U.S. survey disclosed that the industry water usage was about 5 bgd (13,100 m^3/min) of cooling water and 3.5 bgd of process water. Approximately 2 bgd of this water required some form of chemical treatment to make it suitable for use. Of the 3.5 bgd of water that was used in processes, about 65% required some form of treatment before being discharged. As a rough estimate for an integrated mill, 120 tons of water are used for every ton of steel produced (Figure 32.1). A summary of cooling water uses is given in Table 32.1.

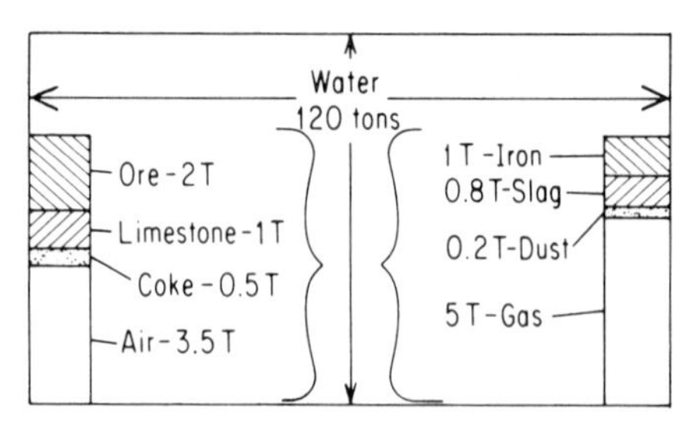

FIG. 32.1 To produce 1 ton (0.907 t) of finished iron, the blast furnace requires 120 tons (109 t) of water.

TABLE 32.1 Cooling Water Use in an Integrated Steel Mill

	gal/ton	m^3/kkg
1. Blast furnace	8000	33
2. Basic oxygen furnace	700	3
3. Open hearth furnace	3600	15
4. Electric arc furnace	2600	11
5. Continuous casting	4200	18
6. Bar mills (hot)	1000	4
7. Plate mills	4080	16
8. Pipe mills	24,800	103
9. Cold sheet mills	3360	14
10. Hot sheet mills	16,000	66
11. By-product coke plant	4500	19
12. Sinter strands	350	1.5

BLAST FURNACE OPERATIONS

The blast furnace is the heart of the iron-making process. Iron ore, coke, and limestone are charged by a skip car to the top of a furnace. Most furnaces range from 16 to 28 ft (4.9 to 8.5 m) in diameter and may be 100 ft (30.5 m) tall. The raw materials, fed in layers to the furnace, are iron ore or pellets, coke, and limestone. In the lower portion of the furnace, hot blast air is injected through a series of circumferential openings called tuyeres, at about 30 to 35 lb/in^2 (2 bars) and 1800°F (985°C). The air flows upward through the burden of raw materials in the furnace, and gas exits the furnace top at 4 to 5 lb/in^2 (0.3 bars).

Coke (carbon) reacts with Fe_3O_4 and Fe_2O_3 in the furnace, releasing the iron and producing CO and CO_2 gas. The iron sinks to the furnace hearth where the original impurities in the charge combine with the lime, forming slag, which floats to the top of the iron. The gas leaving the top of the furnace carries fine dust, which is separated and recovered in collection equipment. The combustible exhaust is used in reheating stoves and in generating steam at boiler houses. The iron and slag are periodically tapped and collected in special railcars every 2 to 6 h, depending upon furnace conditions.

To keep this large furnace working efficiently, much of the equipment connected with the furnace uses cooling water at various points. A modern blast furnace requires 1000 to 15,000 gal/min (63 to 950 L/s) of cooling water. Figure 32.2 illustrates the overall dimensions and general cooling water utilization for a typical blast furnace. This generalized drawing shows that the blast furnace may be constructed in two ways: water sprays may provide cooling water on the outer shell of the furnace, as illustrated on the left side; or, as illustrated on the right side, the furnace may contain stack plates and bosh plates which are hollow passages for cooling water built into the furnace wall. The primary task of the cooling

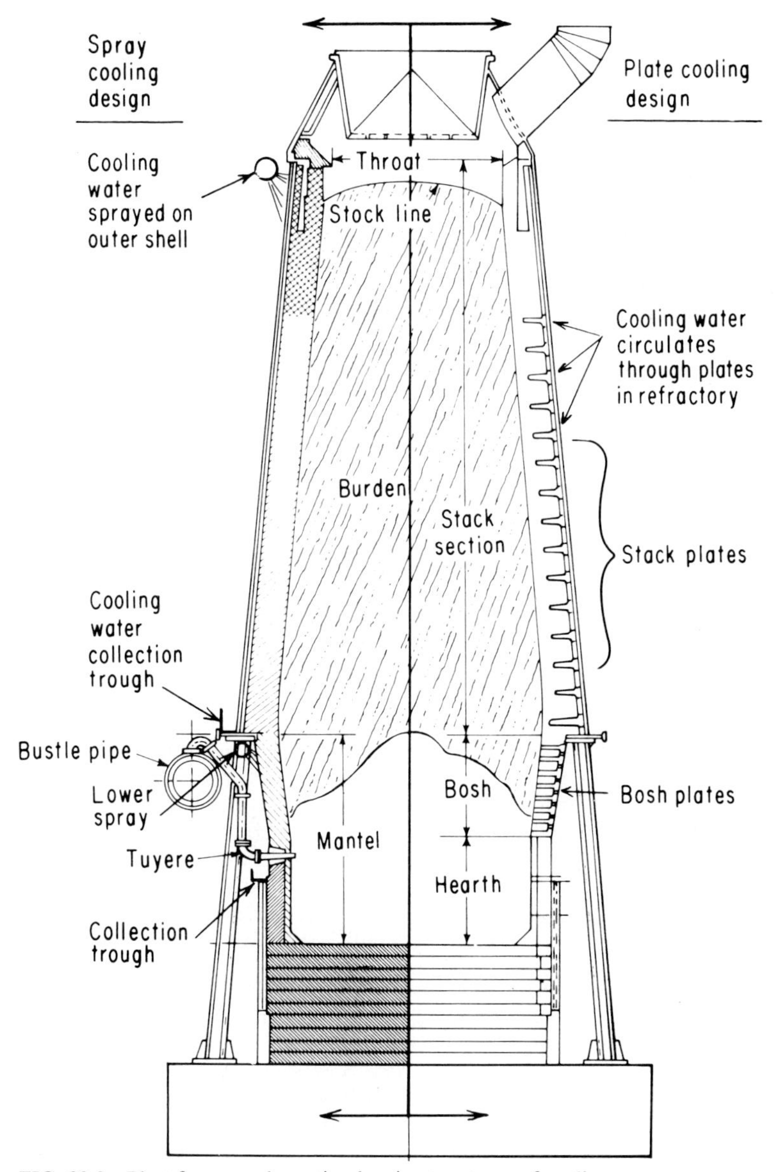

FIG. 32.2 Blast furnace schematic, showing two types of cooling-water systems.

water on the stack area is to prolong the life of the refractory inside the furnace. The stack area uses about one-third of the cooling water through the furnace.

Where plates are used, these are generally connected vertically in a series of four to seven plates with water flow of 15 to 50 gal/min (1 to 3 L/s) in each series. Temperature rises 12 to 25°F (7 to 14°C) throughout the series. The cooling demand for the refractory decreases as height above the ground increases.

Figure 32.3 shows a hearth and bosh section of the blast furnace. Here, too, either sprays or plates may be used for cooling. The bosh area reaches the maximum furnace temperature, ranging up to 3500°F (1927°C). At this point cooling requirements for the furnace are most critical.

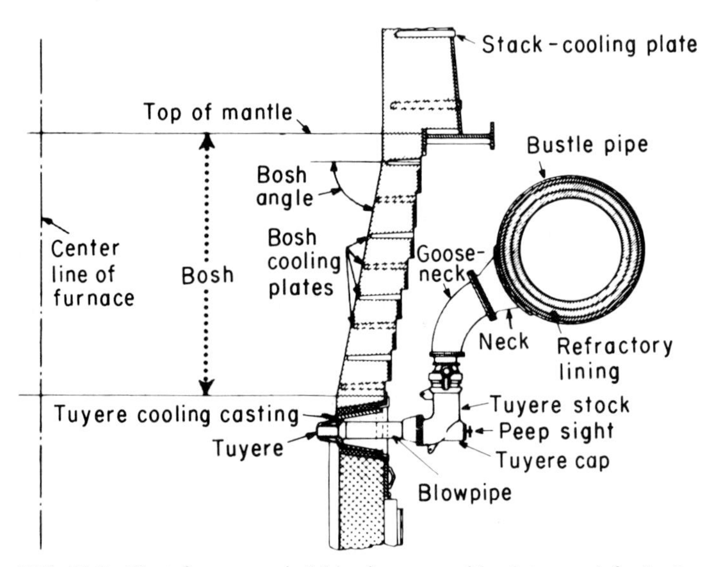

FIG. 32.3 Blast furnace—wind blowing area with plates used for bosh cooling. *(Copyright 1971 by United States Steel Corporation.)*

The left-hand side of Figure 32.2 indicates a carbon or closed bosh arrangement. Here, the water sprays are directed at steel plates containing the carbon brick of the bosh area. It may be seen that the spray water is collected in an annular trough directly above the tuyere zone. Also, a stack trough is indicated just at the top of the mantle line. Cooling water from the blast furnace is generally collected in a circular trough surrounding the furnace itself. On the right-hand side of Figure 32.2, a cooling plate arrangement is indicated. Again, plates in this area may be connected in a series of four to six plates with a temperature rise of 12 to 25°F (7 to 14°C).

The cooling water in the lower section or hearth is not shown on the drawing. Normally, large diameter steel pipes run directly through this area to provide cooling. Stack and bosh cooling equipment is easily accessible and changeable at failure. However, should pipes in the hearth area fail, it is impossible to remove and replace them without major overhaul of the furnace. Figure 32.2 also shows the area in which the molten metal and slag reside. Molten materials are generally not allowed to reach a higher level than just below the tuyeres.

In the tuyere area, air preheated in stoves is blown into the furnace. The tuyeres are copper-jacketed nozzles with cooling water in the jackets. Heat exchange rates are high, so it is important that the cooling system be protected from fouling or plugging to maintain adequate heat transfer.

The valves controlling the flow of hot blast air from the stoves to the furnace must also be cooled to prevent failure or jamming.

The basic problem of water chemistry for all of these systems is protection from corrosion, scale, and fouling from silt or microbiological growths. One of these or any combination may exist within a cooling system at any time.

EXHAUST GAS TREATMENT

Hot air blown through the furnace is changed in composition and expands in volume. The exit gas velocity is high and entrains solids, principally burden fines—ore, coke, and limestone. This dirty gas passes through a dry dust collector, where a majority of the heavier solids are removed, and then proceeds to a wet scrubbing system. Scrubbing water and the dirty gas collide in a venturi or orifice system where almost all suspended solids are removed—usually over 99%. The scrubbed gas has a Btu value of 85 to 100 Btu per standard cubic foot (1950cal/m^3) and is used for heating stoves or firing boiler house furnaces.

The scrubber water containing high concentrations of suspended solids, from 500 to as high as 10,000 mg/L, is normally sent to a thickener or clarifier. Here, the solids are settled, and the effluent water is either sent to the receiving stream, sometimes after additional treatment, or recycled (Figure 32.4).

FIG. 32.4 A clarifier-thickener of this design treats blast furnace gas scrubber water for removal and recovery of solids. *(Courtesy of Inland Steel Company.)*

The heaviest concentrations of solids settled from the water are normally iron, silica, and limestone. Soluble impurities usually include ammonia, phenols, and cyanide. The chemistry of the scrubbing water continually varies as it is being exposed to hot, dusty gas and then cooled and clarified before recycle. Cooling towers are often used to reduce temperature after the water contacts the hot gas.

Evaporation of pure water vapor in the cooling tower concentrates dissolved solids within the water and affects its chemical balance. This must be taken into account when planning a proper cooling water program, as blowdown is required to control the salinity of the recirculating water, and chemical treatment is needed for control of scale, deposits, and corrosion.

Additional water is used for slag granulation or for slag cooling. Where the water is recycled within a slag pit area, there is usually a high potential for deposition occurring within the recycling lines and pumps, and the water chemistry of these systems must be continually monitored.

There are a number of boiler houses in a steel mill complex, and the boiler house in the blast furnace area is one of the most important. Most of these installations have 900-lb/in^2 boilers, but some of the older plants continue to operate a few 450- to 600-lb/in^2 boilers with the higher pressure units. A major use for steam is to operate turbines, which drive the large turbocompressors delivering air to the blast furnaces. A typical turbocompressor discharges 100,000 scfm ft^3/min (standard) (2830 m^3/min) air at a pressure of 30 to 35 lb/in^2 (2 bars), requiring a steam turbine using about 350,000 lb/h (160,000 kg/h) steam at 900 lb/in^2 (60 bars). These turbines operate on a condensing cycle, but some low-pressure steam may be extracted for operation of auxiliaries, such as fans, pumps, and compressors in the utility area.

STEEL PRODUCTION

Steel is manufactured from iron—the blast furnace product—by three different methods: the basic oxygen process, the open hearth process, and the electric arc process. The objective of each is to reduce impurities; for example, the 4% carbon content of the iron is reduced to about 0.2% in the steel product, depending on the metallurgical specifications of individual orders.

Basic Oxygen Process

In the basic oxygen method, the mixture of hot metal from the blast furnace (usually 50 to 60% of the total charge), scrap steel, and slag conditioning materials, such as lime and fluorspar, are charged to the furnace (Figure 32.5). Oxygen at a rate of 15,000 to 20,000 ft^3/min (425 to 566 m^3/min) is injected through a lance lowered into the vessel only inches above the raw materials. The oxygen blowing period continues for 20 to 25 min to melt and burn off impurities.

A typical BOF vessel has a 100 to 300 ton (90 to 270 kkg) capacity and produces steel in about 45 min. The vessel capacity is usually filled before the blowing of oxygen to allow space for the violent reactions to occur. In the Q-BOP system the oxygen is blown through the bottom of the vessel, working through the materials and thus reducing the amount of violent splashing.

There are several water uses in the basic oxygen unit. First, the oxygen lance must be water cooled. In most plants this is a closed recirculating cooling water system. In most of these closed systems, the lance water flows through the shell side of a heat exchanger, with cooling water on the tube side of the exchanger.

Because of the high heat release, gases leaving the furnace hood during the oxygen blow are very hot. The hood is usually cooled with water recirculating through the hood panels. There are several systems where boilers are installed in the hood area for waste heat recovery and cooling of the gases.

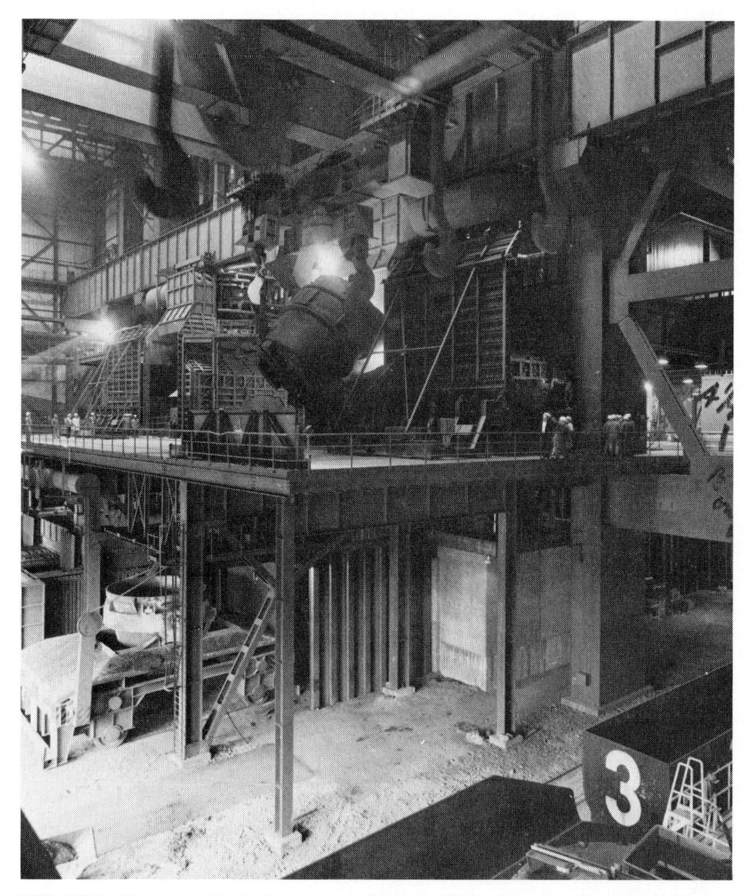

FIG. 32.5 Hot metal is being poured from the ladle into the tilted BOF furnace as the first step in the steel-making process. *(Courtesy of Baumco Gesellschaft für Anlangenstechnik mbH, Essen, West Germany.)*

As the gases leave the hood area, they can be further cooled by a wet scrubber and cooling system which requires large volumes of water (Figure 32.6). This water is then sent to clarifier-thickeners for sedimentation of the solids, and the water can than be recycled or discharged. There is a wide swing in water composition through the entire heat, as shown by a pH record of the effluent. Those systems not using wet gas scrubbers normally have electrostatic precipitators.

Open-Hearth Process

In the open-hearth process (Figure 32.7) the same basic materials used in the BOF process are charged to the open hearth furnace. Hot metal is not as essential to the open hearth as to the basic oxygen unit. These furnaces normally produce 100 to 600 tons (90 to 540 kkg) of steel per heat over a period of 6 to 12 h.

In the open-hearth furnaces, oxygen lance cooling is also required, similar to

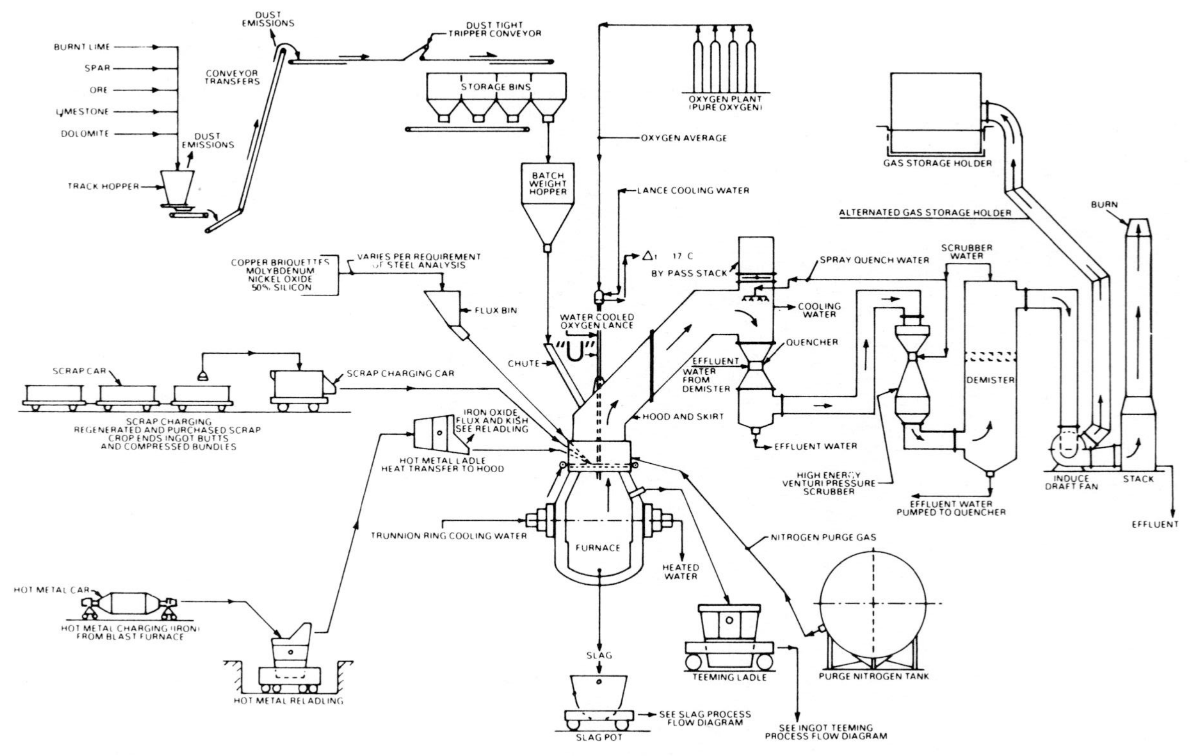

FIG. 32.6 Basic oxygen furnace, showing major water uses for cooling and scrubbing. As different materials are added to the vessel during a heat, the composition of the scrubber water varies. *(From EPA 440/1-74-024a, "Development Document for Effluent Limitations Guidelines and New Source Performance Standards for the Steel Making Segment of the Iron and Steel Manufacturing Point Source Category.")*

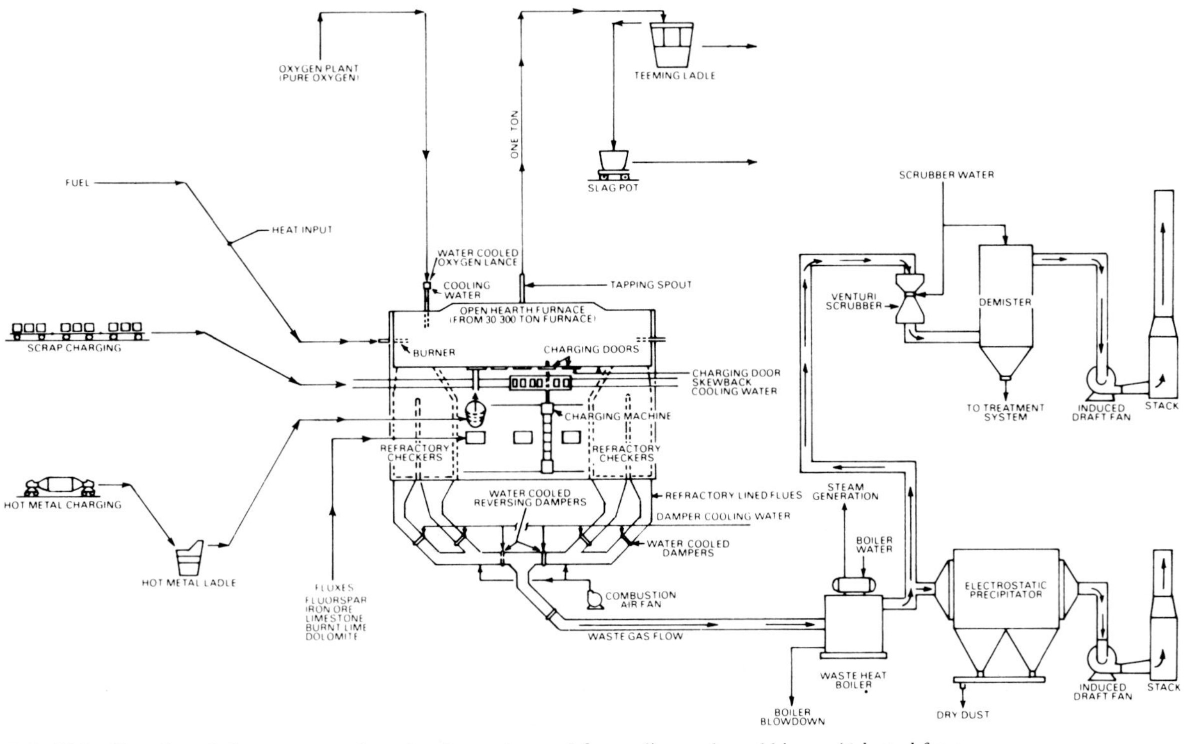

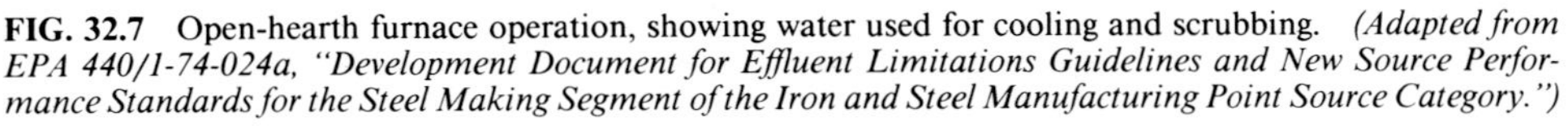
FIG. 32.7 Open-hearth furnace operation, showing water used for cooling and scrubbing. *(Adapted from EPA 440/1-74-024a, "Development Document for Effluent Limitations Guidelines and New Source Performance Standards for the Steel Making Segment of the Iron and Steel Manufacturing Point Source Category.")*

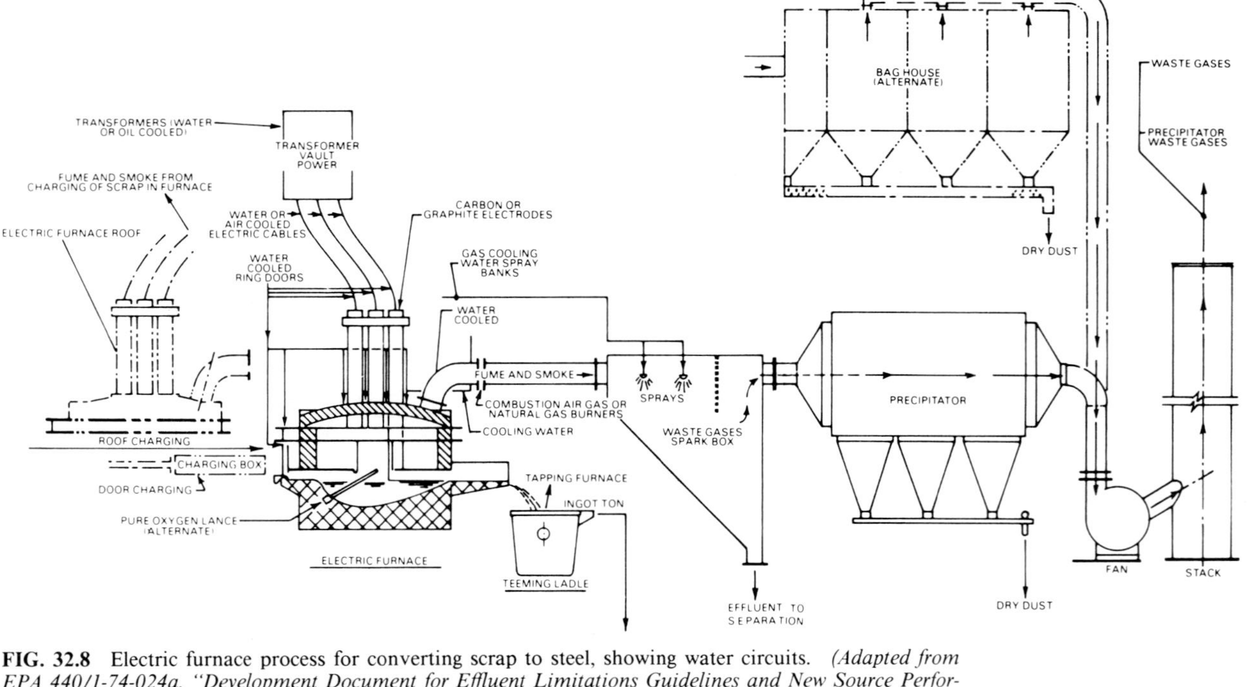

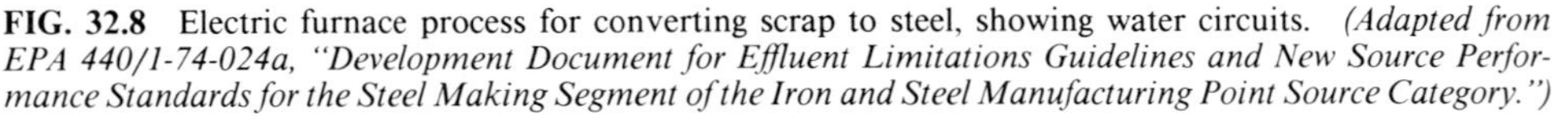

FIG. 32.8 Electric furnace process for converting scrap to steel, showing water circuits. *(Adapted from EPA 440/1-74-024a, "Development Document for Effluent Limitations Guidelines and New Source Performance Standards for the Steel Making Segment of the Iron and Steel Manufacturing Point Source Category.")*

the BOF process. In addition, cooling water in the range of 750 to 1500 gal/min (2.7 to 5.4 m^3/min) is required to cool the skewback channels and doors in the furnace.

Recirculating washwater from open hearth scrubbers is usually acidic so that special materials of construction are required in closing up these systems. However, because the pH of the circulated water is in the range of 2.5 to 3.0, this system can be completely closed without fear of scale. The only water loss is that present in the sludge or filter cake.

The open-hearth process has disappeared from most mills because of the high cost of providing pollution control equipment to handle the acidic dust-laden gases produced.

Electric Furnace Process

The third method of steel making is the electric furnace process (Figure 32.8), which can produce either the common grades of low-carbon steel, or, by charging with alloying materials, special steel such as stainless or tool steel. Electric furnaces normally operate on scrap and have the advantage of being adaptable to almost any part of the country, close to special markets. Because they are not dependent upon the hot iron from a blast furnace for their production, they are not tied down to the traditional steel centers.

Most electric arc furnaces are equipped with water-cooled doors. Water cooling is applied also to the roof ring, the electrode ring, and the electrode clamps. While most electric furnace operations use baghouses to clean their gases, there are plants using wet gas scrubbers.

DIRECT REDUCTION PROCESSES

As true of many developing technologies, a wide variety of processes are currently competing to determine which will be best for a newly developing market, DRI—direct reduced iron. These processes are designed to handle a wide selection of reducing agents from solids (coke, as used in the blast furnace, coal, and lignite) to liquids (oil) as well as gases (coke oven gas, reformed natural gas, and producer gas). The largest DRI production rates are from gas-fired furnaces. The charge may include the solid reductant, with supplemental fuel sometimes added, plus iron ore or pellets, and limestone or dolomite as a sulfur-reducing agent.

There are basically two furnace designs.

1. The horizontal furnace is essentially a tilted kiln (like a lime or cement kiln) with air and fuel fired into the discharge end countercurrent to the flow of solids. Air distribution to the kiln is critical to good temperature control to avoid clinker formation and poor control of carbon content of the sponge iron. Coal can be used directly as a portion of the charge and as fuel, so this type of furnace favors coal-rich nations. However, production can be increased if the secondary fuel is oil or gas instead of coal. Less than 20% of current DRI production worldwide now comes from horizontal furnaces (Figure 32.9).
2. The vertical shaft furnace (Figure 32.10) has been designed for either fixed-bed or moving-bed operation. The reductant and fuel are usually both natural gas; however, coal can be a portion of the reductant, and coal can be converted to

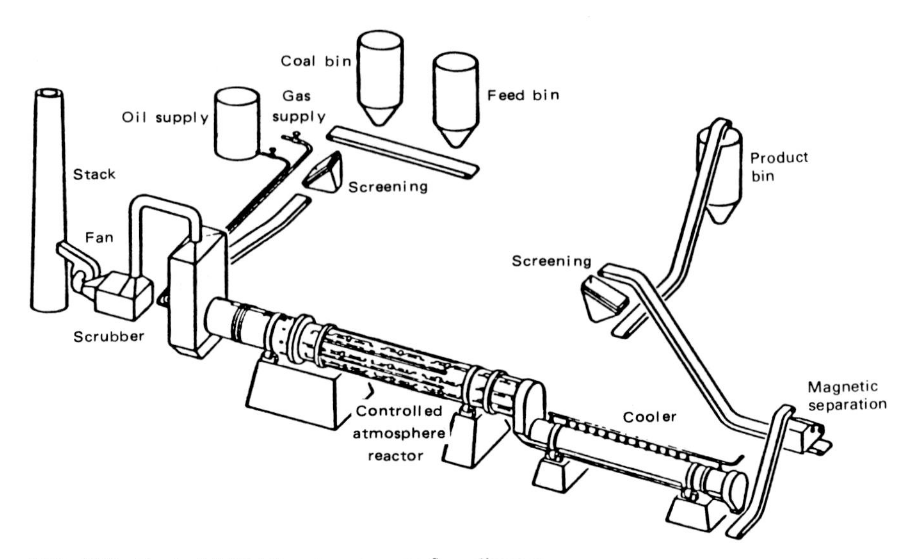

FIG. 32.9 Typical ACCAR system process flow diagram.

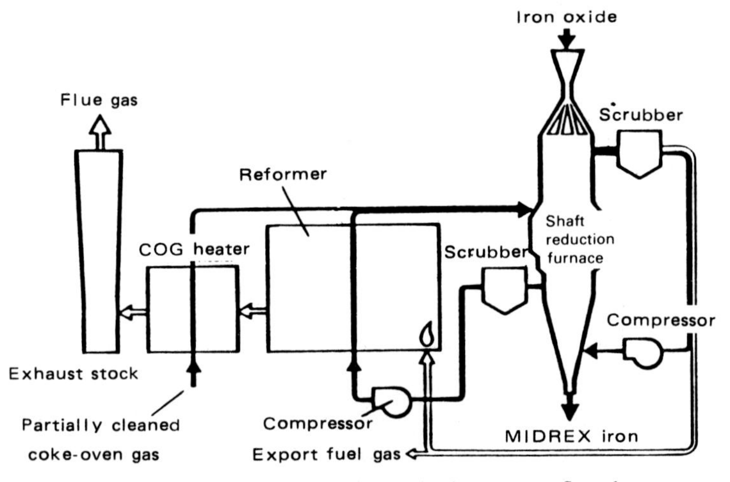

FIG. 32.10 MIDREX DR process—alternate coke-oven gas flow sheet.

producer gas (coal gas), so indirectly coal could be used as both reductant and fuel. Natural gas, however, is most favorable, and it is expected that the OPEC countries will become the major DRI producers, with Venezuela already producing over one-quarter of DRI production worldwide.

The sponge iron discharged from the furnace is crushed, screened, and separated magnetically from slag and char, the latter being recovered in some versions of DRI process. The iron may be briquetted or go directly to electric furnace charging.

Water is used for cooling in a number of operations, such as recycle gas cooling and product cooling or quenching. Water is also used for gas scrubbing, and in the case of coal gas or reformer gas production, for steam generation.

CONTINUOUS CASTING

Continuous casting (Figure 32.11) was developed to reduce the overall cost of steel manufacturing by eliminating several steps in conventional steel preparation such as ingot teeming, soaking, and blooming. Continuous casting is the process of continuously pouring molten metal from a ladle into the complex casting equipment which distributes the liquid, shapes it, cools it, and cuts it to the desired length. The casting is continuous as long as the ladle has available metal.

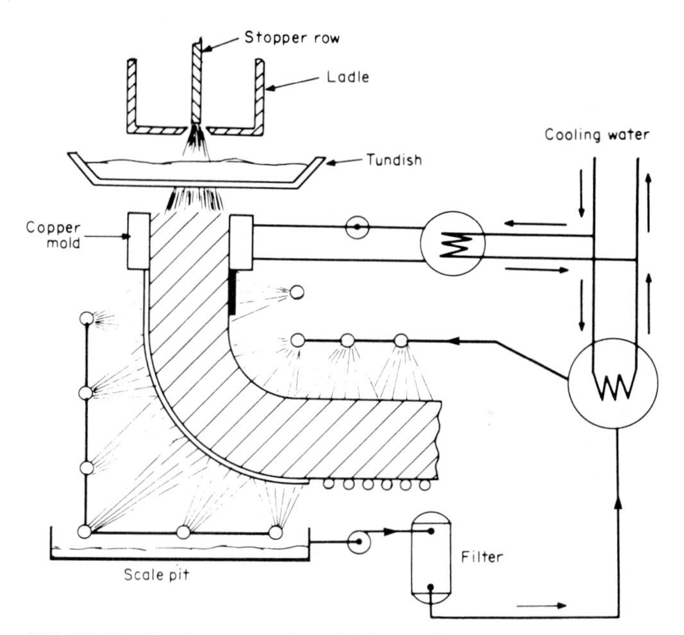

FIG. 32.11 Continuous casting of slabs or billets.

Should one ladle follow another without interruption, the process is called "piggy-backing a cast."

Correct water treatment and distribution is critical to continuous casting. Steel leaving the ladle at about 2800°F (1550°C) is poured into a trough called a tundish. The bottom of the tundish has one or more openings through which the molten steel is distributed to form slabs or billets in the forming area called the mold. The mold is a water-cooled copper jacket providing for high heat exchange rates. At the start of a cast, a dummy bar is moved close to the top of the mold to completely seal the interior. As the cast starts, this bar is slowly lowered to receive molten metal, and the cooling effect of the water-jacketed mold starts the formation of a metal skin. Proceeding through the length of the mold a distance of 30 to 36 in (1 m), the dummy bar and skin-contained metal are exposed to a series of direct-contact water sprays which complete the job of solidifying the steel. As solidification is completed, the dummy bar is cut from the formed metal and removed. The continuously moving, completely formed billet or slab then moves through guides to the straightening rolls and onto the runoff table for cutting to specific lengths.

The crucial point in this process is the copper water-cooled mold which forms the initial skin. Unless the skin is formed quickly and uniformly, a breakout will occur, shutting down the operation. The most reliable cooling water program uses the highest quality water available in a closed loop with a secondary open cooling loop. Condensate, high-purity boiler feed water, or low-hardness waters have been used as makeup. Hardness levels should never exceed 10 mg/L. Since the system is closed, there is little loss and the best corrosion inhibitors and dispersants can be used.

Spray water that contacts the billet or the slab becomes contaminated with iron oxide particles as the hot metal is oxidized. The water is normally processed in a filtration system for solids removal, recirculated through heat exchange equipment, and recycled to the sprays. The sprays must be kept from plugging at all times because the flow of water to the billet or slab being cooled must be uniform at all points.

Auxiliary mechanical equipment of the continuous casting machine is also water cooled. This may have a separate cooling water system, or it may be consolidated with the spray cooling water systems.

THE HOT-MILL ROLLING OPERATION

The hot mill (Figure 32.12) produces such products as sheets, plates, bars, rods, and structural shapes. This operation is the largest user of water in the steel mill, a reasonable use being about 7500 gal per ton of hot metal rolled (31 m^3/kkg).

The first step in rolling is heating the steel billet or slab in a reheat furnace to as high as 2350°F (1300°C). Cooling water must be used to cool the doors and frames in some of the reheat furnaces, and this may be either once-through or recirculated. As the heated billet or slab leaves the furnace, high-pressure descaling water up to 2000 lb/in^2 (135 bars) is blasted onto the surface to remove any oxide scale so that no imperfections are caused by rolling this debris into the metal.

Water is also used for roll cooling and for spraying directly onto the steel prior to its being handled at the end of the machine. A high-speed sheet mill operates in the range of 4000 to 6000 ft/min (1220 to 1830 m/min); a high-speed rod or wire mill operates at 8000 to 9000 ft/min (2440 to 2750 m/min). As water passes over the hot metal from rolling station to rolling station, the oxides washed from the metal are carried to a scale pit. There is a wide particle size range in the operation of descaling as the slab or billet is going through the mill. Larger particles are removed during the initial rolling (the roughing end), and very fine particles are washed off in the final rolling operations (the finishing end).

Much of the scale encountered in the scale pit can be removed with clam-type diggers, electromagnets, or traveling screen grates, but the fine-sized particles are normally separated by coagulation in water clarification equipment. Most new plants recycle the water used on the hot strip mills (Figure 32.13).

When the water is recycled through the mill, attention must be given to potential problems of scale, corrosion, fouling, and microbial activity. This is especially

FIG. 32.12 Hot mill rolling plate for further reduction through a hot strip mill. *(Courtesy of Tippins Machinery Company, Inc.)*

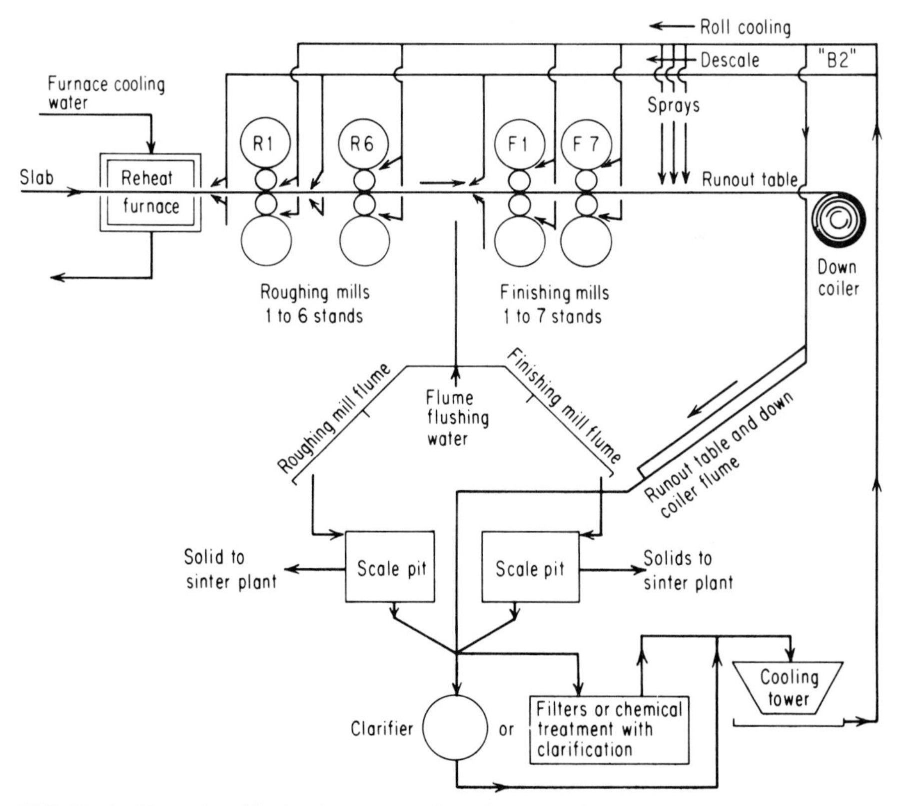

FIG. 32.13 Hot strip mill, showing recycle of clarified water for roll cooling and scale removal.

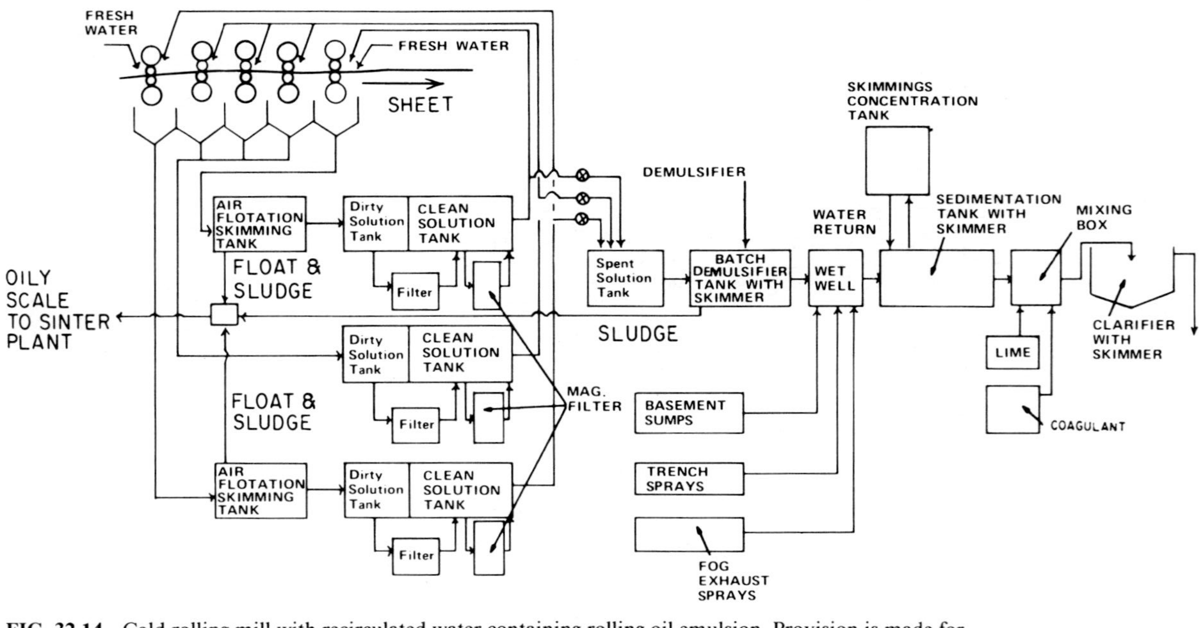

FIG. 32.14 Cold rolling mill with recirculated water containing rolling oil emulsion. Provision is made for periodic treatment of spent rolling oil emulsions.

true for water going to the high-pressure sprays. Heat removal may be required for controlling the work environment.

Considerable amounts of water are used for cooling electric motor systems in many of these mills. The motor-driven rolls keep the product moving to its end point. There can be as many as 300 to 400 motors at an installation.

COLD ROLLING MILLS

Cold rolling mills are divided into two categories: single stand and multistand, where steel is rolled in tandem. Because the steel is cold, it is hard to work, so the cold rolling requires lubricant in water (soluble oil) not only to cool, but to give a good finish to the steel. Water properties to be controlled in this operation include total suspended solids, iron, and oil. There are two systems for feeding lubricant and water in a cold rolling operation: recirculating and once-through.

Recirculating System

In the recirculating system a weak emulsion of oil and water circulates to the roll for cooling and lubrication of the sheet, collects under the roll, and passes through the treatment process. The emulsion is carried from the first stand to the second, third, and fourth. Fresh water is used only on the first and last stands. The spent liquid is normally collected, and an emulsion breaker is used to free the oil. The solids are settled and reclaimed for the iron content, and the oil is reclaimed and reused (Figure 32.14).

Water treatment for these systems consists of sedimentation, flocculation, filtration, and air flotation. The flows from these operations vary from 200 to 1500 gal per ton of steel processed (0.8 to 6 m^3/kkg).

Once-through System

Another cold rolling operation using direct application of oil is the once-through system, used on thin gauge material such as tin plate. The usual treatment system serves a multistand mill, having two to five stands. A 5 to 10% oil-in-water emulsion is applied to the steel at the first four stands, while a detergent solution is applied at the last stand. A once-through system is used for this service since the water must be kept clear. The wastewater, which contains a significant amount of oil, goes to a treatment system that includes an air flotation unit with oil skimmer, chemical treatment, aeration tanks, flocculating tank, and a settling basin or a clarifier.

Emulsion breaking chemicals may be required for efficient treatment of this waste. The oil is reclaimed and reused, while the sludge is disposed of as landfill.

HEAT TREATMENT

To produce special physical properties in certain grades of steel, the metal passes through a series of heat treatment operations including heating in a furnace,

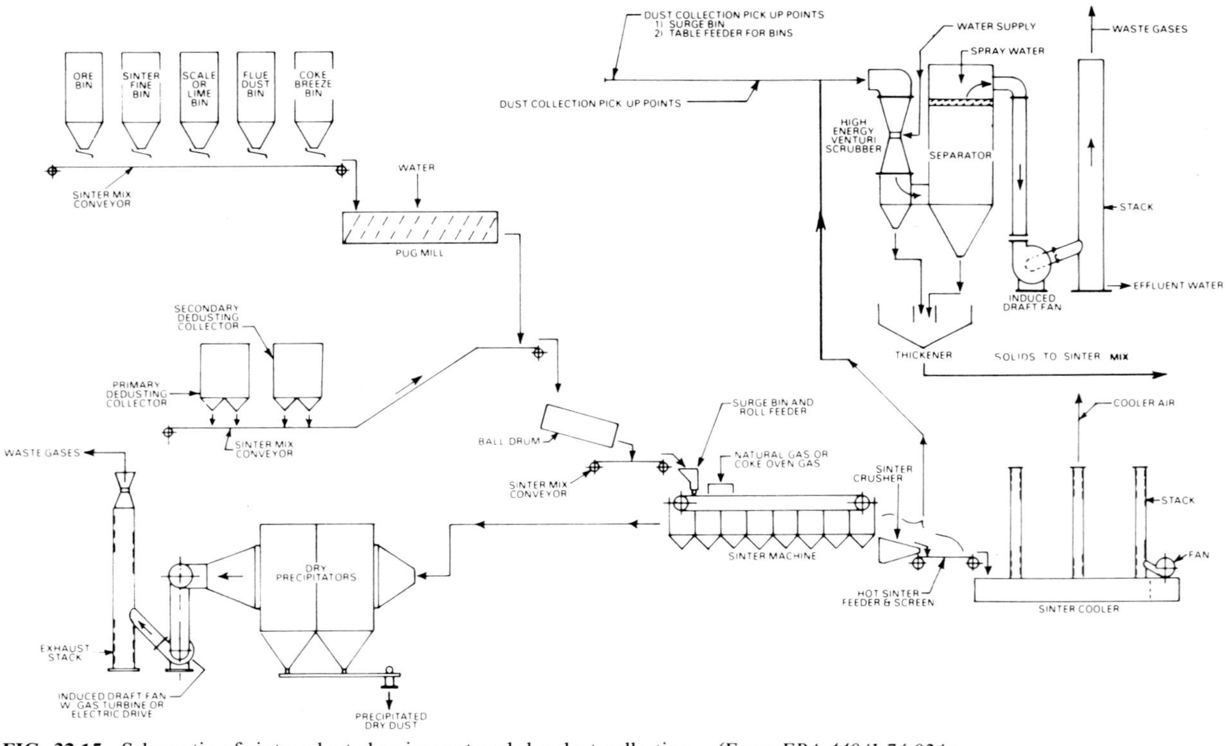

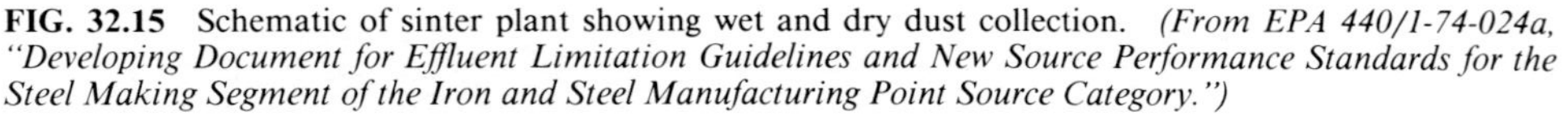

FIG. 32.15 Schematic of sinter plant showing wet and dry dust collection. *(From EPA 440/1-74-024a, "Developing Document for Effluent Limitation Guidelines and New Source Performance Standards for the Steel Making Segment of the Iron and Steel Manufacturing Point Source Category.")*

annealing at a carefully controlled temperature for a specified period of time, quenching in water or oil, and final cooling in air.

Generally, the temperatures in the annealing furnaces are not so high as to require water cooling of the furnace elements, but some cooling may be needed in special cases. The temperature of the quench oil or water quenching tank must be carefully controlled, so the coolant in the quench tank is usually recirculated through a heat exchanger to remove the heat brought into the system from hot metal. Oil discharge, such as could arise by overflowing the quench oil system or by rupture of a tube in the oil-water heat exchanger, must be guarded against. Oil is generally the only likely contaminant in the heat treatment area.

SINTERING

Sintering is a process that recovers solid residues from scrubbers and clarifiers. This process includes collection of useful materials such as iron ore fines, mill scale, limestone, flue dust, and coke fines (Figure 32.15).

The various materials are mixed in controlled proportions with a fixed amount of moisture then distributed onto a permeable grate and passed through to an oil- or gas-fired furnace (ignition furnace). Combustion air is drawn downward through the bed. After a short ignition period the firing of the bed surface is discontinued, and a narrow combustion zone moves downward through the bed, with each layer in turn heating to 2200 to 2250°F (1204 to 1228°C). In advance of the combustion zone, moisture and volatiles vaporize. In the combustion zone bonding of the particles occurs, and a strong agglomerate is formed.

Most of the heat from the combustion zone is absorbed by drying, calcining, and preheating the lower layers of the bed. When the combustion zone reaches the base of the sinter mix, the process is complete. The sinter cake is tipped from the grate and broken up. After screening, the undersize is recycled, and the remaining sinter is sent to the blast furnace.

Water is added to the sinter mix, to control moisture at 5 to 8%. Water is also sprayed for dust control in the plant, and at the many conveyor transfer points where raw materials are moved from storage areas to the sinter line (strand).

Many plants scrub the sinter furnace exhaust gases to remove entrained solids. The scrubber water usually requires treatment by coagulation to remove suspended solids; chemical conditioning to control scale, corrosion, and fouling; and removal of pollutants prior to discharge.

ACID PICKLING

The treatment of steel in an acid bath, known as pickling (Figure 32.16), removes oxide from the metal surface and produces a bright steel stripped down to bare metal and suitable for finishing operations, such as plating, galvanizing, or coating. Both sulfuric and hydrochloric acids are used, with the latter growing in popularity as more by-product hydrochloric acid becomes available from the chemical industry. With either acid, disposal of spent pickle liquor is a serious problem.

Pickling may be either batch or continuous. Usually, the acid is prepared at about 5 to 15% strength, depending upon the work to be processed in the pickle tank and the type of acid used for pickling. As the acid works on the oxide surface, there is a gradual buildup of iron in the pickle solution and a depletion of the acid. When the iron content reaches a level that slows the pickling operation, the bath is either dumped or reprocessed. In some pickling operations, acid is continually withdrawn in order to hold a fairly constant ratio of iron to free acid in the pickle bath, thereby maintaining uniform pickling conditions for all the steel passing through. The metal leaving the pickle bath carries some of the liquor with it

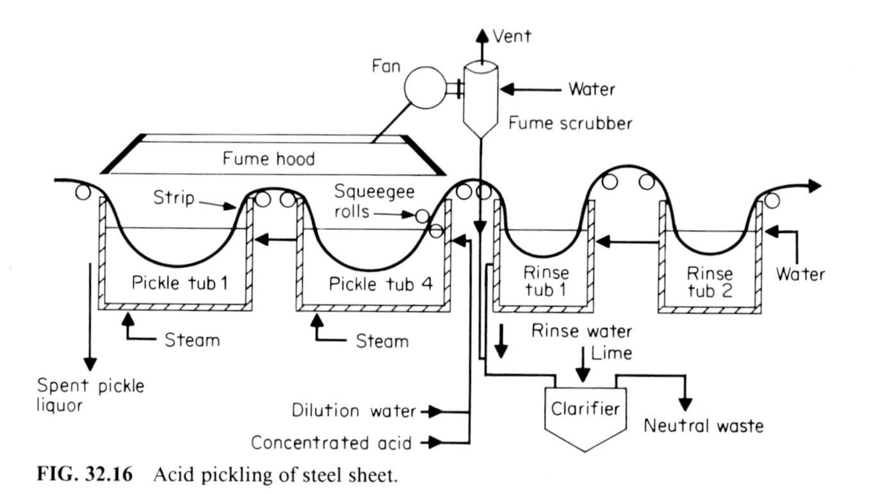

FIG. 32.16 Acid pickling of steel sheet.

into the subsequent rinsing and neutralizing operations. The loss of acid by dragout varies with the type of work, the shape of the products being pickled, and the speed of the operation. It can run as high as 20% of the acid used. Appreciable rinse water must be put into the rinse tank and withdrawn continuously for discharge to a treatment facility.

SLAG PLANT

Various useful products are recovered from slag, some being produced from the slag in a molten form, and others after solidification.

Molten blast furnace slag can be quenched with water to produce lightweight expanded aggregate for the manufacture of cinder block. It can also be spun into mineral wool insulation. Solid air-cooled slag is crushed to various sizes for use as a track ballast, highway foundation, and similar structural material.

In plants handling slag there is normally airborne dust that tends to cake on conveyor belts or interfere with proper operation of mechanical equipment. Water washing of the air is sometimes practiced, or water may be used for washing conveyor belting, creating a high suspended solids wastewater.

BOF, open-hearth, and electric furnace slags are very high in iron. Therefore, they are usually broken up and reclaimed for charging to the blast furnace.

UTILITIES

Because of the useful combustible gases produced at the coke ovens and the blast furnaces, steel mills are able to produce a large percentage of their total power requirements by burning this fuel in boiler houses. In addition to boilers directly fired with these by-product fuels, other boilers reclaim heat from gases discharged from the open-hearth furnaces, the BOF shop, and the engines that operate on coke-oven gas.

The steam generated by these boilers is used throughout the complex for driving turbines, powering presses and forges, and providing heat wherever it may be required. Some of this steam may be treated with steam cylinder oil ahead of steam engines and presses, producing an oily condensate that requires treatment prior to disposal.

In addition to production of power from steam, some by-product gases are used in gas engines to produce electric power or mechanical energy directly for such uses as compressing air for the blast furnaces. Other utilities include air compressors, vacuum pumps, and pump stations to supply the enormous quantities of air and water needed for the operation.

The water requirements for these utilities in the steel mill are similar to those of other industries. High-quality water must be produced for makeup to the steam generators, and these in turn concentrate the water, which is then removed by blowdown at a relatively high salinity level. The water treatment facilities required for producing this high-quality water generate their own wastes, such as lime sludge from lime softening operations, brine from zeolite softening operations, or spent acid and caustic from demineralizer regeneration.

Cooling water is used by these utilities for such purposes as condensing turbine exhaust, cooling compressor jackets, cooling bearings on various types of powerhouse auxiliaries, and conveying ashes from coal-fired furnaces.

SUGGESTED READING

Goodman, R. J.: "Direct Reduction Processing—State of the Art," *Skillings' Mining Review,* **68** (10) (March 10, 1979).

Lepinski, James A.: "The ACCAR System and Its Application to Direct Reduction of Iron Ores," *Iron and Steel Engineer,* December 1980.

Lund, H. F. (ed.): *Industrial Pollution Control Handbook,* McGraw-Hill, New York, 1971.

McGannon, H. E. (ed.): *The Making, Shaping and Treating of Steel,* 8th ed., United States Steel Corporation, 1964.

Sanzenbacher, C. W., Lepinski, J. A., and Jones, M. R.: "Production of Direct Reduced Iron Using Coal," *Iron and Steel Engineer,* October 1982.

Spivey, James J., Carpenter, Ben H., and Westbrook, Clifton W.: "Direct Reduction of Iron Ore: Potential Growth in the U.S.," *Iron and Steel Engineer,* September 1982.

U.S. Environmental Protection Agency: *Development Document for Effluent Limitations Guidelines,* EPA-440/1-74-024a, June 1974.

CHAPTER 33
TEXTILE INDUSTRY

All finished textile products, whether clothing, carpeting, or tire cord, have their origin in wool, cotton, synthetic fibers, or combinations of these. These fibers are processed to make them suitable for their end uses. These processes include: removal of natural impurities from wool and cotton (dirt, grit, grease); removal of process impurities (sizing, metallic contaminants); and finishing, to impart particular qualities of appearance, feel, and durability.

COTTON

The three basic steps in processing cotton (Figure 33.1) are spinning, weaving, and finishing. Cotton spinning and weaving are dry processes. Foreign matter is removed from the raw cotton by opening and cleaning, picking, carding, and combing. The individual fibers are joined, straightened, spun into thread, and wound on spools.

To improve strength and stiffness of lengthwise (warp) yarn, it is passed through a sizing solution that controls abrasion and reduces friction. Starch, polyvinyl acetate (PVA), and carboxymethyl cellulose (CMC) are the sizing agents. This yarn is woven into cloth known as greige goods, which is sent to the finishing mill to process into salable products.

The finishing process, mostly wet, begins with the removal of sizing, natural wax, pectins, alcohols, dirt, oil, and grease to prepare the cloth for the following steps:

Singeing: The cloth passes between heated plates or rollers, or across an open gas flame to burn off loose fibers. Sparks are extinguished as the cloth passes through a water box.

Desizing: Starch is solubilized by enzymes or acid by a 3 to 12 h soaking. Excess liquor is removed. The cloth is then freshwater rinsed and processed through a caustic or penetrant bath.

Caustic scouring: Greige goods are cooked to remove cotton wax, dirt, and grease. The cotton cloth is saturated with liquor consisting of caustic soda, soda ash, pine oil soap, and surfactants and scoured in a steam bath for 1 h. Finally, the cloth is rinsed to remove the scour liquor. This develops a yellow, absorbent pure cellulose fiber.

Bleaching: Peroxide, hypochlorite, or chlorine in combination with sodium silicate and caustic soda are applied as bleach liquor, and the cloth passes into a steam chamber (J-Box) for approximately 1 h. The bleached cloth is rinsed in water and stored for further processing.

Mercerization: Bleached cloth may be mercerized to swell the fiber, thus improving luster, dye affinity, and strength. In mercerization, the cloth is carried through a caustic soda solution (20 to 25% NaOH) while under tension, then through a water rinse, acid dip, and final water wash.

Dyeing: Many different chemicals are used for dyeing:

a. Direct dyes are applied directly to the cloth.

b. Vat dyes and sulfur dyes are applied to cloth in a reduced state and then oxidized.

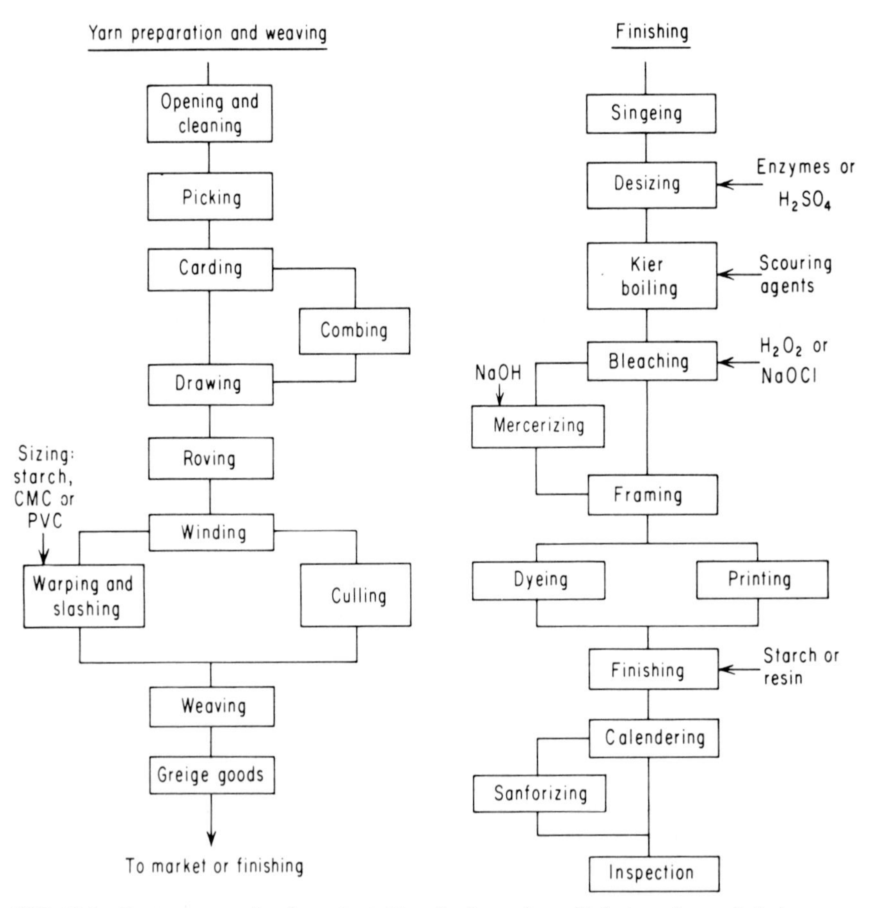

FIG. 33.1 Cotton processing flow sheet. Usually the greige mill that produces cloth is separate from the finishing mill. The finishing mill uses large volumes of water, often as much as 10 mgd, and requires a complex waste treatment system (Figure 33.5). The finishing mill usually has a large boiler house because of its need for steam for process equipment, and often cogenerates power for its machinery requirements.

c. Developed dyes and naphthol dyes are applied to cloth and developed with a secondary chemical.

d. Aniline black dye is oxidized on the cloth by air or steam.

Printing: This process imparts a colored pattern or design to the cloth by a roller or screen print machine. The colors are fixed by steaming or other treatment.

Final finishing: This process involves sizing (starch or resin), waterproofing, fireproofing, or preshrinking.

WOOL

Raw wool, a protein (keratin), contains glandular secretions (suint and wool grease) and feces from the sheep, plus dirt, straw, and vegetable matter. Residues of treatments applied for disease control or for identification of the animal may also be present. Wool is normally insoluble in water, but above 250°F (121°C) some fractions dissolve. Wool fiber expands upon wetting, but contracts to its original size when dried. Being amphoteric, wool is damaged by caustic or acid solutions, so special care must be taken when subjecting it to such treatments in processing.

Wool Processing

Sorting and blending: Raw fibers are sorted into lots according to fineness and length. Fibers from different lots are blended to maintain uniformity.

Scouring and desuinting: Foreign matter is removed by washing with soaps, alkalies, or other chemicals, or by solvent extraction. Scouring the fleece reduces weight by 35 to 65%. The extracted material is processed to recover lanolin.

Washing: The scour is followed by a clean water rinse.

Carding: The wool fibers are disengaged and rearranged into a web.

Oiling: An antistatic agent is applied to the fibers.

Backwashing: Removing oil that has been put into worsted stock in the blending, oiling, and mixing operations.

Gilling: A special procedure for carding which separates the long, choice fibers of the same length from the shorter fibers.

Top dyeing: The application of a tint to the choice, long fibers (called "wool top") used for production of worsted, a tightly spun yarn. The tinting is for identification only.

Roving: Narrow strips of web are gently meshed together and wound onto spools for the spinning frames.

Spinning: The rovings are drawn through small rollers which further extend the web by pulling the fibers apart lengthwise.

Winding: The spun rovings are twisted and wound onto bobbins as finished yarn.

Figure 33.2 outlines the wool-processing flow diagram.

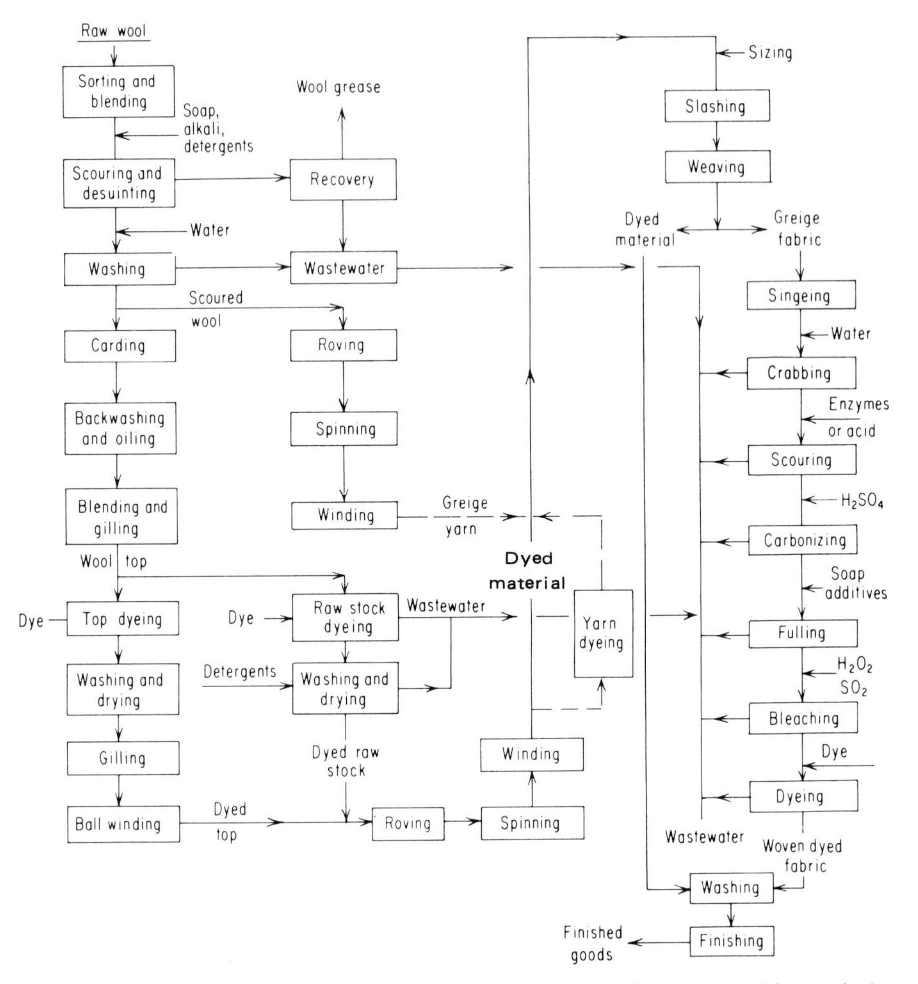

FIG. 33.2 Wool processing flow sheet. This process is not always fully integrated into a single mill. For example, wool scouring may be carried out in a separate plant, and independent dyehouses may process the fiber or fabric.

Scouring (or slashing) prepares the wool for weaving. In detergent scouring, the predominant method, the wool is treated in successive bowls with capacities of 1000 to 3000 gal (3.8 to 10 m^3) each. The first bowl is used for steeping (desuinting); the next two contain soap alkali for grease removal; and the final bowls are for rinsing.

Batch or continuous solvent scouring produces less water pollution than detergent scouring. Grease laden solvent is distilled to recover the solvent. A final detergent washing removes residual solvent and grease.

Dyeing is performed in open or pressure-type machines. The pollution loading is related to the dye used.

Fulling is a process of shrinking the woven fabric by subjecting it to moisture, heat, and friction to produce a feltlike texture. Soap is the felting agent and water is evaporated during the process. Following fulling the wool cloth contains considerable process chemical and must be washed. The cloth passes through a "first soap," is squeezed between rollers, washed in a "second soap," and finally rinsed in a water bath.

In the carbonizing process for removing residual impurities, the wool fabric is impregnated with 4 to 6% sulfuric acid and oven dried at 212 to 220°F (100 to 105°C). The evaporation of the water concentrates the acid, charring organic contaminants. Rollers crush the charred matter, which is then removed by a mechanical dusting machine. The cloth is rinsed, neutralized by soda ash solution, washed again, and dried. Sulfur dioxide or hydrogen peroxide bleaches the natural yellow tint of the wool to white.

SYNTHETICS

The most common synthetic fibers are cellulose base (acetate/rayon) and polymer base (acrylic, nylon, polyester, and orlon). Spun yarn is processed like natural fibers requiring size to impart strength and to provide a protective coating for weaving. Continuous filament yarn requires less sizing.

Static charges build up on synthetic yarn during most processing steps so antistatic oils and lubricants are applied to the fiber before weaving. These include polyvinyl alcohol, styrene-based resins, polyalkylene glycols, gelatin, and polyvinyl acetate.

Synthetic fabric finishing processes are similar to those used with cotton, and include scouring (removal of process chemicals from weaving), initial rinsing, bleaching, second rinsing, dyeing, and final finishing (waterproofing, shrink proofing, etc.).

WATER USES IN THE TEXTILE INDUSTRY

Clean air, free from debris, and controlled at precise temperature and humidity levels is vital to textile processing. The industry is one of the largest users of air-washing equipment to clean and temper air in the processing areas. Tempering requires heating in the winter and cooling in the summer.

Air washers are used throughout cotton mills producing woven fabric; in blending plants where cotton is blended with synthetic staple into yarn and then woven; and in synthetic fiber plants. Air washers also find extensive use in knitting plants, including hosiery and carpet mills. The material removed from the air is transferred to the water, so there are many problems requiring water treatment technology.

The cooling capacity for a typical textile refrigeration unit is 300 to 1200 tons, depending on the size of the plant. Total plant capacity may be from 300 tons in a very small mill up to 18,000 tons. In large plants, the total tonnage may be supplied by a single cooling tower circuit and a single chilled-water system. However, in most cases there are several smaller systems. The tonnage and number of air-conditioning systems operating in a plant depend on the combination of textile processes.

New textile plants are designed to consolidate chilled-water systems into as few independent units as possible to maximize efficiency and reduce maintenance.

Chilled water is piped around the average textile plant for use in air washers, process-heat exchangers, and small office air-conditioning units. By far, the largest user of chilled water is the air washer. The average plant with 1200 to 2400 tons of air-conditioning capacity may have six to eight air washers.

When refrigeration units are operating during warm weather (Figure 33.3), air washer units are supplied with chilled water, 40 to 50°F (5 to 10°C). Most textile chilled-water systems have high-level float switches on the chilled-water sumps. During summer operation, when the chilled water is dehumidifying plant air, the volume of water in the chilled-water system increases as water condenses from the makeup air. When this occurs, the level in the sump rises until it hits a limit switch which diverts excess chilled water (which is essentially condensate) to the cooling tower as makeup. This procedure conserves treatment chemicals, water, and energy, avoiding wasting excess chilled water and increasing the efficiency of the condenser unit.

Chilled-water sumps contain filters for suspended solids removal. Many potential fouling problems in the air-washer systems can be avoided by removing suspended solids in the chilled-water sump in this manner.

In plants where refrigeration is not required during winter months (Figure 33.4), washers operate independently, continually recirculating water from the sump through the spray nozzles. During these months, there is usually evaporation in the air-washer system, eliminating overflow and requiring makeup.

In a typical textile mill air washer, air entering from the plant first passes through a fine mesh screen or drum roll filter to remove lint, dust, oil, and other

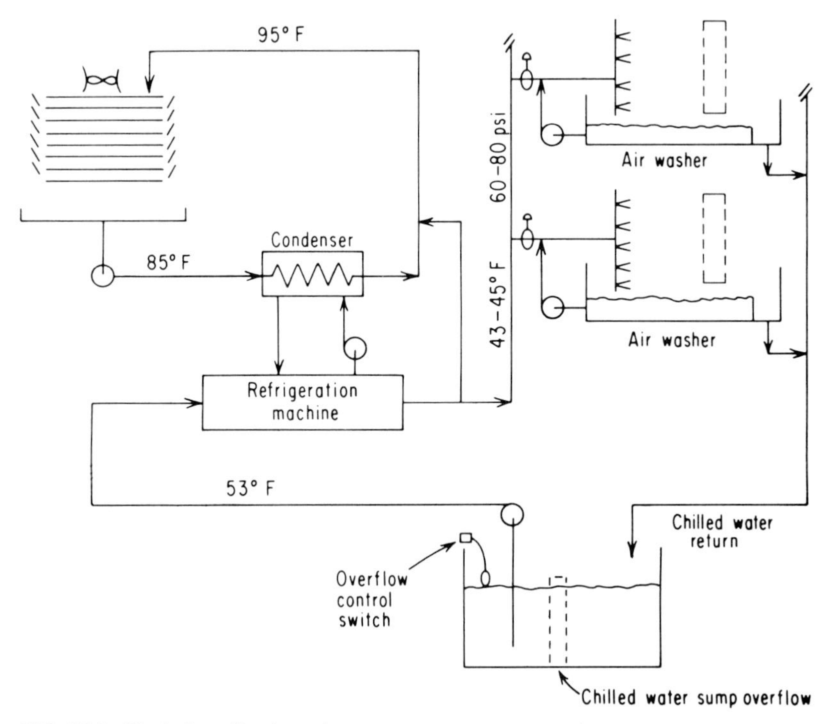

FIG. 33.3 Typical textile air-washer system—summer operation.

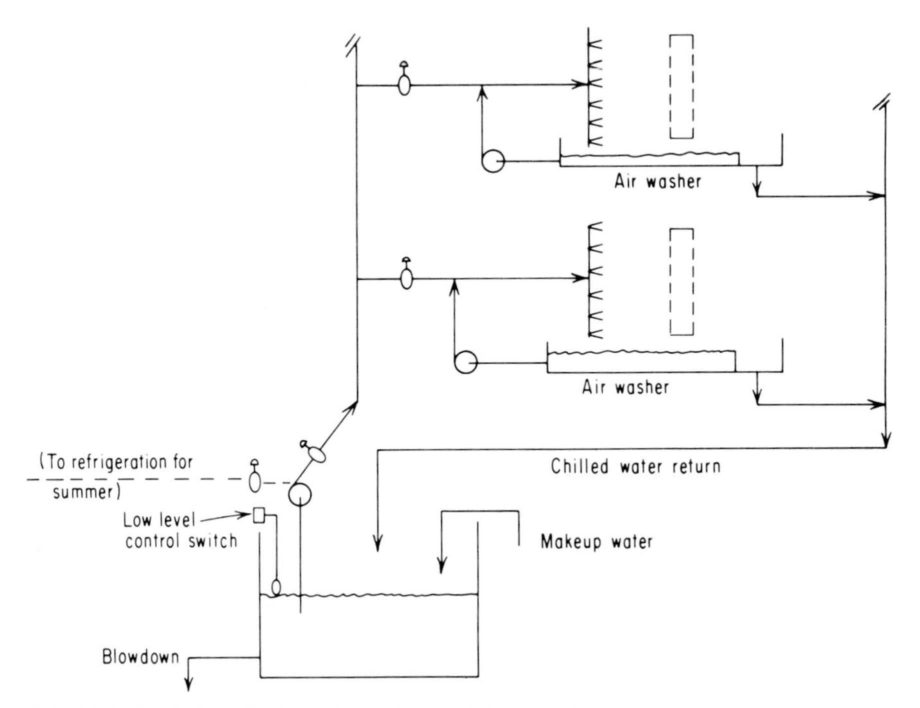

FIG. 33.4 Typical textile air-washer system—winter operation.

debris. Some units have moving paper media while others have stationary synthetic media which are replaced two or three times per year. Some large rotating drum filters have vacuum attachments to continually remove contaminants from the filter medium. A knitting plant or a package dyeing plant with winding operations has oil in the air. However, lint is the major problem in cotton mills.

Most textile air-washer systems automatically blend outside air with in-plant air. The temperature and humidity needs may vary from one department to another in the same plant, requiring separate chilled-water systems.

The mixture of air enters the washing section where several vertical headers with spray nozzles are spaced evenly across the area of air flow. The nozzles spray water against each other so that the incoming air must pass through a barrier of water droplets 2 or 3 ft thick.

The spray nozzle headers are connected to a recirculating pump that has a capacity of 250 to 1200 gal/min (1 to 5 m^3/min). The main pump located at the chilled-water sump continually supplies each individual air-washer sump. Overflow and gravity return the water to the main refrigeration unit and chilled-water sump. Capacities of air-washer sump pans range from 600 to 2500 gal (2 to 10 m^3).

After the spray section of the washer unit, the air passes through mist eliminator blades to remove moisture. Some washers are designed with steam reheat coils to temper the cooled air to suit a particular textile process. Bypass ductwork is sometimes designed into a unit to allow for simple heating of the air without washing. Large fans or blowers take the air from the end of the washer unit and distribute it through the plant ductwork.

Most textile mills use compressed air to control water atomizers and for other purposes. Cooling water must be supplied to the air-compressor heads, oil coolers, and aftercoolers. The cooling water may come from the main cooling tower system in large plants where there are several air conditioning units, or there may be a separate small tower provided for these units.

The few plants set up for once-through cooling water on air-compressor systems conserve this by sending the spent water into one of the cooling towers during summer operations as makeup.

Some air washer units, particularly in synthetics plants, do not use chilled water for the washing process. The eliminator sections of these washers are followed by steam reheat coils, and chilled water coils to adequately control temperature and humidity.

Rotospray systems, similar in principle to the packaged air washer units, are housed in a cylindrical casing slightly larger than the supply ductwork, and are generally located on the roof of the mill. They have stationary spray nozzles and rotating eliminator blades that look like the compressor stages in a gas turbine. There is very little water in these units, and they are somewhat difficult to treat because the dirt is concentrated in a small amount of water.

AIR WASHER MAINTENANCE

Air washers are periodically shut down and washed out to control the severe fouling and deposition problems that occur. The frequency of shutdown and washout depends on the type of textile process being run, and the severity of the problem. It may be weekly in some plants, while in others 5- to 6-week intervals between washouts may be acceptable.

The high degree of recirculation in air washers leads to a variety of water problems including slime formation, deposits, corrosion, and odors. The major part of most deposits is microbial (slime masses). Microbial activity produces a sticky slime that combines with dirt, corrosion products, and crystalline matter to form hard encrusted deposits above the water level and thick slimy masses below the water on metal surfaces inside the washer. Controlling microbial growth in a chilled-water or air-washer system is the key to an effective treatment program. Microbe growth also causes odors, carryover, encrustation, and corrosion under deposits. Oil and other organic matter picked up from the plant air provide food for these microbes. Even though some removal of oil may be accomplished with air filters, residual oil will be present in the washer.

Carryover caused by foaming or by biological growth on eliminator blades disrupts air flow and allows solids to pass into plant ductwork. This can cause a variety of problems in the plant from spotting of the product to disrupting the temperature and humidity controls. Severe damage can be done to the plant ductwork. Most textile plant ductwork contains a mat of lint and fiber. When this becomes wet, it becomes encrusted with the dirt present in the washer water and severe corrosion can result.

Corrosion is most severe in textile air-washer and chilled-water systems during summer operations when the water in the systems deconcentrates. Chiller tube sheets and heads are the most vulnerable areas. Dirt, fiber, oil, and microbiological deposits combine in some cases to totally slime over areas of the tube sheet. Severe pitting occurs under deposits of this nature. Corrosion inside the air washer units themselves may occur on any mild steel structures.

TABLE 33.1 Representative Mill Average Raw Waste Characteristics for Various Textile Operations

Variable	Wool and animal hair scouring	Wool dyeing and finishing	Cotton and synthetic woven fabric finishing	Cotton and synthetic knit fabric finishing	Dyeing and printing of carpet (cotton, wool, syn.)	Cotton and syn. raw stock and yarn dyeing
Water use (gal/lb)	4.3	40	13.5	18	8.3	18
Production (med. size)		2–20	6–180	8–40		
1000 lb/day	80	3–20	7–15	9–60	60	60
BOD_5, range	4740–6220	150–700	250–850	100–650	144–630	75–340
median	5480	300	550	250	340	200
TSS, range	5000–24,500	45–300	45–475	40–485	75–150	25–75
median	7500	130	185	300	120	50
COD, range	29,600–31,300	280–5000	425–1440	450–1440	570–1360	220–1010
median	30,500	1041	850	850	925	524
Oil and grease, range	5000–5600					
median	5340					
Total chromium	0.05	4	0.04	0.05	NI*	0.013
Phenol	1.50	0.5	0.04	0.27	NI	0.12
Sulfide	0.20	0.1	2.72	0.2	NI	NI
Color (ADMI)	2000	500–1700	325	400	600	600
pH (units)	6–9	6–11	7–11	6–9	6–9	7–12

* NI: No information available.

Note: All above, except color and pH, are reported in mg/L; ranges and medians are given.

Source: Lockwood Green Engineers, Inc. "Textile Industry Technology and Costs of Wastewater Control," National Comm. on Water Quality Contract WQSACo-21, June 1975.

DEVELOPING A TREATMENT PROGRAM

A complete equipment survey and a thorough understanding of equipment operations are needed to design an effective water treatment program for air-washer and chilled-water systems.

The selection of an effective biocide and dispersant is generally the starting point in developing a water treatment program for air washers and chilled-water systems. Corrosion control is closely related to the effectiveness of the microbicide and dispersant. There are no easy answers to the selection of a corrosion inhibitor for air-washer and chilled-water systems because of the differences between textile plants. The most important part of controlling corrosion is keeping the system clean.

The static charge on textile fiber as it passes through the various textile processes has a great deal to do with their efficiency. For example, if a carding machine is processing staple that has just been brought in from a cold warehouse,

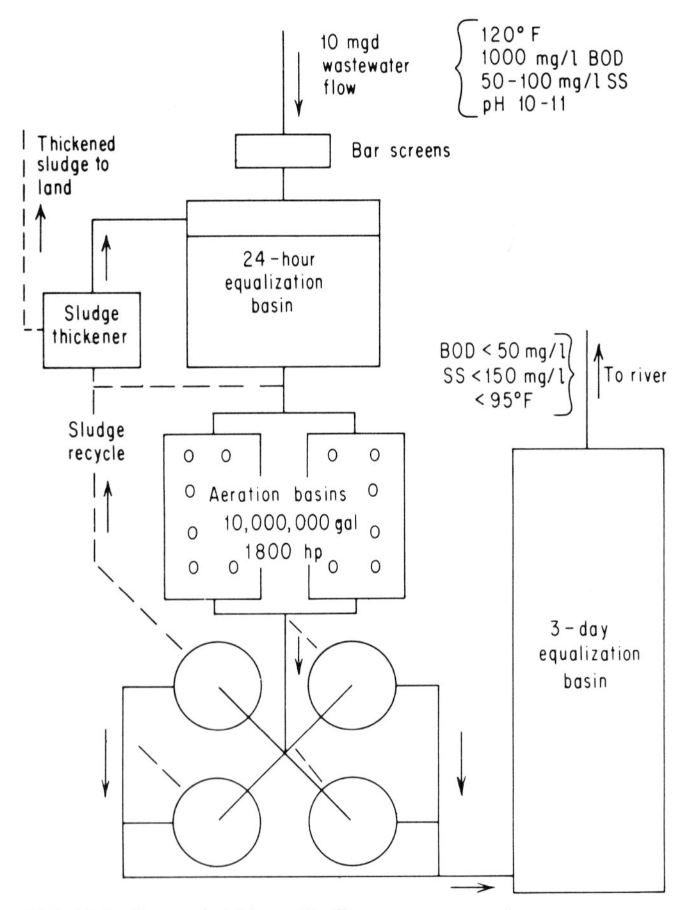

FIG. 33.5 Cotton finishing mill effluent treatment scheme.

the fibers will be negatively charged and will stick to the steel rolls of the card machine rather than smoothly rolling into a sliver of yarn. Fibers that develop too great a negative charge on spinning frames are subject to excessive breaks. Quaternary ammonium compounds may be used to control static charges in air, because they have the added benefits of being good biocides and cleaning agents.

Steam generation facilities in a textile plant will normally be quite simple. Condensate is usually not more than 25% of total feed water. The boilers are generally low pressure and the steam is rarely used for power generation except in the largest integrated mills.

Effluent treatment may be quite complicated because of the residues of processing chemicals present in the raw wastewater. In-plant containment and process modifications are increasingly necessary to meet effluent restrictions. Typical wastewater characteristics are listed in Table 33.1.

A treatment scheme for a large finishing plant is shown in Figure 33.5. The large equalization basin is provided to even out changes in composition and temperature, and to reduce the heat load on the aeration basin. In some plants the presence of strong waste (caustic from mercerizing, dyes, sizing agents) requires segregation and separate treatment of these streams. Activated carbon has been used as part of the treatment where dyes are a problem.

SUGGESTED READING

Shreve, R. N.: *Chemical Process Industries,* 3d ed., McGraw-Hill, New York, 1967.

U.S. Environmental Protection Agency: *Development Document for the Textile Mills,* EPA-440/1-74-022-a, June 1974.

CHAPTER 34
UTILITIES

Steam-electric power generating plants are the largest industrial users of water in the United States. The Federal Power Commission 1970 National Power Survey reported that utilities cooling water constituted 80% of all such water usage, or approximately one-third of the nation's total water withdrawals. About 70% of the water withdrawals by utilities is fresh water. The balance is saline water used by generating stations in coastal regions.

The laws of thermodynamics limit the amount of the energy in fuel that can be converted to work by a steam turbine. More than 50% of the initially available heat remains in the exhaust steam leaving the turbine, and most of this is rejected to cooling water and then lost to the surrounding environment. Cooling water required for condensing the turbine exhaust steam in a 1000 MW (1,000,000 kW) plant is about 400,000 gal/min (1500 m^3/min).

By comparison, other water requirements for steam-electric plants are small. These include replacement of blowdown and other losses from the steam cycle, ash transport (in fossil-fueled plants), equipment cleaning, and domestic uses.

A high percentage of the cooling water is returned directly to its source, because once-through cooling is used in plants having about 70% of the currently installed generating capacity in the United States. Because the condenser outlet water is typically 5 to 25°F (3 to 14°C) warmer than the inlet, heat is the principal concern in discharges of utility cooling water. The goal of U.S. environmental policy to eliminate environmentally harmful thermal discharges to receiving waters by utilities has generally been achieved, with some exceptions.

The difference between total water withdrawal and wastewater discharge is mostly evaporation. Evaporation occurs principally from ponds and cooling towers used in recirculating cooling water systems; there is also some from the surfaces of ash settling ponds. Minor water losses can also occur by seepage into the ground from unlined ponds and in wet ash or water treatment sludges hauled from the plant.

Utility wastewater contaminants include suspended solids, metals such as iron and copper, acids and alkalies from water treating and metal cleaning operations, oil, and grease. Less likely to become a water contaminant, but of special environmental concern, are polychlorinated biphenyl compounds (PCBs) used as transformer fluids, now being replaced by safer materials.

THE PROCESS: ENERGY CONVERSION

In steam-electric power production (Figure 34.1), the chemical energy of fossil fuels or the nuclear energy of fissionable materials is converted to heat for gen-

FIG. 34.1 This utility system incorporates both a nuclear unit (left) and a coal-fired unit (right) at the same site. *(Courtesy of Carolina Power & Light Company.)*

eration of steam that is then converted to mechanical energy in a turbine coupled to an electric generator. Exhaust steam is condensed and returned to the boiler.

Within the fluid (water-steam) cycle of a steam-electric plant, the theoretical efficiency at which heat in the steam can be converted in a turbine to work is shown by:

$$\text{Percent theoretical efficiency} = \frac{T_1 - T_2}{T_1} \times 100$$

where T_1 is the absolute temperature of the turbine inlet steam, and T_2 is the absolute temperature of the turbine exhaust steam.

Thus, the greatest efficiency is achieved by providing the highest inlet steam temperature (T_1) and lowest exhaust temperature (T_2) practical for a given boiler-turbine-condenser system.

In practice, upper temperatures are limited by the strength of metals available for boiler and superheater construction; strength falls off rapidly as temperature exceeds 900°F (480°C, 1360°R). Upper temperatures are generally limited to a maximum of about 1100°F (590°C, 1560°R). A significant temperature difference is required to cause heat to flow from the exhaust steam to the cooling medium circulated through the condenser. The practical effect is that lower steam temperatures (T_2) on condensing turbines typically range 80 to 120°F (27 to 49°C, 540 to 580°R), even with cooling water at 40°F (4°C) in the winter. Within these upper and lower temperatures, then, the best theoretical efficiency would be calculated as:

$$\text{Theoretical efficiency} = \frac{1560 - 540}{1560} \times 100 = 65.4\%$$

Achievable efficiencies are considerably lower than theoretical both for the fluid cycle and for the generating plant as a whole. Significant energy is lost in overcoming fluid and mechanical friction. Sizable amounts of energy are consumed in the operation of the plant's feed water pumps, boiler draft fans, and other auxiliaries. A power plant's pollution control equipment, required to meet air emission and water discharge standards, can consume as much as 10% of the electrical output. Actual steam-electric plant overall efficiencies range from 32 to 39%.

POWER CYCLE

The energy flow through the water-steam phase changes in a steam-electric plant is called the Rankine cycle, having four basic stages: boiler, turbine, condenser, and feed water pump.

The Rankine cycle may be illustrated by several kinds of diagrams. Figure 34.2 shows that cycle, relating the change in temperature of the fluid to its entropy at each stage in the cycle. Entropy is a measure of that portion of the heat received by a cycle that cannot be converted into work because of the random motion of gas (steam) molecules. Entropy is expressed mathematically as the ratio of heat content to temperature.

To achieve the highest possible thermodynamic efficiency, modern fossil-fuel power cycles use steam superheaters and reheaters to raise average temperatures (T_1) of steam flowing through the turbine. Condensers operating at high vacuum permit expansion of steam to the lowest possible exhaust temperature (T_2). And

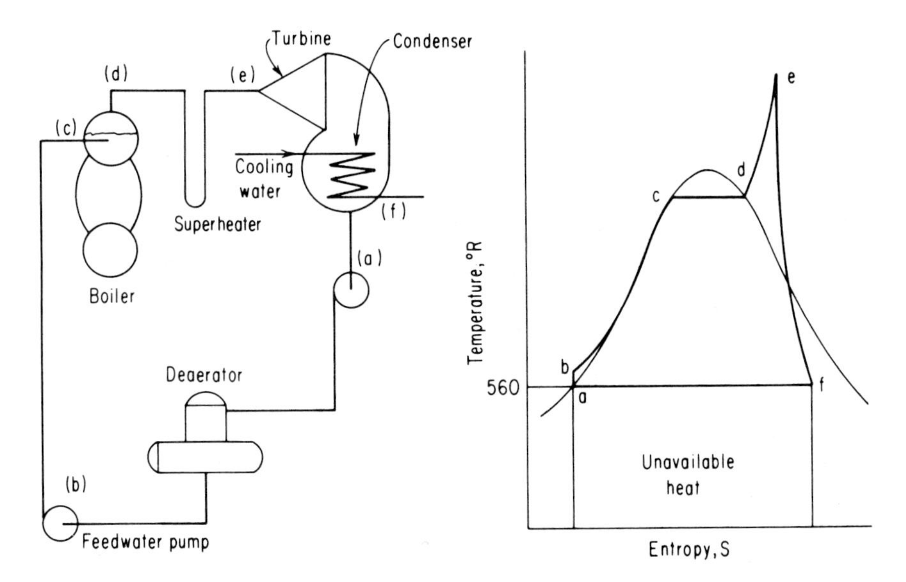

a to b: Liquid heating and compression (feedwater pump).
b to e: Reversible heat addition in feedwater heaters (b to c), boiler (c to d) and superheater (d to e).
e to f: Steam expansion and conversion of heat energy to work.
f to a: Reversible rejection of unavailable heat to cooling water "sink".

FIG. 34.2 Steps in the Rankine cycle from the introduction of water as a liquid into the system, through its phase change to steam which produces work in a turbine, to the condensing of the exhaust vapor to water again for return to the boiler.

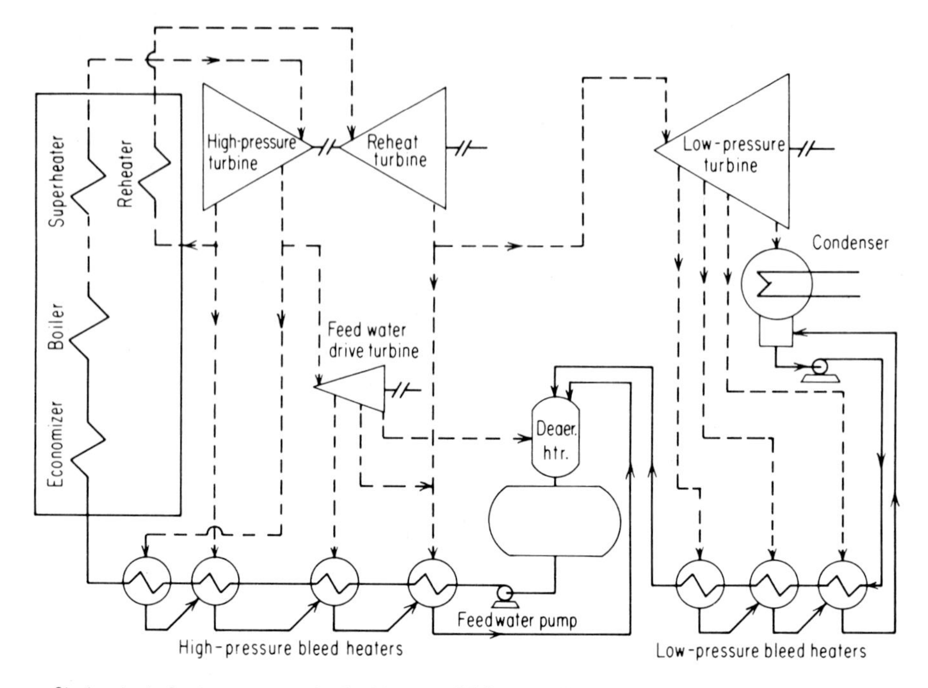

FIG. 34.3 The addition of steam reheat and regenerative feed water heating to increase efficiency.

to further recover as much heat as possible, partially expanded steam is extracted from the turbine at various points for heating feed water (called "regenerative heating"). Figure 34.3 illustrates a cycle having these components, and Figure 34.4 illustrates their effect on a temperature-entropy diagram of the cycle.

Nuclear steam-electric plants of the light water reactor types, typical in U.S. commercial service, cannot deliver steam at the temperatures achieved in fossil-fueled units. Nuclear pressurized water reactors (PWR) (Figure 34.5) employing once-through steam generators achieve only a relatively small degree of steam superheat [30 to 60°F (17 to 34°C)]. But in PWR cycles employing recirculating U-tube steam generators, and in boiling water reactor (BWR) cycles (Figure 34.6) there is no superheat. For this reason, nuclear power cycles are limited to efficiencies of about 32%, compared to 36 to 39% for fossil-fueled cycles, since less heat is supplied to the turbine (at T_1).

In fossil-fueled units, further improvements in thermodynamic efficiency are achieved by economizers and air heaters installed in the path of the exit flue gas between the boiler and stack. Economizers transfer heat from the exit gases to the feed water; air heaters transfer some of the residual stack gas heat to the boiler combustion air supply.

For still higher efficiencies, where suitable fuels are available, some steam-electric plants combine the operation of a combustion gas driven turbine with that of a steam turbine. In such combined cycle generating plants (Figure 34.7), the hot gases of combustion, such as from a jet aircraft engine, are first utilized for spinning the gas-driven turbine; and the turbine exhaust gas, still containing considerable heat, is then used in a boiler-superheater unit to produce steam to drive a steam turbine.

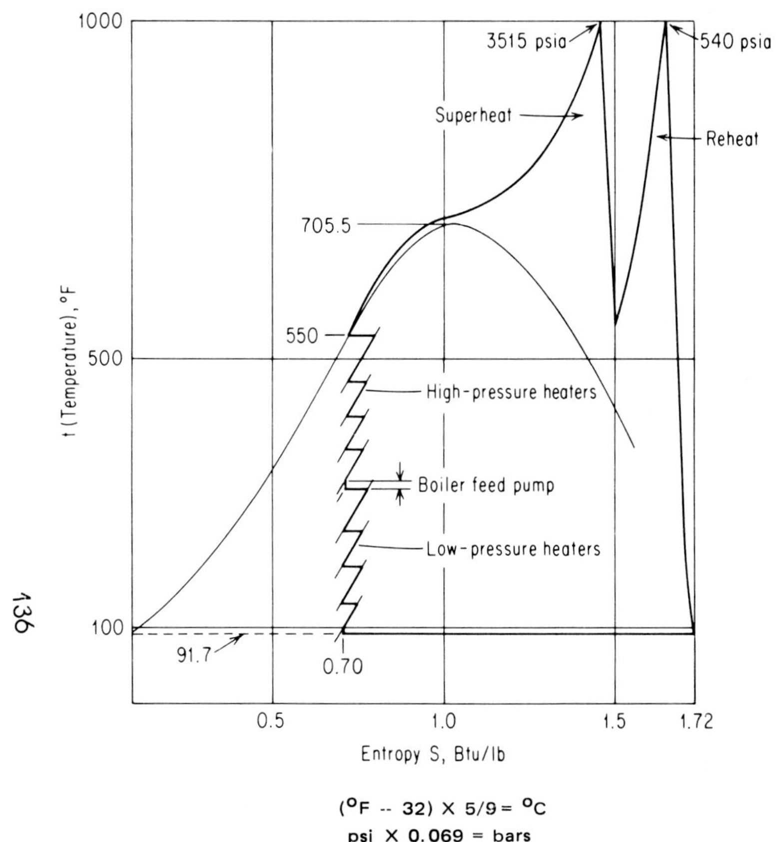

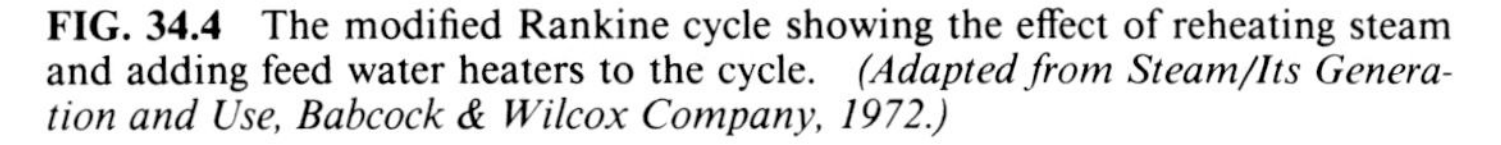

FIG. 34.4 The modified Rankine cycle showing the effect of reheating steam and adding feed water heaters to the cycle. *(Adapted from Steam/Its Generation and Use, Babcock & Wilcox Company, 1972.)*

The heat energy equivalent of one electrical kilowatt hour is 3413 Btu (860 kcal). At 32% overall thermodynamic efficiency, the heat rate for a generating station would be:

$$\frac{3413}{0.32} = 10{,}700 \text{ Btu/kWh or, } \frac{860}{0.32} = 2690 \text{ kcal/kWh}$$

If an efficiency of 40% could be achieved, the station heat rate would be only 8533 Btu/kWh (2150 kcal/kWh)—requiring about 20% less fuel for a given output.

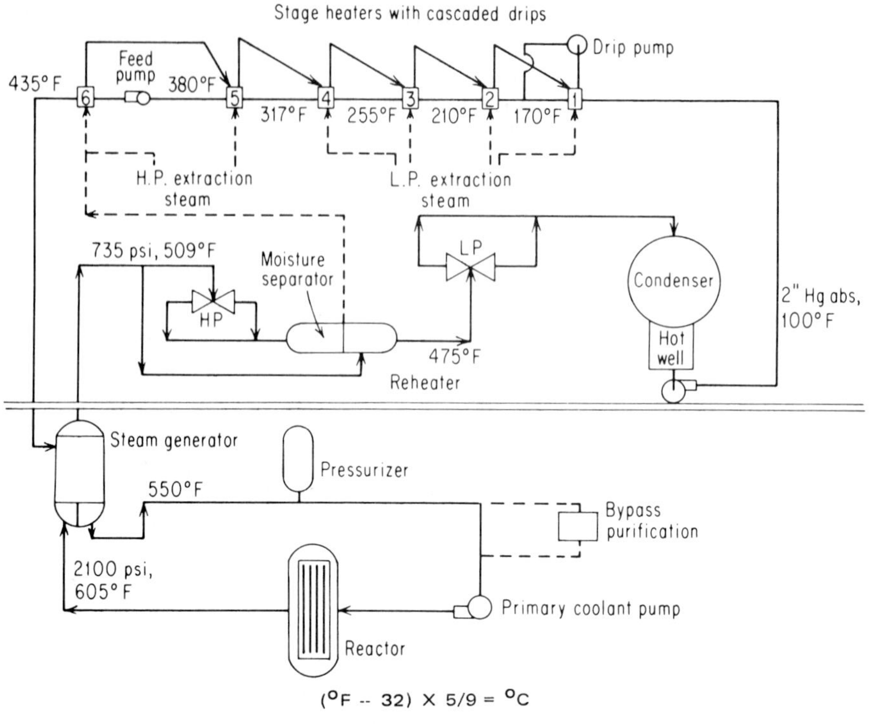

(°F -- 32) X 5/9 = °C

psi X 0.069 = bars

FIG. 34.5 PWR nuclear power cycle showing major components and typical pressures and temperatures. *(Adapted from the Effects of Water Quality on the Performance of Modern Power Plants, Klein and Goldstein, NACE 1968 Conference.)*

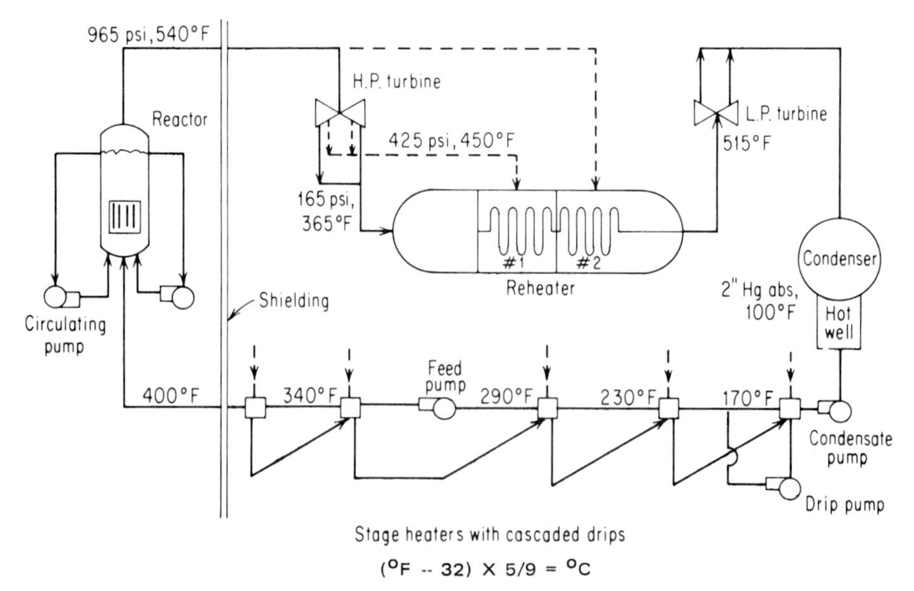

(°F -- 32) X 5/9 = °C

psi X 0.069 = bars

FIG. 34.6 Diagram of a BWR nuclear power cycle showing major components and typical pressures and temperatures. *(Adapted from the Effects of Water Quality on the Performance of Modern Power Plants, Klein and Goldstein, NACE 1968 Conference.)*

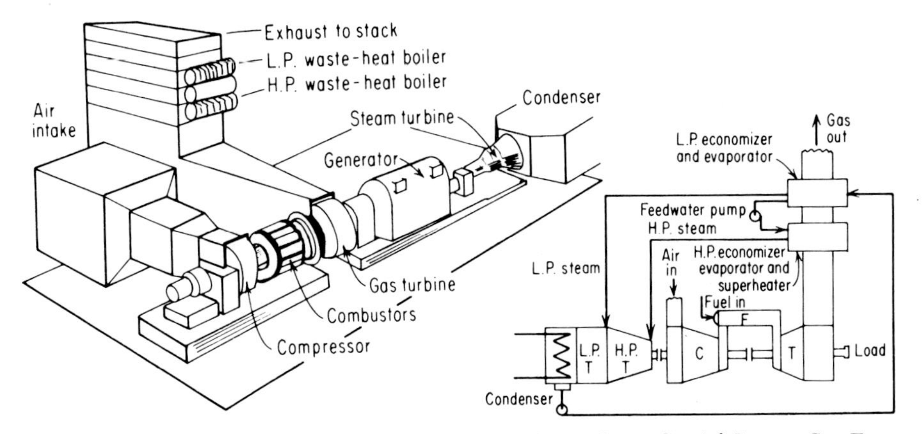

FIG. 34.7 Combined cycle gas turbine-steam plant. *(From Power Special Report, Gas Turbines, December 1963.)*

WATER: THE WORKING FLUID

In steam-electric power cycles, water is subjected to wide variations in temperature and pressure in the cyclic sequences of compression, heating, expansion, and heat rejection. Temperatures may range from 70 to 1200°F (21 to 650°C), with pressures from 0.5 to 5000 psia (0.035 to 345 bars).

As shown by the diagram in Figure 34.8, steam density approaches that of water as pressure increases, until the critical pressure is reached at 3208 lb/in² abs

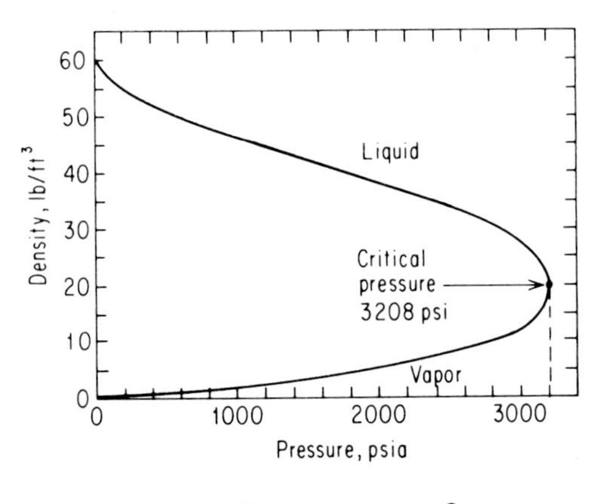

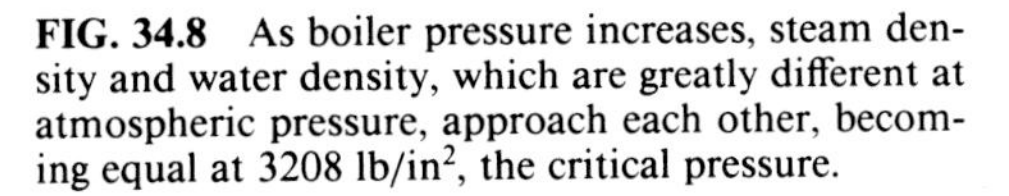

FIG. 34.8 As boiler pressure increases, steam density and water density, which are greatly different at atmospheric pressure, approach each other, becoming equal at 3208 lb/in², the critical pressure.

(220 bars). Above this pressure exists a supercritical fluid which cannot be considered either steam or water. Boilers operating over 3208 lb/in^2 abs are called supercritical units.

In cycles operating at pressures below critical where steam density is less than that of water, phase changes occur from water to steam and then from steam back to water. While there are advantages favoring drum-type boilers (Figure 34.9) for most subcritical pressure cycles, once-through boilers are sometimes used because of lower initial costs.

Drum-type boilers have two major advantages in subcritical steam generation. First, they can produce high-purity steam with less stringent feed water quality requirements. Mechanical disengagement of steam from the recirculating water

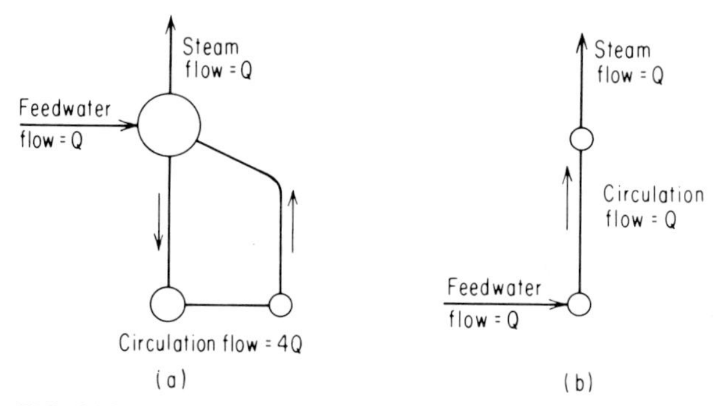

FIG. 34.9 The basic differences between (*a*) drum-type and (*b*) once-through steam generators.

leaves most of the undesirable solids behind in the boiler water. And because boiler internal recirculation rates are about four times the rate of steam flow from the boiler, the fluid rising to the steam drum from the generating tubes will be about 75% water. This means that superheated steam, in which some boiler water impurities would become soluble, cannot exist in the boiler drum. Secondly, since the recirculating mixture in the boiler tubes is primarily water, more efficient heat transfer and tube metal cooling occurs than when steam alone flows through the tubes.

In subcritical once-through boilers, a complete phase change occurs with water entering at one end and superheated steam leaving at the other. There is no recirculation within the unit, nor is there a drum for mechanical steam-water separation. Boiler blowdown is impossible, so to prevent deposition of solids in the system beyond the point of water-steam phase change requires stringent feed water quality requirements—considerably higher than for drum-type boilers.

For cycles operating above critical pressure, no phase change occurs, so no steam-water separation is possible. Thus, at supercritical pressures, all steam generators are once-through. Universal pressure boilers are once-through boilers suitable for use at pressures above or below critical. From the standpoint of tube metal alloy and thickness requirements, they are economically most practical for operation in the range of about 2000 to 3000 lb/in^2 (138 to 207 bars).

In higher subcritical pressure cycles [2750 to 2850 lb/in^2 abs (190 to 197 bars)], the margin between feed water quality requirements of drum and once-through boilers is somewhat lessened. At such pressures, effective mechanical steam-water separation is an advantage provided by drum-type boilers. But the advantage is limited by the increase in the solubility of water contaminants in the steam, as the densities of steam and water approach each other (Figure 34.8). The solids dissolved in steam cannot be removed by mechanical separation devices in a boiler drum. For high-pressure utility boilers, therefore, the concentration of solids in steam is determined to an important extent by their steam-water solubility ratios at the operating pressure. Silica is a notable example (Figure 34.10).

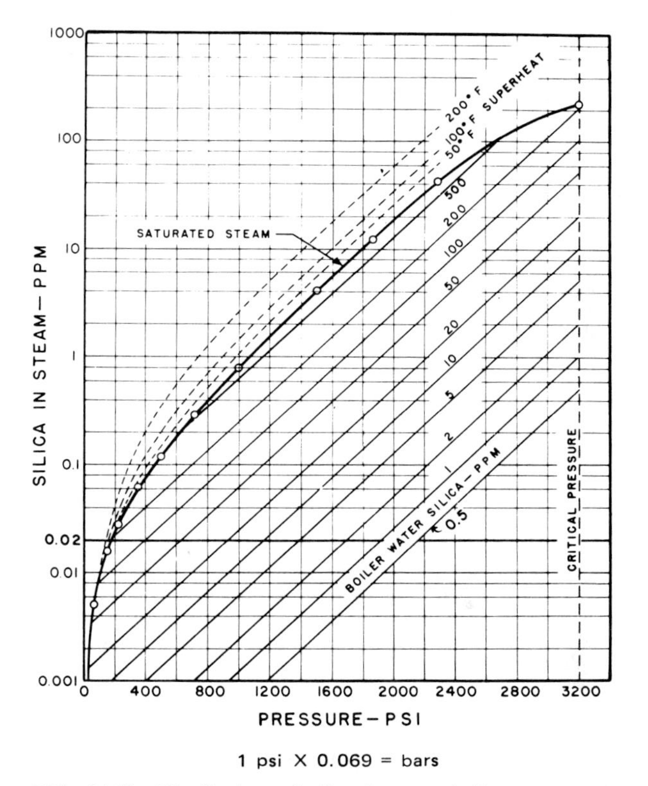

FIG. 34.10 Distribution of silica between boiler water and steam.

If solids concentrations in the steam were to avarage 30 ppb (μg/L), for example, a 200-MW turbine would receive more than 200 lb/yr (91 kg/yr) of potential depositing solids. The portion that would deposit in the turbine and the portion that would remain in the steam to recycle to the boiler is not accurately predictable. But, to avoid excessive turbine fouling, experience has shown the necessity of keeping total salt concentrations in the steam below 50 ppb (μg/L) with silica not to exceed 10 ppb (μg/L). Further, to minimize any potential for stress corrosion cracking of turbine members, the steam solids should be free of sodium hydroxide. The average steam contaminant levels shown in Table 34.1 were collated from a survey of a number of utility plants.

In nuclear power cycles, water purity is important for a different reason.

Although H_2O becomes only slightly radioactive under neutron flux, producing only short-lived isotopes, the response of materials in the water under flux is of concern. Naturally occurring cobalt, for example, would be an extremely undesirable constituent of primary coolant water passing through the reactor core in either BWR or PWR cycles. The product would be highly radioactive, emitting high-energy gamma rays during subsequent decay. So cobalt must be avoided in such systems. Since most other metals also produce radioactive isotopes under neutron flux, the tolerance of water corrosion products is limited in nuclear cycles for operating and maintenance safety.

TABLE 34.1 Typical Contaminant Levels in Utility Plant Steam

Crud constitutents	
Iron, as Fe	17.4 ppb
Copper, as Cu	5.0 ppb
Silica, as SiO_2	17.4 ppb
	39.8 ppb
Cations reported	
Sodium, as Na	6.5 ppb
Potassium, as K	1.5 ppb
Anions reported	
Free OH	0.1 ppb
Phosphate, as PO_4	7.0 ppb
pH	8.55–9.26
OH corresponding to pH 9.0*	500 ppb as $CaCO_3$
Cation conductivity	0.24 μS
Anions corresponding to cation conductivity†	30 ppb

* The significance of pH in this system is not clear; in pure water at pH 7, $[OH]^- = [H]^+$ and as the dissociation constant for water is 10^{-14}, this calculates to 5 ppb H^+ and 5 ppb OH^- as $CaCO_3$. Therefore, at pH 9.0, the OH^- calculates to 500 ppb as $CaCO_3$, or 170 ppb OH^-, which doesn't check with reported anions. The anions must be balanced by the cations, the most prominent of which may be ammonia, morpholine, or hydrazine, none of which was reported.

† After a cation exchange column, the ionic matter should be present as mineral acids, and the conversion factor for them is approximately 8 μS/cm per ppm acidity as $CaCO_3$. Therefore it may be assumed that the 30 ppb is made up of the acids of chloride, sulfate, nitrate, and phosphate.

Source: "Control of Turbine-Steam Chemistry," S. D. Strauss (ed.), *Power,* March 1981.

Although the required degree of purity varies for different cycles, relatively high-purity water is required for all modern power cycles. Many water constituents today must be measured and controlled at parts per billion (ppb, or μg/L) concentration levels compared to their measurement and control at parts per million (mg/L) concentrations in the lower pressure cycles that were once common.

However, make-up water of the purity required can be produced reliably and at reasonable cost by ion exchange, evaporation, and other pretreatment methods. But, because make-up water volumes are usually only 1% or less of total feed water in typical power cycles, the major problem has become keeping the water pure after it has entered the cycle.

The two main sources for contamination of high-purity water are corrosion and inleakage. Corrosion represents not only metal damage, but also contamination of the system water with corrosion products. Low-purity water enters the cycle at points of high vacuum (e.g., the surface condenser), and this inleakage is a second major source of contamination. Careful design and selection of materials of construction minimize the occurrence of these problems.

WATER CHEMISTRY IN FOSSIL-FUELED PLANTS—LIQUID PHASE

Apart from steam quality considerations, a major goal of cycle water chemistry and treatment control is the prevention of tube metal failures in high-pressure steam generators.

Because make-up water is deionized and essentially free of measurable hardness and other solids, tube failure due to water side scale and deposits is far less likely to occur in utility plants than in lower pressure industrial boilers. Instead, in power plant drum-type boilers, tube metal failures are more commonly due to a highly localized "crater" or "ductile gouging" form of water side corrosion, and to water side hydrogen penetration and embrittlement of carbon steel.

The ductile gouging form of corrosion is a chemical attack characterized by irregular craters in the tube wall. The tube fails when wall thickness at the point of wastage is insufficient to withstand the internal pressures (Figure 34.11). The most common chemical agent in this form of attack is sodium hydroxide. Even with a bulk boiler water free of NaOH as determined by usual methods of testing,

FIG. 34.11 Typical chemical-concentrate (ductile gouging)–type attack located downstream of a weld with backing ring.

a porous deposit on the tube wall provides a mechanism for concentrating chemical in contact with the tube surface to several thousand milligrams per liter. Sodium hydroxide concentrates in this manner and reacts first with the magnetite tube surface and then with the elemental iron of the tube to form soluble sodium-iron complexes:

$$4NaOH + Fe_3O_4 \rightarrow 2NaFeO_2 + Na_2FeO_2 + H_2O \quad (1)$$

$$Fe + NaOh + H_2O \rightarrow NaFeO_2 + 3H \quad (2)$$

The parent tube metal is thus exposed to react with water to reform protective magnetite,

$$3Fe + 4H_2O \rightarrow Fe_3O_4 + 8H \quad (3)$$

In either case, localized wastage of tube metal occurs and hydrogen is produced as a reaction product. The atomic hydrogen thus produced is capable of intergranular penetration of steel and of reaction with carbon in the steel to form methane gas:

$$4H + C \rightarrow CH_4 \quad (4)$$

Not only is the steel thus weakened by decarburization along its grain boundaries (Figure 34.12), but the formation of methane develops pressure between grains in the steel microstructure, producing discontinuous intergranular cracks. The end result is failure of the tube wall when such intergranular decarburization and microfissuring have sufficiently reduced its strength. Although some water side wastage of metal may have occurred in the corrosion reaction that produced the hydrogen, hydrogen damage normally results in thick-edged fracture before the wall thickness is reduced to the point that stress rupture would have occurred. In such thick-edged or "brittle" fracture failures, it is not uncommon for an entire irregularly shaped "window" to be blown out of an affected tube.

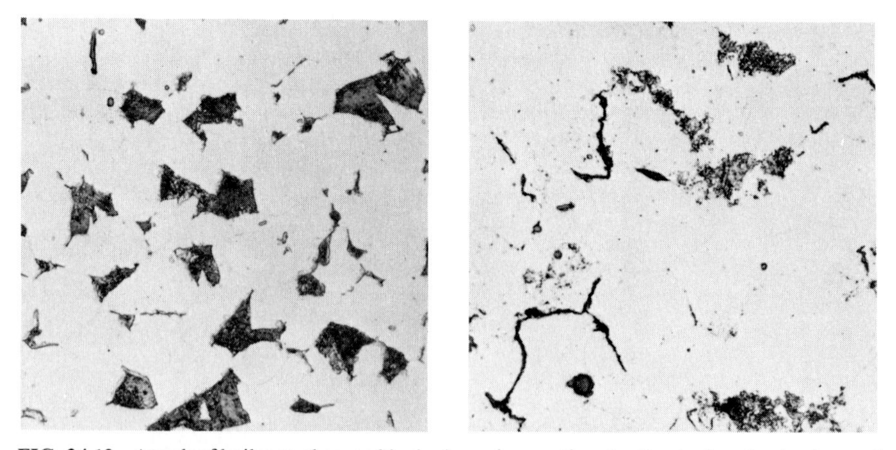

FIG. 34.12 Attack of boiler steel caused by hydrogen generation, leading to decarburization and fissure formation. *(Left)* Typical microstructure of pearlite colonies in a ferrite matrix. Etchant: picral. (500×) *(Right)* Intergranular microcracks and decarburization near fracture surface of actual hydrogen damage failure. Etchant: picral. (500×)

In summary, the major causes of ductile gouging and hydrogen damage of carbon steel tubes are (1) a chemical agent that will be corrosive if sufficiently concentrated and (2) a concentrating mechanism.

There are several mechanisms capable of producing high localized salt concentrations. Any condition of heat input in excess of the cooling capacity of the flow of fluid circulating through a tube can result in film boiling on the metal surface, with evaporative concentration of salts occurring. Localized pressure differentials, as created by high velocity through a tube restriction such as a welding back-up ring, can result in small pockets of "flashing" water, concentrating salts on the downstream side of the restriction. The most common mechanism is preboiler corrosion, with pickup of metals and their subsequent deposition on heated boiler tube surfaces, where they function as porous membranes. Evaporation beneath the deposit draws in boiler water to replace evaporation loss, but prevents solids from leaving the area.

To prevent such corrosion and hydrogen damage, either all concentration mechanisms must be eliminated or potentially corrosive chemical agents must be excluded from the boiler water. Since porous deposits of preboiler corrosion products create the most common mechanism for chemical concentration, strict attention to protection of preboiler metal surfaces is critical.

A significant characteristic of modern utility cycles is the high ratio of preboiler surface area to that in the boiler itself. While a ratio of about 0.3 would be typical in a 750 lb/in^2 (52 bars) system, a ratio of about 1.3 would be typical at 2800 lb/in^2 (93 bars). This is a reflection of the multiple stages of regenerative feed water heating commonly employed to increase cycle efficiency. To the water engineer, the higher ratio means that the area from which metals can be picked up is larger than the area on which they are likely to deposit.

Metals used in fabrication of condensers, low-pressure heaters, and high-pressure heaters are commonly cast iron, stainless steels, and alloys of copper, such as brass and monel. Thus, iron and copper are the principal contaminants likely to be acquired by feed water enroute to the steam generator.

Feed water pH is an important factor in minimizing preboiler corrosion of metals. For iron and its alloys, although a minimum pH of 8.5 might be acceptable, feed water pH values in the range of 9.2 to 9.6 are considered optimum. For copper and its alloys, feed water pH values generally should be in the range of 8.5 to 9.2. A compromise feed water pH control range of 8.8 to 9.2 is usually established where both metals are present. Volatile alkalies such as ammonia and amines that will not contribute dissolved solids are commonly used for pH control in this portion of the cycle.

Feed water oxygen also has a great bearing on preboiler corrosion and contamination; an oxygen level of 5 ppb or less is usually specified. The largest potential for oxygen admission occurs in the high vacuum portions of the cycle. Oxygen can be reduced to low levels mechanically, either in deaerating-type surface condensers or in separate deaerators located ahead of high-pressure feed water pumps. For removal of residual oxygen, organic reducing agents like hydrazine may be injected continuously immediately downstream of the condenser. Hydrazine is a chemical oxygen scavenger that introduces no solids to the cycle. Its concentration is controlled in the range of 10 to 20 ppb in the feed water, usually measured at the economizer inlet. Because of the potential risks to personnel in handling hydrazine, a suspected carcinogen, proprietary substitutes have been developed that react very much as hydrazine. The most widely used substitutes contain carbohydrazide and isoascorbic acid.

Ammonia, an important factor in preboiler pickup of copper, is generally present in power cycles, either due to direct addition or decomposition of chemicals such as hydrazine or amines that have been added to the system. It can also result from thermal decomposition of organic matter in cooling water entering the cycle through condenser leaks.

In sufficient concentrations, ammonia reacts directly with copper to form soluble complexes or it can contribute to stress corrosion cracking of copper alloy tubes. Even at lower concentrations, ammonia can react with protective copper oxides on the water side of condenser and heater tube surfaces, exposing parent metal to wastage by reaction with oxygen in the feed water. To minimize copper attack, most operators limit cycle ammonia concentrations to about 0.1 to 0.3 mg/L. These limits can be met by carefully controlling chemical additions to the cycle and by proper regeneration of condensate polishers. Where ammonia concentrations build up beyond the desired maximum, deconcentration may require temporarily dumping the normally recycled aftercooler drains (Figure 34.13) where ammonia concentrates.

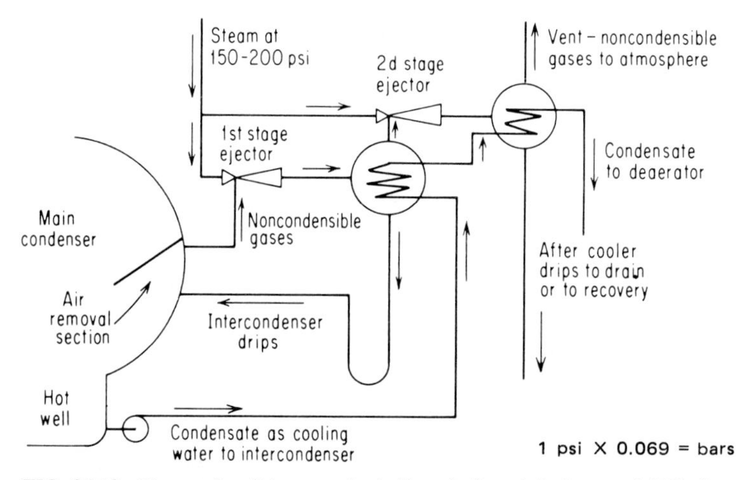

FIG. 34.13 Noncondensible gases, including air from inleakage and NH_3 from steam cycle, are removed by ejection and inter- and after-condensers. Final drips may be reclaimed or discarded if contamination levels are too high. Air discharge (vent) should be limited to 1 ft^3/min (standard) (0.0283 m^3/min) per 100-MW capacity.

Good design calls for careful selection of materials of construction to withstand the environment of the preboiler systems. Condenser tubes in freshwater coolant service are commonly brass and stainless steel. These materials are also common in low-pressure stage heaters, except for supercritical pressure cycles where even minute concentrations of copper can cause serious turbine fouling problems, and the use of carbon steel is required. High-pressure stage heaters are fabricated of copper-nickel alloys, thin-wall stainless steel, and carbon steel.

It is now standard practice to reduce metallic "crud" in utility feed water by polishing the condensate through a filter of mixed cation and anion exchange resins. Filtration removes the insoluble crud and ion exchange removes any dissolved contaminants that have entered the cycle by inleakage.

As shown in Figure 34.14, polishers are generally located between the condenser and the first feed water stage heater, since a large part of the vulnerable preboiler metal surface area is in the condenser and on the steam side of stage heaters cascading drips back to the condenser. This placement protects high-temperature feed water heaters, the steam generator, and the turbine from deposit-forming contaminants. Temperature limitations on ion exchange resin control location of polishers in the cycle: regenerable polishers can withstand only about 120°F (48°C), so must be confined to the condenser; powdered resin units may be placed in higher temperature locations, but still in the lower pressure stage heater train.

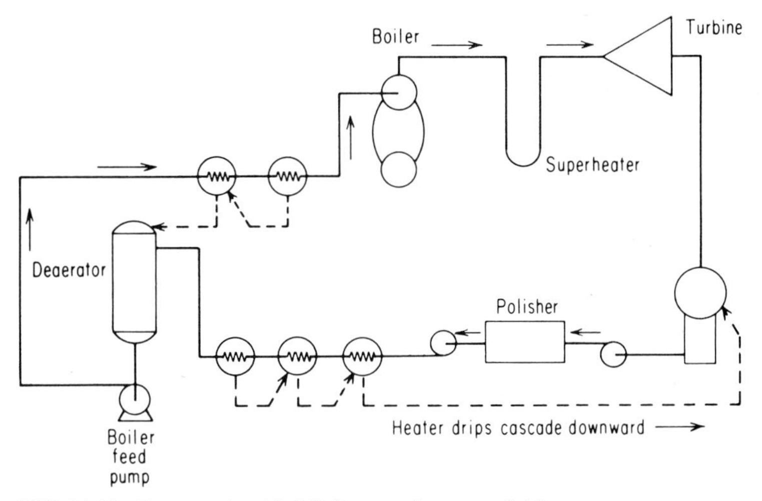

FIG. 34.14 Power cycle with full-flow condensate polishing.

For best cost performance, regenerable mixed bed polishers are commonly operated at service flow rates over 50 gal/min/ft^2 (2 m^3/min/m^2), about 10 times the rate employed in raw water demineralization. To avoid any possibility of contaminating cycle water with regenerant chemicals, resins are removed from the polisher for external regeneration and periodic cleaning.

Nonregenerable, powdered resin polishers have found wide acceptance in utility systems. The finely ground resin particles provide good filtration and excellent kinetics for ion exchange. This type of condensate polisher is typically operated at service flow rates of about 4 gal/min/ft^2 (0.16 m^3/min/m^2). During cycle cleanup, such as during start-up, an inexpensive cellulosic filter aid may be included in the powdered resin mixture. In any case, when the resins become fouled or exhausted, they are discarded, usually every 3 or 4 weeks.

For either regenerable or nonregenerable polishers, the cation exchange resin can be used in ammoniated form to prevent removal of ammonia from the cycle. Anion resins are in the hydroxide form.

Virtually all once-through boilers, because of their especially high water purity requirements, employ polishing of feed water. Many plants polish the total feed water flow; some polish as little as 20% at full load with increasingly higher portions as load is reduced.

In drum-type boiler systems where water purity is less critical, condensate polishing may not be economically justifiable for normal day-to-day operations. But even in these systems, condensate polishing may be worthwhile to protect the boiler and turbine from the high levels of crud and silica common during initial start-up and subsequent restarts. A polisher can greatly shorten the precirculation time and often pay for itself in the cost of purchased power that would otherwise be necessary during the start-up period.

Condensate polishing eliminates most deposit-forming materials that create chemical concentration mechanisms in steam generator tubes, but such measures are never completely effective. Long-term accumulations of such deposits, principally iron and copper oxides, on boiler tube surfaces are removed by periodic cleaning. Utility boilers are generally considered "clean" with deposit weights less than 15 g/ft^2 (1.5 g/m^2) of tube area. Subcritical pressure boilers are considered dirty when accumulations exceed 40 $g/ft.^2$ Supercritical boilers are considered very dirty at accumulations over 25 $g/ft.^2$ Power boilers with normal feed water conditions may be routinely chemically cleaned every 4 years. Longer intervals may allow otherwise soft deposits to harden and become tenacious and difficult to remove.

Corrosion and hydrogen damage of boiler tubes are not likely to occur where rigorous control of preboiler chemistry, condensate polishing, and periodic cleaning keeps boiler surfaces clean. But because clean tube surfaces cannot be assured at all times and conditions other than water side deposits can create chemical concentration mechanisms, it is general utility practice also to exclude sources of potentially aggressive free sodium hydroxide from the boiler water.

Corrosion of boiler steel is also a function of pH (Figure 34.15), and pH values ideally should fall within a general range of 9.0 to 11.0. But a safe value must be achieved without producing measurable residuals of sodium hydroxide. In utility

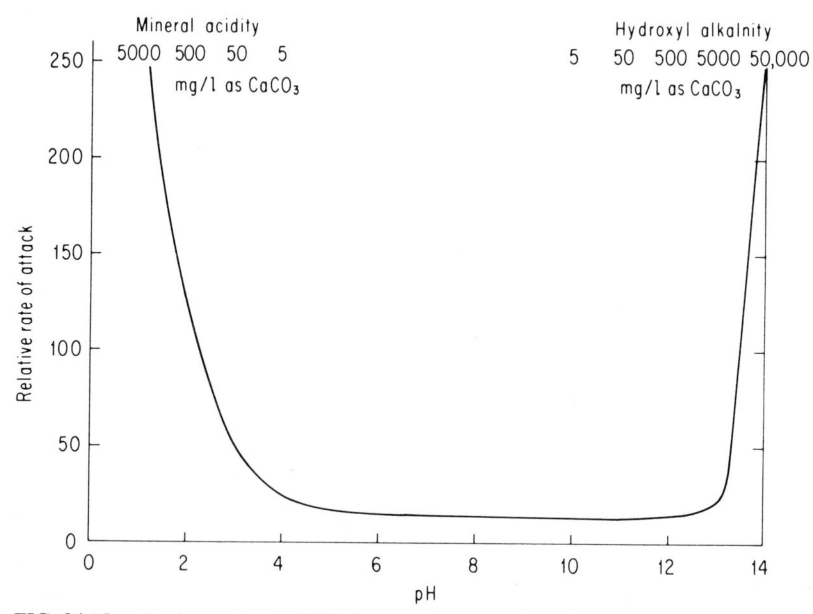

FIG. 34.15 Attack on steel at 590°F (310°C) by water of varying pH.

practice, this is accomplished either with a selection of sodium phosphate salts or with a "zero solids" treatment using all-volatile chemicals.

Sodium phosphates provide the desired pH in boiler water, while controlling the presence of free sodium hydroxide, as shown by these hydrolysis reactions:

$$\underset{\text{trisodium phosphate)}}{Na_3PO_4} \rightleftharpoons 3Na^+ + PO_4^{3-} \tag{5}$$

$$Na_3PO_4 + H_2O \rightleftharpoons Na_2HPO_4 + NaOH \tag{6}$$

$$\underset{\text{disodium phosphate}}{Na_2HPO_4} \rightleftharpoons 2Na^+ + HPO_4^{2-} \;\; \downarrow\!\uparrow \;\; H^+ + PO_4^{3-} \tag{7}$$

$$Na_2HPO_4 + H_2O \rightleftharpoons NaH_2PO_4 + NaOH \tag{8}$$

$$\underset{\text{monosodium phosphate}}{NaH_2PO_4} \rightleftharpoons Na^+ + H_2PO_4\text{'s-} \;\; \downarrow\!\uparrow \;\; H^+ + HPO_4^{2-} \tag{9}$$

The phosphates used are blends of disodium and trisodium salts of phosphoric acid. The species of phosphate present will depend upon pH, as shown in Figure 34.16. This plot shows that phosphates exist almost entirely in the dibasic ion form (HPO_4^{2-}) within the pH range of 9.0 to 11.0 considered optimum for boiler steel. Note that pH is measured on a cooled sample, and equilibrium pH values in the boiler are somewhat different.

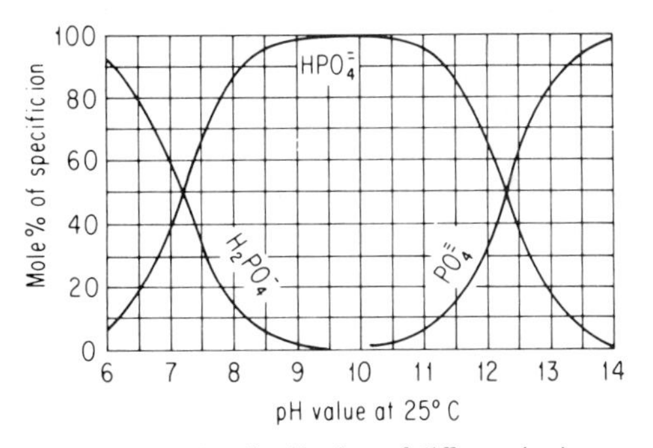

FIG. 34.16 The distribution of different ionic species of phosphate at various pH values.

Where the sodium (Na) to phosphate PO_4) mole ratio in pure water is 3 to 1, for Na_3PO_4, the relationship of pH to phosphate ion concentration will be as shown on the so-called coordinated phosphate curve (Figure 34.17).

Controlling pH solely by sodium phosphate hydrolysis, bulk boiler water pH should never accidently increase to levels aggressive to boiler steel. Reaction (6) shows that the hydrolysis reaction of trisodium phosphate is self-limiting; as pH rises above 10.0, generation of hydroxide ions by hydrolysis decreases. It is theoretically impossible for a solution of Na_3PO_4 to reach a pH much above 12.0. If the Na_3PO_4 solution were to concentrate to 10,000 mg/L, the pH would be 12.

Sodium hydroxide generated solely by this hydrolysis reaction is sometimes called "captive"; it will be "taken back" and revert to Na_3PO_4 at any sites of localized evaporative concentration, so high concentrations of NaOH in a confined area where solids might form is avoided, preventing caustic-gouging–type of metal attack.

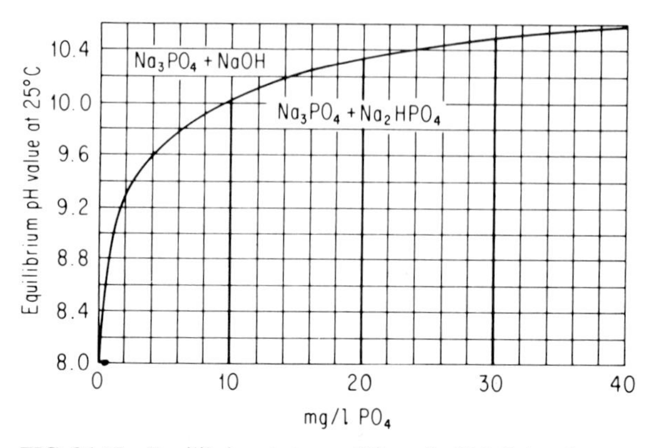

FIG. 34.17 Equilibrium between PO_4 and pH defining the coordinated phosphate boundary. *(Adapted from Combustion, pp. 45–52, October 1962.)*

In reaction (6), if water were to be removed by evaporation, the equilibrium would be forced to the left. Upon evaporation to complete dryness, the residue would contain Na_3PO_4, free of NaOH. But incomplete evaporation, the more likely condition beneath a porous deposit, produces a liquid beneath the deposit rich in sodium hydroxide, especially if incipient localized corrosion were already occurring. Thus, maintaining a 3:1 sodium:phosphate ratio in the boiler water may not provide positive protection against caustic-concentrate–type corrosion damage.

Many utility operators have adopted a modified form of coordinated phosphate control calling for maintenance of a Na:PO_4 ratio not over 2.6, corresponding to a 3:2 blend of trisodium and disodium phosphate. The control diagram shown in Figure 34.18 shows the differing hydrolysis effects of different sodium phosphates in selectively adjusting pH, PO_4, or both to keep coordinates within the desired range.

The all-volatile treatment (AVT) method uses only nitrogen-hydrogen compounds such as ammonia and hydrazine, so no solids are added. This is mandatory in once-through steam generators. Even in drum-type boilers, volatile treatment offers protection against superheater and turbine deposits that might otherwise occur with carryover of boiler water containing dissolved solids.

A disadvantage of AVT for drum-type boilers is that the boiler water is unbuffered and thus subject to extensive and rapid pH excursions in the event of feed water contamination. There is no tolerance for condenser coolant inleakage unless condensate polishing is provided.

Contaminant salts entering the cycle from other sources may produce acids (H^+) or alkalies OH^-) capable of being concentrated locally to corrosive levels. Typical contaminant sources are (1) condenser leakage; (2) treated make-up water

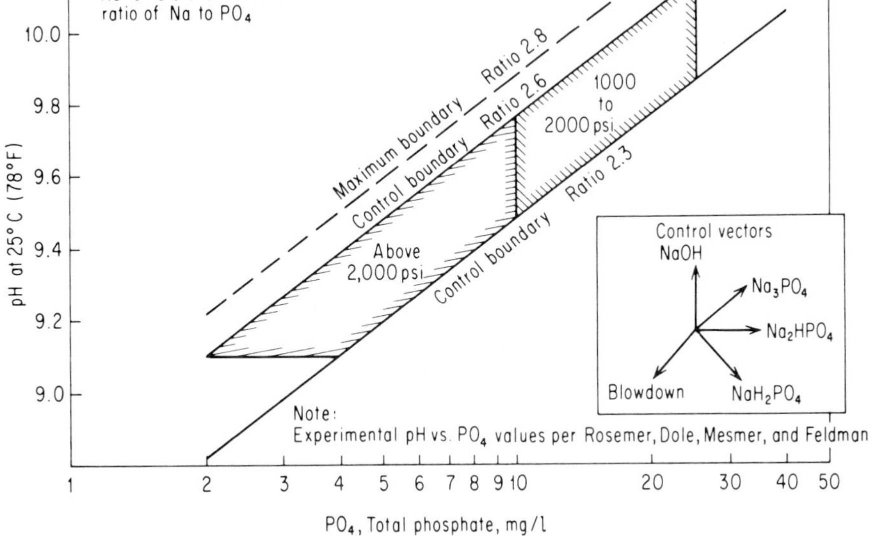

FIG. 34.18 Suggested coordinated phosphate control target for high-pressure boilers.

contamination by evaporator carryover or demineralizer leakage; (3) regenerant contamination from ion exchange units; or (4) incomplete removal of chemical cleaning solvents or alkalies. Of these, condenser leakage is the most significant.

Boiler tube hydrogen damage that results from condenser leakage is particularly likely to occur where the cooling water is brackish. The reactions of contaminants such as calcium chloride and magnesium chloride in boiler water, after depletion of any soluble phosphates, are:

$$MgCl_2 + 2HOH \rightarrow Mg(OH)_2 + 2HCl \tag{10}$$

$$CaCl_2 + 2HOH \rightarrow Ca(OH)_2 + 2HCl \tag{11}$$

In high-purity, unbuffered boiler water, very small cooling water leaks can reduce pH to 4.0 or less. Where the acid is then concentrated locally within a deposit at a tube surface, corrosion and hydrogen damage proceed rapidly.

While protection is afforded by condensate polishers, strict surveillance of condenser leakage is prudent for drum-type boiler cycles without polishers. As a general rule, leaks causing 0.5 mg/L feed water concentrations can usually be accommodated in normal operation without serious difficulty: feed water concentrations of 2.0 mg/L may be tolerated for short periods; but immediate boiler shutdown is usual when a condenser leak produces concentrations greater than 2.0 mg/L. Immediate boiler shutdown is always indicated where acceptable boiler water pH cannot be maintained.

The necessary continuous surveillance of condenser leaks usually relies on so-called cation conductivity monitoring, which effectively senses concentration changes as small as 20 ppb (μg/L). In this method, a condensate sample passes through a column of acid-form cation resin to convert sulfate, chloride, and nitrate salts to more electrically conductive sulfuric, hydrochloric, and nitric acid. The conductivity is then measured in the resin-treated sample.

Monitors for silica and sodium are alternative devices for detection of condenser leaks. Against background concentrations commonly encountered in turbine condensates, available monitoring equipment can reliably detect changes of 5 ppb or less in either of these contaminants.

STEAM PHASE PROBLEMS

In the 1950s, it became apparent that the silica in steam that was causing deposits on the low-pressure stages of utility turbines was not carried out of the boiler in droplets of the boiler salines, but rather it was a gaseous fraction of the high-temperature vapor. (This was a contribution to water chemistry from the field of geology.) Previously, it had been long recognized that carbonates would break down to yield CO_2 in steam, but now there was growing evidence that minerals might actually volatilize and leave the boiler in the gas phase. With development of the once-through boiler, it was apparent that any minerals in the feed water, volatile or not, would be passed along to the turbine and the condenser. This led to the development of polishing demineralizers for the condensate stream to prevent a buildup of minerals in the cycle that would lead to deposits. It also led to development of AVT, or the use of volatile chemicals, such as ammonia and hydrazine, as the only chemical treatment applied to the cycle.

The quality of makeup improved greatly in the succeeding years, demineralizers replacing evaporators not only on the basis of economics, but also on quality; evaporators (as then designed) were unable to produce the low silica levels needed to avoid turbine deposits.

The newer systems began to experience turbine blade cracking, generally occurring on the next-to-last wheel, and it was found that the cracking was caused by caustic. Furthermore, ammonia concentrations and pH were difficult to control, because the polishing demineralizer removed ammonia (usually causing an increase in sodium); and erratic control often led to attack of copper alloy heater tubes.

Because of these problems, it became necessary to study steam-phase conditions throughout the cycle. This required the development of sophisticated and reliable isokinetic sampling systems capable of handling both superheated and wet steam, at high pressure and under vacuum. Analyses were made of the nature and causes of system deposits, and basic research was undertaken in the physical chemistry of the elements and compounds found. One of the surprising conclusions was that NaCl present in superheated steam hydrolyzed, yielding NaOH in dry deposits and HCl in the condensed steam. Because of this, limits were placed on sodium levels because sodium seems to be the major cause of embrittlement-type attacks on turbine blading, a problem more common to AVT systems than to drum-type boilers using phosphate treatment.

There is still more to be learned about vapor-phase chemistry, but present state of the art calls for extensive sampling and continuous analysis of makeup, condenser hotwell water, deaerated water, feed water at the economizer inlet, boiler water, superheated steam, and selected stage heater drains. The analyses include continuous measurement of conductivity, pH, dissolved oxygen, sodium, and silica (on demineralized makeup); Fe, Cu, and Ni on feed water, heater drains, and condensate; these metals plus ammonia before and after condensate polishers;

hydrazine on feed water; and boiler testing (with drum-type boilers). With this type of monitoring, conditions can be continuously scrutinized both during steady load and during shutdown and start-up, so that corrective adjustments can be made in chemical treatment, or operating problems (demineralizer regeneration, control of air inleakage, control of condenser leakage) can be corrected.

The steam-phase chemistry relative to geothermal plants is completely different from conventional plants, as the vapor often contains additional contaminants (e.g., boron and sulfur compounds) at parts per million rather than parts per billion levels. This is a completely new field.

WATER CHEMISTRY IN NUCLEAR FUELED PLANTS

The nuclear pressurized water reactor (PWR) (Figure 34.5) has two major water systems:

1. The primary loop, or reactor coolant system
2. The secondary loop, or steam generator-turbine cycle

Components of the primary loop are the reactor vessel, a pressurizer, steam generators (heat exchangers), and circulating pump. In the primary loop, water temperature is allowed to rise only about 50°F (28°C) passing through the reactor, so recirculation rates are quite high to absorb the amount of heat generated. Pressure of the recirculating water is kept high enough to keep the water from boiling. The pressurizer maintains this pressure and absorbs changes in volume produced by changes in temperature. For maximum corrosion resistance, wetted surfaces in the primary loop are usually stainless steel or nickel-based alloys.

High purity must be maintained in the primary loop water to minimize fouling of reactor and exchanger heat transfer surfaces and to avoid contaminants that could form undesirable radioactive isotopes under neutron flux. Controlled additions and removals of boric acid in the primary loop water provide the correct concentration of neutron-absorbing boron needed to control neutron flux and energy transfer. Chemicals such as lithium hydroxide, forming relatively safe radioisotopes under neutron flux, are used for pH control in the primary loop.

Under nuclear radiation in passage through the reactor, some of the primary loop water is decomposed into hydrogen and oxygen as follows:

$$H_2O \rightarrow H_2 + \tfrac{1}{2}O_2 \tag{12}$$

To suppress oxygen generation and to scavenge oxygen entering the system, hydrogen gas is commonly added. Primary loop water purity is usually maintained by continuously passing a portion of the circulating water through a mixed bed demineralizer. These generally utilize anion resins in the borate form, to avoid removal of boron from the water. As impurities accumulate, these resins tend to become radioactive and regeneration becomes impractical. Consequently, the resins are eventually disposed of as a solid radwaste.

Two basic heat exchanger designs are employed as secondary loop steam generators in PWR systems. The recirculating U-tube–type (Figure 34.19), is analogous to a drum-type boiler in a fossil-fueled system. Having an internal recirculation rate of 3 to 4 times the steam flow, the unit provides mechanical steam-water separation and continuous blowdown. But it differs from fossil-fueled boilers in that there is no provision for superheating the steam.

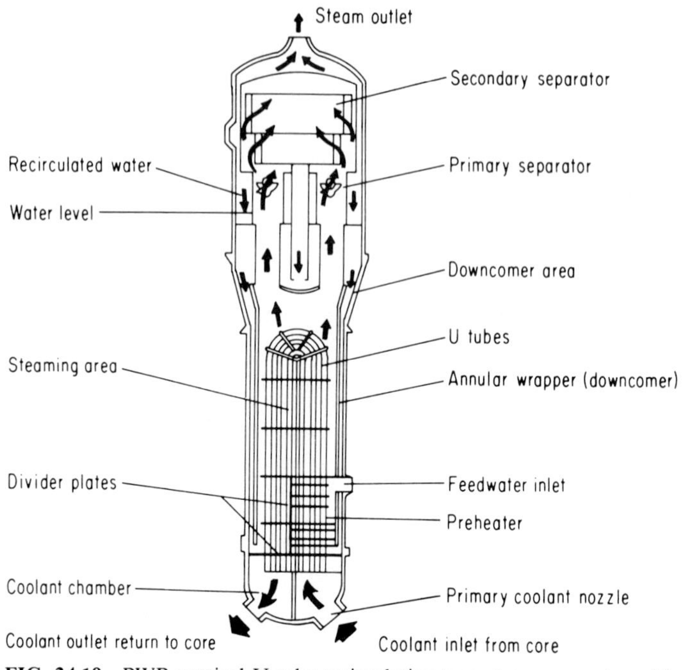

FIG. 34.19 PWR vertical U-tube recirculating-type steam generator with integral feed water preheater.

The other PWR heat exchanger design (Figure 34.20) provides once-through–type steam generation. The heat exchanger is baffled so that incoming feed water first flows down through an outer annulus in which it is heated to saturation temperature, and then it is converted totally to steam and is slightly superheated in a single upward shellside pass through the tube bundle. Depending on load, up to about 60°F (34°C) superheat can be achieved.

Either type steam generator requires maximum corrosion resistance on the primary side, so tube materials are commonly a special nickel-iron-chromium alloy. Carbon steel shells are generally used, although those parts of the shell surface constituting the primary coolant inlet and outlet plenums are clad on the coolant water side.

For recirculating U-tube–type PWR steam generators, the AVT method is preferred to coordinated phosphate treatment. In once-through–type PWR steam generators, of course, AVT is automatically required.

In the nuclear boiling water reactor (BWR) (Figure 34.21), boiling occurs in the reactor itself. The same water serves as the cycle working fluid, the reactor coolant, and the nuclear reaction moderator. Boiling water reactors are comparable to drum-type boilers. A portion of the water passing through the core is converted to steam. The steam-water mixture is then separated, with the steam going to the turbine and boiler water returning through circulating pumps to the core inlet.

An important aspect of BWR operation is the effect on the working fluid of direct exposure to nuclear radiation. Some of the water is decomposed into hydrogen and oxygen. It is not possible to inject hydrogen as in the PWR because of

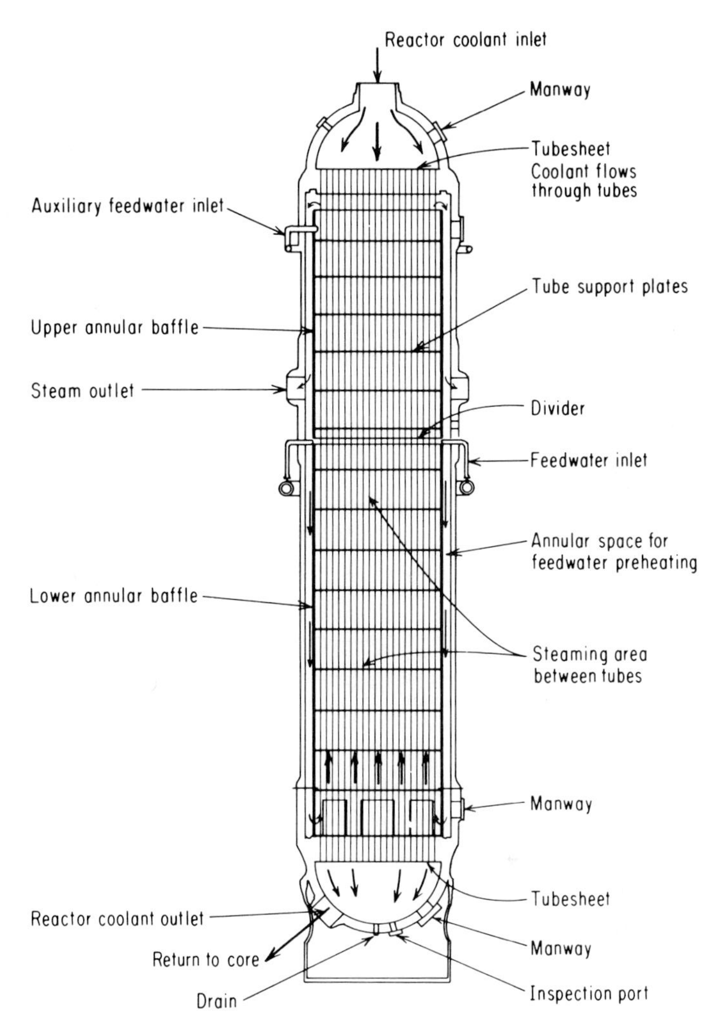

FIG. 34.20 PWR once-through-type steam generator. *(Adapted from Steam/Its Generation and Use, Babcock & Wilcox Company, 1972.)*

the need for continuous removal of all noncondensible gases at the condenser to maintain the vacuum required for turbine efficiency. Thus, steam produced by a BWR reactor contains high concentrations of oxygen, an important factor in corrosion.

Nitrogen entering the cycle by air inleakage or decomposition of nitrogenous compounds can react to form nitric acid:

$$5O_2 + 2N_2 + 2H_2O \rightarrow 4HNO_3 \tag{13}$$

Additives for pH control and oxygen scavenging, such as amines and hydrazine, are not usable in BWR cycles because they are subject to nuclear decomposition. Thus, corrosion control relies primarily on the corrosion resistant materials used throughout the cycle.

Metallic contaminants in the cycle water, either from condenser leakage or from corrosion of system metals, are subject to stringent limitations to avoid formation of hazardous isotopes and to avoid deposition of crud on the core heat transfer surfaces. Copper is specifically limited because of its tendency to foul the

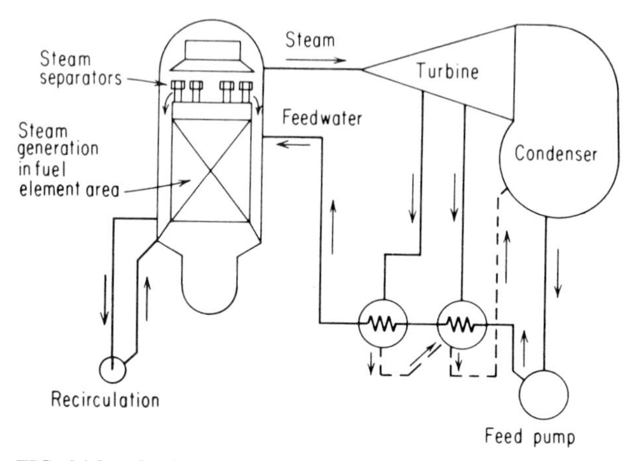

FIG. 34.21 In the BWR nuclear plant, steam is generated in the same vessel that contains the fuel elements, as shown in this simplified diagram.

orifices of the core water distributor. Make-up water is demineralized to less than 1.0 μS/cm specific conductivity and 10 ppb (μg/L) silica. Full flow condensate demineralization is then employed to maintain cycle water purity at limits established by the reactor manufacturer. Typical limits are: 30 ppb total metals, 2 ppb Cu, 2 ppb Cl, and 0.1 μS/cm at 25°C. Resultant feed water is essentially neutral, with pH equal to 7.

Where deep bed demineralizers are used in BWR condensate polishing, resins are usually sluiced to external vessels for scrubbing and regeneration. The regenerant wastes generally contain low-level radioactivity which requires processing in the liquid radwaste system. Where nonregenerable powdered resin units are used for polishing, the spent, sluiced resin must be processed.

CONDENSER COOLING WATER

As previously discussed, the thermal efficiency of modern fossil-fueled steam electric plants averages about 34%. The largest part of the loss, equal to about 50% of the heat released by the fuel, is rejected at the condenser. Nuclear-fueled plants, averaging about 32% efficiency, lose even more heat at the condenser. Heat rejected at the condenser is transferred to cooling water, which subsequently releases this heat to the environment.

Although water circulating through the tubes of a surface condenser is called cooling water, it is important to recognize that there is no cooling of the vapor or condensate. The circulating water simply absorbs the heat of vaporization of steam leaving the turbine, converting it to liquid at the same temperature. For this reason, the term "condenser water" is sometimes preferred to "cooling water."

The primary function of a surface condenser is to maximize cycle efficiency by allowing steam to expand through the turbine to the lowest possible exhaust temperature. Just as in a boiler drum, the temperature of steam and water in a condenser is directly related to pressure. At 102°F (39°C), for example, the corresponding saturated steam pressure is about 1.0 lb/in^2 abs (51.7 mmHg); this is also expressed as a vacuum of about 27.89 in of mercury (705 mmHg). At this vacuum, 1 lb of steam occupies 331.1 ft^3 (20.6 m^3/kg); when condensed it occupies only 0.016 ft^3 (0.001 m^3/kg), about 0.005% of the original steam volume.

Noncondensible gases (air, ammonia) in the condenser reduce the vacuum achievable, so such gases are continuously removed by mechanical vacuum pumps or steam jet eductors (Figure 34.22).

The ability of a surface condenser to provide the lowest possible backpressure for a given load and cooling water temperature is adversely affected by such things as:

1. Waterside scaling and fouling of condenser tube surfaces
2. Partial blockage of cooling water flow through the tubes
3. Inadequate removal of noncondensible gases
4. Faulty distribution of exhaust steam

Waterside scaling and fouling are probably the most common causes of impaired condenser performance.

The effect of a 1.0 in Hg (25.4 mm) rise in backpressure in a 500-MW plant is illustrated in Figure 34.23. At constant steam rate, electrical output would be reduced by about 13 MW, a loss of 2.5%. This 13 MW represents about $1.25

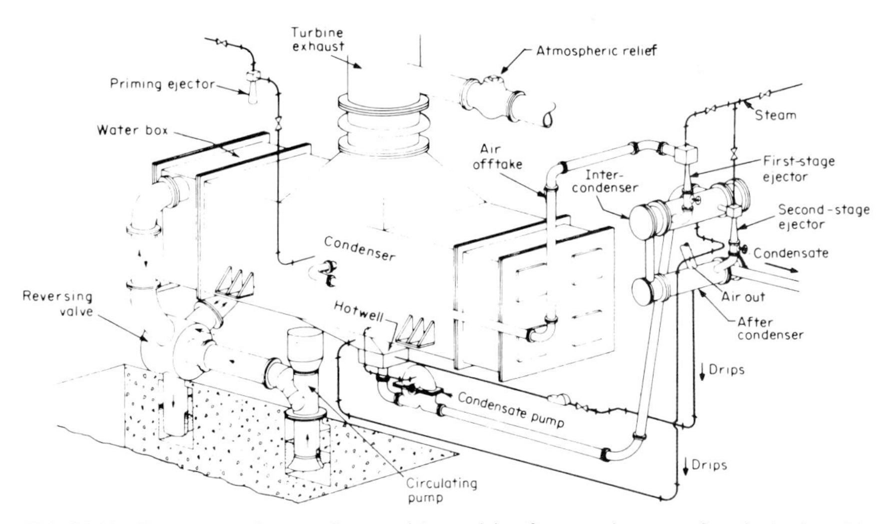

FIG. 34.22 Two-pass surface condenser with provision for reversing water flow for backwashing tubes and waterbox. Inter- and after-condenser drips normally return to the hotwell.

500-MW Generation—Coal Fired (where unit rating permits fixed load maintenance)			
	Backpressure, in mm Hg (abs)	Gross turbine unit	Heat rates, total
	1.5	7600 Btu/kWh	3.8×10^9 Btu/h
	2.5	7800 Btu/kWh	3.9×10^9 Btu/h
A. Rise	1.0	200 Btu/kWh	0.1×10^9 Btu/h
B. Additional fuel Btu (at 90% boiler efficiency) =			0.11×10^9 Btu/h
C. Additional fuel cost (at $1.75/million Btu =			$193/h
			$4620/day
Plus additional fuel cost caused by increased load on auxiliaries. 1 Btu = 0.252 kcal			

FIG. 34.23 The effect of turbine backpressure increase on plant efficiency.

million in plant investment. Effective condenser water treatment is a very important factor in preventing such losses.

Five basic types of condenser water systems are used in utility stations (Note: There are also some all-dry condensing systems, but these are quite rare.):

1. Once-through cooling
2. Cooling lake and cooling pond systems
3. Spray ponds
4. Wet cooling tower systems
5. Wet-dry combination cooling tower systems

In once-through cooling, water passes once through the condenser and returns to its source at a higher temperature. Because residence time and temperature are low, scaling is usually not a problem. The principal fouling problems are usually related to microbial activity and silt deposition. Regularly scheduled biocide control usually achieves a moderate increase in condenser cleanliness for several hours. Application of a biodispersant with each application of chlorine or bromine can increase both the magnitude and duration of this improvement (Figure 34.24). By increasing the effectiveness of chlorine or bromine, such biodispersants also can reduce the dosage of biocide, helping the plant meet discharge regulations.

Corrosion inhibitors are rarely required in freshwater once-through systems. But for protection of copper alloy tubes (e.g., aluminum-brass) in seawater once-through systems, low dosages of ferrous sulfate are sometimes applied intermittently to the condenser inlet water to provide a protective film of iron oxide on the tube internal surfaces. The cost effect of the film in reducing heat transfer must be balanced against the benefit of reduced corrosion.

With the advent of environmental regulations restricting thermal discharges, some once-through systems have been modified so that the condenser outlet water passes over a "helper" cooling tower or through a tempering canal with floating spray modules to dissipate heat before discharge. Since such "helper" additions follow the condenser, biocide-dispersant treatment of the water is the same as for conventional once-through systems.

Treatment*	C.F. improvement	
	Highest percent	Duration
0.05 mg/l Cl_2 residual plus 15 mg/l biodispersant	7.24%	6.5 hours
12.12 mg/l Cl_2 residual plus 5 mg/l biodispersant	2.61%	6.0 hours
12.40 mg/l Cl_2 residual alone no biodispersant	0.55%	2.0 hours

* Short duration treatment applications with system blowdown closed off until chlorine residual gone.

FIG. 34.24 Condenser tube cleanliness factor improvement through the use of a biodispersant. *(Source: Nalco Research Report Serial No. 1642, June 1977, re: Field evaluation data obtained at Duane Arnold Nuclear Center of Iowa Electric Light & Power Company.)*

Cooling water may be discharged to cooling lakes, artificial impoundments created by damming a stream. Condenser water is recirculated through the lake, and heat is dissipated primarily by surface evaporation. The lake receives make-up from rainfall and runoff from its drainage area. Cooling lakes make possible high cooling water withdrawal from streams too small to provide such cooling capacity on a once-through basis. Water treatment practices and discharge limitations are generally the same for these cooling lakes as for once-through cooling.

By contrast, cooling ponds are defined as impoundments that do not impede the flow of a navigable stream. They are usually constructed along a stream from which make-up water may be pumped to meet evaporative losses not replaced by rainfall or runoff from surrounding drainage areas. Cooling ponds contained within earthen levees at levels above the surface water are said to be "perched." The area required for natural evaporation is typically on the order of 1.0 to 1.5 acres (0.4 to 0.6 ha) per megawatt of capacity. A pond typically provides detention of about 5 to 15 days to produce the necessary heat dissipation.

Blowdown from cooling ponds is composed of bottom seepage and controlled overflow to the source stream. Because of the large rainfall-receiving surface areas, concentration ratios are typically quite low, up to about 1.5 times source concentration. But even such low concentration increases may be significant in terms of reduced $CaCO_3$ solubilities, so condenser tubes are more likely to scale on pond systems than on once-through cooling systems. Sulfuric acid may be fed to maintain pH below saturation values (pH_s). Scale inhibitors and dispersants are sometimes required in conjunction with acid feed for complete scale prevention.

A significant aspect of cooling ponds is that they promote "homes" for a variety of aquatic organisms. They often support large fish populations. Specific biodispersants in conjunction with periodic feed for chlorine or bromine may be needed to keep condenser tubes free of biological slimes initiated by the microorganisms of the food chain supported by the pond.

Spray ponds promote evaporation by mechanically dispersing water into sheets and droplets, dissipating heat effectively so that less pond area is required. Evaporation is mechanically promoted, so it is less dependent on climatological conditions. Recirculating water treatment requirements generally correspond more closely to those of evaporative (wet) cooling towers than to cooling ponds or lakes.

Where large cooling ponds or spray ponds are impractical, cooling towers are installed to reduce thermal pollution of receiving waterways. To comply with restrictions on thermal discharge, blowdown is taken from the tower basin, having the lowest temperature of the recirculating system.

An important advantage of wet cooling towers is their ability to significantly reduce raw water withdrawal and wastewater discharge (Table 34.2).

TABLE 34.2 Water Flows for 1000-MW Plant Cooling System
[*Based on 20°F (11°C) temperature rise*]

	Once-through cooling system	Wet cooling tower recirculating system
Condenser flow, gal/min*	400,000	400,000
Evaporative conc. ratio	1×	5×
Make-up water, gal/min	400,000	10,000
Discharge water, gal/min	400,000	2,000

* gpm $\times$ 3.785×10^{-3} = m^3/min

Because drift (entrainment of fine mist) from a cooling tower is part of blowdown, the net discharge to the receiving stream in this example would be about 10 to 200 gal/min (0.038 to 0.76 m^3/min) less than the 2000 gal/min (7.6 m^3/min) shown.

With the constraints of environmental limitations on concentrations of chlorine and other chemicals that may be discharged in blowdown, cooling tower recirculating water treatment practices for stream-electric plants are generally as described elsewhere in this text for industrial plants.

With the increased planning for maximum water conservation, coal-fired plants may substitute cooling tower blowdown for raw water for ash sluicing. Some sluice water is evaporated by the heat of bottom ash, and a small amount is retained by the ash, so the final volume of wastewater for disposal is further reduced.

Zero discharge of tower blowdown may be impractical and unnecessary environmentally in most parts of the country, but strategies for approaching this goal have been devised for arid regions where adequate land is available. In such cases, tower blowdown may be routed to on-site solar evaporation ponds.

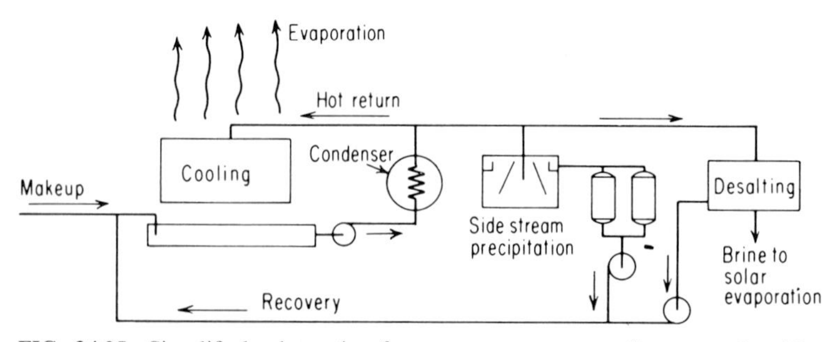

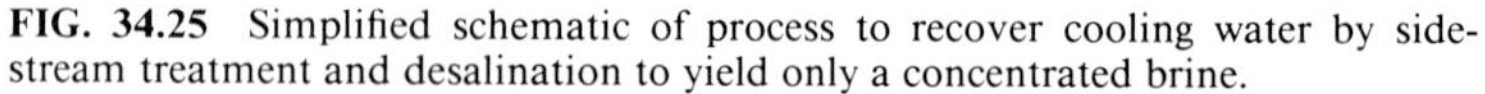
FIG. 34.25 Simplified schematic of process to recover cooling water by side-stream treatment and desalination to yield only a concentrated brine.

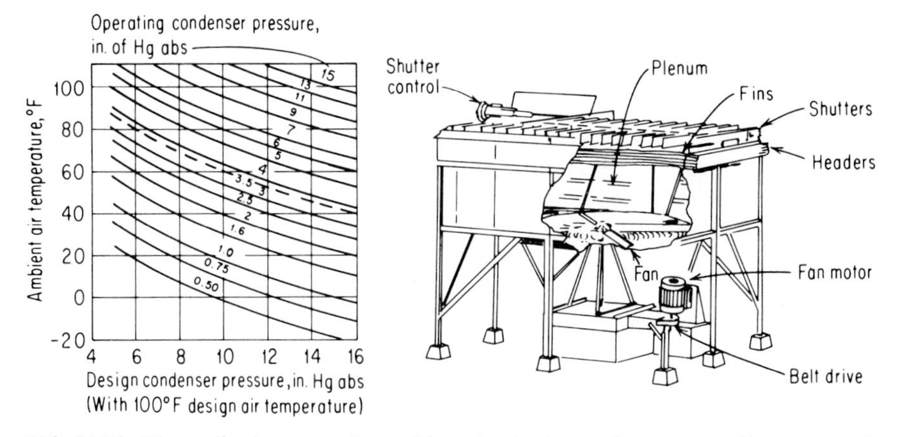

FIG. 34.26 Dry cooling towers can be used in water-short areas, but users usually pay a penalty in energy loss where air temperatures are high. *(From Power Special Report, "Cooling Towers," March 1973.)*

Various processes for side stream treatment of cooling water and for blowdown concentration reduce discharge volume and required evaporation pond areas. One such conceptual system is illustrated in Figure 34.25. Although side stream processing can control concentrations of potentially scale-forming ions, high salinity in the recirculating water is characteristic of these treatment schemes. The effects of high dissolved solids in cooling tower drift and on surounding vegetation and structures must be considered. Also, the significantly greater impact of condenser leaks on the chemistry of the boiler water cycle must be provided for.

Moisture "plumes" emanate from cooling tower stacks like clouds under many weather conditions, literally dominating the atmosphere in the immediate vicinity. One means of eliminating plumes is to use nonevaporative dry cooling towers. In dry cooling towers (Figure 34.26), heat is transferred from the recirculating water to the air by convection rather than by evaporation.

Since dry cooling tower performance is controlled by ambient air dry bulb temperature, rather than by lower wet bulb temperatures, turbine backpressures are higher for dry tower systems than for wet towers, so cycle efficiencies are poorer.

But combinations of wet and dry cooling towers may be used to advantage (Figure 34.27) to alleviate plume problems with minimum loss of cycle efficiency. And, since dry towers handle their part of the heat load without evaporation, overall make-up and blowdown volumes are less for wet-dry cooling tower combinations. Systems may be set up with the recirculating water flowing either in series or parallel through the wet and dry towers. Designs usually provide for a means of totally or partially bypassing either air flow or recirculating water flow selectively around either the wet or dry tower, as may be required to optimize plant efficiency for varying plume abatement and water conservation requirements. Figure 34.28 diagrams a system with provision for water bypass. Figure 34.29 illustrates an integral wet-cooling tower unit with provision for adjusting the balance of air flows through the wet and dry sections. Recirculating water treatment practices for wet-dry cooling tower systems are essentially the same as for straight wet cooling tower systems. As load is shifted from wet to dry cooling, less dirt is washed out of the air to contaminate the recirculating water.

Closed recirculating water systems handle a variety of cooling requirements in

FIG. 34.27 At this utility station a large combination wet-dry tower is used to cool condenser water. *(Courtesy of the Public Service Company of New Mexico.)*

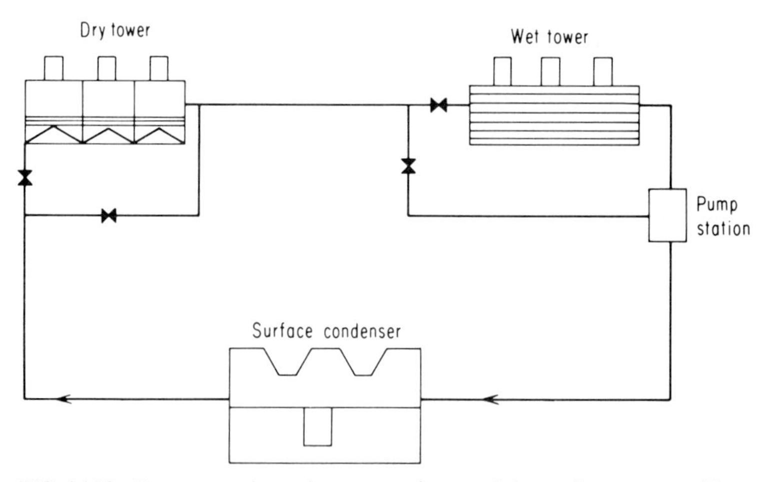

FIG. 34.28 Bypass permits optimum use of wet and dry cooling towers. *(From Combustion, pp. 23–27, October 1976.)*

utility stations, avoiding introduction of outside contaminants that might foul heat exchanger surfaces. Systems of this type are sometimes used for cooling the hydrogen circulated for electric generator cooling. They also may be used for aftercooling of compressed air, jacket cooling for standby diesel engines, and for certain lube oil cooling requirements. In nuclear plants, closed recirculating cooling

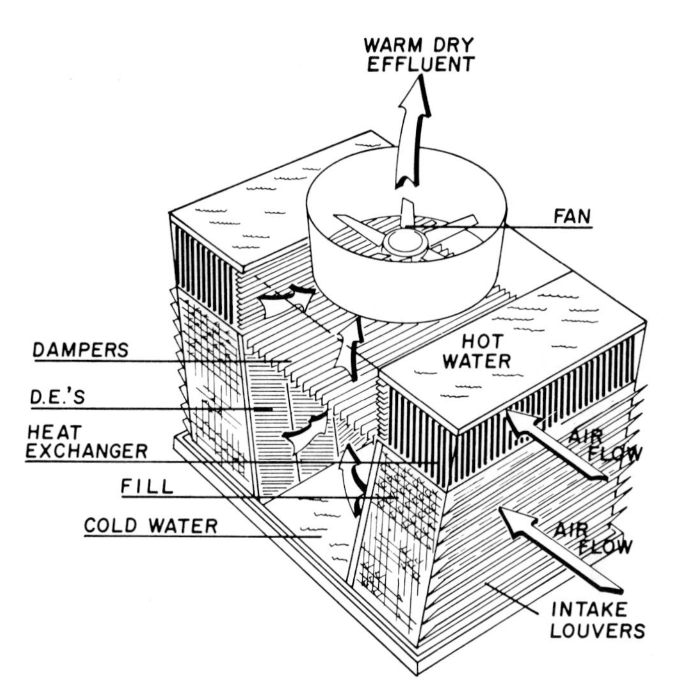

FIG. 34.29 Integral wet/dry cooling tower with damper control (D.E.s—drift eliminators). *(From Reisman, J. I., and Ovard, J. D.: "Cooling Towers and the Environment—An Overview," Proceedings of the American Power Conference, vol. 35, 1973.)*

water systems are commonly used to handle heat loads from the reactor vessel shielding, primary coolant pump seal housings, and other heat loads within the reactor building.

MISCELLANEOUS WATER USES

Another major water use in coal-fired power plants, and to a lesser extent in oil-fired plants, is for ash transport. The solids that fall to the bottom of the boiler furnace (bottom ash) are invariably cooled and conveyed from the boiler by water. The solid products of combustion (fly ash) in the flue gas that do not deposit in the boiler convection passes or air heater, are subsequently removed by electrostatic precipitators, bag filters, and wet scrubbers. Fly ash accumulations are removed from these devices either by water or by mechanical or pneumatic means.

The quantity of water required for ash sluicing varies with design factors such as method of firing and nature of the coal. For a typical 1000-MW coal-fired plant, it amounts to about 10 mgd (26.2 m^3/min).

For molten slag tap or "wet bottom" boiler, where water is used to quench molten ash, heat pick up can be in the range of 10^6 Btu per ton of bottom ash. Suspended solids in bottom ash water are of concern if they are difficult to settle.

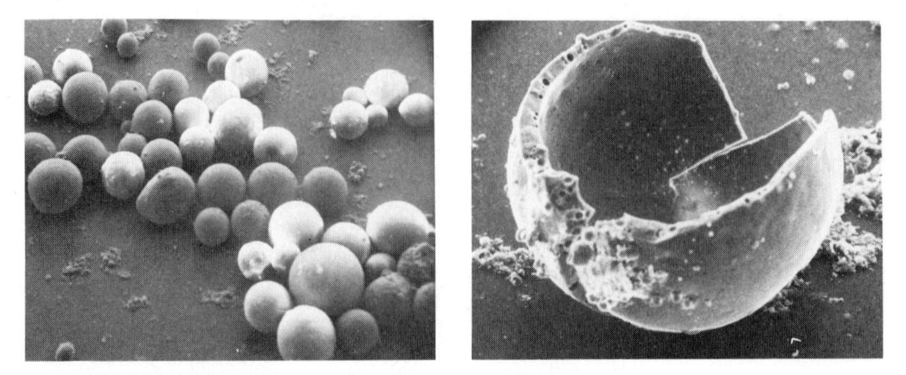

FIG. 34.30 The suspension firing of coal produces slag in a variety of forms, including these microspheres, which float to the surface of wet ash collection ponds. *(Left)* Random spheres; *(right)* broken shell.

Coagulants may be needed to reduce supernatant turbidity. Dissolved solids contribution to the sluice water from the ash is relatively small.

By contrast, fly ash is much more leachable and may contribute on the order of 50 to 800 mg/L dissolved solids to its transport water. Fly ash often contains hollow microspheres (Figure 34.30), which are difficult to separate from sluice water by sedimentation. So, dry transport systems are often used to handle it.

In oil-fired plants, ash is produced in comparatively small quantities. Bottom ash removal from the boiler by fireside washing and sluicing may be needed only occasionally. Oil ash generally does not settle in ponds as well as coal ash. While ash pond overflow waters have many of the same characteristics as in coal-fired plants, high concentrations of vanadium may be present, originating as an ash constituent of some fuel oils. Bottom ash in some oil-fired plants is removed by dry collection rather than by water sluicing.

Where there is a market for the bottom ash, or where insufficient land is available for settling, dewatering bins can be used and the dewatered ash trucked away. Water drainage from dewatering bins or the supernatant from ash ponds can be recycled to the sluice pumps. Make-up to such recycle systems is required for replacement of evaporative losses and moisture that leaves with the dewatered ash. Because of the poor liquids/solids separation of the buoyant fly ash, however, recycle systems are generally not suitable where fly ash is sluiced to the same dewatering point as bottom ash. Chemical coagulants and flocculants are useful in the treatment of ash transport waters, to reduce suspended solids to acceptable levels for recycle or final discharge.

In some utility plants firing high-sulfur coal, water is also used in wet-scrubber–type flue gas desulfurization (FGD) systems. Most FGD systems in U.S. power plants use a wet scrubbing process in which the reactive component of the sorbent is not regenerated. Most commonly, sulfur dioxide in the flue gas is reacted with slurries of lime or limestone to form calcium sulfites and sulfates. Sludge from the process, consisting of fly ash and water in addition to the calcium precipitates, is usually piped to large settling ponds. Supernatant water from the ponds is reclaimed and recycled to the process. The ponded sludge retains a quicksandlike character, limiting any future use of the pond land area. Make-up water requirements for such systems may be in the range of 0.75 to 1.0 gal/min (2.8 to 3.8 l/min) per megawatt of generation rate.

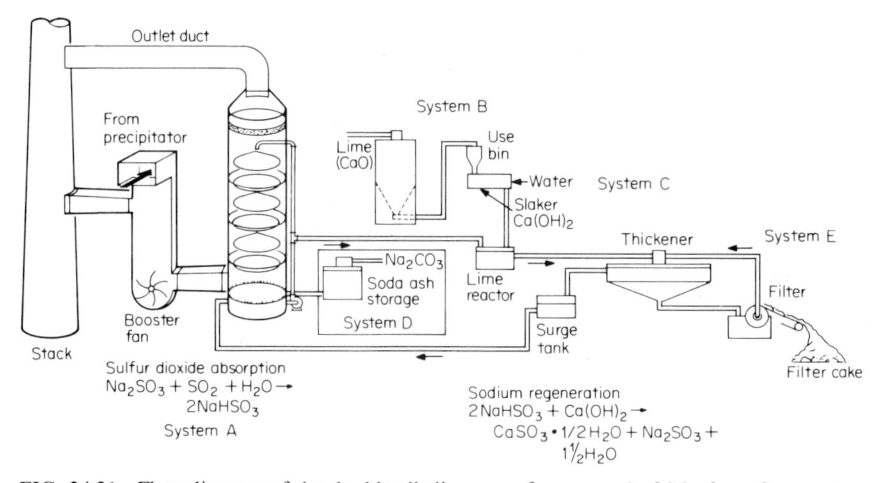

FIG. 34.31 Flow diagram of the double-alkali process for removal of SO_2 from flue gas to produce a disposable sludge cake and recycle soluble alkali.

A smaller number of nonregenerable FGD systems are of the double-alkali type. In this process, flue gas is scrubbed with a solution of sodium hydroxide to form sodium sulfite and sodium bisulfite. The scrubbing solution is then reacted with lime, precipitating sludges similar to those with lime as the primary sorbent, restoring the sodium hydroxide for reuse in the process. Although the absorbing medium is thus "regenerated," this process is still categorized as nonregenerable because the sulfur scrubbed from the flue gas is not recovered (Figure 34.31).

TABLE 34.3 Major Wastewater Sources and Contaminants in Fossil-fueled Steam-Electric Generating Plants

Source	Principal contaminants
Ash sluicing system	TSS,* TDS,† heat, oil
Low-volume wastes‡	TSS, TDS, pH, oil
Metal-cleaning wastes§	TSS, TDS, pH, iron, copper, other metals, oil
Boiler blowdown	TSS, iron, copper, oil
Main condensers, once-through cooling	Heat, residual chlorine
Main condensers, recirculating-water blowdown	Residual chlorine, Zn, Cr, P and other corrosion inhibitors
Area runoff¶	TSS, TDS, pH, oil
Intake water traveling screens and strainers	Solids (debris)

* TSS, total suspended solids.

† TDS, total dissolved solids.

‡ Low volume wastes, collectively including wastes from ion exchange systems regeneration, evaporator blowdown, flue-gas wet scrubbers, floor drainage, cooling-tower basin cleaning, blowdown from recirculating house service water systems.

§ Metal cleaning wastes, including waterborne wastes from boiler tube cleaning, boiler fireside cleaning, air preheater cleaning.

¶ Area runoff, including rainfall runoff from storage of coal, ash and other materials.

Steam-electric power plants are subject to federal, state, and local water pollution control regulations. Where these regulatory bodies differ on specific limitations for a given pollutant, the more stringent regulations generally apply. Some state and local regulations limit certain pollutants not limited by federal guidelines.

A significant feature of pollution control regulations for steam-electric plants is that specific discharge limitations relate to individual processes and sources within the plant rather than to the combined discharge. Thus, no credits accrue from combining or diluting one in-plant waste stream with another. For example, the total weight discharge per day of iron originating in boiler blowdown may not exceed the actual daily flow of boiler blowdown times the permissible concentration limit of iron.

A simplified tabulation of steam-electric plant wastewaters and their contaminant characteristics, by source categories, is shown in Table 34.3.

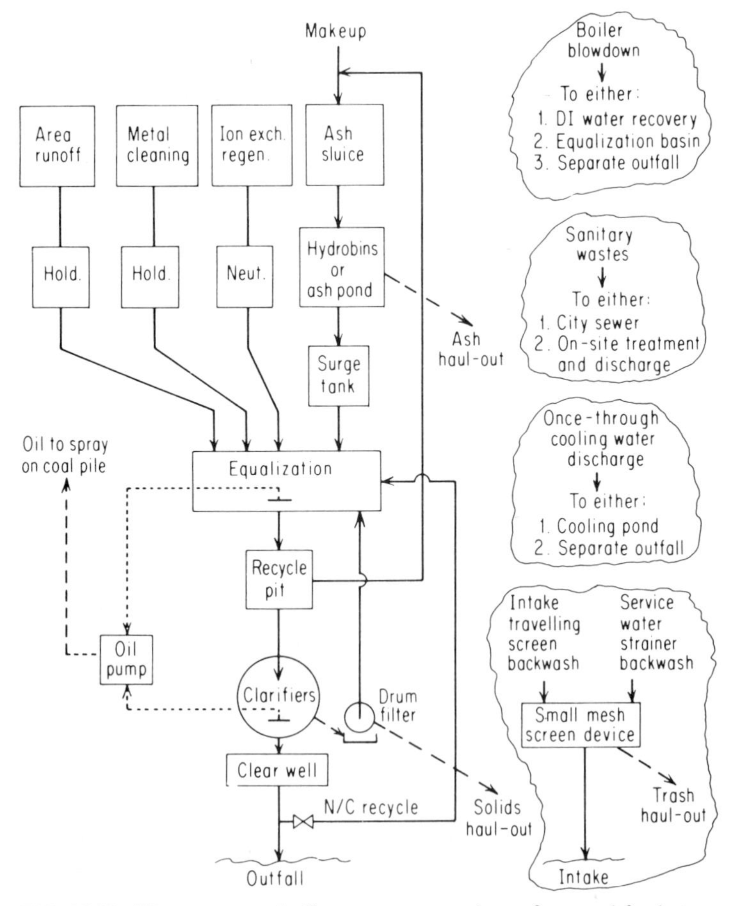

FIG. 34.32 Water reuse and effluent treatment scheme for a coal-fired steam electric plant.

TABLE 34.4 Typical Characteristics of Coal Pile Runoff

Parameter	Concentration, mg/L
Total solids	1500–45000
Total dissolved solids	700–44000
Total suspended solids	20–3300
Total hardness ($CaCO_3$)	130–1850
Alkalinity ($CaCO_3$)	15–80
Acidity ($CaCO_3$)	10–27800
Manganese	90–180
Copper	1.6–3.9
Sodium	160–1260
Zinc	006–23.0
Aluminum	825–1200
Sulfate	130–20000
Phosphorus	0.2–1.2
Iron	0.4–2.0
Chloride	20–480
Nitrate	0.3–2.3
Ammonia	0.4–1.8
BOD	3–10
COD	100–1000
Turbidity (in JTU)	6–605
pH (in units)	2.8–7.8

Source: *Development Document for Proposed Effluent Limitations Guidelines and New Source Performance Standards for Steam Electric Powerplants,* EPA 440-1-73/029, March 1974.

Principles of water recycle, reuse, and effluent treatment for steam-electric plants are generally the same as for other process industries. Figure 34.32 illustrates one reuse and effluent treatment scheme for a coal-fired plant. In this illustration, bottom ash transport water is recycled. Fly ash is separately handled in a dry system. One of the difficult problems is proper handling of coal pile runoff without interfering with coal handling procedures in the large area used for coal inventory. The runoff is often high in heavy metals, alumina, and crud, and may require recycle or direct treatment (Table 34.4).

For nuclear power plants, a principle of special importance is the segregation of dissimilar wastes. High-purity (low-conductivity) wastewaters are kept separate from low-purity (higher conductivity) waters to facilitate reclamation and recycle of the high-purity waters. Chemical wastes, including ion exchange regenerants and chemical cleaning wastes, should be segregated from detergent wastes because of their differing requirements for final treatment before discharge or reuse.

SUGGESTED READINGS (SEE "ENERGY PRIMER," CHAPTER 46)

Auerswald, D. C., and Cutler, F. M.: "Two-Year Study on Condensate Polisher Performance at Southern California Edison," *Proceedings 43rd Internation Water Conference,* October 1982.

Austin, S. M., et al.: "Know Your Condenser," *Power Eng.*, reprint of papers, 1961.

Babcock & Wilcox Company: *Steam/Its Generation and Use,* 1972.

Federal Power Commission: *Steam-Electric Plant Air and Water Quality Control Data,* FPC-S-229, February 1973.

Federal Power Commission: *Steam-Electric Plant Air and Water Quality Control Data,* FPC-S-253, January 1976.

Federal Register: *Steam-Electric Power Generating Point Source Category, Effluent Guidelines and Standards,* **39** Num (196) (October 8, 1974).

Jacklin, C., and Brower, J. R.: Correlation of Silica Carryover and Solubility Studies," *ASME Paper No. 51-A-91.*

Kennedy, George C.: "A Portion of the System Silica-Water," *Econ. Geol.,* **45** (7), 629–52 (1950).

Klein, H. A.: "Use of Coordinated Phosphate Treatment to Prevent Caustic Corrosion in High Pressure Boilers," *Combustion,* October 1962.

Landon, R. D., and Houx, J. R. Jr.: "Plume Abatement And Water Conservation with the Wet-Dry Cooling Tower," *Proc. American Power Conf.,* **35** (1973).

deLorenzi, O. (ed.): *Combustion Engineering,* Riverside, Cambridge, Mass., 1947.

Skrotzky, B. G. A.: *Basic Thermodynamics, Elements of Energy,* McGraw-Hill, New York, 1963.

Strauss, S. D., (ed.): "Control of Turbine-Steam Chemistry," *Power,* **125** (3) March 1981.

U.S. Environmental Protection Agency: *Development Document for Steam Electric Power Generating,* EPA-440/1-74-029a, October 1974.

U.S. Environmental Protection Agency: *Economic Report for Steam Electric,* NTIS-PB-239315/AS.

Van Meter, J. A., Durkin, T. H., Legatski, L. K., and Petkus, R. O.: "Making FGD Work at Sigeco," *Pollut. Eng.,* 29–33 (March 1981).

CHAPTER 35
MUNICIPAL WATER

Communities differ widely in character and size, but all have the common concerns of finding, treating, and distributing water for industrial, commercial, and residential use. Residential uses include washing, transporting wastes, drinking, food preparation, watering lawns and gardens, heating and cooling, and fire protection. It is estimated that public water utilities in the United States deliver over 30 billion gallons of water daily (79×10^3 m^3/min) with approximately 19 bgd (50×10^3 m^3/min) being used for residential purposes. Household use is about 100 gal/day (0.38 m^3/d) per person, of which less than 1% is consumed, yet the water quality established by potable standards is imposed on the entire supply.

Water has been treated for thousands of years by a variety of processes, though procedures that produce safe potable water were not developed until the 19th century, when it became clear that many serious epidemics were related to sewage-contaminated water. It was discovered that chlorine was effective as a disinfectant in destroying pathogenic (disease causing) organisms, and that maintaining a chlorine residual in the distribution system produced a safe water and protected against harmful contamination. Many countries use chlorination today; others use ozone as a disinfecting agent.

Chlorine and other disinfectants can react with trace organics found in many water sources; their by-products may be objectionable in taste or odor, and some may be harmful. Physiologic effects of potentially harmful substances are continually scrutinized by the federal government under authority of the Toxic Substances Act.

Because water is so basic to life, public interest in governing water quality has been strong. This has created a demand for modern treatment plants, trained operators, and careful surveillance of chemical treatment and water quality.

Consumption of water per capita ranges from approximately 100 gal/day (0.38 m^3/d) in very small municipalities, to about 200 gal/day (0.76 m^3/d) in larger systems having greater industrial and commercial demand. Water use primarily depends on the availability of water resources, climatic and seasonal variations, and the cost of the finished water. Where water is scarce, it is used only for essential purposes. Where cost is a prominent factor, it primarily restricts nonessential uses of water, such as lawn watering, since essential uses will be served regardless of cost.

Over the years, water quality standards have become more stringent as a result of public demand and concern for the effects of specific contaminants. In addition to health-related aspects of contaminants, there are also considerations of palatability.

Prior to 1975, the 1962 United States Public Health Service Standards were the quality criteria for the production of potable water. During 1975, the U.S.

TABLE 35.1A U.S. Drinking Water Standards

Constituent	Recommended limit, mg/L	Federal mandatory limit, mg/L
Alkyl benzene sulfonate (ABS)	0.5	—
Arsenic (As)	0.01	0.05
Barium (Ba)	—	1.0
Cadmium (Cd)	—	0.01
Carbon chloroform extract (CCE)	0.2	—
Chloride	250	—
Chromium (total) (Cr)	—	0.05
Copper (Cu)	1.0	—
Cyanide (CN)	0.01	—
Fluoride (F)	0.6–0.9*	1.4–2.4†
Iron (Fe)	0.3	—
Lead (Pb)	—	0.05
Manganese (Mn)	0.05	—
Nitrate (NO_3)	45	45
Phenols	0.001	—
Selenium (Se)	—	0.01
Silver (Ag)	—	0.05
Sulfate (SO_4)	250	—
Total dissolved solids (TDS)	500	—
Zinc (Zn)	5	—
Mercury (Hg)	—	0.002
Turbidity (nephelometric procedure)		1 NTU—monthly average 5 NTU—average of two consecutive days (5 NTU—monthly average may apply at state option)
Radioactivity (natural)	—	Gross Alpha 15 pCi/L
Radioactivity (artificial)	—	Gross Beta 50 pCi/L
		Tritium 20,000 pCi/L
		Strontium 90 8 pCi/L

* Depending upon ambient air temperature.
† Depending upon temperature.

Note: While every effort has been made to ensure the accuracy of the information contained in the above table, it is suggested that the reader verify any information intended to be relied upon with the appropriate governmental authority, and the reader uses the above information at his or her own risk.

Environmental Protection Agency was mandated to institute a new system of quality regulations by the Safe Drinking Water Act (Public Law 93-523). The interim primary (enforceable) standards were then promulgated to become effective in July 1977. Additional regulations may be anticipated if national testing programs detect harmful contaminants for which regulation is necessary. Current standards are listed in Tables 35.1A and B, which include the limits recommended by the World Health Organization.

In addition to the primary (enforceable) regulations, there are secondary (desirable) goals which are to serve as objectives for the industry. Local regulations or objectives may be even more stringent as determined by the individual communities.

TABLE 35.1B World Health Organization Drinking Water Standards

Constituent, mg/L	1950 International			1961 European	
	Permissible limit	Excessive limit	Maximum limit	Recommended limit	Tolerance limit
Ammonia	—	—	—	0.5	—
Anionic detergent	0.2	1.0	—	—	—
Arsenic	—	0.05	—	—	0.2
Cadmium	—	0.01	—	—	0.05
Calcium	75	200	—	—	—
Chloride	200	600	—	350	—
Chromium (+6)	—	—	0.05	—	0.05
Copper	0.05	1.5	—	0.05*	—
Cyanide	—	—	0.05	—	0.01
Fluoride	—	—	—	1.5	—
Iron	0.1	1.0	—	0.1	—
Lead	—	—	0.1	—	0.1
Magnesium†	50	150	—	125†	—
Manganese	0.05	0.5	—	0.1	—
Mercury	—	0.001	—	—	—
Nitrate	—	—	—	50	—
Oil	0.01	0.3	—	—	—
Oxygen	—	—	—	5.0	—
pH range	7–8.5	6.5–9.2	—	—	—
Phenols	0.001	0.002	—	0.001	—
Selenium	—	—	0.01	—	0.05
Sulfate	200	400	—	250	—
Suspended matter	5	25	—	—	—
Total solids	500	1500	—	—	—
Zinc	5.0	15.0	—	5.0	—

* May be higher for new piping.
† If 250 mg/L SO_4 is present, Mg not to exceed 30 mg/L.

Note: While every effort has been made to ensure the accuracy of the information contained in the above table, it is suggested that the reader verify any information intended to be relied upon with the appropriate governmental authority, and the reader uses the above information at his or her own risk.

RAW WATER CHARACTERISTICS

Raw water characteristics vary widely, the major differences being between surface and groundwater, hard and soft water, and river water compared to reservoir water. These differences present varying needs for algae control, turbidity removal, softening, water stabilization, and disinfection. Highly polluted waters also must have organics removed. Certain waters have undesirable inorganic constituents as well.

Generally, water supplies within a defined geologic region are similar. There has been a trend to the use of surface water because of its availability and it minimizes the risks of earth subsidence due to uncontrolled groundwater withdrawal.

A raw water with a constant low turbidity has more treatment options than one of high or variable turbidity. The presence of color in many low alkalinity

TABLE 35.2 Typical Water Impurities and Common Methods of Treatment

	Surface						Ground			Surface or ground				
Typical treatment	Turbidity	Taste and odors	Hardness	Color	Iron/ manganese	Algae	Hardness	Iron/ manganese	H_2S	Bacteria/ virus	Heavy metals	Organics	Corrosion	Scale
Preaeration		X			X			X	X					
Filtration only with or without chemicals	X			X	X	X		X						
Ion exchange			X		X		X	X						
Single or two-stage precipitation softening	X	X	X	X	X	X	X	X						
Chemical clarification and filtration	X	X	X	X	X	X	X	X						
Oxidation		X		X	X	X		X	X	X	X	X		
Chlorination		X		X	X	X		X	X	X	X	X		
Adsorption		X		X								X		
Phosphates													X	X
Zinc phosphates													X	

waters requires special treatment consideration. Corrosion control will become more important as water quality is monitored to meet the secondary standards in the Safe Drinking Water Act.

In some cases, the requirements of the Safe Drinking Water Act will necessitate a change to a more acceptable water source. The implementation of the Safe Drinking Water Act will also result in the consolidation of many small water suppliers into larger regional systems to take advantage of the economy of scale. This will also allow for the funding of more sophisticated monitoring and treatment requirements.

TYPICAL TREATMENT SCHEMES

The type of treatment practiced at a given municipality depends largely on the raw water characteristics. Wherever possible, water resources are acquired and maintained to require a minimum amount of treatment, thereby reducing the capital and operating costs to the municipality. In Table 35.2 are shown the common impurities found in raw water supplies and examples of the general types of treatments that can be employed. Which treatment is optimum depends on local conditions and the level of impurity.

A simple summary of various processes used for treating municipal water is shown in Figure 35.1. By far the largest number of municipalities simply use the scheme shown by flow path 1, with water being taken from a well, sterilized by chlorination, and pumped directly to the distribution system. Most systems like this are small and often operate with only part-time surveillance. Each state is responsible for reviewing the quality of the water both chemically and bacteriologically. In some cases, when this scheme is used a chemical such as polyphosphate may be applied for final stabilization to minimize either scale or corrosion. If the water is undesirable for household use, the individual homeowner then has the responsibility of meeting his or her own needs by installing a zeolite water softener or an activated carbon filter.

The second flow path shows only the addition of a filtering system. In some old plants, the slow sand filter is still used for this purpose. The rapid sand filter is now more commonly used. Because the sand filter is limited in its ability to handle suspended solids, this system is usually applied only to impounded waters of excellent quality where postchlorination can assure a safe potable water. The filter may also be installed to clarify shallow well water where suspended iron may be an occasional problem. There are risks in this kind of system, however, because the filter does no more than remove suspended solids, which are seldom the only problem with a raw water supply.

The third flow path shows the addition of an aerator prior to the filter for removal of tastes and odors and for oxidation of iron. Usually this is followed by lime for pH increase to about 8.0 to 8.5, and a flocculant, since the iron precipitate is usually colloidal.

In modern practice, the minimum equipment is usually considered to be that shown by flow path 4, and it is usually confined to impounded waters where there will be consistently low suspended solids depending on the seasonal presence of algae and where final disinfection can assure a safe finished water for the municipality.

In handling surface waters that contain significant suspended solids, the fast

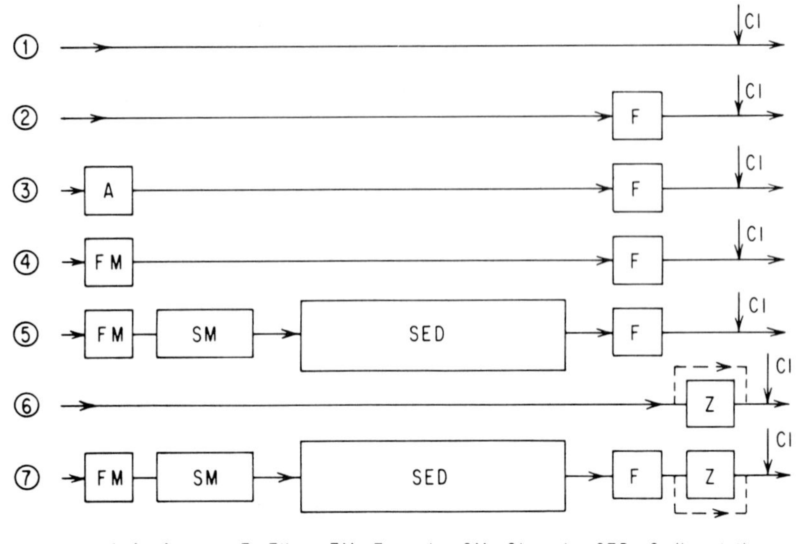

FIG. 35.1 Simplified schematic of municipal water treatment flow sheets.

mix devices of flow path 4 are followed by a slow mix flocculator and a sedimentation basin in flow path 5, providing the detention for settling of most of the flocculated solids so that the load on the filter is measurably reduced. The detention can also provide for disinfection of raw water, with chlorine being added with or ahead of the coagulation chemicals to relieve the final chlorine demand. The prechlorination step may also improve flocculation by destroying some organic contaminants. This also keeps the settled solids from becoming septic and rising to the surface as gas forms.

In some cases where the water is already clear, the major problem may be caused by calcium and magnesium. In flow path 6, this water is softened through zeolite softeners. Some of the water is bypassed so that the effluent may have a hardness controlled at 100 to 150 mg/L. Often the cost of municipal zeolite softening is more than paid for by the savings to the individual homeowner in areas where the water is so hard that most of the homes would be provided with individual zeolite units. A side benefit is reduced discharge of regenerant brine by the more efficient municipal plant.

The final flow pattern 7 includes not only clarification of the surface water, but also final softening by passing a portion of the effluent through a zeolite unit. In this scheme, the sedimentation tank may provide for lime softening as well as clarification, to reduce the load on the zeolite unit.

There are various other combinations of individual treatment processes incorporated into an overall treatment scheme, but the principles are generally little different from those illustrated.

Several municipal water treatment schemes are described below, with a tabulation of water quality and a line diagram of each system.

Case History—Topeka, Kansas (Figure 35.2)

The Topeka water department has demonstrated the value of continuing research in improving water treatment results. Lengthy investigations on the bench and in the plant have produced significant changes in the process for treating the Kansas River water.

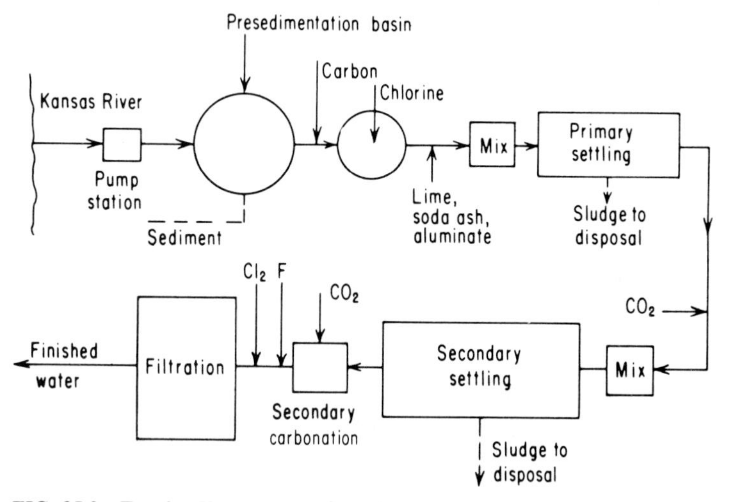

FIG. 35.2 Topeka, Kan., process flow sheet.

The Kansas River, though largely impounded, experiences significant variations in silt content. This previously caused plant upsets and large variations in chemical consumption for clarification, softening, and precipitation. After studying various approaches to reducing high raw water turbidities, the Topeka water

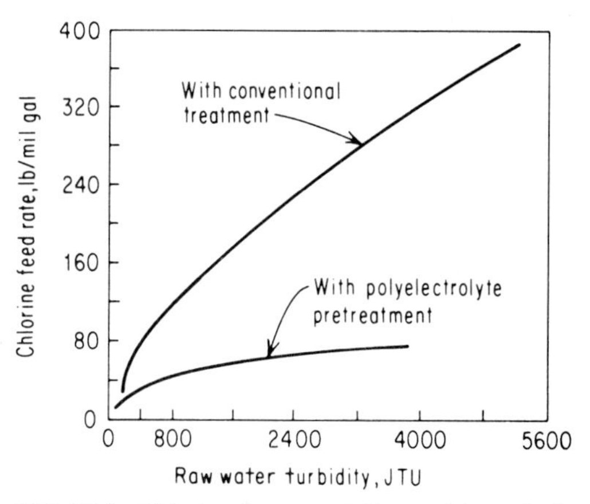

FIG. 35.3 Chlorine dosage as influenced by turbidity and pretreatment program.

TABLE 35.3 Typical Water Characteristics—Topeka, Kansas

	Original process*				Present process			
River turbidity (JTU)	190	605	1020	2000	190	605	1020	2000
River color, APHA	4	5	6	10	4	5	6	10
Polymer, mg/L					1	3	2	4
Presedimentation turbidity	160	370	520	600	13	13	36	40
Primary alum, mg/L	15	27	27	30				3
Activated silica, mg/L	2	2	2	2				2
1st sed. turbidity	3	9	10	20	1	1	2	2
1st sed. color	2	3	3	5	1	1	1	1
Prechlorine, mg/L	7.8	14.4	16.8	21.6	5.4	5.4	6.0	7.2
Postchlorine, mg/L	1.9	2.8	2.6	2.0	1.4	1.6	1.4	1.2
Hardness								
Raw water, mg/L		129				129		
Finished water, mg/L		156				118		

* Stepwise additions to the plant made it possible to compare these processes under identical inlet conditions.

department installed a large presedimentation basin. This basin utilizes a cationic polymer coagulant to remove the incoming silt, allowing its return to the river in a natural state. The significance of this pretreatment shows in reduced chemical usage and sludge production (Figure 35.3). A reduction in the noncarbonate hardness will reduce soda ash requirements correspondingly. Table 35.3 shows raw and finished water analyses.

Case History—Park Forest, Illinois (Figure 35.4)

The village of Park Forest, Illinois, produces potable water in a plant utilizing precipitation softening in conjunction with zeolite softening to give a stable, moderately soft water. In the process, hard well water is first aerated to drive off dissolved CO_2 and reduce lime demand. As the water flows to two sludge blanket softeners, lime and liquid sodium aluminate are added for softening and clarification. The softened water, unstable at this point, is recarbonated with CO_2. Phosphate is also added to stabilize the water and prevent scale formation in the system. Following rapid sand filtration, a portion of the water is softened further by

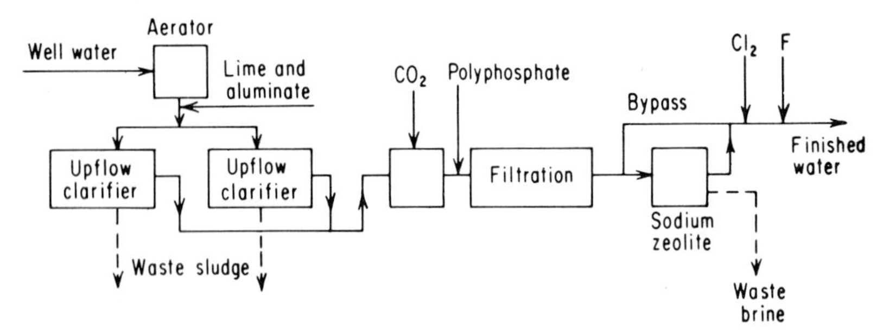

FIG. 35.4 Park Forest, Ill., process flow sheet.

TABLE 35.4 Typical Water Characteristics—Park Forest, Illinois

	Hardness (mg/L, as $CaCO_3$)			Alkalinity			Dissolved solids	pH
	Ca	Mg	Total	P	M	O		
Well water	256	232	488	0	360		630	7.1
Precipitation softened	70	170	240	48	90	6	351	9.8
Zeolite softened			6	0	86		405	8.4
Finished blend	36	100	136	0	80		370	8.1

sodium zeolite. The two water streams are then blended to give the desired finished water characteristics (Table 35.4). Chlorine and fluoride are added before entering the distribution system. Relatively simple control is possible with this system owing to the uniformity of the well water supply with regard to both mineral content and temperature.

Case History—Daytona Beach, Florida (Figure 35.5)

The Daytona Beach water plant uses lime softening to treat a groundwater supply for potable use. Lime and chlorine are applied to the raw water along with a controlled return of filter backwash water. The backwash water helps seed the softening process. A high molecular weight anionic flocculant is used to improve settling. This method replaced the previous use of alum and activated silica because it resulted in lower cost and reduced sludge volume (Table 35.5).

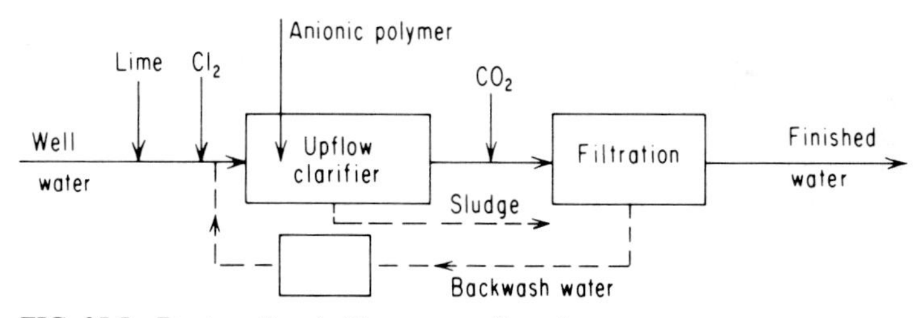

FIG. 35.5 Daytona Beach, Fla., process flow sheet.

TABLE 35.5 Typical Water Characteristics—Daytona Beach, Florida

	Raw	Finished
Alkalinity (as $CaCO_3$) mg/L	290	42
Total hardness (as $CaCO_3$)	324	74
Calcium hardness (as $CaCO_3$)	287	42
Magnesium hardness (as $CaCO_3$)	37	32
CO_2	30	
pH	7.35	8.9
Temperature, °C	23	23

Case History—Niagara Falls, New York (Figure 35.6)

The Niagara Falls, New York, water plant is a conventional clarification plant, typical of those designed to treat low turbidity water sources, such as lake or reservoir supplies. The Niagara River receives low turbidity water from Lake Erie.

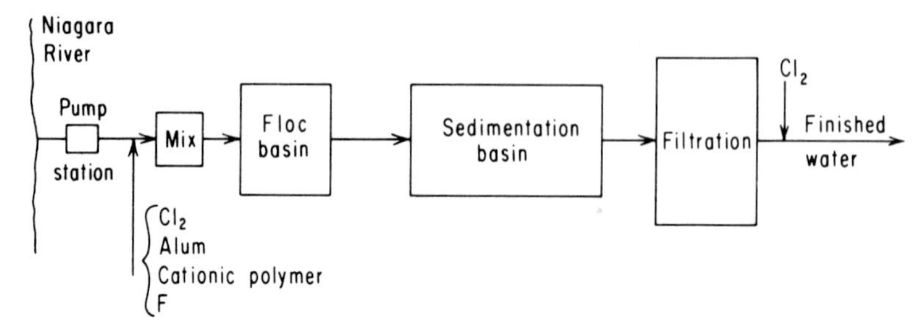

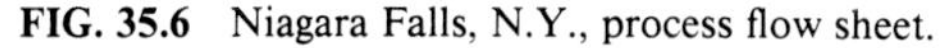
FIG. 35.6 Niagara Falls, N.Y., process flow sheet.

To minimize sludge production, instead of adding alum alone to form the typical large floc structure, the plant restricts the alum dosage to that needed to form a tiny pinpoint floc. The object is to make the solids filterable, not necessarily settleable. A typical dosage is 2 to 4 mg/L alum and 0.25 to 0.5 mg/L polymer coagulant. When the plant operates on alum alone, the dosage is 8 to 12 mg/L. Filter runs of 50 to 72 h are achieved with either program. Floc penetration and solids loading on the filters are greater with alum-polymer than with alum alone, but total sludge volume is less (Table 35.6).

TABLE 35.6 Typical Water Characteristics—Niagara Falls, New York

	Turbidity	pH	M alkalinity, mg/L	Total hardness, mg/L
Raw	3	8.3	92	132
Finished	<.3	7.6	90	132

Case History—San Diego, California (Figure 35.7)

San Diego's municipal water is produced by three plants of conventional design. These treat Colorado River water drawn from impoundments which also receive runoff from the local watershed (Table 35.7). Although there are seasonal changes

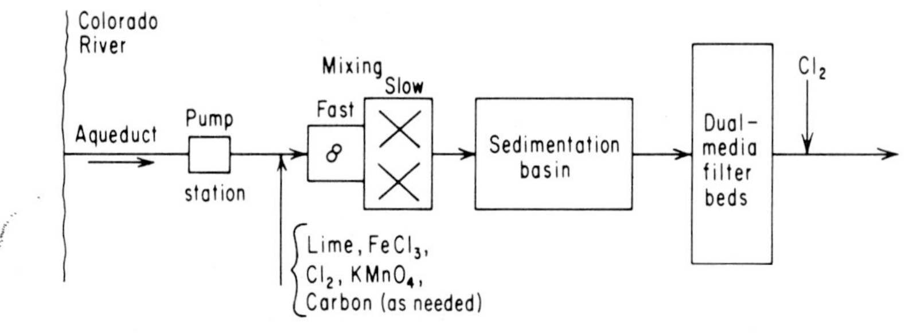

FIG. 35.7 San Diego, Calif., process flow sheet.

TABLE 35.7 Typical Water Characteristics—San Diego, California

	JTU	Alkalinity, mg/L	Hardness, mg/L	pH
Raw water	2.0	119	332	8.1
Finished water	0.2	120	338	8.2

in composition, on a daily basis the raw water is relatively consistent in quality, with the only major variation being temperature. Periodically, copper sulfate is used for algae control in the reservoirs.

WATER TREATMENT PLANT BY-PRODUCT DISPOSAL

In the water treatment process, by-product solids produced by the removal of contaminants from the raw water become a secondary disposal problem. To minimize the costs of sludge handling, the volume of sludge produced should be reduced as much as practicable.

Reduction of the weight of solids removed from the raw water is impossible, of course, if the quality goals are to be achieved. Any weight reduction then must come from the reduction of chemicals added for coagulation and floc formation. In addition to this weight reduction, the type of sludge produced should be controlled, if possible, to produce a minimum volume and to yield a compact sludge that is easy to dewater and dispose of. In general, this may be accomplished by reduction or elimination of inorganic salts which tend to produce a light, fluffy sludge not readily dewatered. This may require partial or complete replacement of the metal salt by organic polymers. In many cases, the addition of a clay nucleating agent allows the complete replacement of the salt, producing an easily dewatered sludge suitable for direct landfill.

Where possible, low turbidity water should be clarified by filtration processes rather than by coagulation/sedimentation processes. This may significantly reduce the amount of sludge produced while maintaining water quality. Many plants with low raw water turbidity practicing conventional coagulation/sedimentation processes apply chemical solids that far exceed the amount of suspended solids present in the raw water to be clarified.

Sludge from precipitation softening processes usually cannot be reduced significantly in weight, since they are primarily a result of the insolubilization of dissolved hardness rather than the addition of unnecessary chemicals. Where softening is achieved by ion exchange, it is possible to reclaim the salt values from spent regeneration solutions while producing a solid waste consisting of calcium and magnesium salts. The recycled salt brine can then be used for subsequent regeneration.

Once the water treatment chemistry has been investigated to determine the optimum process producing the least amount of by-product sludge, consideration then focuses on the type of disposal process needed to handle the solids. Where applicable, the potential use of by-product material by a local industry should be considered. This may include such diverse activities as brick manufacturing, wallboard production, and agriculture. The alkali value of lime sludge may be valuable to neutralize acidic wastes, such as pickle liquor in a steel mill. Finding practical uses for sludge not only reduces the sludge handling cost, but also eliminates the need for landfill of the solid material.

Disposal of sludge by retention in sludge lagoons appears, at first, to be the obvious solution to a sludge disposal problem. Unfortunately, this often is simply a delaying process. Ultimately, the sludge must again be handled if the lagoon must be reclaimed. Lagoons may provide interim storage for a dewatering system which continuously withdraws sludge from the lagoon at a controlled rate, optimizing the design of the dewatering device. This also provides a means to completely flush sedimentation basins in older plants not equipped for continuous sludge withdrawal. Where land is readily available, sludge lagoons may provide an economical method of sludge drying, providing the lagoon depth is not excessive and climatic conditions are favorable.

Sand drying beds have proved acceptable for dewatering water plant sludges. In warmer climates, they may be used year-round; in cold climates, the sand beds are usually used in summer to dewater sludge accumulated throughout the year. Coupled with high molecular weight flocculants, sand bed dewatering provides a practical, inexpensive method of sludge dewatering. It requires little attention by operators and consumes very little energy. Sludge harvesting methods should be considered in designing the beds so that mechanical equipment can be used.

Dewatering by centrifugation has gained popularity because of the compact nature of the equipment and its ability to be operated with relatively little operator attention. Special attention must be given to the details of equipment design to minimize internal abrasion by grit, which leads to costly maintenance. This equipment is able to handle wide variations in feed solids without significant upsets in production.

While vacuum filters have been investigated by some plants, there does not appear to be a widespread use of this equipment to dewater water clarification sludges.

The plate and frame filter press has been successfully adapted to dewatering municipal water plant sludge. To properly assess this process, both capital and operational costs must be compared to the alternative methods. In some cases, very high doses of filler, such as diatomaceous earth, lime, or fly ash, have been required to produce an acceptable cake. Early designs required considerable labor per unit of dry cake production, but newer designs have improved upon this.

The belt press includes several varieties of designs of screens or cloths on a conveyor device to reduce sludge volume. Often the solids are discharged as a paste. In some cases, additional stages use press rolls to squeeze out more water and produce a more concentrated cake for disposal.

Filter backwash water, containing relatively little solids, requires separate treatment. In many cases it is possible to simply pump the backwash water directly to the head of the plant without upsetting the normal clarification process. To facilitate this, it is desirable to add a cationic polymer coagulant directly to the recycle stream during the time the backwash waste is being returned. This agglomerates the solids and prevents them from upsetting the clarification process. The backwash flow rate is often so high that direct return to the sedimentation basin may create a momentary overload. High flow rates can be controlled by an intermediate equalizing basin. Where sedimentation basins are not employed, the use of decant tanks to concentrate the sludge, followed by sand bed dewatering, should prove adequate.

POTABLE WATER OTHER THAN MUNICIPAL SUPPLY

Only a small portion of municipal water, usually less than 1%, is used as potable water, either for drinking, cooking, or for use in beverages. For a variety of reasons, potable water is often derived from other than municipal sources. For example, in isolated areas far from a municipal water plant, institutions, mines, oil fields, and similar entities must supply water for their general uses, and a portion must meet drinking water standards, often requiring a miniature of a typical municipal system. State regulatory agencies require testing—always for bacterial surveillance and sometimes for mineral content. The frequency of analysis usually depends on the number of people drinking the water at each location. Even farms may be required to test and report water quality to health agencies if workers are brought in to harvest crops. Aside from human health considerations, the farmer may need to analyze and treat water for farm animals or for irrigation purposes.

Growing public concern with suspected carcinogens in municipal water, even at trace contamination levels, has led to rapid growth of the bottled water industry. Perhaps this is partly due to a bucolic faith that the artesian or "spring water" that many bottlers offer must be "pure" because it is from a natural source. The analysis of bottled water is seldom given on the label and may not be readily available to the purchaser. In the transition period as regulatory agencies try to develop legislation and surveillance of this industry, the purchaser might consider ways of assuring himself that the analysis of bottled water and its quality are at least comparable to that required of the municipal water treatment plant. As competition develops, there may be increasing installations of commercial treatment and dispensing units in supermarkets. (Figure 35.8). These units process the

FIG. 35.8 Bottled water dispenser with integral treatment units, used in various combinations to produce potable water or low solids water comparable in many respects to distilled water.

municipal water through a combination of filtration, sterilization, activated carbon adsorption, membrane separation, and ion exchange processes.

SUGGESTED READING

American Water Works Association: *Water Quality & Treatment,* 3d ed., McGraw-Hill, 1971.

CHAPTER 36
MUNICIPAL SEWAGE TREATMENT

While municipal sewage treatment is a well-established practice, it is an area of water treatment experiencing revolutionary changes because of increasingly stringent effluent quality limitations and the potential value of treated effluent as a source of water for industry, agriculture, and municipalities. Until a few generations ago, plants were designed to remove 30 to 40% of the undesirable impurities before discharge into the receiving body of water. There were no real limits on effluent quality as long as the discharge was chlorinated and met bacterial count standards. Municipal sewage plant operators had only minimal training, but, fortunately, many of them were able to learn from experience and became knowledgeable in their plants' operations. With the growing sophistication of modern plants designed to meet stringent goals, training programs in sewage plant operation have become available to supplement on-the-job training for upgrading personnel.

Sewage treatment refers to the processing of primarily domestic sewage produced by typical community and household activities. In rural areas the characteristics of raw sewage tend to adhere to this definition. As cities become larger and more industrialized, the waste volume and characteristics of one particular industry may affect sewage composition. In addition, sewage may be from combined or separate sanitary sewer systems. With combined systems, storm water drains into the sewer lines to become part of the total flow to the sewage treatment plant. Such uncontrolled surges in flow can be very disruptive to the process of sewage treatment, severely limiting the plant's ability to remove pollutants. Future practice aims toward segregating storm water flow in sewer systems from those used to collect sanitary waste. This will materially reduce and control the flow entering the treatment plant, thereby increasing its ability to maintain adequate treatment. Storm water retention basins are being built in many cities to even out the severe surges caused by storm water, allowing a more reasonable chance of processing the extra volume.

The flow of sewage to be treated may approximate the flow of municipal water supplied to the same community, being in the range of about 100 gal/day per capita in rural areas to 150 gal in urban areas with industrial users. But, there may be differences between municipal water and sewage flows that could be substantial. Such differences may be caused by evaporation losses in industrial plants (over cooling towers, or as noncondensed steam), storm water or groundwater influx, and local industries that may have their own water sources but discharge into municipal sewers. Sewage is always warmer than the water source, and shows an increase in salinity over potable water of about 50 to 75 mg/L. The most common contaminant additions to filtered municipal water as it is converted to sewage are:

1. Suspended solids, typically about 200 mg/L, of which about two-thirds is organic and two-thirds is settleable.
2. Dissolved organic matter, typically about 150 mg/L as BOD.
3. Nutrients, typically 10 to 30 mg/L phosphate and 10 to 30 mg/L ammonia.

Standards being established for pretreatment of industrial wastes for discharge into municipal sewers are helping to reduce harmful contaminants found at lower concentration levels, such as heavy metals and cyanides from plating operations, oil and grease from metalworking and food-processing plants, and toxic materials (e.g., PCBs, pesticides, solvents) as specified by the U.S. Environmental Protection Agency.

Even with storm water segregation, it appears that the treatment of storm water itself to remove undesirable contaminants may become a requirement for many areas. This is due to its sudden detrimental effect on the receiving streams. In densely populated and highly industrialized areas, storm water may be more polluted than domestic sewage.

Because sewage flow patterns are directly related to residential water use, there are wide variations in inflow during the day, the week, and even the season. Figure 36.1 indicates the degree of variation throughout the day caused by typical residential use patterns. During the week, flows will vary as a result of other factors, such as the traditional Monday washday. Seasonal variations also affect separate sanitary sewers owing to the impact of storm water caused by damaged sewer pipe (infiltration) and sump pump operation. While combined sewer systems are mostly influenced by sudden downpours (inflow), separate systems are affected over a broader period of time because of the slower drainage of water from the saturated soil. After a considerable dry spell, the soil acts as a sponge, reducing the impact of short infrequent showers.

EFFLUENT STANDARDS

The establishment and enforcement of effluent regulations has progressed rapidly as a result of the Clean Water Act (Public Law 92–500). Just as each industrial plant outfall has its National Pollution Discharge Elimination System permit, so do municipal plants. All municipal plants must practice a minimum of secondary sewage treatment prior to discharge into receiving streams. Secondary treatment restricts the effluent to 30 mg/L suspended solids and 30 mg/L BOD, and requires that it be chlorinated to destroy pathogens. While this standard may require some years to achieve, it provides an immediate objective to design engineers upgrading existing plants and installing facilities where none presently exist.

For many waterways, additional, more restrictive standards are imposed because of the particular needs of the receiving stream, defined by its category of usage. A stream used only for navigational purposes may not require elaborate treatment of waste discharges. On the other hand, a receiving body of water used as a source of potable water or for swimming may have very restrictive effluent standards. Regional and local regulations may increase in importance as the Environmental Protection Agency reviews the receiving waters and classifies them

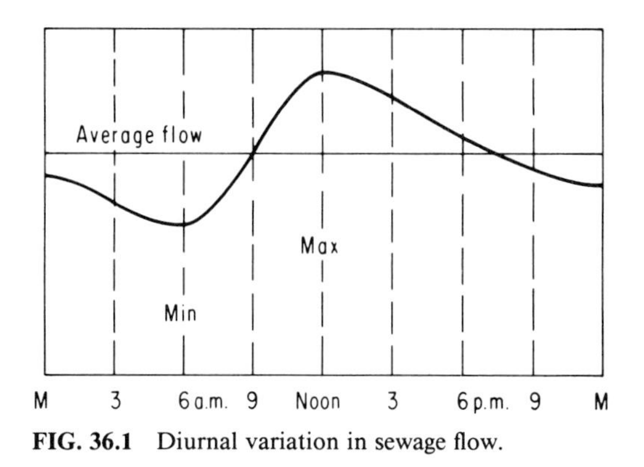

FIG. 36.1 Diurnal variation in sewage flow.

based on present and future needs. Some streams may periodically fail to meet the stream standards set for them, usually in temperature, dissolved oxygen, BOD, and suspended solids. These streams are termed *quality limited.* A plant discharging into a quality limited stream may be required to produce a better effluent than the typical industry standard.

In some cases, the sewage plant effluent may become high enough in quality to permit recycling into the municipal water supply system for reuse. When this is done, safeguards must be instituted to protect the public health. In many cases, sewage plant effluent is used for irrigation of parks, roadways, and agricultural areas because of its nutrient value, further reducing the impact on a receiving stream.

TYPICAL WASTEWATER TREATMENT PROCESSES

Wastewater treatment practices vary in the type of equipment used and in treatment sequences. Most plants fall into a few basic categories, as shown by the simplified flow sheets in Figure 36.2. While these flow sheets show the basic unit operations that make up the total process, they do not include supplemental treatments such as ammonia stripping or selective ion exchange. These additional processes develop from individual plant needs related to sewage characteristics and flow and the effluent limitations to be met. Since the Clean Water Act requires all plants to practice secondary treatment, only those treatment schemes that include secondary treatment are shown. Stabilization ponds, which provide treatment for smaller municipalities, are omitted. Here, the wastewater is contained in a pond for long periods of time during which useful bacteria and algae consume undesirable pollutants. Shown are the most common unit processes of sewage treatment in municipal plants.

Pretreatment

Pretreatment of sewage consists of screening to remove larger solids; or communition, to shred debris. This is followed by grit removal to settle out the heavy grit particles. Often the waste is preaerated to add dissolved oxygen to the wastewater stream and prevent the odors of anaerobic decomposition.

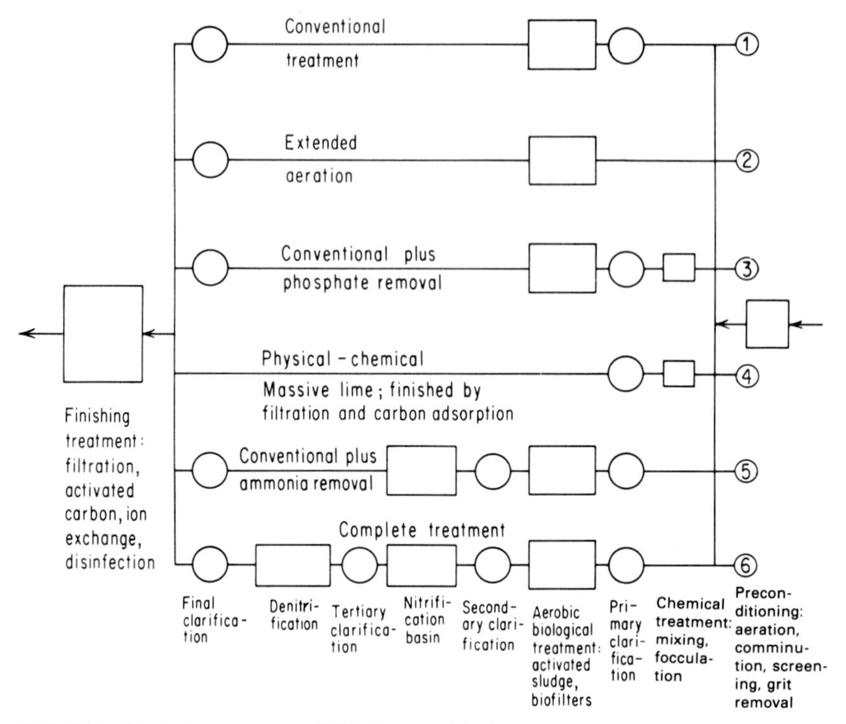

FIG. 36.2 Typical processes available for municipal sewage treatment.

Extended Aeration

This process takes raw sewage directly into an aerated mix tank for 8 h or more to provide bacteria with optimum conditions to consume the BOD present in the wastewater. The effluent from this mix tank goes to a sedimentation tank where the flocculated colonies of organisms are settled to produce a clear overflow. A portion of the settled microbial floc is returned to the headworks and a portion sent to sludge disposal. The clear effluent is then directed to final treatment such as disinfection, perhaps passing through a final polishing filter. This method of treatment is particularly suited to plants that have a low concentration of settleable solids in the raw sewage. It minimizes the number of unit operations involved in smaller plants.

Conventional Primary-Secondary Treatment

In this process, used by most plants in the United States, the raw waste is first directed to a primary clarifier where settleable solids are removed, reducing the load on subsequent treatment units. Colloidal solids and soluble BOD then leave the primary clarifier for the secondary portion of the plant. This secondary unit may be either a trickling filter or, more commonly, an activated sludge basin. Other devices having fixed position bacterial slime in contact with flowing sewage are similar to trickling filters. These include packed beds and rotating disk media.

As in all activated sludge processes, the organisms consume biodegradable organics. After a period of time the organisms and wastewater flow on to the clarifier for separation; with the activated sludge processes, the sludge organisms are returned to the head of the secondary system and a portion wasted to sludge processing. The effluent then goes to final treatment for disinfection and perhaps filtration.

Secondary Treatment with Nitrification

The early stages of this process are similar to conventional secondary treatment. The later stage, nitrification, is the process of converting ammonia and nitrite to nitrate. This is usually done in a separate biological reactor before final clarification. Nitrification may occur in a suspended biological reactor containing specific organisms that use ammonia and nitrite in their metabolism. Following nitrification, a clarifier settles and returns the biological solids to the preceding unit to maintain the nitrification organisms at the proper activity level. Waste sludge is disposed to sludge processing. Disinfection and filtration may also be utilized. Reducing the levels of ammonia and nitrite produces an effluent more compatible to the aquatic life in the receiving stream.

Nitrification / Denitrification

In this process, a continuation of the previous process, nitrates are converted to molecular nitrogen by means of still another biological reactor. This reactor contains organisms that can utilize the combined oxygen from the nitrates and release nitrogen gas. Following clarification the organisms may or may not be returned to the biological reactor depending on the process used. Final filtration and disinfection are utilized as before. The removal of nitrogen can reduce the potential for algal blooms in the receiving water if nitrogen is the limiting nutrient.

Physical-Chemical Treatment

In this process, raw sewage is taken into a rapid mix and flocculation zone where a large dosage of chemicals is added to produce massive chemical coagulation and flocculation. Lime is commonly applied to achieve maximum absorption of impurities in the flocculation process. Following a clarifier to remove the precipitated and coagulated solids, the wastewater is filtered to remove residual suspended solids prior to entering a carbon adsorption column. The carbon adsorption column reduces residual organics to a low level. Following this, the effluent is disinfected before entering the receiving stream. In some cases the lime-treated effluent is air stripped of free ammonia before entering subsequent stages in the process. This process is effective where significant amounts of organics can be settled ahead of the carbon bed.

Nutrient Removal

Removal of nutrient usually refers to phosphorus, but may include nitrogen. The objective is to reduce the content to a level that will inhibit eutrophication of the receiving body of water. It has been determined that when either nitrogen or phosphorus is reduced to low levels, algae growth can be controlled to prevent the massive algal blooms that give rise to eutrophication. A study may be necessary to determine which nutrient is limiting, reducing the cost of waste treatment, since only that nutrient need be removed.

Phosphorus is removed by precipitation with calcium at high pH to precipitate calcium phosphate, or with aluminum or iron salts to precipitate phosphorus as the metal phosphate. In either case, phosphorus can be removed to a level that prevents troublesome amounts from entering the receiving stream. Consideration must be given to the form of phosphate present in the wastewater. Phosphorus in the orthophosphate form reacts completely with the precipitant; polyphosphate is not completely removed. Some plants have been provided with clarifiers specifically designed for the purpose of precipitating phosphorus, although this often is unnecessary. Where lime treatment is used, it may be added to the primary clarifier if sufficient flocculation time is available to complete the chemical reactions.

Underflow pumps must be sized so that the solids precipitated may be removed effectively from the primary clarifier. Metal salts may be added either ahead of the primary clarifier or into the activated sludge tank prior to its discharge into the biological clarifier. The contained metal hydroxide and metal phosphate sludge does not inhibit biological reactions, providing the pH is in the 7 to 10 range.

The most efficient performance of metal salts is usually in the activated sludge tanks, where polyphosphate has had time for conversion to the ortho form. Regardless of the chemical used or selection of a reaction site, it is helpful to add an anionic polymer flocculant to the stream as it enters the clarifier for more complete removal of suspended materials and reduction of chemical dosage for effective clarification. This will minimize the volume of waste chemical sludge generated. Without anionic flocculants, significantly more metal salt is required to achieve acceptable results.

Ammonia removal may be practiced by nitrification/denitrification. Other ways to remove ammonia nitrogen include: lime treatment to achieve a high pH followed by air stripping of the ammonia from the waste stream; and ion exchange, where the stream is passed through a bed of clinoptilolite clay, which has a specific affinity for ammonia.

DISINFECTION

While disinfection may take various forms, chlorine application to a detention basin providing at least 15 min reaction time has usually been used for this purpose. In some cases, dechlorination may follow to remove any trace amounts of residual chlorine that might be toxic to stream organisms. Because of concern for the production of chlorinated organics, which might also be toxic, other methods of disinfection are gaining popularity. Ozonation is becoming more practical as improvements in equipment design continue to reduce costs. Further research is needed to define the potential hazards of by-products of the oxidation of organic materials with chlorine and ozone.

Several plants shown in Figures 36.3 and 36.4 are examples of conventional systems which were upgraded to meet increasingly stringent effluent quality goals.

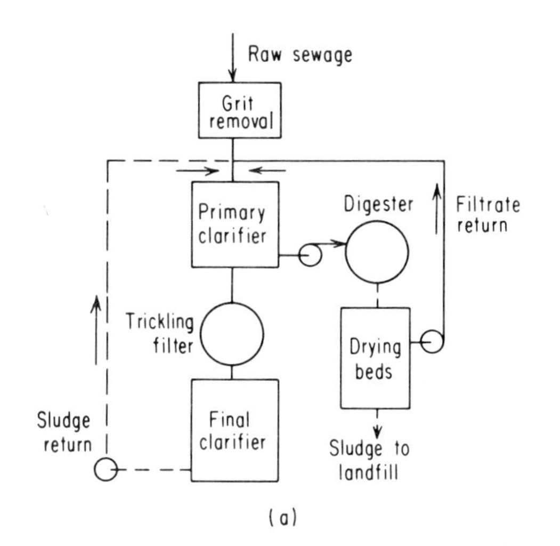

(a)

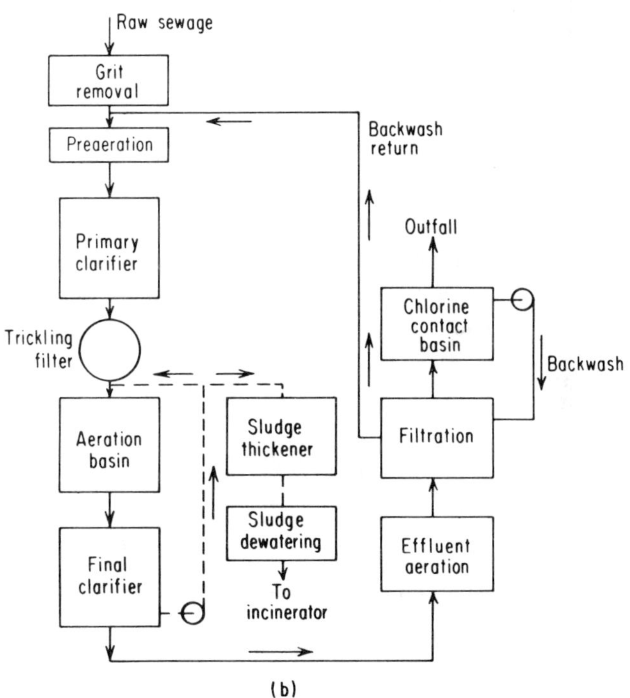

(b)

FIG. 36.3 (*a*) South Buffalo Creek—original flow sheet. (*b*) Current flow sheet showing improvements in basic process and addition of tertiary treatment.

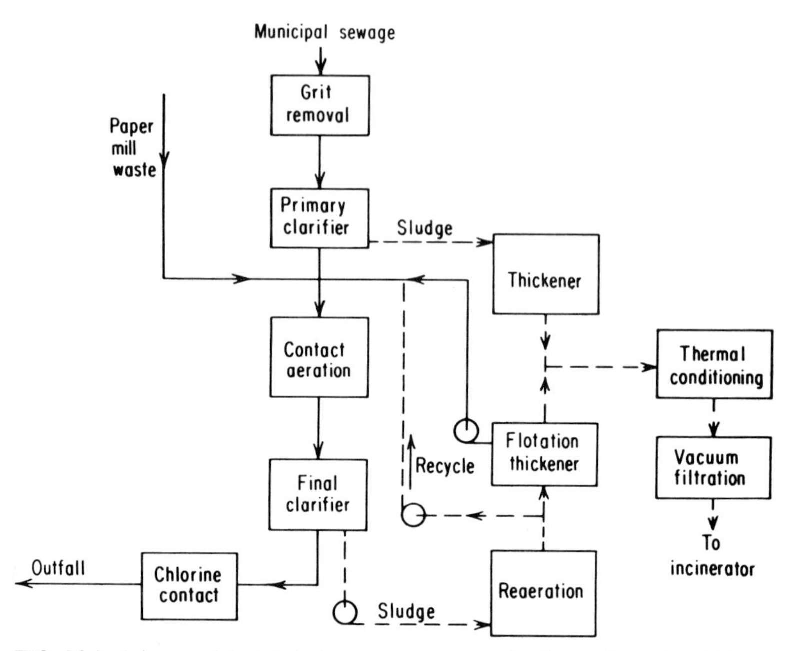

FIG. 36.4 Joint municipal-industrial waste treatment facility at Green Bay, Wisc.

PLANT PROCESS CHANGES CAUSED BY GROWTH AND INDUSTRIALIZATION

The South Buffalo Creek treatment plant in Greensboro, N.C. (Figure 36.3*a*) is an example of the impact of industrial and urban growth on both plant size and preferred treatment processes. When originally designed in the early 1930s, domestic sewage was by far the largest portion of the raw sewage. The design flow of 3.25 mgd (8.5 m^3/min) was sufficient for an ultimate population of 37,000. The plant performance was adequate during the next few years with 90% BOD removal. By the late 1950s industrial discharges had begun to change the character of the raw waste. The industries discharging to the system included textile, meat packing, chemical processing, poultry processing, and metal plating. By the mid-1960s, the industrial BOD contribution had risen to over 65%. Although treatment plant improvements had been made, they were generally expansions of existing equipment. In the mid-1960s, a major change in the method of treatment was initiated. This included conversion of trickling filters to activated sludge aeration basins. However, some of the trickling filters were retained to serve as roughing filters ahead of the activated sludge basins. This allowed more effective handling of load surges and also protected the activated sludge organisms critical to effective plant performance. By these changes, plant efficiency of 70% BOD removal was increased above 90% again.

Additional plant improvements (Figure 36.3*b*) will achieve the increased removal efficiencies required for local stream standards while at the same time improving odor control around the plant. Effluent BOD of 15 mg/L and suspended solids of 10 mg/L represent over 98% removal. Phosphorus removal is also practiced.

This is a rather typical case history as it illustrates the steady growth in water consumption, the continual improvements required to meet new standards, and the relatively long time needed to implement changes because of problems of funding, design, and construction.

Another example of retrofitting an older plant to meet today's effluent quality standards involves the conversion of the aeration basin to a pure oxygen digester. The Boar Tusk Creek plant in Convers, Ga., typically handles a raw waste with a BOD of 225 mg/L and TSS of 150 mg/L. Originally designed for a loading of about 0.5 mgd, the plant discharge contained 50 mg/L BOD and 125 mg/L TSS at the increased load of 0.75 mgd. There was no space available for conventional facilities needed to handle a projected need for treating over 1.0 mgd. Installation of a specially designed pure O_2 system and larger final clarifiers has enabled the plant to meet state standards of 20 mg/L BOD, 30 mg/L TSS, and 10 mg/L ammonia. Oxygen application rate averages about 275 mg/L.

MUNICIPALITY AND INDUSTRY JOIN FORCES TO TREAT WASTES MORE ECONOMICALLY

In Green Bay, Wisconsin, the cooperation and planning of industry and government has produced a unique municipal-industrial waste treatment plant. In working out the project, the Green Bay Metropolitan Sewerage District and two paper mills found that it was less expensive and more efficient to treat a complex combined waste than the separate individual plant wastes. In the process (Figure 36.4), the municipal and paper mill wastes are collected and proportioned into the plant process by separate pumping systems. The municipal wastes undergo preliminary treatment to remove grit and settleable solids before entering the secondary treatment phase. Mill wastes, however, do not contain appreciable settleable material so they go directly to secondary treatment. Hydraulic design for the plant is based on a nominal flow of 52 mgd (137 m^3/min) with a peak flow of 172 mgd (450 m^3/min). To treat this complex and variable waste, the process flow sheet was kept as flexible as possible. The plant can segregate the secondary portion into four separate treatment systems. While the contact stabilization process was chosen for its adaptability to the anticipated changes, the contact basin can receive feed in either slug flow, step feed, or complete mix modes. The flows within the plant are monitored and can be proportioned by a computer in the main control room.

Nutrient removal is practiced by using alum and anionic flocculant for optimum settling. While phosphorus nutrient should generally be available somewhat in excess of that required for biological metabolism, provisions were made to allow phosphorus addition if required because of the nutrient-deficient nature of the paper mill wastes.

Excess activated sludge from the plant is thickened by air flotation and then thermally conditioned to give a sterilized sludge. Currently, this sludge is vacuum filtered and incinerated, producing steam from a waste heat boiler. The sludge may also be disposed of on land depending on prevailing conditions.

SEWAGE PLANT EFFLUENT IS FED TO WATER FACTORY 21

An example of more complete purification of municipal wastewater is found in Orange County, California, Water Factory 21. This easily expandable plant combines physical/chemical treated sewage with desalted ocean water to give a high-quality product water. This is used as injection water, both to prevent seawater intrusion and to recharge the groundwater aquifer.

In this plant, the first step is treatment of the secondary effluent of the Orange County Sanitation District. The process (Fig. 36.5) involves chemical clarification

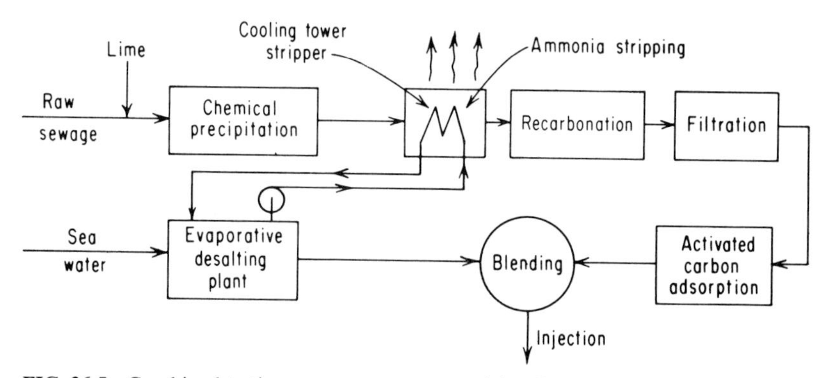

FIG. 36.5 Combined tertiary sewage treatment and desalination plant for underground injection at Orange County, Calif.

with lime to precipitate solids and to liberate ammonia-nitrogen, which is then air stripped. The stripping towers were designed to enhance the removal of ammonia and also provide cooling for the adjacent desalting plant. The absorbed heat improves ammonia stripping efficiency.

Following ammonia stripping and recarbonation, the wastewater is filtered prior to entering carbon absorption columns. The water is then chlorinated to produce a water acceptable for drinking, although it is high in dissolved solids. To remedy this deficiency the water factory also processes an equal amount of

TABLE 36.1 Quality of Reclaimed Wastewater in Orange County, California

	Concentrations, mg/L			
Constituent	Wastewater before treatment	Reclamation plant product	Desalted seawater	Blended product water
Total dissolved solids	1300	1100	25	540
Total hardness	400	200	1	110
Sodium	220	220	7	110
Sulfate	250	250	1	125
Chloride	240	240	10	120

Source: Chandler, Courtney R.: "Water Factory 21," *NWSA J.*, July 1974, p. 26.

seawater using an advanced vertical tube evaporator. The desalted water is also of excellent quality except for its corrosive character due to very low dissolved solids and lack of hardness. When blended (see Table 36.1) the two streams produce a water that is acceptable for drinking at a cost comparable to development of natural water sources.

When further expanded to full capacity this facility will provide the flow of water necessary to maintain an effective hydraulic barrier against seawater intrusion.

WORLDS'S LARGEST TREATMENT PLAN

The largest sewage treatment plant in the world is the West-Southwest plant of the Chicago, Ill., Metropolitan Sanitary District with a capacity of 1.44 bgd (3800 m^3/min). This plant processes its own sludge plus that produced at another location. The flow diagram for this plant is shown in Figure 36.6. Table 36.2 compares certain characteristics of Chicago's municipal water supply, which originates in Lake Michigan, and the sewage plant effluent. The treated effluent discharges to the sanitary and barge canal, which joins the Des Plaines River, which eventually drains into the Mississippi River.

The Metropolitan Sanitary District of Greater Chicago (MSDGC) has undertaken an extensive strip mine reclamation project in north-central Illinois using waste sludge from the West-Southwest plant. This project has significantly advanced the knowledge of land applications of municipal sewage sludge.

The MSDGC has also moved forward in the area of storm water retention by designing an extensive network of large diameter collection mains under the city. These tunnels are to store runoff during rain to be pumped later for processing in the existing treatment facilities.

BY-PRODUCT WASTES

In the treatment of municipal wastewater, some contaminants are actually consumed, but a significant portion simply are converted to solid wastes. When they are removed from the water, they become by-product sludge, which must be disposed of.

The first step in disposal is to stabilize the sludge to eliminate odors and pathogenic organisms. Most plants use aerobic or anerobic digesters to accomplish this. Following stabilization, the sludge is concentrated and dewatered prior to final disposal in a landfill or by incineration. New processes are being continually investigated to use this waste sludge. It has been used for years as a lawn and agricultural soil supplement. It also finds use as a fuel and as a raw material for conversion to activated carbon.

Sludge is often concentrated before dewatering to permit more efficient utilization of the final dewatering equipment. This is particularly true in the case of waste activated sludge, which may contain only 0.5 to 1.5% solids. The position of the sludge concentrator will vary, though when used to concentrate waste activated sludge, it is generally ahead of the digester to provide a concentrated feed. In some cases this is the final sludge processing step prior to landfill of digested

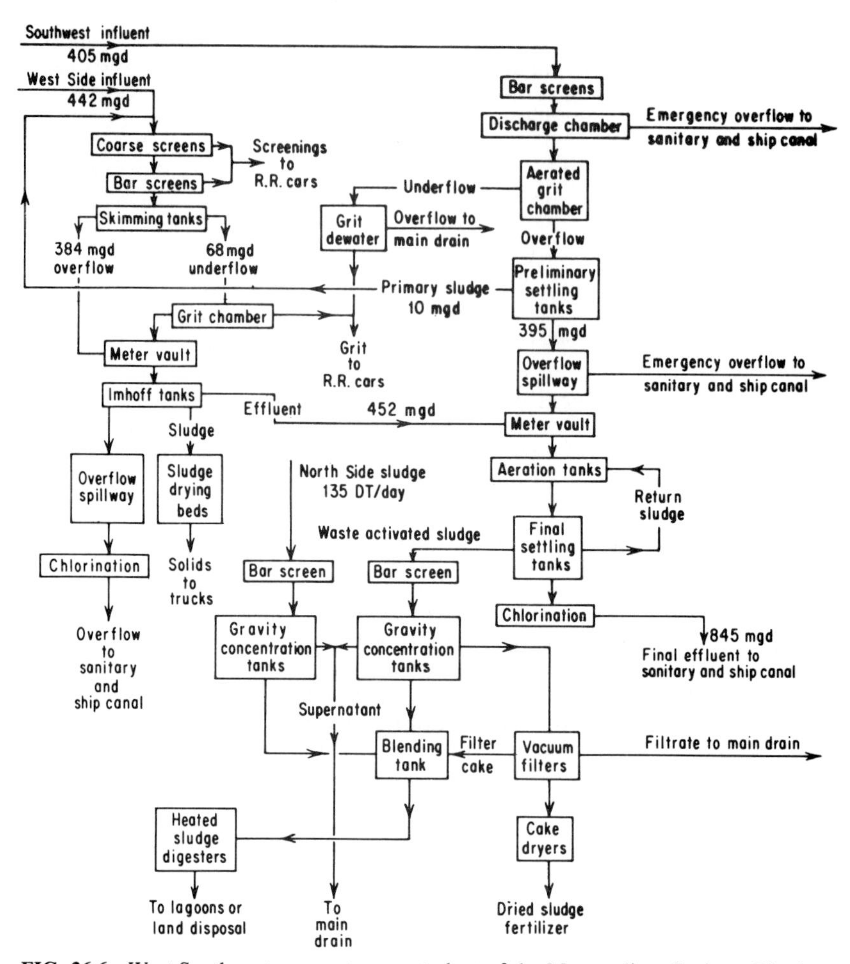

FIG. 36.6 West-Southwest sewage treament plant of the Metropolitan Sanitary District of Greater Chicago—schematic of wastewater flow.

TABLE 36.2 Water Quality Changes from Urban Use

Constituents	Chicago tap water	Raw sewage*	Final effluent*
Total solids, mg/L	164	644	
Suspended solids	Nil	294	9.0
BOD		181	7.0
Ammonia nitrogen		6.8	1.8
Oil/grease, hexane soluble		60	19
pH	7.8	7.3	7.6

* Averages for December 1976.

sludge where the higher moisture does not represent significant haulage costs. The equipment used may vary greatly from plant to plant, though it would be one of the following.

Thickeners

The common thickener is a gravity sedimentation tank with very slowly moving rakes surmounted by pickets. Sludge is fed to a centerwell and the movement of the pickets assists in sludge compaction. The purpose is to increase sludge density, thereby minimizing sludge volume and increasing detention in the sludge digester. Usually the thickener is used to densify activated sludge prior to digestion. Thickeners can be very troublesome because the overflow goes back to the primary, and is often the largest source of solids fed to the primary clarifier. The thickener is likely to carry over badly if there are unusual temperature changes or if the sludge becomes septic and floats because of gasification. Chemical treatment is often helpful in controlling solids carryover.

Elutriation

This sludge washing process is sometimes used to reduce the ammonia and organic content of the sludge, which increases the effectiveness of chemicals used for subsequent dewatering. Sludge passes in one direction through two stages of thickeners, with final effluent moving in the opposite direction to wash it. About 1 to 2 parts of washwater are mixed with one part of sludge in each thickener, and the sludge is settled before passing on to the next unit.

Flotation

This process is an improvement over gravity thickening because it provides a greater difference in density between water and rising air-entrained sludge than is usually possible between water and sinking sludge floc. Usually final effluent is pressurized with air at 60 to 75 lb/in^2 (4.2 to 5.3 kg/cm^2) and then mixed with sludge.

Pressure on the mixture is suddenly released, allowing minute air bubbles to attach to the sludge particles and cause them to rise as a blanket of froth. The froth is scraped from the surface, flows down a ramp, and is then pumped at 4 to 6% solids to the digester or directly to a dewatering system. Chemicals aid in forming a cohesive froth and in improving the quality of the separated water, which is returned to the headworks.

Drainage Concentrators

There are two types of equipment available in this area. One is a twin-drum–type unit with sludge fed to the inside allowing water to drain through a cloth covering on the drum. Others are simple top-loaded belt screens which allow drainage of water by gravity, later assisted by pressure rolls.

Digester

Digestion applies to the bacterial system of sludge destruction, traditionally anaerobic sludge digestion, although there is a growing use of aerobic sludge digesters because they usually have fewer operating problems. In many cases, though, aerobic digesters produce a very thin sludge (less than 1% solids), which is hard to dispose of. The anaerobic digester is a large fermentation vat. Several units may be used for flexibility and to keep within practical size limits. The digester stores and ferments sludge during an approximately 30-day detention period, and yields a stabilized sludge—that is, it will no longer support active bacterial growth, so odor-production is negligible.

Anaerobic digested sludge is drawn off periodically for final disposal, generally at a concentraton of 4 to 7% solids. The sludge is almost black and has a musty or earthy odor. The anaerobic activity generates gas, a mixture of chiefly ammonia (NH_3), methane (CH_4), carbon dioxide (CO_2), and some sulfide (H_2S). This combustible gas is burned in furnaces to heat sludge recirculated through the furnace from the digester. In some plants this is an indirect exchange of heat, the fuel heating water in a closed loop, which in turn heats the sludge. This maintains the anaerobic process at optimum temperature—usually about 90°F (32°C). To maintain a uniform temperature, the digester must be insulated.

Some systems use several digesters in series, the first unit operated at 90°F (32°C) and actively undergoing digestion and the secondary unit operating at ambient temperature, usually 70 to 80°F (21 to 27°C), so that the sludge no longer ferments, thus allowing clearer separation of sludge and supernatant.

To make room for the fresh sludge being fed to the digester, thickened digested sludge is withdrawn for disposal and supernatant must also be drawn off. The drawdown of supernatant is usually returned to the raw sewage line intermittently. The supernatant contains a high level of suspended solids and is also highly alkaline, since digestion produces NH_3 and CO_2 which form NH_4HCO_3, producing an alkalinity of 1200 to 1500 mg/L. Therefore, this intermittent return of highly alkaline supernatant can be a disruptive factor in a good chemical coagulation program in the primary clarifier.

Vacuum Filter

This consists of a vat, which receives the sludge feed, and a horizontal drum with a filtering surface partially submerged in the sludge. An internal piping system draws filtrate through the filter surface and discharges it to a receiver and then draws air through the cake formed on the filter surface. A "doctor blade" or similar device scrapes, blows, or otherwise disengages the cake from the drum and puts the cake onto a conveyor. Considerable art is involved in filter operation, usually requiring an operator to be on hand at all times for adjustment as sludge density or quality changes. Chemicals are usually required to properly condition the sludge for efficient vacuum filter operation.

A pretreatment device flocculates the sludge before it is fed to the filter vat or pan. Variables are many, including feed sludge density, sludge flow rate, drum submergence, drum speed, flocculating drum detention time and agitation, vat agitation, and vacuum. Filter cloth—including type of fabric and weave—is another variable, though not one that is readily changed. Chemicals are used to increase production, clarify filtrate, or increase cake solids. The filtrate returns to the headworks. Cake is disposed of by incineration or hauling, so maximum density is desired to reduce final disposal costs. The cake may contain as little as 14% solids (typical of thickened waste activated) to as much as 25% (typical of mixtures of primary and digested sludges). In smaller plants, filters may be operated intermittently, often only on the day shift.

Centrifuge

This dewatering device is generally a horizontal, solid bowl centrifugal separator operating at 1000 to 6000 rev/min and having an internal screw conveyor (scroll) to separate and withdraw the centrifuged sludge from the slurry pool. Variables include bowl speed, differential speed between bowl and scroll, feed solids and flow rate, pool depth, and chemical application point. Chemicals needed to control both cake density and centrate clarity are fed internally through a special injection pipe or to the sludge feed itself or both. In small plants, centrifuges, like filters, may be operated only intermittently.

Roll Presses

Many of these units are primarily designed for sludge concentration. Some can be considered as dewatering devices because of their ability to provide cake solids in excess of 12 to 13%. This is done by a series of opposing rollers which squeeze additional moisture out of the cake.

Filter Presses

This equipment has demonstrated its ability to provide high solids cake. The sludge is processed through a series of plate-and-frame sections with filter cloth inserted over the plate. A filter aid such as lime or fly ash is used, and the sludge is pumped at 100 to 150 lb/in^2 (7 to 10.5 kg/cm^2) until the frame is full of compressed cake. The plates and frames are then separated to allow the cake to fall out.

Sludge Beds

Sludge beds are a series of rectangular earthen basins underlain with a tile collector, above which is a graded gravel drainage and collection layer 12 in (30 cm) deep, topped by sand 9 in (23 cm) deep. Sludge is applied by gravity, usually through a simple standpipe. The sludge fills the bed to a level of about 9 to 12 in (23 to 30 cm) above the sand surface. After filling, the bed is quiescent until water drains into the underdrain system, usually 18 to 24 h. Moisture then evaporates leaving a cracked surface within 3 to 4 weeks, depending on weather, sludge conditioning, and initial sludge density. A relatively dry cake (24% solids) is finally harvested for disposal as land dressing. Chemical conditioning speeds drainage and facilitates drying, reducing time between harvesting. Sludge beds are used extensively by small plants.

SLUDGE DISPOSAL

Sludge disposal is becoming more important each year because of the amount of sludge generated by improved and expanded treatment processes. Many smaller towns simply pile the sludge near the plant and allow local residents to take it as a fertilizer. Some cities process the sludge to salable fertilizer. In an increasing number of cases, land is purchased for large scale disposal of sludge with the growing of crops being incidental to the operation. Long-term effects on the soil must be studied for each new situation.

Many cities use incinerators to dispose of sludge generated from sewage. Sometimes sludge is combined with garbage for incineration. Because of energy considerations, emphasis is being placed on power generation in newer units.

CHAPTER 37

COMMERCIAL, INSTITUTIONAL, AND RESIDENTIAL WATER TREATMENT

This chapter covers the special requirements in treating water for nonindustrial facilities such as office buildings, hospitals, apartments, schools, universities, department store chains, hotels/motels, shopping centers, and commercial enterprises such as laundries. The systems in such establishments that require water treatment are heating, air conditioning, and domestic water.

In some respects the water problems encountered in buildings and institutions are similar to those in industrial plants. But there are several important differences. First, the consequences of improper water treatment may be even more critical to a hospital or living complex than to a large industrial plant, since human health, comfort, and even survival may be at stake. Second, the utility services are often required intermittently on an instantaneous demand basis, so the system may be severely taxed physically. This is particularly true during spring and fall heating seasons when heat may be required at night but not during the day; a steam heating system under these conditions is subject to oxygen corrosion during the idle daylight hours when condensate receivers draw in air, often aggravated by CO_2 corrosion when heat is again called for and cold condensate returns to the steam plant.

Designers of modern commercial, institutional, and residential facilities are well aware of the importance of water treatment. They provide sophisticated chemical feeding and control systems for these installations (Figure 37.1), which are designed to use the same kinds of chemicals as industrial systems. The operating personnel in charge are skilled in the operation of mechanical systems, and are well trained in proper chemical control testing.

Older, smaller facilities may lack good feed systems and skilled personnel. This requires some modification of the chemical program: simple feeders are devised to apply multipurpose chemical formulas, containing a variety of ingredients for controlling any of the problems that might be encountered (Figure 37.2).

HEATING SYSTEMS

Heating systems using water are of two basic designs: (1) steam boilers and (2) hot water "boilers." Steam boilers are usually package-type firetube or watertube

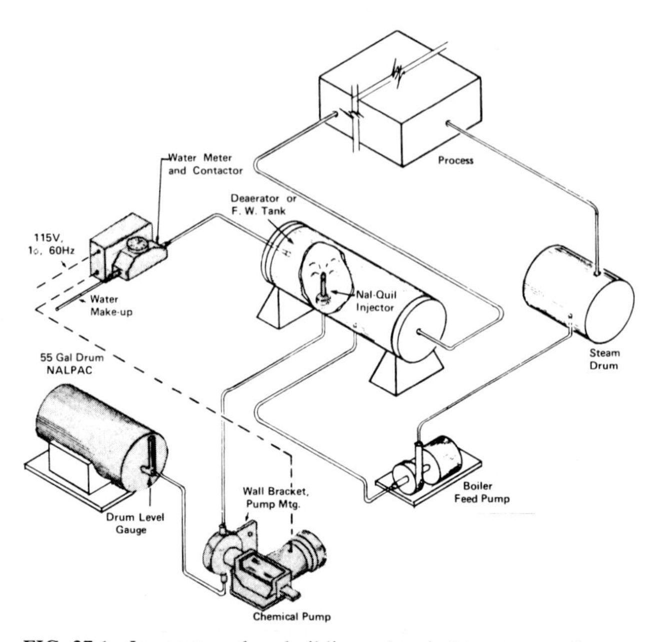

FIG. 37.1 In most modern buildings, chemical treatment of water for heating or cooling is automated for simple, reliable control.

FIG. 37.2 Where there are multiple units to be treated, a simple feeder charged with briquetted chemical is often more practical than individual mechanical feeders.

FIG. 37.3 A typical packaged steam generator with complete fireside and waterside controls. *(Courtesy of Cleaver Brooks, division of Aqua-Chem, Inc.)*

design operating at less than 200 lb/in^2 (13.8 bars) (Figure 37.3). However, in some of the more sophisticated installations, such as an energy supply complex for a consortium of medical centers or a large university campus, there is a trend toward cogeneration and high-pressure (900 to 1200 lb/in^2) boilers. These are being installed as topping plants to produce electrical energy from turbines that then discharge steam at perhaps 100 lb/in^2 to the campus central heating system.

Most smaller institutional boilers are usually gas- or oil-fired, but occasionally an electrically heated boiler is found. Fuel oils can range from No. 2 to No. 6, including mixtures of several grades. Many boilers contain burners designed to burn either gas or fuel oil, a useful feature in areas where shortages of either fuel can occur. Coal-fired boilers in commercial buildings and institutions are rare.

In heating applications the steam generated in these boilers is used in one of two ways: (1) directly; i.e., circulated throughout the building, where the heat is extracted through radiators and the condensate is returned to the boiler to be used as feedwater; (2) indirectly; i.e., passed through a heat exchanger located relatively close to the boiler, with the condensate returning to the boiler. The other side of the heat exchanger is part of a closed heating water loop which extends throughout the building. The water heated by the steam is pumped around the loop, giving up its heat through radiators, mixing boxes, and other types of heat exchangers.

The steam generated in these boilers can have other uses, such as humidification, dishwashing, food preparation, and sterilization in autoclaves (as in hospitals and research laboratories). A careful study must be made of the uses to which the steam is put so that acceptable, safe treatment chemicals can be used. If any part of the steam contacts food, for example, FDA-approved chemicals are required. If the steam is used for humidification, OSHA guidelines for condensate corrosion inhibitors must be followed and care must be taken that condensate corrosion inhibitors be safe and odor-free. In some university, medical, or research centers, pure steam is needed in small quantities for these and other uses. Often the quantity needed at a research site or a hospital work station may be only 100 to 200 lb/h, a very tiny fraction of the 200,000 to 500,000 lb/h generated by the central boiler house. It is essential that the multimillion dollar steam-condensate system be thoroughly protected against corrosion, yet the corrosion inhibitors in the steam may be objectionable for these small critical uses. In that case, the institution must install a "pure steam generator," where needed, to convert plant steam to research-quality steam free of the inhibitors.

Water treatment in these steam plants is similar to that used in most industrial boilers. The chemical products used should control scale, corrosion, and foaming in the boiler, and corrosion in the condensate system. The types of products used are similar to those used in industrial systems; however, they are often blended and packaged so as to be convenient and easy to handle, feed, and control.

The condensate return system of a university complex is subject to unusual corrosion problems because of the stress of seasonal changes. Because of this, in addition to having a central chemical treatment station in the boiler house, satellite systems are sometimes installed to supply booster shots of steam treatment to correct for seasonal upsets. The quality of the condensate must be under constant surveillance, because often the steam heating coils in the individual building's hot water supply system may fail, and the city water is then diverted back through the condensate return lines to the boiler plant. It is common practice to install ion-exchange–type condensate polishers in the boiler house to handle such failures and prevent having to dump large amounts of valuable hot condensate before the source of coil failure is found. Monitoring of conductivity at significant condensate return junction points helps to locate and isolate the failure promptly.

Hot water "boilers" are really misnamed since they do not boil water and produce no steam. They can heat water to 500°F (260°C), but typically operate at temperatures between 180° to 250°F in a vessel almost identical to a steam boiler. The heated water is circulated through the building to various heat exchangers and radiators and back to the boiler.

In addition to hot water boilers, which use gas or fuel oil for heat, there are also electric boilers. These use either enclosed, clad-type immersion heaters, or resistance heaters, which operate by a flow of current between electrodes in the boiler water, which is conductive because of its mineral content.

These systems are designed to be closed, so theoretically there should be no makeup water required, but normally a small amount is required to replace that lost by leakage of liquid or vapor. In these systems, the primary concern is prevention of corrosion. Scale formation, although not usually a major factor, can also occur in hot water boilers. This may be caused by a combination of very hard water, used to fill the system initially and to provide makeup, and the high temperature of the boiler tube surfaces. The gradual accumulation of calcium and alkalinity precipitates $CaCO_3$.

In addition to corrosion and scaling, foaming can also occur. This can lead to severe cavitation-erosion in recirculating pump impellers, caused by the impingement of foam bubbles on the metal at high velocity. The preferred chemical treatment in hot water boilers is a borate/nitrite-based product. Such products contain a mixture of corrosion inhibitors to protect steel, aluminum, copper, and admiralty; a scale control agent; and high- and low-temperature antifoams. This treatment is initially charged at high dosage levels for maximum protection. Since hot water boiler systems have little makeup and the chemicals are relatively stable, only small additions of treatment are required thereafter to maintain protection.

Occasionally, hot water systems require high makeup, caused by excessive leaks of water or vapor from the system. In some cases these systems are very extensive, such as in a university complex with a major portion of the system buried underground. This makes repair of leaks costly compared to the cost of makeup water, so the system has to be treated accordingly. Maintaining high level dosages of a borate/nitrite-based inhibitor becomes prohibitive economically. Under these conditions less expensive treatments are used. These still provide acceptable protection, but need more attention for control of chemicals. Sodium silicate treatments, combination hydrazine/amine programs and glassy phosphate treatments fall into this category.

AIR-CONDITIONING SYSTEMS

Most large buildings, shopping centers, and similar installations use water-cooled air-conditioning units. These fall into three categories:

1. Reciprocating compressor units, usually electrically driven and using Freon* as the refrigerant (Figure 37.4).

FIG. 37.4 Packaged water chiller, using a reciprocating-type compressor. *(Courtesy of The Trane Company.)*

2. Centrifugal compressor units, driven by electric motor, steam turbine, or gas engine, also use Freon as refrigerant (Figure 37.5).
3. Absorption refrigeration units, in which the refrigerant (water) is absorbed in concentrated lithium bromide solution and then evaporated by steam (Figure 37.6).

In the reciprocating and centrifugal systems, the hot compressed refrigerant vapor first passes through a water-cooled condenser (usually a shell-and-tube–

* Registered trademark of the E. I. du Pont de Nemours & Co.

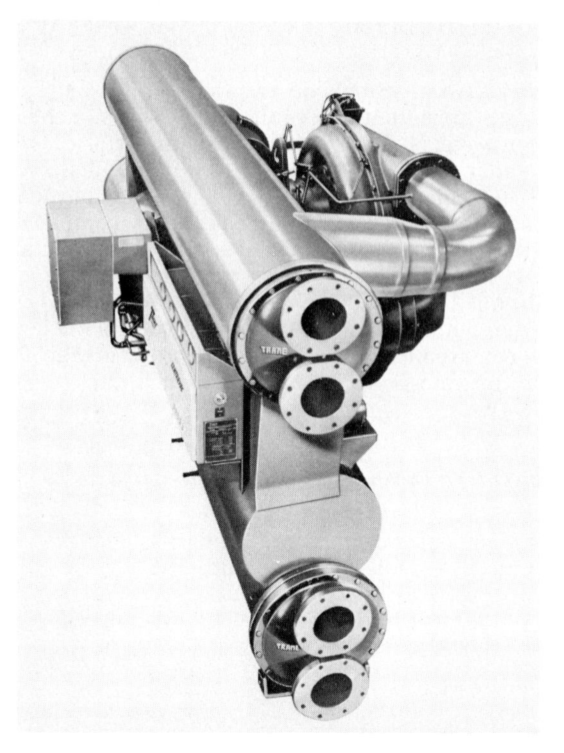

FIG. 37.5 A centrifugal-type compressor is used in this package refrigeration unit. *(Courtesy of The Trane Company.)*

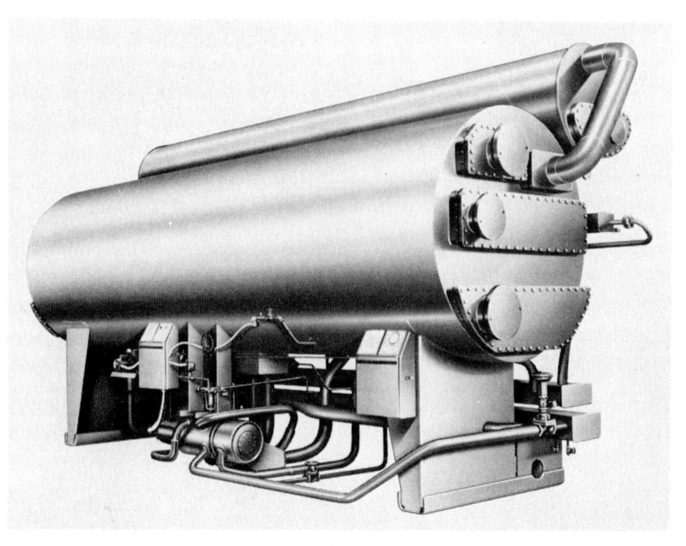

FIG. 37.6 An absorption-type refrigeration system. *(Courtesy of The Trane Company.)*

type heat exchanger) where it is cooled and condensed into a liquid. From there it passes into an evaporator (commonly called a chiller) where it boils under reduced pressure into a cool vapor. The boiling liquid is used to extract heat and the vapor is then compressed, repeating the cycle (Figure 37.7).

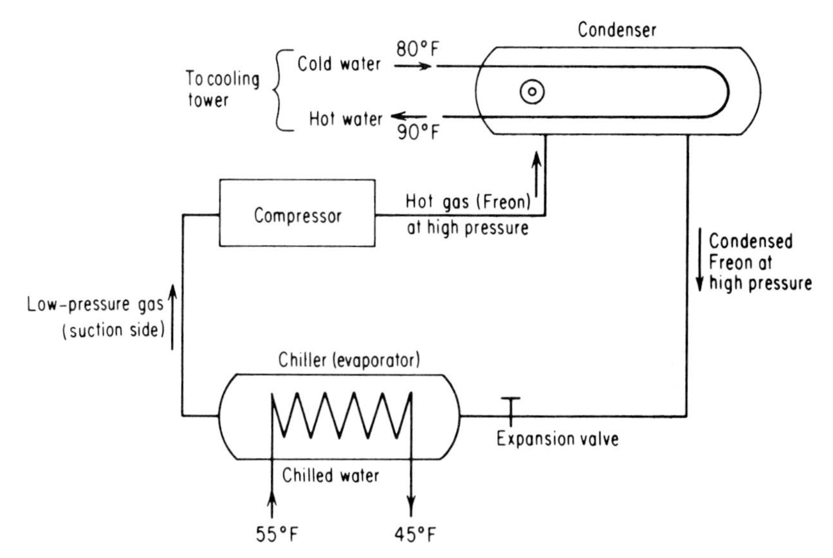

FIG. 37.7 Typical air-conditioner schematic.

The water used to cool the refrigerant in the condenser is itself cooled in a typical cooling tower, and then recycled back to the condenser. Typically, water enters the condenser at 80 to 90°F (28 to 32°C), and leaves at 90 to 100°F (32 to 38°C). This is known as the open recirculating portion of the cooling water system, since the cooling tower is open to the atmosphere.

The evaporator, or chiller, also a shell-and-tube–type heat exchanger, provides chilled water for air-conditioning systems. When the condensed refrigerant leaves the condenser, it has been cooled, but is still at a high pressure. In the tube side of the chiller, the Freon vaporizes by extracting heat from the relatively warm [55°F (13°C)] chilled water entering the shell side. This cools the chilled water to about 45°F (8°C) for circulation throughout the building to provide air conditioning. The air being cooled warms this water back to 55°F (13°C) for return to the chiller. The chilled water system is not open to the atmosphere and is known as a closed recirculating system (Figure 37.8).

In simple terms, the heat within a building is absorbed by the chilled water, which then gives up this heat to the refrigerant in the chiller. The refrigerant is then compresed and passed into the condenser, giving up the heat to the open recirculating cooling water. The cooling water then gives up this heat to the atmosphere as it evaporates while passing through the cooling tower. An air-conditioning system is nothing more than a mechanism designed to move heat from inside the building to the outside.

Steam absorption units are more complicated in their operation. They use water as a refrigerant, a lithium bromide solution as refrigerant absorber, and

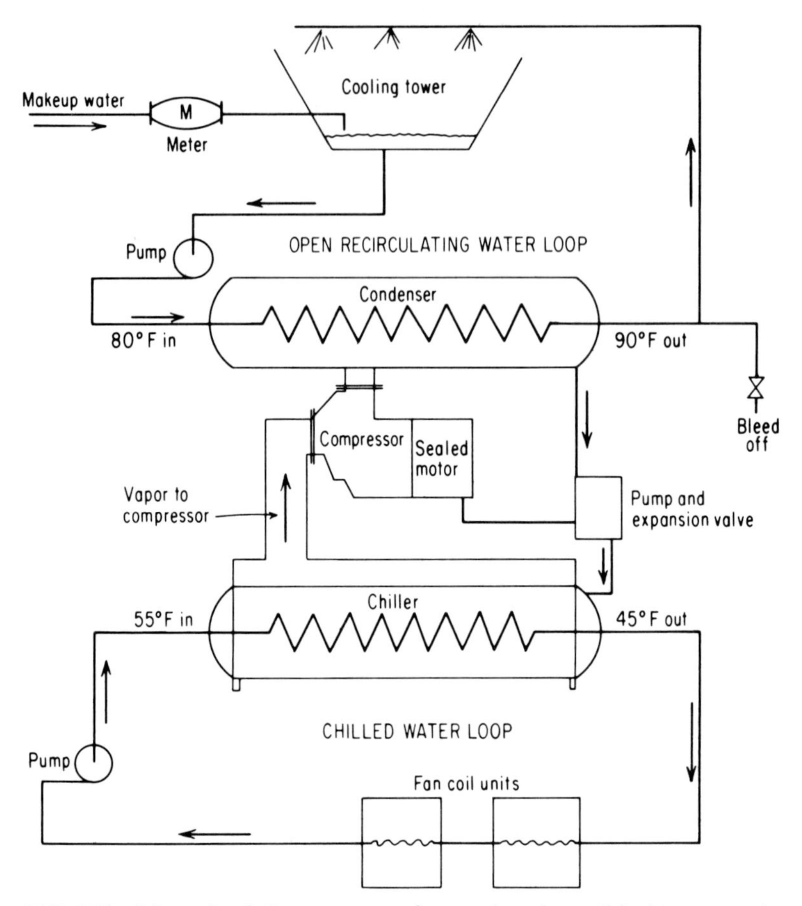

FIG. 37.8 Schematic of all components of a complete air-conditioning system. At certain times of the year, the cooled water in the cooling tower basic can bypass the refrigerant system and go directly through the chilled water loop ("free cooling") at a significant cost savings.

steam as a heat source to evaporate the water (Figure 37.9). Absorption units use condensers, cooled by water from a cooling tower, and chillers to provide chilled water for air conditioning. From the water treatment point of view, the operating principles of the open recirculating and chilled water systems are the same in all of the above systems, except that much higher temperatures and heat fluxes are involved.

In order for water-cooled air-conditioning systems to operate at design efficiency, careful attention must be paid to the condition of both the open recirculating water and the chilled water. The chemical makeup of water varies widely, depending on the source. Even water from the same metropolitan area can have wide variation. For example, Chicago's municipal water, obtained from Lake Michigan, has approximately 120 mg/L hardness, but many suburban communities outside Chicago obtain their water from wells, typically with hardness in the 300 to 500 mg/L range. In some areas of the country, such as the eastern

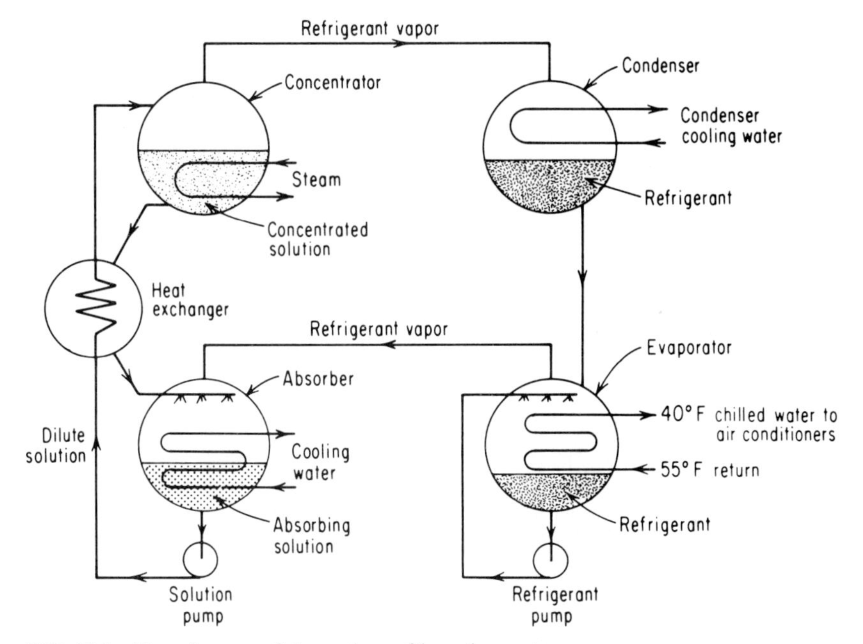

FIG. 37.9 Flow diagram of absorption refrigeration system.

seaboard and the Pacific Northwest, waters of exceptionally low hardness are common.

Problems that can occur on the water side of air-conditioning systems are:

1. Formation of scale, especially on heat transfer tubes
2. Corrosion of the metals used in the system
3. Microbial fouling
4. Fouling from dirt and silt accumulation

As with industrial heat exchangers, scale and fouling deposits cause poor heat transfer through the exchanger tubes, reducing efficiency and increasing energy costs. The problem is worse in air-conditioning systems than in most applications because the temperature driving force is much lower than in most other heat transfer applications (see Table 37.1). Corrosion causes failure of the system parts, resulting in expensive repair or replacement.

Water treatment procedures for these systems are the same as for industrial systems. Scale and corrosion control in open recirculating systems are maintained either by: (1) keeping the water in a nonscaling condition by using acid to lower the pH, and adding chromate to prevent corrosion or (2) keeping the water in an alkaline condition, and preventing scale by the addition of organophosphate scale modifying agents. To provide the required copper corrosion protection, organic nitrogen compounds are added.

In systems where concentration results in a pH of 8.5 or above and sufficient alkalinity is present, the addition of corrosion inhibitors may not be necessary. Corrosion coupons at critical spots in the system can be inserted to provide the data to determine whether inhibitors are needed.

TABLE 37.1 The Cost of Energy Loss Caused by Scale in an Air-Conditioner Condenser*

	Scale thickness		
A/C tonnage	Light 0.001 fouling factor (1/64 in)	Moderate 0.002 fouling factor (1/42 in)	Heavy 0.003 fouling factor (1/32 in)
500	$ 5,100	$ 11,100	$ 15,600
1000	10,200	22,200	31,200
2500	25,500	55,500	78,000
5000	51,000	111,000	156,000

* Assumptions: (1) unit operating 8 h/day, 240 days/yr, (2) one ton of A/C consumes 0.7 kWh, (3) electricity costs 7.5¢ per kWh.

Some cooling towers are located in areas where appreciable quantities of airborne dirt and silt are drawn into the recirculating water. These impurities become suspended in the water, and tend to settle out in areas of low velocity, such as in heat exchangers. In addition to impeding heat transfer, settled sludge causes corrosion and encourages microbial activity underneath these deposits. A dispersant may be added to keep the particulates in suspension, until they are eventually removed from the system with the blowdown water. Dispersants are available as separate products, or they can be included in the formulation of multipurpose products, which also contain scale and corrosion inhibitors.

Problems can be caused by aerobic slime-forming microorganisms, anaerobic bacteria, molds, yeasts, and algae. Slime-forming bacteria cause loss of heat transfer efficiency in condensers; in severe cases they can completely block the flow of water through the tubes.

Anaerobic bacteria generally grow under deposits in areas where no oxygen is available. These produce corrosive by-products such as hydrogen sulfide, which can eat holes in piping quite rapidly. Algae growth on tower decks and other areas exposed to sunlight can become so severe as to plug the drain holes on the tower deck. The application of microbial control chemicals is essential to correct these problems.

Closed chilled water recirculating systems, like closed hot water systems, are generally not susceptible to scale formation, since little or no evaporation occurs in a properly operating system. However, leaks and evaporation can result in scale formation. The primary goals in treating these systems are to prevent metal corrosion, to control anaerobic microorganisms, and to prevent foaming and iron fouling.

The preferred chemical treatment for corrosion, scale, and foaming is the same as for hot water boilers—a borate/nitrite-based formulation containing scale control agents and antifoams.

Life cannot exist in hot water systems, because of the high temperatures involved; but microbes can and do grow in chilled water systems. Typical examples are nitrate-reducing and sulfate-reducing bacteria. The end result of the activity of these microbes is corrosion of the system metal, which can become severe. Proper control is maintained by periodic analysis of the water and by the application of biocides when necessary.

In colder climates, the addition of ethylene glycol antifreeze solutions to both hot and chilled water systems is common. The chemical treatment selected must be compatible with the antifreeze. For example, borate/nitrate-based inhibitors are usually compatible, while chromate-based inhibitors are not.

FIG. 37.10 These two large centrifugal chilling machines provide for the air conditioning of an entire hospital. *(Courtesy of Carrier Corporation.)*

FIG. 37.11 An automated feed and control system for an air-conditioner cooling tower.

An example of an institutional air-conditioning system is shown in Figure 37.10, representing a typical hospital installation. The completely automated chemical feed system for the open cooling tower is shown in Figure 37.11.

ENERGY STORAGE

The public utility practice of offering a special discount on electrical costs to customers able to use off-peak energy has brought a new factor into the economics of air-conditioning operations called "load-leveling." One way to take advantage of off-peak electrical rates for comfort cooling is to manufacture ice during the cheaper off-peak hours and to melt it during peak periods. Another method is to provide storage of refrigerated water in large, insulated reservoirs which may hold a 12-h supply of chilled water. Taking advantage of this method of electrical cost saving, however, requires special consideration of the chemical treatment program to prevent microbial attack and concentration cell corrosion beneath the sediment that invariably forms in large, relatively quiescent reservoirs.

DOMESTIC WATER

Domestic water is that used for drinking, cooking, bathing, washing of dishes and clothes, toilet flushing, and lawn or garden sprinkling. It is usually municipal water but may be from private wells. Because of the large size of modern buildings, hot water is often recirculated throughout the building from large holding tanks.

Domestic cold water problems are limited to corrosion of the water piping and holding tanks. Domestic hot water problems include both scale and corrosion. Tastes and odors may be the cause of occasional complaints, but these are not major problems because they do not involve equipment damage and they apply to a very small percentage of the total water used.

The corrosion in both hot and cold systems is caused by (1) water that is naturally corrosive as supplied to the building, and (2) the use of a zeolite softener, which replaces Mg and Ca ions with Na ion, creating a more corrosive soft water.

Scale in hot water systems occurs in the hot water heater tubes, and is caused by the precipitation of $CaCO_3$, which becomes less soluble as the temperature rises. The EPA has responsibility for granting approval of those chemical products

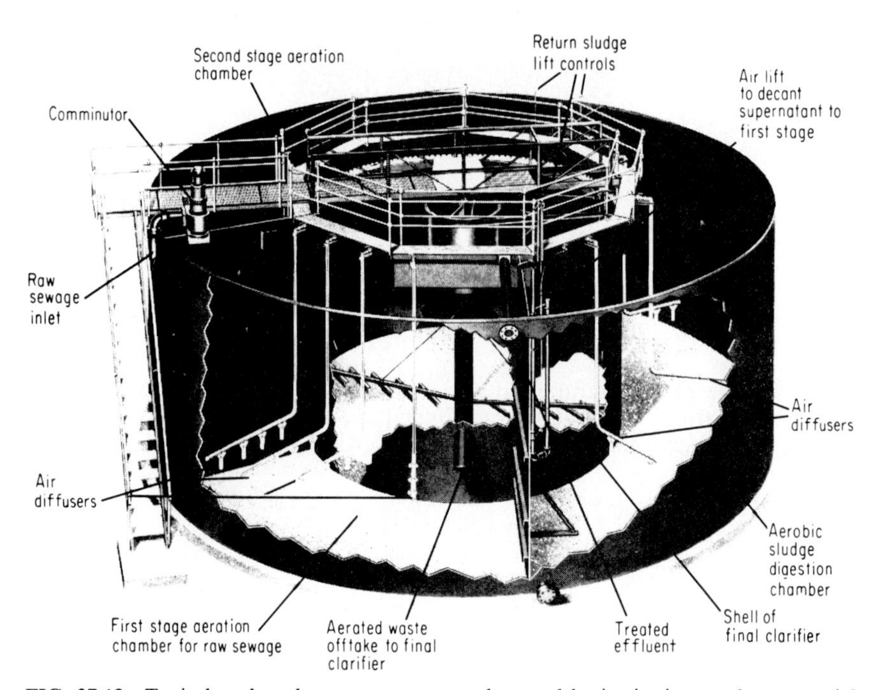

FIG. 37.12 Typical packaged sewage treatment plan used by institutions and commercial centers not served by a municipal sewage plant. *(Courtesy of Smith & Loveless Division, Ecodyne Corporation.)*

that are safe to use in potable systems. In potable water systems within buildings, the accepted practice is to use only those chemicals that are approved for use in municipal water supplies. These chemicals are usually polyphosphates, sodium silicates, or combinations thereof. These products provide both scale and corrosion protection. These chemicals are applied through the simple feeder shown in Figure 37.2.

WASTE TREATMENT

In the building and institutional field, most individual installations discharge their wastes into the municipal sewer system. In some water-short areas, such as Los Angeles, noncontaminated wastes may be segregated from contaminated wastes, such as domestic sewage, for recovery and return to the potable water source.

There are certain commercial operations, such as laundries, where strong wastes are given special treatment before discharge. In the case of laundries, the cost of special treatment is partly offset by heat recovery and use of rinse water in the washing cycle. The recovered water contains both valuable heat and unspent detergent. Some institutions, such as hospitals and hotels/motels, operate their own laundries, but dilution with other wastewater is usually adequate to produce a combined effluent meeting local discharge standards, which usually apply to oils, pH, BOD, and suspended solids.

Some shopping centers, resort hotels, hospitals, and similar establishments may not have access to municipal sewers. These are required to install waste treatment facilities to meet the same standards applied to municipal sewage plant effluents. An illustration of a complete packaged sewage treatment facility serving such an installation is shown in Figure 37.12.

P • A • R • T • 4

SPECIALIZED WATER TREATMENT TECHNOLOGIES

CHAPTER 38
COOLING WATER TREATMENT

Most of the water employed for industrial purposes is used for cooling a product or process. The availability of water in most industrialized areas and its high heat capacity have made water the favored heat transfer medium in industrial and utility type applications. Direct air cooling is finding increasing use, particularly in water-short areas but is still far behind water in total numbers of applications and total heat transfer loading.

During recent years, the use of water for cooling has come under increasing scrutiny from both environmental and conservational points of view and as a result, cooling water use patterns are changing and will continue to do so. For example, many systems pass cooling water through the plant system only once and return it to the watershed. This creates a high water withdrawal rate and adds heat to the receiving stream. On the other hand, cooling towers permit reusing water to such a large extent that most modern evaporative cooling systems reduce stream withdrawal rates by over 90% compared to once-through cooling. This substantially reduces the heat input to the stream but not to the environment, since the heat is transferred to the air.

These changes in cooling water system design and operation have a profound impact on the chemistry of water as it influences corrosion, deposition, and fouling potential in the system. This chapter reviews the industrial operations which use water for cooling purposes, the problems of corrosion, scale, and fouling in these systems and how these problems affect plant production through loss of heat transfer, equipment failures, or both. In addition, various cooling water treatment concepts are examined and the control procedures required for their success are discussed.

HEAT TRANSFER

Heat transfer is simply the movement of heat from one body to another, the hotter being the source and the cooler the receiver. In cooling water systems, the product or process being cooled is the source and cooling water the receiver.

Cooling water usually does not contact the source directly; the materials are usually both fluids, separated by a barrier that is a good conductor of heat, usually a metal. The barrier that allows heat to pass from the source to the receiver is called the heat transfer surface, and the assembly of barriers in a containment vessel is a heat exchanger.

In many industrial heat exchangers both the source and receiver are liquids. If the source is steam or other vapor that is liquefied, the heat exchanger is called a condenser; if the receiver is a liquid that is vaporized, the exchanger is called an evaporator.

The simplest type of heat exchanger consists of a tube or pipe located concentrically inside another—the shell. This is called a double pipe exchanger (Figure 38.1). In this simple exchanger, process liquid flows through the inner tube and cooling water through the annulus between the tubes. Heat flows across the metal wall separating the fluids. Since both fluids pass through the exchanger only once, the arrangement is called a single-pass heat exchanger. If both liquids flow

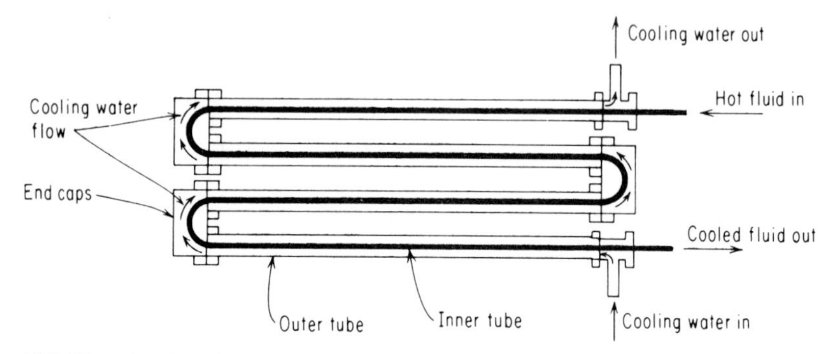

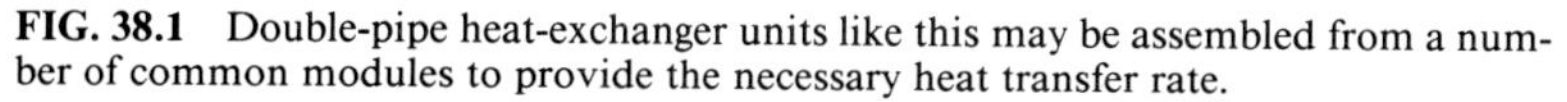

FIG. 38.1 Double-pipe heat-exchanger units like this may be assembled from a number of common modules to provide the necessary heat transfer rate.

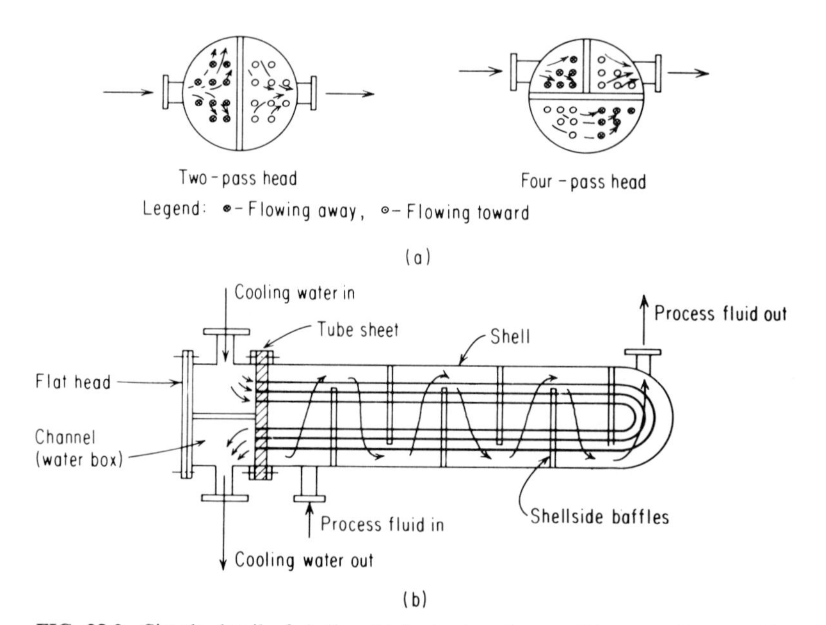

FIG. 38.2 Simple detail of shell-and-tube heat exchanger. The water box may be designed for as many as eight passes, and a variety of configurations of shell-side baffles may be used to improve heat transfer. (*a*) Several water box arrangements for tube-side cooling. (*b*) Assembly of simple two-pass exchanger with U-tubes.

in the same direction, the exchanger is parallel or cocurrent flow; if they move in opposite directions, the exchanger is a countercurrent type.

Progressing from this exchanger, more sophisticated units are designed to improve the efficiency of the heat exchange process. Figure 38.2 shows a shell-and-tube exchanger. Process fluid and cooling water could be located on either side of the barrier.

Another simple heat exchange device is the jacketed vessel, with cooling water passing through the space between the double walls of a chemical reaction vessel, removing heat from the process. This design is like a thermos bottle, but in this case, the double wall is used for heat removal instead of insulation. Plate-type heat exchangers, somewhat resembling plate-and-frame filters, are used in many chemical process industries because of their compact design and availability in a wide range of materials of construction.

Removing Undesirable Heat

Once the water completes its job and cools the source, it contains heat that must be dissipated. This is accomplished by transferring heat to the environment. In once-through systems cool water is withdrawn, heated, and returned to a receiving stream, which subsequently becomes warmer. In this type of system each pound (0.454 kg) of cooling water is heated 1°F (0.56°C) for each Btu (0.252 cal) removed from the source.

In open recirculating systems, water is evaporated; this phase change from liquid to gas discharges heat to the atmosphere instead of to a stream. Evaporating water dissipates about 1000 Btu per pound (555 cal/kg) of water converted to vapor. When evaporation is used in the cooling process, it can dissipate 50 to 100 times more heat to the environment per unit of water than a nonevaporative system. (This is explained in more detail in a later section of this chapter.)

Sensible Heat Transfer

The two most common ways heat is transferred from process fluid to cooling water in the heat exchange process are conduction and convection. Heat flows from a hot fluid through a heat exchange surface to the other side by conduction. Heat is then removed from this hot surface by direct contact with cooling water i.e., by conduction. Subsequently this heated water then mixes with other cooler water in a heat transfer process called convection.

The five factors controlling conductive heat transfer are:

1. The heat transfer characteristics (thermal conductivity) of the barrier.
2. The thickness of the heat transfer barrier.
3. The surface area of the barrier.
4. The temperature difference between the source and the cooling water (the driving force).
5. Insulating deposits on either side of the barrier.

Of these five factors, the first three are inherent in the design of the exchanger. Items 4 and 5 are operational characteristics that change depending on the conditions of service. Deposits on either side of a metal barrier have a lower thermal conductivity than the metal itself, so the rate of heat conduction is reduced by any deposit. For example, a buildup of 0.1 in. (0.25 cm) of calcium carbonate scale on a heat exchanger tube wall may reduce the rate of heat transfer by as much as 40%.

This reduction means that the cooling water may not remove sufficient heat from the process. Therefore, production must be slowed or the flow of cooling water must be increased to maintain the same cooling rate that prevailed before fouling developed. Frequently the latter is not possible, and the productivity of the process unit or the entire plant is reduced.

Heat Exchanger Design

In the simple heat exchanger noted earlier (Figure 38.1), both fluids pass through the exchanger only once, so this is known as a single-pass exchanger. The countercurrent arrangement in which source and receiver liquids flow in opposite directions is superior to the cocurrent design because it provides a greater driving force (measured as a mean temperature difference) for the same terminal temperatures; less surface area can transfer the same amount of heat. Counterflow shell-and-tube exchangers are thus more commonly found because more heat can be transferred for a given set of conditions.

The principal disadvantage of the double-pipe exchanger is the small heat transfer surface provided by a single tube relative to the total space required for installation. To offset this limitation, modern exchangers increase the effective heat transfer surface area by using multiple tubes in the shell. This allows more intimate contact between the source and receiver. Also, most industrial heat exchangers employ multiple pass designs. Both of these techniques combine efficient heat transfer with practical equipment sizing. Figure 38.2 shows a double-tube pass, single-shell pass exchanger. In almost all units of this kind, water flows inside the tubes, with process fluids on the shell side, outside the tubes. However, there are notable exceptions, and any investigation of heat exchanger performance must begin with identification of the shell side and tube side fluids. For convenience, this text concentrates on conventional flow, with cooling water in the tubes and process fluid in the shell.

In the design of heat exchangers, the engineer obtains from handbooks heat transfer rates in Btu per hour per square foot of surface, per inch of barrier thickness, per degree of temperature difference, or U_1 = Btu/h/ft^2/in/°F (cal/h/m^2/cm/°C). For a given heat exchange tube size, which may be standardized in specific designs such as condensers, since the wall thickness of the tube is known, the transfer rate may be shortened to

$$U_2 = \text{Btu/h/ft}^2/°\text{F}$$

$$= (\text{cal/h/m}^2/°\text{C})$$

The expression may be further reduced for fixed process temperature conditions to

$$U_3 = \text{Btu/h/ft}^2$$

$$= (\text{cal/h/m}^2)$$

And finally, a given exchanger in a fixed application which shows little change in temperature conditions may simply be rated in total heat transfer in Btu/h (cal/

h). If the heat flux in Btu/h falls off, then an investigation is needed to determine if this is because of changed temperature conditions, insulating deposits, plugged tubes, or some other factor. These problems can occur on either the water or process side.

In most process heat exchangers, heat flux averages 5000 to 6000 Btu/h/ft^2 (120 to 150 cal/h/m^2). However, heat flux in some exchangers exceeds 30,000 Btu/h/ft^2 (750 cal/h/m^2) when vapor is condensed on the process side. Transfer rates also vary considerably with water flow velocities (Figure 38.3).

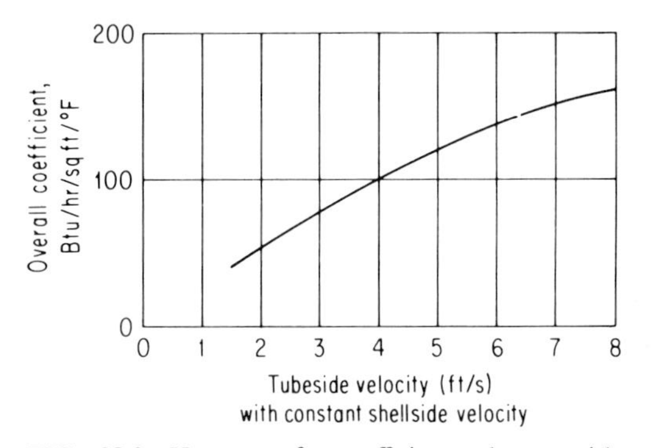

FIG. 38.3 Heat transfer coefficients change with velocity, both on shell side and tube side. Uniform shell-side velocity cannot be achieved because of limitations in geometry.

Cooling water temperature varies across the cross section of a tube, the hottest water being that contacting the tube wall. The temperature of the tube wall, called the skin temperature, is important in designing chemical treatment programs. In fact, skin temperature is the most important variable controlling corrosion and deposition. The individual contributing factors of water velocity, heat flux, and water and process temperatures all combine to define skin temperature. For example, where high skin temperatures occur [above 200°F (93°C)] the plant can find the most probable locations for scale formation and corrosion. Many compounds found in water-formed deposits are less soluble at increased temperatures, and corrosion reactions proceed faster at elevated temperatures.

Cooling water can be on either the tube side (inside the tubes) or shell side (surrounding the tubes) of an exchanger. From a water treatment perspective, there are significant advantages to having tube-side water. With this type of exchanger, water velocity is usually maintained above 2 to 3 ft/s (0.6 to 0.9 m/s) to as high as 7 to 8 ft/s (2.1 to 2.4 m/s) to help keep the tube walls free of suspended solids deposition. Lower velocities encourage tube deposits to form by the settling out of suspended solids. In a bundle of tubes, perhaps only one tube may have a low velocity due to plugging or poor distribution.

Occasionally, high pressures on the process side make it more economical to design the exchanger with cooling water on the shell side. One major problem in such exchangers is the low flow velocities frequently encountered around baffles, tube supports, and tube sheets even when the average flow velocity through the shell appears acceptable. These low velocity areas influence skin temperature and greatly increase the potential for deposits and rapid metal deterioration. For example, mild steel exchangers with water on the shell side have been known to fail from perforations in as little as 3 months, even in the presence of a strong corrosion inhibitor like chromate.

COOLING WATER SYSTEM: PROBLEMS & TREATMENT

The principal water side problems encountered in cooling systems are:

Corrosion is a function of water characteristics and the metals in the system. Corrosion causes premature metal failures; deposits of corrosion products reduce both heat transfer and flow rates.

Scale is caused by precipitation of compounds that become insoluble at higher temperatures, such as calcium carbonate. Scale interferes with heat transfer and reduces flow.

Fouling results from the settling out of suspended solids, build up of corrosion products, and growth of microbial masses. Fouling has the same effect on the system as scaling, but fouling also promotes severe corrosion under deposits.

The treatment of cooling water follows the same basic principles for all types of cooling systems. The first step is to properly identify the problem as scale, corrosion, fouling, or combinations of these factors. The next step is a thorough survey to understand both the process and water side of the system. This establishes the system design, operating characteristics, and water chemistry, important considerations for selecting and applying a reliable, economical treatment program. Special considerations are given to systems restricted to specific treatments; the potential for cross-contamination of water with process or product may not permit employing the most effective treatment. There are three basic types of cooling water systems: once-through, closed recirculating (nonevaporative), and open recirculating (evaporative).

Once-through Cooling

Once-through water is taken from the plant supply, passed through the cooling system, and returned to the receiving body of water. Heat has been picked up from the source. The chief characteristic of once-through water systems is the relatively large quantity of water that is usually used for cooling. A simple flow diagram for a once-through cooling water system is shown in Figure 38.4. Some once-through systems use plant water for drinking as well as cooling, thereby requiring chemicals that are safe for potable use.

A typical chemical treatment program for corrosion control may use various types of inorganic phosphates alone or synergized with zinc. When applied at the low levels required for economical treatment of once-through systems, these materials form no visible film on the metal surface; nevertheless, they can reduce the corrosion rate by as much as 90% over nontreated systems. Corrosion protection is provided because the chemicals act at the point of potential metal loss, hindering the corrosion reaction and thereby reducing the amount of metal removed from the surface.

Where scale is a problem, it is most often calcium carbonate resulting from a change in the stability index of the water.

Polyphosphates are typically used for scale control in potable water systems.

In nonpotable applications, phosphonates, specific acrylate polymers or phosphonate/acrylate combinations are more effective scale inhibitors. These inhibitors function in two ways to prevent calcium carbonate scale on heat transfer surfaces and in distribution lines:

1. They interfere with the potentially scaling ions and prevent crystal growth. Inorganic polyphosphates and organophosphorus compounds are normally used alone or together for this purpose (threshold treatment). Occasionally acid is used to adjust the stability index of the water thereby preventing $CaCO_3$ scale. Acid will not control iron and manganese scales. Usually it is not the most economical method for treating high-volume once-through systems for prevention of $CaCO_3$ scale.
2. They condition crystal nuclei to prevent their growth on heat transfer surfaces and transmission lines. This process of crystal modification uses various polymers and phosphate compounds—both organic and inorganic—and sometimes natural organics.

Fouling, the deposition of particulate matter, iron, manganese, or microbial masses, is a complex mechanism governed by variables such as particle size and charge; water velocity, composition, and temperature; and bacterial populations.

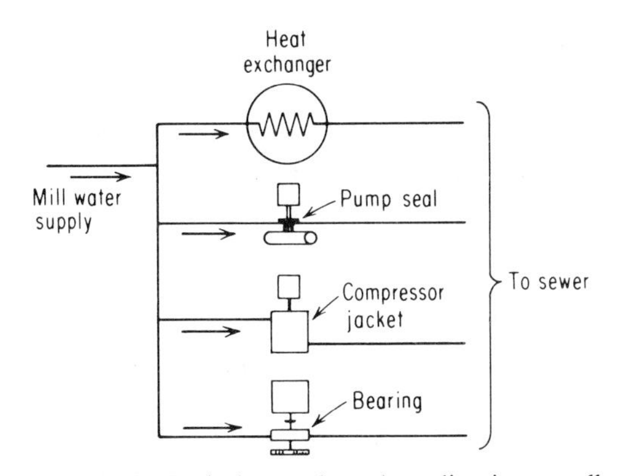

FIG. 38.4 Typical once-through cooling in a small industrial plant.

One approach to handling this problem is to condition foulants such as iron and manganese as they develop by continuously applying specific polymers so that the conditioned material will be carried out of the system. The success of this approach depends on adequate water velocities throughout the system. Low velocity areas, such as in shell side exchangers, reactor jackets, and compressor jackets, are likely to accumulate some sludge and may not be amenable to protection.

A second approach involves dispersing the suspended solids into very tiny particles thereby preventing their agglomeration into sufficiently large particles that would readily settle out of the water. These small particles can be more easily carried through the system. Chemicals frequently used include surfactants and low molecular weight polymers. The choice of the best dispersant depends on the problem to be solved. Polymers can be tailor-made to optimize dispersant performance for specific foulants. This is especially true for foulants such as iron and manganese.

Most fouling problems in all types of cooling systems are complicated by microbial activity. Slime deposits on tubes not only interfere with efficient heat transfer, but act as a trap to enmesh suspended solids, further impeding heat transfer. In addition, by-products of bacterial metabolism influence water chemistry, including the tendency for scale to form or metal to corrode. Proper use of biocides and biodispersants can be a major step toward solving a once-through fouling problem.

Rarely do corrosion, scale, and fouling occur independently of one another. Usually two or all three develop together to cause loss of heat transfer and premature metal loss. For example, microbial fouling can cause scaling and corrosion to occur; corrosion can contribute to iron fouling and encourage more corrosion to occur. To break this cycle, proper problem identification is important for selecting and applying a practical, economical solution to any deposit problem.

Closed Recirculating Systems

A closed recirculating system is one in which the water is circulated in a closed loop with negligible evaporation or exposure to the atmosphere or other influences that would affect the chemistry of the water in the system. These systems usually require high chemical treatment levels, and, since water losses are negligible, these levels are economical. High-quality makeup water is generally used for best system operation. These systems are frequently employed for critical cooling applications, such as continuous casters in the steel industry where the slightest deposit from any source could cause equipment failure.

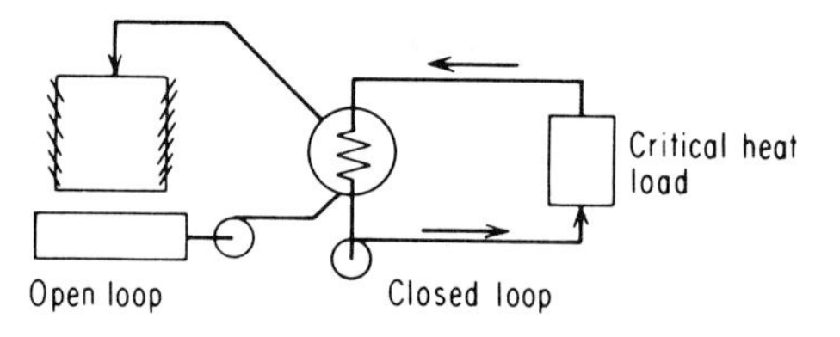

FIG. 38.5 For many critical applications of heat transfer, water in a closed loop is used for fail-safe chemical control, and this water is cooled by an open system.

Figure 38.5 shows a simplified closed recirculating system. Heat is transferred to the closed cooling water loop by typical heat exchange equipment and is removed from the closed system loop by a second exchange of heat from the closed loop to a secondary cooling water cycle. The secondary loop could use either evaporative or once-through water cooling, or air cooling.

Velocity of water in closed systems is generally in the 3 to 5 ft/s (0.9 to 1.5 m/s) range. Temperature rise usually averages 10 to 15°F (6 to 9°C), although some systems can exceed this substantially. Generally closed systems require little or no makeup water except for pump seal leaks, expansion tank overflows, and surface evaporation from system vents. This periodic makeup requires regular analysis for control of correct treatment chemical residuals.

Closed systems usually contain a combination of different metals which provide a high potential for galvanic corrosion. The potential for dissolved oxygen attack is generally quite low in closed systems because of the small amount of makeup water—the main oxygen source. However, in systems that require substantial makeup because of loss of water from leaks, oxygen is continually supplied and oxygen corrosion presents a serious problem. Oxygen can, at elevated temperatures or at points of high heat transfer, cause severe pitting corrosion.

Since relatively little makeup is added to most closed recirculating systems, it is practical and desirable to maintain the system in a corrosion-free condition. This is normally achieved by applying chromates, nitrite/nitrate-based inhibitors, or soluble oil-type treatments at rather high concentrations.

Theoretically, scale should be a minor problem in a closed system since the water is not concentrated by evaporation. In a tightly closed system, none of the common scale-forming constituents deposit on metal surfaces to interfere with heat transfer or encourage corrosion.

With high makeup rates, however, additional scale forms with each new increment of water added so that in time, scale becomes significant. In addition, there is opportunity for sludge, rust, and suspended solids to drop out at low flow points and bake on heat transfer surfaces to form a hard deposit. Therefore, scale retardants and dispersants are usually included as part of a closed system treatment program where makeup rates are high. Often soft water or condensate is used for makeup to closed systems depending on the characteristics of the system being protected.

Because water circulating through a closed system is not exposed to the atmosphere, fouling by airborne silt and sand is rare. However, fouling by microbial masses may occur in closed systems where makeup rate is significant or process leaks encourage bacterial growths. These are controlled with biological control agents formulated to be compatible with the chemical treatments and operating conditions found in closed systems.

It is desirable as a part of routine maintenance to flush closed water systems with high-pressure high-velocity water to remove accumulated debris if makeup rates are high.

Open Recirculating Systems

An open recirculating system incorporates a cooling tower or evaporation pond to dissipate the heat it removes from the process or product. An open recirculating system (Figure 38.6) takes water from a cooling tower basin or pond, passes it through process equipment requiring cooling, and then returns the water through the evaporation unit, where the water evaporated cools the water that remains. The open recirculating system repeats this process of reuse, taking in sufficient freshwater makeup to balance the water evaporated and that blown down from the system to control the chemical character of the recirculating water. This greatly reduces water demand (e.g., withdrawal from a river) and discharge, as shown by Figure 38.7.

The following definitions are used to explain the operation of an evaporative

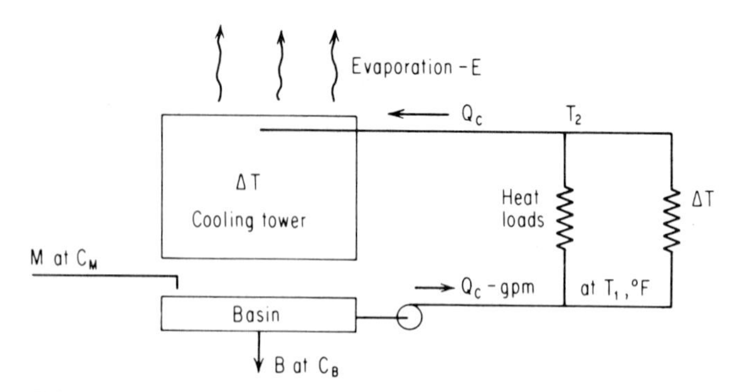

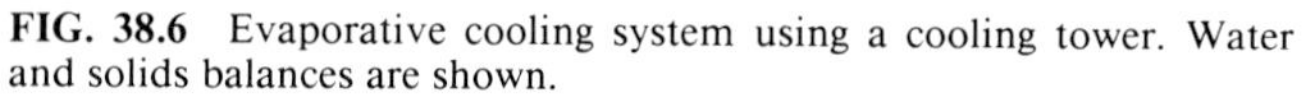

FIG. 38.6 Evaporative cooling system using a cooling tower. Water and solids balances are shown.

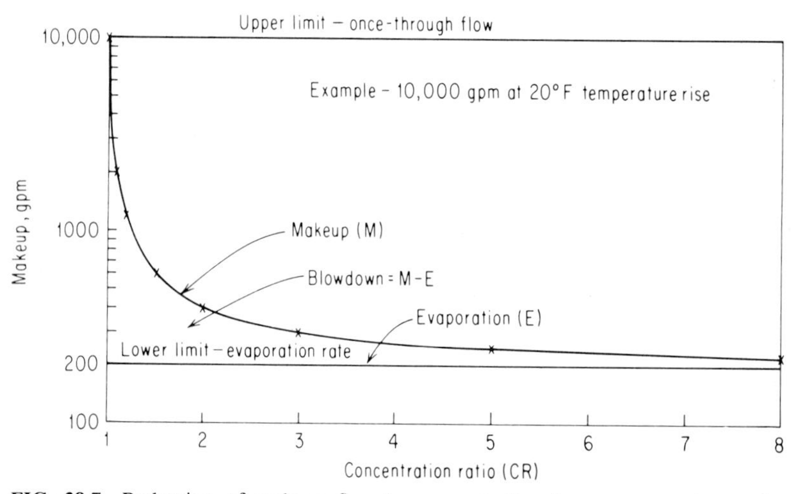

FIG. 38.7 Reduction of makeup flow by concentration in an evaporative cooling system.

system and to permit the plant operator to calculate performance. Figure 38.6 helps to clarify these definitions.

1. *Recirculation rate* (Q_c): This is the flow of cooling water being pumped through the entire plant cooling loop, usually cooling a number of exchangers. Q_c can usually be estimated from the recirculating pump nameplate data; however, actual measurements are more accurate. The actual recirculation is seldom more than the nameplate data and frequently may be 10 to 20% less. A pump curve, usually available from the manufacturer, plots recirculating flow against head; a pressure gauge on the pump discharge should provide a reasonably accurate estimate of flow.
2. *Temperature differential or range* (ΔT): This term refers to the difference between the average water temperature returning to the tower from the plant

exchangers (T_2) and the average water temperature following evaporation (T_1) (tower basin).

3. *Evaporation (E):* E is the water lost to the atmosphere in the cooling process [gallons per minute (m^3/min)]. The evaporation rate is dependent on the amount of water being cooled (Q_c) and the temperature differential, ΔT. As a rule of thumb, for each 10°F (5.6°C) temperature drop across the evaporation process, 1% of the recirculation rate (Q_c) is evaporated. Therefore, a 20°F (11.2°C) ΔT across a cooling tower produces an evaporation loss of 2% of the recirculation rate ($0.02\ Q_c = E$).

$$E = Q_c \times \frac{(T_2 - T_1)}{1000}$$

$$E = Q_c \times \frac{(T_2 - T_1)}{560} \qquad \text{(metric)}$$

The amount of evaporation that can take place over a given tower is limited primarily by the relative humidity of the air. Relative humidity is determined by measuring the wet and dry bulb temperatures of the air. E can be as low as 0.75% per 10°F of ΔT in high humidity areas like the Gulf Coast in the United States. Conversely, E may be as high as 1.2% per 10°F ΔT in regions of very low humidity, such as the arid regions of the southwestern United States. E depends to a lesser degree on the liquid-to-gas flow ratio, and conductive heat losses in other areas of cooling systems.

4. *Makeup (M):* The input of water required to replace the water lost by evaporation plus that being lost through blowdown, tower drift, and other miscellaneous losses. It is usually measured by a flow meter; if not it may be calculated as shown below:

$$M = E \times \left(\frac{\text{CR}}{\text{CR} - 1}\right)$$

5. *Concentration ratio (CR):* Makeup to a recirculating cooling water system contains dissolved impurities. The evaporating water produces pure H_2O vapor, leaving behind these impurities. The ratio of the concentrations of salts in the circulating water (C_B) to those of the makeup (C_M) is the concentration ratio.

$$\text{CR} = C_B/C_M$$

Since the input solids must equal the output solids,

$$M \times C_M = B \times C_B$$

where M is the makeup flow and B represents loss of concentrated water. Therefore, the concentration ratio is also

$$\text{CR} = M/B$$

The CR should be calculated for several individual components of the water to determine if the system is "in balance." In the ideal case, the system is in balance when the CRs of all ions in the water (Ca, Mg, alkalinity, etc.) are equal. If the concentration ratios are not equal, it can indicate that some mineral ($CaCO_3$, SiO_2, etc.) is precipitating from the recirculating water. For example, if the CR for calcium or alkalinity is more than 0.5 below the CR for magnesium, then $CaCO_3$ is probably precipitating in the cooling system. By knowing what may precipitate, the CR can be a valuable indicator that a problem is occurring.

The concentration ratios of some ions will be affected by chemicals added to the cooling system. The CR for SO_4^{2-} would be increased when sulfuric acid is added or where the plant atmosphere contains SO_2. In these cases, the CR for alkalinity would be decreased because alkalinity is destroyed by acid added to the tower. Chlorination of the cooling water will increase the CR for Cl^-. A CR based on conductivity will also be increased by some of the treatment chemicals added.

As shown in Figure 38.7, even a small degree of concentration enormously reduces water demand, and the greater the CR, the lower the demand as the evaporation rate is approached as a limit.

6. *Blowdown (B):* Since pure water vapor is discharged by evaporation, the dissolved and suspended solids left behind concentrate. If there were no water loss other than evaporation, these solids would concentrate to brine, causing massive scale and corrosion. To balance this, regulated flow is bled from the circulating system. This blowdown (B_R) is calculated and controlled to remove solids at the same rate at which they are introduced by the makeup.

   There are other uncontrolled losses from the system. One is drift (B_D); the other is leakage (B_L), sometimes deliberate, but usually accidental. These are included in the total blowdown calculation,

$$B = B_R + B_D + B_L$$

   Blowdown is related to other factors thus:

$$B = M - E$$

   and

$$B = M/\mathrm{CR}$$

7. *Drift (B_D):* Even though evaporating water is pure, some water droplets escape as mist through the evaporation equipment. In modern cooling towers, very intricate mist and drift eliminators may be added to reduce this droplet loss to about 0.0005% of the recirculation rate. A more usual drift loss in conventional cooling towers is in the range of 0.05 to 0.2% loss based on the recirculation rate. Since drift contains dissolved solids it is really a portion of the blowdown. In the absence of a controlled blowdown, as when the blowdown valve is deliberately closed, drift establishes the maximum concentration ratio in the absence of other system losses.

8. *System losses (B_L):* Circulating water may be lost in the plant through pump or valve leaks; by tap-off for once-through cooling of pump glands, compressor jackets, or bearings; or draw-off for such uses as equipment or floor area washup when the cooling water line happens to be close to where water is needed. In many plants, miscellaneous draw-off of recirculating cooling water is so great that it is impossible to build up the concentration ratio over 1.2 to 1.5. This severely limits the selection of an economical chemical treatment program and prevents effective conservation of water.

9. *Holding capacity of system (V):* Usually most of the water in a system is contained in the cooling tower basin or spray pond. An approximation of the holding capacity can be obtained by calculating the volume of water in the

basin and adding an extra 20 to 30% for the water contained in the lines and equipment. Additional increases may be required if the system has an unusually large number of open box condensers, jacketed vessels, or furnaces with substantial water holding capacity.

10. *Time/cycle (t):* One cycle is defined as the time required for water to make one trip around the circulating loop. This time is a function of the holding capacity and the recirculation rate.

$$t = V/Q_c$$

11. *Holding time index (HTI):* The holding time index is an expression of the half-life of a treatment chemical added to an evaporative cooling system. Mathematically, this index represents the time required to dilute an added chemical to 50% of its original concentration after the chemical addition is discontinued. It is also the time required to concentrate the makeup solids by a factor of 2. This is an important factor in setting control limits where chemical feed may be interrupted. It is also important for establishing an effective dosage for biological control agents, which are slug fed into the system. A dilution curve for a chemical slug fed into a cooling tower system is illustrated by Figure 38.8.

Calculation of Holding Time Index

The half-life of a system depends on the capacity and the rate at which the components are leaving the system. For a cooling tower, the half-life depends primarily on the system capacity and the blowdown rate. In its simplest form, the equation to calculate holding time index is

$$HTI = 0.693 \times \frac{\text{capacity (gal)}}{B\ \text{(gal/min)}}$$

where 0.693 = ln 2.0, a number derived from standard half-life equations.

To illustrate how the holding time index depends on other elements of a cooling system, the factors used to calculate blowdown (B) can be substituted:

$$HTI = 0.693 \times \frac{\text{capacity} \times (\text{CR} - 1)}{E}$$

or

$$HTI = 0.693 \times \frac{\text{capacity} \times (\text{CR} - 1)}{Q_c \times \Delta T \times 0.001}$$

This illustrates that a change in several factors (Q_c, ΔT, CR, or capacity) will all affect the *HTI*.

Another method of calculating holding time index is illustrated in the following example.

1. Calculate the time per cycle, t.
2. From Table 38.1 determine the number of cycles required to reach the prevailing concentration ratio, based on the temperature drop through the tower or spray pond.

3. Multiply this by the time per cycle to get the holding time index expressed in minutes, then divide by 60 or 1440 to convert to hours or days, respectively.

The following example illustrates this calculation:

1. *Recirculation rate:* Pump data show a recirculation rate of 5500 gal/min (21 m^3/min). Use 5000 gal/min (19 m^3/min) as a good estimate of actual recirculation.

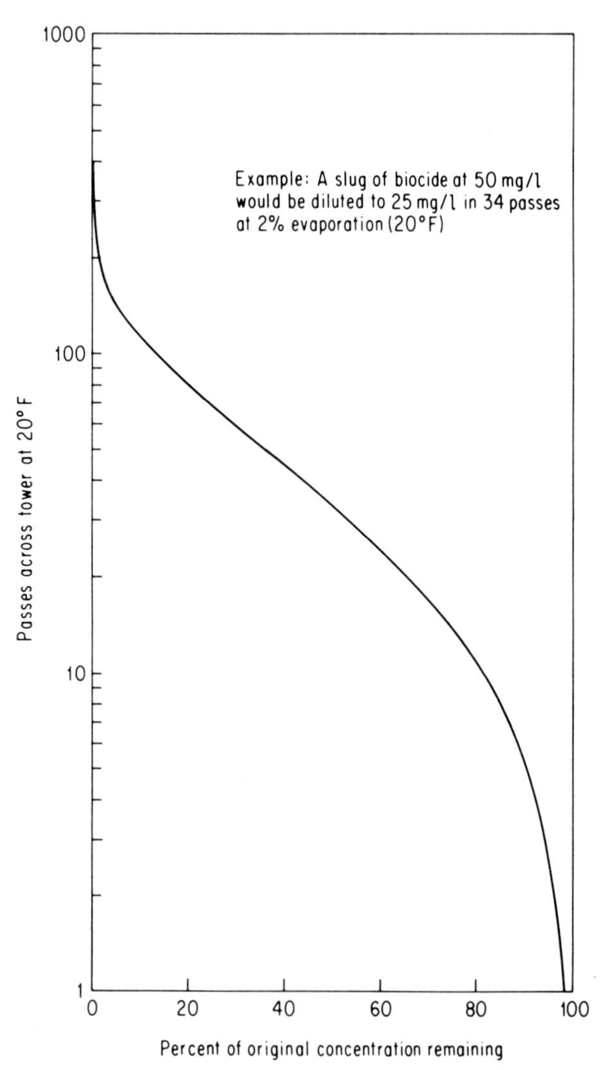

FIG. 38.8 The effect of time and makeup dilution on slug application of a chemical to an evaporative cooling water system.

TABLE 38.1 Cycles Required to Concentrate in the Absence of Blowdown

Concentration ratio	Temperature drop through tower		
	10°	20°	30°
1.1	6.5 cycles	3.2 cycles	2.1 cycles
1.3	20.2	10.2	7.0
1.5	34.5	17.2	11.7
1.8	55	28	19
2.0	69	34	23
2.5	105	52	35
3.0	138	70	47
3.5	174	87	58
4.0	208	104	70
4.5	241	120	81
5.0	324	162	93

2. *Temperature drop:*

$$105°\text{F} - 85°\text{F} = 20°\text{F}$$

$$(41°\text{C} - 30°\text{C} = 11°\text{C})$$

3. *Evaporation loss:* 20°F (11°C) is equivalent to 2% evaporation loss:

$$0.02 \times 5000 = 100 \text{ gal/min evaporation loss}$$

4. *Concentration ratio* (see analyses on Figure 38.9): Tests on the makeup and recirculating water show concentration ratios of 2.8 to 8.3.

The approximate concentration ratio is 3.0 based on magnesium and silica, since magnesium and silica remain soluble at the prevailing pH and concentration conditions. Chlorine and sulfuric acid are both being added to the system and thereby eliminate the use of Cl^- and SO_4^{2-} as valid CR indicators.

5. *Makeup:*

$$M = E \times \left(\frac{\text{CR}}{\text{CR} - 1}\right)$$

$$M = 100 \times \frac{3}{2} = 150 \text{ gal/min}$$

$$\left(M = 0.38 \times \frac{3}{2} = 0.57 \text{ m}^3\text{/min}\right)$$

6. *Holding capacity of the system:* Basin contains 72,000 gal (284 m³). The total holding capacity of the system is estimated to be 100,000 gal (379 m³).

Identification of Analyses Tabulated Below:

A. Raw water
B. Recirculated cooling water
C. Concentration ratios
D.
E.
F.

Constituent	As	A	B	C	D	E	F
Calcium	$CaCO_3$	182	526	2.8			
Magnesium	"	78	240	3.1			
Sodium (by difference)	"	40	120				
Total Electrolyte	$CaCO_3$	305	886	2.9			
Bicarbonate	$CaCO_3$	190	80				
Carbonate	"	0	0				
Hydroxyl	"	0	0				
Sulfate	"	84	696	8.3*			
Chloride	"	29	94	3.2*			
Nitrate	"	2	6				
Chromate		0	10	--			
M Alk.	$CaCO_3$	190	80				
P Alk.	"	TR	0				
Carbon Dioxide		TR					
pH		8.3	7.5				
Silica		22	68	3.1			
Iron	Fe	0.3					
Turbidity		10					
TDS		360	ND				
Color		Nil					

* Not valid, since Cl_2 and H_2SO_4 are being fed.

FIG. 38.9 Concentration ratios in an evaporative cooling system.

7. *Time/cycle:*

$$t = \frac{100{,}000}{5000} = 20 \text{ min}$$

$$\left(t = \frac{379 \text{ m}^3}{19 \text{ m}^3/\text{min}} = 20 \text{ min} \right)$$

8. *Holding time index:* If CR = 3, and ΔT = 20°F (11°C), Table 38.1 shows average cycles at 70; at 70 cycles,

$$\text{HTI} = 70 \times 20 = 1400 \text{ min, or } 23 \text{ h}$$

COOLING TOWERS

Cooling towers are designed to evaporate water by intimate contact of water with air. Cooling towers are classified by the method used to induce air flow (natural or mechanical draft) and by the direction of air flow (either counterflow or crossflow relative to the downward flow of water).

In natural draft towers, air flow depends on the surrounding atmosphere, which establishes the difference in densities between the warmer air inside the tower and the external atmosphere; wind velocity also affects performance. Most natural draft towers in modern utility service are of hyperbolic design (Figure 38.10), which has been used for many years in European installations. These tall towers provide cooling without fan power, and they also minimize plume problems and drift.

FIG. 38.10 Hyperbolic towers cooling condenser water in a utility station. *(Courtesy of The Marley Company.)*

Mechanical draft cooling towers use fans to move air instead of depending on natural draft or wind. This speeds the cooling process and increases the efficiency of the tower by increasing the air velocity over droplets of water falling through the tower. Mechanical towers can, therefore, evaporate much more water than natural draft towers of the same size.

There are two designs of mechanical draft towers, forced and induced draft. In forced draft towers (Figure 38.11) fans mounted on the side of the tower force air through the tower packing, producing intimate mixing of air with the falling water.

Induced draft cooling towers (Figure 38.12) are either counterflow or crossflow with fans on top pulling cooling air up through or horizontally across the falling water. The choice between forced draft and induced draft is based on engineering considerations that take prevailing weather patterns into account. A major consideration is to avoid recirculation of the warm air discharge, which would greatly

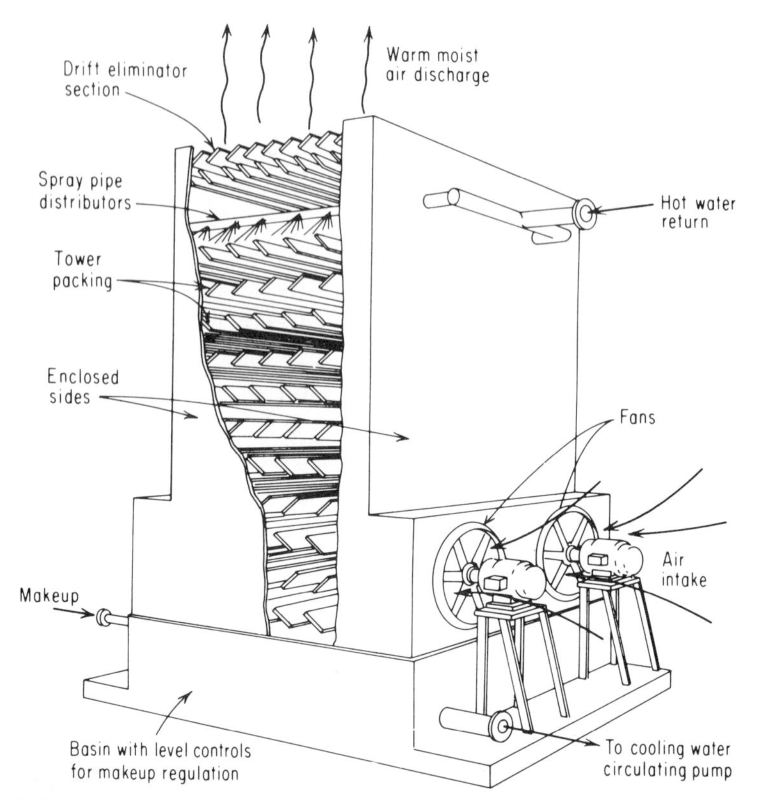

FIG. 38.11 Forced-draft tower design. This design was widely used prior to development of the induced-draft design. Fans and motors are conveniently located for maintenance.

FIG. 38.12 Induced-draft tower. Air enters the tower louvers and is well distributed before exiting at top.

reduce tower performance. The main advantage of a counterflow tower is that the coldest water contacts the driest air, providing the most efficient evaporation sequence. A complete survey is shown in Figure 38.14.

COOLING WATER TREATMENT AND CONTROL

Every cooling water system presents a unique combination of equipment, water chemistry, contaminants, blowdown, and control considerations. Proper selection of a sound cooling water treatment program requires collecting a considerable amount of information. This is often a painstaking task because of the complexity of the mechanical equipment involved and the variations encountered in operating conditions. Figure 38.13 and Table 38.2 show an example of a system sur-

1. COOLING SYSTEM DATA

METALS	Tubes	CS, Adm, SS
	Tube Sheets	CS, SS
	Lines	CS
	Coils	CS
VELOCITY Ft/Sec	Min	2.5
	Max	7 +
	Avg	3.0
MAX. TEMP °F.	Process Side	510°F
	Water Side	120°F
NUMBER OF EXCHANGERS	Tube & Shell	35
	Barometric	
	Annular Space	
	Open Box	
	Other	
No Exch with Water Shell Side		6
TYPE TOWER	Deck Open or Closed	Open
	Nat or Mech Draft	Mech.
Side Stream Filter Yes/No		No
Recirc. Rate GPM		46,000
Holding Capacity Gal		600,000
TOWER TEMPS	Max Return °F	115°F
	Max Supply °F	85°F
	ΔT °F	30°F
Avg No Cycles		5.7
Holding Time Index in Hours		28.9
Bleed-Off GPM		359
Make-Up GPM		1436

2. COOLING WATER CHEMISTRY

PPM	MAKE UP	RECIRC WATER
TDS	330	1680
SS (circled) or JTU	22	86
H as $CaCO_3$	182	960
HCa as $CaCO_3$	102	590
Mg as $CaCO_3$	80	370
P as $CaCO_3$	0	0
M as $CaCO_3$	130	50
Cl as NaCl	116	620
SO_4 as Na_2SO_4		
Total PO_4	0.4	2.0
Ortho PO_4	0.4	2.0
SiO_2	2-10	10-50
Total Fe	0.1	0.6
Sol Fe	0.1	0.4
Total Al_2O_3	0.4	2.0
Sol Al_2O_3	0.4	1.8
pH	7.3	7.1

MAKE UP WATER PRETREATMENT
Include: Source, Pretreatment Method, All Chemicals, Results
Des Plaines River with traveling screens

CONTAMINATION OF RECIRCULATING WATER
Include: Source, Type, Amount, and Frequency
Hydrocarbons, H_2S (periodic leaks)

3. HEAT TRANSFER DATA

MONITORING	PLANT DATA	Include: Temperature, Pressures, Vacuum, Cleanliness Factors, heat transfer coefficients, etc. Temperature, pressure
	TOOLS USED	Include: Methods, Location of Tools (Temperature, Pressure, Velocity), Typical Results, etc.
CONTROL	METHOD	Include: Throttling of Cooling Water, Throttling of Process, By Passes – Used to Control Temperature or Heat Transfer
PRESENT	CONDITIONS	Include: Condition of Heat Transfer Equipment, Amount, Type and Location of Deposits, Condition of Tower, Etc. Fouling present in heat exchangers. Moderate tuberculation in last passes of exchangers and water boxes.
CLEANING	PROCEDURE	Include: On-Stream and Turnaround Cleaning Methods, and Frequency Annual turnaround. Refinery has been replacing admiralty bundles with mild steel due to process side corrosion.

4. PLANT EFFLUENT CONSIDERATIONS

Where is Bleed-Off Discharged	To API separator and waste
Present Waste Treatment	treatment plant.
CrO_4, Zn, PO_4 Removal Capabilities of present waste water facilities	Combining blowdown with other effluent streams reduces chromate
Plant Eff. Limits on CrO_4, Zn, PO_4	No limits

5. PRESENT TREATMENT AND CONTROL METHODS

pH		CORROSION AND SCALE INHIBITOR		DISPERSANT		BIOCIDE		CHLORINE		CONTROLS		
Type Acid & Strength	66°	Product		Product		Product	None	Form: Gas, Liq, Pulv	Gas	Cycles of Concentration		
Slug/Cont. or Auto	H_2SO_4	Slug/Cont or Auto	Cont.	Slug/Cont or Auto	Slug	Slug/Cont. or Auto		Slug/Cont or Auto	Slug	Actual No. Desired No.	5.7	6
Desired pH Range	6-7	Desired ppm Range	40-45	Desired ppm Range	50 ppm	Frequency		Frequency	m/w/f		Makeup	Bleed
Actual pH Range	5-8	Actual ppm Range	20-50	Actual ppm Range	3 hrs	Residual Time		Residual Time	1 ppm	Metered Yes/No	Yes	Yes
Lbs/Day	Slug	Lbs/Day	255	Lbs/Day	104	Lbs/Appli		Lbs/Day	for 4 hours	Controled Yes/No	Yes	Yes

FIG. 38.13 A condensed summary of data required in making a cooling system survey. A complete survey is much more extensive (Figure 38.14) and includes flow diagrams and tabulation of individual heat exchanger data (Table 38.2).

TABLE 38.2 Performance Data for Individual Heat Exchangers Tabulated on General Survey (Figure 38.13)

Exc. number	Nomenclature FCC	General: Shell side or tube side cooling water	General: Heat flux, Btu, $ft^2 \cdot h$	General: Number of passes, ft/s	Cooling water: Velocity	Cooling water: Max exit temp, °F	Cooling water: ΔT across exc.	Cooling water: Pressure, lb/in^2	Process side: Product nomenclature	Process side: Max. inlet temp, °F	Process side: Design outlet temp, °F	Process side: Pressure, lb/in^2	Materials of construction: Tubes	Materials of construction: Channel	Materials of construction: Head	Materials of construction: Sheet
307	Compressor inlet stage cooler	Tube	9,475	1	3	100	5	35	Hydrocarbon vapor	262	154	70	Adm	CS	CS	Adm
309	Compressor seal oil cooler	Tube	6,000	2	5.5	120	10	35	Oil	245	159	25	Adm	CS	CS	Adm
314	Compressor seal oil cooler	Tube	3,780	2	6.5	120	8	40	Oil	245	159	25	SS	SS	SS	SS
316*	Bearing lube oil cooler	Shell	4,600	2	4.5	120	10	35	Oil	230	159	30	SS	SS	SS	SS
322	Main overhead trim cooler	Tube	12,000	4	3.0	135	25	30	Hydrocarbon vapor	510	154	120	CS	CS	CS	CS
330	Main air blower lube system	Tube	8,750	2	6.5	118	12	40	Oil	225	154	40	CS	CS	CS	CS
333	Main air blower turbine exhaust	Shell	10,250	1	2.5	125	10	40	Air	320	115	65	Adm	CS	CS	CS
337	Main air blower turbine gland cooler	Shell	8,000	1	5.0	115	5	40	Air	200	110	40	CS	CS	CS	CS

* Information was taken from the most critical exchanger in a series of similar exchangers.

vey, divided into five major categories: (1) cooling system data, (2) cooling water chemistry, (3) heat transfer data, (4) effluent considerations, and (5) present treatment and control methods.

Cooling System Data

This section identifies physical aspects of the system such as number and type of heat exchangers, materials of construction of exchangers and piping, type of tower, maximum temperatures of the water and process, tower operating temperatures, and system characteristics such as velocities, makeup, bleed-off, and holding time index. This section should be supplemented with the process flow and water flow diagram as well as complete specifications on critical heat exchangers.

Cooling Water Chemistry

This section identifies the chemical environment of the system. The information is divided into the chemistry of the makeup and recirculating water, including a

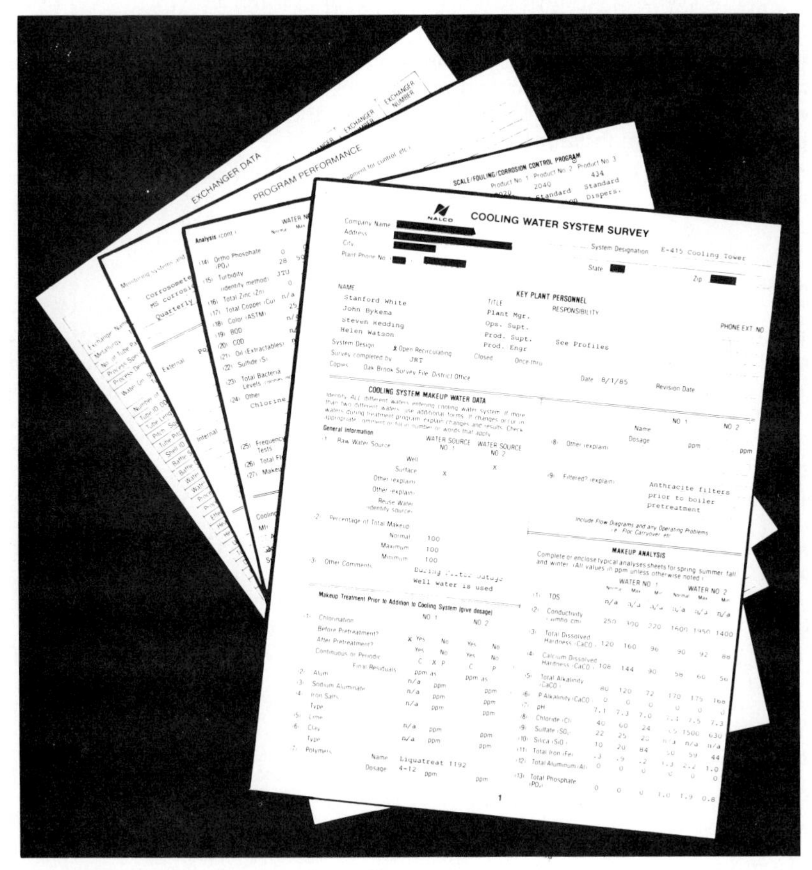

FIG. 38.14 Complete survey of a cooling water system.

description of pretreatment processes and sources and types of contamination of the recirculating water. Causes of poor makeup water quality and persistent sources of contamination should be examined; these are critical to the treatment program, so the possibility of correction is a decisive factor in program selection.

Heat Transfer Data

This survey section is organized into four parts.

1. *Results monitoring:* Defines how heat transfer is evaluated, including the use of corrosion coupons and test heat exchangers; data on plant heat exchangers, permitting calculation and monitoring of heat transfer rates.
2. *Control methods:* Indicates how heat transfer is controlled. For example, a common method in many plants is to throttle cooling water entering certain heat exchangers in the winter to prevent overcooling of the process. However, throttling reduces velocities and promotes fouling which leads to a loss of heat transfer that cannot always be recovered by reopening the throttled valve. Alternate control methods, such as water recycle or process stream bypass, should be considered in preference to throttling water flow.
3. *Present conditions:* Defines the physical conditions of the heat exchange equipment inspected during the survey, supplemented by the analysis of significant deposits. This information provides a basis for recommendations for cleaning, preconditioning of metal surfaces, and application of chemicals for proper system maintenance.
4. *Cleaning procedure:* This includes mechanical and chemical cleaning procedures currently employed.

Plant Effluent Considerations

Some cooling systems are bled off directly into a receiving stream; others are discharged to various kinds of waste treatment processes; and some discharge to municipal sewage systems. Each imposes considerations on the choice and application of a chemical treatment program.

Control Monitoring and Followup

Chemical control, monitoring of results, and corrective action are required for an effective cooling water treatment program. A wide variety of analytical tools and monitoring devices are available to aid in developing and maintaining a chemical program that will provide an efficient operation.

The goal of analysis and monitoring is to identify potential problems before they occur. The major diagnostic tools include:

1. Water analyses (on-site and laboratory)
2. Deposit analyses (organic, inorganic, and microbial)
3. Corrosion and deposition monitoring devices
4. Metallographic analyses
5. Microbial analyses

TABLE 38.3 Important Cooling Water Variables

Variable	Effects
Ca, Mg	Define scaling tendency of water
M, pH, T	Define concentrations of carbonate and bicarbonate, and solubility of calcium carbonate
SO_4, SiO_2	Must be controlled to prevent sulfate and silicate scales
Suspended solids	Cause fouling, require dispersants
Contaminants: Hydrocarbons, glycols, NH_3, SO_2, H_2S	Cause fouling and microbial growth, high chlorine demand, precipitate chemical treatments

Table 38.3 lists some of the more important variables that must be controlled in cooling systems. Calcium and magnesium hardness define the scaling tendency of the water. Total alkalinity, pH, and temperature define the concentrations of carbonate and bicarbonate ions in the water, and also the solubility of calcium carbonate. All of these must be controlled within acceptable ranges for each system to ensure scale-free operation. Concentrations of sulfate and silica must also be controlled at reasonable levels to prevent formation of gypsum and silica scale.

Many systems contain suspended solids which concentrate in the tower and cause fouling. Dispersants may be used to control this once the problem has been properly defined. Some of the solids may come from the makeup, some may be washed from the air, and some may be precipitation products or microbial masses. If the suspended solids are excessive, as evidenced by plugged tubes, a filter system must be added to the circuit to filter a portion of the circulating water, approximately equal to the evaporation rate, Figure 38.15. This sidestream filter must be designed to avoid excessive backwash, as this represents

FIG. 38.15 This 15-ft-diameter automatic valveless gravity unit filters a cooling tower sidestream at a Texas natural gas plant.

TABLE 38.4 Chemical Components of Cooling Water Treatments

	Problems			
Chemical treatments	Corrosion	Scaling	Fouling	Microbes
Chromates	x			
Zinc	x			
Molybates	x			
Silicates	x			
Polyphosphates	x	x		
Polyol esters		x		
Phosphonates	x	x		
All-organics	x	x	x	
Natural organics			x	
Synthetic polymers		x	x	
Nonoxidizing biocides			x	x
Chlorine/bromine				x
Ozone				x

uncontrolled blowdown loss. These filters can also become incubators for microbial growths if the biocidal treatments are not assiduously applied.

Table 38.4 lists some of the major chemical components available for cooling water treatment. In using this chart note that no one of these components is useful by itself. For example, good corrosion control is difficult or impossible in a dirty, scaled system. Good scale and fouling control to maintain clean surfaces minimizes the dosage of corrosion inhibitors. Many chemicals help to solve more than one problem, as shown in Table 38.4. Whether a particular chemical is the best choice for a given system depends on the specific conditions in that system.

AMBIENT AIR EFFECTS

Cooling towers scrub the air passing through them to provide the evaporative conditions, handling about 200 ft^3 of air per gallon of water (about 1500 m^3 air per cubic meter of water). It is not surprising, then, that the atmospheric environment has profound effects on the performance of the cooling system.

In some areas, the air contains a large mass of dust, as in arid sections of the country, especially where dust storms are common. The western and southwestern areas of the United States are prone to this problem, and cooling systems there cannot work effectively without side-stream filtration. In a complex industrial plant, solids may become airborne from dirt on roads, open areas between plant buildings, or from open storage of solids (e.g., ore or coal), and this self-generated source of particulates is as damaging as silt, also requiring side-stream filtration.

A more subtle, difficult problem is the presence of acidic or alkaline gases in the atmosphere. These gases affect the pH of the system, a critical control factor in the chemical treatment program that has a direct bearing on the scale-forming or corrosive tendencies of the water. An unusual, but pertinent, example is an ammonia plant cooling tower located between the ammonia process and the nitric acid process. When the wind came from one direction, ammonia in the atmosphere raised the pH of the system water; when the wind came from the opposite direction, acidic nitrogen-oxide gases reduced the pH. Although this is an exaggerated case, it makes the point that in most plants, the pH of the system may be affected by wind direction.

The most prominent of the atmospheric gases are acidic, and chief among them is carbon dioxide, which occurs at an average concentration of about 0.03% by volume in the atmosphere. The amount of CO_2 supported in water at about 68°F (20°C) by this partial pressure is less than 1 mg/L. However, the actual CO_2 level in a cooling tower system varies considerably from one plant to another because of local atmospheric conditions, such as the presence of industrial stack gas discharges.

There is a definite relationship between CO_2, alkalinity, and pH. (See Chapter 4.) If CO_2 is variable, alkalinity must be varied to maintain a pH control point. This makes it difficult to predict the alkalinity concentration that will be required to achieve a specific control pH value. It must be done empirically.

If experience shows that the atmospheric CO_2 concentration is constant, then the correct alkalinity for a selected pH can be chosen. The pH will then vary as a logarithmic function: i.e., if the alkalinity doubles, the pH will increase by the log of 2, or a value of 0.3; similarly, if the atmospheric CO_2 doubles with a fixed alkalinity, the pH will be reduced by the log of 2, or 0.3 pH unit. The expected equilibrium pH of any new system can only be established empirically, unless there is a record of previous experience in the vicinity of the new cooling tower or strong evidence that the average atmospheric CO_2 concentration of 0.03% applies.

CORROSION AND SCALE CONTROL

Corrosion in recirculating cooling water systems is controlled by employing either inorganic or organic inhibitors. The four major inorganic inhibitors are chromate, zinc, orthophosphate, and polyphosphate. Minor supplements include molybdate, nitrite, nitrate, various organic nitrogen compounds, silicate, and natural organics.

The earliest chemicals for treating recirculating cooling waters were inorganic polyphosphates and natural organic materials. The concept was to add a small amount of acid to control the stability index to a slightly scale-forming value. Organic corrosion inhibitors include organic phosphorus compounds, specific synthetic polymers, organic nitrogen compounds, and long-chain carboxylic acids.

Polyphosphate and natural organic materials were added to the program to provide both corrosion protection and scale inhibition. The scale inhibition stemmed from the use of the polyphosphate as a threshold treatment. In addition, the polyphosphate combined with calcium to form a cathodic inhibitor that reduced the corrosion rate. The natural organic material tended to keep the metal surface relatively clean and aid the inhibitor in establishing a protective film. It also dispersed suspended solids, and modified calcium carbonate and tricalcium phosphate precipitates if they tended to develop on hot surfaces.

The greatest disadvantage of this treatment approach is the reversion of polyphosphate to orthosphosphate, which can combine with calcium to form calcium phosphate scale. For this reason, this type of program has evolved into the stabilized phosphate program. In this treatment, both ortho- and polyphosphate are used as corrosion inhibitors. To prevent calcium phosphate deposition, the pH is generally controlled at 7.0 and specific synthetic polymers are added to disperse and stabilize calcium phosphate.

The next cooling water treatment was chromate, an exceptionally reliable corrosion inhibitor. Initially, chromate was applied at very high dosages, frequently in the range of 200 to 300 mg/L as CrO_4. Acid was added to the system to lower the pH to between 6 and 7, preventing calcium carbonate from precipitating. This treatment was quite effective in both scale inhibition and corrosion protection, but one shortcoming was that pitting attack tended to occur if the chromate residual became low. It was found that if chromate were combined with other inhibitors, particularly cathodic types (e.g., zinc and polyphosphate), the chromate level could be reduced to 20 to 30 mg/L CrO_4 with better results than obtained at 200 to 300 mg/L CrO_4 used alone. The synergized chromate approach also employed acid, frequently controlling the pH to 6 to 7. An additional advantage of synergized chromate was the margin of safety provided against pitting attack should the chromate be momentarily underfed.

These synergized chromate formulations are still considered among the best corrosion inhibitors in use today. However, increasing environmental pressures are forcing the development of innovative synergized chromate formulations that permit carrying chromate levels in a recirculating system substantially below 10 mg/L CrO_4 while continuing to provide acceptable corrosion protection. To achieve results with this approach the system pH must be controlled precisely, and dispersants and biocides used to keep the system clean. An obvious limitation to this approach is that the reservoir of protection available with the higher CrO_4 levels does not exist. Therefore, process contamination, uncontrolled microbial activity, fouling, and deposition will disrupt the system much more quickly than at the more traditional 20 to 30 mg/L CrO_4 levels.

Although chromate has done an outstanding job for years, increasing environmental concerns have brought pressure on research into new corrosion inhibitors with potentially less environmental impact. An early result of such research was the development of organozinc combinations. Since zinc, a cathodic inhibitor, has a lower film strength than chromate, the pH of the system for an organozinc program was increased to between 7 and 8 to make the water less corrosive, allowing the zinc to form a satisfactory inhibitor barrier. The organic portion of the treatment was a dispersant to keep the system free of deposits, thereby encouraging formation of an adequate zinc film. In addition to dispersancy, certain types of organics increased zinc solubility at the higher pH required for this method of treatment. These programs were adequate in many industrial plants, but because the inhibitor film at the operating pH was not as effective as a chromate film, these programs did not substantially replace traditional chromate-type treatments.

Subsequently, an innovative concept in cooling water chemistry arrived with the introduction of organophosphorus compounds. Like inorganic polyphosphates, these prevent scale formation by the threshold effect. However, there the similarity ends; inorganic polyphosphates easily revert to orthophosphates, with increasing holding time, temperature, and microbiological attack. Organophosphorus compounds do not revert under normal cooling tower conditions except under severe microbiological attack. Further, unlike the inorganic polyphosphates, the organophosphorus compounds are generally able to inhibit precipitation of calcium carbonate and other scale-forming species at a higher pH and alkalinity than tolerated by the inorganic polyphosphates. This development opened the door to what is now known as the alkaline approach to treating cooling water systems.

The basic treatment concept is to raise the pH of the operating system to 7.5 to 9.0, thereby substantially reducing the natural corrosivity of the recirculating water. Experience has shown that although the higher pH provides a less corrosive water, frequently this reduction is not of sufficient magnitude to protect all mild steel systems, especially mild steel heat exchangers with high heat flux or low flow velocities. Thus a specific all-organic inhibitor package is required to control corrosion and scale. In general, all-organic inhibitors combine organic phosphorus compounds, synthetic polymers, and aromatic azoles. These combinations provide corrosion control for steel and copper alloys, scale control, and deposit control.

Another approach to alkaline treatment involves the use of modern scale and deposit control agents along with more traditional corrosion inhibitors. Organic phosphorus compounds and polymers can be supplemented with inorganics like chromate or zinc. These programs can provide the performance of an all-organic program at a lower cost, where chromate or zinc can be used.

The significant advantage provided by alkaline operation over earlier treatments is the buffer capacity provided by the water that reduces the impact of system upsets on performance. Another particular advantage of the alkaline concept of treatment is the substantial reduction or occasional elimination of acid feed. This, of course, depends on the chemistry of the system.

FOULING CONTROL

Deposit control in cooling water systems is absolutely essential for maintenance of heat transfer rates. However, control of deposits is often more difficult in alkaline systems than in lower pH systems. The makeup water may contain dissolved solids, organic matter, and suspended solids, any of which can contribute to fouling. A system may become grossly contaminated with microbes; for example, makeup water with a high BOD, such as a recycled municipal or industrial effluent, is particularly susceptible to fouling from slime-forming bacteria.

Table 38.5 shows some sources of foulants in a typical recirculating system. The raw water and air inoculate a system with colloidal organic matter, silt, soluble iron, and microbes. Hydrogen sulfide, sulfur dioxide, and ammonia may enter from the plant atmosphere.

The selection of the proper dispersant for any operating system is based on actual analysis of a deposit. Synthetic organics, including polymers and surface-active agents, are generally applied for dispersing microbial and organic deposits.

TABLE 38.5 Sources of Fouling Deposits

Raw water	Airborne	Recirculating water
Colloidal organics	Dirt	Scale: $CaCO_3$, $CaSO_4$, $MgSiO_3$
Silt, dirt	Reactive gases—H_2S, SO_2, NH_3	Corrosion products: Fe_2O_3
Soluble iron		Process leaks—hydrocarbons, sulfides
Microbial contamination	Microbial contamination	Microbial deposits

Synthetic polymers such as polyacrylates or polyacrylamides are dispersants for silt, sand, iron, and other inorganic deposits. These polymers can be tailor-made by varying the components and molecular weights to maximize dispersant performance on specific foulants. Organophosphorus compounds, including polyol esters and phosphonates, are inhibitors for calcium carbonate and calcium sulfate precipitates. However, once deposits form, any scale removing action by these dispersants takes place slowly, so the best approach is to prevent the scale from forming in the first place.

MICROBIAL CONTROL

Microbial deposits present a special case of fouling. Treatment often requires biocides to kill microbe colonies and dispersants to loosen and wash them away. The most common biocide employed in all systems is chlorine. In general, chlorine is the only biocide required in most systems. If applied continuously at a residual of 0.2 to 0.4 ppm it will provide effective control at all cooling water pH values. At alkaline pH, the continuous presence of chlorine species in the water will provide the required microbial killing power because of the infinite contact time available. In intermittent chlorination, such as utility cooling systems, the chlorine contacts the microbial organisms for short periods of time. In this case pH can be more important. Sterilization studies have shown that chlorine kills faster at pH 7 than above pH 8. This may be due to the greater amount of HOCl present in the hypochlorite equilibrium at pH 7. Thus slug chlorination may be more effective at neutral pH because HOCl has a faster killing power than OCl^-.

There are problems associated with the use of chlorine. It can react with some organic materials, particularly phenolic compounds, to form reaction products that are nonbiodegradable or refractory, presenting potential effluent problems. Generally speaking, chlorine can be applied to most recirculating systems without danger of tower lumber delignification if free chlorine residuals do not exceed 1 mg/L. It is seldom necessary to continually carry a free chlorine residual over 0.2 to 0.3 mg/L to control microbial growths in most systems. Bromine is often a more practical treatment than chlorine because it remains effective at higher pH values and avoids formation of the kinds of halogenated by-products resulting from chlorination.

Although chlorine and bromine are excellent killing agents, their performance can be significantly improved by the use of biodispersants. Biodispersants aid the toxicant by breaking loose the biofilms and enabling them to contact more microbial organisms. In cases of gross contamination or loss of toxicant feed, a contingency nonoxidizing biocide may be required (See Chapter 22).

CHAPTER 39
BOILER WATER TREATMENT

Of the many uses for energy in the United States today—in industry, in transportation, in homes and commercial buildings—the largest portion of total use is directed toward producing steam through the combustion of fossil fuels. Utilities account for the greatest share of this, but industrial plants also produce enormous quantities of steam for process uses, often generating electric power through turbines as a by-product (Cogeneration).

The treatment of water for steam generation is one of the most sophisticated branches of water chemistry. An understanding of the fundamentals of boiler water chemistry is essential to the power engineer who continually strives to increase the efficiency of the boilers and steam-using equipment.

The pressure and design of a boiler determine the quality of water it requires for steam generation. Municipal or plant water of good quality for domestic use is seldom good enough for boiler feed water. These sources of makeup are nearly always treated to reduce contaminants to acceptable levels; in addition, corrective chemicals are added to the treated water to counteract any adverse effects of the remaining trace contaminants. The sequence of treatment depends on the type and concentration of contaminants found in the water supply and the desired quality of the finished water to avoid the three major boiler system problems—deposits, corrosion, and carryover.

DEPOSITS

Deposits, particularly scale, can form on any water-washed equipment surface—especially on boiler tubes—as the equilibrium conditions in the water contacting these surfaces are upset by an external force, such as heat. Each contaminant has an established solubility in water and will precipitate when it has been exceeded. If the water is in contact with a hot surface and the solubility of the contaminant is lower at higher temperatures, the precipitate will form on the surface, causing scale. The most common components of boiler deposits are calcium phosphate, calcium carbonate (in low-pressure boilers), magnesium hydroxide, magnesium silicate, various forms of iron oxide, silica adsorbed on the previously mentioned precipitates, and alumina (see Table 39.1). If phosphate salts are used to treat the boiler water, calcium will preferentially precipitate as the phosphate before precipitating as the carbonate, and calcium phosphate becomes the most prominent feature of the deposit.

At the high temperatures found in a boiler, deposits are a serious problem, causing poor heat transfer and a potential for boiler tube failure. In low-pressure

TABLE 39.1 Expected Composition of Boiler Sludge

Constituent	Coagulation-type treatment	PO_4 residual treatment
Calcium carbonate	High	Usually less than 5%
Calcium phosphate	Usually less than 15%	High
Calcium silicate	Usually less than 3%	Trace or none
Calcium sulfate	None	None
Calcium hydroxide	None	None
Loss on ignition	Usually less than 5%	Usually 8–12% except higher in very pure feed waters
Magnesium phosphate	None	Usually less than 5% except in some high-pressure boilers
Magnesium hydroxide	Moderate	Moderate
Magnesium silicate	Moderate	Moderate
Silica	Usually less than 10%	Usually less than 10%
Alumina	Less than 10%	Usually less than 10%
Oil	None	None
Iron oxide	Usually less than 5%	Usually less than 5% except in high-purity feed waters
Sodium salts	Usually less than 1.5%	Usually less than 1.5%
Copper	Trace	Usually low
Other metals	Trace	Low

boilers with low heat transfer rates, deposits may build up to a point where they completely occlude the boiler tube.

In modern intermediate and higher pressure boilers with heat transfer rates in excess of 200,000 Btu/ft^2/h (5000 cal/m^2/hr), the presence of even extremely thin deposits will cause a serious elevation in the temperature of tube metal. The deposit coating retards the flow of heat from the furnace gases into the boiler water. This heat resistance results in a rapid rise in metal temperature to the point at which failure can occur. The action that takes place in the blistering of a tube by deposit buildup is illustrated by Figure 39.1. For simplification, no tempera-

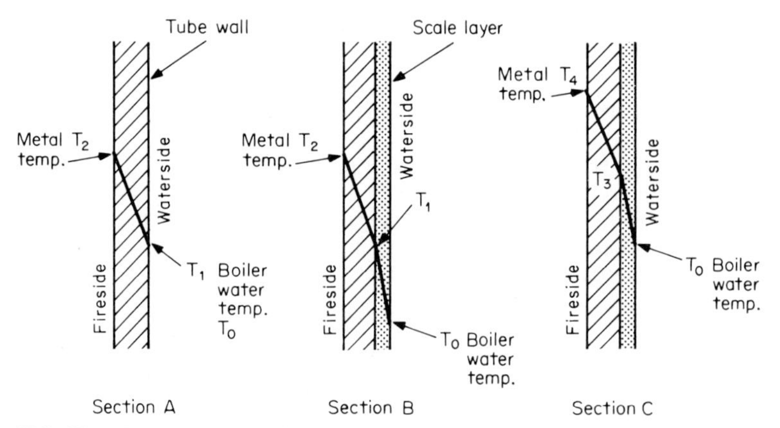

FIG. 39.1 Temperature profile across clean tube and tube having a water-side deposit.

ture drops through gas or water films have been shown. Section A shows a cross section of the tube metal with a completely deposit-free heating surface. There is a temperature drop across the tube metal from the outside metal (T_2) to the metal in contact with boiler water (T_1). Section B illustrates this same tube after the development of a heat-insulating deposit layer. In addition to the temperature drop from T_2 to T_1, there would be an additional temperature drop through the deposit layer from T_1 to T_0. This condition would, of course, result in a lower boiler water temperature T_0. However, boiler water temperature is fixed by the operating pressure, and operating conditions require that the same boiler water temperature be maintained as before the development of the deposit layer. Section C illustrates the condition that actually develops. Starting at the base boiler water temperature of T_0, the increase through the scale layer is represented by the line from T_0 to T_3. The further temperature increase through the tube wall is represented by the line from T_3 to T_4. The outside metal temperature T_4 is now considerably higher than the temperature T_2, which was the outside metal temperature prior to the formation of deposit on the tube surfaces. If continued deposition takes place, increasing the thickness of the heat-insulating deposits, further increases will take place in the tube metal temperature until the safe maximum temperature of the tube metal is exceeded. Usually this maximum temperature is 900 to 1000°F (480 to 540°C). At higher heat transfer rates, and in high-pressure boilers, the problem is more severe: at temperatures in the 900 to 1350°F (482 to 732°C) range, carbon steel begins to deteriorate. Figure 39.2 shows the normal structure of carbon steel boiler tubes, and Figure 39.3 illustrates the spheroidization of carbon and successive changes in structure, which begin to take place above 800°F (427°C), weakening the metal. Temperatures within the boiler furnace are considerably above this critical temperature range. Water circulating

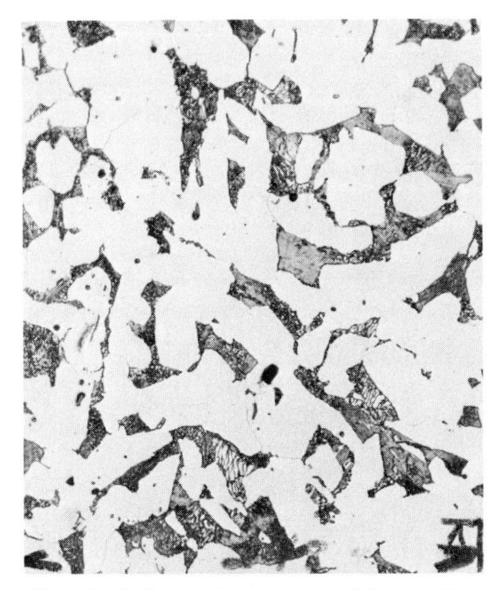

FIG. 39.2 Normal structure of low-carbon boiler steel. The dark is formed by an alternation of cementite (Fe_3C) and ferrite platelets, collectively called pearlite, in the larger matrix of light colored ferrite.

FIG. 39.3 Above 800°F, the carbon begins to form spheroids and grain growth develops.

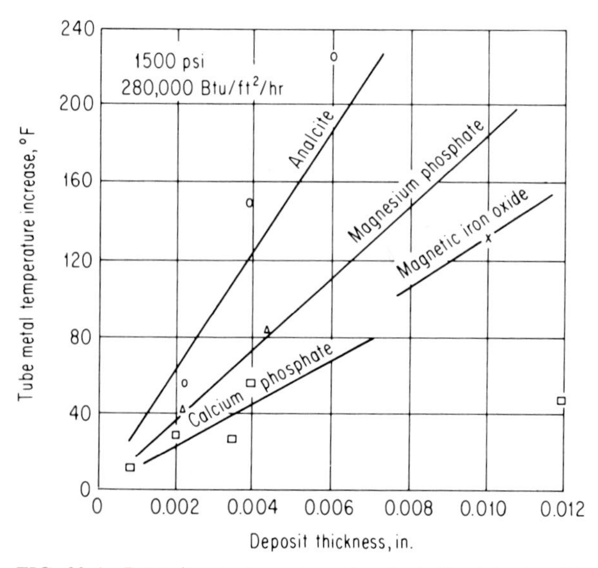

FIG. 39.4 Deposits on the water-side of a boiler tube insulate the metal from the cooling effect of water flow. The metal on the fireside may then become overheated. *(Adapted from Jacklin, C.: "Deposits in Boilers," Ind. Eng. Chem., May 1954.)*

through the tubes normally conducts heat away from the metal, preventing the tube from reaching this range. Deposits insulate the tube, reducing the rate at which this heat can be removed (Figure 39.4); this leads to overheating and eventual tube failure (Figure 39.5). If the deposit is not thick enough to cause such a failure, it can still cause a substantial loss in efficiency and disruption of the heat transfer load in other sections of the boiler.

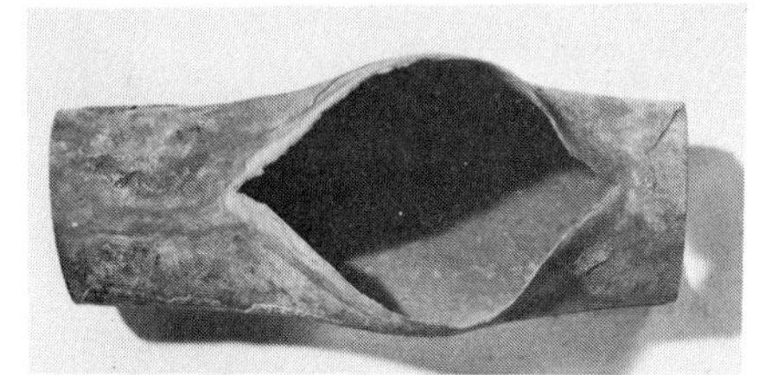

FIG. 39.5 A typical tube failure caused by overheating.

Deposits may be scale, precipitated in situ on a heated surface, or previously precipitated chemicals, often in the form of sludge. These drop out of water in low-velocity areas, compacting to form a dense agglomerate similar to scale, but retaining the features of the original precipitates. In the operation of most industrial boilers, it is seldom possible to avoid formation of some type of precipitate at some time. There are almost always some particulates in the circulating boiler water which can deposit in low-velocity sections, such as the mud drum. The exception would be high-purity systems, such as utility boilers, which remain relatively free of particulates except under conditions where the system may become temporarily upset.

CORROSION

The second major water-related boiler problem is corrosion, the most common example being the attack of steel by oxygen. This occurs in water supply systems, preboiler systems, boilers, condensate return lines, and in virtually any portion of the steam cycle where oxygen is present. Oxygen attack is accelerated by high temperature and by low pH. A less prevalent type of corrosion is alkali attack, which may occur in high-pressure boilers where caustic can concentrate in a local area of steam bubble formation because of the presence of porous deposits.

Some feed water treatment chemicals, such as chelants, if not properly applied, can corrode feed water piping, control valves, and even the boiler internals.

While the elimination of oxygen from boiler feed water is the major step in controlling boiler corrosion, corrosion can still occur. An example is the direct attack by steam of the boiler steel surface at elevated temperatures, according to the following reaction:

$$4H_2O + 3Fe \rightarrow Fe_3O_4 + 4H_2\uparrow \quad (1)$$

This attack can occur at steam-blanketed boiler surfaces where restricted boiler water flow causes overheating. It may also occur in superheater tubes subjected to overheating. Since this corrosion reaction produces hydrogen, a device for analyzing hydrogen in steam, Figure 39.6, is useful as a corrosion monitor.

The third major problem related to boiler operations is carryover from the boiler into the steam system. This may be a mechanical effect, such as boiler water spraying around a broken baffle; it may be caused by the volatility of certain boiler

FIG. 39.6 This instrument is used to monitor corrosion-produced hydrogen concentrations in steam.

water salts, such as silica and sodium compounds; or it may be caused by foaming. Carryover is most often a mechanical problem, and the chemicals found in the steam are those originally present in the boiler water, plus the volatile components that distill from the boiler even in the absence of spray.

There are three basic means for keeping these major problems under control.

1. *External treatment:* Treatment of water—makeup, condensate, or both, before it enters the boiler, to reduce or eliminate chemicals (such as hardness or silica), gases or solids.
2. *Internal treatment:* Treatment of the boiler feed water, boiler water, steam, or condensate with corrective chemicals.
3. *Blowdown:* Control of the concentration of chemicals in the boiler water by bleeding off a portion of the water from the boiler.

EXTERNAL TREATMENT

Most of the unit operations of water treatment (Table 39.2) can be used alone or in combination with others to adapt any water supply to any boiler system. The suitability of the processes available is judged by the results they produce and the costs involved.

The boiler treatment program aims at control of seven broad classifications of impurities: suspended solids, hardness, alkalinity, silica, total dissolved solids (TDS), organic matter, and gases. The extent to which each of the unit processes applicable to boiler makeup treatment, as described in earlier chapters, reduces or removes these impurities is summarized by Table 39.2.

Suspended Solids

The removal of suspended solids is accomplished by coagulation/flocculation, filtration, or precipitation. Other unit processes, except direct reaction, usually require prior removal of solids. For example, water to be processed by ion exchange should contain less than 10 mg/L suspended solids to avoid fouling of the exchanger and operating problems.

Hardness

A number of unit operations remove calcium and magnesium from water, as summarized by Table 39.2. Sodium exchange removes hardness and does nothing else; other processes provide additional benefits. Figure 39.7 compares these softening processes, showing the additional reduction of other impurities that may occur. Differences between softening processes are summarized in Table 39.3.

Alkalinity

It is desirable to have some alkalinity in boiler water, so complete removal of alkalinity from boiler makeup is seldom practiced except in demineralization. Some alkalinity is also needed to provide optimum pH in the feed water to prevent corrosion of piping and equipment.

The makeup alkalinity may be present as HCO_3^-, CO_3^{2-}, or OH^-. If the makeup is city water that has been zeolite softened, alkalinity is usually in the bicarbonate (HCO_3^-) form; if lime softened, it is mostly carbonate (CO_3^{2-}), but the water may also contain some hydroxide (OH^-). When bicarbonates and carbonates are exposed to boiler temperatures, they break down to release CO_2:

$$2NaHCO_3 \rightarrow Na_2CO_3 + H_2O + CO_2\uparrow \qquad (2)$$

The sodium carbonate then breaks down further to caustic:

$$Na_2CO_3 + H_2O \rightarrow 2NaOH + CO_2\uparrow \qquad (3)$$

The carbon dioxide gas redissolves when the steam condenses, producing corrosive carbonic acid:

$$CO_2 + H_2O \rightleftharpoons H_2CO_3 \rightleftharpoons H^+ + HCO_3^- \qquad (4)$$

The amount of CO_2 generated is proportional to the alkalinity. For a given alkalinity twice as much CO_2 is formed from HCO_3^- as from CO_3^{2-} because the bicar-

TABLE 39.2 Boiler Makeup Treatment Processes

Impurity to be removed	Direct addition[a]	Coagulation/ flocculation	Solids/ liquid separation	Precipi- tation	Adsorption	Ion exchange	Evapora- tion	Degasi- fication	Membrane separation
Suspended solids	NA NA	10 mg/L	<1 mg/L	10 mg/L	← These processes require pretreatment for suspended solids removal →				
Hardness	NA	NA	NA	Partial removal[b]	NA	0–2%[c]	↑	*e*	↑
Alkalinity	Can be decreased to 0 or increased as needed	*g*	NA	Partial removal[b]	NA	*h*		NA	
Silica	NA	Slight removal	Slight removal	Partial removal[i]	Partial removal[i]	*j*	*d*	NA	*f*
Dissolved solids (TDS)	May increase[k]	NA	NA	Partial removal[b]	NA	*l*		NA	
Organics	*m*	Partial removal[m]	NA	Partial removal[m]	5–10%	Partial removal[n]		NA	
Gases	Can be decreased to 0	May increase[h]	NA	*o*	*p*	*h*	↓	Note *q*	↓

[a] Direct addition is the application of a chemical directly to water, where the by-products remain in solution.

[b] In the precipitation process, hardness can be reduced by a controlled amount, depending on the lime dosage. In partial lime softening, only the Ca is removed. This is accomplished by adding just enough lime to react with bicarbonate alkalinity plus free CO_2. The Ca hardness after treatment is approximately 35 mg/L cold or 15 to 20 mg/L hot, if the raw water alkalinity exceeds the calcium; if the raw water calcium exceeds the alkalinity, then the CO_3 concentration of the treated water is 35 mg/L cold, 15 to 20 mg/L hot. The reduction of calcium equals the reduction of alkalinity, since $CaCO_3$ forms as a precipitate. More complete hardness removal requires adding enough lime to react with the magnesium as well as the bicarbonate alkalinity and the free CO_2. In this case, if the alkalinity of the raw water exceeds the total hardness, the Ca residual is approximately 35 mg/L cold, 15 to 20 mg/L hot; if the hardness exceeds the alkalinity, then the CO_3 after treatment is 35 mg/L cold, 15 to 20 mg/L hot. The residual magnesium is approximately 30 ppm cold, 2 to 3 mg/L hot, when excess lime is added to produce a hydroxide alkalinity of 10 mg/L. The reduction of total solids equals the hardness reduction with coincident reduction of silica and organic matter.

[c] The residuals after ion exchange vary with the water analysis, the regenerant dosage and application method, and the arrangement of the units in the system. Typical leakages through a sodium exchanger regenerated with brine are shown in the chapter on ion exchange.

[d] Makeup to an evaporator is usually treated either ahead of the evaporator or in the evaporator body; the concentrated water is similar to boiler water, and the chemical treatment program is like boiler treatment.

[e] If water is degasified for CO_2 removal, the chemical balance may be disturbed and hardness may precipitate in the degasifier.

[f] In the membrane process, incoming solids remain in the saline stream unless removed by pretreatment. The degree of purification varies with water analysis and membrane characteristics, such as ion selectivity.

[g] Alkalinity may be reduced by coagulation with alum or iron salts; it may be increased by aluminate; it may remain constant if the dosages of alum and aluminate are balanced.

[h] Alkalinity is reduced by blending the effluents of sodium and hydrogen zeolite units (split stream treatment). It can be reduced to any desired residual. The alkalinity reduction produces an equivalent concentration of CO_2 gas, which can be degassed to 5 to 10 mg/L at ambient temperature or to zero at 212°F (100°C).

[i] The residual silica can be predicted from the water analysis and the dosage of adsorbent applied in the treatment process. Residuals range from 90% in cold process to as little as 5% in hot process, depending on the adsorbent added or on the magnesium precipitated by lime softening.

[j] Silica is removed in ion exchange processes only by strong base anion resins regenerated with caustic. If the anion resin unit follows a sodium zeolite, the residual may be 10% of the feed; if the anion unit is part of a demineralizer, residuals as low as 0.01 mg/L can be achieved.

[k] Some reactions, such as that between sodium sulfite and oxygen, produce a soluble by-product (sodium sulfate) which increases TDS.

[l] The reduction of dissolved solids by ion exchange varies with different cation and anion exchange processes. With sodium zeolite treatment, TDS is unchanged on a $CaCO_3$ equivalent basis; with split stream hydrogen zeolite, H_2X, TDS reduction equals the alkalinity reduction; with a demineralizer, removal is essentially complete and the residuals depend on the combination of units in the system.

[m] Organic matter is reduced by both coagulation/flocculation or lime softening. Organic color over 50 interferes with lime softening unless counteracted by adsorption on carbon or by chemical oxidation. It is usual to obtain about 30% removal of organic matter in cold clarification or lime softening, and 50% removal in hot process precipitation. These removal efficiencies vary with the selection of coagulants, chlorine, permanganate, or powdered activated carbon.

[n] Special ion exchange resins (like activated carbon) can be used to remove color. However, organic matter tends to accumulate irreversibly on anion resins, presenting serious operating problems. Organic matter should be removed ahead of a demineralizing system. Organic matter also interferes with membrane processes.

[o] Carbon dioxide is removed by lime in cold process softeners and by degasification in the spray section of hot process softeners, which also reduce dissolved oxygen to a residual of about 0.5 mg/L. H_2S may be removed by adding a heavy metal precipitant such as an iron or zinc salt.

[p] Taste and odor producing gases can be removed by activated carbon. Carbon also removes excess chlorine, through a chemical reaction.

[q] Degasifiers using air for stripping reduce CO_2 and H_2S by 90 to 95% if the pH is kept below 7. Vacuum deaerators do as well on CO_2 and H_2S and also reduce dissolved O_2 to less than 1.0 mg/L. Steam-heated deaerators will remove all free CO_2 and reduce dissolved O_2 to 0.005 mg/L.

Identification of Analyses Tabulated Below:

A. Raw Water
B. Simple Na_2Z Softening
C. Na_2Z + Acid + Degasifier
D. Split-Stream + Degasifier
E. Partial Lime-Cold
F. Hot Lime-Zeolite

Constituent	As	A	B	C	D	E	F
Calcium	$CaCO_3$	115	Nil	Nil	Nil	35	Nil
Magnesium	"	45	Nil	Nil	Nil	40	Nil
Sodium	"	60	220	220	105	60	115
Total Electrolyte	$CaCO_3$	220	220	220	105	135	115
Bicarbonate	$CaCO_3$	135	135	20	20	15	0
Carbonate	"	0	0	0	0	35	20
Hydroxyl	"	0	0	0	0	0	10
Sulfate	"	65	65	180	65	65	65
Chloride	"	20	20	20	20	20	20
Nitrate	"	0	0	0	0	0	0
M Alk.	$CaCO_3$	135	135	20	20	50	30
P Alk.	"	0	0	0	0	18	20
Carbon Dioxide		10	10	5	5	0	0
pH		7.3	7.3	6.8	6.8	9.8	10.3
Silica		10	10	10	10	9	1
Iron	Fe	0	0	0	0	0	0
Turbidity		Nil	Nil	Nil	Nil	Nil	Nil
TDS (estimated)		275	290	290	170	190	155
Color		0	0	0	0	0	0
Total Hardness		160	Nil	Nil	Nil	75	Nil

FIG. 39.7 A comparison of water-softening processes.

TABLE 39.3 Water Softening Process Summary
(Summary of Figure 39.6)

Process	Residuals, mg/L			
	Hardness	Alkalinity	Silica	TDS
Original water	160	135	10	275
Sodium exchange	Nil	135	10	290
Sodium exchange and acid	Nil	20	10	290
Split stream	Nil	20	10	170
Partial lime (cold)	75	50	9	190
Hot lime–zeolite	Nil	30	1	155

bonate breakdown is the sum of both reactions (2) and (3) above. The carbonic acid is usually neutralized by chemical treatment of the steam—either directly or indirectly through the boiler—to produce a condensate pH in the range of 8.5 to 9.0. Reduction of feed water alkalinity is desirable, then, to minimize CO_2 formation and reduce chemical treatment costs.

The hydroxide produced by the breakdown of HCO_3^- and CO_3^{2-} is beneficial for precipitation of magnesium, to provide a good environment for sludge conditioning, and to minimize SiO_2 carryover. However, too high an excess of caustic can be corrosive, particularly if localized concentration can occur. The breakdown of HCO_3^- is complete, but not all the CO_3^{2-} converts to caustic. The conversion varies from one boiler to another and increases with temperature. As a general rule, at 600 lb/in^2 65 to 85% of boiler water alkalinity is NaOH, the remainder Na_2CO_3. (This is based on the equilibrium in the cooled sample of boiler water.)

Identification of Analyses Tabulated Below:

A. Raw Water
B. Direct H_2SO_4 + Degasifier
C. Split-Stream + Degasifier
D. Na_2Z + Chloride Anion Exchanger
E. Partial Lime-Cold
F. Hot Lime Zeolite

Constituent	As	A	B	C	D	E	F
Calcium	$CaCO_3$	115	115	Nil	Nil	35	Nil
Magnesium	"	45	45	Nil	Nil	40	Nil
Sodium	"	60	60	105	220	60	115
Total Electrolyte	$CaCO_3$	220	220	105	220	135	115
Bicarbonate	$CaCO_3$	135	20	20	20	15	0
Carbonate	"	0	0	0	0	35	20
Hydroxyl	"	0	0	0	0	0	10
Sulfate	"	65	180	65	5	65	65
Chloride	"	20	20	20	195	20	20
Nitrate	"	0	0	0	0	0	0
M Alk. (Total Alk.)	$CaCO_3$	135	20	20	20	50	30
P Alk.	"	0	0	0	0	18	20
Carbon Dioxide		10	5	5	5	0	0
pH		7.3	6.8	6.8	6.8	9.8	10.3
Silica		10	10	10	10	9	1
Iron	Fe	0	0	0	0	0	0
Turbidity		Nil	Nil	Nil	Nil	Nil	Nil
TDS (estimated)		275	290	170	290	190	165
Color		0	0	0	0	0	0

FIG. 39.8 A comparison of processes for alkalinity reduction.

The degree of alkalinity reduction is therefore dictated by boiler water control limits and steam quality goals. The best unit process for alkalinity reduction may be chosen for the other benefits it provides as well as its efficiency in alkalinity reduction. Figure 39.8 compares alkalinity reduction processes and their supplemental benefits. The major differences between these processes are summarized in Table 39.4.

TABLE 39.4 Alkalinity Reduction Processes
(Summary of Figure 39.7)

Process	Residuals, mg/L			
	Alkalinity	Hardness	Silica	TDS
Original water	135	160	10	275
Direct acid addition	20	160	10	290
Split stream	20	Nil	10	170
Na_2X and anion (Cl)	20	Nil	10	290
Cold lime	50	75	9	190
Hot lime–zeolite	30	Nil	1	155

Silica

The permissible concentrations of silica in boiler water at various operating pressures are given in Table 39.5. Silica reduction is not always necessary, especially in the absence of a condensing turbine. Low concentrations of silica can sometimes produce sticky sludge in low-pressure boilers treated with phosphate. A

TABLE 39.5 Silica Concentration in Boiler Water

Drum pressure, lb/in^2 gage	Silica concentration, mg/L	
	Recommended*	To produce 0.02 mg/L†
0–300	150	150
301–450	90	90
451–600	40	55
601–750	30	35
751–900	20	20
901–1000	8	15

* Recommended limits. Consensus on operating practices for control of feed water and boiler water quality in modern industrial boilers (ASME, 1979).

† In saturated steam at upper pressure limit.

makeup treatment process may be selected to provide just the proper degree of silica reduction required by the steam system. Figure 39.9 shows the treatment results achieved by the various silica removal processes; the influence of each of these processes on other feed water contaminants is summarized in Table 39.6.

Identification of Analyses Tabulated Below:

A. Raw Water
B. Cold Lime with $FeCl_3$
C. Cold Lime-Soda
D. Lime-Soda @ 220°F
E. Na_2Z + A.OH + H_2SO_4
F. Demineralization

Constituent	As	A	B	C	D	E	F
Calcium	$CaCO_3$	115	70	35	15	Nil	Nil
Magnesium	"	45	40	30	2	Nil	Nil
Sodium	"	60	60	85	108	220	1-2
Total Electrolyte	$CaCO_3$	220	170	150	125	220	1-2
Bicarbonate	$CaCO_3$	135	0	0	0	0	1-2
Carbonate	"	0	35	35	30	20	0
Hydroxyl	"	0	0	30	10	0	0
Sulfate	"	65	65	65	65	180	0
Chloride	"	20	70	20	20	20	0
Nitrate	"	0	0	0	0	0	0
M Alk.	$CaCO_3$	135	35	65	40	20	1-2
P Alk.	"	0	18	48	25	10	0
Carbon Dioxide		10	0	0	0	0	Nil
pH		7.3	10.0	10.8	10.3	10.0	7-8
Silica		10	6	7-8	1	0.5-1	0.05
Iron	Fe	0	0	0	0	0	0
Turbidity		Nil	Nil	Nil	Nil	Nil	Nil
TDS (estimated)		275	225	210	145	290	1-2
Color		0	0	0	0	0	0

FIG. 39.9 A comparision of processes for silica reduction.

TABLE 39.6 Silica Reduction Processes
(Summary of Figure 39.9)

	Residuals, mg/L			
Process	Silica	Hardness	Alkalinity	TDS
Original water	10	160	135	275
Cold lime and iron salts	6	110	35	225
Cold lime–soda	7–8	65	65	210
Hot lime–soda	1	17	40	145
Na_2X + anion exchanger	1	Nil	20	290
Demineralization	0.05	Nil	1–2	1–2

Total Dissolved Solids

Some treatment processes increase dissolved solids by adding soluble by-products to water; sodium zeolite softening increases solids by adding an ion (sodium) having a higher equivalent weight (23) than the calcium (20) or magnesium (12.2) removed from the raw water. Processes to reduce dissolved solids achieve various degrees of success. Usually, reduction of dissolved solids is accomplished by a reduction of several individual contaminants. Table 39.7 summarizes the analyses of effluents produced by processes which reduce dissolved solids.

Organic Matter

Organic matter as a general classification is only a qualitative term. It includes a wide variety of compounds that are seldom analyzed as specific materials. Problems in boiler systems attributed to organic matter have often been traced to organic materials from plant processes in returned condensate, rather than makeup water contaminants. However, in high-pressure utility systems, organic matter is the major impurity in makeup and can result in formation of organic acids.

Dissolved Gases

Degasifiers are commonly used to remove gas mechanically rather than chemically. Blower types are used for CO_2 removal at ambient temperatures following acid or hydrogen-exchange units. Vacuum degasifiers provide the same extent of CO_2 removal, but also reduce O_2 to less than 0.5 to 1.0 mg/L, offering corrosion protection, especially if the vacuum degasifier is part of a demineralizing system. Steam-scrubbing degasifiers, called deaerating heaters, usually produce an effluent free of CO_2 with O_2 concentrations in the range of 0.005 to 0.01 mg/L. Direct reaction of this low residual with catalyzed sulfite, hydrazine, or hydrazine substitutes (all-volatile oxygen-reducing compounds) eliminates O_2 completely to prevent preboiler corrosion.

CONDENSATE RETURNS

In addition to makeup treatment, acceptable feed water quality may require cleanup of condensate to protect the boiler system, particularly if there is process condensate containing oil. Boilers requiring high-quality demineralized water also demand high-quality condensate. Some plants operate both high- and low-pressure boilers; high-quality feed water for the high-pressure boilers may be provided entirely by a demineralizer, with lower quality condensate segregated for return to the low-pressure boilers.

Septum filters are usually selected for oily condensate treatment. A cellulose-type filter aid (processed wood pulp) is applied both as a precoat and a body feed. The temperature should be less than 200°F (93°C) to avoid degradation of the filter aid. Anthracite filters precoated with a floc produced from alum and sodium aluminate are also effective. However, the pH of the condensate must be controlled in the range of 7 to 8 to avoid solubilizing the alumina floc. Condensate contaminated with corrosion products and inleakage of hard water is cleaned up through specially designed, high flow rate sodium exchangers (Figure 39.10). They have been used to process condensate at temperatures up to 300°F (149°C). One

TABLE 39.7 Reduction of Total Dissolved Solids
(Summary of Tables 39.3, 39.4, and 39.6)

	Residuals, mg/L			
Process	TDS	Hardness	Alkalinity	Silica
Original water	275	160	135	10
Split stream	170	Nil	20	10
Partial cold lime	190	75	50	9
Hot lime–soda	145	17	40	1
Hot lime–zeolite	155	Nil	30	1
Demineralization	1–2	Nil	1–2	0.05

serious limitation of the simple sodium exchanger is its ability to pick up neutralizing amines such as morpholine (present in the condensate as morpholine bicarbonate) and exchange this for sodium. This causes excessive use of amines, but a more serious problem arises if the condensate is returned to a high-pressure boiler where the presence of sodium may be objectionable in deterioration of steam quality. Special regeneration procedures would then be needed.

Heavily contaminated process condensates, such as those produced in kraft pulp mills and petroleum refineries, present special problems in use as boiler feed water. Their composition is usually variable and may include complex organic compounds and unusual ions such as cyanide, thiocyanate, and sulfide. The treatment program cannot be selected simply on the basis of the condensate analysis; research on the bench and by pilot plant operation may be required, but the recovery of condensate pays good dividends in both heat saving and reduced cost of makeup and treatment chemicals.

INTERNAL TREATMENT

Scale formation within a boiler is controlled by one of four chemical programs: coagulation (carbonate), phosphate residual, chelation, or coordinated phosphate.

Coagulation Program

In this process, sodium carbonate, sodium hydroxide, or both are added to the boiler water to supplement the alkalinity supplied by the makeup, which is not softened. The carbonate causes deliberate precipitation of calcium carbonate under favorable, controlled conditions, preventing deposition at some subsequent point as scale. Under alkaline conditions, magnesium and silica are also precipitated as magnesium hydroxide and magnesium silicate. There is usually a fairly high concentration of suspended solids in the boiler water, and the precipitation occurs on these solids. This method of treatment is used only with boilers (usually firetube design) using high-hardness feed water and operating below 250 lb/in^2 (17 bars). This type of treatment must be supplemented by some form of sludge conditioner. Even with a supplemental sludge conditioner, heat transfer is hindered by deposit formation, and blowdown rates are excessive because of high suspended solids. Coagulation programs are becoming obsolete as pretreatment systems become more common and competitive with the high internal treatment cost.

(*a*)

(*b*)

FIG. 39.10 An ion exchange system designed for treating chemical plant condensate: (*a*) exchange units; (*b*) control panel. *(Courtesy of Dow Chemical Company.)*

Phosphate Program

Where the boiler pressure is above 250 lb/in^2, high concentrations of sludge are undesirable. In these boilers, feed water hardness should be limited to 60 mg/L, and phosphate programs are preferred. Phosphate is also a common treatment below 250 lb/in^2 with soft makeup.

A sodium phosphate compound is fed either to the boiler feed water or to the boiler drum, depending on water analysis and the preboiler auxiliaries, to form an insoluble precipitate, principally hydroxyapatite, $Ca_{10}(PO_4)_6(OH)_2$. Magnesium and silica are precipitated as magnesium hydroxide, magnesium silicate (often combined as $3MgO \cdot 2SiO_2 \cdot 2H_2O$), or calcium silicate. The alkalinity of the makeup is usually adequate to produce the OH^- for the magnesium precipitation. Phosphate residual programs which produce high suspended solids require the addition of a sludge conditioner/dispersant. Because these programs restrict heat transfer, owing to the deposition of calcium and magnesium salts, precipitation programs of this type are often replaced with solubilizing treatments such as chelants and polymer/dispersants.

Chelant Programs

A chelate is a molecule similar to an ion exchanger; it is low in molecular weight and soluble in water. The sodium salts of ethylene diamine tetraacetic acid (EDTA) and nitrilotriacetic acid (NTA) are the chelating agents most commonly used for internal boiler treatment. These chelate (form complex ions with) calcium and magnesium. Because the resulting complex is soluble, this treatment is advantageous in minimizing blowdown. The higher cost compared to phosphate usually limits the use of chelates to feed waters having low hardness. There is the risk that breakdown of the organic molecule at higher temperatures could create a potential problem of control that could result in corrosion, so chelate programs are usually limited to boilers operating below 1500 lb/in^2 (100 bars). The addition of polymers as scale control agents increases the effectiveness of chelate programs.

It also reduces the corrosion potential by reducing the chelant dosage below theoretical requirements, so that there is no chelant residual in the boiler water.

Chelates can react with oxygen under boiler water conditions, which can increase the cost of a chelate program substantially. Overfeed of chelates and concentration mechanisms in the boiler can lead to severe localized corrosion and subsequent unit failure.

Coordinated Phosphate Program

In high-pressure, high heat transfer rate boilers, the internal treatment program must contribute little or no solids. The potential for caustic attack of boiler metal increases with increasing pressure, so free caustic alkalinity must be minimized. The coordinated phosphate program is chosen for these conditions. This differs from the standard program in that the phosphate is added to provide a controlled pH range in the boiler water as well as to react with calcium if hardness should enter the boiler. Trisodium phosphate hydrolyzes to produce hydroxide ions:

$$Na_3PO_4 + H_2O \rightleftharpoons 3Na^+ + OH^- + HPO_4^{2-} \tag{5}$$

This cannot occur with the ionization of disodium and monosodium phosphate:

$$Na_2HPO_4 \rightleftharpoons 2Na^+ + HPO_4^{2-} \quad (6)$$

$$NaH_2PO_4 \rightleftharpoons Na^+ + H^+ + HPO_4^{2-} \quad (7)$$

The program is controlled by feeding combinations of disodium phosphate with trisodium or monosodium phosphate to produce an optimum pH without the presence of free OH^-. To successfully control a coordinated phosphate program, the feed water must be extremely pure and of consistent quality. Coordinated phosphate programs do not reduce precipitation; they simply cause precipitation of less adherent calcium phosphate in the absence of caustic. A dispersant must be added to condition deposits that would otherwise reduce the heat transfer rate. The coordinated phosphate program was first developed for high-pressure utility boilers, and most experience with this program has been gained in this field. More details on control of the program can be found in Chapter 34, "Utilities."

COMPLEXATION/DISPERSION

The newest addition to internal treatment technology is the use of synthetic organic polymers for complexation and dispersion. This type of program can be used to 1500 lb/in^2 (100 bars) and is economical in all low-hardness feed water systems typical of those produced by ion exchange. Heat transfer rates are maximized because these polymers produce the cleanest tube surfaces of any of the available internal treatment programs. This treatment solubilizes calcium, magnesium, and aluminum, and maintains silica in solution while avoiding corrosion potential side effects as determined by hydrogen levels in the steam. Iron particulates returned from the condensate system are likewise dispersed for removal via blowdown. A simple measure of ion transport is used to demonstrate on-line performance of this program.

PROGRAM SUPPLEMENTS

In addition to controlling scale and deposits, internal treatment must also control carryover, defined as entrainment of boiler water into the steam. Boiler salts carried as a mist may subsequently deposit in the superheater, causing tube failures, or deposit on the blades of a turbine. They may also contaminate a process in which the steam is used. Since a high percentage of carryover is caused by foaming, this problem is usually solved by the addition of antifoam to the boiler feed water.

Sludge in boiler water may settle to form deposits, which are as serious a problem as scale. Chemicals are used to condition boiler water particulates so that they do not form large crystalline precipitates; smaller particles will remain dispersed at the velocities encountered in the boiler circuit. At lower pressures both the coagulation and phosphate residual programs incorporate sludge conditioning agents for this purpose. A variety of natural organic materials are used, including starches, tannins, and lignins. Figure 39.11 shows the effect of tannin in stunting the growth of $CaCO_3$ crystals; Figure 39.12 illustrates the effectiveness of tannin

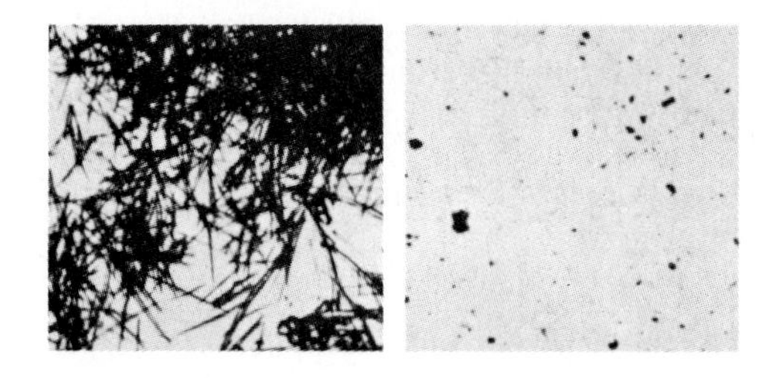

FIG. 39.11 The effect of organic treatment—in this case tannin—on the nature of $CaCO_3$ crystals.

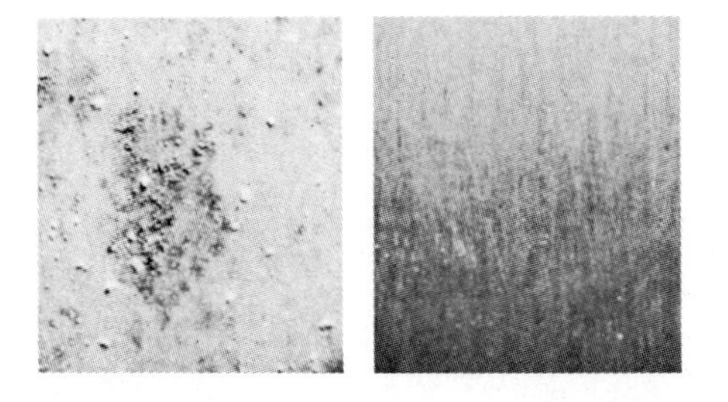

FIG. 39.12 In early experimental boiler studies, the value of tannin in controlling deposits on boiler surfaces was clearly shown.

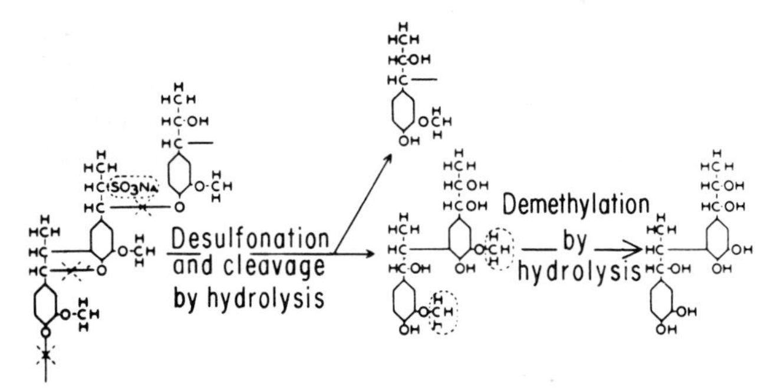

FIG. 39.13 Because tannins lose effectiveness at higher temperatures, other organics were developed. This is the structure of lignin processed for deposit control in higher pressure boilers.

in preventing $CaCO_3$ scale in 250 lb/in^2 (17 bars) experimental boilers by this ability to control crystal growth and disperse the precipitated $CaCO_3$.

At intermediate pressures, chemically reacted lignins have been widely used, though synthetic polymers are replacing them. Figure 39.13 shows the approximate molecular configuration of a lignin processed for high-temperature stability.

Its effectiveness in controlling calcium phosphate scale and magnetic iron oxide deposits at 1500 lb/in^2 (100 bars) is shown in Figures 39.14 and 39.15.

At pressures up to 1800 lb/in^2 (120 bars), heat-stable polymers such as anionic carboxylates and their derivatives are used as effective dispersants. An alkaline environment generally increases their effectiveness. Lignin-type dispersants and

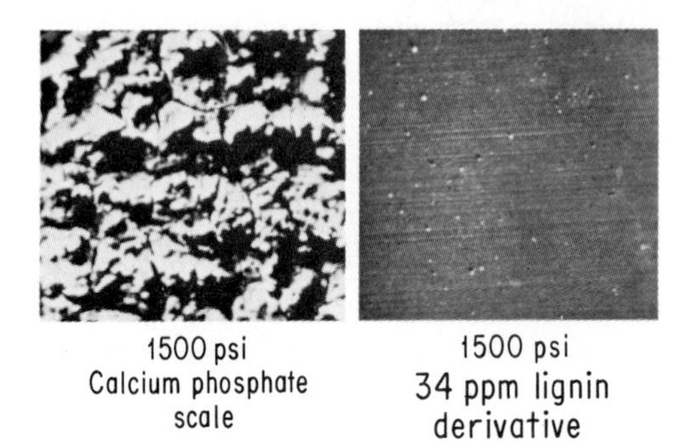

FIG. 39.14 The effectiveness of a lignin derivative in controlling calcium phosphate scale at 1500 lb/in^2 is shown here. The effectiveness of organic scale control agents has been greatly expanded with synthetic polymer combinations.

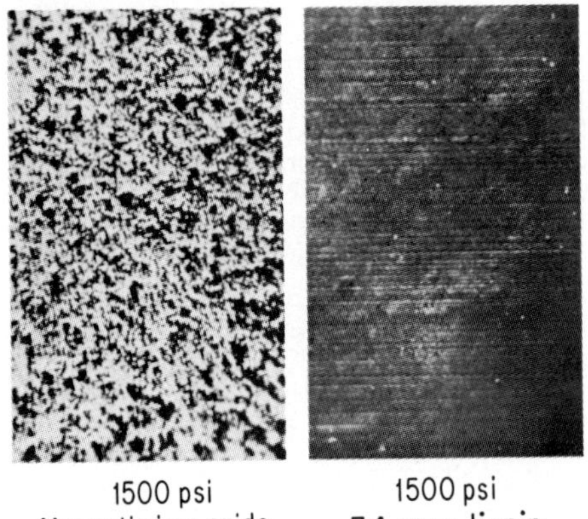

FIG. 39.15 The effectiveness of a lignin derivative in controlling iron deposits at 1500 lb/in^2 (*Left*) no treatment; (*right*) with organic treatment.

other natural organic derivatives are being replaced by these more effective synthetic organic polymers. These dispersants have been designed for specific dispersion problems, with tailored molecules for magnesium silicate, calcium phosphate, and iron particulates being available.

Somewhat related to carryover, in that steam quality is affected, is the discharge of contaminants that volatilize under boiler operating conditions. The major volatiles are CO_2, created by the breakdown of carbonate and bicarbonate mentioned earlier, and SiO_2. Although the CO_2 can be neutralized, it is prudent to reduce feed water alkalinity to minimize its formation. For all practical purposes, external treatment for silica reduction and blowdown are the only means to avoid excessive SiO_2 discharges for protection of turbine blading. Hydroxyl alkalinity helps reduce silica volatility.

Oxygen is the chief culprit in boiler systems corrosion. Deaeration reduces this to a low concentration in the preboiler system, but does not completely eliminate it. Application of sulfite, hydrazine, or hydrazine-like (all-volatile) compounds after deaeration scavenges the remaining O_2 and maintains a reducing condition in the boiler water. An advantage of hydrazine is that it is discharged into the steam to become available in the condensate as protection against oxygen corrosion in the return system. If oxygen is present, ammonia can attack copper alloys in condensers and stage heaters. The removal of NH_3 by external treatment may be necessary. The corrosive aspects of CO_2 have already been mentioned in relation to condensate systems. The beneficial and detrimental aspects of NaOH in the boiler circuit in relation to corrosion control have also been discussed earlier.

BLOWDOWN

Boiler feed water, regardless of the type of treatment used to process the makeup, still contains measurable concentrations of impurities. In some plants, contaminated condensate contributes to feed water impurities. Internal boiler water treatment chemicals also add to the level of solids in the boiler water.

When steam is generated, essentially pure H_2O vapor is discharged from the boiler, leaving the solids introduced in the feed water to remain in the boiler circuits. The net result of impurities being continuously added and pure water vapor being withdrawn is a steady increase in the level of dissolved solids in the boiler water. There is a limit to the concentration of each component of the boiler water. To prevent exceeding these concentration limits, boiler water is withdrawn as blowdown and discharged to waste. Figure 39.16 illustrates a material balance for a boiler, showing that the blowdown must be adjusted so that solids leaving the boiler equal those entering and the concentration is maintained at the predetermined limits.

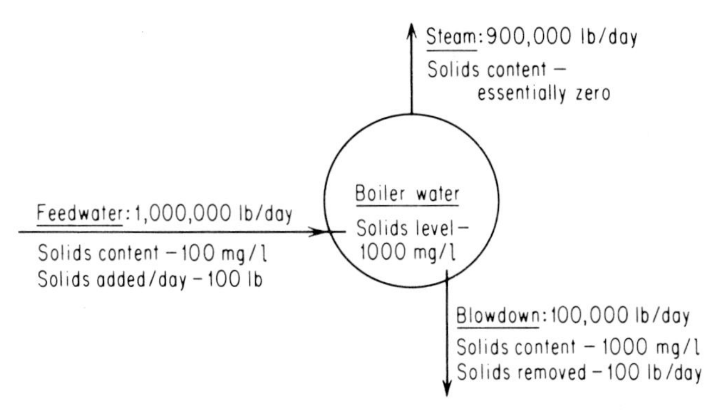

FIG. 39.16 How boiler water solids are controlled by blowdown.

Of course it is apparent that the substantial heat energy in the blowdown represents a major factor detracting from the thermal efficiency of the boiler, so minimizing blowdown is a goal in every steam plant. There are ways to reclaim this heat that will be examined later in the chapter.

One way of looking at boiler blowdown is to consider it a process of diluting boiler water solids by withdrawing boiler water from the system at a rate that induces a flow of feed water into the boiler in excess of steam demand.

There are two separate blowdown points in every boiler system. One accommodates the blowdown flow that is controlled to regulate the dissolved solids or other factors in the boiler water. The other is an intermittent or mass blowdown, usually from the mud drum or waterwall headers, which is operated intermittently at reduced boiler load to rid the boiler of accumulated settled solids in relatively stagnant areas. The following discussion of blowdown will be confined only to that used for adjusting boiler water dissolved solids concentrations.

Blowdown may be either intermittent or continuous. If intermittent, the boiler is allowed to concentrate to a level acceptable for the particular boiler design and pressure. When this concentration level is reached, the blowdown valve is opened for a short period of time to reduce the concentration of impurities, and the boiler is then allowed to reconcentrate until the control limits are again reached. In continuous blowdown, on the other hand, the blowdown valve is kept open at a fixed setting to remove water at a steady rate, maintaining a relatively constant boiler water concentration. Since the average concentration level in a boiler blown down intermittently is substantially less than that maintained by continuous blowdown, intermittent blowdown is less efficient—more costly—than continuous blowdown.

Figure 39.17 is a schematic diagram of a typical industrial boiler plant that discharges steam to a turbine, with part of the steam being condensed in the condenser and the remainder extracted for a process use where the steam may be lost or the condensate become so contaminated that it must be wasted. With reference

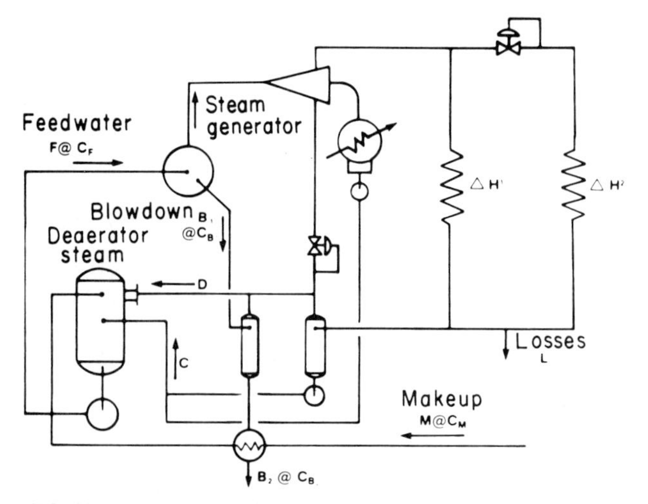

FIG. 39.17 Schematic of industrial boiler system.

to this diagram, the following relationships apply in determining blowdown losses:

$$M \times C_M = F \times C_F = B_1 \times C_{B1}$$

$$M + C + D = F = S + B \qquad \text{(all in lb/h) (or kg/h)}$$

$$M = L + B_2 \qquad \text{(all in lb/h) (or kg/h)}$$

$$\text{CR}_F = \frac{C_{B1}}{C_F} = \frac{F}{B_1} \qquad \text{CR}_M = \frac{C_{B1}}{C_M} = \frac{M}{B_1}$$

$$F = S \times \frac{\text{CR}_F}{\text{CR}_F - 1}$$

$$B_1 = F/\text{CR}_F$$

$$\text{Percent blowdown} = \frac{100}{\text{CR}}$$

where B_1 = boiler blowdown flow containing C_{B1} mg/L solids
B_2 = blowdown flow from flash tank
CR = concentration ratio; CR_F is concentration ratio based on feed water; CR_M is based on makeup
C = returned condensate flow, assumed to be free of solids
D = steam flow used for deaeration
F = feed water flow containing C_F mg/L solids
L = loss of steam or condensate
M = makeup flow containing C_M mg/L solids
S = steam flow

It is common to express blowdown as a percentage of feed water. However, this may give the utilities engineer a false sense of security. If the plant has 80% condensate return and 20% makeup, a 5% blowdown would appear satisfactory, but it indicates that the makeup is being concentrated only 4 times—of the four units of makeup entering the boiler, one unit is being thrown away. Perhaps that is as much usage as can be made of that particular quality makeup, but the operator should be aware of it.

Since the main purpose of blowdown control is to reach the maximum permissible concentrations for best boiler efficiency without exceeding concentrations that would harm the system, the first step in developing a blowdown control program is to establish allowable limits. The conventional limits recommended to provide boiler cleanliness and adequate steam quality are shown in Tables 39.8 and 39.9. These limits cover most situations encountered in industrial boiler operations, but not the coagulation treatment used in low-pressure boilers. With the coagulation treatment, total dissolved solids are usually limited to 3500 mg/L, and adequate alkalinity is maintained to provide the carbonate for calcium precipitation and the hydroxide for magnesium precipitation. These levels can be established only after the nature of the makeup treatment system has been considered.

As shown by Tables 39.8 and 39.9, the limits on such things as total dissolved solids, silica, and alkalinity are basically related to the amounts of these materials

TABLE 39.8 Optimum Boiler Water Control Limits*

Drum-type boilers using softened (not deionized) feed waters

	Pressure, lb/in^2					
	150	300	600	900	1200	1500
TDS (max.)	4000	3500	3000	2000	500	300
Phosphate (as PO_4)†	30–60	30–60	20–40	15–20	10–15	5–10
Hydroxide (as $CaCO_3$)	300–400	250–300	150–200	120–150	100–120	80–100
Sulfite	30–60	30–40	20–30	15–20	10–15	5–10
Silica (as SiO_2) mas.‡	100	50	30	10	5	3
Total iron (as Fe) max.	10	5	3	2	2	1
Organics	70–100	70–100	70–100	50–70	50–70	50–70

* Heat release below 150,000 Btu/h/ft.^{2}

† Where conditions warrant, chelants (EDTA or NTA) may be used in place of phosphates to achieve a hardness-solubilizing rather than a hardness-precipitating effect. In certain programs, phosphate and chelant are used together. Where chelant residuals are to be maintained, recommended boiler water control limits are: (1) boiler pressure below 400 lb/in^2, 4 to 8 mg/L; (2) 401 to 600 lb/in^2, 3 to 6 mg/L; (3) 601 to 1000 lb/in^2, 3 to 5 mg/L (all residuals are as $CaCO_3$).

‡ See Table 39.5 for intermediate pressure.

entering with the makeup water; these concentrations can be adjusted by blowdown, but also by some adjustment in the makeup treatment system if that flexibility is provided. On the other hand, such constituents as phosphate, organics, and sulfite are introduced as internal treatment chemicals, and their concentration can be adjusted both by blowdown and by rate of application.

For purposes of illustrating the calculation of boiler blowdown related to the concentrations, a 900-lb/in^2 (40 bars) boiler system in a paper mill is used as an example. The steam goes both to a condensing turbine and a back pressure turbine, with 50% condensate return. The makeup is treated by hot lime–zeolite softening, and, after treatment, has a total dissolved solids concentration of 150 mg/L, silica of 3 mg/L, and total alkalinity of 20 mg/L. Table 39.10 summarizes the conditions established in this example.

With a silica concentration of 1.5 mg/L in the feed water and an allowable limit of only 10 mg/L in the boiler water, silica is the controlling factor and sets the concentration ratio (based on feed water) at 6.7. Since the water could be concentrated to a factor of 10 based on total dissolved solids, there is incentive for additional silica reduction in the hot process unit. If the addition of dolomitic lime would permit a reduction from 3 mg/L shown to less than 2 mg/L in the makeup, the blowdown rate could be reduced from 15 to 10%.

A second example explores the use of a simple sodium zeolite softener to treat city water as makeup for a 300-lb/in^2 boiler. The water analyses in Figure 39.18 show the results of treating the city water through a zeolite softener, and the allowable concentrations in a 300-lb/in^2 (20 bars) boiler. The concentration ratio is calculated for each of the constituents to be controlled; the lowest CR determines the blowdown rate. In this example, the lowest ratio is 2.5, applying to alkalinity. So the blowdown rate, controlled by alkalinity, would be:

$$\text{Percent blowdown} = \frac{100}{2.5} = 40\%$$

This is a high blowdown loss, expressed as a percentage of makeup. However, in a small plant that generates less than 50,000 lb/h (22,700 kg/h) of steam with

TABLE 39.9 Optimum Boiler Water Control Limits*

Drum-type boilers using high purity (deionized) feed waters

	Pressure, lb/in^2					
	Up to 600	900†	1200†	1500	1800	2400
TDS (max.)	Use same optimum limits as for soft (not deionized) feed waters	500	300	200	100	50
Phosphate (as PO_4)		15–25	15–25	5–10	5–10	5–10
pH		9.8–10.2	9.8–10.2	9.4–9.7	9.4–9.7	9.4–9.7
Silica (aş SiO_2)		10	5	2	1	0.25
Total iron (as Fe)		2	2	1	0.5	0.2
Hydrazine (as N_2H_4)		0.04–0.06‡	0.04–0.06‡	0.04–0.06‡	0.04–0.06‡	0.04–0.06‡

* Heat release below 150,000 Btu/h/ft^2.

† For most boilers installed before 1950, and for boilers installed since but not having waterwall heat transfer rates conducive to DNB (departure from nucleate boiling) under anticipated operating conditions, control limits applicable for softened (not deionized) feed waters may be used.

‡ Hydrazine residuals in feed water just ahead of boiler, e.g., at economizer inlet.

less than 10 to 20% makeup, this process might be acceptable just for its simplicity and low cost. Larger plants would find this high blowdown loss unacceptable because of the high energy loss and the cost of preparing and treating makeup that is concentrated to such a limited degree.

TABLE 39.10 Summary of Controls
(All concentrations in mg/L)

Factor	Makeup	Feed water	Boiler limit*	Max CR_F
TDS	150	75	2000	26.7
SiO_2	3	1.5	10	6.7
Alkalinity	20	10	150†	15

* From Table 39.8.

† Assuming 90% of boiler alkalinity is hydroxyl, an average hydroxyl alkalinity of 135 mg/L (Table 39.8) corresponds to an average total alkalinity of 150 mg/L. Since silica is controlling, the blowdown rate is:

$$\text{Blowdown} = \frac{100}{CR_F} = 15\% \text{ of feed water}$$

Two processes are explored to modify the sodium zeolite system to reduce alkalinity, Figure 39.19 shows these two modifications, sodium zeolite plus acid and split-stream treatment. Both of these significantly reduce alkalinity and blowdown. The first process increases the critical CR to 12.5, so blowdown would be controlled by alkalinity at a level close to the optimum TDS. Further reduction in blowdown is achieved by using split-stream treatment, since TDS is reduced as well as alkalinity. At these low levels, silica becomes a controlling factor at a blowdown of 6% of makeup. Capital cost and operating cost figures are needed to decide whether the reduction in blowdown from 8% achieved with the first process to 6%, which can be reached with the split-stream treatment, is justified. The split-stream process is more costly and it creates a secondary problem—disposal of spent regenerant acid.

These examples show that concentration ratios are determined by chemical analyses. Since blowdown rate is never measured, but most plants meter both makeup and feed water as well as steam, the chemical determination of concentration ratio is the most accurate means of determining blowdown loss. It is apparent that careful sampling of both the feed water and makeup is required to properly control blowdown and be able to determine blowdown rate. The boiler water must be cool before it can be analyzed, and leakage in the cooler could affect the composition of the boiler water. The boiler water sample is usually taken from the blowdown collection pipe in the boiler drum, and if this is not properly designed, the blowdown sample may be nonrepresentative. An example is the accumulation of steam bubbles within the blowdown line which is then condensed through the sample cooler and dilutes the boiler water.

Although one of several boiler water constituents may determine the required blowdown rate—for example, silica—it is general practice to determine all of the critical concentrations in the boiler on a regular basis. Each of these can then be related to the total dissolved solids as measured by a conductivity instrument,

Identification of Analyses Tabulated Below:

A. City Water
B. Na_2Z Effluent
C. Boiler Limits
D. Concentration Ratios
E. Blowdown, % of Makeup
F.

Constituent	As	A	B	C	D	E	F
Calcium	$CaCO_3$	75	Nil				
Magnesium	"	45	Nil				
Sodium	"	20	140				
Total Electrolyte	$CaCO_3$	140	140				
Bicarbonate	$CaCO_3$	100	100				
Carbonate	"	0	0				
Hydroxyl	"	0	0	225			
Sulfate	"	10	10				
Chloride	"	30	30				
Nitrate	"	0	0				
M Alk.	$CaCO_3$	100	100	250	2.5	40%	
P Alk.	"	Tr	Tr	238			
Carbon Dioxide		0	0				
pH		8.2	8.2				
Silica		3	3	50	16.7	6%	
Iron	Fe	0	0				
Turbidity		0	0				
TDS		160	185	2000	10.8	9.2	
Color		0	0				

FIG. 39.18 Simple zeolite softening for a 300-lb/in^2 boiler.

and the actual control of blowdown can then be related to conductivity for simplicity of control. The chloride test is another simple test to use for controlling blowdown.

The continuous blowdown withdrawal pipe should be located in the boiler drum in the area where the risers return to release steam behind baffles (Figure 39.20). It should never be located where it can remove feed water and fresh chemicals that have not reacted. The holes in the collecting pipe should face upward so that the pipe cannot become steam-bound, restricting the rate of blowdown withdrawal and interfering with testing.

The valve controlling boiler blowdown is usually calibrated so that the operator can make a simple adjustment if tests indicate that a change in blowdown rate is required. Because the boiler operates at constant pressure and the blow-

Identification of Analyses Tabulated Below:

A. City Water
B. Na_2Z + Acid + Degasifier
C. Split-Stream + Degasifier
D. Boiler Limits @ 300 psi
E. CR for "B" @ 300 psi
F. CR for "C" @ 300 psi

Constituent	As	A	B	C	D	E	F
Calcium	$CaCO_3$	75	Nil	Nil			
Magnesium	"	45	Nil	Nil			
Sodium	"	20	140	55			
Total Electrolyte	$CaCO_3$	140	140	55			
Bicarbonate	$CaCO_3$	100	20	15			
Carbonate	"	0	0	0			
Hydroxyl	"	0	0	0	225		
Sulfate	"	10	90	10			
Chloride	"	30	30	30			
Nitrate	"	0	0	0			
M Alk.	$CaCO_3$	100	20	15	250	12.5	16.7
P Alk.	"	Tr	0	0	238		
Carbon Dioxide		Tr	5	5			
pH		8.2	7.0	6.6			
Silica		3	3	3	50	16.7	16.7
Iron	Fe	0	0	0			
Turbidity		0	0	0			
TDS		160	160	75	2000	12.5	26.6
Color		0	0	0			

FIG. 39.19 Reducing alkalinity to reduce blowdown.

down discharges to constant pressure, this is a very reliable and reproducible method of control. The control valve can be designed for automatic actuation by a conductivity recorder.

Most plants have installed systems to recover valuable heat from boiler drum blowdown. In smaller plants, the blowdown may go directly through a heat exchanger, where the heat of the blowdown water is transferred to makeup ahead of the deaerating heater. In larger plants, blowdown is directed to a flash tank equalized to a process steam line, for example, operating at 15 lb/in^2 gage (1 bar). In rare cases, a high-pressure boiler may be blown down into a low-pressure boiler to obtain additional concentrations and steam, or it may be blown down to an evaporator in a utility station. In some plants, more than one flash tank is used, depending on the heat balance and the optimum recovery achievable in that plant.

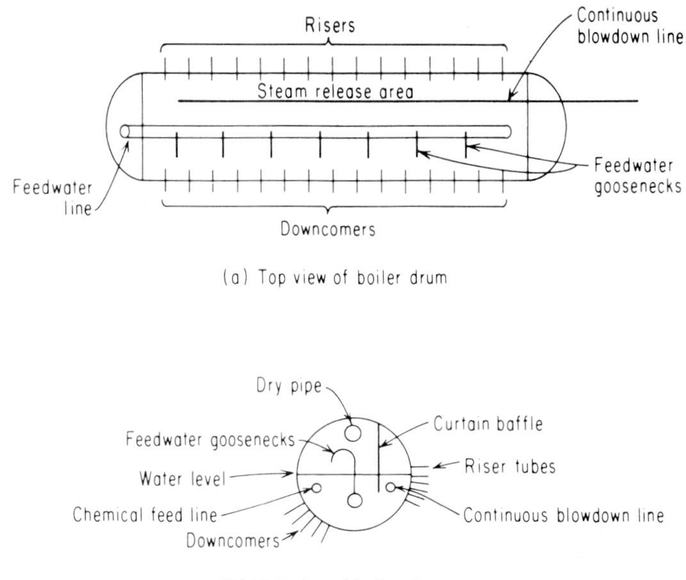

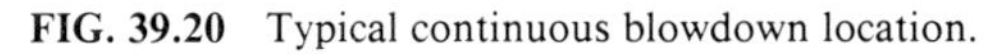

FIG. 39.20 Typical continuous blowdown location.

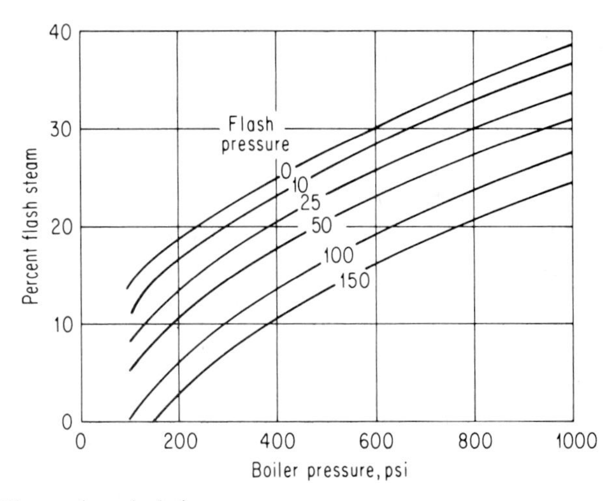

For precise calculations:

$$\% \text{ flash} = \frac{h_{f(1)} - h_{f(2)}}{h_{fg(2)}}$$

where

$h_{f(1)}$ = enthalpy of boiler water at boiler pressure 1

$h_{f(2)}$ = enthalpy of blowdown water at flash pressure 2

$h_{fg(2)}$ = enthalpy added to flash steam as it converts from water to steam at flash pressure

FIG. 39.21 The production of flash steam from boiler blowdown.

Many industrial plants produce an excess of low-pressure exhaust steam because of a variety of processes that operate at temperatures below 250 to 300°F (118 to 149°C) and have a fluctuating demand for this exhaust steam (so-called because it is steam exhausted from a bleed point, or extraction point, of a turbine or from a pressure reading station). The excess low-pressure steam may be a regular operating situation or may be intermittent. Since the flash steam from the blowdown flash tank usually goes to this low-pressure steam line which, in turn, supplies steam to the deaerator, this type of blowdown system does not recover energy in plants having excess low-pressure steam; the heat recovered at the heat exchanger simply means that less steam is condensed in the deaerator and more exhaust steam is vented to the atmosphere. However, some type of blowdown system may still be required to cool water going to the sewer or to condense the exhaust steam as a high-purity source of makeup.

The amount of steam produced by blowdown to a flash tank can be calculated by using the chart illustrated in Figure 39.21. Several typical blowdown arrangements are shown in Figures 39.22 and 39.23. Where the heat is recovered through a heat exchanger, the exchanger is normally designed to reduce the blowdown temperature to within 10 to 20°F (6 to 11°C) of the cooling water, which is usually makeup water.

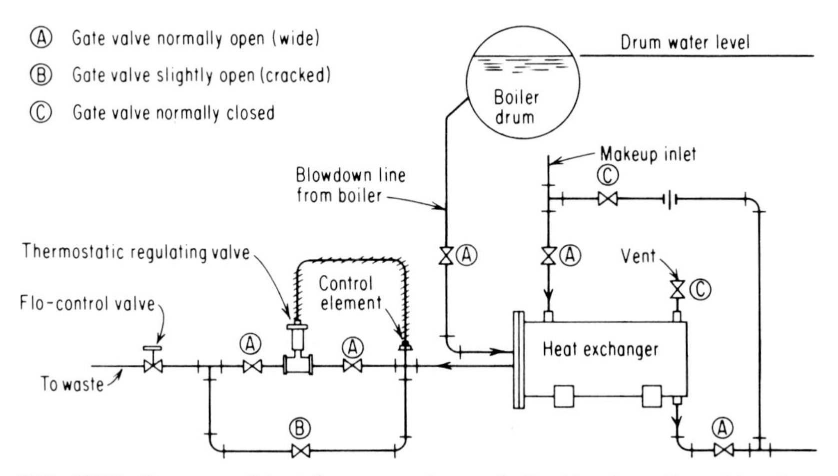

FIG. 39.22 Recovery of heat from a continuous boiler blowdown line with a heat exchanger. *(Courtesy of Cochrane Environmental Systems Division, Crane Company.)*

BOILER TYPES

A boiler is a vessel in which water is continuously vaporized into steam by the application of heat. A primary objective in designing a boiler is to provide for the greatest possible efficiency in absorption of heat. Other objectives are production of pure steam and safe, reliable operation.

Variations in the design of steam generators are almost without limit. This is partly because each new development in improving the quality of boiler makeup

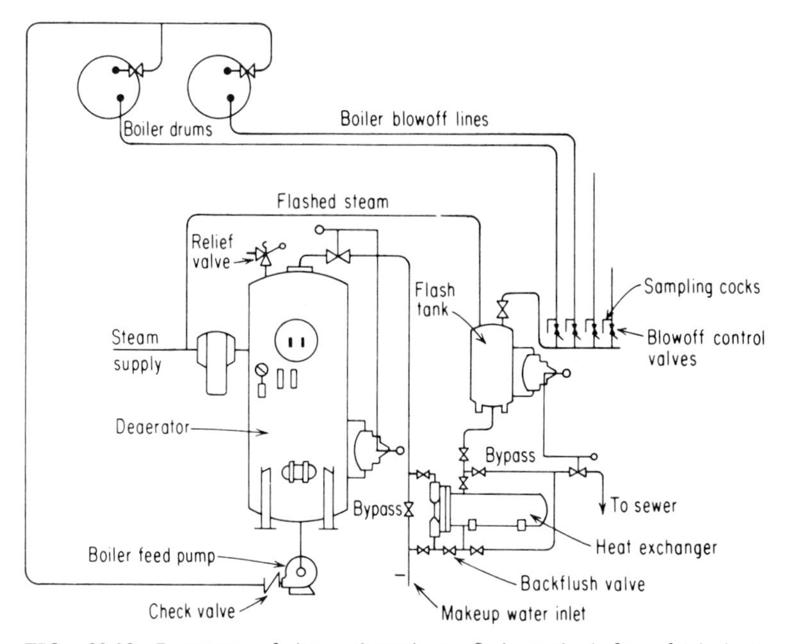

FIG. 39.23 Recovery of heat through a flash tank before final heat exchange. *(Courtesy of Cochrane Environmental Systems Division, Crane Company.)*

has influenced the boiler designer to make the steam generator more compact and efficient.

Design variations occur because of the numerous factors involved in selection and operation of a steam generator: these include capacity, types of fuel available, burner design, pressure and temperature conditions, feed water quality, load variations anticipated, and space available for the installation.

Each of these factors affects the way steam bubbles form and the mechanism of boiler water concentration at the metal surface. Figure 39.24 shows how scale initially forms as a ring at the point of bubble formation and how this ring completely fills with scale if the chemical environment is not properly controlled.

Boilers are of two general designs, firetube and watertube. In firetube boilers the flame and hot gases are confined within tubes arranged in a bundle within a water drum. Water circulates on the outside of these tubes (Figure 39.25). As the water changes to steam, it rises to the top of the boiler drum and exits through a steam header. Firetube boilers are efficient steam generators for steam requirements below 150,000 lb/h (68,000 kg/h) and 150 lb/in^2 (10 bars). Higher pressure and greater capacity require thicker plates and tube walls in this design, so watertube boilers are more economical for conditions above these limits.

The watertube boiler differs from the firetube in that the flame and hot combustion gases flow across the outside of the tubes and water is circulated within the tubes. Combustion of the fuel occurs in a furnace and some of the water tubes usually form the furnace walls.

In a simple watertube circuit (Figure 39.26) steam bubbles form on the heated side of the tubes. The resulting steam-water mixture has a density below that of

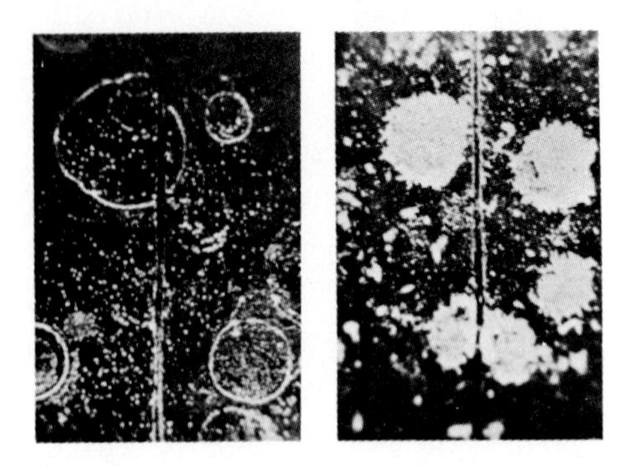

FIG. 39.24 The progression of initial scale formation.

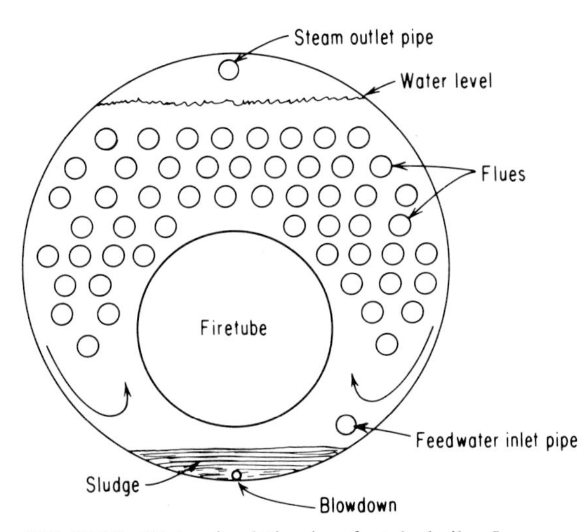

FIG. 39.25 Water circulation in a firetube boiler. Low-pressure boilers of this design were often treated by coagulation (carbonate) and were, in effect, hot process softeners. This is similar to the steam locomotive boiler.

the cooler water on the unheated side and rises, creating a circulation through the system. The steam bubbles rise until they reach the steam drum where steam is released from the water into the vapor space.

Natural circulation boilers—where circulation is induced by density differences—usually have many parallel circuits. Those sections of tubes in which heated water rises to the steam drum are called risers and those through which the cooler water descends are downcomers. Circulation usually occurs between several drums. The steam drum at the top separates steam from water; the mud drum at the bottom separates suspended solids and sludge from water. A sche-

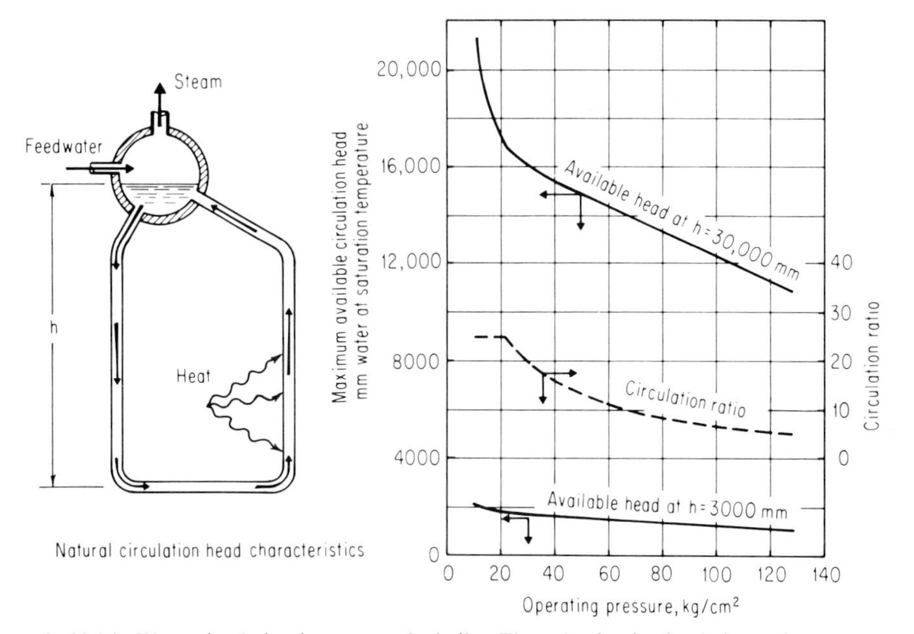

FIG. 39.26 Water circulation in a watertube boiler. The reduction in circulation ratio as boiler pressure increases is shown on the graph. Note also the reduction in available head for a short package boiler (3000 mm = 10 ft high) compared to a tall field erected boiler 100 ft high. *(Courtesy of Combustion Engineering.)*

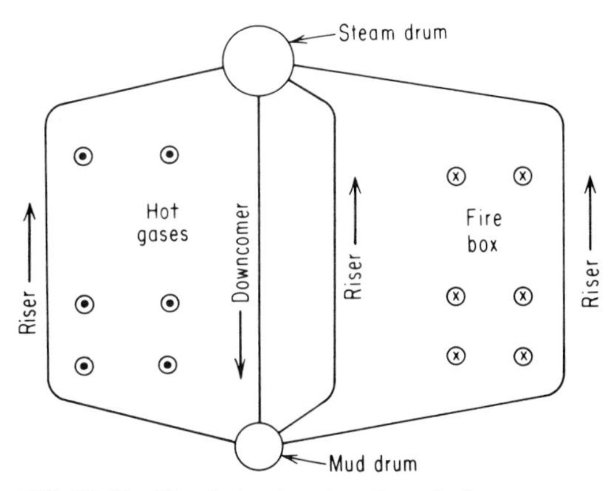

FIG. 39.27 Circulation in a two-drum boiler.

matic of such a boiler design showing risers, downcomers, and the location of the steam, and mud drums is given in Figure 39.27. These basic items are found in all natural circulation boilers regardless of the details of boiler design.

In forced circulation boilers, a pump provides the water circulation (Figure 39.28). This is a more positive control of the circulation pattern than that created simply by density difference. This is important if the boiler may operate over a wide capacity range. If space requirements limit the height of a boiler installation, the density differences available to create natural circulation become small and forced circulation becomes advantageous. It is also advantageous at high pressures as the difference in density between steam and water diminishes. Once-through boilers are special designs for utility operation.

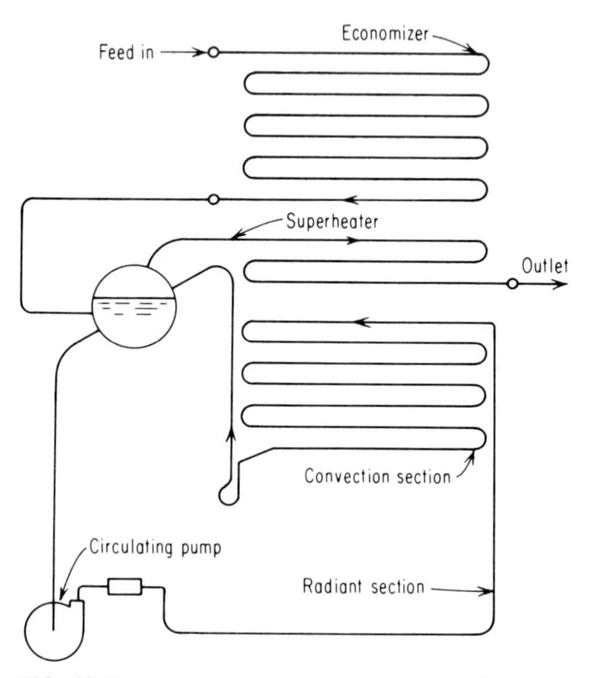

FIG. 39.28 Pumps assist in circulating water through a boiler when natural circulation is restricted by pressure or elevation.

Increased boiler efficiency can be obtained by bringing water into close contact with the source of heat. Tubes are built into the boiler furnace to absorb the greatest amount of heat possible. Usually the entire firebox, called the radiant section, is surrounded by waterwall tubes, through which water circulates (Figure 39.29). Connections between various tube sections of the boiler are accomplished by headers, named for their location, such as waterwall headers or drum headers.

Steam quality is of paramount importance in most operations if high turbine performance and long equipment life are to be achieved. Boiler water is separated from steam by cyclone separators and steam scrubbers located in the steam drum (Figure 39.30). Steam entering the steam drum is directed first toward the cyclone separator by baffles. The cyclones force the steam into spiral motion on its path toward the exit at the top of the separator, and centrifugal force separates the water from the steam. The purified steam exits at the top of the separator, while the heavier water drains out the bottom and reenters the boiler drum.

After passing through the cyclone separators, final removal of entrained water from the steam is accomplished by secondary steam scrubbers. They consist of baffles that change the direction of the steam so that water impinging on them drains back to the steam drum. The final steam should have less than 0.1 to 0.5% entrained water, depending on the design and effectiveness of the separators. High-pressure boilers perform much better than this because of their sophisticated designs required by low-steam sodium specifications.

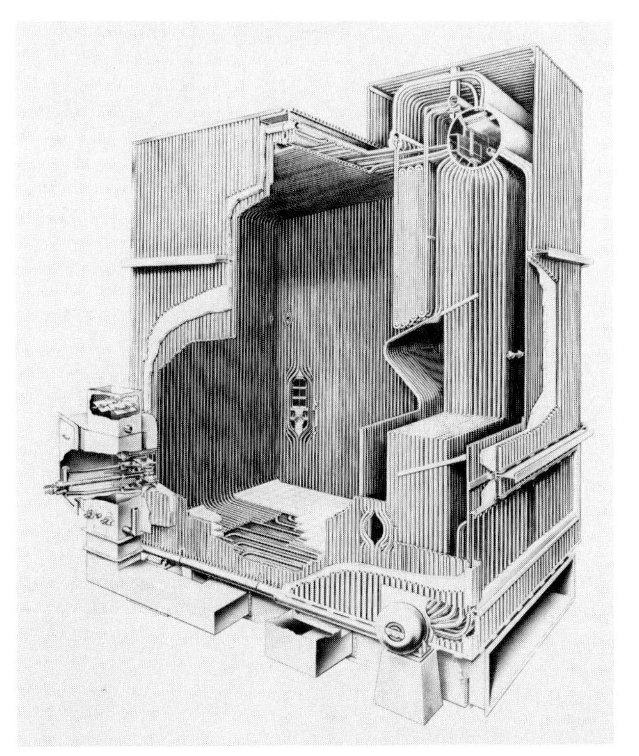

FIG. 39.29 In this boiler, the radiant section of the furnace is cooled by wall tubes in which steam is generated. In this design, the tangential location of burners produces a turbulence that promotes efficient combustion. *(Courtesy of Combustion Engineering.)*

The presence of entrained moisture determines steam quality. A steam quality of 95% means the steam contains 5% moisture. This may be measured thermodynamically by a device called a throttling calorimeter, but can be determined much more accurately by detecting the solids in the entrained moisture. Since the bulk of the boiler solids are sodium salts, an ion electrode specifically measuring sodium is used. This measures the sodium content of a sample of condensed steam.

Figure 39.31 shows a plot of sodium content in steam leaving a 250 lb/in^2 (17 bars) boiler, and how the sodium level, which reflects entrained boiler water, is influenced by boiler operating conditions. A plot of steam demand shows the sudden surges which produced gross carryover. Assuming the boiler water contains

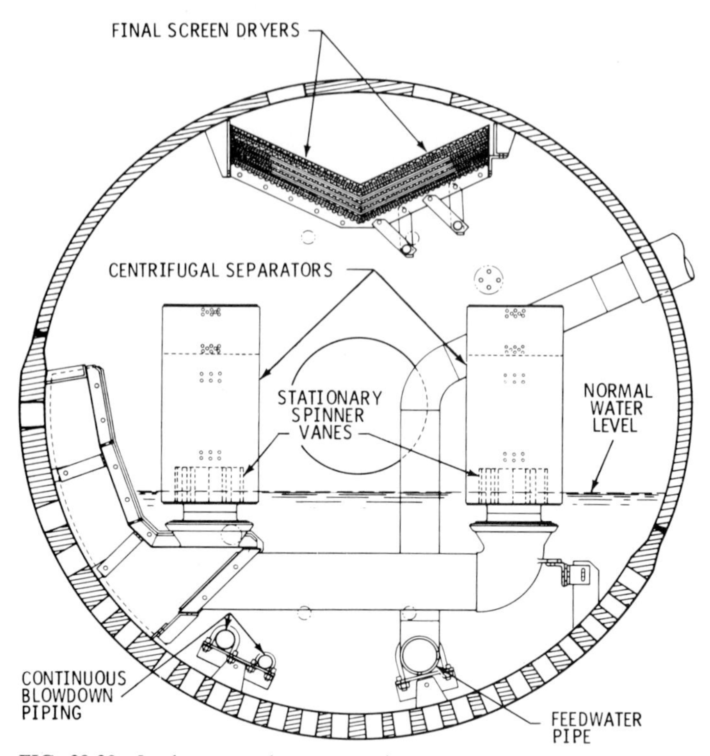

FIG. 39.30 In the steam drum, water is removed by cyclone separators followed by screen dryers. *(Courtesy of Combustion Engineering.)*

500 mg/L Na, a sodium content of 50 μg/L Na in the steam indicates the following moisture content of the steam:

$$\text{Percent moisture} = \frac{\mu\text{g/L Na in steam}}{\text{mg/L Na in boiler}} \times 0.1$$

$$= \frac{50 \times 0.1}{500} = 0.01\%$$

The average efficiency of a watertube boiler producing saturated steam is about 85%. Most of the heat loss is via hot stack gases and radiation. Improved efficiency can be achieved by adding heat recovery devices.

Table 39.11 shows a typical survey of energy losses in a simple industrial boiler plant. This is a 250 lb/in^2 oil-fired boiler delivering saturated steam for a process having no heat recovery devices. In this example, the plant has already achieved its optimum water conditions with a blowdown of only 3.5%, and increased efficiency with the existing design is achievable only by a change in firing conditions. However, appreciable energy could be recovered if the plant were retrofitted with heat recovery auxiliaries.

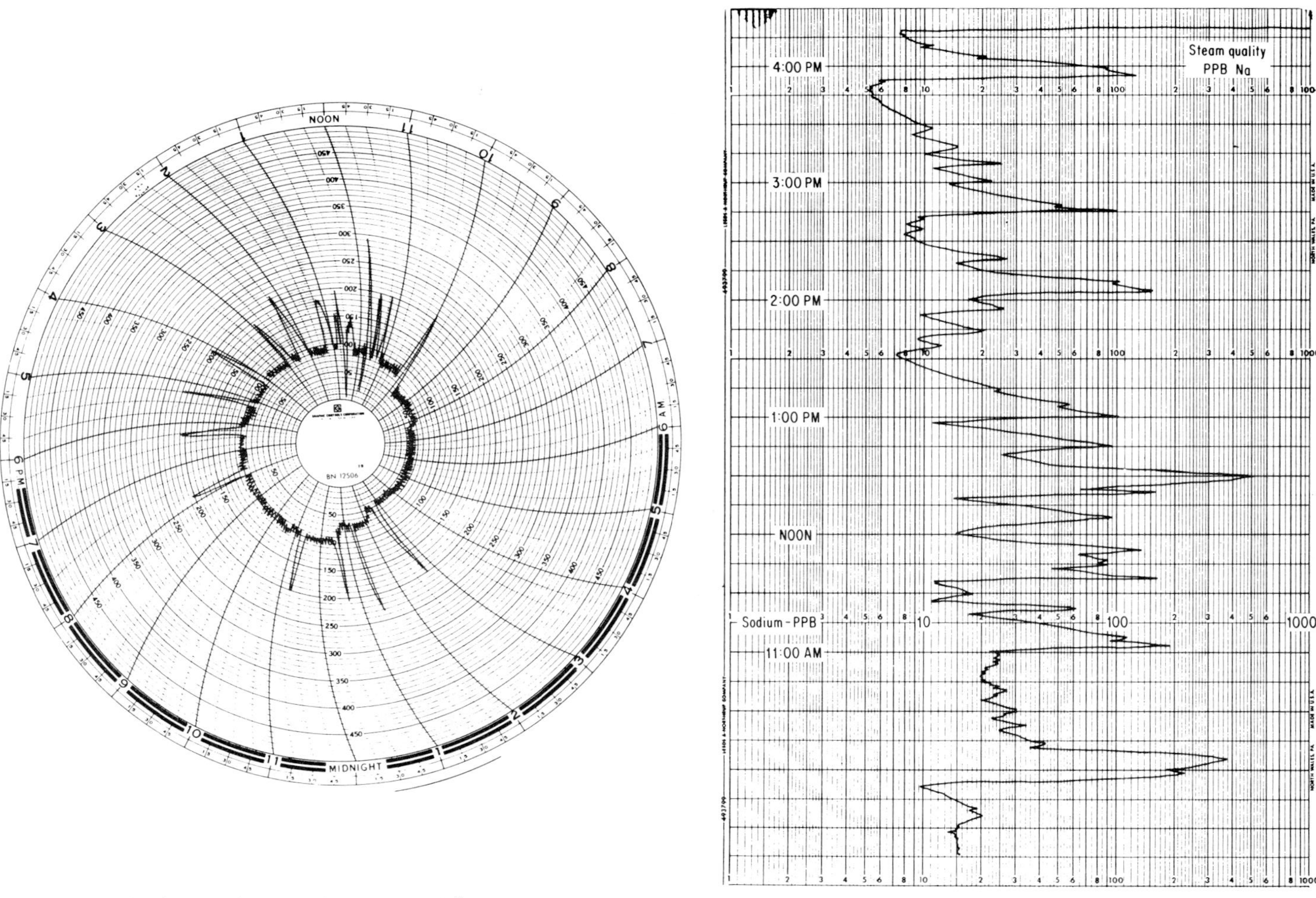

FIG. 39.31 Test of steam discharged from a 250-lb/in^2 boiler, showing increases in steam solids with corresponding sudden steam demand.

TABLE 39.11 Distribution of Boiler Losses
(All figures in millions of Btu/day)

	Actual loss	Potential	Saving
Blowdown	170.0	170.0	none
Stack heat	287.0	275.5	11.5
Excess air	96.9	49.8	47.1
Total	401.0	342.4	58.6

SUPERHEATERS AND OTHER AUXILIARIES

A thermodynamic gain in efficiency can be obtained by raising the temperature of the steam above its saturation point. This is accomplished by passing the steam through a series of tubes, called superheaters, located in the radiant section of the boiler (Figure 39.32). Similarly, steam that has only a portion of its heat content removed and is at a reduced pressure can be reheated to a temperature close to its initial temperature and reused. This is accomplished in a bundle of tubes called reheaters, also located in the radiant section.

A substantial gain in efficiency can be achieved by raising the temperature of the feed water entering the boiler to that approaching boiler water temperature by recovering heat from the hot flue gases before they are discharged up the stack to the atmosphere. This heating of the feed water, as shown in Figure 39.33, occurs in the economizer.

Similarly, air needed for combustion in the furnace can reclaim additional heat from the flue gas by passing through a series of tubes called air heaters. They are usually located immediately after the economizer. The relative increase in heat absorption gained by adding these heat recovery auxiliaries is shown by Figure 39.34.

Watertube boilers may be of the horizontal straight tube design or the bent tube design. The horizontal straight tube boiler is made up of banks of tubes, usually staggered as shown in Figure 39.35 and inclined at an angle of about 15° to promote circulation. The ends of the tubes are expanded into headers, which provide for circulation between the tubes and the steam drum. The drum may either be longitudinal or crosswise with reference to the axis of the tube bank. The advantages of the straight tube boilers include: visibility through the tubes for inspection, ease of tube replacement, low head room, and accessibility of all components for inspection and gas-side cleaning by hand if necessary during operation. Some of the inherent disadvantages are: difficult access because of the many hand hole caps and gaskets that must be removed, replaced, and tightened; limited steam disengaging surfaces with poor separation of steam and water at high steaming rates; and limited steaming rates caused by relatively low circulation and poor distribution.

The horizontal straight tube boiler is limited to production of approximately 10,000 lb/h of steam per foot of boiler width.

The bent tube design is a multidrum boiler with the mud drum at the bottom, the additional drums, all called steam drums, being located at the top. Figures 39.36 and 39.37 show a two-drum and a three-drum bent water tube boiler. Tubes in the bent tube boiler are either inclined or arranged in vertical banks within the

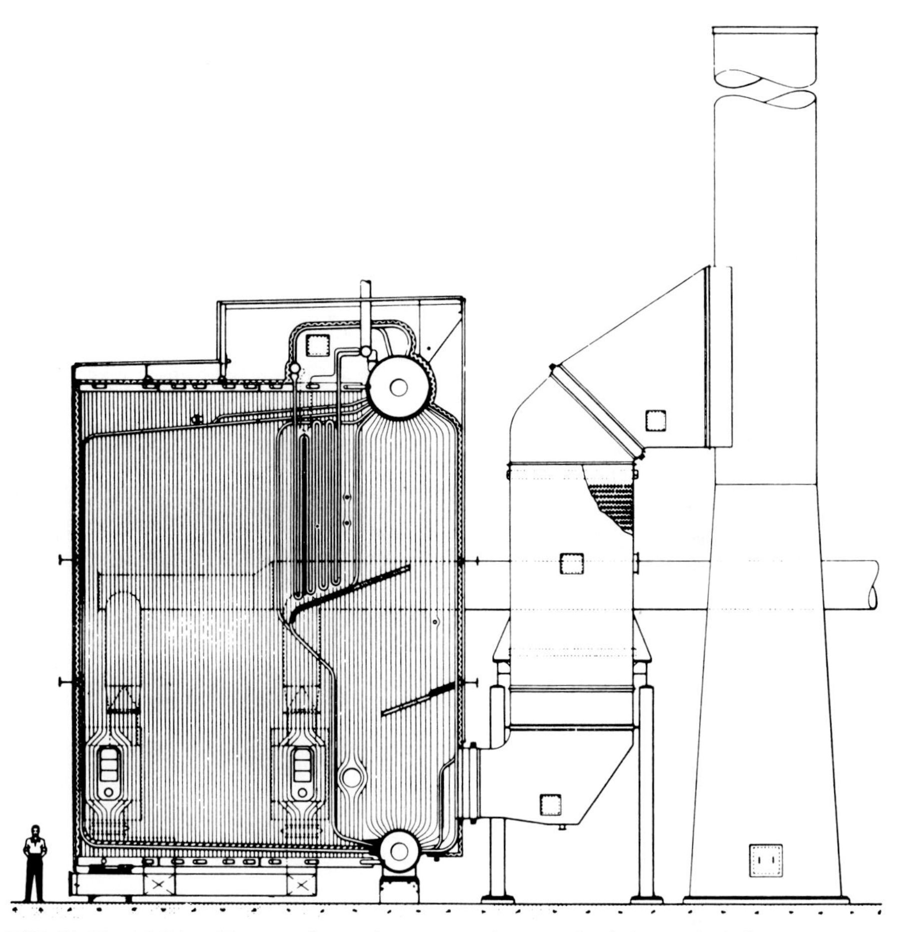

FIG. 39.32 Additional heat can be put into saturated steam after it leaves the boiler drum by a heat exchanger called a superheater, shown here in a furnace designed to burn CO produced in oil refineries as a by-product of catalytic cracking. *(Courtesy of Combustion Engineering.)*

combustion space or made up as a waterwall assembly, backed with refractories. The bent tube boiler is a rapid steamer, with quick response to fluctuating loads because of its small water volume relative to its generating capacity. Advantages of the bent tube boiler include: greater economy in fabrication and operation, accessibility for inspection, cleaning, and maintenance, and ability to operate at higher steaming rates and deliver drier steam.

Bent tube boilers are usually classified further into the number of drums and the arrangement of the tubing within the boiler (Figure 39.38). The three major classifications are:

Type A: A type A unit consists of one steam drum and two mud drums or headers arranged in an A-pattern with the steam drum at the apex and the headers at the bottom. Bottom blowdown is required from both headers.

FIG. 39.33 Heat in the stack gases is transferred to feed water in the economizer and also may be recovered by preheating combustion air. The heat-exchange bundle (*left*) is mounted in the stack (*right*). *(Courtesy of General Resource Corporation.)*

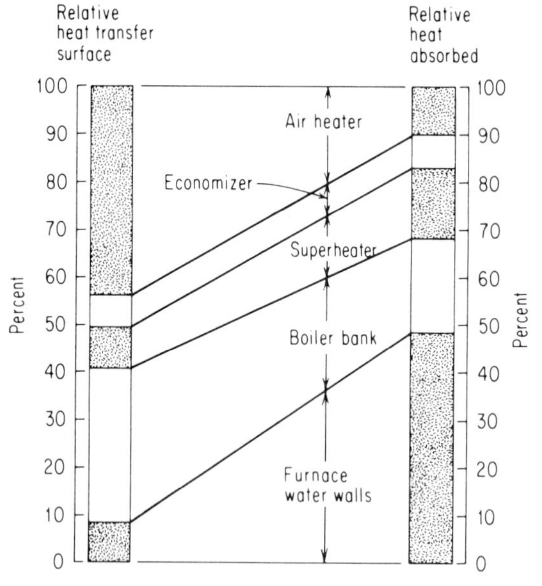

FIG. 39.34 This diagram shows the relative heat transfer duties of different sections of a steam generator. *(Reprinted with permission from Power Special Report, "Steam Generation," June 1964.)*

FIG. 39.35 Inclined horizontal tubes are used for steam generation in this boiler design with box headers at each end connected to the boiler drum. This unit was designed for marine service. *(Courtesy of Babcock & Wilcox Company.)*

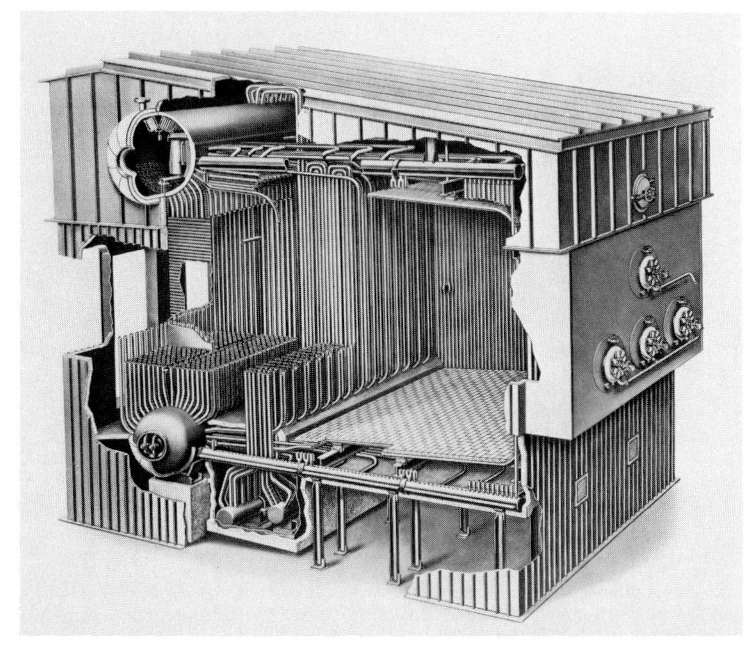

FIG. 39.36 Two-drum bent-tube boiler with waterwall tubes absorbing heat in the radiant section of the furnace. *(Courtesy of Babcock & Wilcox Company.)*

Type D: The steam drum and mud drum are located directly above one another and off to one side of the furnace in a D-pattern. A series of tubes run vertically between the mud drum and the steam drum. The rest of the tubes extend horizontally from the steam and mud drums to the furnace wall, at which point they become waterwall tubes.

Type O: In this arrangement the steam drum is also located directly above the mud drum, but both are located in the center of the boiler, and the connecting tubes are in an O-pattern.

Depending on the size and the complexity of the boiler, it may be erected on-site, or it may be preassembled at the manufacturer's plant. Because the package

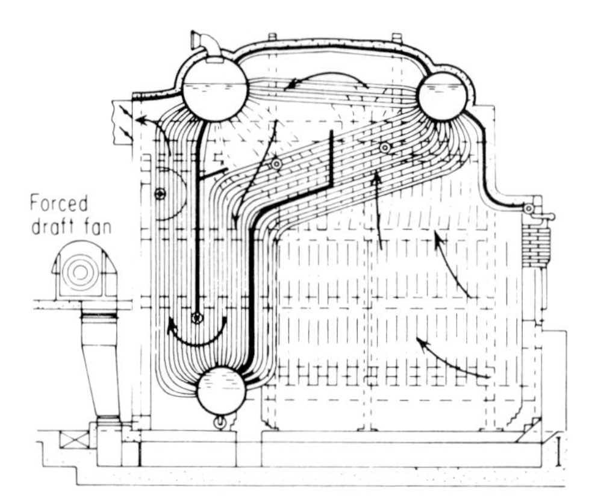

FIG. 39.37 Three-drum bent-tube boiler, common in older plants and widely used before the development of waterwall tube designs.

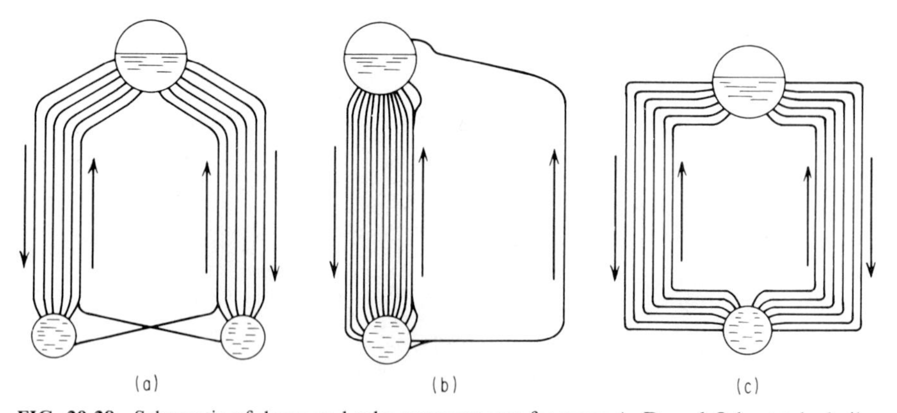

FIG. 39.38 Schematic of drum and tube arrangements for types A, D, and O bent-tube boiler designs. (*a*) A-type, with large upper drum for effective water-steam separation and lower drums (headers) for circulation. (*b*) D-type, with steaming tubes discharging near water line, provide more effective feed water and blowdown piping locations. (*c*) O-type, similar to A-type in circulation pattern with risers entering center of steam drum.

boilers are usually smaller, they employ high heat transfer rates to obtain the required steaming capacity. For this reason, they may require closer control of boiler water treatment and operation than the field-erected units.

Environmental factors and fuel availability are certain to influence boiler designs of the future. For example, fluidized bed combustion is one possibility for solving the environmental problems of burning coal. It may present new problems in the steam generation section of the system, but even so, the basic principles of boiler water chemistry will still apply.

The fluidized bed boiler is the newest of boiler designs, used both for industrial and utility service. There are variations in the basic concept, but the fundamental principles are the same: to burn the fuel in the presence of a bed of limestone so that sulfur and nitrogen oxides are fixed on the alkaline bed to keep emissions within acceptable limits. Figure 39.39 shows the basic elements of a circulating fluidized bed (CFB). Varying the ratio of primary to secondary air provides fine-tuning of combustion conditions that affect the formation of NO_x. Most of these units are being designed to burn high-sulfur coal but can be modified to accommodate other common fuels or waste products, including sewage plant sludges.

For a typical emission limitation of 1.2 lb of SO_2 per million Btu, about 6 lb of limestone is needed per pound of sulfur in the coal to produce about 80% sulfur retention. The combination of solids from coal ash and limestone usually requires recovery of the heat in the spent solids to achieve acceptable thermal efficiency.

With certain techniques of fluidized bed combustion, existing stoker-fired fur-

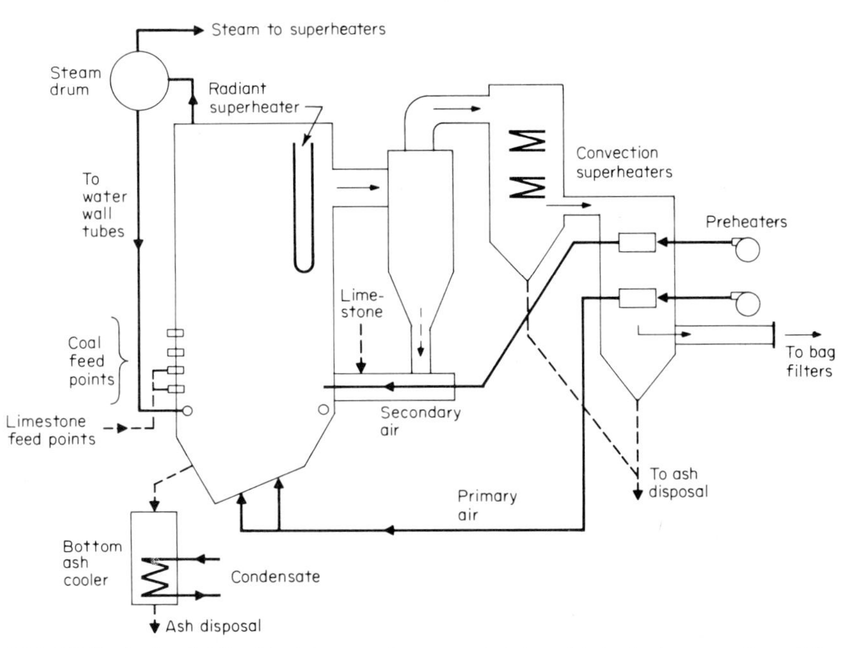

FIG. 39.39 In the fluidized bed furnace, limestone is injected into the lower section of the furnace with coal or other fuel to capture the acid gases, SO_2 and NO_x. Cyclones remove limestone fines from the flue gas for recycle to improve capture. This is a schematic of a circulating fluidized bed unit, one of several designs available.

naces can be retrofitted as limestone fluidized bed units. This new design appears to be ideal for the current trend toward cogeneration by industrial steam plants.

There are a number of aspects of boiler design that influence the selection and control of the internal treatment program. The first involves the mixing and distribution of chemicals and feed water. In many boilers, location of feed water piping, chemical distribution, and blowdown lines have been more a matter of convenience in design and fabrication than of purpose. It is essential for boiler water, feed water, and treatment chemicals to mix before the cooler water descends into the downcomers and is then heated in the radiant section. Improper mixing may produce scale in the downcomers and hinder sludge conditioning.

CONTAMINANTS IN RETURNED CONDENSATE

With increasing use of demineralized water, even for intermediate pressure boilers (600 to 900 lb/in^2), the major impurities in the feed water are no longer introduced by the makeup but rather by the returned condensate, principally as corrosion products.

These corrosion products are not solubilized by chelates and are difficult to disperse. If they deposit, boiler salts may concentrate under them, because they are relatively porous and permit boiler water to enter with only steam escaping. This may lead to caustic attack. To prevent this, boilers should be cleaned at a set frequency.

The likelihood of alkalinity attack is increased by variable loads and firing conditions, which cause flexing and cracking of the normal dense magnetite film on the boiler metal, exposing fresh metal to attack.

In forced circulation boilers, there is a tendency for deposits to form on the downstream side of flow control orifices. To avoid this, feed water should be free of deposit-forming material, especially corrosion products, so that a chemical program can be applied successfully. Condensate polishing may be required since the condensate is the source of the corrosion products. Steam separators are essential to production of acceptable steam, and their deterioration will depreciate steam quality.

Boilers are often damaged by corrosion during out-of-service periods. Idle boilers are very vulnerable to attack when air contacts moist metal surfaces. To prevent this corrosion, the boiler metal must be protected either by (1) keeping the surfaces completely dry or by (2) excluding all air from the boiler by completely filling it with properly treated water. Because of variations in boiler design, there is no single, detailed procedure that covers all steps in boiler lay-up, including both chemical and mechanical aspects. The basic principles in protecting boilers against corrosion are simple.

There are two basic ways of laying up boilers—wet and dry. In storing a boiler dry, trays of moisture absorbing chemicals, such as quicklime, are distributed on trays in the boiler drum (or drums) and the boiler is sealed. The alternate method, wet storage, involves forcing air out of the boiler by completely filling it to overflowing with water which has been specially treated. Nitrogen gas under slight pressure can also be used to displace air and blanket the boiler surfaces. Special consideration must be given to protecting superheaters during lay-up, particularly, the nondrainable type.

The choice between the wet and dry methods of lay-up depends to a great extent on how long the boiler is to be out of service. Dry lay-up is preferable for long outages; the wet method has the advantage of permitting the boiler to be returned to service on reasonably short notice. It is a good idea to drain, flush, and inspect a boiler prior to any lay-up. When time does not permit this, the boiler may be stored wet without first draining it. In this case the chemical treatments for lay-up—including catalyzed sulfite, caustic, and organic dispersants—are injected into the boiler just before it comes off the line.

TYPES OF STEAM-USING EQUIPMENT

Steam is a convenient form in which to package energy because it has a high heat content and can be generated at one point and distributed to many different energy-using units throughout a plant. The steam may be used for its heating effect or for motive power.

In heating applications heat transfer may be by direct contact of steam with a fluid, or with a solid object, as in steam cleaning of oily parts. The most common application of direct contact heating is the deaerating feed water heater in a boiler plant. In this case, the steam serves not only to heat the water at the most efficient point in the feed water system, but it also acts as an inert gas to strip oxygen and carbon dioxide from the feed water and carry these noncondensible gases out of the system; the deaerating heater is a sophisticated degasifier. Steam is used in a more primitive way for sparging and heating a variety of liquids, from food products to wastewaters.

Another example of direct contact of steam with another fluid is the thermocompressor, where high-pressure steam may be used to boost low-pressure steam to a higher pressure where it can be used. For example, an industrial plant will often have an excess of low-pressure steam at 5 to 15 lb/in^2. This plant may need 50 to 100 lb/in^2 steam for dryers, reboilers, or some other use where higher temperature is needed. A thermocompressor could boost the 5 to 15 lb/in^2 steam to 50 to 100 lb/in^2 by using live steam directly from the boiler on the jet of the thermocompressor.

Most heating applications are indirect, with a heat transfer surface between the steam and the fluid to be heated. These units are called surface heat exchangers and include a wide variety of devices such as condensers, stage heaters, process heat exchangers, reboilers, evaporators, and space heaters.

The most common surface heat exchanger, the shell-and-tube type, transfers heat from steam or another hot fluid to a colder fluid principally by conduction across a metal tube wall separating the two. The rate of heat transfer depends on the temperature difference between the fluids, the metal surface area and thickness, and the conductivity of the metal itself. Shell-and-tube heat exchangers can be of several designs. The more complex designs are usually more efficient, but may be more costly and more difficult to clean and maintain.

Closed feed water heaters, also called stage heaters, are heat exchangers used in steam plants to increase thermal efficiency by raising the temperature of the feed water as close to boiler temperature as possible with extracted or exhausted steam. A utility system may have as many as 12 feed water heaters, although it is more common to find eight, the choice being based on the most practical heat balance. Figure 39.40 shows the straight condensing feed water heater; steam is always on the shell side, with feed water passing through the tubes. Carbon-steel is used for construction of shells, heads, and tube sheets in most feed water heaters. In high-pressure heaters, tubes may be constructed of carbon steel, monel, or

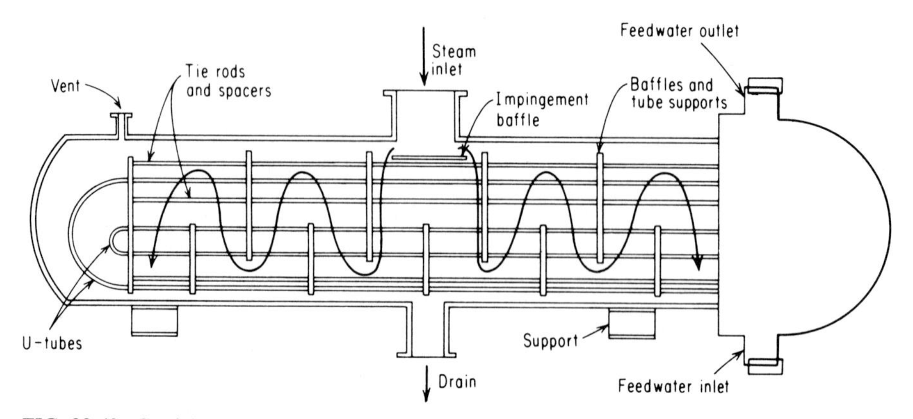

FIG. 39.40 Straight condensing feed water heater.

cupronickel; low-pressure heaters are commonly fabricated with carbon steel, stainless steel, admiralty, or 90-10 cupronickel tubes.

The surface condenser is a special design of shell-and-tube heat exchanger operating under a vacuum created by condensing steam exhausted from a turbine or from process equipment such as multiple-effect evaporators in bauxite processing plants, kraft pulp mills, and sugar refineries. Air ejectors or mechanical pumps remove noncondensible gases, such as O_2, CO_2, and NH_3, from the vapor space. Utility condensers usually include a degassing or deaerating section in the hotwell to assist in the removal of noncondensibles. Figure 39.41 illustrates a typical utility-type surface condenser.

The cooling water flowing through the condenser tubes may or may not be makeup or feed water. In a utility, there is no way to reclaim this heat because the cooling water requirements are so much greater than makeup needs. However, many industrial plants, such as steel mills, pulp/paper mills, need relatively large volumes of makeup, and condenser cooling water may be used for this purpose if

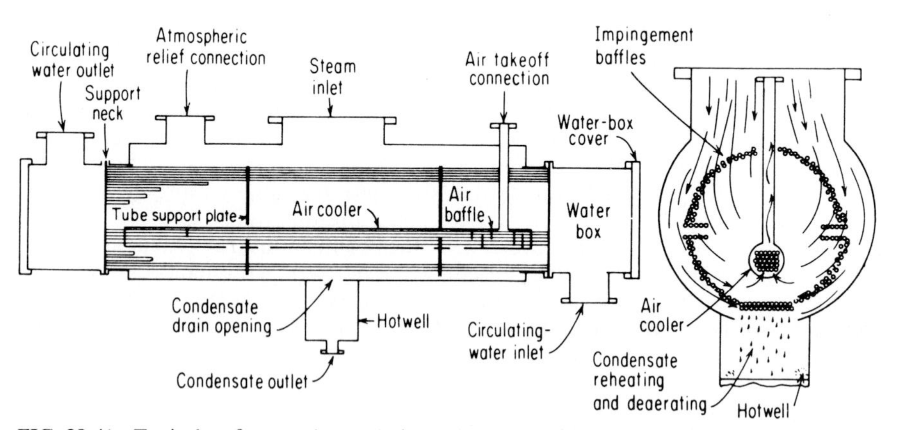

FIG. 39.41 Typical surface condenser design. *(Courtesy of Basco Division, American Precision Industries.)*

the recovered heat is usable; it may be recoverable in the winter, but not in the summer, which is often the case for pulp/paper mills.

Reboilers are tubular heat exchangers commonly used in petroleum refineries, usually inserted in the bottom of a distillation column where the tubes are surrounded with hydrocarbon being processed, and steam is supplied to a headbox from which it passes into the tubes. This is one of several process applications where steam flows inside rather than outside the tube bundle.

Space heaters are widely used for comfort heating of offices, shops, and plants. If steam is available, it is often used in tubes within the space heater.

There is a large variety of steam-jacketed processing equipment in industry, including kettles in the food industry, reactors in the chemical industry, rotary drum dryers, and evaporators. The latter usually comprise a series of vessels containing reboiler elements, with a condenser and a heat exchanger for optimum thermal efficiency. Many of these process vessels are operated intermittently, but some of the more sophisticated, such as the continuous digester in the pulp industry, operate on a continuous basis. The practice of superheating steam to increase cycle efficiency in plants generating power through the turbines was mentioned earlier. However, superheat is objectionable in process steam, principally because it creates a lower heat transfer rate than does saturated steam and equipment must therefore be larger if superheated steam is used. (The optimum heat transfer rate occurs with "dropwise condensation," where droplets of condensate repeatedly break the stagnant film on the heat transfer surface.)

Because of this it is often necessary to desuperheat steam for process use. High-purity water is injected into the desuperheater, and the evaporation of this water extracts heat from the steam; the injection rate is controlled automatically by temperature sensors that modulate the flow of injection water. In some respects related to the desuperheater, the attemperator is a device used to control superheat temperature at the boiler itself, particularly where steam load is variable and other superheat temperature controllers tend to cycle or "hunt."

It is unusual to find deposits in most steam process equipment, particularly if the equipment is in continuous, steady operation; but deposits do occasionally occur. They usually develop because of inleakage into the steam side from the process side of the vessel, caused by the vacuum that develops during shutdown, drawing in contaminants through a leaky tube. Gross carryover of boiler solids with periodic wetting and drying in the process equipment may also cause deposits as the solids concentrate.

A much more common problem with process equipment is corrosion, often due to inleakage of oxygen caused by vacuum during shutdown or under throttle steam conditions. Concentration of CO_2 also occurs with improper devices for temperature or pressure control. This shows up quickly as a localized drop in condensate pH where the CO_2 is allowed to concentrate. Ammonia attacks copper alloys, in areas of O_2 inleakage, particularly where conditions permit its concentration beyond the initial NH_3 concentration in the steam itself.

Once condensate has formed in steam-using equipment it must be drained to avoid flooding and reduction of heat transfer surface exposed to steam. (In some cases, however, controlled flooding is a preferred method of product temperature control instead of throttling the steam admission valve.) This is the job of the trap and subsequent condensate collection and forwarding systems. There are many designs of traps, the major types being (a) thermostatically operated, (b) float- or bucket-operated, and (c) throttling orifice controlled. All of them are expected to be effective in (a) venting air (or noncondensible gases)—especially important where the trap handles condensate from a cyclic operation, such as a cooking kettle in a food-processing plant; (b) draining condensate with a minimum of restriction; and (c) promptly sensing the presence of steam and restricting its loss from the system. Each of these designs has its special niche, but it has taken the energy crisis to educate industrial steam users to the excessive energy losses that occur when the wrong type of trap (or a faulty or damaged trap) is used in a specific application. As a result, in many modern plants, an operator reporting to an "energy czar" is trained as a trap expert, responsible for selection of the correct trap for each job and for its regular inspection.

The trap, of course, does not convey the condensate; it is a throttling device, and the condensate flows because of pressure drop or level gradient. One of the problems of condensate handling is the flashing that occurs with pressure drop, resulting in an increase in the specific volume of the flowing stream. Another common problem is air inleakage that occurs at low loads when a steam admission valve is throttled. An example is the common problem of controlling condensate pH in plants that use steam-heated space heaters during spring and fall weather changes.

STEAM USED FOR POWER

The major use for steam is motive power, since this is its sole use in the utility industry, which is the largest consumer of energy in most industrialized countries. Even in moderate-sized industrial plants where the chief reason for generating steam is for process requirements, the steam may be put through a turbine at high pressure, with the exhaust steam then being used for the process needs. Turbines are the major prime movers because they are compact and efficient and require little maintenance; however, reciprocating piston steam engines are still found in many older plants throughout the world. These prime movers may be used for direct drive of pumps, compressors, and other mechanical devices, or they may be connected to a generator to supply electric energy for motor-driven devices and other electrical accessories.

The development of the modern steam turbine was a monumental engineering achievement—first, for the designer who was able to translate the complex mathematics of thermodynamics into blueprints, and second, for the builder who could convert the intricate design into precise shapes for manufacture almost on a mass production basis. In the simple, single-stage turbine, steam is directed through a series of carefully shaped nozzles onto blades, or buckets, fixed at the circumference of the disk attached to the turbine shaft. This is known as the impulse stage. As the steam leaves the rotating blades, a second set of stationary nozzles may be used to collect it and redirect it onto a second set of blades, in which case the set of nozzles and blades is called a reaction stage. Most turbines contain multiple stages, some impulse and some reaction, depending entirely on the duty required of it and the balance of cost and performance. A simple single-stage turbine is shown in Figure 39.42, and a multiple-stage utility-type turbine is shown in Figure 39.43.

There are two basic categories of steam turbine operation: condensing, where exhaust steam discharges to a condenser at subatmospheric pressure; and noncondensing, where steam exhausts into process steam headers under pressure (e.g., 100 lb/in^2 is common in the paper industry). Turbine operation and design are further classified as to the flow of steam through the turbine: flow patterns include straight flow, reheat, automatic extraction, and nonautomatic extraction.

FIG. 39.42 This simple, solid wheel turbine is commonly used to drive auxiliary equipment such as boiler feed pumps and rotary compressors. Cutaway (*top*); plant installation (*bottom*). *(From Power Special Report, "Steam Turbines," June 1962. Courtesy of Terry Corporation.)*

FIG. 39.43 This large turbogenerator is typical of nuclear plant units. The cutaway section shows high-pressure and low-pressure turbine blades. *(From Power Special Report, "Steam Turbines," June 1962. Courtesy of General Electric Company.)*

In straight flow systems (Figure 39.44), steam at full throttle is directed by nozzles through the entire bank of turbine stages to exhaust. Reheat turbines (Figure 39.45) gain efficiency by discharging steam at an intermediate stage for reheat in the boiler furnace, with the higher temperature steam being returned to the next stage of the turbine for further work. In automatic expansion, steam is extracted from the main flow at one or more points for process use, with the balance passing through to exhaust (Figure 39.46). Nonautomatic turbines bleed steam at one or more stages for feed water heating (Figure 39.47).

The most troublesome type of turbine problem related to water treatment and boiler operation is the accumulation of material on the turbine blading. This may be either inorganic, which is most common, or organic. Some of the volatile materials released from the boiler may condense as the pressure and temperature drop as the steam passes through the turbine. Silica has been mentioned as one of the components of boiler water that can volatilize, and it is one of the principal contributors to turbine deposits; it tends to deposit on lower pressure stages of the turbine, usually at 50 lb/in^2 or below. The silica limits set for boiler water assume that the steam will be passed through turbines. If the turbine exhausts at 100 lb/in^2, there may be no problem in carrying higher silica concentrations than those

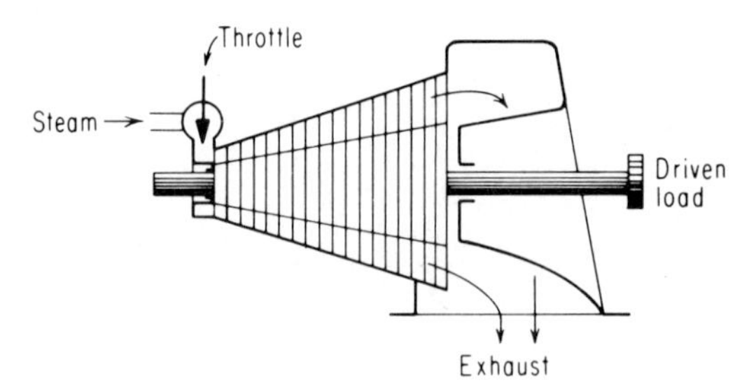

FIG. 39.44 Straight-flow turbine. *(From Power Special Report, "Steam Turbines," June 1962.)*

recommended in Tables 39.8 and 39.9. The relationship between silica in boiler water and silica in steam has been carefully studied, and it is possible to predict the point at which silica will sublime.

Turbine deposits may originate from organic materials in the boiler steam as well as inorganic materials. Usually they are caused by process leaks into the condensate system, but they may also originate in the raw water either from naturally occurring organic materials or from organic wastes discharged into the water source from industrial plants or from municipal sewage treatment plants.

There are numerous additional uses for steam in industrial operation. Many fabrication shops use steam for operation of forges, hammers, and presses. The

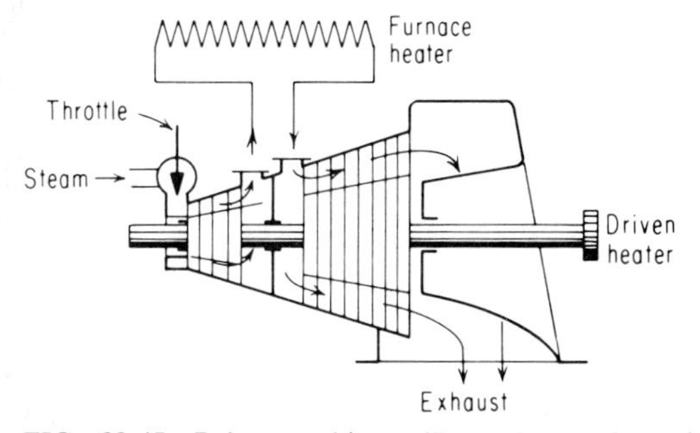

FIG. 39.45 Reheat turbine. *(From Power Special Report, "Steam Turbines," June 1962.)*

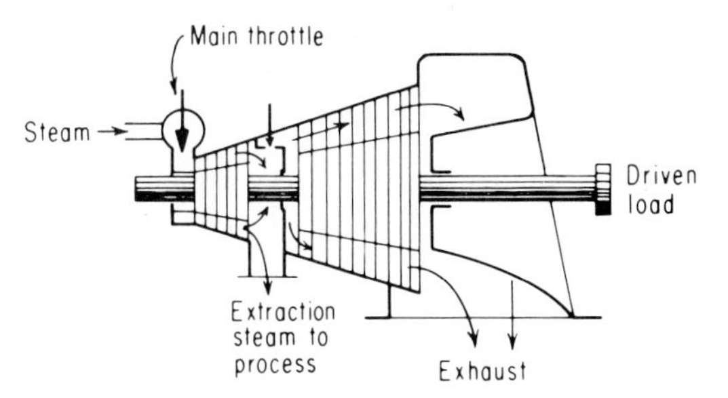

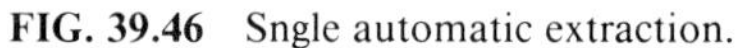

FIG. 39.46 Sngle automatic extraction.

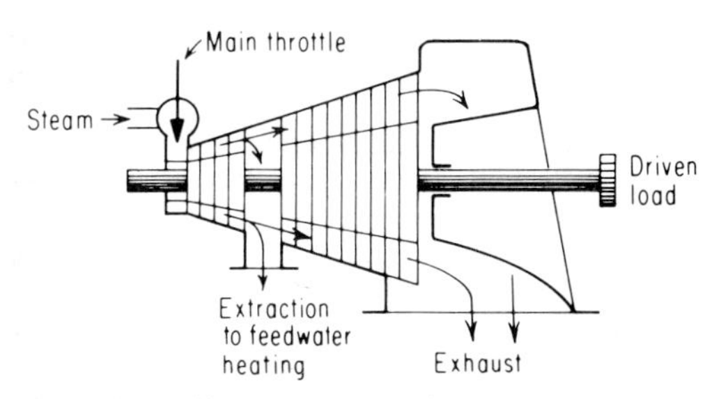

FIG. 39.47 Single nonautomatic extraction.

steam cylinders usually require saturated or slightly wet steam and are lubricated with specially compounded oil, so in plants of this kind it is common to find oil in the exhaust steam and condensate. Steam jet ejectors with barometric condensers are frequently used to produce vacuum for such diverse purposes as degassing steel and distilling heavy hydrocarbon fractions.

CONDENSATE RETURN SYSTEMS

Condensate produced when steam is used in any kind of process is seldom cooled measurably below the steam temperature. Because it is hot, close to the steam temperature, the collection system and piping used to handle it must be carefully selected. The piping is generally larger than that used for cold water because pressure drop will cause steam to flash from the flowing condensate, choking the pipe and restricting flow. Condensate may be picked up by a pump for boosting to the best point of return, or it may be delivered to a point of lower pressure simply by the pressure gradient.

Turbine condensate is normally collected in a hotwell, and a level is maintained in the hotwell so that the pump transferring the turbine condensate to the deaerator will have adequate net positive suction head. The level in the hotwell is maintained by returning some of the pump discharge, depending on load fluctuations.

Process condensate is also usually returned by pumps to the deaerating heater. Condensate receivers collect the process condensate and maintain level control so that the pump handling the hot condensate will have adequate suction head. Even so, special designs of centrifugal pumps are usually required for handling hot condensate. Condensate systems can become quite sophisticated, as in a paper mill, where condensate from high-pressure dryer rolls may be flashed to lower pressure steam in the condensate system and then routed to another dryer section. Stage heater condensates are normally handled by gravity, returning to a lower pressure stage heater, the deaerator, or even the condenser hotwell.

Once the condensate has been collected, the proper point of return must be decided upon. In the utility station, the entire flow of condensate may be polished through some type of ion exchange system before being returned to the deaerating heater. In industrial plants, if the condensate is contaminated it will be sent to a treatment plant before returning to the deaerator. For the most part, condensates are returned to the deaerating element itself, as they may contain dissolved oxygen and other gases; however, high-pressure returns free of dissolved oxygen may be sent directly to the storage section of the deaerating heater to flash and supply steam for the deaerating operation.

For the most part, the condensate handling system is of ordinary carbon steel construction, although pump impellers, valve trim, and heat exchange tubes are usually of copper alloys. Because condensate is usually hot, if corrosive agents are present, the rate of corrosion will usually be greater than what would be expected in cold water. The principal agents of corrosion are carbon dioxide and oxygen. The CO_2 is normally produced by the breakdown of alkalinity in the boiler and the oxygen may be drawn into the system by inleakage of air or of water containing dissolved oxygen (pump sealing water, for example). Inspection of the condensate piping provides a good clue as to the cause of corrosion, as shown in Figures 39.48 and 39.49. The principal cause of corrosion of copper alloys is ammonia in systems containing O_2 (Figure 39.50).

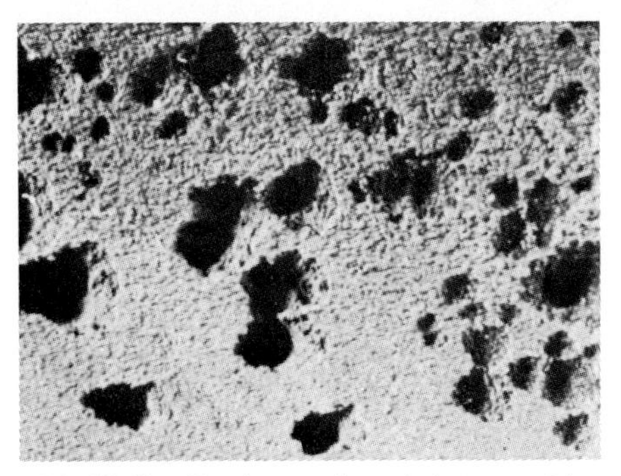

FIG. 39.48 Condensate line pitting caused by oxygen and carbon dioxide.

FIG. 39.49 Uniform CO_2 attack below water line in a condensate pipe.

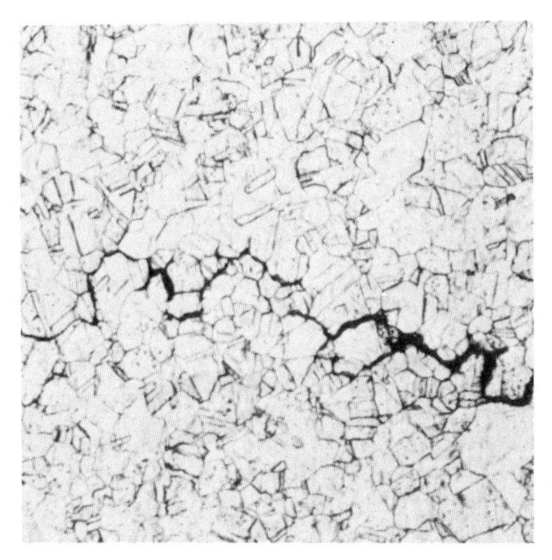

FIG. 39.50 Attack of copper alloy by ammonia in a stage heater.

Without adequate treatment of the steam, the pH of condensate would normally be low because of the presence of carbon dioxide. The application of volatile alkaline amines will control attack by neutralizing carbonic acid, thus raising the pH value of the system. In a tight system, the neutralizing amine is often adequate for the complete corrosion control program. However, many systems are operated intermittently or under throttling (flow-restricting) conditions where oxygen inleakage can occur. At these higher levels of oxygen, neutralization is inadequate as the sole protective measure against corrosion of steel piping. In such cases, volatile filming amines are added to the steam, which upon condensation produce a waxy substance on the metal and provide a barrier between the flowing condensate and the pipe wall so that corrosion cannot occur. Figure 39.51 illustrates the action of filming corrosion inhibitors in a condensate system.

FIG. 39.51 FIlming inhibitors coat condensate piping with a nonwettable film, protecting the metal surface from corrosive attack.

Hydrazine and other all-volatile oxygen scavengers may be used both for pH correction and oxygen scavenging, but it becomes uneconomical when high levels of carbon dioxide and oxygen occur, a common condition in most industrial operations.

In plants where gross contamination occurs, the source and cause should be located and corrected. An example of such an occurrence is the use of steam for producing hot water through a heat exchanger. The industrial operation may require the hot water at a specific temperature, 150°F for example, and a thermostatic element is installed in the water line to regulate the steam flow into the heat exchanger according to water flow and exit temperature. At low water flows where the steam demand is low, the steam admission valve may be so fully throttled that there is actually a vacuum in the vapor space. Most of these systems are designed for pressure operation, and under vacuum, air leakage is common. If a neutralizing amine is used in a system of this kind, it may be easy to locate inleakage of this sort because the air in the industrial atmosphere will contain enough CO_2 to drop the pH of the condensate at that particular point. So a way to find the source of air inleakage in a complex industrial plant is to sample condensate at all sources and compare the pH of the sample condensate with the pH of a condensed steam sample or a sample of condensate known to be free of atmospheric contamination.

Where attack of copper alloy has been found to be caused by ammonia, filming inhibitors will usually prevent further attack by preventing O_2 from reaching the surface. If the ammonia concentration is high, reduction of ammonia in the pretreatment system should be considered.

Neutralizing Amines

The most commonly used neutralizing inhibitors are amines such as morpholine, cyclohexylamine, and diethylaminoethanol. The ability of each product mentioned to enter the condensate or water phase is indicated by its vapor-to-liquid distribution ratio. This ratio compares the concentration of amine in the vapor phase to the concentration in the water phase.

Product	Vapor to liquid distribution ratio*
Morpholine	0.4 to 1
Cyclohexylamine	4.0 to 1
Diethylaminoethanol	1.7 to 1

* At atmospheric pressure.

In order to neutralize carbonic acid, the amine must be present in the water phase. The distribution ratio indicates the preference of an amine for the water phase or the vapor phase. An amine such as morpholine, preferring the water phase, will be present in the initially formed condensate at high temperatures. On the other hand, cyclohexylamine tends to remain with the steam to enter the condensate as the temperatures decrease.

Because of their differing vapor-to-liquid distribution ratios, two or more such amines may be used together to provide effective neutralization programs for complex systems.

Neutralizing amines are fed to the feed water after deaeration, boiler steam drum, or steam header. They are controlled by monitoring the returned condensate pH from samples taken at the beginning, middle, and end of a condensate system.

Filming Inhibitors

Inhibitors used to film condensate systems are amines with chainlike molecules. One end of each molecule is hydrophilic (loves water), and the other end is hydrophobic (hates water). The hydrophilic end attaches to the metal, leaving the other end to repel water. As the molecules accumulate, the surface becomes nonwettable. The film, therefore, provides a barrier against metal attack by water containing carbon dioxide, oxygen, or ammonia. Since the molecules also repel each other, they don't tend to build up layers or thick films. Instead they remain a monomolecular, protective film.

A film one molecule thick actually improves heat transfer in condensers, dryers, and other heat exchange equipment. By promoting dropwise condensation, an insulating water film between the water and metal surface is prevented.

Good distribution of filming inhibitors is of prime importance in preventing condensate corrosion. Protection depends on the maintenance of a continuous film. Since steam and condensate can wash away the film, it must be constantly repaired by continuous feed of the inhibitor.

Octadecylamine and certain of its salts were the first chemicals to be used as filming inhibitors in steam-condensate systems. However, because of their waxlike nature (whether supplied in flake or emulsion form), it was difficult to put these chemicals into uniform solutions for feeding. A relatively narrow condensate pH range of 6.5 to 8.0 is required for the octadecylamine to form a film and to remain on the metal surface. To overcome these limitations molecules formulated specifically for boiler plant conditions have been developed as alternatives to octadecylamine.

Most filming inhibitors are normally fed to the steam header but may be fed to the feed water or boiler drum; the latter feed points result in some loss of product to blowdown. If only the process equipment needs protection, the inhibitor may be conveniently fed into the desuperheating water at the process steam header. Regardless of the feedpoint, however, the inhibitor should be fed continuously for best results. Dosages are not based on oxygen or carbon dioxide content of the steam. The amount of inhibitor required is set according to the system's surface area. Creation of an effective film is a physical process, highly dependent on flow rates, feeding, and testing techniques.

EVALUATING RESULTS

There are several good ways of finding out how much corrosion is occurring in a system and how effective a prevention program is. Because this requires extensive monitoring, it is essential that sampling be done at the significant points of the system and with adequate facilities. There must be a quill installed in the line that projects into the flowing stream; a sample taken along the pipe wall is meaningless. Sample lines must be of stainless-steel tubing.

1. pH monitoring: This involves checking the pH throughout a system, necessary to make sure that:

   a. Sufficient amine is being fed to neutralize carbon dioxide

   b. The proper amine is used to give total system protection

   c. Process contamination or air inleakage is not occurring

   When taking condensate samples, to avoid carbon dioxide flashing off and giving false results, the sample must be cooled prior to contact with the atmosphere. This means using a sample cooler attached to the condensate line. The sample may be throttled at the outlet but not at the inlet. This is done to prevent a vacuum from occurring in the coil and drawing in air, which will give false results. Failure to set up a cooler properly will give inaccurate and misleading results.

2. Conductivity monitoring: The conductivity of the returning condensate can be an indicator of process contaminants and corrosion.

3. Carbon dioxide testing: By actually measuring the carbon dioxide content of the condensate, the problem of corrosion from carbon dioxide can be directly evaluated.

4. Hardness: Similarly to conductivity measurement, this can be an indicator of process contaminants, particularly from cooling water leakage.

5. Test nipples: The use of test nipples, installed in steam or condensate lines, permits both visual inspection of system conditions and a measure of corrosion.

The nipple can be weighed before and after the test interval to determine weight loss caused by corrosion.

6. Test coupons: These have also been used to evaluate condensate corrosion conditions. Preweighed coupons in holders are inserted in condensate lines. After an arbitrary time interval (usually at least 30 days), the coupons are removed, cleaned, and reweighed. The difference in original and final weights, when coupon surface area and exposure time are known, gives an indication of rate of metal loss owing to corrosion. It should be noted, however, that such test coupons usually cannot identify bottom "grooving" and threaded-joint corrosion as they occur in actual piping.

7. Testing for corrosion products (iron and copper) in condensate is a preferred and widely used means of observing corrosion trends. Since metal corrosion products in condensate are mainly present as insoluble particulates rather than in dissolved form, methods of sampling that insure representative and proportional collection of particulates are important.

A general level of corrosion products in the 0 to 50 μg/L range indicates the system is in control. Because of the problem of sampling error when looking for this small quantity of particulates, a relatively large amount of sample needs to be run. Two accepted methods used are:

a. Visual estimation of iron concentration based on the degree of discoloration of a membrane filter through which a known volume of condensate is passed. Estimation is made by comparing the membrane pad, after filtration of the sample, with prepared standard pads having color equivalents of specific iron values (Figure 39.52).

b. Very precise quantitative determinations of total iron and copper values are easily made by the Analex method. In this procedure, a sample stream of condensate is allowed to flow through a small plastic cartridge containing a high-purity filter of ion exchange materials for a period of 7 to 30 days. Particulates are captured by filtration, while dissolved solids are retained by ion exchange. By a unique laboratory process, the total weight of each metal is calculated. A series of successive Analex analyses gives a complete and accurate picture of condensate corrosion trends. Normally, when inhibitors are used in a system previously untreated, the corrosion products observed in the condensate will increase temporarily. Most inhibitors have a "detergent effect" and tend to slough off old oxides. This must be taken into consideration when evaluating test results. Normally a plant uses more than one of these monitoring methods (e.g., pH monitoring and corrosion coupons). Although it may seem inconvenient when initially setting up a pro-

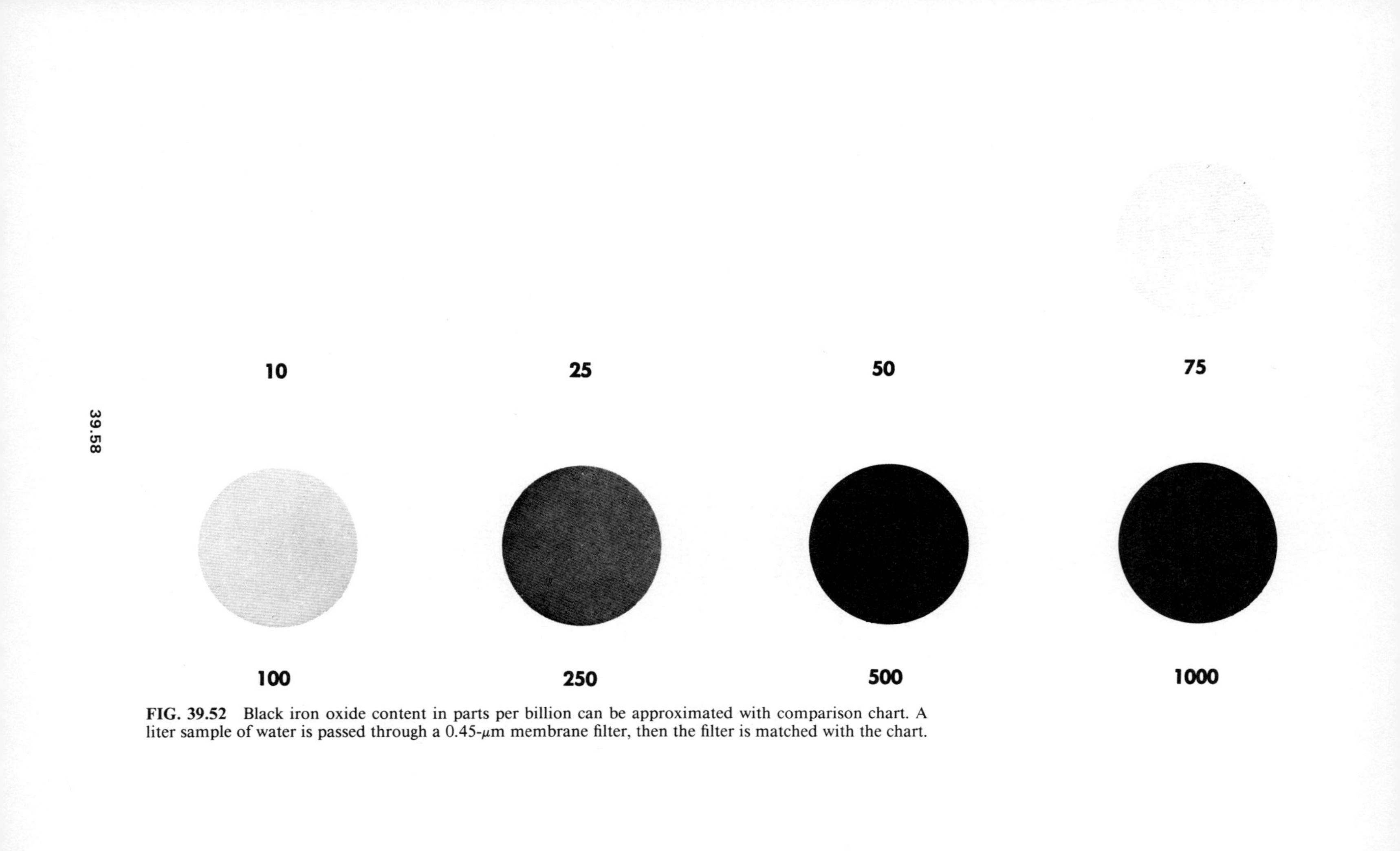

FIG. 39.52 Black iron oxide content in parts per billion can be approximated with comparison chart. A liter sample of water is passed through a 0.45-μm membrane filter, then the filter is matched with the chart.

gram to monitor the condensate, the results are more than worth the time invested in avoiding downtime and gaining energy savings in the high-heat value and high-purity of the returned condensate.

SURVEYING A BOILER SYSTEM

The purpose of surveying a boiler system is to develop physical and chemical data defining the system and its relationship to plant operations with the goal of operating at peak efficiency with minimum maintenance. Physical or mechanical data cover the various types of equipment in the system water supply and treatment equipment, steam-consuming units, condensate handling, preboiler auxiliaries, and the boiler itself. Heat balance data are also included. The chemical data include analyses of all water streams of significance—raw water, condensate streams, treated water, boiler feed water, boiler water, and wastes. This extensive data gathering is necessary because each boiler system is unique and the optimum performance standards change as new equipment is added, water supplies are changed, and relative costs of fuel, capital equipment, and labor shift.

Making a complete power plant survey requires review of six specific sources of information:

1. Utilities system flow diagram
2. Descriptions of the mechanical equipment and water systems, including physical dimensions and capacities
3. Analyses and flow records of steam and all water sources in the boiler system, including wastes
4. Description of all treatment chemicals, use rates, and details of injection points
5. Deposit/corrosion analyses from past equipment inspections
6. Detailed boiler inspection report of both water and fire sides
7. Metallographic analyses of any system metal failures
8. Operator log sheets and service reports

Power Plant Flow Diagram

The flow diagram showing the utility system and all its parts traces water from its source through the pretreatment and then into the feed water system, where it joins the condensate. It shows the path of the feed water to the boilers, and the distribution of steam to turbines and process. The flow diagram incorporates all steam-using process equipment and the returned condensate flow. Figure 39.53 shows an example of a utilities diagram. A diagram like this should be prepared or checked by the investigator by touring the plant and interviewing operating personnel. Blueprints may not give a current picture because many of these are prepared at the time of construction and may not have been revised as changes were made after startup.

While preparing the flow diagram of the system, data related to that system should be collected. Figure 39.54 shows a typical plant survey form that is used in recording data of the utility system shown in Figure 39.53.

The survey now requires updated water analyses of all the sources contributing to the boiler feed water, including raw water, clarified and filtered water, zeolite-treated water, individual condensate returns, and the feed water itself.

Discretion must be used when deciding how complete an analysis is required of each of these samples. For example, if the water supply is colored, but there is no evidence of problems having developed in the plant from organic materials, there is no need to determine organic matter using such tests as BOD, COD, or TOC on any of the sampling points along the makeup path. If there is occasional hardness leakage into the condensate, the only analyses which may be needed at the condensate sampling points would be total hardness, iron, pH, and conductivity. The troublesome point in a condensate system can often be detected by filtering the individual samples through membrane filters and comparing the filter pads based on the filtration of a uniform sample volume, 500 mL, for example.

On the other hand, condensate being sewered may require rather sophisticated analysis to determine its suitability for boiler feed water. It may contain chemical contamination from processing operations, oil, high concentrations of dissolved gases such as NH_3 and HCN, and high levels of various suspended solids such as corrosion products or debris.

Determining the flows of the many streams usually requires ingenuity on the part of the investigator, because for the most part, flow meters are found only on

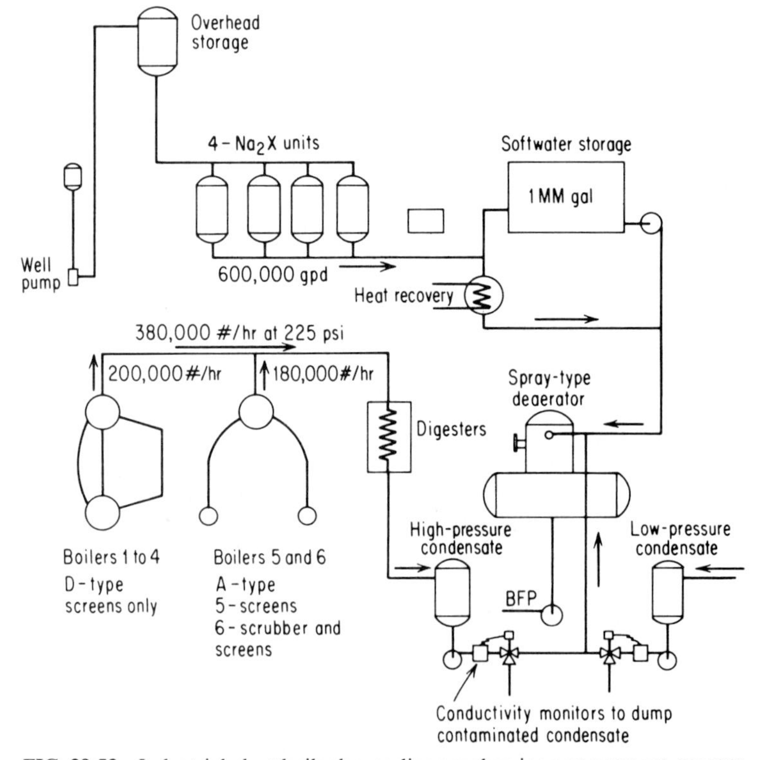

FIG. 39.53 Industrial plant boiler house diagram showing pretreatment, process steam use, and condensate return.

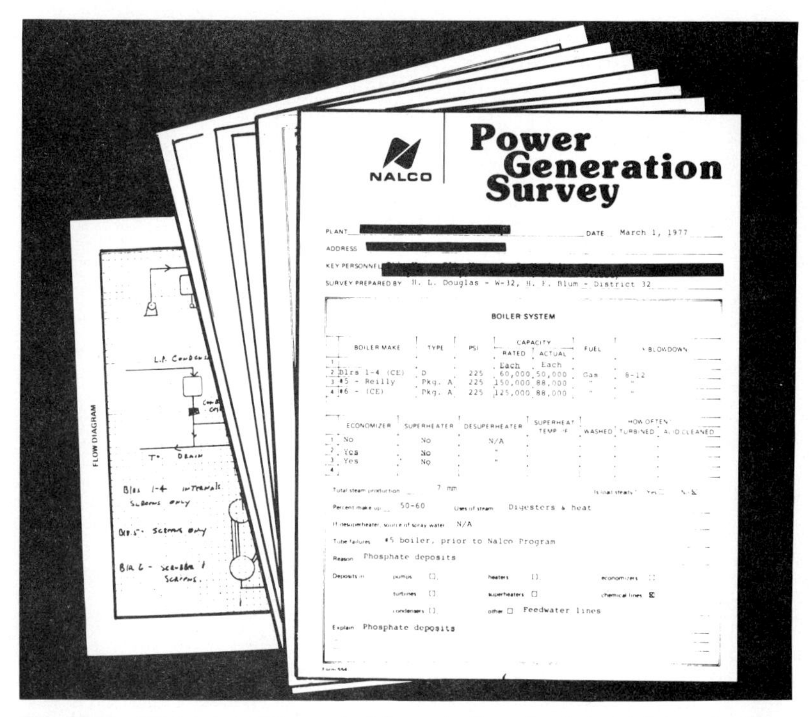

FIG. 39.54 Plant survey data required for evaluation of an industrial steam/condensate system.

the clarifier inlet, the individual zeolite units, total feed water to the boiler, and total steam production. Determining the ratio of condensate to makeup is relatively easy, since the condensate in most plants is quite pure compared to the makeup, and conductivity is a simple means of determining the ratio. However, where the plant has a demineralizer, metered flows must be used since the makeup is then as high in quality as the condensate. In determining the ratio of various condensate streams to one another, sometimes temperature differences provide a handy means of making the calculation. For example, in the illustration used here, the temperature of the turbine condensate will probably be about 100°F (38°C), with the process condensate coming back at about 200°F (93°C). The actual temperatures can be measured, and the temperatures of the combined condensates determined for calculating the ratios.

STEAM SAMPLING

The most difficult sampling is the steam itself, because a representative sample is hard to obtain without making special provisions for it. Figure 39.55 indicates the recommended procedure for sampling steam and condensing this for analysis. The analysis may only require a determination of pH to establish the level pro-

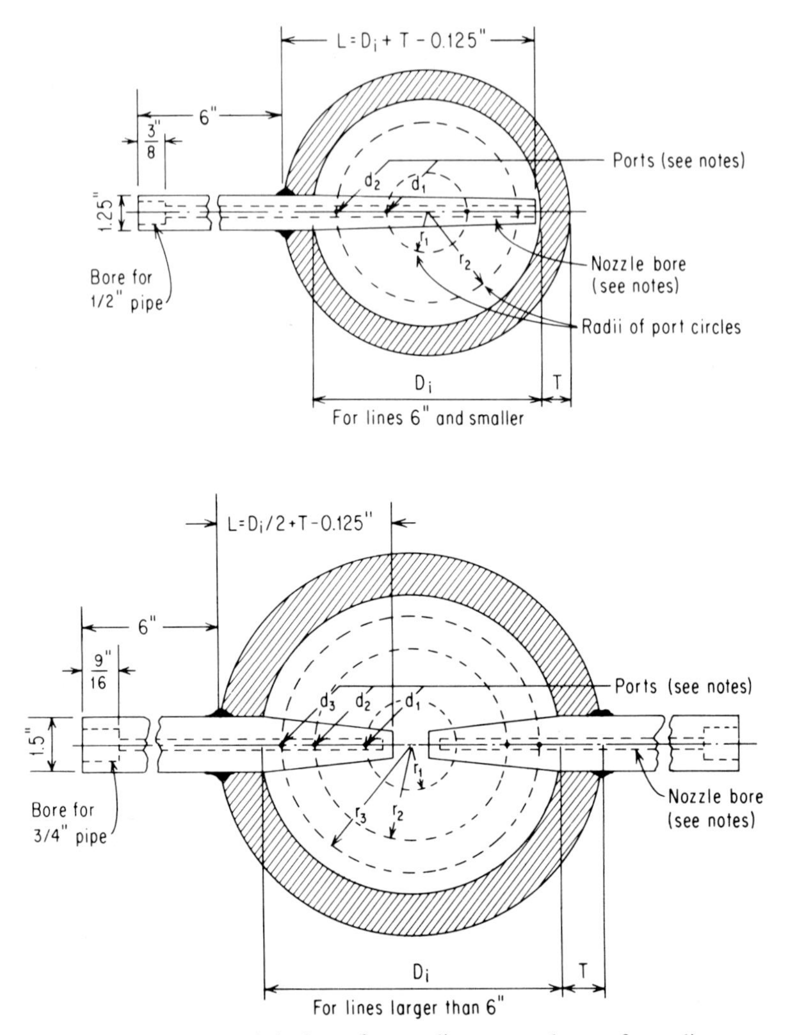

FIG. 39.55 Recommended scheme for sampling saturated steam for quality monitoring. Superheated steam requires a modified nozzle to recirculate a portion of the condensed sample for desuperheating. *(From ASTM Standard Method D 1066-69.)*

duced by the application of a neutralizing amine as a benchmark for interpreting the pH of samples of condensate. On the other hand, steam sampling is extremely valuable as a means of locating sources of trouble in a system.

Hydrogen analysis of steam will help determine boiler corrosion rate with respect to steaming load. Hydrogen evolution results from active corrosion cells or chemical decomposition. Experienced consultants should conduct these tests so that correct interpretations can be made. Figure 39.56 shows data obtained from the use of a hydrogen analyzer (Figure 39.6) to determine the hydrogen level in a steam produced from a boiler having process-contaminated feed water. The corrosion action within the boiler caused by destruction of the protective film of magnetite produced hydrogen directly from the attack of the boiler steel, which is easily observed on the hydrogen analyzer chart.

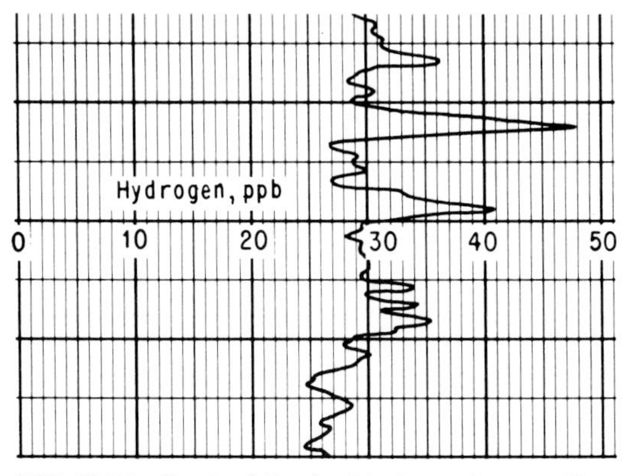

FIG. 39.56 Graph of dissolved hydrogen in steam from a 250-lb/in^2 industrial boiler. A careful study detected periodic contamination of condensate by perchloroethylene that hydrolyzed to HCl in the boiler generating hydrogen.

A common method for determining steam quality is by use of the specific ion electrode for sodium. Since the salts dissolved in the boiler water are sodium salts, the presence of sodium in the steam sample is a direct indication of carryover. Figure 39.31, discussed earlier, shows the chart of such a steam analysis in a plant where periodic carryover was occurring.

STEAM PLANT EFFICIENCY

If the survey is being conducted to improve the efficiency of the steam plant, special attention must be given to two major water-related operations, the boiler blowdown losses and condensate losses. The boiler blowdown can be calculated from chemical ratios. Condensate losses will then be determined as the difference between makeup and feed water, which are usually metered quantities. Of course, a complete study of the boiler plant is needed to establish the actual operating efficiency, the water-related factors being only part of the energy losses. A typical energy survey for a small plant has been shown earlier in Table 39.11.

Recommendations for improvement may include such steps as changing the combustion conditions in the boiler to reduce excess air; reduction in blowdown; and a plan for increasing the amount of condensate return by either condensate treatment, or simply a repiping of condensate that had been lost to a sewer.

DEPOSIT AND FAILURE ANALYSIS

The next item on the survey agenda is the collection of deposit analyses from past boiler inspections. Again, discretion is needed in deciding how extensive these analyses need be. Typical analyses have been given earlier in this chapter. For the most part, chemical analyses are adequate for evaluating the problem, but in some cases the crystallographic structure, which must be determined by x-ray, will also be needed. Table 39.12 shows the analysis of a deposit as determined both by chemical analysis and by x-ray diffraction. A great deal can be gained by having the observation of the person collecting the sample recorded as part of the data. For example, the physical appearance of the sample as taken from the site can be valuable information. Many deposits show a laminar structure, indicating periodic changes in the water environment forming the deposit. If the problem should require such painstaking investigation, the individual layers can be analyzed to piece together a story of the causes of variable environmental conditions. If a deposit has been found in the superheater tube, it is often enough simply to know whether the deposit is water soluble or not. So there is considerable judgment required on the part of the investigator to decide what is really needed in analyzing deposits found in various parts of the water circuit.

Metallographic evaluation of specimens of failed tubes, pump parts, or process equipment is a valuable technique for determining the cause of failure so that the program can be modified or changed to correct the cause of failure. A complete history of the failed section is needed, showing its location in the boiler circuit or in the process equipment and the conditions prevailing at the time of failure. As

TABLE 39.12 Elemental and X-Ray Diffraction Analyses of Brown Deposit from a Steam Drum

Elemental analysis			
Inorganics in dried sample (scaled to 100%)		Major calculated combinations (weight percent)	
Iron (Fe_2O_3)	64	Net loss at 800°C	19
Magnesium (MgO)	10	Magnetic iron oxide	51
Calcium (CaO)	8	Magnesium silicate	13
Silicon (SiO_2)	8	Basic calcium phosphate	8
Phosphorus (P_2O_5)	4	Calcium sulfate	4
Sulfur (SO_3)	4	Magnesium sulfate	2
Manganese (MnO_2)	1	Magnesium oxide	1
Carbonate (CO_2)	0		

The following elements were absent (below detection limit):
Na, Al, Cl, K, Ti, V, Cr, Co, Ni, Cu, Zn, Sr, Sn, Ba, Pb.

Loss at 800°C: 19%; $CHCl_3$ extractables: 0%.

X-ray analysis	
X-ray diffraction indicates the presence of the following:	
Fe_3O_4	Magnetite
Fe_2O_3	Hematite
MgO	Periclase (probably)
$Ca(OH)_2{:}3Ca_3(PO_4)_2$	Hydroxylapatite (possibly)

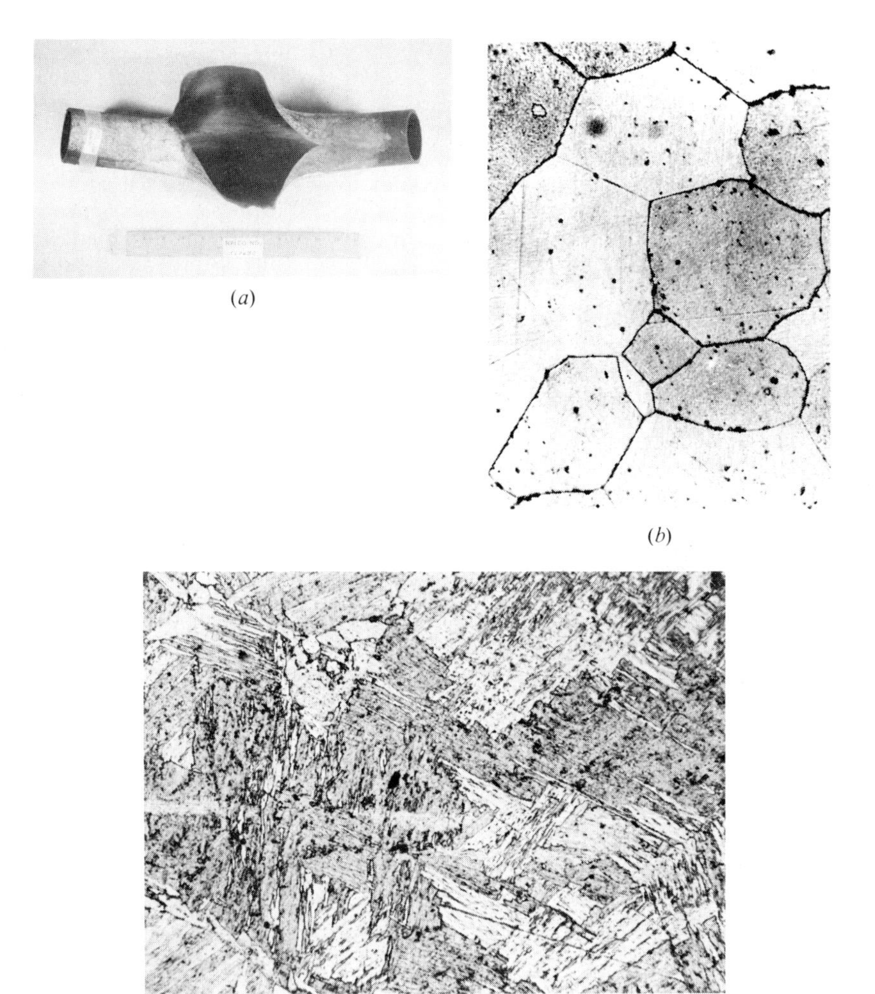

(*a*)

(*b*)

(*c*)

FIG. 39.57 (*a*) Failure of a superheater tube. (*b*) The normal pearlitic structure (Fig. 39.2) was changed by overheating. (*c*) The bursting of the tube cooled the overheated metal to produce the quenched structure.

part of the metallographic study, any deposits found on the specimen are analyzed as additional clues to the cause of failure. Figure 39.57 illustrates the section of a failed superheater tube. The section of the tube opposite the failure was of normal structure (pearlitic). The change in metal structure under conditions of overheat are seen in the second section of this illustration, and the nature of the metal at final failure is shown in the third panel, where the overheated metal was actually cooled as the steam rushed through the rupture. This rapid cooling caused quenching of the overheated metal to produce the crystalline structure seen in the third panel. This failure was caused by gradual accumulation of deposits within the superheater tube owing to improper layup procedures when the boiler was shut down. During layup, corrosion of the internal tube surfaces occurred, and the corrosion products led to ultimate failure of the tube by insulating the tube surface from normal heat transfer rates.

Finally, the investigator should review all operating log sheets and service reports related to the utility system. There will be found the record of normal and abnormal conditions which were observed day-to-day by the plant operators, and such items as repeated failure of a chemical pump, for example, can be extremely valuable information in pinpointing the cause of plant problems. If a plant has a boiler water consultant, service calls are usually carefully recorded and provide another source of information for investigation. A service of this type not only serves to check on the accuracy of the chemical tests conducted by the operators on a routine basis, but it should provide recommendations for improvement in operations based on the experience and judgment of a person who has serviced many plants and can provide recommendations for continual upgrading of the plant based on knowledge of progress made in the water treatment industry at a variety of locations.

CHAPTER 40
EFFLUENT TREATMENT OPTIMIZATION

In the United States, effluent water quality criteria are no longer matters of local agreements between regional regulatory agencies and industries; they are established by state and federal statutes. In considering effluent treatment for discharge into a receiving stream or lake, each plant outfall must be considered individually as each requires a separate discharge permit from the Environmental Protection Agency (EPA). This is because the effluent characteristics vary widely; even in the same industry, plants having the same capacity and processes may produce effluents of widely different quality and quantity. This may be due to differences in raw material composition, sources of plant water supply, water use and recycle, geographic location, or age of the facility.

SOURCES OF EFFLUENT

Data on raw water consumption may not provide accurate information as to the quantity of effluent that will require treatment. Instantaneous plant effluent flows can be as small as 10% or over 200% of the incoming supply rate. Even on a daily average, the effluent may be only 20% of the raw water supply because of evaporation losses, or may exceed the raw water flow on a rainy day (Figure 40.1). To obtain information regarding type of treatment and equipment size requirements, all sources of water must be examined.

Although normally containing little contamination after the initial flow, storm water runoff can contribute significant volume to the plant effluent. Along the Gulf Coast, where precipitation may exceed 60 in (150 cm) per year, plants must consider runoff in their treatment plans. In other areas, climatic conditions may provide a means to eliminate storm water discharge by favorable evaporation rates. Storm water may be collected and impounded as a good plant water source—often lower in dissolved solids than the normal supply; instead of being a discharge load, the storm water can become a benefit. For example, a refinery on a 500-acre (200-ha) site in an area where annual rainfall is 36 in (90 cm) discharges an average storm water flow of 1.5 mgd (2.62 m^3/min). A significant portion of this may be reclaimable if provisions are made to collect it during the rainy season.

The quantity of wastewater flow can be influenced by infiltration of groundwater into the sewer system, especially during the rainy season. Sewers located

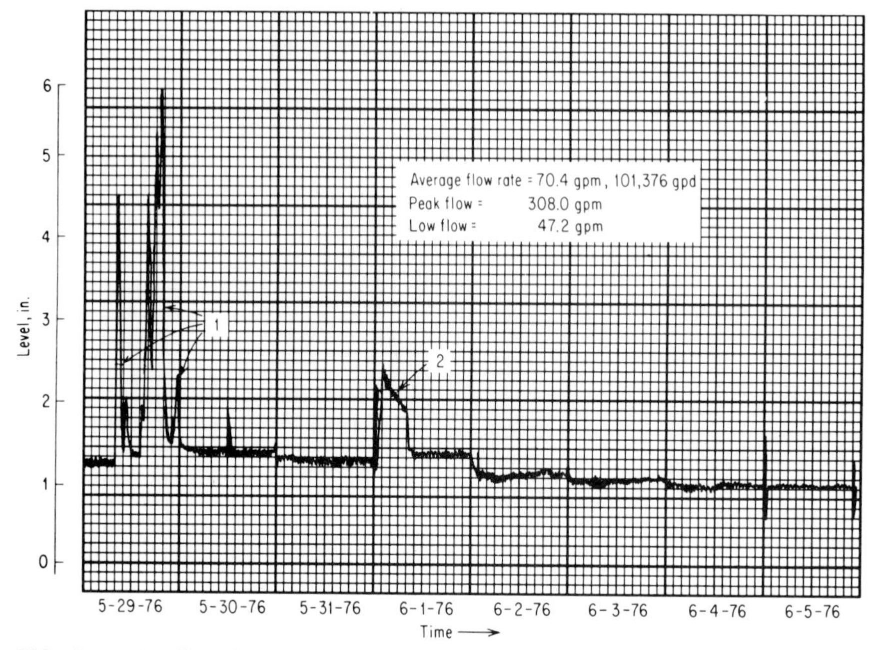

FIG. 40.1 The effect of storm water on plant effluent flow rate. 1 and 2 are peaks caused by rainfall.

below the groundwater table or near stream beds may have continual infiltration entering through poorly constructed joints and leaking manholes or manhole covers. Along the seacoast, the rate of infiltration may be influenced by tides. Rates of infiltration range from 15,000 to 50,000 gal/day/mi (35 to 120 m^3/km) of sewer.

PROCESS WATER USE AND CONTAMINATION

Discharge patterns are normally determined by examining individual process water uses. Process waters may be required for washing, rinsing, direct contact cooling, solution makeup, chemical reactions, process condensation, and gas scrubbing operations. Spent process waters normally generate the largest load of contamination in plant effluents on a daily average basis. This is usually caused by inadvertent leakage of process liquors into the water system. Individual process water uses within a plant may also produce contaminant discharges that may be incompatible in a combined industrial-municipal treatment system. Table 40.1 lists some process waters in several major industries where salinity increase and industry-specific pollutants occur.

Contamination may occur from spills and leaks of process or product liquors from manufacturing operations. The volume from spills is usually low in comparison to total water used, but owing to high concentration, its loss to the sewer often produces a significant impact on the waste treatment operation. Similarly, storage and transport of raw material or product may require attention as a potential source of contamination. In analyzing wastewaters, it should be assumed that

TABLE 40.1 Processes that Increase Effluent Salinity

Industry	Process
Pulp/paper	Bleaching, pulp washing
Petroleum	Crude-oil desalting
Steel	Blast-furnace gas scrubbing
Coke	Condensation of moisture from coal
Textile	Washing of fiber or yarn
Food	Peeling, washing, cleanup
Machined parts	Cleaning, paint-spray control
All industries	Water treatments producing strong liquid wastes (e.g., ion exchange, reverse osmosis)

raw materials and products in the process area may somehow find their way into the sewer, and these should be sought in the analysis if they are objectionable contaminants.

Maintenance operations from cleanup and repair of equipment, tankage, and tank trucks or cars also produce concentrated wastes.

UTILITY SYSTEMS

Additional wastes are generated in preparing raw water for use in the plant's utility system. The water treatment plant must dispose of suspended or dissolved solids removed from the supply water as well as chemicals added for treatment. These wastes may discharge on a continuous or batch basis to the plant sewer from clarifiers, filters, softeners, and demineralizers.

The plant's utility system discharges less water than it receives. Steam is lost to the atmosphere, but the greatest evaporation loss occurs over cooling towers, when these provide the main source of heat removal. To control the buildup of undesirable materials in its water systems caused by evaporation, the utilities plant removes a small portion of water from the boilers or cooling systems as blowdown. The blowdown may be only 5 to 10% of boiler and cooling tower makeup, accounting for the large "shrinkage" between raw water and effluent flows. This blowdown contributes some salinity to the effluent since it has been concentrated by evaporation and also contains chemicals that have been added to control corrosion, scaling, and fouling. A typical water balance of a refinery having an evaporative cooling system shows a 60 to 80% loss of water because of evaporation in the plant.

Finally, plant washrooms and dining facilities generate sanitary wastes. If these are not sent to a municipal treatment system, on-site treatment must be provided for disposal.

PLANT SURVEY

Industrial waste treatment planning begins with a review of plant plot plan and sewer diagram. The plot plan provides general information showing plant size, storage of raw materials, intermediates and finished products, manufacturing operations, materials flow, water supply, and effluent disposal sites. The sewer drawings show sewer and manhole locations, floor drains, separated waste collection systems, downspouts and storm drains, and any existing effluent treatment facilities. A review of these drawings and available rainfall data will give a picture of sewer loadings, including storm water and plant wastewater.

A valuable reference to possible contaminant discharges is a review of purchase order records, chemical inventories used in conjunction with manufacturing operations, and final production. Discrepancies between input and output provide insights into the types and probable quantities of contaminants that may be present in the plant effluent (Table 40.2). In reviewing the list of chemicals used in the plant and potentially present at certain times in the effluent, the survey team should be on the alert for toxic pollutants cited by the EPA as "priority pollutants." If these are present in the plant, their analysis will be required in plant wastewaters. Any treatment program proposed for the plant will have to reduce these pollutants to acceptable levels—and that may be a level not detectable by current analytical methods. Table 40.3 lists the EPA priority toxic pollutants to be aware of in making a pollution survey.

The potential for disposing of spent chemicals by dumps to the plant sewer or other careless practices must be considered and fail-safe procedures developed to prevent such occurrences.

Inspection of the manufacturing facilities should follow the review of drawings. This tour should focus on water use and resulting waste discharges from each process to identify problems not readily evident by data review alone. Excess water usage, faulty control, poor housekeeping, and inadequate equipment are common examples of factors contributing to unnecessary contaminant discharges. The inspection offers the opportunity of correcting and preventing contaminant discharges at the source rather than removing the contaminant "after the fact" in the plant effluent. Critical to reducing pollution in a plant is a program to make employees as aware of pollution as they are of safety. As with a successful plant safety program, employee awareness requires continual communication by posters and training sessions.

TABLE 40.2 Raw Materials List for a Small Plant Producing Textile Chemicals*

Acids (sulfuric, stearic, oleic)
Aluminum stearate
Alcohols (isopropyl, methyl, butyl)
Butyl carbitol
Chlorinated phenols
Formaldehyde
Kerosene
Mineral oil
Monoethanolamine
Oils (pine, peanut, castor)
Polyethylene
Sodium chloride, perborate, tripolyphosphate, hydroxide
Soap flakes
Tallow
Tapioca flour
Waxes (petroleum and synthetic)

* These chemicals include detergents, softeners, wetting agents, and sizes. All of them are potential wastewater pollutants.

TABLE 40.3 EPA Priority Toxic Pollutants*

1. Elements and their compounds (organic and inorganic)	DDT and metabolites
Antimony	Dichlorobenzinine
Arsenic	Dichloroethylenes
Beryllium	Dichloropropane and propene
Cadmium	Diphenyl hydrazine
Chromium	Endosulfan and metabolites
Copper	Endrin and metabolites
Lead	Ethylbenzene
Mercury	Fluoranthene
Nickel	Haloethers
Selenium	Halomethanes
Silver	Heptachlor and metabolites
Thallium	Hexachlorobutadiene
Zinc	Hexachlorocyclohexane
2. Inorganic compounds	Hexachlorocyclopentadiene
Asbestos (mineral)	Isophorone
Cyanides	Naphthalene and chlorinated naphthalene
3. Organic compounds	Nitrosamines
Acenaphthene	Phenols, chlorinated phenols, nitrophenols, and 2,4-dimethylphenol
Acrolein	Phthalate esters
Acrylonitrile	Polychlorinated biphenyls (PCBs)
Aldrin/dieldrin	Polynuclear aromatic hydrocarbons
Benzene, chlorinated benzene, and nitrobenzenes	2,3,7,8-tetrachlorodibenzo-p-dioxin (TCDD)
Benzidene	Tetrachloroethylene
Carbon tetrachloride	Toluene and dinitrotoluene
Chlordane and metabolites	Toxaphene
Chlorinated ethanes	Trichloroethylene
Chloroalkyl ethers	Vinyl chloride
Chloroform	

* This list is being continually updated and must be checked for current use.

The plant tour should also include an inspection of existing waste treatment facilities to obtain operating data and equipment sizes. In some plants, night operations may differ from day operations, so inspection of each shift is usually required.

ASSESSING WASTE FLOWS AND QUALITY

Of prime importance is the record of volumes of water requiring treatment. Flow of total plant discharge as well as individual process wastes must be measured. Many plants have permanent flow monitoring devices to meter total plant effluent. However, individual process area flow measurements are also necessary. To provide realistic data on flows to be treated, measurements should be taken over as long a period as practical to show what flow variations must be accommodated in a waste treatment unit (Figure 40.2). (See Chapter 7, Flow Measurement).

Chemical characteristics of the wastewater and quantity of contaminants present must be established. These are dependent on the raw water supply itself, the

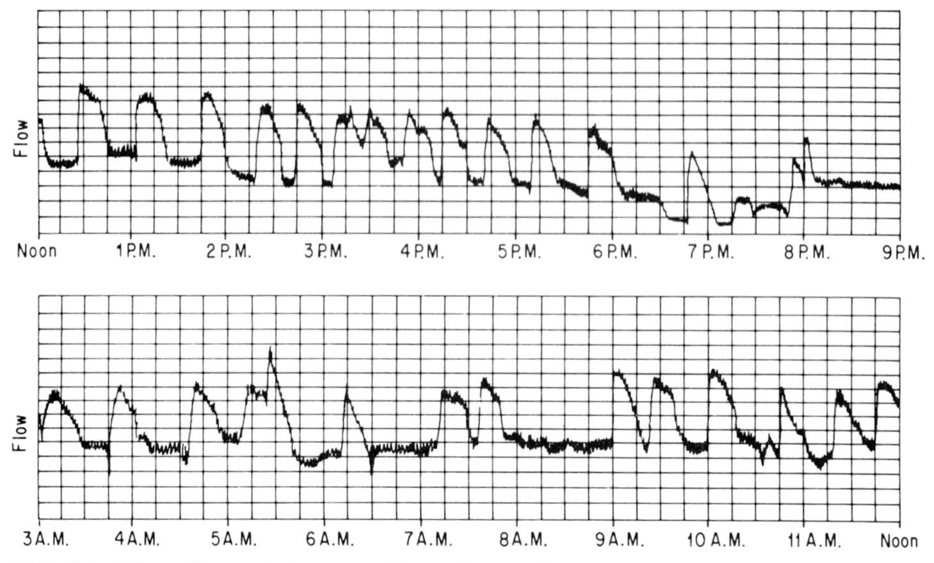

FIG. 40.2 Effluent flow variations caused by cyclic operations.

water treatment processes, the concentration through utility use, contaminants added in the process operation, or chemicals lost from manufacturing and inventory. Besides the probable contaminants identified through raw materials use, additional types of contaminants may be present as by-products from the plant processes or as trace contamination in raw materials. Each industry has its own wastewater profile. For example, petrochemical production, petroleum refining, and iron and steel manufacturing produce contaminants such as cyanide, phenols, ammonia, and sulfides at different concentrations and ratios. A guide to determining probable contaminants in industrial wastes may be found in numerous EPA-funded studies that were used to establish criteria for effluent compliance by each major industry. Some of this data is summarized by Table 40.4.

SAMPLING AND ANALYSIS

After determining the contaminants to be analyzed for, sampling of the plant streams, both in selected manufacturing areas and in plant outfalls, can proceed to define contaminant loading. The sampling schedule should uncover average contaminant discharges as well as peaks from cyclic or batch operations and process variables. In conjunction with the sampling, continuous monitoring of pH, temperature, and conductivity further define variations in wastewater characteristics (Figure 40.3). Existing waste treatment facilities should also be sampled to establish unit loadings, operating conditions, and contaminant removal efficiencies.

A summary of the raw data may be prepared to show a breakdown of contaminants and flows making up the total plant discharge, focusing on contributions from the various processes and manufacturing areas.

TABLE 40.4 Raw Wastewater Profiles for Selected Industries

	Industry*									
Contaminants	Aluminum	Automotive	Chemical	Coke	Food	Mining	Paper	Petroleum	Steel	Textile
Suspended solids	M	M	C	M	M	M	M	C	M	C
Salinity	C	C	C	M	V	M	C	C	C	C
pH variations	C	M	M	M	V	M	M	C	C	C
Oil and grease	C	M	M	M	V	V	I	M	M	C
Settleable solids	M	M	C	M	M	M	M	C	M	C
BOD	V	M	M	M	M	C	M	M	V	M
COD	V	M	M	M	M	I	M	M	V	M
Heat	C	I	V	C	C	V	C	C	M	M
Color	V	I	V	M	C	V	M	I	I	M
Odor	I	I	C	C	C	I	C	C	I	C
Heavy metals	C	C	V	I	I	M	V	V	M	I
Cyanides	I	V	V	M	I	V	I	C	V	I
Thiocyanates	I	I	V	M	I	I	I	C	V	I
Chromates	C	C	C	I	I	I	I	C	V	V
Phosphates	I	C	V	I	V	V	V	V	I	V
Fluorides	C	I	V	V	I	V	I	V	V	I
Ammonia	C	I	V	M	V	I	V	C	V	I
Organics (general)	I	M	M	M	M	I	M	M	V	M
Phenolics	I	I	V	M	I	I	V	C	V	I
Pesticides, biocides	I	I	V	I	I	I	C	C	I	I
Surfactants	V	M	V	I	C	V	V	C	I	C

* M, major factor; C, contributes to the problem; I, insignificant; V, varies in the industry, may contribute.

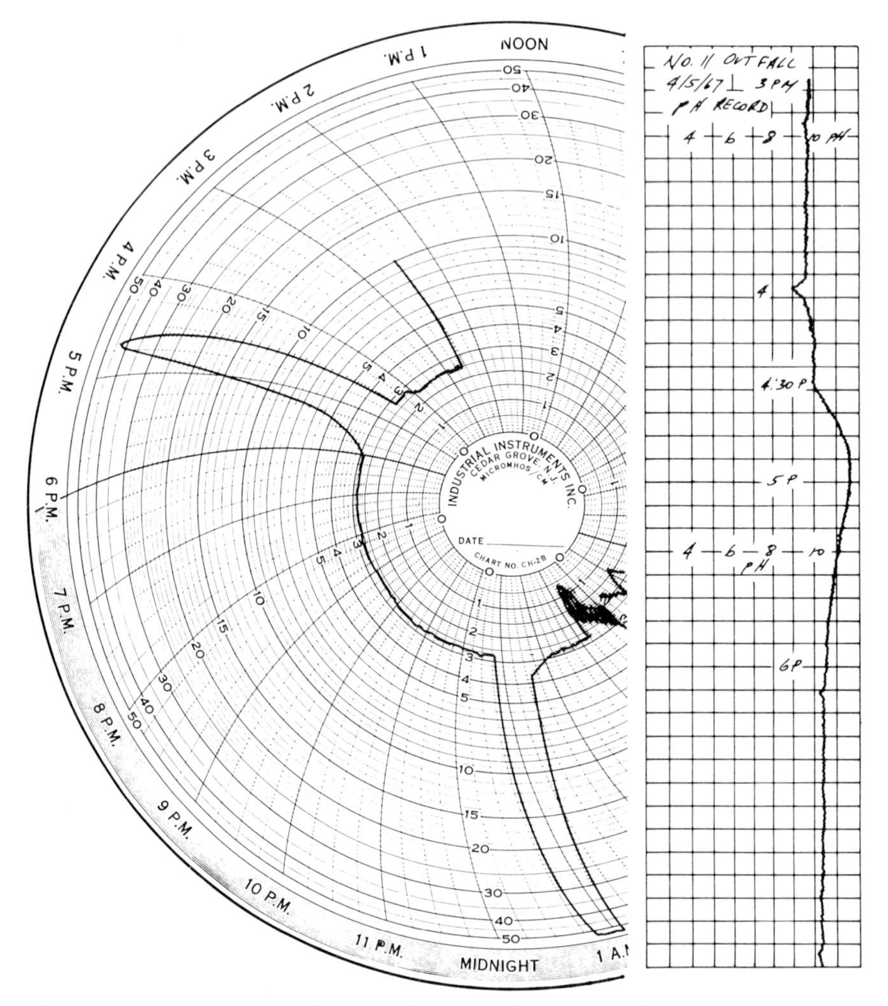

FIG. 40.3 Conductivity and pH record of a mill outfall, showing discharges at about 1 a.m. and 4 to 5 p.m., events repeated each day. This is an old city sewer passing through the mill, and investigation disclosed dumps of spent alkali by a neighboring plant.

EQUALIZATION

One of the most common practical first steps to take in planning waste treatment programs is to provide a large basin for equalization of flow, concentration, and in some cases, temperature. It is clear that no physical plant, especially one using a biological process, can successfully cope with the storm water surges shown in Figure 40.1 or the cyclical process discharges of Figure 40.2, nor can it easily deal with the concentration changes represented by Figure 40.3.

After all precautions have been taken to minimize the range of swings in flow and concentration, new data must be obtained to establish the variations still present. Then a holding basin must be designed to receive and blend various wastes to produce an outflow of strength and rate that subsequent waste treatment equipment can handle.

Sizing the equalization basin volume requires a knowledge of the tolerance levels of the process equipment to flow and concentration changes. Solids/liquid separation devices are subject to upset by rate of change in flow and temperature, while biodigestion devices are more disrupted by change in waste concentration and temperature. Sometimes maximum flow may occur at the same time as maximum waste strength, so enough data must be obtained to establish the probability of this. Then, these data must be plotted, the areas integrated, and average, maximum, and minimum loadings established against time to provide a reasonable time in detention.

In some cases, the equalization of a widely variable tributary can minimize the size of the mainstream equalization basin. Diversion basins are also valuable to deal with storm water and accidental dumps. But, in every case, the equalization basin itself, despite the complications of its design formulas, is a simple, practical pretreatment structure for any industrial wastewater treatment program.

CONSERVATION TO MINIMIZE EFFLUENT LOADING

At this point if it appears that a relatively large volume of water may require treatment, consideration is given to reducing plant water use. The first area for water conservation is usually process water where water use often exceeds equipment requirements. Secondly, waters identified as mildly contaminated but acceptable for direct reuse in a process not requiring high quality should be segregated for recovery. The third area considered is substitution of recirculating water systems for once-through indirect or direct contact cooling waters. In the steel and petroleum industries, these can make up as much as 90% of total water use. Figure 40.4 is a general schematic of water usage that is typical of many industrial plants. The major water sources are storm water plus a conventional raw water supply (which may be surface or well water). The potential opportunities for flow reduction shown by Figure 40.4 are: (1) the elimination of unnecessary or wasteful flows, (2) water recovery and diversion to secondary uses, and (3) water recycled within the system. Figure 40.5 shows the reduction achieved in going from a once-through to an evaporative cooling system.

By simply providing additional rinse tanks with a countercurrent flow of rinse water, metal plating and finishing plants may reduce rinse water requirements by over 75% (Figures 40.6 and 40.7). Installation of water pretreatment or sidestream treatment to remove objectionable contaminants from recirculating cooling systems can extend cooling water use and reduce blowdown.

Evaluation of the survey data may show that a single process or manufacturing area discharges the bulk of the most significant contaminants present in the total plant discharge. Individualized in-plant treatment of that particular source may be warranted. In plating shops this is usually required because treatment of individual contaminants such as cyanide or hexavalent chromium requires chemical reactions incompatible with one another. Poor maintenance practices may cause unnecessary contamination; improved practices and installation of containments

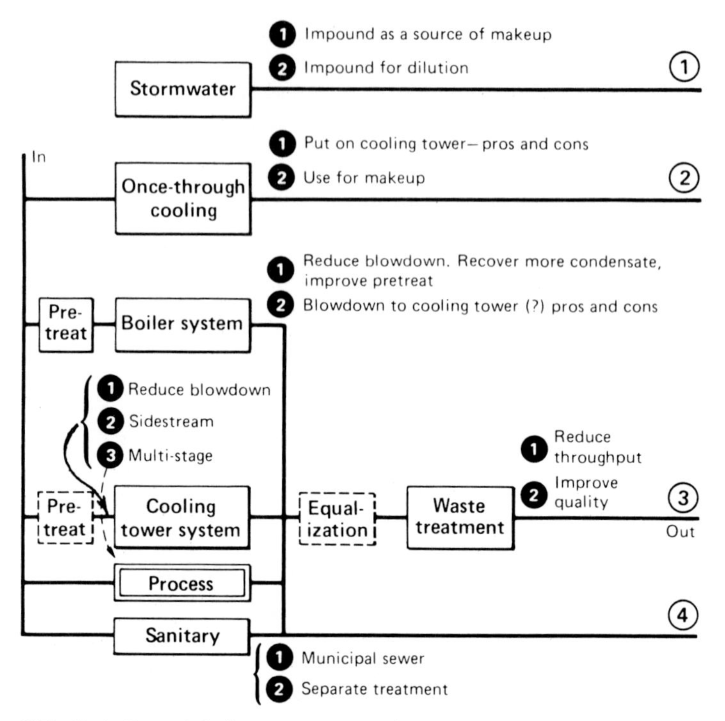

FIG. 40.4 Potentials for water conservation.

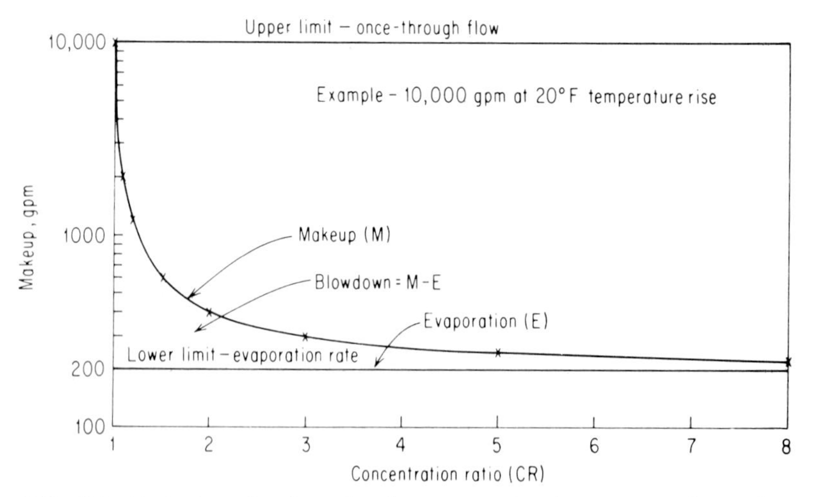

FIG. 40.5 Reduction of makeup flow by concentration in an evaporative cooling system.

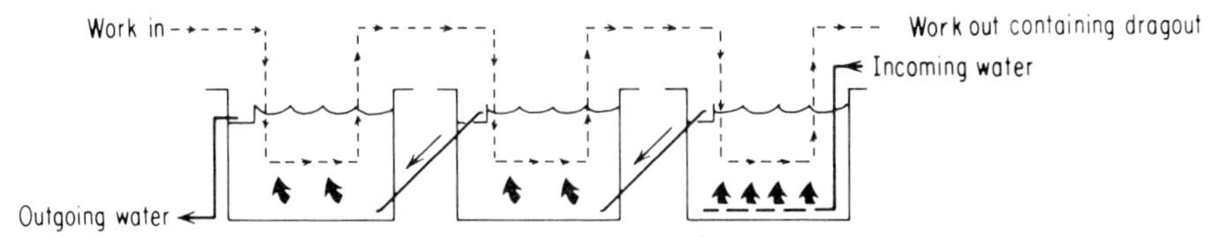

FIG. 40.6 Counterflow rinsing in a metal finishing shop saves water.

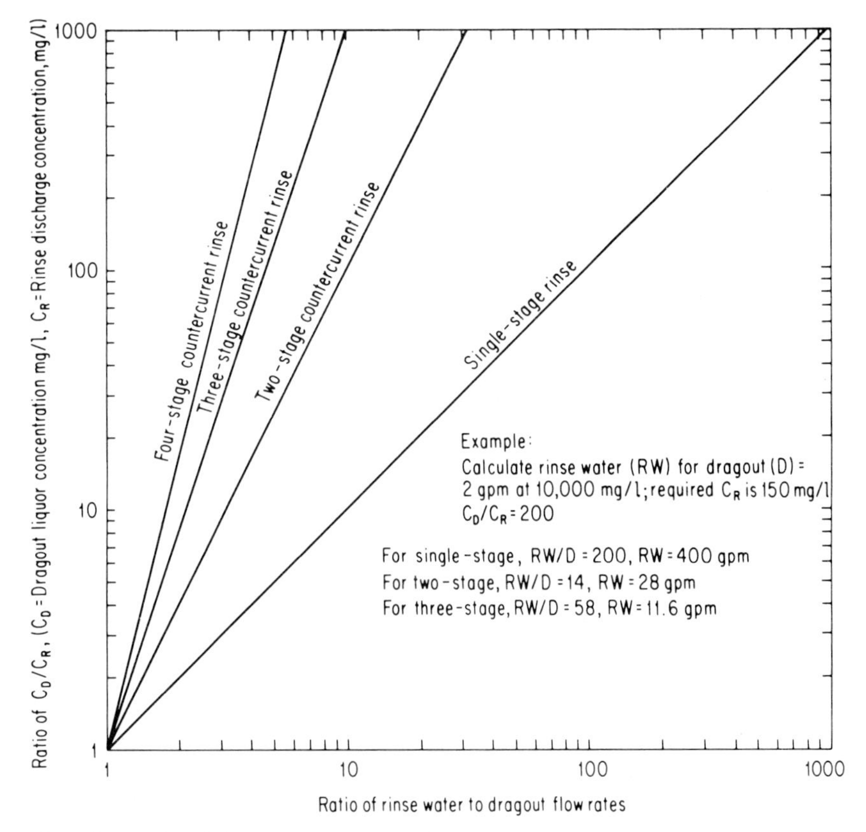

FIG. 40.7 Ratio of rinse water to dragout flow rates.

for spills and leaks can reduce contamination. Figure 40.8 shows a scheme for containment of strong wastes for recovery or separate disposal.

If possible, sanitary wastes should be collected and sent to a municipal treatment plant. Alternatively, septic tanks for the sanitary waste and discharge into a drainage field could be provided if volume is small and ground conditions permit. The septic tank discharge may also be chlorinated and then mixed with the plant's process wastes.

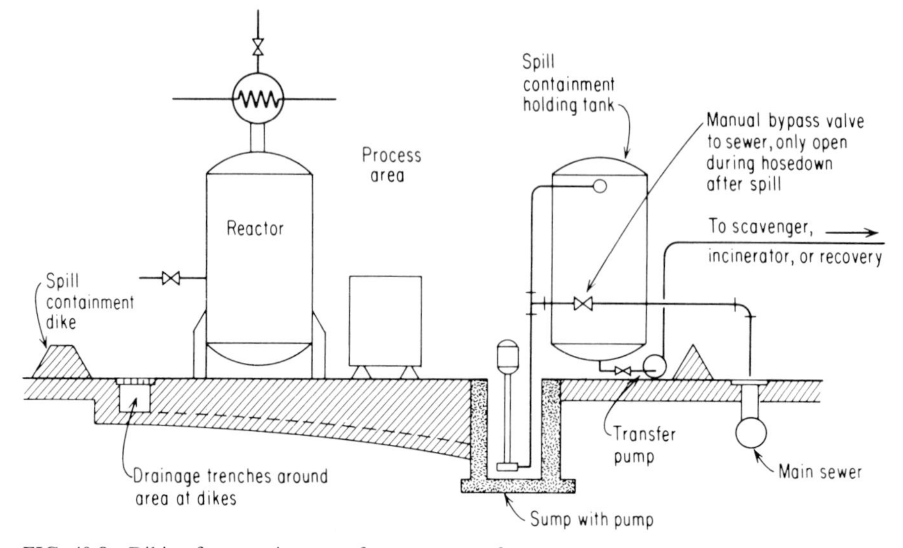

FIG. 40.8 Diking for containment of strong wastes for separate treatment.

REVIEW OF EXISTING FACILITIES

Some industrial discharges contain colloidal solids or emulsified oils that are difficult to remove and may carry through the waste treatment unit. Coagulants are required to destabilize these suspensions and allow the particles to float or sink. A common problem in waste treatment units is short-circuiting of the wastewater flow, caused by poor distribution of inflow and discharge, tilted overflow weirs, or other poor design features. Installation of baffles or redesign of distributor and collector manifolds may alleviate the problem and allow effective use of the total unit. (See Chapters 8 and 9.)

Not to be overlooked in the plant survey is the adequacy of existing equipment for treating the average wastewater flow. If clarifiers are found to have a high hydraulic loading, reduction in water usage, the addition of properly sized equipment, or additional equalization volume may be required.

Biological waste treatment units are very sensitive to wastewater conditions. Ineffective contaminant removals can result from one or several factors such as variable flow; too high or low an organic loading; low dissolved oxygen levels; inadequate mixing; insufficient nitrogen or phosphorus nutrient levels; presence of toxic compounds; erratic changes in waste composition, pH, or temperature; and growth of undesirable filamentous bacteria or fungi. The ineffective treatment may also be related to inadequate design, such as insufficient aeration capacity or detention time and inadequate solids/liquid separation facilities. A biosystem operation may be optimized by careful review of all operating variables and institution of proper controls. (See Chapter 23, Biological Digestion.)

Although biological waste treatment can effectively reduce organics, microbial growths occurring in other types of waste treatment units may hinder overall treatment efficiency. These growths can produce nonflocculating solids or odorous gases that prevent effective sedimentation and cause fouling or plugging of equipment. They may be controlled by preventing accumulations in dead spaces, elimination or control of their source of food, and application of biocides.

Through the plant tour and review of survey data, it is often possible to reduce both the volume of water and quantity of contaminants that require final treatment. Plants having existing waste treatment systems may find that the in-plant contaminant reduction, segregation, or water reuse studies provide the means for sufficient improvement to achieve current effluent compliance. The improvements may be realized by reduction of water use with resultant increase in residence time.

Control or replacement of chemicals at the process source may eliminate the need for a particular waste treatment process, such as activated carbon for removal of refractory organics. Segregation or in-plant treatment of certain contaminants may prevent adverse effects that can occur with mixtures of different process streams, such as the complexing of metals from one stream with cyanides from another, and solids density loss by mixing a turbid stream with an oily stream.

PROCESS MODIFICATION

One alternative for eliminating or reducing process wastes involves the modification or elimination of steps producing these wastes. For example, the peeling process is one of the greatest sources of waste in most fruit and vegetable processing plants. Research has been directed toward modifying the peeling process so the peel waste can be removed without excessive use of water. One such process modification is the "dry" caustic peeling process for potatoes. In the conventional steam or hot lye peeling processes, potato peels may contribute up to 80% of the total effluent BOD. The new dry caustic peeling method collects peels and caustic as a solid residue, preventing their entrance into plant wastewaters.

In a chemical plant, additional stages of separation to improve product yield from evaporators, washers, filters, or crystallizers may reduce loss of product to the sewer. Often these additional stages are not economical in terms of product recovery costs, but they pay out in reducing final effluent treatment costs.

If possible, the in-plant or waste treatment improvements should be accomplished prior to making final plans for additional facilities. With completion of this program, the total plant discharge should be recharacterized in terms of both flow and contaminant concentration to reveal the extent of the improvements as well as the contaminants still requiring reduction. Since the wastewater is likely to contain an array of contaminants of various concentrations, a single treatment process usually will not effectively remove each contaminant to the required levels for discharge to a receiving stream. If the plant has no existing waste treatment facilities, a system of two or more stages may be required.

The purpose of the first stage (referred to as primary) or an intermediate stage treatment process would be to remove or reduce contaminants that would interfere or overload a subsequent treatment unit operation (secondary or advanced waste treatment stage).

With the final wastewater characteristics and effluent quality requirements established, the selection of the treatment program may be initiated. The first step is to define the progressive stages of treatment required to obtain the final waste quality goals. The designer should prepare a preliminary schematic flow sheet listing each sequential process treatment step. Perhaps available technology may offer several choices of methods as alternate schematic flow sheets for comparison.

Preliminary evaluation of the possible treatment schemes in terms of estimated installed cost, system limitations, complexity of controls, area and labor requirements, and flexibility for future needs may decide the final treatment selection or simplify the choices considerably.

PILOT STUDIES ESSENTIAL

A pilot study to simulate the waste process is required to generate the necessary design information. This may be performed by laboratory bench tests or pilot-scale equipment. Bench scale studies are quicker and easier to perform, and may be used to screen and define chemical treatment requirements, and determine contaminant removal efficiency. Obtaining a representative sample is critical for reliable results. The bench test should be performed on composites and various grab samples to determine reliability and establish the ability to obtain repetitive results. (See Chapter 7, Sampling.)

Pilot plant studies should be undertaken where influent waste fluctuations are present and must be considered in the process performance. The pilot studies should include a continuous flow evaluation, preferably on a slipstream of the

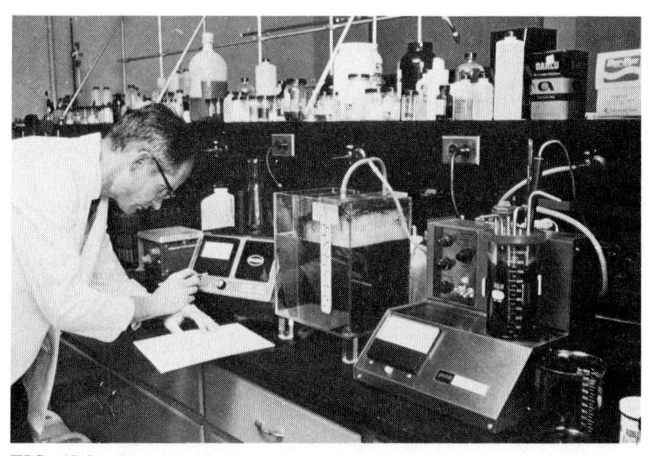

FIG. 40.9 Bench-scale biological studies require frequent attention to observation and recording of data during and following the period of acclimatization.

actual waste flow. The study should continue for a long enough time to include all the major variables encountered in the plant and to provide sufficient data for design purposes. In some pilot scale studies, such as biological treatment, an acclimatization period is required to obtain reliable, steady-state conditions (Figure 40.9).

These studies determine the quantity of solid wastes generated by the proposed treatment. In many industrial wastewater treatment applications, from simple operations such as food processing to the complex processes of a petrochemical plant, solid waste handling and disposal can present a formidable technical and economic problem. In some cases a chemical wastewater treatment program may be modified to reduce solids generation. For example, this could involve replacing inorganic coagulants with organic polyelectrolytes to minimize sludge formation. Sources of excess solids may be located and brought under control. Reduction of chemical losses in the process operations may significantly reduce sludge production. This is particularly true in metallic waste treatment where lime is often used for neutralization and precipitation.

If solids are difficult to remove in a wastewater clarifier, they will more than likely be difficult to dewater. These sludges will often require chemical sludge conditioning and specialized processing such as precoat vacuum or pressure filtration. Inorganic sludge conditioning chemicals and precoat materials contribute to operating cost not only from a raw materials standpoint, but also from rehandling, since their use contributes to the final volume, thereby increasing ultimate disposal or haul-out costs. A process of growing interest is solids stabilization, or conversion to a calcareous, cementlike solid that will resist leaching of its inorganic toxic components.

In dealing with in-plant or "end-of-pipe" treatment, consideration must be given to ultimate disposal of the contaminants removed from the wastewater. Volatile ingredients may cause air pollution, and sludges may contain materials that would not be acceptable for landfill because of possible water or soil contamination.

USING PLANT EFFLUENTS

The question often arises as to whether plant effluent can be considered a source of plant water. It may, though probably not without additional treatment. Perhaps it is useful simply because it is water—it may be usable for irrigation, for example. It may be used in the plant strictly for wash-up if separate piping can be installed to accommodate it. But in most plants, it has already become concentrated by evaporation in boilers or in evaporative cooling towers, and it may be even further concentrated by influx of high-salinity wastes from ion exchange regeneration or from sour water strippers, for example. The effluent probably will have become concentrated beyond the tolerance levels of the cooling system or the boiler system. Additional treatment of the effluent would then be necessary to upgrade its quality by reducing critical limiting factors such as water hardness, alkalinity, dissolved solids, or silica (Figure 40.10).

The location of points of use relative to the waste treatment plant, the cost of separate piping to reclaim the water, and the balance between further treatment of the effluent versus more complete treatment of the raw water source must be carefully studied before a decision can be made on the potential of reusing plant effluent as a water source. Most studies show that improved treatment of a water supply to a boiler or cooling system is preferred to recycling of the final effluent from a waste treatment plant as the best method of optimizing the total water balance. Often the effluent liquid contains organic matter that would interfere with the proposed chemical treatment and control.

It is fundamental that where an evaporative system is limited by the concentration of salinity, adding wastewater with a higher salinity than the control limit set for the evaporative system is counterproductive, as it actually increases the demand for low-salinity makeup.

FIG. 40.10 Electrodialysis system processing wastewater for concentration and final disposal. *(Courtesy of HPD, Inc., Naperville, Ill.)*

Because of this increasing salinity with water use—which is multiplied by recycle—and because of both legal and technical limitations, it is apparent that the major goal of recycle is not minimum water consumption, but optimum water usage. However, these goals coincide when water is scarce or where zero discharge is a realistic possibility. In water-short areas of the west and southwest, zero discharge may be an EPA-permit requirement attached to the location of a plant. Some utility plants in such areas have installed specially designed evaporators to process all wastewaters (cooling tower blowdown, spent regenerants, sluice waters, etc.). These evaporators produce distillate for recycle plus precipitated salts as waste solids. Such equipment is costly and can be justified only because of the National Pollutant Discharge Elimination System requirement of zero discharge. In some areas, solar evaporation ponds achieve zero discharge at the expense of land area instead of energy. Underground disposal, where permitted, may compete with evaporation.

SURVEILLANCE OF PROGRAM

To ensure consistent compliance with effluent restrictions, not only a well-designed but also a well-operated facility is necessary. In dealing with chemical waste treatment, over- or underfeed of chemicals will result in inefficiency and poor quality. Chemical feed rates should be verified periodically by jar or laboratory tests. Also of importance with chemical feeds is correct application point and uninterrupted feed.

Most regulatory agencies require routine scheduled analyses for effluent compliance purposes. The critical effluent analyses should also be checked on the raw waste flow to adjust unit operations accordingly. The frequency of such tests is based on influent waste variation.

To keep waste discharges under control, a schedule of in-plant waste surcharges, based on BOD, flow, and suspended solids from each operating area (like municipal surcharges) keeps area supervisors aware of the performance of their departments.

In many instances, on-line monitors are desirable to provide a continuous record of effluent quality and to adjust chemical additions and equipment performance as necessary to maintain optimum conditions. These devices can warn the operator if a test is out of its control range. The monitoring devices should be cleaned and calibrated as directed by the equipment supplier and the schedule adjusted through experience.

In addition to control analyses normally required on waste treatment effluents, regulatory agencies may require biological monitoring for discharges to receiving streams or bodies of water designated for wildlife or recreational activities. One form of this monitoring may be performance of toxicity measurements. These may expose indigenous fish species to controlled dilutions of waste or, in cages, to the receiving stream below the plant outfall. Total counts of various aquatic organisms above and below the outfall are sometimes preferable. A change in the variety of species and the populations of each provides information on the influence of the plant wastewater on the receiving stream.

The industrial discharge may comply with applicable criteria but still contain compounds harmful to the aquatic organisms or wildlife exposed to the receiving stream. Another form of monitoring related to biological activity of discharges into surface waters is dissolved oxygen sag measurements downstream of discharges. The temperature and residual organic and dissolved solids content may produce a reduction in oxygen levels essential to fish life. (See Chapter 5.)

ZERO DISCHARGE—POSSIBILITIES AND REALITIES

As designers of industrial plants and utilities have attempted to minimize the cost of wastewater disposal, most have almost completely eliminated once-through cooling in favor of evaporative cooling. A significant result is the discharge of water from the plant as vapor rather than as liquid. For example, a typical plant that used once-through cooling of 50,000 gal/min of water with a temperature rise of 20°F (a heat duty of 8.3 million Btu/min), converting to a recirculating, evaporative system would still require a recirculation rate of 50,000 gal/min with a 20°F temperature rise, but would evaporate about 8300 lb/min or 1000 gal/min, requiring about 1200 gal/min of makeup water. At these rates, the plant would return only 200 gal/min to the local environment (as blowdown), with 1000 gal/min lost as vapor to the atmosphere to return to the surface many miles away—perhaps in a different watershed or at sea. So, even though the trend has been to abandon once-through cooling, this may not always have been the best plan for the local environment, which in this example has lost 1000 gal/min from the watershed and has received 200 gal/min of concentrated blowdown in place of the 1200 gal/min withdrawn for makeup.

In the United States, substantial reductions in the rate of water withdrawal and wastewater discharge have been made in all industrial plants through water reuse. The establishment in each plant of a hierarchy of water uses based on quality requirements often results in the spent water from the highest water quality user being of adequate quality for users of lower quality. Cascading water from high to low quality uses has achieved high rates of reuse and almost reached a point of diminishing returns.

Although there has been strong interest in zero water discharge—and this practice has actually been mandated in certain areas where the environment seemed to require it—the cost is prohibitive in most cases; i.e., the cost would make production at the specific plant site uneconomical. Zero discharge requires separation of water from its dissolved solids by some combination of freeze concentration, evaporation, and crystallization, preceded by electrodialysis or some such membrane process (see Figure 40.10). Solar evaporation may be practical in arid regions, but in most cases energy must be applied, and it is this cost that makes zero discharge uneconomical. Even where zero water discharge is actually practiced, there is the requirement for some adequate disposal of the solids removed from the water, most of which are soluble and therefore unsuited to outside storage where storm water would redissolve them.

SUGGESTED READING

Arbuckle, J. G., and Vanderver, T. A., Jr.: "Water Pollution Control," Chapter 3, in *Environmental Law Handbook,* 7th ed., Government Institute, Inc., Rockville, Md., 1983.

Athavaley, A. S., Funk, R. J., Sweet, R. G., and Coffey, W. A.: "Deepwell Injection of Industrial Wastes," *Industrial Wastes,* May/June 1981.

Kemmer, F. N.: "Optimizing Water Supply, Treatment and Recycle Practices," *Chem. Eng.,* October 6, 1980.

Lund, Herbert, F. (ed): *"Industrial Pollution Control Handbook,"* McGraw-Hill, New York, 1971.

Rubin, Alan, J.: *"Chemistry of Wastewater Technology,"* Ann Arbor Science, Ann Arbor, Mich., 1979.

CHAPTER 41
WET GAS SCRUBBING

Wet scrubbers use a liquid to remove solid, liquid, or gaseous contaminants from a gas stream. The scrubbing liquid performs this separation by dissolving, trapping, or chemically reacting with the contaminant.

Scrubbers are used extensively to control air polluting emissions. So many different scrubber configurations have been used that there is some confusion as to whether they all belong in the same category. In some references, for example, the definition of a scrubber may be restricted to certain design criteria, such as whether the units are open or packed. In this text, any device fitting the definition of the first sentence is a wet scrubber.

Scrubber systems can be designed to remove entrained particulate materials such as dust, fly ash, or metal oxides, or to remove gases, such as oxides of sulfur (SO_x), from a flue gas stream to meet air emission standards.

PARTICLE COLLECTION CONCEPTS

In scrubbing particulate matter from gases, the principal concern is usually removal of particles smaller than than 10 μm. Larger particles are relatively easy to separate. The successful design and operation of wet scrubbers depends on knowing the size, composition, and derivation of the particles to be collected.

Figure 41.1 shows estimated size for some common pollutants. Just as fine particles in water (colloids) carry a charge of static electricity, so do colloidal particles in the fumes and dust, defined as aerosols. If these particles carry no charge, they may be deliberately charged to assist removal by special separators called electrostatic precipitators.

Among the particulates (term for the suspended solid materials) collected by wet scrubbers are dispersion aerosols from processes such as grinding, solid and liquid atomization, and transport of suspended powders by air currents or vibration. Dispersion aerosols are usually coarse and contain a wide range of particle sizes. Dispersion aerosols consisting of individual or slightly aggregated, irregularly formed particles are called dusts.

Condensation aerosols are formed when supersaturated vapors condense or when gases react chemically, forming a nonvolatile product. These aerosols are usually smaller than 1 μm. In condensed aerosols, solid particles are often loose aggregates of a large number of primary particles of crystalline or spherical form. Condensation aerosols with a solid dispersed phase or a solid-liquid dispersion phase are classed as smokes or fumes. Aerosols which include a liquid dispersion

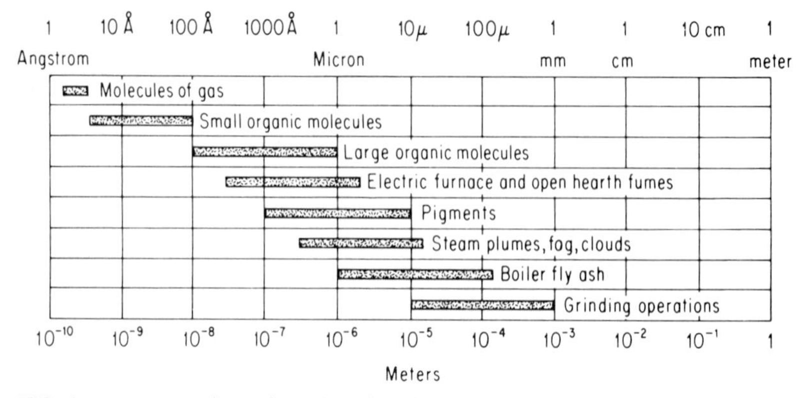

FIG. 41.1 A comparison of the size of particles present in air emissions.

phase are called mists. This classification usually applies regardless of particle size, and differentiation is sometimes difficult.

In practice, a combination of dispersion and condensation aerosols is encountered. Different size particles behave differently because of such physical properties as light scattering, evaporation rates, and particle movement. The choice of the best device for particle removal from gas is affected by these differences. Particle sizes, volumes, and weights may be obtained by microscopic sizing and density estimation.

PARTICULATE EMISSIONS

Limits on particulate emissions (smoke, mist, dust) are usually established in four ways:

1. *Emission rate:* The maximum weight that can be legally emitted in pounds per hour (kg/h). This may be expressed as the rate for a specific industry in production terms, e.g., pounds per hour per ton (kg/h/kkg) of pulp.
2. *Maximum concentration:* Maximum amount of particulate matter in the gas stream released, e.g., g/m^3, grains/cubic foot, or lb/1000 lb gas.
3. *Maximum opacity:* Maximum opacity of the gas stream emitted, usually measured by observation and comparison to empirical standards (Ringlemann numbers).
4. *Corrected emission rate.* Corrected emission rate is tied to an air quality standard by a formula based on atmospheric dispersion considerations.

Often, several particular emission regulations are enforced simultaneously. If all four types of restrictions are employed, a plant might pass on emission rate and concentration, but fail on opacity. This is an understandable situation, since large particles are the major contributors to weight while smaller particles, in the 0.1 to 2.0 μm range, are the major contributors to opacity.

The addition of chemical additives to the scrubber water to capture particles in the 0.1 to 2.0 μm range is often an economical way to meet air quality standards, particularly when compared to the cost of modifications and additions to equipment.

The wet scrubbers discussed in this chapter use water to remove particulates, gases, or both from industrial gas streams or stacks. Water chemistry is often extremely complicated in these scrubber systems because of the variety of operations occurring simultaneously in the scrubber environment.

1. *Heat transfer:* The gas and water are often at different temperatures, so heat will be transferred in the scrubbing process.
2. *Evaporation/condensation:* The gas may be hot and saturated with water vapor. Contact with colder water will dehumidify the gas, and the scrubbing water will be diluted with condensate. If the stack gas is hot and dry, the scrubber water will evaporate, as in a cooling tower, and become concentrated.
3. *Mass transfer:* The gas may contain water-soluble solids or gases that will dissolve in the scrubber water. The water may transfer gases to the gas stream also. For example, the water may be recycled over a cooling tower becoming saturated with O_2 and N_2, later releasing them to the gas stream.
4. *Scaling:* As the scrubber water is heated or increases in pH, alkalinity, or sulfate-sulfite content, precipitation of $CaCO_3$, $CaSO_4$, or $CaSO_3$ may occur and the scrubber may become scaled.
5. *Corrosion:* A common and troublesome problem encountered in most wet scrubbers.
6. *Fouling:* Fouling may occur from the coagulation of the particulate being removed or from microbial activity. Many industrial gas streams contain organics that supply food to microbes.

PRINCIPLES OF OPERATION

Scrubber manufacturers offer a bewildering array of products. Scrubbers are available in a wide range of designs, sizes, and performance capabilities. Some are designed primarily for collection of particles and others for mass transfer (gas removal by a chemical reaction). As good liquid-gas contact is needed for both operations, all scrubbers can collect both particles and gases to some extent. The degree to which the particle collection and mass transfer characteristics of a scrubber can be utilized determines the applicability of the scrubber for each specific purification problem. Figure 41.2 shows the commonly accepted domain of wet

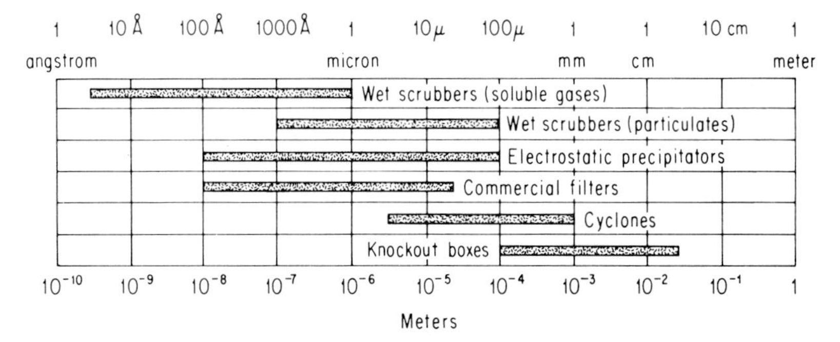

FIG. 41.2 General scheme of application of wet scrubbers compared to competitive devices.

scrubbers, based on particle size, relative to other competitive devices. Figure 41.3 shows the relative particulate removal efficiency of the more common types.

Particle size is one of the most important factors affecting removal efficiency, larger particles being much more easily removed. Submicron particles (1 μm = 10^{-6} m) are the most difficult to remove.

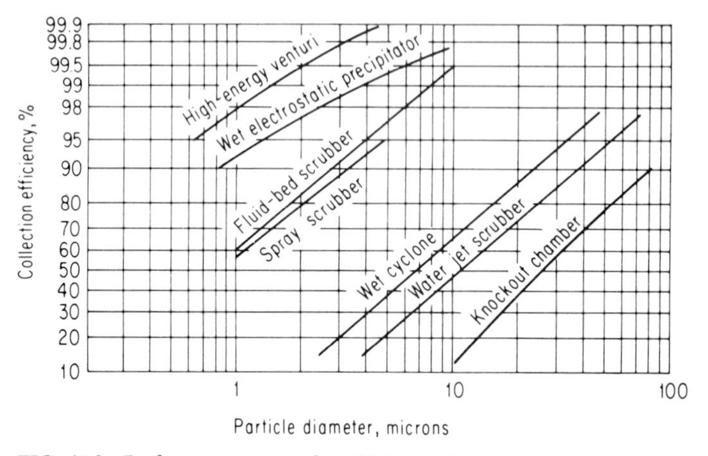

FIG. 41.3 Performance range of scrubbing devices.

All wet particle scrubbers operate on the same basic aerodynamic principle. A simple analogy: If water droplets of basketball size were projected to collide with gas-stream particles the size of BBs, the statistical chances of collision would be small. As the size of the droplets is reduced to more nearly the size of the particles, the chances of collison improve. Studies have shown that a surface film surrounding a water droplet has an approximate thickness of 1/200 of its diameter. A BB (the particle in flight) having a diameter less than 1/200 the diameter of the basketball will flow through the streamline film around the basketball without collision (Figure 41.4). But if the droplet were a baseball instead of basketball, collision would occur. A 0.5-μm fume particle requires water droplets smaller than 100 μm (200 $\times$ 0.5) for adequate collection. Efficient scrubbing, therefore, requires atomizing the liquid to a fineness related to particle size to afford maximum contact with the particles to be captured.

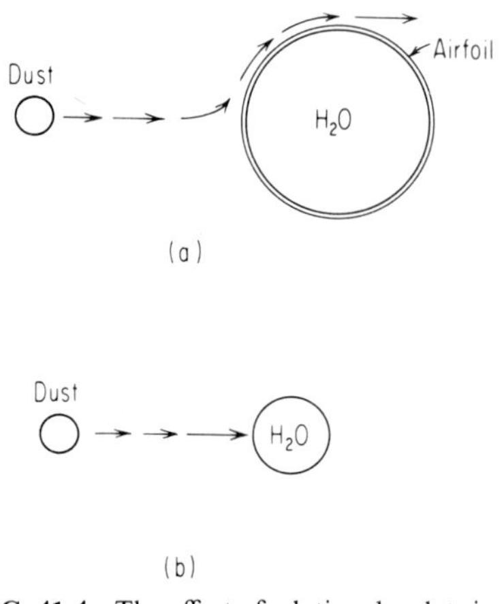

FIG. 41.4 The effect of relative droplet size to dust size in the process of particulate capture.

The probability of a droplet hitting the dust particles is proportional to the dust concentration; a ball would be less likely to hit a single BB than a swarm of them.

To equalize these factors, scrubbers are regulated as to the volume of gas to be scrubbed (measured by pressure drop of the gas stream), and water to be sprayed (measured by hydraulic pressure at the spray nozzles).

The scrubbing chamber's height and diameter are also tailored to the known characteristics of the gas.

CATEGORIZING WET SCRUBBERS

Wet scrubbers differ principally in their methods of effecting contact between the recirculating liquid and the gas stream. Techniques employed include injecting the liquid into collection chambers as a spray, flowing the liquid into chambers over weirs, bubbling gas through trays or beds containing the liquid, and atomization of the liquid by injection into a rapidly moving gas stream.

One way of categorizing wet scrubbers is by their energy requirements. Some require high energy to perform their task while others require very little. Generally speaking, low-energy scrubbers are used for removal of large particulate matter and gaseous contaminants. They rely on high liquid/gas ratios and contact time in the scrubber to increase removal efficiency. High-energy scrubbers are used for the removal of very small particulates (1 μm and less). They depend on high gas velocity for atomization to form small liquid droplets, with maximum impact between water droplets and particulate matter.

A second way of categorizing wet scrubbers is based on their selectivity toward either gaseous contaminants or particulate matter. Scrubbers designed primarily for removal of gas are called mass transfer scrubbers or gas absorbers; those designed for removal of particulate are called wet particle scrubbers.

GAS ABSORPTION SCRUBBERS

Gas absorbers, the first category, are designed to maximize contact time and surface area between the scrubbing liquid and the gas. This provides maximum opportunity for liquid/gas chemical reactions to occur. Absorption scrubbers usually have low energy requirements. The types most commonly used are the packed bed, moving bed, impingement, and plate-type scrubbers. Although wet particle scrubbers will also provide mass transfer removal of some gases, these four types of absorption scrubbers will do the job more completely and with greater efficiency.

Mass transfer (gas absorption) reactions require long residence times because the contaminants must first be absorbed by the scrubbing liquor and then react chemically to form a product that remains in the liquid phase.

PACKED TOWER

The packed tower (packed bed) consists of a vertical vessel containing packing materials such as rings, saddles, or tellerettes (Figure 41.5). Water is sprayed across the top of the bed and trickles through the packing material. Gas enters near the bottom and contaminants are removed as the gas stream moves upward through the water-washed packing.

The cleaned gas stream passes through a mist eliminator near the top where entrained moisture is removed prior to discharge. Scrubbing liquor is collected at the bottom. A portion is usually recycled to the inlet, and the balance discharged to the sewer.

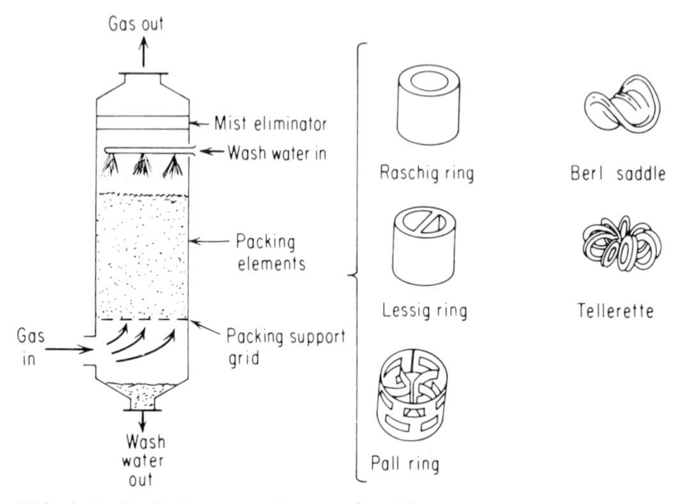

FIG. 41.5 Packed tower and types of packing.

Although flows can also be cocurrent or crosscurrent, the countercurrent type is most widely used. Packed beds have long been used for gas absorption operations because they are able to reduce odor and pollutant gases to low residual concentrations. The limiting factor is economics. As better separation is called for, beds require greater packing depth and operate with higher pressure drops. Gases entering a packed bed should not be heavily laden with solid particles as these cause clogging of the packing material. Pressure drop is typically 0.5 in of H_2O per foot of packing (4 cm H_2O/m).

Typical applications include rendering plants, food-processing plants, sewage treatment plants, and metal pickling plants.

MOVING BED SCRUBBERS

The moving bed wet scrubbers are well suited for high heat transfer and mass transfer rates (Figure 41.6). They are able to handle viscous liquids and heavy slurries without plugging. They accomplish this by using lightweight sphere packing that is free to move between upper and lower retaining grids. Countercurrent gas and liquid flows cause the spheres to move in a random, turbulent motion, causing intimate mixing of the liquid and gas.

In addition to excellent gas/liquid contact, the turbulence provides continuous cleaning of the moving spheres to minimize plugging or channeling of the bed.

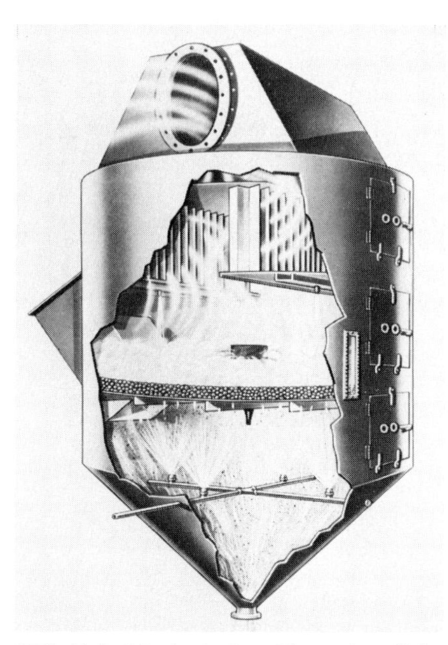

FIG. 41.6 Moving bed scrubber, using a lightweight packing of spheres which continually shift, preventing plugging with deposits. *(Courtesy of The Ducon Company, Inc.)*

Moving bed scrubbers are useful for absorbing gas and removing particulates simultaneously. This type of scrubber is especially suited for use with gases containing viscous or gummy substances, which would result in plugging of conventional packed bed scrubbers.

Efficiency is good for collection of particles larger than 1 μm. Both particle collection and gas absorption efficiency may be increased by employing several stages in series. Pressure drop is typically 0.2 to 0.5 in H_2O per stage.

TURBULENT CONTACT ABSORBER (TCA)

The turbulent contact absorber was developed as an extension of the moving bed scrubber, the difference being the increased turbulence of the TCA unit, resulting from using fewer spheres per unit volume. The TCA enhances the beneficial characteristics of the moving bed scrubber and permits high liquid and gas flows.

Plate Scrubbers

A plate scrubber consists of a tower having plates (trays) mounted inside (Figure 41.7). Liquid introduced at the top flows successively across each plate as it moves downward. Gas passing upward through the openings in each plate mixes with the liquid flowing over it. The gas/liquid contact causes gas absorption or particle

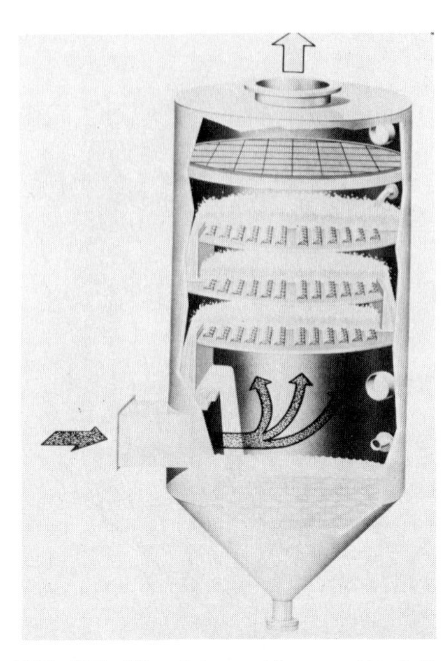

FIG. 41.7 The plate scrubber provides intimate gas/liquid contact. The flat plates are kept relatively free of deposits in most applications by turbulence. *(Courtesy of Koch Engineering Company, Inc.)*

removal. A plate scrubber is named for the type of plates it contains: if the plates are sieves, it is called a sieve plate tower.

Impingement Scrubbers

In some designs, impingement baffles are placed a short distance above each perforation on a sieve plate to form an impingement plate to increase turbulence and enhance gas/particle/liquid interaction (Figure 41.8). The impingement baffles are below the liquid level. Pressure drop is about 1 to 2 in H_2O for each plate.

WET PARTICLE SCRUBBERS

The four basic factors determining the efficiency of wet particle scrubbers are:

1. Water surface area
2. Liquid/gas ratio
3. Particle size and scrubber energy
4. Particulate affinity for water (wettability)

Anything mechanical or chemical that causes the water spray nozzles to form smaller water droplets with a larger surface area increases the collision rate

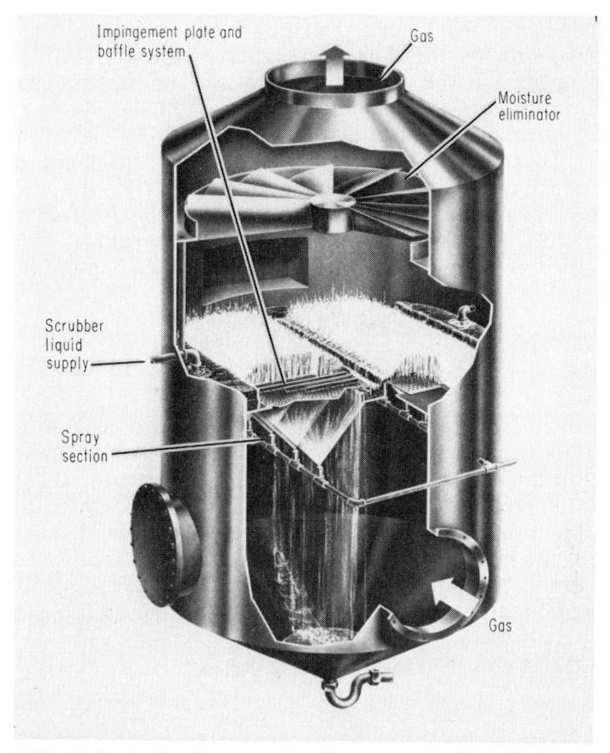

FIG. 41.8 An impingement scrubber uses a perforated plate to promote gas/liquid contact with impingement baffles over the perforations to atomize the water. *(Courtesy of W. W. Sly Manufacturing Company.)*

between the particulate in the gas phase and the water, resulting in increased particulate removal.

The second way to increase the collision rate is to pump more water through the scrubber. Increasing the liquid/gas ratio is an inefficient means of increasing the surface area of water. Usually, it is more economical to increase the effective liquid/gas ratio by causing smaller droplets to form mechanically or chemically.

The force with which particulate matter strikes the water is a third factor in scrubber performance. Since a liquid film barrier separates particulate in the gas phase from the water, a particle must have enough energy to force its way into the water droplet to be captured. Smaller particles require more energy than larger ones because they have a lower mass and strike the barrier with less momentum than larger particles moving at the same velocity. So, higher energy is required for scrubbing small particles than for large particles.

A higher energy must also be expended to scrub a particle that is hydrophobic (repelled by water) than a similar size particle that is attracted to water—and most particulates in the gas stream are hydrophobic. Increasing the mechanical energy of a wet scrubber increases its ability to remove smaller, and more hydrophobic, particulates.

Another means for increasing the removal of these particles is chemical reduction of surface tension, increasing the wetting power of the water.

Either a high-energy or low-energy scrubber may be used for removal of particulates from gas streams, the choice depending on the size of the particles. Low-energy scrubbers, such as the spray tower and wet cyclone, may be used for particles over 5 μm. For particles smaller than 5 μm, a high-energy scrubber such as a venturi or a venturi ejector provides more complete removal. Many wet scrubbing systems, such as those cleaning steel mill blast furnace gas, employ a low-energy scrubber followed by a high-energy scrubber. The function of the low-energy scrubber is to cool the gas (reduce volume) and remove large particles, thereby reducing the load on the high-energy venturi scrubber. This also reduces the size and power of the induced draft fan, because the cooling effect reduces gas volume.

Spray Towers

The spray tower collects particles or gases on liquid droplets produced by spray nozzle atomization. The characteristics of the droplets are determined by the design of the nozzle. The sprays are directed into a chamber shaped to conduct the gas to the atomized droplets. Spray towers can be used for both mass transfer and particle collection. Their low pressure drop, 1 to 2 in H_2O, and high water rate make them the least expensive of the mass transfer scrubbers. Spray chambers are most applicable for removal of large particles, gas cooling, humidification or dehumidification, and the removal of gases with high liquid solubilities.

Wet Cyclones

Wet cyclone scrubbers (Figure 41.9) are effective for removing dusts and liquid aerosols. A finely atomized water spray contacts the gas stream, which enters tangentially at the bottom to pursue a spiral path upward. The atomized droplets are

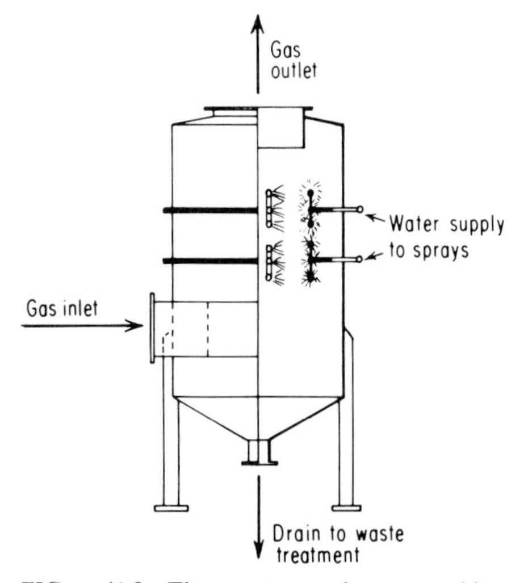

FIG. 41.9 The wet cyclone scrubber. *(Courtesy of Ceilcote Compay, a unit of General Signal.)*

caught in the spinning gas stream and swept by centrifugal force across to the walls of the cylinder, colliding with, absorbing, and collecting the dust or fume particles en route. The scrubbing liquid and particles drain down the wall to the bottom, and clean gas leaves through the top. The higher pressure drop, 6 to 8 in of H_2O, increases energy costs over those for a spray tower.

Venturi Scrubbers

Venturi scrubbers (Figure 41.10) are best suited for removal of 0.05 to 5 μm particulates such as those created by condensation of a liquid or metallic vapor or by a chemical reaction forming a mist or fume. Typical examples are ammonium chloride fumes from steel galvanizing, phosphorus pentoxide fumes from phosphoric acid concentration, mists from dry ice plants, and zinc oxide fumes from reverberatory furnaces.

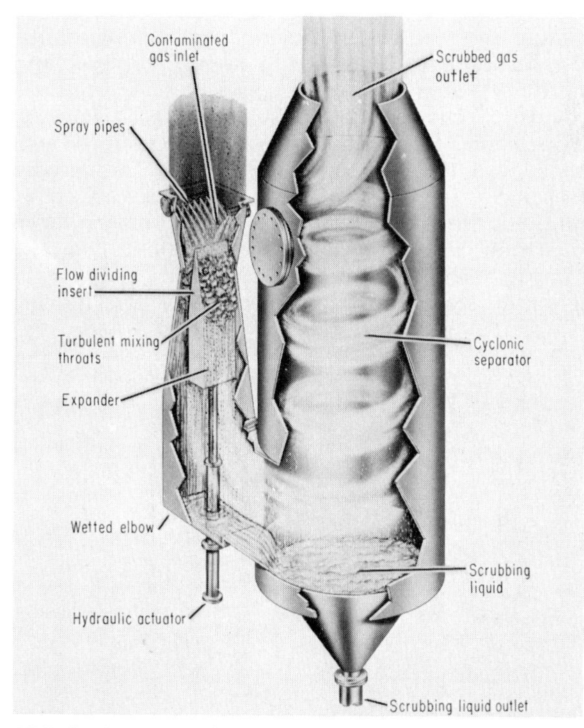

FIG. 41.10 A venturi scrubber with a variable throat to accommodate changes in gas flow. *(Courtesy of FMC Corporation.)*

These aerosols are removed by passing the gas and water streams cocurrently through the extremely small throat section of a venturi. As the velocity is accelerated in the throat, the liquid breaks up into extremely fine drops. High gas velocities, ranging from 200 to 400 ft/s, make the relative velocity between gas and liquid high enough to cause good liquid atomization and particle collection.

The liquid drops collide with and remove the particles in the gas stream and the drops then agglomerate for separation from the gas. The cleaned gas stream then passes thrugh a separator to eliminate entrained liquid.

Venturi scrubbers require high pressure drops (5 to 100 in H_2O). Pressure drop must be increased as the particle size becomes smaller to ensure adequate removal.

Venturi scrubbers can also be used for removing soluble gases. However, such applications are limited to situations where small particulates are also present, because the high energy requirements for operating venturi scrubbers make them costly for controlling gaseous pollutants.

Several modifications of the basic venturi scrubber are available to meet specific requirements of the size and type of particle to be removed. Low-, medium-, or high-energy venturi scrubbers are available, with energy requirements directly related to the pressure drop needed for removal of submicron particulates.

Many venturi scrubbers have a variable throat to allow for change in load. Also, at a fixed load, as the throat is decreased in size, velocity increases, resulting in increased pressure drop and better efficiency in removing submicron particles.

Venturi scrubbers use several methods to atomize the scrubbing water. In the most common, liquid is sprayed through jets across the venturi throat (Figure 41.10). This provides effective removal of submicron dust, fume, and mist particles, and is the first choice for the majority of applications. In another common venturi, the flooded-wall–type, the scrubbing liquid is introduced tangentially at the top, as shown in Figure 41.11. It spirals down the converging walls to the throat in a continuous film. At the entrance of the throat, it forms a curtain of liquid in the gas stream. The impaction of the gas into this curtain atomizes the liquid. Further impaction and agglomeration occur in the diverging section.

This type of venturi scrubber is recommended for hard-to-handle situations:

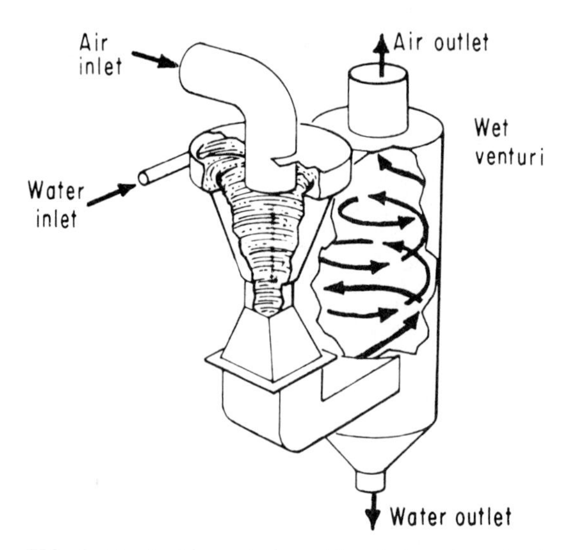

FIG. 41.11 In this venturi scrubber design, the water is atomized and it flows down a spiral path toward the venturi throat.

removal of "sticky" solids from gases; recycling of dirty water where water supplies are limited; and recovery of process materials in concentrated form.

Another common type is the flooded disk scrubber, in which liquid is introduced to a disk slightly upstream of the venturi throat. This liquid flows to the edge of the disk and is atomized by the high-velocity gas stream.

Figure 41.12 shows an ejector venturi which uses high-pressure spray nozzles to collect particulates, absorb gases, and move the gas. It derives its energy from the high-pressure liquid, while the regular venturi derives most of its energy from the high gas velocity produced by the induced draft fan. An induced draft fan may or may not be required, depending on the ejector venturi being used.

There is a high velocity difference between the liquid droplets and the gas; this affects particle separation. Collection efficiency is generally high for solid particles larger than 1 μm. Mass transfer is affected by the cocurrent flow of the gas-liquid. Energy consumption is relatively high because of pumping costs.

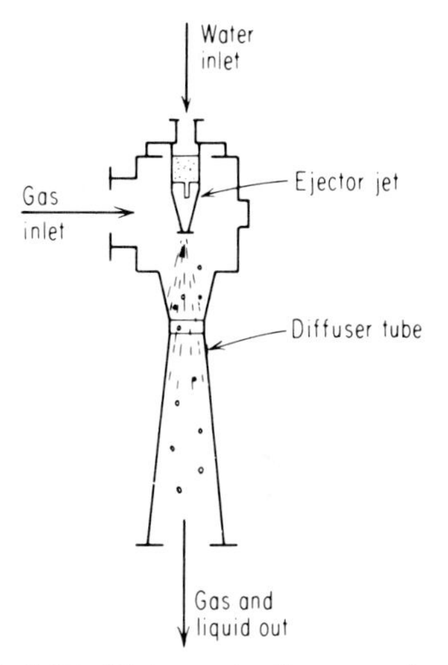

FIG. 41.12 Water creates the energy for this ejector venturi scrubber.

Ejector venturis may be used alone or as the first stage of a more complex system. The principal collection mechanism in this type of scrubber is inertial impaction, which is effected by liquid drops. Particle adherence upon striking the droplets is dependent upon the wettability of the particles. Because scrubbing liquid is usually recirculated, nozzles must be capable of handling a high solids concentration.

WET ELECTROSTATIC PRECIPITATORS

Water can also be used with electrostatic precipitation to improve removal of particulates. In this type of system, water is continually recirculated over the plates and discharged to an ash sluice pond or thickener to be clarified for reuse.

WATERSIDE PROBLEMS

Operating conditions in the scrubber may produce a severely corrosive or scaling water, depending on the gas stream being scrubbed and on the nature of any chemicals being added to the water. Many systems use lime/limestone slurries to react with SO_2, forming insoluble compounds to be removed in a thickener. The problems encountered in wet scrubber operation parallel those found in an open recirculating cooling water system, but the scrubber water is often more saline.

Chemical treatment programs for wet scrubber systems are designed to:

1. Maintain clean nozzles and collection surfaces, preventing deposit or scale buildup, thus helping to maintain unit efficiency.
2. Improve particulate capture or, for gas removal, mass transfer.
3. Control corrosion in the scrubber and recirculating water system.

Water problems in scrubbers range from scale and deposits to corrosion and waste disposal. Depending on the moisture level or dewpoint of the gas stream, gas cooling can result in evaporation or condensation, leading to concentration or dilution of scrubber water.

If the gas stream is above the dewpoint, recycling water for wet scrubbing results in evaporation and concentration of the scrubber water, adding to corrosion and scale problems inherent with the gases or particulates removed from the gas stream. However, in some systems the gas contains substantial water vapor, so condensation with resultant dilution of the recycle water may occur. This tends to lessen the potential for scale, deposits, and corrosion. Dilution is less common than concentration.

Each scrubbing system must be considered individually because of the wide variety of construction materials available, including mild and stainless steels, copper and nickel alloys, fiberglass, PVC, ceramics, lead, and refractories, to name a few. Many manufacturers construct scrubbers of alloyed metals and nonmetallic materials to avoid corrosion problems but still encounter the problems of scale and deposition. The same techniques and principles used for corrosion and deposit control in cooling systems apply to scrubbers.

Scrubber systems that operate under low pH conditions usually have deposits

TABLE 41.1 Typical Deposit Analyses*,†

	Sugar mill scrubber supply line	Electric utility demister scrubber	Process scrubber—calcium carbide	Coke gas scrubber—calcium carbide	Blast furnace scrubber—return to furnace
Calcium (as CaO)	60	49	36.4	33	52
Iron (as Fe_2O_3)	1		2.6	5	9
Carbonate (as CO_2)	2	16		0	21
Sulfur (as SO_3)	36	32		44	4
Silicon (as SiO_2)	1	2	2	4	6
Aluminum (as Al_2O_3)	1	1		1	3
Chlorine (as Cl)	0	0		2	
Sodium (as Na_2O)	0	0		5	
Magnesium (as MgO)	0	0	58.5	5	1
Loss at 800°C	14	16	60	67	23
Carbon	0	0	5	62	5
Hydrocarbons	0	0	19		

* Inorganics in dried sample scaled to 100%.
† All figures in %.

that are primarily particulates removed from the gas. Those systems that operate at a pH greater than 7.0 will normally yield scale deposits produced from water reactions, or scale and suspended solids combined.

However, even high-pH systems of this type will sometimes yield deposits that are primarily particulates removed from the gas. It is difficult to generalize regarding composition of deposits because of the wide variability of scrubber designs, process gases, and water characteristics. The most commonly encountered deposits are calcium carbonate, calcium sulfate, lime [$Ca(OH)_2$], iron oxide, carbon black (soot), oils and greases, aluminosilicates (clays), and metal sulfides. Deposits, like corrosion, can plague virtually every section of the scrubber system—from the inlet gas ports to the induced draft fan and stack. Deposits usually form at the venturi throat, trays, and packing in gas absorbers, liquid recycle lines and pumps, mist eliminators, induced draft fans, and clarifier supply lines.

The deposit analyses shown in Table 41.1 illustrate the primary components of deposits removed from different types of scrubbers to illustrate the wide diversity from one industry to another. These analyses show that deposit compositions vary widely, with calcium occurring most frequently.

Chemical treatment for scale and deposit control is effective in controlling the majority of deposits, but it must be individualized for each scrubbing system owing to variation in deposition problems.

WASTE TREATMENT

Since the basic function of the wet scrubber is to remove contaminants from process and combustion gases, once the scrubber liquid has done its job, the disposal of contaminants transferred to the water must be considered. For the small plant, this may entail merely discharging a bleed-off of recycle water to the sanitary sewer. The larger plant may be required to install an in-plant clarification system. There are two basic types: The first is the full-flow, inline clarifier which clarifies all scrubber water after it has made one pass through the scrubber. Water is recycled from the clarifier to the scrubber for further reuse. Problems are usually less severe with this type of unit because suspended solids are usually maintained at a fairly low level. This type of system is in many respects comparable to a once-through system.

The second basic type of clarifier is that used for clarification of a blowdown sidestream from the recycle system prior to discharge. Problems are normally more severe with this type of system.

Scrubbing liquid that contains high BOD, heavy metals, or toxic matter may require additional treatment, such as biological oxidation, prior to discharge.

AUXILIARY EQUIPMENT

Although the wet scrubber is the heart of the gas cleaning system, auxiliary equipment is required to help the scrubber work efficiently. Auxiliary equipment may be categorized as follows:

1. *Dust catchers* remove gross solids to prevent overloading the scrubber.
2. *Gas quenchers* cool high-temperature (over 1000°F) gas and reduce evaporation in the scrubber.

3. *Entrainment separators* (demisters) reduce water droplets in the exit gas.
4. *Gas cooling towers* reduce plume discharge.
5. *Water cooling towers, cooling ponds,* or *spray ponds* facilitate optimum water recycle and minimize makeup.
6. *Induced draft fan* moves gas from the scrubber to the discharge vent or stack.
7. *Forced draft fans* move gas to the scrubber and through to the vent or stack.
8. *Gas reheat system* reduces plume discharge by raising the dewpoint of the gas.
9. *Clarifiers, thickeners, and settling ponds* facilitate recycle and recover solids for disposal or reuse.
10. *Sludge dewatering devices* consolidate recovered solids.

SELECTED GAS SCRUBBING SYSTEMS

This section presents examples of gas scrubbing systems used in the electric utility, steel, and paper industries, including the basic types of scrubbers and auxiliary systems used in these applications and the nature of the waterside problems.

Removal of Particulate (Fly Ash) and SO_2 from Flue Gas

Although several different scrubbing systems are available for this application in the utility industry, the process using limestone has been selected to illustrate many of the aspects of wet scrubbing previously discussed.

Figure 41.13 shows the limestone scrubbing process for removal of SO_2 from boiler flue gas. Two venturi/absorber scrubber modules, the heart of the system, are used to treat the entire flow of flue gas. This system is designed to be added to an existing conventional stack-gas cleaning system.

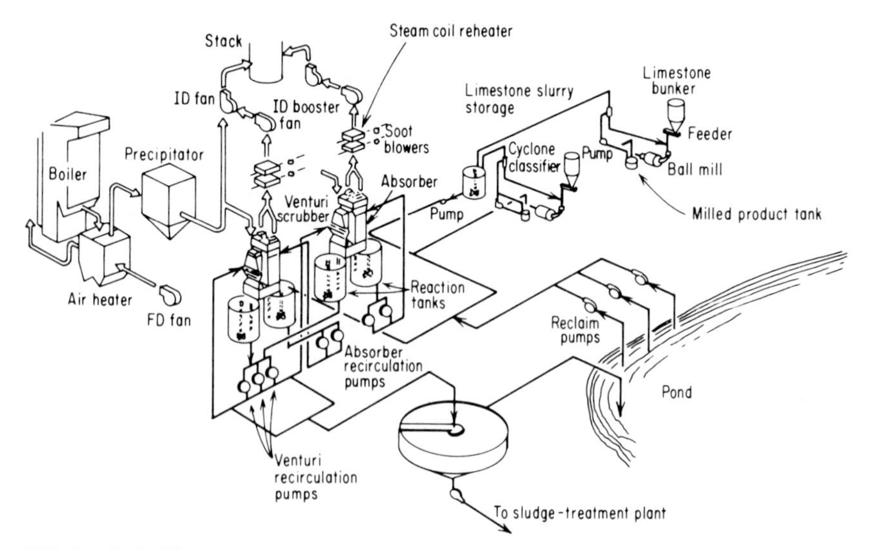

FIG. 41.13 Utility stack gas scrubbing system. *(From Power, September 1974.)*

Limestone stored in silos discharges through a gravimetric feeder, a wet ball mill, and classification system. Limestone slurry at 20% solids leaves the milling system and is stored for transfer to the scrubber modules as needed.

Each scrubber unit consists of (1) a variable-throat venturi that removes fly ash and provides an initial stage of SO_2 removal, followed by (2) a countercurrent tray absorber. Sprays in the venturi and the absorber drain into separate reaction tanks, where chemical reactions are allowed to go to completion before the slurry is returned to the scrubber.

Spent slurry from the scrubber, principally calcium sulfite ($CaSO_3$), is pumped to a thickener. Clarified water from this unit is discharged to a pond and returned to the reaction tanks. Thickener underflow is pumped to a sludge-treatment plant, where fly ash and other dry additives are blended to modify the gell-like sludge to produce a stable landfill.

Cleaned flue gases leaving each absorber pass through a bare-tube steam-coil reheater and then to a booster fan. Both booster fans discharge into ID fan inlets. An electrostatic precipitator ahead of the scrubbers serves to remove the bulk of fly ash initially. In case of scrubber system malfunction, a bypass of the precipitator discharge goes directly to the stack.

Major control functions, including limestone feed rate, venturi spray liquor rate, venturi differential pressure, slurry solids concentration, and milling system operation, are handled automatically from the boiler control room.

Primary waterside problems are scale and deposits in the venturi scrubber, gas absorbers, and venturi and absorber recirculation lines and pumps.

Steel Mill Flue Gas Scrubbers

Wet scrubbers are commonly employed by the steel industry for scrubbing gases produced by the blast furnace and basic oxygen furnace (BOF). A diagram and description of each will be presented to show how the basic scrubber and auxiliary equipment may be combined to meet effluent gas requirements.

Blast furnace gas contains CO and is used for combustion in boilers. This requires the effluent gas to be clean and cooled to reduce gas volumes and moisture content prior to combustion. Prior cooling and reduction in gas volume results in substantial savings in delivery costs through the extensive distribution system throughout the mill. Since the primary objective of cleaning blast furnace gas is to produce dust-free, cooled gas to be used as fuel for the boilers, the scrubbing system is designed as shown in Figure 41.14.

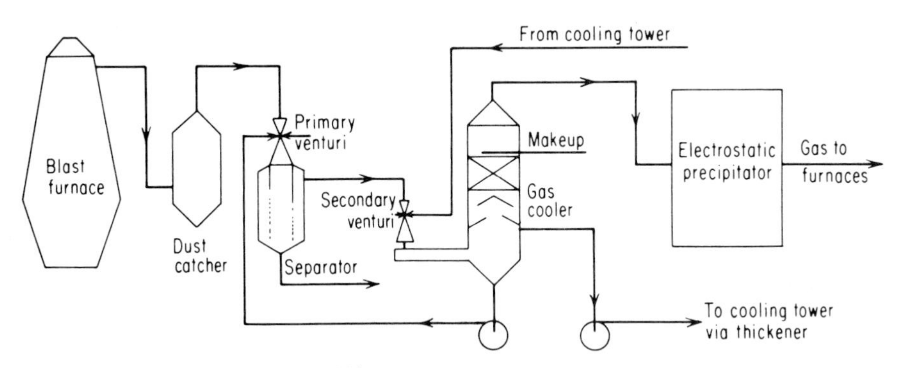

FIG. 41.14 Blast furnace gas scrubbing system.

Effective removal of a mixture of coarse and fine dust from a very dusty gas necessitates the use of a dust catcher and a multiventuri scrubbing system. Effective cooling requires the use of a gas cooling tower prior to effluent gas discharge to the boiler.

The dust catcher is merely a settling chamber to remove large particles and reduce loading on the venturi scrubbers. The gas passes through both a primary venturi (with separator) and a secondary venturi for even more effective particulate removal. Then the gas passes through the entrainment separator/gas cooling tower combination. The cleaned, cooled gas is then sent to furnaces. Adequate cooling is required to reduce the moisture level of the gas to avoid problems in distribution lines and furnaces, especially in winter.

The recycle water collected from the first venturi, containing a high level of particulates, is sent directly to a clarifier-thickener. The recycle water collected from the separator is recirculated to the first venturi scrubber. Makeup water is added at the clarifier, and the combined overflow is recycled to the secondary venturi. So the cleanest water contacts the cleanest gas, and works its way back to the first venturi and then to the clarifier. Water can be recycled from the thickener to the scrubber or may be used for some other purpose such as slag quenching.

A conventional cooling tower is normally used for removal of heat from the gas cooling tower water or scrubber water. The design of the tower makes it possible to keep the cooling water and the venturi scrubber water separate.

The scrubber water generally contains considerable hardness and alkalinity from the lime fines in the burden in the blast furnace. Consequently, scale is frequently encountered. Deposits of iron oxide and unburned carbon are also a concern in many systems. Deposition problems are most frequently encountered in the primary venturi nozzles and throat region, where the gas contains the highest level of particulates, and in the lines and pumps going to and from the thickener. However, deposition can occur in both venturi scrubbers, the separators, the gas

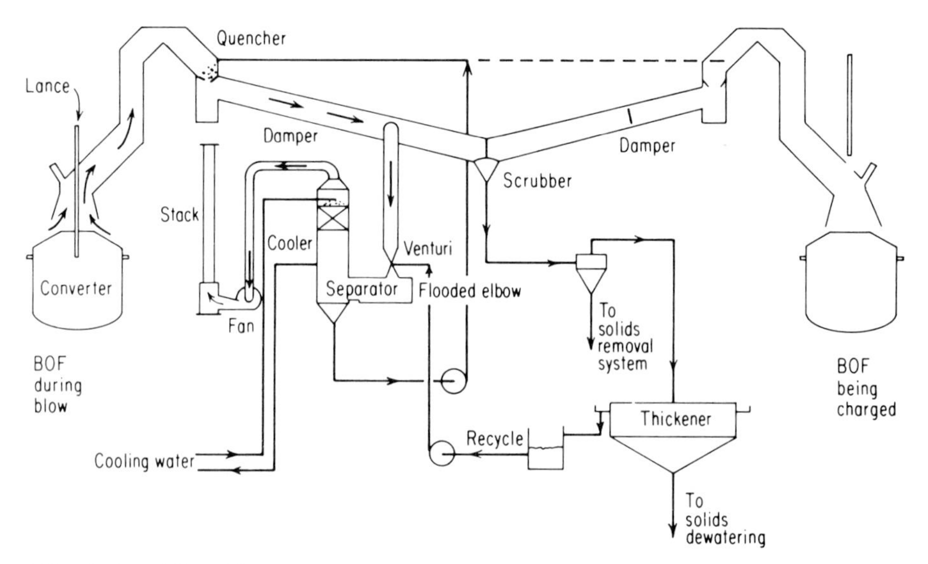

FIG. 41.15 BOF scrubbing systems.

cooling tower, or the scrubber recycle lines and pumps. Clarification is another major problem area since inadequate liquids/solids separation results in poor water quality of the clarifier overflow.

The objective of BOF gas cleaning is to produce effluent acceptable for discharge into the environment, since the cleaned BOF gas cannot be reused. As a result, the gas scrubber system for a BOF, shown in Figure 41.15, is somewhat different from the system used for blast furnace gas cleaning. The blast furnace dust catcher is replaced by a quenching system and the primary and secondary venturi replaced by a single venturi scrubber.

The function of the quencher is to substantially reduce gas temperatures, by 1000 to 1500°F (560 to 840°C), and to reduce gas volumes and resultant fan size and power requirements. A secondary function is to remove dust and particulates greater than 10 μm. This particulate is collected in the scrubber and sent to the solids removal system and thickener. The humidified, cooled gas then passes through the venturi, flooded elbow, separator, and gas cooler prior to discharge up the stack.

The BOF recycle water from the clarifier is pumped to the venturi scrubber, collected in the separator, pumped to the quencher, collected in the scrubber, and transferred to the thickener system for clarification and reuse. A cooling tower is then used for removal of waste heat from the gas cooling tower water.

Scale is normally the primary problem in most BOF scrubber systems. Calcium and alkalinity levels vary, but are usually not as high as in a blast furnace scrubbing system. Scaling potential is severe in a BOF scrubber because pH occa-

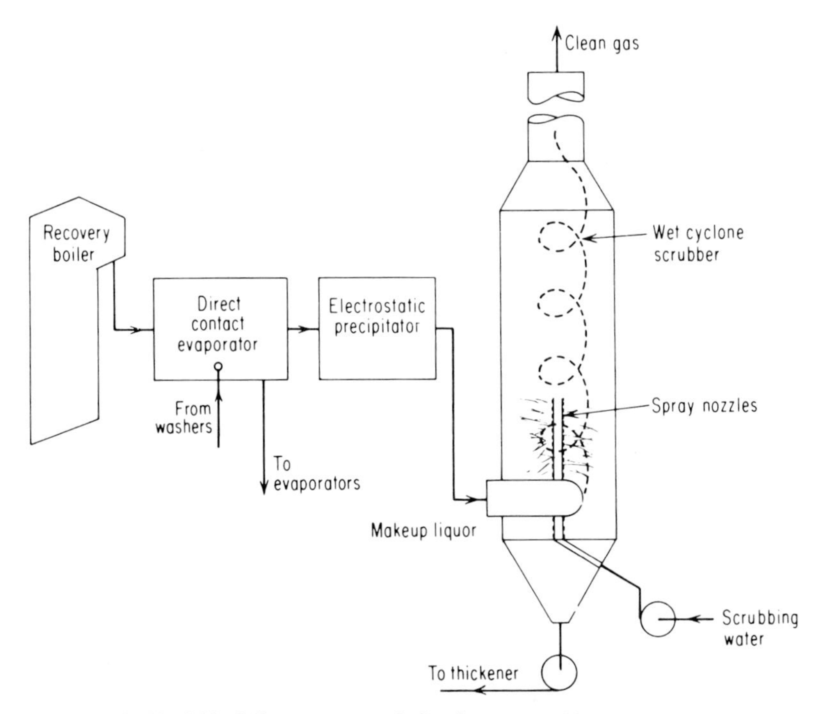

FIG. 41.16 Kraft black liquor recovery boiler flue-gas scrubber.

sionally gets as high as 11.0 during lime addition to the converter during a blow. The pH in the thickener is generally more consistent than in the scrubbing water because of equalization in the thickener.

Problem locations in the scrubber system are similar to the blast furnace circuit.

Pulp/Paper Wet Scrubbers

The most common applications for wet scrubbers in the pulp and paper industry are for coal-fired, bark-fired (hog fuel), and recovery boiler flue gases, and lime kiln recovery operations. The type of wet scrubber that finds the greatest application is the venturi with cyclonic separator combination, or electrostatic/wet cyclone combination. The recovery boiler flue gas scrubber uses a wet cyclonic separator in combination with an electrostatic precipitator (Figure 41.16), while the other processes use the venturi/cyclone combination.

The venturi/cyclone or cyclone scrubbers are very basic systems that suffer from the standard problems associated with wet scrubbing systems—deposition and scaling. Calcium carbonate scaling potential is most severe in the lime kiln scrubber because of the CaO and $CaCO_3$ particulate in the effluent gas. Problems in thickener operation are similar to those encountered in the utility and steel industries.

CHAPTER 42
AGRICULTURAL USES OF WATER

More fresh water is used for agriculture in the United States than for all industrial purposes, including utility cooling water. By far the largest portion of this is for irrigation, the principal focus of this chapter. Farm uses include water supply for livestock and domestic use, and water for farm equipment, particularly for diesel engine cooling, and preparation of agricultural chemical solutions, such as fertilizers and herbicides.

Both surface and underground sources are drawn on for agricultural water supply; the greatest volume is from surface sources, but the largest number of installations use well water. An example of surface water use is western irrigation with Colorado River water. Land flooding for leaching and irrigation in that area has become so widespread that it has affected the salinity of the Colorado River, requiring the United States to desalinate the river to meet its obligations to its downstream neighbor, Mexico.

CROP IRRIGATION

Although some crops along the eastern seaboard are irrigated, the major applications of irrigation water are in states where both rainfall and water are relatively scarce, i.e., west of the Mississippi River. Because of this, there is growing use of reclaimed wastewater for crop irrigation. For many years, the entire discharge of the sewage treatment plant at Lubbock, Texas, has been used for irrigation of cotton. But the use of municipal sewage plant effluent for irrigation of food crops is a more sensitive problem and requires careful evaluation of such potential problems as the heavy metals content of the wastewater, the presence of pathogenic organisms, and the effect of each of these on the crops.

Irrigation water must be pumped from the source and passed through some type of distribution system, either open or closed. Many distribution systems are open ditches and flumes that develop problems of biological growths, plugging with sediment, seepage into the ground (which is usually the highest portion of the water lost en route to the point of use), and evaporation loss.

The problems in closed distribution systems are discussed elsewhere in this text; they are categorized as corrosion, scaling, and deposits caused by sediment or microbial activity. Precipitation may occur in the distribution system if ammonia is added to the irrigation water as a fertilizer.

Many irrigation systems use plastic pipe because of its light weight and easy assembly. Corrosion in such pipelines is not a problem. After reaching the point of use, the water is fed into an irrigation system which may be of the gravity,

FIG. 42.1 Overhead irrigation systems distribute water over broad expanses of cropland, slowly moving across the fields to optimize water used. *(Courtesy of Valmont Industries, Inc.)*

sprinkler, or drip type. With the gravity system, fields may be completely flooded, as for rice cultivation; or the irrigation water may flow across a sloping field. The bulk of irrigation water today is handled by gravity.

Sprinkler irrigation systems include overhead application, which is commonly used on crops like alfalfa. Overhead spray systems are designed to cover single fields having a diameter of as much as 2500 ft (750 m) (Figure 42.1). The second basic type of spray system uses low-trajectory sprinklers. Citrus crops may use this design to confine wetting to the soil, leaving the foliage dry. There is considerable evaporation loss in the use of spray irrigation because of the large water surface area produced by the spray mechanisms. This evaporation also produces cooling, which may present a problem of thermal stress to the plant.

The final category of irrigation, drip irrigation, has been developed to meet the needs of ariculture in water-short areas where evaporation losses cannot be tolerated. A controlled volume of water is delivered to each unit under cultivation at predetermined rates,using the device shown in Figure 42.2. Since the flow rate

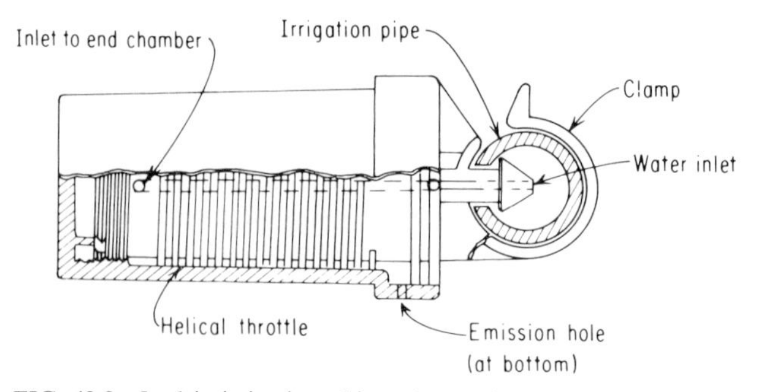

FIG. 42.2 In drip irrigation, this emitter reduces water pressure by frictional resistance through a long helical path so the emission hole can be as large as possible to avoid plugging.

per unit is small, this type of system must be kept completely free of silt or slimes produced by microbial activity to prevent clogging orifices that control the rate of flow to each site.

THE NATURE OF SOIL

Soil is a miraculous mixture of inorganic materials produced by the weathering of rocks and clays, and organic material produced by the decomposition of vegetation. Physical stresses, produced by diurnal temperature changes or seasonal freezing and thawing assisted by erosion from wind and storms, cause exfoliation of exposed rocks, breaking them down into fine particles. The solvent action of water produces a similar effect by selectively dissolving oxides, carbonates, and feldspar, releasing the particles of less soluble material. The effect of both of these weathering reactions is to produce silt and sand, which is washed into the valleys to deposit with other soil-forming materials.

Clays are broken down in a somewhat similar fashion, yet they still bring their swelling properties and their ion exchange capacity to the soil mixture. Finally, high molecular weight organic materials, such as humic acid, fulvic acid, and polysaccharides, with lower molecular weight materials such as protein, are broken down by bacterial action. Just as with the organic components in colored water, these colloidal materials carry a charge that is neutralized as their ion exchange sites take up calcium, magnesium, and iron, causing flocculation and sedimentation of the organic material with the sand and silt.

In this soil mixture, the clay is usually of small particle size, less than 5 μm. Silt is normally in the range of 5 to 50 μm, and sand may be found in particle sizes from 50 μm to 2 mm.

The properties of several soils are shown in Table 42.1, and a diagram that defines the various descriptive categories of soil is shown in Figure 42.3. Voids containing air and moisture occupy about 50% of the soil volume, an important fact in understanding the environment of soil bacteria and the capillary effects that not only retain soil moisture, but at times provide water to the root zone from the water table far below the plant.

TABLE 42.1 Selected Soil Analyses

Property	Sample A	Sample B	Sample C
Organic matter, %	0.9	1.0	1.4
Ca, ppm*	1600	700	2000
Mg, ppm	150	160	440
K, ppm	210	71	397
P, ppm	103	9	135
pH	7.9	6.2	7.4
H+, meq/100 g	0.0	0.7	0.0
Cation exchange cap., meq/100 g	9.8	5.7	14.7
Sand, %	68.0	74.0	58.8
Silt, %	27.2	17.2	33.2
Clay, %	4.8	8.8	8.8

* mg/kg of soil.

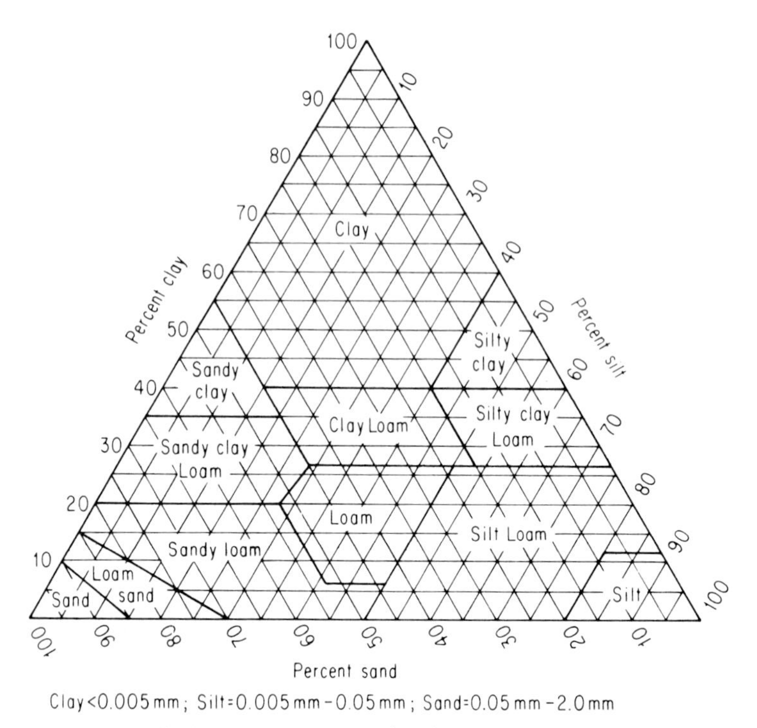

FIG. 42.3 United States Department of Agriculture soil classification.

The moisture content of the soil is measured by a tensiometer. If the void space is saturated with water, the instrument reads zero; as the soil drains and air occupies an increasing portion of the voids a suction effect is produced, and the instrument dial (a vacuum gauge) shows increasing levels of vacuum. The dial is calibrated in units of barometric pressure (bars), and a reading of 70 centibars usually warns of the soil becoming too dry and requiring irrigation. A tensiometer may be used to automatically control the irrigation system; it may initiate water flow at 30 to 40 centibars and discontinue flow at 10 to 20 centibars, for example.

MEASURABLE SOIL CHARACTERISTICS

Although many properties of soil are classified subjectively, such as friability, many other properties can be measured quite precisely. These include the permeability of the soil, its swelling properties, its ion exchange capacity, and the conductivity of the water extract. The pH of the water extract is also an important property as this influences the selection and dosage of chemicals used for soil conditioning.

The plant growing in any soil, properly irrigated and exposed to sunlight, combines carbon dioxide and water to produce carbohydrates and oxygen according to the following reaction.

$$CO_2 + H_2O \xrightarrow[\text{chlorophyll}]{\text{sunlight}} CH_2O + O_2 \tag{1}$$

This reaction shows that 60% of the weight of carbohydrate originated with water. In other words, the water required to produce the carbohydrate in a bushel of wheat, weighing about 60 lb (27 kg) is about 40 lb (18 kg). Considerably more water is applied to the plant than is converted to carbohydrate, because of transpiration (the plant's equivalent of perspiration) from the foliage, and drainage of excess water past the root zone. The transpiration process takes pure water vapor from the plant's system, leaving behind a more concentrated fluid. In other words, the plant acts as an evaporator, the evaporation providing cooling to the skin just as perspiration does for animals.

Another important fact about the photosynthesis reaction is that it requires chlorophyll as a catalyst. Magnesium is needed to produce the chlorophyll molecule, and this must be taken into the plant fluids through the roots by some process of selection. The plant also requires many other materials for cell synthesis, including phosphorus, potassium, nitrogen, and sulfur. Depending on the type of plant, certain trace elements are needed for nutrition, and these too must come from the water-soil environment. Plants irrigated with municipal sewage plant effluent show uptakes of heavy metals, including potentially toxic materials such as lead and zinc. But these may become concentrated in the stalk, the leaves, or the roots and may not become a part of the fruit. Each of these problems has to be investigated selectively to determine the effect of heavy metals in municipal sewage on specific food crops.

Soil bacteria are partly responsible for solubilizing the required nutrients for transfer across the outer skin of the roots into the plant fluids. For the most part, these are aerobic organisms, so the creation of anaerobic conditions which could be caused by poor drainage or by irrigation with wastewaters high in organic food substances could be detrimental to the plant.

CONCENTRATION EFFECTS

Since the plant foliage evaporates water, the plant fluids become quite concentrated relative to groundwater. Osmotic pressure then causes water to pass across the root surfaces from the soil, and this leaves a more concentrated water in the root zone. If the salinity of the water in the root zone becomes too high, the osmotic flow may reverse, and the plant can become dehydrated as it passes water into the soil. The relative salt tolerance of a scattered grouping of crops is shown by Table 42.2, based on the conductivity of a paste prepared from a soil sample.

In addition to the concentration effects caused by transpiration of water from the plant and osmotic flow, there are other factors affecting the composition of water held in the root zone. A prominent force changing this composition is the capillary action of the soil, causing water to rise to the surface crust to be evaporated to the atmosphere. As water is concentrated, whether by transpiration, osmosis, or evaporation, two important indexes of water quality are affected: the stability index changes (either the Langelier index or the Stiff-Davis index), and the sodium adsorption ratio (SAR) changes. This ratio is defined as

$$\text{SAR} = \frac{\text{Na}}{\sqrt{\dfrac{\text{Ca} + \text{Mg}}{2}}}$$

where ionic concentrations are in equivalents per million, or milliequivalents/liter (milligrams per liter divided by equivalent weight).

TABLE 42.2 Tolerance of Certain Crops to Salinity

High tolerance (10,000–16,000 μs)	Moderate tolerance (4000–10,000 μs)	Low tolerance (below 4000 μs)
Barley	Alfalfa	Celery
Sugar beet	Rye	Green beans
Cotton	Wheat	Field beans
Spinach	Oats	Citrus crops
	Rice	
	Corn	
	Potatoes	
	Carrots	

Source: *Water Quality Criteria*, 1972. EPA publication R.3-73-033. Original data by Salinity Laboratory Staff, USDA 1953.

The clay component of soil has ion exchange properties which play an important role in holding trace metals for the plant to use as they may be needed. The clay component may also include swelling clays, such as bentonite. A high SAR value affects the plant itself, the clay minerals with ion exchange capacities (causing the trace metals to be regenerated off the clay particle), and dispersion and swelling of the clay, reducing its permeability (drainage ability).

There are some general rules about the suitability of irrigation water for certain crops. For example, the SAR limit for certain fruits may be only four, whereas grain and alfalfa may tolerate an SAR as high as 18. However, it is not the irrigation water itself that is of greatest concern, but rather the nature of the water at the root zone, which has been affected by concentration and often by precipitation of calcium carbonate. Figure 42.4 is an illustration of what may happen to irrigation water in concentrating by a factor of four in the root zone. The SAR of the Colorado River at Parker Dam is given as 2.52, and the Langelier index is +1.05. In the second column of the tabulation, this water has been concentrated four times, which is a typical concentration ratio for soil water about 12 in below the ground level. The Langlier index has increased to +2.3 and the SAR is now 5.04. Several processes are going on in the root zone which affect the actual composition, including precipitation of calcium carbonate and respiration of soil organisms to produce CO_2. In the third column, it is assumed that the Langelier index has reached equilibrium by precipitation of calcium carbonate and by the production of 65 mg/L of CO_2 by bacterial respiration. Because of the precipitation of calcium carbonate, the sodium adsorption ratio has increased almost to six.

A modification of the Langelier index has been proposed to provide a means of estimating the sodium adsorption ratio of the soil moisture based on the SAR of the irrigation water and the concentration ratio. These relationships are shown below:

$$pH_c = pK_2 - pK_c + p(Ca + Mg) + p\,Alk \tag{1}$$

$$PI\ (\text{precipitation index}) = pH_{se} - pH_c \tag{2}$$

$$SAR_{se} = SAR_w \sqrt{CR} \cdot (1 + PI) \tag{3}$$

where pH_c = equilibrium of pH of irrigation water at $CaCO_3$ saturation
pK_2 = $-\log K_2$, second dissociation constant of H_2CO_3

pK_c = − log K_c, equilibrium constant for $CaCO_3$
p(Ca + Mg) = − log (Ca + Mg), the total hardness
p Alk = − log Alk, total alkalinity
CR = concentration ratio
se = soil extract
w = irrigation water

As shown above, this modified index assumes the precipitation of magnesium as well as calcium in the soil as the imbibed water in the root zone concentrates. It also assumes the ability to accurately measure the pH of the water in the root zone.

Of course, one means of controlling the concentration effect is to leach the soil by applying an excess of water to dilute the concentrate. But, in many areas where

Identification of Analyses Tabulated Below:

A. Colorado River at Parker Dam D.

B. "A" at 4 concentrations E.

C. "B" assuming $CaCO_3$ precipitation F.

Constituent	As	A	B	C	D	E	F
Calcium	$CaCO_3$	198	792	512			
Magnesium	"	105	420	420			
Sodium	"	220	880	880			
Total Electrolyte	$CaCO_3$	523	2092	1812			
Bicarbonate	$CaCO_3$	113	452	200			
Carbonate	"	7	28	0			
Hydroxyl	"	0	0	0			
Sulfate	"	302	1208	1208			
Chloride	"	100	400	400			
Nitrate	"	Nil	Nil	Nil			
Fluoride		1	4	4			
M Alk.	$CaCO_3$	120	480	200			
P Alk.	"	4	16	0			
Carbon Dioxide		0	0	65			
pH		8.40	8.4	6.65			
pH_s @ 100°F		7.35	6.1	6.65			
Silica							
Iron	Fe						
Langelier Index		+1.05	+2.3	0.0			
Turbidity							
TDS		661	2644	2364			
Color							
SAR		2.52	5.04	5.8			

FIG. 42.4 An example of changes in SAR as groundwater becomes concentrated.

irrigation is practiced, there is a shortage of water and concentration control by leaching may be impractical. If soil drainage is poor, excess water may simply flood the voids and restrict air supply to soil organisms.

In some areas where irrigation is practiced, there are times when groundwater may rise from the water table to supply the root zone. The compatibility of the concentrated irrigation water with the groundwater may become a problem if precipitation occurs and reduces soil permeability.

CONDITIONING IRRIGATION WATER

Quality standards have been recommended for irrigation that are as complete as those issued for potable water. Table 42.3 lists recommended maximum levels of trace constituents in irrigation water.

In addition to these general recommendations, further preferred limits for salinity and important elements such as boron have been proposed for specific types of crops.

TABLE 42.3 Recommended Maximum Concentrations of Trace Elements in Irrigation Waters*

Element	For waters used continuously on all soils, mg/L	For use up to 20 years on fine textured soils of pH 6.0 to 8.5, mg/L
Aluminum	5.0	20.0
Arsenic	0.10	2.0
Beryllium	0.10	0.50
Boron	0.75	2.0
Cadmium	0.010	0.050
Chromium	0.10	1.0
Cobalt	0.050	5.0
Copper	0.20	5.0
Fluoride	1.0	15.0
Iron	5.0	20.0
Lead	5.0	10.0
Lithium	2.5†	2.5†
Manganese	0.20	10.0
Molybdenum	0.010	0.050†
Nickel	0.20	2.0
Selenium	0.020	0.020
Tin†		
Titanium‡		
Tungsten§		
Vanadium	0.10	1.0
Zinc	2.0	10.0

* These levels will normally not adversely affect plants or soils.

† Recommended maximum concentration for irrigating citrus is 0.075 mg/L.

‡ See *Water Quality Criteria*, EPA Publication R.3-73-033, 1972, pp. 337–353, for a discussion of these elements.

§ For only fine textured soils or acid soils with relatively high iron oxide contents.

Because of the large volume of water used for crop irrigation generally, water treatment for removal of specific constituents has historically been considered uneconomical. For the most part, the only treatment being practiced is threshold treatment for (1) scale control in distribution systems and (2) to prevent calcium carbonate precipitation in flume water to which ammonia has been added as a fertilizer. The addition of ammonia to irrigation water with a positive Langelier index may cause calcium carbonate to precipitate, but this can be controlled by the use of polyphosphates. In the newer drip-irrigation systems, where very small volumes of water are brought to individual plants, because of the small flow through the individual distribution units, plugging is an ever-present threat. Clarification, treatment with dispersants, application of biocides, or a combination of these may be necessary to keep the system functioning.

The use of municipal sewage plant effluent for irrigation of croplands can serve both the municipality and agricultural interests. Spray irrigation has proved to be an effective tertiary treatment for both municipal and certain industrial wastes—such as food processing wastes. There will undoubtedly be increased use of waste effluents for irrigation, and the savings to the waste plant operator in not having to install exotic teriary treatment facilities may justify additional treatment of some kind in the waste plant if the quality of the effluent must be upgraded to meet limitations on such constituents as heavy metals.

The desalination of the Colorado River to upgrade its quality to meet the needs of the Mexican farmers suggests that there will be growing consideration of desalination processes as water quality in water-short areas continues to depreciate and the value of crops increases.

SUGGESTED READING

Bower, C. A., Walcox, L. V., Akin G. W., and Keyes, M. C.: "Precipitation and Solution of Calcium Carbonate in Irrigation Operations," Soil Science Society of America, Proceedings 29 (1) and 93–94 (1965).

Croneis, C., and Krumbein, W. C.: *The Crust of the Earth,* in "Rocks and Minerals", Rapport, S., and Wright, H. (eds.), New American Library, New York, 1955.

Environmental Studies Board: Water Quality Criteria—1972. A Report of the Committee on Water Quality, EPA Publication R3-73-033, March 1973.

Salinity Laboratory Staff: U. S. Dept. of Agriculture: "Diagnosis and Improvement of Saline and Alkaline Soils," Handbook 60, U. S. Government Printing Office, Washington, D.C., 1954.

Shoji, K.: "Drip Irrigation," *Scientific American.* **237**,5 (1977).

Stumm, W., and Morgan, J.: *Aquatic Chemistry,* Wiley, New York, 1970.

CHAPTER 43
OIL FIELD WATER TECHNOLOGY

THEORY OF OIL FORMATION

The organic theory, the most generally accepted hypothesis on the origin of oil, presumes that microscopic plant and animal life from the sea and tidal areas provided oil's raw materials, hydrogen and carbon. A less prevalent theory proposes that methane and possibly other related hydrocarbons originate in the earth's mantle and work their way to faults and traps in the crust. This theory has gained support because of fairly recent findings of hydrocarbons in meteorites and in the more distant planets.

According to the prevailing theory of the biological origin of oil and gas, through millions of years, rivers transported great volumes of mud, sand, and products of surface erosion to the sea floor, to be spread by tides and currents. Under the increasing weight of this accumulating debris, the ocean floors slowly sank and were compressed to form the sedimentary rocks which contain petroleum—sandstone, shale, porous limestone, and dolomite, a mixture of calcium and magnesium carbonates.

The organic components of trapped microscopic organisms were changed over millions of years to petroleum, through chemical, physical and biological influences (Figure 43.1).

Naturally occurring petroleum is complex and variable in chemical composition. Its color ranges from light greenish brown to black. It may be fluid or so viscous as to be nearly solid. The petroleum refiners classify crudes according to their base as follows:

1. *Paraffin-base:* High in wax and lube oil fractions, containing small amounts of naphthenes or asphalt; low in sulfur, nitrogen, and oxygen compounds.
2. *Asphalt base:* High in pitch, asphalt, and heavy fuel oil.
3. *Mixed-base:* Characteristics midway between paraffin and asphalt bases.
4. *Aromatic-base:* High in low molecular weight aromatic compounds and naphthene, with smaller portions of asphalt and lube oils.

Crudes are commonly identified by API gravity, a figure which is inversely related to specific gravity (sp. gr.).

$$\text{API gravity} = \frac{141.5}{\text{S.G.}} - 131.5^\circ$$

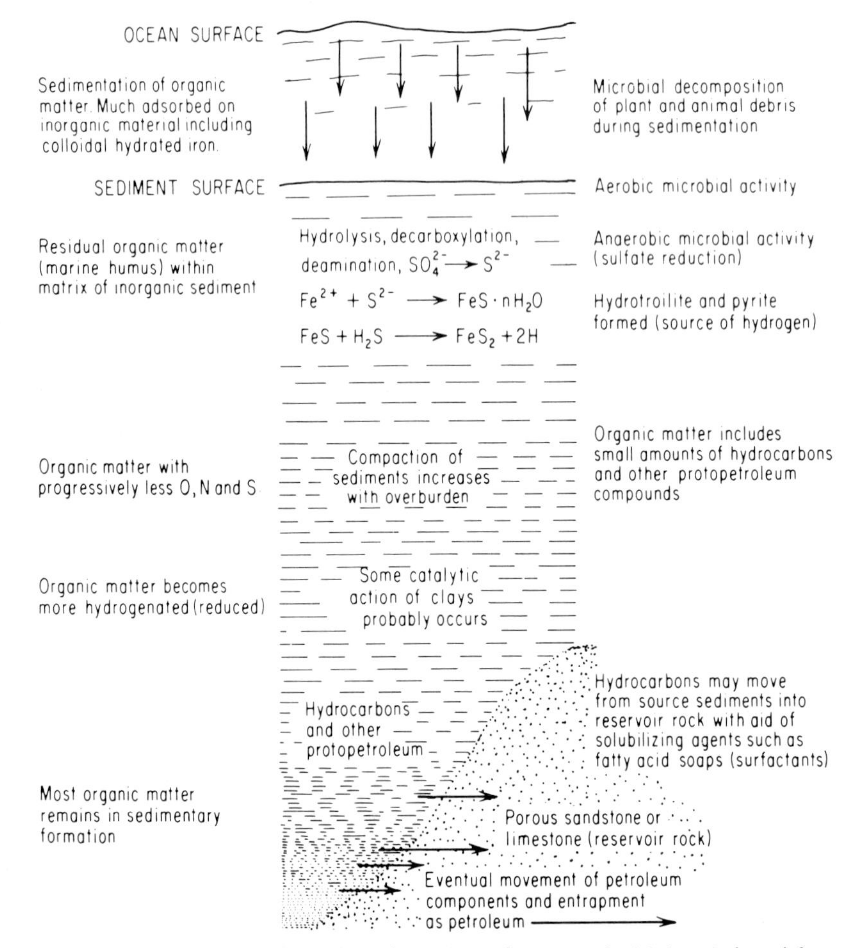

FIG. 43.1 Events in the formation of petroleum from organic debris. *(Adapted from Davis, J. B.: Petroleum Microbiology, Elsevier, Amsterdam, 1967.)*

API gravity can be confusing to those unfamiliar with petroleum terminology. In the oil industry, a "high-gravity crude" is rich in volatile materials and has a low specific gravity (0.75 to 0.84), so the API gravity is in the range 35 to 55°. Some "heavy crudes" have a specific gravity close to that of water, and the API gravity may be as low as 15° at the standard temperature of 60°F (16°C).

THE OIL-BEARING RESERVOIR

Petroleum, found only in porous sedimentary rock, migrates laterally and upward reaching some local structure or trap having a caprock seal which contains the oil, creating the reservoir.

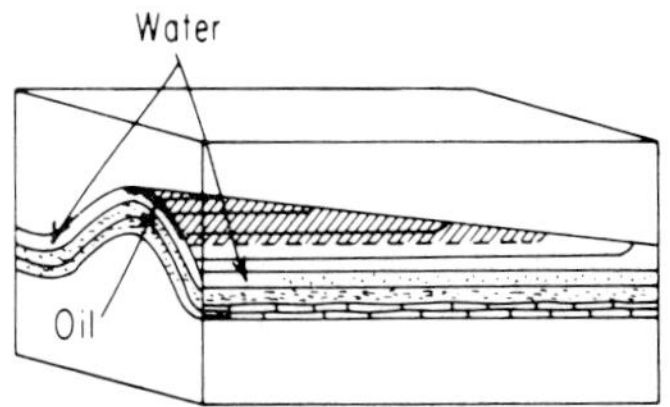

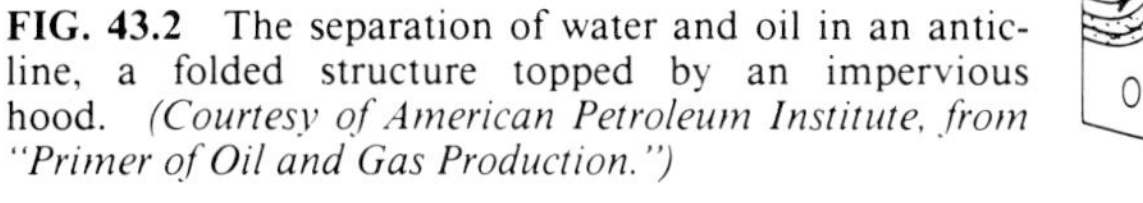

FIG. 43.2 The separation of water and oil in an anticline, a folded structure topped by an impervious hood. *(Courtesy of American Petroleum Institute, from "Primer of Oil and Gas Production.")*

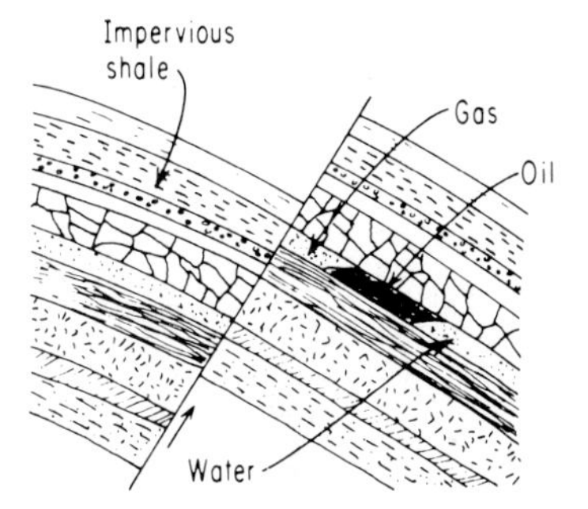

FIG. 43.3 The separation of water and oil in sections of a fault, the displacement of layers along a slip plane. *(Courtesy of American Petroleum Institute, from "Primer of Oil and Gas Production.")*

Many different shapes, sizes, and types of geologic structures form reservoirs. These include: (1) domes and anticlines, (2) fault traps, (3) unconformities, (4) dome and plug traps, (5) lens-type traps, and (6) combination traps. An anticlinal type of folded structure is shown in Figure 43.2, and a trap resulting from faulting is shown in Figure 43.3.

The length and width of a reservoir can vary from one to several miles, and the depth from a few feet to several hundred. A 1 acre (0.405 ha) reservoir with a depth of 10 ft (3.05 m) would contain 10 acre-ft, a common unit of measure. To estimate its petroleum content, the total pore volume (porosity) and percentage oil saturation (10 to 99%) must also be known. The remaining fluid is interstitial (or connate) water. The U.S. standard of measurement of volume in the petroleum industry is the barrell (42 gal, 5.62 ft^3, 0.16 m^3).

A 10 acre-ft reservoir contains 435,600 ft^3 (12,327 m^3). With a porosity of 20% and an oil saturation of 80% this would contain:

$$435{,}600 \times 0.2 \times 0.8 = 69{,}696 \text{ ft}^3 \text{ of oil } (1972 \text{ m}^3)$$

$$= 12{,}424 \text{ barrels}$$

PETROLEUM PRODUCTION

When the drill penetrates the reservoir, oil and gas are forced to the surface by natural reservoir pressure. The period of time during which oil is produced by natural reservoir pressure is referred to as primary production, a period of a few months or several years.

The flowing well is constructed of "strings" of concentric vertical pipes called casings and a smaller pipe, usually 2 to 3 in (5 to 7.5 cm) in diameter, called

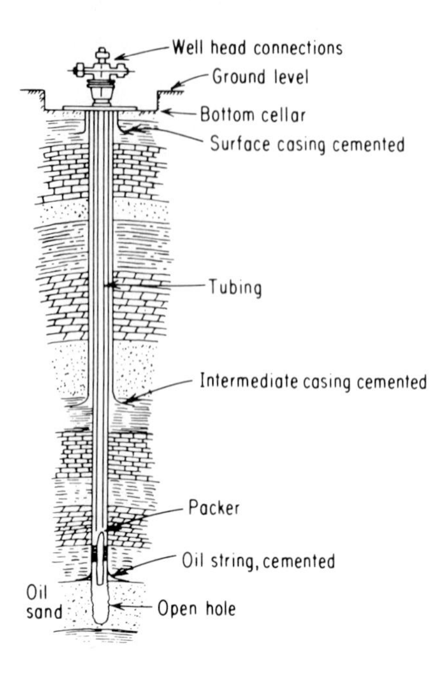

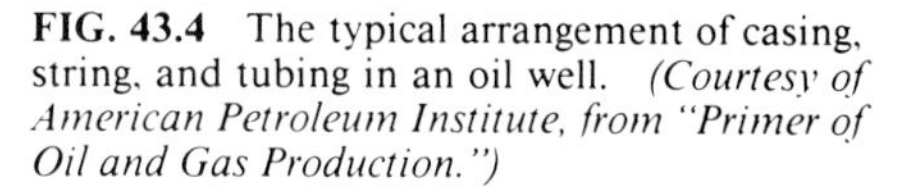

FIG. 43.4 The typical arrangement of casing, string, and tubing in an oil well. *(Courtesy of American Petroleum Institute, from "Primer of Oil and Gas Production.")*

FIG. 43.5 This large "Christmas tree" is made up of 4-in valves and fittings designed for 10,000 lb/in^2 working pressure. *(Courtesy of McEvoy Oilfield Equipment Company.)*

FIG. 43.6 This oil well pump is a walking beam unit. Submersible centrifugal units are also commonly used.

tubing, through which produced fluid flows. The largest diameter casing (the surface string) extends to a depth of 200 to 1500 feet (61 to 460 m); the intermediate string may reach a depth of up to 5000 feet (1530 m); a third casing (the oil string) may reach the producing zone. Some producing zones are at depths of 20,000 feet (6100 m) or more. The tubing into the producing formation is secured by a packer which seals the space between the tubing and the final casing (Figure 43.4). Occasionally intermediate level strings are perforated, allowing production to flow simultaneously from shallower producing zones.

A series of valves and flanges at the wellhead (called a "Christmas tree") includes a small orifice plate (called a "choke") (Figure 43.5) to control flow.

When the natural reservoir energy subsides, some method of pumping is employed to maintain production (Figure 43.6).

Oil Dehydration

Oil leaving the producing well is a mixture of liquid petroleum, natural gas, and formation water (Figure 43.7). During early primary production, water may be insignificant. Most production, however, contains sizable proportions of produced water (up to 90%). This must be separated from the oil since pipeline specifications stipulate maximum water content—as low as 1%, but up to 3 to 4% in some localities.

The initial separation vessel in a modern treating plant is called a free-water-knockout (FWKO) (Figure 43.8). Free water, defined as that which separates within 5 min, is drawn off to a holding tank, to be clarified prior to reinjection or discharge. The remaining oil usually contains emulsified water and must be further processed to break the emulsion, usually assisted by heat, electric energy, or both.

Heater treaters (Figure 43.9) are vertical or horizontal vessels in which the water-in-oil emulsion is resolved, invariably with the assistance of emulsion-

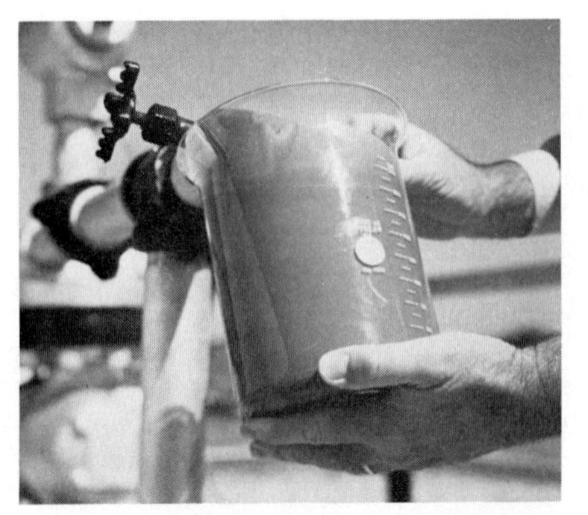

FIG. 43.7 Oil is almost always produced as an emulsion, very differnt in appearance from finished oil products.

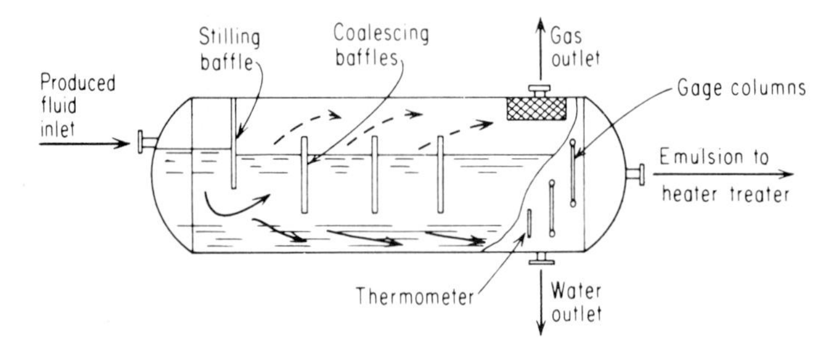

FIG. 43.8 Free water and gas are separated from the produced fluid in this gravity separation vessel, simplifying emulsion breaking. *(Courtesy of American Petroleum Institute from "Treating Oil Field Emulsions.")*

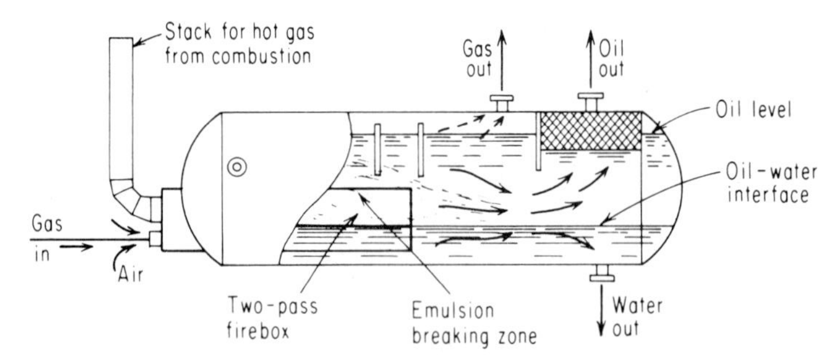

FIG. 43.9 This heater-treater uses the heat of combustion of gas or oil to heat the emulsion from the free water knockout drum, assisting chemical treatment to yield an oil of minimum water content. *(Courtesy of American Petroleum Institute, from "Treating Oil Field Emulsions.")*

breaking chemicals. The electrostatic treater employs heat, but also uses high-voltage alternating current to charge the water droplets, accelerating the process of coalescing smaller droplets into larger drops. The demulsified crude oil flows to a stock tank for pipeline shipment to a refinery.

Enhanced Recovery

Enhanced recovery describes any additional production after primary production that results from the introduction of artificial energy into the reservoir. This includes waterflooding, gas injection, and other processing involving fluid or energy injection whether for secondary or tertiary oil recovery.

Secondary recovery is any enhanced recovery first undertaken in the reservoir. Usually, it follows primary production, but may be conducted concurrently with it. Waterflooding is the most common method of secondary recovery.

Tertiary recovery is enhanced recovery undertaken following secondary operations designed for total recovery of the remaining petroleum. As much as 50% of the original oil may remain in place after primary and secondary processes are terminated.

Secondary Recovery by Waterflooding

Waterflooding is injection of water as a uniform barrier through the producing formation from a series of injection wells toward the producing well. Such injection wells can either be converted producing wells or new wells drilled specifically for injection of a flooding water.

Proper spacing of the injection wells is important. Most reservoirs are flooded through wells distributed uniformly throughout the reservoir. Others may be flooded by peripheral injectors.

A typical arrangement, called a "five-spot pattern," is shown in Figure 43.10. The density of well spacing (the area enclosed by the perimeter of the five-spot pattern) may be 3-acre, 5-acre, or whatever density is determined to be most efficient by the reservoir engineer.

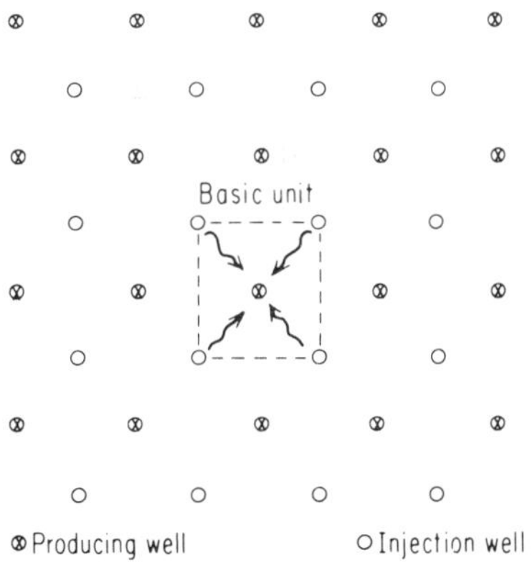

FIG. 43.10 Typical 5-spot well arrangement.

The permeability of the reservoir rock has a great bearing on its suitability for waterflooding. Rock is considered permeable if a significant fluid flow will pass through it in a short time; it is impermeable if the rate of passage is negligible.

The unit of permeability is the *darcy,* standardized by the American Petroleum Institute as follows: "A porous medium has a permeability of one darcy when a single phase fluid of one centipoise viscosity that completely fills the voids (or pores) of the medium will flow through it at a rate of one centimeter per second per square centimeter of cross-sectional area under a pressure or equivalent hydraulic gradient of one atmosphere (760 mm of Hg) per centimeter."

The permeabilities of formation cores are generally in the range of 5 to 1000 millidarcies (md) (1 md = 0.001 d). A rough practical example of 1d would be 1 ft^3 (0.0283 m^3) of sandstone passing approximately one barrel of oil (0.16 m^3) per day with a pressure drop of 1 lb/in^2 (0.068 bars, 0.0703 kg/cm^2).

Permeability and porosity vary greatly both laterally and vertically in the typical reservoir rock. A rock whose permeability is 5 md or less is called a "tight sand" or a "dense limestone" according to its composition. The following are rough permeability ratings:

1 to 10 md	Fair permeability
10 to 100 md	Good permeability
100 to 1000 md	Very good permeability

It is imperative for efficient flooding that the water be totally compatible with the reservoir formation. A desirable water for this purpose is that produced from

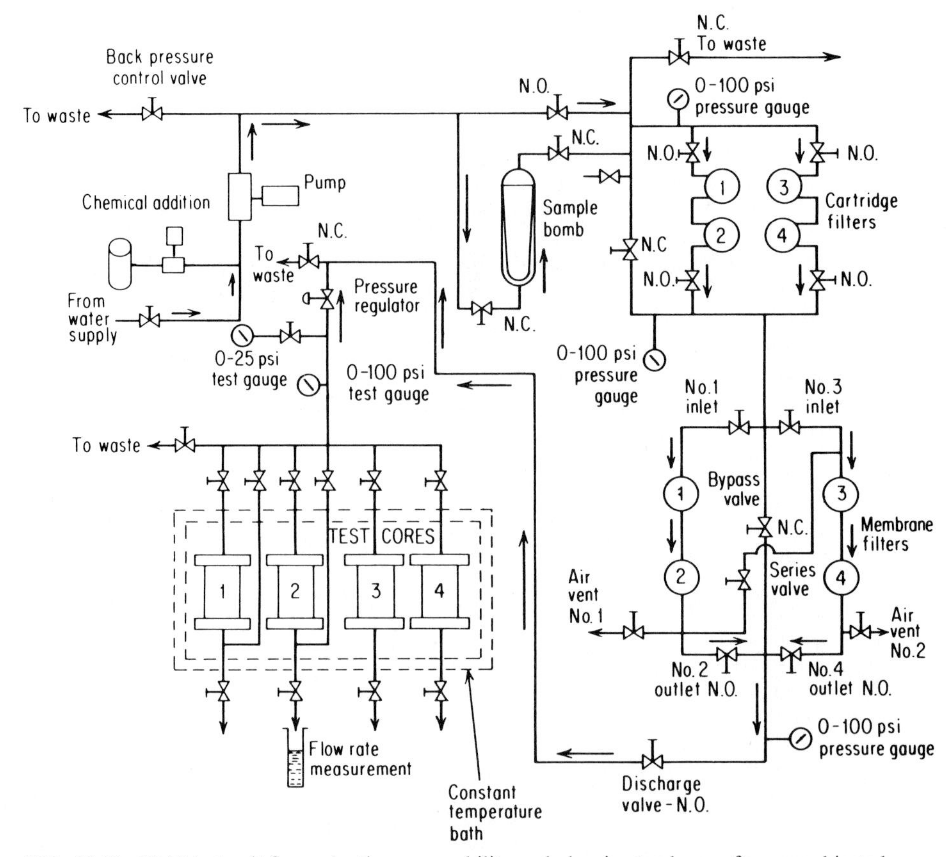

FIG. 43.11 Field test unit for evaluating permeability and plugging tendency of a core subjected to a specific flood water. *(From McCune, C. C.: "On Site Testing to Define Injection Water Quality Requirements," Society of Petroleum Engineers Publication 5865.)*

the formation, after separation from the oil and gas. However, the amount of water required for flooding far exceeds the volume produced, so supplemental water is needed.

Produced water and supplemental surface waters must be clarified to remove residual oil accumulations, sand, or dirt washed from the producing formation, oxidized inorganic or organic suspended solids, or corrosion products. Quality required for waterflooding is dictated by the permeability restriction of the reservoir. Currently, the most popular equipment for clarification is the flotation cell employing gas diffusion to produce clear effluent.

Where injection water must include surface or well water of lower salinity than the original formation water, chemical incompatibility with the formation rock may result. This may cause swelling of clay in the reservoir rock. Historically, the suitability of injection water has been determined through analyses of the waters in question, a membrane filtration, or laboratory work with formation cores. All of these procedures have the disadvantage that the waters have been removed from the actual reservoir environment and may have changed.

A field unit has been developed for defining injection water quality standards (Figure 43.11). The injection water does not undergo any change during sampling, handling or shipment, and storage. Formation rock is used in the study of the effect of chemical additions and filtration conditions.

FIG. 43.12 In this waterflooding operation, both sand filters and septum-type filters are used to prevent underground plugging. *(Courtesy L*A Water Conditioning Company.)*

Most injection waters are passed through some type of filter, the media and design depending on the reservoir permeability. Types of filters include mixed media beds, individual well cartridges, or septum filters using diatomaceous earth as a filter aid (Figure 43.12). In some cases more sophisticated treatment, such as clarification or lime softening, may be needed. Deaeration may be required to reduce corrosion or to prevent iron oxidation (Figure 43.13).

FIG. 43.13 Vacuum degasifiers remove dissolved oxygen from water used for secondary oil recovery on this offshore rig. *(Courtesy of Infilco Degremont Inc.)*

SECONDARY RECOVERY BY STEAM FLOODING

Water flooding is not effective for secondary recovery of low API-gravity oil from relatively shallow formations at temperatures below about 120°F. This situation responds, however, to steam flooding. Sometimes the steam is used to stimulate an oil well: The well is taken out of service and steam is injected for a period of several weeks to heat the oil-bearing rock; then the oil well is returned to service until stimulation may be required again. (This is known as the "huff-and-puff" method of steaming). The most common method of steaming, however, is introduction of the steam-water mixture from a once-through steam generator into one of a group of centrally located displacement wells, with the steam and hot water then radiating outward toward the peripheral oil wells served by each of the injection wells.

In the Imperial Valley of central California, crude oil having an API gravity of about 11 to 13, and occurring in a very loose formation at depths of 500 to 1500 ft, is recovered by the displacement method of steam injection. In the United States, this area is the largest producer of oil by steaming; the total California production by this process is about 375,000 bpd. About 30% of the crude produced is used as fuel in the steam generators, although some generators are gas-fired, either continuously or intermittently, as may be required to meet air emission standards.

As time passes at any steam-flood site, there is a gradual increase in temperature of the produced fluids to about 160 to 180°F (71 to 82°C). The wells produce 6 to 8 bbl of water per barrel of oil. The oil and water are separated in dehydration tanks using emulsion breakers, and the water must then be processed to render it suitable for feed to the steam generators and for disposal. The salinity and chemical characteristics of the produced water vary greatly from one site to another. Table 43.1 includes the analysis of several produced waters after the dehydration tanks. Even though these waters are softened, chemically deaerated, and chelated, it is still surprising that such high-salinity waters could be suitable feed water for steam generators that sometimes operate at pressures over 1000 lb/in^2.

Figure 43.14 shows a typical flow diagram of a water treatment plant servicing a steam-flooding facility. After dehydration, oil is removed in dispersed-air flotation units (see Chapter 9), assisted by cationic emulsion breakers. The skimmings are recovered and treated water is discharged to storage. The flotation unit and storage tanks are gas blanketed to avoid pickup of oxygen. An oxygen scavenger is usually applied at this point. If the Stiff-Davis index is strongly scaling, a stabilizing inhibitor is also applied here to prevent problems owing to calcium carbonate scale. Gypsum (calcium sulfate) may also be a problem. The water is then polished through septum-type filters, Figure 43.12. (See also Figure 11.11.) A variety of filter aids can be used, including specially processed cellulose, diatomaceous earth (DE), and vermiculite. The filter is first precoated by recirculation; then it is put onstream, with a continuous application of filter aid (body feed) of 2 to 3 ppm per ppm oil.

The filtered water is then softened through zeolite softeners (Figure 43.15). The high salinity of the produced water makes softening difficult, so almost universally, a polishing softener follows the primary unit to assure zero hardness in the finished water. The softened water is then sent to the various steam generators in different areas of the oil field—sometimes several miles away.

The steam generator is unusual in that it is a once-through unit (Figure 43.16). Feed water is preheated (to avoid acid deposit attack of the economizer), and then flows through the economizer to recover heat from the stack gas. The feed water then passes through the radiant and convection sections of the steam generator where steam is generated. However, unlike conventional drum-type boilers, the fluid leaving the generator has a quality of 80%; that is, it contains 80% steam and 20% water by weight. Because the steam volume at 900 lb/in^2 is 0.49 ft^3/lb, where the specific volume of water is only 0.021 ft^3/lb, the discharged fluid contains 99% steam by volume and only 1% water by volume.

However, the water in the discharged steam contains all of the solids contained in the feed water. At 80% quality, 5 lb of feed water produces 4 lb of steam and 1 lb of water, so the solids have been concentrated 5 times. The relationship between steam quality and concentration is:

$$\text{CR (concentration ratio)} = \frac{100}{100\text{-quality}}$$

Depending on the depth of the injection well and the permeability of the formation, the steam generators may operate at a pressure as low as 300 lb/in^2 and as high as 1000 (68 bars). This pressure is maintained by the boiler feed pumps.

TABLE 43.1* Analyses of Several Formation Waters

			Constituents, mg/L						
State	County	Formation	Na	Ca	Mg	Cl	SO_4	HCO_3	TDS
Oklahoma	Kingfisher	Oswego	56,250	8,300	260	98,300	180	50	166,652
Kansas	Ellis	Arbuckle	16,800	2,630	690	30,500	2,880	315	54,072
New Mexico	Lea	San Andreas	9,150	1,500	500	17,800	2,000	1,000	32,329
Texas	Hopkins	Paluxy	5,640	630	40	8,350	120	500	15,417
California	Kern†		3,936	124	66	5,800	216	73	10,159
California	Kern		6,725	235	115	10,714	216	52	18,382
California	Kern (Kern River)		184	25	5	170	65	234	620

* The form for water analysis reporting recommended by the American Petroleum Institute is shown in Figure 43.14.

† These brines are treated and used as feed water to 800 to 1000 lb/in^2 steam generators in steam flooding applications.

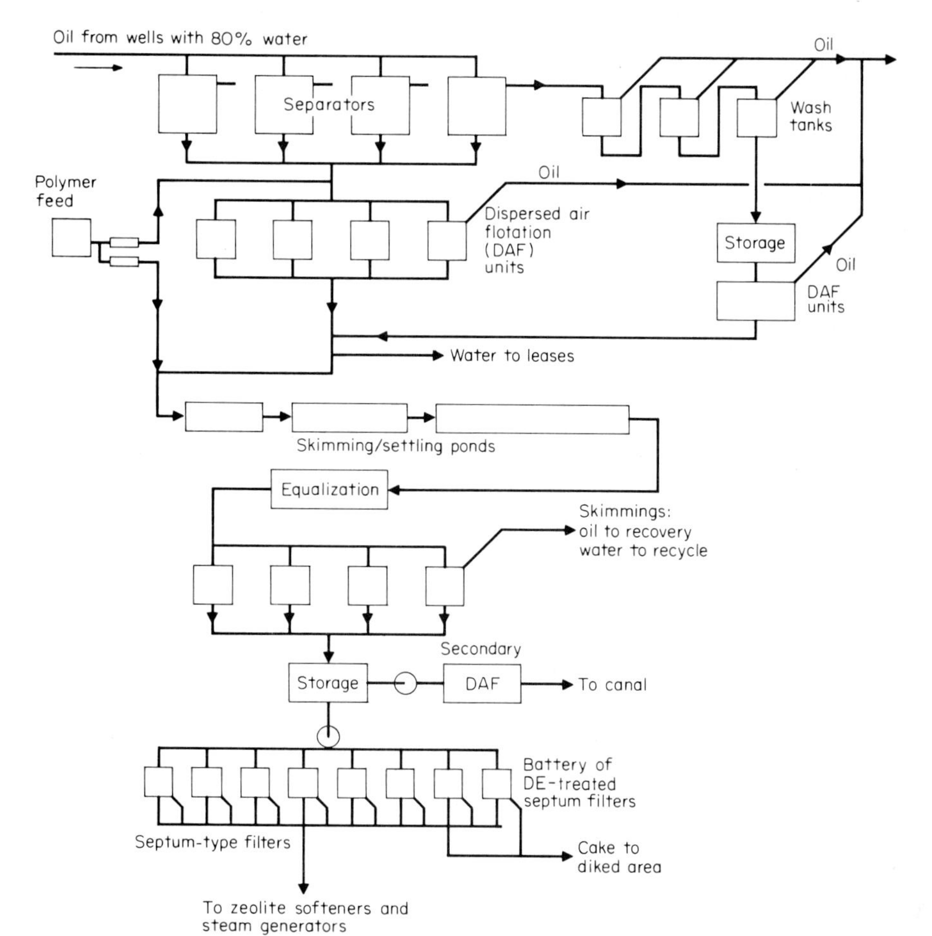

FIG. 43.14 Treatment of produced water for steam flooding.

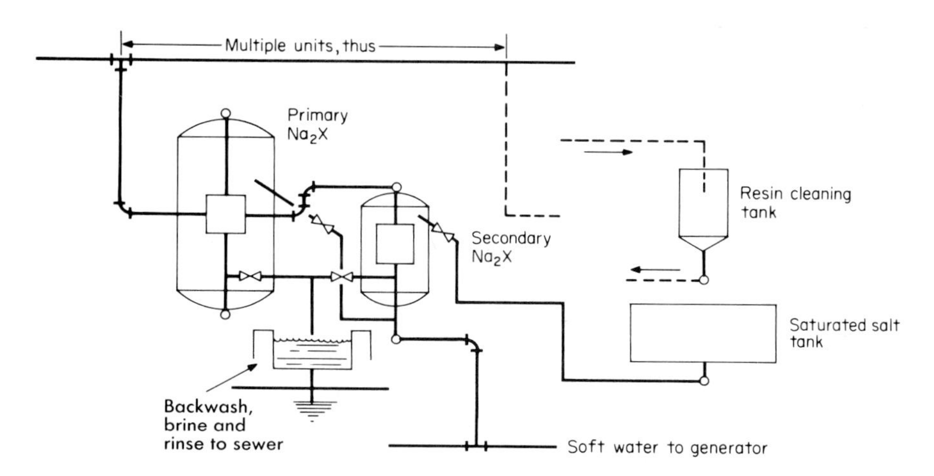

FIG. 43.15 Two-stage sodium zeolite softening of deoiled, filtered, produced water to render water suitable for high-pressure boiler.

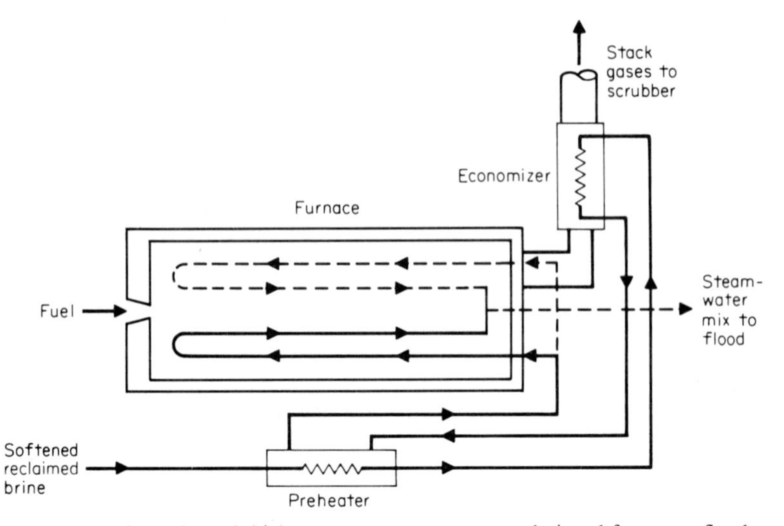

FIG. 43.16 Once-through high-pressure steam generator designed for steam-flooding and recovery of the low-gravity, viscous oils of lower California.

Since produced crude is usually burned in the steam generators, the stack gases are scrubbed in wet scrubbers for removal of SO_2 and NO_x. Makeup to the scrubbers becomes concentrated and is usually treated with an antifoam and dispersant to minimize scaling, carryover, and restriction of gas flow that would interfere with furnace firing conditions.2

THE NATURE OF OIL FIELD WATERS

After establishing injection water quality, steps must be taken to eliminate or control scaling, corrosion, and fouling of surface equipment, tanks, and lines to prevent interruption of water injection or a decrease in oil production.

It is necessary to conduct several types of analyses to identify and evaluate these potential problems. Analyses must be run at points throughout the water system, because changes through the system provide data for proper control. The job of injection system control starts at the producing well, for it is here that water begins to change. As water enters the production tubing, a pressure drop occurs that in itself could cause precipitation either in the tubing string or in the formation. As the produced fluid approaches the surface, further pressure drop causes loss of gases and a reduction in temperature.

Substances in oil field waters are classified as: (1) dissolved ionic solids, (2) suspended solids, (3) dissolved gases, (4) problem-causing bacteria, and (5) residual oil.

Total Dissolved Solids (TDS)

In oil field brines, TDS ranges from less than 10,000 mg/L to over 350,000 mg/L, of which (NaCl) constitutes 80% or more. Troublesome cations found in oil

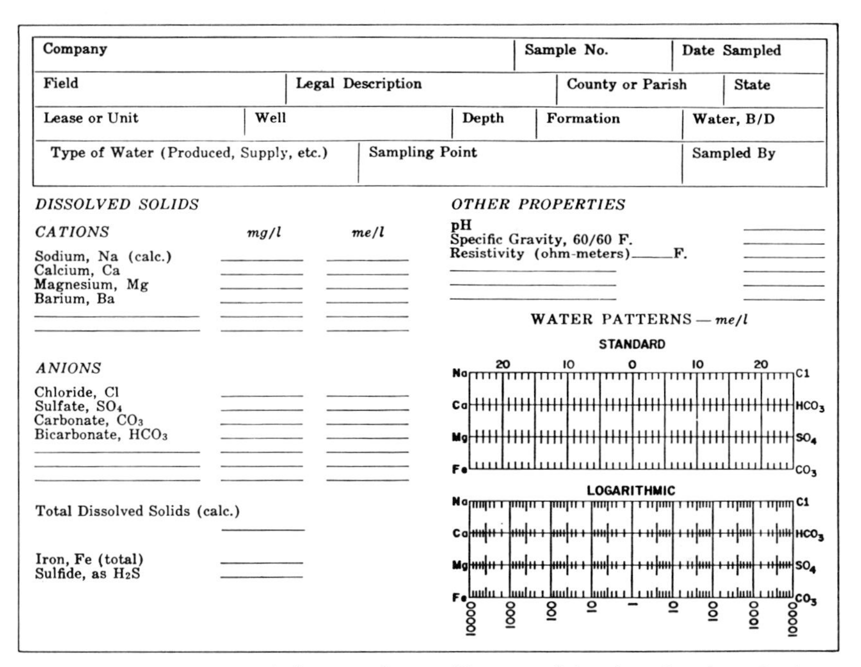

Company			Sample No.	Date Sampled
Field	Legal Description		County or Parish	State
Lease or Unit	Well	Depth	Formation	Water, B/D
Type of Water (Produced, Supply, etc.)	Sampling Point			Sampled By

DISSOLVED SOLIDS

CATIONS	*mg/l*	*me/l*
Sodium, Na (calc.)		
Calcium, Ca		
Magnesium, Mg		
Barium, Ba		

ANIONS		
Chloride, Cl		
Sulfate, SO_4		
Carbonate, CO_3		
Bicarbonate, HCO_3		

Total Dissolved Solids (calc.)

Iron, Fe (total)

Sulfide, as H_2S

OTHER PROPERTIES

pH

Specific Gravity, 60/60 F.

Resistivity (ohm-meters)____F.

WATER PATTERNS — *me/l*

FIG. 43.17 API water analysis report form. *(Courtesy of American Petroleum Institute, from API RP 45, "APT Recommended Practice for Analysis of Oil-Field Waters.")*

field waters are calcium (Ca); magnesium (Mg); barium (Ba); strontium (Sr); and ferrous iron (Fe). Commonly encountered anions are chloride (Cl), sulfate (SO_4), bicarbonate (HCO_3), and sulfide (S).

Potassium (K), lithium (Li), boron (B), zinc (Zn), and copper (Cu) are also frequently measured. A few typical brine combinations are shown in Table 43.1. The "as calcium carbonate" designation, valuable to engineers working with surface waters, is virtually unknown in oil field water work. The standardized API report (Figure 43.17) includes both tabular and graphical forms.

Suspended Solids

Various inorganic and organic solids are found in petroleum waters. These may be particles of metal oxides from well casings, or oxidized iron or manganese originally in the water. Other suspended solids may be silt, sand, clay, or bacterial bodies. These particulates are collected on a 0.45-μm membrane filter for identification. The filtered solids are analyzed separately (Figure 43.18). A high concentration of Freon-soluble extractables usually indicates the need for better demulsification or clarification of the water to be injected. Hydrochloric acid soluble materials include carbonates of calcium, magnesium, and iron, as well as oxides and sulfides of iron. Residue remaining after treatment could include silica, barium sulfate, calcium sulfate, or heavy hydrocarbons such as asphaltenes, which are insoluble in most solvents. The residue could also include bacteria.

NALCO

ANALYTICAL SERVICE LABORATORY REPORT

FROM: [redacted]

DATE: 1-4-78
ANALYSIS NO: 76V-2495,B
SAMPLING DATE: 12-28-77
DATE SAMP. REC'D: 12-28-77

SAMPLE MARKED: Injection Water

PRESSURE (Psig)		20
THROUGHPUT (milliliters)		1000
EXTRACTABLES (mg/l)		
Water Soluble		56.4
Freon Soluble		12.2
Hydrochloric Acid Soluble		3.8
Iron (Fe)	0.5	
Calcium (Ca)	0.3	
RESIDUE (mg/l)		
Total Residue		5.8
SiO_2 and/or $BaSO_4$	2.2	
TOTAL FILTERABLE SOLIDS (mg/l)		78.2

FIG. 43.18 Analysis of suspended solids.

Dissolved Gases

The gases of greatest concern are hydrogen sulfide (H_2S), carbon dioxide (CO_2), and oxygen (O_2).

Produced waters containing *hydrogen sulfide* are called sour waters. Oil reservoirs can become sour through the activity of sulfate-reducing bacteria in the producing formation. H_2S concentrations can reach several hundred milligrams per liter. In many areas of the country elaborate mechanical systems are used to remove this gas from the produced water before reinjection.

As H_2S is extremely poisonous, all sour oil field operators post conspicuous signs at tanks or vessels cautioning against inspection without the use of breathing equipment.

H_2S in contact with iron produces iron sulfide, which can accelerate corrosion or act as a serious plugging agent. The sulfides of most metals are insoluble in water.

H_2S can also be produced in the water handling system by sulfate-reducing bacteria. Any increase in H_2S concentration through the water system not caused by blending is considered an indication of bacterial activity.

H_2S in contact with dissolved oxygen can produce elemental sulfur, also a serious plugging agent.

$$H_2S + \tfrac{1}{2}O_2 \rightarrow = H_2O + S^{\circ} \tag{1}$$

Carbon dioxide (CO_2) is an ionizable gas, forming weak carbonic acid when dissolved in water. It is one of the greatest contributors to production well and waterflood system corrosion. Production well concentrations of CO_2 can exceed 200 mg/L, much of which is lost to the atmosphere when the produced fluids leave the well. Carbon dioxide is also responsible for dissolving limestone reservoir rock, increasing hardness and alkalinity.

Dissolved oxygen (DO) is rarely present in produced fluids coming from the reservoir, unless entrained through leakage, but it is perhaps the most serious corrodent participating in oil field water corrosion mechanisms.

It is also responsible for creating plugging agents through oxidation of ferrous iron and hydrogen sulfide. It is important to try to exclude oxygen from all vessels, casings, and surface supply lines.

Problem-Causing Bacteria

Of several types of bacteria responsible for corrosion or production of plugging type solids, the most serious offender is the anaerobic sulfate-reducing bacterium, *Desulfovibrio desulfricans.* Historically these bacteria have been identified through the serial dilution or "extinction" dilution technique. The disadvantage of this technique is that a time lapse of 7 to 30 days might result before positive determination is made.

Using a new technique, bacteria filtered onto a pad from a water sample (0.45-μm filter) are immediately immersed in boiling water containing a buffer solution that destroys the enzymes which would normally consume ATP (adenosine triphosphate) upon cessation of life. The preserved ATP is reacted with luciferin and luciferase enzymes in a photometric cell, and the photons of light produced indicate total bacteria population within minutes after sampling.

Residual Oil

For a qualitative determination of residual oil, Freon extraction followed by weighing is recommended. This procedure requires time, limiting its use for monitoring.

A field procedure involving extraction of crude with chloroform gives good information concerning changes in oil concentration throughout the water system. In many cases, qualitative correlation with laboratory extraction methods is quite close. The chloroform-extracted hydrocarbon is compared to prepared standards in a simple photometer. This method is restricted to use with crudes having a definite color.

MINERAL SCALES

The mineral scales of greatest concern to oil producers are calcium carbonate, calcium sulfate, and barium sulfate. Strontium sulfate, a less common scale, can also cause problems. Several iron compounds are also of concern, related to corrosion or to oxidation of ferrous iron as a consequence of oxygen intrusion. Table 43.2 lists common scales with system variables that affect their occurrence.

Mineral scales can form in many areas; supersaturation can occur in the formation face as the produced fluids enter the production tubing; scale can form throughout the length of the production tubing, on sucker rods, and in downhole pumps; it can form in surface vessels and on the heating surfaces in the heater treater. In the water handling system, scale can form in injection water pumps, surface lines leading to injection wells, and rock surfaces in the injection formation.

Some scales can be removed by chemical treatment, an expensive, time consuming process. Scales such as barium sulfate, which cannot be removed chemically, may require replacement of surface lines or abandonment of the well.

Calcium Carbonate

Calcium carbonate equilibria are frequently upset in oil field waters, because of changes in temperature, pressure, and pH. Because calcium carbonate solubility decreases with increasing temperature, injection from the surface to a warm formation increases the chance of calcium carbonate deposition. Its solubility

TABLE 43.2 Most Common Scales

Name	Chemical formula	Primary variables
Calcium carbonate (calcite)	$CaCO_3$	Partial pressure of CO_2, temperature, total dissolved salts
Calcium sulfate		
Gypsum (most common)	$CaSO_4 \cdot 2H_2O$	Temperature, total dissolved salts, pressure
Anhydrite	$CaSO_4$	
Barium sulfate	$BaSO_4$	Temperature, total dissolved salts
Strontium sulfate	$SrSO_4$	
Iron compounds		
Ferrous carbonate	$FeCO_3$	Corrosion, dissolved gases, pH
Ferrous sulfide	FeS	
Ferrous hydroxide	$Fe(OH)_2$	
Ferric hydroxide	$Fe(OH)_3$	
Ferric oxide	Fe_2O_3	

increases as the total dissolved solids content increases. For instance, the addition of 200,000 mg/L NaCl increases $CaCO_3$ solubility by over 100%. Prediction of calcium carbonate scale potential has been the subject of much research. The Langelier saturation index attempted to relate calcium and alkalinity concentrations to pH, temperature, and total dissolved solids. But this had limited value in oil field brines.

The Stiff-Davis index is now widely accepted for predicting calcium carbonate deposition in oil field systems. To be useful, the analytical data must be determined on freshly drawn samples; analyses determined in a laboratory are not reliable since the character of the water may change substantially. The Stiff-Davis index ranges from -2.0 to $+2.0$. Positive numbers indicate a progressively severe supersaturation; negative numbers indicate an undersaturated solution. The validity of the index depends on the accuracy of the analysis.

Calcium Sulfate

Most calcium sulfate deposits found in the oil field are gypsum ($CaSO_4 \cdot 2H_2O$), the predominant form at temperatures below 100°F (38°C); above this temperature anhydrite ($CaSO_4$) may be found. Gypsum solubility increases with temperature up to about 100°F, and then decreases with increasing temperature. Sodium chloride increases the solubility of anhydrite, as it does for calcium carbonate, up to a salt concentration of approximately 150,000 mg/L. Higher salt concentrations decrease calcium sulfate solubility. The addition of 150,000 mg/L NaCl to distilled water triples the solubility of gypsum.

One of the more recent predictive indexes for gypsum is that developed by Skillman, McDonald, and Stiff. Calcium sulfate precipitation usually results from the mixing of two waters, one of which has a high calcium or sulfate concentration. A solubility graph relating calcium and sulfate concentration to total brine concentration is shown in Figure 43.19.

Barium Sulfate

Barium sulfate solubility is the lowest of the usual scales, approximately 2.3 mg/L in distilled water. The solubility of barium sulfate is also increased by increasing salt concentration. The addition of 100,000 mg/L NaCl to distilled water increases the solubility of barium sulfate from 2.3 mg/L to approximately 30 mg/L at 77°F (25°C).

The solubility of barium sulfate increases with temperature so that the combined effect of temperature and sodium chloride concentration can increase solubility to approximately 65 mg/L at 203°F (95°C).

Barium sulfate deposition usually results from the mixing of a barium-rich water with a sulfate-rich water. Such a combination should be avoided, but where mixing is unavoidable, chemical inhibitors may control deposits.

Scale Prevention

The first scale inhibitors were the inorganic polyphosphates, but their limitations were quickly discovered: above 140°F (60°C) they revert to orthophosphate, as they do at acidic pH or over a long residence time. The orthophosphate form does

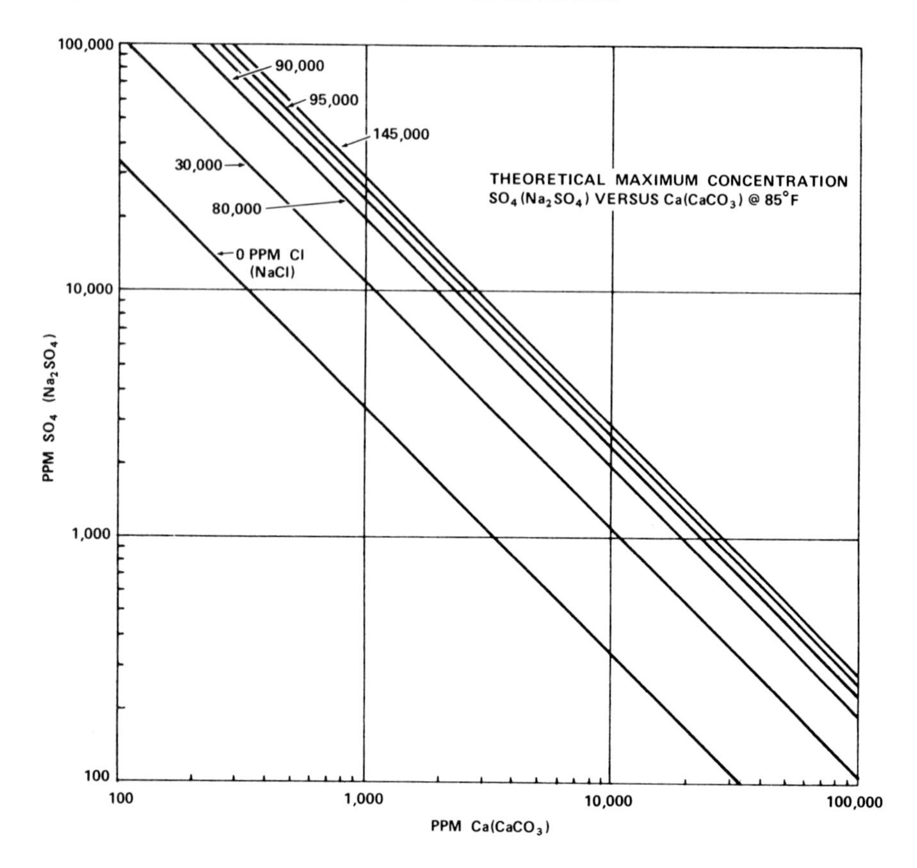

Instructions for using the graph:

1. Determine concentrations of Ca (as $CaCO_3$), SO_4 (as Na_2SO_4) and Cl (as NaCl).
2. Plot the intersecting point of Ca and SO_4 concentrations on the graph.
3. If point lies above the diagonal line for NaCl level, the water will tend to precipitate $CaSO_4$. If point falls below the NaCl diagonal, the water is undersaturated and should not precipitate $CaSO_4$.

FIG. 43.19 Calcium sulfate solubility graph at various brine levels.

not inhibit scale. Present day oil field scale inhibitors are of three types: (1) esters of polyphosphoric acid, (2) phosphonates, and (3) organic polymers, such as polymers or copolymers of acrylic or methacrylic acid.

The phosphate ester formulations are versatile for most oil field water applications. Phosphonates and polymer products have definite advantages where water temperatures exceed 200°F (90°C). Dosages vary with temperature and the concentration of suspended solids, since solids adsorb the inhibitor, requiring higher concentrations for effective inhibition. Dosage is also directly proportional to the degree of supersaturation.

When applying scale inhibitors to producing wells the "squeeze" technique is frequently used. This involves injection of a scale inhibitor into the producing formation through the production tubing string under pressure. The scale inhibitor molecules adsorb to the formation, to be gradually released with the produced fluids. Scale inhibitor squeeze applications can last for periods up to 6 months. The need for renewed treatment is determined through tests for inhibitor residual in the produced fluids.

CORROSION

The primary corrodents in oil field water systems are carbon dioxide (CO_2), hydrogen sulfide (H_2S), and oxygen (O_2).

One reason oxygen is corrosive, even at low concentrations, is its participation in creating differential cells beneath deposits on metal surfaces, which become anodic to adjacent deposit-free areas. Control of oxygen corrosion in oil field water systems requires a conscientious effort to exclude air from all surface tanks and vessels and from the casings of producing wells. Field gas is used to maintain a positive gas blanket in these areas whenever possible.

After establishing control over oxygen exposure, it is then practical to use an oxygen scavenger to react with trace quantities of remaining oxygen. Several types of sulfite are used, including sodium bisulfite ($NaHSO_3$), ammonium bisulfate (NH_4HSO_3), sodium sulfite (Na_2SO_3), and sulfur dioxide (SO_2).

The chemistry is the same with all of these: the sulfite reacts with oxygen to form a sulfate ion.

$$SO_3^{2-} + \tfrac{1}{2}O_2 \rightarrow SO_4 \tag{2}$$

The sulfite-oxygen reaction is influenced by temperature and pH. Optimum pH for quick reaction is above 7. Most oil field waters will range in pH from 6.0 to 7.5.

Catalysis of the reaction is necessary. Sometimes naturally occurring metal ions in the produced water will do the job, but usually the catalyst is provided in the sulfite formulation.

In high-oxygen freshwater systems, passivating inhibitors such as chromate supplemented by zinc and phosphate may provide an alternative to oxygen scavenging. However, the presence of H_2S or certain easily oxidized hydrocarbons in injection systems reduces chromate and precipitates zinc, so these alternatives are seldom applicable in waterfloods.

Hydrogen sulfide is corrosive because it ionizes to form a weak acid:

$$H_2S \rightarrow H + HS^- \tag{3}$$

Whenever H_2S is present, iron sulfide deposits. These deposits are cathodic to base metal, so it usually follows that severe pitting occurs beneath iron sulfide deposits, all as a consequence of the H_2S. Where O_2 intrudes into a sulfide system, the rate of corrosion can become uncontrollable. Carbon dioxide mixed with H_2S is also much more aggressive than either of the gases alone. Where H_2S is present, an efficient film-forming inhibitor must be used to prevent severe, localized corrosion.

Invariably, corrosion inhibitors used in oil field water work are organic film formers. The molecules adsorb onto metal surfaces to shield the metal from the corrodents. Since the film-forming corrosion inhibitors must be added to a wide range of brine concentrations and in many cases to a mixture of water and hydrocarbon, inhibitors with a wide range of solubilities must be available. In the producing well where a mixture of crude oil and brine must be treated, an inhibitor that is oil soluble and only slightly water dispersible is often required. This will film metal from the oil phase, providing long-term persistency to the metal surface; that is, the film will be maintained even when the inhibitor feed is discontinued. This allows batch treatment of producing wells in many instances. One of the most common methods of treating producing wells is using a treating truck to pump inhibitor down the tubing or into the annulus, followed by an overflush of produced water. The application is repeated periodically at intervals depending on the aggressiveness of the corrodents.

In injection water systems where water is the predominant phase, inhibitors must be either totally water soluble or highly water dispersible to carry through the surface line and tubing system. These inhibitors are not persistent, and it is necessary to feed continuously, always maintaining a residual in the system. Loss of inhibitor residuals results in desorption of inhibitor film and loss of protection.

In monitoring the chemical program, inhibitor residuals are determined through field extraction tests for close control.

Corrosion monitoring is usually accomplished through a combination of metal weight-loss specimens, corrosion meters, pipe spools or nipples (which have the advantage of duplicating flow conditions), and "iron counts." Residual iron concentrations are valid in systems without hydrogen sulfide. Where H_2S is present, iron is deposited as iron sulfide.

BACTERIA CONTROL

Control of bacteria is important in oil field water operations because they can cause plugging of the injection water formation or serious corrosion.

The most troublesome of these organisms is the anaerobic sulfate-reducing bacterium, *Desulfovibrio desulfricans.* Present in many oil-bearing formations, it is involved in the chemical changes which occur during the formation of petroleum. It reduces inorganic sulfate (SO_4^{2-}) to sulfide (S^{2-}), which leads to iron sulfide precipitates.

Clostridia, another anaerobic sulfate-reducing organism, is not as prevalent as *Desulfovibrio.* This organism is reported to be thermophilic, preferring a temperature range of 55 to 70°C.

Several species of aerobic bacteria are also important, most of which are found in surface waters used to supplement produced water in a water flood. Some of the more important organisms are *Pseudomonas, Flavobacterium, Aerobacter, Escherichia,* and *Bacillus.* These form slime masses, which can cause plugging or which shield the anaerobic *Desulfovibrio.*

Iron bacteria are occasionally found in some water floods, the two major species being *Sphaerotilus* and *Gallionella.* These oxidize the ferrous ion (Fe^{2+}) to ferric ion (Fe^{3+}). Iron bacteria are identifiable through a microscopic analysis following staining.

Beggiatoa oxidize sulfide to sulfate. *Beggiatoa* usually grow in open produced water systems where a gathering line dumps into an open pit. They are often responsible for filter plugging.

Microbes are controlled by chemical application. The concentration of chemical and the period of contact varies among systems. Normally, chemical is fed at a selected dosage, such as 50 to 100 mg/L, for a period of 4 to 8 h. Following application, samples are withdrawn from the system and microbial population is determined. The frequency of biocide application is dictated by such control tests.

In many situations, physical system cleanup is necessary to gain proper microbial control. Low spots in lines or tanks with bottom accumulations of sediment or debris protect bacteria from chemical contact. In these cases, the chemical program must be supplemented by a thorough "house cleaning."

SUGGESTED READING

American Petroleum Institute: *Primer of Oil and Gas Production,* 1983.

American Petroleum Institute: *Treating Oil Field Emulsions,* 3d ed., 1974.

Cole, Frank W.: *Reservoir Engineering Manual*, Gulf, Houston, 1969.

Collins, A. Gene: *Geochemistry of Oilfield Waters*, Elsevier, Amsterdam, 1975.

Levorsen, A. I.: *Geology of Petroleum*, Freeman, San Francisco, 1966.

Uren, L. C.: *Petroleum Production Engineering*, McGraw-Hill, New York.

CHAPTER 44
PRODUCTION OF ULTRAPURE WATER

One of the lessons learned early in the Space Age was that the normal work environment was almost universally contaminated by invisible crud. Fine debris appeared to surround all human industry and to infiltrate sensitive instruments and controls, to lower performance, cast doubt on instrumental data, and occasionally interfere with such critical processes as controlling the flow of rocket fuel, where failure of the fuel control valve could abort a mission or destroy a vehicle in flight. The aerospace industry developed clean-room techniques that required use of equipment to continually circulate and filter fine particles from the air and procedures for personnel to follow in clothing themselves properly to avoid introducing debris into the work area. Research and development are still progressing in this field. Recent studies have shown that humans emit particulates continually, and some newer manufacturing processes, in the electronics field especially, cannot tolerate this debris and will require that robots replace humans in certain supercritical manufacturing operations to avoid contamination.

It is not surprising that water chemists have focused their attention on a similar problem—particulates, or "crud," in water. Water used in such sensitive operations as the preparation of pharmaceuticals, washing cathode ray tubes, printed circuits and transistors, and makeup for nuclear power plants must be ultrapure. Ionic impurities can be reduced to such low concentrations—at the parts per billion (micrograms per liter) level—that measurement of significant conductivity differences is virtually impossible. Yet particulate impurities undetectable by online analytical devices may remain to damage products washed with this water. For example, linewidths of less than 1 μm on a transistor chip make it obvious that the deposition of particulates of even submicron dimension on the surface during rinsing of an integrated circuit could ruin the device, so final filtration through submicron filters is essential. In the nuclear power industry, the presence of organic acids in steam generators treated with only a few selected all-volatile chemicals has shown the need for more complete removal from makeup water of the organic matter that produces these acids than was anticipated earlier in the history of these critical plants. Sources of this organic matter include not only the raw water, but also ion exchange resins, regenerant solutions, and metabolites from microbes lodged in various parts of the system.

Defining the specifications for the ultrapure water needed by various industries has been complicated by failure of trade associations to keep pace with the rapid changes in their industries. For example, the American Society for Testing and Materials (ASTM) standards on reagent grade water (currently ASTM D 1193-77)

TABLE 44.1 Abstract of Proposed ASTM Standards for Electronic Grade Water

Measurement	Type E-I	Type E-II	Type E-III	Type E-IV
Resistivity, min MΩ-cm at 25°C	18(90% of time) 17 minimum	15(90% of time) 12 minimum	2	0.5
Particle count,* max per mL	2	2	100	500
Viable bacteria, max per mL	less than 1	10	50	100
TOC, max ppb	50	200	1000	5000
Total solids, max ppb	10	50	500	2000
Na, max ppb	less than 1	10	200	1000
Cl, max ppb	2	10	100	1000

* Particles larger than 1 μm.

have been recognized as inadequate for the electronics industry, and a new standard will list four types of electronic grade water. The interesting thing about the new standard is its separate specifications for (1) analytical limits for each of the four grades and (2) the unit operations required for their production.

As an example, Table 44.1 lists certain chemical quality limits: specific resistivity varies from only 0.5 MΩ for the lower quality E-IV water to 18 MΩ for the highest quality E-I water, requiring this quality to be produced 90% of the time the treatment system is in operation. It is incumbent on the plant engineers to select their own limits based on the nature of their specific manufacturing operations. Likewise, the production of the lowest grade water (E-IV) requires only a primary section in the treatment system, a demineralizing unit—ion exchange, reverse osmosis, or distillation device—plus preconditioning filters, the choice of which is based on the raw water quality and the E-IV specifications. Secondary and tertiary devices are added to the specifications to meet the more stringent quality requirements of water types E-III to E-I.

For the pharmaceutical industry, Table 44.2 lists current standards for USP XX. Here, too, methods of production are included in the specifications. A second

TABLE 44.2 Abstract of USP-XX Specifications

Measurement	Purified water	Water for injection
pH (with KCl)	5.0–7.0	5.0–7.0
Ca and heavy metals	None by test*	None by test*
Sulfate, chloride, CO_2	None by test*	None by test*
Oxidizables, max ppm	0.3	0.3
Total solids, max ppm	10	10
Pyrogenicity	Not specified	None by test
Microbiological purity	Distillation, reverse osmosis, ion exchange, other	Distillation, reverse osmosis
Aerobic organisms† colony-forming units per ml	100	50

* Test methods not specified.

† Not USP, but suggested by the Pharmacopoeial Forum.

standards group, the Pharmacopoeial Forum, has set suggested guidelines (added to Table 44.2) assuming that equipment standards alone may be inadequate for the required microbiological purity.

In addition to these trade association standards, both the Food and Drug Administration (FDA) and the Environmental Protection Agency (EPA) have jurisdiction over water quality standards for food and drug applications. For example, the FDA has established so-called good manufacturing practices (GMP) for the manufacture, processing, packing, or holding of parenterals produced by the drug industry. Three significant requirements not covered by other standards-setting agencies for parenterals are:

1. *Microbial populations:* Microbial populations are restricted to 50 per 100 mL for cleansing and initial rinsing of drug product containers and closures; and to 10 per 100 mL for water used for manufacturing; both are based on evaluation of three consecutive samples from the same location.
2. *Holding temperature:* If water for manufacture is to be held in storage for longer than 24 h, it must be held at 80°C or higher. If it is held for less than 24 h, it may be stored at ambient temperature, but the storage vessel must be completely drained every 24 h. Obviously this specification has a large influence on the choice of evaporation systems (Chapter 18) selected for production of this water, since vapor compression stills produce water at slightly over ambient temperature, whereas multiple effect evaporators can be designed to produce water at over 80°C.
3. *Filters:* Filters are excluded from use in any part of the piping systems for water for manufacture or final rinse. This is difficult to interpret because "filter" has not been defined. It has been found that certain kinds of disposable cartridge filters are unreliable and may actually introduce contamination, but ultrafilters are commonly used in these systems.

The EPA is involved in pure water quality because of their concern for the presence of halogenated organics, particularly total trihalomethanes in potable water, and therefore, in food products and drugs using pure water.

From these examples, it is clear that all users of ultrapure water must be aware not only of the quality standards they need for manufacture of acceptable products, but also the jurisdiction that governmental regulatory agencies may have in control of manufacturing practices and product quality.

In putting together a set of process units to meet high purity standards, the water chemist is faced with selecting equipment that will reduce each of the following:

1. Soluble matter—minerals, organic matter, gases
2. Particulate matter*—filterable by a 0.22-μm membrane
3. Colloidal matter*—smaller than 0.22 μm, usually including SiO_2 and heavy metal oxides
4. Biological matter—microbes, viruses, metabolites

The unit operations available for meeting the goals of contaminant removal have been presented in earlier sections of this *Handbook,* but it is important to

* In this respect, the users of ultrapure water have revised the long-standing specification of 0.45 μm as the definition of the boundary between particulate matter and colloidal matter.

recognize at this stage that the selection of the best process set requires a knowledge of how each process unit relates to the others as well as to the tasks each performs in contaminant reduction. For example, in some cases a membrane unit (ultrafiltration, reverse osmosis, or electrodialysis) may be best located ahead of an ion exchange unit—to reduce its ion loading or to protect it from organic fouling—and in some cases it may be best following ion exchange (to eliminate the organics that may be present from deteriorating the ion exchange resins, or to filter out bacterial debris that may slough from the ion exchange beds).

Another factor to be considered in the flow sheet is the effect of flow rate on water quality. Two contributors to flow-affected quality are the ion exchange units, which should not be kept on-line at zero or low flows (less than about 20% of rated capacity) and the materials of construction of the storage and distribution system, which may contaminate water at low flow because of leaching or organics from plastics or corrosion of metallic alloys. A brief list of acceptable materials for use in ultrapure water systems is given in Table 44.3. In many ultrapure water systems a provision for final recycle minimizes the depreciation of water quality from these two potential sources of flow-sensitive contamination. Table 44.4 lists the various unit operations of water treatment and their function in an ultrapure system.

TABLE 44.3 Suggested Materials of Construction for Distribution and Storage of Ultrapure Water

(Arranged in order of preference)

Teflon (tetrafluoroethylene)	Polypropylene
Tefzel (trifluoromonethylethylene)	Borosilicate glass
Tantalum	304 L stainless steel
Titanium	304 stainless steel
Pure block tin	316 L stainless steel
High-purity quartz	316 stainless steel
PVDF (polyvinylidene fluoride)	321 stainless steel
Polyethylene	PVC (polyvinyl chloride)

Source: V. C. Smith, Vaponics, Inc.

An example of an ultrapure water system used in the microelectronics industry is shown in Figure 44.1. The function of each of the units shown on this flowsheet is described below.

1. *Water clarifier:* This is required if the water source is surface water containing suspended solids. Coagulants and flocculants may be applied to the clarifier for removal of these solids, and chlorine may be applied for disinfection (Chapters 8 and 9). If the water is hard (usually over 100 mg/L hardness), the clarifier can be treated with lime for partial softening (Chapter 10)—and partial demineralization—of the supply. The resulting high pH assists in the removal of heavy metals and in disinfection. If the raw water is taken from a municipal distribution system, the clarifier as such is unnecessary, but may be replaced by a zeolite softener (Chapter 12) to protect the downstream reverse osmosis unit from fouling, since it is generally controlled to produce a negative Langelier index in the reject saline stream.

TABLE 44.4 Performance of Unit Operations of Water Treatment in an Ultrapure Water System*

Process units (with handbook chapter references)	Soluble matter			Particulate matter	Colloidal matter	Biological matter
	Ionic	Organic	Gas	Larger than 0.45 μm	Smaller than 0.45 μm	Living cells or cell debris
Solids/liquid separation (Chapters 8 and 9)	0	1	0	2	2	1†
Precipitation (Chapter 10)	1	1	0	1	1	1
Ion exchange (Chapter 12)						
Softening	0	0	0	1	0	0
Demineralization	2	1	0	1	0	0
Disinfection (Chapter 22)						
Chemical	0	0	0	0	0	2–3†
Ultraviolet	0	2	0	0	0	2–3†
Membrane separations (Chapter 15)						
Ultrafiltration	0	1	0	3	2	2
Reverse osmosis	2	2	0	3	2	2–3
Degasification (Chapter 14)	0	0	1–3	0	0	0
Distillation (Chapter 18)	2–3	2–3	2–3	2–3	3	3
Adsorption (Chapter 17)	0	1	0	1–2	0	0–1

* Legend: 0, no removal; 1, partial removal, 2, removal to low levels, 3, removal below current levels of analytical detection

† Living organisms may be killed, but cellular material may remain.

Note: Where individual process units may reduce contaminants to the 2-level, the same units in series may reduce contaminants to the 3-level.

2. *Depth filter* (see Chapter 9): This unit polishes the clarifier or lime softener effluent and reduces turbidity to below 0.1 NTU. It provides for removal of many types of microbes; even viruses may be removed if consolidated into a filterable mass by flocculation. It is unnecessary if the first unit is a zeolite softener.
3. *Carbon filter* (see Chapter 17): This filter serves two purposes: it eliminates residual chlorine by direct reduction and thus provides protection for the ion exchange resins downstream, and it adsorbs certain nonpolar organics from the raw water supply. This latter function reduces dissolved solids, but also helps to protect anion exchangers downstream from organic fouling. This may be substituted by a resin trap, if the chlorine is destroyed first by sulfite. The removal of chlorine makes the reverse osmosis (RO) membranes susceptible to biological attack, so ultraviolet sterilization of the carbon filtrate may be necessary. Certainly, frequent membrane cleaning is essential.

4. *Membrane separation* (see Chapter 15): This process may include an RO unit that provides (a) reduction of dissolved ions, (b) reduction of dissolved organic matter, and (c) filtration of colloidal matter, microbes, and microbial debris. The first benefit, reduction of dissolved ions, greatly reduces the ion loading on downstream ion exchange units; the second, reduction of organics, improves the final quality and, at the same time, helps to protect the downstream ion exchangers from organic fouling. The third, filtration of biological materials, protects the downstream equipment from either direct microbial attack or from becoming an incubator of microbial life; it also removes larger colloidal silica fractions. The RO unit may be replaced by a membrane ultrafilter that has all its advantages, except ion reduction, at a lower energy cost of operation and with lower loss of bleed-off water.
5. *Primary demineralizer:* This is usually a two-bed ion exchanger (see Chapter 12), especially if the membrane unit contains reverse osmosis membranes. However, this may be an evaporator, which provides for organic removal, disinfection, and degasification, operating effectively on an influent that has been processed by reverse osmosis while providing a benefit in low levels of dissolved gases.
6. *Storage tank:* This unit is provided to allow for continuous recycle of downstream water to minimize leaching and degradation of water quality. A vapor-space protection device, nitrogen blanketing, or both must be provided. They are, in fact, process units in themselves—designed to protect water quality from depreciation by airborne debris, especially microbes, as the storage tank breathes. Ultraviolet sterilizing lights may be mounted on this storage tank.
7. *Polishing demineralizer.* This is usually a mixed-bed ion exchanger, but could be an RO unit or an evaporator. In some plants the polishing demineralizer is a nonregenerable unit packed with specially processed, nuclear grade ion exchange resins.
8. *Ultraviolet sterilizer:* UV irradiation using a wavelength of 254 nm provides for final nonchemical destruction of any microbes that may have entered the system after storage and survived the earlier treatment stages. The persistence of microbial contamination through successive physical barriers is an accepted fact to designers and operators of ultrapure water systems. Conventional chemical biocides (e.g., chlorine) are obviously to be avoided as a source of contamination. Since ultraviolet radiation oxidizes organic matter to produce carbon dioxide, which would reduce resistivity, the sterilizer cannot aways be used in this location if the water contains excessive organic contamination at this point.
9. *Submicron filter:* This filter removes the residual debris, such as fragments of microbial cells, before the point of final use.
10. *Point of use:* There are often disposable mixed-bed demineralizer cartridges containing special nuclear grade resin followed by submicron filters at each workstation at the final point of use. Since the excess water used in the rinse operation is often quite pure, it is reclaimed, reprocessed, and returned to the main storage tank. A special ultraviolet unit may be installed here for oxidation of organic matter at a wavelenth of 185 nm.

Figure 44.2 shows a 100-gal/min system of this type producing ultrapure water for electronics manufacture.

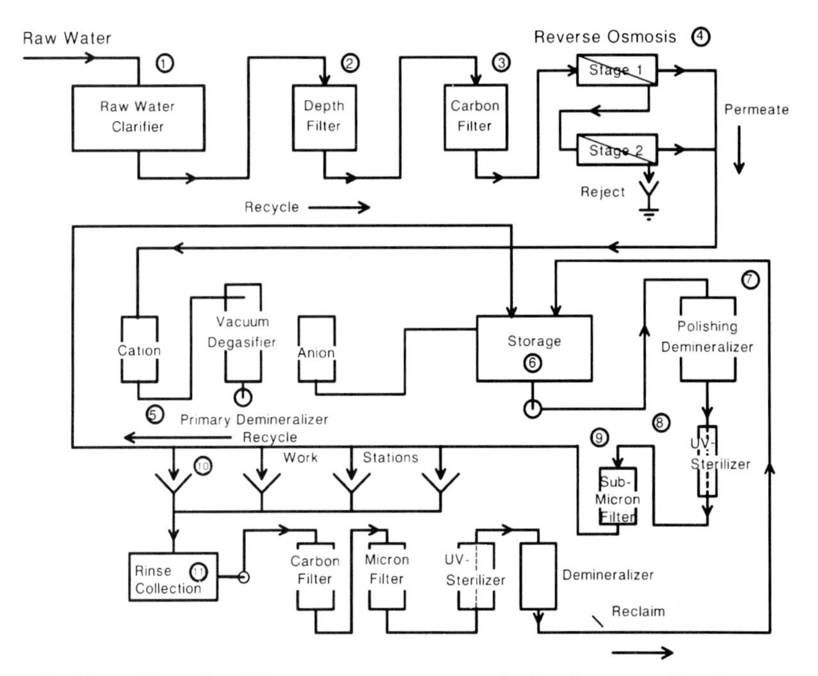

FIG. 44.1 Schematic of ultrapure water system producing rinse water for microprocessor manufacture. The rinse water after use is generally of such good quality that it is reclaimed for partial reprocessing.

FIG. 44.2 A 100-gal/min ultrapure water system including the following components: sodium zeolite softener (center background), depth filter, activated carbon filter, reverse osmosis unit (on right), storage tank with vent protection (left), a final loop containing ultraviolet sterilizers and mixed-bed polishers, and submicron filters at each workstation. *(Courtesy of Culligan USA.)*

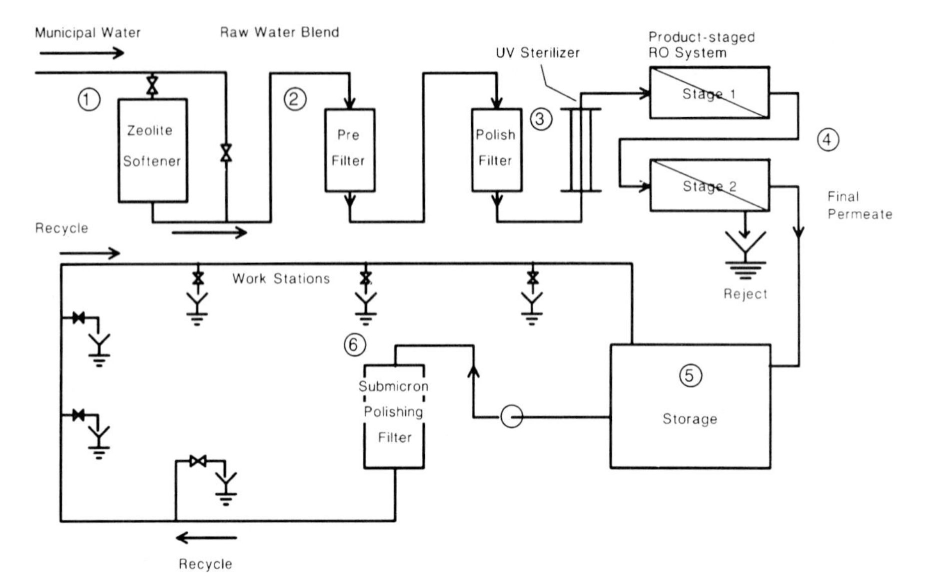

FIG. 44.3 System designed to produce pyrogen-free USP XX water for washing of surgical implant devices.

A second example is the flow sheet of a system producing USP XX pyrogen-free water used in the manufacture of implantable medical devices (Figure 44.3). This is a simpler system because the water quality specifications relative to resistivity are less strict than for most electronic water uses. The system includes:

1. *A zeolite softener:* The softener has a controlled bypass to adjust the final stability index for protection of the downstream RO units from deposits. This plant processes municipal water: if another source were used, such as well water or surface water, additional pretreatment would be needed.
2. *Cartridge filters:* These filters are in series, the first being rated at 50 μm and the second at 10 μm.
3. *Ultraviolet sterilizer:* The clear filtrate is easily sterilized to prevent microbial activity from fouling downstream units and reducing flow or depreciating quality.
4. *Two-step, product-staged RO.* This unit reduces ionic matter by over 98%, so a raw water of 500 mg/L total dissolved solids would be reduced to less than 10 mg/L required by USP XX specifications. This unit also removes organic matter of molecular weight greater than 300 and discharges a sterile effluent.
5. *Storage:* Product is held in a storage tank with the vapor space protected by an air filtration device against ingress of airborne debris as levels change and the tank breathes.
6. *Recycle and final filtration:* Finished water is continuously recycled through a 0.2 μm absolute filter to various workstations. Note that this type filter may be forbidden in the FDA GMP regulations for parenterals; so its future use may be questioned in this operation.

A final example of an ultrapure water system is shown in Figure 44.4, having an evaporator as the central purification device. One of the advantages of this system is that the product is gas-free, so bubbles will not form if the water is heated for final use. In ultrapure systems a deaerating device may be needed for the same purpose if the water supply is saturated with air and likely to release bubbles on heating or passage through a high pressure drop throttling valve.

1. Steam is required as the major source of water. In most plants, steam has been treated for corrosion control, so any steam additives must be removed.
2. The evaporator is started up with deionized feed water. Steam from the stripper and the evaporator coil is condensed to become feed water after startup.
3. Cooled condensate is passed through a cation exchanger to remove the usual major steam anticorrosion additives, morpholine or cyclohexylamine.
4. Decationized water is polished through a mixed bed deionizer.
5. Organics, including resin breakdown products, are taken out by activated carbon or a resin trap for organics.
6. The overhead vapor is condensed and steam-stripped to remove noncondensible gases and volatile organics.
7. The stripped condensate is cooled for process use.

It is obvious in each of these three systems that ultrapure water may be produced using a variety of building blocks, and as much care must then go into the

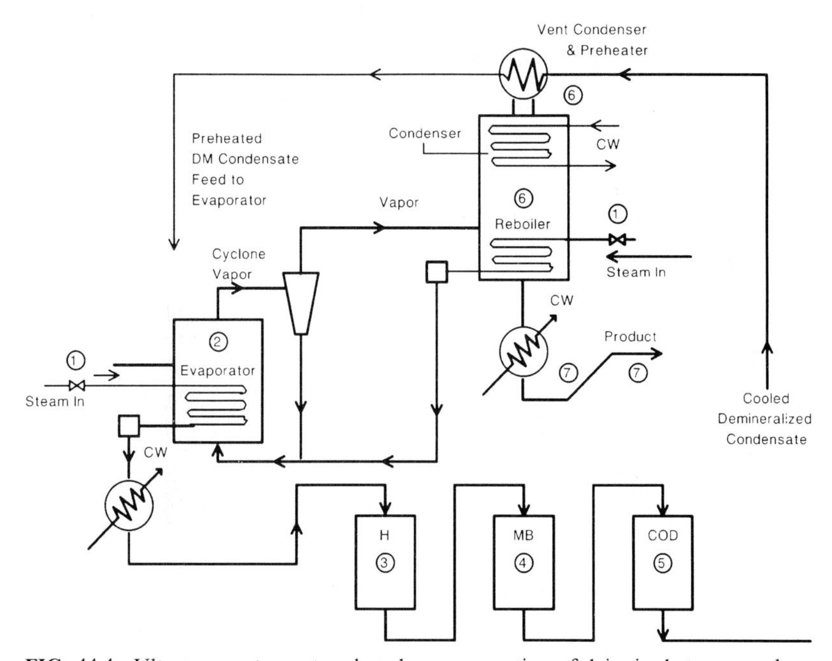

FIG. 44.4 Ultrapure water system based on evaporation of deionized steam condensate. *(Courtesy of Vaponics, Incorporated.)*

piping, storage, and distribution system as into the engineering of the process units to preserve ultrapure quality to the point of final use.

SUGGESTED READING

ASTM: *Aemrican Society for Testing and Materials Annual Book of ASTM Standards, Part 23,* "Water and Atmospheric Analysis," 1977.

"Cleaner Rooms for Better Wafers," *Sci. News,* **124,** 358 (1983).

Ford, Donald: "Water Purification," *Circuits Manufact.,* January 1983.

Human Drugs: Current Good Manufacturing Practice in Manufacture, Processing, Packing, or Holding of Large Volume Parenterals, Department of Health, Education and Welfare; Food and Drug Administration, Federal Register, 21 CFR Part 212; **41**(104) June 1, 1974.

Jones, G. D., Gagnon, S. R., and Kaszyski, M. J.: "The Role of Ultrafiltration in the Production of Ultra-High Purity Water" 41st Annual Meeting, International Water Conference 1980.

Montgomery, Greg: "Producing Ultra-Pure Water for Microelectronics Manufacturing," *Microelectronic Manufac. Testing,* August 1982.

National Interim Primary Drinking Water Regulations: Trihalomethanes, Environmental Protection Agency; Part III, Federal Register, 40 CFR Part 142; **48**(40) February 28, (1983).

Noyce, R. N.: "Microelectronics," *Sci. Am.,* September 1977.

Stadnisky, Walt: "Product-Staged Reverse Osmosis System," *Pharm Eng,* May–July 1981.

Stasche, A. W., Jr.: "Ultrapure Water Needs (and Headaches) in the Pharmaceutical Industry," *Ultrapure Water,* July/August 1984.

CHAPTER 45
CHEMICAL FEED SYSTEMS

Understanding the chemistry of water treatment processes is the first step toward mastering the ability to develop a chemical treatment program that produces water of acceptable quality—meeting predetermined specifications for hardness, dissolved solids, silica, and other controlled impurities—at reasonable cost. The next important step is selection of a reliable chemical feeding system—sensors and instruments, controls, and feeders—to consistently and accurately apply the chemicals needed by each treatment process.

There are many varieties of chemical feeders available for this purpose. Systems may be required to handle dry products like lime or liquid products like alum or caustic soda. A 1000 ton/day (908 t/day) kraft pulp/paper mill using 26 mgd (98,400 m^3/day) of water would require 25,000 lb/day (11,300 kg/day) of lime if the water is softened with lime at 120 mg/L. The same plant may generate 5,000,000 lb/day (2,270,000 kg/day) of high-pressure steam, requiring the application of hydrazine at 20 ppb, or only 0.1 lb/day (0.04 kg/day). The design engineer must exercise the same care in designing the system for feeding 12.5 tons of lime per day as the one for feeding 0.1 lb/day of hydrazine. The problems of inventory, of course, are also greatly different in the storage and handling of heavy chemicals shipped in bulk from those with specialty chemicals shipped in pails, drums, or bags. The storage properties of common water treatment chemicals are listed in Table 45.1. This chapter summarizes the selection process for the most common types of feed systems for liquid and dry chemicals used in water treatment, for feed rates ranging from a few pounds per day to many tons per day.

LIQUID FEED SYSTEMS

There are feeders for application of liquid chemicals by drip feed ranging from a constant head device used for feeding tiny amounts of liquid (as in intravenous feeding) to large flows of acid from bulk storage tanks. However, for the most part, the need for accuracy in chemical treatment is such that this chapter will deal only with systems using (1) metering or flow-controlled pumps, of which there is a wide choice based on delivery rate and application specifications (chemical to be fed, concentration, pressure, temperature, etc.); or (2) motor-controlled decanting devices combined with receivers and nonmetering delivery pumps.

TABLE 45.1 Common Chemicals Used to Treat Water

Chemical	Common name	Typical specs	Equiv. weight	Bulk density, lb/ft or lb/gal	Approx. pH 1% solution	Solubility
Aluminum sulfate	Alum	Lump—17% Al_2O_3	100*	60	3.4	4.2 lb/gal at 60°F
[$Al_2(SO_4)_3 \cdot 14H_2O$]		Liquid—8.5% Al_2O_3		11		
Bentonitic clay	Bentonite			60	—	Insoluble
Calcium carbonate ($CaCO_3$)	Limestone	96% $CaCO_3$	50	80	9	Insoluble
Calcium hydroxide [$Ca(OH)_2$]	Hydrated lime, slaked lime	96% $Ca(OH)_2$	40*	40	12	Insoluble
Calcium hypochlorite	HTH	70% Cl_2	103	55	6–8	3% at 60°F
[$C(OCl)_2 \cdot 4H_2O$]						
Calcium oxide (CaO)	Burned lime, quicklime	96% CaO	30*	60	12	Slake at 10–20%
Calcium sulfate ($CaSO_4 \cdot 2H_2O$)	Gypsum	98% Gypsum	86*	55	5–6	Insoluble
Chlorine (Cl_2)	Chlorine	Gas—99.8% Cl_2	35.5	gas	—	0.07 lb/gal at 60°F
Copper sulfate ($CuSO_4 \cdot 5H_2O$)	Blue vitriol	98% Pure	121*	75	5–6	2 lb/gal at 60°F
Dolomitic lime [$Ca(OH)_2 \cdot MgO$]	Dolomitic lime	36–40% MgO	67†	40	12.4	Insoluble
Ferric chloride ($FeCl_3 \cdot 6H_2O$)	Iron chloride	Lump—20% Fe	91*	70	3–4	45% at 60°F
		Liquid—20% Fe		13		
Ferric sulfate [$Fe_2(SO_4)_3 \cdot 3H_2O$]	Iron sulfate	18.5% Fe	51.5*	70	3–4	30% at 60°F
Ferrous sulfate ($FeSO_4 \cdot 7H_2O$)	Copperas	29% Fe	139*	70	3–4	1 lb/gal at 60°F
Hydrochloric acid (HCl)	Muriatic acid	30% HCl	120*	9.6	1–2	35% at 60°F
		20° Baumé				
Sodium aluminate ($NaAlO_2$)	Aluminate	Flake—46% Al_2O_3	100*	50	11–12	40% at 60°F
		Liquid—26% Al_2O_3		13		
Sodium chloride (NaCl)	Rock salt, salt	98% Pure	58.5	60	6–8	2.6 lb/gal at 60°F
Sodium carbonate ($NaCo_3$)	Soda ash	98% Pure				
		58% Pure Na_2O	53	60	11	1.5 lb/gal at 60°F
Sodium hydroxide (NaOH)	Caustic,	Flake—99% NaOH		65	12.8	70% at 60°F
	Lye	Liquid—50–70%	40	12		
Sodium phosphate (Na_2HPO_4)	Disodium phsophate	49% P_2O_5	47.3	55	9	20% at 60°F
Sodium metaphosphate ($NaPO_3$)	Hexametaphosphate	66% P_2O_5	34	47	5–6	1 lb/gal at 60°F
Sulfuric acid (H_2SO_4)	Oil of vitriol	94–96%				
		66° Baumé	50*	15	1–2	Infinite

* Effective equivalent weight of commercial product.
† Effective equivalent weight based on $Ca(OH)_2$ content.

Positive Displacement Pumps

In a positive displacement pump, the liquid is first drawn into a cavity, then forced through an outlet port into the discharge line. The discharge volume is relatively independent of the discharge head, and the discharge is never deliberately throttled. Pressure relief valves must be installed to protect these pumps against accidental shut-off of the discharge line. A separate pump should be provided for each product and for each feed point. A common error is to attempt to use a single pump to supply a chemical to several feed points (e.g., a chelate solution to three boiler drums in the same steam plant). It is impossible to control the split flow by throttling valves in the three feed lines; separate pumps are needed for each point of feed. Spare pumps should be provided to allow for maintenance without interruption of chemical feeding.

The reciprocating pump, Figure 45.1, has a piston that moves in a cylinder, and each discharge stroke delivers a volume close to the product of the cross-sectional area of the piston and the stroke length. Both suction and discharge pass through check valves, so there is little backflow or slippage; therefore, these pumps are also called "metering pumps." Discharge pressure varies from typical plant water main pressure in the range of 50 to 100 lb/in^2 (3.4 to 6.9 bars), to boiler pressure of 3000 lb/in^2 (207 bars). It is essential to have some back pressure on the discharge to ensure that the check valves close quickly and tightly. If the system pressure is too low to do this, it is common practice to pump through a pressure relief valve set for the minimum effective pressure to cause seating. The discharge flow has a pulsating characteristic, which must sometimes be compensated for to ensure even chemical application.

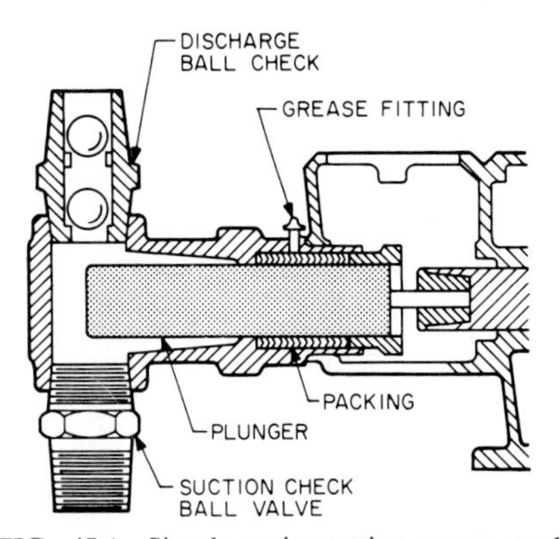

FIG. 45.1 Simple reciprocating pump used for relatively small flow chemical applications, for discharge against normal water supply pressure up to boiler pressure. *(Courtesy Milton Roy Company.)*

Some examples of reciprocating pump applications are:

1. *Heavy chemical usage:* Chemical fed directly from bulk liquid storage tanks, such as:

   Caustic soda, 50% NaOH: Most commonly used for demineralizer regeneration, where the average flow may be from 2 gal/h to 10 gal/min (7.6 L/h to 37.9 L/min). The discharge is diluted to about 5% in such a way that fluctuation in delivery rate is modulated, reaching the resin bed at a relatively uniform concentration.

   Alum, 50% liquid: Commonly used for coagulation, where the average delivery rate may be in the range of 2 gal/h to 2 gal/min (7.6 L/h to 7.6 L/min). Fluctuating flow is no problem when alum is applied to a clarifier.

   Sulfuric acid, 94% H_2SO_4: Commonly used for demineralizer regeneration in the range of 2 gal/h to 20 gal/min (7.6 L/h to 76 L/min). Pulsations are modulated in the final dilution system. Also commonly used for cooling tower alkalinity and pH control, in the range of 2 gal/h to 20 gal/min.

Note that in these examples, the cost of a piston-type reciprocating pump for use in the gallon per minute (or liter per minute) flow range is so high that centrifugal pumps or gear pumps are generally more practical and economical.

2. *Specialty chemical usage:* In these applications, total usage is often less than 2 gal/h (7.6 L/h). Where usage is extremely small, e.g., less than 5 gal/day (18.9 L/day), the product is commonly diluted to 5 to 10% of product strength before feeding to improve the precision of feed rate control.

There are three principal techniques used to control the delivery rate of reciprocating pumps:

1. *Stroke adjustment:* A crank is imposed between the rotating drive shaft and the reciprocating piston. The crank arm pivot can be positioned at the end of the arm, providing maximum stroke, or at the center of the drive shaft, producing zero reciprocating motion (Figure 45.2). Stroke adjustment is often done manually with the drive motor operating. However, special devices are available for making the stroke adjustment from a distant location in a central control room, either manually or instrumentally, such as by a pH controller.

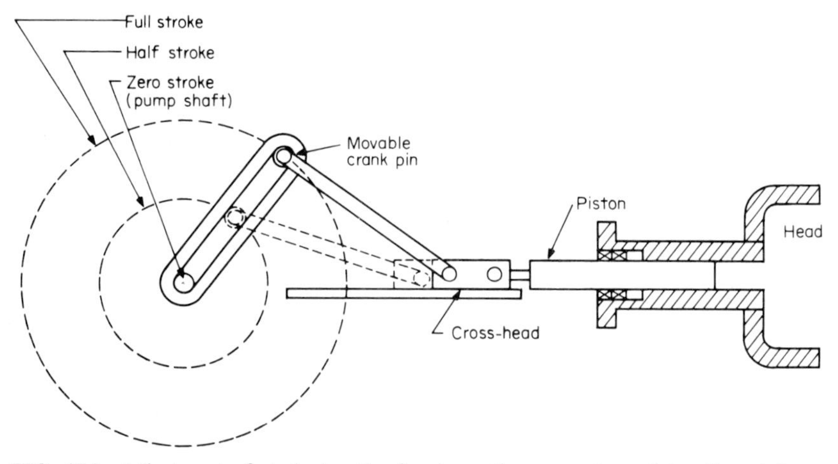

FIG. 45.2 Adjustment of stroke length of reciprocating pump is used to adjust delivery rate.

2. *Pump speed:* The drive can be a dc motor adjustable in speed by a rheostat or a variable-speed gear reducer, thus providing a simple means for control of delivery rate. Some pumps are controlled by a solenoid rather than a motor. The frequency of actuating the solenoid is varied to change pump delivery rate.
3. *Stroke interruption:* A device is built into the pump cylinder to interrupt the piston displacement even though the crosshead makes a full stroke.

There are several modifications to the simple reciprocating pump that keep the piston from contacting the fluid being pumped, usually to avoid corrosion or to avoid the nuisance of fluid leakage on the floor from packing glands. The diaphragm pump is commonly used for this purpose. The piston is sometimes directly connected to the diaphragm, but the diaphragm is more commonly positioned by oil sealed in the cylinder head (Figure 45.3).

The reciprocating pump follows the laws of simple harmonic motion, so the maximum rate of displacement occurs when the crank arm is at right angles to the linear movement of the crosshead. At this point, the fluid velocity into and

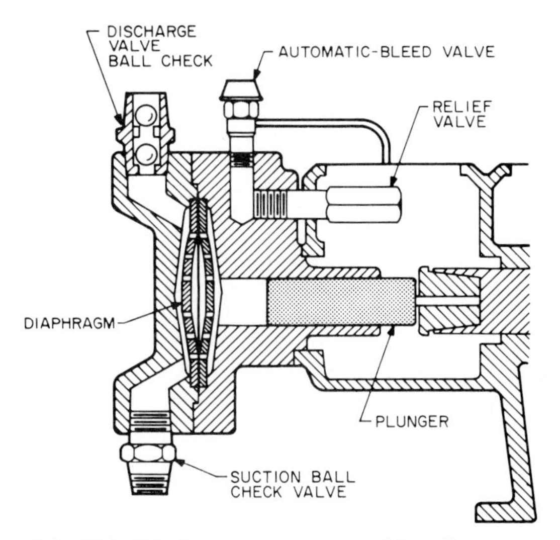

FIG. 45.3 Diaphragm pump actuated by oil pressure developed by the pump piston moving in an oil reservoir. *(Courtesy Milton Roy Company.)*

out of the pump head is 3.14 times the average flow. Piping must be sized to take this into account. Furthermore, the acceleration of the fluid from 0 to 3.14 times average flow subtracts a substantial head from atmospheric pressure at the suction, which is often the only pressure available to fill the pump cavity. Therefore, it is essential that these pumps have a flooded suction—despite the temptation to simplify the system by mounting on a chemical tank. For similar reasons, pumping viscous products causes problems; handling viscous chemicals should be referred to the pump specialist. Table 45.2 shows the viscosities of certain common fluids; and Figure 45.4 shows the relationship between several methods of expressing kinematic viscosity.

The second major design category of positive displacement pumps is the rotary pump, which delivers a continuous rather than a pulsating flow. This type is available in two styles, the progressing cavity design (Figure 45.5) and the gear pump (Figure 45.6). This style of pump is affected somewhat more by throttling or change in discharge pressure than the reciprocating pump, but this is not a procedure used deliberately to control feed rate, because it will either burn out the motor drive or rupture the pump casing. Unlike the reciprocating pump, the rotary pump works best on viscous fluids—in fact, it is seldom recommended for handling water or solutions below a viscosity of about 50 SSU, whether aqueous or organic. The following are examples of rotary positive displacement pump applications.

1. Heavy chemical usage directly from bulk liquid storage tanks, such as:

   Caustic soda, 50% NaOH: Where the delivery rate exceeds 1 gal/min (3.8 L/min), as in demineralizer regeneration.

   Alum, 50% liquid: Where the flow exceeds 1 gal/min (3.8 L/min), as in large coagulation-flocculation systems.

   Sulfuric acid, 94% H_2SO_4: Where the flow rate exceeds 1 gal/min (3.8 L/min), as in demineralizer regeneration and application of acid to cooling tower systems.

   Lime slurries have been fed with this type of pump, but because of the tendency for settling, the pump should be used at full flow rate.

2. Specialty chemical usage, such as:

   High molecular weight polymer solutions used for flocculation. The pump is used to prepare the stock solution (0.5 to 2.0%) held in a day tank and sometimes to feed the stock solution to a dilution line enroute to the point of use, if the rate of feed exceeds 1 gal/min (3.8 L/in).

   Emulsion-type high molecular weight polymers as supplied. This type pump is used to deliver product from inventory to a special dilution-feeding system.

TABLE 45.2 Viscosity of Some Common Fluids*

[*Based on measurements at 20°C (68°F)*]

Liquid	Sp. gr.	Viscosity cP	Kinematic viscosity, cSt	Viscosity English units, ft^2/s ($\times 10^{-5}$)	Viscocity Saybolt, SSU
Water	1.00	1.0	1.0	1.076	31.0
Ethanol	0.79	1.2	1.52	1.64	31.7
Saturated brine	1.19	3.0	2.5	2.69	34.0
HCl, 31.5%	1.05	2.0	1.9	2.04	33.0
NaOH, 30%	1.33	13.3	10.0	10.76	59
H_2SO_4, 66° Be	1.84	26.7	14.5	15.6	75
Propylene glycol	1.04	54	52	56	245
Motor oils					
10W	0.90	90	100	108	470
30W	0.90	315	350	377	1650
Glycerine	1.26	781	620	667	2950

* Viscosity relationships:

Metric:

$$\text{centistokes} = \frac{\text{absolute viscosity (cP)}}{\text{density (g/cm}^3)}$$

English: $ft^2/s = cSt \times 1.076 \times 10^{-5}$

Saybolt seconds universal: the time required for a measured volume of fluid to pass through a standardized orifice at a controlled temperature. Because the test is conducted under fluid flow conditions, SSU is related to kinematic viscosity rather than absolute viscosity.

Note: For effect of temperature on water viscosity, see Figure 1.6.

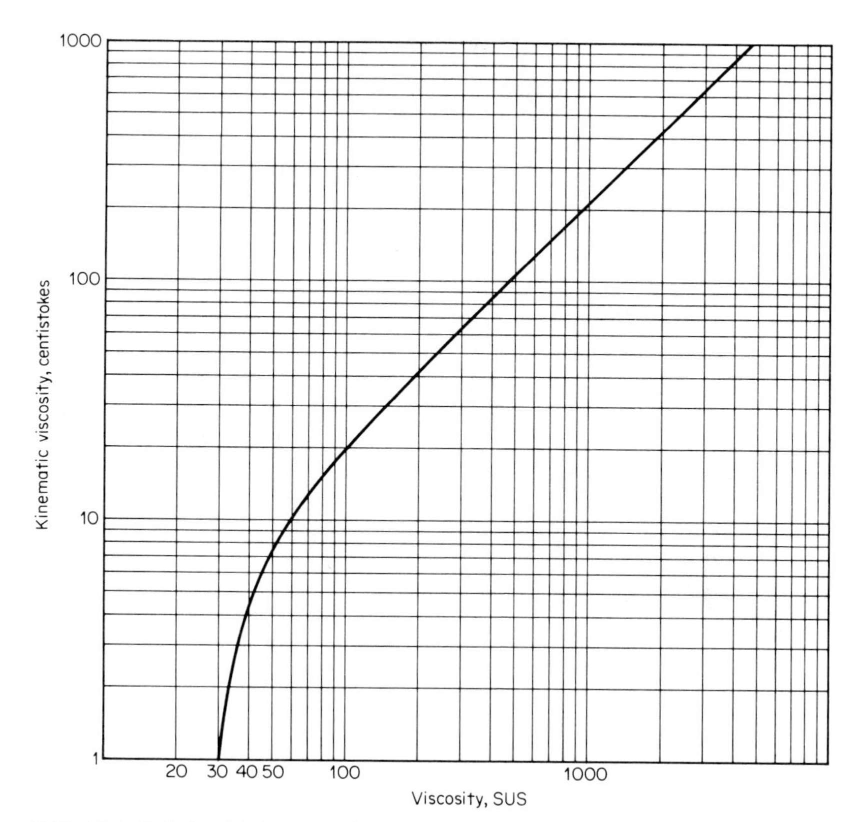

FIG. 45.4 Relationship between kinematic viscosity and the empirical Saybolt viscosity measurement at 20°C (68°F).

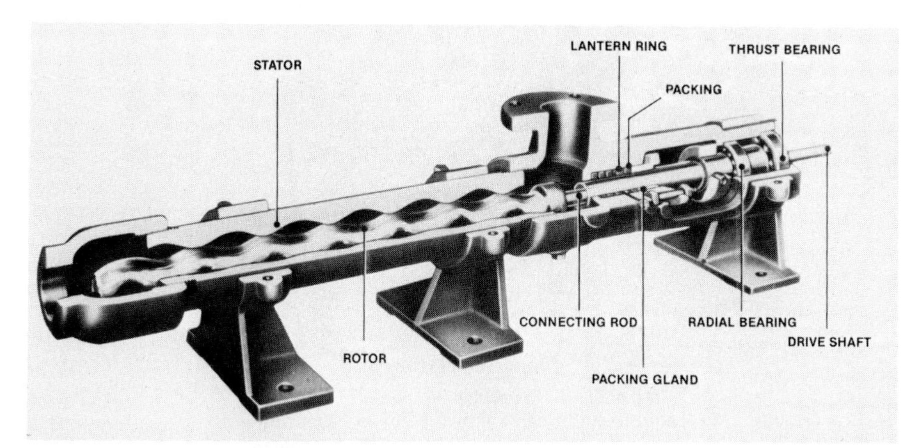

FIG. 45.5 A screwlike shaft rotating in a resilient stator of matched form produces positive displacement of progressing cavities in this pump, delivering a continuous flow at the discharge end. *(Courtesy Fluids Handling Division of Robbins & Myers, Inc.)*

FIG. 45.6 In the gear pump, cavities created between gear teeth and housing are moved from suction to discharge, and these are sealed from each other by the meshing of one gear with another. One gear is driven, and the other is an idler. *(Courtesy Liquiflo Equipment Co.)*

FIG. 45.7 In the peristaltic pump, the fluid is squeezed through a flow tube by external rollers. *(Courtesy Waukesha/Bredel Peristaltic Hosepump, Waukesh Div., Abex.)*

Two techniques are used to control the delivery rate from this type of rotary pump, only one of which actually controls the pump itself. The pump discharge rate may be controlled by speed variation, or the flow can be intermittently interrupted by a timer-controlled diversion valve, shunting the pump discharge back to the pump supply tank.

Another version of the progressive cavity pump is the peristaltic pump (Figure 45.7). Small pumps like this are used in a variety of instruments for sampling, automatic titration, and colorimetric measurements. They are used in many medical applications. They develop only limited head pressure, and the peristaltic action applied to the flexible tube requires close attention so that the tubing can be replaced before it fails. They can handle low-viscosity fluids. Flow rate is controlled by change of speed of the rollers that "milk" the tubing or by substitution of different sized tubing.

Rotary positive displacement pumps require large suction piping of short length, with limited valving and fittings, plus ample static head, because of the viscosity of the fluids being handled. The fluid acts as a lubricant for the moving parts, so continuous flow is essential. This type of pump must never be kept in operation with the discharge shut off or the suction line empty.

Centrifugal Pumps

In this type of pump, an impeller (a disk having radial vanes) rotates at high speed—usually 1800 or 3600 r/min—in a housing. Liquid enters at the center of the impeller, and centrifugal force throws the liquid at an accelerating velocity to the periphery of the housing, where the velocity is converted to head pressure. Liquid discharges tangentially from the peripheral passage of the housing.

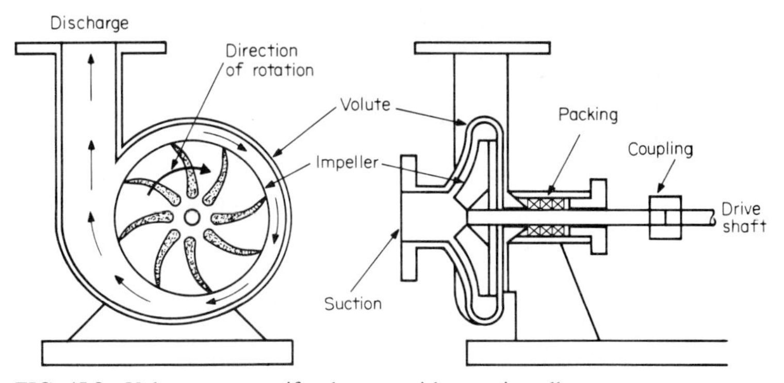

FIG. 45.8 Volute-type centrifugal pump with open impeller.

The volute-type centrifugal pump (Figure 45.8) is the most common design. It has impeller vanes with spiral passages, the vanes beginning at a point near the center and advancing to the edge of the impeller. In the open impeller design, the vanes are visible when the pump casing is removed; with the closed impeller, a more efficient design, the vanes are encased like a sandwich between the impeller disk and the thin-walled cover, having an opening at the eye for entrance of the fluid. The open impeller (Figure 45.9) is generally used for pumping slurries; the closed impeller (Figure 45.10) is used for clear liquids, especially for large water supply pump stations. Close clearances must be maintained to prevent internal recirculation and loss of efficiency. Throttling the pump discharge is the common

FIG. 45.9 Open impeller centrifugal pump used for pumping slurries such as lime or clay. *(Courtesy Ingersoll-Rand.)*

FIG. 45.10 Closed impeller centrifugal pump used for pumping clear liquids. It provides higher efficiency than the open impeller design. *(Courtesy Ingersoll-Rand.)*

method of controlling delivery rate. Examples of the use of the volute-type centrifugal pump are as follows:

Hydrated lime, 5 to 10% slurry: In many plants, lime is fed as a slurry from a day tank in which the slurry is prepared. In the example given earlier, a plant feeding 25,000 lb/day (11,000 kg/day) of lime would pump about 250,000 lb/day (110,000 kg/day) of slurry, requiring an average flow of about 20 gal/min (75 L/min) of lime slurry. To prevent deposition in the lime feed lines, a recirculation system is recommended, especially for such large volumes of lime feed. The recirculation pump may feed 50 gal/min to the circuit, with intermittent "blips" of 40 gal/min (150 L/min) to the point of feed, 50% of the on-time (Figure 45.11).

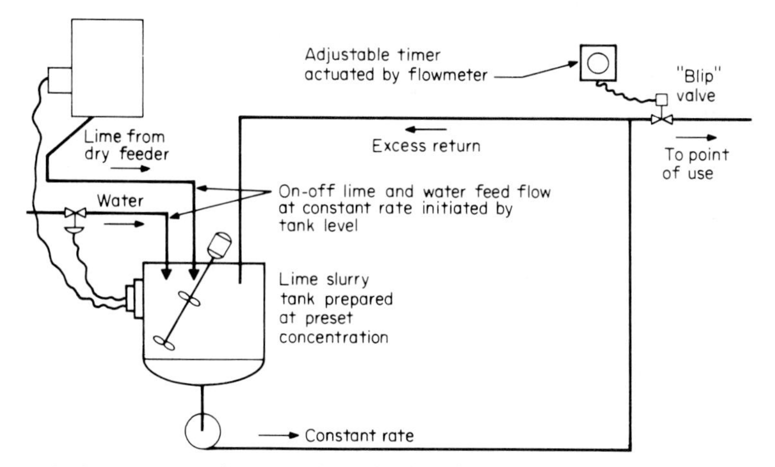

FIG. 45.11 Scheme for proportional feeding of lime with meter-actuated timer dosage control for large volume requirements.

Brine (saturated NaCl solution): In many municipal zeolite softening plants, salt is delivered in bulk to a wet salt storage basin (see Chapter 12). During regeneration, the 26% brine is usually delivered directly from the saturation basin to the zeolite unit at rates up to 50 gal/min (190 L/min), usually through a brine meter. Centrifugal pumps are ideal for this service.

Because delivery rate changes with discharge pressure, as shown by a typical centrifugal pump curve, Figure 45.12, control of flow is important because line

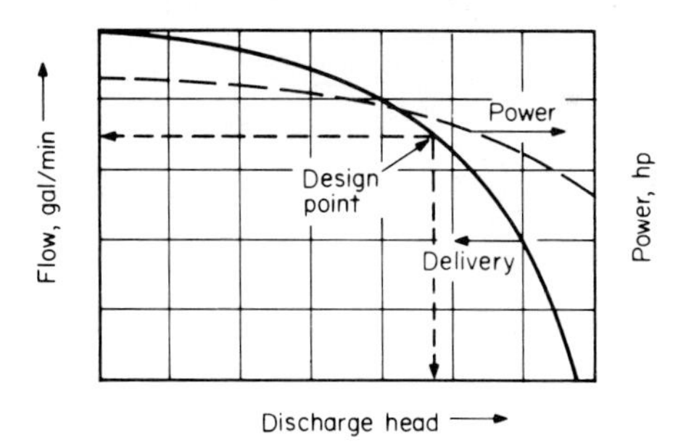

FIG. 45.12 Typical performance curve for an open impeller centrifugal pump. The design point allows for control of flow over a reasonable range by throttling discharge pressure. Maximum power is required at zero discharge head.

pressure may change and upset the initial flow setting. Because of this, a rate controller actuated by a flow meter is often used.

One of the major limitations of volute-type pumps is viscosity. Highly viscous fluids are not easily handled by the volute-type pump. Another limitation is the handling of abrasive slurries, especially lime. This requires the use of lantern glands on the pump shaft to flush the packing into the pump housing. The flushing water must be nonreactive to the lime slurry—even zeolite-softened water is not acceptable. Any dilution water that reacts with the lime will scale the pump and piping.

The pump housing and the impeller are usually of cast construction, using iron, steel, bronze, or brass. Some alloys are not easily or economically fabricated by casting, so where these alloys are needed, the volute pump design may be ruled out.

A much less common type of centrifugal pump is the turbine design (Figure 45.13). In this pump, pressure builds up progressively as fluid enters vanes machined on the circumference of the impeller disk, and recirculates in stages as the vanes move through 360° from inlet to outlet openings. This type of pump is more like a positive displacement pump in its characteristics than a volute pump;

FIG. 45.13 Although it operates as a centrifugal pump, the turbine pump is in many respects a cross between a positive displacement pump and a volute pump. *(Courtesy Burks Pumps, Decatur Pump Co.)*

flow should not be controlled by throttling the discharge. Since the pump casing and impeller can be machined, this pump can be made of a variety of alloys as required by the chemical to be fed. It is not recommended for handling viscous fluids.

As mentioned earlier, meter-actuated flow controllers either of the loss-of-head type or the variable area (rotameter) design can be used to control the delivery rate of centrifugal pumps where large volumes of liquid [over 10 gal/min (38 L/min)] are fed from bulk storage tanks. A much more common liquid feed device, used for applying 5 to 10% chemical solutions or slurries to clarifiers or lime softeners, where the chemical feed application rate is usually less than about 500 lb/day (225 kg/day), is the decanter, Figure 45.14. This comprises a chemical solution tank holding either a slurry (e.g. lime or clay) or a solution (e.g., alum, caustic soda, or soda ash) fitted with a motor-operated decanting pipe that usually lowers the pipe into the liquid at a fixed rate in the range of 0.1 to 0.3 in/min (0.25 to 0.75 cm/min). The lowering motor is operated through a timer. The decanted fluid flows into a receiver, where it is diluted, and the final solution is then delivered by centrifugal pump to the point of use. Change of dosage is easily handled by changing the timer setting, supplemented by change of solution strength where practical. In this system, the centrifugal pump is the device that delivers the solution to the point of use, but its operation does not play a part in dosage rate or changes in feed rate.

The limitation of this device in feeding solutions of dry chemicals as its need for manual charging, usually from 50-lb (23-kg) bags. When the application rate demands too much labor, then the plant may elect to feed the dry chemicals directly from storage silos through dry feeders.

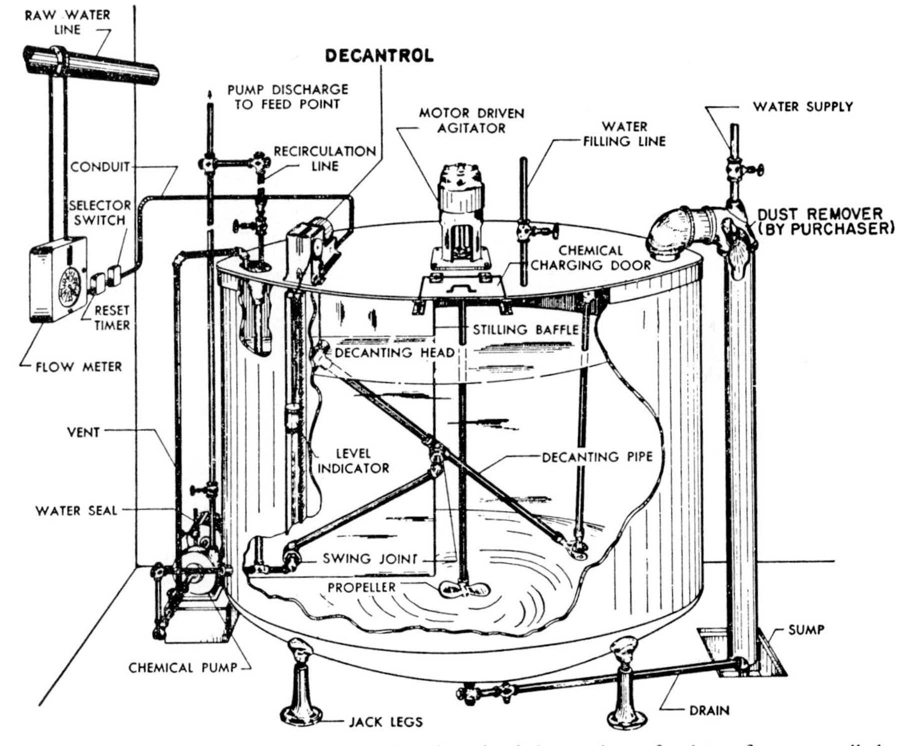

FIG. 45.14 In the decanting feeder, a skimming pipe is lowered at a fixed rate for a controlled period of time to withdraw chemical from the tank into a feed reservoir for delivery to the point of use. *(Courtesy Crane Company, Cochrane Division.)*

DRY FEED SYSTEMS

Dry chemical feeders are selected over wet feeders for handling large volume requirements of chemicals available in dry form and in bulk. In feeding lime as a slurry through a wet feeder, the delivery of 1 ft^3 of slurry made up as a 5% concentration would apply only about 3 lb (1.4 kg) of lime to the point of use. One cubic foot (0.028 m^3) of lime fed through a dry feeder, on the other hand, would deliver about 40 lb (18 kg) of lime, or more than 10 times that of the liquid feeder. This illustrates the basic advantage of the dry feeder in being more compact, less costly, and able to handle materials delivered to the feeder in bulk.

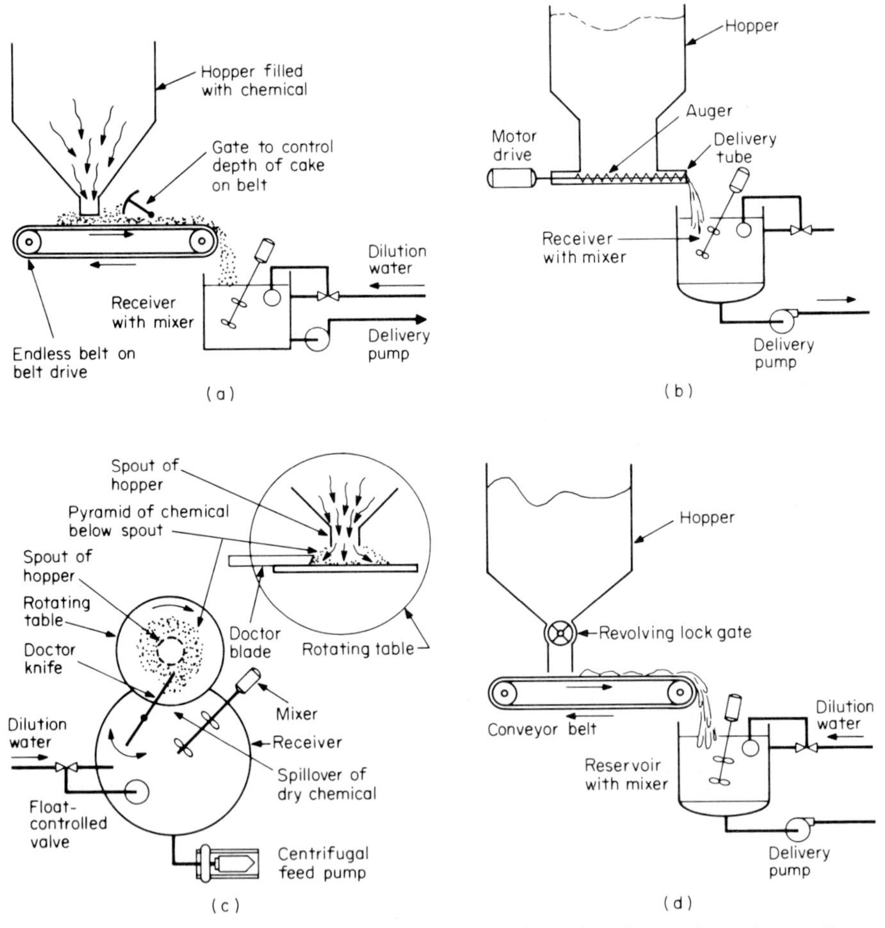

FIG. 45.15 (*a*) Endless belt with gate that sets the dimensions of the ribbon of chemical. Delivery rate may be changed by varying speed of belt drive or adjusting the control gate height. (*b*) Auger delivers chemical through tube to receiver. Delivery controlled by adjusting auger revolutions per minute. (*c*) Rotating table with adjustable plow. Delivery volume is changed by speed of table and pitch of the plow on the table. (*d*) Revolving lock gate delivers chemical onto conveyor belt. Dosage is controlled by speed of revolution.

Dry feeders are widely used in industries other than the water treatment industry, handling high-tonnage flows of coal in coal-burning steam plants and of chemicals in the chemical and mining industries.

The two basic categories of dry chemical feeders are volumetric and gravimetric. The volumetric feeder is simpler and less costly, with a volumetric accuracy of about 98 to 99%; but because the bulk density of the dry chemical leaving the silo or hopper varies, the accuracy on a delivered weight basis is reduced to 94 to 95%. The gravimetric feeder is more sophisticated and actually weighs out the delivered chemical with an accuracy greater than 99%.

Volumetric Feeders (Figure 45.15)

Dry chemical stored in a hopper is fed by gravity into the feeder mechanism. The device for displacing the chemical at a controlled rate from the feed bin to the reservoir, or solution tank, may be:

1. A traveling belt, with some type of gate to control the depth and width of the band of chemical leaving the spout
2. A screw or auger, turning on its axis in a tube
3. A rotating table, or disk, directly below the hopper spout, with an adjustable doctor blade to deflect a controlled volume of chemical from the table into a receiver
4. A rotating "paddle wheel" lock valve, similar in some respects to the liquid gear pump, delivering the measured volume contained in each compartment onto a conveyor belt as the lock valve slowly revolves

Each of these devices can deliver chemical at a rate proportional to the flow of water to be treated by using a flow meter signal either directly—to control belt speed, for example—or indirectly, through a timer.

Gravimetric Feeders (Figure 45.16)

In the gravimetric feeder, chemical is actually weighed out of the hopper onto a conveyor belt supported on a scale. The belt delivers the material to a feed reservoir, common to almost all dry feed systems. If the rate of delivery falls off or increases, material either collects on the belt or is fed too rapidly, upsetting the balance and initiating corrective action, e.g., adjustment of the control valve or the speed of the revolving gate rotor.

With either design of dry feeder, there are some common variables to consider to make sure that performance is consistent and reliable:

1. The characteristics of the chemical, including
   - *a.* Tendency to flood, requiring a rotary lock in the hopper spout
   - *b.* Tendency to cake, requiring protection from moisture, and vibrators or similar accessory devices to promote material flow
   - *c.* Angle of repose, for proper design of hopper and chutes
   - *d.* Particle size distribution, so that dusty materials can be confined and kept from damaging electric devices
   - *e.* Method of storage and transfer from inventory

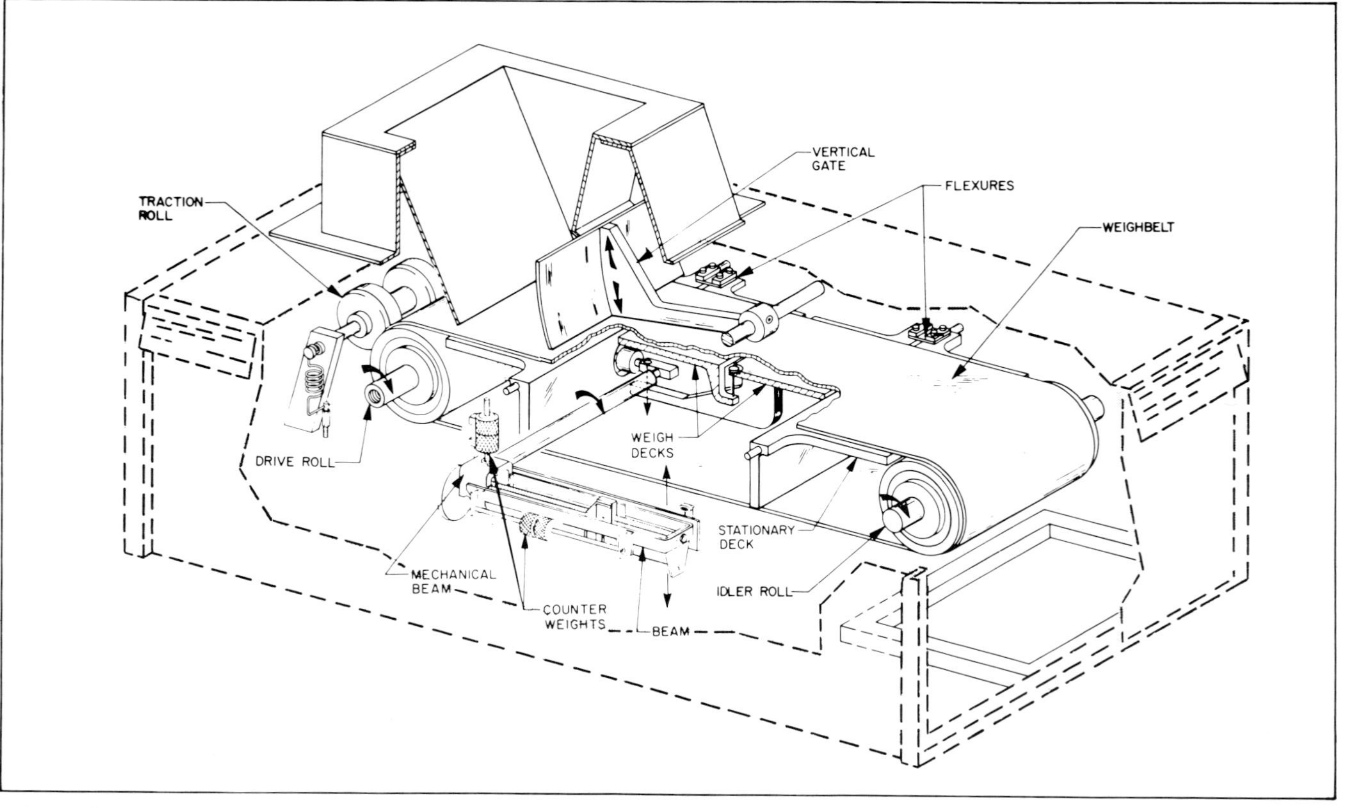

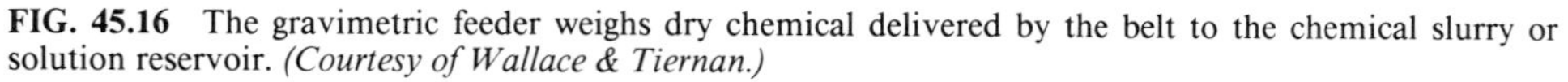
FIG. 45.16 The gravimetric feeder weighs dry chemical delivered by the belt to the chemical slurry or solution reservoir. *(Courtesy of Wallace & Tiernan.)*

TABLE 45.3 Typical Control Programs for Chemical Feeders

1. Method of initiation
   A. Nonproportional to flow
      (1) On-off manual control
      (2) Clock-type program timer, on-off*
   B. Proportional to flow
      (1) Meter delivers signal from integrator
      (2) Meter delivers a digital signal
2. Method of dosage control
   A. Timer setting, initiated by integrator signal (e.g., timer controls on-time of pump, delivery valve, etc.)
   B. Control of motor speed by digital signal
   C. Gear reducer speed change (usually manual)
   D. Change of mechanical linkage (e.g., pump stroke, gate setting, etc.)
   E. Recycle of discharge to chemical tank†

* Nonproportional control is commonly used for such services as (a) ion exchange regerneration, where the dosage and concentration are seldom changed and the dilution water flow rate is constant and (b) polymer feed to a sludge flowing at a constant rate to a dewatering device.

† pH controller can be used alone or with flowmeter to produce a signal for dosage correction.

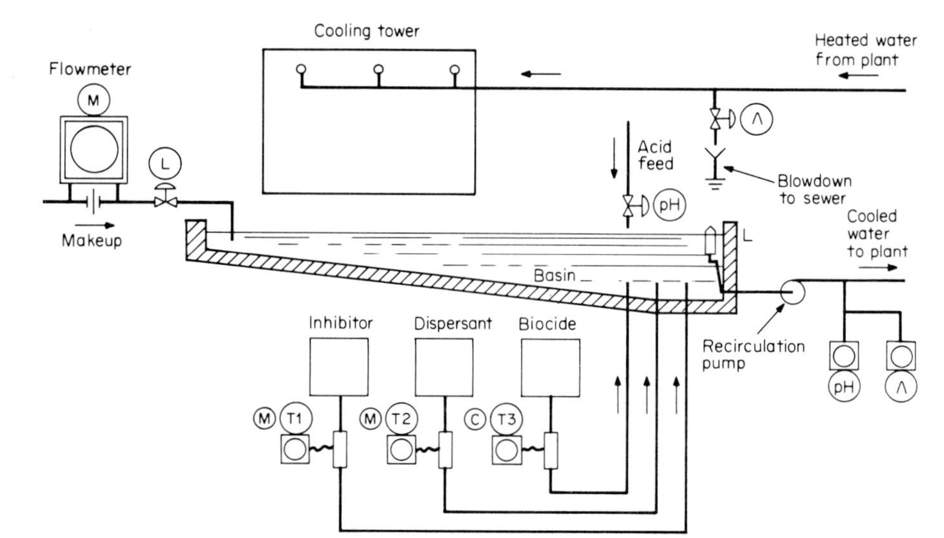

FIG. 45.17 Adaptation of instruments and controllers to chemical feed equipment serving a cooling system. (1) Flow meter (M) controls feed of inhibitor and dispersant proportional to water makeup rate. Dosage controlled by individual timers or by pump. Biocide timer is clock controlled (C). (2) Makeup valve is level controlled (L). (3) Blowdown valve is conductivity controlled ($\wedge$). (4) Acid feed is pH controlled (pH). In some systems the acid feed may be meter controlled with a pH override.

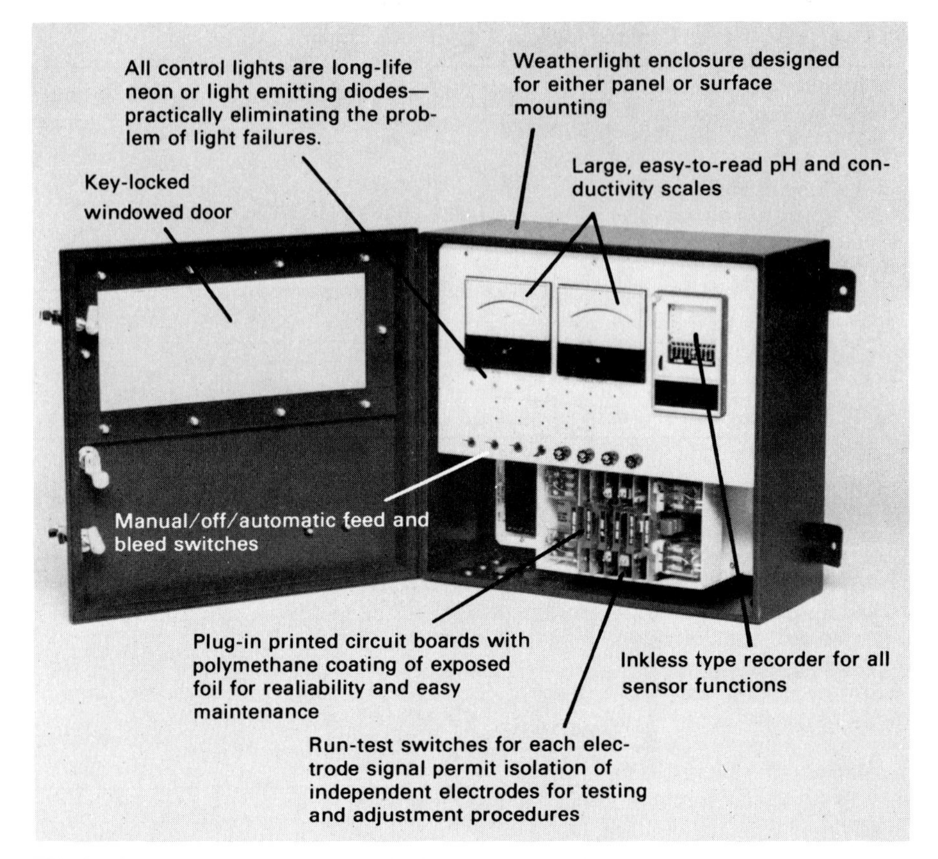
All control lights are long-life neon or light emitting diodes—practically eliminating the problem of light failures.
Weatherlight enclosure designed for either panel or surface mounting
Large, easy-to-read pH and conductivity scales
Key-locked windowed door
Manual/off/automatic feed and bleed switches
Plug-in printed circuit boards with polymethane coating of exposed foil for realiability and easy maintenance
Inkless type recorder for all sensor functions
Run-test switches for each electrode signal permit isolation of independent electrodes for testing and adjustment procedures

FIG. 45.18 A compact control panel with instruments used for control of the system illustrated in Figure 45.17.

2. The design of the silo, chutes, and hoppers to prevent size segregation and variability in density of the material entering the feed device
3. The operating environment of the feeder, such as ranges of temperature and humidity, and dust loading in the air

There are many methods of controlling chemical feed devices from simple on-off operation of a pump to meter-proportioned control of a feeder with a pH-controlled override. The more common of these various methods are listed in Table 45.3.

A typical example of a chemical treatment control system used with open, evaporative cooling systems is shown in Figure 45.17. A meter on the makeup water line controls the application of most treatment chemicals (corrosion or scale inhibitors, dispersants), while other chemicals (biocides, acid) may be controlled by a program timer or pH controller. The blowdown from the system to hold mineral solids content of the water in an acceptable range is actuated by a conductivity meter with high- and low-conductivity contacts. The assembly of these units into a single control panel is shown in Figure 45.18.

CHAPTER 46
ENERGY PRIMER

Energy is defined as the ability to do work. Although energy is abundant, it is seldom in a form that is easily convertible into useful work. In fact, it is often in a state that cannot be used at all. For example, a block sliding down an incline is heated by friction, but this heat, even though a form of energy, cannot do the work of returning the block to its original position.

Energy exists in many forms, the most familiar being chemical, mechanical, electric, thermal, radiant, and nuclear. Of immediate concern is the growing shortage of the chemical energy stored in fossil fuels—coal, gas, and petroleum. Their accessibility, transportability, high energy content, and easy conversion into heat have made them widely used in modern society.

Mechanical energy available from water power is another traditional source. However, this is confined to limited geographic areas of ample rainfall and mountainous terrain. There is considerable energy available in the ocean tides and waves, but these, too, are geographically restricted—and practical technology for capturing this energy has not yet been developed.

Nuclear energy remains a promising souce for the future. However, its development is in its infancy and has been held back by public concern for long-range safeguards against potential radiation hazards.

Geothermal energy in the earth's interior is evidenced by occasional spectacular and often catastrophic volcanic eruptions. Here, too, is a technology in its infancy, with sources presently limited to a few scattered geographic sites.

Since energy means the ability to do work, the term "work" needs to be examined to focus more precisely on energy. In popular language, work applies to any form of labor; but work also has a specific technical definition. If a person lifts a weight from the floor to the top of a table, he or she not only has exerted force, but maintains this force in moving the weight a definite distance. Work (W) is defined as the product of a force (F) exerted uniformly through a given distance (d), or mathematically,

$$W = F \times d$$

In this example, the person moving the weight possessed energy and converted it to work. Steam can be shown to possess energy because it is able to push a piston within a cylinder or rotate a turbine wheel. The amount of energy a body possesses is equal to the amount of work it can do.

What becomes of the energy that has left the body? It is transferred to the object being worked on. This gives rise to the first law of thermodynamics: *Energy cannot be created or destroyed.*

Since energy is never destroyed, how can there be an energy shortage? The answer is that energy always runs downhill. This is most easily seen in the transfer of heat energy. Absolute zero—the absence of all heat—is −460°F (−273°C). Everything above absolute zero contains some heat energy. A piece of ice at 32°F (0°C) contains considerable heat energy; but this energy cannot do work on a pan of warm water. On the contrary, the warm water does work on the ice, melting it and becoming cooler itself. The pan of water now contains heat energy equal to the total of the original warm water and the ice. This exemplifies the problems facing a society dependent on energy—finding energy at a high enough temperature level so that some can be transformed to work before reaching a temperature (or other energy stage) where it is no longer available to do work.

ENERGY MEASUREMENTS

The example of a person moving a weight shows that energy and work are interchangeable. Since work is the product of force and distance, it is measured in units called foot-pounds in the English system (one example of a metric system work unit is dyne-centimeters). If the person has lifted a 10-lb weight 3 ft off the floor, the energy expended on the weight is equal to 30 ft-lb. The weight now contains this 30 ft-lb of potential energy because of its new elevation.

While the foot-pound is useful for measuring mechanical energy, it is not commonly used to measure heat energy. For energy in the thermal or chemical form, it is common to use the British thermal unit, or Btu (the calorie is the analogous metric unit). A Btu is 1/180 of the energy needed to heat 1 lb of water from 32°F to 212°F; so the average energy required to heat 1 lb of water 1°F is 1 Btu (to raise 1 g of water 1°C, a calorie is required). In an ingenious experiment in the early part of the nineteenth century, Joule heated water in a closed, insulated vessel by mechanical agitation alone, thus establishing the heat equivalent of work as 778 ft-lb per Btu.

For energy in the electrical form, the unit of measurement in the English system is the kilowatt-hour. Since each of these forms of energy is convertible to another, their units are also convertible:

$$1 \text{ Btu} = 778 \text{ ft-lb} = 0.000293 \text{ kWh}$$

$$1 \text{ kWh} = 3413 \text{ Btu} = 2,655,000 \text{ ft-lb}$$

These conversion factors are equivalencies that are unrelated to the efficiency of working process. For example, although 1 Btu is equivalent to 778 ft-lb of work, heat does not actually produce that much useful work, because the conversion is never 100% efficient.

Power is the speed of doing work or the rate at which energy is expended. The size of an engine is defined not by the total amount of work to be done, but by the rate at which it can do work. A boiler generating 100,000 lb/h of steam has twice the power output of a boiler generating 50,000 lb/h. The relationships between the common energy and power units are:

Energy units	*Power units*
778 ft-lb = 1 Btu	550 ft-lb/s = 1 hp
1 kWh = 3413 Btu	3413 Btu/h = 1 kW
1 hph = 2545 Btu	1 kW = 1.34 hp
1 J = 0.239 cal	1 hp = 0.75 kW
	1 kcal/min = 4186 J/min

It is not uncommon to find inaccurate references that confuse energy and power terms; the most common error is to find electric energy expressed as kilowatts instead of kilowatt-hours. A 10,000-kW turbine produces no energy if the steam is shut off, but it is still a 10,000-kW turbine; it will produce 10,000 kWh of energy (work) each hour that the throttle valve is open.

The efficiency of energy conversion into useful work can be very poor. In many cases nothing can be done to improve it beyond a certain point, limited partly by the laws of physics and partly by economics.

In mechanical work much of the energy expended is used to overcome friction, which comes from many sources—rolling friction from bearings or drag caused by air resistance. The burning of fuels, which is the transformation of chemical energy into heat energy, is beset with inefficiencies because of the difficulty of getting complete combustion. But the conversion of heat into mechanical work is an even more difficult transformation.

In the nineteenth century, Carnot proposed a theoretical cycle through which a gas could be heated and expanded against a piston, converting heat energy into mechanical work. Although his theoretical engine is impossible to use practically, it defines the most efficient transformation attainable. Carnot found that thermal efficiency of a heat engine depends only on the absolute temperature of the source (T_s) and of the receiver (T_r). In equation form:

$$\text{Thermal efficiency} = \frac{T_s - T_r}{T_s} \times 100\%$$

where T is in absolute temperature units, degrees Rankine, °R (°R = °F + 460).

This law states that 100% conversion efficiency would be possible only if the exhaust gas (receiver) were without heat (at zero absolute temperature). In the real world, the ambient temperature of the typical receiver may be 70°F (530°R), so 100% efficiency is impossible. This fact is stated as the second law of thermodynamics: *Not all energy received as heat by a heat engine cycle can be converted into mechanical work.*

To examine the thermal efficiency encountered under practical applications, consider a boiler supplying 1050°F steam to a turbine at 1000 lb/in^2. The turbine exhausts to a condenser at 1 lb/in^2 abs (pounds per square inch absolute), corresponding to approximately 102°F. The maximum thermal efficiency is:

$$\text{Efficiency} = \frac{(1050 + 460) - (102 + 460)}{1050 + 460} \times 100\% = 63\%$$

In practice, even the most efficient steam electric power plants operate below this figure—about 38% for fossil fuels and 32% for nuclear fuels. There are losses of energy in the turbine, through friction across the blades, and impingement because of condensation, so the best turbines may be only 80% efficient. Other losses occur in the condenser needed to provide vacuum, and in the generator that converts mechanical energy to electrical.

The Carnot formula for thermal efficiency shows that lower steam temperatures or the absence of a condenser would seriously reduce efficiency. This is the most important reason why the long-range trend has been toward higher pressures and temperatures in steam boilers, within the limits of stress and strain of the alloys available to withstand these conditions. It is one reason for the demise of the steam locomotive; without a condenser, its engine had a T_r of 212°F, 672°R.

Efficiency to a large extent is limited by the high ambient temperature of the earth (approximately 530°R); heat energy below this temperature is unavailable for work because the lowest temperature of the receiver, T_r (exhaust), is limited to the ambient temperature of its surroundings. To attain temperatures below ambient would be impractical, uneconomical, or both, because additional energy would be used to do the work of the refrigeration that subambient temperatures would require.

STEAM PROPERTIES

This background of theoretical work cycles and efficiencies can be applied in studying the use of vapor as the medium for transformation of heat energy into work. Water is the working fluid—both as vapor and as liquid—in most power cycles.

As heat is supplied to water, the temperature increases until boiling occurs (212°F at atmospheric pressure). At the boiling point, the temperature remains constant until all the water is converted to vapor. Although it requires only 1 Btu for each degree of temperature increase to the boiling point, 970 Btu's are required to change 1 lb of water at 212°F to steam—about 6 times the energy necessary to heat the water from 60°F to 212°F.

If the pressure of the boiler water is increased, as it is, for example, in a pressure cooker, the boiling point also increases, requiring the application of still more Btu's both to raise the temperature and to vaporize the water. Steam at higher pressure, then, has a higher temperature and contains more energy than steam at atmospheric pressure. Another important change is the reduction in volume as the pressure increases.

Steam tables have been developed to relate these properties of steam and water at various temperatures and pressures. Table 46.1 illustrates the type of information found in these tables. Although a complete discussion of the steam tables is beyond the scope of this handbook, some basic orientation is necessary for understanding power generation processes.

ENTHALPY

The term *enthalpy* denotes the heat content of water or steam expressed in Btu/ per pound. The table shows 32°F as the zero point for enthalpy, so strictly speaking, enthalpy is not the total energy content. However, cycle calculations involve "changes" in enthalpy, which can then be related to equal changes in energy.

Here is a simple example: How much energy is required to raise the temperature of water in a heat exchanger from 60°F to 120°F when the water flow is 100 gal/min? Since 1 lb of water at 32°F has zero enthalpy and absorbs 1 Btu in warming each 1°F, water at 60°F will have an enthalpy of 60° − 32° = 28 Btu/lb; and 1 lb of water at 120°F will have an enthalpy of 120°F − 32°F = 88 Btu/lb. Therefore, the difference in energy between the water at the two temperatures will be the difference in enthalpies or 88 − 28 = 60 Btu/lb, which is numerically equal to the difference in temperatures. To find the rate of energy addition to heat the water in the example, the calculation is:

$$60 \text{ Btu/lb} \times 100 \text{ gal/min} \times 8.3 \text{ lb/gal} = 49{,}800 \text{ Btu/min}$$

The same procedure is used in reverse for cooling. Although the enthalpies of water at 60°F and 120°F were calculated in this example, they could have been found in the steam tables opposite the temperature and under the column heading, "Enthalpy, Saturated Liquid, h_f."

In the same example, how much steam would condense if steam were the heat source for heating water in the exchanger by 60°F? Under the enthalpy column for condensation (or evaporation) at 212°F we find that 970 Btu/lb are required to change water to vapor (shown as h_{fg} or "fluid-to-gas enthalpy change"). Under the enthalpy for saturated vapor (steam), H_g, 1150 Btu/lb is the total heat content of the steam, but it is only the heat of condensation, h_{fg}, that is used in this process (assuming the condensate formed is not further cooled, which is essentially true). So the steam required for this process would be:

$$49{,}800 \text{ Btu/min} \div 970 \text{ Btu/lb} = 51.3 \text{ lb/min}$$

STEAM QUALITY

The terms "saturated," "dry," and "moist" are often used in describing steam. What the steam tables call "saturated steam" could also be called "dry and saturated." In other words, it is vapor in equilibrium with water at the boiling temperature, containing no liquid. Steam that contains water droplets is called "wet steam." Steam issuing from the spout of the teakettle on the stove is colorless or invisible, just as it leaves the spout; this is dry saturated steam. A short distance from the spout it turns white, because some of the vapor has given up its heat and condensed into tiny droplets of water; this "cloudy steam" contains less heat per pound and is called "wet steam."

Wet steam always has less heat per pound than dry steam; consequently, it does less work. The term quality is often used to denote the dryness of steam: 100% quality means "dry," and 90% quality means the steam contains 10% by weight of moisture. The properties of dry steam are shown in Tables 46.1 and 46.2.

After vaporization, additional heat may be added to the steam, increasing its temperature and energy; this is called "superheated steam." Superheating not only increases the energy content, but also permits transporting the steam through pipelines with less condensation occurring. Condensation decreases the efficiency of heat engines and causes heat loss and water hammer in steam lines. The properties of superheated steam are shown in Table 46.3.

FUEL AND BOILER EFFICIENCIES

The potential energy (heating value) of fuel varies from one source to another (Table 46.4).

In burning any fuel, considerable energy is lost to a lower energy level—the atmosphere, in this case—rendering it unavailable for use. Most of this lost heat is in the stack gases leaving the furnace. The lower the temperature of the exit gas the higher the efficiency will be. Since a major portion of the fuel burned in industry goes into producing steam, the efficiency of various boilers is of considerable significance in all industrialized countries.

TABLE 46.1 Saturated Steam: Temperature Table (32 to 705.47°F)

Temp Fahr t	Abs Press. Lb per Sq In. p	Specific Volume Sat. Liquid v_f	Evap v_{fg}	Sat. Vapor v_g	Enthalpy Sat. Liquid h_f	Evap h_{fg}	Sat. Vapor h_g	Entropy Sat. Liquid s_f	Evap s_{fg}	Sat. Vapor s_g	Temp Fahr t
32.0*	0.08859	0.016022	3304.7	3304.7	−0.0179	1075.5	1075.5	0.0000	2.1873	2.1873	32.0*
34.0	0.09600	0.016021	3061.9	3061.9	1.996	1074.4	1076.4	0.0041	2.1762	2.1802	34.0
36.0	0.10395	0.016020	2839.0	2839.0	4.008	1073.2	1077.2	0.0081	2.1651	2.1732	36.0
38.0	0.11249	0.016019	2634.1	2634.2	6.018	1072.1	1078.1	0.0122	2.1541	2.1663	38.0
40.0	0.12163	0.016019	2445.8	2445.8	8.027	1071.0	1079.0	0.0162	2.1432	2.1594	40.0
42.0	0.13143	0.016019	2272.4	2272.4	10.035	1069.8	1079.9	0.0202	2.1325	2.1527	42.0
44.0	0.14192	0.016019	2112.8	2112.8	12.041	1068.7	1080.7	0.0242	2.1217	2.1459	44.0
46.0	0.15314	0.016020	1965.7	1965.7	14.047	1067.6	1081.6	0.0282	2.1111	2.1393	46.0
48.0	0.16514	0.016021	1830.0	1830.0	16.051	1066.4	1082.5	0.0321	2.1006	2.1327	48.0
50.0	0.17796	0.016023	1704.8	1704.8	18.054	1065.3	1083.4	0.0361	2.0901	2.1262	50.0
52.0	0.19165	0.016024	1589.2	1589.2	20.057	1064.2	1084.2	0.0400	2.0798	2.1197	52.0
54.0	0.20625	0.016026	1482.4	1482.4	22.058	1063.1	1085.1	0.0439	2.0695	2.1134	54.0
56.0	0.22183	0.016028	1383.6	1383.6	24.059	1061.9	1086.0	0.0478	2.0593	2.1070	56.0
58.0	0.23843	0.016031	1292.2	1292.2	26.060	1060.8	1086.9	0.0516	2.0491	2.1008	58.0
60.0	0.25611	0.016033	1207.6	1207.6	28.060	1059.7	1087.7	0.0555	2.0391	2.0946	60.0
62.0	0.27494	0.016036	1129.2	1129.2	30.059	1058.5	1088.6	0.0593	2.0291	2.0885	62.0
64.0	0.29497	0.016039	1056.5	1056.5	32.058	1057.4	1089.5	0.0632	2.0192	2.0824	64.0
66.0	0.31626	0.016043	989.0	989.1	34.056	1056.3	1090.4	0.0670	2.0094	2.0764	66.0
68.0	0.33889	0.016046	926.5	926.5	36.054	1055.2	1091.2	0.0708	1.9996	2.0704	68.0
70.0	0.36292	0.016050	868.3	868.4	38.052	1054.0	1092.1	0.0745	1.9900	2.0645	70.0
72.0	0.38844	0.016054	814.3	814.3	40.049	1052.9	1093.0	0.0783	1.9804	2.0587	72.0
74.0	0.41550	0.016058	764.1	764.1	42.046	1051.8	1093.8	0.0821	1.9708	2.0529	74.0
76.0	0.44420	0.016063	717.4	717.4	44.043	1050.7	1094.7	0.0858	1.9614	2.0472	76.0
78.0	0.47461	0.016067	673.8	673.9	46.040	1049.5	1095.6	0.0895	1.9520	2.0415	78.0
80.0	0.50683	0.016072	633.3	633.3	48.037	1048.4	1096.4	0.0932	1.9426	2.0359	80.0
82.0	0.54093	0.016077	595.5	595.5	50.033	1047.3	1097.3	0.0969	1.9334	2.0303	82.0
84.0	0.57702	0.016082	560.3	560.3	52.029	1046.1	1098.2	0.1006	1.9242	2.0248	84.0
86.0	0.61518	0.016087	227.5	527.5	54.026	1045.0	1099.0	0.1043	1.9151	2.0193	86.0
88.0	0.65551	0.016093	496.8	496.8	56.022	1043.9	1099.9	0.1079	1.9060	2.0139	88.0
90.0	0.69813	0.016099	468.1	468.1	58.018	1042.7	1100.8	0.1115	1.8970	2.0086	90.0
92.0	0.74313	0.016105	441.3	441.3	60.014	1041.6	1101.6	0.1152	1.8881	2.0033	92.0
94.0	0.79062	0.016111	416.3	416.3	62.010	1040.5	1102.5	0.1188	1.8792	1.9980	94.0
96.0	0.84072	0.016117	392.8	392.9	64.006	1039.3	1103.3	0.1224	1.8704	1.9928	96.0
98.0	0.89356	0.016123	370.9	370.9	66.003	1038.2	1104.2	0.1260	1.8617	1.9876	98.0

100.0	0.94924	0.016130	350.4	350.4	67.999	1037.1	1105.1	0.1295	1.8530	1.9825	**100.0**
102.0	1.00789	0.016137	331.1	331.1	69.995	1035.9	1105.9	0.1331	1.8444	1.9775	**102.0**
104.0	1.06965	0.016144	313.1	313.1	71.992	1034.8	1106.8	0.1366	1.8358	1.9725	**104.0**
106.0	1.1347	0.016151	296.16	296.18	73.99	1033.6	1107.6	0.1402	1.8273	1.9675	**106.0**
108.0	1.2030	0.016158	280.28	280.30	75.98	1032.5	1108.5	0.1437	1.8188	1.9626	**108.0**
110.0	1.2750	0.016165	265.37	265.39	77.98	1031.4	1109.3	0.1472	1.8105	1.9577	**110.0**
112.0	1.3505	0.016173	251.37	251.38	79.98	1030.2	1110.2	0.1507	1.8021	1.9528	**112.0**
114.0	1.4299	0.016180	238.21	238.22	81.97	1029.1	1111.0	0.1542	1.7938	1.9480	**114.0**
116.0	1.5133	0.016188	225.84	225.85	83.97	1027.9	1111.9	0.1577	1.7856	1.9433	**116.0**
118.0	1.6009	0.016196	214.20	214.21	85.97	1026.8	1112.7	0.1611	1.7774	1.9386	**118.0**
120.0	1.6927	0.016204	203.25	203.26	87.97	1025.6	1113.6	0.1646	1.7693	1.9339	**120.0**
122.0	1.7891	0.016213	192.94	192.95	89.96	1024.5	1114.4	0.1680	1.7613	1.9293	**122.0**
124.0	1.8901	0.016221	183.23	183.24	91.96	1023.3	1115.3	0.1715	1.7533	1.9247	**124.0**
126.0	1.9959	0.016229	174.08	174.09	93.96	1022.2	1116.1	0.1749	1.7453	1.9202	**126.0**
128.0	2.1068	0.016238	165.45	165.47	95.96	1021.0	1117.0	0.1783	1.7374	1.9157	**128.0**
130.0	2.2230	0.016247	157.32	157.33	97.96	1019.8	1117.8	0.1817	1.7295	1.9112	**130.0**
132.0	2.3445	0.016256	149.64	149.66	99.95	1018.7	1118.6	0.1851	1.7217	1.9068	**132.0**
134.0	2.4717	0.016265	142.40	142.41	101.95	1017.5	1119.5	0.1884	1.7140	1.9024	**134.0**
136.0	2.6047	0.016274	135.55	135.57	103.95	1016.4	1120.3	0.1918	1.7063	1.8980	**136.0**
138.0	2.7438	0.016284	129.09	129.11	105.95	1015.2	1121.1	0.1951	1.6986	1.8937	**138.0**
140.0	2.8892	0.016293	122.98	123.00	107.95	1014.0	1122.0	0.1985	1.6910	1.8895	**140.0**
142.0	3.0411	0.016303	117.21	117.22	109.95	1012.9	1122.8	0.2018	1.6534	1.8852	**142.0**
144.0	3.1997	0.016312	111.74	111.76	111.95	1011.7	1123.6	0.2051	1.6759	1.8810	**144.0**
146.0	3.3653	0.016322	106.58	106.59	113.95	1010.5	1124.5	0.2084	1.6684	1.8769	**146.0**
148.0	3.5381	0.016332	101.68	101.70	115.95	1009.3	1125.3	0.2117	1.6610	1.8727	**148.0**
150.0	3.7184	0.016343	97.05	97.07	117.95	1008.2	1126.1	0.2150	1.6536	1.8686	**150.0**
152.0	3.9065	0.016353	92.66	92.68	119.95	1007.0	1126.9	0.2183	1.6463	1.8646	**152.0**
154.0	4.1025	0.016363	88.50	88.52	121.95	1005.8	1127.7	0.2216	1.6390	1.8606	**154.0**
156.0	4.3068	0.016374	84.56	84.57	123.95	1004.6	1128.6	0.2248	1.6318	1.8566	**156.0**
158.0	4.5197	0.016384	80.82	80.83	125.96	1003.4	1129.4	0.2281	1.6245	1.8526	**158.0**
160.0	4.7414	0.016395	77.27	77.29	127.96	1002.2	1130.2	0.2313	1.6174	1.8487	**160.0**
162.0	4.9722	0.016406	73.90	73.92	129.96	1001.0	1131.0	0.2345	1.6103	1.8448	**162.0**
164.0	5.2124	0.016417	70.70	70.72	131.96	999.8	1131.8	0.2377	1.6032	1.8409	**164.0**
166.0	5.4623	0.016428	67.67	67.68	133.97	998.6	1132.6	0.2409	1.5961	1.8371	**166.0**
168.0	5.7223	0.016440	64.78	64.80	135.97	997.4	1133.4	0.2441	1.5892	1.8333	**168.0**
170.0	5.9926	0.016451	62.04	62.06	137.97	996.2	1134.2	0.2473	1.5822	1.8295	**170.0**
172.0	6.2736	0.016463	59.43	59.45	139.98	995.0	1135.0	0.2505	1.5753	1.8258	**172.0**
174.0	6.5656	0.016474	56.95	56.97	141.98	993.8	1135.8	0.2537	1.5684	1.8221	**174.0**
176.0	6.8690	0.016486	54.59	54.61	143.99	992.6	1136.6	0.2568	1.5616	1.8184	**176.0**
178.0	7.1840	0.016498	52.35	52.36	145.99	991.4	1137.4	0.2600	1.5548	1.8147	**178.0**

*The states shown are metastable

TABLE 46.1 Saturated Steam: Temperature Table (32 to 705.47°F) (Continued)

Temp Fahr	Abs Press. Lb per Sq In.	Specific Volume			Enthalpy			Entropy			Temp Fahr
t	p	Sat. Liquid v_f	Evap v_{fg}	Sat. Vapor v_g	Sat. Liquid h_f	Evap h_{fg}	Sat. Vapor h_g	Sat. Liquid s_f	Evap s_{fg}	Sat. Vapor s_g	t
180.0	7.5110	0.016510	50.21	50.22	148.00	990.2	1138.2	0.2631	1.5480	1.8111	**180.0**
182.0	7.850	0.016522	48.172	18.189	150.01	989.0	1139.0	0.2662	1.5413	1.8075	**182.0**
184.0	8.203	0.016534	46.232	46.249	152.01	987.8	1139.8	0.2694	1.5346	1.8040	**184.0**
186.0	8.568	0.016547	44.383	44.400	154.02	986.5	1140.5	0.2725	1.5279	1.8004	**186.0**
188.0	8.947	0.016559	42.621	42.638	156.03	985.3	1141.3	0.2756	1.5213	1.7969	**188.0**
190.0	9.340	0.016572	40.941	40.957	158.04	984.1	1142.1	0.2787	1.5148	1.7934	**190.0**
192.0	9.747	0.016585	39.337	39.354	160.05	982.8	1142.9	0.2818	1.5082	1.7900	**192.0**
194.0	10.168	0.016598	37.808	37.824	162.05	981.6	1143.7	0.2848	1.5017	1.7865	**194.0**
196.0	10.605	0.016611	36.348	36.364	164.06	980.4	1144.4	0.2879	1.4952	1.7831	**196.0**
198.0	11.058	0.016624	34.954	34.970	166.08	979.1	1145.2	0.2910	1.4888	1.7798	**198.0**
200.0	11.526	0.016637	33.622	33.639	168.09	977.9	1146.0	0.2940	1.4824	1.7764	**200.0**
204.0	12.512	0.016664	31.135	31.151	172.11	975.4	1147.5	0.3001	1.4697	1.7698	**204.0**
208.0	13.568	0.016691	28.862	28.878	176.14	972.8	1149.0	0.3061	1.4571	1.7632	**208.0**
212.0	14.696	0.016719	26.782	26.799	180.17	970.3	1150.5	0.3121	1.4447	1.7568	**212.0**
216.0	15.901	0.016747	24.878	24.894	184.20	967.8	1152.0	0.3181	1.4323	1.7505	**216.0**
220.0	17.186	0.016775	23.131	23.148	188.23	965.2	1153.4	0.3241	1.4201	1.7442	**220.0**
224.0	18.556	0.016805	21.529	21.545	192.27	962.6	1154.9	0.3300	1.4081	1.7380	**224.0**
228.0	20.015	0.016834	20.056	20.073	196.31	960.0	1156.3	0.3359	1.3961	1.7320	**228.0**
232.0	21.567	0.016864	18.701	18.718	200.35	957.4	1157.8	0.3417	1.3842	1.7260	**232.0**
236.0	23.216	0.016895	17.454	17.471	204.40	954.8	1159.2	0.3476	1.3725	1.7201	**236.0**
240.0	24.968	0.016926	16.304	16.321	208.45	952.1	1160.6	0.3533	1.3609	1.7142	**240.0**
244.0	26.826	0.016958	15.243	15.260	212.50	949.5	1162.0	0.3591	1.3494	1.7085	**244.0**
248.0	28.796	0.016990	14.264	14.281	216.56	946.8	1163.4	0.3649	1.3379	1.7028	**248.0**
252.0	30.883	0.017022	13.358	13.375	220.62	944.1	1164.7	0.3706	1.3266	1.6972	**252.0**
256.0	33.091	0.017055	12.520	12.538	224.69	941.4	1166.1	0.3763	1.3154	1.6917	**256.0**
260.0	35.427	0.017089	11.745	11.762	228.76	938.6	1167.4	0.3819	1.3043	1.6862	**260.0**
264.0	37.894	0.017123	11.025	11.042	232.83	935.9	1168.7	0.3876	1.2933	1.6808	**264.0**
268.0	40.500	0.017157	10.358	10.375	236.91	933.1	1170.0	0.3932	1.2823	1.6755	**268.0**
272.0	43.249	0.017193	9.738	9.755	240.99	930.3	1171.3	0.3987	1.2715	1.6702	**272.0**
276.0	46.147	0.017228	9.162	9.180	245.08	927.5	1172.5	0.4043	1.2607	1.6650	**276.0**
280.0	49.200	0.017264	8.627	8.644	249.17	924.6	1173.8	0.4098	1.2501	1.6599	**280.0**
284.0	52.414	0.01730	8.1280	8.1453	253.3	921.7	1175.0	0.4154	1.2395	1.6548	**284.0**
288.0	55.795	0.01734	7.6634	7.6807	257.4	918.8	1176.2	0.4208	1.2290	1.6498	**288.0**
292.0	59.350	0.01738	7.2301	7.2475	261.5	915.9	1177.4	0.4263	1.2186	1.6449	**292.0**
296.0	63.084	0.01741	6.8259	6.8433	265.6	913.0	1178.6	0.4317	1.2082	1.6400	**296.0**

300.0	67.005	0.01745	6.4483	6.4658	269.7	910.0	1179.7	0.4372	1.1979	1.6351	**300.0**
304.0	71.119	0.01749	6.0955	6.1130	273.8	907.0	1180.9	0.4426	1.1877	1.6303	**304.0**
308.0	75.433	0.01753	5.7655	5.7830	278.0	904.0	1182.0	0.4479	1.1776	1.6256	**308.0**
312.0	79.953	0.01757	5.4566	5.4742	282.1	901.0	1183.1	0.4533	1.1676	1.6209	**312.0**
316.0	84.688	0.01761	5.1673	5.1849	286.3	897.9	1184.1	0.4586	1.1576	1.6162	**316.0**
320.0	89.643	0.01766	4.8961	4.9138	290.4	894.8	1185.2	0.4640	1.1477	1.6116	**320.0**
324.0	94.826	0.01770	4.6418	4.6595	294.6	891.6	1186.2	0.4692	1.1378	1.6071	**324.0**
328.0	100.245	0.01774	4.4030	4.4208	298.7	888.5	1187.2	0.4745	1.1280	1.6025	**328.0**
332.0	105.907	0.01779	4.1788	4.1966	302.9	885.3	1188.2	0.4798	1.1183	1.5981	**332.0**
336.0	111.820	0.01783	3.9681	3.9859	307.1	882.1	1189.1	0.4850	1.1086	1.5936	**336.0**
340.0	117.992	0.01787	3.7699	3.7878	311.3	878.8	1190.1	0.4902	1.0990	1.5892	**340.0**
344.0	124.430	0.01792	3.5834	3.6013	315.5	875.5	1191.0	0.4954	1.0894	1.5849	**344.0**
348.0	131.142	0.01797	3.4078	3.4258	319.7	872.2	1191.1	0.5006	1.0799	1.5806	**348.0**
352.0	138.138	0.01801	3.2423	3.2603	323.9	868.9	1192.7	0.5058	1.0705	1.5763	**352.0**
356.0	145.424	0.01806	3.0863	3.1044	328.1	865.5	1193.6	0.5110	1.0611	1.5721	**356.0**
360.0	153.010	0.01811	2.9392	2.9573	332.3	862.1	1194.4	0.5161	1.0517	1.5678	**360.0**
364.0	160.903	0.01816	2.8002	2.8184	336.5	858.6	1195.2	0.5212	1.0424	1.5637	**364.0**
368.0	169.113	0.01821	2.6691	2.6873	340.8	855.1	1195.9	0.5263	1.0332	1.5595	**368.0**
372.0	177.648	0.01826	2.5451	2.5633	345.0	851.6	1196.7	0.5314	1.0240	1.5554	**372.0**
376.0	186.517	0.01831	2.4279	2.4462	349.3	848.1	1197.4	0.5365	1.0148	1.5513	**376.0**
380.0	195.729	0.01836	2.3170	2.3353	353.6	844.5	1198.0	0.5416	1.0057	1.5473	**380.0**
384.0	205.294	0.01842	2.2120	2.2304	357.9	840.8	1198.7	0.5466	0.9966	1.5432	**384.0**
388.0	215.220	0.01847	2.1126	2.1311	362.2	837.2	1199.3	0.5516	0.9876	1.5392	**388.0**
392.0	225.516	0.01853	2.0184	2.0369	366.5	833.4	1199.9	0.5567	0.9786	1.5352	**392.0**
396.0	236.193	0.01858	1.9291	1.9477	370.8	829.7	1200.4	0.5617	0.9696	1.5313	**396.0**
400.0	247.259	0.01864	1.8444	1.8630	375.1	825.9	1201.0	0.5667	0.9607	1.5274	**400.0**
404.0	258.725	0.01870	1.7640	1.7827	379.4	822.0	1201.5	0.5717	0.9518	1.5234	**404.0**
408.0	270.600	0.01875	1.6877	1.7064	383.8	818.2	1201.9	0.5766	0.9429	1.5195	**408.0**
412.0	282.894	0.01881	1.6152	1.6340	388.1	814.2	1202.4	0.5816	0.9341	1.5157	**412.0**
416.0	295.617	0.01887	1.5463	1.5651	392.5	810.2	1202.8	0.5866	0.9253	1.5118	**416.0**
420.0	308.780	0.01894	1.4808	1.4997	396.9	806.2	1203.1	0.5915	0.9165	1.5080	**420.0**
424.0	322.391	0.01900	1.4184	1.4374	401.3	802.2	1203.5	0.5964	0.9077	1.5042	**424.0**
428.0	336.463	0.01906	1.3591	1.3782	405.7	798.0	1203.7	0.6014	0.8990	1.5004	**428.0**
432.0	351.00	0.01913	1.30266	1.32179	410.1	793.9	1204.0	0.6063	0.8903	1.4966	**432.0**
436.0	366.03	0.01919	1.24887	1.26806	414.6	789.7	1204.2	0.6112	0.8816	1.4928	**436.0**
440.0	381.54	0.01926	1.19761	1.21687	419.0	785.4	1204.4	0.6161	0.8729	1.4890	**440.0**
444.0	397.56	0.01933	1.14874	1.16806	423.5	781.1	1204.6	0.6210	0.8643	1.4853	**444.0**
448.0	414.09	0.01940	1.10212	1.12152	428.0	776.7	1204.7	0.6259	0.8557	1.4815	**448.0**
452.0	431.14	0.01947	1.05764	1.07711	432.5	772.3	1204.8	0.6308	0.8471	1.4778	**452.0**
456.0	448.73	0.01954	1.01518	1.03472	437.0	767.8	1204.8	0.6356	0.8385	1.4741	**456.0**

TABLE 46.1 Saturated Steam: Temperature Table (32 to 705.47°F) (Continued)

Temp Fahr t	Abs Press. Lb per Sq In. p	Specific Volume Sat. Liquid v_f	Specific Volume Evap v_{fg}	Specific Volume Sat. Vapor v_g	Enthalpy Sat. Liquid h_f	Enthalpy Evap h_{fg}	Enthalpy Sat. Vapor h_g	Entropy Sat. Liquid s_f	Entropy Evap s_{fg}	Entropy Sat. Vapor s_g	Temp Fahr t
460.0	466.87	0.01961	0.97463	0.99424	441.5	763.2	1204.8	0.6405	0.8299	1.4704	**460.0**
464.0	485.56	0.01969	0.93588	0.95557	446.1	758.6	1204.7	0.6454	0.8213	1.4667	**464.0**
468.0	504.83	0.01976	0.89885	0.91862	450.7	754.0	1204.6	0.6502	0.8127	1.4629	**468.0**
472.0	524.67	0.01984	0.86345	0.88329	455.2	749.3	1204.5	0.6551	0.8042	1.4592	**472.0**
476.0	545.11	0.01992	0.82958	0.84950	459.9	744.5	1204.3	0.6599	0.7956	1.4555	**476.0**
480.0	566.15	0.02000	0.79716	0.81717	464.5	739.6	1204.1	0.6648	0.7871	1.4518	**480.0**
484.0	587.81	0.02009	0.76613	0.78622	469.1	734.7	1203.8	0.6696	0.7785	1.4481	**484.0**
488.0	610.10	0.02017	0.73641	0.75658	473.8	729.7	1203.5	0.6745	0.7700	1.4444	**488.0**
492.0	633.03	0.02026	0.70794	0.72820	478.5	724.6	1203.1	0.6793	0.7614	1.4407	**492.0**
496.0	656.61	0.02034	0.68065	0.70100	483.2	719.5	1202.7	0.6842	0.7528	1.4370	**496.0**
500.0	680.86	0.02043	0.65448	0.67492	487.9	714.3	1202.2	0.6890	0.7443	1.4333	**500.0**
504.0	705.78	0.02053	0.62938	0.64991	492.7	709.0	1201.7	0.6939	0.7357	1.4296	**504.0**
508.0	731.40	0.02062	0.60530	0.62592	497.5	703.7	1201.1	0.6987	0.7271	1.4258	**508.0**
512.0	757.72	0.02072	0.58218	0.60289	502.3	698.2	1200.5	0.7036	0.7185	1.4221	**512.0**
516.0	784.76	0.02081	0.55997	0.58079	507.1	692.7	1199.8	0.7085	0.7099	1.4183	**516.0**
520.0	812.53	0.02091	0.53864	0.55956	512.0	687.0	1199.0	0.7133	0.7013	1.4146	**520.0**
524.0	841.04	0.02102	0.51814	0.53916	516.9	681.3	1198.2	0.7182	0.6926	1.4108	**524.0**
528.0	870.31	0.02112	0.49843	0.51955	521.8	675.5	1197.3	0.7231	0.6839	1.4070	**528.0**
532.0	900.34	0.02123	0.47947	0.50070	526.8	669.6	1196.4	0.7280	0.6752	1.4032	**532.0**
536.0	931.17	0.02134	0.46123	0.48257	531.7	663.6	1195.4	0.7329	0.6665	1.3993	**536.0**
540.0	962.79	0.02146	0.44367	0.46513	536.8	657.5	1194.3	0.7378	0.6577	1.3954	**540.0**
544.0	995.22	0.02157	0.42677	0.44834	541.8	651.3	1193.1	0.7427	0.6489	1.3915	**544.0**
548.0	1028.49	0.02169	0.41048	0.43217	546.9	645.0	1191.9	0.7476	0.6400	1.3876	**548.0**
552.0	1062.59	0.02182	0.39479	0.41660	552.0	638.5	1190.6	0.7525	0.6311	1.3837	**552.0**
556.0	1097.55	0.02194	0.37966	0.40160	557.2	632.0	1189.2	0.7575	0.6222	1.3797	**556.0**
560.0	1133.38	0.02207	0.36507	0.38714	562.4	625.3	1187.7	0.7625	0.6132	1.3757	**560.0**
564.0	1170.10	0.02221	0.35099	0.37320	567.6	618.5	1186.1	0.7674	0.6041	1.3716	**564.0**
568.0	1207.72	0.02235	0.33741	0.35975	572.9	611.5	1184.5	0.7725	0.5950	1.3675	**568.0**
572.0	1246.26	0.02249	0.32429	0.34678	578.3	604.5	1182.7	0.7775	0.5859	1.3634	**572.0**
576.0	1285.74	0.02264	0.31162	0.33426	583.7	597.2	1180.9	0.7825	0.5766	1.3592	**576.0**
580.0	1326.17	0.02279	0.29937	0.32216	589.1	589.9	1179.0	0.7876	0.5673	1.3550	**580.0**
584.0	1367.7	0.02295	0.28753	0.31048	594.6	582.4	1176.9	0.7927	0.5580	1.3507	**584.0**
588.0	1410.0	0.02311	0.27608	0.29919	600.1	574.7	1174.8	0.7978	0.5485	1.3464	**588.0**
592.0	1453.3	0.02328	0.26499	0.28827	605.7	566.8	1172.6	0.8030	0.5390	1.3420	**592.0**
596.0	1497.8	0.02345	0.25425	0.27770	611.4	558.8	1170.2	0.8082	0.5293	1.3375	**596.0**

600.0	1543.2	0.02364	0.24384	0.26747	617.1	550.6	1167.7	0.8134	0.5196	1.3330	**600.0**
604.0	1589.7	0.02382	0.23374	0.25757	622.9	542.2	1165.1	0.8187	0.5097	1.3284	**604.0**
608.0	1637.3	0.02402	0.22394	0.24796	628.8	533.6	1162.4	0.8240	0.4997	1.3238	**608.0**
612.0	1686.1	0.02422	0.21442	0.23865	634.8	524.7	1159.5	0.8294	0.4896	1.3190	**612.0**
616.6	1735.9	0.02444	0.20516	0.22960	640.8	515.6	1156.4	0.8348	0.4794	1.3141	**616.0**
620.0	1786.9	0.02466	0.19615	0.22081	646.9	506.3	1153.2	0.8403	0.4689	1.3092	**620.0**
624.0	1839.0	0.02489	0.18737	0.21226	653.1	496.6	1149.8	0.8458	0.4583	1.3041	**624.0**
628.0	1892.4	0.02514	0.17880	0.20394	659.5	486.7	1146.1	0.8514	0.4474	1.2988	**628.0**
632.0	1947.0	0.02539	0.17044	0.19583	665.9	476.4	1142.2	0.8571	0.4364	1.2934	**632.0**
636.0	2002.8	0.02566	0.16226	0.18792	672.4	465.7	1138.1	0.8628	0.4251	1.2879	**636.0**
640.0	2059.9	0.02595	0.15427	0.18021	679.1	454.6	1133.7	0.8686	0.4134	1.2821	**640.0**
644.0	2118.3	0.02625	0.14644	0.17269	685.9	443.1	1129.0	0.8746	0.4015	1.2761	**644.0**
648.0	2178.1	0.02657	0.13876	0.16534	692.9	431.1	1124.0	0.8806	0.3893	1.2699	**648.0**
652.0	2239.2	0.02691	0.13124	0.15816	700.0	418.7	1118.7	0.8868	0.3767	1.2634	**652.0**
656.0	2301.7	0.02728	0.12387	0.15115	707.4	405.7	1113.1	0.8931	0.3637	1.2567	**656.0**
660.0	2365.7	0.02768	0.11663	0.14431	714.9	392.1	1107.0	0.8995	0.3502	1.2498	**660.0**
664.0	2431.1	0.02811	0.10947	0.13757	722.9	377.7	1100.6	0.9064	0.3361	1.2425	**664.0**
668.0	2498.1	0.02858	0.10229	0.13087	731.5	362.1	1093.5	0.9137	0.3210	1.2347	**668.0**
672.0	2566.6	0.02911	0.09514	0.12424	740.2	345.7	1085.9	0.9212	0.3054	1.2266	**672.0**
676.0	2636.8	0.02970	0.08799	0.11769	749.2	328.5	1077.6	0.9287	0.2892	1.2179	**676.0**
680.0	2708.6	0.03037	0.08080	0.11117	758.5	310.1	1068.5	0.9365	0.2720	1.2086	**680.0**
684.0	2782.1	0.03114	0.07349	0.10463	768.2	290.2	1058.4	0.9447	0.2537	1.1984	**684.0**
688.0	2857.4	0.03204	0.06595	0.09799	778.8	268.2	1047.0	0.9535	0.2337	1.1872	**688.0**
692.0	2934.5	0.03313	0.05797	0.09110	790.5	243.1	1033.6	0.9634	0.2110	1.1744	**692.0**
696.0	3013.4	0.03455	0.04916	0.08371	804.4	212.8	1017.2	0.9749	0.1841	1.1591	**696.0**
700.0	3094.3	0.03662	0.03857	0.07519	822.4	172.7	995.2	0.9901	0.1490	1.1390	**700.0**
702.0	3135.5	0.03824	0.03173	0.06997	835.0	144.7	979.7	1.0006	0.1246	1.1252	**702.0**
704.0	3177.2	0.04108	0.02192	0.06300	854.2	102.0	956.2	1.0169	0.0876	1.1046	**704.0**
705.0	3198.3	0.04427	0.01304	0.05730	873.0	61.4	934.4	1.0329	0.0527	1.0856	**705.0**
705.47*	3208.2	0.05078	0.00000	0.05078	906.0	0.0	906.0	1.0612	0.0000	1.0612	**705.47***

*Critical temperature

Source: Combustion Engineering, Inc.

TABLE 46.2 Saturated Steam: Pressure Table (0.08865 to 3208.2 lb per Square Inch Absolute Pressure)

Abs Press. Lb/Sq In. p	Temp Fahr t	Specific Volume: Sat. Liquid v_f	Specific Volume: Evap v_{fg}	Specific Volume: Sat. Vapor v_g	Enthalpy: Sat. Liquid h_f	Enthalpy: Evap h_{fg}	Enthalpy: Sat. Vapor h_g	Entropy: Sat. Liquid s_f	Entropy: Evap s_{fg}	Entropy: Sat. Vapor s_g	Abs Press. Lb/Sq In. p
0.08865	32.018	0.016022	3302.4	3302.4	0.0003	1075.5	1075.5	0.0000	2.1872	2.1872	**0.08865**
0.25	59.323	0.016032	1235.5	1235.5	27.382	1060.1	1087.4	0.0542	2.0425	2.0967	**0.25**
0.50	79.586	0.016071	641.5	641.5	47.623	1048.6	1096.3	0.0925	1.9446	2.0370	**0.50**
1.0	101.74	0.016136	333.59	333.60	69.73	1036.1	1105.8	0.1326	1.8455	1.9781	**1.0**
5.0	162.24	0.016407	73.515	73.532	130.20	1000.9	1131.1	0.2349	1.6094	1.8443	**5.0**
10.0	193.21	0.016592	38.404	38.420	161.26	982.1	1143.3	0.2836	1.5043	1.7879	**10.0**
14.696	212.00	0.016719	26.782	26.799	180.17	970.3	1150.5	0.3121	1.4447	1.7568	**14.696**
15.0	213.03	0.016726	26.274	26.290	181.21	969.7	1150.9	0.3137	1.4415	1.7552	**15.0**
20.0	227.96	0.016834	20.070	20.087	196.27	960.1	1156.3	0.3358	1.3962	1.7320	**20.0**
30.0	250.34	0.017009	13.7266	13.7436	218.9	945.2	1164.1	0.3682	1.3313	1.6995	**30.0**
40.0	267.25	0.017151	10.4794	10.4965	236.1	933.6	1169.8	0.3921	1.2844	1.6765	**40.0**
50.0	281.02	0.017274	8.4967	8.5140	250.2	923.9	1174.1	0.4112	1.2474	1.6586	**50.0**
60.0	292.71	0.017383	7.1562	7.1736	262.2	915.4	1177.6	0.4273	1.2167	1.6440	**60.0**
70.0	302.93	0.017482	6.1875	6.2050	272.7	907.8	1180.6	0.4411	1.1905	1.6316	**70.0**
80.0	312.04	0.017573	5.4536	5.4711	282.1	900.9	1183.1	0.4534	1.1675	1.6208	**80.0**
90.0	320.28	0.017659	4.8779	4.8953	290.7	894.6	1185.3	0.4643	1.1470	1.6113	**90.0**
100.0	327.82	0.017740	4.4133	4.4310	298.5	888.6	1187.2	0.4743	1.1284	1.6027	**100.0**
110.0	334.79	0.01782	4.0306	4.0484	305.8	883.1	1188.9	0.4834	1.1115	1.5950	**110.0**
120.0	341.27	0.01789	3.7097	3.7275	312.6	877.8	1190.4	0.4919	1.0960	1.5879	**120.0**
130.0	347.33	0.01796	3.4364	3.4544	319.0	872.8	1191.7	0.4998	1.0815	1.5813	**130.0**
140.0	353.04	0.01803	3.2010	3.2190	325.0	868.0	1193.0	0.5071	1.0681	1.5752	**140.0**
150.0	358.43	0.01809	2.9958	3.0139	330.6	863.4	1194.1	0.5141	1.0554	1.5695	**150.0**
160.0	363.55	0.01815	2.8155	2.8336	336.1	859.0	1195.1	0.5206	1.0435	1.5641	**160.0**
170.0	368.42	0.01821	2.6556	2.6738	341.2	854.8	1196.0	0.5269	1.0322	1.5591	**170.0**
180.0	373.08	0.01827	2.5129	2.5312	346.2	850.7	1196.9	0.5328	1.0215	1.5543	**180.0**
190.0	377.53	0.01833	2.3847	2.4030	350.9	846.7	1197.6	0.5384	1.0113	1.5498	**190.0**
200.0	381.80	0.01839	2.2689	2.2873	355.5	842.8	1198.3	0.5438	1.0016	1.5454	**200.0**
210.0	385.91	0.01844	2.16373	2.18217	359.9	839.1	1199.0	0.5490	0.9923	1.5413	**210.0**
220.0	389.88	0.01850	2.06779	2.08629	364.2	835.4	1199.6	0.5540	0.9834	1.5374	**220.0**
230.0	393.70	0.01855	1.97991	1.99846	368.3	831.8	1200.1	0.5588	0.9748	1.5336	**230.0**
240.0	397.39	0.01860	1.89909	1.91769	372.3	828.4	1200.6	0.5634	0.9665	1.5299	**240.0**
250.0	400.97	0.01865	1.82452	1.84317	376.1	825.0	1201.1	0.5679	0.9585	1.5264	**250.0**
260.0	404.44	0.01870	1.75548	1.77418	379.9	821.6	1201.5	0.5722	0.9508	1.5230	**260.0**
270.0	407.80	0.01875	1.69137	1.71013	383.6	818.3	1201.9	0.5764	0.9433	1.5197	**270.0**
280.0	411.07	0.01880	1.63169	1.65049	387.1	815.1	1202.3	0.5805	0.9361	1.5166	**280.0**
290.0	414.25	0.01885	1.57597	1.59482	390.6	812.0	1202.6	0.5844	0.9291	1.5135	**290.0**
300.0	417.35	0.01889	1.52384	1.54274	394.0	808.9	1202.9	0.5882	0.9223	1.5105	**300.0**
350.0	431.73	0.01912	1.30642	1.32554	409.8	794.2	1204.0	0.6059	0.8909	1.4968	**350.0**
400.0	444.60	0.01934	1.14162	1.16095	424.2	780.4	1204.6	0.6217	0.8630	1.4847	**400.0**

450.0	456.28	0.01954	1.01224	1.03179	437.3	767.5	1204.8	0.6360	0.8378	1.4738	**450.0**
500.0	467.01	0.01975	0.90787	0.92762	449.5	755.1	1204.7	0.6490	0.8148	1.4639	**500.0**
550.0	476.94	0.01994	0.82183	0.84177	460.9	743.3	1204.3	0.6611	0.7936	1.4547	**550.0**
600.0	486.20	0.02013	0.74962	0.76975	471.7	732.0	1203.7	0.6723	0.7738	1.4461	**600.0**
650.0	494.89	0.02032	0.68811	0.70843	481.9	720.9	1202.8	0.6828	0.7552	1.4381	**650.0**
700.0	503.08	0.02050	0.63505	0.65556	491.6	710.2	1201.8	0.6928	0.7377	1.4304	**700.0**
750.0	510.84	0.02069	0.58880	0.60949	500.9	699.8	1200.7	0.7022	0.7210	1.4232	**750.0**
800.0	518.21	0.02087	0.54809	0.56896	509.8	689.6	1199.4	0.7111	0.7051	1.4163	**800.0**
850.0	525.24	0.02105	0.51197	0.53302	518.4	679.5	1198.0	0.7197	0.6899	1.4096	**850.0**
900.0	531.95	0.02123	0.47968	0.50091	526.7	669.7	1196.4	0.7279	0.6753	1.4032	**900.0**
950.0	538.39	0.02141	0.45064	0.47205	534.7	660.0	1194.7	0.7358	0.6612	1.3970	**950.0**
1000.0	544.58	0.02159	0.42436	0.44596	542.6	650.4	1192.9	0.7434	0.6476	1.3910	**1000.0**
1050.0	550.53	0.02177	0.40047	0.42224	550.1	640.9	1191.0	0.7507	0.6344	1.3851	**1050.0**
1100.0	556.28	0.02195	0.37863	0.40058	557.5	631.5	1189.1	0.7578	0.6216	1.3794	**1100.0**
1150.0	561.82	0.02214	0.35859	0.38073	564.8	622.2	1187.0	0.7647	0.6091	1.3738	**1150.0**
1200.0	567.19	0.02232	0.34013	0.36245	571.9	613.0	1184.8	0.7714	0.5969	1.3683	**1200.0**
1250.0	572.38	0.02250	0.32306	0.34556	578.8	603.8	1182.6	0.7780	0.5850	1.3630	**1250.0**
1300.0	577.42	0.02269	0.30722	0.32991	585.6	594.6	1180.2	0.7843	0.5733	1.3577	**1300.0**
1350.0	582.32	0.02288	0.29250	0.31537	592.3	585.4	1177.8	0.7906	0.5620	1.3525	**1350.0**
1400.0	587.07	0.02307	0.27871	0.30178	598.8	576.5	1175.3	0.7966	0.5507	1.3474	**1400.0**
1450.0	591.70	0.02327	0.26584	0.28911	605.3	567.4	1172.8	0.8026	0.5397	1.3423	**1450.0**
1500.0	596.20	0.02346	0.25372	0.27719	611.7	558.4	1170.1	0.8085	0.5288	1.3373	**1500.0**
1550.0	600.59	0.02366	0.24235	0.26601	618.0	549.4	1167.4	0.8142	0.5182	1.3324	**1550.0**
1600.0	604.87	0.02387	0.23159	0.25545	624.2	540.3	1164.5	0.8199	0.5076	1.3274	**1600.0**
1650.0	609.05	0.02407	0.22143	0.24551	630.4	531.3	1161.6	0.8254	0.4971	1.3225	**1650.0**
1700.0	613.13	0.02428	0.21178	0.23607	636.5	522.2	1158.6	0.8309	0.4867	1.3176	**1700.0**
1750.0	617.12	0.02450	0.20263	0.22713	642.5	513.1	1155.6	0.8363	0.4765	1.3128	**1750.0**
1800.0	621.02	0.02472	0.19390	0.21861	648.5	503.8	1152.3	0.8417	0.4662	1.3079	**1800.0**
1850.0	624.83	0.02495	0.18558	0.21052	654.5	494.6	1149.0	0.8470	0.4561	1.3030	**1850.0**
1900.0	628.56	0.02517	0.17761	0.20278	660.4	485.2	1145.6	0.8522	0.4459	1.2981	**1900.0**
1950.0	632.22	0.02541	0.16999	0.19540	666.3	475.8	1142.0	0.8574	0.4358	1.2931	**1950.0**
2000.0	635.80	0.02565	0.16266	0.18831	672.1	466.2	1138.3	0.8625	0.4256	1.2881	**2000.0**
2100.0	642.76	0.02615	0.14885	0.17501	683.8	446.7	1130.5	0.8727	0.4053	1.2780	**2100.0**
2200.0	649.45	0.02669	0.13603	0.16272	695.5	426.7	1122.2	0.8828	0.3848	1.2676	**2200.0**
2300.0	655.89	0.02727	0.12406	0.15133	707.2	406.0	1113.2	0.8929	0.3640	1.2569	**2300.0**
2400.0	662.11	0.02790	0.11287	0.14076	719.0	384.8	1103.7	0.9031	0.3430	1.2460	**2400.0**
2500.0	668.11	0.02859	0.10209	0.13068	731.7	361.6	1093.3	0.9139	0.3206	1.2345	**2500.0**
2600.0	673.91	0.02938	0.09172	0.12110	744.5	337.6	1082.0	0.9247	0.2977	1.2225	**2600.0**
2700.0	679.53	0.03029	0.08165	0.11194	757.3	312.3	1069.7	0.9356	0.2741	1.2097	**2700.0**
2800.0	684.96	0.03134	0.07171	0.10305	770.7	285.1	1055.8	0.9468	0.2491	1.1958	**2800.0**
2900.0	690.22	0.03262	0.06158	0.09420	785.1	254.7	1039.8	0.9588	0.2215	1.1803	**2900.0**
3000.0	695.33	0.03428	0.05073	0.08500	801.8	218.4	1020.3	0.9728	0.1891	1.1619	**3000.0**
3100.0	700.28	0.03681	0.03771	0.07452	824.0	169.3	993.3	0.9914	0.1460	1.1373	**3100.0**
3200.0	705.08	0.04472	0.01191	0.05663	875.5	56.1	931.6	1.0351	0.0482	1.0832	**3200.0**
3208.2*	705.47	0.05078	0.00000	0.05078	906.0	0.0	906.0	1.0612	0.0000	1.0612	**3208.2***

*Critical pressure

Source: Combustion Engineering, Inc.

TABLE 46.3 Superheated Steam (1 to 15,500 lb per Square Inch Absolute Pressure)

Abs Press. Lb/Sq In. (Sat. Temp)		Sat. Water	Sat. Steam	Temperature — Degrees Fahrenheit 200	250	300	350	400	450	500	600	700	800	900	1000	1100	1200
1 (101.74)	Sh			98.26	148.26	198.26	248.26	298.26	348.26	398.26	498.26	598.26	698.26	798.26	898.26	998.26	1098.26
	v	0.01614	333.6	392.5	422.4	452.3	482.1	511.9	541.7	571.5	631.1	690.7	750.3	809.8	869.4	929.0	988.6
	h	69.73	1105.8	1150.2	1172.9	1195.7	1218.7	1241.8	1265.1	1288.6	1336.1	1384.5	1433.7	1483.8	1534.9	1586.8	1639.7
	s	0.1326	1.9781	2.0509	2.0841	2.1152	2.1445	2.1722	2.1985	2.2237	2.2708	2.3144	2.3551	2.3934	2.4296	2.4640	2.4969
5 (162.24)	Sh			37.76	87.76	137.76	187.76	237.76	287.76	337.76	437.76	537.76	637.76	737.76	837.76	937.76	1037.76
	v	0.01641	73.53	78.14	84.21	90.24	96.25	102.24	108.23	114.21	126.15	138.08	150.01	161.94	173.86	185.78	197.70
	h	130.20	1131.1	1148.6	1171.7	1194.8	1218.0	1241.3	1264.7	1288.2	1335.9	1384.3	1433.6	1483.7	1534.7	1586.7	1639.6
	s	0.2349	1.8443	1.8716	1.9054	1.9369	1.9664	1.9943	2.0208	2.0460	2.0932	2.1369	2.1776	2.2159	2.2521	2.2866	2.3194
10 (193.21)	Sh			6.79	56.79	106.79	156.79	206.79	256.79	306.79	406.79	506.79	606.79	706.79	806.79	906.79	1006.79
	v	0.01659	38.42	38.84	41.93	44.98	48.02	51.03	54.04	57.04	63.03	69.00	74.98	80.94	86.91	92.87	98.84
	h	161.26	1143.3	1146.6	1170.2	1193.7	1217.1	1240.6	1264.1	1287.8	1335.5	1384.0	1433.4	1483.5	1534.6	1586.6	1639.5
	s	0.2836	1.7879	1.7928	1.8273	1.8593	1.8892	1.9173	1.9439	1.9692	2.0166	2.0603	2.1011	2.1394	2.1757	2.2101	2.2430
14.696 (212.00)	Sh				38.00	88.00	138.00	188.00	238.00	288.00	388.00	488.00	588.00	688.00	788.00	888.00	988.00
	v	.0167	26.799		28.42	30.52	32.60	34.67	36.72	38.77	42.86	46.93	51.00	55.06	59.13	63.19	67.25
	h	180.17	1150.5		1168.8	1192.6	1216.3	1239.9	1263.6	1287.4	1335.2	1383.8	1433.2	1483.4	1534.5	1586.5	1639.4
	s	.3121	1.7568		1.7833	1.8158	1.8459	1.8743	1.9010	1.9265	1.9739	2.0177	2.0585	2.0969	2.1332	2.1676	2.2005
15 (213.03)	Sh				36.97	86.97	136.97	186.97	236.97	286.97	386.97	486.97	586.97	686.97	786.97	886.97	986.97
	v	0.01673	26.290		27.837	29.899	31.939	33.963	35.977	37.985	41.986	45.978	49.964	53.946	57.926	61.905	65.882
	h	181.21	1150.9		1168.7	1192.5	1216.2	1239.9	1263.6	1287.3	1335.2	1383.8	1433.2	1483.4	1534.5	1586.5	1639.4
	s	0.3137	1.7552		1.7809	1.8134	1.8437	1.8720	1.8988	1.9242	1.9717	2.0155	2.0563	2.0946	2.1309	2.1653	2.1982
20 (227.96)	Sh				22.04	72.04	122.04	172.04	222.04	272.04	372.04	472.04	572.04	672.04	772.04	872.04	972.04
	v	0.01683	20.087		20.788	22.356	23.900	25.428	26.946	28.457	31.466	34.465	37.458	40.447	43.435	46.420	49.405
	h	196.27	1156.3		1167.1	1191.4	1215.4	1239.2	1263.0	1286.9	1334.9	1383.5	1432.9	1483.2	1534.3	1586.3	1639.3
	s	0.3358	1.7320		1.7475	1.7805	1.8111	1.8397	1.8666	1.8921	1.9397	1.9836	2.0244	2.0628	2.0991	2.1336	2.1665
25 (240.07)	Sh				9.93	59.93	109.93	159.93	209.93	259.93	359.93	459.93	559.93	659.93	759.93	859.93	959.93
	v	0.01693	16.301		16.558	17.829	19.076	20.307	21.527	22.740	25.153	27.557	29.954	32.348	34.740	37.130	39.518
	h	208.52	1160.6		1165.6	1190.2	1214.5	1238.5	1262.5	1286.4	1334.6	1383.3	1432.7	1483.0	1534.2	1586.2	1639.2
	s	0.3535	1.7141		1.7212	1.7547	1.7856	1.8145	1.8415	1.8672	1.9149	1.9588	1.9997	2.0381	2.0744	2.1089	2.1418
30 (250.34)	Sh					49.66	99.66	149.66	199.66	249.66	349.66	449.66	549.66	649.66	749.66	849.66	949.66
	v	0.01701	13.744			14.810	15.859	16.892	17.914	18.929	20.945	22.951	24.952	26.949	28.943	30.936	32.927
	h	218.93	1164.1			1189.0	1213.6	1237.8	1261.9	1286.0	1334.2	1383.0	1432.5	1482.8	1534.0	1586.1	1639.0
	s	0.3682	1.6995			1.7334	1.7647	1.7937	1.8210	1.8467	1.8946	1.9386	1.9795	2.0179	2.0543	2.0888	2.1217

Abs press. (Sat. temp.)		Sat. water	Sat. steam												
35 (259.29)	Sh			40.71	90.71	140.71	190.71	240.71	340.71	440.71	540.71	640.71	740.71	840.71	940.71
	v	0.01708	11.896	12.654	13.562	14.453	15.334	16.207	17.939	19.662	21.379	23.092	24.803	26.512	28.220
	h	228.03	1167.1	1187.8	1212.7	1237.1	1261.3	1285.5	1333.9	1382.8	1432.3	1482.7	1533.9	1586.0	1638.9
	s	0.3809	1.6872	1.7152	1.7468	1.7761	1.8035	1.8294	1.8774	1.9214	1.9624	2.0009	2.0372	2.0717	2.1046
40 (267.25)	Sh			32.75	82.75	132.75	182.75	232.75	332.75	432.75	532.75	632.75	732.75	832.75	932.75
	v	0.01715	10.497	11.036	11.838	12.624	13.398	14.165	15.685	17.195	18.699	20.199	21.697	23.194	24.689
	h	236.14	1169.8	1186.6	1211.7	1236.4	1260.8	1285.0	1333.6	1382.5	1432.1	1482.5	1533.7	1585.8	1638.8
	s	0.3921	1.6765	1.6992	1.7312	1.7608	1.7883	1.8143	1.8624	1.9065	1.9476	1.9860	2.0224	2.0569	2.0899
45 (274.44)	Sh			25.56	75.56	125.56	175.56	225.56	325.56	425.56	525.56	625.56	725.56	825.56	925.56
	v	0.01721	9.399	9.777	10.497	11.201	11.892	12.577	13.932	15.276	16.614	17.950	19.282	20.613	21.943
	h	243.49	1172.1	1185.4	1210.4	1235.7	1260.2	1284.6	1333.3	1382.3	1431.9	1482.3	1533.6	1585.7	1638.7
	s	0.4021	1.6671	1.6849	1.7173	1.7471	1.7748	1.8010	1.8492	1.8934	1.9345	1.9730	2.0093	2.0439	2.0768
50 (281.02)	Sh			18.98	68.98	118.98	168.98	218.98	318.98	418.98	518.98	618.98	718.98	818.98	918.98
	v	0.01727	8.514	8.769	9.424	10.062	10.688	11.306	12.529	13.741	14.947	16.150	17.350	18.549	19.746
	h	250.21	1174.1	1184.1	1209.9	1234.9	1259.6	1284.1	1332.9	1382.0	1431.7	1482.2	1533.4	1585.6	1638.6
	s	0.4112	1.6586	1.6720	1.7048	1.7349	1.7628	1.7890	1.8374	1.8816	1.9227	1.9613	1.9977	2.0322	2.0652
55 (287.07)	Sh			12.93	62.93	112.93	162.93	212.93	312.93	412.93	512.93	612.93	712.93	812.93	912.93
	v	0.01733		7.945	8.546	9.130	9.702	10.267	11.381	12.485	13.583	14.677	15.769	16.859	17.948
	h	256.43		1182.9	1208.9	1234.2	1259.1	1283.6	1332.6	1381.8	1431.5	1482.0	1533.3	1585.5	1638.5
	s	0.4196		1.6601	1.6933	1.7237	1.7518	1.7781	1.8266	1.8710	1.9121	1.9507	1.987	2.022	2.055
60 (292.71)	Sh			7.29	57.29	107.29	157.29	207.29	307.29	407.29	507.29	607.29	707.29	807.29	907.29
	v	0.01738	7.174	7.257	7.815	8.354	8.881	9.400	10.425	11.438	12.446	13.450	14.452	15.452	16.450
	h	262.21	1177.6	1181.6	1208.0	1233.5	1258.5	1283.2	1332.3	1381.5	1431.3	1481.8	1533.2	1585.3	1638.4
	s	0.4273	1.6440	1.6492	1.6934	1.7134	1.7417	1.7681	1.8168	1.8612	1.9024	1.9410	1.9774	2.0120	2.0450
65 (297.98)	Sh			2.02	52.02	102.02	152.02	202.02	302.02	402.02	502.02	602.02	702.02	802.02	902.02
	v	0.01743	6.653	6.675	7.195	7.697	8.186	8.667	9.615	10.552	11.484	12.412	13.337	14.261	15.183
	h	267.63	1179.1	1180.3	1207.0	1232.7	1257.9	1282.7	1331.9	1381.3	1431.1	1481.6	1533.0	1585.2	1638.3
	s	0.4344	1.6375	1.6390	1.6731	1.7040	1.7324	1.7590	1.8077	1.8522	1.8935	1.9321	1.9685	2.0031	2.0361
70 (302.93)	Sh				47.07	97.07	147.07	197.07	297.07	397.07	497.07	597.07	697.07	797.07	897.07
	v	0.01748	6.205		6.664	7.133	7.590	8.039	8.922	9.793	10.659	11.522	12.382	13.240	14.097
	h	272.74	1180.6		1206.0	1232.0	1257.3	1282.2	1331.6	1381.0	1430.9	1481.5	1532.9	1585.1	1638.2
	s	0.4411	1.6316		1.6640	1.6951	1.7237	1.7504	1.7993	1.8439	1.8852	1.9238	1.9603	1.9949	2.0279
75 (307.61)	Sh				42.39	92.39	142.39	192.39	292.39	392.39	492.39	592.39	692.39	792.39	892.39
	v	0.01753	5.814		6.204	6.645	7.074	7.494	8.320	9.135	9.945	10.750	11.553	12.355	13.155
	h	277.56	1181.9		1205.0	1231.2	1256.7	1281.7	1331.3	1380.7	1430.7	1481.3	1532.7	1585.0	1638.1
	s	0.4474	1.6260		1.6554	1.6868	1.7156	1.7424	1.7915	1.8361	1.8774	1.9161	1.9526	1.9872	2.0202

Sh = superheat, F
v = specific volume, cu ft per lb
h = enthalpy, Btu per lb
s = entropy, Btu per F per lb

TABLE 46.3 Superheated Steam (1 to 15,500 lb per Square Inch Absolute Pressure) (Continued)

Abs Press. Lb/Sq In. (Sat. Temp)		Sat Water	Sat Steam	Temperature – Degrees Fahrenheit 350	400	450	500	550	600	700	800	900	1000	1100	1200	1300	1400
80 (312.04)	Sh			37.96	87.96	137.96	187.96	237.96	287.96	387.96	487.96	587.96	687.96	787.96	887.96	987.96	1087.96
	v	0.01757	5.471	5.801	6.218	6.622	7.018	7.408	7.794	8.560	9.319	10.075	10.829	11.581	12.331	13.081	13.829
	h	282.15	1183.1	1204.0	1230.5	1256.1	1281.3	1306.2	1330.9	1380.5	1430.5	1481.1	1532.6	1584.9	1638.0	1692.0	1746.8
	s	0.4534	1.6208	1.6473	1.6790	1.7080	1.7349	1.7602	1.7842	1.8289	1.8702	1.9089	1.9454	1.9800	2.0131	2.0446	2.0750
85 (316.26)	Sh			33.74	83.74	133.74	183.74	233.74	283.74	383.74	483.74	583.74	683.74	783.74	883.74	983.74	1083.74
	v	0.01762	5.167	5.445	5.840	6.223	6.597	6.966	7.330	8.052	8.768	9.480	10.190	10.898	11.604	12.310	13.014
	h	286.52	1184.2	1203.0	1229.7	1255.5	1280.8	1305.8	1330.6	1380.2	1430.3	1481.0	1532.4	1584.7	1637.9	1691.9	1746.8
	s	0.4590	1.6159	1.6396	1.6716	1.7008	1.7279	1.7532	1.7772	1.8220	1.8634	1.9021	1.9386	1.9733	2.0063	2.0379	2.0682
90 (320.28)	Sh			29.72	79.72	129.72	179.72	229.72	279.72	379.72	479.72	579.72	679.72	779.72	879.72	979.72	1079.72
	v	0.01766	4.895	5.128	5.505	5.869	6.223	6.572	6.917	7.600	8.277	8.950	9.621	10.290	10.958	11.625	12.290
	h	290.69	1185.3	1202.0	1228.9	1254.9	1280.3	1305.4	1330.2	1380.0	1430.1	1480.8	1532.3	1584.6	1637.8	1691.8	1746.7
	s	0.4643	1.6113	1.6323	1.6646	1.6940	1.7212	1.7467	1.7707	1.8156	1.8570	1.8957	1.9323	1.9669	2.0000	2.0316	2.0619
95 (324.13)	Sh			25.87	75.87	125.87	175.87	225.87	275.87	375.87	475.87	575.87	675.87	775.87	875.87	975.87	1075.87
	v	0.01770	4.651	4.845	5.205	5.551	5.889	6.221	6.548	7.196	7.838	8.477	9.113	9.747	10.380	11.012	11.643
	h	294.70	1186.2	1200.9	1228.1	1254.3	1279.8	1305.0	1329.9	1379.7	1429.9	1480.6	1532.1	1584.5	1637.7	1691.7	1746.6
	s	0.4694	1.6069	1.6253	1.6580	1.6876	1.7149	1.7404	1.7645	1.8094	1.8509	1.8897	1.9262	1.9609	1.9940	2.0256	2.0559
100 (327.82)	Sh			22.18	72.18	122.18	172.18	222.18	272.18	372.18	472.18	572.18	672.18	772.18	872.18	972.18	1072.18
	v	0.01774	4.431	4.590	4.935	5.266	5.588	5.904	6.216	6.833	7.443	8.050	8.655	9.258	9.860	10.460	11.060
	h	298.54	1187.2	1199.9	1227.4	1253.7	1279.3	1304.6	1329.6	1379.5	1429.7	1480.4	1532.0	1584.4	1637.6	1691.6	1746.5
	s	0.4743	1.6027	1.6187	1.6516	1.6814	1.7088	1.7344	1.7586	1.8036	1.8451	1.8839	1.9205	1.9552	1.9883	2.0199	2.0502
105 (331.37)	Sh			18.63	68.63	118.63	168.63	218.63	268.63	368.63	468.63	568.63	668.63	768.63	868.63	968.63	1068.63
	v	0.01778	4.231	4.359	4.690	5.007	5.315	5.617	5.915	6.504	7.086	7.665	8.241	8.816	9.389	9.961	10.532
	h	302.24	1188.0	1198.8	1226.6	1253.1	1278.8	1304.2	1329.2	1379.2	1429.4	1480.3	1531.8	1584.2	1637.5	1691.5	1746.4
	s	0.4790	1.5988	1.6122	1.6455	1.6755	1.7031	1.7288	1.7530	1.7981	1.8396	1.8785	1.9151	1.9498	1.9828	2.0145	2.0448
110 (334.79)	Sh			15.21	65.21	115.21	165.21	215.21	265.21	365.21	465.21	565.21	665.21	765.21	865.21	965.21	1065.21
	v	0.01782	4.048	4.149	4.468	4.772	5.068	5.357	5.642	6.205	6.761	7.314	7.865	8.413	8.961	9.507	10.053
	h	305.80	1188.9	1197.7	1225.8	1252.5	1278.3	1303.8	1328.9	1379.0	1429.2	1480.1	1531.7	1584.1	1637.4	1691.4	1746.4
	s	0.4834	1.5950	1.6061	1.6396	1.6698	1.6975	1.7233	1.7476	1.7928	1.8344	1.8732	1.9099	1.9446	1.9777	2.0093	2.0397
115 (338.08)	Sh			11.92	61.92	111.92	161.92	211.92	261.92	361.92	461.92	561.92	661.92	761.92	861.92	961.92	1061.92
	v	0.01785	3.881	3.957	4.265	4.558	4.841	5.119	5.392	5.932	6.465	6.994	7.521	8.046	8.570	9.093	9.615
	h	309.25	1189.6	1196.7	1225.0	1251.8	1277.9	1303.3	1328.6	1378.7	1429.0	1479.9	1531.6	1584.0	1637.2	1691.4	1746.3
	s	0.4877	1.5913	1.6001	1.6340	1.6644	1.6922	1.7181	1.7425	1.7877	1.8294	1.8682	1.9049	1.9396	1.9727	2.0044	2.0347

	Sh			8.73	58.73	108.73	158.73	208.73	258.73	358.73	458.73	558.73	658.73	758.73	858.73	958.73	1058.73
120	v	0.01789	3.7275	3.7815	4.0786	4.3610	4.6341	4.9009	5.1637	5.6813	6.1928	6.7006	7.2060	7.7096	8.2119	8.7130	9.2134
(341.27)	h	312.58	1190.4	1195.6	1224.1	1251.2	1277.4	1302.9	1328.2	1378.4	1428.8	1479.8	1531.4	1583.9	1637.1	1691.3	1746.2
	s	0.4919	1.5879	1.5943	1.6286	1.6592	1.6872	1.7132	1.7376	1.7829	1.8246	1.8635	1.9001	1.9349	1.9680	1.9996	2.0300
	Sh			2.67	52.67	102.67	152.67	202.67	252.67	352.67	452.67	552.67	652.67	752.67	852.67	952.67	1052.67
130	v	0.01796	3.4544	3.4699	3.7489	4.0129	4.2672	4.5151	4.7589	5.2384	5.7118	6.1814	6.6486	7.1140	7.5781	8.0411	8.5033
(347.33)	h	318.95	1191.7	1193.4	1222.5	1249.9	1276.4	1302.1	1327.5	1377.9	1428.4	1479.4	1531.1	1583.6	1636.9	1691.1	1746.1
	s	0.4998	1.5813	1.5833	1.6182	1.6493	1.6775	1.7037	1.7283	1.7737	1.8155	1.8545	1.8911	1.9259	1.9591	1.9907	2.0211
	Sh				46.96	96.96	146.96	196.96	246.96	346.96	446.96	546.96	646.96	746.96	846.96	946.96	1046.96
140	v	0.01803	3.2190		3.4661	3.7143	3.9526	4.1844	4.4119	4.8588	5.2995	5.7364	6.1709	6.6036	7.0349	7.4652	7.8946
(353.04)	h	324.96	1193.0		1220.8	1248.7	1275.3	1301.3	1326.8	1377.4	1428.0	1479.1	1530.8	1583.4	1636.7	1690.9	1745.9
	s	0.5071	1.5752		1.6085	1.6400	1.6686	1.6949	1.7196	1.7652	1.8071	1.8461	1.8828	1.9176	1.9508	1.9825	2.0129
	Sh				41.57	91.57	141.57	191.57	241.57	341.57	441.57	541.57	641.57	741.57	841.57	941.57	1041.57
150	v	0.01809	3.0139		3.2208	3.4555	3.6799	3.8978	4.1112	4.5298	4.9421	5.3507	5.7568	6.1612	6.5642	6.9661	7.3671
(358.43)	h	330.65	1194.1		1219.1	1247.4	1274.3	1300.5	1326.1	1376.9	1427.6	1478.7	1530.5	1583.1	1636.5	1690.7	1745.7
	s	0.5141	1.5695		1.5993	1.6313	1.6602	1.6867	1.7115	1.7573	1.7992	1.8383	1.8751	1.9099	1.9431	1.9748	2.0052
	Sh				36.45	86.45	136.45	186.45	236.45	336.45	436.45	536.45	636.45	736.45	836.45	936.45	1036.45
160	v	0.01815	2.8336		3.0060	3.2288	3.4413	3.6469	3.8480	4.2420	4.6295	5.0132	5.3945	5.7741	6.1522	6.5293	6.9055
(363.55)	h	336.07	1195.1		1217.4	1246.0	1273.3	1299.6	1325.4	1376.4	1427.2	1478.4	1530.3	1582.9	1636.3	1690.5	1745.6
	s	0.5206	1.5641		1.5906	1.6231	1.6522	1.6790	1.7039	1.7499	1.7919	1.8310	1.8678	1.9027	1.9359	1.9676	1.9980
	Sh				31.58	81.58	131.58	181.58	231.58	331.58	431.58	531.58	631.58	731.58	831.58	931.58	1031.58
170	v	0.01821	2.6738		2.8162	3.0288	3.2306	3.4255	3.6158	3.9879	4.3536	4.7155	5.0749	5.4325	5.7888	6.1440	6.4983
(368.42)	h	341.24	1196.0		1215.6	1244.7	1272.2	1298.8	1324.7	1375.8	1426.8	1478.0	1530.0	1582.6	1636.1	1690.4	1745.4
	s	0.5269	1.5591		1.5823	1.6152	1.6447	1.6717	1.6968	1.7428	1.7850	1.8241	1.8610	1.8959	1.9291	1.9608	1.9913
	Sh				26.92	76.92	126.92	176.92	226.92	326.92	426.92	526.92	626.92	726.92	826.92	926.92	1026.92
180	v	0.01827	2.5312		2.6474	2.8508	3.0433	3.2286	3.4093	3.7621	4.1084	4.4508	4.7907	5.1289	5.4657	5.8014	6.1363
(373.08)	h	346.19	1196.9		1213.8	1243.4	1271.2	1297.9	1324.0	1375.3	1426.3	1477.7	1529.7	1582.4	1635.9	1690.2	1745.3
	s	0.5328	1.5543		1.5743	1.6078	1.6376	1.6647	1.6900	1.7362	1.7784	1.8176	1.8545	1.8894	1.9227	1.9545	1.9849
	Sh				22.47	72.47	122.47	172.47	222.47	322.47	422.47	522.47	622.47	722.47	822.47	922.47	1022.47
190	v	0.01833	2.4030		2.4961	2.6915	2.8756	3.0525	3.2246	3.5601	3.8889	4.2140	4.5365	4.8572	5.1766	5.4949	5.8124
(377.53)	h	350.94	1197.6		1212.0	1242.0	1270.1	1297.1	1323.3	1374.8	1425.9	1477.4	1529.4	1582.1	1635.7	1690.0	1745.1
	s	0.5384	1.5498		1.5667	1.6006	1.6307	1.6581	1.6835	1.7299	1.7722	1.8115	1.8484	1.8834	1.9166	1.9484	1.9789
	Sh				18.20	68.20	118.20	168.20	218.20	318.20	418.20	518.20	618.20	718.20	818.20	918.20	1018.20
200	v	0.01839	2.2873		2.3598	2.5480	2.7247	2.8939	3.0583	3.3783	3.6915	4.0008	4.3077	4.6128	4.9165	5.2191	5.5209
(381.80)	h	355.51	1198.3		1210.1	1240.6	1269.0	1296.2	1322.6	1374.3	1425.5	1477.0	1529.1	1581.9	1635.4	1689.8	1745.0
	s	0.5438	1.5454		1.5593	1.5938	1.6242	1.6518	1.6773	1.7239	1.7663	1.8057	1.8426	1.8776	1.9109	1.9427	1.9732

Sh = superheat, F
v = specific volume, cu ft per lb
h = enthalpy, Btu per lb
s = entropy, Btu per F per lb

TABLE 46.3 Superheated Steam (1 to 15,500 lb per Square Inch Absolute Pressure) (Continued)

Abs Press. Lb/Sq In. (Sat. Temp)		Sat Water	Sat Steam	Temperature — Degrees Fahrenheit 400	450	500	550	600	700	800	900	1000	1100	1200	1300	1400	1500
	Sh			14.09	64.09	114.09	164.09	214.09	314.09	414.09	514.09	614.09	714.09	814.09	914.09	1014.09	1114.09
210	v	0.01844	2.1822	2.2364	2.4181	2.5880	2.7504	2.9078	3.2137	3.5128	3.8080	4.1007	4.3915	4.6811	4.9695	5.2571	5.5440
(385.91)	h	359.91	1199.0	1208.02	1239.2	1268.0	1295.3	1321.9	1373.7	1425.1	1476.7	1528.8	1581.6	1635.2	1689.6	1744.8	1800.8
	s	0.5490	1.5413	1.5522	1.5872	1.6180	1.6458	1.6715	1.7182	1.7607	1.8001	1.8371	1.8721	1.9054	1.9372	1.9677	1.9970
	Sh			10.12	60.12	110.12	160.12	210.12	310.12	410.12	510.12	610.12	710.12	810.12	910.12	1010.12	1110.12
220	v	0.01850	2.0863	2.1240	2.2999	2.4638	2.6199	2.7710	3.0642	3.3504	3.6327	3.9125	4.1905	4.4671	4.7426	5.0173	5.2913
(389.88)	h	364.17	1199.6	1206.3	1237.8	1266.9	1294.5	1321.2	1373.2	1424.7	1476.3	1528.5	1581.4	1635.0	1689.4	1744.7	1800.6
	s	0.5540	1.5374	1.5453	1.5808	1.6120	1.6400	1.6658	1.7128	1.7553	1.7948	1.8318	1.8668	1.9002	1.9320	1.9625	1.9919
	Sh			6.30	56.30	106.30	156.30	206.30	306.30	406.30	506.30	606.30	706.30	806.30	906.30	1006.30	1106.30
230	v	0.01855	1.9985	2.0212	2.1919	2.3503	2.5008	2.6461	2.9276	3.2020	3.4726	3.7406	4.0068	4.2717	4.5355	4.7984	5.0606
(393.70)	h	368.28	1200.1	1204.4	1236.3	1265.7	1293.6	1320.4	1372.7	1424.2	1476.0	1528.2	1581.1	1634.8	1689.3	1744.5	1800.5
	s	0.5588	1.5336	1.5385	1.5747	1.6062	1.6344	1.6604	1.7075	1.7502	1.7897	1.8268	1.8618	1.8952	1.9270	1.9576	1.9869
	Sh			2.61	52.61	102.61	152.61	202.61	302.61	402.61	502.61	602.61	702.61	802.61	902.61	1002.61	1102.61
240	v	0.01860	1.9177	1.9268	2.0928	2.2462	2.3915	2.5316	2.8024	3.0661	3.3259	3.5831	3.8385	4.0926	4.3456	4.5977	4.8492
(397.39)	h	372.27	1200.6	1202.4	1234.9	1264.6	1292.7	1319.7	1372.1	1423.8	1475.6	1527.9	1580.9	1634.6	1689.1	1744.3	1800.4
	s	0.5634	1.5299	1.5320	1.5687	1.6006	1.6291	1.6552	1.7025	1.7452	1.7848	1.8219	1.8570	1.8904	1.9223	1.9528	1.9822
	Sh				49.03	99.03	149.03	199.03	299.03	399.03	499.03	599.03	699.03	799.03	899.03	999.03	1099.03
250	v	0.01865	1.8432		2.0016	2.1504	2.2909	2.4262	2.6872	2.9410	3.1909	3.4382	3.6837	3.9278	4.1709	4.4131	4.6546
(400.97)	h	376.14	1201.1		1233.4	1263.5	1291.8	1319.0	1371.6	1423.4	1475.3	1527.6	1580.6	1634.4	1688.9	1744.2	1800.2
	s	0.5679	1.5264		1.5629	1.5951	1.6239	1.6502	1.6976	1.7405	1.7801	1.8173	1.8524	1.8858	1.9177	1.9482	1.9776
	Sh				45.56	95.56	145.56	195.56	295.56	395.56	495.56	595.56	695.56	795.56	895.56	995.56	1095.56
260	v	0.01870	1.7742		1.9173	2.0619	2.1981	2.3289	2.5808	2.8256	3.0663	3.3044	3.5408	3.7758	4.0097	4.2427	4.4750
(404.44)	h	379.90	1201.5		1231.9	1262.4	1290.9	1318.2	1371.1	1423.0	1474.9	1527.3	1580.4	1634.2	1688.7	1744.0	1800.1
	s	0.5722	1.5230		1.5573	1.5899	1.6189	1.6453	1.6930	1.7359	1.7756	1.8128	1.8480	1.8814	1.9133	1.9439	1.9732
	Sh				42.20	92.20	142.20	192.20	292.20	392.20	492.20	592.20	692.20	792.20	892.20	992.20	1092.20
270	v	0.01875	1.7101		1.8391	1.9799	2.1121	2.2388	2.4824	2.7186	2.9509	3.1806	3.4084	3.6349	3.8603	4.0849	4.3087
(407.80)	h	383.56	1201.9		1230.4	1261.2	1290.0	1317.5	1370.5	1422.6	1474.6	1527.1	1580.1	1634.0	1688.5	1743.9	1800.0
	s	0.5764	1.5197		1.5518	1.5848	1.6140	1.6406	1.6885	1.7315	1.7713	1.8085	1.8437	1.8771	1.9090	1.9396	1.9690
	Sh				38.93	88.93	138.93	188.93	288.93	388.93	488.93	588.93	688.93	788.93	888.93	988.93	1088.93
280	v	0.01880	1.6505		1.7665	1.9037	2.0322	2.1551	2.3909	2.6194	2.8437	3.0655	3.2855	3.5042	3.7217	3.9384	4.1543
(411.07)	h	387.12	1202.3		1228.8	1260.0	1289.1	1316.8	1370.0	1422.1	1474.2	1526.8	1579.9	1633.8	1688.4	1743.7	1799.8
	s	0.5805	1.5166		1.5464	1.5798	1.6093	1.6361	1.6841	1.7273	1.7671	1.8043	1.8395	1.8730	1.9050	1.9356	1.9649

	Sh			35.75	85.75	135.75	185.75	285.75	385.75	485.75	585.75	685.75	785.75	885.75	985.75	1085.75
290	v	0.01885	1.5948	1.6988	1.8327	1.9578	2.0772	2.3058	2.5269	2.7440	2.9585	3.1711	3.3824	3.5926	3.8019	4.0106
(414.25)	h	390.60	1202.6	1227.3	1258.9	1288.1	1316.0	1369.5	1421.7	1473.9	1526.5	1579.6	1633.5	1688.2	1743.6	1799.7
	s	0.5844	1.5135	1.5412	1.5750	1.6048	1.6317	1.6799	1.7232	1.7630	1.8003	1.8356	1.8690	1.9010	1.9316	1.9610
	Sh			32.65	82.65	132.65	182.65	282.65	382.65	482.65	582.65	682.65	782.65	882.65	982.65	1082.65
300	v	0.01889	1.5427	1.6356	1.7665	1.8883	2.0044	2.2263	2.4407	2.6509	2.8585	3.0643	3.2688	3.4721	3.6746	3.8764
(417.35)	h	393.99	1202.9	1225.7	1257.7	1287.2	1315.2	1368.9	1421.3	1473.6	1526.2	1579.4	1633.3	1688.0	1743.4	1799.6
	s	0.5882	1.5105	1.5361	1.5703	1.6003	1.6274	1.6758	1.7192	1.7591	1.7964	1.8317	1.8652	1.8972	1.9278	1.9572
	Sh			29.64	79.64	129.64	179.64	279.64	379.64	479.64	579.64	679.64	779.64	879.64	979.64	1079.64
310	v	0.01894	1.4939	1.5763	1.7044	1.8233	1.9363	2.1520	2.3600	2.5638	2.7650	2.9644	3.1625	3.3594	3.5555	3.7509
(420.36)	h	397.30	1203.2	1224.1	1256.5	1286.3	1314.5	1368.4	1420.9	1473.2	1525.9	1579.2	1633.1	1687.8	1743.3	1799.4
	s	0.5920	1.5076	1.5311	1.5657	1.5960	1.6233	1.6719	1.7153	1.7553	1.7927	1.8280	1.8615	1.8935	1.9241	1.9536
	Sh			26.69	76.69	126.69	176.69	276.69	376.69	476.69	576.69	676.69	776.69	876.69	976.69	1076.69
320	v	0.01899	1.4480	1.5207	1.6462	1.7623	1.8725	2.0823	2.2843	2.4821	2.6774	2.8708	3.0628	3.2538	3.4438	3.6332
(423.31)	h	400.53	1203.4	1222.5	1255.2	1285.3	1313.7	1367.8	1420.5	1472.9	1525.6	1578.9	1632.9	1687.6	1743.1	1799.3
	s	0.5956	1.5048	1.5261	1.5612	1.5918	1.6192	1.6680	1.7116	1.7516	1.7890	1.8243	1.8579	1.8899	1.9206	1.9500
	Sh			23.82	73.82	123.82	173.82	273.82	373.82	473.82	573.82	673.82	773.82	873.82	973.82	1073.82
330	v	0.01903	1.4048	1.4684	1.5915	1.7050	1.8125	2.0168	2.2132	2.4054	2.5950	2.7828	2.9692	3.1545	3.3389	3.5227
(426.18)	h	403.70	1203.6	1220.9	1254.0	1284.4	1313.0	1367.3	1420.0	1472.5	1525.3	1578.7	1632.7	1687.5	1742.9	1799.2
	s	0.5991	1.5021	1.5213	1.5568	1.5876	1.6153	1.6643	1.7079	1.7480	1.7855	1.8208	1.8544	1.8864	1.9171	1.9466
	Sh			21.01	71.01	121.01	171.01	271.01	371.01	471.01	571.01	671.01	771.01	871.01	971.01	1071.01
340	v	0.01908	1.3640	1.4191	1.5399	1.6511	1.7561	1.9552	2.1463	2.3333	2.5175	2.7000	2.8811	3.0611	3.2402	3.4186
(428.99)	h	406.80	1203.8	1219.2	1252.8	1283.4	1312.2	1366.7	1419.6	1472.2	1525.0	1578.4	1632.5	1687.3	1742.8	1799.0
	s	0.6026	1.4994	1.5165	1.5525	1.5836	1.6114	1.6606	1.7044	1.7445	1.7820	1.8174	1.8510	1.8831	1.9138	1.9432
	Sh			18.27	68.27	118.27	168.27	268.27	368.27	468.27	568.27	668.27	768.27	868.27	968.27	1068.27
350	v	0.01912	1.3255	1.3725	1.4913	1.6002	1.7028	1.8970	2.0832	2.2652	2.4445	2.6219	2.7980	2.9730	3.1471	3.3205
(431.73)	h	409.83	1204.0	1217.5	1251.5	1282.4	1311.4	1366.2	1419.2	1471.8	1524.7	1578.2	1632.3	1687.1	1742.6	1798.9
	s	0.6059	1.4968	1.5119	1.5483	1.5797	1.6077	1.6571	1.7009	1.7411	1.7787	1.8141	1.8477	1.8798	1.9105	1.9400
	Sh			15.59	65.59	115.59	165.59	265.59	365.59	465.59	565.59	665.59	765.59	865.59	965.59	1065.59
360	v	0.01917	1.2891	1.3285	1.4454	1.5521	1.6525	1.8421	2.0237	2.2009	2.3755	2.5482	2.7196	2.8898	3.0592	3.2279
(434.41)	h	412.81	1204.1	1215.8	1250.3	1281.5	1310.6	1365.6	1418.7	1471.5	1524.4	1577.9	1632.1	1686.9	1742.5	1798.8
	s	0.6092	1.4943	1.5073	1.5441	1.5758	1.6040	1.6536	1.6976	1.7379	1.7754	1.8109	1.8445	1.8766	1.9073	1.9368
	Sh			10.39	60.39	110.39	160.39	260.39	360.39	460.39	560.39	660.39	760.39	860.39	960.39	1060.39
380	v	0.01925	1.2218	1.2472	1.3606	1.4635	1.5598	1.7410	1.9139	2.0825	2.2484	2.4124	2.5750	2.7366	2.8973	3.0572
(439.61)	h	418.59	1204.4	1212.4	1247.7	1279.5	1309.0	1364.5	1417.9	1470.8	1523.8	1577.4	1631.6	1686.5	1742.2	1798.5
	s	0.6156	1.4894	1.4982	1.5360	1.5683	1.5969	1.6470	1.6911	1.7315	1.7692	1.8047	1.8384	1.8705	1.9012	1.9307

Sh = superheat, F
v = specific volume, cu ft per lb
h = enthalpy, Btu per lb
s = entropy, Btu per F per lb

TABLE 46.3 Superheated Steam (1 to 15,500 lb per Square Inch Absolute Pressure) (Continued)

Abs Press. Lb/Sq In. (Sat. Temp)		Sat. Water	Sat. Steam	Temperature — Degrees Fahrenheit 450	500	550	600	650	700	800	900	1000	1100	1200	1300	1400	1500
	Sh			5.40	55.40	105.40	155.40	205.40	255.40	355.40	455.40	555.40	655.40	755.40	855.40	955.40	1055.40
400	v	0.01934	1.1610	1.1738	1.2841	1.3836	1.4763	1.5646	1.6499	1.8151	1.9759	2.1339	2.2901	2.4450	2.5987	2.7515	2.9037
(444.60)	h	424.17	1204.6	1208.8	1245.1	1277.5	1307.4	1335.9	1363.4	1417.0	1470.1	1523.3	1576.9	1631.2	1686.2	1741.9	1798.2
	s	0.6217	1.4847	1.4894	1.5282	1.5611	1.5901	1.6163	1.6406	1.6850	1.7255	1.7632	1.7988	1.8325	1.8647	1.8955	1.9250
	Sh			.60	50.60	100.60	150.60	200.60	250.60	350.60	450.60	550.60	650.60	750.60	850.60	950.60	1050.60
420	v	0.01942	1.1057	1.1071	1.2148	1.3113	1.4007	1.4856	1.5676	1.7258	1.8795	2.0304	2.1795	2.3273	2.4739	2.6196	2.7647
(449.40)	h	429.56	1204.7	1205.2	1242.4	1275.4	1305.8	1334.5	1362.3	1416.2	1469.4	1522.7	1576.4	1630.8	1685.8	1741.6	1798.0
	s	0.6276	1.4802	1.4808	1.5206	1.5542	1.5835	1.6100	1.6345	1.6791	1.7197	1.7575	1.7932	1.8269	1.8591	1.8899	1.9195
	Sh				45.97	95.97	145.97	195.97	245.97	345.97	445.97	545.97	645.97	745.97	845.97	945.97	1045.97
440	v	0.01950	1.0554		1.1517	1.2454	1.3319	1.4138	1.4926	1.6445	1.7918	1.9363	2.0790	2.2203	2.3605	2.4998	2.6384
(454.03)	h	434.77	1204.8		1239.7	1273.4	1304.2	1333.2	1361.1	1415.3	1468.7	1522.1	1575.9	1630.4	1685.5	1741.2	1797.7
	s	0.6332	1.4759		1.5132	1.5474	1.5772	1.6040	1.6286	1.6734	1.7142	1.7521	1.7878	1.8216	1.8538	1.8847	1.9143
	Sh				41.50	91.50	141.50	191.50	241.50	341.50	441.50	541.50	641.50	741.50	841.50	941.50	1041.50
460	v	0.01959	1.0092		1.0939	1.1852	1.2691	1.3482	1.4242	1.5703	1.7117	1.8504	1.9872	2.1226	2.2569	2.3903	2.5230
(458.50)	h	439.83	1204.8		1236.9	1271.3	1302.5	1331.8	1360.0	1414.4	1468.0	1521.5	1575.4	1629.9	1685.1	1740.9	1797.4
	s	0.6387	1.4718		1.5060	1.5409	1.5711	1.5982	1.6230	1.6680	1.7089	1.7469	1.7826	1.8165	1.8488	1.8797	1.9093
	Sh				37.18	87.18	137.18	187.18	237.18	337.18	437.18	537.18	637.18	737.18	837.18	937.18	1037.18
480	v	0.01967	0.9668		1.0409	1.1300	1.2115	1.2881	1.3615	1.5023	1.6384	1.7716	1.9030	2.0330	2.1619	2.2900	2.4173
(462.82)	h	444.75	1204.8		1234.1	1269.1	1300.8	1330.5	1358.8	1413.6	1467.3	1520.9	1574.9	1629.5	1684.7	1740.6	1797.2
	s	0.6439	1.4677		1.4990	1.5346	1.5652	1.5925	1.6176	1.6628	1.7038	1.7419	1.7777	1.8116	1.8439	1.8748	1.9045
	Sh				32.99	82.99	132.99	182.99	232.99	332.99	432.99	532.99	632.99	732.99	832.99	932.99	1032.99
500	v	0.01975	0.9276		0.9919	1.0791	1.1584	1.2327	1.3037	1.4397	1.5708	1.6992	1.8256	1.9507	2.0746	2.1977	2.3200
(467.01)	h	449.52	1204.7		1231.2	1267.0	1299.1	1329.1	1357.7	1412.7	1466.6	1520.3	1574.4	1629.1	1684.4	1740.3	1796.9
	s	0.6490	1.4639		1.4921	1.5284	1.5595	1.5871	1.6123	1.6578	1.6990	1.7371	1.7730	1.8069	1.8393	1.8702	1.8998
	Sh				28.93	78.93	128.93	178.93	228.93	328.93	428.93	528.93	628.93	728.93	828.93	928.93	1028.93
520	v	0.01982	0.8914		0.9466	1.0321	1.1094	1.1816	1.2504	1.3819	1.5085	1.6323	1.7542	1.8746	1.9940	2.1125	2.2302
(471.07)	h	454.18	1204.5		1228.3	1264.8	1297.4	1327.7	1356.5	1411.8	1465.9	1519.7	1573.9	1628.7	1684.0	1740.0	1796.7
	s	0.6540	1.4601		1.4853	1.5223	1.5539	1.5818	1.6072	1.6530	1.6943	1.7325	1.7684	1.8024	1.8348	1.8657	1.8954
	Sh				24.99	74.99	124.99	174.99	224.99	324.99	424.99	524.99	624.99	724.99	824.99	924.99	1024.99
540	v	0.01990	0.8577		0.9045	0.9884	1.0640	1.1342	1.2010	1.3284	1.4508	1.5704	1.6880	1.8042	1.9193	2.0336	2.1471
(475.01)	h	458.71	1204.4		1225.3	1262.5	1295.7	1326.3	1355.3	1410.9	1465.1	1519.1	1573.4	1628.2	1683.6	1739.7	1796.4
	s	0.6587	1.4565		1.4786	1.5164	1.5485	1.5767	1.6023	1.6483	1.6897	1.7280	1.7640	1.7981	1.8305	1.8615	1.8911

560 (478.84)	Sh			21.16	71.16	121.16	171.16	221.16	321.16	421.16	521.16	621.16	721.16	821.16	921.16	1021.16
	v	0.01998	0.8264	0.8653	0.9479	1.0217	1.0902	1.1552	1.2787	1.3972	1.5129	1.6266	1.7388	1.8500	1.9603	2.0699
	h	463.14	1204.2	1222.2	1260.3	1293.9	1324.9	1354.2	1410.0	1464.4	1518.6	1572.9	1627.8	1683.3	1739.4	1796.1
	s	0.6634	1.4529	1.4720	1.5106	1.5431	1.5717	1.5975	1.6438	1.6853	1.7237	1.7598	1.7939	1.8263	1.8573	1.8870
580 (482.57)	Sh			17.43	67.43	117.43	167.43	217.43	317.43	417.43	517.43	617.43	717.43	817.43	917.43	1017.43
	v	0.02006	0.7971	0.8287	0.9100	0.9824	1.0492	1.1125	1.2324	1.3473	1.4593	1.5693	1.6780	1.7855	1.8921	1.9980
	h	467.47	1203.9	1219.1	1258.0	1292.1	1323.4	1353.0	1409.2	1463.7	1518.0	1572.4	1627.4	1682.9	1739.1	1795.9
	s	0.6679	1.4495	1.4654	1.5049	1.5380	1.5668	1.5929	1.6394	1.6811	1.7196	1.7556	1.7898	1.8223	1.8533	1.8831
600 (486.20)	Sh			13.80	63.80	113.80	163.80	213.80	313.80	413.80	513.80	613.80	713.80	813.80	913.80	1013.80
	v	0.02013	0.7697	0.7944	0.8746	0.9456	1.0109	1.0726	1.1892	1.3008	1.4093	1.5160	1.6211	1.7252	1.8284	1.9309
	h	471.70	1203.7	1215.9	1255.6	1290.3	1322.0	1351.8	1408.3	1463.0	1517.4	1571.9	1627.0	1682.6	1738.8	1795.6
	s	0.6723	1.4461	1.4590	1.4993	1.5329	1.5621	1.5884	1.6351	1.6769	1.7155	1.7517	1.7859	1.8184	1.8494	1.8792
650 (494.89)	Sh			5.11	55.11	105.11	155.11	205.11	305.11	405.11	505.11	605.11	705.11	805.11	905.11	1005.11
	v	0.02032	0.7084	0.7173	0.7954	0.8634	0.9254	0.9835	1.0929	1.1969	1.2979	1.3969	1.4944	1.5909	1.6864	1.7813
	h	481.89	1202.8	1207.6	1249.6	1285.7	1318.3	1348.7	1406.0	1461.2	1515.9	1570.7	1625.9	1681.6	1738.0	1794.9
	s	1.6828	1.4381	1.4430	1.4858	1.5207	1.5507	1.5775	1.6249	1.6671	1.7059	1.7422	1.7765	1.8092	1.8403	1.8701
700 (503.08)	Sh				46.92	96.92	146.92	196.92	296.92	396.92	496.92	596.92	696.92	796.92	896.92	996.92
	v	0.02050	0.6556		0.7271	0.7928	0.8520	0.9072	1.0102	1.1078	1.2023	1.2948	1.3858	1.4757	1.5647	1.6530
	h	491.60	1201.8		1243.4	1281.0	1314.6	1345.6	1403.7	1459.4	1514.4	1569.4	1624.8	1680.7	1737.2	1794.3
	s	0.6928	1.4304		1.4726	1.5090	1.5399	1.5673	1.6154	1.6580	1.6970	1.7335	1.7679	1.8006	1.8318	1.8617
750 (510.84)	Sh				39.16	89.16	139.16	189.16	289.16	389.16	489.16	589.16	689.16	789.16	889.16	989.16
	v	0.02069	0.6095		0.6676	0.7313	0.7882	0.8409	0.9386	1.0306	1.1195	1.2063	1.2916	1.3759	1.4592	1.5419
	h	500.89	1200.7		1236.9	1276.1	1310.7	1342.5	1401.5	1457.6	1512.9	1568.2	1623.8	1679.8	1736.4	1793.6
	s	0.7022	1.4232		1.4598	1.4977	1.5296	1.5577	1.6065	1.6494	1.6886	1.7252	1.7598	1.7926	1.8239	1.8538
800 (518.21)	Sh				31.79	81.79	131.79	181.79	281.79	381.79	481.79	581.79	681.79	781.79	881.79	981.79
	v	0.02087	0.5690		0.6151	0.6774	0.7323	0.7828	0.8759	0.9631	1.0470	1.1289	1.2093	1.2885	1.3669	1.4446
	h	509.81	1199.4		1230.1	1271.1	1306.8	1339.3	1399.1	1455.8	1511.4	1566.9	1622.7	1678.9	1735.7	1792.9
	s	0.7111	1.4163		1.4472	1.4869	1.5198	1.5484	1.5980	1.6413	1.6807	1.7175	1.7522	1.7851	1.8164	1.8464
850 (525.24)	Sh				24.76	74.76	124.76	174.76	274.76	374.76	474.76	574.76	674.76	774.76	874.76	974.76
	v	0.02105	0.5330		0.5683	0.6296	0.6829	0.7315	0.8205	0.9034	0.9830	1.0606	1.1366	1.2115	1.2855	1.3588
	h	518.40	1198.0		1223.0	1265.9	1302.8	1336.0	1396.8	1454.0	1510.0	1565.7	1621.6	1678.0	1734.9	1792.3
	s	0.7197	1.4096		1.4347	1.4763	1.5102	1.5396	1.5899	1.6336	1.6733	1.7102	1.7450	1.7780	1.8094	1.8395
900 (531.95)	Sh				18.05	68.05	118.05	168.05	268.05	368.05	468.05	568.05	668.05	768.05	868.05	968.05
	v	0.02123	0.5009		0.5263	0.5869	0.6388	0.6858	0.7713	0.8504	0.9262	0.9998	1.0720	1.1430	1.2131	1.2825
	h	526.70	1196.4		1215.5	1260.6	1298.6	1332.7	1394.4	1452.2	1508.5	1564.4	1620.6	1677.1	1734.1	1791.6
	s	0.7279	1.4032		1.4223	1.4659	1.5010	1.5311	1.5822	1.6263	1.6662	1.7033	1.7382	1.7713	1.8028	1.8329

Sh = superheat, F
v = specific volume, cu ft per lb
h = enthalpy, Btu per lb
s = entropy, Btu per F per lb

TABLE 46.3 Superheated Steam (1 to 15,500 lb per Square Inch Absolute Pressure) (Continued)

Abs Press. Lb/Sq In. (Sat. Temp)		Sat. Water	Sat. Steam	Temperature – Degrees Fahrenheit 550	600	650	700	750	800	850	900	1000	1100	1200	1300	1400	1500
950 (538.39)	Sh			11.61	61.61	111.61	161.61	211.61	261.61	311.61	361.61	461.61	561.61	661.61	761.61	861.61	961.61
	v	0.02141	0.4721	0.4883	0.5485	0.5993	0.6449	0.6871	0.7272	0.7656	0.8030	0.8753	0.9455	1.0142	1.0817	1.1484	1.2143
	h	534.74	1194.7	1207.6	1255.1	1294.4	1329.3	1361.5	1392.0	1421.5	1450.3	1507.0	1563.2	1619.5	1676.2	1733.3	1791.0
	s	0.7358	1.3970	1.4098	1.4557	1.4921	1.5228	1.5500	1.5748	1.5977	1.6193	1.6595	1.6967	1.7317	1.7649	1.7965	1.8267
1000 (544.58)	Sh			5.42	55.42	105.42	155.42	205.42	255.42	305.42	355.42	455.42	555.42	655.42	755.42	855.42	955.42
	v	0.02159	0.4460	0.4535	0.5137	0.5636	0.6080	0.6489	0.6875	0.7245	0.7603	0.8295	0.8966	0.9622	1.0266	1.0901	1.1529
	h	542.55	1192.9	1199.3	1249.3	1290.1	1325.9	1358.7	1389.6	1419.4	1448.5	1505.4	1561.9	1618.4	1675.3	1732.5	1790.3
	s	0.7434	1.3910	1.3973	1.4457	1.4833	1.5149	1.5426	1.5677	1.5908	1.6126	1.6530	1.6905	1.7256	1.7589	1.7905	1.8207
1050 (550.53)	Sh				49.47	99.47	149.47	199.47	249.47	299.47	349.47	449.47	549.47	649.47	749.47	849.47	949.47
	v	0.02177	0.4222		0.4821	0.5312	0.5745	0.6142	0.6515	0.6872	0.7216	0.7881	0.8524	0.9151	0.9767	1.0373	1.0973
	h	550.15	1191.0		1243.4	1285.7	1322.4	1355.8	1387.2	1417.3	1446.6	1503.9	1560.7	1617.4	1674.4	1731.8	1789.6
	s	0.7507	1.3851		1.4358	1.4748	1.5072	1.5354	1.5608	1.5842	1.6062	1.6469	1.6845	1.7197	1.7531	1.7848	1.8151
1100 (556.28)	Sh				43.72	93.72	143.72	193.72	243.72	293.72	343.72	443.72	543.72	643.72	743.72	843.72	943.72
	v	0.02195	0.4006		0.4531	0.5017	0.5440	0.5826	0.6188	0.6533	0.6865	0.7505	0.8121	0.8723	0.9313	0.9894	1.0468
	h	557.55	1189.1		1237.3	1281.2	1318.8	1352.9	1384.7	1415.2	1444.7	1502.4	1559.4	1616.3	1673.5	1731.0	1789.0
	s	0.7578	1.3794		1.4259	1.4664	1.4996	1.5284	1.5542	1.5779	1.6000	1.6410	1.6787	1.7141	1.7475	1.7793	1.8097
1150 (561.82)	Sh				39.18	89.18	139.18	189.18	239.18	289.18	339.18	439.18	539.18	639.18	739.18	839.18	939.18
	v	0.02214	0.3807		0.4263	0.4746	0.5162	0.5538	0.5889	0.6223	0.6544	0.7161	0.7754	0.8332	0.8899	0.9456	1.0007
	h	564.78	1187.0		1230.9	1276.6	1315.2	1349.9	1382.2	1413.0	1442.8	1500.9	1558.1	1615.2	1672.6	1730.2	1788.3
	s	0.7647	1.3738		1.4160	1.4582	1.4923	1.5216	1.5478	1.5717	1.5941	1.6353	1.6732	1.7087	1.7422	1.7741	1.8045
1200 (567.19)	Sh				32.81	82.81	132.81	182.81	232.81	282.81	332.81	432.81	532.81	632.81	732.81	832.81	932.81
	v	0.02232	0.3624		0.4016	0.4497	0.4905	0.5273	0.5615	0.5939	0.6250	0.6845	0.7418	0.7974	0.8519	0.9055	0.9584
	h	571.85	1184.8		1224.2	1271.8	1311.5	1346.9	1379.7	1410.8	1440.9	1499.4	1556.9	1614.2	1671.6	1729.4	1787.6
	s	0.7714	1.3683		1.4061	1.4501	1.4851	1.5150	1.5415	1.5658	1.5883	1.6298	1.6679	1.7035	1.7371	1.7691	1.7996
1300 (577.42)	Sh				22.58	72.58	122.58	172.58	222.58	272.58	322.58	422.58	522.58	622.58	722.58	822.58	922.58
	v	0.02269	0.3299		0.3570	0.4052	0.4451	0.4804	0.5129	0.5436	0.5729	0.6287	0.6822	0.7341	0.7847	0.8345	0.8836
	h	585.58	1180.2		1209.9	1261.9	1303.9	1340.8	1374.6	1406.4	1437.1	1496.3	1554.3	1612.0	1669.8	1727.9	1786.3
	s	0.7843	1.3577		1.3860	1.4340	1.4711	1.5022	1.5296	1.5544	1.5773	1.6194	1.6578	1.6937	1.7275	1.7596	1.7902
1400 (587.07)	Sh				12.93	62.93	112.93	162.93	212.93	262.93	312.93	412.93	512.93	612.93	712.93	812.93	912.93
	v	0.02307	0.3018		0.3176	0.3667	0.4059	0.4400	0.4712	0.5004	0.5282	0.5809	0.6311	0.6798	0.7272	0.7737	0.8195
	h	598.83	1175.3		1194.1	1251.4	1296.1	1334.5	1369.3	1402.0	1433.2	1493.2	1551.8	1609.9	1668.0	1726.3	1785.0
	s	0.7966	1.3474		1.3652	1.4181	1.4575	1.4900	1.5182	1.5436	1.5670	1.6096	1.6484	1.6845	1.7185	1.7508	1.7815

1500	Sh			3.80	53.80	103.80	153.80	203.80	253.80	303.80	403.80	503.80	603.80	703.80	803.80	903.80
(596.20)	v	0.02346	0.2772	0.2820	0.3328	0.3717	0.4049	0.4350	0.4629	0.4894	0.5394	0.5869	0.6327	0.6773	0.7210	0.7639
	h	611.68	1170.1	1176.3	1240.2	1287.9	1328.0	1364.0	1397.4	1429.2	1490.1	1549.2	1607.7	1666.2	1724.8	1783.7
	s	0.8085	1.3373	1.3431	1.4022	1.4443	1.4782	1.5073	1.5333	1.5572	1.6004	1.6395	1.6759	1.7101	1.7425	1.7734
1600	Sh				45.13	95.13	145.13	195.13	245.13	295.13	395.13	495.13	595.13	695.13	795.13	895.13
(604.87)	v	0.02387	0.2555		0.3026	0.3415	0.3741	0.4032	0.4301	0.4555	0.5031	0.5482	0.5915	0.6336	0.6748	0.7153
	h	624.20	1164.5		1228.3	1279.4	1321.4	1358.5	1392.8	1425.2	1486.9	1546.6	1605.6	1664.3	1723.2	1782.3
	s	0.8199	1.3274		1.3861	1.4312	1.4667	1.4968	1.5235	1.5478	1.5916	1.6312	1.6678	1.7022	1.7347	1.7657
1700	Sh				36.87	86.87	136.87	186.87	236.87	286.87	386.87	486.87	586.87	686.87	786.87	886.87
(613.13)	v	0.02428	0.2361		0.2754	0.3147	0.3468	0.3751	0.4011	0.4255	0.4711	0.5140	0.5552	0.5951	0.6341	0.6724
	h	636.45	1158.6		1215.3	1270.5	1314.5	1352.9	1388.1	1421.2	1483.8	1544.0	1603.4	1662.5	1721.7	1781.0
	s	0.8309	1.3176		1.3697	1.4183	1.4555	1.4867	1.5140	1.5388	1.5833	1.6232	1.6601	1.6947	1.7274	1.7585
1800	Sh				28.98	78.98	128.98	178.98	228.98	278.98	378.98	478.98	578.98	678.98	778.98	878.98
(621.02)	v	0.02472	0.2186		0.2505	0.2906	0.3223	0.3500	0.3752	0.3988	0.4426	0.4836	0.5229	0.5609	0.5980	0.6343
	h	648.49	1152.3		1201.2	1261.1	1307.4	1347.2	1383.3	1417.1	1480.6	1541.4	1601.2	1660.7	1720.1	1779.7
	s	0.8417	1.3079		1.3526	1.4054	1.4446	1.4768	1.5049	1.5302	1.5753	1.6156	1.6528	1.6876	1.7204	1.7516
1900	Sh				21.44	71.44	121.44	171.44	221.44	271.44	371.44	471.44	571.44	671.44	771.44	871.44
(628.56)	v	0.02517	0.2028		0.2274	0.2687	0.3004	0.3275	0.3521	0.3749	0.4171	0.4565	0.4940	0.5303	0.5656	0.6002
	h	660.36	1145.6		1185.7	1251.3	1300.2	1341.4	1378.4	1412.9	1477.4	1538.8	1599.1	1658.8	1718.6	1778.4
	s	0.8522	1.2981		1.3346	1.3925	1.4338	1.4672	1.4960	1.5219	1.5677	1.6084	1.6458	1.6808	1.7138	1.7451
2000	Sh				14.20	64.20	114.20	164.20	214.20	264.20	364.20	464.20	564.20	664.20	764.20	864.20
(635.80)	v	0.02565	0.1883		0.2056	0.2488	0.2805	0.3072	0.3312	0.3534	0.3942	0.4320	0.4680	0.5027	0.5365	0.5695
	h	672.11	1138.3		1168.3	1240.9	1292.6	1335.4	1373.5	1408.7	1474.1	1536.2	1596.9	1657.0	1717.0	1771.1
	s	0.8625	1.2881		1.3154	1.3794	1.4231	1.4578	1.4874	1.5138	1.5603	1.6014	1.6391	1.6743	1.7075	1.7389
2100	Sh				7.24	57.24	107.24	157.24	207.24	257.24	357.24	457.24	557.24	657.24	757.24	857.24
(642.76)	v	0.02615	0.1750		0.1847	0.2304	0.2624	0.2888	0.3123	0.3339	0.3734	0.4099	0.4445	0.4778	0.5101	0.5418
	h	683.79	1130.5		1148.5	1229.8	1284.9	1329.3	1368.4	1404.4	1470.9	1533.6	1594.7	1655.2	1715.4	1775.7
	s	0.8727	1.2780		1.2942	1.3661	1.4125	1.4486	1.4790	1.5060	1.5532	1.5948	1.6327	1.6681	1.7014	1.7330
2200	Sh				.55	50.55	100.55	150.55	200.55	250.55	350.55	450.55	550.55	650.55	750.55	850.55
(649.45)	v	0.02669	0.1627		0.1636	0.2134	0.2458	0.2720	0.2950	0.3161	0.3545	0.3897	0.4231	0.4551	0.4862	0.5165
	h	695.46	1122.2		1123.9	1218.0	1276.8	1323.1	1363.3	1400.0	1467.6	1530.9	1592.5	1653.3	1713.9	1774.4
	s	0.8828	1.2676		1.2691	1.3523	1.4020	1.4395	1.4708	1.4984	1.5463	1.5883	1.6266	1.6622	1.6956	1.7273
2300	Sh					44.11	94.11	144.11	194.11	244.11	344.11	444.11	544.11	644.11	744.11	844.11
(655.89)	v	0.02727	0.1513			0.1975	0.2305	0.2566	0.2793	0.2999	0.3372	0.3714	0.4035	0.4344	0.4643	0.4935
	h	707.18	1113.2			1205.3	1268.4	1316.7	1358.1	1395.7	1464.2	1528.3	1590.3	1651.5	1712.3	1773.1
	s	0.8929	1.2569			1.3381	1.3914	1.4305	1.4628	1.4910	1.5397	1.5821	1.6207	1.6565	1.6901	1.7219

Sh = superheat, F
v = specific volume, cu ft per lb
h = enthalpy, Btu per lb
s = entropy, Btu per F per lb

TABLE 46.3 Superheated Steam (1 to 15,500 lb per Square Inch Absolute Pressure) (Continued)

Abs Press. Lb/Sq In. (Sat. Temp)		Sat Water	Sat Steam	Temperature — Degrees Fahrenheit 700	750	800	850	900	950	1000	1050	1100	1150	1200	1300	1400	1500
	Sh			37.89	87.89	137.89	187.89	237.89	287.89	337.89	387.89	437.89	487.89	537.89	637.89	737.89	837.89
2400	v	0.02790	0.1408	0.1824	0.2164	0.2424	0.2648	0.2850	0.3037	0.3214	0.3382	0.3545	0.3703	0.3856	0.4155	0.4443	0.4724
(662.11)	h	718.95	1103.7	1191.6	1259.7	1310.1	1352.8	1391.2	1426.9	1460.9	1493.7	1525.6	1557.0	1588.1	1649.6	1710.8	1771.8
	s	0.9031	1.2460	1.3232	1.3808	1.4217	1.4549	1.4837	1.5095	1.5332	1.5553	1.5761	1.5959	1.6149	1.6509	1.6847	1.7167
	Sh			31.89	81.89	131.89	181.89	231.89	281.89	331.89	381.89	431.89	481.89	531.89	631.89	731.89	831.89
2500	v	0.02859	0.1307	0.1681	0.2032	0.2293	0.2514	0.2712	0.2896	0.3068	0.3232	0.3390	0.3543	0.3692	0.3980	0.4259	0.4529
(668.11)	h	731.71	1093.3	1176.7	1250.6	1303.4	1347.4	1386.7	1423.1	1457.5	1490.7	1522.9	1554.6	1585.9	1647.8	1709.2	1770.4
	s	0.9139	1.2345	1.3076	1.3701	1.4129	1.4472	1.4766	1.5029	1.5269	1.5492	1.5703	1.5903	1.6094	1.6456	1.6796	1.7116
	Sh			26.09	76.09	126.09	176.09	226.09	276.09	326.09	376.09	426.09	476.09	526.09	626.09	726.09	826.09
2600	v	0.02938	0.1211	0.1544	0.1909	0.2171	0.2390	0.2585	0.2765	0.2933	0.3093	0.3247	0.3395	0.3540	0.3819	0.4088	0.4350
(673.91)	h	744.47	1082.0	1160.2	1241.1	1296.5	1341.9	1382.1	1419.2	1454.1	1487.7	1520.2	1552.2	1583.7	1646.0	1707.7	1769.1
	s	0.9247	1.2225	1.2908	1.3592	1.4042	1.4395	1.4696	1.4964	1.5208	1.5434	1.5646	1.5848	1.6040	1.6405	1.6746	1.7068
	Sh			20.47	70.47	120.47	170.47	220.47	270.47	320.47	370.47	420.47	470.47	520.47	620.47	720.47	820.47
2700	v	0.03029	0.1119	0.1411	0.1794	0.2058	0.2275	0.2468	0.2644	0.2809	0.2965	0.3114	0.3259	0.3399	0.3670	0.3931	0.4184
(679.53)	h	757.34	1069.7	1142.0	1231.1	1289.5	1336.3	1377.5	1415.2	1450.7	1484.6	1517.5	1549.8	1581.5	1644.1	1706.1	1767.8
	s	0.9356	1.2097	1.2727	1.3481	1.3954	1.4319	1.4628	1.4900	1.5148	1.5376	1.5591	1.5794	1.5988	1.6355	1.6697	1.7021
	Sh			15.04	65.04	115.04	165.04	215.04	265.04	315.04	365.04	415.04	465.04	515.04	615.04	715.04	815.04
2800	v	0.03134	0.1030	0.1278	0.1685	0.1952	0.2168	0.2358	0.2531	0.2693	0.2845	0.2991	0.3132	0.3268	0.3532	0.3785	0.4030
(684.96)	h	770.69	1055.8	1121.2	1220.6	1282.2	1330.7	1372.8	1411.2	1447.2	1481.6	1514.8	1547.3	1579.3	1642.2	1704.5	1766.5
	s	0.9468	1.1958	1.2527	1.3368	1.3867	1.4245	1.4561	1.4838	1.5089	1.5321	1.5537	1.5742	1.5938	1.6306	1.6651	1.6975
	Sh			9.78	59.78	109.78	159.78	209.78	259.78	309.78	359.78	409.78	459.78	509.78	609.78	709.78	809.78
2900	v	0.03262	0.0942	0.1138	0.1581	0.1853	0.2068	0.2256	0.2427	0.2585	0.2734	0.2877	0.3014	0.3147	0.3403	0.3649	0.3887
(690.22)	h	785.13	1039.8	1095.3	1209.6	1274.7	1324.9	1368.0	1407.2	1443.7	1478.5	1512.1	1544.9	1577.0	1640.4	1703.0	1765.2
	s	0.9588	1.1803	1.2283	1.3251	1.3780	1.4171	1.4494	1.4777	1.5032	1.5266	1.5485	1.5692	1.5889	1.6259	1.6605	1.6931
	Sh			4.67	54.67	104.67	154.67	204.67	254.67	304.67	354.67	404.67	454.67	504.67	604.67	704.67	804.67
3000	v	0.03428	0.0850	0.0982	0.1483	0.1759	0.1975	0.2161	0.2329	0.2484	0.2630	0.2770	0.2904	0.3033	0.3282	0.3522	0.3753
(695.33)	h	801.84	1020.3	1060.5	1197.9	1267.0	1319.0	1363.2	1403.1	1440.2	1475.4	1509.4	1542.4	1574.8	1638.5	1701.4	1763.8
	s	0.9728	1.1619	1.1966	1.3131	1.3692	1.4097	1.4429	1.4717	1.4976	1.5213	1.5434	1.5642	1.5841	1.6214	1.6561	1.6888
	Sh				49.72	99.72	149.72	199.72	249.72	299.72	349.72	399.72	449.72	499.72	599.72	699.72	799.72
3100	v	0.03681	0.0745		0.1389	0.1671	0.1887	0.2071	0.2237	0.2390	0.2533	0.2670	0.2800	0.2927	0.3170	0.3403	0.3628
(700.28)	h	823.97	993.3		1185.4	1259.1	1313.0	1358.4	1399.0	1436.7	1472.3	1506.6	1539.9	1572.6	1636.7	1699.8	1762.5
	s	0.9914	1.1373		1.3007	1.3604	1.4024	1.4364	1.4658	1.4920	1.5161	1.5384	1.5594	1.5794	1.6169	1.6518	1.6847

3200	Sh			44.92	94.92	144.92	194.92	244.92	294.92	344.92	394.92	444.92	494.92	594.92	694.92	794.92
(705.08)	v	0.04472	0.0566	0.1300	0.1588	0.1804	0.1987	0.2151	0.2301	0.2442	0.2576	0.2704	0.2827	0.3065	0.3291	0.3510
	h	875.54	931.6	1172.3	1250.9	1306.9	1353.4	1394.9	1433.1	1469.2	1503.8	1537.4	1570.3	1634.8	1698.3	1761.2
	s	1.0351	1.0832	1.2877	1.3515	1.3951	1.4300	1.4600	1.4866	1.5110	1.5335	1.5547	1.5749	1.6126	1.6477	1.6806
	Sh															
3300	v			0.1213	0.1510	0.1727	0.1908	0.2070	0.2218	0.2357	0.2488	0.2613	0.2734	0.2966	0.3187	0.3400
	h			1158.2	1242.5	1300.7	1348.4	1390.7	1429.5	1466.1	1501.0	1534.9	1568.1	1632.9	1696.7	1759.9
	s			1.2742	1.3425	1.3879	1.4237	1.4542	1.4813	1.5059	1.5287	1.5501	1.5704	1.6084	1.6436	1.6767
	Sh															
3400	v			0.1129	0.1435	0.1653	0.1834	0.1994	0.2140	0.2276	0.2405	0.2528	0.2646	0.2872	0.3088	0.3296
	h			1143.2	1233.7	1294.3	1343.4	1386.4	1425.9	1462.9	1498.3	1532.4	1565.8	1631.1	1695.1	1758.5
	s			1.2600	1.3334	1.3807	1.4174	1.4486	1.4761	1.5010	1.5240	1.5456	1.5660	1.6042	1.6396	1.6728
	Sh															
3500	v			0.1048	0.1364	0.1583	0.1764	0.1922	0.2066	0.2200	0.2326	0.2447	0.2563	0.2784	0.2995	0.3198
	h			1127.1	1224.6	1287.8	1338.2	1382.2	1422.2	1459.7	1495.5	1529.9	1563.6	1629.2	1693.6	1757.2
	s			1.2450	1.3242	1.3734	1.4112	1.4430	1.4/09	1.4962	1.5194	1.5412	1.5618	1.6002	1.6358	1.6691
	Sh															
3600	v			0.0966	0.1296	0.1517	0.1697	0.1854	0.1996	0.2128	0.2252	0.2371	0.2485	0.2702	0.2908	0.3106
	h			1108.6	1215.3	1281.2	1333.0	1377.9	1418.6	1456.5	1492.6	1527.4	1561.3	1627.3	1692.0	1755.9
	s			1.2281	1.3148	1.3662	1.4050	1.4374	1.4658	1.4914	1.5149	1.5369	1.5576	1.5962	1.6320	1.6654
	Sh															
3800	v			0.0799	0.1169	0.1395	0.1574	0.1729	0.1868	0.1996	0.2116	0.2231	0.2340	0.2549	0.2746	0.2936
	h			1064.2	1195.5	1267.6	1322.4	1369.1	1411.2	1450.1	1487.0	1522.4	1556.8	1623.6	1688.9	1753.2
	s			1.1888	1.2955	1.3517	1.3928	1.4265	1.4558	1.4821	1.5061	1.5284	1.5495	1.5886	1.6247	1.6584
	Sh															
4000	v			0.0631	0.1052	0.1284	0.1463	0.1616	0.1752	0.1877	0.1994	0.2105	0.2210	0.2411	0.2601	0.2783
	h			1007.4	1174.3	1253.4	1311.6	1360.2	1403.6	1443.6	1481.3	1517.3	1552.2	1619.8	1685.7	1750.6
	s			1.1396	1.2754	1.3371	1.3807	1.4158	1.4461	1.4730	1.4976	1.5203	1.5417	1.5812	1.6177	1.6516
	Sh															
4200	v			0.0498	0.0945	0.1183	0.1362	0.1513	0.1647	0.1769	0.1883	0.1991	0.2093	0.2287	0.2470	0.2645
	h			950.1	1151.6	1238.6	1300.4	1351.2	1396.0	1437.1	1475.5	1512.2	1547.6	1616.1	1682.6	1748.0
	s			1.0905	1.2544	1.3223	1.3686	1.4053	1.4366	1.4642	1.4893	1.5124	1.5341	1.5742	1.6109	1.6452
	Sh															
4400	v			0.0421	0.0846	0.1090	0.1270	0.1420	0.1552	0.1671	0.1782	0.1887	0.1986	0.2174	0.2351	0.2519
	h			909.5	1127.3	1223.3	1289.0	1342.0	1388.3	1430.4	1469.7	1507.1	1543.0	1612.3	1679.4	1745.3
	s			1.0556	1.2325	1.3073	1.3566	1.3949	1.4272	1.4556	1.4812	1.5048	1.5268	1.5673	1.6044	1.6389

Sh = superheat, F
v = specific volume, cu ft per lb

h = enthalpy, Btu per lb
s = entropy, Btu per F per lb

TABLE 46.3 Superheated Steam (1 to 15,500 lb per Square Inch Absolute Pressure) (Continued)

Abs Press. Lb/Sq In. (Sat. Temp)		Sat. Water	Sat. Steam	Temperature — Degrees Fahrenheit 750	800	850	900	950	1000	1050	1100	1150	1200	1250	1300	1400	1500
	Sh																
4600	v			0.0380	0.0751	0.1005	0.1186	0.1335	0.1465	0.1582	0.1691	0.1792	0.1889	0.1982	0.2071	0.2242	0.2404
	h			883.8	1100.0	1207.3	1277.2	1332.6	1380.5	1423.7	1463.9	1501.9	1538.4	1573.8	1608.5	1676.3	1742.7
	s			1.0331	1.2084	1.2922	1.3446	1.3847	1.4181	1.4472	1.4734	1.4974	1.5197	1.5407	1.5607	1.5982	1.6330
	Sh																
4800	v			0.0355	0.0665	0.0927	0.1109	0.1257	0.1385	0.1500	0.1606	0.1706	0.1800	0.1890	0.1977	0.2142	0.2299
	h			866.9	1071.2	1190.7	1265.2	1323.1	1372.6	1417.0	1458.0	1496.7	1533.8	1569.7	1604.7	1673.1	1740.0
	s			1.0180	1.1835	1.2768	1.3327	1.3745	1.4090	1.4390	1.4657	1.4901	1.5128	1.5341	1.5543	1.5921	1.6272
	Sh																
5000	v			0.0338	0.0591	0.0855	0.1038	0.1185	0.1312	0.1425	0.1529	0.1626	0.1718	0.1806	0.1890	0.2050	0.2203
	h			854.9	1042.9	1173.6	1252.9	1313.5	1364.6	1410.2	1452.1	1491.5	1529.1	1565.5	1600.9	1670.0	1737.4
	s			1.0070	1.1593	1.2612	1.3207	1.3645	1.4001	1.4309	1.4582	1.4831	1.5061	1.5277	1.5481	1.5863	1.6216
	Sh																
5200	v			0.0326	0.0531	0.0789	0.0973	0.1119	0.1244	0.1356	0.1458	0.1553	0.1642	0.1728	0.1810	0.1966	0.2114
	h			845.8	1016.9	1156.0	1240.4	1303.7	1356.6	1403.4	1446.2	1486.3	1524.5	1561.3	1597.2	1666.8	1734.7
	s			0.9985	1.1370	1.2455	1.3088	1.3545	1.3914	1.4229	1.4509	1.4762	1.4995	1.5214	1.5420	1.5806	1.6161
	Sh																
5400	v			0.0317	0.0483	0.0728	0.0912	0.1058	0.1182	0.1292	0.1392	0.1485	0.1572	0.1656	0.1736	0.1888	0.2031
	h			838.5	994.3	1138.1	1227.7	1293.7	1348.4	1396.5	1440.3	1481.1	1519.8	1557.1	1593.4	1663.7	1732.1
	s			0.9915	1.1175	1.2296	1.2969	1.3446	1.3827	1.4151	1.4437	1.4694	1.4931	1.5153	1.5362	1.5750	1.6109
	Sh																
5600	v			0.0309	0.0447	0.0672	0.0856	0.1001	0.1124	0.1232	0.1331	0.1422	0.1508	0.1589	0.1667	0.1815	0.1954
	h			832.4	975.0	1119.9	1214.8	1283.7	1340.2	1389.6	1434.3	1475.9	1515.2	1552.9	1589.6	1660.5	1729.5
	s			0.9855	1.1008	1.2137	1.2850	1.3348	1.3742	1.4075	1.4366	1.4628	1.4869	1.5093	1.5304	1.5697	1.6058
	Sh																
5800	v			0.0303	0.0419	0.0622	0.0805	0.0949	0.1070	0.1177	0.1274	0.1363	0.1447	0.1527	0.1603	0.1747	0.1883
	h			827.3	958.8	1101.8	1201.8	1273.6	1332.0	1382.6	1428.3	1470.6	1510.5	1548.7	1585.8	1657.4	1726.8
	s			0.9803	1.0867	1.1981	1.2732	1.3250	1.3658	1.3999	1.4297	1.4564	1.4808	1.5035	1.5248	1.5644	1.6008
	Sh																
6000	v			0.0298	0.0397	0.0579	0.0757	0.0900	0.1020	0.1126	0.1221	0.1309	0.1391	0.1469	0.1544	0.1684	0.1817
	h			822.9	945.1	1084.6	1188.8	1263.4	1323.6	1375.7	1422.3	1465.4	1505.9	1544.6	1582.0	1654.2	1724.2
	s			0.9758	1.0746	1.1833	1.2615	1.3154	1.3574	1.3925	1.4229	1.4500	1.4748	1.4978	1.5194	1.5593	1.5960

Pressure															
6500	Sh														
	v	0.0287	0.0358	0.0495	0.0655	0.0793	0.0909	0.1012	0.1104	0.1188	0.1266	0.1340	0.1411	0.1544	0.1669
	h	813.9	919.5	1046.7	1156.3	1237.8	1302.7	1358.1	1407.3	1452.2	1494.2	1534.1	1572.5	1646.4	1717.6
	s	0.9661	1.0515	1.1506	1.2328	1.2917	1.3370	1.3743	1.4064	1.4347	1.4604	1.4841	1.5062	1.5471	1.5844
7000	Sh														
	v	0.0279	0.0334	0.0438	0.0573	0.0704	0.0816	0.0915	0.1004	0.1085	0.1160	0.1231	0.1298	0.1424	0.1542
	h	806.9	901.8	1016.5	1124.9	1212.6	1281.7	1340.5	1392.2	1439.1	1482.6	1523.7	1563.1	1638.6	1711.1
	s	0.9582	1.0350	1.1243	1.2055	1.2689	1.3171	1.3567	1.3904	1.4200	1.4466	1.4710	1.4938	1.5355	1.5735
7500	Sh														
	v	0.0272	0.0318	0.0399	0.0512	0.0631	0.0737	0.0833	0.0918	0.0996	0.1068	0.1136	0.1200	0.1321	0.1433
	h	801.3	889.0	992.9	1097.7	1188.3	1261.0	1322.9	1377.2	1426.0	1471.0	1513.3	1553.7	1630.8	1704.6
	s	0.9514	1.0224	1.1033	1.1818	1.2473	1.2980	1.3397	1.3751	1.4059	1.4335	1.4586	1.4819	1.5245	1.5632
8000	Sh														
	v	0.0267	0.0306	0.0371	0.0465	0.0571	0.0671	0.0762	0.0845	0.0920	0.0989	0.1054	0.1115	0.1230	0.1338
	h	796.6	879.1	974.4	1074.3	1165.4	1241.0	1305.5	1362.2	1413.0	1459.6	1503.1	1544.5	1623.1	1698.1
	s	0.9455	1.0122	1.0864	1.1613	1.2271	1.2798	1.3233	1.3603	1.3924	1.4208	1.4467	1.4705	1.5140	1.5533
8500	Sh														
	v	0.0262	0.0296	0.0350	0.0429	0.0522	0.0615	0.0701	0.0780	0.0853	0.0919	0.0982	0.1041	0.1151	0.1254
	h	792.7	871.2	959.8	1054.5	1144.0	1221.9	1288.5	1347.5	1400.2	1448.2	1492.9	1535.3	1615.4	1691.7
	s	0.9402	1.0037	1.0727	1.1437	1.2084	1.2627	1.3076	1.3460	1.3793	1.4087	1.4352	1.4597	1.5040	1.5439
9000	Sh														
	v	0.0258	0.0288	0.0335	0.0402	0.0483	0.0568	0.0649	0.0724	0.0794	0.0858	0.0918	0.0975	0.1081	0.1179
	h	789.3	864.7	948.0	1037.6	1125.4	1204.1	1272.1	1333.0	1387.5	1437.1	1482.9	1526.3	1607.9	1685.3
	s	0.9354	0.9964	1.0613	1.1285	1.1918	1.2468	1.2926	1.3323	1.3667	1.3970	1.4243	1.4492	1.4944	1.5349
9500	Sh														
	v	0.0254	0.0282	0.0322	0.0380	0.0451	0.0528	0.0603	0.0675	0.0742	0.0804	0.0862	0.0917	0.1019	0.1113
	h	786.4	859.2	938.3	1023.4	1108.9	1187.7	1256.6	1318.9	1375.1	1426.1	1473.1	1517.3	1600.4	1679.0
	s	0.9310	0.9900	1.0516	1.1153	1.1771	[illegible]320	1.2785	1.3191	1.3546	1.3858	1.4137	1.4392	1.4851	1.5263
10000	Sh														
	v	0.0251	0.0276	0.0312	0.0362	0.0425	0.0495	0.0565	0.0633	0.0697	0.0757	0.0812	0.0865	0.0963	0.1054
	h	783.8	854.5	930.2	1011.3	1094.2	1172.6	1242.0	1305.3	1362.9	1415.3	1463.4	1508.6	1593.1	1672.8
	s	0.9270	0.9842	1.0432	1.1039	1.1638	1.2185	1.2652	1.3065	1.3429	1.3749	1.4035	1.4295	1.4763	1.5180
10500	Sh														
	v	0.0248	0.0271	0.0303	0.0347	0.0404	0.0467	0.0532	0.0595	0.0656	0.0714	0.0768	0.0818	0.0913	0.1001
	h	781.5	850.5	923.4	1001.0	1081.3	1158.9	1228.4	1292.4	1351.1	1404.7	1453.9	1500.0	1585.8	1666.7
	s	0.9232	0.9790	1.0358	1.0939	1.1519	1.2060	1.2529	1.2946	1.3371	1.3644	1.3937	1.4202	1.4677	1.5100

Sh = superheat, F
v = specific volume, cu ft per lb
h = enthalpy, Btu per lb
s = entropy, Btu per F per lb

TABLE 46.3 Superheated Steam (1 to 15,500 lb per Square Inch Absolute Pressure) (Continued)

Abs Press. Lb/Sq In. (Sat. Temp)		Sat. Water	Sat. Steam	Temperature – Degrees Fahrenheit 750	800	850	900	950	1000	1050	1100	1150	1200	1250	1300	1400	1500
11000	v			0.0245	0.0267	0.0296	0.0335	0.0386	0.0443	0.0503	0.0562	0.0620	0.0676	0.0727	0.0776	0.0868	0.0952
	h			779.5	846.9	917.5	992.1	1069.9	1146.3	1215.9	1280.2	1339.7	1394.4	1444.6	1491.5	1578.7	1660.6
	s			0.9196	0.9742	1.0292	1.0851	1.1412	1.1945	1.2414	1.2833	1.3209	1.3544	1.3842	1.4112	1.4595	1.5023
11500	v			0.0243	0.0263	0.0290	0.0325	0.0370	0.0423	0.0478	0.0534	0.0588	0.0641	0.0691	0.0739	0.0827	0.0909
	h			777.7	843.8	912.4	984.5	1059.8	1134.9	1204.3	1268.7	1328.8	1384.4	1435.5	1483.2	1571.8	1654.7
	s			0.9163	0.9698	1.0232	1.0772	1.1316	1.1840	1.2308	1.2727	1.3107	1.3446	1.3750	1.4025	1.4515	1.4949
12000	v			0.0241	0.0260	0.0284	0.0317	0.0357	0.0405	0.0456	0.0508	0.0560	0.0610	0.0659	0.0704	0.0790	0.0869
	h			776.1	841.0	907.9	977.8	1050.9	1124.5	1193.7	1258.0	1318.5	1374.7	1426.6	1475.1	1564.9	1648.8
	s			0.9131	0.9657	1.0177	1.0701	1.1229	1.1742	1.2209	1.2627	1.3010	1.3353	1.3662	1.3941	1.4438	1.4877
12500	v			0.0238	0.0256	0.0279	0.0309	0.0346	0.0390	0.0437	0.0486	0.0535	0.0583	0.0629	0.0673	0.0756	0.0832
	h			774.7	838.6	903.9	971.9	1043.1	1115.2	1184.1	1247.9	1308.8	1365.4	1418.0	1467.2	1558.2	1643.1
	s			0.9101	0.9618	1.0127	1.0637	1.1151	1.1653	1.2117	1.2534	1.2918	1.3264	1.3576	1.3860	1.4363	1.4808
13000	v			0.0236	0.0253	0.0275	0.0302	0.0336	0.0376	0.0420	0.0466	0.0512	0.0558	0.0602	0.0645	0.0725	0.0799
	h			773.5	836.3	900.4	966.8	1036.2	1106.7	1174.8	1238.5	1299.6	1356.5	1409.6	1459.4	1551.6	1637.4
	s			0.9073	0.9582	1.0080	1.0578	1.1079	1.1571	1.2030	1.2445	1.2831	1.3179	1.3494	1.3781	1.4291	1.4741
13500	v			0.0235	0.0251	0.0271	0.0297	0.0328	0.0364	0.0405	0.0448	0.0492	0.0535	0.0577	0.0619	0.0696	0.0768
	h			772.3	834.4	897.2	962.2	1030.0	1099.1	1166.3	1229.7	1291.0	1348.1	1401.5	1451.8	1545.2	1631.9
	s			0.9045	0.9548	1.0037	1.0524	1.1014	1.1495	1.1948	1.2361	1.2749	1.3098	1.3415	1.3705	1.4221	1.4675
14000	v			0.0233	0.0248	0.0267	0.0291	0.0320	0.0354	0.0392	0.0432	0.0474	0.0515	0.0555	0.0595	0.0670	0.0740
	h			771.3	832.6	894.3	958.0	1024.5	1092.3	1158.5	1221.4	1283.0	1340.2	1393.8	1444.4	1538.8	1626.5
	s			0.9019	0.9515	0.9996	1.0473	1.0953	1.1426	1.1872	1.2282	1.2671	1.3021	1.3339	1.3631	1.4153	1.4612
14500	v			0.0231	0.0246	0.0264	0.0287	0.0314	0.0345	0.0380	0.0418	0.0458	0.0496	0.0534	0.0573	0.0646	0.0714
	h			770.4	831.0	891.7	954.3	1019.6	1086.2	1151.4	1213.8	1275.4	1332.9	1386.4	1437.3	1532.6	1621.1
	s			0.8994	0.9484	0.9957	1.0426	1.0897	1.1362	1.1801	1.2208	1.2597	1.2949	1.3266	1.3560	1.4087	1.4551
15000	v			0.0230	0.0244	0.0261	0.0282	0.0308	0.0337	0.0369	0.0405	0.0443	0.0479	0.0516	0.0552	0.0624	0.0690
	h			769.6	829.5	889.3	950.9	1015.1	1080.6	1144.9	1206.8	1268.1	1326.0	1379.4	1430.3	1526.4	1615.9
	s			0.8970	0.9455	0.9920	1.0382	1.0846	1.1302	1.1735	1.2139	1.2525	1.2880	1.3197	1.3491	1.4022	1.4491
15500	v			0.0228	0.0242	0.0258	0.0278	0.0302	0.0329	0.0360	0.0393	0.0429	0.0464	0.0499	0.0534	0.0603	0.0668
	h			768.9	828.2	887.2	947.8	1011.1	1075.7	1139.0	1200.3	1261.1	1319.6	1372.8	1423.6	1520.4	1610.8
	s			0.8946	0.9427	0.9886	1.0340	1.0797	1.1247	1.1674	1.2073	1.2457	1.2815	1.3131	1.3424	1.3959	1.4433

Sh = superheat, F
v = specific volume, cu ft per lb
h = enthalpy, Btu per lb
s = entropy, Btu per F per lb

Source: Combustion Engineering, Inc.

TABLE 46.4 Fuel Heating Values

Fuel	Heating value
Bituminous coal	12,000–14,000 Btu/lb (6700–7800 kcal/kg)
No. 6 fuel oil	18,000–19,000 Btu/lb (10,000–10,500 kcal/kg) (about 150,000 Btu/gal) (10×10^6 kcal/m^3)
Natural gas	1000–1200 Btu/ft^3 (8900–10,700 kcal/m3)

Unless special heat-recovery equipment is provided, the efficiency of the boiler itself ranges from 65 to 80% (Table 46.5). Very large boilers with sophisticated economizers and air heaters can sometimes reach 90%, but such auxiliaries are generally restricted to high-pressure boilers used by utilities. In a 1000 lb/in^2 (69 bars) boiler, the water temperature is approximately 550°F (288°C). Assuming that the temperature difference between the furnace combustion gas and the boiler water is about 400°F (204°C) to provide the driving force to transfer heat, the exit gas temperature would be 950°F (510°C).

Addition of an economizer to this boiler (to heat the feed water) and an air heater (to heat the combustion air) would probably reduce the exit gas temperature to 400°F (204°C). This results in a 12 to 13% increase in efficiency. The total surface area of these heat recovery devices may be twice the heating surface of the boiler itself!

An illustration of the addition of an economizer to a low-pressure package boiler is shown in Figure 46.1. This economizer can be installed at the time of erection or later as an add-on unit. A rule of thumb sometimes used is that a 1% gain in boiler efficiency can be achieved with each 40°F reduction in stack temperature accomplished by the economizer.

Another contributor to energy loss and corresponding reduction in efficiency is the use of excess air beyond that needed for combustion. It is practically impossible to burn a fuel in a boiler using exactly the theoretical volume of air for com-

TABLE 46.5 Typical Range of Boiler Efficiency

Fuel	Type of steam generating equipment	Service	Average full-load efficiency, %
Coal	Horizontal and vertical return tubular	Heat	65 to 75
Coal	Low head—watertube	Heat	70 to 75
Oil or gas	Low head—watertube	Heat	75
Coal	Watertube—no heat recovery	Power	75 to 77
Oil or gas	Boiler only	Power	75 to 80
Stoker coal (low grade)	Boiler and heat recovery	Power	80 to 83
Stoker coal (high grade)	Boiler and heat recovery	Power	83 to 86
Pulverized fuel—coal (low grade)	Boiler and heat recovery	Power	82 to 85
Pulverized fuel—coal (high grade)	Boiler and heat recovery	Power	85 to 90
Natural gas and oil (high grade)	Boiler and heat recovery	Power	82 to 85

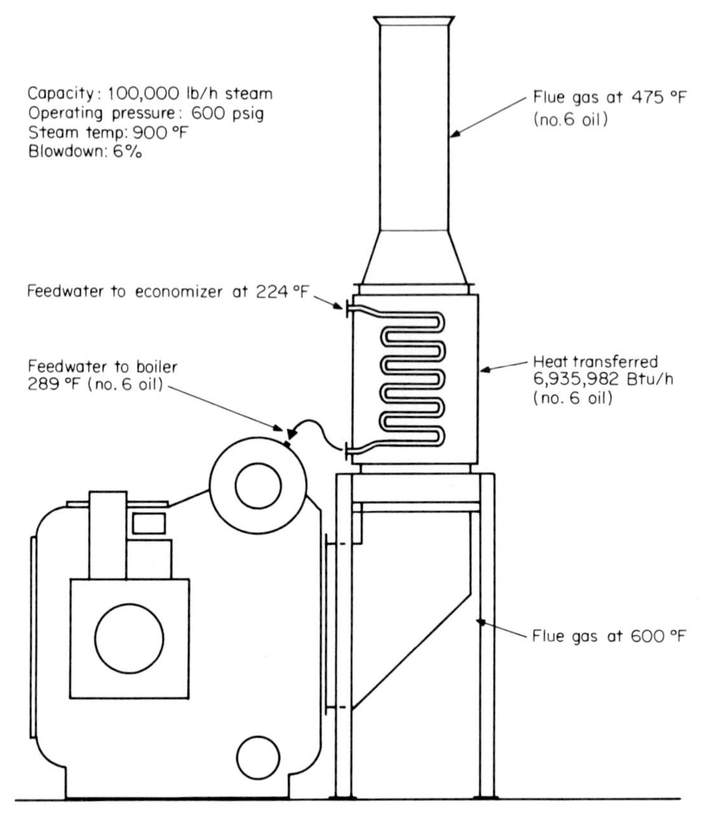

FIG. 46.1 Illustration of an economizer added to a package boiler with fuel savings of up to $200,000 per year.

bustion. Excess air, above the theoretical requirement, is always used, because the driving force of the excess oxygen ensures that all of the fuel is burned rapidly and efficiently before it leaves the furnace. But, excess air takes energy from the fuel because it has to be heated to the temperature of the other furnace gases. So this heat acquired by the excess air becomes another loss (increasing unavailable energy) in the exit stack gas.

IMPROVING BOILER EFFICIENCY

A boiler generating 100,000 lb/h (45,400 kg/h) of steam at a pressure of 600 lb/in^2 (41.4 bars) and temperature of 900°F (482°C) (superheated steam) produces steam with an enthalpy of 1461 Btu/lb (812 kcal/kg) (see Table 46.3). The *power* output of the boiler will be 100,000 lb/h × 1461 Btu/lb = 146 × 10^6 Btu/h (45,400 kg/h × 810 kcal/kg = 36.9 × 10^6 kcal/h).

This is equivalent to

$$\frac{146 \times 10^6 \text{ Btu/h}}{3413 \text{ Btu/kWh}} = 42{,}778 \text{ kW}$$

To see how much fuel is needed to produce this steam if the boiler is 80% efficient, the Btu/h output is divided by the efficiency (as a decimal) to give the Btu/h in fuel consumed:

$$\frac{146 \times 10^6}{0.80} = 182.5 \times 10^6 \text{ Btu/h } (46 \times 10^6 \text{ kcal/h})$$

This is equivalent to approximately 7.6 ton/h (6.9 t/h) of coal or 1200 gal/h (4.5 M^3/h) of No. 6 oil. The cost would be approximately $365 per hour, or the cost of steam would be $2 per million Btu ($8 per million kcal). This boiler uses $32 million worth of fuel per year. A steam production of only 100,000 lb/h is typical of a relatively small industrial plant; for comparison, the steam plant in a large integrated pulp/paper mill might produce 15 to 20 times as much steam.

Depending on the pulping process, very few pulp/paper mills generate all of their steam with fossil fuel. Kraft mills have considerable by-product fuel in the form of bark and black liquor. However, even this fuel costs money to prepare and handle, so it is quite common to have an overall fuel cost of $25 million/year.

There are substantial rewards in cost savings from seemingly small gains in efficiency anywhere in the cycle. In the small boiler example above, a 1% increase in boiler efficiency results in a savings of $32,000 per year; for the paper mill this 1% increased efficiency could mean $250,000 a year.

MECHANICAL HEAT RECOVERY

The points in the heat cycle where energy is most vulnerable to loss are readily apparent. To achieve good combustion of fuel in the furnace requires a certain minimum of excess air to thoroughly mix the fuel molecules with the necessary oxygen molecules. Since air is 80% nitrogen and only 20% oxygen, it is immediately apparent that the nitrogen is a major parasite in the system, contributing nothing while robbing energy as it is heated in the furnace and discharged at a temperature of perhaps 400°F (204°C) above ambient.

The percentage of excess air required varies with the fuel and the burner or stoker employed. For instance, coal firing may require 20 to 50% excess air, depending upon the type of furnace, while gas or oil may require only 5 to 20% excess air. Although this excess air is necessary for mixing, it does rob energy from the fuel, carrying heat out with the stack gases and increasing the volume of stack gas that must be pulled through the boiler by the induced draft fan.

Poor equipment operation is a major cause of inefficiency. If an oil-fired boiler requires 20% excess air for optimum combustion, but actually operates at 40% excess air because of poor control (a common situation), the total stack gas volume increases by 16%. If the 100,000 lb/h (45,400 kg/h) boiler of the earlier example is designed for 80% efficiency and 20% excess air, about 15% of the efficiency loss may be due to the heat in the stack gas. The fuel equivalent of this 15% efficiency loss would be: $0.15 \times 182.5 \times 10^6 = 27.4 \times 10^6$ Btu/h (6.9×10^6 kcal/h). The increase in gas volume of 16% with excess air at 40% instead of 20% adds the following heat loss to the expected design performance.

$$0.16 \times 27.4 = 10^6 = 4.38 \times 10^6 \text{ Btu/h } (1.1 \times 10^6 \text{ kcal/h})$$

This is over 2% of the total fuel burned, or a waste of approximately $64,000 a year. As a rule of thumb, each 10% of excess air results in a 1% loss of boiler efficiency.

In conventional boiler firing controls, pressure deviation in the steam header regulates the flow of fuel and air to the furnace. The fuel/air ratio varies with steam load. On start-up, this ratio is established and the correct mechanical linkages are set by the control manufacturer. However, as the boiler ages, the original ratios change because the linkages wear and the pressure loss through boiler passes changes. In practice, the operator adjusts the air/fuel ratio by a manual override to maintain a certain fire quality that the operator has become used to; in other words, boiler operation is subjective and varies considerably among operators.

To correct inefficiencies caused by subjective operator control, additional maintenance or instrumentation may be required. For instance, most boiler furnaces operate at negative pressures, so perforations that develop in the boiler casing admit excess air, upsetting the fuel/air ratio and decreasing efficiency. This requires regular inspection and maintenance of the furnace casing. Analytical instruments are needed to monitor and optimize the fuel/air ratio. The most common are a smoke density meter and O_2 and CO_2 meters, because there is a direct relationship between these gases and excess air in the stack.

Since stack gas temperature so markedly affects efficiency, the decision on investing in cold-end heat recovery equipment (economizers and air heaters) is usually made when a boiler is designed. With fuel costs continually increasing, these heat recovery devices that provide an increase of 1% efficiency for every 40°F (22°C) reduction in stack gas temperature have become common on all but the smallest boilers.

In addition to mechanical heat recovery, the companions to keeping stack temperature down are soot-blowing devices, combustion aid chemicals, and close operator attention.

FEED WATER TEMPERATURE

On the waterside of the boiler, each 10°F (6°C) increase in feed water temperature usually results in a 1% fuel savings in typical industrial steam plants having an excess of exhaust steam. Savings are only about a third of this where steam is bled off from a turbine at various stages for preheating water and there is no excess exhaust steam.

The deaerating heater is one example of a feed water heating device. Most of them operate at about 220°F (104°C) water temperature, but there are exceptions. The steam table shows that 220°F corresponds to 17.2 lb/in^2 abs (2.5 lb/in gage) (multiply pounds per square inch by 0.069 to get bars). If the heater has been designed for 10 lb/in^2 gage [25 lb/in^2 abs, equivalent to 240°F (116°C)] and if the steam going to the heater is 20 lb/in^2 gage (35 lb/in^2 abs) before the pressure reducing valve, it is practical to adjust the controls to maintain 240°F and thus avoid energy loss across the pressure control valve. This would increase efficiency 2% [for the 20°F (11°C) temperature increase], amounting to $64,000 per year in the small boiler (Figure 46.1).

STEAM PRESSURE

The goal of the utilities engineer is to use steam at its highest energy level. Whether the 20 lb/in^2 gage steam supplied to the deaerating heater is condensed at 10 or 2.5 lb/in^2 gage, it still transfers approximately the same amount of Btu's per pound. However, condensing at 10 lb/in^2 gage produces a temperature of approximately 240°F, while condensing at 2.5 lb/in^2 gage yields a temperature of only 220°F. In this example, the water at 240°F has 20 Btu/lb more energy than water at 220°F and thus requires slightly more steam to attain this temperature. To gain this 20 Btu/lb in the higher temperature water only requires 0.02 lb (9 g) additional steam having an enthalpy of about 1150 Btu/lb (640 cal/g). This almost negligible increase in steam has produced a savings of about 2% in fuel consumption.

DECREASING BLOWDOWN

The steam tables show that in a boiler operating at 600 lb/in^2 gage (615 lb/in^2 abs) the boiler water enthalpy is approximately 475 Btu/lb (264 cal/g).

Assume that feed water is being heated from 220°F (104°C) [enthalpy 188 Btu/lb (105 cal/g)] to the enthalpy of the boiler water, 475 Btu/lb (264 cal/g), an addition of 287 Btu/lb (160 cal/g). For the example boiler making 100,000 lb/h (45,400 kg/h) steam, 1% blowdown for practical purposes is about 1000 lb/h (454 kg/h). The energy required to heat this 1000 lb/h of blowdown is:

$$1000 \text{ lb/h} \times 287 \text{ Btu/lb} = 287{,}000 \text{ Btu/h}$$

This is about 0.16% of our total energy input, costing approximately $5000/year. In this example then, a reduction in blowdown of 1% (e.g., from 10 to 9%) results in approximately 0.2% fuel savings. Additional savings accrue from reduced chemical treatment, since the treatment is in the blowdown. Decreases in blowdown can be accomplished by improved pretreatment of feed water, by operating the boiler at a higher concentration of dissolved solids, or by returning more condensate. The solids level in the boiler can be optimized through steam purity testing.

RECLAIMING BLOWDOWN HEAT

All boilers will require blowdown, the quantity depending upon the purity of the feed water, the steam pressure, the allowable concentration of dissolved solids, and boiler design. Extracting heat from blowdown to approximately 120°F (49°C) is common practice—unless, of course, the plant has an excess of exhaust steam not readily recoverable. In that case, increasing the temperature of water going to the deaerator simply reduces the amount of exhaust steam that can be condensed in the deaerator.

Blowdown water from a 600-lb/in^2 gage (41.4 bars) boiler has an enthalpy of 475 Btu/lb (264 cal/g); water at 120°F (49°C) has an enthalpy of 88 Btu/lb (49 cal/g); so an energy recovery of 387 Btu/lb (215 cal/g) from the boiler blowdown is possible.

Assuming 10,000 lb/h (4540 kg/h) of blowdown, equal to approximately 9% in the boiler example, 387 Btu/lb (215 cal/g) recovery yields a total energy savings of 3.87×10^6 Btu/h (1×10^6 kcal/h), or a savings of $186/day at $2 per million Btu.

Transferring the boiler blowdown heat through an exchanger directly to 60°F (16°C) makeup water would be a misuse of high-level energy because of the approximately 430°F (221°C) temperature difference between the blowdown and the makeup. It is much more efficient to flash the blowdown water at some intermediate pressure, thereby producing steam at this energy level where it could do

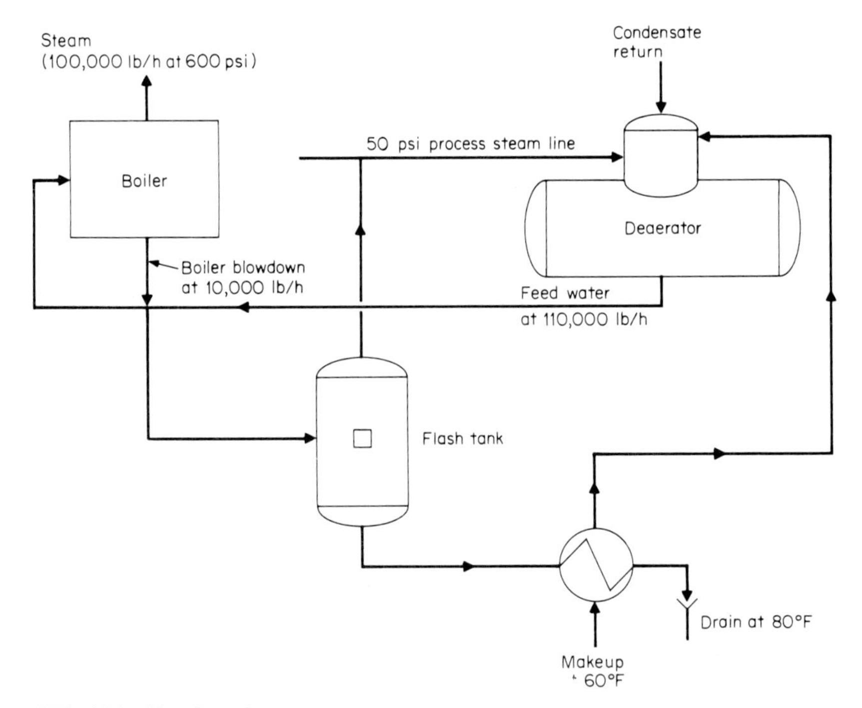

FIG. 46.2 Blowdown heat recovery system.

work at higher temperature. The remaining water after flashing could then be heat exchanged at the lower level of makeup water temperature. A typical system using a blowdown flash tank followed by a heat exchanger is shown in Figure 46.2

If the plant has a 50-lb/in^2 gage low-pressure steam system used for process heating, the blowdown is commonly flashed at this pressure. Using the enthalpy of 600-lb/in^2 gage (41.4 bars) water, 50-lb/in^2 gage (3.5 bars) steam, and 50-lb/in^2

gage water, a simple algebraic equation shows how much of the blowdown is converted to steam at 50 lb/in² gage:

$$W_F = \frac{W_B(h_p - h_R)}{L_R}$$

where W_F = flash steam, lb/h
W_B = blowdown, lb/h
h_p = enthalpy of boiler water, Btu/lb
h_R = enthalpy of liquid at lower pressure, Btu/lb
L_R = heat of vaporization, Btu/lb, at flash pressure

In the example the calculation would be as follows:

$$W_F = \frac{10{,}000\ (475 - 267)}{912} = 2281\ \text{lb/h}$$

Using the sensible heat in the boiler water, the drop in pressure at the flash tank converts some blowdown to steam with over twice the amount of energy per pound as the boiler water and at a temperature level of 298°F (148°C), allowing it to be used in many high-temperature processes. The remaining blowdown water, having had its energy value decreased, is still available for heat exchange with the 60°F (16°C) makeup water. The blowdown water going to the sewer will be considerably below 120°F (49°C), which was the temperature on which the 387 Btu/lb potential energy recovery calculated above was based. A terminal difference of 20°F (11°C) between cold makeup and cooled blowdown is the usual design factor.

BOILER LOADING

A boiler generally produces its maximum efficiency at about 80 to 85% of rating. Therefore, a powerhouse with a number of boilers in operation gains by judicious loading of each unit. However, a boiler with full heat recovery may have such an advantage in efficiency over one with no heat recovery that it would be better in the overall picture to run the more efficient boiler even at lower ratings than to run the less efficient at its optimum rating. Optimizing performance requires experience and study.

TURBINE OPERATION

Power plants that operate condensing turbines can justify careful attention to maintaining a clean condenser. Good heat transfer here means better vacuum and more heat energy available to the turbine. As a good rule of thumb, a loss of 1 in (25.4 mm) Hg in condenser vacuum results in a 1% loss of efficiency in the turbine.

In industrial turbines practicing extraction, it is most efficient to operate at maximum extraction and minimum condensing. The cost of the fuel being burned, whether fossil or waste, determines the amount of steam going to the condenser in an extraction turbine. An industrial plant burning oil, for example, in most cases finds it cheaper to buy power than to try to produce it by operating a turbine where the steam must go to a condenser. However, that power generated with extraction or in noncondensing turbines usually costs less than purchased power. This is the essence of the trend toward cogeneration, or "topping" a process need for steam with a turbine exhausting to the process steam line.

Figure 46.3 shows the operating constraints on a fictional noncondensing turbine generating electric power and exhausting the used steam to a process. The process requirements have a direct bearing on how much power can be generated. Generally, process requirements are rated in Btu/h—easily converted to pounds-per-hour steam based on the enthalpy of the steam.

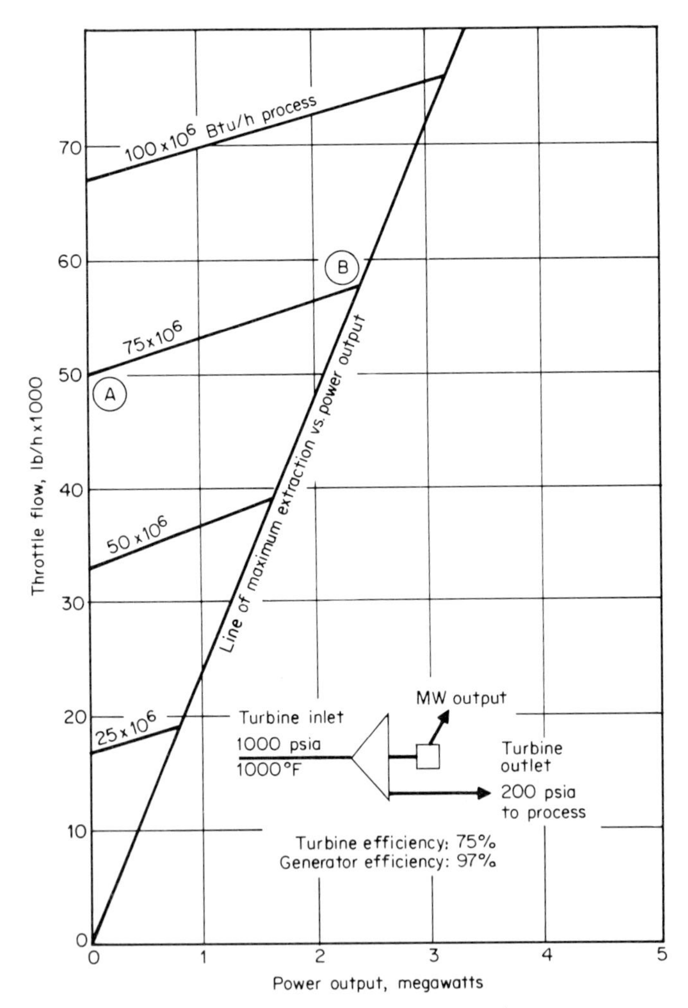

FIG. 46.3 Throttle flow versus power output at various process steam requirements for a fictional noncondensing turbine. Point A represents no electrical load, point B maximum electrical load.

If the process requires 75 × 10^6 Btu/h (19 × 10^6 kcal/h), the power plant could theoretically generate zero power (Figure 46.3, point A), or progressively up to approximately 2.4 mW (megawatts; 1 mW = 100 kW) (Figure 46.3, point B), while at the same time supplying the required 75 × 10^6 Btu/h to process. However, in going from 0 to 2.4 mW, the turbine throttle flow increases from 50,000 to 57,000 lb/h (22,700 to 25,900 kg/h). Even though the steam flow increased, the Btu's per hour to the process remains constant because the enthalpy in the exhaust steam decreases a proportionate amount.

The turbine throttle valve on this fictional machine is modulated by a governor shaft designed to control steam flow to maintain a shaft speed of 3600 r/min, or 60 r/s, corresponding to the operation of 60-cycle current from the generator. Pressure variations in the 200-lb/in^2 abs (14 bars) exhaust line also signal for throttle adjustments.

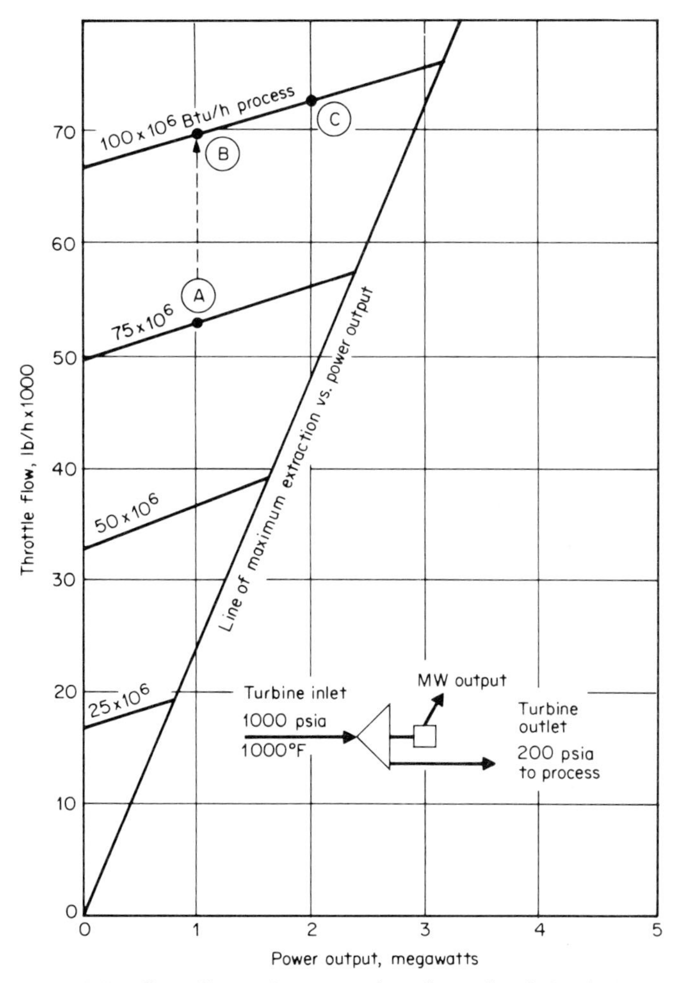

FIG. 46.4 Effect of increasing process heat demand and electrical load on throttle flow, shown by points A, B, and C.

Generally, a plant loads the electric generator to some optimum point by regulating the amount of current being fed to the electrical network. If the electrical system draws a 1-mW load from the generator at a process steam demand of 75 $\times$ 10^6 Btu/h (19 $\times$ 10^6 kcal/h), this operating condition is represented in Figure 46.4, by point A. The plant conditions are: (1) throttle flow is 53,000 lb/h (24,000 kg/h), (2) generator output is 1 mW, (3) process steam demand is 75 $\times$ 10^6 Btu/h (19 $\times$ 10^6 kcal/h).

If the process department now puts another unit into service by opening up a steam valve and increasing the overall process steam requirement to 100 $\times$ 10^6 Btu/h (25 $\times$ 10^6 kcal/h), this additional steam demand causes the pressure in the 1200-lb/in^2 abs (14 bars) line to drop. Immediately a pressure sensor detects this and signals the turbine throttle valve to open to maintain the 200-lb/in^2 abs (14

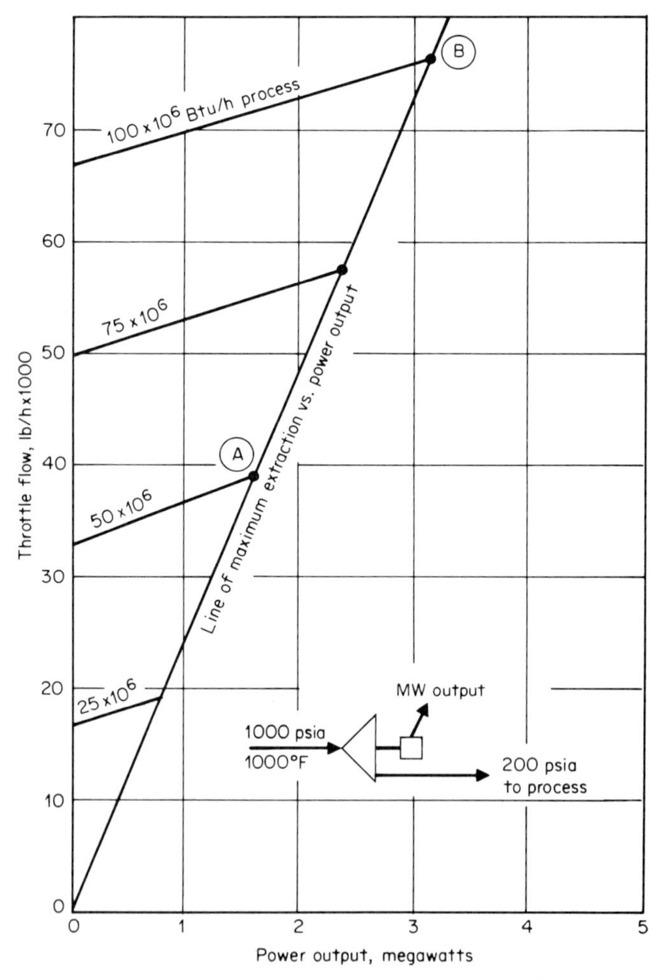

FIG. 46.5 Maximum power output varies with throttle flow and with amount of process steam being used.

bars) pressure. The operating point during this change goes from point A (Figure 46.4) vertically up the 1-mW line (keeping the same electric load) until it reaches point B. The new set of conditions are: (1) throttle flow is 69,500 lb/h (31.500 kg/h), (2) generator output is still 1 mW, (3) process steam demand is 100 × 10^6 Btu/h (25 × 10^6 kcal/h).

The power plant operator may suddenly need to increase the electric generator output to 2 mW to meet the plant's electric demands, without a need for increasing the heat load of 100 × 10^6 Btu/h (25 × 10^6 kcal/h). As the power plant operator gradually increases the load on the generator, it in turn produces a torque on the turbine drive shaft which tends to slow it down. Immediately the turbine governor senses the drop in revolutions per minute and signals the throttle valve to open. The operating point now moves from Figure 46.4, point B, upward along

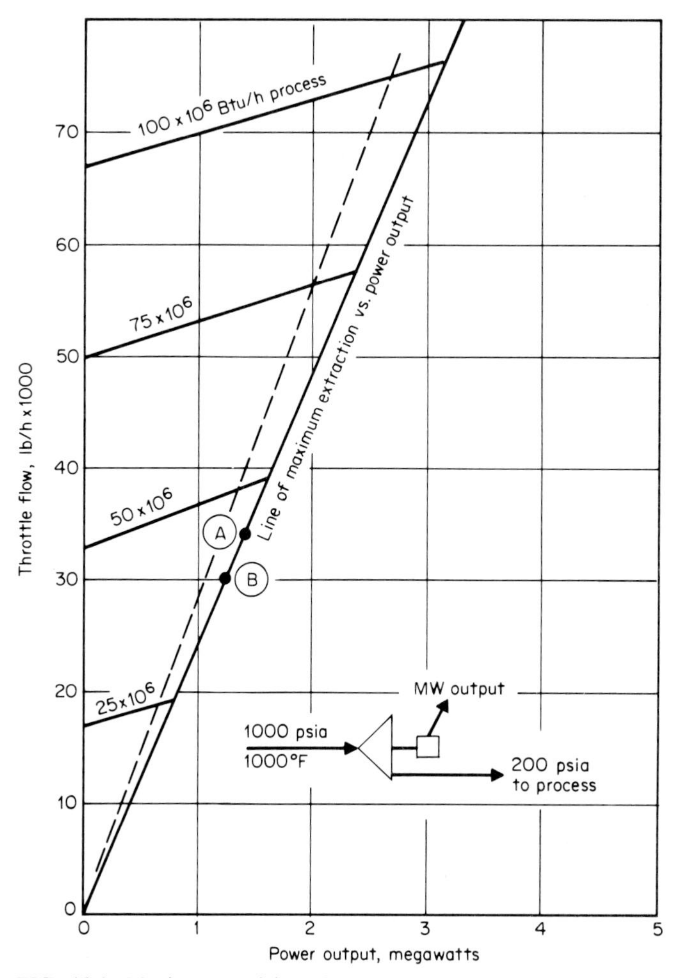

FIG. 46.6 Maximum anticipated power output at 34,000 lb steam/h (point A) is 1.4 mW; at 30,000 lb steam per hour it is 1.2 mW.

the 100 × 10^6 Btu/h (25 × 10^6 kcal/h) line until it intersects the vertical 2-mW line at point C. After equilibrium, the new set of conditions is: (1) throttle flow is 72,600 lb/h (33,000 kg/h), (2) generator output is now 2 mW, (3) process load is still 100 × 10^6 Btu/h (25 × 10^6 kcal/h).

A third condition will arise if the power plant operator needs maximum electric generation. Figure 46.5 shows that the maximum power output line varies with throttle flow as might be expected. However, maximum power output also depends upon the amount of process steam being used. If the process requires 50 × 10^6 Btu/h (12.6 × 10^6 kcal/h), the maximum generator output will be at point A, or 1.6 mW; on the other hand, if the process demand is 100,000 Btu/h (25 × 10^6 kcal/h), the maximum generator output will be at point B, or 3.1 mW.

There are a few basic problems that interfere with the use of diagrams of this kind:

1. Boiler water carryover causes turbine deposits. When this occurs the efficiency of the turbine decreases and the line of maximum power output moves to a position similar to the dotted line on Figure 46.6, and the plant can't obtain as much power as before.
2. If the steam flow meter were in error—reading 34,000 lb/h (15,400 kg/h) when it should be reading 30,000 lb/h (13,600 kg/h)—the diagram would anticipate an output of 1.4 mW (point A, Figure 46.6), when in actuality the maximum would be 1.2 mW (point B, Figure 46.6)
3. The throttle inlet pressure and temperature readings may be in error. If reading higher than actual conditions, there would be less enthalpy in the steam and therefore lower output from the turbine. If for some reason the 200 lb/in^2 abs (14 bars) regulator was holding a slightly higher back pressure in the process line, there would be less output from the generator.
4. Another factor might be worn turbine interstage or shaft seals which would lower the turbine's efficiency.

STEAM TRAP

A very important mechanical item for maintaining peak efficiency in steam sysems is the steam trap. A large industrial plant has hundreds of steam traps in its complex.

A steam trap is used to remove and recover low-energy condensate from a steam-heating system, while at the same time preventing the loss of high-energy steam. If steam traps are not maintained, condensate may be retained in the system, severely limiting heat transfer. The traps may also stick open, allowing high-energy steam to blow through and be wasted.

Heat tracing of process or fuel pipelines is required in colder climates where these liquids must be kept hot and fluid for transport. The tracing involves taping ¼-or ⅜-in copper tubing to the pipeline beneath the pipe insulation. Steam is then run through this tubing to maintain heat in the line. It is common to allow this steam to blow to the atmosphere at the end of the tracing line, resulting in substantial steam loss. Steam traps should be used on these tracer lines to eliminate loss of steam to the atmosphere.

APPENDIX

SIGN CONVENTIONS IN ELECTROCHEMISTRY

In representing electrochemical processes, all half-cell reactions (those reactions that occur at each electrode) are written as reduction reactions, thus:

$$Na^+ + e \rightarrow Na \quad E = -2.71 \text{ V}$$

$$Fe^{2+} + 2e \rightarrow Fe \quad E = 0.44 \text{ V}$$

The addition of electrons to the system should be shown on the left side of the reaction. By definition these are cathodic reactions: reduction always occurs at the cathode and oxidation at the anode.

In the electrolytic cell, the anode may be shown with a minus charge and the cathode with a positive charge. When the half-cells are coupled, electrons flow from the anode through the conductor to the cathode.

In electrolysis, an external source of electron flow (a battery, rectifier, or direct-current generator) is connected to the half-cells. The half-cell receiving electrons is the cathode, because that is where cations are reduced. The cathode may be shown with a positive sign, but this is unnecessary and may be confusing. Where a battery is used as the direct-current source, the negative pole of the battery is connected to the cathode, which may lead to a common and erroneous impression that the cathode of the electrolytic cell is negatively charged. In fact, the battery is itself an electrolytic cell, and its negative pole is the anode of that cell and is supplying electrons to the cathode of the cell being electrolyzed.

Confusion is avoided by noting the half-cell reactions being studied and identifying the cathode as that surface where reduction is occurring, with the anode being that surface where oxidation is occurring.

TABLE A.1 English/Metric Conversions

English unit	Multiplied by	Equals	Metric unit
Acres (43,560 ft^2)	0.405		Hectares (ha)
(640 acres/m^2)			
Acre-feet (acre-ft)	1234		Cubic meters (m^3)
(325,851 gal/acre-ft)			
Barrel (bbl) liquid	0.16		m^3
(42 gal/bbl)			
British thermal units (Btu)	0.252		Kilocalories (kcal)
Btu/lb	0.556		kcal/kg (cal/g)
Btu/h/ft^2	2.69		kcal/h/m^2
Cubic feet (ft^3)	28.32		Liters (L)
(1 $ft^3 H_2O$ = 62.4 lb)			(1 L = 1000 mL = 1000 cm^3)
(1 ft^3 = 7.48 gal)	0.0283		m^3
Cubic feet per second (ft^3/s)	1.7		m^3/min
(1 ft^3/s = 449 gal/min)	2.45×10^3		m^3/day
Fahrenheit degrees (F°)	0.556		Celsius degrees (C°)*
(absolute zero = −460°F or 0° Rankine)			(absolute zero = − 273°C or 0 kelvin)
Feet (ft)	30.48		Centimeters (cm)
(1 ftH_2O = 0.434 lb/in^2)	0.305		Meters (m)
Feet per second (ft/s)	0.305		m/sec
Gallons (gal)	3.785		Liters (L)
(1 gal H_2O = 8.345 lb)	0.00379		m^3
(7.48 gal/ft^3)	3.785×10^{-3}		m^3
Gallons per minute (gal/min)	5.45		m^3/day
(695 gal/min = 1 mgd)			
gal/min/ft^2	40.6		L/min/m^2
(1 gal/min/ft^2 = 1.6 in/min)	4.06		cm/min
	0.0406		m^3/min/m^2
	0.0406		m/min
Grains per gallon (gr/gal)	17.1		Milligrams per liter (mg/L)
(1 gr/gal = 17.1 ng/L)			
Horsepower (hp)	0.746		Kilowatts (kW)
Inches (in)	2.54		Centimeters (cm)
Mils (0.01 in)	25.4		Microns (10^{-6} m)
Miles	1.61		Kilometers (km)
Million gallons daily (mgd)	3785		m^3/day
(695 gal/min = 1 mgd)	2.62		m^3/min

TABLE A.1 English/Metric Conversions (*Continued*)

English unit	Multiplied by	Equals	Metric unit
Ounces (oz)	28.4		g
Ounces (fluid)	29.6		Milliliters (mL)
Pounds (lb)	0.454		Kilograms (kg)
	453.6		Grams (g)
Pounds per cubic foot (lb/ft^3)	0.016		g/cm^3
	16.02		kg/m^3
Pounds per square inch (lb/in^2)	0.068		Atmospheres (atm) or bars
($1\ lb/in^2 = 2.31$ ft H_2O)	703		kg/m^2
	0.0703		kg/cm^2
	51.7		mmHg
Square feet (ft^2)	0.093		m^2
Square inches (in^2)	6.452		cm^2
Square miles (mi^2)	260		ha
Tons, long (t, long)	1016		kg
Tons, short (t, short)	0.907		Tons (t)
Tons of refrigeration (12,000 Btu/h)	3024		kcal/h
Yards	0.9144		m

* For temperature conversions: (°F − 32) × 0.555 = °C.

TABLE A.2 Metric Prefixes

mega (M)	10^6	milli (m)	10^{-3}
kilo (k)	10^3	micro (μ)	10^{-6}
deci (d)	10^{-1}	nano (n)	10^{-9}
centi (c)	10^{-2}	pico (p)	10^{-12}

GLOSSARY

Absorption: Assimilation of molecules or other substances into the physical structure of a liquid or solid without chemical reaction.

Acidity: Theoretically, in water, an excess of H^+ ions over OH^- ions that occurs at a pH below 7. In water analysis, an excess of H^+ ions that is measurable by titration where the pH is less than 4.2 to 4.4, where M alkalinity disappears (at the methyl orange endpoint).

Activated sludge: An aerobic biological process for conversion of soluble organic matter to solid biomass, removable by gravity or filtration.

Adsorption: Physical adhesion of molecules or colloids to the surfaces of solids without chemical reaction.

Aerobic organism: An organism that requires oxygen for its respiration.

Aerosol: A colloidal system involving liquid or solid particulates dispersed in air.

Agglomerate: To gather fine particulates together into a larger mass.

Algae: Simple plants containing chlorophyll. Many are microscopic, but under conditions favorable for their growth they grow in colonies and produce mats and similar nuisance masses.

Alkalinity: By definition, total alkalinity (also called M alkalinity) is that which will react with acid as the pH of the sample is reduced to the methyl orange endpoint—about pH 4.2. Another significant expression is P alkalinity, which exists above pH 8.2 and is that which reacts with acid as the pH of the sample is reduced to 8.2.

Amphoteric: Capable of reacting in water either as a weak acid or as a weak base. For example, aluminum salts hydrolyze in water to produce a compound that may be considered a weak base, $Al(OH)_3$, or a weak acid, H_3AlO_3. A property of certain oxides makes them reactive both with acids and bases. Al_2O_3 is an example:

1. $Al_2O_3 + 3H_2SO_4 \rightarrow Al_2(SO_4)_3 + 3H_2O$
2. $Al_2O_3 + 2NaOH \rightarrow Na_2Al_2O_4 + H_2O$

Anaerobic organism: An organism that can thrive in the absence of oxygen.

Anion: A negatively charged ion resulting from dissociation of salts, acids, or alkalies in aqueous solution.

Anionic: The condition of a polymer, colloid, or large particle having exchangeable cations on its surface and an opposite, negative charge on the substrate.

Anode: In electrolysis or electrochemical corrosion, a site where metal goes into solution as a cation leaving behind an equivalent of electrons to be transferred to an opposite electrode, called a cathode.

Anodizing: The treatment of a metal surface whereby the metal is made anodic.

API gravity: An index of specific gravity defined by the American Petroleum Institute.

API separator: A simple gravity separator meeting the design standards of the American Petroleum Institute for separation of oil and solids from wastewater.

Aquifier: A porous, subsurface geological structure carrying or holding water, such as a well.

Avogadro's number: The number of molecules in a gram-molecular weight of any substance, 6.02×10^{23}.

Bacteria: Microscopic single-cell plants which reproduce by fission or by spores, identified by their shapes: coccus, spherical; bacillus, rod-shaped; and spirillum, curved.

Base: An alkaline substance.

Biocide: A chemical used to control the population of troublesome microbes.

Biota: All living organisms of a region or system.

Black liquor: Kraft cooking liquor recovered from brown stock washers in the pulp mill.

Blast furnace: A furnace producing iron from ore by reduction wtih coke.

Blowdown: The withdrawal of water from an evaporating water system to maintain a solids balance within specified limits of concentration of those solids.

Blowpit: The vessel receiving cooked wood pulp from the digester.

BOD: Biochemical oxygen demand of a water, being the oxygen required by bacteria for oxidation of the soluble organic matter under controlled test conditions.

Broke: Trim or excess sheet from paper manufacture returned to a pulping device for recovery.

BS&W (bottom sediment and water): A measure of oil quality based on the volume percent of sediment and water that can be centrifuged from a sample.

Buffer: A substance in solution which accepts hydrogen ions or hydroxyl ions added to the solution as acids or alkalies, minimizing a change in pH.

Bulk density: The measured density/volume ratio for a solid including or not corrected for the voids contained in the bulk of material, in lb/ft^3 or kg/m^3.

Bulking: Production of a light, fluffy biomass, usually due to the presence of filamentous organisms.

Cake: A term applied to a dewatered residue from a filter, centrifuge, or other dewatering device.

Carbonate hardness: That hardness in a water caused by bicarbonates and carbonates of calcium and magnesium. If alkalinity exceeds total hardness, all hardness is carbonate hardness; if hardness exceeds alkalinity, the carbonate hardness equals the alkalinity.

Carryover: The presence of boiler water in steam caused by foaming or entrainment.

Catalysis: Addition of a material (catalyst) that does not take a direct part in a chemical reaction but increases the rate of the reaction.

Cathode: In electrolysis or electrochemical corrosion, a site on a surface where cations in solution are neutralized by electrons to become elements that either plate out on the surface or react with water to produce a secondary reaction.

Cation: A positively charged ion resulting from dissociation of molecules in solution.

Cationic: The condition of a polymer, colloid, or large particle having exchangeable anions on its surface and an opposite, positive charge on the substrate.

Caustic soda: A common water treatment chemical, sodium hydroxide (lye).

Centrate: The liquid remaining after removal of solids as a cake in a centrifuge.

Chelating agents: Organic compounds having the ability to withdraw ions from their water solutions into soluble complexes.

Coagulation: The neutralization of the charges on colloidal matter (sometimes also considered to be flocculation).

Coalescence: The gathering together of coagulated colloidal liquid particles into a single continuous phase.

COD: Chemical oxygen demand, a measure of organic matter and other reducing substances in water.

Coliform bacteria: Bacteria found in the intestinal tract of warm-blooded animals and used as indicators of pollution if found in water.

Colloids: Matter of very fine particle size, usually in the range of 10^{-5} to 10^{-7} cm in diameter.

Concentration: The process of increasing the dissolved solids per unit volume of solution, usually by evaporation of the liquid; also, the amount of material dissolved in a unit volume of solution.

Concentration cell: The connection of two solutions of the same composition but different concentrations by a metal conductor to produce current flow through the circuit.

Concentration ratio: In an evaporating water system, the ratio of the concentration of a specific substance in the makeup to its concentration in the evaporated water, usually measured in the blowdown.

Condensate: Water obtained by evaporation and subsequent condensation.

Conduction: The transfer of heat through a body by molecular motion.

Conductivity: The ability of a substance to conduct heat or electricity. Electrical conductivity is usually expressed in microsiemans per centimeter.

Connate water: Fossil water produced with oil.

Consistency: In the pulp/paper industry, a term for the density in percent by weight dry matter, of a slurry of pulp.

Contaminant: Any foreign component present in another substance; e.g., anything in water that is not H_2O is a contaminant.

Convection: The transfer of heat through a fluid by circulating currents.

Coordinated phosphate: A boiler treatment scheme using phosphate buffers to avoid the presence of hydroxyl alkalinity.

Cracking: An oil-refining process that breaks large molecules into smaller ones.

Critical pressure: The pressure at the critical temperature above which the fluid no longer has the properties of a liquid, regardless of further increase in pressure.

Cupola: A furnace for melting scrap or pig iron with coke.

Cycles of concentration: Concentration ratio.

Dealkalization: Any process for reducing the alkalinity of water.

Decantation: An elutriation process, where the supernatant liquor contains recoverable leaching chemical.

Deinking: The process of removing ink from secondary fibers.

Deionization: Any process removing ions from water, but most commonly an ion exchange process where cations and anions are removed independently of each other.

Demineralization: Any process used to remove minerals from water; however, commonly the term is restricted to ion exchange processes.

Desalination: The removal of inorganic dissolved solids from water.

Desalting: The removal of salt from crude oil.

Detackify: Treatment of solids from a paint spray booth to eliminate their sticky properties.

Dewater: To separate water from sludge to produce a cake that can be handled as a solid.

Dialysis: A separation process that depends on differences in diffusion rates of solutes across a permeable membrane.

Diatoms: Organisms related to algae, having a brown pigmentation and a siliceous skeleton.

Disinfection: Application of energy or chemical to kill pathogenic organisms.

Dispersant: A chemical which causes particulates in a water system to remain in suspension.

Donnan effect: The rejection of diffusion of external ions by a semipermeable membrane because of a high internal concentration of ions of the same charge.

Drift: Entrained water in the stack discharge of a cooling tower.

Economizer: A heat exchanger in a furnace stack that transfers heat from the stack gas to the boiler feedwater.

EDTA: Ethylenediaminetetraacetic acid. The sodium salt is the usual form of this chelating material.

Electrolyte: A substance that dissociates into two or more ions when it dissolves in water.

Elution: The process of extracting one solid from another. Often used incorrectly to describe the regeneration of an ion exchanger.

Elutriation: The washing of a sludge with water to free it of its mother liquor.

Emulsion: A colloidal dispersion of one liquid in another.

Endothermic: Absorbing heat.

Enthalpy: The total heat content of a body.

Entrainment: The transport of water into a gas stream. In a boiler, this is carryover; in a cooling tower, drift.

Entropy: A mathematical expression applying to the limits to the availability of energy; a measure of the random motion of matter.

Enzyme: As applied to water, a chemical produced by living cells having the ability to reduce large organic molecules to units small enough to diffuse through the cell membrane.

EPA: Environmental Protection Agency.

Equalization: Minimization of variations in flow and composition by means of a storage reservoir.

Equivalent weight: The weight in grams of a substance which combines with or displaces one gram of hydrogen; it is usually obtained by dividing the formula weight by the valence.

Eutrophication: Enrichment of water, causing excessive growth of aquatic plants and an eventual choking and deoxygenation of the water body.

Exothermic: Evolving heat.

Facultative organisms: Microbes capable of adapting to either aerobic or anaerobic environments.

FDA: Food and Drug Administration.

Fermentation: The conversion of organic matter to CO_2, CH_4, and similar low-molecular weight compounds by anaerobic bacteria.

Filler: Clay, calcium carbonate, or other minerals added to cellulose fiber in the production of certain grades of paper or board.

Filtrate: The liquid remaining after removal of solids as a cake in a filter.

Filtration: The process of separating solids from a liquid by means of a porous substance through which only the liquid passes.

Fission: In biology, the process of reproduction by cell splitting.

Flash: The portion of a superheated fluid converted to vapor when its pressure is reduced.

Flocculation: The process of agglomerating coagulated particles into settleable flocs, usually of a gelatinous nature.

Flotation: A process for separating solids from water by developing a froth in a vessel in such fashion that the solids attach to air particles and float to the surface for collection.

Flume: A raceway or channel constructed to carry water or to permit flow measurements.

F/M ratio: Food-to-mass or food-to-microorganism ratio used to predict the phase of growth being experienced by the major microbial populations in a biological digestion process.

Fourdrinier: A design of paper machine using a continuous wire for forming the sheet.

Freundlich isotherm: The plot of test data related to the removal of colloidal matter from water showing the process to be adsorption.

Fumes: An aerosol with solids as the dispersed colloids.

Fungi: As applied to water, simple, one-celled organisms without chlorophyll, often filamentous. Molds and yeasts are included in this category.

Galvanic couple: The connection of two dissimilar metals in an electrolyte that results in current flow through the circuit.

Gangue: The earthy material remaining from ore beneficiation.

Grains per gallon: A unit of concentration. 1 gr/gal = 17.1 mg/L.

Green liquor: The liquor resulting from dissolving molten smelt from the kraft recovery furnace in water.

Hardness: The concentration of calcium and magnesium salts in water. Hardness is a term originally referring to the soap-consuming power of water; as such it is sometimes also taken to include iron and manganese. "Permanent hardness" is the excess of hardness over alkalinity. "Temporary hardness" is hardness equal to or less than the alkalinity. These are also referred to as "noncarbonate" or "carbonate" hardness, respectively.

Heat rate: An expression of heat-conversion to power, given in Btu/kWh. Theoretical conversion is 3413 Btu/kWh.

Henry's law: An expression for calculating the solubility of a gas in a fluid based on temperature and partial pressure.

Hindered settling: A stage of settling where the accumulated settled solids have compacted to an extent that egress of water from the mass is hindered and, therefore, settling is slowed.

Humidification: The addition of water vapor to air.

Hydrophilic: Having an affinity for water. Its opposite, non-water-wettable, is hydrophobic.

Infiltration: Inleakage of groundwater into sewage piping.

Inhibitor: A chemical that interferes with a chemical reaction, such as corrosion or precipitation.

Ion: An atom or radical in solution carrying an integral electric charge, either positive (cation) or negative (anion).

Ion exchange: A process by which certain undesired ions of given charge are absorbed from solution within an ion-permeable absorbent, being replaced in the solution by desirable ions of similar charge from the absorbent.

Ionic strength: A measure of the strength of a solution based on both the concentrations and valences of the ions present.

Kraft: An alkaline chemical pulping process, using salt cake as makeup.

Langelier index: A means of expressing the degree of saturation of a water as related to calcium carbonate solubility.

Leakage: The presence in the effluent of a species of ions in the feed to an ion exchanger.

Lignin: The major noncellulose constituent of wood.

Lime: A common water treatment chemical. Limestone, $CaCO_3$, is burned to produce quicklime, CaO, which is mixed with water to produce slaked, or hydrated, lime, $Ca(OH)_2$.

Lipophilic: Having an affinity for oil. The opposite of hydrophilic (i.e., hydrophobic).

Membrane: A barrier, usually thin, that permits the passage only of particles up to a certain size or of special nature.

Metabolize: To convert food, such as soluble organic matter, to cellular matter and gaseous by-products by a biological process.

Microorganism: Organisms (microbes) observable only through a microscope; larger, visible types are called *macroorganisms.*

Mineral: Any inorganic or fossilized organic material having a definite chemical composition and structure found in a natural state.

Miscibility: The ability of two liquids, not mutually soluble, to mix.

Mist: An aerosol with liquids as the dispersed colloids.

Mole: A unit weight or volume of a chemical corresponding to its molecular weight. A mole of water weighs 18 g, and its vapor occupies 22.4 L at standard temperature and pressure.

Monomer: A molecule, usually an organic compound, having the ability to join with a number of identical molecules to form a polymer.

Neutralization: Most commonly, a chemical reaction that produces a resulting environment that is neither acidic nor alkaline. Also, the addition of a scavenger chemical to an aqueous system in excess concentration to eliminate a corrosive factor, such as dissolved oxygen.

Noncarbonate hardness: Hardness in water caused by chlorides, sulfates, and nitrates of calcium and magnesium.

Noncondensibles: Gaseous material not liquefied when associated water vapor is condensed in the same environment.

NPDES permit: The National Pollution Discharge Elimination System permit required by and issued by EPA.

NSSC: The neutral sulfite, semichemical pulping process.

NTA: Nitrilotriacetic acid, a chelant with the sodium salt being the usual form.

Occlusion: An absorption process by which one solid material adheres strongly to another, sometimes occurring by coprecipitation.

Opacity: The percentage of light transmission through a plume.

Ore: A mineral containing useful substances which can be extracted.

Orifice: An opening through which a fluid can pass; a restriction placed in a pipe to provide a means of measuring flow.

ORP: Oxidation Reduction Potential. See "Redox potential."

Osmosis: The passage of water through a permeable membrane separating two solutions of different concentrations; the water passes into the more concentrated solution.

Oxidation: A chemical reaction in which an element or ion is increased in positive valence, losing electrons to an oxidizing agent.

Packing: The fill in a confined space in a stripping vessel, ranging from simply shaped units such as rocks or slats to complex shapes that provide large surface area per unit volume.

Pasteurization: A process for killing pathogenic organisms by heat applied for a critical period of time.

Pathogens: Disease-producing microbes.

Periodic chart: An arrangement of the elements in order of increasing atomic number that illustrates the repetition (or periodicity) of key characteristics.

Permeability: The ability of a body to pass a fluid under pressure.

pH: A means of expressing hydrogen ion concentration in terms of the powers of 10; the negative logarithm of the hydrogen ion concentration.

Photosynthesis: The process of converting carbon dioxide and water to carbohydrates, activated by sunlight in the presence of chlorophyll, liberating oxygen.

Pickle liquor: Acid used in treating steel for removal of oxide scale.

Plankton: Small organisms with limited powers of locomotion, carried by water currents from place to place.

Polarize: In corrosion, to develop a barrier on the anodic or cathodic surface, disrupting the corrosion process.

Pollutant: A contaminant at a concentration high enough to endanger the aquatic environment or the public health.

Polyelectrolyte: A polymeric material having ion exchange sites on its skeleton.

Polymer: A chain of organic molecules produced by the joining of primary units called *monomers.*

Polyphosphate: Molecularly dehydrated orthophosphate.

Precipitate: An insoluble reaction product; in an aqueous chemical reaction, usually a crystalline compound that grows in size to become settleable.

Protozoa: Large, microscopic single-cell organisms higher on the food chain than bacteria, which consume bacteria.

Pulp: Fibrous matter.

Radiation: In a furnace, the transfer of heat by energy waves, much like other forms of electromagnetic waves (e.g., light and radio waves).

Rag: Debris that accumulates at an oil-water interface.

Rankine cycle: The successive changes in heat content and temperature as water is converted to steam, expands through a prime mover, condenses, and returns to the boiler.

Recovery furnace: A furnace which burns black liquor from the kraft pulping process, to recover the cooking chemicals as smelt.

Red mud: The gangue from bauxite processing.

Redox potential: Reduction-oxidation potential measured against a standard electrode.

Reduction: A chemical reaction in which an element or compound gains electrons, being reduced in positive valence.

Regenerative heating: In utility stations, a scheme for reducing heat losses to the main condenser in the cycle by using steam extracted from the turbine to heat feedwater. In engineering designs, the use of a heat exchanger to preheat the feed to a process by extracting heat from the product.

Reheater: A heat exchanger located in a furnace to increase the temperature of steam extracted from a turbine for reinjection.

Resolution: The breaking of an emulsion into its individual components.

Reverse osmosis: A process that reverses (by the application of pressure) the flow of water in the natural process of osmosis so that it passes from the more concentrated to the more dilute solution.

Reversion: The return of molecularly dehydrated phosphate (polyphosphate) to its hydrated origin (orthophosphate).

Ringlemann test: A method of comparing the opacity of a stack plume to an arbitrary set of standard disks of increasing degrees of discoloration.

Salinity: The presence of soluble minerals in water.

Salt-splitting: The ability of an anion exchanger to convert a salt solution to caustic; the ability of a cation exchanger to convert a salt solution to acid.

Saturation index: The relation of calcium carbonate to the pH, alkalinity, and hardness of a water to determine its scale-forming tendency.

Saveall: A general term for several designs of devices used to recover fiber from white water and clarify the water for reuse.

Scale: The precipitate that forms on surfaces in contact with water as the result of a physical or chemical change.

Scale pit: A collection chamber alongside a rolling mill that receives roll cooling water containing metallic scale.

Scouring: The removal of surface debris from raw textile fibers.

Secondary fibers: In the paper industry, fibers reclaimed from waste paper.

Sedimentation: Gravitational settling of solid particles in a liquid system.

Seed: A particle or particles, usually crystalline, added to a supersaturated solution to induce precipitation.

Selectivity: The order of preference of an ion exchange material for each of the ions in the surrounding aqueous environment.

Sensible heat: Heat measurable by temperature alone.

Sequester: To form a stable, water-soluble complex.

Sewage: Waste fluid in a sewer.

Silt density index: A measure of the tendency of a water to foul a reverse osmosis membrane, based on timed flow through a membrane filter at constant pressure.

Sinter: A clinker-like material produced in a special furnace from a mixture of coal and recovered iron-bearing materials, such as scale pit solids, used as charge for a blast furnace.

Sizing: A surface finish, such as starch, applied to paper and textile fibers.

Slag: In metallurgical processing, the impurities separated from molten metal during refining; in boiler furnaces, the noncombustible ash which has reached fusion temperatures.

Slop oil: A general term in oil refining applying to tramp oil discharge to the oily sewer during shutdown and startup or through abnormal operation.

Sludge volume index: An inverse measure of sludge density.

Slurry: A water containing a high concentration of suspended solids, usually over 5000 mg/L.

Smelt: Molten slag; in the pulp industry, the cooking chemicals tapped from the recovery boiler as molten material and dissolved in the smelt tank as green liquor.

Soda ash: A common water-treatment chemical, sodium carbonate.

Sodium absorption ration (SAR): In irrigation water, a relationship between sodium and hardness used to predict acceptability for both the plant and soil being irrigated.

Softening: The removal of hardness (calcium and magnesium) from water.

Sour water: Waste waters containing malodorous materials, usually sulfur compounds.

Spore: A reproductive cell, or seed, of algae, fungi, or protozoa.

Stability index: An empirical modification of the saturation index used to predict scaling or corrosive tendencies in water systerms.

Stickwater: The distillate produced in the cooking of meat or in the rendering of fat and scraps.

Stiff-Davis index: An index used to predict the stability of brackish waters, such as those used in waterflooding.

Stoichiometric: The ratio of chemical substances reacting in water that corresponds to their combining weights in the theoretical chemical reaction.

Stokes' law: An expression for calculating the rate of fall of particles through a fluid based on densities, viscosity, and particle size.

Superheater: A heat exchanger located in a furnace to increase the temperature of steam leaving the boiler drum.

Supernate: The liquid overlying the sludge layer in a sedimentation vessel.

Surfactant: A surface active agent: usually an organic compound whose molecules contain a hydrophilic group at one end and a lipophilic group at the other.

Synergism: The combined action of several chemicals which produces an effect greater than the additive effects of each.

Synfuels: Liquid or gaseous fuels produced from coal, lignite, or other solid carbon sources.

Tailings: The residue from separation of useful values from an ore.

Thermocline: The layer in a lake dividing the upper, current-mixed zone, from the cool lower stagnant zone.

Threshold treatment: The control of scale or deposits by application of sub-stoichiometric dosage of treatment chemical.

Transpiration: Respiration of plants.

Tuberculation: A corrosion process that produces hard mounds of corrosion products on the metal surface, increasing friction and reducing flow in a water distribution system.

Turbidity: A suspension of fine particles that obscures light rays but requires many days for sedimentation because of the small particle size.

Turnover: The mixing of lower and upper layers in a lake in spring and fall caused by temperature and density equalization.

USDA: United States Department of Agriculture.

USGS: United States Geological Survey, Department of the Interior.

Venturi: A device for measuring fluid flow, including a short converging cone succeeded on the same axis by a long diverging cone. This device is also used in gas scrubbing.

Waterflooding: A process of displacing oil from underground formations with water and returning it to the surface for recovery.

Weir: A spillover device used to measure or control water flow.

White liquor: Cooking liquor from the kraft pulping process produced by recausticizing green liquor with lime.

White water: The filtrate from a paper- or board-forming machine, usually recycled for density control.

Yield: The rate of production of cake from a dewatering device.

Zeta potential: The difference in voltage between the surface of the diffuse layer surrounding a colloidal particle and the bulk liquid beyond.

INDEX

INDEX